Azeotropic Data—III

Compiled by Lee H. Horsley

The Dow Chemical Co.

Midland, Mich.

ADVANCES IN CHEMISTRY SERIES **116**

AMERICAN CHEMICAL SOCIETY

WASHINGTON, D. C.   1973

QD
1
A355
no. 116

Library of Congress Catalog Card 73-75991

ISBN 8412-0166-8

ISBN 0-8412-0533-7 (paperback)

ADCSAJ 116 1-628 (1973)

Paperback printing 1979

3 3001 00702 0356

Advances in Chemistry Series

Robert F. Gould, *Editor*

FOREWORD

ADVANCES IN CHEMISTRY SERIES was founded in 1949 by the American Chemical Society as an outlet for symposia and collections of data in special areas of topical interest that could not be accommodated in the Society's journals. It provides a medium for symposia that would otherwise be fragmented, their papers distributed among several journals or not published at all. Papers are referred critically according to ACS editorial standards and receive the careful attention and processing characteristic of ACS publications. Papers published in ADVANCES IN CHEMISTRY SERIES are original contributions not published elsewhere in whole or major part and include reports of research as well as reviews since symposia may embrace both types of presentation.

CONTENTS

PREFACE

This volume is a complete revision of Azeotropic Data I and II published as ADVANCES IN CHEMISTRY SERIES No. 6 and No. 35, American Chemical Society. It includes revised data on systems in the original tables plus new data on azeotropes, nonazeotropes, and vapor-liquid equilibria collected since 1962.

No attempt has been made to evaluate the accuracy of the data. Where appreciable differences occur in the data from a single investigator, only the most recent data are recorded. Where differences in values from two different sources occur, both sets of data are recorded. To aid the reader in evaluating data on a given system, however, all references to that system are included.

In general, data have been obtained from the original literature. Where the original litereature was not available, data have been taken from *Chemical Abstracts*. In a few instances, data have been taken from collections of azeotropic data in handbooks, review articles, and so forth.

The tables are arranged in the same manner as the previous volumes. This is based on empirical formula as in *Chemical Abstracts* except that all inorganic compounds are listed first, alphabetically by empirical formula.

For a given binary system the lower order component according to empirical formula is chosen as the A-component, and under each A-component the B-components are also arranged according to empirical formula. For ternary and quarternary systems, the same arrangement is used, using the lowest order formula as A-component, the next lowest order as B-component, and so on.

Abbreviations

max. b.p.　Maximum boiling point azeotrope (negative azeotrope)
min. b.p.　Minimum boiling point azeotrope (positive azeotrope)
atm.　　　Pressure in standard atmospheres
mm.　　　Pressure in millimeters of Hg
p.s.i.a.　　Pressure in pounds per square inch absolute
p.s.i.g.　　Pressure in pounds per square inch gage
v–l　　　　Vapor-liquid equilibrium data are given in the original reference
v.p.　　　Vapor pressure

vol. % Azeotropic concentration is given in volume per cent. Unless so indicated, all concentration data are in weight %

~ Approximate

>, < Greater than, less than

While all volumes in this series, "Azeotropic Data," are collections of data, mostly from the literature, the supplement that was published as No. 35 in the Advances in Chemistry Series included previously unpublished data from industrial files, of which most was from Union Carbide Chemicals Co., assembled by William S. Tamplin. Additional contributions came from Commercial Solvents Corp., Eastman Chemical Products, Inc., Farbenwerke Hoechst, Imperial Chemical Industries Ltd., and Minnesota Mining and Manufacturing Co. These data are continued in the present volume.

Special thanks go to David C. Young for his invaluable work on this volume.

The Dow Chemical Co.
Midland, Mich.
November 1972

Lee H. Horsley

Tables of
Azeotropes and Nonazeotropes

Table I. Binary Systems

No.	Formula	Name	B.P., °C	B.P., °C	Wt.%A		Ref.
		B-Component		**Azeotropic Data**			
A =	**Ar**	**Argon**	**—186**				
1	CO₂	Carbon dioxide					
		—50°–20°C.		Nonazeotrope		v-l	433c
1a	N₂	Nitrogen 1.36 kg./cm²		Nonazeotrope		v-l	685c
		" 500-1500 mm.	—195	Nonazeotrope		v-l	395,685
		" 80°–105°K.				v-l	961c
2	O₂	Oxygen 1.36 kg./cm²		Nonazeotrope		v-l	685c
		"	—183	Nonazeotrope		v-l	208
		" 1-15 atm.				v-l	686
		" 90°–96°K.				v-l	1007
		" 0-10 atm.		Nonazeotrope		v-l	36
3	C₅F₁₂	Perfluoropentane, 25°C.		Nonazeotrope		v-l	695
A =	**AgCl**	**Silver Chloride**	**1550**				
4	Cl₂Pb	Lead chloride	954	Nonazeotrope			575
A =	**AlCl₃**	**Aluminum Chloride**	**183**				
4a	Cl₅Ta	Tantalum chloride	242	235	9.6	v-l	770e
A =	**AsCl₃**	**Arsenic Chloride**	**130**				
4b	Cl₂OSe	Selenyl chloride	179	Nonazeotrope		v-l	738a
4c	Cl₃P	Phosphorus trichloride 0.4–4 atm.		Nonazeotrope		v-l	27f
5	SbCl₃	Antimony chloride 0.3-4 atm.		Nonazeotrope		v-l	701
6	Cl₄Ge	Germanium chloride	86.5	Nonazeotrope		v-l	859,760c
6a	Cl₄Sn	Tin tetrachloride	114.1	Nonazeotrope		v-l	700c
A =	**AsH₃**	**Arsine**	**55**				
6b	ClH	Hydrochloric acid —80°C.		Nonazeotrope v.p. curve			206c
6c	H₂S	Hydrogen sulfide		— 75	78.5	v-l	206c
		"		— 85	78.0		206c
		"		— 95	74.7		206c
6d	H₃P	Phosphine —80°C.		Nonazeotrope v.p. curve			206c
A =	**BCl₃**	**Boron Chloride**	**11.5**				
7	B₂H₆	Boron hydride	— 92.5	Nonazeotrope			638
A =	**BF₃**	**Boron Fluoride**	**—100**				
8	B₂H₆	Boron hydride	— 92	—106	77.2		638
9	H₂O	Water 100 mm.			62		591
		"	100		60		591
		" 1 mm.		46	65		951
10	H₃N	Ammonia	— 33	180	80		951
11	CH₂O₂	Formic acid 11 mm.		43	42		951

3

No.		B-Component			Azeotropic Data		
	Formula	Name		B.P., °C	B.P., °C	Wt.%A	Ref.
A =	BF₃	Boron Fluoride (continued)		−100			
12	CH₄O	Methanol	4 mm.		58	52	951
13	C₂H₃N	Acetonitrile		81.6	101	62	951
14	C₂H₄O₂	Acetic acid	15 mm.		70	47	591
		"		118.1	150	36	591
		"	746 mm.	118	140		951
		"	13 mm.		59	36	951
15	C₂H₄O₂	Methyl formate		31.9	91	53	951
16	C₂H₅ClO	2-Chloroethanol	2 mm.		59	30	943
17	C₂H₆O	Ethyl alcohol	15 mm.		51	42	951
18	C₂H₆O	Methyl ether		− 21	127	60	951
19	C₃H₆O₂	Ethyl formate		54.1	102	48	951
20	C₃H₆O₂	Methyl acetate		57.1	110	48	951
21	C₃H₆O₂	Propionic acid	17 mm.		62	31	951
22	C₃H₆O₃	Methyl glycolate	3 mm.		60	43	951
23	C₃H₈O	Ethyl methyl ether		10.8	127	53	951
24	C₃H₈O	Propyl alcohol	2 mm.		56	36	951
25	C₃H₉N	Trimethylamine		3.5	230	53	951
26	C₄H₆O₂	Crotonic acid	12.5 mm.		81	28	951
27	C₄H₈O₂	Butyric acid	11 mm.		64	28	951
28	C₄H₈O₂	Ethyl acetate		77.05	119	44	951
29	C₄H₁₀O	Butyl alcohol	3 mm.		64.5	31	951
30	C₄H₁₀O	Ethyl ether		34.5	125	48	951
31	C₅H₅N	Pyridine		115.5	300	46	951
32	C₅H₁₀O₂	Ethyl propionate		99.15	116	40	951
33	C₅H₁₀O₂	Propyl acetate		101.6	127	40	951
34	C₆H₁₄O	Amyl methyl ether					
			10 mm.		55	40	951
35	C₆H₁₄O	Isopropyl ether	98 mm.		61	40	951
A =	B₂H₆	Boron Hydride		− 92.5			
36	BrH	Hydrobromic acid		− 67	Nonazeotrope		638
37	ClH	Hydrochloric acid		− 85	− 94	64	638
		"	205 mm.	−106	−115	68	638
38	C₂H₆	Ethane	100-760 mm.	− 88	Nonazeotrope		638
39	C₄H₁₀O	Ethyl ether					
		25-100 p.s.i.g.			Nonazeotrope	v-l	602
40	C₆H₁₅B	Triethylborane				v-l	660
		-25 to 25°C.					
A =	BeF₂	Beryllium Fluoride					
40a	FLi	Lithium fluoride					
		900°–1050°C.				v-l	896c
41	FNa	Sodium fluoride					
		509°-1061°C.				v-l	861
A =	BrF₃	Bromine Trifluoride		135			
42	BrF₅	Bromine pentafluoride			Nonazeotrope	v-l	593
43	Br₂	Bromine	1760 mm.		75	84.4	v-l	275
		"	3800 mm.		100	81.5	v-l	275

TABLE I. *Binary Systems*   5

No.	Formula	B-Component Name	B.P., °C	B.P., °C	Wt.%A		Ref.
				Azeotropic Data			
A =	BrF$_3$	**Bromine Trifluoride** (*continued*)	135				
44	FH	Hydrogen fluoride	19.4	Azeotropic			238
45	F$_6$U	Uranium hexafluoride	56	Nonazeotrope			238
A =	BrF$_5$	**Bromine Pentafluoride**					
46	FH	Hydrogen fluoride	19.4	20	56		9,238
		" 4 atm.			79		238
47	F$_6$U	Uranium hexafluoride	56	Min. b.p.			
					82		238
		" 3 atm.		Nonazeotrope		v-l	597
		" 70°		Nonazeotrope		v-l	597
		" 90°		Nonazeotrope		v-l	597
		" 100 p.s.i.a			62.5		238
A =	BrH	**Hydrobromic Acid**	— 73				
48	H$_2$O	Water	100	126	47.5		563
		" 100 mm.		74.12	49.80 ⎤		
		" 500 mm.		112.94	48.19 ⎪		69,483,
		" 900 mm.		129.13	47.40 ⎬		820
		" 1200 mm.		137.34	47.03 ⎦		
49	H$_2$S	Hydrogen sulfide					
		480 mm.	— 70	— 70	60.5	v-l	915
50	SO$_2$	Sulfur dioxide	— 10	Nonazeotrope		v-l	915
A =	Br$_2$	**Bromine**	58.9				
51	FH	Hydrogen fluoride	19.4	Azeotropic			238
52	F$_6$U	Uranium hexafluoride	56	Azeotropic			238
53	H$_2$O	Water low Br$_2$ conc.				v-l	423
54	I$_2$	Iodine	185.3	Nonazeotrope			563
55	CCl$_4$	Carbon tetrachloride					
		735 mm.	76	57.7	89	v-l	908
56	C$_2$Br$_2$F$_4$	1,2-Dibromotetrafluoroethane		43	29.8		214
57	C$_2$Cl$_3$F$_3$	1,1,2-Trichloro-1,2, 2-trifluoroethane	47.6	41	40.8	v-l	909
58	C$_2$Cl$_4$F$_2$	1,1,1,2-Tetrachloro- 2,2-difluoroethane	91.6	57.8	89.5	v-l	909
59	C$_2$HCl$_3$F$_2$	1,2,2-Trichloro-1,1- difluoroethane, 736 mm.	71.1	54.1	73.5	v-l	910
60	C$_2$H$_2$Cl$_2$F$_2$	1,1-Dichloro-2,2- difluoroethane, 735 mm.	59	49.6	62	v-l	910
61	C$_3$Cl$_3$F$_5$	1,2,2-Trichloro-1,1, 3,3,3-pentafluoro- propane	72.5	49.1	60.5	v-l	909
62	C$_7$H$_5$F$_3$	α,α,α-Trifluorotoluene	103.9	58.1	97	v-l	909
A =	Br$_3$P	**Phosphorus Tribromide**	175.3				
63	C$_n$H$_{2n+2}$	Paraffin hydrocarbons		Min. b.p.			691
A =	Br$_4$Sn	**Tin Bromide**	206.7				
64	I$_4$Sn	Tin iodide	346.0⁻	Nonazeotrope		v-l	778

No.	Formula	B-Component Name	B.P., °C	B.P., °C	Wt.%A		Ref.
A =	Br₄Sn	**Tin Bromide** *(continued)*	**206.7**				
65	C₇H₁₂O₄	Ethyl malonate	198.9	Reacts			563
66	C₁₀H₈	Naphthalene	218.1	Nonazeotrope			563
A =	C	**Graphite**	**2300/0.01**				
67	MnS	Manganese sulfide		1375/0.01 mm.			753
A =	CClN	**Cyanogen Chloride**	**12.5**				
68	HCN	Hydrocyanic acid					
		15 mm.		Nonazeotrope		v-l	332
68a	H₂O	Water	100	Nonazeotrope		v-l	40e
A =	CCl₂O	**Phosgene**	**8.2**				
69	FH	Hydrofluoric acid					
		3000 mm.		21	77		48
70	C₂H₄Cl₂	1,2-Dichloroethane	83.45	Nonazeotrope			575
A =	CF₂O	**Carbonyl Fluoride**					
71	CF₄O	Trifluoromethyl					
		hypofluorite	— 94.2	— 97.0	10		448
A =	CHN	**Hydrocyanic Acid**	**26**				
72	H₂O	Water	100	V.P. curve			563
73	CH₄O	Methanol	64.7	Nonazeotrope			563
74	C₂H₄O₂	Methyl formate	31.7	24.0	52		575
75	C₂H₅NO₂	Ethyl nitrite	17.4	16.5	15		575
76	C₃H₃N	Acrylonitrile 400 mm.		Nonazeotrope			905
		" 400 mm.				v-l	905c
		"	77.3	Nonazeotrope			905
		" 200–760 mm.		Nonazeotrope		v-l	905f
76a	C₃H₄O	Acrolein 200–760 mm.		Nonazeotrope		v-l	905f
A =	CO	**Carbon Monoxide**	**—192**				
76b	CO₂	Carbon dioxide	— 78	Nonazeotrope		v-l	
							433c, 434c
77	CH₄	Methane -125 to 255°F.		Nonazeotrope		v-l	967
A =	CO₂	**Carbon Dioxide**	**— 79.1**				
78	ClH	Hydrochloric acid	— 82	Nonazeotrope			563
79	Cl₂	Chlorine	— 37.6	Nonazeotrope			575
80	H₂	Hydrogen -40 to 25°C				v-l	434
		" 219°–290°K.				v-l	905t
		" 90°–220°K.		ideal		v-l	43c
81	H₂O	Water	100	Nonazeotrope			563
82	H₂S	Hydrogen sulfide					
		20-80 atm.	— 59.6	Nonazeotrope		v-l	56,57
82a	He	Helium 220°–290°K.				v-l	601c
		" 253°–293°K.		Nonazeotrope		v-l	104c
83	N₂	Nitrogen —40 to 25°C				v-l	434
84	N₂O	Nitrous oxide	— 90.7	Min. b.p.		v-l	829
85	O₂	Oxygen —40 to 25°C				v-l	434
		" — 5.0°–10°C.				v-l	289c

TABLE I. *Binary Systems* 7

No.	Formula	B-Component Name	B.P.. °C	B.P., °C	Azeotropic Data Wt.%A		Ref.
A =	**CO₂**	**Carbon Dioxide** *(continued)*	**— 79.1**				
86	SO₂	Sulfur dioxide	— 10	Nonazeotrope			563
87	CS₂	Carbon disulfide	46.2	Nonazeotrope			563
88	CH₃Cl	Chloromethane	— 23.7	Nonazeotrope			499
88a	CH₄	Methane	—161	Nonazeotrope		v-l	434c
89	C₂H₂	Acetylene	— 84	Nonazeotrope		v-l	829
		" Crit. press.		Nonazeotrope		v-l	829
90	C₂H₄	Ethylene <4 atm.		Nonazeotrope		v-l	829
		" 12 atm.			28.9	v-l	829
		" Crit. prcss.			51	v-l	829
91	C₂H₅Cl	Chloroethane	12.4	Nonazeotrope			559
92	C₂H₆	Ethane	— 93	Max. v.p. mixture			499
		Ethane	— 88.6		59.5	v-l	829
		" Crit. press.			77.5	v-l	829
		"			77.2	v-l	453
93	C₂H₆O	Methyl ether	— 23.65	Nonazeotrope			575
93a	C₃H₆	Propylene 0°C.				v-l	1042e
		" — 43.5°C.				v-l	1042e
A =	**ClF₃**	**Chlorine Trifluoride**					
94	ClH	Hydrogen chloride	— 85	Reacts			9
95	FH	Hydrogen fluoride	19.4	Azeotropic			238
		" 1183 mm.		20	93	v-l	237
		" 90 p.s.i.g.			94.5		238
		" 125 p.s.i.g.			94		238
		" 143 p.s.i.g.			93.8		238
		" 148 p.s.i.g.			93.7		238
96	F₆U	Uranium hexafluoride	56	Nonazeotrope		v-l	235,238
		" 2600 mm.		Nonazeotrope		v-l	29c
A =	**ClH**	**Hydrogen Chloride**	**— 85**				
97	Cl₂	Chlorine 350 mm.	— 44	Nonazeotrope			9
98	Cl₄Ge	Germanium tetrachloride	86.5		79.8	v-l	10
99	H₂O	Water 50 mm.		48.724	23.42 ⎫		
		" 250 mm.		81.205	21.883 ⎪		70,118,
		" 760 mm.	100	108.584	20.222 ⎬		483,821
		" 1220 mm.		122.98	19.358 ⎭		
		" 100 p.s.i.g.	169	177	14.8		981
		" 520 p.s.i.g.	244	250	6.5		981
		" 860 p.s.i.g.	275	280	2.8		981
		" 1360 p.s.i.g.	306	310	0.6		981
		" 1815 p.s.i.g.	328	330	0.1		981
100	SO₂	Sulfur dioxide	— 10	Nonazeotrope at —35°C.			915
101	CH₄O	Methanol	64.7	Max. b.p.			215
102	C₂H₆	Ethane 48 atm.		15	56		563
		"		25.4	59		563
103	C₂H₆O	Methyl ether	— 22	— 2	38		295
		"		Azeotropic to critical point			500
		"	— 23.65	— 1.5	60		563
104	C₆H₇N	Aniline	184.35	244.8	~27.5		563

No.	Formula	B-Component Name	B.P., °C	B.P., °C	Wt.%A	Ref.
A =	ClHO$_4$	Perchloric Acid	110			
105	H$_2$O	Water	100	203	71.6	563
A =	Cl$_2$	Chlorine	— 34.6			
106	FH	Hydrogen fluoride				
		350 mm.	3.0	— 47	92	9
	"		19.4	— 35		9
107	H$_2$O	Water	100	Nonazeotrope		563
108	SO$_2$	Sulfur dioxide				
		400–760 mm.			v-l	326c
	"		— 9.7	— 34.7	89	134
	"	7 atm.		18	80	134
	"	20 atm.		57.5	75.5	134
A =	Cl$_2$Cu	Cupric Chloride				
109	Cl$_2$Pb	Lead chloride	954	Min. b.p.		575
110	Cl$_2$Zn	Zinc chloride	732	Min. b.p.		575
A =	Cl$_2$OSe	Selenyl Chloride	179			
110a	Cl$_4$Sn	Tin tetrachloride	114	Nonazeotrope	v-l	738a
A =	Cl$_2$O$_2$S	Sulfuryl Chloride	69.1			
111	Cl$_3$OP	Phosphorus oxychloride	107.2		96.5/0°C.	575
112	CCl$_4$	Carbon tetrachloride	76.75	Nonazeotrope	v-l	991
113	C$_2$Cl$_6$	Hexachloroethane	184.8	Nonazeotrope	v-l	991
114	C$_2$H$_2$Cl$_4$	1,1,2,2-Tetra-				
		chloroethane	146.2	Nonazeotrope	v-l	991
115	C$_2$H$_4$Cl$_2$	1,2-Dichloroethane	83.45	Nonazeotrope	v-l	991
A =	Cl$_2$Pb	Lead Chloride	954			
116	Cl$_2$Zn	Zinc chloride	732	Nonazeotrope		575
A =	Cl$_2$S$_2$	Sulfur Monochloride	138			
116a	CHCl$_3$S	Perchloromethyl mercaptan		Nonazeotrope	v-l	668h
		15–150 mm.				
A =	Cl$_3$Fe	Ferric Chloride	315			
116b	Cl$_3$HSi	Trichlorosilane	32	Nonazeotrope	v-l	66c
116c	Cl$_4$Si	Silicon tetrachloride	56.9	Nonazeotrope	v-l	66c
A =	Cl$_3$HSi	Trichlorosilane	32			
116d	Cl$_3$OP	Phosphorus oxychloride				
		300–1520 mm.		Nonazeotrope	v-l	522c
117	Cl$_3$P	Phosphorus trichloride	76	Nonazeotrope	v-l	522,
						66c,229c
118	CCl$_4$	Carbon tetrachloride	76.75	Nonazeotrope	v-l	67
118a	C$_4$H$_8$Cl$_2$O	Bis-2-chloroethyl ether	178.5	Nonazeotrope	v-l	521c
118b	C$_4$H$_8$O$_2$	Dioxane	101.1	Nonazeotrope	v-l	522e
119	C$_6$H$_6$	Benzene 30°–40°C.		Nonazeotrope		718e
				b.p. curve		718e
120	C$_6$H$_{14}$O	Isopropyl ether	69	Nonazeotrope	v-l	700
121	C$_6$H$_{14}$O	Propyl ether	85	Nonazeotrope	v-l	521
122	C$_8$H$_{18}$O	Butyl ether	142.4	Nonazeotrope	v-l	699
123	C$_{10}$H$_{22}$O	Amyl ether	185	Nonazeotrope	v-l	521
124	C$_{10}$H$_{22}$O	Isoamyl ether	170	Nonazeotrope	v-l	521

TABLE I. *Binary Systems* 9

	B-Component			Azeotropic Data		
No.	Formula	Name	B.P., °C	B.P., °C	Wt.%A	Ref.
A =	Cl₃OP	Phosphorus Oxychloride	107.2			
125	Cl₃OV	Vanadium oxychloride	126.5	Nonazeotrope	v-l	702
125a	Cl₃P	Phosphorus trichloride	76	Nonazeotrope	v-l	412c
126	Cl₄Ti	Titanium tetrachloride	136.5	143.2 46.6	v-l	702
126a	CH₃Cl₃Si	Trichloromethylsilane	66.4	Nonazeotrope	v-l	450c
126b	C₄Cl₆	Hexachlorobutadiene		Nonazeotrope	v-l	68f
A =	Cl₃OV	Vanadium Oxychloride	126.5			
126c	Cl₄Ge	Germanium tetrachloride	83.5	Nonazeotrope		718e
				b.p. curve		718e
126d	Cl₄Si	Silicon tetrachloride	56.8	Nonazeotrope		718e
				b.p. curve		718e
126e	Cl₄Sn	Tin tetrachloride	113.8	Nonazeotrope		718e
				b.p. curve		718e
127	Cl₄Ti	Titanium tetrachloride	136.5	Nonazeotrope	v-l	702,1075
	"		136	b.p. curve		718e
127a	Cl₄V	Vanadium tetrachloride	153	Nonazeotrope		718c
				b.p. curve		718c
127b	CCl₄	Carbon tetrachloride	76.5	Nonazeotrope		718c
				b.p. curve		718e
A =	Cl₃P	Phosphorus Trichloride	76			
128	Cl₄Si	Silicon tetrachloride	56.7	Nonazeotrope	v-l	63,522,66c
128a	CH₃Cl₃Si	Trichloromethylsilane	66.4	Nonazeotrope	v-l	328c,450c
129	C₆H₁₂	Cyclohexane	80.75	Nonazeotrope		693
130	C₆H₁₄	Hexane	68.8	68.7 8 vol. %		691
131	C₇H₁₆	2,2-Dimethylpentane	79.1	Min. b.p.		691,693
132	C₇H₁₆	2,3-Dimethylpentane	89.8	74.5 98.8 vol. %		691
133	C₇H₁₆	2,4-Dimethylpentane	80.5	74.2 73		691
134	C₇H₁₆	2,2,3-Trimethylbutane	80.9	74.5 77		691
A =	Cl₃Sb	Antimony Chloride				
135	CₙH₂ₙ₊₂	Paraffins	200-220	Min. b.p.		196,938
136	CₙH₂ₙ₋₆	Aromatics	200-220	Nonazeotrope		196,938
A =	Cl₄Ge	Germanium Tetrachloride	86.5			
136a	Cl₄Si	Silicon tetrachloride	56.9	Nonazeotrope	v-l	205c
136b	CCl₄	Carbon tetrachloride	75.9	Nonazeotrope	v-l	205c
136c	CHCl₃	Chloroform	61	Nonazeotrope	v-l	205c
136d	CH₂Cl₂	Dichloromethane	41.5	Nonazeotrope	v-l	205c
136e	C₂H₄Cl₂	1,1-Dichloroethane	57.4	Nonazeotrope	v-l	205c
136f	C₂H₄Cl₂	1,2-Dichloroethane	83.45	Nonazeotrope	v-l	205c
136g	C₂H₅Cl	Chloroethane	13	Nonazeotrope	v-l	205c
A =	Cl₄Si	Silicon Tetrachloride	56.9			
137	Cl₄Ti	Titanium tetrachloride	136.5	Nonazeotrope	v-l	575,987
138	CCl₄	Carbon tetrachloride	76.75	Nonazeotrope	v-l	987,1031
139	CHCl₃	Chloroform	61	55.6 70		844
140	CH₃SiCl₃	Methyl trichloro-silane, 20°-66°		Nonazeotrope, v.p. curve		485
141	CH₃NO₂	Nitromethane	101	53.8 94		841
142	CH₄SiCl₂	Methyl dichloro-silane, 20°-66°		Nonazeotrope, v.p. curve		485

		B-Component		Azeotropic	Data		
No.	Formula	Name	B.P., °C	B.P., °C	Wt.%A		Ref.
A =	Cl₄Si	**Silicon Tetrachloride**	**56.9**				
		(continued)					
143	C₂H₃N	Acetonitrile	81.5	49.1	90.4	v-l	484
		"	82	49.0	90.6		841,843
144	C₂H₄Cl₂	1,1-Dichloroethane	57.4	52.7	63.5		844
145	C₂H₄Cl₂	1,2-Dichloroethane	83.7	Azeotropic?			844
146	C₃H₃N	Acrylonitrile	79	51.2	89		841,843
147	C₃H₅N	Propionitrile	97	55.6	92		841
148	C₃H₉SiCl	Chlorotrimethylsilane		Azeotrope composition			
				independent of			
				pressure			840,843
		"	57.5	54.7	64.8		841
		" 2247 mm.		90	66.2		484
		" 760 mm.	57.7	54.5	64.8	v-l	484
		" 83 mm.		0	58.8		484
		" 17 mm.		30.0	55.1		484
148a	C₄H₈Cl₂O	Bis-2-chloroethyl ether	178.5	Nonazeotrope		v-l	521c
		"		b.p. curve			886c
148b	C₄H₈O₂	Dioxane	101.1	Nonazeotrope		v-l	522e
149	C₆H₁₄	3-Methylpentane	63.3	Nonazeotrope			844
150	C₆H₁₄	2-Methylpentane	60.4	Nonazeotrope			844
151	C₆H₁₄O	Isopropyl ether	69	Nonazeotrope		v-l	700
151a	C₆H₁₄O	Propyl ether	91	Nonazeotrope		v-l	521c
152	C₈H₁₈O	Butyl ether	142.4	Nonazeotrope		v-l	699
152a	C₁₀H₂₂O	Isoamyl ether	173	Nonazeotrope		v-l	521c
152b	C₁₀H₂₂O	Amyl ether	190	Nonazeotrope		v-l	521c
A =	Cl₄Sn	**Tin Chloride**	**113.85**				
153	Cl₄Ti	Titanium chloride	136	Nonazeotrope			575
154	CCl₄	Carbon tetrachloride	76.8	Nonazeotrope		v-l	124
155	C₃H₅ClO	Epichlorohydrin	116.45	Reacts			563
156	C₅H₅N	Pyridine	115.5	Reacts			563
157	C₆H₆	Benzene	80.2	Nonazeotrope			563
158	C₆H₁₂	Cyclohexane	80.75	Nonazeotrope			575
159	C₆H₁₂O₂	Ethylbutyrate	119.9	Reacts			563
160	C₇H₈	Toluene	110.7	109.15	52		563
161	C₇H₁₄	Methylcyclohexane	101.15	<100.8	>15		562
		"		101.1	Nonazeotrope		545
162	C₈H₁₀	Ethylbenzene	136.15	Nonazeotrope			575
163	C₈H₁₆	1,3-Dimethylcyclohexane	120.7	112.5	80		562
164	C₈H₁₈	2,5-Dimethylhexane	109.4	107.5	40		548,575
165	C₈H₁₈	Octane	125.75	<113.2	>80		545,562
		"		125.7	Nonazeotrope	v-l	124
A =	Cl₄Ti	**Titanium Chloride**	**136**				
166	CCl₄	Carbon tetrachloride	76.75	Nonazeotrope		v-l	
							575,987,1076
167	CS₂	Carbon disulfide	46.25	Nonazeotrope		v-l	987
167a	CH₃Cl₃Si	Trichloromethylsilane	66.4	Nonazeotrope		v-l	522g
168	C₂Cl₄O	Trichloroacetyl chloride	118	Nonazeotrope		v-l	
							868,869,1075

TABLE I. *Binary Systems* 11

No.		B-Component			Azeotropic Data		
	Formula	Name	B.P., °C	B.P., °C	Wt.%A		Ref.
A =	Cl₄Ti	**Titanium Chloride** *(continued)*	136				
169	C₂H₂Cl₂O	Chloroacetyl chloride	106	105	20	v-l	868
		"	105	Nonazeotrope		v-l	869
170	C₂H₂Cl₄	1,1,2,2-Tetrachloro-					
		ethane, 740 mm.	146.6	135.4	91.7	v-l	115
A =	Cu	**Copper**	2310				
171	Pb	Lead	1525	Azeotropic			575
172	Sn	Tin	2275	Max. b.p.			563
A =	DH	**Deuterium Hydride**					
173	D₂	Deuterium, 18°-28°K.	—249.7	Nonazeotrope		v-l	696
174	H₂	Hydrogen, 18°-28°K.	—252.7	Nonazeotrope		v-l	696
A =	D₂	**Deuterium**	—249.7				
175	H₂	Hydrogen, 18°-28°K.	—252.7	Nonazeotrope		v-l	696
A =	D₂O	**Deuterium Oxide**					
176	H₂O	Water		221	53		1055
		Azeo. comp. varies					
		from 0-100% H₂O					
		from 220-222°C.					1055
		Nonazeo. below 220°C					
A =	FH	**Hydrogen Fluoride**	19.4				
177	F₅Sb	Antimony pentafluoride	142.7	Nonazeotrope		v-l	870
178	F₆U	Uranium hexafluoride					
		" 85 p.s.i.g.			22		238
		" 110 p.s.i.g.			18		238
		" 132 p.s.i.g.			15		238
		" 145 p.s.i.g.			14		238
179	H₂O	Water	100	111.35	35.6 ⎱		289,820
		"		B.p. curve ⎰			" "
		" 750 mm.	100	112.0	38.26	v-l	670
180	SO₂	Sulfur dioxide	— 10	Nonazeotrope		v-l	433
		" 2280 mm.	28	Nonazeotrope		v-l	433
181	CCl₂F₂	Dichlorodifluoromethane		20	8		
				(Under pressure)			48
		" 150 p.s.i.g.	48	39	7.5		981
182	CHClF₂	Chlorodifluoro-					
		methane 70 p.s.i.g.	7	< 7	3		981
		" 150 p.s.i.g.	29	24	2.7		981
		" 230 p.s.i.g.	45	36	2.8		981
		"			1-2.2		48
183	C₂HF₃O₂	Trifluoroacetic acid		Nonazeotrope		v-l	659
184	C₄H₁₀	Butane	0	Min. b.p.			291,325
185	C₄H₁₀	2-Methylpropane	— 10	Min. b.p.			291,325
186	C₄H₁₀O	Ethyl ether	34.5	74	40		135
A =	F₃Sb	**Antimony Fluoride**	319				
187	F₅Sb	Antimony pentafluoride	155	390	62		563
		"	155	384	80		563

No.	Formula	Name	B.P., °C	B.P., °C	Wt.%A		Ref.
	B-Component			Azeotropic Data			
A =	F₅Nb	Niobium Pentafluoride					
187a	F₆U	Uranium hexafluoride					
		5.9 atm.		Nonazeotrope		v-l	756c
A =	F₆S	Sulfur Hexafluoride					
188	C₅F₁₂	Perfluoropentane, 25°C.		Nonazeotrope		v-l	695
A =	F₆U	Uranium Hexafluoride	56				
188a	F₆W	Tungsten hexafluoride					
		1520–2600 mm.		Nonazeotrope		v-l	768c
188b	C₂Cl₂F₄	1,2-Dichlorotetrafluoroethane					
		2600 mm.		Nonazeotrope		v-l	131c
A =	F₆W	Tungsten Hexafluoride	25/1019; 45/1982 mm.				
189	C₅F₁₀	Perfluorocyclo-					
		pentane	25/833	25/1035	85.4	v-l	819
	"		45/1642	45/2010	83.4	v-l	819
190	C₅F₁₂	Perfluoro-					
		pentane, 1140 mm.	40.86	28.11	93.2	v-l	41
A =	HI	Hydriodic Acid	— 34				
191	H₂O	Water 744 mm.	100	127	57		483,820
				18	60.5		820
				100	58.2		820
192	H₂S	Hydrogen sulfide 60°C.		Nonazeotrope/		v-l	915
A =	HNO₃	Nitric Acid	86				
193	H₂O	Water 735 mm.	100	120.5	68		820
	"	75 mm.			66.7		820
	"	1200 mm.			68.7		820
	"	7 mm.		25	66.2	v-l	278
	"	30 mm.				v-l	629c
	"	50 mm.				v-l	629c
	"	100 mm.				v-l	629c
	"	50 mm.	37	57.8	65.8	v-l	72
	"	100 mm.	51.6	72.4	66.0	v-l	72
	"	200 mm.	66.5	86.4	66.4	v-l	72
	"	400 mm.	83.0	103.2	66.8	v-l	72
	"	760 mm.	100	120.7	67.4	v-l	72
		See also H₂O-N₂O₅ below					
194	CHCl₃	Chloroform	61	47.5	15		738
A =	HO₄Re	Rhenic Acid					
194a	H₂O	Water	100	Nonazeotrope		v-l	68c
A =	H₂	Hydrogen	—252.7				
194b	He	Helium 14°–15.5°K.				v-l	342c
	"	15.5°–29.8°K.				v-l	900c
195	Ne	Neon			28.6	v-l	927
	"	0-26 atm.		Azeotropic		v-l	372
	"			26°K	33	v-l	372
	"			30°K	27.2	v-l	372

TABLE I. *Binary Systems*   13

No.	Formula	B-Component Name	B.P., °C	B.P., °C	Wt.%A		Ref.
A·=	**H₂**	**Hydrogen** *(continued)*	**−252.7**				
195a	C₂H₄	Ethylene	−104	Nonazeotrope		v-l	386m
A =	**H₂O**	**Water**	**100**				
196	H₂O₂	Hydrogen peroxide	152.1	Nonazeotrope		v-l	326
		"				v-l	281
197	H₂SO₄	Hydrogen sulfate		330	1.7		981
		" 200 mm.			1.6		981
198	H₂S	Hydrogen sulfide	− 63.5	Nonazeotrope			563
199	H₃N	Ammonia	− 33.5	Nonazeotrope			563
200	H₄N₂	Hydrazine	113.5	120	28.5		563
		"	113.5	120	30		402
		" 20-760 mm.				v-l	986
		" 124.8 mm.	66.8	74.2	33.2	v-l	107
		" 281.8 mm.	86.5	93.3	32.3	v-l	107
		" 411.2 mm.	96.8	103.6	31.0	v-l	107
		" 560.4 mm.	105.2	111.3	31.4	v-l	107
		" 700.6 mm.	111.7	117.6	32.6	v-l	107
		" 760 mm.	113.8	120	32.3	v-l	107
201	N₂O₅	Nitrogen pentoxide		Max. b.p.	40	v-l	588
		"		Min. b.p.	14.3		588
		"		Max. b.p.	12.5		588
202	O₂S	Sulfur dioxide	− 10	Nonazeotrope			563
203	O₃S	Sulfur trioxide	47	338	~ 19		563
		"			17.2		820
204	O₁₀P₄	Phosphorus pentoxide 104 mm.		694	8.9		954
		" 753 mm.		869	7.9		954
205	CCl₄	Carbon tetrachloride	76.75	66	4.1		215,689
		"	75.9	66	4		623c
206	CS₂	Carbon disulfide	46.25	42.6	2.8		201,606
207	CHCl₃	Chloroform	61	56.1	2.8		171,803,1000
208	CH₂Cl₂	Dichloromethane	43.5	38.1	1.5		36,620
209	CH₂O	Formaldehyde	− 21	Nonazeotrope		v-l	756
210	CH₂O₂	Formic acid	100.75	107.2	22.6		803
		"			22.5		260
		" 45 p.s.i.a.		139	15		260
		" 175 mm.		63	35		260
		" 15 mm.			40		260
		" 40-760 mm.				v-l	145
		"	100.75	107.65	25.5	v-l	171,649,1058
		"	100.7	107.6	22.3		112
		" 50 mm.				v-l	507c
211	CH₃NO₂	Nitromethane	101.2	83.59	23.6	v-l	168,285,806,857
		"				v-l	856c
212	CH₃NO₃	Methyl nitrate	64.8	< 61.5	< 16		560

No.	Formula	B-Component Name	B.P., °C	B.P., °C	Azeotropic Data Wt.%A	Ref.
A =	H₂O	Water *(continued)*	100			
213	CH₄O	Methanol				
		0-150 p.s.i.g.		Nonazeotrope		728
				v-l data	198,483,728,1046	
214	CH₅N	Methylamine	— 6	Nonazeotrope		575
215	CH₆N₂	Methylhydrazine	88	105.2	52.7	402
216	C₂Cl₃F₃	1,1,2-Trichlorotri-				
		fluoroethane	47.5	44.5	1.0	981
217	C₂Cl₄	Tetrachloroethylene	121	88.5	17.2	981
		"	121	87.7	15.8	466
218	C₂HCl₃	Trichloroethylene	86.2	73.4	7.02	497
		"	86.2-86.6	73.6	5.4	337,803
219	C₂HCl₃O	Chloral	97.75	95	7	813
220	C₂HCl₅	Pentachloroethane	162.0	95.9		575
221	C₂HF₃O₂	Trifluoroacetic acid		105	21	659
222	C₂H₂Cl₂	*cis*-1,2-Dichloroethylene	60.2	55.3	3.35	148
223	C₂H₂Cl₂	*trans*-1,2-Dichloroethylene	48.35	45.3	1.9	148
224	C₂H₃-Cl₂NO₂	Methyl *N,N*-dichlo-rocarbamate		93	50 vol. %	142
225	C₂H₃Cl₃	1,1,2-Trichloro-ethane	113.8	86.0	16.4	981
226	C₂H₃N	Acetonitrile	81.6		16.5 v-l	997f
		" 10 mm.	— 15	< -16	2.6	983
		" 50 mm.	13	< 12	5.8	983
		" 760 mm.	80.1	76.5	16.3	983
		"	81.5	76.0	14.2 v-l	173,583,
		" 300 mm.	54.4	51.1	10.5 v-l⎫	727,763,
		" 150 mm.	36.7	34.1	7.2 v-l⎬	959
		"	81.5	76	14.2 v-l	58
		" 750 mm.		76.1	15.85	700
		"		76.2	17.1	997c
		" 100-3168 mm.		% H₂0 increases with pressure	v-l	150,742
227	C₂H₄Cl₂	1,2-Dichlorethane	83.5	72.28	9.2	112
		"	83.5	71.6	8.2	981
		"		75.5	8.2	806
		" 150 mm.		33.5	4.9	806
		" 75 mm.		19.0	4.9	806
		"	84	72	7.9	33,357
228	C₂H₄Cl₂O	Bis(chloromethyl) ether	106	Min. b.p.		723
229	C₂H₄O	Acetaldehyde	20.2	Nonazeotrope	v-l	167
		" 5 atm.		Nonazeotrope	v-l	41c
230	C₂H₄O	Ethylene oxide	10	Nonazeotrope	v-l	167,980
231	C₂H₄O₂	Acetic acid 50 mm.		Nonazeotrope	v-l	507e
		"	118	Nonazeotrope	v-l	171,483,901
232	C₂H₄O₂	Methyl formate	31.9	Nonazeotrope		359

TABLE I.   *Binary Systems*   15

No.	Formula	B-Component Name	B.P., °C	B.P., °C	Azeotropic Data Wt.%A	Ref.
A =	H₂O	Water *(continued)*	100			
233	C₂H₅Br	Bromoethane	38.4	37	1.3 vol.	820
234	C₂H₅BrO	2-Bromoethanol				
		150 mm.	100	58	55.7	215
235	C₂H₅ClO	2-Chloroethanol				
		50 mm.	60	37.1	60.2	981
		" 100 mm.	75	51.1	59.3	981
		" 748 mm.	128.7	97.75	58	38,76,462
		" 50 mm.		35—36	60	38,76,462
		" 200 mm.		65.46	58.6	481a
		" 200–760 mm.			56 v-l	1c
236	C₂H₅I	Iodoethane	70	66	3—4 vol.	754
237	C₂H₅IO	1-Iodo-2-ethanol	176	98.7	77	215
238	C₂H₅NO	Acetamide	221.2	Nonazeotrope		529
239	C₂H₅NO	N-Methylformamide				
		200 mm.		Nonazeotrope	v-l	373
		" 500 mm.		Nonazeotrope	v-l	373
240	C₂H₅NO₂	Nitroethane	114.07	87.22	28.5	168,806
241	C₂H₅NO₃	Ethyl nitrate	87.68	74.35	22	538
241a	C₂H₆	Ethane crit. region			v-l	190c
242	C₂H₆O	Ethyl alcohol	78.3	78.174	4.0	34,131,
						337,706,803,1046
				Effect of pressure		471,1001
		" 150°-350°C.			v-l	42
		" 250-2500 mm.			v-l	719
243	C₂H₆SO	Dimethyl sulfoxide		Nonazeotrope	v-l	671
244	C₂H₆O₂	Ethylene glycol				
		76-760 mm.		Nonazeotrope	v-l	
						183,201,529
245	C₂H₆SO₄	Methyl sulfate	189.1	98.6	73	575
246	C₂H₇N	Dimethylamine	7.3	Nonazeotrope		575
247	C₂H₇N	Ethylamine	16.55	Nonazeotrope		575
248	C₂H₇NO	2-Aminoethanol	170.5	Nonazeotrope		981
		" 100 mm.	112	Nonazeotrope		981
249	C₂H₈N₂	1,1-Dimethylhydra-				
		zine 102 mm.		Max. b.p. 82.5		132
		"		Nonazeotrope		132
250	C₂H₈N₂	1,2-Ethylenediamine	116		18.0 v-l	215
		"	116	118.5	20 v-l	386c
		" 1.77 atm.		Nonazeotrope	v-l	386c
		" >3400 mm.		Nonazeotrope		215
		"		116.9	119 18.4	981
		"		116	118 20-25	173
251	C₃HF₅O₂	Pentafluoropropionic acid		109	10	659

No.	Formula	B-Component Name	B.P., °C	Azeotropic Data B.P., °C	Wt.%A		Ref.
A =	**H₂O**	**Water** *(continued)*	**100**				
251a	C₃H₂F₆O	1,1,1,3,3,3-Hexafluoro- 2-propanol 100–300 mm.		Nonazeotrope		v-l	674c
252	C₃H₃N	Acrylonitrile	77.2	70.6	14.3		981
		"		70	13		959
		"	77.3	71	12		215
		"			12.7	v-l	997c
		" 200 mm.		33	9.0	v-l	905i
		" 400 mm.		52.1	10.5	v-l	905i
		"	77.3	70.9	12.0	v-l	905i
253	C₃H₃NS	Thiazole 695.5 mm.		90	34.8	v-l	653
		" 750 mm.	111.5	92.1	35.3	v-l	653
254	C₃H₄O	Acrolein	52.8	52.4	2.6		981
		" 200 mm.		18.35	1.28	v-l	469
		" "		18.6	1.02	v-l	905i
		" 400 mm.		34.4	1.74	v-l	905i
		"	52.5	52.3	2.48	v-l	905i
255	C₃H₄O	Propargyl alcohol	113	96	65		172
		"	115	97	54.5		264
		"				v-l	885
256	C₃H₄O₂	Acrylic acid 90 mm.	84	Nonazeotrope		v-l	157
		"	141.2	Nonazeotrope			981
		" 750 mm.		99.85	99.6	v-l	301c
		"		% acid increases with pressure			301c
257	C₃H₄O₃	Ethylene carbonate		Nonazeotrope			981
258	C₃H₅Cl	3-Chloropropene	44.9	43.0	2.2		981
259	C₃H₅Cl	Methylvinyl chloride		33	0.9		981
260	C₃H₅ClO	1-Chloro-2-propanone	121	Min. b.p.			723
261	C₃H₅ClO	α-Chloropropionaldehyde	86	80.5-81			712
262	C₃H₅ClO	Epichlorohydrin	115.2	88.5	26		980
		"	117	88	25		265
263	C₃H₅ClO₂	Methyl chloroacetate	131.4	92.7	36.15		121
264	C₃H₅I	3-Iodopropene	102.0	80.7	10		563
265	C₃H₅N	Propionitrile	97	81.5-83	24		959
266	C₃H₅NO	Hydracrylonitrile	229.7	Nonazeotrope			981
267	C₃H₆Cl₂	1,2-Dichloropropane	97	78	12		333
268	C₃H₆Cl₂O	2,3-Dichloro-1-propanol	183.8	99.4	87		981
269	C₃H₆O	Acetone	56.1	Nonazeotrope		v-l	726
		" 50 p.s.i.a.		Nonazeotrope		v-l	726
		" 100 p.s.i.a.		125.4	5.2	v-l	726
		" 200 p.s.i.a.		157.6	7.2	v-l	726
		" 250 p.s.i.a.		168.4	9.4	v-l	726
		" 500 p.s.i.a.		206.0	14.3	v-l	726
		" 0-35 p.s.i.g.		Nonazeotrope			371,728,
		" 85 p.s.i.g.	125.1	124.1	3		803,854
		" 0-185 p.s.i.g.				v-l	

TABLE I.   *Binary Systems*   17

No.	Formula	B-Component Name	B.P., °C	Azeotropic Data B.P., °C	Wt.% A		Ref.
A =	H₂O	Water *(continued)*	100				
270	C₃H₆O	Allyl alcohol	96.90	88.89	27.7	v-l	358,875, 1004,1024
	"	"		89.14	29.2	v-l	338
	"	" 752 mm.	96.90			v-l	364
271	C₃H₆O	Propionaldehyde	47.9	47.5	2		981
		"		47.5	2.5		227
272	C₃H₆O	Propylene oxide	35	Nonazeotrope		v-l	183
		"	34.1	33.8	1.0		575
		"	35	Nonazeotrope			215
		" 60 p.s.i.g.	88	86.5	0.2	v-l	980
		" 30 p.s.i.g.	69	69	0.1	v-l	980
		"	34	Nonazeotrope		v-l	980
273	C₃H₆O₂	1,3-Dioxolane	75	70-73	6.7		344
		"	75.6	71.9	7		344,981, 148a
274	C₃H₆O₂	Ethyl formate	54.2	52.6	5		981
		"	54.1	Nonazeotrope			359
275	C₃H₆O₂	Methoxyacetaldehyde, 770 mm.	92.3	88.8	20		216
		" 770 mm.	92	88.5	12.8		217
276	C₃H₆O₂	Methyl acetate, <10 p.s.i.a		Nonazeotrope			366
		"	56.3	56.1	5		981
		" 265 mm.	30	30	1.5		981
		"	57	56.4	3.2-3.7		303
		"	57	Nonazeotrope			359
		"	57	56.5		v-l	633
		"		55.8	2.6		36g
277	C₃H₆O₂	Propionic acid	141.4	99.1	82.2	v-l	483,722, 563
		"	141.1	99.2	83.7	v-l	26,190
		"		99.9	82.3	v-l	812
		" 400 mm.		82.9	83.9	v-l	812
		" 200 mm.		66.3	85.5	v-l	812
		" 100 mm.		Nonazeotrope		v-l	812
		" 149.8 mm.		60.0	85	v-l	80
		" 92.6 mm.		50.0	90.0	v-l	80
		" 55.3 mm.		40.0	94.5	v-l	80
278	C₃H₆O₃	Methyl carbonate	90.25	77.5	11		575
279	C₃H₆O₃	Trioxane	114.5	91.4	30		1003
		" 754.5 mm.	113.82	91.32	30.1	v-l	867,225c
279a	C₃H₇Br	1-Bromopropane	71.0	63.2	4.0		498i
280	C₃H₇Cl	1-Chloropropane	46.6	44	2.2		981
		"	46.4	43.4	1.0		215
281	C₃H₇Cl	2-Chloropropane	36.5	33.6	1.2		215
		"	36.5	35.0	1		981

No.	Formula	B-Component Name	B.P., °C	B.P., °C	Wt.%A		Ref.
A =	H₂O	Water *(continued)*	100				
282	C₃H₇ClO	1-Chloro-2-propanol	127		49		130
		"	127.4	95.4	45.8		128
		743 mm.		96	50.9		462
283	C₃H₇ClO	2-Chloro-1-propanol	133.7	96	50.9		575
284	C₃H₇N	Allylamine	52.9	Nonazeotrope			878
285	C₃H₇NO	Dimethylformamide	153	Nonazeotrope		v-l	207
		" 50-760 mm.		Nonazeotrope		v-l	655
		" 500 mm.	138	Nonazeotrope			981
		" 200-760 mm.		Nonazeotrope		v-l	934
286	C₃H₇NO	Propionamide	222.1	Nonazeotrope			535
287	C₃H₇NO₂	Isopropyl nitrite	40.1	Nonazeotrope			575
288	C₃H₇NO₂	Propyl nitrite	47.75	Nonazeotrope			575
289	C₃H₇NO₂	1-Nitropropane	131.18	91.63	36.5		168,806
290	C₃H₇NO₂	2-Nitropropane	120.25	88.55	29.4		168,806
291	C₃H₇NO₃	Propyl nitrate	110.5	84.8	25		538,560
292	C₃H₈O	Isopropyl alcohol	82-82.3	80.3	12.6	v-l	164,525,
							834,857
							1046
		" 67 mm.		30	12	v-l	979c
		" 153 mm.		45	12	v-l	979c
		" 319 mm.		60	12	v-l	979c
		" 95 mm.		36	13	v-l	1027
		" 190 mm.		49.33	12.8	v-l	1027
		" 380 mm.		63.90	12.6	v-l	1027
		" 760 mm.	82.5	80.10	12.0	v-l	1027,
							1042g
		" 3087 mm.		120.45	11.7	v-l	1027
		" 150°-300° C.				v-l	42
		"		Evaporation data			577
				Effect of dissolved			
				salt		v-l	764
		"		80.3			1042
293	C₃H₈O	Propyl alcohol,					
		" 47 mm.			31.8		322
		" 200 mm.		56.68	29.6	v-l	674,890
		" 400 mm.		71.92	29.0	v-l	890
		" 600 mm.		81.68	28.5	v-l	890
		"		87.65	28.3	v-l	890
		"		In 1.5M CaCl₂			
				Solution		v-l	209
		" 740 mm.	97.3	87	28.3		584
		" 1790 mm.		110	27.8		584
		" 2830 mm.		124	27.5		584
		" 3860 mm.		135	27.2		584
		" 5930 mm.		151	26.7		584
		"	97.3	87.76	29.1,	v-l	285,483,
							760,1046

TABLE I. *Binary Systems*   19

		B-Component		Azeotropic	Data	
No.	Formula	Name	B.P., °C	B.P., °C	Wt.%A	Ref.
A =	H₂O	**Water** *(continued)*	**100**			
294	C₃H₈O₂	2-Methoxyethanol	124.5	99.9	77.8	129,556
		" 100 mm.		51.5	80.5 v-l	421
		" 752 mm.		99.2	81 v-l	421
		" 150 mm.	79.2	Nonazeotrope		981
		" 760 mm.	124.6	99.9	84.7	981
		" 100 p.s.i.g.	212	169	73.3	981
		" 750 mm.		99.5	80.0	311
		" 268 mm.		71	88	311
		" 212 mm.		67.6	91	311
		" 133 mm.		57.2	94	311
		" 60 mm.		42	99	311
		" 40 mm.		Nonazeotrope		311
295	C₃H₈O₂	Dimethoxymethane	42.3	41.9	0.65	581c
			42.3	42.05	1.4	324
			42.25	Nonazeotrope		563
296	C₃H₈O₂	1,2-Propanediol	187.8	Nonazeotrope		575
		1,2-Propanediol	188	Nonazeotrope	v-l	183
297	C₃H₈O₂	1,3-Propanediol	214.8	Nonazeotrope	v-l	981
298	C₃H₉N	Propylamine	47.8	Nonazeotrope		981
299	C₃H₉N	Trimethylamine	3.2	75.5	10 v-l	414
		" 0°–100°C.			v-l	414
		"	3.5	Nonazeotrope		575
300	C₃H₉NO	1-Amino-2-propanol	159.9	Nonazeotrope		981
301	C₃H₁₀N₂	1,2-Propanediamine	119.7	Nonazeotrope		128
302	C₄HF₇O₂	Perfluorobutyric acid	122.0	97	71	659
303	C₄H₄N₂	Pyrazine	114-115	95.5	40	751
304	C₄H₄O	1-Butyn-3-one	85	74	35	902
305	C₄H₄O	Furan	31.7	Nonazeotrope		575
306	C₄H₄S	Thiophene	84	Min. b.p.		1017
307	C₄H₅N	3-Butenenitrile	118.9	89.4	34	981
308	C₄H₅N	*cis*- and *trans*-crotononitrile	107.5-120.5	85		136
309	C₄H₅N	Methacrylonitrile		76.5	16	767
310	C₄H₅N	Pyrrol	129.8	93-93.5		39
311	C₄H₆ClN	2-Chloro-2-methyl-propionitrile	116	87	22	767
312	C₄H₆O	1-Butene-3-one				
		" 743 mm.		75.5	14.1 v-l	420
		" 300 mm.		51.1	14.1	420
		" 150 mm.		32	11.9 v-l	420
313	C₄H₆O	3-Butyn-1-ol	128.9	Min. b.p.		267
314	C₄H₆O	3-Butyn-2-ol	109	92		172
315	C₄H₆O	Crotonaldehyde,				
		" 111 mm.	84.9	40	19	981
		" 273 mm.	112.3	60	22	981

		B-Component		Azeotropic Data			
No.	Formula	Name	B.P., °C	B.P.. °C	Wt.%A		Ref.
A =	H₂O	Water *(continued)*	100				
		Crotonaldehyde					
		" 412 mm.	126.4	70	23		981
		"	102.4	84	24.8		266,398, 981
316	C₄H₆O	Methacrylaldehyde	68.0	63.6	7.7		767,981
317	C₄H₆O₂	Biacetyl	87-88	78.5			119
318	C₄H₆O₂	3-Butenoic acid		Nonazeotrope			981
319	C₄H₆O₂	*trans*-Crotonic acid	185	Nonazeotrope			981
320	C₄H₆O₂	Crotonic acid		99.9	97.8		227
321	C₄H₆O₂	Butyrolactone	204.3	Nonazeotrope			981
322	C₄H₆O₂	Dioxene 753 mm.	94	79.3	13.7		215
		" 200 mm.		46.5	12.3		215
323	C₄H₆O₂	Methacrylic acid	160.5	99.3	76.9	v-l	299
		" " 25 mm.		26.0	84.0		215
		" " 50 mm.		37.3	82.4		215
		" " 100 mm.		50.5	81.1		215
324	C₄H₆O₂	Methyl acrylate	80	71	7.2		800
325	C₄H₆O₂	Vinyl acetate	72.7	66	7.3		981
326	C₄H₆O₃	Propylene carbonate	242.1	Nonazeotrope			981
327	C₄H₇Cl	1-Chloro-2-methyl-1-propene	68.1	61.9	7.5		105
328	C₄H₇ClO	α-3-Chloro-2-buten-1-ol	164	98.1			367
329	C₄H₇ClO	β-3-Chloro-2-buten-1-ol	166	98.8			367
330	C₄H₇ClO	2-Chloroethyl vinyl ether	109.1	84	17		981
331	C₄H₇ClO₂	4-Chloromethyl 1,3-dioxolane, 40 mm.	67	99			860
332	C₄H₇ClO₂	Ethyl chloroacetate	143.5	95.2	45.12		121
333	C₄H₇N	Butyronitrile	118	87.5	31		959
		"	117.6	88.7	32.5		981
334	C₄H₇N	Isobutyronitrile	103	82.5	23		959
335	C₄H₇NO	2-Hydroxyisobutyro-nitrile, 30 mm.		Nonazeotrope			981
		" 50 mm.		Nonazeotrope			981
336	C₄H₈Cl₂O	Bis(2-chloroethyl) ether	179.2	98	65.5		981
337	C₄H₈Cl₂O	1,3-Dichloro-2-methyl-2-propanol	174	98.3	64.8		105
338	C₄H₈O	2-Butanone	79.6	73.41	11.3	v-l	552,633, 876,877
		" 768-1243 mm		Effect of pressure			363
		"	79.6	73.4	11.0		218
		" 3.5 p.s.i.g.		79.3	12.1		218
		" 9.2 p.s.i.g.		88.0	12.5		218
		" 30 p.s.i.g.		111	15.8		218
		" 60 p.s.i.g.		125	18.3		218
		" 14.7 p.s.i.a.		73.3	11.6	v-l	726
		" 50 p.s.i.a.		112.2	15.9		726

TABLE I.   *Binary Systems*   21

No.	Formula	B-Component Name	B.P., °C	Azeotropic Data B.P., °C	Wt.%A		Ref.
A =	H$_2$O	**Water** *(continued)*	**100**				
		2-Butanone 100 p.s.i.a.		139.0	19.3		726
		" 250 p.s.i.a.		180.7	23.4		726
		" 500 p.s.i.a.		216.1	26.4		726
		"	79.6	73.77	12.7	v-l	15
		" 600 mm.		67.11	12.1	v-l	15
		" 400 mm.		56.81	11.0	v-l	15
		" 200 mm.		40.16	8.3	v-l	15
339	C$_4$H$_8$O	1-Buten-3-ol	97.4	85	26		172
		"	96-97	Azeotropic			575
340	C$_4$H$_8$O	Butyraldehyde	74	68	6		519
		"		Evaporation behavior			577
		"	74.8	68.0	9.7		981
		"		67.8	6.7		227
341	C$_4$H$_8$O	Crotonyl alcohol	119-120		60		575
342	C$_4$H$_8$O	Ethyl vinyl ether	35.5	34.6	1.5		981
343	C$_4$H$_8$O	Isobutyraldehyde	63.5	64.3	6.7		227
		"	63.3	60.1	9.6		665
344	C$_4$H$_8$O	Methyl propenyl ether	46.3	46.3	0.5		981
345	C$_4$H$_8$O	Tetrahydrofuran					
		100 p.s.i.g.			12	v-l	224
		"	66	64	5.3	v-l	224
		Tetrahydrofuran		63.8	4.6	v-l	158
346	C$_4$H$_8$OS	2-Methylthiopropional-					
		dehyde, 85 mm.		48	64		215
		" 412 mm.		82	60		215
		" 753 mm.		97.5	68		215
		" 759 mm.		97.5	63		215
347	C$_4$H$_8$OS	1,4-Oxathiane	149.2	95.6	48		981
348	C$_4$H$_8$O$_2$	Butyric acid	163.5	99.4	97	v-l	26
		" "	162.45	99.4	81.5		} 483,545,
					80	v-l	} 722,757
349	C$_4$H$_8$O$_2$	Dioxane 50 mm.		26.9	10.5	v-l	481g
		" 100 mm.		39.9	12.1	v-l	481g
		"	101.3	87.8	17.6	v-l	481g
		"	101.32	87.82	18	v-l	128, 201,898
		"		Effect of dissolved			
				salt		v-l	641
		" 260 mm.		60	15.4		183
350	C$_4$H$_8$O$_2$	1,3-Dioxane	104-105	86.5			860
351	C$_4$H$_8$O$_2$	Ethoxyacetaldehyde	105	90	21.8		217
		"	103	90	24.4		881
352	C$_4$H$_8$O$_2$	Ethyl acetate,					
		" 25 mm.	2.51	— 1.90	3.60		}
		" 250 mm.	46.87	42.55	6.28		} 650,
		" 760 mm.	77.15	70.38	8.47		} 877,
		" 1441 mm.	97.80	89.08	9.94		} 1033

No.	Formula	B-Component Name	B.P., °C	Azeotropic Data B.P., °C	Wt.%A		Ref.
A =	H₂O	**Water** *(continued)*	**100**				
		Ethyl acetate		50		v-l	452c
		"		60		v-l	452c
		"		70		v-l	452c
		"		80		v-l	452c
353	C₄H₈O₂	2-Hydroxybutyraldehyde, 80 mm.		Nonazeotrope			981
354	C₄H₈O₂	Isobutyric acid	154.5	98.8	71.8		227
		"	154.35	99.3	79		563
355	C₄H₈O₂	Isopropyl formate	68.8	65.0	3		575
356	C₄H₈O₂	3-Methoxypropionaldehyde, 100 mm.		45	30		981
357	C₄H₈O₂	2-Methyl-1,3-dioxolane	82.5	75	8		215
358	C₄H₈O₂	Methyl propionate	79.7	71.0	8.2		227
		" "	79.85	71.4	3.9		531
359	C₄H₈O₂	Propyl formate	80.9	71.6	2.3		359, 531,804
360	C₄H₈O₂	2-Vinyloxyethanol	143	98	65		263
361	C₄H₈O₃	Methyl lactate	144.8	99	80		852
361a	C₄H₉Br	1-Bromo-2-methylpropane	91.4	75.3	7.3		498i
362	C₄H₉Cl	1-Chlorobutane	77.9	68.1	6.6		215
363	C₄H₉Cl	1-Chloro-2-methylpropane	68.8	61.6	3.3		215
364	C₄H₉ClO	1-Chloro-2-methyl-2-propanol	126.7	93-94	34		105
365	C₄H₉I	1-Iodo-2-methylpropane	122.5	95-96	21 vol.%		754
366	C₄H₉N	Methallylamine	78.7	78.4	4.1		878
367	C₄H₉NO	Morpholine	128.3	Nonazeotrope			981
368	C₄H₉NO₂	Butyl nitrite	78.2	70.0	~7		575
369	C₄H₉NO₂	Isobutyl nitrite	67.1	63.2	8		575
370	C₄H₉NO₂	N-(2-Hydroxyethyl)acetamide		Nonazeotrope			981
371	C₄H₉NO₃	Isobutyl nitrate	122.9	88.5	28		559
371a	C₄H₁₀	Butane crit. region				v-l	190c
372	C₄H₁₀O	Butyl alcohol	117.4	92.7	42.5	v-l	943,452c 128,535, 755,497, 877,896, 943
		"		Evaporation behavior			577
		" 250-2500 mm.				v-l	719
		" 30 mm.	48	28	52.4		982
		" 685 mm.		88.7	43.5	v-l	992
		"	118	92.6	42.2	v-l	380
		" 1485 mm.	139	111.2	42.2	v-l	380
		" 3690 mm.	171.2	141.8	41.8	v-l	380
		" 5900 mm.	191.8	158.8	41.8	v-l	380
		" 8110 mm.	206.1	172.3	38.6	v-l	380

TABLE I. *Binary Systems*

No.	Formula	B-Component Name	B.P., °C	B.P., °C	Wt.%A		Ref.
A =	H₂O	**Water** *(continued)*	**100**				
		Butyl alcohol 100 mm.		48	49.8		266
		" 270 mm.		70	46.6		266
		" 755 mm.		92.4	42.8		266
373	C₄H₁₀O	*sec*-Butyl alcohol	99.5	87.0	26.8	v-l	1039
		"		Evaporation behavior			577
		"	99.5	88.5	32		981
		" 20 mm.	27.3	16.0	32.2		413
		"	99.5	87.35	28.4		14
		" 600 mm.		81.50	28.8		14
		" 400 mm.		72.07	29.3		14
		" 200 mm.		56.95	29.5		14
		" 60-80°C.				v-l	14
		"	99.4	87.5	27.3		164,877
		"		87.5	26	v-l	109
374	C₄H₁₀O	*tert*-Butyl alcohol 100 mm.			9.4	v-l	771c
		"	82.5	79.9	11.76		877,1046
		" 8120 mm.				v-l	771c
		"		79.85	14.0		581c
		"	82.9			v-l	937c
375	C₄H₁₀O	Ethyl ether	34.5	34.15	1.26		877
		" 11 atm		114	4.5		775
		" 20 p.s.i.g.	62	60	2.0		981
376	C₄H₁₀O	Isobutyl alcohol, 745 mm.				v-l	982
		"	107.0	89.8	33.0	v-l	483,776, 921,1046
		100-130°C.		Effect of pressure			117
377	C₄H₁₀O	Methyl propyl ether	38.9	~ 38.7	~ 2		563
378	C₄H₁₀O₂	*l*-2,3-Butanediol, 14-75 p.s.i.g.		Nonazeotrope		v-l	965
379	C₄H₁₀O₂	*meso*-2,3-Butanediol 200-760 mm.	183-184	Nonazeotrope		v-l	731
380	C₄H₁₀O₂	1,1-Dimethoxyethane	64.3	61.3	3.6		45
381	C₄H₁₀O₂	1,2-Dimethoxyethane	83	76	10.5		129,417
		"	35		6		23
		"	85.2		10.4		23
		"	85	77.4	10.1		215
382	C₄H₁₀O₂	2-Ethoxyethanol, <100 mm.		Nonazeotrope			71
		" 200 mm.		66.4	70	v-l	71
		" 400 mm.		82.4	79	v-l	71
		"	134	98.2	87	v-l	71
		" 200 mm.	96.5	66.4	85		981
		" 400 mm.	115.6	82.4	76		981
		"	135.6	99.4	71.2		981
		"	135.1	99.4	71.2		129,526
		"			70.0	v-l	34,210

No.	Formula	B-Component Name	B.P., °C	Azeotropic Data B.P., °C	Wt.%A		Ref.
A =	H₂O	**Water** *(continued)*	100				
383	C₄H₁₀O₂	Ethoxymethoxymethane	65.91	61.25	4.4		1035
384	C₄H₁₀O₂	1-Methoxy-2-propanol	118	96	~ 48.5		215
		" "	118	97.5	35		797
385	C₄H₁₀O₂	2-Methoxy-1-propanol	130	98	67		797
386	C₄H₁₀O₃	Diethylene glycol	245.5	Nonazeotrope			556
		" "		Nonazeotrope		v-l	183
		" " 10 mm.		Nonazeotrope		v-l	183
387	C₄H₁₁N	Butylamine, 575 mm.	69	69	1.3		981
		" 20 p.s.i.g.	106		6.5		981
		"	77.8	Nonazeotrope		v-l	481c
388	C₄H₁₁N	Diethylamine		Effect of NaOH		v-l	416
		"	55.5	Nonazeotrope		v-l	415
389	C₄H₁₁N	Isobutylamine	68	Nonazeotrope		v-l	787
390	C₄H₁₁NO	2-Dimethylamino-ethanol, 27 p.s.i.g.	174		90.2		981
		" 744 mm.	133.9	99	92.6		981
		" 540 mm.	123.4	91	95.2		981
		" 250 mm.	100.7	71	98.2		981
391	C₄H₁₁NO	3-Methoxypropylamine	116		~95		17
392	C₄H₁₁NO₂	2,2'-Iminodiethanol	268	Nonazeotrope			981
393	C₄H₁₂N₂	Tetramethylhydrazine	74.3	69	3		401
394	C₅H₄O₂	Furfural	161.7			v-l	972
		" 600 mm.		91.3	65	v-l	972
		" 300 mm.		74.1	65.5	v-l	972
		" 150 mm.		58.2	66.6	v-l	972
		" 55 mm.		40.1	68.6	v-l	972
		"	161.45	97.85	65		556
		" 100-200°F.				v-l	743
		" 1-18 atm.	161.7			v-l	648
395	C₅H₅N	Pyridine 230 mm.		62	45		215
		" 490 mm.		79	45		215
		"	115	94	43		44,553
		"	115.3	93.6	41.3		413
		" 120 mm.			46.2		322
		" 758 mm.			40.5		322
		" <760 mm.		30	40.7	v-l	412
		" <760 mm.		50	—	v-l	412
		" <760 mm.		80	40.7		412,1058
		"		Effect of dissolved salt			
						v-l	764
396	C₅H₆N₂	2-Methylpyrazine,	133	97	55		1036
		" 737 mm.	130	92.6	36 vol. %		789
397	C₅H₆O	2-Methylfuran	63.7	58.2	—		769
		" 740 mm.	62.7	57.3	3.4	v-l	897
398	C₅H₆O₂	Furfuryl alcohol	169.35	98.5	80		545
399	C₅H₇N	3-Methyl-3-butene-nitrile	137.0	93.0	43.2		981

TABLE I. *Binary Systems*   25

No.	Formula	B-Component Name	B.P., °C	Azeotropic Data B.P., °C	Wt.%A		Ref.
A =	H₂O	Water *(continued)*	100				
400	C₅H₇NO	Furfurylamine	144	99	74		920
400a	C₅H₈	Isoprene	34.1	32.4	0.14		581c
401	C₅H₈O	Allyl vinyl ether	67.4	60	5.4		981
402	C₅H₈O	Cyclopentanone	130.8	94.6	42.4	v-l	981
		" 740 mm.	130	92.6	36 vol. %		789
403	C₅H₈O	1-Methoxy-1,3-buta- diene	90.7	76.2	12.7		981
404	C₅H₈O	3-Methyl-3-butene- 2-one	97.9	81.5	18.4		981
		" 735 mm.	98.5	82	—		79
		" 100 mm.	45-46	34-35	—		79
405	C₅H₈O	2-Methyl-3-butyne-2-ol, 100 mm.		Min. b.p.			895
		"	103	90	26		172,900
		"	104.4	91.0	29	v-l	169
406	C₅H₈O	4-Pentenal	106	84.3	21		981
407	C₅H₈O	3-Penten-2-one	123.5	92	28.6		981
408	C₅H₈O₂	Allyl acetate	104.1	83	16.7		981
		" "	105	Azeotropic			723
409	C₅H₈O₂	Ethyl acrylate	100	98.3	—		804
		" "	99.5	81.1	15		807
		" 195 mm.	61	48	12		981
410	C₅H₈O₂	Isopropenyl acetate, 200 mm.	60.2	48	11		981
		"	97.4	79.3	13.4		981
411	C₅H₈O₂	Methyl methacrylate	100.8	83	14		807
		" " 200 mm.		49	11.6		1032
		" "	99.5	86-92	—		1032
412	C₅H₈O₂	2,3-Pentanedione	109	86	—		119
413	C₅H₈O₂	2,4-Pentanedione	140.6	94.4	41		981
414	C₅H₈O₂	Δ-Valerolactone		Nonazeotrope			981
415	C₅H₈O₂	Vinyl propionate	95.0	79	13		981
416	C₅H₉ClO₂	Propyl chloroacetate	162.3	97.1	57.5		121
416a	C₅H₉NO	N-Methylpyrrolidone 200 mm.				v-l	742c
		" 760 mm.				v-l	742c
417	C₅H₁₀Cl₂O₂	Bis(2-chloroethoxy) methane	218.1	99.4	86.8		981
418	C₅H₁₀N₂	3-Dimethylamino- propionitrile	174.5	99.6	84		981
419	C₅H₁₀O	*cis*-1-Butenyl methyl ether	72.0	64	6.1		981
420	C₅H₁₀O	*trans*-1-Butenyl methyl ether	76.7	67	7.2		981
421	C₅H₁₀O	Cyclopentanol	140.85	96.25	58		575
422	C₅H₁₀O	Isopropenyl ethyl ether	61.9	58	2		981

No.	Formula	B-Component Name	B.P., °C	Azeotropic Data B.P., °C	Wt.%A		Ref.
A =	H₂O	Water *(continued)*	100				
423	C₅H₁₀O	Isopropyl vinyl ether	55.7	51.8	2.7		981
424	C₅H₁₀O	Isovaleraldehyde	—		17.9		665
		"	92.5	77	12		215
424a	C₅H₁₀O	3-Methyl-2-buten-1-ol	140	96.4	69.0		581c
424b	C₅H₁₀O	3-Methyl-3-buten-1-ol	130.0	95.0	57.9		581c
424c	C₅H₁₀O	2-Methyl-3-buten-2-ol	97.0	86.5	23.5		581c
425	C₅H₁₀O	2-Methyltetrahydrofuran	77	Min. b.p.			350
426	C₅H₁₀O	3-Methyl-2-butanone	94	~ 79	~13		563
427	C₅H₁₀O	2-Pentanone	102.3	83.3	19.5		552,877
428	C₅H₁₀O	3-Pentanone	102.05	82.9	14		552
		"		82.9	18.8		497
429	C₅H₁₀O	Propyl vinyl ether	65.1	59	5		981
430	C₅H₁₀O	Tetrahydropyran	88	71	8.5		223
		"		Min. b.p.			81
431	C₅H₁₀O	Valeraldehyde	103.3	83	19		981
432	C₅H₁₀O	Valeraldehydes (isomers)	98.6	80	17		981
433	C₅H₁₀O₂	Butyl formate	106.6	83.8	16.5		359,538 498c
434	C₅H₁₀O₂	4,5-Dimethyl-1,3-dioxolane		Min. b.p.			860
435	C₅H₁₀O₂	3-Ethoxy-1,2-epoxypro-pane	124-126	90-91			265
436	C₅H₁₀O₂	Ethyl propionate	99.15	81.2	10		531,804
		" 350 mm.	76.0	61	13.3		981
437	C₅H₁₀O₂	3-Hydroxy-3-methyl-2-butanone	141.0	98.6	61.0	v-l	169
438	C₅H₁₀O₂	Isobutyl formate	98.3	79.5	18.9		359,1034
		"	98.3	79.6	12.4		498i
439	C₅H₁₀O₂	Isopropyl acetate	88.6	76.6	10.6		531,877
		"		76.6	10.4		497
440	C₅H₁₀O₂	Isovaleric acid	176.5	99.5	81.6		563
441	C₅H₁₀O₂	3-Methoxybutyraldehyde,					
		" 100 mm.		50	37		981
		" 200 mm.		64	37		981
		"	131	>92	35		981
442	C₅H₁₀O₂	Methyl butyrate	102.65	82.7	11.5		531
443	C₅H₁₀O₂	Methyl isobutyrate	92.3	77.7	6.8		531
444	C₅H₁₀O₂	Propyl acetate	101.6	82.4	14		333,359
		" 200-700 mm.				v-l	891
445	C₅H₁₀O₂	Tetrahydrofurfuryl alcohol				v-l	197
446	C₅H₁₀O₂	Valeric acid	185.5	99.8	89		981
447	C₅H₁₀O₂	Valeric acid (isomers)	183.2	99.6	85		981
448	C₅H₁₀O₂	1-Vinyloxy-2-propanol		~100	75		263
449	C₅H₁₀O₂	3-Vinyloxy-1-propanol		~100	75		263
450	C₅H₁₀O₃	β-Ethoxypropionic acid	219.2	Nonazeotrope			981
451	C₅H₁₀O₃	γ-Methoxybutyric acid		Nonazeotrope			981

TABLE I.   *Binary Systems*   27

No.	Formula	B-Component Name	B.P., °C	B.P., °C	Wt.%A	Ref.
A =	H₂O	Water *(continued)*	100	Azeotropic Data		
452	C₅H₁₀O₃	Ethyl carbonate	126.5	91	30	575
453	C₅H₁₀O₃	2-Methoxyethyl acetate	144.6	97.0	51.5	526
		" 45 mm.		58 v-l		920c
454	C₅H₁₀O₃	Methoxymethyl propionate		95	56	981
455	C₅H₁₀O₃	Methyl β-methoxypropionate, 100 mm.	84	Azeotropic		87
456	C₅H₁₁Cl	1-Chloropentane	108.35	82		409,723
457	C₅H₁₁N	Piperidine	105.8	92.8	35	914
458	C₅H₁₁NO	4-Methylmorpholine	115.6	94.2	24	981
459	C₅H₁₁NO	Tetrahydrofurfurylamine	153	Nonazeotrope		920
460	C₅H₁₁NO₂	Isoamyl nitrite	97.15	<80.6	<15	575
461	C₅H₁₁NO₃	Isoamyl nitrate	149.75	95.0	40	560
462	C₅H₁₂	Pentane	36.1	34.6	1.4	981
463	C₅H₁₂N₂	1-Methylpiperazine	138.0	Nonazeotrope		981
464	C₅H₁₂O	*n*-Amyl alcohol	137.8	95.8	54.4	324,359, 545,633, 755
465	C₅H₁₂O	*tert*-Amyl alcohol	102.25	87.35	27.5	545
466	C₅H₁₂O	*tert*-Butyl methyl ether	55	52.6	4	256
		"	55.0	52.1	2.1	581c
467	C₅H₁₂O	Ethyl propyl ether	63.6	59.5	4	538
468	C₅H₁₂O	Isoamyl alcohol	132.05	95.15	49.60	760,1046
469	C₅H₁₂O	3-Methyl-2-butanol	112.9	91.0	33	576
470	C₅H₁₂O	2-Pentanol	119.3	91.7	36.5	545,877
471	C₅H₁₂O	3-Pentanol	115.4	91.7	36.0	545
472	C₅H₁₂O₂	1,1-Diethoxymethane	87.5	75.2	10	324,668 970
473	C₅H₁₂O₂	1,2-Dimethoxypropane	92-93	80		417
		"	92	80	11	981
474	C₅H₁₂O₂	1-Ethoxy-2-propanol	132.2	97.3	50.1	981
475	C₅H₁₂O₂	3-Methoxy-1-butanol	161.1	98.5	80	981
475a	C₅H₁₂O₂	3-Methyl-1,3-butanediol	203	Nonazeotrope		581c
476	C₅H₁₂O₂	1,5-Pentanediol	242.5	Nonazeotrope		981
477	C₅H₁₂O₂	2-Propoxyethanol	151.5	98.8	70	981
		"	151.35	98.75	72	556
478	C₅H₁₂O₃	2-(2-Methoxyethoxy)ethanol	192.95	Nonazeotrope		556
479	C₅H₁₂O₃	1,1,2-Trimethoxyethane	126	93	30	346
480	C₅H₁₃N	*N*-Methylbutylamine	91.1	82.7	15	981
481	C₅H₁₃NO	1-Ethylamino-2-propanol	159.4	Nonazeotrope		981
482	C₅H₁₃NO	3-Ethoxypropylamine		80		17
483	C₅H₁₄N₂	*N,N*-Dimethyl-1,3-propanediamine	134.9	Nonazeotrope		981
484	C₆H₅Cl	Chlorobenzene	131.8	90.2	28.4	762
484a	C₆H₅ClO	*o*-Chlorophenol				
		120 mm.		54	72	215
		" 750 mm.		97.5	67	215
485	C₆H₅NO₂	Nitrobenzene	210.85	98.6	88 vol.%	689

No.	Formula	B-Component Name	B.P., °C	B.P., °C	Azeotropic Data Wt.%A		Ref.
A =	H₂O	Water (continued)	100				
486	C₆H₆	Benzene 25–400 p.s.i.a				v-l	103c
		"	80.2	69.25	8.83		689,877, 1043
		" 537.0 mm.		60	8.35		979
		" 295.0 mm.		45	6.36		979
		" 150.5 mm.		30	5.98		979
		"		69.25	8.95		497
		"		69.25			1042
487	C₆H₆O	Phenol <24°C		Nonazeotrope			25
		"		160	89.1		25
		"		275	94.4		25
		" 200 mm.		66.4	96.5		472
		"		15		v-l	627
		" 127 mm.		56.3	94.5		853
		" 294 mm.		75.0	92.8		853
		" 531 mm.		90.0	91.71		853
			182	99.52	90.79		762,877
488	C₆H₇N	Aniline		41	86.6	v-l	853
				56.3	84	v-l	853
		" 742 mm.		98.6	80.8	v-l	403
		" 6 atm.		155	76.6	v-l	403
		" 11 atm.		182	76.2	v-l	403
		" 16.4 atm.		200	77.4	v-l	403
		"		75	81.8	v-l	853
				90	80.5	v-l	853
489	C₆H₇N	2-Picoline	129.5	93.5	48		39,43
490	C₆H₇N	3-Picoline, 700 mm.		94.1	61.4		39,141, 175,585
		"	144.1	97	60	v-l	1063
		"		96.7			160c,585
491	C₆H₇N	4-Picoline	144.3	97.35	62.8	v-l	1063
				97.4	63.5		981
		" 700 mm.		94.6	63.5		39,141, 175,585
		"		97.4	63.5		981,160c
492	C₆H₈	1,4-Cyclohexadiene	85.6	71.3			575
493	C₆H₈	1,3-Cyclohexadiene	80.8	68.9	9		563
494	C₆H₈N₂	2,5-Dimethylpyrazine	154	98.5	65		1036
495	C₆H₈N₂	2-Ethylpyrazine		98-99	>0.18		215
496	C₆H₈N₂	Phenylhydrazine	243	Nonazeotrope			563
497	C₆H₈O	2,5-Dimethylfuran	93.3	77.0	11.7		981
498	C₆H₈O	2,4-Hexadienal	171	98.0	70		981
499	C₆H₈O₂	1,3-Butadienyl acetate	138.5	93	35.6		981
500	C₆H₈O₂	Vinyl crotonate	133.9	92	31		981
		"	132.7	91.0	24.2		873
501	C₆H₈O₄	Methyl fumarate	193.25	98.85	74.5		575
502	C₆H₉N₃	3,3'-Iminodipropionitrile		Nonazeotrope			981

TABLE I.   *Binary Systems*   29

No.	Formula	B-Component Name	B.P., °C	Azeotropic Data B.P., °C	Wt.%A		Ref.
A =	H₂O	Water *(continued)*	100				
503	C₆H₁₀	Cyclohexene	82.75	70.8	10		563
		"		70.8	8.93		497
504	C₆H₁₀	2-Ethyl-1,3-butadiene	66.9	60.2	5.3		981
505	C₆H₁₀	4-Methyl-1,3-pentadiene		67.0	7.5		849
506	C₆H₁₀O	Cyclohexa-none, <760 mm.		90	—	v-l	334
		"	155.6	96.3	55	v-l	334
		"	155.4	95	61.6		981
507	C₆H₁₀O	2-Ethylcrotonaldehyde	135.3	92.7	38		981
508	C₆H₁₀O	2-Hexenal	149	95.1	48.6		981
509	C₆H₁₀O	1-Hexen-5-one	129	Min. b.p.			723
		"	128.9	92.1	35.3		981
510	C₆H₁₀O	Mesityl oxide	129.5	91.8	34.8		723,877
		"	128	91.3	29 vol%		724
510a	C₆H₁₀O	Methyldihydropyran	118.5	87.6	20.5		581c
		" <760 mm.		60	18.4	v-l	581g
511	C₆H₁₀O	2-Methyl-2-pentenal	138.2	93.5	40		981
512	C₆H₁₀O₂	Crotonyl acetate	129	Min. b.p.			723
513	C₆H₁₀O₂	Ethyl crotonate	137.8	93.5	38		981
513a	C₆H₁₀O₂	4-Vinyl-1,3-dioxane	144.9	94.5	57.5		581c
514	C₆H₁₀O₂	Vinyl butyrate	116.7	87.2	20.4		981
515	C₆H₁₀O₂	Vinyl isobutyrate	105.4	83.5	17		981
516	C₆H₁₀O₄	Ethylene glycol diacetate	190.8	99.7	84.6		981
517	C₆H₁₁ClO₂	Butyl chloroacetate	181.9	98.12	75.49		121
518	C₆H₁₁ClO₂	Isobutyl chloroacetate	174.4	97.8	64.18		121
519	C₆H₁₁N	Diallylamine	110.4	87	22-23		878
			110.5	87.2	24		981
520	C₆H₁₁NO	Caprolactam 30 mm.		Nonazeotrope		v-l	617c
		" 80 mm.		Nonazeotrope		v-l	617c
		" 180 mm.		Nonazeotrope		v-l	617c
		" 760 mm.		Nonazeotrope		v-l	617c,905r
		" 50-760 mm.				v-l	976
521	C₆H₁₁NO₃	2-Methyl-2-nitropropyl vinyl ether, 10 mm.	77-78	—	8.6		1008
522	C₆H₁₂	Cyclohexane	80.8	69.5	8.4		413
		"	80.75	68.95	9		563
		" 25-400 p.s.i.a.				v-l	103c
523	C₆H₁₂	1-Hexene	63.84	57.7	5.7		806
524	C₆H₁₂	4-Methyl-2-pentene	56.7	53.5	3.5		981
525	C₆H₁₂Cl₂O	Bis(chloroisopropyl) ether	187.0	98.5	62.6		981
526	C₆H₁₂Cl₂O₂	1,2-Bis(2-chloro-ethoxy)ethane	240.9	99.7	94.0		981
527	C₆H₁₂O	Butyl vinyl ether	94.2	77.5	11.6		981
528	C₆H₁₂O	Cyclohexanol, 42 mm.		35	86		1072

| | | B-Component | | Azeotropic | Data | |
No.	Formula	Name	B.P., °C	B.P., °C	Wt.%A	Ref.
A =	H₂O	**Water** *(continued)*	**100**			
		Cyclohexanol, 57 mm.		40	84.8	1072
		" 95 mm.		50	82.5	1072
		" 158 mm.		60	80.2	1072
		" 252 mm.		70	77.8	1072
		" 385 mm.		80	75.2	1072
		" 570 mm.		90	72.6	1072
		" 684 mm.		95	70.7	1072
		"	160.65	97.8	69.5	1072
		" <760 mm.		90	74 v-l	335
		"	160.65	~ 97.8	~80	563
529	C₆H₁₂O	2,2-Dimethyltetrahy-drofuran	90	Min. b.p.		350
530	C₆H₁₂O	2,5-Dimethyltetrahy-drofuran	90	78	13	172
531	C₆H₁₂O	2-Ethylbutyraldehyde	116.7	87.5	23.7	981
532	C₆H₁₂O	Hexaldehyde	128.3	91.0	31.3	981
533	C₆H₁₂O	2-Hexanone	127	90.5	26 vol.%	552,723, 724,877
534	C₆H₁₂O	3-Hexanone	124	Min. b.p.		723
535	C₆H₁₂O	Isobutyl vinyl ether	83.4	70.5	7.8	981
536	C₆H₁₂O	2-Methylpentanal	118.3	88.5	23	981
537	C₆H₁₂O	4-Methyl-2-pentanone	115.9	87.9	24.3	877
538	C₆H₁₂O	2-Methyl-2-pentene-4-ol		94.6	40.8	849
539	C₆H₁₂O	Pinacolone	106	~ 85	~14.5	563
540	C₆H₁₂OS	2-Ethylthioethyl vinyl ether	169.7	97.8	61	981
541	C₆H₁₂O₂	Amyl formate	132	91.6	28.4	359,723
542	C₆H₁₂O₂	Butyl acetate	126.2	90.2	28.7	359,723,877
543	C₆H₁₂O₂	*sec*-Butyl acetate	112.4	87	22.5	723,877
543a	C₆H₁₂O₂	4,4-Dimethyl-1,3-dioxane	133.4	92.8	35	581c
		" <760 mm.		60	28.2 v-l	581g
544	C₆H₁₂O₂	Ethyl butyrate	120.1	87.9	21.5	531
545	C₆H₁₂O₂	2-Ethylbutyric acid	194.2	99.7	87	981
546	C₆H₁₂O₂	Ethyl isobutyrate	110.1	85.2	15.2	531
547	C₆H₁₂O₂	2-Ethyl-2-methyl-1,3-dioxolane	117.6	88.5	20	981
548	C₆H₁₂O₂	Hexanoic acid	205.7	99.8	92.1	981
549	C₆H₁₂O₂	4-Hydroxy-4-methyl-2-pentanone 50 mm.		Nonazeotrope	v-l	352
		" 100 mm.		Nonazeotrope	v-l	352
		" 200 mm.	123.5	66.4	97 v-l	352
		" 400 mm.	143	82.6	90 v-l	352
		" 760 mm.	161	99.5	85 v-l	352
		"	169.2	99.6	87	982
		"	166	98.8	87.3	128,877
550	C₆H₁₂O₂	Isoamyl formate	124.2	90.2	21	359,531,723
551	C₆H₁₂O₂	Isobutyl acetate	117.2	87.4	16.5	359,498c, 531,723
552	C₆H₁₂O₂	Isopropyl propionate	110.3	85.2	19.9	575

TABLE I. *Binary Systems*   31

No.	Formula	Name	B.P., °C	B.P., °C	Wt.%A	Ref.
		B-Component		**Azeotropic Data**		
A =	H₂O	**Water** *(continued)*	**100**			
552a	C₆H₁₂O₂	4-Methyl-4-hydroxy-tetrahydropyran	188	Nonazeotrope		581c
553	C₆H₁₂O₂	Methyl isovalerate	116.3	87.2	19.2	531
554	C₆H₁₂O₂	2-Methylpentanoic acid	196.4	99.4	87.9	981
555	C₆H₁₂O₂	Propyl propionate	122.1	88.9	23	531,723
556	C₆H₁₂O₂	4-Vinyloxy-1-butanol		Min. b.p.		263
557	C₆H₁₂O₂	Tetrahydropyran-2-methanol	187.2	Nonazeotrope		981
558	C₆H₁₂O₃	2,2-Dimethoxy-3-butanone	145	94		119
559	C₆H₁₂O₃	2-Ethoxyethyl acetate	156.8	97.4	45	526
		" "	156.2	97.5	55.6	981
560	C₆H₁₂O₃	Methyl 3-ethoxy-propionate, 50 mm.		37	50	981
561	C₆H₁₂O₃	Paraldehyde	124	90	28.5	563,975
562	C₆H₁₂O₃	Trioxane	114.5	91.4	30	1003
563	C₆H₁₂O₃	2-(2-Vinyloxyethoxy)ethanol		~ 100	97-8	263
		"	207.6	Nonazeotrope		981
564	C₆H₁₃Cl	1-Chlorohexane	134.5	91.8	29.7	981
565	C₆H₁₃N	Cyclohexylamine,				
		" 40 mm.	51.4	31.7	69.0	139
		" 70 mm.		41.9	66.0	139
		" 100 mm.	72	49.0	64.1	139
		" 200 mm.	90.9	63.6	60.7	139
		" 300 mm.	102.5	72.7	59.1	139
		" 500 mm.	118.9	85.3	57.0	139
		" 760 mm.	134.5	96.4	55.8	139
		"			v-l	718c
566	C₆H₁₃N	Hexamethyleneimine	138	95.5	49.5	222,224
		"	134.5	96.5	55.78	413
567	C₆H₁₃NO	2,6-Dimethyl-morpholine	146.6	99.6	70	981
568	C₆H₁₃NO	4-Ethylmorpholine	138.3	96.7	46.2	981
569	C₆H₁₃NO₂	4-Morpholineethanol	225.5	Nonazeotrope		981
570	C₆H₁₄	Hexane	68.7	61.6	5.6	563,981
		" 25–400 p.s.i.a			v-l	103c
		"		61.5	5.4	623c
571	C₆H₁₄N₂	2,5-Dimethylpiperazine	164	Nonazeotrope		981
572	C₆H₁₄N₂O	4-(2-Aminoethyl)morpholine	204.7	Nonazeotrope		981
573	C₆H₁₄N₂O	1-Piperazineethanol	246.3	Nonazeotrope		981
574	C₆H₁₄O	*tert*-Amyl methyl ether	86	73.8	9	256
575	C₆H₁₄O	Butyl ethyl ether	92.2	76.6	11.9	981
		" " "		76.7	10.8	497
576	C₆H₁₄O	*tert*-Butyl ethyl ether	73	65.2	6	256
577	C₆H₁₄O	2-Ethyl-1-butanol	148.9	96.7	58.7	128
		" "	147.0	96.7	58	982

No.	Formula	B-Component Name	B.P., °C	B.P., °C	Azeotropic Data Wt.%A	Ref.
A =	H₂O	Water *(continued)*	100			
578	C₆H₁₄O	Hexyl alcohol	157.1	97.8	67.2	982
		" "	157.85	97.8	75	545
579	C₆H₁₄O	Isopropyl ether	69	62.2	4.5	196,877
		"			v-l	1042g
		" 131 mm.	22.47	20.0	2.6	981
		" 297 mm.	41.82	38.0	3.4	981
		" 481 mm.	54.75	50.0	4.0	981
		" 1520 mm.	92	88	7.6	981
580	C₆H₁₄O	2-Methyl-1-pentanol	148	97.2	60	981
581	C₆H₁₄O	4-Methyl-2-pentanol	131.8	94.3	43.3	981
582	C₆H₁₄O	Propyl ether	90.7	75.4		760
583	C₆H₁₄O₂	Acetal	103.6	82.6	14.5	45,563
584	C₆H₁₄O₂	2-Butoxyethanol	171.2	98.8	79.2	129,527
585	C₆H₁₄O₂	1,2-Diethoxyethane	123	89.4	25	129,563
586	C₆H₁₄O₂	1,3-Dimethoxybutane	120.3	89.6	30	981
587	C₆H₁₄O₂	1,1-Dimethoxy-2-methylpropane	104.7	83.9	14.3	981
588	C₆H₁₄O₂	Ethoxypropoxymethane	113.7	85.90	18.4	1035
589	C₆H₁₄O₂	2-Methyl-1,5-pentanediol	242.4	Nonazeotrope		981
590	C₆H₁₄O₂	3-Methyl-1,5-pentanediol	248.4	Nonazeotrope		981
591	C₆H₁₄O₂	Pinacol	174.35	Nonazeotrope		526
592	C₆H₁₄O₃	Bis(2-methoxyethyl) ether, 100 mm.	103		89.5	23
		" 760 mm.	162		80.2	23
		" 800 mm.	164		80	23
		" 760 mm.	164	99.55	78 v-l	215
593	C₆H₁₄O₃	2-(2-Ethoxyethoxy)ethanol	202.8	Nonazeotrope		981
593a	C₆H₁₄O₃	3-Methyl-1,3,5-pentanetriol	295	Nonazeotrope		581c
594	C₆H₁₄O₄	Triethylene glycol	288.7	Nonazeotrope		526
595	C₆H₁₅N	Diisopropylamine 38.8 mm.		10	2.4 v-l	194c
		" 51.2 mm.		15	2.9 v-l	194c
		" 67.2 mm.		20	3.8 v-l	194c
		" 86.8 mm.		25	4.3 v-l	194c
		" 112.2 mm.		30	4.3 v-l	194c
		" 180.8 mm.		39.95	5.3 v-l	194c
		"	83.86	74.1	9.2	878
		"	84.1	74.1	9	981
596	C₆H₁₅N	1,3-Dimethylbutylamine	108.5	89.5	28.6	981
597	C₆H₁₅N	3,3-Dimethyl-1-butylamine	112.8	92.9		406
598	C₆H₁₅N	Dipropylamine 15.4 mm.		10	16.2 v-l	194c
		" 30.0 mm.		20	19.1 v-l	194c
		" 55.5 mm.		30	v-l	194c

TABLE I.   *Binary Systems*   33

	B-Component			Azeotropic Data		
No.	Formula	Name	B.P., °C	B.P., °C	Wt.%A	Ref.
A =	H_2O	Water *(continued)*	**100**			
		Dipropylamine				
		97.0 mm.		39.95		v-l 194c
		"	109	86.7		144
599	$C_6H_{15}N$	N-Ethylbutylamine	111.2	87.5	43.6	981
		" 13.95 mm.		10	~16	v-l 194f
		" 27.1 mm.		20	~20	v-l 194f
		" 50.3 mm.		30	~30	v-l 194f
		" 88.75 mm.		40	~50	v-l 194f
599a	$C_6H_{15}N$	N-Ethyl-*sec*-butylamine				
		22 mm.		10	~5	v-l 194f
		" 30 mm.		15	~6	v-l 194f
		" 40 mm.		20	~7.1	v-l 194f
		" 54 mm.		25	~8	v-l 194f
		" 72 mm.		30	~9	v-l 194f
		" 122 mm.		40	~10	v-l 194f
600	$C_6H_{15}N$	Hexylamine	132.7	95.5	49	981
601	$C_6H_{15}N$	Triethylamine				
		75°–100°C.				v-l 156c
		"	89.4			v-l 156c,923
		"	89.4	75	10	977
602	$C_6H_{15}NO$	2-Butylaminoethanol	199.3	Nonazeotrope		981
603	$C_6H_{15}NO$	2-Diethylaminoethanol	162.1	98.9	74.4	981
		"	162	Azeotropic		22
604	$C_6H_{15}NO$	1-Isopropylamino-2-propanol	164.5	99.8	86	981
605	$C_6H_{15}NO$	3-Isopropoxypropylamine	147		67	17
606	$C_6H_{15}N_3$	4-(2-Aminoethyl)piperazine	222.0	Nonazeotrope		981
607	$C_6H_{16}N_2$	N,N-Diethylethylenediamine	144.9	99.8	79.5	981
608	$C_6H_{16}N_2$	N,N,N',N'-Tetramethylethylenediamine	119-22	95.6	30	816
609	C_7H_7Cl	p-Chlorotoluene	163.5	95		101
610	C_7H_8	Toluene	110.7	84.1	13.5	563,689,877
		"	110.6	84.6	18	935
		"	110.7	Evaporation behavior		577
		"	110.6	85	20.2	981
611	C_7H_8O	Anisole	153.85	95.5	40.5	531
612	C_7H_8O	Benzyl alcohol	205.2	99.9	91	535
		"	204.7			v-l 935
613	$C_7H_8O_2$	Guaiacol	205.0	99.5	87.5	266
614	$C_7H_8O_2$	m-Methoxyphenol	244.7	99.25	80	575
615	$C_7H_9ClO_4$	2-Chloroallylidene diacetate	212.1	99.7	85	981
616	C_7H_9N	2,6-Lutidine	142	96.02	51.8	v-l 1063
		" 700 mm.		93.3	51.5	39,141,175,585

No.	Formula	Name	B.P., °C	B.P., °C	Wt.%A	Ref.
		B-Component		**Azeotropic Data**		
A =	**H₂O**	**Water** *(continued)*	**100**			
617	C₇H₉N	*o*-Toluidine	199.7		15.4	413
618	C₇H₉N	*p*-Toluidine	200.4		13.8	413
619	C₇H₉N	Tetrahydrobenzonitrile	195.1	98.8	78.3	981
620	C₇H₁₀O	1,2,3,6-Tetrahydro-benzaldehyde	164.2	96.9	60	981
621	C₇H₁₀O₄	Allylidene diacetate		98.7	71	981
622	C₇H₁₂	2,4-Dimethyl-1,3-pen-tadiene, 750.6 mm.	93.3	76.8	13	981
623	C₇H₁₂O	3-Hepten-2-one	162.9	96	55.7	981
624	C₇H₁₂O₂	Butyl acrylate	147	94.5	40	215
		" "		94.3	38	981
		" 100 mm.		47.8	41	981
625	C₇H₁₂O₂	2-Ethoxy-3,4-dihydro-1,2-pyran	142.9	93.6	34.9	981
625a	C₇H₁₂O₂	4-Methyl-4-vinyl-1,3-dioxane	151	95		581c
626	C₇H₁₂O₄	Pimelic acid, 100 mm.	272	Nonazeotrope		981
627	C₇H₁₃ClO₂	Isoamyl chloroacetate	195.2	98.95	77.76	121
628	C₇H₁₄	Methylcyclohexane	101.15	81.0		275
629	C₇H₁₄	1-Heptene	93.64	77.0	14.8	806
630	C₇H₁₄O	Butyl isopropenyl ether	114.8	86.3	18.8	981
631	C₇H₁₄O	2-Heptanone	149	95	48	724,725
632	C₇H₁₄O	3-Heptanone	147.6	94.6	42.2	981
633	C₇H₁₄O	4-Heptanone	143.7	94.3	40.5	981
		"	143	94		723,724
634	C₇H₁₄O	2-Methylcyclohexanol	168.8	98.4	80	576
635	C₇H₁₄O	5-Methyl-2-hexanone	144	94.7	44	981
		"		93.0	75	227
636	C₇H₁₄O₂	Amyl acetate (isomers)	146	94	36.2	981
637	C₇H₁₄O₂	Amyl acetate	148.8	95.2	41	359,723,725
638	C₇H₁₄O₂	*sec*-Amyl acetate	133.5	92.0	33.2	877
639	C₇H₁₄O₂	Butyl propionate	146.8	94.8	41	575
		" "	137	Min. b.p.		723
640	C₇H₁₄O₂	Enanthic acid	222.0	Heteroazeotrope		575
641	C₇H₁₄O₂	Ethyl isovalerate	134.7	92.2	30.2	531
642	C₇H₁₄O₂	Ethyl valerate	145.45	94.5	40	575
643	C₇H₁₄O₂	Isoamyl acetate	142	93.8	36.3	359,531,723
644	C₇H₁₄O₂	Isobutyl propionate	136.85	92.75	52.2	531
645	C₇H₁₄O₂	Isopropyl isobutyrate	120.8	88.4	23	575
646	C₇H₁₄O₂	Methyl caproate	149.8	95.3	41	575
647	C₇H₁₄O₂	Propyl butyrate	142.8	94.1	36.4	531,723
648	C₇H₁₄O₂	Propyl isobutyrate	133.9	92.15	30.8	531
649	C₇H₁₄O₃	1,3-Butanediol methyl ether acetate	171.75	97.8	60	575
650	C₇H₁₄O₃	2,2-Dimethoxy-3-pentanone	162.5	95.5		119

TABLE I. *Binary Systems* 35

		B-Component		Azeotropic Data		
No.	Formula	Name	B.P., °C	B.P., °C	Wt.%A	Ref.
A =	H₂O	Water *(continued)*	100			
651	C₇H₁₄O₃	Ethyl 3-ethoxy-propionate	170.1	97	63	981
		" 100 mm.	107.8	50.5	71	981
651a	C₇H₁₄O₃	4-Hydroxyethyl-4-methyl-1,3-dioxane		Nonazeotrope		581c
651b	C₇H₁₄O₃	4-Hydroxy-3-methylol-4-methyltetrahydropyran		Nonazeotrope		581c
652	C₇H₁₄O₃	3-Methoxybutyl acetate	171.3	96.5	65.4	981
653	C₇H₁₄O₄	2-(2-Methoxyethoxy) ethyl acetate	208.9	Nonazeotrope		981
654	C₇H₁₆	Heptane	98.4	79.2	12.9	981
		"	98.4	80.0		575
655	C₇H₁₆O	Amyl ethyl ether	120	Min. b.p.		723
656	C₇H₁₆O	*tert*-Amyl ethyl ether	101	81.2	13	256
657	C₇H₁₆O	Heptyl alcohol	176.15	98.7	83	545
658	C₇H₁₆O	5-Methyl-2-hexanol		96.5	59.1	227
659	C₇H₁₆O₂	1-Butoxy-2-methoxy-ethane	149.9	95.6	42	981
660	C₇H₁₆O₂	1-Butoxy-2-propanol	170.1	98.6	72	981
661	C₇H₁₆O₂	Diisopropoxymethane		79-80	12	970
662	C₇H₁₆O₂	Dipropoxymethane	137.2	92.2	40.3	324,970
663	C₇H₁₆O₂	2-Ethyl-1,5-pentane-diol	253.3	Nonazeotrope		981
664	C₇H₁₆O₃	1-(2-Ethoxyethoxy)-2-propanol	198.1	Nonazeotrope		981
665	C₇H₁₆O₃	2-Ethoxyethyl 2-methoxyethyl ether		99.5	82	981
666	C₇H₁₆O₃	2-(2-Propoxyethoxy) ethanol	215.8	Nonazeotrope		981
667	C₇H₁₇NO	1-Diethylamino-2-propanol	159.5	97.2	55	981
668	C₇H₁₈N₂	3-Diethylaminopropylamine	169.4	99.8	93	981
669	C₈H₈	Styrene	145.1	93.9	40.9	981
		"	145	93		688
670	C₈H₈Cl₂O₂	2-(2,4-Dichloro-phenoxy)ethanol		~100	~99.6	981
671	C₈H₈O	Acetophenone	202.0	98	81.7	581c,981
672	C₈H₈O	(Epoxyethyl)benzene	194.2	99.2	77.6	981
673	C₈H₈O₂	Benzyl formate	202.3	99.2	80	538
674	C₈H₈O₂	Methyl benzoate	199.45	99.08	79.2	531
675	C₈H₈O₂	Phenyl acetate	195.7	98.9	75.1	531
676	C₈H₁₀	Ethylbenzene, 60 mm.	60.5	33.5	33	53,688
		"	136.2	92	33.0	981
		"		91	30.6	227
677	C₈H₁₀	*m*-Xylene	139.1	94.5	40	981
		"	139	92	35.8	689,803

No.	B-Component Formula	Name	B.P., °C	B.P., °C	Azeotropic Data Wt.%A	Ref.
A =	H_2O	Water *(continued)*	100			
678	$C_8H_{10}O$	α-Methylbenzyl alcohol	203.4	99.7	89	982
679	$C_8H_{10}O$	Phenetole	170.4	97.3	59	531
680	$C_8H_{10}O_2$	Veratrole	205.5	99.0	76.5	531
681	$C_8H_{11}N$	s-Collidine	171	Min. b.p.		813
682	$C_8H_{11}N$	N-Ethylaniline	204.8	99.2	83.9	981
683	$C_8H_{11}N$	α-Methylbenzylamine	188.6	99.4	83.8	981
684	$C_8H_{11}N$	2-Methyl-5-ethyl-pyridine	178.3	98.4	72	981
685	$C_8H_{11}N$	ar-Methyl-1,2,3,6-tetrahydrobenzo-nitrile	205.4	99.1	82.6	981
686	$C_8H_{12}O$	2-Methyl-1,2,3,6-tetrahydrobenz-aldehyde	176.4	97.7	92.2	981
687	$C_8H_{12}O_2$	3,4-Dihydro-2,5-dimethyl-2H-pyran-2-carboxaldehyde	170.9	97.4	56	981
688	$C_8H_{12}O_4$	Diethyl fumarate	218.1	99.5	87.5	981
689	$C_8H_{12}O_4$	Ethyl maleate	223.3	99.65	88.2	575
690	C_8H_{14}	Diisobutylene	101	81	87	877
		"	102.3	82	12	981
691	$C_8H_{14}O$	Bicyclo[2.2.1]-heptane-2-methanol	203.9	99.7	91	981
692	$C_8H_{14}O$	Diisobutylene oxide		94	37	981
693	$C_8H_{14}O$	2-Ethyl-2-hexenal	176	97.6	60.9	981
694	$C_8H_{14}O$	2-Methallyl ether	134.6	92.5	31.0	879
695	$C_8H_{14}O$	2-Octenal		99.2	76.2	981
696	$C_8H_{14}O_2$	1,1-Diallyloxyethane	150.9	95.3	41	981
697	$C_8H_{14}O_2$	2-Ethyl-3-hexenoic acid	231.8	99.9	97.4	981
698	$C_8H_{14}O_2$	Vinyl 2-methyl-valerate	148.8	95	38	981
699	$C_8H_{14}O_3$	Bis(2-vinyloxyethyl) ether	198.7	99.4	82	981
700	$C_8H_{14}O_3$	Butyl acetoacetate	213.9	99.4	84.1	981
701	$C_8H_{14}O_4$	Diethyl succinate	216.2	99.9	91	981
702	$C_8H_{15}N$	2-(Aminomethyl)bicyclo[2.2.1]heptane	185.9	99	82	981
703	$C_8H_{15}N$	Dimethallylamine	149.0	94.1	40.3	878
704	C_8H_{16}	1-Octene	121.28	88.0	28.7	806
705	$C_8H_{16}O$	Allyl isoamyl ether	120	Min. b.p.		723
706	$C_8H_{16}O$	2-Ethylhexaldehyde	163.6	96.4	51.6	981
707	$C_8H_{16}O$	2,2,5,5-Tetramethyltetrahydro-furan	115	Min. b.p.		350
708	$C_8H_{16}O$	2,4,4-Trimethyl-1,2-epoxypentane	140.9	93.4	33	981

TABLE I. *Binary Systems*   37

No.	B-Component Formula	B-Component Name	B.P., °C	Azeotropic Data B.P., °C	Wt.% A	Ref.
A =	**H₂O**	**Water** *(continued)*	**100**			
709	C₈H₁₆O	2,4,4-Trimethyl-2,3-epoxypentane	127.3	91	25	981
710	C₈H₁₆OS	2-Butylthioethyl vinyl ether	210.5	99.3	80	981
711	C₈H₁₆O₂	2-Butoxyethyl vinyl ether		97.0	52.8	981
712	C₈H₁₆O₂	Butyl butyrate	165.7	97.2	53	538
713	C₈H₁₆O₂	2,3-Epoxy-2-ethyl-hexanol		100	99.5	981
714	C₈H₁₆O₂	2-Ethylbutyl acetate	162.3	97.0	52.4	981
715	C₈H₁₆O₂	Ethyl caproate	166.8	97.15	54	538
716	C₈H₁₆O₂	2-Ethylhexanoic acid	227.6	99.9	96.4	981
				99.5	97.6	227
717	C₈H₁₆O₂	Hexyl acetate	171.0	97.4	61	981
718	C₈H₁₆O₂	Isoamyl propionate	160.3	96.55	48.5	531
719	C₈H₁₆O₂	Isobutyl butyrate	156.8	96.3	46	531
720	C₈H₁₆O₂	Isobutyl isobutyrate	147.3	95.5	39.4	531
721	C₈H₁₆O₂	Iso-octanoic acid (isomers)	220	99.9	96	981
722	C₈H₁₆O₂	4-Methyl-2-pentyl acetate	146.1	94.8	36.7	981
723	C₈H₁₆O₂	Propyl isovalerate	155.8	96.2	45.2	531
724	C₈H₁₆O₃	2-Butoxyethyl acetate	192.2	98.8	71.9	981
725	C₈H₁₆O₃	2,2-Diethoxy-3-butanone	163.5	95-96		119
726	C₈H₁₆O₃	2,5-Diethoxytetra-hydrofuran	173.0	98	60	981
727	C₈H₁₆O₃	2-Ethoxyethyl-2-vinyl-oxyethyl ether	194.0	99.3	82.3	981
728	C₈H₁₆O₄	2-(2-Ethoxyethoxy) ethyl acetate	217.4	Nonazeotrope		981
		"	218.5	99.2	76	575
729	C₈H₁₇Cl	1-Chloro-2-ethyl-hexane	173	97.3	55	981
730	C₈H₁₇N	N-Ethylcyclohexyl-amine	164.9	97.1	58	981
731	C₈H₁₇N	5-Ethyl-2-methyl-piperidine	163.4	97.1	57.0	981
732	C₈H₁₇N	ar-Methylcyclo-hexylmethylamine		99.0	79	981
733	C₈H₁₇NO	4-Ethyl-2,6-dimethyl-morpholine	158.1	97.5	49	981
734	C₈H₁₈	Octane	125.7	89.6	25.5	981
		"	124.75	89.4		575
734a	C₈H₁₈	Isooctane	118	78.8	11.1	498i
735	C₈H₁₈O	Butyl ether	142.6	92.9	33	723,725,760
736	C₈H₁₈O	sec-Butyl ether	121	Min. b.p.		723

		B-Component		Azeotropic	Data	
No.	Formula	Name	B.P., °C	B.P., °C	Wt.%A	Ref.

No.	Formula	Name	B.P., °C	B.P., °C	Wt.%A	Ref.
A =	H₂O	**Water** *(continued)*	**100**			
737	C₈H₁₈O	2-Ethylhexanol	183.5	99.1	80	128
738	C₈H₁₈O	Ethyl hexyl ether	143-144	92.9	29 vol.%	724
739	C₈H₁₈O	Isobutyl ether	122.2	88.6	23	538,723
740	C₈H₁₈O	Iso-octyl alcohol (isomers)	186.5	99.8	82	981
741	C₈H₁₈O	Octyl alcohol	195.15	99.4	90	550
742	C₈H₁₈O	*sec*-Octyl alcohol	178.7	98	73	563
743	C₈H₁₈O₂	Acetaldehyde dipropyl acetal	147.7	94.7	36.6	45
744	C₈H₁₈O₂	2-Ethyl-1,3-hexanediol	243.1	Nonazeotrope		981
745	C₈H₁₈O₂	1-Butoxy-2-ethoxy-ethane	164.2	96.8	50	981
746	C₈H₁₈O₂	1,1-Diethoxybutane	146.3	94.2	34.5	981
747	C₈H₁₈O₂	5-Ethoxy-3-methyl-pentanol	211.7	99.9	97	981
748	C₈H₁₈O₂	2-Ethyl-3-methyl-1,5-pentanediol	265.5	Nonazeotrope		981
749	C₈H₁₈O₂	2-Hexyloxyethanol	208.1	99.7	91	981
750	C₈H₁₈O₂	2-(2-Methylpentyloxy)ethanol	197.1	99.6	86	981
751	C₈H₁₈O₃	2-(2-Butoxyethoxy)ethanol	230.6	Nonazeotrope		981
752	C₈H₁₈O₃	Bis(2-ethoxyethyl)ether	188.4	99.4	69	981
		"		98.4	78.5	129
753	C₈H₁₈O₄	1,2-Bis(2-methoxyethoxy)ethane		Nonazeotrope		23
754	C₈H₁₉N	Dibutylamine	159.6	97	50.5	475,981
		"		96.9	51.7 v-l	481c
755	C₈H₁₉N	2-Ethylhexylamine	169.1	98.2	64	981
756	C₈H₁₉N	1,1,3,3-Tetramethylbuty-lamine	140	86	35	817
757	C₈H₁₉NO	2-Diisopropylamino-ethanol	190.9	99.2	85	981
758	C₈H₁₉NO₂	2,2'-Butyliminodi-ethanol		Nonazeotrope		981
759	C₈H₁₉NO₂	1,1'-Ethyliminodi-2-propanol	238.9	Nonazeotrope		981
760	C₉H₇N	Quinoline	237.3		96.6 v-l	603
		"	Azeo. composition independent of press.			603
761	C₉H₈O₂	Vinyl benzoate		99.3	82.6	981
762	C₉H₁₀O₂	Benzyl acetate	214.9	99.60	87.5	531
763	C₉H₁₀O₂	1,2-Epoxy-3-phenoxypropane	244.4	99.8	96.1	981
764	C₉H₁₀O₂	Ethyl benzoate	212.4	99.40	84.0	531,689
765	C₉H₁₀O₂	Methyl α-toluate	215.3	99.6	88	575

TABLE I.   *Binary Systems*   39

No.	Formula	B-Component Name	B.P., °C	Azeotropic Data B.P., °C	Wt.%A	Ref.
A =	H₂O	**Water** *(continued)*	**100**			
766	C₉H₁₁N	5-Ethyl-2-vinyl-pyridine		99.4	85	981
767	C₉H₁₂	Cumene	152.4	95	43.8	981
768	C₉H₁₂	Mesitylene	164.6	96.5		575
768a	C₉H₁₂O	α,α-Dimethylbenzyl alcohol		98.1	82	581c
769	C₉H₁₂O	Phenyl propyl ether	190.2	98.5	66	538
770	C₉H₁₂O₂	Bicyclo[2.2.1]hept-5-ene-2-ol acetate	188.6	98.6	70	981
771	C₉H₁₃NO	5-Ethyl-2-pyridine-ethanol		Nonazeotrope		981
772	C₉H₁₄O	Isophorone	215.2	99.5	83.9	981
		" 25 p.s.i.g.	251	130	86.5	981
773	C₉H₁₅O	1-Methyl-2,5-endomethylene-cyclohexane-1-methanol	211.1	99.7	90.6	981
774	C₉H₁₅N	Triallylamine	151.1	95	38	981
775	C₉H₁₆O	5-Ethyl-3-hepten-2-one	193.5	98.7	73.4	981
776	C₉H₁₆O₄	Dimethyl pimelate	248.9	99.9	96.8	981
777	C₉H₁₈	1-Nonene	146.87	94.5	46.3	806
778	C₉H₁₈O	2,6-Dimethyl-4-heptanone	169.4	97.0	51.9	981
779	C₉H₁₈O₂	Butyl isovalerate	177.6	98.0	63	575
780	C₉H₁₈O₂	Ethyl enanthate	188.7	98.5	72	575
781	C₉H₁₈O₂	2-Heptyl acetate	176.4	97.8	58.9	981
782	C₉H₁₈O₂	3-Heptyl acetate	173.8	97.5	57.6	981
783	C₉H₁₈O₂	Isoamyl butyrate	178.5	98.05	63.5	531
784	C₉H₁₈O₂	Isoamyl isobutyrate	168.9	97.35	56.0	531
785	C₉H₁₈O₂	Isobutyl isovalerate	168.7	97.4	55.8	531
786	C₉H₁₈O₂	Methyl caprylate	192.9	98.8	74	575
787	C₉H₁₈O₃	β-(2-Ethylbutoxy)propionic acid		100	> 99	981
788	C₉H₁₈O₃	Isobutyl carbonate	190.3	98.6	74	575
789	C₉H₂₀	Nonane	150.7	94.8	82	968
		"	150.8	95	39.8	981
		"	150.8	94.6	53.2	806
790	C₉H₂₀O	2,6-Dimethyl-4-heptanol	178.1	98.5	70.4	982
791	C₉H₂₀O₂	Dibutoxy methane	181.8	98.2	62	324
792	C₉H₂₀O₂	Diisobutoxymethane	163.8	97.2	47.5	324,970
793	C₉H₂₀O₂	2-Ethyl-2-butyl-1,3-propanediol		Nonazeotrope		981
794	C₉H₂₀O₃	1-(2-Butoxyethoxy)-2-propanol	230.3	99.9	95	981
795	C₉H₂₀O₃	2-Methoxymethyl-2,4-dimethyl-1,5-pentanediol		Nonazeotrope		981

| | | B-Component | | Azeotropic | Data | |
No.	Formula	Name	B.P., °C	B.P., °C	Wt.%A	Ref.
A =	H₂O	Water *(continued)*	100			
796	C₉H₂₀O₃	1,1,3-Triethoxy-propane		99	70	981
797	C₉H₂₁N	N-Methyldibutylamine	163.1	96.5	48.0	981
798	C₉H₂₁N	Tripropylamine	156	94.3		144
799	C₉H₂₁NO₂	1,1'-Isopropylimino-di-2-propanol	248.6	Nonazeotrope		981
800	C₉H₂₁NO₄	2-(2-[2-(3-Amino-propoxy)ethoxy]-ethoxy)ethanol		Nonazeotrope		981
801	C₁₀H₈	Naphthalene	218	98.8	84	689
802	C₁₀H₁₀O₂	Isosafrole	252.0	99.8	96.0	538
803	C₁₀H₁₀O₂	Methyl cinnamate	261.9	99.9	95.5	538
804	C₁₀H₁₀O₂	Safrol	235.9	99.72	92.3	531
805	C₁₀H₁₀O₄	Methyl phthalate	283.2	99.95	97.5	575
		" "	282.9	100	98.9	981
806	C₁₀H₁₂	1,2,3,4-Terahydronaph-thalene		99.1	80	266
807	C₁₀H₁₂O	Anethole	235.7	99.7	92	575
808	C₁₀H₁₂O	Estragole	215.6	99.3	82	538
809	C₁₀H₁₂O₂	Ethyl α-toluate	228.75	99.73	91.3	531
810	C₁₀H₁₂O₂	Propyl benzoate	230.85	99.70	90.9	531
811	C₁₀H₁₂O₃	2-Phenoxyethyl acetate	260.6	99.9	97.4	981
812	C₁₀H₁₄	Dicyclopentadiene	172	98	67.7	981
813	C₁₀H₁₄N₂	Nicotine, 110 mm.		Nonazeotrope	v-l	284
		" 478 mm.			99.70 v-l	284
		" 572 mm.			99.02 v-l	284
		" 624 mm.			98.50 v-l	284
		" 760 mm.		99.85	97.48 v-l	284
		"		99.988	2.5	894
814	C₁₀H₁₄O₂	m-Diethoxybenzene	235.0	99.7	91	538
815	C₁₀H₁₄O₂	Ethyl bicyclo[2.2.1]hept-5-ene-2-carboxylate	198	99.2	80	981
816	C₁₀H₁₄O₃	2-(2-Phenoxyethoxy)ethanol	297.9	Nonazeotrope		981
817	C₁₀H₁₅N	N-Butylaniline	240.4	99.8	94.4	981
818	C₁₀H₁₅N	N-Ethyl-α-ethyl-benzylamine	201.2	99.2	80	981
819	C₁₀H₁₅N	N,N,α-Trimethyl-benzylamine	195.8	98.6	74.8	981
820	C₁₀H₁₅NO	2-(α-Mehylbenzyl-amino)ethanol		Nonazeotrope		981
821	C₁₀H₁₆	Camphene	159.6	96.0		575
822	C₁₀H₁₆O	Dicyclopentenol		100	96.6	981
823	C₁₀H₁₆O	Trimethyltetrahydro-benzaldehyde	204.5	99.0	77.0	981

TABLE I.   *Binary Systems*   41

No.	B-Component Formula	Name	B.P., °C	Azeotropic Data B.P., °C	Wt.%A	Ref.
A =	**H₂O**	**Water** *(continued)*	**100**			
824	C₁₀H₁₆O₄	Diisopropyl maleate	228.7	99.9	93	981
825	C₁₀H₁₈O	Cineol	176.35	99.55	57.0	531
826	C₁₀H₁₈O	Linalool	199	~99.7		563
827	C₁₀H₁₈O₂	Vinyl 2-ethylhexanoate	185.2	98.6	68	981
828	C₁₀H₁₈O₂	Vinyl octanoate (isomers)		99.1	74	981
829	C₁₀H₂₀	1-Decene	170.57	96.7	64.2	806
830	C₁₀H₂₀O	2-Ethylhexyl vinyl ether	177.7	97.8	59.1	981
831	C₁₀H₂₀O₂	2-Ethylbutyl butyrate	199.6	98.6	74.9	981
832	C₁₀H₂₀O₂	2-Ethylhexyl acetate	198.4	99.0	73.5	981
833	C₁₀H₂₀O₂	Ethyl caprylate	208.35	99.25	82	575
834	C₁₀H₂₀O₂	Isoamyl isovalerate	193.5	98.8	74.1	531
835	C₁₀H₂₀O₂	4-Methyl-2-pentyl butyrate	182.6	98.2	60.8	981
836	C₁₀H₂₀O₂	Methyl pelargonate	213.8	99.45	85	575
837	C₁₀H₂₀O₃	2-Butoxyethyl 2-vinyl-oxyethyl ether	226.7	99.8	90	981
838	C₁₀H₂₀O₃	2,2-Dipropoxy-3-butanone	196-7	98.5		119
839	C₁₀H₂₀O₄	2-(2-Butoxyethoxy) ethyl acetate	245.3	99.8	92	575
840	C₁₀H₂₁Cl	Chlorodecane (isomers)	210.6	99.7	84	981
841	C₁₀H₂₁N	N-Butylcyclohexyl-amine	209.5	99.5	81	981
842	C₁₀H₂₂	Decane	173.3	97.2		575
843	C₁₀H₂₂	2,7-Dimethyloctane	160.1	96.1		575
844	C₁₀H₂₂O	Amyl ether	190	98.4		760
845	C₁₀H₂₂O	Decyl alcohol (isomers)	217.3	100	94.8	982
846	C₁₀H₂₂O	2-Ethyloctanol	220.5	99.9	94.0	981
847	C₁₀H₂₂O	Isoamyl ether	172.6	97.4	54	538,760
848	C₁₀H₂₂O	2-Propylheptanol	217.9	99.8	92	981
849	C₁₀H₂₂O₂	Acetaldehyde dibutyl acetal	188.8	98.7	66.3	45,895
850	C₁₀H₂₂O₂	Acetaldehyde diisobutyl acetal	171.3	97.4	52.5	45
851	C₁₀H₂₂O₂	1,2-Dibutoxyethane	203.6	99.1	76.8	981
852	C₁₀H₂₂O₃	2-(2-Hexyloxyethoxy) ethanol	259.1	100	98.1	981
853	C₁₀H₂₂O₄	1,2-Bis(2-ethoxy-ethoxy)ethane	246.9	Nonazeotrope		981
854	C₁₀H₂₂O₅	Bis[2-(2-methoxy-ethoxy)ethyl] ether		Nonazeotrope		23,981
855	C₁₀H₂₃N	Decylamine (isomers)	203.7	99.5	82	981
856	C₁₀H₂₃N	Diamylamine (isomers)	190	99.3	76	981

No.	B-Component		B.P., °C	Azeotropic Data		
	Formula	Name	B.P., °C	B.P., °C	Wt.%A	Ref.
A =	H₂O	Water *(continued)*	100			
857	$C_{10}H_{23}N$	N,N-Dimethyl-2-ethyl-hexylamine	176.1	98.2	58	981
858	$C_{10}H_{23}NO$	2-Dibutylaminoethanol	228.7	99.9	91.0	981
859	$C_{11}H_{10}$	1-Methylnaphthalene	245	99.8	94	266
860	$C_{11}H_{12}O_2$	Ethyl cinnamate	272	99.93	97	575
861	$C_{11}H_{14}O_2$	1-Allyl-3,4-dimethoxybenzene	255.0	99.85	96.2	538
862	$C_{11}H_{14}O_2$	Butyl benzoate	249.8	99.88	94	538
863	$C_{11}H_{14}O_2$	1,2-Dimethoxy-4-propenylbenzene	270.5	99.95	98.8	575
864	$C_{11}H_{14}O_2$	Isobutyl benzoate	242.15	99.82	92.6	531
865	$C_{11}H_{14}O_3$	Butyl salicylate	268.2	99.9	95.8	981
866	$C_{11}H_{14}O_3$	Ethyl 6-formylbicyclo[2.2.1]hept-5-en-2-carboxylate		100	97	981
867	$C_{11}H_{16}O_3$	Allyl 6-methyl-3,4-epoxycyclohexane-carboxylate	251.4	100	98.1	981
868	$C_{11}H_{18}O_2$	Isopropyl 6-methyl-3-cyclohexene-carboxylate	215.2	99.7	84	981
869	$C_{11}H_{20}O$	5-Ethyl-3-nonen-2-one	226.4	99.6	92	981
870	$C_{11}H_{20}O$	Isobornyl methyl ether	192.2	98.55	68	538
871	$C_{11}H_{20}O$	Methyl terpineol ether	216.2	99.3	83	575
872	$C_{11}H_{20}O_4$	Diethyl pimelate	268.1	100	98.3	981
873	$C_{11}H_{22}O$	5-Ethyl-2-nonanone	222.9	99.6	87.1	981
874	$C_{11}H_{22}O_2$	2,6-Dimethyl-4-heptyl acetate	192.2	98.7	67.6	981
875	$C_{11}H_{22}O_2$	Ethyl pelargonate	227	99.6	88	575
876	$C_{11}H_{22}O_3$	Isoamyl carbonate	232.2	99.75	91	575
877	$C_{11}H_{22}O_3$	4-Methoxy-2,6-dipropyl-1,3-dioxane	223.6	99.6	88.1	981
878	$C_{11}H_{24}$	Undecane	194.5	98.85	96	968
879	$C_{11}H_{24}O$	5-Ethyl-2-nonanol	225.4	99.7	89.1	981
880	$C_{11}H_{24}O_2$	Diamyloxymethane	221.6	99.2	93	324
881	$C_{11}H_{24}O_2$	Diisoamyloxymethane	207	99.3	78.8	45,970
882	$C_{11}H_{24}O_2$	2,2-Dibutoxypropane		98.9	69.6	981
883	$C_{11}H_{24}O_2$	2,6-Dimethyl-4-heptyloxyethanol	225.5	99.9	91	981
884	$C_{11}H_{24}O_4$	1,1,3,3-Tetraethoxy-propane	220.1	99.8	87.4	981
885	$C_{11}H_{25}NO$	1-Dibutylamino-2-propanol	229.1	99.8	88.4	981
886	$C_{12}H_{10}O$	o-Phenylphenol		99.95	98.75	266

TABLE I. *Binary Systems* 43

No.	Formula	B-Component Name	B.P., °C	B.P., °C	Wt.%A		Ref.
				Azeotropic Data			
A =	H$_2$O	**Water** *(continued)*	**100**				
887	C$_{12}$H$_{10}$O	Phenyl ether	259.3	99.33	96.75		531
888	C$_{12}$H$_{14}$O$_4$	Ethyl phthalate	298.5	99.98	98		575
889	C$_{12}$H$_{16}$O$_2$	Isoamyl benzoate	262.3	99.9	95.6		531
890	C$_{12}$H$_{18}$O	Triisobutylene oxide		99.3	72		981
891	C$_{12}$H$_{19}$N	*N*-Butyl-α-methyl-benzylamine	239.3	99.8	92		981
892	C$_{12}$H$_{20}$O$_2$	Bornyl acetate	227.6	99.62	87.3		531
893	C$_{12}$H$_{20}$O$_2$	*sec*-Butyl-6-methyl-3-cyclohexene-carboxylate		100	92		981
894	C$_{12}$H$_{20}$O$_4$	Dibutyl fumarate	285.2	99.9	98.5		981
895	C$_{12}$H$_{20}$O$_4$	Dibutyl maleate	280.6	99.9	98.4		981
896	C$_{12}$H$_{22}$O	Ethyl isobornyl ether	203.8	98.9	75		575
897	C$_{12}$H$_{22}$O$_2$	2-Ethylhexyl crotonate	241.2	99.9	93.4		981
898	C$_{12}$H$_{22}$O$_2$	Vinyl decanoate (isomers)		99.9	88		981
899	C$_{12}$H$_{22}$O$_4$	Diethyl 2-ethyl-3-methylglutarate	255.8	100	97.1		981
900	C$_{12}$H$_{23}$N	Dicyclohexylamine	255.8	Nonazeotrope			139
901	C$_{12}$H$_{24}$O	2,6,8-Trimethyl-4-nonanone	218.2	99	84		981
902	C$_{12}$H$_{24}$O$_2$	2-Ethylbutyl 2-ethylbutyrate	222.6	99.6	85.6		981
903	C$_{12}$H$_{24}$O$_2$	2-Ethylbutyl hexanoate	236.2	99.7	91.2		981
904	C$_{12}$H$_{24}$O$_2$	Hexyl 2-ethylbutyrate	230.3	99.7	88.8		981
905	C$_{12}$H$_{24}$O$_2$	Hexyl hexanoate	245.2	99.8	93.3		981
906	C$_{12}$H$_{24}$O$_3$	2,2-Dibutoxy-3-butanone	228-230	97-8			119
907	C$_{12}$H$_{24}$O$_3$	2,2-Diisobutoxy-3-butanone	214-215	98			119
908	C$_{12}$H$_{26}$	Dodecane	214.5	99.45	98		968
909	C$_{12}$H$_{26}$O	2-Butyl-1-octanol	253.4	99.9	97.5		981
910	C$_{12}$H$_{26}$O	2,6,8-Trimethyl-4-nonanol	225.5	99.6	89.7		982
911	C$_{12}$H$_{26}$O$_2$	Acetaldehyde diamyl acetal	225.3	99.8	85.5		45
912	C$_{12}$H$_{26}$O$_2$	Acetaldehyde diisoamyl acetal	213.6	99.3	78.8		45
913	C$_{12}$H$_{26}$O$_2$	1,1-Diethoxy-2-ethylhexane	207.8	99.3	78.6		981
914	C$_{12}$H$_{26}$O$_2$	3-Ethoxy-4-ethyl-octanol	249.2	100	98		981
915	C$_{12}$H$_{26}$O$_3$	Bis(2-butoxyethyl) ether	254.6	99.8	94.7		981
916	C$_{12}$H$_{26}$O$_3$	1,1,3-Triethoxyhexane		99.6	85		981
917	C$_{12}$H$_{27}$N	Dihexylamine	239.8	99.8	92.8		981
918	C$_{12}$H$_{27}$N	Tributylamine	213.9	99.65	79.7	v-l	481c

No.	Formula	Name	B.P., °C	B.P., °C	Wt.%A	Ref.
		B-Component		Azeotropic Data		
$A =$	H_2O	**Water** *(continued)*	**100**			
919	$C_{12}H_{27}O_4P$	Tributyl phosphate		100	99.4	981
920	$C_{13}H_{24}O_2$	Decyl acrylate (isomers)		99.9	94.9	981
921	$C_{14}H_{22}O$	2-(Ethylhexyl)phenol	297.0	100	>99	981
922	$C_{14}H_{23}N$	N-(Ethylhexyl)aniline		100	99.3	981
923	$C_{14}H_{24}$	1,3,6,8-Tetramethyl-1,6-cyclodecadiene	220.5	99.5	82.3	981
924	$C_{14}H_{26}O_4$	Dibutyl adipate		100	>99	981
925	$C_{14}H_{28}O$	Trimethylnonyl vinyl ether	223.4	99.6	84.3	981
926	$C_{14}H_{28}O_2$	2-Ethylbutyl 2-ethyl-hexanoate	261.5	99.9	95.8	981
927	$C_{14}H_{28}O_2$	2-Ethylhexyl 2-ethyl-butyrate	252.8	99.9	94.8	981
928	$C_{14}H_{28}O_2$	2-Ethylhexyl hexanoate	267.2	99.9	96.4	981
929	$C_{14}H_{28}O_2$	Hexyl 2-ethyl-hexanoate	254.3	99.9	94.6	981
930	$C_{14}H_{29}N$	N-(2-Ethylhexyl)cyclohexylamine		100	99.7	981
931	$C_{14}H_{30}O$	7-Ethyl-2-methyl-4-undecanol	264.3	99.9	96.3	981
932	$C_{14}H_{30}O_2$	2-(2,6,8-Trimethyl-4-nonyloxy)ethanol		100	99.0	981
933	$C_{15}H_{28}O_4$	Dibutyl pimelate		100	>99.5	981
934	$C_{15}H_{32}O$	2,8-Dimethyl-6-isobutyl-4-nonanol	265.4	99.9	97.2	981
935	$C_{16}H_{18}O$	Bis(α-methylbenzyl)ether	286.7	100	98.7	981
936	$C_{16}H_{28}O_4$	Bis(4-methyl-2-pentyl) maleate		100	99	981
937	$C_{16}H_{30}O_2$	Tridecyl acrylate		100	98.8	981
938	$C_{16}H_{31}N$	Bis(methylcyclohexyl-methyl)amine		100	99.45	981
939	$C_{16}H_{32}O_2$	2-Ethylhexyl 2-ethylhexanoate	280.4	99.9	97.9	981
940	$C_{16}H_{34}O$	Bis(2-ethylhexyl)ether	269.8	99.8	96.4	981
941	$C_{16}H_{35}N$	Bis(2-ethylhexyl)amine	280.7	100	97.6	981
942	$C_{17}H_{36}O$	3,9-Diethyl-6-tridecanol	309	100	>99	981
943	$C_{18}H_{24}N_2$	Bis(α-methylbenzyl)ethylenediamine		100	>99.9	981
944	$C_{18}H_{38}O_2$	1,1-Bis(2-ethylhexyloxy)ethane		99.0	99.9	981
945	$C_{18}H_{39}NO$	2-[Bis(2-ethylhexyl)amino]ethanol		100	>99.5	981

TABLE I.   *Binary Systems*   45

No.	Formula	B-Component Name	B.P., °C	Azeotropic Data B.P., °C	Wt.%A		Ref.
A =	**H₂O**	**Water** *(continued)*	**100**				
946	C₂₀H₃₆O₄	Bis(2-ethylhexyl) fumarate		100	>99.9		981
947	C₂₀H₃₆O₄	Bis(2-ethylhexyl) maleate		100	>99.9		981
948	C₂₀H₄₀O₃	2-Ethylhexyl 3- (2-ethylhexyloxy) butyrate		100	>99.5		981
949	C₂₀H₄₂O	Decyl ether (isomers)		100	99.6		981
950	C₂₀H₄₂O	Eicosanol (isomers)		100	99.8		981
951	C₂₀H₄₃N	Didecylamine (isomers)		100	99.6		981
952	C₂₁H₃₈O₃	Allyl 9,10-epoxystearate		Nonazeotrope			981
953	C₂₄H₅₂O₄Si	Tetra(2-ethylbutoxy) silane		100	99.9		981
954	C₃₁H₅₈O₆	Tri(2-ethylhexyl) 1,2,4-butane- tricarboxylate		100	99.8		981
A =	**H₂S**	**Hydrogen Sulfide**	**— 59.6**				
955	C₂H₆	Ethane 200 p.s.i.g.		— 21.6	7.9	v-l	443
		" 300 p.s.i.g.		— 6.5	11.6	v-l	443
		" 400 p.s.i.g.		5	14.5	v-l	443
		" 500 p.s.i.g.		15	17.1	v-l	443
		" 600 p.s.i.g.		23.5	19.6	v-l	443
956	C₃H₈	Propane 200 p.s.i.g.		7.8	75.2	v-l	445
		" 400 p.s.i.g.		37.1	82	v-l	445
		" 600 p.s.i.g.		56	83.7	v-l	445
		" 800 p.s.i.g.		72	87.2	v-l	445
		" 1000 p.s.i.g.		84.2	89.9	v-l	445
		" 1200 p.s.i.g.		95	92.7	v-l	445
A =	**H₃N**	**Ammonia**	**— 33.5**				
956a	H₄N₂	Hydrazine —40°–50°C.		Nonazeotrope		v-l	31h
957	CH₅N	Methylamine	— 6.32	Nonazeotrope			818
958	C₂H₂	Acetylene 15-65°C.		Nonazeotrope		v-l	524
959	C₂H₃Cl	Chloroethylene 15 atmos.		38.6	66.5	v-l	138
960	C₂H₆O	Methyl ether		— 37.5	42.6	v-l	770
		"		— 60	38.6	v-l	770
		"	— 23	— 37	42.5		378
		" 11 atm.		25	56		378
961	C₂H₇N	Dimethylamine	6.88	Nonazeotrope			818
962	C₂H₇N	Ethylamine, (0°-30° C.)		Nonazeotrope			981
963	C₃H₄	Propadiene	— 32	— 45	44.3		356
964	C₃H₄	Propyne	— 23	— 35	75		215
		" 498 mm.		— 43.9	54.6	v-l	871
		" 248 mm.		— 56.7	46.3	v-l	871
		" 94 mm.		— 70.5	47.5	v-l	871

No.	Formula	B-Component Name	B.P., °C	Azeotropic Data B.P., °C	Wt.%A		Ref.
A =	**H₃N**	**Ammonia** *(continued)*	**—33.5**				
965	C₃H₅F	2-Fluoropropene	— 24	— 40.5	34		356
966	C₃H₆	Cyclopropane	— 31.5	— 44	20		215
967	C₃H₆	Propene 1200 mm.	— 34.2	— 40	10-15		215
968	C₃H₈	Propane	— 42	— 44	5-10		215
969	C₃H₈O	Propyl alcohol	97.2	Nonazeotrope			813
970	C₃H₉N	Trimethylamine	2.87	— 34	73		20,818
		" 210 p.s.i.g.			82		818
971	C₄H₂	Diacetylene -35 to -45°		Nonazeotrope		v-l	871
972	C₄H₄	Vinylacetylene -35 to -55°		Nonazeotrope		v-l	871
973	C₄H₆	1,3-Butadiene	— 4.5	— 37	55		215
974	C₄H₆	1-Butyne	7	Nonazeotrope			215
975	C₄H₈	1-Butene	— 6	— 37.5	45		215
976	C₄H₈	2-Methylpropene	— 6	— 38.5	45		215
977	C₄H₁₀	Butane	— 0.5	— 37.1	45		215
		"		— 37.0	54		378
		" 375 p.s.i.g.		55.5	57		195
		" 300 p.s.i.g.		43	56.8	v-l	444
		" 500 p.s.i.g.		66	59.0	v-l	444
		" 700 p.s.i.g.		81	60.9	v-l	444
		" 900 p.s.i.g.		94	62.1	v-l	444
		" 1100 p.s.i.g.		104	63.4	v-l	444
978	C₄H₁₀	2-Methylpropane	— 10	— 38.4	35		215
		" 12 atm.		25	45		378
979	C₅H₁₂	2-Methylbutane	27.6	— 34.5	65		215
980	C₈H₁₈	Iso-octane,					
		" 200-1600 p.s.i.g.				v-l	446
		" >1400 p.s.i.g.		Min. b.p. 98-100%		v-l	446
A =	**H₄N₂**	**Hydrazine**	**113.5**				
981	C₂H₈N₂	1,1-Dimethylhydra- zine 250-760 mm.		Nonazeotrope		v-l	656,736
A =	**H₄Si**	**Silane**	**—111.86**				
982	C₂H₄	Ethylene	—103.7	v.p. curve			1074
A =	**He**	**Helium**	**—268.9**				
982a	Ne	Neon 26.95°–41.90°K.				v-l	372c
983	CH₄	Methane 5-170 atm.		Nonazeotrope		v-l	450
		" 95°–185°K.				v-l	372e
A =	**Kr**	**Krypton**	**—152**				
983a	C₃H₆	Propylene	— 48			v-l	63e
A =	**NO**	**Nitric Oxide**	**—153.6**				
984	NO₂	Nitrogen peroxide	26	Nonazeotrope			575
A =	**N₂**	**Nitrogen**	**—196**				
984a	N₂O	Nitrous oxide 4–45 atm.				v-l	850c
985	Ne	Neon 66.13-120°K				v-l	926

TABLE I.   *Binary Systems*   47

		B-Component			Azeotropic Data		
No.	Formula	Name	B.P., °C	B.P., °C	Wt.%A		Ref.
A =	N₂	**Nitrogen** *(continued)*	**—196**				
986	O₂	Oxygen	—183	Nonazeotrope			813
		" 1.36 kg./cm²				v-l	685c
987	CH₄	Methane	—164	Nonazeotrope			575
A =	N₂O	**Nitrous Oxide**	**15**				
988	C₂H₆	Ethane, 45 atm.	28	12.8	80		563
				Min. b.p.	85.5		499
A =	O₂S	**Sulfur Dioxide**	**— 10**				
988a	O₃S	Sulfur trioxide					
		760 mm.				v-l	200a
		" 850 mm.				v-l	200a
988b	CH₄O	Methanol 20°–40°C.		Nonazeotrope		v-l	44c
989	CH₄S	Methanethiol	6.8	Nonazeotrope			575
990	C₂H₄	Ethylene	— 103.9	Nonazeotrope			349
991	C₂H₆	Ethane	— 83.3	Min. b.p.			349
992	C₂H₆O	Methyl ether					
		6.6 atm.		56.1	60		82
		" 12.1 atm.		77.1	60		82
		" 26.7 atm.		108.7	60		82
		"	— 23.6	0	65		279
		"		0.4	65.8	v-l	770c
993	C₃H₆	Propene	— 48	Nonazeotrope			349,985
993a	C₃H₆O	Acetone 20°–40°C.		Nonazeotrope		v-l	44c
993b	C₃H₆O₂	Methyl acetate					
		20°–40°C.		Nonazeotrope		v-l	44c
994	C₃H₈	Propane, 7 kg./cm²			22		349,985
		"		Azeotropic at all pressures			985
995	C₄H₈	1-Butene	— 6.7	—16	61		292,636
		" 2.37 atm.		3	62		636
996	C₄H₈	2-Methylpropene	— 6.7	—14	59		636
		" 0.46 atm.		—30	57		636
		" 2.40 atm.		3	66		636
997	C₄H₈	*trans*-2-Butene	1.0	—14	71		636
		" 0.46 atm.		—29	70		636
		" 2.05 atm.		3	75		636
998	C₄H₈	*cis*-2-Butene	3.7	—13	72		636
		" 2.05 atm.		3	75		636
999	C₄H₁₀	Butane	— 0.6	—18	63.3		636
		" 0.46 atm.		—35	62		636
		" 2.65 atm.		3	66		636
1000	C₄H₁₀	2-Methylpropane	— 12.4	—24			636
		" 3.17 atm.		3	57.4		636
1001	C₅H₁₀	2-Methyl-1-butene	32.0	Min. b.p.			292
1002	C₅H₁₀	3-Methyl-1-butene	21.2	Min. b.p.			292
1003	C₅H₁₀	2-Methyl-2-butene	37.7	Min. b.p.			292
1004	C₅H₁₀	1-Pentene	30.2	Min. b.p.			292
1005	C₅H₁₀	2-Pentene	35.8	Min. b.p.			292
1006	C₅H₁₂	2-Methylbutane	27.9	Min. b.p.			292
1007	C₅H₁₂	Pentane	36.2	Min. b.p.			292

	B-Component			Azeotropic Data		
No.	Formula	Name	B.P., °C	B.P., °C	Wt.%A	Ref.
A =	**Pb**	**Lead**	**1525**			
1008	Sn	Tin	2275	Nonazeotrope		563
A =	**S**	**Sulfur**	**444.6**			
1009	Se	Selenium	688	Compound formation		
					v-l	13,206
1009a	Te	Tellurium		Nonazeotrope	v-l	486c
A =	**Se**	**Selenium**				
1010	Te	Tellurium 50-760 mm.		Nonazeotrope	v-l	486c
A =	**Xe**	**Xenon**	**—109**			
1010a	C₃H₆	Propylene	— 48	Nonazeotrope	v-l	63e
A =	**CCl₂F₂**	**Dichlorodifluoromethane**	**— 29.8**			
1011	CHClF₂	Chlorodifluoro-				
		methane	— 40.8	—41.4	25	231
		" 4.93 atm.	0.04	0.00	2.1	745
		" 2059 mm.		Nonazeotrope		1019
		" —60°–70°C.		effect of press.	v-l	491c
1012	CH₃Cl	Chloromethane,				
		5380 mm.	33.5	25.0	78	796
1013	C₂H₂F₄	1,1,2,2-Tetrafluoro-				
		ethane	— 19.7	—33.1	72	508
1014	C₂H₄F₂	1,1-Difluoroethane				
		52.72 p.s.i.g.		0	74,	v-l	746
		"		—30.5	77.55	744
		"		0.00	73.80	744
		"		24.90	71.22	744
		"		40.08	69.31	744
		" 60 p.s.i.a.		4.44	76.2	795
		" 112 p.s.i.a.		25	74	795
1015	C₂H₆O	Methyl ether				
		2340 mm.	6	0	90	796
1016	C₃F₆	Hexafluoropropene				
		2059 mm.	—6.1	—7.1	46.3	v-l	1019
1017	C₃HF₇	Heptafluoropropane				
		2328 mm.	17	0.00	86.5	745
1018	C₄F₈	Perfluorocyclobutane				
		2059 mm.	21	Nonazeotrope		1019
A =	**CCl₃F**	**Trichlorofluoro-**				
		methane	**24.9**			
1019	C₂H₄O	Acetaldehyde	20.2	15.6	55	279
1020	C₂H₄O₂	Methyl formate	32	20	82	279
1021	C₅H₁₂	2-Methylbutane	27	23.16	92	225
A =	**CCl₃NO₂**	**Trichloronitromethane**	**111.9**			
1022	CHBrCl₂	Bromodichloromethane	90.1	Nonazeotrope		554
1023	CH₂Br₂	Dibromomethane	97.0	Nonazeotrope		554
1024	CH₂O₂	Formic acid	100.75	91		543
1025	CH₃NO₂	Nitromethane	101.2	<100.4	<15	575

TABLE I.   *Binary Systems*   49

No.		B-Component			Azeotropic Data		
	Formula	Name	B.P., °C	B.P., °C	Wt.%A		Ref.
A =	**CCl₃NO₂**	**Trichloronitromethane**	**111.9**				
		(continued)					
1026	CH₄O	Methanol	64.65	Nonazeotrope			554
1027	C₂Cl₄	Tetrachloroethylene	121.1	Nonazeotrope			554
1028	C₂H₄O₂	Acetic acid	118.1	107.65	80.5		554
1029	C₂H₅ClO	2-Chloroethanol	128.6	108.9	85		554
1030	C₂H₆O	Ethyl alcohol	78.32	77.5	34		554
1031	C₃H₅ClO	Epichlorohydrin	116.45	~106			563
1032	C₃H₅I	3-Iodopropene	101.8	Nonazeotrope			554
1033	C₃H₆Cl₂	1,3-Dichloropropane	129.8	Nonazeotrope			554
1034	C₃H₆O	Allyl alcohol	96.85	94.2	56		554,875
1035	C₃H₆O₂	Propionic acid	141.3	Nonazeotrope			554
1036	C₃H₇ClO	1-Chloro-2-propanol	127.0	<110.8	<96		554
1037	C₃H₇I	1-Iodopropane	102.4	Nonazeotrope			554
1038	C₃H₈O	Isopropyl alcohol	82.4	81.95	35		554
1039	C₃H₈O	Propyl alcohol	97.2	94.05	58.5		554
1040	C₃H₈O₂	2-Methoxyethanol	124.5	<110.5	<82		554
1041	C₄H₈O₂	Dioxane	101.35	Nonazeotrope			554
1042	C₄H₈O₂	Isobutyric acid	154.6	Nonazeotrope			554
1043	C₄H₈S	Tetrahydrothiophene	118.8	Nonazeotrope			566
1044	C₄H₉Br	1-Bromobutane	101.5	Nonazeotrope			554
1045	C₄H₉I	1-Iodo-2-methylpropane	120.8	Nonazeotrope			554
1046	C₄H₁₀O	*n*-Butyl alcohol	117.8	106.65	80		554
1047	C₄H₁₀O	*sec*-Butyl alcohol	99.5	96.1	60		554
1048	C₄H₁₀O	*tert*-Butyl alcohol	82.45	82.25	37		554
1049	C₄H₁₀O	Isobutyl alcohol	108.0	102.05	68		554
1050	C₄H₁₀S	Butanethiol	97.5	Nonazeotrope			575
1051	C₅H₅N	Pyridine	115.4	Nonazeotrope			553
1052	C₅H₁₀O	Isovaleraldehyde	92.1	Nonazeotrope			554
1053	C₅H₁₀O	3-Pentanone	102.05	Nonazeotrope			575
1054	C₅H₁₀O₂	Ethyl propionate	99.1	Nonazeotrope			554
1055	C₅H₁₀O₂	Methyl butyrate	102.65	Nonazeotrope			554
1056	C₅H₁₀O₂	Propyl acetate	101.6	Nonazeotrope			532
1057	C₅H₁₁Br	1-Bromo-3-methylbutane	120.65	Nonazeotrope			554
1058	C₅H₁₁Cl	1-Chloro-3-methylbutane	99.4	Nonazeotrope			554
1059	C₅H₁₂O	*tert*-Amyl alcohol	102.35	98.9	65		554
1060	C₅H₁₂O	Isoamyl alcohol	131.9	111.15	93		554
1061	C₅H₁₂O	3-Methyl-2-butanol	112.9	<106.5	<80		554
1062	C₅H₁₂O	2-Pentanol	119.8	108.0	83		554
1063	C₅H₁₂O	3-Pentanol	116.0	<107.3	<82		554
1064	C₆H₆	Benzene	80.15	Nonazeotrope			554
1065	C₆H₁₀	Cyclohexene	82.75	Nonazeotrope			554
1066	C₆H₁₂	Cyclohexane	80.75	Nonazeotrope			554
1067	C₆H₁₂O	Cyclohexanol	160.8	Nonazeotrope			554
1068	C₆H₁₂O	3-Hexanone	123.3	Nonazeotrope			575
1069	C₆H₁₂O₂	Ethyl isobutyrate	110.1	Nonazeotrope			532
1070	C₆H₁₂O₂	Isobutyl acetate	117.4	Nonazeotrope			554
1071	C₆H₁₂O₂	Methyl isovalerate	116.5	Nonazeotrope			554
1072	C₆H₁₄O	*n*-Hexyl alcohol	157.85	Nonazeotrope			554

		B-Component		Azeotropic Data		
No.	Formula	Name	B.P., °C	B.P., °C	Wt.%A	Ref.
A =	CCl₃NO₂	**Trichloronitromethane**	**111.9**			
		(continued)				
1073	C₆H₁₄O₂	Acetal	103.55	Nonazeotrope		554
1074	C₆H₁₄S	Isopropyl sulfide	120.5	Nonazeotrope		566
1075	C₇H₈	Toluene	110.75	Nonazeotrope		554
1076	C₇H₁₄	Methylcyclohexane	101.15	100.8	27	554
1077	C₇H₁₄O	2-Methylcyclohexanol	168.5	Nonazeotrope		554
1078	C₇H₁₆	n-Heptane	98.4	98.32	7	554
1079	C₇H₁₆O	n-Heptyl alcohol	176.15	Nonazeotrope		554
1080	C₈H₁₀	Ethylbenzene	136.15	Nonazeotrope		554
1081	C₈H₁₀	m-Xylene	139.2	Nonazeotrope		554
1082	C₈H₁₆	1,3-Dimethylcyclohexane	120.7	111.0	80	554
1083	C₈H₁₈	2,5-Dimethylhexane	109.3	<107.5	<55	554
1084	C₈H₁₈O	Isobutyl ether	122.3	Nonazeotrope		554

A =	CCl₄	**Carbon Tetrachloride**	**76.75**			
1085	CS₂	Carbon disulfide	46.25	Nonazeotrope	v-l	734
1086	CHCl₃	Chloroform	61.2	Nonazeotrope	v-l	371,930c
1087	CH₂O₂	Formic acid	100.7	66.65	81.5	563
1087a	CH₃Cl₃Si	Trichloromethylsilane				
		65°–74°C.		Nonazeotrope	v-l	522g
1088	CH₃NO₂	Nitromethane	101.2	71.3	83	554
		" 303 mm.		45	89.4	91
1089	CH₃NO₃	Methyl nitrate	64.8	<63.5		560
1090	CH₄O	Methanol	64.7	55.7	79.44	563,903,
						1045
		"	64.7		v-l	710
1091	C₂Cl₄	Tetrachloroethylene	120.8	Nonazeotrope,	v-l	641
1092	C₂F₄N₂O₄	1.1,2,2-Terafluorodinitro-				
		ethane		62	23.4	288
1093	C₂HCl₃	Trichloroethylene	86.2	Ideal system	v-l	501
1094	C₂H₃Cl₃O₂	Chloral hydrate	97.5	~76		563
1095	C₂H₃N	Acetonitrile	81.6	65.1	83	527
		" 371.2 mm		45	84.5	v-l 91
		"		65.9	83.8	v-l 635c
1095a	C₂H₃NO	Methyl isocyanate	37.9	Nonazeotrope	v-l	45c
1096	C₂H₄Br₂	1,2-Dibromoethane	131.5	Nonazeotrope		563
1097	C₂H₄Cl₂	1,1-Dichloroethane	57		v-l	435
1098	C₂H₄Cl₂	1,2-Dichloroethane	82.85	75.3	78.4,	v-l 465,1047
		"	83.45	75.5	80	581
1099	C₂H₄O₂	Acetic acid				
		" <50 mm.		Nonazeotrope		368
		" 90 mm.		18.7	99.28	368
		" 340 mm.		51.5	99.42	368
		" 530 mm.		64.6	99	368
		" 760 mm.	118.1	76	98.46	368
		" 1080 mm.		90	97.7	368
		" 1400 mm.			97.0	368
		"	118.5	76.55	97	542
		" 20°C.		Nonazeotrope	v-l	587c

TABLE I. *Binary Systems*   51

No.	Formula	Name	B.P., °C	B.P., °C	Wt.%A		Ref.
		B-Component		**Azeotropic Data**			
A =	**CCl₄**	**Carbon Tetrachloride**	**76.75**				
		(continued)					
1100	C₂H₅Br	1-Bromoethane	38.4	Nonazeotrope			575
1101	C₂H₅ClO	Chloromethyl methyl ether	59.5	Nonazeotrope			556
1102	C₂H₅I	Iodoethane	72.3	Nonazeotrope			549
		"	72.3	Min. b.p.			814
1103	C₂H₅NO₂	Nitroethane 25°C		Nonazeotrope		v-l	845
1104	C₂H₅NO₃	Ethyl nitrate	87.68	74.95	84.5		536
1105	C₂H₆O	Ethyl alcohol	78.3	65.04	84.2	v-l	385
		"	78.3	65	84	v-l	386
		"	78.3	65.08	84.15		574
		" 200 mm.		32.1	90		371,
		" 380 mm.		47.0	85		382,
		" 760 mm.		64.9	80		854,
				Vapor pressure curves			855,978
		" 600 mm.		58.4	86	v-l	73c
		" 752 mm.		64.6	83	v-l	73c
		" 900 mm.		69.4	81	v-l	73c
1106	C₃H₃N	Acrylonitrile	77.3	66.2	79		215
1107	C₃H₅ClO	Epichlorohydrin	116.4	Nonazeotrope			556
1108	C₃H₆O	Acetone	56.15	56.08	11.5		29,371,551
		" 513.2 mm.		45	9	v-l	92
		" 300 mm.	31.29	31.22	9.03	v-l	32
		" 450 mm.	41.56	41.47	11.80	v-l	32
		" 600 mm.	49.36	49.26	12.48	v-l	32
		" 760 mm.	56.08	55.98	12.6	v-l	32
1109	C₃H₆O	Allyl alcohol	97.1	72.3	88.5		981
		"		72.6	79.6	v-l	1
		"	96.9	72.5	91.15		358,563, 875
1110	C₃H₆O₂	Methyl acetate	57.0	Nonazeotrope			563
		"	57.1			v-l	682
1111	C₃H₆O₃	Methyl carbonate	90.25	75.75	88		527
		"	90.35	Nonazeotrope			547
1112	C₃H₇Br	1-Bromopropane	71.0	Nonazeotrope			549
1113	C₃H₇NO₂	1-Nitropropane 25°		Nonazeotrope		v-l	845
1114	C₃H₇NO₂	2-Nitropropane 25°		Nonazeotrope		v-l	845
1115	C₃H₈O	Isopropyl alcohol	82.5	68.65	81.9	v-l	1037, 1049
		"	82.45	68.95	82		572, 1051
1116	C₃H₈O	Propyl alcohol	97.25	73.4	92.1,	v-l	133, 389,574, 1051
		" 20-40°C				v-l	629
		"	97.2	72.8	88.5		981
1117	C₃H₉SiCl	Chlorotrimethylsilane	57.5	Nonazeotrope			844
1118	C₃H₉BO₃	Methyl borate	68.7	Nonazeotrope			547
1119	C₄H₄S	Thiophene	84	Nonazeotrope			527

No.	Formula	B-Component Name	B.P., °C	Azeotropic Data B.P., °C	Wt.%A		Ref.
A =	**CCl₄**	**Carbon Tetrachloride**	**76.75**				
		(continued)					
1119a	C₄H₆	1,3-Butadiene					
		25°C.				v-l	1030c
1120	C₄H₆O₂	Allyl formate	80.0	74.3	66		562
1121	C₄H₈O	2-Butanone	79.6	73.8	71		29,552
		"				v-l	598c
		" 342 mm.		50.0	84.3	v-l	287
		"	79.6	73.7	81.6	v-l	287,501
1122	C₄H₈O	Isobutyraldehyde	63.5	Nonazeotrope			575
1123	C₄H₈O₂	Butyric acid	163.5	Nonazeotrope			678
1124	C₄H₈O₂	Dioxane	101.35	Nonazeotrope			559
		" 20-30°				v-l	916
1125	C₄H₈O₂	Ethyl acetate	76.7	74.8	57		981
		" 789.2 mm.		76.15	68.7		858
		" 583.7 mm.		66.72	73.0		858
		" 484.5 mm.		61.32	75.4		858
		" 385.2 mm.		55.22	78.6		858
		" 285.7 mm.		47.36	82.2		858
		" 685.0 mm.		71.56	70.9,	v-l	858,978
		" 20-30°C				v-l	916
1126	C₄H₈O₂	Isopropyl formate	68.8	68.0	12		562
1127	C₄H₈O₂	Methyl propionate	79.85	76.0	~75		573
1128	C₄H₈O₂	Propyl formate	80.8	74.6	60		572
1129	C₄H₉Br	2-Bromo-2-methylpropane	73.3	Nonazeotrope			563
1130	C₄H₉ClO	2-Chloroethyl ethyl ether	98.5	Nonazeotrope			575
1131	C₄H₉NO₂	Butyl nitrite	78.2	75.3	70		550
1132	C₄H₉NO₂	Isobutyl nitrite	67.1	Nonazeotrope			550
1133	C₄H₁₀O	Butyl alcohol	117.75	76.55	97.5		574
		"	117.75	76.55	97.6	v-l	386,385
1134	C₄H₁₀O	*sec*-Butyl alcohol	99.5	74.6	92.4		215
		"	99.4	74	92		623c
1135	C₄H₁₀O	*tert*-Butyl alcohol	82.55	71.1	83		29,527
1136	C₄H₁₀O	Ethyl ether	34.6	Nonazeotrope		v-l	978
1137	C₄H₁₀O	Isobutyl alcohol	108	75.8	94.5		563,1051
1138	C₄H₁₀S	Ethyl sulfide	92.2	Nonazeotrope			532
1139	C₄H₁₁N	Diethylamine 20-40°C		Nonazeotrope		v-l	457
1140	C₅H₄O₂	2-Furaldehyde	162	Nonazeotrope		v-l	1029
1141	C₅H₅N	Pyridine	115.5	Nonazeotrope			563
1141a	C₅H₁₀	Cyclopentane	49.3	Nonazeotrope		v-l	815c,304c
1142	C₅H₁₀O	Isovaleraldehyde	92.1	Nonazeotrope			575
1143	C₅H₁₀O	3-Methyl-2-butanone	95.4	Nonazeotrope			552
1143a	C₅H₁₀O	2-Methyl-3-buten-2-ol	97	75.3	94.2		581c
1144	C₅H₁₀O₂	Isobutyl formate	98.2	Nonazeotrope			575
1145	C₅H₁₀O₂	Isopropyl acetate	90.8	Nonazeotrope			547
1146	C₅H₁₀O₂	Methyl isobutyrate	92.3	Nonazeotrope			563
1147	C₅H₁₁NO₂	Isoamyl nitrite	97.15	Nonazeotrope			550
1147a	C₅H₁₂O	Amyl alcohol					
		744 mm.				v-l	213c

TABLE I.   *Binary Systems*   53

No.	Formula	B-Component Name	B.P., °C	Azeotropic Data B.P., °C	Wt.%A	Ref.
A =	CCl₄	**Carbon Tetrachloride** *(continued)*	**76.75**			
1148	C₅H₁₂O	*tert*-Amyl alcohol	102.25	76.57 95.5		525
1149	C₅H₁₂O	Isoamyl alcohol	131.3	Nonazeotrope		527
1150	C₅H₁₂O	2-Pentanol	119.8	Nonazeotrope		575
1151	C₅H₁₂O	3-Pentanol	116.0	Nonazeotrope		575
1152	C₆H₅Cl	Chlorobenzene	131.8	Nonazeotrope		563
1153	C₆H₅NO₂	Nitrobenzene	210.75	Nonazeotrope		554
		" 25°		Nonazeotrope	v-l	845
1154	C₆H₆	Benzene, <280 mm.	80.1	Azeotropic		113
		" 100 mm.		51.93 99		113
		"	80.12	Nonazeotrope,	v-l	123, 389,855, 1045,815c,304c
		"	80.1	Min. b.p. 98	v-l	710
		" 40°C.		Nonazeotrope	v-l	286
		" 760 mm.	80.1	Nonazeotrope	v-l	286
		" >1800 mm.		Min. b.p.		286
		" 15-35°C		Nonazeotrope	v-l	628
		"		Nonazeotrope	v-l	598
1155	C₆H₆O	Phenol 24-50°C.		Nonazeotrope	v-l	149
1156	C₆H₈	1,3-Cyclohexadiene	80.8	Azeotrope doubtful		563
1157	C₆H₈	1,4-Cyclohexadiene	85.6	Nonazeotrope		563
1158	C₆H₁₀	Cyclohexene	82.75	Nonazeotrope		563
		"	82.75	Nonazeotrope	v-l	815c
1159	C₆H₁₂	Cyclohexane, 40-70°C.	80.75	Nonazeotrope,	v-l	847
		"	80.75	76.5		562
		"	80.7	Nonazeotrope	v-l	598,1049, 815c
		" 20-40°C		Nonazeotrope	v-l	457
1159a	C₆H₁₂	1-Hexene	63.6	Nonazeotrope	v-l	815c
1160	C₆H₁₂	Methylcyclopentane	72.0	<71.6 <32		562
		"	72.0	Nonazeotrope	v-l	815c
1161	C₆H₁₂O₂	Butyl acetate	126.2	Nonazeotrope	v-l	108
1161a	C₆H₁₄	2,2-Dimethylbutane	49.7	Nonazeotrope	v-l	815c
1161b	C₆H₁₄	2,3-Dimethylbutane	58.0	Nonazeotrope	v-l	815c
1162	C₆H₁₄	Hexane	68.95	Azeotrope doubtful		563
		"	68.9	Nonazeotrope	v-l	815c
1162a	C₆H₁₄	2-Methylpentane	60.4	Nonazeotrope	v-l	815c
1162b	C₆H₁₄	3-Methylpentane	63.3	Nonazeotrope	v-l	815c
1163	C₆H₁₄O	Isopropyl ether 685–2280 mm.		Nonazeotrope	v-l	995,994c
1164	C₆H₁₄O	Propyl ether	90.55	Nonazeotrope		548
1165	C₆H₁₄O₂	Acetal	104.5	Nonazeotrope		563
1166	C₇H₈	Toluene	110.3	Nonazeotrope		388
				B.p. curve		388
		"	110.7	Nonazeotrope	v-l	815c
1166a	C₇H₁₄	Methylcyclohexane	101.6	Nonazeotrope	v-l	815c
1166b	C₇H₁₆	2,4-Dimethylpentane	80.8	76.3 86	v-l	815c

No.	Formula	Name	B.P., °C	B.P., °C	Wt.%A		Ref.
		B-Component		Azeotropic Data			

No.	Formula	Name	B.P., °C	B.P., °C	Wt.%A		Ref.
A =	**CCl₄**	**Carbon Tetrachloride**	**76.75**				
		(continued)					
1167	C₇H₁₆	Heptane	98.4	Nonazeotrope			899
				Vapor pressure data			
		"	98.4	Nonazeotrope		v-l	815c
1167a	C₈H₁₀	Ethylbenzene	136	Nonazeotrope		v-l	815c
1167b	C₈H₁₀	m-Xylene	139	Nonazeotrope			815c
1167c	C₈H₁₀	o-Xylene	143.6	Nonazeotrope			815c
1167d	C₈H₁₀	p-Xylene	138.4	Nonazeotrope			815c
1167e	C₈H₁₆	1-Octene	121.6	Nonazeotrope			815c
1168	C₈H₁₈	2,5-Dimethylhexane	109.4	Nonazeotrope			575
1168a	C₈H₁₈	Octane	125.7	Nonazeotrope			815c
1168b	C₈H₁₈	2,2,4-Trimethylpentane	99.3	Nonazeotrope			815c
1168c	C₉H₂₀	2,2,5-Trimethylhexane	120	Nonazeotrope			815c
A =	**CF₄**	**Carbon Tetrafluoride**					
1168d	CHF₃	Trifluoromethane					
		50°–200°F.				v-l	751c
1168e	C₃H₆	Propylene 145°K.		v.p. curve			63c
A =	**CS₂**	**Carbon Disulfide**	**46.25**				
1169	CHCl₃	Chloroform	61.2	Nonazeotrope			527,834
1170	CH₂Cl₂	Dichloromethane	40	35.7	35		215
1171	CH₂O₂	Formic acid	100.75	42.55	83		555
1172	CH₃I	Iodomethane	42.55	46	94		981
		"	42.55	41.5	40		527
1173	CH₃NO₂	Nitromethane	101.2	41.2	18.6	v-l	320
1174	CH₃NO₃	Methyl nitrate	64.8	44.25	90		555
1175	CH₄O	Methanol	64.7	39.8	71		560
1176	C₂Cl₆	Hexachloroethane	184.8	37.65	86		29,555
1177	C₂H₃Br	1-Bromoethylene	15.8	Nonazeotrope			813
1178	C₂H₄Cl₂	1,1-Dichloroethane	57.25	Nonazeotrope			566
		"	57.2	44.75	72		527
1179	C₂H₄Cl₂O	Bis(chloromethyl) ether	104	43.1	75		555
		"	105.5	Nonazeotrope			566
1180	C₂H₄O₂	Acetic acid	118.5	Nonazeotrope			834
1181	C₂H₄O₂	Methyl formate	31.7	24.75	33		555
1182	C₂H₅Br	Bromoethane	38.4	37.85	33		563,834
1183	C₂H₅Cl	Chloroethane	13	Nonazeotrope			531
1184	C₂H₅ClO	Chloromethyl methyl ether	59.15	43.1	75		555
1185	C₂H₅I	Iodoethane	72.3	Nonazeotrope			834
1186	C₂H₅NO₂	Ethyl nitrite	17.4	16.5	~5		538
1187	C₂H₅NO₂	Nitroethane	114.2	Nonazeotrope			554
1188	C₂H₅NO₃	Ethyl nitrate	87.7	Nonazeotrope			527
1189	C₂H₆O	Ethyl alcohol	78.3	42.6	91		54,555,834
1190	C₂H₆S	Methyl sulfide	37.4	Nonazeotrope			575
1191	C₃H₄O	Acrolein	52.45	<42.5	<71		566
1192	C₃H₅Br	3-Bromopropene	70.5	Nonazeotrope			566
1193	C₃H₅Cl	3-Chloropropene	45.15	41.2	50		566

TABLE I.   *Binary Systems*   55

No.	Formula	Name	B.P., °C	B.P., °C	Wt.%A	Ref.
		B-Component		**Azeotropic Data**		
A =	**CS₂**	**Carbon Disulfide** (*continued*)	**46.25**			
1194	C₃H₆O	Acetone	56.15	39.25	67	555,943
		" 1 kg./sq. cm.			66	834,964
		" 16.5 kg./sq. cm.			62.6	964
		" 32.5 kg./sq. cm.			59.4	964
		" 42 kg./sq. cm.			55.5	964
1195	C₃H₆O	Allyl alcohol	96.95	Nonazeotrope		563
		"	96.85	45.25	93.5	527
1196	C₃H₆O	Propionaldehyde	48.7	40.0	60	566
1197	C₃H₆O₂	Ethyl formate	54.15	39.35	63	555
1198	C₃H₆O₂	Methyl acetate	57	39.55	70	555,834
1199	C₃H₆O₂	Propionic acid	141.3	Nonazeotrope		566
1200	C₃H₆O₃	Methyl carbonate	90.25	45.72	91	527
1201	C₃H₇Br	1-Bromopropane	71.0	Nonazeotrope		566
1202	C₃H₇Br	2-Bromopropane	59.4	46.08	89.5	527
1203	C₃H₇Cl	1-Chloropropane	46.65	42.05	55.5	527
		"	46.6	45.2	55	981
1204	C₃H₇Cl	2-Chloropropane	35.0	33.5	~20	548
1205	C₃H₇NO₂	Isopropyl nitrite	40.1	35.5	42	527
1206	C₃H₇NO₂	Propyl nitrite	47.75	40.15	62	527
1207	C₃H₈O	Ethyl methyl ether	10.8	Nonazeotrope		215
1208	C₃H₈O	Isopropyl alcohol	82.45	44.22	92.4	555,834
1209	C₃H₈O	Propyl alcohol	97.1	45.65	94.5	324,555
1210	C₃H₈O₂	Methylal	42.25	37.25	46	555
1211	C₃H₉BO₃	Methyl borate	68.7	Nonazeotrope		527
		"	68.7	44.0	~ 84	548
1212	C₄H₄O	Furan	31.7	Nonazeotrope		566
1213	C₄H₅NS	Allylisothiocyanate	152.05	Nonazeotrope		575
1213a	C₄H₆	1,3-Butadiene	—4.6		v-l	1030c
1214	C₄H₆O₂	Biacetyl	87.5	Nonazeotrope		563
1215	C₄H₇N	Pyrroline	90.9	Nonazeotrope		575
1216	C₄H₈O	2-Butanone	79.6	45.85	84	552
1217	C₄H₈O	Butyraldehyde	75.2	Nonazeotrope		566
1218	C₄H₈O	Isobutyraldehyde	63.5	44.7	86	569
1219	C₄H₈O₂	Butyric acid	164.0	Nonazeotrope		566
1220	C₄H₈O₂	Ethyl acetate	77.1	46.02	92.7	555,834
		"	76.7	46.1	97	981
1221	C₄H₈O₂	Isobutyric acid	154.6	Nonazeotrope		575
1222	C₄H₈O₂	Isopropyl formate	68.8	43.0	~82	548
1223	C₄H₈O₂	Methyl propionate	79.85	Nonazeotrope		527
1224	C₄H₈O₂	Propyl formate	80.8	Nonazeotrope		527
1225	C₄H₉Br	2-Bromo-2-methylpropane	73.25	Nonazeotrope		566
1226	C₄H₉Cl	1-Chlorobutane	78.5	Nonazeotrope		566
1227	C₄H₉Cl	2-Chlorobutane	68.25	Nonazeotrope		566
1228	C₄H₉Cl	1-Chloro-2-methylpropane	68.85	Nonazeotrope		531
1229	C₄H₉Cl	2-Chloro-2-methylpropane	50.8	43.5	62	527
1230	C₄H₉ClO	2-Chloroethyl ethyl ether	98.5	Nonazeotrope		566

No.	Formula	B-Component Name	B.P., °C	B.P., °C	Wt.%A	Ref.
A =	CS₂	**Carbon Disulfide** *(continued)*	**46.25**			
1231	$C_4H_9NO_2$	Butyl nitrite	78.2	Nonazeotrope		527
1232	$C_4H_9NO_2$	Isobutyl nitrite	67.1	45.55	86	555
1233	$C_4H_{10}O$	Butyl alcohol	116.9	Nonazeotrope		527
1234	$C_4H_{10}O$	*tert*-Butyl alcohol	82.45	44.9	93	555
1235	$C_4H_{10}O$	Ethyl ether	34.6	34.5	13?	555,834
		"	34.6	34.4	1	981
1236	$C_4H_{10}O$	Isobutyl alcohol	107.85	Nonazeotrope		527
1237	$C_4H_{10}O$	Methyl propyl ether	38.8	36.2	~18	563
1238	$C_4H_{10}O_2$	Acetaldehyde dimethyl acetal	64.3	<45.9		575
1239	$C_4H_{11}N$	Diethylamine	55.9	Nonazeotrope		531
1240	C_5H_6O	2-Methylfuran	63.8	Nonazeotrope		566
1241	C_5H_8	Isoprene	34.3	<34.15	<7	566
1242	C_5H_{10}	Cyclopentane	49.4	44.0	67	566
1243	C_5H_{10}	2-Methyl-2-butene	37.15	36.5	~17	563
1244	C_5H_{10}	3-Methyl-1-butene	20.6	Nonazeotrope		566
1245	$C_5H_{10}O$	Cyclopentanol	140.85	Nonazeotrope		575
1246	$C_5H_{10}O$	3-Methyl-2-butanone	95.4	Nonazeotrope		552
1247	$C_5H_{10}O$	2-Pentanone	102.35	Nonazeotrope		552
1248	$C_5H_{10}O$	3-Pentanone	102.05	Nonazeotrope		552
1249	$C_5H_{10}O_2$	Ethyl propionate	99.1	Nonazeotrope		566
1250	$C_5H_{10}O_2$	Isobutyl formate	98.2	Nonazeotrope		566
1251	$C_5H_{10}O_2$	Isopropyl acetate	89.5	Nonazeotrope		575
1252	$C_5H_{10}O_2$	Isovaleric acid	176.5	Vapor pressure data		563
1253	$C_5H_{10}O_2$	Propyl acetate	101.6	Nonazeotrope		566
1254	$C_5H_{11}NO_2$	Isoamyl nitrite	97.15	Nonazeotrope		527
1255	C_5H_{12}	2-Methylbutane	27.95	Nonazeotrope		538
1256	C_5H_{12}	Pentane	36.15	35.7	11	555
1257	$C_5H_{12}O$	Amyl alcohol	138.2	Nonazeotrope		566
1258	$C_5H_{12}O$	*tert*-Amyl alcohol	102.35	Nonazeotrope		527
1259	$C_5H_{12}O$	Ethyl propyl ether	63.85	Nonazeotrope		566
1260	$C_5H_{12}O$	Isoamyl alcohol	131.9	Nonazeotrope		527
1261	$C_5H_{12}O$	3-Methyl-2-butanol	112.9	Nonazeotrope		575
1262	$C_5H_{12}O$	2-Pentanol	119.8	Nonazeotrope		575
1263	$C_5H_{12}O$	3-Pentanol	116.0	Nonazeotrope		575
1264	$C_5H_{12}O_2$	Diethoxymethane	87.95	Nonazeotrope		566
1265	C_6H_6	Benzene	80.2	Nonazeotrope	v-l	89,834
1266	C_6H_8	1,3-Cyclohexadiene	80.4	Nonazeotrope		575
1267	C_6H_{10}	Cyclohexene	82.75	Nonazeotrope		566
1268	C_6H_{10}	Methylcyclopentene	75.85	Nonazeotrope		575
1269	C_6H_{12}	Cyclohexane	80.75	Nonazeotrope		563
1270	C_6H_{12}	Methylcyclopentane	72.0	Nonazeotrope		566
1271	$C_6H_{12}O$	Pinacolone	106.2	Nonazeotrope		566
1272	$C_6H_{12}O$	4-Methyl-2-pentanone	116.05	Nonazeotrope		552
1273	C_6H_{14}	2,3-Dimethylbutane	58.0	<46.15	<97	566
1274	C_6H_{14}	Hexane	68.95	Nonazeotrope		563

TABLE I.   *Binary Systems*   57

No.		B-Component			Azeotropic Data		
	Formula	Name	B.P., °C	B.P., °C	Wt.%A	Ref.
A =	CS₂	**Carbon Disulfide** *(continued)*	**46.25**			
1275	C₆H₁₅N	Triethylamine	89.35	Nonazeotrope		575
1276	C₇H₈	Toluene	110.7	Nonazeotrope		563
1277	C₇H₁₄	Methylcyclohexane	101.15	Nonazeotrope		566
1278	C₇H₁₆	Heptane	98.45	Nonazeotrope		575
A =	CHBrCl₂	**Bromodichloromethane**	**90.2**			
1279	CH₂O₂	Formic acid	100.7	78.15	~76	563
1280	CH₃NO₂	Nitromethane	101.2	87.3	75	554
1281	CH₃NO₃	Methyl nitrate	64.8	Nonazeotrope		560
1282	CH₄O	Methanol	64.7	63.8	60	563
1283	C₂HCl₃	Trichloroethylene	86.9	Nonazeotrope		549
		"	86.95	86.7	22	528
1284	C₂HCl₃O	Chloral	97.75	90.1	97.5	572
1285	C₂H₄O₂	Acetic acid	118.5	Nonazeotrope		563
1286	C₂H₅BrO	2-Bromoethanol	150.2	Nonazeotrope		575
1287	C₂H₅ClO	2-Chloroethanol	128.6	Nonazeotrope		564
1288	C₂H₅NO₂	Nitroethane	114.2	Nonazeotrope		554
1289	C₂H₅NO₃	Ethyl nitrate	90.1	86.85	35	527
1290	C₂H₆O	Ethyl alcohol	78.3	75.5	72	563
1291	C₂H₆O₂	Glycol	197.4	Nonazeotrope		573
1292	C₃H₆O	Acetone	56.15	Nonazeotrope		552
1293	C₃H₆O	Allyl alcohol	96.95	85.85	82.5	563,875
1294	C₃H₆O₃	Methyl carbonate	90.35	91.95	64.5	572
1295	C₃H₇ClO	1-Chloro-2-propanol	127.0	Nonazeotrope		575
1296	C₃H₇I	2-Iodopropane	89.45	90.7	<50	549
1297	C₃H₈O	Isopropyl alcohol	82.45	79.4	62	572
1298	C₃H₈O	Propyl alcohol	97.2	86.4	80.5	563
1299	C₃H₉BO₃	Methyl borate	68.7	Nonazeotrope		547
1300	C₄H₈O	2-Butanone	79.6	90.85	89.5	569
1301	C₄H₈O₂	Ethyl acetate	77.1	90.55	88	572
1302	C₄H₈O₂	Methyl propionate	79.85	91.2	~85	538
1303	C₄H₈O₂	Propyl formate	80.85	90.9	82	573
1304	C₄H₉Br	2-Bromobutane	91.2	91.65	45	562
1305	C₄H₉Br	1-Bromo-2-methylpropane	91.4	91.8	45	549
1306	C₄H₉NO₂	Butyl nitrite	78.2	Nonazeotrope		550
1307	C₄H₁₀O	Butyl alcohol	117.75	Nonazeotrope		527
1308	C₄H₁₀O	*sec*-Butyl alcohol	99.6	87.5		563
1309	C₄H₁₀O	*tert*-Butyl alcohol	82.55	79.0	~65	532
1310	C₄H₁₀O	Isobutyl alcohol	108	89.3	89	563
1311	C₄H₁₀O₂	Acetaldehyde dimethyl acetal	64.3	Nonazeotrope		559
1312	C₄H₁₀S	Ethyl sulfide	92.2	96.7	~58	531
1313	C₅H₁₀O	3-Methyl-2-butanone	95.4	97.2	50	552
1314	C₅H₁₀O	2-Pentanone	102.35	102.85	35	552
1315	C₅H₁₀O	3-Pentanone	102.05	102.65	36	552
1316	C₅H₁₀O₂	Butyl formate	106.7	Nonazeotrope		547
1317	C₅H₁₀O₂	Ethyl propionate	99.15	100.6	35	538

		B-Component		Azeotropic Data		
No.	Formula	Name	B.P., °C	B.P., °C	Wt.%A	Ref.
A =	**CHBrCl₂**	**Bromodichloromethane**	**90.2**			
		(*continued*)				
1318	C₅H₁₀O₂	Isobutyl formate	97.9	98.7	40	573
1319	C₅H₁₀O₂	Isopropyl acetate	90.8	96.0	55	547
1320	C₅H₁₀O₂	Methyl butyrate	102.65	103.5	25	538
1321	C₅H₁₀O₂	Methyl isobutyrate	92.3	93.8	58	573
1322	C₅H₁₀O₂	Propyl acetate	101.6	102.3	29.5	572
1323	C₅H₁₁NO₂	Isoamyl nitrite	97.15	Nonazeotrope		550
1324	C₅H₁₂O	*tert*-Amyl alcohol	102.0	~88.8	~92	535
1325	C₅H₁₂O	3-Methyl-2-butanol	112.6	Nonazeotrope		575
1326	C₆H₁₂O₂	Diethoxymethane	87.9	94.05	74	568
1327	C₆H₆	Benzene	80.2	Nonazeotrope		528
1328	C₆H₁₀	Cyclohexene	82.75	82		563
1329	C₆H₁₂	Cyclohexane	80.75	Nonazeotrope		575
1330	C₆H₁₂	Methylcyclopentane	72.0	Nonazeotrope		575
1331	C₆H₁₂O	4-Methyl-2-pentanone	116.05	Nonazeotrope		527
1332	C₆H₁₂O	Pinacolone	106.2	Nonazeotrope		552
1333	C₆H₁₄	Hexane	68.8	Nonazeotrope		575
1334	C₆H₁₄O	Propyl ether	90.55	97.0	54	559
1335	C₆H₁₄O₂	Acetal	103.55	Nonazeotrope		572
1336	C₇H₁₄	Methylcyclohexane	101.15	Nonazeotrope		575
1337	C₇H₁₆	Heptane	98.4	<90.0		575
A =	**CHBr₃**	**Bromoform**	**149.5**			
1338	CH₂O₂	Formic acid	100.75	97.4	52	568
1339	C₂H₂Cl₄	1,1,2,2-Tetrachloroethane	146.2	145.5	45	549
1340	C₂H₃BrO₂	Bromoacetic acid	205.1	Nonazeotrope		575
1341	C₂H₃ClO₂	Chloroacetic acid	189.35	148.5	96.9	564
1342	C₂H₄Br₂	1,2-Dibromoethane	129.8	Nonazeotrope, b.p. curve		388,563
1343	C₂H₄Cl₂O	2,2-Dichloroethanol	146.2	<143.0	<55	575
1344	C₂H₄O₂	Acetic acid	118.5	118.3	18	542
1345	C₂H₅ClO	2-Chloroethanol	128.6	127.4	46	564
1346	C₂H₅NO	Acetamide	221.15	Nonazeotrope		527
		"	221.2	149	98	535
1347	C₂H₆O₂	Glycol	197.4	146.75	~93.5	532
1348	C₂H₇NO	2-Aminoethanol	170.8	Reacts		527
1349	C₃H₅ClO₂	Methyl chloroacetate	130.0	Nonazeotrope		532
1350	C₃H₆Cl₂O	1,3-Dichloro-2-propanol	175.8	Nonazeotrope		575
1351	C₃H₆O	Allyl alcohol	96.95	Nonazeotrope		532
1352	C₃H₆O₂	Propionic acid	140.9	138.0	63	527
1353	C₃H₇ClO	1-Chloro-2-propanol	127.0	Nonazeotrope		575
1354	C₃H₇NO	Propionamide	222.2	Nonazeotrope		527
1355	C₃H₇NO₂	Ethyl carbamate	185.25	149.25	97.5	564
1356	C₃H₈O	Propyl alcohol	97.2	Nonazeotrope		532
1357	C₃H₈O₂	2-Methoxyethanol	124.5	Nonazeotrope		526
1358	C₄H₆O₄	Methyl oxalate	164.45	Nonazeotrope		575
1359	C₄H₇BrO₂	Ethyl bromoacetate	~158.2	Nonazeotrope		538
1360	C₄H₇ClO₂	Ethyl chloroacetate	143.55	143.52	4	527
1361	C₄H₇Cl₃O	Ethyl 1,1,2-trichloroethyl ether	172.5	Nonazeotrope		575

TABLE I. *Binary Systems*   59

No.	Formula	B-Component Name	B.P., °C	Azeotropic Data B.P., °C	Wt.%A	Ref.
A =	**CHBr₃**	**Bromoform** *(continued)*	**149.5**			
1362	C₄H₈Cl₂O	Bis(2-chloroethyl) ether	178.65	Nonazeotrope		527
1363	C₄H₈Cl₂O	1,2-Dichloroethyl ethyl ether	145.5	151.3	91	575
1364	C₄H₈O₂	Butyric acid	162.45	146.8	93.2	389,526
1365	C₄H₈O₂	Isobutyric acid	154.35	145.5	81	541
1366	C₄H₈O₃	Methyl lactate	144.8	~152		563
1367	C₄H₉I	1-Iodo-2-methylpropane	120.4	Nonazeotrope		389
1368	C₄H₁₀O	Butyl alcohol	117.75	Nonazeotrope		527
1369	C₄H₁₀O	Isobutyl alcohol	107.85	Nonazeotrope		532
1370	C₄H₁₀O₂	2-Ethoxyethanol	135.3	Nonazeotrope		556
1371	C₅H₄O₂	2-Furaldehyde	161.45	Nonazeotrope		527
1372	C₅H₉ClO₂	Propyl chloroacetate	162.5	Nonazeotrope		575
1373	C₅H₁₀O₂	Isovaleric acid	176.5	148.7	96	527
		"	176.5	Nonazeotrope		542
1374	C₅H₁₀O₂	Valeric acid	186.35	Nonazeotrope		575
1375	C₅H₁₁I	1-Iodo-3-methylbutane	147.65	Nonazeotrope		531
1376	C₅H₁₁NO₃	Isoamyl nitrate	149.75	144.8	57	560
1377	C₅H₁₂O	Isoamyl alcohol	129	Nonazeotrope		389
		"	130.8	131.35	43	527
1378	C₅H₁₂O₂	2-Propoxyethanol	151.35	147.15	84	527
1379	C₆H₅Br	Bromobenzene	156.1	Nonazeotrope		549
1380	C₆H₅Cl	Chlorobenzene	131.75	Nonazeotrope		575
1381	C₆H₅ClO	o-Chlorophenol	176.8	Nonazeotrope		575
1382	C₆H₅NO₂	Nitrobenzene	210.75	Nonazeotrope		554
1383	C₆H₆O	Phenol	182.2	Nonazeotrope		542
1384	C₆H₇N	Aniline	184.35	Nonazeotrope		575
1385	C₆H₁₀O	Cyclohexanone	155.6	158.5	~52	573
1386	C₆H₁₀O	Mesityl oxide	129.45	Nonazeotrope		527
1387	C₆H₁₀O₄	Ethylidene diacetate	168.5	Nonazeotrope		527
1388	C₆H₁₀S	Allyl sulfide	139.35	>150.5		566
1389	C₆H₁₂O	Cyclohexanol	160.7	Nonazeotrope		532
		"		149.5?	95?	545
1390	C₆H₁₄O	Hexyl alcohol	157.85	147.7	86	542
1391	C₆H₁₄O₂	2-Butoxyethanol	171.15	Nonazeotrope		556
1392	C₆H₁₄S	Propyl sulfide	140.8	151.5	90	555
1393	C₇H₈	Toluene	110.65	Nonazeotrope		563
1394	C₇H₈O	Anisole	153.85	Nonazeotrope		563
1395	C₇H₈O	o-Cresol	191.1	Nonazeotrope		575
1396	C₇H₁₄O	4-Heptanone	143.55	151.0	77	552
1397	C₆H₁₄O	3-Methylcyclohexanol	168.5	Nonazeotrope		575
1398	C₇H₁₄O₂	Amyl acetate	148.8	>154.0	<65	562
1399	C₇H₁₄O₂	Ethyl isovalerate	134.7	Nonazeotrope		547
1400	C₇H₁₄O₂	Ethyl valerate	145.45	>152.7		575
1401	C₇H₁₄O₂	Isoamyl acetate	142.1	150.2	82	389,573
1402	C₇H₁₄O₂	Isobutyl propionate	137.5	150.0		575
1403	C₇H₁₆O	Heptyl alcohol	176.15	Nonazeotrope		575

No.	Formula	Name	B.P., °C	B.P., °C	Wt.%A	Ref.
		B-Component		**Azeotropic Data**		

A = CHBr₃   Bromoform *(continued)*   **149.5**

No.	Formula	Name	B.P., °C	B.P., °C	Wt.%A	Ref.
1404	$C_7H_{16}O_3$	Ethyl orthoformate	145.75	Nonazeotrope		559
1405	C_8H_{10}	Ethylbenzene	136.15	Nonazeotrope		575
1406	C_8H_{10}	m-Xylene	139.0	Nonazeotrope		538
1407	$C_8H_{10}O$	Benzyl methyl ether	167.8	Nonazeotrope		559
1408	$C_8H_{16}O_2$	Butyl butyrate	166.4	Nonazeotrope		557
1409	$C_8H_{16}O_2$	Isoamyl propionate	160.7	>161.0	>18	575
1410	$C_8H_{16}O_2$	Isobutyl butyrate	156.8	157.7	35	573
1411	$C_8H_{16}O_2$	Isobutyl isobutyrate	147.3	151	75	573
1412	$C_8H_{18}O$	Butyl ether	142.2	Nonazeotrope		548
1413	$C_8H_{18}O$	Isobutyl ether	122.3	Nonazeotrope		559
1414	C_9H_{12}	Propylbenzene	158.9	Nonazeotrope		538
1415	$C_{10}H_{16}$	Camphene	159.6	~ 148.5	~ 95	535
1416	$C_{10}H_{16}$	α-Pinene	155.8	146.5	75	528
1417	$C_{10}H_{16}$	Nopinene	163.8	<149.0	>91	575
1418	$C_{10}H_{22}$	2,7-Dimethyloctane	160.25	Nonazeotrope		573

A = CHClF₂   Chlorodifluoromethane   — 40.8

No.	Formula	Name	B.P., °C	B.P., °C	Wt.%A	Ref.
1419	C_2ClF_5	Chloropentafluoroethane	— 38.5	— 45.6	48.7	47
1420	$C_2H_2Cl_2F_2$	1,2-Dichloro-1,2-difluoroethane, 755 mm.	29.8	— 41.4	87.6	266
1421	C_3F_6	Hexafluoropropene, 2059 mm.	— 6.1	— 17.3	69.7	1019
		" 9.6 atm.		20	77	615
		"		Min. b.p.		954f
1422	C_3F_8	Perfluoropropane, 6.064 atm.	12.5	0	46	745
1423	C_3H_8	Propane, 86.2 p.s.i.a.		0	68	794
		" 6.002 atm.	8.6	0	68.3	745
1424	C_4F_8	Perfluorocyclobutane, 2059 mm.	21.0	Nonazeotrope		1019
				Min. b.p.		922

A = CHCl₂F   Dichlorofluoromethane   7.63/723 mm.

No.	Formula	Name	B.P., °C	B.P., °C	Wt.%A	Ref.
1425	$C_2Cl_2F_4$	1,2-Dichloro-1,1,2,2-tetrafluoroethane, 723 mm.	2.22	0.00	25	745

A = CHCl₃   Chloroform   61.2

No.	Formula	Name	B.P., °C	B.P., °C	Wt.%A		Ref.
1426	CH_2Cl_2	Dichloromethane	41.5	Nonazeotrope			261
1427	CH_2O_2	Formic acid	100.7	59.15	85		537
		"	100.75			v-l	171
1428	CH_3I	Iodomethane	42.5	Nonazeotrope			575
1429	CH_3NO_2	Nitromethane	101.15	Nonazeotrope			548

TABLE I. *Binary Systems*   61

No.	Formula	Name	B.P., °C	B.P., °C	Wt.%A		Ref.
		B-Component		**Azeotropic Data**			
A =	**CHCl₃**	**Chloroform** *(continued)*	**61.2**				
1430	CH₄O	Methanol	64.7	53.43	87.4	v-l	679,978
		"		20	91.7		371,464
		"		35	89.7 }		563,
		"		49	87.8 }		834
		"	64.7			v-l	110
		"	400 mm.	36.3	88.9	v-l	687
		"	500 mm.	41.6	88.4	v-l	687
		"	600 mm.	46.2	87.9	v-l	687
1431	C₂Cl₄	Tetrachloroethylene	121.1	Nonazeotrope		v-l	187
1432	C₂H₃Cl₃O₂	Chloral hydrate	97.5	Nonazeotrope			563
1433	C₂H₃N	Acetonitrile	81.6	Nonazeotrope		v-l	635c,981
1433a	C₂H₃NO	Methylisocyanate	37.9	Nonazeotrope		v-l	931c
1434	C₂H₄Cl₂	1,1-Dichloroethane	57.3			v-l	435
1435	C₂H₄Cl₂	1,2-Dichloroethane	83.28	Nonazeotrope		v-l	463
1436	C₂H₄Cl₂O	Bis(chloromethyl) ether	59.15	>63.9	<80		575
1437	C₂H₄O₂	Acetic acid	118.1	Nonazeotrope			575
		"	118.1	Nonazeotrope		v-l	171
1438	C₂H₄O₂	Methyl formate	31.9	Nonazeotrope			563
1439	C₂H₅Br	Bromoethane	38.4	Nonazeotrope			834
1440	C₂H₅Cl	Chloroethane	13.3	Nonazeotrope			563
1441	C₂H₅I	Iodoethane	72.3	Nonazeotrope			834
1442	C₂H₆O	Ethyl alcohol	78.3	59.35	93		1000
		"		35	95.9	v-l }	371,834
		"		45	94.8	v-l }	846,
		"		55	93.7	v-l }	960
		"	20 p.s.i.g.	101.7	82	89	982
1443	C₃H₆O	Acetone,	101 mm.	15	74.3		815
		"	129 mm.	20	75.0		815
		"	202 mm.	30	76.1		815
		"	250 mm.	35	76.3		815
		"	308 mm.	40	76.7		815
		"	455 mm.	50	77.1		815
		"	546 mm.	55	77.3		815
		"	56.5	64.4	78.1		498
		"	547 mm.	55	78.0	v-l	498
		"	380 mm.	45	77.9	v-l	498
		"	251 mm.	35	77.8	v-l	498
		"	56.10	64.43	78.5	v-l	802
				64-65	80	{ 282,371 803,834,903 960,978	
1444	C₃H₆O	Allyl alcohol	96.95	Nonazeotrope			563
1445	C₃H₆O	Propionaldehyde	50	Max. b.p.			262
1446	C₃H₆O	Propylene oxide	35	Nonazeotrope			262
1447	C₃H₆O₂	Ethyl formate	54.15	62.7	87		563,579,720
1448	C₃H₆O₂	Methyl acetate	57.1	64.74	64.35	v-l	110,579
		"	57.05	64.8	77		563

		B-Component		Azeotropic	Data	
No.	Formula	Name	B.P., °C	B.P., °C	Wt.%A	Ref.
A =	CHCl₃	**Chloroform** *(continued)*	**61.2**			
		Methyl acetate	57	64-65	78	834,903
		"		20	52.9	814
		"		40	50.3	814
		"		63.3	46.6	814
1449	C₃H₇Br	1-Bromopropane	71.0	Nonazeotrope		549
1450	C₃H₇Br	2-Bromopropane	59.4	62.2	65	570,579,720
1451	C₃H₇NO₂	Isopropyl nitrite	40.1	Nonazeotrope		550
1452	C₃H₇NO₂	Propyl nitrite	47.75	Nonazeotrope		550
1453	C₃H₈O	Isopropyl alcohol	82.45	Nonazeotrope		834
		"	82.45	60.8	95.5	563
		"	82.5	Nonazeotrope	v-l	680
1454	C₃H₈O	Propyl alcohol	97.2	Nonazeotrope		563
1455	C₃H₈O₂	Methylal	42.3	61.8	92.5	262,527
1456	C₃H₈S	Ethyl methyl sulfide				
		700 mm.	64.0	66.6	52.3	v-l 635
1457	C₃H₈S	Propanethiol	67.5	Nonazeotrope		563
1458	C₃H₉BO₃	Methyl borate	68.7	> 70		575
1459	C₃H₉SiCl	Chlorotrimethylsilane	57.5	Nonazeotrope		844
1460	C₄H₈O	2-Butanone	79.6	Nonazeotrope		527,1025
		"	79.6	Nonazeotrope	v-l	492,501
		"	79.6	79.9	17	981
1461	C₄H₈O	Butyraldehyde	76	Max. b.p.		262
		"	76	Nonazeotrope		575
1462	C₄H₈O	Isobutyraldehyde	63	Max. b.p.		262
		"	63	Nonazeotrope		575
1463	C₄H₈O	Isobutylene oxide	50	Max. b.p.		262
1464	C₄H₈O	Tetrahydrofuran	66	72.5	65.5	224
1465	C₄H₈O₂	Dioxane	101	Nonazeotrope		262
1466	C₄H₈O₂	Ethyl acetate	76	Nonazeotrope		834
		"	77.1	77.8	28.1	679
1467	C₄H₈O₂	Isopropyl formate	68.8	70.0	>14	562
		"	68.8	70	13	579
1468	C₄H₈O₂	Methyl propionate	79.85	Nonazeotrope		575
1469	C₄H₈O₂	Propyl formate	80.85	Nonazeotrope		575
1470	C₄H₉NO₂	Butyl nitrite	78.2	Nonazeotrope		550
1471	C₄H₉NO₂	Isobutyl nitrite	67.1	Nonazeotrope		550
1472	C₄H₁₀O	sec-Butyl alcohol	99.5	Nonazeotrope		575
1473	C₄H₁₀O	tert-Butyl alcohol	82.55	Nonazeotrope		531,581c
1474	C₄H₁₀O	Ethyl ether	35	Nonazeotrope		262,834
		"	34.5	Nonazeotrope	v-l	481
1475	C₄H₁₀O	Isobutyl alcohol	108.0	Nonazeotrope		575
1476	C₄H₁₀O	Methyl propyl ether	38.9	Nonazeotrope		559
1477	C₄H₁₀O₂	Ethoxymethoxymethane	65.9	> 67.5	20	559
1478	C₄H₁₀O₂	Acetaldehyde dimethyl				
		acetal	64.3	67.2	32	559
1479	C₄H₁₀S	Ethyl sulfide	92.2	Nonazeotrope		531

TABLE I.   *Binary Systems*   63

No.	Formula	Name	B.P., °C	B.P., °C	Wt.%A	Ref.
		B-Component		**Azeotropic Data**		
A =	CHCl₃	**Chloroform** *(continued)*	**61.2**			
1480	C₅H₄O₂	Furfuraldehyde	162	Nonazeotrope	v-l	2
1480a	C₅H₅N	Pyridine 50°–63.5°C.		Nonazeotrope	v-l	274c
1481	C₅H₁₀	2-Methyl-2-butene	37.15	Nonazeotrope		563
1481a	C₅H₁₀O	2-Methyl-3-buten-2-ol	97.0	Nonazeotrope		581c
1482	C₅H₁₂	Pentane	36.15	Nonazeotrope		563
1483	C₅H₁₂O	Ethyl propyl ether	63.85	> 69.0	>35	559
		" 400 mm.	44.6	49.05	58	v-l	148f
		" 650 mm.	58.2	62.45	56	v-l	148f
		" 698 mm.	60.6	64.6	54.7	v-l	635
1484	C₆H₅Cl	Chlorobenzene	131.8	Nonazeotrope		563
1485	C₆H₅NO₂	Nitrobenzene	210.75	Nonazeotrope		554
1486	C₆H₆	Benzene	79.90	Nonazeotrope,	v-l	262,
						646,802,834,978
1487	C₆H₈	1,3-Cyclohexadiene	80.8	Nonazeotrope		563
1488	C₆H₁₀	Biallyl	60.2	~ 55		563
1489	C₆H₁₀	Cyclohexene	82.75	Nonazeotrope		563
1490	C₆H₁₂	Cyclohexane	80.75	Nonazeotrope		563
1491	C₆H₁₂	Methylcyclopentane	72.0	60.5	80	575
		"	71.9	Azeotropic		929
1492	C₆H₁₂O	4-Methyl-2-pentanone	115.9	Nonazeotrope	v-l	438
1493	C₆H₁₂O₂	Butyl acetate	126.2	Nonazeotrope	v-l	186
1494	C₆H₁₄	2,3-Dimethylbutane	58.0	55.5	47	562
		"		55.5	55.3	638c
		"	58.0	56	~53	v-l	1026c
1495	C₆H₁₄	Hexane	68.7	60.4	83.5	v-l	498
		" 637 mm.		55	82.6	v-l	498
		" 454 mm.		45	82.3	v-l	498
		" 310 mm.		35	82.1	v-l	498
		"	68.95	59.95	72	563
1495a	C₆H₁₄O	Butyl ethyl ether				
		" 250 mm.	59.0	Nonazeotrope	v-l	148f
		" 650 mm.	86.6	Nonazeotrope	v-l	148f
1496	C₆H₁₄O	Isopropyl ether	68	70.5	36	v-l	261,262
		" 246 mm.	36.5	39.6	44	v-l	148f
		" 577 mm.	59.3	61.5	44	v-l	148f
1496a	C₆H₁₄O	Propyl ether 400 mm.	70	Nonazeotrope	v-l	148f
		" 650 mm.	84	Nonazeotrope	v-l	148f
1497	C₆H₁₅N	Triethylamine	89	Nonazeotrope		262
1498	C₇H₈	Toluene	110.65	Nonazeotrope		563
1499	C₇H₁₄	Methylcyclohexane	101.15	Nonazeotrope		575
1500	C₇H₁₆	Heptane	98.45			563
1501	C₈H₁₈	2,5-Dimethylhexane	109.4	Nonazeotrope		575
1501a	C₈H₁₈O	Butyl ether 390 mm.	118.4	Nonazeotrope	v-l	148f
		" 650 mm.	132.0	Nonazeotrope	v-l	148f
1502	C₉H₁₀O₂	Ethyl benzoate	213.3	Nonazeotrope		981
A =	CHF₃	**Fluoroform**				
1503	C₂H₆	Ethane	— 88	— 96	58	120

		B-Component		Azeotropic	Data	
No.	Formula	Name	B.P., °C	B.P., °C	Wt.%A	Ref.
A =	CH₂ClBr	Bromochloromethane	69			
1504	CH₂Cl₂	Dichloromethane	40.7	Nonazeotrope		266
A =	CH₂Br₂	Dibromomethane	97.0			
1505	CH₄O	Methanol	64.65	64.25	52	575
	"		64.7	Azeotrope doubtful		563
1506	C₂H₄O₂	Acetic acid	118.1	94.8	84	562
1507	C₂H₅ClO	2-Chloroethanol	128.6	Nonazeotrope		575
1508	C₂H₅NO₃	Ethyl nitrate	87.7	Nonazeotrope		560
1509	C₂H₆O	Ethyl alcohol	78.3	76	62	573
1510	C₃H₆O	Allyl alcohol	96.95	~ 86.5	~80	563
1511	C₃H₆O₂	Propionic acid	141.3	Nonazeotrope		575
1512	C₃H₆O₃	Methyl carbonate	90.35	Nonazeotrope		547
1513	C₃H₇ClO	1-Chloro-2-propanol	127.0	Nonazeotrope		575
1514	C₃H₈O	Isopropyl alcohol	82.4	< 81.0	>32	575
1515	C₃H₈O	Propyl alcohol	97.2	< 90.5	>74	567
1516	C₄H₉Br	1-Bromobutane	101.5	Nonazeotrope		575
1517	C₄H₉ClO	2-Chloroethyl ethyl ether	98.5	< 96.0	<72	575
1518	C₄H₁₀O	Isobutyl alcohol	108.0	94.8	82	567
1519	C₄H₁₀S	Butanethiol	97.5	< 95.5	<72	575
1520	C₅H₁₀O	3-Methyl-2-butanone	95.4	98.0	70	552
1521	C₅H₁₀O	2-Pentanone	102.35	Nonazeotrope		552
1522	C₅H₁₀O	3-Pentanone	102.05	Nonazeotrope		552
1523	C₅H₁₀O₂	Isopropyl acetate	90.8	Nonazeotrope		547
1524	C₅H₁₀O₂	Methyl isobutyrate	92.3	92		547
1525	C₅H₁₀O₂	Propyl acetate	101.6	Nonazeotrope		547
1526	C₅H₁₁NO₂	Isoamyl nitrite	97.15	96.5		550
1527	C₅H₁₂O	Isoamyl alcohol	131.9	Nonazeotrope		575
1528	C₅H₁₂O₂	Diethoxymethane	87.95	Nonazeotrope		527
1529	C₆H₆	Benzene	80.15	Nonazeotrope		575
1530	C₆H₁₄O	Propyl ether	90.1	Nonazeotrope		559
1531	C₆H₁₄O₂	Acetal	103.55	Nonazeotrope		559
1532	C₇H₈	Toluene	110.75	Nonazeotrope		575
1533	C₇H₁₄	Methylcyclohexane	101.15	< 96.4	<75	575
1534	C₇H₁₆	Heptane	98.4	< 95.5	>58	562
A =	CH₂ClF	Chlorofluoromethane				
1534a	C₂Cl₂F₄	1,1-Dichlorotetrafluoro-ethane 45 p.s.i.a.		18	55	885c
A =	CH₂ClNO₂	Chloronitromethane	122.5			
1535	C₂Cl₄	Tetrachloroethylene	121.1	115.2	45	562
1536	C₅H₅N	Pyridine	115.4	Nonazeotrope		553
1537	C₅H₁₁Br	1-Bromo-3-methylbutane	120.65	115.5	40	562
1538	C₅H₁₁Cl	1-Chloro-3-methylbutane	99.4	Nonazeotrope		575
1539	C₆H₁₄S	Isopropyl sulfide	120.5	<119.7	20	554
1540	C₇H₈	Toluene	110.75	Nonazeotrope		575
1541	C₈H₁₈	Octane	125.75	<121.0	<80	575

TABLE I. *Binary Systems* 65

No.	Formula	Name	B.P., °C	B.P., °C	Wt.%A		Ref.
		B-Component		**Azeotropic Data**			
A =	**CH$_2$Cl$_2$**	**Dichloromethane**	**40.0**				
1542	CH$_3$I	1-Iodomethane	42.5	39.8	79		527
1543	CH$_3$NO$_3$	Methyl nitrate	64.8	Nonazeotrope			527
1544	CH$_4$O	Methanol	64.65	37.8	92.7		261,527
1545	C$_2$Cl$_3$F$_3$	1,1,2-Trichlorotri-fluoroethane	47.6	37	48		75
1546	C$_2$H$_3$N	Acetonitrile	81.6	Nonazeotrope			565
1547	C$_2$H$_4$O	Acetaldehyde	20.65	Nonazeotrope			575
1548	C$_2$H$_4$O$_2$	Methyl formate	32	Nonazeotrope			262
			31.9	~ 30.8	~20		563
1549	C$_2$H$_5$Br	1-Bromoethane	38.4	38.1	20		549
1550	C$_2$H$_5$ClO	Chloromethyl methyl ether	59.15	Nonazeotrope			835
1551	C$_2$H$_6$O	Ethyl alcohol	78.3	< 39.85	>95		527
		" 680 mm.		Nonazeotrope		v-l	434f
1552	C$_2$H$_6$S	Ethanethiol	36.2	Nonazeotrope			563
1553	C$_3$H$_6$O	Acetone	56.	Nonazeotrope			261,262
		" 680 mm.		Nonazeotrope		v-l	434f
1554	C$_3$H$_6$O	Propionaldehyde	50	Max. b.p.			262
			50	Nonazeotrope			575
1555	C$_3$H$_6$O	Propylene oxide	34.1	40.6	77		262,559
1556	C$_3$H$_6$O$_2$	Ethyl formate	54	Nonazeotrope			262
		"	54.15	41	92		547
1557	C$_3$H$_6$O$_2$	Methyl acetate	57.0	Nonazeotrope			527
1558	C$_3$H$_6$O$_3$	Trioxane	114.5			v-l	866
1559	C$_3$H$_7$NO$_2$	Isopropyl nitrite	40.1	39.45	53		527
1560	C$_3$H$_7$NO$_2$	Propyl nitrite	47.75	Nonazeotrope			527
1561	C$_3$H$_8$O	Isopropyl alcohol	82.4	Nonazeotrope			527
1562	C$_3$H$_8$O$_2$	Methylal	42.3	45.0	41		262,559
1563	C$_4$H$_4$O	Furan	31.7	Nonazeotrope			527
1563a	C$_4$H$_6$	1,3-Butadiene	—4.6	Nonazeotrope		v-l	1030c
1564	C$_4$H$_8$O	2-Butanone	79.6	Nonazeotrope		v-l	620
1565	C$_4$H$_8$O	Isobutylene oxide	50	Max. b.p.			262
1566	C$_4$H$_{10}$O	Ethyl ether	34.6	40.8	70		262,527
		"	34.6	40.2	68	v-l	996
1567	C$_4$H$_{10}$O	Methyl propyl ether	38.9	44.8	57		527
1568	C$_5$H$_{10}$	Cyclopentane	49.3	38.0	70		527
1569	C$_5$H$_{10}$	2-Methyl-2-butene	37.1	< 36.5	< 52		527
1570	C$_5$H$_{12}$	2-Methylbutane	27.95	26.0	27		527
1571	C$_5$H$_{12}$	Pentane	36.15	< 35.5	<49		575
1572	C$_6$H$_{12}$	Methylcyclopentane	72.0	Nonazeotrope			527
1573	C$_6$H$_{14}$	2,2-Dimethylbutane, 742 mm.	49.74	35.6	53 vol. %		692
		"	49.7	Min. b. p.			59
1574	C$_6$H$_{14}$	2,3-Dimethylbutane	58	Min. b. p.			59
		"	58	39.0	83		527
1575	C$_6$H$_{14}$	Hexane	68.8	Nonazeotrope			527
A =	**CH$_2$I$_2$**	**Diiodomethane**	**181**				
1576	C$_2$H$_4$O$_2$	Acetic acid	118.1	Nonazeotrope			565

	B-Component			Azeotropic Data		
No.	Formula	Name	B.P., °C	B.P., °C	Wt.%A	Ref.
A =	**CH₂I₂**	**Diiodomethane** *(continued)*	**181**			
1577	C₂H₆O₂	Glycol	197.4	168.7	86	569
1578	C₂H₇NO	2-Aminoethanol	170.8	Reacts		527
1579	C₃H₆O₂	Propionic acid	141.3	140.65	27	527
1580	C₃H₇NO₂	Ethyl carbamate	185.25	169.35	75	527
1581	C₄H₈O₂	Butyric acid	164.0	159.1	60	569
1582	C₄H₈O₂	Isobutyric acid	154.6	151.8	47	527
1583	C₄H₁₀O₂	2-Ethoxyethanol	135.3	Nonazeotrope		575
1584	C₅H₆O₂	Furfuryl alcohol	169.35	165.8	55	565
1585	C₅H₁₀O₂	Isovaleric acid	176.5	168.5	75	565
1586	C₅H₁₂O	Isoamyl alcohol	131.9	Nonazeotrope		565
1587	C₅H₁₂O₂	2-Propoxyethanol	151.35	Nonazeotrope		575
1588	C₆H₄Cl₂	*p*-Dichlorobenzene	174.4	171.3	48	549
1589	C₆H₅Br	Bromobenzene	156.1	Nonazeotrope		565
1590	C₆H₁₀O₄	Ethylidene diacetate	168.5	164.15	44	569
1591	C₆H₁₂O₃	2-Ethoxyethyl acetate	156.8	Nonazeotrope		575
1592	C₆H₁₄O₂	2-Butoxyethanol	171.15	167.15	58	527
1593	C₇H₇Cl	*o*-Chlorotoluene	159.2	Nonazeotrope		565
1594	C₇H₁₄O₂	Butyl propionate	146.8	Nonazeotrope		565
1595	C₇H₁₆O	Hepyl alcohol	176.15	169.8	62	527
1596	C₈H₁₆	*m*-Xylene	139.2	Nonazeotrope		575
1597	C₈H₁₆O₂	Butyl butyrate	166.4	164.0	38	565
1598	C₈H₁₆O₂	Isoamyl propionate	160.7	159.5	22	565
1599	C₈H₁₈O	*sec*-Octyl alcohol	180.4	174.0	72	565
1600	C₉H₁₈O₂	Isobutyl isovalerate	171.2	167.9	52	565
1601	C₁₀H₁₈O	Cineole	176.35	169.6	60	559
1602	C₁₀H₂₂O	Isoamyl ether	173.2	166.5	55	559
A =	**CH₂O₂**	**Formic Acid**	**100.75**			
1603	CH₃I	Iodomethane	42.6	42.1	6	527,541
1604	CH₃NO₂	Nitromethane	101.22	97.05	45.5	554
1605	C₂Cl₄	Tetrachloroethylene	121.1	88.15	50.0	538
1606	C₂HCl₃	Trichloroethylene	86.95	74.1	25	563
1607	C₂HCl₅	Pentachloroethane	161.95	Nonazeotrope		541
1608	C₂H₂Cl₄	1,1,2,2-Tetrachloroethane	146.25	99.25	68	538
1609	C₂H₃Br	Bromoethylene	15.8	Nonazeotrope		542
1609a	C₂H₃N	Acetonitrile	81.6	Nonazeotrope	v-l	635e
1610	C₂H₄Br₂	1,2-Dibromoethane	131.65	94.65	51.5	538
1611	C₂H₄Cl₂	1,1-Dichloroethane	57.25	56.0	5	542
1612	C₂H₄Cl₂	1,2-Dichloroethane	83.7	77.4	14	537
		"	83.7	77.02	25.77	112
1613	C₂H₄O₂	Acetic acid 50 mm.		Nonazeotrope	v-l	507e
		" 200 mm.		Nonazeotrope	v-l	507e
		"	118.1	Nonazeotrope,	v-l	11,26,171
1614	C₂H₅Br	Bromoethane	38.40	38.23	3	538
1615	C₂H₅Cl	Chloroethane	13.1	Nonazeotrope		541
1616	C₂H₅ClO	Chloromethyl methyl ether	59.5	Nonazeotrope		563
1617	C₂H₅I	Iodoethane	72.3	65.6	22	537

TABLE I. *Binary Systems* 67

No.	Formula	B-Component Name	B.P., °C	Azeotropic Data B.P., °C	Wt.%A		Ref.
A =	CH₂O₂	**Formic Acid** *(continued)*	**100.75**				
1618	C₂H₅NO₂	Nitroethane	114.2	Nonazeotrope			554
1619	C₂H₆S	Methyl sulfide	37.4	Nonazeotrope			566
1620	C₃H₅Br	3-Bromopropene	70.5	64.5	~22		562
1621	C₃H₅Cl	2-Chloropropene	22.65	Nonazeotrope			575
1622	C₃H₅Cl	3-Chloropropene	45.7	45.0	7.5		542
1623	C₃H₅ClO	1-Chloro-2-propanone	119	Nonazeotrope			563
1624	C₃H₅ClO	Epichlorohydrin	116.45	Nonazeotrope			563
1625	C₃H₅I	3-Iodopropene	102	85	~35		563
1626	C₃H₆Cl₂	2,2-Dichloropropane	70.4	< 66.0	<25		575
1627	C₃H₆O	Acetone	56.15	Nonazeotrope			552
1628	C₃H₆O₂	Propionic acid	141.1	Nonazeotrope		v-l	26,507g
1629	C₃H₇Br	1-Bromopropane	71.0	64.7	27		537
1630	C₃H₇Br	2-Bromopropane	59.35	56.0	14		541
1631	C₃H₇Cl	1-Chloropropane	46.4	45.7	8		555
1632	C₃H₇Cl	2-Chloropropane	34.8	34.7	1.5		541
1633	C₃H₇I	1-Iodopropane	102.4	82	36		541
1634	C₃H₇I	2-Iodopropane	89.45	75.2	29		562
1635	C₃H₇NO	*N, N*-Dimethylformamide	153.0	153.2	1.2		832
		"		158.8			616
		" 100 mm.	90	98.5		v-l	616
		" 200 mm.	107.9	117.0		v-l	616
		" 760 mm.	153	158.8		v-l	616
		" 757 mm.	153	153.2	1.2		224
		" 50 mm.		85	33		224
1636	C₃H₈O₂	Methylal	42.15	Nonazeotrope			556
1637	C₃H₉N	Trimethylamine, azeotrope composition independent of pressure	3.5	179	~24.5		563
1638	C₄H₄S	Thiophene	84	Min. b.p.			1017
1639	C₄H₆O	Crotonaldehyde	102.15	~95			563
1640	C₄H₈O	2-Butanone	79.6	Nonazeotrope			526
1641	C₄H₈O₂	Butyric acid	163.5	Nonazeotrope			26
1642	C₄H₈O₂	Dioxane	101.35	113.35	43		556
1642a	C₄H₈O₂	Propyl formate	80.9	Nonazeotrope		v-l	8c
1643	C₄H₈S	Tetrahydrothiophene	118.8	< 94.5	<73		566
1644	C₄H₉Br	1-Bromobutane	101.5	81.4	35		562
1645	C₄H₉Br	1-Bromo-2-methylpropane	91.3	76.7	30		555
1646	C₄H₉Br	2-Bromo-2-methylpropane	73.3	66.2	22		541
1647	C₄H₉Cl	1-Chlorobutane	78.5	69.4	25		562
1648	C₄H₉Cl	1-Chloro-2-methylpropane	68.85	62.95	19		563
1649	C₄H₉Cl	2-Chloro-2-methylpropane	51.6	50.0	11.2		541
1650	C₄H₉I	1-Iodobutane	130.4	92.6	52		562
1651	C₄H₉I	1-Iodo-2-methylpropane	120.4	89.5	45		542
1652	C₄H₁₀O	Ethyl ether	34.6	Nonazeotrope			563
1653	C₄H₁₀S	Ethyl sulfide	92.2	82.2	35		555
1654	C₅H₅N	Pyridine	115.5	150-151	63.5		360,563
		"	115.5	107.43	61.4	v-l	1058

No.	Formula	B-Component Name	B.P., °C	Azeotropic Data B.P., °C	Wt.%A	Ref.
A =	CH₂O₂	**Formic Acid** *(continued)*	**100.75**			
1655	C₅H₁₀	Cyclopentane	49.3	46.0	16	562
1656	C₅H₁₀	2-Methyl-2-butene	37.15	35.0	10.5	541
1657	C₅H₁₀	3-Methyl-1-butene	22.5	~ 22.2	~ 2	537
1658	C₅H₁₀O	3-Methyl-2-butanone	95.4	>102.15	<85	552
1659	C₅H₁₀O	2-Pentanone	102.35	105.5	32	552
1660	C₅H₁₀O	3-Pentanone	102.05	105.25	33	552
1660a	C₅H₁₀O₂	Butyl formate	106.6	99.7	38.4	8c
1661	C₅H₁₀O₂	Isobutyl formate	98.3	Nonazeotrope		1023
1661a	C₅H₁₀O₂	Valeric acid				
		50°–100°C.		Nonazeotrope	v-l	471c
1662	C₅H₁₁Br	1-Bromo-3-methylbutane	120.3	90.5	47	541
1663	C₅H₁₁Cl	1-Chloro-3-methylbutane	99.4	80.0	33.5	542
1664	C₅H₁₁I	1-Iodo-3-methylbutane	147.65	97.0	62	562
1665	C₅H₁₂	2-Methylbutane	27.95	27.2	4	537
1666	C₅H₁₂	Pentane	36.15	34.2	10	537
1667	C₅H₁₂O	Ethyl propyl ether	63.6	Azeotrope doubtful		563
1668	C₆H₄Cl₂	p-Dichlorobenzene	174.6	Nonazeotrope		542
1669	C₆H₅Br	Bromobenzene	156.1	98.1	68	568
1670	C₆H₅Cl	Chlorobenzene	131.75	93.7	59	568
1671	C₆H₅F	Fluorobenzene	84.9	73.0	27	562
1672	C₆H₆	Benzene	80.2	71.05	31	563
		" 521.4 mm.		60	30.69 v-l	979
		" 296.8 mm.		45	30.02 v-l	979
		" 157.7 mm.		30	29.35 v-l	979
1673	C₆H₇N	Aniline	184.35	Nonazeotrope		263
1674	C₆H₇N	2-Picoline	129	158	25	263
1675	C₆H₇N	3-Picoline, 200 mm.		100-125		174,810
		" 100 mm.		98-110		174,810
1676	C₆H₇N	4-Picoline, 200 mm.		100-175		174,810
		" 100 mm.		98-110		174,810
1677	C₆H₈	1,3-Cyclohexadiene	80.8	~ 71	30	563
1678	C₆H₁₀	Biallyl	60.2	~ 46		541
1679	C₆H₁₀	Cyclohexene	82.75	71.5	21	563
1680	C₆H₁₀S	Allyl sulfide	139.35	97.5	80	566
1681	C₆H₁₂	Cyclohexane	80.75	70.7	30	563
1682	C₆H₁₂	Methylcyclopentane	72.0	63.3	29	562
1683	C₆H₁₂O	Pinacolone	106.2	>107.1	<24	552
1684	C₆H₁₄	2,3-Dimethylbutane	58.0	52.5	22	542
1685	C₆H₁₄	Hexane	68.95	60.6	28	537
1686	C₆H₁₄S	Isopropyl sulfide	120.5	93.5	62	566
1687	C₆H₁₄S	Propyl sulfide	141.5	98.0	83	566
1688	C₇H₇Cl	o-Chlorotoluene	159.3	100.2	83	541
1689	C₇H₇Cl	p-Chlorotoluene	162.4	100.5	88	541
1690	C₇H₈	Toluene	110.7	85.8	50	563
1691	C₇H₉N	2,6-Lutidine, 200 mm.		100-125		174,810
		" 100 mm.		98-110		174,810

TABLE I. *Binary Systems* 69

No.	Formula	Name	B.P., °C	B.P., °C	Wt.%A	Ref.
		B-Component		**Azeotropic Data**		
A =	CH₂O₂	**Formic Acid** *(continued)*	**100.75**			
1692	C₇H₁₄	Methylcyclohexane	101.1	80.2	46.5	541
1693	C₇H₁₆	*n*-Heptane	98.45	78.2	56.5	527,804
1694	C₈H₈	Styrene	145.8	95.75	73	17
1695	C₈H₁₀	Ethylbenzene	136.15	~ 94.0	68	541
	"	" 200 mm.		60	69	53
1696	C₈H₁₀	*m*-Xylene	139	92.8	71.8	526,803
1697	C₈H₁₀	*o*-Xylene	143.6	95.5	74	541
1698	C₈H₁₀	*p*-Xylene	138.4	~ 95	70.0	541
1699	C₈H₁₁N	Dimethylaniline	194.05	Nonazeotrope		563
1700	C₈H₁₆	1,3-Dimethylcyclohexane	120.7	89.0	51	562
1701	C₈H₁₈	2,5-Dimethylhexane	109.4	83.2	48	562
1702	C₈H₁₈	Octane	125.8	90.5	63	541
1703	C₈H₁₈O	Butyl ether	141	Nonazeotrope		537
1704	C₉H₇N	Quinoline	237.3	Nonazeotrope		727
1705	C₉H₈	Indene	182.4	Nonazeotrope		543
1706	C₉H₁₂	Cumene	152.8	97.2	< 88	562
1707	C₉H₁₂	Mesitylene	164.6	< 99.7	< 96	575
1708	C₉H₁₂	Propylbenzene	159.3	< 98.8	< 93	575
1709	C₁₀H₁₆	Camphene	159.6	Nonazeotrope		538
1710	C₁₀H₁₆	*d*-Limonene	177.8	Nonazeotrope		538
1711	C₁₀H₁₆	Thymene	179.7	Reacts		542
1712	C₁₀H₂₂	2,7-Dimethyloctane	160.1	< 98.5	< 93	575
A =	CH₃Br	**Bromomethane**	**3.65**			
1713	CH₄O	Methanol	64.7	3.55	99.45	1025
1714	C₂H₄O	Acetaldehyde	20.2	Nonazeotrope		563
1715	C₂H₄O₂	Methyl formate	31.75	Nonazeotrope		547
1716	C₂H₅NO₂	Ethyl nitrite	17.4	Nonazeotrope		550
1717	C₃H₇NO₂	Isopropyl nitrite	40.1	Nonazeotrope		550
1718	C₄H₆	Butadiene	— 4.5	Nonazeotrope		215
1719	C₄H₈	1-Butene	— 6.5	Nonazeotrope		215
1720	C₄H₁₀	Butane	— 0.6	— 4.4	57.3	375
A =	CH₃Cl	**Chloromethane**	**— 23.7**			
1721	C₂H₆O	Methyl ether	— 23.65	Azeotropic		559
1722	C₄H₁₀	2-Methylpropane	— 10	Azeotropic		161
A =	CH₃Cl₃Si	**Trichloromethylsilane**	**66.8**			
1723	C₂H₆SiCl₂	Dichlorodimethylsilane				
		40-760 mm.		Nonazeotrope	v-l	229
1724	C₃H₉SiCl	Chlorotrimethylsilane	57.5	V.p. curves,		485
				Nonazeotrope		
1724a	C₄H₈Cl₂O	Bis(2-chloroethyl) ether	178	Nonazeotrope		886c
A =	CH₃I	**Iodomethane**	**42.5**			
1725	CH₃NO₃	Methyl nitrate	64.8	Nonazeotrope		527
1726	CH₄O	Methanol	64.7	37.8	95.5	389,527
1727	C₂H₄O₂	Methyl formate	31.9	31	~ 17	563
1728	C₂H₆O	Ethyl alcohol	78.3	41.2	96.8	573

No.	Formula	B-Component Name	B.P., °C	B.P., °C	Azeotropic Data Wt.%A	Ref.
A =	**CH₃I**	**Iodomethane** *(continued)*	**42.5**			
1729	C₃H₆O	Acetone	56.15	42.4	95	527
1730	C₃H₆O₂	Ethyl formate	54.1	Nonazeotrope		538
	"	"	54.15	41.8		571c
1731	C₃H₆O₂	Methyl acetate	56.95	Nonazeotrope		527
1732	C₃H₇Cl	1-Chloropropane	46.65	42.1	85	562
1733	C₃H₇NO₂	Isopropyl nitrite	40.1	39.5	> 30	527
1734	C₃H₈O	Isopropyl alcohol	82.4	42.4	98.2	531
1735	C₃H₈O	Propyl alcohol	97.2	Nonazeotrope		371
1736	C₃H₈O₂	Methylal	42.2	39.45	57	527
1737	C₄H₈O₂	Ethyl acetate	77.1	Nonazeotrope		527
1738	C₄H₁₀O	Ethyl ether	34.6	Nonazeotrope		527
1739	C₄H₁₀O	Methyl propyl ether	38.8	Nonazeotrope		527
1740	C₄H₁₂Si	Tetramethylsilane	26.64	26.1	28.8	28
1741	C₅H₁₀	Cyclopentane	49.3	< 42.0	> 66	575
1742	C₅H₁₀	2-Methyl-2-butene	37.1	< 36.2	> 40	575
1743	C₅H₁₀	3-Methyl-1-butene	22.5	Azeotrope doubtful		563
1744	C₅H₁₂	2-Methylbutane	27.95	< 25.0	> 20	562
1745	C₅H₁₂	Pentane	36.2	< 33.8	> 38	527
1746	C₆H₅NO₂	Nitrobenzene	210.75	Nonazeotrope		554
1747	C₆H₁₄	Hexane	68.85	Nonazeotrope		527
A =	**CH₃NO₃**	**Methyl Nitrite**	**— 16**			
1748	C₄H₆	Butadiene	— 4.7	Nonazeotrope		642
1749	C₄H₈	1-Butene	— 6	— 16		642
1750	C₄H₈	2-Methylpropene	— 6	— 16		642
1751	C₄H₁₀	Butane	— 0.6	— 20		642
1752	C₄H₁₀	2-Methylpropane	— 11	— 20		642
A =	**CH₃NO₃**	**Nitromethane**	**101.2**			
1753	CH₄O	Methanol		64.4	9.1	v-l	684g
	"	"	64.51	64.33	12.2	806, 554
1754	C₂Cl₄	Tetrachloroethylene	121.1	95.0	80?	554
1755	C₂HCl₃	Trichloroethylene	86.2		v-l	1048
	"	"	86.9	81.4	20	554
1756	C₂HCl₃O	Chloral	97.75	93	35	572
1757	C₂H₃N	Acetonitrile, 60°C.		Nonazeotrope	v-l	91
	"	"	81.6	Nonazeotrope	v-l	635e
1758	C₂H₄O₂	Acetic acid	118.1	101.2	96	554
1759	C₂H₄S	Ethylene sulfide	55.7	Nonazeotrope		575
1760	C₂H₅ClO	2-Chloroethanol	128.6	Nonazeotrope		554
1761	C₂H₅I	Iodoethane	72.3	71.2	10	554
1762	C₂H₅NO₃	Ethyl nitrate	87.70	87.68	1.2	527
1763	C₂H₆O	Ethyl alcohol	78.3	75.95	26.8	554
	"	"	78.32	76.05	29.0	806
1764	C₂H₆O₂	Glycol	197.4	Nonazeotrope		576

TABLE I. *Binary Systems* 71

No.	B-Component Formula	Name	B.P., °C	Azeotropic Data B.P., °C	Wt.%A		Ref.
A =	CH₃NO₃	**Nitromethane** *(continued)*	**101.2**				
1765	C₂H₆S	Ethanethiol	35.8	Nonazeotrope			554
1766	C₂H₆S	Methyl sulfide	37.4	Nonazeotrope			554
1767	C₂H₇NO	2-Aminoethanol	170.8	Nonazeotrope			575
1768	C₃H₅Br	3-Bromopropene	70.0	< 69.8	> 4		554
1769	C₃H₅I	3-Iodopropene	101.8	89.0			548,554
1770	C₃H₆O	Acetone, 45°C.		Nonazeotrope		v-l	91
1771	C₃H₆O	Allyl alcohol	96.85	89.3	43		554
1772	C₃H₆O₂	Propionic acid	141.3	Nonazeotrope			554
1773	C₃H₇Br	1-Bromopropane	71.0	70.6	7		554
1774	C₃H₇Br	2-Bromopropane	59.2	Nonazeotrope			548
1775	C₃H₇Cl	1-Chloropropane	46.4	Nonazeotrope			555
1776	C₃H₇ClO	1-Chloro-2-propanol	127.0	Nonazeotrope			554
1777	C₃H₇ClO	2-Chloro-1-propanol	133.7	Nonazeotrope			554
1778	C₃H₇I	1-Iodopropane	102.4	89.2	> 42		554
1779	C₃H₇I	2-Iodopropane	89.45	82.0	33		554
1780	C₃H₇NO₃	Propyl nitrate	110.5	100.2	75		554
1781	C₃H₈O	Isopropyl alcohol	82.0	79.3	28.2	v-l	156,857
		" "	82.40	79.33	27.6		806
1782	C₃H₈O	Propyl alcohol	97.25	Nonazeotrope		v-l	383
		"	97.15	89.09	48.4		806
		"	97.2	89.3	47.5		285,554
1783	C₃H₉SiCl	Chlorotrimethylsilane	57.7	Nonazeotrope			841
1783a	C₄H₆	1,3-Butadiene	—4.6	Nonazeotrope		v-l	1030c
1784	C₄H₆O	Crotonaldehyde	102.15	99			563
1785	C₄H₈O	2-Butanone	79.6	Nonazeotrope			552
1786	C₄H₈O₂	Dioxane	101.35	100.55	56.5		527
		" 722 mm.		98	30.7	v-l	912
		" 398 mm.		80	30.7	v-l	912
		" 103 mm.		46	30.7	v-l	912
		" 48 mm.		30	30.7	v-l	912
1787	C₄H₈O₂	Ethyl acetate	77.1	Nonazeotrope			554
1788	C₄H₈O₂	Methyl propionate	79.85				554,1038
1789	C₄H₈O₂	Propyl formate	80.85	Nonazeotrope			554
1790	C₄H₉Br	1-Bromobutane	101.5	90.0	50		554
1791	C₄H₉Br	1-Bromo-2-methylpropane	91.4	84.0	34		554
1792	C₄H₉Br	2-Bromo-2-methylpropane	73.25	72.2	9		554
1793	C₄H₉Cl	1-Chlorobutane	78.5	75.5	16		554
1794	C₄H₉Cl	1-Chloro-2-methylpropane	68.85	68.35	6		554
1795	C₄H₉I	1-Iodobutane	130.4	99.8	~90		548,554
1796	C₄H₉I	1-Iodo-2-methylpropane	120.8	96.7	>60		554
1797	C₄H₁₀O	Butyl alcohol	117.8	97.8	70		554
		" "	117.73	97.99	71.4		806
1798	C₄H₁₀O	*sec*-Butyl alcohol	99.53	91.14	45.8		806
		" "	99.5	91.1	46		554
1799	C₄H₁₀O	*tert*-Butyl alcohol	82.45	79.4	32		554
		" "	82.41	80.04	21.2		806

No.	Formula	B-Component Name	B.P., °C	Azeotropic Data B.P., °C	Wt.%A		Ref.
A =	CH_3NO_3	**Nitromethane** *(continued)*	**101.2**				
1800	$C_4H_{10}O$	Isobutyl alcohol	107.89	94.46	57.6		806
		" "	108.0	94.6	56.5		554
1801	$C_4H_{10}O_2$	2-Ethoxyethanol	135.3	Nonazeotrope			554
1802	$C_4H_{10}S$	1-Butanethiol	97.5	< 93.2			554
1803	$C_4H_{10}S$	Ethyl sulfide	92.1	85.0	30		554
1804	C_5H_5N	Pyridine	115.4	<100.5	>85		575
1805	C_5H_8	Isoprene	34.1	Nonazeotrope		v-l	716
1806	C_5H_8	1,3-Pentadiene	42.3	Nonazeotrope		v-l	716
1807	C_5H_{10}	2-Methyl-1-butene	31.1	Nonazeotrope		v-l	716
1808	C_5H_{10}	2-Methyl-2-butene	38.5	38	5.0	v-l	716
1809	C_5H_{10}	3-Methyl-1-butene	20.6	Nonazeotrope			554
1810	C_5H_{10}	Cyclopentane	49.3	<47.5	> 9		554
1811	$C_5H_{10}O$	Cyclopentanol	140.85	Nonazeotrope			554
1812	$C_5H_{10}O$	3-Methyl-2-butanone	95.4	<94.8			575
1813	$C_5H_{10}O$	2-Pentanone	102.35	99.15	56		552
1814	$C_5H_{10}O$	3-Pentanone	102.05	99.1	55		552
1815	$C_5H_{10}O_2$	Butyl formate	106.8	< 98.7	<60		554
1816	$C_5H_{10}O_2$	Ethyl propionate	99.1	96.0	35		554
1817	$C_5H_{10}O_2$	Isobutyl formate	98.2	94.7	32		554
1818	$C_5H_{10}O_2$	Isopropyl acetate	89.5	< 89.3			554
1819	$C_5H_{10}O_2$	Methyl butyrate	102.65	97.95	50		554
1820	$C_5H_{10}O_2$	Methyl isobutyrate	92.5	91.2			554
1821	$C_5H_{10}O_2$	Propyl acetate	101.6	97.6	45		554
1822	$C_5H_{11}Br$	1-Bromo-3-methylbutane	120.65	97.5			554
1823	$C_5H_{11}Cl$	1-Chloro-3-methylbutane	99.4	88.2	48		554
1824	$C_5H_{11}NO_2$	Isoamyl nitrite	97.15	94.2			554
1825	C_5H_{12}	2-Methylbutane	27.95	Nonazeotrope			554
		"	27.9	Nonazeotrope		v-l	716
1826	C_5H_{12}	Pentane	36.07	35	1		806
1827	$C_5H_{12}O$	*tert*-Amyl alcohol	102.35	93.1	49.5		554
1828	$C_5H_{12}O$	Isoamyl alcohol	131.9	100.6	88		554
1829	$C_5H_{12}O$	3-Methyl-2-butanol	112.9	96.4	63		554
1830	$C_5H_{12}O$	2-Pentanol	119.8	98.5	73		554
1831	$C_5H_{12}O$	3-Pentanol	116.0	97.4	68		554
1832	$C_5H_{12}O_2$	2-Propoxyethanol	151.35	Nonazeotrope			554
1833	C_6H_5Cl	Chlorobenzene	131.75	Nonazeotrope			575
1834	$C_6H_5NO_2$	Nitrobenzene	210.9	Nonazeotrope		v-l	963
1835	C_6H_6	Benzene 228 mm.		45	9.6		91
		"	80.15	79.15	14		554
		" 97.7 mm.		25	6.4	v-l	845
		"	80.1	25	12.7	v-l	1010
1836	C_6H_{10}	Cyclohexene	82.75	< 74.5	<31		554
1837	C_6H_{10}	Biallyl	60.1	< 57.5	<23		554
1838	$C_6H_{10}S$	Allyl sulfide	139.35	Nonazeotrope			554
1839	C_6H_{12}	Cyclohexane	80.75	70.2	28		554
		"	80.75	69.5	26.5	v-l	1010

TABLE I. *Binary Systems* 73

No.	Formula	Name	B.P., °C	B.P., °C	Wt.%A	Ref.
		B-Component			**Azeotropic Data**	
A =	**CH₃NO₃**	**Nitromethane** *(continued)*	**101.2**			
1840	C_6H_{12}	1-Hexene	63.84	59.7	7.4	806
1841	C_6H_{12}	Methylcyclopentane	72.0	64.2	23	554
1842	$C_6H_{12}O$	Cyclohexanol	160.8	Nonazeotrope		554
1843	$C_6H_{12}O$	4-Methyl-2-pentanone	116.05	Nonazeotrope		552
1844	$C_6H_{12}O$	Pinacolone	106.2	<100.5		552
1845	$C_6H_{12}O_2$	Ethyl butyrate	121.5	Nonazeotrope		554
1846	$C_6H_{12}O_2$	Ethyl isobutyrate	110.1	100.1	72	554
1847	$C_6H_{12}O_2$	Isobutyl acetate	117.2	Nonazeotrope		548
1848	$C_6H_{12}O_2$	Methyl isovalerate	116.5	Nonazeotrope		548
1849	C_6H_{14}	2,3-Dimethylbutane	58.0	< 54.5	<26	554
1850	C_6H_{14}	n-Hexane	68.8	62.0	21	554
		"	68.74	61.7	18.5	806
1851	$C_6H_{14}O$	n-Hexyl alcohol	157.85	Nonazeotrope		554
1852	$C_6H_{14}O_2$	Acetal	104.5	95	~65	563
1853	$C_6H_{14}S$	Isopropyl sulfide	120.5	< 99.5	>85	575
1854	$C_6H_{14}S$	Propyl sulfide	141.5	Nonazeotrope		566
1855	$C_6H_{15}NO$	2-(Diethylamino)ethanol	162.2	Nonazeotrope		575
1856	C_7H_8	Toluene	110.75	96.5	55	321,554
1857	C_7H_{14}	1-Heptene	93.64	79.4	31.1	806
1858	C_7H_{14}	Methyl cyclohexane	101.15	81.25	39.5	554
		" "	100.1	81.7	39.5	1038
1859	C_7H_{16}	Heptane 748 mm.		79.70	35.6	613
		"	98.43	79.9	34.7	806
		"	98.4	80.2	37	554
1860	C_8H_8	Styrene	145.8	Nonazeotrope		554
1861	C_8H_{10}	Ethylbenzene	136.15	Nonazeotrope		554
1862	C_8H_{10}	m-Xylene	139.2	Nonazeotrope		514
1863	C_8H_{10}	o-Xylene	144.3	Nonazeotrope		554
1864	C_8H_{16}	1,3-Dimethylcyclohexane	120.7	90.2	50	554
1865	C_8H_{16}	1-Octene	121.28	91.2	52.3	806
1866	C_8H_{18}	2,5-Dimethylhexane	109.4	85.5	43	554
1867	C_8H_{18}	n-Octane	125.75	92.0	53	554
		" 748 mm.		90.23	55.2	613
		"	125.66	90.6	54.0	806
1868	$C_8H_{18}O$	Isobutyl ether	122.3	Nonazeotrope		554
1869	C_9H_{12}	Cumene	152.8	Nonazeotrope		554
1870	C_9H_{12}	Mesitylene	164.6	Nonazeotrope		554
1871	C_nH_{2n+2}	Paraffins	90-118	25-90		321
1872	C_9H_{18}	1-Nonene	146.87	97.1	72.3	806
1873	C_9H_{20}	Nonane	150.85	96.5	70.2	806
		" 748 mm.		96.14	71.6	613
1874	$C_{10}H_{20}$	1-Decene	170.57	99.8	84.2	806
1875	$C_{10}H_{22}$	Decane	174.12	99.1	84.1	806
		" 748 mm.		98.81	83.9	613
1876	$C_{11}H_{24}$	Undecane 748 mm.		100.01	90.7	613

No.	B-Component Formula	Name	B.P., °C	Azeotropic Data B.P., °C	Wt.%A	Ref.
A =	CH₃NO₃	Nitromethane *(continued)*	101.2			
1877	C₁₂H₂₆	Dodecane 748 mm.		100.60	95.8	613
1878	C₁₃H₂₈	Tridecane 748 mm.		Nonazeotrope		613
1879	C₁₆H₃₄	Hexadecane 748 mm.		Nonazeotrope		613
A =	CH₃NO₃	Methyl Nitrate	64.8			
1880	CH₄O	Methanol	64.65	52.5	73	560
1881	C₂H₄Cl₂	1,2-Dichloroethane	83.45	Nonazeotrope		560
1882	C₂H₅Br	Bromoethane	38.4	Nonazeotrope		560
1883	C₂H₅I	Iodoethane	72.3	< 63.5	<72	560
1884	C₂H₆O	Ethyl alcohol	78.3	< 59.5	> 64	560
1885	C₃H₅Br	3-Bromopropene	70.5	62.8	68	560
1886	C₃H₅Cl	3-Chloropropene	45.3	Nonazeotrope		560
1887	C₃H₇Br	1-Bromopropane	71.0	63.0	70	560
1888	C₃H₇Br	2-Bromopropane	59.4	57.3	32	560
1889	C₃H₇Cl	1-Chloropropane	46.65	Nonazeotrope		560
1890	C₃H₈O	Isopropyl alcohol	82.42	< 62.5	>78	560
1891	C₃H₈O₂	Methylal	42.3	Nonazeotrope		557
1892	C₄H₄S	Thiophene	84.7	Nonazeotrope		560
1893	C₄H₉Br	2-Bromo-2-methylpropane	73.25	63.8	<80	560
1894	C₄H₉Cl	1-Chlorobutane	78.5	Nonazeotrope		560
1895	C₄H₉Cl	2-Chlorobutane	68.25	< 62.0	<64	560
1896	C₄H₉Cl	1-Chloro-2-methylpropane	68.85	61.2	61	560
1897	C₄H₁₀O	*tert*-Butyl alcohol	82.45	63.2	84	560
1898	C₄H₁₀O	Ethyl ether	34.6	Nonazeotrope		557
1899	C₄H₁₀O₂	Ethoxymethoxymethane	65.8	< 63.9		557
1900	C₅H₁₀	Cyclopentane	49.4	< 47.2	>20	560
1901	C₅H₁₂	Pentane	36.15	< 35.5	<10	560
1902	C₅H₁₂O	Ethyl propyl ether	63.85	< 61.5		557
1903	C₆H₅F	Fluorobenzene	84.9	Nonazeotrope		560
1904	C₆H₆	Benzene	80.15	Nonazeotrope		560
1905	C₆H₁₂	Cyclohexane	80.75	61.0	77	560
1906	C₆H₁₂	Methylcyclopentane	72.0	57.8	60	560
1907	C₆H₁₄	2,3-Dimethylbutane	58.0	51.0	38	560
1908	C₆H₁₄	Hexane	68.8	56.0	56	560
1909	C₇H₁₆	Heptane	98.4	Nonazeotrope		560
A =	CH₄	Methane	—164			
1910	C₂H₆	Ethane	— 88	Nonazeotrope	v-l	766
1910a	C₃H₆	Propylene	— 48	Nonazeotrope		63c
1911	C₃H₈	Propane	— 44	Nonazeotrope	v-l	766
A =	CH₄Cl₂Si	Dichloromethysilane				
1912	C₃H₉ClSi	Chlorotrimethyl-silane, 30°-40°		V.p. curve, nonazeotrope		872
A =	CH₄O	Methanol	64.7			
1913	C₂Cl₃F₃	1,1,2-Trichlorotri-fluoroethane	47.5	39.9	6	46,982

TABLE I. *Binary Systems* 75

No.	Formula	Name	B.P., °C	B.P., °C	Wt.%A		Ref.
		B-Component		**Azeotropic Data**			
A =	**CH₄O**	**Methanol** *(continued)*	**64.7**				
1914	C₂Cl₄	Tetrachloroethylene	121.1	63.75	63.5		574
1915	C₂HCl₃	Trichloroethylene	87	59.3	38	v-l	276,297
1916	C₂H₂BrCl	1-Bromo-2-chloroethylene	106.7	Nonazeotrope			575
1917	C₂H₂Br₂	*cis*-1,2-Dibromoethylene	112.5	Nonazeotrope			563
1918	C₂H₂Br₂	*trans*-1,2-Dibromoethylene	108	Nonazeotrope			575
		"	108	~ 64.1	~72		563
1919	C₂H₂Cl₂	1,1-Dichloroethylene	31	27.5-28	6 vol. %		956
1920	C₂H₂Cl₂	*cis*-1,2-Dichloroethylene	60.25	51.5	~13		563
		"	60.3	51.5	15.1	v-l	12
1921	C₂H₂Cl₂	*trans*-1,2-Dichloroethylene	48.3	41.9	9.02	v-l	12
1922	C₂H₃Br	Bromoethylene	15.8	< 15.7			575
1923	C₂H₃Cl₃	1,1,1-Trichloroethane	74.1	56	21.7		215
1924	C₂H₃Cl₃	1,1,2-Trichloroethane	113.65	Nonazeotrope			575
		"	114	~ 64.5	97		563
1925	C₂H₃N	Acetonitrile	81.6	63.45	19		563
1926	C₂H₄BrCl	1-Bromo-2-chloroethane	106.7	64.5?			563
1927	C₂H₄Br₂	1,1-Dibromoethane	109.5	Nonazeotrope			575
		"	~110	64.2	~82		563
1928	C₂H₄Br₂	1,2-Dibromoethane	131.65	Nonazeotrope			574
1929	C₂H₄Cl₂	1,1-Dichloroethane	57.3	49.05	11.5		563
1930	C₂H₄Cl₂	1,2-Dichloroethane	83.7	60.95	32		572
		"			40	v-l	282
		"	83.5	59.5	35		982
1931	C₂H₄O	Acetaldehyde	20.2	Nonazeotrope		v-l	467,575
1932	C₂H₄O	Ethylene oxide	10.75	Nonazeotrope			575
1933	C₂H₄O₂	Acetic acid	118.1	Nonazeotrope		v-l	601,811
1934	C₂H₄O₂	Methyl formate	31.9	Nonazeotrope			563
1935	C₂H₄S	Ethylene sulfide	55.7	< 47.0	<21		575
1936	C₂H₅Br	Bromoethane	38	35	5		563,834
		"		34.9	5.3		497
1937	C₂H₅Cl	Chloroethane	13.5	Nonazeotrope			563
1938	C₂H₅ClO	Chloromethyl methyl ether	59.5	56	~35		563
1939	C₂H₅I	Iodoethane	72.3	55	17		563,834
1940	C₂H₅NO	Acetamide	220.9	Nonazeotrope			527
1941	C₂H₅NO₂	Nitroethane	114.2	Nonazeotrope			576
1942	C₂H₅NO₃	Ethyl nitrate	87.68	61.77	57		560
1943	C₂H₆	Ethane	—93	Nonazeotrope			563
1944	C₂H₆O	Ethyl alcohol	78.3	Nonazeotrope		v-l	114,563, 889
1945	C₂H₆O₂	Ethylene glycol	197.4	Nonazeotrope		v-l	35,859c
1946	C₂H₆S	Ethanethiol	36.2	Nonazeotrope			563
1947	C₂H₆S	Methyl sulfide	37.3	< 34.5	<13		566
1948	C₃H₃N	Acrylonitrile	77.3	61.4	61.3		215
		" 175 mm.	37	29	47		982
1949	C₃H₄Cl₂	1,2-Dichloro-1-propene	76.8	56.5	25		410

		B-Component		Azeotropic Data			
No.	Formula	Name	B.P., °C	B.P., °C	Wt.%A		Ref.
A =	**CH₄O**	**Methanol** *(continued)*	**64.7**				
1950	C₃H₄O	Acrolein	52.45	Nonazeotrope			575
1951	C₃H₅Br	*trans*-1-Bromopropene	63.25	50.8	15		563
1952	C₃H₅Br	*cis*-1-Bromopropene	57.8	48	12		563
1953	C₃H₅Br.	2-Bromopropene	48.35	42.7	11		563
1954	C₃H₅Br	3-Bromopropene	70.5	54.0	20.5		567
1955	C₃H₅Cl	2-Chloropropene	22.65	22.0	3		573
1956	C₃H₅Cl	3-Chloropropene	45.15	39.85	10		567
1957	C₃H₅ClO	Epichlorohydrin	116.4	Nonazeotrope			556
1958	C₃H₅ClO₂	Methyl chloroacetate	131.4	Nonazeotrope			121
1959	C₃H₅I	3-Iodopropene	102.0	63.5	~62		563
1960	C₃H₅N	Propionitrile	97.1	Nonazeotrope			563
1961	C₃H₆Cl₂	1,2-Dichloropropane	96.8	62.9	53		276
1962	C₃H₆Cl₂	2,2-Dichloropropane	69.8	55.5	21		573
1963	C₃H₆O	Acetone	56.15	55.5	12		261,371,
							527,834
		" 100 mm.		Nonazeotrope		v-l	282
		" 182 mm.		20	9.5	v-l	44c
		" 283 mm.		30	8.5	v-l	44c
		" 426 mm.		40	<9.5	v-l	44c
		" 758 mm.		55	14	v-l	290c
		"		55	12.9	v-l	621
		"		45	9.2	v-l	621
		"		35	4.7	v-l	621
		" 752 mm.		55.07	14.8	v-l	16,364
		"	56.1	Nonazeotrope			518
		" 4.56 atm.	108	102	32		982
		" 7.82 atm.	132	124	46		982
		" 11.6 atm.	150	140	56		982
1964	C₃H₆O	Propionaldehyde	48.7	Nonazeotrope			575
1965	C₃H₆OS	Methyl thioacetate	95.5	Nonazeotrope			575
1966	C₃H₆O₂	Ethyl formate	54.15	50.95	16		563
1967	C₃H₆O₂	Methyl acetate					
		100 mm.		8.5	11.6		861c
		" 200 mm.		21.5	13.6	v-l	861c,36c
		" 400 mm.		37.2	15.6		861c
		" 760 mm.		53.5	19		861c,36c
		" 4 atm.		97.0	27	v-l	36c
		" 8 atm.		123.2	33	v-l	36c
		"		53.8	17.7		359,834
		"	57.1	53.9	17.7	v-l	110,179
		" 4.4 atm.	107	99	29		982
		" 7.8 atm.	132	120	34.6		982
		" 11.2 atm.	149	135	40.4		982
1968	C₃H₆O₃	Methyl carbonate	90.35	62.7	~70		536
1969	C₃H₇Br	1-Bromopropane	71.0	54.5	21		389,535
1970	C₃H₇Br	2-Bromopropane	59.4	49.0	14.5		527
1971	C₃H₇Cl	1-Chloropropane	46.6	40.6	10		555
1972	C₃H₇Cl	2-Chloropropane	36.25	33.4	6		573

TABLE I.   *Binary Systems*   77

No.	Formula	B-Component Name	B.P., °C	B.P., °C	Wt.%A	Ref.
				Azeotropic Data		
A =	CH₄O	Methanol *(continued)*	64.7			
1973	C₃H₇I	1-Iodopropane	102.4	63.1	50	527
1974	C₃H₇I	2-Iodopropane	89.35	61.0	38	573
1975	C₃H₇NO	Propionamide	222.1	Nonazeotrope		531
1976	C₃H₇NO₂	1-Nitropropane	131.18	Nonazeotrope		806
1977	C₃H₇NO₂	2-Nitropropane	120.25	Nonazeotrope		806
1978	C₃H₈	Propane	— 44	Min. b.p.		563
1978a	C₃H₈O	Isopropyl alcohol 55°C.		Nonazeotrope	v-l	290c
		"	82.3	Nonazeotrope	v-l	290c
1979	C₃H₈O₂	2-Methoxyethanol	124	Nonazeotrope	v-l	884
		" 752 mm.		Nonazeotrope	v-l	982
		" 800 mm.		Nonazeotrope	v-l	982
1980	C₃H₈O₂	Methylal	42.3	41.82	7.85	140,324,545
		"		41.7	4.4	581c
1981	C₃H₈S	Propanethiol	67.3	< 58.0	<35	566
1982	C₃H₉BO₃	Methyl borate	68.7	54.6	32	574
		"	68.0	54.0	27	318,982
		" 60 p.s.i.g.		100	33	982
		" 30 p.s.i.g.		84	29	982
		" 200 mm.		25	22	982
		Methyl borate	69	54.18	25 v-l	865
1983	C₄H₄Cl₂	2,3-Dichloro-1,3-butadiene	98	61.5	50.0	1023
		" 200 mm.		31.65	47 v-l	158c
		" 275 mm.		36.0	53.5	1023
		" 475 mm.		50.0	52.0	1023
		" 1000 mm.		70		1023
1984	C₄H₄N₂	Pyrazine	114	Nonazeotrope		751
1985	C₄H₄O	Furan	31.7	< 30.5	< 7	575
1986	C₄H₄S	Thiophene	84	< 59.55	<55	527
		"		59.71	16.4 v-l	605
1987	C₄H₅NS	Allyl isothiocyanate	152.05	Nonazeotrope		575
1988	C₄H₆O	Crotonaldehyde	102	Nonazeotrope		241
1989	C₄H₆O₂	Biacetyl	87.5	< 62.0	<75	552
1990	C₄H₆O₂	Methyl acrylate	80	62.5	54	799,800
1991	C₄H₆O₂	Vinyl acetate	72.7	58.8	36.6	982
		"	72.6	59.05	36.6	266
		"		Azeo. comp. independent		
				of press.		830c
1992	C₄H₇N	Pyrroline	90.9	Nonazeotrope		575
1993	C₄H₈O	2-Butanone	79.6	63.5	70	83
				Effect of pressure	v-l	83
		"	79.6	64.5	70 v-l	383
1994	C₄H₈O	Butyraldehyde	74.8	Nonazeotrope		132
1995	C₄H₈O	Isobutyraldehyde	63.5	62.7	40	575
1996	C₄H₈O	Tetrahydrofuran,				
		" 740 mm.	65	59.1	31.1	318
		"	66	60.7	31.0	224
		" 349 mm.		40	25.4 v-l	859c
1997	C₄H₈O₂	1,2-Dimethoxyethylene	102	63-64	90	345

	B-Component			Azeotropic	Data	
No.	Formula	Name	B.P., °C	B.P., °C	Wt.%A	Ref.
A =	CH₄O	**Methanol** (*continued*)	**64.7**			
1998	C₄H₈O₂	Dioxane	101.05	Nonazeotrope		v-l	241,733
1999	C₄H₈O₂	Ethyl acetate	77.1	62.25	44	572,834
		"	77.1	62.3	45.8	v-l	7
		"	77.1	62.4	47.1	v-l	679
		" 40°-60°		% alcohol increases with pressure		v-l	674
		"	76.7	62.1	48.6	982
		" 730 mm.		62.2	47.5	v-l	684f
2000	C₄H₈O₂	Isopropyl formate	68.8	57.2	33	567
2001	C₄H₈O₂	Methyl propionate	79.8	62.45	47.5	572
		" "	79.7	62.0	50	227
2002	C₄H₈O₂	Propyl formate	80.8	61.9	50.2	572
2003	C₄H₈S	Tetrahydrothiophene	118.8	Nonazeotrope		566
2004	C₄H₉Br	1-Bromobutane	101.5	63.5	59	527
2005	C₄H₉Br	2-Bromobutane	91.2	61.5	41.5	567
2006	C₄H₉Br	1-Bromo-2-methyl-propane	91.0	61.55	42	389,555
2007	C₄H₉Br	2-Bromo-2-methyl-propane	73.3	55.6	~24	563
2008	C₄H₉Cl	1-Chlorobutane	78.05	57.2	28.5	555
2009	C₄H₉Cl	2-Chlorobutane	68.25	52.7	20	567
2010	C₄H₉Cl	1-Chloro-2-methyl-propane	68.9	53.05	23	563
2011	C₄H₉Cl	2-Chloro-2-methyl-propane	51.6	43.75	10	532
2012	C₄H₉I	1-Iodobutane	130.4	Nonazeotrope		575
2013	C₄H₉I	2-Iodobutane	120.0	< 64.60	>65	575
2014	C₄H₉I	1-Iodo-2-methyl-propane	120.4	Nonazeotrope		532
		"	119	64	<70	834
2015	C₄H₁₀O	Butyl alcohol, crit. region	117.75	Nonazeotrope		v-l	213,383
2015a	C₄H₁₀O	*tert*-Butyl alcohol	82.5	Nonazeotrope		581c
2016	C₄H₁₀O	Ethyl ether	34.6	Nonazeotrope		834
2017	C₄H₁₀O	Methyl propyl ether	39	38	11.94	73
		" " "	38.95	38.85	10	545
2018	C₄H₁₀O₂	Acetaldehyde dimethyl acetal	64.3	57.5	24.2	45
2019	C₄H₁₀O₂	Ethoxymethoxymethane	65.90	57.1	25.3	1035
2020	C₄H₁₀S	Ethyl sulfide	92.2	61.2	62	555
2020a	C₄H₁₁N	Butylamine 730 mm.		Nonazeotrope		v-l	684e
2021	C₄H₁₁N	Diethylamine 730 mm.		67.3	58	v-l	684e
		"	55.9	Nonazeotrope		545
		" 740 mm.	54.7	66.2	40	982
2022	C₄H₁₁N	Isobutylamine	68.0	Reacts		545
2023	C₄H₁₂SiO	Methoxytrimethylsilane	57.0		14-16	839

TABLE I. *Binary Systems* 79

No.	Formula	Name	B.P., °C	B.P., °C	Wt.%A		Ref.
		B-Component		Azeotropic Data			
A =	CH₄O	**Methanol** *(continued)*	**64.7**				
2024	C₅H₅N	Pyridine	115.4	Nonazeotrope		v-l	684g,553
2025	C₅H₆O	2-Methylfuran	63.7	51.5	22.3		769
2026	C₅H₈	Cyclopentene	43	37	20 vol.%		878,1015
2027	C₅H₈	Isoprene	34.8	~ 29.5			563
		"	34.1	30.0	7.92	v-l	651
		"		30.7	8.0		476
		"	34.3	29.57	5.2		946
		"	34.3	30.45	4.1		581c,946
2028	C₅H₈	3-Methyl-1,2-butadiene	40.8	34.7	8.5		946
		"	40.8	~ 35	~10		563
2029	C₅H₈	*cis*-1,3-pentadiene	42	37.5	16.7 vol. %		1015
		"	42.3	37.6	13		476
		"	44.0	38.1	16 vol. %		826
2030	C₅H₈	*trans*-1,3-Pentadiene	42.0	36.5	15 vol. %		826
		"		36.5	12.9	v-l	715
2031	C₅H₈O	1,3-Butadienyl methyl ether	90.7	62	57.5		982
2032	C₅H₈O₂	Ethyl acrylate		64.5	84.4		799,800
2033	C₅H₈O₂	Methyl methacrylate	99.5	64.2	82	v-l	1032
		" 200 mm.	61.5	34.5	82	v-l	1032
		"		64.5	89.3	v-l	300
2034	C₅H₁₀	Cyclopentane	49.4	38.8	14		567
2035	C₅H₁₀	2-Methyl-1-butene	32	27.4	8.1	v-l	715
		"	31.1	27.6	6.5		476
2036	C₅H₁₀	2-Methyl-2-butene	38.5	33	10.5		476
		"	37.7	33.1	11.2	v-l	715
		"		33.1	11.4		497
			37.15	31.75	7		563
2037	C₅H₁₀	3-Methyl-1-butene	22.5	19.8	3		537
		"	21.2	17.9	4.28	v-l	715
		"	20.1		4.0		476
2038	C₅H₁₀	1-Pentene	30.0	26.8	8.5		476
		"	29.92	26.4	13 vol. %		826
		"	30.1	26.3	8.92	v-l	715
2039	C₅H₁₀	*cis*-2-Pentene	37.1	31.8	7 vol. %		826
2040	C₅H₁₀	2-Pentene	37	31.6	10		476
		"	35.8	31.5	12 vol. %		1015
2041	C₅H₁₀O	3-Methyl-2-butanone	95.4	Nonazeotrope			552
2042	C₅H₁₀O	3-Pentanone	102.2	Nonazeotrope		v-l	383
2043	C₅H₁₀O₂	Butyl formate	106.8	Nonazeotrope			575
2044	C₅H₁₀O₂	Ethyl propionate	99.15	Nonazeotrope			537
2045	C₅H₁₀O₂	Isobutyl formate	97.9	64.6	~95		536
2046	C₅H₁₀O₂	Isopropyl acetate	91.0	64.5	80		536
		"	88.7	64.0	70.2		982
2047	C₅H₁₀O₂	Methyl butyrate	102.65	Nonazeotrope			536
2048	C₅H₁₀O₂	Methyl isobutyrate	92.3	64.0	75		536
		" "	92.3	Nonazeotrope			563

| | | B-Component | | | Azeotropic | Data | |
No.	Formula	Name	B.P., °C	B.P., °C	Wt.%A		Ref.
A =	**CH₄O**	**Methanol** *(continued)*	**64.7**				
2049	C₅H₁₀O₂	Propyl acetate	101.6	Nonazeotrope			536
2049a	C₅H₁₀O₃	2-Methoxyethyl acetate	144.6	Nonazeotrope		v-l	920c
2050	C₅H₁₁Br	1-Bromo-3-methylbutane	118.2	Nonazeotrope			389
				B.p. curve			389
2051	C₅H₁₁Cl	1-Chloropentane	108.35	Nonazeotrope			409
2052	C₅H₁₁Cl	1-Chloro-3-methylbutane	99.4	62.0	57		527
2053	C₅H₁₁N	Piperidine	106.4	Nonazeotrope			575
2054	C₅H₁₂	2-Methylbutane	27.95	24.5	~ 4		563
		"	27.6	24.62	4		946
		"		24.2	6.98	v-l	715
		"	27.9	24.7	6.3		476
2055	C₅H₁₂	Pentane	36	30.6	7.6		476
		"	36.1	30.6	15 vol. %		1015
		"	37.15	30.8	9		538
		"	36.15	30.85	7		946
2056	C₅H₁₂O	Amyl alcohol	137.8	Nonazeotrope		v-l	383
2057	C₅H₁₂O	Butyl methyl ether	71	56.3	35.35		73
2058	C₅H₁₂O	*tert*-Butyl methyl ether	55	50.6	10		581c,256
2059	C₅H₁₂O	Ethyl propyl ether	63.6	55.5	24		556
2060	C₅H₁₂O₂	Diethoxymethane	87.95	63.2	65		527
2061	C₅H₁₂O₂	2,2-Dimethoxypropane	80	61-62	45		594
2062	C₅H₁₄SiO	Methoxymethyl-trimethylsilane	83	60	36 vol. %		907
2063	C₆H₅Cl	Chlorobenzene	132.0	Nonazeotrope		v-l	574,684g
2064	C₆H₅F	Fluorobenzene	85.15	59.7	32		545
2065	C₆H₅NO₂	Nitrobenzene	210.75	Nonazeotrope			554
2066	C₆H₆	Benzene	80.1	57.50	39.1	v-l	1026
		" 770 mm.		58	38.4		297,814,
		" 400 mm.		40	36.8		834,903,1044
		" 223 mm.		25	33.1		
		"	80.1	58	38	v-l	1026,406a
		" 64.7 p.s.i.a.	108	102	49	v-l	783
		" 112.7 p.s.i.a.	128	123	54	v-l	783
		" 159.7 p.s.i.a.	141	138	58	v-l	783
		" 209.7 p.s.i.a.	152	148	62	v-l	783
		" 259.7 p.s.i.a.	161	159	65	v-l	783
		" to crit. pt.		Azeotropic		v-l	888
2067	C₆H₈	1,3-Cyclohexadiene	80.8	56.38	38.8		563
2068	C₆H₈	1,4-Cyclohexadiene	85.6	58	42.5		563
2069	C₆H₈O	2,5-Dimethylfuran	93.3	61.5	51		982
2070	C₆H₁₀	Biallyl	60.2	47.05	22.5		563
2071	C₆H₁₀	Cyclohexene	82.75	55.9	40		563
2072	C₆H₁₀	2,3-Dimethyl-1,3-butadiene	68.9	52	25		78
2073	C₆H₁₀	1,3-Hexadiene	72.9	< 58	~40		221
2074	C₆H₁₀	2,4-Hexadiene	82	~ 58	~40		221
2075	C₆H₁₀	Methylcyclopentene	75.85	53.0	35		567
2076	C₆H₁₀	3-Methyl-1,3-pentadiene	77	~ 58	~40		221

TABLE I.   *Binary Systems*   81

No.	Formula	B-Component Name	B.P., °C	Azeotropic Data B.P., °C	Wt.%A		Ref.
A =	**CH₄O**	**Methanol** *(continued)*	**64.7**				
2076a	C₆H₁₀O	Methyldihydropyran	118.5	Nonazeotrope			581c
2077	C₆H₁₀O₂	Isopropyl acrylate		Min. b.p.			799
2078	C₆H₁₀O₂	Propyl acrylate		Min. b.p.			799
2079	C₆H₁₂	Cyclohexane	80	54	38		276,563
		"		35	34.8	v-l	623
		"		55	37.6	v-l	623
		"	520 mm.	45	35.6	v-l	623
		"		53.9	36.4	v-l	604
		"		54.28	36.7	v-l	623
		"	527 mm.	45	36.4	v-l	604
		"		55	36.4	v-l	604
		"	<760 mm.	27.5	34		925
		"	<760 mm.	30	32.6		925
		"	<760 mm.	38	31.6		925
		"	<760 mm.	42	26.8		925
2079a	C₆H₁₂	1-Hexene	200 mm.	20	18.4	v-l	498m
		"	400 mm.	32.6	19.1	v-l	498m
		"	600 mm.	42.0	19.6	v-l	498m
		"	760 mm.	63.6	48.2	20.8 v-l	498m
2080	C₆H₁₂	*cis*-3-Hexene	66.4	49.6	26 vol. %		826
2081	C₆H₁₂	Hexenes	68	49.5	27 vol. %		1015
2082	C₆H₁₂	Methylcyclopentane	72.0	51.3	32		567
2083	C₆H₁₂O	Butyl vinyl ether	94.2	62	52		982
2084	C₆H₁₂O	4-Methyl-2-pentanone	116.2	Nonazeotrope		v-l	383
2085	C₆H₁₄	2,2-Dimethylbutane	49.74	39.6	17 vol. %		826
2086	C₆H₁₄	2,3-Dimethylbutane	58.0	45.0	20		567
		"	58.15	44.5	19.2	v-l	461c
2087	C₆H₁₄	Hexane	68	50	26 vol. %		541,1015
		"		49.9	26.4		497
		"	68.95	49.5	26.4		478
		"	68.95	50.57	28		946
		"	210.3 mm.	20	22.9	v-l	850f
		"	267.5 mm.	25	23.5	v-l	850f
		"	336.7 mm.	30	24.2	v-l	850f
2088	C₆H₁₄	2-Methylpentane	60.27	45.6	21 vol. %		826
2089	C₆H₁₄	3-Methylpentane	63.28	47.1	20 vol. %		826
2090	C₆H₁₄O	*tert*-Amyl methyl ether	86-7	62.3	50		256
2091	C₆H₁₄O	Butyl ethyl ether	92.2	62.6	56		982
2091a	C₆H₁₄O	Isopropyl ether					
			730 mm.	57.0	24	v-l	684f
2092	C₆H₁₄O	Propyl ether	90.4	63.8	72		545
2093	C₆H₁₄O₂	Acetal	103.55	Nonazeotrope			556
2094	C₆H₁₄O₂	1,1-Dimethoxybutane	114	Nonazeotrope			981
2095	C₆H₁₄O₂	2,2-Dimethoxybutane	106-7	64.5	81.5		594
2096	C₆H₁₄S	Isopropyl sulfide	120.5	Nonazeotrope			566
2097	C₆H₁₅N	Triethylamine	89.35	Nonazeotrope			575
			730 mm.	Nonazeotrope		v-l	684e

		B-Component		Azeotropic Data		
No.	Formula	Name	B.P., °C	B.P., °C	Wt.%A	Ref.
A =	CH₄O	**Methanol** *(continued)*	**64.7**			
2098	C₇H₈	Toluene	110.7	63.8	69	50
		"		0.5	71.6	537,814,834
		"		25	73.0	537,814,834
		"		50	74.0	537,814,834
		"		62.5	75.0	537,814,834
		"	110.6	63.5	72.5	662
		" 708 mm.		62.0	72.0	662
		" 613 mm.		58.0	71.6	662
		" 512 mm.		53.5	71.0	662
		" 412 mm.		48.0	70.0	662
		"	110.7	63.6	70.8 v-l	595
2098a	C₇H₈O	Anisole	154	Nonazeotrope v-l		684g
2099	C₇H₁₄	*trans*-1,3-Dimethyl-				
		cyclopentane	90.77	57.3	45 vol. %	826
		"	90.7		~45	928
2100	C₇H₁₄	Methylcyclohexane	100.8	59.2	54	50,572
		" 127.5 mm.		20	45.1 v-l	850f
		" 209.8 mm.		30	46.0 v-l	850f
		" 334.3 mm.		40	46.8 v-l	850f
2101	C₇H₁₆	*n*-Heptane	98.45	59.1	51.5	572
		"	98.4	58.8	46.1	477
		" 406 mm.		43.83		1064
2102	C₇H₁₆	2-Methylhexane	90.05	57.1	44 vol. %	826
		"	90.0		~40	928
2103	C₇H₁₆	3-Methylhexane	91.8		~40	928
		"	91.85	57.6	44 vol. %	826
2104	C₇H₁₆	2,2,3-Trimethylbutane	80.88	54.1	38 vol. %	826
2105	C₈H₈	Styrene	145.8	64.2		545
2106	C₈H₁₀	Ethylbenzene	136.15	Nonazeotrope		537
2107	C₈H₁₀	*m*-Xylene	139.0	Nonazeotrope		537
2108	C₈H₁₀	*o*-Xylene	143.6	Nonazeotrope		541
2109	C₈H₁₀	*p*-Xylene	138.3	Nonazeotrope		540
		"	138.35	64.5	99.5	227
2110	C₈H₁₄O	2-Ethyl-2-hexenal	176	Nonazeotrope		981
2111	C₈H₁₆	1,3-Dimethyl-				
		cyclohexane	120.7	< 62.5		575
2112	C₈H₁₈	2,5-Dimethylhexane	109.2	61.0	60	575
2113	C₈H₁₈	Octane	125.6	63.0	72	537
		"	125.75	62.75	67.5	477
		" 406 mm.		47.65		1064
2114	C₈H₁₈	2,2,4-Trimethylpentane	99.3	59.4	53	575
2115	C₉H₁₂	Cumene	152.8	Nonazeotrope		575
2116	C₉H₁₂	Mesitylene	164.6	Nonazeotrope		537
2117	C₉H₁₂	Propylbenzene	159.3	Nonazeotrope		575
2118	C₉H₁₈O	2,6-Dimethyl-4-				
		heptanone	169.4	Nonazeotrope		981

TABLE I. *Binary Systems* 83

No.	Formula	B-Component Name		B.P., °C	Azeotropic Data B.P., °C	Wt.%A		Ref.
A =	**CH₄O**	**Methanol** *(continued)*		**64.7**				
2119	C_9H_{20}	Nonane,	406 mm.		48.93			1064
		"		150.7	64.1	83.4		477
2120	$C_{10}H_{14}$	Cymene		176.7	Nonazeotrope			537
2121	$C_{10}H_{16}$	Camphene		159.6	64.67?	98.8?		574
2122	$C_{10}H_{16}$	d-Limonene		177.8	64.63	99.2		572
2123	$C_{10}H_{16}$	α-Pinene		155.8	64.55	90.7		528
2124	$C_{10}H_{16}$	Thymene		179.7	Nonazeotrope			537
2125	$C_{10}H_{22}$	2,7-Dimethyloctane		160.1	< 64.6	> 3		575
2126	$C_{10}H_{22}$	Decane	406 mm.		Nonazeotrope			1064
		"		171.8	Nonazeotrope		v-l	715
2127	$C_{11}H_{24}$	Undecane	406 mm.		Nonazeotrope			1064
A =	**CH₄S**	**Methanethiol**		**6.8**				
2128	$C_2H_4O_2$	Methyl formate		31.7	Nonazeotrope			575
2129	C_2H_5Cl	Chloroethane		12.4	Nonazeotrope			575
2130	$C_2H_5NO_2$	Ethyl nitrite		17.4	< 6.4	>82		575
2131	C_4H_8	2-Methyl- propene	95 p.s.i.a.		53	19.5		408
2132	C_4H_{10}	Butane		0.6	— 0.5	25		575
2133	C_4H_{10}	2-Methylpropane		—11.70	— 13.00	4.9		88
2134	C_5H_{12}	2-Methylbutane		27.95	Nonazeotrope			575
A =	**CH₅N**	**Methylamine**		**— 6.32**				
2135	C_2H_7N	Dimethylamine		6.88	Nonazeotrope			818
2136	C_3H_9N	Trimethylamine						
		"	60 p.s.i.g.		36	85		818
		"	210 p.s.i.g.		75	90-92		818
		"	370 p.s.i.g.		Nonazeotrope			818
		"		3.5	— 5	70		20,21
		"	1040 mm.		0	68	v-l	353
2137	C_4H_4	1-Butene-3-yne		5.0	— 6.8	97.5	v-l	84
2138	C_4H_6	1,3-Butadiene		— 4.5	— 9.5	41.4	v-l	84
		"		— 4.5	— 10.4			215
		"				58		85
		"	5 atm.			74		85
		"	20 atm.			96		85
		"	4 kg./cm.²			49.4	v-l	407
		"		— 4.6	— 9.2	41.3	v-l	305
2139	C_4H_8	1-Butene	4 kg./cm.²		62.5		v-l	407
		"		— 6.3	— 13.6	36.5	v-l	305
		"				50		85
		"	5 atm.			64		85
		"	20 atm.			74		85
		"		— 5.6	— 13	22.2	v-l	84
		"		— 6.0	— 13.8			215
2140	C_4H_8	cis-2-Butene		3.5	— 9.6	47.5	v-l	84
		"		3.7	— 10.4	56.5	v-l	305

	B-Component		B.P., °C	Azeotropic Data			
No.	Formula	Name		B.P., °C	Wt.%A		Ref.
A =	CH₅N	**Methylamine** *(continued)*	— 6.32				
2141	C₄H₈	trans-2-Butene	0.9	— 11.6	55.0	v-l	305
		"	0.9	— 10.4	48.5	v-l	84
2142	C₄H₈	2-Methylpropene	— 6.0	— 14.3	32	v-l	84
		"	— 6.9	— 15.5	47.5		305
2143	C₄H₁₀	Butane	— 0.5	— 16.0	43.7		305
		"	— 0.6	Min. b.p.			195
		Butane	1.0	— 14.0	37.6	v-l	84
2144	C₄H₁₀	2-Methylpropane	—10.0	— 19.9	25.5	v-l	84
		"	—11.7	— 24.0	27.0		305
		% CH₃NH₂ increases with pressure in above azeotropes					305,407
2145	C₅H₈	Isoprene	34	Min. b.p.			818
2146	C₅H₁₀	Amylenes		Min. b.p.			252
A =	C₂Br₂Cl₂	**1,2-Dibromo-1,2,- dichloroethylene**	172				
2147	C₂H₆O	Ethyl alcohol	78.3	Nonazeotrope			533
2148	C₄H₁₀O	Butyl alcohol	117.75	Nonazeotrope			533
A =	C₂ClF₅	**Chloropentafluoro- ethane**	— 38.5				
2149	C₂H₄F₂	1,1-Difluoroethane	— 24.7	— 41.3	83.8		582
A =	C₂Cl₂F₄	**1,2-Dichlorotetra- fluoroethane**					
2150	C₄H₁₀	Butane	— 0.5	— 2.2	59		279
A =	C₂Cl₃F₃	**1,1,2-Trichlorotri- fluoroethane**	47.5				
2151	C₂Cl₄F₂	1,1,2,2-Tetrachloro- difluoroethane	92.4	Nonazeotrope			266,981
2152	C₂H₆O	Ethyl alcohol	78.3	43.8	96.2		982
		" "	78.3	Nonazeotrope			46
2153	C₃H₆O	Acetone	56.5	45	87.5		232
A =	C₂Cl₃N	**Trichloroacetonitrile**					
2154	C₂H₃N	Acetonitrile	82	75.6	71		437
A =	C₂Cl₄	**Tetrachloroethylene**	121.1				
2155	C₂HCl₃O	Chloral	97.5	Nonazeotrope			575
2156	C₂H₃Cl₃	1,1,2-Trichloroethane	112.4	112	57		215
		"	113.65	112.9	26		581
2157	C₂H₄Cl₂O	2,2-Dichloroethanol	146.2	<119.5	<96		575
2158	C₂H₄O₂	Acetic acid	118.5	107.35	61.5		563
2159	C₂H₅BrO	2-Bromoethanol	150.2	116.5	85		575
2160	C₂H₅ClO	2-Chloroethanol	128.6	110.0	75.7		568
2161	C₂H₅NO	Acetamide	221.2	120.45	97.4		574
2162	C₂H₆O	Ethyl alcohol	78.3	76.75	~37		574

TABLE I. *Binary Systems* 85

No.	Formula	Name	B.P., °C	B.P., °C	Wt.%A		Ref.
		B-Component		**Azeotropic Data**			
A =	**C₂Cl₄**	**Tetrachloroethylene**	**121.1**				
		(continued)					
2163	C₂H₆O₂	Glycol	197.4	119.1	94		574
2164	C₃H₅BrO	Epibromohydrin	138.5	<119.5	<92		575
2165	C₃H₅ClO	1-Chloro-2-propanone	119	118			563
2166	C₃H₅ClO	Epichlorohydrin	116.45	110.12	48.5		563
2167	C₃H₅ClO₂	Methyl chloroacetate	129.95	120.8	94		575
2168	C₃H₆O	Acetone	56.1	Nonazeotrope		v-l	187
2169	C₃H₆O	Allyl alcohol	96.95	93.15	55		527,875
2170	C₃H₆O₂	Propionic acid	140.9	119.1	91.5		527
2171	C₃H₇ClO	1-Chloro-2-propanol	127.0	113.0	72		567
2172	C₃H₇ClO	2-Chloro-1-propanol	133.7	115.0	87		567
2173	C₃H₇NO	Propionamide	222.1	Nonazeotrope			527
2174	C₃H₇NO₂	Ethyl carbamate	185.25	<120.8	<96		564
2175	C₃H₇NO₃	Propyl nitrate	110.5	109.6	18		560
2176	C₃H₈O	Isopropyl alcohol	82.4	81.7	30		535
	" "		82.3	81.7	19		981
2177	C₃H₈O	Propyl alcohol	97.25	94.05	52		535
2178	C₃H₈O₂	2-Methoxyethanol	124.5	109.7	75.5		569
2179	C₄H₅N	Pyrrol	130.0	113.35	80.5		553
2180	C₄H₈O₂	Butyric acid	164.0	121.0	98.8		527
			162.45	Nonazeotrope			541
2181	C₄H₈O₂	Dioxane	101.35	Nonazeotrope			527
2182	C₄H₈O₂	Isobutyric acid	154.35	120.5	~97		541
2183	C₄H₈O₃	Methyl lactate	143.8	120.0	90		575
2184	C₄H₉I	1-Iodo-2-methylpropane	120.8	119.2	40		549
2185	C₄H₉NO₃	Isobutyl nitrate	122.9	117.45	70		560
2186	C₄H₁₀O	Butyl alcohol	117.75	108.95	71		574
	" "		117.7	110	68		981
2187	C₄H₁₀O	*sec*-Butyl alcohol	99.5	97.0	43		567
2188	C₄H₁₀O	*tert*-Butyl alcohol	82.45	Nonazeotrope			575
2189	C₄H₁₀O	Isobutyl alcohol	108	103.05	60		563
2190	C₄H₁₀O₂	2-Ethoxyethanol	135.3	116.25	83.5		527
2191	C₅H₄O₂	2-Furaldehyde	161.45	Nonazeotrope			527
2192	C₅H₅N	Pyridine	115.4	112.85	51.5		553
	"	126 mm.		60	59	v-l	292c
	"	270 mm.		80	58	v-l	292c
	"	527 mm.		100	56	v-l	292c
2193	C₅H₆O₂	Furfuryl alcohol	169.35	Nonazeotrope			575
2194	C₅H₈O	Cyclopentanone	130.65	120.1	86		552
2195	C₅H₉N	Isovaleronitrile	130.5	113.5	72		562
2196	C₅H₁₀O	Cyclopentanol	140.85	118.8	92		567
2197	C₅H₁₀O₂	Isovaleric acid	176.5	Nonazeotrope			527
2198	C₅H₁₀O₃	Ethyl carbonate	126.0	118.55	74		547
2199	C₅H₁₀O₃	2-Methoxyethyl acetate	144.6	120.9	96		556
2200	C₅H₁₁Br	1-Bromo-3-methylbutane	120.65	119.25	48		549
2201	C₅H₁₂O	Amyl alcohol	138.2	117.0	85		567
2202	C₅H₁₂O	*tert*-Amyl alcohol	102.35	101.4	27		567
2203	C₅H₁₂O	Isoamyl alcohol	131.3	116.2	81		527
2204	C₅H₁₂O	2-Pentanol	119.8	113.2	66		567

No.	Formula	B-Component Name	B.P., °C	Azeotropic Data B.P., °C	Wt.%A		Ref.
A =	C_2Cl_4	**Tetrachloroethylene** *(continued)*	**121.1**				
2205	$C_5H_{12}O_2$	2-Propoxyethanol	151.35	120.6	95		527
2206	$C_6H_{10}O$	Mesityl oxide	129.45	119.8	83.5		527
2207	$C_6H_{10}S$	Allyl sulfide	139.35	Nonazeotrope			566
2207a	C_6H_{12}	1-Hexene	60	Nonazeotrope		v-l	362c
2208	$C_6H_{12}O$	3-Hexanone	123.3	118.15	55		552
2209	$C_6H_{12}O$	4-Methyl-2-pentanone	116.05	113.85	48		569
2210	$C_6H_{12}O_2$	Butyl acetate	126.0	120.1	79		527
		" "	125.0	120.5			547
2211	$C_6H_{12}O_2$	Ethyl butyrate	119.9	119.5	57		547
2212	$C_6H_{12}O_2$	Ethyl isobutyrate	110.1	Nonazeotrope			563
2213	$C_6H_{12}O_2$	Isoamyl formate	123.6	117.9	65		547
2214	$C_6H_{12}O_2$	Isobutyl acetate	117.2	115.5	47		573
2215	$C_6H_{12}O_2$	Methyl isovalerate	116.5	Nonazeotrope			575
2216	$C_6H_{12}O_2$	Propyl propionate	122.5	120.0			547
2217	$C_6H_{12}O_3$	Paraldehyde	124	118.75	68		563
2217a	C_6H_{14}	Hexane 60°C.		Nonazeotrope		v-l	362c
2218	$C_6H_{14}O_2$	2-Butoxyethanol	171.25	Nonazeotrope			526
2219	$C_6H_{15}BO_3$	Ethyl borate	118.6	117.5	48		538
2220	C_7H_8	Toluene	110.75	Nonazeotrope			538
2221	$C_7H_{14}O_2$	Ethyl isovalerate	134.7	Nonazeotrope			575
2222	$C_7H_{14}O_2$	Isobutyl propionate	136.9	Nonazeotrope			547
2223	$C_7H_{14}O_2$	Isopropyl isobutyrate	120.8	119.0	45		547
2224	$C_7H_{14}O_2$	Propyl isobutyrate	134.0	Nonazeotrope			547
2225	C_8H_{10}	Ethylbenzene	136.15	Nonazeotrope			575
2226	C_8H_{16}	1,3-Dimethylcyclohexane	~120.5	118			563
2227	C_8H_{18}	Octane	125.75	<120.5	<92		575
2228	$C_8H_{18}O$	Isobutyl ether	122.2	~119.5	~65		548
A =	C_2Cl_6	**Hexachloroethane**	**185**				
2229	$C_2HCl_3O_2$	Trichloroacetic acid	196	181	85		563
2230	$C_2H_3ClO_2$	Chloroacetic acid	189.35	171.2	75		529
2231	C_2H_3NO	Acetamide	221.2	Nonazeotrope			535
2232	$C_2H_6O_2$	Glycol	197.4	Nonazeotrope			530
2233	$C_2H_6SO_4$	Methyl sulfate	189.1	<181.5	<72		575
2234	C_3H_7NO	Propionamide	222.1	Nonazeotrope			535
2235	$C_4H_6O_4$	Methyl oxalate	164.2	Nonazeotrope			547
2236	$C_4H_8O_2$	Butyric acid	162.45	162.0			542
2237	$C_5H_4O_2$	2-Furaldehyde	161.45	Nonazeotrope			556
2238	$C_5H_8O_4$	Methyl malonate	181.4	176.0	45		538
2239	$C_5H_{10}O_2$	Isovaleric acid	176.5	172.6	63		527
2240	$C_5H_{10}O_2$	Valeric acid	186.35	179.0	70		562
2241	C_6H_6O	Phenol	182.2	173.7	70		574
2242	C_6H_7N	Aniline	184.35	176.75	66		551
2243	$C_6H_{10}O_3$	Ethyl acetoacetate	180.4	172.5	51		552
2244	$C_6H_{10}O_4$	Ethylidene diacetate	168.5	Nonazeotrope			575
2245	$C_6H_{10}O_4$	Ethyl oxalate	185.65	178.6	57		574
2246	$C_6H_{10}O_4$	Methyl succinate	195.5	<184.0			547

TABLE I. *Binary Systems* 87

No.	Formula	B-Component Name	B.P., °C	Azeotropic Data B.P., °C	Wt.%A	Ref.
A =	C₂Cl₆	**Hexachloroethane** *(continued)*	**185**			
2247	C₆H₁₂O	Cyclohexanol	160.65	Nonazeotrope		531
2248	C₇H₆O	Benzaldehyde	179.2	Nonazeotrope		563
2249	C₇H₇Br	*p*-Bromotoluene	185	~183.5	~70	530
2250	C₇H₈O	Benzyl alcohol	205.15	182.0	88	529
2251	C₇H₈O	*m*-Cresol	202.2	183.2	92	575
		"	202.2	Nonazeotrope		542
2252	C₇H₈O	*o*-Cresol	191.1	181.3	72	538
2253	C₇H₈O	*p*-Cresol	201.7	183.0	90	562
2254	C₇H₁₂O₄	Ethyl malonate	199.2	Nonazeotrope		547
2255	C₈H₈O₂	Phenyl acetate	195.7	Nonazeotrope		547
2256	C₉H₁₈O₂	Butyl isovalerate	177.6	Nonazeotrope		547
2257	C₉H₁₈O₂	Isoamyl butyrate	178.5	Nonazeotrope		538
2258	C₉H₁₈O₂	Isobutyl isovalerate	171.2	Nonazeotrope		575
2259	C₉H₁₈O₃	Isobutyl carbonate	190.3	184.0	<80	547
2260	C₁₀H₁₄	Butylbenzene	183.1	Nonazeotrope		575
2261	C₁₀H₁₄	Cymene	176.7	Nonazeotrope		575
2262	C₁₀H₁₆	*d*-Limonene	177.8	Nonazeotrope		530
2263	C₁₀H₁₆	*α*-Terpinene	173.4	Nonazeotrope		575
2264	C₁₀H₁₆	Terpinolene	~185	~182.5		563
2265	C₁₀H₁₆	Thymene	179.7	Nonazeotrope		530
2266	C₁₀H₁₆O	Fenchone	193	Nonazeotrope		563
2267	C₁₀H₁₈O	Cineol	176.35	Nonazeotrope		548
2268	C₁₀H₁₈O	Linalool	198.6	Nonazeotrope		532
2269	C₁₀H₂₀O₂	Isoamyl isovalerate	192.7	Nonazeotrope		547
A =	C₂F₆	**Hexafluoroethane**	**— 78**			
2270	C₂H₆	Ethane	— 88	— 92	69	120
A =	C₂HBrClF₃	**2-Bromo-2-chloro-1,1,1-trifluoroethane**				
2271	C₄H₁₀O	Ethyl ether	34.6	52.7	83.2	355
A =	C₂HBrCl₂	*cis*-**1-Bromo-1,2-dichloroethylene**	113.8			
2272	C₂H₆O	Ethyl alcohol	78.3	77.4	30.9	988
A =	C₂HBrCl₂	*trans*-**1-Bromo-1,2-dichloroethylene**				
2273	C₂H₆O	Ethyl alcohol	78.3	74.9	65.5	988
A =	C₂HBrCl₂	**1-Bromo-2,2-dichloroethylene**	**107-108**			
2274	C₂H₆O	Ethyl alcohol	78.3	77.25	39.5	988
A =	C₂HBr₂Cl	**1,2-Dibromochloroethylene**	**140**			
2275	C₂H₆O	Ethyl alcohol	78.3	74.9	65.5	533
2276	C₄H₁₀O	Butyl alcohol	117.75	117.0		533
A =	C₂HBr₃O	**Bromal**	**174**			
2277	C₅H₁₀O₂	Isovaleric acid	176.5	~170.3		563

		B-Component		Azeotropic Data			
No.	Formula	Name	B.P., °C	B.P., °C	Wt.%A		Ref.
A =	**C₂HClF₂**	**Chlorodifluoroethylene**					
2278	C₃F₆	Hexafluoropropene					
		5170 mm.		20°	27.7	v-l	615
A =	**C₂HClF₄**	**Tetrafluorochloroethane**	**— 10**				
2279	C₄F₈	Octafluorocyclobutane	— 4	— 12	80 vol. %		49
		" —20° to 80°		v.p. curve			748
		"		— 13	60		748,922
		" 30 kg./cm.²		107			922
A =	**C₂HCl₃**	**Trichloroethylene**	**86.9**				
2280	C₂H₃N	Acetonitrile 778 mm.	81.6	74.6	71	v-l	763,635c
2281	C₂H₄Cl₂	1,2-Dichloroethane	83.7	82.1	43.5		801
		"	83.45	82.6	18		549
		"	83.45	82.2	39		581
2282	C₂H₄O₂	Acetic acid	117.9	Nonazeotrope			981
		" "	118.5	86.5	96.2		545
2283	C₂H₅BrO	2-Bromoethanol	150.2	Nonazeotrope			575
2284	C₂H₅ClO	2-Chloroethanol	128.6	Nonazeotrope			564
		"	128.6	86.55	97.5		527
2285	C₂H₅NO₃	Ethyl nitrate	87	83.5	62		527
2286	C₂H₆O	Ethyl alcohol	78.3	70.9	72.5		297,803,73c,
							837
2287	C₂H₆O₂	Glycol	197.4	Nonazeotrope			575
2288	C₃H₆Cl₂	1,2-Dichloropropane	96.3	Nonazeotrope			981
2289	C₃H₆O	Acetone	56.15	Nonazeotrope			552
2290	C₃H₆O	Allyl alcohol	96.9	80.9	84.4		358,563
2291	C₃H₆O₃	Methyl carbonate	90.35	85.95	90		527
2292	C₃H₇ClO	1-Chloro-2-propanol	127.0	Nonazeotrope			575
2293	C₃H₇ClO	2-Chloro-1-propanol	133.7	Nonazeotrope			575
2294	C₃H₇I	2-Iodopropane	89.45	< 86.5	<88		575
2295	C₃H₈O	Isopropyl alcohol	82.45	75.5	70		573
2296	C₃H₈O	Propyl alcohol	97.2	81.75	83		563
2297	C₃H₉BO₃	Methyl borate	68.7	Nonazeotrope			575
2298	C₄H₄S	Thiophene	84	Nonazeotrope			563
2299	C₄H₈O	2-Butanone	79.6	Nonazeotrope			501,552
2300	C₄H₈O₂	Butyric acid	162.5	Nonazeotrope			678
2301	C₄H₈O₂	Dioxane	101.35	Nonazeotrope			527
2302	C₄H₈O₂	Ethyl acetate	77.05	Nonazeotrope			563
		" "					
		" 700-760 mm.		Nonazeotrope		v-l	782
2303	C₄H₈O₂	Methyl propionate	79.85	Nonazeotrope			547
2304	C₄H₈O₂	Propyl formate	80.85	79.5	20		547
2305	C₄H₉NO₂	Butyl nitrite	78.2	Nonazeotrope			550
2306	C₄H₁₀O	Butyl alcohol	117.75	86.65	97		527
2307	C₄H₁₀O	sec-Butyl alcohol	99.5	84.2	85		567
2308	C₄H₁₀O	tert-Butyl alcohol	82.55	75.8	~ 67		532
2309	C₄H₁₀O	Isobutyl alcohol	108	85.4	91		563

TABLE I. *Binary Systems* 89

No.	Formula	Name	B.P., °C	B.P., °C	Wt.%A		Ref.
		B-Component		Azeotropic Data			
A =	C_2HCl_3	**Trichloroethylene** *(continued)*	**86.9**				
2310	$C_4H_{10}S$	Diethyl sulfide 701 mm.	89.5	Nonazeotrope		v-l	635
2311	$C_4H_{10}S$	2-Methyl-1-propanethiol	88	Nonazeotrope			563
2312	$C_5H_{10}O$	Isovaleraldehyde	92.1	Nonazeotrope			575
2313	$C_5H_{10}O$	3-Methyl-2-butanone	95.4	Nonazeotrope			552
2314	$C_5H_{10}O$	3-Pentanone	102.05	Nonazeotrope			552
2315	$C_5H_{10}O_2$	Ethyl propionate	99.1	Nonazeotrope			575
2316	$C_5H_{10}O_2$	Isobutyl formate	98.2	Nonazeotrope			547
2317	$C_5H_{10}O_2$	Isopropyl acetate	89.5	Nonazeotrope			575
2318	$C_5H_{10}O_2$	Methyl isobutyrate	92.5	Nonazeotrope			547
2319	$C_5H_{11}NO_2$	Isoamyl nitrite	97.15	Nonazeotrope			550
2320	$C_5H_{12}O$	*tert*-Amyl alcohol	102.25	86.67	92.5		545
2321	$C_5H_{12}O$	Isoamyl alcohol	131.3	Nonazeotrope			527
2322	$C_5H_{12}O$	3-Methyl-2-butanol	112.6	Nonazeotrope			575
2323	$C_5H_{12}O$	2-Pentanol	119.8	Nonazeotrope			575
2324	$C_5H_{12}O$	3-Pentanol	116.0	Nonazeotrope			575
2325	$C_5H_{12}O_2$	Diethoxymethane	87.9	89.2	53.5		568
2326	C_6H_6	Benzene	80.2	Nonazeotrope			563
		"	80.1	Nonazeotrope		v-l	780
2327	C_6H_{10}	Cyclohexene	82.75	Azeotrope doubtful			563
2328	C_6H_{12}	Cyclohexane	80.75	Nonazeotrope			563
		"	80.7	80.5	16.6	v-l	780
2328a	C_6H_{12}	1-Hexene 60°C.		Nonazeotrope		v-l	362c
2329	C_6H_{12}	Methylcyclopentane	72.0	Nonazeotrope			575
2330	C_6H_{14}	Hexane 60°C.		Nonazeotrope		v-l	362c
		"	68.8	Nonazeotrope			575
2331	$C_6H_{14}O$	Isopropyl ether	68.3	Nonazeotrope			559
2332	$C_6H_{14}O$	Propyl ether 699 mm.	87.2	Nonazeotrope		v-l	635
2333	$C_6H_{14}O_2$	Acetaldehyde diethyl acetal	103.55	Nonazeotrope			559
2334	C_7H_{14}	Methylcyclohexane	101.15	Nonazeotrope			575
2335	C_7H_{16}	Heptane	98.45	Nonazeotrope			563
A =	C_2HCl_3O	**Chloral**	**97.75**				
2336	$C_2H_4Cl_2$	1,2-Dichloroethane	83.75	Nonazeotrope			532
2337	$C_2H_5NO_2$	Nitroethane	114.2	Nonazeotrope			575
2338	C_2H_6O	Ethyl alcohol	78.3	116.2			563
2339	C_3H_5I	3-Iodopropene	101.8	~ 97.0	~80		548
2340	$C_3H_6O_3$	Methyl carbonate	90.35	~ 98.0	~85		548
2341	C_3H_7I	1-Iodopropane	102.4	~ 97.3			563
2342	C_3H_7I	2-Iodopropane	89.45	Nonazeotrope			548
2343	C_4H_8O	2-Butanone	79.6	Nonazeotrope			563
2344	$C_4H_8O_2$	Propyl formate	80.85	Nonazeotrope			548
2345	C_4H_9Br	1-Bromobutane	101.5	96.5			575
2346	C_4H_9Br	1-Bromo-2-methylpropane	91.4	Azeotrope doubtful			563
2347	C_4H_9Cl	1-Chlorobutane	78.5	Nonazeotrope			575
2348	C_4H_9I	1-Iodo-2-methylpropane	120.8	Nonazeotrope			575

		B-Component			Azeotropic Data		
No.	Formula	Name	B.P., °C	B.P., °C	Wt.%A	Ref.	
A =	**C₂HCl₃O**	**Chloral** *(continued)*	**97.75**				
2349	C₄H₁₀O	Isobutyl alcohol	108	~138		563	
2350	C₅H₁₀O	3-Pentanone	102.2	102.9	~23	563	
2351	C₅H₁₀O₂	Butyl formate	106.8	Nonazeotrope		545	
2352	C₅H₁₀O₂	Ethyl propionate	99.15	100.8		545	
2353	C₅H₁₀O₂	Isobutyl formate	97.9	100.1	~60	528	
2354	C₅H₁₀O₂	Isopropyl acetate	90.8	98.2	~85	548	
2355	C₅H₁₀O	Methyl butyrate	102.65	103.3	45	545	
2356	C₅H₁₀O	Methyl isobutyrate	92.3	98.2	~90	545	
2357	C₅H₁₀O₂	Propyl acetate	101.6	102.55	50.5	528	
2358	C₅H₁₁Br	1-Bromo-3-methylbutane	120.65	Nonazeotrope		575	
2359	C₅H₁₁Cl	1-Chloro-3-methylbutane	99.8	< 97.0	<85	548	
2360	C₆H₆	Benzene	80.2	Nonazeotrope		528	
2361	C₆H₁₂	Cyclohexane	80.75	Nonazeotrope		545	
2362	C₆H₁₂	Methylcyclopentane	72.0	Nonazeotrope		575	
2363	C₆H₁₂O₂	Ethyl isobutyrate	110.1	Nonazeotrope		548	
2364	C₆H₁₄	Hexane	68.8	Nonazeotrope		575	
2365	C₇H₈	Toluene	110.75	Nonazeotrope		548	
2366	C₇H₁₄	Methylcyclohexane	100.95	94.45	57	572	
2367	C₇H₁₆	Heptane	98.45	93	53	545	
2368	C₈H₁₆	1,3-Dimethylcyclohexane	120.7	Nonazeotrope		575	
2369	C₈H₁₈	2,5-Dimethylhexane	109.3	< 97.0	<90	545	
2370	C₈H₁₈	Octane	125.75	Nonazeotrope		583	
A =	**C₂HCl₃O₂**	**Trichloroacetic Acid**	**197.55**				
2371	C₂HCl₅	Pentachloroethane	161.95	161.8	3.5	574	
2372	C₂H₃ClO₂	Chloroacetic acid	189.35	Nonazeotrope		530	
2373	C₂H₄O₂	Methyl formate	31.9	Nonazeotrope		563	
2374	C₄H₁₀O	Ethyl ether	34.6	Nonazeotrope		563	
2375	C₆H₄BrCl	*p*-Bromochlorobenzene	196.4	<191.5	<47	575	
2376	C₆H₄Cl₂	*p*-Dichlorobenzene	174.35	174.0	~12	530	
2377	C₆H₅Br	Bromobenzene	156.1	Nonazeotrope		535	
2378	C₆H₅I	Iodobenzene	188.55	~181	~25	563	
2379	C₆H₅NO₂	Nitrobenzene	210.75	Nonazeotrope		554	
2380	C₆H₁₂O₂	Caproic acid	205.15	210.4?	45?	575	
		" "	204.5	Nonazeotrope		563	
2381	C₇H₇Br	*o*-Bromotoluene	181.45	180.0	~18	535	
2382	C₇H₇Cl	α-Chlorotoluene	179.3	~178.2	~14	530	
2383	C₇H₇Cl	*o*-Chlorotoluene	159.2	Nonazeotrope		575	
2384	C₇H₇I	*p*-Iodotoluene	214.5	<196.8		575	
2385	C₇H₈O	*m*-Cresol	202.8	Nonazeotrope		563	
2386	C₇H₈O	*o*-Cresol	190.8	Nonazeotrope		563	
2387	C₇H₈O	*p*-Cresol	201.7	Reacts		535	
2388	C₇H₈O₂	Guaiacol	205.05	Reacts		535	
2389	C₈H₅O	Acetophenone	202.05	Nonazeotrope		530	
2390	C₁₀H₈	Naphthalene	218.05	Nonazeotrope		530	
2391	C₁₀H₁₄	Butylbenzene	183.1	181.3	20	562	
2392	C₁₀H₁₄	Cymene	176.7	176.0?		575	
2393	C₁₃H₁₅	*d*-Limonene	177.8	171		563	

TABLE I. *Binary Systems*   91

No.	Formula	B-Component Name	B.P., °C	Azeotropic Data B.P., °C	Wt.% A	Ref.
A =	$C_2HCl_3O_2$	**Trichloroacetic Acid**	**197.55**			
		(continued)				
2394	$C_{11}H_{20}O$	Terpineol methyl ether	216.2	Nonazeotrope		537
A =	C_2HCl_5	**Pentachloroethane**	**162.0**			
2395	$C_2H_3BrO_2$	Bromoacetic acid	205.1	Nonazeotrope		527
2396	$C_2H_3ClO_2$	Chloroacetic acid	189.35	158.65	90.1	530
2397	$C_2H_4O_2$	Acetic acid	118.5	Nonazeotrope		542
2398	C_2H_5ClO	2-Chloroethanol	128.6	Nonazeotrope		526
2399	C_2H_5NO	Acetamide	221.2	160.5	97	574
2400	$C_2H_6O_2$	Glycol	197.4	154.5	~85	528
2401	$C_2H_6SO_4$	Methyl sulfate	189.1	Nonazeotrope		547
2402	$C_3H_5BrO_2$	α-Bromopropionic acid	205.8	Nonazeotrope		575
2403	$C_3H_6Cl_2O$	1,3-Dichloro-2-propanol	175.1	159.7	77.5	529
2404	$C_3H_6O_2$	Propionic acid	140.7	Nonazeotrope		563
2405	C_3H_7NO	Propionamide	222.1	Nonazeotrope		527
2406	$C_3H_7NO_2$	Ethyl carbamate	185.25	159.8	91	527
2407	$C_4H_6O_4$	Methyl oxalate	163.3	157.55	68	563
2408	$C_4H_7BrO_2$	Ethyl bromoacetate	158.9	158.5	30	516
2409	$C_4H_7ClO_2$	Ethyl chloroacetate	143.55	Nonazeotrope		580
2410	$C_4H_8O_2$	Butyric acid	163.5	156.75	74	565
2411	$C_4H_8O_2$	Isobutyric acid	154.35	152.9	57	563
2412	$C_4H_8O_3$	Methyl lactate	143.8	Nonazeotrope		573
2413	$C_4H_{10}O_2$	2-Ethoxyethanol	135.3	Nonazeotrope		526
2414	$C_5H_4O_2$	2-Furaldehyde	161.4	156.75	60	556
		"	161.4	155.15	50	528
2415	$C_5H_8O_3$	Methyl acetoacetate	169.5	<159.4	>40	563
2416	$C_5H_8O_4$	Methyl malonate	181.5	Nonazeotrope		547
2417	$C_5H_9ClO_2$	Propyl chloroacetate	162.5	160.5	60	575
2418	$C_5H_{10}O_2$	Isovaleric acid	176.5	160.25	91	527
2419	$C_5H_{10}O_2$	Valeric acid	186.35	161.5	97.2	527
2420	$C_5H_{10}O_3$	Ethyl lactate	153.9	153.45	35	529
2421	$C_5H_{10}O_3$	2-Methoxyethyl acetate	144.6	Nonazeotrope		575
2422	$C_5H_{11}NO_3$	Isoamyl nitrate	~149.6	Nonazeotrope		541
2423	$C_5H_{12}O$	Isoamyl alcohol	131.3	Nonazeotrope		527
2424	$C_5H_{12}O_2$	2-Propoxyethanol	151.35	Nonazeotrope		526
2425	C_6H_5Br	Bromobenzene	156.1	Nonazeotrope		575
2426	C_6H_5ClO	o-Chlorophenol	176.8	Nonazeotrope		575
		"	175.5	160		563
2427	$C_6H_5NO_2$	Nitrobenzene	210.75	Nonazeotrope		554
2428	C_6H_6O	Phenol	181.5	160.85	90.5	563
2429	C_6H_7N	Aniline	184.35	Nonazeotrope		551
2430	$C_6H_{10}O$	Cyclohexanone	155.7	165.0	73	552
2431	$C_6H_{10}O_3$	Ethyl acetoacetate	180.4	Nonazeotrope		552
2432	$C_6H_{10}O_4$	Ethyl oxalate	185.65	Nonazeotrope		527
2433	$C_6H_{11}BrO_2$	Ethyl α-bromoisobutyrate	178	Nonazeotrope		532
2434	$C_6H_{11}ClO_2$	Isobutyl chloroacetate	174.5	Nonazeotrope		575
2435	$C_6H_{12}O$	Cyclohexanol	160.65	157.9	64	563
2436	$C_6H_{12}O_2$	Isocaproic acid	199.5	Nonazeotrope		575

No.	Formula	B-Component Name	B.P., °C	Azeotropic Data B.P., °C	Wt.%A	Ref.
A =	C₂HCl₅	**Pentachloroethane** *(continued)*	**162.0**			
2437	C₆H₁₂O₃	2-Ethoxyethyl acetate	156.8	Nonazeotrope		556
2438	C₆H₁₂O₃	Propyl lactate	171.7	Nonazeotrope		563
2439	C₆H₁₄O	Hexyl alcohol	157.95	155.75	54	538
2440	C₆H₁₄O₂	2-Butoxyethanol	171.15	Nonazeotrope		575
2441	C₆H₁₄O₂	Pinacol	174.35	158.8	~84	529
2442	C₇H₆O	Benzaldehyde	179.2	Nonazeotrope		536
2443	C₇H₇Cl	o-Chlorotoluene	159.2	Nonazeotrope		575
2444	C₇H₇Cl	p-Chlorotoluene	161.3	Nonazeotrope		563
2445	C₇H₈O	Anisole	153.85	Nonazeotrope		529
2446	C₇H₈O	o-Cresol	190.8	Nonazeotrope		563
2447	C₇H₁₄O	Heptaldehyde	155	Max. b.p.		262
2448	C₇H₁₄O₂	Amyl acetate	148.8	Nonazeotrope		575
2449	C₇H₁₄O₂	Ethyl valerate	145.45	Nonazeotrope		575
2450	C₇H₁₄O₂	Isoamyl acetate	142.1	Nonazeotrope		575
2451	C₇H₁₄O₂	Methyl caproate	149.8	Nonazeotrope		575
2452	C₇H₁₄O₂	Propyl butyrate	143.7	Nonazeotrope		575
2453	C₇H₁₄O₃	1,3-Butanediol methyl ether acetate	171.75	Nonazeotrope		575 575
2454	C₇H₁₆O	Heptyl alcohol	176.15	Nonazeotrope		575
2455	C₇H₁₆O₃	Ethyl orthoformate	145.75	Nonazeotrope		559
2456	C₈H₈	Styrene	145.8	Nonazeotrope		575
2457	C₈H₁₀	o-Xylene	144.3	Nonazeotrope		575
2458	C₈H₁₀O	Benzyl methyl ether	167.8	Nonazeotrope		559
2459	C₈H₁₀O	p-Methylanisole	177.05	Nonazeotrope		559
2460	C₈H₁₀O	Phenetole	170.35	Nonazeotrope		530
2461	C₈H₁₄O	Methyl heptenone	173.2	Nonazeotrope		552
2462	C₈H₁₆O	2-Octanone	174.1	Nonazeotrope		573
2463	C₈H₁₆O₂	Ethyl caproate	167.8	Nonazeotrope		547
2464	C₈H₁₆O₂	Hexyl acetate	171.5	Nonazeotrope		575
2465	C₈H₁₆O₂	Isoamyl propionate	160.3	158.7	50	563
2466	C₈H₁₆O₂	Isobutyl butyrate	157	<156.5		547
2467	C₈H₁₆O₂	Propyl isovalerate	155.7	Nonazeotrope		547
2468	C₈H₁₈O	sec-Octanol	179.0	Nonazeotrope		529
2469	C₈H₂₀SiO₄	Ethyl silicate	168.8	Nonazeotrope		575
2470	C₉H₁₂	Cumene	152.8	Nonazeotrope		575
2471	C₉H₁₂	Mesitylene	164.6	166.0	40	529
2472	C₉H₁₂	Pseudocumene	168.2	>168.35	<22	575
		"	169	Nonazeotrope		563
2473	C₉H₁₃N	N,N-Dimethyl-o-toluidine	185.3	Nonazeotrope		551
2474	C₉H₁₈O	2,6-Dimethyl-4-heptanone	168.0	169.0	35	552
2475	C₉H₁₈O₂	Isoamyl isobutyrate	169.8	Nonazeotrope		575
2476	C₉H₁₈O₂	Isobutyl isovalerate	171.35	Nonazeotrope		538
2477	C₁₀H₁₄	Cymene	176.7	Nonazeotrope		575
2478	C₁₀H₁₆	Camphene	159.6	159.5	3	529
2479	C₁₀H₁₆	Dipentene	177.7	Nonazeotrope		575
2480	C₁₀H₁₆	α-Pinene	155.8	155.6	11	529

TABLE I.   *Binary Systems*   93

No.	Formula	Name	B.P., °C	B.P., °C	Wt.%A	Ref.
		B-Component		Azeotropic	Data	

No.	Formula	Name	B.P., °C	B.P., °C	Wt.%A	Ref.
A =	**C₂HCl₅**	**Pentachloroethane** *(continued)*	**162.0**			
2481	C₁₀H₁₆	Nopinene	163.8	160.7	>62	562
		"	163.8	~166	~42	563
2482	C₁₀H₁₆	α-Terpinene	173.4	Nonazeotrope		575
2483	C₁₀H₁₈	*m*-Menthene-8	170.8	Nonazeotrope		575
2484	C₁₀H₁₈O	Cineol	176.4	Nonazeotrope		528
2485	C₁₀H₂₂	Decane	173.3	Nonazeotrope		575
2486	C₁₀H₂₂O	Isoamyl ether	173.5	Nonazeotrope		548
A =	**C₂HF₃O₂**	**Trifluoroacetic Acid**				
2486a	C₄F₁₀	Perfluorobutane				
		crit. press.			v-l	1049c
A =	**C₂H₂**	**Acetylene**	**— 84**			
2487	C₂H₄	Ethylene	—103.9	Min. b.p.		156
		"	—103.7		18	829
		" Crit. press.			19	829
		" —35°, 0°, 40°F.			v-l	387
2488	C₂H₆	Ethane	— 88.3		39	829
		" Crit. press.			44	829
		" —35°, 0°, 40°F.			v-l	387
		"	— 88.3	— 94.5	40.75	499,646
2489	C₃H₄	Propyne,				
		—50° to 35°C.		Nonazeotrope	v-l	103
A =	**C₂H₂BrCl**	*cis*-**1-Bromo-2-**				
		chloroethylene	**106.7**			
2490	C₂H₆O	Ethyl alcohol	78.3	72.4	73.3	533
2491	C₅H₁₀O₂	Ethyl propionate	99.1	Nonazeotrope		575
2492	C₅H₁₀O₂	Propyl acetate	101.6	Nonazeotrope		575
2493	C₆H₁₂O₂	Ethyl isobutyrate	110.1	Nonazeotrope		575
A =	**C₂H₂BrCl**	*trans*-**1-Bromo-2-**				
		chloroethylene	**75.3**			
2494	C₂H₆O	Ethyl alcohol	78.3	66.3	82	533
A =	**C₂H₂BrI**	*cis*-**1-Bromo-2-iodoethylene**	**149.05**			
2495	C₂H₄O₂	Acetic acid	118.1	115.6	40.5	575
2496	C₃H₆O₂	Propionic acid	141.3	135.3	65.2	575
2497	C₄H₁₀O	Butyl alcohol	117.8	117.3	32.4	575
2498	C₈H₁₆O₂	Butyl Butyrate	165.8	141.5	55	575
A =	**C₂H₂Br₂**	*cis*-**1,2-Dibromoethylene**	**112.5**			
2499	C₂H₆O	Ethyl alcohol	78.3	77.7	32.5	563
A =	**C₂H₂Br₂**	*trans*-**1,2-Dibromoethylene**	**108**			
2500	C₂H₆O	Ethyl alcohol	78.3	75.6	64	563
A =	**C₂H₂ClI**	*cis*-**1-Chloro-2-**				
		iodoethylene	**116-117**			
2501	C₃H₈O	Propyl alcohol	97.20	93.6	55.6	988
2502	C₄H₁₀O	Butyl alcohol	117.8	108.5	75	575

No.	Formula	Name	B.P., °C	B.P., °C	Wt.%A	Ref.
		B-Component		**Azeotropic Data**		

No.	Formula	Name	B.P., °C	B.P., °C	Wt.%A		Ref.
A =	**C₂H₂ClI**	*trans*-1-Chloro-2-					
		iodoethylene	**113-114**				
2503	C₃H₈O	Propyl alcohol	97.20	87.5	96		988
A =	**C₂H₂Cl₂**	*cis*-1,2-Dichloroethylene	**60.25**				
2504	C₂H₆O	Ethyl alcohol	78.3	57.7	90.20		148
		" "	78.3	Calculated		v-l	12
2505	C₃H₆O	Acetone	56.4	61.9	73	v-l	12
2506	C₃H₆O₂	Ethyl formate	54.0	Nonazeotrope		v-l	12
2507	C₃H₆O₂	Methyl acetate	57.2	61.7	73	v-l	12
2508	C₃H₈O₂	Methylal	42.6	Nonazeotrope		v-l	280
2509	C₄H₈O	2-Butanone	79.6	Nonazeotrope		v-l	12
2510	C₄H₈O	Tetrahydrofuran	66.1	69.8	44.5	v-l	280
		"	66	69.9	46.0		224
2511	C₆H₆	Benzene	80.15	Nonazeotrope			575
2512	C₆H₁₄O	Isopropyl ether	68.0	Nonazeotrope		v-l	280
A =	**C₂H₂Cl₂**	*trans*-1,2-Dichloroethylene	**48.35**				
2513	C₂H₆O	Ethyl alcohol	78.3	46.5	94.0		148
		" "	78.3	Calculated		v-l	12
2514	C₃H₆O	Acetone	56.4	Nonazeotrope		v-l	12
2515	C₃H₆O₂	Ethyl formate	54.0	Nonazeotrope		v-l	12
2516	C₃H₆O₂	Methyl acetate	57.2	Nonazeotrope		v-l	12
2517	C₃H₈O₂	Methylal	42.6	48.6	79.3	v-l	280
2518	C₄H₈O	2-Butanone	79.6	Nonazeotrope		v-l	12
2519	C₄H₈O	Tetrahydrofuran	66.1	Nonazeotrope		v-l	280
2520	C₆H₁₄O	Isopropyl ether	68.0	Nonazeotrope		v-l	280
A =	**C₂H₂Cl₂O₂**	**Dichloroacetic Acid**	**190**				
2521	C₂H₄O₂	Methyl formate	31.9	Nonazeotrope			563
2522	C₄H₁₀O	Ethyl ether	34.6	Nonazeotrope			563
2523	C₆H₅NO₂	Nitrobenzene	210.75	Nonazeotrope			554
2524	C₆H₆O	Phenol	181.5	Nonazeotrope			563
2525	C₇H₇Br	*o*-Bromotoluene	181.75	175.5	25		563
2526	C₇H₈O	*m*-Cresol	202.2	Nonazeotrope			544
2527	C₇H₈O	*o*-Cresol	190.8	~189			563
2528	C₇H₈O₂	Guaiacol	205.05	Reacts			575
A =	**C₂H₂Cl₄**	**1,1,2,2-Tetrachloroethane**	**146.25**				
2529	C₂H₃ClO₂	Chloroacetic acid	189.35	Nonazeotrope			575
		" "	189.35	146.25	98.2		530
2530	C₂H₄Cl₂O	2,2-Dichloroethanol	146.2	<144.0	52		575
2531	C₂H₄O₂	Acetic acid	118.5	Nonazeotrope			563
2532	C₂H₅BrO	2-Bromoethanol	150.2	141.5			575
2533	C₂H₅ClO	2-Chloroethanol	128.6	128.2	31		564
2534	C₂H₅IO	2-Iodoethanol	176.5	Nonazeotrope			575
2535	C₂H₅NO	Acetamide	221.2	Nonazeotrope			527
2536	C₂H₆O₂	Glycol	197.4	144.9	93		526
2537	C₃H₅ClO₂	Methyl chloroacetate	130.0	Nonazeotrope			532

TABLE I. *Binary Systems* 95

No.	B-Component			Azeotropic Data		
	Formula	Name	B.P., °C	B.P., °C	Wt.%A	Ref.
A =	**C₂H₂Cl₄**	**1,1,2,2-Tetrachloroethane**	**146.25**			
		(continued)				
2538	C₃H₆Cl₂O	1,3-Dichloro-2-propanol	174.5	Nonazeotrope		563
2539	C₃H₆O₂	Propionic acid	140.7	140.4	40	379,527
2540	C₃H₇ClO	1-Chloro-2-propanol	127.0	Nonazeotrope		575
2541	C₃H₇NO	Propionamide	222.1	Nonazeotrope		527
2542	C₃H₇NO₂	Ethyl carbamate	185.25	Nonazeotrope		564
2543	C₃H₈O	Propyl alcohol	97.25	Nonazeotrope		574
2544	C₃H₈O₂	2-Methoxyethanol	124.5	Nonazeotrope		526
2545	C₄H₆O₄	Methyl oxalate	164.2	Nonazeotrope		547
2546	C₄H₇BrO₂	Ethyl bromoacetate	158.2	Nonazeotrope		532
2547	C₄H₇ClO₂	Ethyl chloroacetate	143.6	147.45	73	528
2548	C₄H₈O₂	Butyric acid	162.45	145.65	96.2	527
2549	C₄H₈O₂	Isobutyric acid	154.35	144.8	93	563
2550	C₄H₈O₃	Methyl lactate	143.8	Nonazeotrope		572
2551	C₄H₁₀O	Butyl alcohol	117.75	Nonazeotrope		527
2552	C₄H₁₀O	Isobutyl alcohol 80°C.		Nonazeotrope	v-l	791c
		" 95°C.		Nonazeotrope	v-l	791c
		"	107	Nonazeotrope	v-l	296,791c
2553	C₄H₁₀O₂	2-Ethoxyethanol	135.3	Nonazeotrope		556
2554	C₅H₄O₂	2-Furaldehyde	161.45	161.55	3	556
2555	C₅H₆O₂	Furfuryl alcohol	169.35	Nonazeotrope		575
2556	C₅H₈O₃	Methyl acetoacetate	169.5	Nonazeotrope		563
2557	C₅H₉ClO₂	Propyl chloroacetate	162.5	Nonazeotrope		575
2558	C₅H₁₀O₂	Isovaleric acid	176.5	Nonazeotrope		527
2559	C₅H₁₀O₂	Valeric acid	186.35	Nonazeotrope		527
2560	C₅H₁₀O₃	Ethyl carbonate	126.0	Nonazeotrope		547
2561	C₅H₁₀O₃	Ethyl lactate	153.9	Nonazeotrope		573
2562	C₅H₁₀O₃	2-Methoxyethyl acetate	144.6	150.9	63	527
2563	C₅H₁₁I	1-Iodo-3-methylbutane	147.65	Nonazeotrope		572
2564	C₅H₁₁NO₃	Isoamyl nitrate	149.75	Nonazeotrope		560
2575	C₅H₁₂O	Isoamyl alcohol	131.3	Nonazeotrope		527
		" "	131.3	131.25	2	529
2566	C₅H₁₂O₂	2-Propoxyethanol	151.35	Nonazeotrope		556
2567	C₆H₅Br	Bromobenzene	156.1	Nonazeotrope		575
2568	C₆H₅ClO	o-Chlorophenol	175.5	Nonazeotrope		563
2569	C₆H₅NO₂	Nitrobenzene	210.75	Nonazeotrope		554
2570	C₆H₆O	Phenol	181.5	Nonazeotrope		563
2571	C₆H₇N	Aniline	184.35	Nonazeotrope		551
2572	C₆H₁₀O	Cyclohexanone	155.7	159.0	45	552
2573	C₆H₁₀O	Mesityl oxide	129.4	147.5	85	573
2574	C₆H₁₀S	Allyl sulfide	139.35	>148.5		556
2574a	C₆H₁₂	1-Hexene 60°C.		Nonazeotrope	v-l	362c
2575	C₆H₁₂O	Cyclohexanol	160.7	Nonazeotrope		532
2576	C₆H₁₂O₂	Isobutyl acetate	117.4	Nonazeotrope		575
2577	C₆H₁₂O₃	2-Ethoxyethyl acetate	156.8	158.25	26	556
2577a	C₆H₁₄	Hexane 60°C.		Nonazeotrope	v-l	362c
2578	C₆H₁₄O	Hexyl alcohol	157.85	Nonazeotrope		575
2579	C₆H₁₄O₂	2-Butoxyethanol	171.25	Nonazeotrope		526

No.	Formula	B-Component Name	B.P., °C	B.P., °C	Wt.%A	Ref.
A =	$C_2H_2Cl_4$	**1,1,2,2-Tetrachloroethane**	**146.25**			
		(*continued*)				
2580	$C_6H_{14}S$	Propyl sulfide	141.5	>150.0	82	562
2581	C_7H_8	Toluene	110.75	Nonazeotrope		575
2582	C_7H_8O	Anisole	153.85	Nonazeotrope		572
2583	$C_7H_{14}O$	Heptaldehyde	155	Max. b.p.		262
2584	$C_7H_{14}O$	4-Heptanone	143.55	148.5		552
2585	$C_7H_{14}O_2$	Amyl acetate	148.8	153.1	40	562
2586	$C_7H_{14}O_2$	Butyl propionate	146.5	152.5	55	547
2587	$C_7H_{14}O_2$	Ethyl isovalerate	134.7	147.0		538
2588	$C_7H_{14}O_2$	Isoamyl acetate	142.1	150.1	68	530
		" "	138.8	Nonazeotrope		563
2589	$C_7H_{14}O_2$	Isobutyl propionate	136.9	>148.5	90	572
		" "	136.9	Nonazeotrope		563
2590	$C_7H_{14}O_2$	Methyl caproate	149.7	153	50	573
2591	$C_7H_{14}O_2$	Propyl butyrate	142.8	150.2	66	538
2592	$C_7H_{14}O_3$	1,3-Butanediol methyl ether acetate	171.75	Nonazeotrope		575
2593	$C_7H_{16}O_3$	Ethyl orthoformate	145.75	151.5	61	568
2594	C_8H_8	Styrene	145.7	~143.5	~55	563
2595	C_8H_{10}	Ethylbenzene	136.15	Nonazeotrope		575
2596	C_8H_{10}	m-Xylene	139.2	Nonazeotrope		527
		"	139.1		99.94	592
2597	C_8H_{10}	o-Xylene	144.4	147	70.2	592
		"	144.3	Nonazeotrope		575
2598	C_8H_{15}	1,3-Dimethylcyclohexane	120.7	Nonazeotrope		575
2599	$C_8H_{16}O_2$	Butyl butyrate	166.4	Nonazeotrope		546
2600	$C_8H_{16}O_2$	Isoamyl propionate	160.7	Nonazeotrope		575
2601	$C_8H_{16}O_2$	Isobutyl butyrate	156.8	158.0	~88	538
2602	$C_8H_{16}O_2$	Isobutyl isobutyrate	147.3	151.5	65	573
2603	$C_8H_{18}O$	Butyl ether	142.2	148.0	70	559
2604	C_9H_{12}	Cumene	152.8	Nonazeotrope		575
A =	$C_2H_2F_4$	**1,1,2,2-Tetrafluoroethane**	**— 23**			
2605	$C_2H_2F_4$	1,1,1,2-Tetrafluoroethane	— 26.1	— 29		182
A =	C_2H_3Br	**Bromoethylene**	**15.8**			
2606	$C_2H_4O_2$	Methyl formate	31.9	Nonazeotrope		563
2607	C_2H_5Cl	Chloroethane	13.3	Nonazeotrope		563
2608	$C_2H_5NO_2$	Ethyl nitrite	17.4	< 14.8	>64	550
2609	C_2H_6O	Ethyl alcohol	78.3	Nonazeotrope		542
2610	$C_3H_7NO_2$	Isopropyl nitrite	40.1	Nonazeotrope		550
2611	C_5H_8	Isoprene	34.3	Nonazeotrope		575
2612	C_5H_{10}	3-Methyl-1-butene	20.6	< 15.0	<78	575
2613	C_5H_{12}	2-Methylbutane	27.95	< 13.0	75	575
2614	C_5H_{12}	Pentane	36.15	Nonazeotrope		575
A =	$C_2H_3BrO_2$	**Bromoacetic Acid**	**205.1**			
2615	$C_6H_4Br_2$	p-Dibromobenzene	220.25	<201.5	>55	527
2616	$C_6H_4Cl_2$	o-Dichlorobenzene	179.5	177.0	16	575
2617	$C_6H_4Cl_2$	p-Dichlorobenzene	174.4	172.8	13	527

TABLE I.   *Binary Systems*   97

No.	Formula	B-Component Name	B.P., °C	Azeotropic Data B.P., °C	Wt.%A	Ref.
A =	$C_2H_3BrO_2$	**Bromoacetic Acid** *(continued)*	**205.1**			
2618	C_6H_5Br	Bromobenzene	156.1	Nonazeotrope		527
2619	C_6H_5I	Iodobenzene	188.45	<184.3	20	575
2620	$C_6H_5NO_2$	Nitrobenzene	210.75	202.25	63	554
2621	$C_6H_{12}O_2$	Caproic acid	205.15	204.4		575
2622	C_7H_7Br	*m*-Bromotoluene	184.3	181.2		527
2623	C_7H_7Br	*o*-Bromotoluene	181.5	179.0	18	527
2624	C_7H_7I	*p*-Iodotoluene	214.5	<198.0	54	575
2625	C_7H_8O	*p*-Cresol	201.8	Nonazeotrope		563
2626	$C_7H_8O_2$	Guaiacol	205.05	203.7	40	527
			205.1	Nonazeotrope		563
2627	$C_7H_{14}O_2$	Enanthic acid	222.0	Nonazeotrope		575
2628	C_8H_8O	Acetophenone	202.0	206.5	70	527
2629	$C_8H_8O_2$	Methyl benzoate	199.4	Nonazeotrope		527
2630	$C_8H_{10}O_2$	*o*-Ethoxyphenol	216.5	Nonazeotrope		527
2631	$C_9H_{10}O_2$	Ethyl benzoate	212.5	Nonazeotrope		575
2632	C_9H_{12}	Mesitylene	164.6	Nonazeotrope		575
2633	C_9H_{12}	Propylbenzene	159.3	Nonazeotrope		575
2634	$C_{10}H_8$	Naphthalene	218.0	<201.3	>72	562
2635	$C_{10}H_{14}$	Butylbenzene	183.1	179.5	25	562
2636	$C_{10}H_{14}$	Cymene	176.7	174.7	15	562
2637	$C_{11}H_{10}$	2-Methylnaphthalene	241.15	Nonazeotrope		575
2638	$C_{11}H_{20}O$	Terpineol methyl ether	216	Reacts		563
2639	$C_{12}H_{18}$	1,3,5-Triethylbenzene	215.5	<199.0	<76	562
A =	C_2H_3Cl	**Vinyl Chloride**	**— 13.4**			
2639a	C_2H_4	Ethylene crit. region		v-l		221c
2640	$C_2H_4Cl_2$	1,2-Dichloroethane	83.5	Nonazeotrope		981
		"	83.7	Nonazeotrope	v-l	951c
2641	C_3H_6O	Acetone	56.1	Nonazeotrope		981
2642	C_4H_6	1,3-Butadiene	— 4.5	Nonazeotrope		215
		" —30° to —60°C.		v.p. curves		351c
2643	C_4H_8	1-Butene	— 6	Nonazeotrope		215
A =	$C_2H_3ClO_2$	**Chloroacetic Acid**	**189.35**			
2644	$C_2H_4Br_2$	1,2-Dibromoethane	131.65	Nonazeotrope		535
2645	$C_2H_4O_2$	Methyl formate	31.9	Nonazeotrope		563
2646	$C_2H_6SO_4$	Methyl sulfate	189.1	194.5?		575
2647	$C_3H_5Cl_3$	1,2,3-Trichloropropane	156.85	154.5	10	574
2648	C_3H_8O	Propyl alcohol	97.2	Nonazeotrope		575
2649	$C_4H_6O_4$	Methyl oxalate	164.45	Nonazeotrope		575
2650	$C_4H_{10}O$	Ethyl ether	34.6	Nonazeotrope		563
2651	$C_5H_{10}O_2$	Isovaleric acid	176.5	Nonazeotrope		563
2652	$C_5H_{10}O_2$	Valeric acid	186.35	186.33	3	527
2653	$C_5H_{11}I$	1-Iodo-3-methylbutane	147.65	147.4		545
2654	$C_6H_3Cl_3$	1,3,5-Trichlorobenzene	208.4	<185.0	<72	575
2655	C_6H_4BrCl	*p*-Bromochlorobenzene	196.4	<181.5	<58	575
2656	$C_6H_4Br_2$	*p*-Dibromobenzene	220.25	186.3	74	535
2657	$C_6H_4Cl_2$	*o*-Dichlorobenzene	179.5	170.8	28	545

| | B-Component | | | Azeotropic | Data | |
No.	Formula	Name	B.P., °C	B.P., °C	Wt.%A	Ref.
A =	$C_9H_3ClO_2$	**Chloroacetic Acid** *(continued)*	**189.35**			
2658	$C_6H_4Cl_2$	*p*-Dichlorobenzene	174.1	167.55	24.5	529
2659	C_6H_5Br	Bromobenzene	156.1	154.3	11	573
2660	C_6H_5Cl	Chlorobenzene	132.0	Nonazeotrope		573
2661	C_6H_5I	Iodobenzene	188.55	175.3	~35	563
2662	$C_6H_5NO_2$	Nitrobenzene	210.75	Nonazeotrope		554
2663	C_6H_6O	Phenol	181.5	Nonazeotrope		563
2664	$C_6H_8O_4$	Methyl fumarate	193.25	195.7	42	569
2665	$C_6H_8O_4$	Methyl maleate	204.05	Nonazeotrope		575
2666	$C_6H_{10}O_4$	Methyl succinate	195.5	197.0	28	562
2667	$C_6H_{10}O_4$	Ethyl oxalate	185.65	190.25	70	568
2668	$C_6H_{12}O_2$	Caproic acid	205.15	Nonazeotrope		575
2669	$C_7H_6Cl_2$	α,α-Dichlorotoluene	205.2	189.1	97	538
			205.1	Nonazeotrope		563
2670	C_7H_6O	Benzaldehyde	179.2	Azeotrope doubtful		563
2671	C_7H_7Br	α-Bromotoluene	198.5	~183	~82	563
2672	C_7H_7Br	*m*-Bromotoluene	183.8	174	30	527
2673	C_7H_7Br	*o*-Bromotoluene	181.75	172.95	32	563
2674	C_7H_7Br	*p*-Bromotoluene	185.0	174.1	34	574
2675	C_7H_7Cl	α-Chlorotoluene	179.3	173.8	25	530
2676	C_7H_7Cl	*o*-Chlorotoluene	159.3	156.8	12	545
2677	C_7H_7Cl	*p*-Chlorotoluene	162.4	159.3	14	545
2678	C_7H_7I	*p*-Iodotoluene	214.5	<184.8	<78	575
2679	C_7H_8O	*m*-Cresol	202.2	Nonazeotrope		575
2680	C_7H_8O	*o*-Cresol	191.1	187.5	~54	535
			191.8	Nonazeotrope		563
2681	C_7H_8O	*p*-Cresol	201.7	Nonazeotrope		544
2682	$C_7H_8O_2$	Guaiacol	205.05	Nonazeotrope		575
2683	$C_7H_{13}ClO_3$	Isoamyl chloroacetate	195	Nonazeotrope		563
2684	C_8H_8	Styrene	145.8	144.8	14	562
2685	$C_8H_8O_2$	Methyl benzoate	199.4	Nonazeotrope		575
2686	$C_8H_8O_2$	Phenyl acetate	195.7	Nonazeotrope		575
2687	C_8H_{10}	Ethylbenzene	136.15	Nonazeotrope		575
2688	C_8H_{10}	*o*-Xylene	144.3	143.5	12	562
2689	C_8H_{10}	*m*-Xylene	139.2	139.05	7	527
2690	C_8H_{10}	*p*-Xylene	138.45	138.35	4?	575
2691	$C_8H_{10}O$	Phenetole	171.5	Nonazeotrope		563
2692	$C_8H_{12}O_4$	Ethyl maleate	223.3	195.7	42	526
2693	$C_8H_{16}O_2$	Butyl butyrate	166.4	Nonazeotrope		575
2694	$C_8H_{16}O_2$	Ethyl caproate	167.7	Nonazeotrope		575
2695	$C_8H_{16}O_2$	Hexyl acetate	171.5	Nonazeotrope		575
2696	C_8H_{18}	Octane	125.75	Nonazeotrope		575
2697	C_9H_8	Indene	182.5	174.5		575
2698	C_9H_{12}	Cumene	152.8	150.8	21	562
2699	C_9H_{12}	Mesitylene	164.6	162	17	573
2700	C_9H_{12}	Propylbenzene	158.9	156.0		545
2701	C_9H_{12}	Pseudocumene	168.2	162.8	34	562

TABLE I. *Binary Systems*   99

No.	Formula	B-Component Name	B.P., °C	Azeotropic Data B.P., °C	Wt.%A		Ref.
A =	C₂H₃ClO₂	**Chloroacetic Acid** *(continued)*	**189.35**				
2702	C₉H₁₈O₂	Butyl isovalerate	177.6	Nonazeotrope			575
2703	C₉H₁₈O₂	Ethyl enanthate	188.7	185.5	48		562
2704	C₉H₁₈O₂	Isoamyl butyrate	181.05	Nonazeotrope			575
2705	C₉H₁₈O₂	Isobutyl isovalerate	171.2	Nonazeotrope			575
2706	C₉H₁₈O₂	Methyl caprylate	192.9	187.5	67		562
2707	C₉H₁₈O₃	Isobutyl carbonate	190.3	192.5	40		562
2708	C₁₀H₈	Naphthalene	218.05	187.1	78		530
			218.05	Nonazeotrope			528
2709	C₁₀H₁₄	Butylbenzene	183.1	172.8	52		562
2710	C₁₀H₁₄	Cymene	175.3	166	~35		563
		"	176.7	169.0	42		562
2711	C₁₀H₁₆	Camphene	159.6	~154.7	~15		530
2712	C₁₀H₁₆	d-Limonene	177.8	167.8	34		563
2713	C₁₀H₁₆	Nopinene	163.8	157.6	30		562
2714	C₁₀H₁₆	α-Phellandrene	171.5	~163.5	~20		563
2715	C₁₀H₁₆	α-Pinene	155.8	152.0			545
2716	C₁₀H₁₆	α-Terpinene	173.4	166.0			575
2717	C₁₀H₁₆	Terpinolene	185	~ 173	~ 47		563
2718	C₁₀H₁₆	Terpinene	180.5	170	~38		563
		"	181.5	170			545
2719	C₁₀H₁₈O	Cineol	176.4	Nonazeotrope			556
2720	C₁₀H₂₀O₂	Isoamyl isovalerate	192.7	187.7	65		564
2721	C₁₀H₂₂	Decane	173.3	165.2	42		564
2722	C₁₀H₂₂	2,7-Dimethyloctane	160.1	155.7	28		575
2723	C₁₀H₂₂O	Amyl ether	187.5	<184.3	<50		575
2724	C₁₀H₂₂O	Isoamyl ether	173.2	171.95	16		556
		" "	172.6	Nonazeotrope			537
2725	C₁₁H₁₀	2-Methylnaphthalene	241.15	Nonazeotrope			575
2726	C₁₁H₂₀O	Isobornyl methyl ether	192.2	Reacts			563
2727	C₁₂H₁₈	1,3,5-Triethylbenzene	215.5	185.5	75		562
A =	C₂H₃Cl₃	**1,1,1-Trichloroethane**	**74.1**				
2728	C₂H₃Cl₃	1,1,2-Trichloroethane	113.5	Nonazeotrope		v-l	58,240
2729	C₂H₄Cl₂	1,2-Dichloroethane	83.7	Nonazeotrope		v-l	58
2730	C₂H₄Cl₂	1,1-Dichloroethane	57.4	Nonazeotrope			266
2730a	C₆H₁₂	1-Hexene	68	Nonazeotrope		v-l	362c
2730b	C₆H₁₄	Hexane 581 mm.		60	28.9	v-l	362c
A =	C₂H₃Cl₃	**1,1,2-Trichloroethane**	**113.5**				
2731	C₂H₄Cl₂	1,2-Dichloroethane	83.7	Nonazeotrope		v-l	58
2732	C₂H₄O₂	Acetic acid	118.1	106.0	70		562
2733	C₂H₅NO	Acetamide	221.2	Nonazeotrope			535
2734	C₂H₆O	Ethyl alcohol	78.3	77.8	30		532
2735	C₂H₆O₂	Glycol	197.4	Nonazeotrope			575
2736	C₃H₆O	Acetone	56.1	Nonazeotrope		v-l	969
2737	C₃H₆O₂	Propionic acid	141.3	Nonazeotrope			575
2738	C₄H₈O₂	Butyric acid	164.0	Nonazeotrope			575
2739	C₄H₈O₂	Dioxane	101	Max. b.p.			262

No.	Formula	Name	B.P., °C	B.P., °C	Wt.%A		Ref.
		B-Component		**Azeotropic Data**			
A =	$C_2H_3Cl_3$	**1,1,2-Trichloroethane**	**113.5**				
		(continued)					
2740	$C_4H_{10}O$	Isobutyl alcohol	108.0	<103.8	>62		575
2741	C_5H_5N	Pyridine	115	Max. b.p.			262
2742	$C_5H_{10}O_2$	Propyl acetate	101.6	Nonazeotrope			575
2742a	C_6H_{12}	1-Hexene 60°C.		Nonazeotrope		v-l	362c
2743	$C_6H_{12}O_2$	Ethyl butyrate	121	Max. b.p.			262
2744	$C_6H_{12}O_2$	Ethyl isobutyrate	110.1	Nonazeotrope			547
2744a	C_6H_{14}	Hexane 60°C.		Nonazeotrope		v-l	362c
2745	C_7H_{14}	Methylcyclohexane	101.15	Nonazeotrope			575
A =	$C_2H_3Cl_3O$	**Methyl Trichloromethyl**					
		Ether	**131.2**				
2746	$C_3H_8O_2$	2-Methoxyethanol	124.5	123.0	75?		575
2747	C_4H_5N	Pyrrol	130.0	<127.5			575
2748	C_5H_8O	Cyclopentanone	130.65	<130.2			575
2749	$C_6H_{12}O$	2-Hexanone	127.2	Nonazeotrope			575
2750	$C_6H_{14}S$	Isopropyl sulfide	120.5	Nonazeotrope			575
2751	C_8H_{10}	Ethylbenzene	136.15	Nonazeotrope			575
A =	$C_2H_3Cl_3O$	**2,2,2-Trichloroethanol**					
2752	$C_4H_{11}PO_3$	Diethylphosphite 3mm.		60	49.5		24
A =	$C_2H_3Cl_3O_2$	**Chloral Hydrate**	**97.5**				
2753	$C_4H_8O_2$	Ethyl acetate	77.05	Nonazeotrope			563
2754	$C_5H_{10}O_2$	Propyl acetate	101.55	~ 96.5			563
2755	C_6H_5Cl	Chlorobenzene	131.8	Nonazeotrope			563
2756	C_6H_{12}	Cyclohexane	80.75	76	~22		563
A =	$C_2H_3F_3O$	**2,2,2-Trifluoroethanol**					
2757	C_2H_6O	Ethyl alcohol	78.3	81.75	57.65		669
A =	C_2H_3N	**Acetonitrile**	**81.6**				
2758	$C_2H_4Cl_2$	1,2-Dichloroethane	83.15	~ 79.1	49	v-l	765
2758a	$C_2H_4O_2$	Acetic acid	118.1	Nonazeotrope		v-l	635e
2759	C_2H_5I	Iodoethane	72.3	< 64.2			565
2760	C_2H_6O	Ethyl alcohol	78.3	72.5	44		563
2761	C_3H_3N	Acrylonitrile	77.3	Nonazeotrope		v-l	768,63
		"	77.3	ideal		v-l	997c
		" 200–760 mm.		Nonazeotrope		v-l	905f
2762	C_3H_6O	Acetone	56.4	Nonazeotrope		v-l	763,94
		" 30°–60°C.		ideal		v-l	989c
2763	$C_3H_6O_2$	Methyl acetate	56.95	Nonazeotrope			565
2764	C_3H_7Br	1-Bromopropane	71.0	63.0	22		565
2765	C_3H_7NO	N, N-Dimethylform-					
		amide	153	Nonazeotrope			981
		" 100–500 mm.		Nonazeotrope			981
2766	C_3H_8	Propane, 280 p.s.i.a.		55	2.2		437
2767	C_3H_8O	Isopropyl alcohol	82.5	74.5	52		527
2768	C_3H_8O	Propyl alcohol	97.2	81.2	~72		563
2769	C_3H_9SiCl	Chlorotrimethylsilane	57.5	56	7.4		841,843
		"	57.9	56.5	7.33	v-l	484
2769a	$C_4H_8O_2$	Dioxane	101.3	Nonazeotrope		v-l	635e

TABLE I. *Binary Systems* 101

No.	Formula	B-Component Name	B.P., °C	B.P., °C	Wt.%A		Ref.
					Azeotropic Data		
A =	C₂H₃N	**Acetonitrile** *(continued)*	**81.6**				
2770	C₄H₈O₂	Ethyl acetate	77.1	74.8	23		527
		"		76.1	18.6	v-l	635e
2771	C₄H₈O₂	Methyl propionate	79.85	76.2	30		565
2772	C₄H₈O₂	Propyl formate	80.85	76.5	33		565
2773	C₄H₉Br	1-Bromobutane	101.5	< 79.0			565
2774	C₄H₉Br	1-Bromo-2-methylpropane	91.4	< 74.5			565
2775	C₄H₉Cl	1-Chlorobutane	78.5	67.2	33		565
2776	C₄H₉Cl	1-Chloro-2-methylpropane	68.85	62.0	20		565
2777	C₄H₉NO₂	Butyl nitrite	78.2	< 77.0			575
2778	C₄H₁₀O	Ethyl ether	34.6	Nonazeotrope			575
2779	C₄H₁₀O	Isobutyl alcohol	108.0	Nonazeotrope			565
2779a	C₅F₁₁IO	Heptafluoroisopropyl 2-iodo-tetrafluoroethyl ether	86	69.5	60		21c
2779b	C₅H₅N	Pyridine	115.3	Nonazeotrope		v-l	635e
2780	C₅H₈	Isoprene	**39.1**	33.7	2.4	v-l	716,742
2780a	C₅H₈	Piperylene 30°–40°C.		Nonazeotrope		v-l	347c
2781	C₅H₁₀	Cyclopentane	49.3	< 44.5	<14		565
2782	C₅H₁₀	β-Isoamylene		35.5	8.2		742
2783	C₅H₁₀	γ-Isoamylene		29.7	5.4		742
2784	C₅H₁₀	2-Methyl-2-butene	38.5	36.1	8.4	v-l	716
2785	C₅H₁₀	1-Pentene	**30.0**	28.7	5.0	v-l	716
2786	C₅H₁₀	2-Pentene (mixed isomers)	37	34.2		v-l	716
2787	C₅H₁₀O₂	Ethyl propionate	99.1	Nonazeotrope			565
2788	C₅H₁₀O₂	Isopropyl acetate	89.5	79.5	60		565
2789	C₅H₁₀O₂	Propyl acetate	101.55	Nonazeotrope			565
2790	C₅H₁₁Br	1-Bromo-3-methylbutane	120.65	Nonazeotrope			565
2791	C₅H₁₂	2-Methylbutane	27.9	27.4		v-l	716
		"		25.3	6.4		742
2792	C₅H₁₂	Pentane 24 p.s.i.g.	65	58	13		981
		"	36		10		437
		" 60°–120°C.				v-l	1049e
2793	C₅H₁₂O	*tert*-Amyl alcohol	102.35	Nonazeotrope			565
2794	C₅H₁₂O	3-Pentanol	116.0	Nonazeotrope			565
2794a	C₆H₅Cl	Chlorobenzene	131	Nonazeotrope		v-l	635c
2795	C₆H₆	Benzene	80.1	73	34		62,569,763
		" 278 mm.		45	30.7	v-l	91
		"	80.1	73.7	31.8	v-l	635c,62
2796	C₆H₇N	2-Picoline	129	Nonazeotrope		v-l	713
2797	C₆H₁₂	Cyclohexane	80.8	62.2	33 vol. %		62,565
2798	C₆H₁₂	Methylcyclopentane	72.0	< 60.5			565
2799	C₆H₁₄	2,3-Dimethylbutane	58.0	48	13		527
2800	C₆H₁₄	Hexane	68.8	56.8	25 vol. %		62,437,565
2800a	C₆H₁₅N	Triethylamine	89.4	70.9	37	v-l	635e
2801	C₇H₈	Toluene	110.7	81.1	78 vol. %		62,565
		"	110.7	81.4	80	v-l	635c
2802	C₇H₁₄	Methylcyclohexane	100.8	71.1	51 vol. %		62
2803	C₇H₁₆	Heptane	98.4	69.4	44 vol. %		62,437,565
		"	98.4	69.55			1066

No.	Formula	B-Component Name	B.P., °C	B.P., °C	Wt.%A	Ref.
A =	C₂H₃N	**Acetonitrile** (*continued*)	**81.6**			
2804	C₈H₁₀	Ethylbenzene	136.2	Nonazeotrope		62
2805	C₈H₁₀	Mixed xylenes	138-144	Nonazeotrope		62
2805a	C₈H₁₂	1,3-*trans*-6-*cis*-Octatriene	132	81	95	269c
2805b	C₈H₁₂	1,3,7-Octatriene	125	80.5	90.5	269c
2805c	C₈H₁₂	4-Vinylcyclohexene	127	79	89	269c
2806	C₈H₁₆	1-Octene	121.6	78.0	60 vol. %	62
2807	C₈H₁₆	2-Octene	125.2	78.0	62 vol. %	62
2808	C₈H₁₈	2,5-Dimethylhexane	109.4	< 75.5		565
2809	C₈H₁₈	2-Methyl-3-ethylpentane	114	65	55	437
2810	C₈H₁₈	Octane	125.6	77.2	64 vol. %	62
		"	125.75	76.7		1066
2811	C₈H₁₈	2,2,4-Trimethylpentane	99.2	68.9	38 vol. %	62
2812	C₉H₂₀	Nonane	150.7	79.82		1066
2813	C₉H₂₀	2,2,5-Trimethylhexane	120.1	76.1	58 vol. %	62
2814	C₁₀H₂₀	1-Decene	172.0	81.6	95 vol. %	62
2815	C₁₀H₂₂	Decane	173.3	81.45		1066
2816	C₁₁H₂₄	Undecane	195.4	Nonazeotrope		1066
A =	C₂H₃NO	**Hydroxyacetonitrile**				
2817	C₄H₁₁PO₃	Diethylphosphite				
		1.5 mm.		66	38.3	24
A =	C₂H₃NO	**Methylisocyanate**	**37.9**			
2817a	C₂H₄Cl₂	1,2-Dichloroethane	83.7	Nonazeotrope	v-l	996i
2817b	C₆H₅Cl	Chlorobenzene	132	Nonazeotrope	v-l	996i
2817c	C₇H₈	Toluene	109.9	Nonazeotrope	v-l	45c
2817d	C₇H₁₆	Heptane	98.4	Nonazeotrope	v-l	931c
A =	C₂H₃NS	**Methyl Thiocyanate**	**132.5**			
2818	C₄H₈Cl₂O	1,2-Dichloroethyl ethyl ether	145.5	Nonazeotrope		575
A =	C₂H₄	**Ethylene**	**—103.9**			
2819	C₂H₆	Ethane	— 88.3	Nonazeotrope		156
		" —35°, 0°, 40°F.	— 88.3		v-l	333
		" 0°, —40°, —100°F.		Nonazeotrope	v-l	362
A =	C₂H₄BrCl	**1-Bromo-2-chloroethane**	**106.7**			
2820	C₂H₄O₂	Acetic acid	118.5	~102	~87	563
2821	C₂H₆O	Ethyl alcohol	78.3	~ 76.5	~50	563
2822	C₂H₆O₂	Glycol	197.4	Nonazeotrope		575
2823	C₃H₅ClO	Epichlorohydrin	116.45	103.5	83	563
2824	C₃H₆O₂	Propionic acid	141.3	Nonazeotrope		575
2825	C₄H₁₀O	Isobutyl alcohol	108	100		563
2826	C₅H₁₀O	2-Pentanone	102.25	Nonazeotrope		563
2827	C₅H₁₀O	3-Pentanone	102.2	Nonazeotrope		563
2828	C₆H₁₂O₂	Ethyl isobutyrate	110.1	Nonazeotrope		563
2829	C₆H₁₂O₂	Methyl isovalerate	116.3	Nonazeotrope		563
2830	C₆H₁₄O₂	Acetal	103.55	108.5	65	559
2831	C₇H₁₄	Methylcyclohexane	101.15	<100.8	> 8	575

TABLE I. *Binary Systems* 103

No.	Formula	B-Component Name	B.P., °C	Azeotropic Data B.P., °C	Wt.%A	Ref.
A =	$C_2H_4Br_2$	**1,1-Dibromoethane**	**109.5**			
2832	$C_2H_4O_2$	Acetic acid	118.1	103.7	75	562
2833	C_2H_5ClO	2-Chloroethanol	128.6	108.5	?	575
2834	C_2H_5NO	Acetamide	221.15	Nonazeotrope		575
2835	C_2H_6O	Ethyl alcohol	78.3	77	46	563
2836	$C_2H_6O_2$	Glycol	197.4	Nonazeotrope		575
2837	$C_3H_6O_2$	Propionic acid	141.3	Nonazeotrope		575
2838	$C_3H_7NO_3$	Propyl nitrate	110.5	<109.2	>58	560
2839	C_3H_8O	Isopropyl alcohol	82.4	< 82.0		575
2840	C_3H_8O	Propyl alcohol	97.2	< 94.0	>57	567
2841	C_4H_5N	Pyrrol	130.0	Nonazeotrope		553
2842	$C_4H_8O_2$	Butyric acid	164.0	Nonazeotrope		575
2843	$C_4H_{10}O$	Butyl alcohol	117.8	104.5	80	567
2844	$C_4H_{10}O$	Isobutyl alcohol	108	101		563
2845	$C_5H_{10}O$	3-Pentanone	102.05	Nonazeotrope		552
2846	$C_5H_{10}O_2$	Methyl butyrate	102.75	Nonazeotrope		563
2847	$C_5H_{10}O_2$	Propyl acetate	101.6	Nonazeotrope		547
2848	$C_5H_{11}NO_2$	Isoamyl nitrite	97.15	Nonazeotrope		550
2849	$C_5H_{12}O$	*tert*-Amyl alcohol	102.35	<101.3	>45	575
2850	$C_5H_{12}O$	Isoamyl alcohol	131.9	Nonazeotrope		575
2851	$C_6H_{12}O$	4-Methyl-2-pentanone	116.05	Nonazeotrope		552
2852	$C_6H_{12}O_2$	Methyl isovalerate	116.3	Nonazeotrope		563
2853	$C_6H_{14}O_2$	Acetal	103.55	Nonazeotrope		559
2854	C_7H_8	Toluene	110.75	Nonazeotrope		575
2855	C_7H_{16}	Heptane	98.4	Nonazeotrope		575
2856	$C_8H_{18}O$	Isobutyl ether	122.3	Nonazeotrope		559
A =	$C_2H_4Br_2$	**1,2-Dibromoethane**	**131.5**			
2857	$C_2H_4Cl_2$	1,2-Dichloroethane	83.7	Nonazeotrope		563
2858	$C_2H_4O_2$	Acetic acid	118.5	114.35	45	563
2859	C_2H_5BrO	2-Bromoethanol	150.2	130.5	90	575
2860	C_2H_5ClO	2-Chloroethanol	128.6	122.3	66.5	526
2861	C_2H_5NO	Acetamide	221.2	Nonazeotrope		527
2862	C_2H_6O	Ethyl alcohol	78.3	Nonazeotrope		574
2863	$C_2H_6O_2$	Glycol	197.4	130.85	96.5	574
2864	C_3H_5BrO	Epibromohydrin	138.5	<128.8	<80	575
2865	$C_3H_5ClO_2$	Methyl chloroacetate	129.95	127.7	56	572
2866	$C_3H_6Br_2$	1,2-Dibromopropane	141	134	50	563
2867	C_3H_6O	Allyl alcohol	96.85	< 96.7		575
		" "	96.95	Nonazeotrope		532
2868	$C_3H_6O_2$	Propionic acid	140.7	127.75	82.5	563
2869	C_3H_7ClO	1-Chloro-2-propanol	127.0	<124.8	>38	575
2870	C_3H_7ClO	2-Chloro-1-propanol	133.7	128.0	67	566
2871	C_3H_7NO	Propionamide	222.1	Nonazeotrope		535
2872	$C_3H_7NO_2$	Ethyl carbamate	185.25	Nonazeotrope		527
2873	$C_3H_7NO_2$	1-Nitropropane				
			613 mm.	75	73	511
			133 mm.	120	72	511

		B-Component		Azeotropic	Data	
No.	Formula	Name	B.P., °C	B.P., °C	Wt.%A	Ref.
A =	**C₂H₄Br₂**	**1,2-Dibromoethane** (continued)	**131.5**			
2874	C₃H₈O	Isopropyl alcohol	82.4	Nonazeotrope		532
2875	C₃H₈O	Propyl alcohol	97.2	Nonazeotrope		573
2876	C₃H₈O₂	2-Methoxyethanol	124.5	120.55	63.5	527
2877	C₄H₅N	Pyrrol	130.0	126.5	67	553
2878	C₄H₇ClO₂	Ethyl chloroacetate	143.6	Nonazeotrope		532
2879	C₄H₈O₂	Butyric acid	162.45	131.1	96.5	527
2880	C₄H₈O₂	Isobutyric acid	154.35	130.5	93.5	541
2881	C₄H₈O₃	Methyl lactate	143.8	130.0	82	567
2882	C₄H₉I	1-Iodobutane	130.4	129.0	65	562
2883	C₄H₉I	1-Iodo-2-methylpropane	120.4	Nonazeotrope B.p. curve		389
2884	C₄H₁₀O	Butyl alcohol	117.75	114.75	56	574
2885	C₄H₁₀O	sec-Butyl alcohol	99.5	Nonazeotrope		575
2886	C₄H₁₀O	Isobutyl alcohol	108	105	38	573,834
2887	C₄H₁₀O₂	2-Ethoxyethanol	135.3	127.75	77	556
2888	C₅H₄O₂	2-Furaldehyde	161.45	Nonazeotrope		527
2889	C₅H₅N	Pyridine	115.5	Nonazeotrope		548
2890	C₅H₉N	Valeronitrile	141.3	<129.5	<83	575
2891	C₅H₁₀O₂	Isovaleric acid	176.5	Nonazeotrope		527
2892	C₅H₁₀O₃	Ethyl carbonate	125.9	Nonazeotrope		552
2893	C₅H₁₀O₃	2-Methoxyethyl acetate	144.6	Nonazeotrope		556
2894	C₅H₁₂O	Amyl alcohol	138.2	<127.3	<78	567
2895	C₅H₁₂O	tert-Amyl alcohol	102.35	Nonazeotrope		575
2896	C₅H₁₂O	Isoamyl alcohol	131.8	124.15	69.5	527,834
2897	C₅H₁₂O	2-Pentanol	119.8	<119.0	<47	567
2898	C₅H₁₂O₂	2-Propoxyethanol	151.35	Nonazeotrope		556
2899	C₆H₅Br	Bromobenzene	152	Nonazeotrope		389
2900	C₆H₅Cl	Chlorobenzene				
		" 128 mm.		75	61.6	v-l	511
		" 311 mm.		100	63.3	v-l	511
		"	131.75	130.05	59	527
2901	C₆H₅NO₂	Nitrobenzene	210.75	Nonazeotrope		554
2902	C₆H₆	Benzene	80.2	Nonazeotrope		563
2903	C₆H₆O	Phenol	182.2	Nonazeotrope		575
2904	C₆H₇N	Aniline	184.35	Nonazeotrope		575
2905	C₆H₁₀	Cyclohexene	82.75	Nonazeotrope		563
2906	C₆H₁₀O	Mesityl oxide	129.45	Nonazeotrope		527
2907	C₆H₁₂	Cyclohexane	80.75	Nonazeotrope		563
2908	C₆H₁₂O₂	Butyl acetate	124.8	Nonazeotrope		527
2909	C₆H₁₂O₂	Ethyl butyrate	121.5	Nonazeotrope		547
2910	C₆H₁₂O₂	Isoamyl formate	123.8	123.7	~12	531
2911	C₆H₁₂O₂	Propyl propionate	122.5	Nonazeotrope		547
2912	C₆H₁₂O₃	Paraldehyde	124	Nonazeotrope		563
2913	C₆H₁₄	Hexane	68.95	Nonazeotrope		563
2914	C₆H₁₄O	Hexyl alcohol	157.85	Nonazeotrope		575
2915	C₆H₁₄O₂	2-Butoxyethanol	171.25	Nonazeotrope		526
2916	C₇H₈	Toluene	110.7	Nonazeotrope		563

ACS ADVANCES IN CHEMISTRY SERIES 116

"Azeotropic Data—III"

―――――――――――

ERRATUM

―――――――――――

On page 455, for compound 15917, under the column "Azeotropic Data," the Wt. % B should be 95.5 and the Wt. % C should be 3.9.

TABLE I. *Binary Systems*   105

		B-Component		Azeotropic Data		
No.	Formula	Name	B.P., °C	B.P., °C	Wt.%A	Ref.
A =	**C₂H₄Br₂**	**1,2-Dibromoethane** *(continued)*	**131.5**			
2917	C₇H₁₄O	4-Heptanone	143.55	Nonazeotrope		552
2918	C₇H₁₄O₂	Ethyl isovalerate	134.7	Nonazeotrope		573
2919	C₇H₁₄O₂	Isoamyl acetate	137.5	Nonazeotrope		388
2920	C₇H₁₄O₂	Isobutyl propionate	136.9	Nonazeotrope		563
2921	C₇H₁₄O₂	Propyl isobutyrate	134.0	Nonazeotrope		575
2922	C₇H₁₆	Heptane	98.4	Nonazeotrope		527
2923	C₈H₈	Styrene 60 mm.	68	Nonazeotrope		53
2924	C₈H₁₀	Ethylbenzene	136.15	131.1	90	563
		60 mm.	60.5	57	87	53
2925	C₈H₁₀	*m*-Xylene	139.0	Nonazeotrope		315,527
2926	C₈H₁₀	*p*-Xylene	138.4	131.0	94	315
		"	138.45	Nonazeotrope		575
		"	138.25	131.3	~97	563
2927	C₈H₁₆	1,3-Dimethylcyclohexane	120.7	Nonazeotrope		575
2928	C₈H₁₈	2,5-Dimethylhexane	109.4	Nonazeotrope		575
2929	C₈H₁₈O	Butyl ether	142.4	Nonazeotrope		559
2930	C₉H₁₂	Mesitylene	164	Nonazeotrope		563
2931	C₁₀H₁₄	Cymene	175.3	Nonazeotrope		563
A =	**C₂H₄Cl₂**	**1,1-Dichloroethane**	**57.3**			
2932	C₂H₄Cl₂	1,2-Dichloroethane	83.7		v-l	435
2933	C₂H₅ClO	Chloromethyl methyl ether	59.5	<54?	<80	563
2934	C₂H₆O	Ethyl alcohol	78.3	54.6	88.5	573
2935	C₃H₆O	Acetone	56.15	57.55	70	552
2936	C₃H₆O	Allyl alcohol	96.85	Nonazeotrope		527
2937	C₃H₆O	Propionaldehyde	50	Nonazeotrope		262
2938	C₃H₆O₂	Ethyl formate	54.15	Nonazeotrope		563
2939	C₃H₆O₂	Methyl acetate	57	~ 56		262,563
2940	C₃H₇Br	2-Bromopropane	59.4	Nonazeotrope		575
2941	C₃H₇NO₂	Propyl nitrite	47.75	Nonazeotrope		550
2942	C₃H₈O	Isopropyl alcohol	82.45	56.6	~92	573
2943	C₃H₈O	Propyl alcohol	97.2	Nonazeotrope		573
2944	C₃H₈O₂	Methylal	42.3	Nonazeotrope		559
2945	C₃H₉BO₃	Methyl borate	65	Nonazeotrope		563
2946	C₃H₉SiCl	Chlorotrimethylsilane	57.7	56.4		844
2947	C₄H₈O	2-Butanone	79.6	Nonazeotrope		552
2948	C₄H₈O	Isobutylene oxide	50	Max. b.p.		262
2949	C₄H₈O	Isobutyraldehyde	63	Nonazeotrope		262
2950	C₄H₉NO₂	Isobutyl nitrite	67.1	Nonazeotrope		550
2951	C₄H₁₀O	*tert*-Butyl alcohol	82.55	57.1	~94	532
2952	C₄H₁₁N	Diethylamine	56	52	~45	563
2053	C₅H₁₀	Cyclopentane	49.3	Nonazeotrope		575
2054	C₅H₁₂O	Ethyl propyl ether	63.6	Nonazeotrope		548
2955	C₆H₁₀	Biallyl	60.2	56.5	~77	563
2956	C₆H₁₂	Methylcyclopentane	72.0	Nonazeotrope		575
2957	C₆H₁₄	2,3-Dimethylbutane	58.0	< 56.0	<58	562

No.	Formula	B-Component Name	B.P., °C	Azeotropic Data B.P., °C	Wt.%A	Ref.
A =	$C_2H_4Cl_2$	**1,1-Dichloroethane** *(continued)*	**57.3**			
2958	C_6H_{14}	Hexane	68.85	Nonazeotrope		538
2959	$C_6H_{14}O$	Isopropyl ether	68	Nonazeotrope		262
A	$C_2H_4Cl_2$	**1,2-Dichloroethane**	**83.45**			
2960	C_2H_4O	Ethylene oxide	10.75	Nonazeotrope		559
2961	$C_2H_4O_2$	Acetic acid	118.1	Nonazeotrope	v-l	722
2962	C_2H_5ClO	2-Chloroethanol	128.6	Nonazeotrope		564
2963	$C_2H_5NO_3$	Ethyl nitrate	87.68	Nonazeotrope		527
2964	C_2H_6O	Ethyl alcohol	78.3	70.5	63	572
		" 550–800 mm.			v-l	168c
2965	$C_3H_6Cl_2$	1,2-Dichloropropane	96.3	Nonazeotrope		981
2966	C_3H_6O	Acetone	56.25	Nonazeotrope	v-l	282,552, 832
2967	C_3H_6O	Allyl alcohol	96.9	80.9	85.5	358,532, 875
2968	$C_3H_6O_3$	Methyl carbonate	90.35	Nonazeotrope		572
2969	C_3H_7ClO	1-Chloro-2-propanol	127.0	Nonazeotrope		575
2970	C_3H_8O	Isopropyl alcohol	82.45	74.7	56.5	572
		" "	82.3	72.7	60.8	981
2971	C_3H_8O	*n*-Propyl alcohol	97.2	80.65	~81	572
2972	$C_3H_9BO_3$	Methyl borate	68.7	Nonazeotrope		547
2973	C_4H_4S	Thiophene	84	83.5		563
2974	$C_4H_6O_2$	Allyl formate	80.0	83.55		575
2975	$C_4H_6O_2$	Methacrylic acid	160.5	Nonazeotrope	v-l	301
2976	$C_4H_8Cl_2O$	Bis(2-chloro ethyl)- ether	179.2	Nonazeotrope		981
		" 100 mm.		Nonazeotrope	v-l	427c
		" 760 mm.	178	Nonazeotrope	v-l	427c
2977	C_4H_8O	2-Butanone	80	Max. b.p.		262
			79.6	Nonazeotrope		527
2978	$C_4H_8O_2$	Butyric acid	162	Nonazeotrope		678
2979	$C_4H_8O_2$	Dioxane	101.35	Nonazeotrope		527
2980	$C_4H_8O_2$	Ethyl acetate	77	Nonazeotrope		572
2981	$C_4H_8O_2$	Propyl formate	80.8	84.05	~90	572
2982	C_4H_9ClO	2-Chloroethyl ethyl ether	98.5	Nonazeotrope		575
2983	$C_4H_9NO_2$	Butyl nitrite	77.8	Nonazeotrope		547
2984	$C_4H_{10}O$	Butyl alcohol	117.75	Nonazeotrope		527,574
		" 723 mm.		Nonazeotrope	v-l	930h
2985	$C_4H_{10}O$	*sec*-Butyl alcohol	99.5	< 82.2	88	575
2986	$C_4H_{10}O$	*tert*-Butyl alcohol	82.45	< 76.5	<78	575
2987	$C_4H_{10}O$	Ethyl ether	34.6	Nonazeotrope		559
2988	$C_4H_{10}O$	Isobutyl alcohol	107.85	83.45	93.5	574
2989	$C_4H_{10}S$	2-Methyl-1-propanethiol	88	Nonazeotrope		568
2990	$C_5H_{10}O$	Isovaleraldehyde	92.1	Nonazeotrope		575
2991	$C_5H_{10}O$	3-Methyl-2-butanone	95.4	Nonazeotrope		552
2991a	$C_5H_{10}O$	2-Methyl-3-buten-2-ol	97	82.7	93.5	581c
2992	$C_5H_{10}O_2$	Isopropyl acetate	90.8	Nonazeotrope		547
2993	$C_5H_{11}NO_2$	Isoamyl nitrite	97.15	Nonazeotrope		550
2994	$C_5H_{12}O$	*tert*-Amyl alcohol	102.35	Nonazeotrope		575

TABLE I.   *Binary Systems*   107

No.	Formula	B-Component Name	B.P., °C	B.P., °C	Azeotropic Data Wt.%A		Ref.
A =	C₂H₄Cl₂	**1,2-Dichloroethane** *(continued)*	**83.45**				
2995	C₅H₁₂O	Isoamyl alcohol	131.9	Nonazeotrope			575
2996	C₅H₁₂O	3-Methyl-2-butanol	112.6	Nonazeotrope			575
2997	C₅H₁₂O	3-Pentanol	116.0	Nonazeotrope			575
2998	C₅H₁₂O₂	Diethoxymethane	87.95	88.95	22		527
2999	C₆H₆	Benzene	80.2	Nonazeotrope			465,734
		"	80.1	80.1	15 vol. %		692
3000	C₆H₁₀	Cyclohexene	82.75	Azeotrope doubtful			563
3001	C₆H₁₂	Cyclohexane		74.4	49.6	v-l	276,282, 563,734
		"	80.75	74.7	38 vol. %		692
3002	C₆H₁₂	Methylcyclopentane	72.0	Nonazeotrope			575
3003	C₆H₁₄	Hexane	68.95	Nonazeotrope			563
3004	C₆H₁₄O	Isopropyl ether	68.3	Nonazeotrope			559
3005	C₆H₁₄O	Propyl ether	90.55	Nonazeotrope			573
3006	C₇H₈	Toluene, 25°C.		Nonazeotrope		v-l	282,429, 563
		"	110.7	Nonazeotrope		v-l	11
		" 200 mm.	69.5	Nonazeotrope		v-l	236
3007	C₇H₁₄	Methylcyclohexane	101.15	Nonazeotrope			575
3008	C₇H₁₆	2-4-Dimethylpentane	80.8	73.7	35 vol. %		692
3009	C₇H₁₆	*n*-Heptane	98.4	81	75.8		269
		"	98.45	Nonazeotrope			527
3010	C₈H₁₈	2-5-Dimethylhexane	109.4	Nonazeotrope			575
A =	C₂H₄Cl₂O	**Bis(chloromethyl) Ether**	**105.5**				
3011	C₃H₆Cl₂	2,2-Dichloropropane	70.4	Nonazeotrope			575
3012	C₃H₇Cl	1-Chloropropane	46.4	Nonazeotrope			555
3013	C₃H₈O₂	2-Methoxyethanol	124.5	Nonazeotrope			575
3014	C₄H₅N	Pyrrol	130.0	Nonazeotrope			572
3015	C₄H₁₀S	Ethyl sulfide	92.1	Nonazeotrope			566
3016	C₅H₇N	1-Methylpyrrol	112.8	<104.8			566
3017	C₆H₁₄O	Propyl ether	90.1	89.0	10		575
3018	C₆H₁₄S	Isopropyl sulfide	120.5	Nonazeotrope			575
3019	C₇H₈	Toluene	110.75	Nonazeotrope			566
3020	C₇H₁₆	Heptane	98.4	Nonazeotrope			575
A =	C₂H₄Cl₂O	**2,2-Dichloroethanol**	**146.2**				
3021	C₃H₆Br₂	1,3-Dibromopropane	166.9	Nonazeotrope			575
3022	C₃H₇I	1-Iodopropane	102.4	Nonazeotrope			575
3023	C₄H₉Br	1-Bromobutane	101.5	Nonazeotrope			575
3024	C₄H₉I	1-Iodobutane	130.4	128.0	15		565
3025	C₄H₉I	1-Iodo-2-methylpropane	120.8	<120.5			575
3026	C₅H₁₁Cl	1-Chloro-3-methylbutane	99.4	Nonazeotrope			575
3027	C₅H₁₁I	1-Iodo-3-methylbutane	147.65	138.5	50		567
3028	C₆H₄Cl₂	*p*-Dichlorobenzene	174.4	Nonazeotrope			575
3029	C₆H₅Br	Bromobenzene	156.1	142.5	70		567
3030	C₆H₅Cl	Chlorobenzene	131.75	130.0	20		567
3031	C₇H₈	Toluene	110.75	Nonazeotrope			575
3032	C₇H₈O	Anisole	153.85	145.5			575

No.	Formula	B-Component Name	B.P., °C	Azeotropic Data B.P., °C	Wt.%A		Ref.
A =	$C_2H_4Cl_2O$	**2,2-Dichloroethanol** *(continued)*	**146.2**				
3033	C_8H_8	Styrene	145.8	<140.0			575
3034	C_8H_{10}	*m*-Xylene	139.2	<136.0	>32		575
3035	C_8H_{10}	*o*-Xylene	144.3	139.0	50		567
3036	$C_8H_{10}O$	Phenetole	170.45	Nonazeotrope			575
3037	$C_8H_{18}O$	Butyl ether	142.4	136.0	45		567
3038	C_9H_8	Indene	182.6	Nonazeotrope			575
3039	C_9H_{12}	Cumene	152.8	142.0	65		567
3040	C_9H_{12}	Mesitylene	164.6	<145.0			575
3041	C_9H_{12}	Propylbenzene	159.3	143.5	75		575
3042	$C_{10}H_{14}$	Butylbenzene	183.1	Nonazeotrope			575
3043	$C_{10}H_{16}$	Camphene	159.6	139.0	75		567
3044	$C_{10}H_{16}$	Dipentene	177.7	143.0	80		567
3045	$C_{10}H_{22}O$	Isoamyl ether	173.2	<145.5	>85		575
A =	C_2H_4O	**Acetaldehyde**	**20.4**				
3046	C_2H_4O	Ethylene oxide	10.4	Nonazeotrope		v-l	217
3047	$C_2H_4O_2$	Methyl formate	31.9	Nonazeotrope			537
3048	C_2H_5Br	Bromoethane	38.4	Nonazeotrope			563
3049	C_2H_5Cl	Chloroethane	14.0	$<\ 9$	<32		563
		"	12.3	11	9.5		266
3050	C_2H_6O	Ethyl alcohol	78.3	Nonazeotrope			563
3051	C_3H_6O	Acetone	56.15	Nonazeotrope			552
3052	C_3H_6O	Propylene oxide, 35 p.s.i.g.	73	Nonazeotrope			981
3052a	$C_3H_6O_2$	Methyl acetate	47	Nonazeotrope		v-l	985c
3053	C_3H_7Cl	2-Chloropropane	34.9	Nonazeotrope			575
3054	$C_3H_7NO_2$	Isopropyl nitrite	40.0	Nonazeotrope			548
3055	C_4H_6	1,3-Butadiene	— 4.5	5.0	5.2	v-l	111
3056	$C_4H_6O_2$	Vinyl acetate	72.5	Nonazeotrope			266
3057	C_4H_8O	Ethyl vinyl ether	35.5	Nonazeotrope			981
3058	C_4H_{10}	Butane	— 0.5	— 7	16		981
3059	$C_4H_{10}O$	Ethyl ether	34.5	18.9	76.5		707
3060	$C_5H_4O_2$	Furfuraldehyde	161.7	Nonazeotrope		v-l	722
3061	C_5H_{12}	2-Methylbutane	27.95	~ 17			563
3062	C_5H_{12}	Pentane	36.15	Azeotrope doubtful			563
3063	C_6H_6	Benzene	80.1	Nonazeotrope		v-l	722
3064	$C_6H_{12}O_3$	Paraldehyde	124	Nonazeotrope			563
3065	$C_6H_{14}O$	Isopropyl ether	68.3	Nonazeotrope			981
3066	C_7H_8	Toluene	110.8	Nonazeotrope		v-l	722
A =	C_2H_4O	**Ethylene Oxide**	**10.75**				
3067	$C_2H_4O_2$	Methyl formate	31.7	Nonazeotrope			575
3068	C_3H_6O	Propylene oxide	34.1	Nonazeotrope			575
3069	C_4H_6	1,3-Butadiene	— 5.3	Nonazeotrope			215
3070	C_4H_8	1-Butene	— 6.5	— 7			215,292
3071	C_4H_8	*cis*-2-Butene	3.6	Min. b.p.			292
3072	C_4H_8	*trans*-2-Butene	0.9	Min. b.p.			292
3073	C_4H_8	2-Methylpropene	— 7.5	Min. b.p.			292

TABLE I.   *Binary Systems*   109

No.	B-Component			Azeotropic Data		
	Formula	Name	B.P., °C	B.P., °C	Wt.%A	Ref.
A =	**C_2H_4O**	**Ethylene Oxide** *(continued)*	**10.75**			
3074	C_4H_{10}	*n*-Butane	0.6	< 0.0	> 5	292,558
	"	"	— 0.5	— 6.2	22	980
	"	"	— 0.5	— 6.5	22	981
3075	C_4H_{10}	2-Methylpropane	— 12.2	Min. b.p.		292
3076	C_5H_{10}	2-Methyl-1-butene	32.0	Min. b.p.		292
3077	C_5H_{10}	3-Methyl-1-butene	21.2	Min. b.p.		292
3078	C_5H_{10}	2-Methyl-2-butene	37.7	Min. b.p.		292
3079	C_5H_{10}	1-Pentene	30.2	Min. b.p.		292
3080	C_5H_{10}	2-Pentene	35.8	Min. b.p.		11
3081	C_5H_{12}	2-Methylbutane	27.9	Min. b.p.		292
			27.95	Nonazeotrope		558
3082	C_5H_{12}	Pentane	36.2	Min. b.p.		292
A =	**C_2H_4OS**	**Thioacetic Acid**	**89.5**			
3083	C_6H_6	Benzene	80.15	Nonazeotrope		575
3084	C_6H_{12}	Cyclohexane	80.75	Nonazeotrope		575
3085	C_6H_{12}	Methylcyclopentane	72.0	Nonazeotrope		575
A =	**$C_2H_4O_2$**	**Acetic Acid**	**118.5**			
3086	C_2H_5I	Iodoethane	72.3	Nonazeotrope		572
3087	C_2H_5NO	Acetamide	222.0	Nonazeotrope	v-l	722
3088	$C_2H_5NO_2$	Nitroethane	114.2	112.4	30	554
3089	$C_2H_5NO_3$	Ethyl nitrate	87.7	Nonazeotrope		527
3090	C_2H_6O	Ethyl alcohol	78.3	Nonazeotrope	v-l	525,811
3091	C_2H_6S	Methyl sulfide	37.4	Nonazeotrope		566
3092	C_3H_5Br	3-Bromopropene	70.5	Nonazeotrope		575
3093	C_3H_5BrO	Epibromohydrin	138.5	Nonazeotrope		556
3094	C_3H_5ClO	Epichlorohydrin	116.4	115.05	34.5	556
3095	$C_3H_5ClO_2$	Methyl chloroacetate	129.95	Nonazeotrope		575
3096	$C_3H_5Cl_3$	1,2,3-Trichloropropane	156.85	Nonazeotrope		541
3097	C_3H_5I	3-Iodopropene	101.8	97.2	15	562
3098	$C_3H_6Br_2$	1,2-Dibromopropane	140.5	116.0	70	555
3099	$C_3H_6Br_2$	1,3-Dibromopropane	166.9	Nonazeotrope		575
3100	$C_3H_6Cl_2$	2,2-Dichloropropane	70.4	Nonazeotrope		575
3101	C_3H_6O	Acetone	56.1	Nonazeotrope	v-l	722
3101a	$C_3H_6O_2$	Methyl acetate	57.1	Nonazeotrope	v-l	36e
3102	$C_3H_6O_2$	Propionic acid	140.7	Ideal system	v-l	26,154
3103	$C_3H_6O_3$	Methyl carbonate	90.35	Nonazeotrope		575
3104	C_3H_7Br	1-Bromopropane	71.0	Nonazeotrope		575
3105	C_3H_7Br	2-Bromopropane	59.4	Nonazeotrope		527
3106	C_3H_7I	1-Iodopropane	102.4	99.2	20	541
3107	C_3H_7I	2-Iodopropane	89.2	88.3	9	545
3108	$C_3H_7NO_3$	Propyl nitrate	110.5	107.5	23	560
3109	C_3H_8O	Propyl alcohol	97.25	Nonazeotrope	v-l	525,811
3110	C_3H_9N	Trimethyl-amine 37 mm.		80-81	20	563
			3.5	148-150	80	360

		B-Component		Azeotropic Data		
No.	Formula	Name	B.P., °C	B.P., °C	Wt.%A	Ref.
A =	C₂H₄O₂	**Acetic Acid** *(continued)*	118.5			
3111	C₄H₆O	Crotonaldehyde	102.2	Nonazeotrope		575
3112	C₄H₆O₂	Biacetyl	88.0	Nonazeotrope	v-l	722
3113	C₄H₆O₂	Vinyl acetate	72.6	Nonazeotrope	v-l	127
3114	C₄H₆O₃	Acetic anhydride 92°C.			v-l	954c
			139.6	Nonazeotrope	v-l	428
		" 100 mm.		Nonazeotrope	v-l	428
3115	C₄H₆O₃	Methyl pyruvate	137.5	Nonazeotrope		552
3116	C₄H₈Cl₂O	1,2-Dichloroethyl ethyl ether	145.5	Nonazeotrope		575
3117	C₄H₈O	2-Butanone	79.6	Nonazeotrope		552
		"	79.6	Nonazeotrope	v-l	331
3118	C₄H₈O₂	Butyric acid	163.5	Nonazeotrope	v-l	26
		" '	163.5	Vapor pressure data		563
3119	C₄H₈O₂	Dioxane	101.35	119.5	77	556
		"		119.4	79.5	486
3120	C₄H₈O₂	Ethyl acetate	77.1	Nonazeotrope	v-l	331
3121	C₄H₈O₂	Propyl formate	80.85	Nonazeotrope		575
3122	C₄H₈S	Tetrahydrothiophene	118.8	<113.5	< 4.7	568
3123	C₄H₉Br	1-Bromobutane	100.35	97.6	18	541
3124	C₄H₉Br	2-Bromobutane	91.2	89.0	13	562
3125	C₄H₉Br	1-Bromo-2-methylpropane	91.3	90.2	12	541
		"	91.6	Nonazeotrope		563
3126	C₄H₉Br	2-Bromo-2-methyl-propane	73.25	73.2?	2?	575
3127	C₄H₉Cl	1-Chlorobutane	78.05	Nonazeotrope		541
3128	C₄H₉Cl	2-Chlorobutane	68.25	Nonazeotrope		575
3129	C₄H₉Cl	1-Chloro-2-methyl-propane	68.85	Nonazeotrope		575
3130	C₄H₉I	1-Iodobutane	130.4	112.4	47	562
3131	C₄H₉I	2-Iodobutane	120.0	110.7	30	562
3132	C₄H₉I	1-Iodo-2-methyl-propane	120.4	109.5	37	541
3133	C₄H₉NO	N,N-Dimethylacetamide	165	170.8	21.1	832
3134	C₄H₉NO₃	Isobutyl nitrate	123.5	114.2	50	560
3135	C₄H₁₀O	Butyl alcohol	117.1	120.3	43	525,811
3136	C₄H₁₀O	Ethyl ether	34.6	Nonazeotrope		563
3137	C₄H₁₀S	Ethyl sulfide	92.1	91.5	10	566
3138	C₅H₄O₂	Furfuraldehyde	161.45	Nonazeotrope	v-l	722
		"	161.7	Nonazeotrope	v-l	971
3139	C₅H₅N	Pyridine	115	138.36	51.1	307
		"	115.5	138.1	51.1	1071
		" Crit. press.	345	348	20.2	945
		"	115.5	139-141	53	360
3140	C₅H₈O	Cyclopentanone	130.65	Nonazeotrope		552
3140a	C₅H₈O₄	Methylene diacetate 92°C.			v-l	954c
3141	C₅H₁₀	Cyclopentane	49.3	Nonazeotrope		575
3142	C₅H₁₀O	Isovaleraldehyde	92.1	Nonazeotrope		575

TABLE I.   *Binary Systems*   111

No.	Formula	B-Component Name	B.P., °C	Azeotropic Data B.P., °C	Wt.%A		Ref.
A =	C₂H₄O₂	**Acetic Acid** *(continued)*	**118.5**				
3143	C₅H₁₀O	3-Pentanone	102.05	Nonazeotrope			552
3144	C₅H₁₀O₂	Butyl formate	106.8	Nonazeotrope			575
3145	C₅H₁₀O₂	Ethyl propionate	99.1	Nonazeotrope			575
3146	C₅H₁₀O₂	Isobutyl formate	98.3	Nonazeotrope			563
3147	C₅H₁₀O₂	Isopropyl acetate	88.7	Nonazeotrope			981
3148	C₅H₁₀O₂	Methyl butyrate	102.65	Nonazeotrope			575
3149	C₅H₁₀O₂	Methyl isobutyrate	92.5	Nonazeotrope			575
3150	C₅H₁₀O₂	Propyl acetate	101.6	Nonazeotrope		v-l	722
3151	C₅H₁₀O₂	Valeric acid	187	Nonazeotrope		v-l	26
3152	C₅H₁₀O₃	2-Methoxyethyl acetate	144.6	Nonazeotrope			526
3153	C₅H₁₁Br	1-Bromo-3-methylbutane	120.65	108.65	38		568
3154	C₅H₁₁Cl	1-Chloro-3-methylbutane	99.8	97.2	18.5		541
3155	C₅H₁₁I	1-Iodo-3-methylbutane	147.65	117.65	80		541
3156	C₅H₁₂	Pentane	36.15	Nonazeotrope			504
3156a	C₅H₁₂O	Isoamyl alcohol					
		20 mm.		54.5	18.5	v-l	496c
		"	132	133	16	v-l	496c
3157	C₅H₁₂O₂	Diethoxymethane	87.95	Nonazeotrope			575
3158	C₆H₄Cl₂	*p*-Dichlorobenzene	174.4	Nonazeotrope			575
3159	C₆H₅Br	Bromobenzene	156.1	118.35	95		541
			156.15	Nonazeotrope			563
3160	C₆H₅Cl	Chlorobenzene	131.8	114.65	58.5		563
3161	C₆H₅F	Fluorobenzene	84.9	Nonazeotrope			575
3162	C₆H₅NO₂	Nitrobenzene	210.85	Nonazeotrope			563
3163	C₆H₆	Benzene	80.2	80.05	2		575
			80.2	Nonazeotrope			834
		" 288 mm.				v-l	997
		" 99 mm.	26.1	Nonazeotrope		v-l	997
		" 20°C.		Nonazeotrope		v-l	1018
		"		79.6	1.7	v-l	370c
		"	80.1			v-l	659f
3164	C₆H₇N	Aniline	184.35	Nonazeotrope			563
3165	C₆H₇N	2-Picoline	129	145	49		360
		"	129	144.12	40.4		1067
3166	C₆H₇N	3-Picoline	144	152.5	30.4		
		212 mm.		114.5	35.0		174,175,810
3167	C₆H₇N	4-Picoline	145.3	154.3	30.3		
		212 mm.		116.5	36.1		
3168	C₆H₈	1,3-Cyclohexadiene	80.4	80.0	2		575
3169	C₆H₈	1,4-Cyclohexadiene	85.6	84.0	6		562
3170	C₆H₁₀	Cyclohexene	82.75	81.8	6.5		541
3171	C₆H₁₀O	Mesityl oxide	129.45	Nonazeotrope			552
3172	C₆H₁₀S	Allyl sulfide	139	116.55	78.5		527
3173	C₆H₁₂	Cyclohexane	80.75	79.7	2		563
		"		78.8	9.6	v-l	40b
3174	C₆H₁₂	Methylcyclopentane	72.0	Nonazeotrope			575
3175	C₆H₁₂O	4-Methyl-2-pentanone	115.80	Nonazeotrope		v-l	722

No.	Formula	Name	B.P., °C	B.P., °C	Wt.%A		Ref.
		B-Component			Azeotropic Data		
A =	C₂H₄O₂	Acetic Acid *(continued)*	118.5				
3176	C₆H₁₂O	Pinacolone	106.2	Nonazeotrope			552
3177	C₆H₁₂O₂	Butyl acetate	125	Nonazeotrope		v-l	722
3178	C₆H₁₂O₂	Ethyl butyrate	121.5	Nonazeotrope			575
3179	C₆H₁₂O₂	Ethyl isobutyrate	110.1	Nonazeotrope			541
3180	C₆H₁₂O₂	Isoamyl formate	123.8	Nonazeotrope			575
3181	C₆H₁₂O₂	Propyl propionate	123.0	Nonazeotrope			575
3182	C₆H₁₃Br	1-Bromohexane	156.5	117.5	92		575
3183	C₆H₁₄	2,3-Dimethylbutane	58.0	Nonazeotrope			575
3184	C₆H₁₄	Hexane	68.8	67.5?	5		575
		"	68.7	68.35	5.7	v-l	331
		"	68.7	Nonazeotrope			981
		"	68.60	68.25	6.0		504,507
3185	C₆H₁₄O	Isopropyl ether	68.3	Nonazeotrope			981
3186	C₆H₁₄O	Propyl ether	90.55	Nonazeotrope			537
3187	C₆H₁₄O₂	Acetal	104.5	Azeotrope doubtful			563
3188	C₆H₁₄S	Isopropyl sulfide	120	111.5	48		555
3189	C₆H₁₄S	Propyl sulfide	141.5	116.9	83		566
3190	C₆H₁₅N	Triethylamine	89	163	67	v-l	989
				162	81.3		360
		" 40 mm.		91-92			360
		"		162	65.3	v-l	394
		" 380 mm.		142	65.3	v-l	394
		" 190 mm.		120	65.3	v-l	394
		" 95 mm.		102	65.3	v-l	394
		" 47.5 mm.		85	65.3	v-l	394
3191	C₇H₇Cl	α-Chlorotoluene	179.3	Nonazeotrope			575
3192	C₇H₇Cl	o-Chlorotoluene	159.3	Nonazeotrope			541
3193	C₇H₇Cl	p-Chlorotoluene	162.4	Nonazeotrope			575
3194	C₇H₈	Toluene	110.8	100.6	28.1	v-l	722,725, 884
		"	110.6	104.6	34.9	v-l	370
		" 288 mm.				v-l	997
		" 50 mm.	36.4	34	21.8	v-l	997
		" 600 mm.		96.4	32.6	v-l	672c
		" 760 mm.	110.7	104.8	34.4	v-l	672c
3195	C₇H₈O	Anisole	153.85	Nonazeotrope			556
3196	C₇H₉N	2,6-Lutidine, 212 mm.		110-111	34.4		174,175,810
		"	144	148	27.8		174,175,810
		"	143.41	147.28	24	v-l	1056
		"	144	148.1	22.9		1062
				162.3	19.5		1068
3197	C₇H₁₂O	Methylcyclohexanone	165.0	Nonazeotrope		v-l	722
3198	C₇H₁₄	Methycyclohexane	101.1	963	31		541,571
3199	C₇H₁₄O	2-Heptanone	149	Nonazeotrope		v-l	722
3200	C₇H₁₄O₂	Amyl acetate	149	Nonazeotrope			725
3201	C₇H₁₄O₂	Ethyl isovalerate	134.7	Nonazeotrope			575

TABLE I. *Binary Systems* 113

No.	Formula	B-Component Name	B.P., °C	Azeotropic Data B.P., °C	Wt.%A		Ref.
A =	**C₂H₄O₂**	**Acetic Acid** *(continued)*	**118.5**				
3202	C₇H₁₄O₂	Isoamyl acetate					
		20 mm.	52.3	Nonazeotrope		v-l	496e
		"	142.2	Nonazeotrope		v-l	496e,722
	C₇H₁₄O₂	"	142.1	Nonazeotrope		v-l	722
3203	C₇H₁₄O₂	Propyl isobutyrate	133.9	Nonazeotrope			543
3204	C₇H₁₆	*n*-Heptane	98.4	95	17		571,725
		"	98.25	91.72	33		504,507, 1052,1071
		" 41.2 mm.		20	22	v-l	1018
3205	C₈H₈	Styrene	145.2	116.8	85.7		369
		" 90.1 mm.		60	88.3		369
		" 50 mm.		46.3	91		369
		"	145.8	116.0	83		545
3206	C₈H₁₀	Ethylben- zene 60 mm.	60.5	48	75		53
		"	136.15	114.65	66		563
3207	C₈H₁₀	*m*-Xylene	139.0	115.35	72.5		527,834
3208	C₈H₁₀	*o*-Xylene	143.6	116.0	76		541
		"	143.6	116.6	78		1070
		" 27 mm.	51.5	33.6	76	v-l	997
3209	C₈H₁₀	*p*-Xylene	138.4	115.25	72		307
		"	138.4	115.25	72		542
3210	C₈H₁₀	Xylene	138.8	115.2	70.9	v-l	722
3211	C₈H₁₀O	Benzyl methyl ether	167.8	Nonazeotrope			575
3212	C₈H₁₁N	Dimethylaniline	194.05	Nonazeotrope			563
3213	C₈H₁₄O₂	Cyclohexyl acetate	177.0	Nonazeotrope		v-l	722
3214	C₈H₁₄O₄	*meso*-2,3-butanediol diacetate, 150-760 mm.	190-193	Nonazeotrope		v-l	731
3215	C₈H₁₆	1,3-Dimethylcyclohexane	120.7	109.0	45		575
3216	C₈H₁₆	Ethylcyclohexane	131.8	107.9			1020
3217	C₈H₁₆O₂	Methyl isoamyl acetate		Nonazeotrope		v-l	722
3218	C₈H₁₈	2,5-Dimethylhexane	109.2	100.0	35		545,571
3219	C₈H₁₈	*n*-Octane	125.5	105.1	52.5	v-l	541,725, 850
		"	125.75	105.7	53.7	v-l	507, 1052,1057
3220	C₈H₁₈O	Butyl ether	141	Nonazeotrope			537
3221	C₈H₁₈O	Isobutyl ether	122.3	113.5?	48?		575
3222	C₈H₁₉NO	α-Diethylaminobutane- γ-ol 7 mm.		83.5	43.6		974
3223	C₉H₇N	Quinoline	237.3	Nonazeotrope			553
3224	C₉H₁₂	Cumene	152.3	116.8			30,1021
		"	152.8	116	84		215
3225	C₉H₁₂	Mesitylene	164.6	Nonazeotrope			575
3226	C₉H₁₂	Propylbenzene	158.9	Nonazeotrope			542
3227	C₉H₁₈	Nonanaphthene	136.7	109.6			1020
3228	C₉H₁₈O	2,6-Dimethyl-4-heptanone	164	Nonazeotrope		v-l	722

No.	Formula	Name	B.P., °C	B.P., °C	Wt.%A	Ref.
		B-Component		**Azeotropic Data**		
A =	C₂H₄O₂	**Acetic Acid** *(continued)*	**118.5**			
3229	C₉H₂₀	2-Methyloctane	135.2	108.8		1020
3230	C₉H₂₀	Nonane	150.7	112.6		1021
		"	150.2	112.8	69	504,507,1052
		"	150.2	113.25	69.6	946
		"	150.8	112.9	69.0	307
3231	C₁₀H₁₄	Cymene	176.7	Nonazeotrope		575
3232	C₁₀H₁₆	Camphene	159.6	118.2	97	541
		" 100 mm.		60.6	90	v-l 830
3233	C₁₀H₁₆	α-Pinene	155.8	117.2	83	541
3234	C₁₀H₁₆	α-Terpinene	173.4	Nonazeotrope		575
3235	C₁₀H₁₆O	Fenchone	193.0	Nonazeotrope	v-l	722
3236	C₁₀H₂₂	2,7-Dimethyloctane	160.1	117.0	94	562
		"	160.25	Nonazeotrope		563
3237	C₁₀H₂₂	Decane	173.3	116.75	79.5	504,507, 1052
		"	173.3	117.2	79	946
		"		116.10	87	v-l 1056
3238	C₁₁H₂₄	Undecane	194.5	117.72	95	504,507 1052
		"	194.5	117.17	78	946
3239	C₁₂H₂₀O₂	Isobornyl acetate	225.8	Ideal system	v-l	830
3240	C₁₂H₂₆	Dodecane	216	Nonazeotrope		504
A =	C₂H₄O₂	**Methyl Formate**	**31.7**			
3241	C₂H₄S	Ethylene sulfide	55.7	Nonazeotrope		566
3242	C₂H₅Br	Bromoethane	38.4	29.85	>66	555
		"		29.9	16.5	497
3243	C₂H₅Cl	Chloroethane	13.3	Nonazeotrope		563
3244	C₂H₅ClO	Chloromethyl methyl ether	59.5	Nonazeotrope		563
3245	C₂H₅NO₂	Ethyl nitrite	17.4	Nonazeotrope		549
3246	C₂H₆O	Ethyl alcohol	78.3	Nonazeotrope		536
3247	C₂H₆S	Ethanethiol	36.2	27	~ 30	563
3248	C₂H₆S	Methyl sulfide	37.2	29.0	62	555
3249	C₃H₅Cl	2-Chloropropene	22.65	<22.0	<13	575
3250	C₃H₅Cl	3-Chloropropene	46.15	Nonazeotrope		547
3251	C₃H₆O	Acetone	56.15	Nonazeotrope		552
3252	C₃H₇Cl	1-Chloropropane	46.65	Nonazeotrope		555
3253	C₃H₇Cl	2-Chloropropane	35.0	Nonazeotrope		555
3254	C₃H₇NO	Dimethylformamide 100-760 mm.		Nonazeotrope	v-l	827
3255	C₃H₇NO₂	Isopropyl nitrite	40.1	Nonazeotrope		549
3256	C₃H₇NO₂	Propyl nitrite	47.75	Nonazeotrope		549
3257	C₃H₈O₂	Methylal	42.25	Nonazeotrope		557
3258	C₄H₄O	Furan	31.7	<28.6		557
3259	C₄H₈	1-Butene	— 6.5	Min. b.p.		292

TABLE I.   *Binary Systems*   115

No.	Formula	Name	B.P., °C	B.P., °C	Wt.%A	Ref.
		B-Component		**Azeotropic Data**		
A =	**C₂H₄O₂**	**Methyl Formate** *(continued)*	**31.7**			
3260	C₄H₈	*cis*-2-Butene	3.6	Min. b.p.		292
3261	C₄H₈	*trans*-2-Butene	0.9	Min. b.p.		292
3262	C₄H₈	2-Methylpropene	— 7.5	Min. b.p.		292
3263	C₄H₈O₂	Butyric acid	163.5	Nonazeotrope		563
3264	C₄H₉Cl	2-Chloro-2-methylpropane	51.6	Nonazeotrope		563
3265	C₄H₁₀	*n*-Butane	— 0.6	Min. b.p.		292
3266	C₄H₁₀	2-Methylpropane	— 12.2	Min. b.p.		292
3267	C₄H₁₀O	Ethyl ether	34.6	28.2	56	557
		"		28.4	55	497
3268	C₄H₁₀O	Methyl propyl ether	38.9	<31.2	<88	557
3269	C₅H₆	Cyclopentadiene	41.0	Min. b.p.		259
3270	C₅H₈	Cyclopentene	43	Min. b.p.		1013
3271	C₅H₈	Isoprene	34.1	22.5	50	259,563,
						1013
3272	C₅H₈	3-Methyl-1,2-butadiene	40.8	26.5	~68	563
3273	C₅H₈	Piperylene	42.5	Min. b.p.		259
3274	C₅H₁₀	Cyclopentane	49.3	26.0	60 vol. %	562,788
3275	C₅H₁₀	2-Methyl-1-butene	31.05	Min. b.p.		259,292,
						1013
3276	C₅H₁₀	3-Methyl-1-butene	20.1	Min. b.p.		259,292
3277	C₅H₁₀	2-Methyl-2-butene	37.15	24.3	54	259,292,
						563
3278	C₅H₁₀	1-Pentene	30.1	Min. b.p.		259,292
3279	C₅H₁₀	2-Pentene	36.4	Min. b.p.		259,292,
						1013
3280	C₅H₁₂	2-Methylbutane	27.95	17.05	47	292,563
		"		18.4	44.6	497
3281	C₅H₁₂	Pentane		21.7	52.8	497
		"	36.15	21.8	53	285,563
3282	C₆H₁₀	Biallyl	60.2	Nonazeotrope		563
3283	C₆H₁₂	Methylcyclopentane	72.0	Nonazeotrope		575
3284	C₆H₁₄	2,2-Dimethylbutane	49.7	25.4	55 vol. %	788
		"	49.7	Min. b.p.		59
3285	C₆H₁₄	2,3-Dimethylbutane	58	Min. b.p.		59
		2,3-Dimethylbutane	58.0	30.5	85	562
3286	C₆H₁₄	*n*-Hexane	69.0	Nonazeotrope		546
A =	**C₂H₄S**	**Ethylene Sulfide**	**55.7**			
3287	C₂H₅Br	Bromoethane	38.4	Nonazeotrope		566
3288	C₂H₅ClO	Chloromethyl methyl ether	59.15	Nonazeotrope		575
3289	C₃H₆O	Acetone	56.15	51.5	57	566
3290	C₃H₆O₂	Ethyl formate	54.15	50.5	53	566
3291	C₃H₇NO₂	Isopropyl nitrite	40.1	Nonazeotrope		566
3292	C₄H₈O	2-Butanone	79.6	Nonazeotrope		566
3293	C₄H₁₀O	Ethyl ether	34.6	Nonazeotrope		566
3294	C₅H₁₂	Pentane	36.15	Nonazeotrope		575
3295	C₆H₁₄	2,3-Dimethylbutane	58.0	54.0	65	575
3296	C₆H₁₄	Hexane	68.8	Nonazeotrope		575

No.	Formula	B-Component Name	B.P., °C	Azeotropic Data B.P., °C	Wt.%A	Ref.
A =	**C₂H₅Br**	**Bromoethane**	**38.4**			
3297	C₂H₅ClO	Chloromethyl methyl ether	59.15	Nonazeotrope		556
3298	C₂H₅I	Iodoethane	72.3	Nonazeotrope		563,899
3299	C₂H₅NO₂	Ethyl nitrite	17.4	Nonazeotrope		550
3300	C₂H₆O	Ethyl alcohol	78.3	37	97	563,834
3301	C₂H₆S	Ethanethiol	36.2	Nonazeotrope		563
3302	C₂H₆S	Methyl sulfide	37.4	< 37.0	<46	566
3303	C₃H₆O	Acetone	56.1	Nonazeotrope		552,834
3304	C₃H₆O	Propionaldehyde	48.7	Nonazeotrope		575
3305	C₃H₆O	Propylene oxide	34.1	Nonazeotrope		559
3306	C₃H₆O₂	Ethyl formate	54.1	Nonazeotrope		531
3307	C₃H₆O₂	Methyl acetate	57.0	Nonazeotrope		563
3308	C₃H₇Cl	2-Chloropropane	34.9	Nonazeotrope		575
3309	C₃H₇NO₂	Isopropyl nitrite	40.1	37.7	68	550
3310	C₃H₇NO₂	Propyl nitrite	47.75	Nonazeotrope		550
3311	C₃H₈O	Isopropyl alcohol	82.4	Nonazeotrope		575
		"	82.45	38.35?	99?	563
3312	C₃H₈O	Propyl alcohol	97.2	Nonazeotrope		575
3313	C₃H₈O₂	Methylal	42.2	Nonazeotrope		555
3314	C₄H₈O₂	Butyric acid	163.5	Nonazeotrope, vapor pressure data		563
3315	C₄H₈O₂	Ethyl acetate	77.1	Nonazeotrope		575
3316	C₄H₁₀O	sec-Butyl alcohol	99.5	Nonazeotrope		575
3317	C₄H₁₀O	Ethyl ether	34.6	Nonazeotrope		563
3318	C₄H₁₀O	Methyl propyl ether	38.8	Nonazeotrope		563
3319	C₄H₁₁N	Diethylamine	55.9	Nonazeotrope		531
3320	C₅H₈	Isoprene	34.1	32	<35	563
3321	C₅H₈	3-Methyl-1,2-butadiene	40.8	~ 36		563
3322	C₅H₁₀	Cyclopentane	49.3	< 37.5	<80	575
3323	C₅H₁₀	2-Methyl-2-butene	37.15	35.0	<59	555
		"		35.4	62	497
3324	C₅H₁₂	2-Methylbutane		27.4	29.7	497
		"	27.95	23.7	70	555
3325	C₅H₁₂	Pentane	36.15	~ 33	~50	563
3326	C₆H₆	Benzene	80.2	Nonazeotrope v-l		563,978
3327	C₆H₁₀	Biallyl	60.2	Nonazeotrope		563
3328	C₆H₁₂	Methylcyclopentane	72.0	Nonazeotrope		575
3329	C₆H₁₄	2,3-Dimethylbutane	58.0	Nonazeotrope		575
3330	C₆H₁₄	Hexane	68.85	Nonazeotrope		538
3331	C₇H₁₆	Heptane, 30° C.	98.4	Vapor pressure data		899
A =	**C₂H₅BrO**	**2-Bromoethanol**	**150.2**			
3332	C₃H₆Br₂	1,2-Dibromopropane	140.5	137.0		575
3333	C₃H₈O₂	2-Methoxyethanol	124.5	Nonazeotrope		526
3334	C₄H₇ClO₂	Ethyl chloroacetate	143.55	Nonazeotrope		575
3335	C₄H₉Br	1-Bromobutane	101.5	Nonazeotrope		575
3336	C₄H₉Br	1-Bromo-2-methylpropane	91.4	Nonazeotrope		575
3337	C₄H₁₀O₂	2-Ethoxyethanol	135.3	Nonazeotrope		526

TABLE I. *Binary Systems* 117

No.	Formula	Name	B.P., °C	B.P., °C	Wt.%A	Ref.
		B-Component		**Azeotropic Data**		
A =	**C_2H_5BrO**	**2-Bromoethanol** *(continued)*	**150.2**			
3338	$C_5H_6O_2$	Furfuryl alcohol	169.35	Nonazeotrope		575
3339	$C_5H_{10}O$	Cyclopentanol	140.85	Nonazeotrope		575
3340	$C_5H_{10}O_3$	Ethyl carbonate	126.5	Nonazeotrope		575
3341	$C_5H_{10}O_3$	2-Methoxyethyl acetate	144.6	Nonazeotrope		575
3342	$C_5H_{11}Br$	1-Bromo-3-methylbutane	120.65	<119.5	> 7	575
3343	$C_5H_{12}O$	Amyl alcohol	138.2	Nonazeotrope		575
3344	$C_5H_{12}O$	Isoamyl alcohol	131.9	Nonazeotrope		527
3345	C_6H_5Cl	Chlorobenzene	131.75	128.7	20	575
3346	$C_6H_{10}O$	Cyclohexanone	155.7	Nonazeotrope		552
3347	$C_6H_{10}O$	Mesityl oxide	129.45	Nonazeotrope		575
3348	$C_6H_{10}S$	Allyl sulfide	139.35	135.5	20	566
3349	$C_6H_{12}O_2$	Butyl acetate	126.0	Nonazeotrope		575
3350	$C_6H_{14}O_2$	2-Butoxyethanol	171.15	Nonazeotrope		575
3351	$C_6H_{14}S$	Isopropyl sulfide	120.5	Nonazeotrope		566
3352	$C_7H_{14}O$	4-Heptanone	143.55	Nonazeotrope		575
3353	$C_7H_{14}O_2$	Butyl propionate	146.8	146.6	50	575
3354	$C_7H_{14}O_2$	Isoamyl acetate	142.1	Nonazeotrope		575
3355	C_7H_{16}	Heptane	98.4	<97.5		575
3356	C_8H_{10}	Ethylbenzene	136.15	131.5	40	575
3357	C_8H_{10}	*m*-Xylene	139.2	133.5	43	575
3358	C_8H_{10}	*p*-Xylene	138.45	133.0	42	575
3359	C_8H_{16}	1,3-Dimethylcyclohexane	120.7	<117.0		575
3360	$C_8H_{18}O$	Butyl ether	142.4	<138.0		575
3361	$C_8H_{18}S$	Butyl sulfide	185.0	Nonazeotrope		566
3362	$C_9H_{18}O$	2,6-Dimethyl-4-heptanone	168.0	Nonazeotrope		552
A =	**C_2H_5BrO**	**Bromomethyl Methyl Ether**	**87.5**			
3363	C_6H_6	Benzene	80.15	Nonazeotrope		575
3364	C_6H_{14}	Hexane	68.8	Nonazeotrope		575
A =	**C_2H_5Cl**	**Chloroethane**	**12.4**			
3365	$C_2H_5NO_2$	Ethyl nitrite	17.4	< 12.2	>85	550
3366	C_2H_6O	Ethyl alcohol	78.3	Nonazeotrope		573
3367	$C_3H_7NO_2$	Isopropyl nitrite	40.1	Nonazeotrope		550
3368	C_4H_4O	Furan	31.7	Nonazeotrope		559
3369	C_4H_{10}	*n*-Butane	0		20	460
		"	— 0.5		15.6	732
		Butane 738.6 mm.	— 0.5	-1.4	20.2 v-l	749
3370	C_5H_{10}	3-Methyl-1-butene	20.6	< 11.5	<73	575
3371	C_5H_{12}	2-Methylbutane	27.95	~12	95	563
3372	C_5H_{12}	Pentane	36.15	Nonazeotrope		563
A =	**C_2H_5ClO**	**2-Chloroethanol**	**128.6**			
3373	C_2H_5NO	Acetamide	221.15	Nonazeotrope		527
3374	$C_2H_5NO_2$	Nitroethane	114.2	Nonazeotrope		554
3375	$C_2H_6O_2$	Glycol	197.4	Nonazeotrope		526
3376	$C_3H_5ClO_2$	Methyl chloroacetate	129.95	<128.0	<85	575

No.	Formula	B-Component Name	B.P., °C	B.P., °C	Azeotropic Data Wt.%A	Ref.
A =	C₂H₅ClO	2-Chloroethanol *(continued)*	**128.6**			
3377	C₃H₅I	3-Iodopropene	101.8	100.2	8	564
3378	C₃H₆Br₂	1,2-Dibromopropane	140.5	126.0		555
3379	C₃H₆Br₂	1,3-Dibromopropane	166.9	Nonazeotrope		564
3380	C₃H₇I	1-Iodopropane	102.4	99.7	15	567
3381	C₃H₇I	2-Iodopropane	89.45	< 88.5	> 8	575
3382	C₃H₇N	Propionamide	222.2	Nonazeotrope		526
3383	C₃H₈O₂	2-Methoxyethanol	124.5	130.0	69	568
3384	C₄H₇ClO	2-Chloroethyl vinyl ether, 120 mm.		55.62	14	1008
3385	C₄H₇ClO₂	Ethyl chloroacetate	143.55	Nonazeotrope		575
3386	C₄H₈Cl₂O	Bis(2-Chloroethyl)ether	177.4	128.2	86.3 v-l	901
		" 50 mm.	96	Nonazeotrope		981
		"	179.2	Nonazeotrope		981
3387	C₄H₈O₂	Dioxane	101.35	Nonazeotrope		527
3387	C₄H₈S	Tetrahydrothiophene	118.8	115.0	~28	575
3389	C₄H₉Br	1-Bromobutane	101.5	100.1	10	564
3390	C₄H₉Br	1-Bromo-2-methyl-propane	91.4	90.2		526
3391	C₄H₉I	Iodobutane	130.4	119.0	38	564
3392	C₄H₉I	1-Iodo-2-methyl-propane	120.8	112.5	30	567
3393	C₄H₁₀O	Butyl alcohol	117.2	Nonazeotrope v-l		901
3394	C₄H₁₀O	Isobutyl alcohol	107.5	Nonazeotrope v-l		901
3395	C₄H₁₀O₂	2-Ethoxyethanol	135.3	135.65	15	568
3396	C₄H₁₀S	Butanethiol	97.5	Nonazeotrope		575
3397	C₄H₁₀S	Ethyl sulfide	92.1	Nonazeotrope		566
3398	C₅H₄O₂	2-Furaldehyde	161.45	Nonazeotrope		575
3399	C₅H₅N	Pyridine	115.4	Nonazeotrope (reacts)		575
3400	C₅H₆O₂	Furfuryl alcohol	169.35	Nonazeotrope		575
3401	C₅H₁₀O	Cyclopentanol	140.85	Nonazeotrope		526
3402	C₅H₁₀O₂	Isobutyl formate	97.9	Nonazeotrope		575
3403	C₅H₁₀O₂	Methyl butyrate	102.65	Nonazeotrope		575
3404	C₅H₁₀O₂	Propyl acetate	101.6	Nonazeotrope		575
3405	C₅H₁₀O₃	Ethyl carbonate	126.5	<125.7	>28	575
3406	C₅H₁₀O₃	2-Methoxyethyl acetate	144.6	Nonazeotrope		526
3407	C₅H₁₁Br	1-Bromo-3-methylbutane	120.3	113.5	30	564
3408	C₅H₁₁Cl	1-Chloro-3-methylbutane	99.4	98.5	8	564
3409	C₅H₁₁I	1-Iodo-3-methylbutane	147.65	125.0	55	564
3410	C₅H₁₂O	Amyl alcohol	138.2	Nonazeotrope		526
3411	C₅H₁₂O	Isoamyl alcohol	131.9	127.8	75	527
3412	C₅H₁₂O	2-Pentanol	119.8	Nonazeotrope		526
3413	C₅H₁₂O	3-Pentanol	116.0	Nonazeotrope		575
3414	C₅H₁₂O₂	2-Propoxyethanol	151.35	Nonazeotrope		526
3415	C₅H₁₃ClSiO	2-Chloroethoxytrimethyl-silane	134.3	120-122		842
3416	C₆H₄Cl₂	o-Dichlorobenzene	179.5	Nonazeotrope		564

TABLE I. *Binary Systems* 119

No.	Formula	Name	B.P., °C	B.P., °C	Wt.%A		Ref.
		B-Component		**Azeotropic Data**			
A =	C₂H₅ClO	**2-Chloroethanol** *(continued)*	**128.6**				
3417	C₆H₄Cl₂	*p*-Dichlorobenzene	174.6	Nonazeotrope			564
3418	C₆H₅Br	Bromobenzene	156.1	127.45	68		564
3419	C₆H₅Cl	Chlorobenzene	131.75	119.95	42		564
3420	C₆H₅F	Fluorobenzene	84.9	Nonazeotrope			575
3421	C₆H₆	Benzene	80.0	Nonazeotrope		v-l	901
3422	C₆H₆O	Phenol	182.2	Nonazeotrope			575
3423	C₆H₆S	Benzenethiol	169.5	Nonazeotrope			566
3424	C₆H₁₀	Cyclohexene	82.75	81.0	11		575
3425	C₆H₁₀O	Mesityl oxide	129.45	130.2	33		527
3426	C₆H₁₀O₄	Ethylidene diacetate	168.5	Nonazeotrope			527
3427	C₆H₁₀S	Allyl sulfide	139.35	124.5	61		555
3428	C₆H₁₂	Cyclohexane	80.75	78.5	10		575
3429	C₆H₁₂	Methylcyclopentane	72	< 71.4			575
3430	C₆H₁₂O	2-Hexanone	127.2	129.0	75		552
3431	C₆H₁₂O	3-Hexanone	123.3	Nonazeotrope			552
3432	C₆H₁₂O	4-Methyl-2-pentanone	116.05	Nonazeotrope			527
3433	C₆H₁₂O	Pinacolone	106.2	Nonazeotrope			552
3434	C₆H₁₂O₂	Butyl acetate	126.0	125.6	31		527
3435	C₆H₁₂O₂	Isoamyl formate	123.8	123.15	21		526
3436	C₆H₁₂O₂	Isobutyl acetate	117.2	Nonazeotrope			526
3437	C₆H₁₂O₂	Methyl isovalerate	116.3	Nonazeotrope			526
3438	C₆H₁₂O₂	Propyl propionate	123.0	122.7			575
3439	C₆H₁₂O₃	Paraldehyde	124.35	Reacts			526
3440	C₆H₁₃Br	1-Bromohexane	156.5	126.5			571
3441	C₆H₁₄	Hexane	68.8	< 68.0	<13		575
3442	C₆H₁₄O	Isopropyl ether	68.4	Nonazeotrope		v-l	901
3443	C₆H₁₄O	Propyl ether	90.1	Nonazeotrope			575
3444	C₆H₁₄O₂	2-Butoxyethanol	171.25	Nonazeotrope			526
3445	C₆H₁₄S	Isopropyl sulfide	120	115.5	30		555
3446	C₆H₁₄S	Propyl sulfide	141.5	125.5	67		566
3447	C₇H₇Br	*m*-Bromotoluene	184.3	Nonazeotrope			564
3448	C₇H₇Br	*o*-Bromotoluene	181.5	Nonazeotrope			564
3449	C₇H₇Br	*p*-Bromotoluene	185.0	Nonazeotrope			564
3450	C₇H₇Cl	*o*-Chlorotoluene	159.2	128.0	75		564
3451	C₇H₇Cl	*p*-Chlorotoluene	162.4	Nonazeotrope			575
3452	C₇H₈	Toluene	110.6	106.9	24.4	v-l	571,901
3453	C₇H₈O	Anisole	153.85	128.55	97.5		556
3454	C₇H₁₄	Methylcyclohexane	101.15	96.5	30		564
3455	C₇H₁₄O	4-Heptanone	143.55	Nonazeotrope			552
3456	C₇H₁₄O	5-Methyl-2-hexanone	144.2	Nonazeotrope			552
3457	C₇H₁₄O₂	Amyl acetate	148.8	Nonazeotrope			575
3458	C₇H₁₄O₂	Isoamyl acetate	142.1	Nonazeotrope			526
3459	C₇H₁₄O₂	Isobutyl propionate	137.5	Nonazeotrope			526
3460	C₇H₁₄O₂	Propyl butyrate	143.7	Nonazeotrope			575
3461	C₇H₁₄O₂	Propyl isobutyrate	134.0	<128.3	<94		575
3462	C₇H₁₆	Heptane	98.4	92.0	25		564

	B-Component			Azeotropic Data		
No.	Formula	Name	B.P., °C	B.P., °C	Wt.%A	Ref.
A =	C₂H₅ClO	**2-Chloroethanol** (continued)	**128.6**			
3463	C₇H₁₆O₃	Ethyl orthoformate	145.75	Reacts		526
3464	C₈H₆	Phenylacetylene	142	min. b. pt.		245
3465	C₈H₆	Styrene	144.7	min. b. pt.		245
	"	"	145.8	123.0		564
3466	C₈H₁₀	Ethylbenzene	136.15	121.0	62	564
	"	"	136.2	min. b. pt.		245
3467	C₈H₁₀	Xylene	140	min. b. pt.		245
3468	C₈H₁₀	m-Xylene	139.2	121.9	55.5	526
3469	C₈H₁₀	o-Xylene	144.3	123.2	68	564
3470	C₈H₁₀	p-Xylene	138.45	121.5	54	567
3471	C₈H₁₀O	Benzyl methyl ether	167.8	Nonazeotrope		575
3472	C₈H₁₀O	Phenetole	170.45	Nonazeotrope		556
3473	C₈H₁₆	1,3-Dimethylcyclohexane	120.7	109.5	42	567
3474	C₈H₁₆O₂	Isobutyl butyrate	156.9	Nonazeotrope		575
3475	C₈H₁₈	2,5-Dimethylhexane	109.2	100.5	33	526
3476	C₈H₁₈	Octane	125.75	115.0		564
3477	C₈H₁₈O	Butyl ether	141.7	123.0	56.8 v-l	901
3478	C₈H₁₈O	Isobutyl ether	122.3	<117.0	<42	567
3479	C₈H₁₈S	Isobutyl sulfide	172.0	Nonazeotrope		566
3480	C₉H₈	Indene	182.6	Nonazeotrope		564
3481	C₉H₁₂	Cumene	152.8	125.35	70	526
3482	C₉H₁₂	Mesitylene	164.6	<128.0		575
3483	C₁₀H₈	Naphthalene	218.0	Nonazeotrope		564
3484	C₁₀H₁₄	Butylbenzene	183.1	Nonazeotrope		564
3485	C₁₀H₁₄	Cymene	176.7	Nonazeotrope		575
3486	C₁₀H₁₆	Camphene	159.6	125.5	80	564
3487	C₁₀H₁₆	α-Terpinene	173.4	<127.0	<85	575
3488	C₁₀H₂₂	2,7-Dimethyloctane	160.1	123.5	68	526
3489	C₁₀H₂₂O	Isoamyl ether	173.2	Nonazeotrope		556
A =	C₂H₅ClO	**Chloromethyl Methyl Ether**	**59.5**			
3490	C₂H₆O	Ethyl alcohol	78.3	58.4	~84	565
3491	C₂H₆S	Ethanethiol	35.8	Nonazeotrope (reacts)		573
3492	C₂H₆S	Methyl sulfide	37.4	Nonazeotrope		566
3493	C₃H₅Br	3-Bromopropene	70.0	Nonazeotrope		548
3494	C₃H₅Cl	3-Chloropropene	45.15	Nonazeotrope		548
3495	C₃H₆O	Acetone	56.15	55.9	13	552
3496	C₃H₆O₂	Ethyl formate	54.1	Nonazeotrope		531
3497	C₃H₆O₂	Methyl acetate	56.25	Nonazeotrope		556
3498	C₃H₇Br	1-Bromopropane	71.0	Nonazeotrope		556
3499	C₃H₇Br	2-Bromopropane	59.4	< 57.1	>45	575
3500	C₃H₇Cl	1-Chloropropane	46.65	Nonazeotrope		556
3501	C₃H₇NO₂	Propyl nitrite	47.75	Nonazeotrope		575
3502	C₃H₈O	Propyl alcohol	97.2	Nonazeotrope		563
3503	C₃H₈O₂	Methylal	42.3	Nonazeotrope		548
		" 128-720 mm.			v-l	634
3504	C₃H₉BO₃	Methyl borate	68.75	Nonazeotrope		531

TABLE I.   *Binary Systems*   121

No.	Formula	Name	B.P., °C	B.P., °C	Wt.%A	Ref.
		B-Component		**Azeotropic Data**		
A =	**C₂H₅ClO**	**Chloromethyl Methyl Ether**	**59.5**			
		(continued)				
3505	C₄H₈O	2-Butanone	79.6	Nonazeotrope		552
3506	C₄H₈O₂	Ethyl acetate	77.05	Nonazeotrope		563
3507	C₄H₈O₂	Isopropyl formate	68.8	Nonazeotrope		548
3508	C₄H₉Cl	1-Chlorobutane	78.5	Nonazeotrope		556
3509	C₄H₉Cl	1-Chloro-2-methyl-propane	68.85	Nonazeotrope		556
3510	C₄H₉Cl	2-Chloro-2-methyl-propane	50.8	Nonazeotrope		548
3511	C₄H₁₀O	Ethyl ether	34.6	Nonazeotrope		548
3512	C₅H₁₀	Cyclopentane	49.4	Nonazeotrope		575
3513	C₅H₁₀	2-Methyl-2-butene	37.15	Nonazeotrope		563
3514	C₅H₁₂O	Ethyl propyl ether	63.6	Nonazeotrope		548
3515	C₆H₆	Benzene	80.15	Nonazeotrope		575
3516	C₆H₁₀	Biallyl	60.1	~55.5	55	548
3517	C₆H₁₄	2,3-Dimethylbutane	58.0	56.0	42	562
3518	C₆H₁₄	Hexane	68.85	~58.5	~90	548
A =	**C₂H₅I**	**Iodoethane**	**72.3**			
3519	C₂H₆O	Ethyl alcohol	78.3	63	86	563,834
3520	C₃H₅Br	3-Bromopropene	70.5	Nonazeotrope		547
3521	C₃H₆O	Acetone	56.2	55-56	40	552,834
3522	C₃H₆O	Allyl alcohol	96.85	69.4	88	567
3523	C₃H₆O₂	Ethyl formate	54.1	Nonazeotrope		538
3524	C₃H₆O₂	Methyl acetate	56.95	Nonazeotrope		538
3525	C₃H₆O₃	Methyl carbonate	90.35	Nonazeotrope		547
3526	C₃H₇Br	1-Bromopropane	71.0	Nonazeotrope		549
3527	C₃H₈O	Isopropyl alcohol	82.45	66	87	573,834
3528	C₃H₈O	Propyl alcohol	97.2	70	93	563,834
3529	C₃H₉BO₃	Methyl borate	68.7	67.8	48	538
3530	C₄H₄S	Thiophene	84.7	Nonazeotrope		527
3531	C₄H₆O₂	Allyl formate	80.0	< 71.5		575
3532	C₄H₈O	2-Butanone	79.6	< 71.5	>75	552
3533	C₄H₈O₂	Ethyl acetate	77.1	70	78	573,834
3534	C₄H₈O₂	Isopropyl formate	68.8	< 66.5	>38	575
3535	C₄H₈O₂	Methyl propionate	79.85	Nonazeotrope		547
3536	C₄H₈O₂	Propyl formate	80.85	72.0	90	547
3537	C₄H₉Br	2-Bromo-2-methyl-propane	73.25	Nonazeotrope		549
3538	C₄H₁₀O	Butyl alcohol	117.8	Nonazeotrope		527
3538a	C₄H₁₀O	*tert*-Butyl alcohol	82.45	68.5		571c
3539	C₄H₁₀O	Isobutyl alcohol	108	Nonazeotrope		834
3540	C₄H₁₀O₂	Acetaldehyde dimethyl acetal	64.3	Nonazeotrope		559
3541	C₅H₁₀O₂	Methyl isobutyrate	92.3	Nonazeotrope		563
3542	C₅H₁₂O	*tert*-Amyl alcohol	102.35	Nonazeotrope		575
3543	C₅H₁₂O	Isoamyl alcohol	131.8	Nonazeotrope		527

No.	Formula	B-Component Name	B.P., °C	Azeotropic Data B.P., °C	Wt.%A	Ref.
A =	C₂H₅I	**Iodoethane** (continued)	72.3			
3544	C₆H₆	Benzene	80.2	Nonazeotrope		563
	"	"		74-75	80	834
3545	C₆H₁₂	Cyclohexane	80.75	Nonazeotrope		575
3546	C₆H₁₄	Hexane	68.95	68	76	563
3547	C₆H₁₄O	Isopropyl ether	68.3	Nonazeotrope		559
3548	C₇H₁₆	Heptane	98.4	V.p. data		
				Nonazeotrope		899
A =	C₂H₅IO	**2-Iodoethanol**	176.5			
3549	C₃H₆Br₂	1,2-Dibromopropane	140.5	Nonazeotrope		575
3550	C₅H₁₁I	1-Iodo-3-methylbutane	147.65	145.8	23	575
3551	C₆H₅Br	Bromobenzene	156.1	153.5	25	567
3552	C₆H₆O	Phenol	182.2	Nonazeotrope		575
3553	C₇H₇Cl	o-Chlorotoluene	159.2	155.5	29	567
3554	C₇H₈	Toluene	110.75	Nonazeotrope		575
3555	C₈H₁₀	o-Xylene	144.3	<143.5	>10	575
3556	C₈H₁₀O	Benzyl methyl ether	167.8	164.0	40	575
3557	C₈H₁₀O	Phenetole	170.45	166.0	38	567
3558	C₈H₁₈O	Butyl ether	142.4	Nonazeotrope		575
3559	C₉H₁₂	Mesitylene	164.6	158.5	35	567
3560	C₉H₁₂	Propyl benzene	159.3	155.0	30	567
3561	C₉H₁₈O₂	Isoamyl isobutyrate	169.8	Nonazeotrope		575
3562	C₁₀H₂₂O	Isoamyl ether	173.2	166.5	50	567
A =	C₂H₅NO	**Acetamide**	221.2			
3563	C₂H₆O₂	Glycol	197.4	Nonazeotrope		529
3564	C₂H₇NO	2-Aminoethanol	170.8	Nonazeotrope		551
3565	C₃H₅Br₃	1,2,3,-Tribromopropane	220	200	~17	535
3566	C₃H₅Cl₃	1,2,3-Trichloropropane	156.85	154.5	7.5	535
3567	C₃H₆Br₂	1,2-Dibromopropane	140.5	Nonazeotrope		527
3568	C₃H₆Br₂	1,3-Dibromopropane	166.9	<165.5	<11	575
3569	C₃H₆Cl₂O	1,3-Dichloro-2-propanol	175.8	<175.5		575
3570	C₃H₇I	1-Iodopropane	102.4	Nonazeotrope		527
3571	C₃H₇NO	Propionamide	222.2	220.9	72	538
3572	C₃H₇NO₂	Ethyl carbamate	185.25	Nonazeotrope		527
3573	C₃H₈O₂	1,2-Propanediol	187.8	Nonazeotrope		575
3574	C₃H₈O₃	Glycerol	290	Nonazeotrope		564
3575	C₄H₅NS	Allyl isothiocyanate	152.0	Nonazeotrope		575
3576	C₄H₆O₄	Methyl oxalate	164.2	Nonazeotrope		535
3577	C₄H₇BrO₂	Ethyl bromoacetate	158.8	Nonazeotrope		527
3578	C₄H₇ClO₂	Ethyl chloroacetate	143.55	Nonazeotrope		575
3579	C₄H₈Cl₂O	1,2-Dichloroethyl ethyl ether	145.5	Nonazeotrope		556
3580	C₄H₈Cl₂O	Bis(2-chloroethyl) ether	178.65	178.25	3	527
3581	C₄H₈O₃	Glycol monoacetate	190.9	190.7	5	527
3582	C₄H₉I	1-Iodobutane	130.4	130.1	~3	575
3583	C₄H₉I	1-Iodo-2-methylpropane	120.8	120.5	1.5	575
3584	C₄H₁₀O	Butyl alcohol	117.75	Nonazeotrope		527

TABLE I. *Binary Systems*   123

No.	Formula	Name	B.P., °C	B.P., °C	Wt.%A	Ref.
		B-Component		**Azeotropic Data**		
A =	C_2H_5NO	**Acetamide** *(continued)*	**221.2**			
3585	$C_4H_{10}O_3$	Diethylene glycol	245.5	Nonazeotrope		527
3586	$C_4H_{11}NO_2$	2,2'-Iminodiethanol	268.0	Nonazeotrope		551
3587	$C_5H_4O_2$	2-Furaldehyde	161.45	Reacts		535
3588	$C_5H_6O_2$	Furfuryl alcohol	169.35	Nonazeotrope		575
3589	$C_5H_8O_3$	Levulinic acid	252	Nonazeotrope		527
3590	$C_5H_9ClO_2$	Propyl chloroacetate	163.5	Nonazeotrope		575
3591	$C_5H_{10}O_2$	Isovaleric acid	176.5	Nonazeotrope		575
3592	$C_5H_{10}O_3$	Ethyl carbonate	126.5	Nonazeotrope		575
3593	$C_5H_{11}Br$	1-Bromo-3-methylbutane	120.65	Nonazeotrope		527
		"	120.3	120.0	~1	535
3594	$C_5H_{11}I$	1-Iodo-3-methylbutane	147.65	146	5	535
3595	$C_5H_{12}O$	Amyl alcohol	138.2	Nonazeotrope		575
3596	$C_5H_{12}O$	Isoamyl alcohol	131.3	Nonazeotrope		527
3597	$C_5H_{12}O$	2-Pentanol	119.8	Nonazeotrope		575
3598	$C_5H_{12}O_2$	2-Propoxyethanol	151.35	Nonazeotrope		526
3599	$C_5H_{12}O_3$	2-(2-Methoxyethoxy) ethanol	192.95	Nonazeotrope		527
3600	$C_6H_4Br_2$	*p*-Dibromobenzene	220.25	199.35	18	574
3601	C_6H_4BrCl	*p*-Bromochlorobenzene	196.4	<187.0		562
3602	C_6H_4ClNO	*m*-Chloronitrobenzene	235.5	212.5	50	554
3603	C_6H_4ClNO	*o*-Chloronitrobenzene	246.0	216.0	60	554
3604	C_6H_4ClNO	*p*-Chloronitrobenzene	239.1	213.6	55	527
3605	$C_6H_4Cl_2$	*o*-Dichlorobenzene	179.2	174.0	10	564
3606	$C_6H_4Cl_2$	*p*-Dichlorobenzene	174.35	169.9	10	574
3607	C_6H_5Br	Bromobenzene	156.1	154.85	4.2	527
3608	C_6H_5BrO	*o*-Bromophenol	194.8	223.0	50	562
3609	C_6H_5Cl	Chlorobenzene	132.0	~131.85	~3	574
3610	C_6H_5ClO	*o*-Chlorophenol	175.8	Nonazeotrope		575
3611	C_6H_5ClO	*p*-Chlorophenol	219.75	231.7	33	574
3612	C_6H_5I	Iodobenzene	188.5	180	13	527
3613	$C_6H_5NO_2$	Nitrobenzene	210.75	201.95	24	554
3614	$C_6H_5NO_3$	*o*-Nitrophenol	217.25	207.7	24.2	527
3615	C_6H_6O	Phenol	182.2	Nonazeotrope		527
		"	182.2	221.3	~98	529
3616	$C_6H_6O_2$	Pyrocatechol	245.9	Nonazeotrope		527
3617	$C_6H_6O_2$	Resorcinol	281.4	Nonazeotrope		541
3618	C_6H_7N	Aniline	184.35	Nonazeotrope		551
3619	$C_6H_8N_2$	*o*-Phenylenediamine	258.6	Nonazeotrope		527
3620	$C_6H_8O_4$	Methyl maleate	204.05	201.9	11	570
3621	$C_6H_{10}O$	Cyclohexanone	155.7	Nonazeotrope		552
3622	$C_6H_{10}O_4$	Ethylidene diacetate	168.5	Nonazeotrope		527
3623	$C_6H_{10}O_4$	Ethyl oxalate	185.65	185.3	4.2	574
3624	$C_6H_{10}O_4$	Glycol diacetate	186.3	Nonazeotrope		575
3625	$C_6H_{10}S$	Allyl sulfide	139.35	Nonazeotrope		566
3626	$C_6H_{11}NO_2$	Nitrocyclohexane	205.3	<200	<22	575

No.	Formula	B-Component Name	B.P., °C	B.P., °C	Wt.%A	Ref.
A =	**C₂H₅NO**	**Acetamide** *(continued)*	**221.2**			
3627	C₆H₁₂O	Cyclohexanol	160.7	Nonazeotrope		527
3628	C₆H₁₂O₂	Butyl acetate	126.0	Nonazeotrope		527
3629	C₆H₁₂O₂	Caproic acid	205.15	<202.8		575
3630	C₆H₁₂O₂	Isoamyl formate	123.8	Nonazeotrope		575
3631	C₆H₁₂O₂	Propyl propionate	123.0	Nonazeotrope		575
3632	C₆H₁₂O₃	2-Ethoxy ethyl acetate	156.8	Nonazeotrope		526
3633	C₆H₁₂O₃	Propyl lactate	171.7	Nonazeotrope		575
3634	C₆H₁₃Br	1-Bromohexane	156.5	154.5	7.5	562
3635	C₆H₁₄O	Hexyl alcohol	157.8	Nonazeotrope		527
3636	C₆H₁₄O₂	2-Butoxyethanol	171.15	Nonazeotrope		527
3637	C₆H₁₄O₂	Pinacol	174.3	Nonazeotrope		535
3638	C₆H₁₅NO	2-Diethylaminoethanol	162.2	Nonazeotrope		551
3639	C₇H₅Cl₃	α,α,α-Trichlorotoluene	220.9	Reacts		535
3640	C₇H₆Cl₂	α,α-Dichlorotoluene	205.15	190.8	15.5	534
3641	C₇H₆O	Benzaldehyde	179.2	178.6	6.5	527
3642	C₇H₆O₂	Benzoic acid	250.5	Nonazeotrope		527
3643	C₇H₇Br	*m*-Bromotoluene	184.3	177.05	11.0	527
3644	C₇H₇Br	*o*-Bromotoluene	181.45	175	11.0	527
3645	C₇H₇Br	*p*-Bromotoluene	185.0	178.0	12	535
3646	C₇H₇BrO	*o*-Bromoanisole	217.7	<207.7		575
3647	C₇H₇Cl	α-Chlorotoluene	179.3	173.7	11	534
3648	C₇H₇Cl	*o*-Chlorotoluene	159.3	157.8	8	535
3649	C₇H₇Cl	*p*-Chlorotoluene	162.4	160.0	8.5	527
3650	C₇H₇ClO	*o*-Chloroanisole	195.7	191.0	20	562
3651	C₇H₇ClO	*p*-Chloroanisole	197.8	<193.0	<26	575
3652	C₇H₇I	*p*-Iodotoluene	212	195	17	535
3653	C₇H₇NO₂	*m*-Nitrotoluene	230.8	210.8	42	564
3654	C₇H₇NO₂	*o*-Nitrotoluene	221.75	206.45	32.5	564
3655	C₇H₇NO₂	*p*-Nitrotoluene	238.9	213.4	48	527
3656	C₇H₈	Toluene	110.75	Nonazeotrope		527,529
3657	C₇H₈O	Anisole	153.85	Nonazeotrope		527
3658	C₇H₈O	Benzyl alcohol	205.1	Nonazeotrope		527
3659	C₇H₈O	*m*-Cresol	202.1	Nonazeotrope		527
3660	C₇H₈O	*o*-Cresol	191.1	Nonazeotrope		527
3661	C₇H₈O	*p*-Cresol	201.7	Nonazeotrope		527
3662	C₇H₈O₂	Guaiacol	205.05	204.55	7.5	574
3663	C₇H₈O₂	*m*-Methoxyphenol	244	220.8	~80	535
3664	C₇H₉N	Methylaniline	196.25	193.8	14	551
3665	C₇H₉N	*m*-Toluidine	203.1	200.95	14	551
3666	C₇H₉N	*o*-Toluidine	200.35	198.55	12	527
3667	C₇H₉N	*p*-Toluidine	200.55	198.7	12	551
3668	C₇H₁₃ClO₂	Isoamyl chloroacetate	195.0	<194.1		575
3669	C₇H₁₄	Methylcyclohexane	101.15	Nonazeotrope		527
3670	C₇H₁₄O	4-Heptanone	143.55	Nonazeotrope		552
3671	C₇H₁₄O	2-Methylcyclohexanol	168.5	Nonazeotrope		575
3672	C₇H₁₄O	5-Methyl-2-hexanone	144.2	Nonazeotrope		552
3673	C₇H₁₄O₂	Amyl acetate	148.8	Nonazeotrope		575

TABLE I. *Binary Systems*   125

No.	Formula	Name	B.P., °C	B.P., °C	Wt.%A	Ref.
		B-Component		**Azeotropic Data**		
A =	**C₂H₅NO**	**Acetamide** *(continued)*	**221.2**			
3674	C₇H₁₄O₂	Butyl propionate	146.8	Nonazeotrope		527
3675	C₇H₁₄O₂	Enanthic acid	222.0	<216.5		575
3676	C₇H₁₄O₂	Ethyl isovalerate	134.7	Nonazeotrope		527
3677	C₇H₁₄O₂	Ethyl valerate	145.45	Nonazeotrope		527
3678	C₇H₁₄O₂	Isoamyl acetate	142.1	Nonazeotrope		527
3679	C₇H₁₄O₂	Isobutyl propionate	137.5	Nonazeotrope		575
3680	C₇H₁₄O₂	Propyl butyrate	143.7	Nonazeotrope		575
3681	C₇H₁₄O₃	1,3-Butanediol methyl ether acetate	171.75	Nonazeotrope		575
3682	C₇H₁₄O₃	Isobutyl lactate	182.15	<181.5	<12	575
3683	C₇H₁₆	Heptane	98.4	Nonazeotrope		527
3684	C₇H₁₆O	Heptyl alcohol	176.35	Nonazeotrope		527
3685	C₇H₁₆O₃	Ethyl orthoformate	145.75	Nonazeotrope		527
3686	C₇H₁₆O₄	2-[2-(2-Methoxyethoxy)-ethoxy]ethanol	245.25	Nonazeotrope		527
3687	C₈H₇N	Indole	253.5	Nonazeotrope		527
3688	C₈H₈	Styrene	145.8	144	12	574
3689	C₈H₈O	Acetophenone	202.0	197.45	16.3	527
3690	C₈H₈O₂	Benzyl formate	203.0	193.0	22	564
3691	C₈H₈O₂	Methyl benzoate	199.45	193.8	15	574
3692	C₈H₈O₂	Phenyl acetate	195.7	~194.5	~7	574
3693	C₈H₈O₂	α-Toluic acid	266.5	Nonazeotrope		527
3694	C₈H₈O₃	Methyl salicylate	222.3	205.8	29	528
3695	C₈H₉BrO	p-Bromophenetole	234.2	212.0	35	562
3696	C₈H₁₀	Ethylbenzene	136.15	135.6	~8	535
3697	C₈H₁₀	o-Xylene	144.3	142.6	11	562
3698	C₈H₁₀	m-Xylene	139.0	138.4	10	527
3699	C₈H₁₀	p-Xylene	138.2	137.75	8	527
3700	C₈H₁₀O	Benzyl methyl ether	167.8	166.0	10	575
3701	C₈H₁₀O	p-Methyl anisole	177.05	174.2	11	556
3702	C₈H₁₀O	Phenethyl alcohol	219.5	214.05	35	528
3703	C₈H₁₀O	Phenetole	170.5	168.3	10.8	574
3704	C₈H₁₀O	2,4-Xylenol	210.5	Nonazeotrope		575
3705	C₈H₁₀O	3,4-Xylenol	226.8	221.1	96	527
3706	C₈H₁₀O₂	m-Dimethoxybenzene	214.7	199.0	25	535
3707	C₈H₁₀O₂	o-Ethoxyphenol	216.5	<215.0		575
3708	C₈H₁₀O₂	Veratrol	205.5	193.5	23	535
3709	C₈H₁₁N	Dimethylaniline	194.15	186.95	17.5	551
3710	C₈H₁₁N	2,4-Xylidine	214.0	<209.5	21	551
3711	C₈H₁₁N	3,4-Xylidine	225.5	<213.5	<29	551
3712	C₈H₁₁N	Ethylaniline	205.5	199.0	18	551
3713	C₈H₁₁NO	o-Phenetidine	232.5	216.0	55	527
3714	C₈H₁₁NO	p-Phenetidine	249.9	Nonazeotrope		551
3715	C₈H₁₂O₄	Ethyl fumarate	217.85	205.5	26.7	527
3716	C₈H₁₂O₄	Ethyl maleate	223.3	210.15	32	527
3717	C₈H₁₄O	Methyl heptenone	173.2	Nonazeotrope		552

No.	Formula	B-Component Name	B.P., °C	Azeotropic Data B.P., °C	Wt.%A	Ref.
A =	**C₂H₅NO**	**Acetamide** (*continued*)	**221.2**			
3718	C₈H₁₆O	2-Octanone	172.85	Nonazeotrope		552
3719	C₈H₁₆O₂	Butyl butyrate	166.4	164.5	7	564
3720	C₈H₁₆O₂	Caprylic acid	283.5	<219.5		575
3721	C₈H₁₆O₂	Hexyl acetate	171.5	169.5	10	562
3722	C₈H₁₆O₂	Isoamyl propionate	160.7	159.8	4	564
3723	C₈H₁₆O₂	Isobutyl butyrate	156.8	Nonazeotrope		535
3724	C₈H₁₆O₂	Isobutyl isobutyrate	148.6	Nonazeotrope		575
3725	C₈H₁₆O₂	Propyl isovalerate	155.7	<155.3	>3	575
3726	C₈H₁₆O₃	Isoamyl lactate	202.4	<196.0	<28	575
3727	C₈H₁₈	2,5-Dimethylhexane	109.4	Nonazeotrope		527
3728	C₈H₁₈	Octane	125.7	125.6	~1	535
3729	C₈H₁₈O	Butyl ether	142.4	<142.0	<10	575
3730	C₈H₁₈O	Octyl alcohol	195.2	194.45	9.5	564
3731	C₈H₁₈O	*sec*-Octyl alcohol	179.0	Nonazeotrope		527
3732	C₈H₁₈S	Butyl sulfide	185.0	180.0	8	566
3733	C₈H₁₈S	Isobutyl sulfide	172.0	<170.5	<7	566
3734	C₉H₇N	Quinoline	237.3	Nonazeotrope		527
3735	C₉H₈	Indene	183.0	177.2	17.5	527
3736	C₉H₈O	Cinnamaldehyde	253.5	Nonazeotrope		527
3737	C₉H₁₀O	Cinnamyl alcohol	257	Nonazeotrope		564
3738	C₉H₁₀O	*p*-Methyl acetophenone	226.35	209.8	38.3	552
3739	C₉H₁₀O	Propiophenone	217.7	204.0	31	527
3740	C₉H₁₀O₂	Benzyl acetate	~214.9	204.8	27.5	574
3741	C₉H₁₀O₂	Ethyl benzoate	212.6	200.85	24	574
3742	C₉H₁₀O₂	Methyl α-toluate	215.3	203.0	30	562
3743	C₉H₁₀O₃	Ethyl salicylate	233.7	209.2	40.2	536
3744	C₉H₁₂	Cumene	152.8	<150.8	>8	**575**
3745	C₉H₁₂	Mesitylene	164.6	~160.0	~15	573
3746	C₉H₁₂	Pseudocumene	168.2	<164.8		575
3747	C₉H₁₂O	Benzyl ethyl ether	185.0	179.0	17	562
3748	C₉H₁₂O	3-Phenyl propanol	235.6	Nonazeotrope		527
3749	C₉H₁₂O	Phenyl propyl ether	190.2	183.5	20	535
3750	C₉H₁₂O₂	2-Benzyloxyethanol	265.2	Nonazeotrope		575
3751	C₉H₁₃N	N,N-Dimethyl-*o*-toluidine	185.3	177.95	16.5	551
3752	C₉H₁₃N	N,N-Dimethyl-*p*-toluidine	210.2	194.0	22	551
3753	C₉H₁₄O	Phorone	197.8	194.8	12	552
3754	C₉H₁₈O	2,6-Dimethyl-4-heptanone	168.0	Nonazeotrope		552
3755	C₉H₁₈O₂	Ethyl enanthate	188.7	183.0	16	562
3756	C₉H₁₈O₂	Isoamyl butyrate	178.5	174.75	11.8	536
3757	C₉H₁₈O₂	Isoamyl isobutyrate	169.8	167.5	9	564
3758	C₉H₁₈O₂	Isobutyl isovalerate	171.35	169.3	10.5	541
3759	C₉H₁₈O₂	Methyl caprylate	192.9	186.0	15	564
3760	C₁₀H₇Br	1-Bromonaphthalene	281.8	217.35	56.5	527
3761	C₁₀H₇Cl	1-Chloronaphthalene	~262.7	213.9	52.2	527
3762	C₁₀H₈	Naphthalene	218.05	199.55	27	527
3763	C₁₀H₈O	1-Naphthol	288	Nonazeotrope		527

TABLE I. *Binary Systems* 127

No.	Formula	Name	B.P., °C	B.P., °C	Wt.%A	Ref.
		B-Component		**Azeotropic Data**		
A =	**C₂H₅NO**	**Acetamide** *(continued)*	**221.2**			
3764	$C_{10}H_8O$	2-Naphthol	290	Nonazeotrope		544
3765	$C_{10}H_9N$	1-Naphthylamine	300.8	Nonazeotrope		551
3766	$C_{10}H_9N$	Quinaldine	246.5	Nonazeotrope		575
3767	$C_{10}H_{10}O_2$	Isosafrol	252.1	214.0	47	574
3768	$C_{10}H_{10}O_2$	Methyl cinnamate	261.95	219.1	62	527
3769	$C_{10}H_{10}O_2$	Safrol	235.9	208.8	32	527
3770	$C_{10}H_{10}O_4$	Methyl phthalate	283.7	Nonazeotrope		542
3771	$C_{10}H_{12}O$	Anethole	235.7	208.0	38	527
3772	$C_{10}H_{12}O$	Estragole	215.8	~199.8	~24	574
3773	$C_{10}H_{12}O_2$	Ethyl α-toluate	228.75	209.6	35.5	536
3774	$C_{10}H_{12}O_2$	Eugenol	255	220.8	88	527
3775	$C_{10}H_{12}O_2$	Isoeugenol	268.8	Nonazeotrope		527
3776	$C_{10}H_{12}O_2$	Propyl benzoate	230.85	209.0	38	527
3777	$C_{10}H_{14}$	Butylbenzene	183.1	<176.0		575
3778	$C_{10}H_{14}$	Cymene	176.7	170.5	19	527
3779	$C_{10}H_{14}O$	Carvacrol	237.85	<220.8		575
3780	$C_{10}H_{14}O$	Carvone	231	210.65	42.5	527
3781	$C_{10}H_{14}O$	Thymol	232.8	219.9	70.5	527
3782	$C_{10}H_{14}O_2$	*m*-Diethoxybenzene	235.0	208.5	34	535
3783	$C_{10}H_{15}N$	Diethylaniline	217.05	198.05	24	551
3784	$C_{10}H_{16}$	Camphene	159.6	155.5	12	527
3785	$C_{10}H_{16}$	Dipentene	177.7	169.15	18	527
3786	$C_{10}H_{16}$	*d*-Limonene	177.8	169.2	16	529
3787	$C_{10}H_{16}$	Nopinene	163.8	159.5	18	527
3788	$C_{10}H_{16}$	α-Pinene	155.8	152.5	13	536
3789	$C_{10}H_{16}$	α-Terpinene	173.4	167.5	18	562
3790	$C_{10}H_{16}$	γ-Terpinene	183	175.0	~20	575
3791	$C_{10}H_{16}$	Terpinolene	184.6	176.5	20	562
3792	$C_{10}H_{16}$	Thymene	179.7	169.8	18	535
3793	$C_{10}H_{16}O$	Camphor	209.1	199.8	23	552
3794	$C_{10}H_{16}O$	Carvenone	234.5	233.0	44	552
3795	$C_{10}H_{16}O$	Fenchone	193.6	<192.8	> 5	552
3796	$C_{10}H_{16}O$	Pulegone	223.8	205.9	36	527
3797	$C_{10}H_{17}Cl$	Bornyl chloride	207.5	<195.0		575
3798	$C_{10}H_{18}O$	Borneol	213.4	205.4	26	527
3799	$C_{10}H_{18}O$	Cineol	176.35	170.9	17	574
3800	$C_{10}H_{18}O$	Citronellal	208.0	199	Reacts	575
3801	$C_{10}H_{18}O$	Geraniol	229.6	213.5	45	527
3802	$C_{10}H_{18}O$	Linalool	198.6	<198.0	< 12	575
3803	$C_{10}H_{18}O$	α-Terpineol	217.8	205.2	28	529
3804	$C_{10}H_{18}O$	β-Terpineol	210.5	203.0	22	567
3805	$C_{10}H_{20}O$	Citronellol	224.4	209.5	40	567
3806	$C_{10}H_{20}O$	Menthol	216.4	204.45	27	564
3807	$C_{10}H_{20}O_2$	Ethyl caprylate	208.35	196.0	24	562
3808	$C_{10}H_{20}O_2$	Isoamyl isovalerate	192.7	184.85	16	564
3809	$C_{10}H_{20}O_2$	Isoamyl valerate	192.7	184.85	16	541
3810	$C_{10}H_{20}O_2$	Methyl pelargonate	213.7	268.5	28	564

No.	Formula	B-Component Name	B.P., °C	Azeotropic Data B.P., °C	Wt.%A	Ref.
A =	C$_2$H$_5$NO	Acetamide *(continued)*	**221.2**			
3811	C$_{10}$H$_{22}$	2,7-Dimethyloctane	160.1	155.5	15	562
3812	C$_{10}$H$_{22}$O	*n*-Decyl alcohol	232.9	211.1	49	529
3813	C$_{10}$H$_{22}$O	Amyl ether	187.5	178.0	20	556
3814	C$_{10}$H$_{22}$O	Isoamyl ether	173.4	166.95	14.5	556
3815	C$_{10}$H$_{22}$S	Isoamyl sulfide	214.8	199.5	17	566
3816	C$_{11}$H$_{10}$	1-Methylnaphthalene	245.1	209.8	43.8	574
3817	C$_{11}$H$_{10}$	2-Methylnaphthalene	241.15	208.25	40	527
		"	241.1		55	270
3818	C$_{11}$H$_{12}$O$_2$	Ethyl cinnamate	272.5	271.5	70	564
3819	C$_{11}$H$_{14}$O$_2$	1-Allyl-3,4-dimethoxy-benzene	255.2	216.85	50	527
3820	C$_{11}$H$_{14}$O$_2$	Butyl benzoate	251.2	214.0	49	535
3821	C$_{11}$H$_{14}$O$_2$	1,2-Dimethoxy-4-propenyl-benzene	270.5	219.55	69	574
3822	C$_{11}$H$_{14}$O$_2$	β-phenethyl propionate	248.1	215.5	48	562
3823	C$_{11}$H$_{14}$O$_2$	Isobutyl benzoate	241.9	211.2	42	570
3824	C$_{11}$H$_{20}$O	Isobornyl methyl ether	192.4	<185.5	<23	562
3825	C$_{11}$H$_{20}$O	Methyl α-terpineol ether	216.2	<200.5	<28	562
3826	C$_{11}$H$_{22}$O$_3$	Isoamyl carbonate	232.2	205.65	32	564
3827	C$_{12}$H$_{10}$	Acenaphthene	277.9	217.1	64.2	527
3828	C$_{12}$H$_{10}$	Biphenyl	255.9	212.95	50.5	527
3829	C$_{12}$H$_{10}$O	Phenyl ether	259.3	214.55	52	574
3830	C$_{12}$H$_{14}$O$_4$	Ethyl phthalate	295	Nonazeotrope		527
3831	C$_{12}$H$_{16}$O$_2$	Isoamyl benzoate	262.05	215.4	55	574
3832	C$_{12}$H$_{16}$O$_3$	Isoamyl salicylate	277.5	220.0	70	575
3833	C$_{12}$H$_{18}$	1,3,5-Triethylbenzene	215.5	198.0	27	574
3834	C$_{12}$H$_{20}$O$_2$	Bornyl acetate	227.6	205.0	32	527
3835	C$_{12}$H$_{22}$O	Ethyl isobornyl ether	203.8	<193.0	<25	575
3836	C$_{12}$H$_{22}$O$_4$	Isoamyl oxalate	268.0	~217	~60	542
3838	C$_{13}$H$_{10}$	Fluorene	295	<219.7	>72	575
3839	C$_{13}$H$_{10}$O$_2$	Phenyl benzoate	315	Nonazeotrope		527
3839	C$_{13}$H$_{12}$	Diphenyl methane	265.6	215.15	56.5	574
3840	C$_{13}$H$_{12}$O	Benzyl phenyl ether	286.5	<220.8	>92	575
3841	C$_{14}$H$_{12}$	Stilbene	306.5	220.5	~88	575
3842	C$_{14}$H$_{14}$	1,2-Diphenylethane	284	218.2	68	537
3843	C$_{14}$H$_{14}$O	Benzyl ether	297	Nonazeotrope		575
A =	C$_2$H$_5$NO$_2$	**Ethyl Nitrite**	**17.4**			
3844	C$_2$H$_6$S	Methyl sulfide	37.4	Nonazeotrope		550
3845	C$_3$H$_5$Cl	2-Chloropropene	22.65	Nonazeotrope		550
3846	C$_3$H$_7$Cl	2-Chloropropane	34.9	Nonazeotrope		550
3847	C$_3$H$_8$O	Isopropyl alcohol	82.35	Min. b.p.		576
3848	C$_3$H$_8$O$_2$	Methylal	42.3	Nonazeotrope		550
3849	C$_4$H$_4$O	Furan	31.7	Nonazeotrope		550
3850	C$_4$H$_{10}$	Butane	0.6	Nonazeotrope		550
3851	C$_4$H$_{10}$O	Ethyl ether	34.6	Nonazeotrope		550
3852	C$_4$H$_{10}$O	Methyl propyl ether	38.95	Nonazeotrope		550
3853	C$_5$H$_{10}$	Cyclopentane	49.3	Nonazeotrope		550

TABLE I. *Binary Systems*   129

No.	Formula	B-Component Name	B.P., °C	Azeotropic Data B.P., °C	Wt.%A	Ref.
A =	**C₂H₅NO₂**	**Ethyl Nitrite** *(continued)*	**17.4**			
3854	C₅H₁₀	3-Methyl-1-butene	20.6	15.5	60	550
3855	C₅H₁₀	2-Methyl-2-butene	37.15	Nonazeotrope		550
3856	C₅H₁₂	2-Methylbutane	27.95	16.7	90	550
3857	C₅H₁₂	Pentane	36.15	Nonazeotrope		550
A =	**C₂H₅NO₂**	**Nitroethane**	**114.2**			
3858	C₂H₅NO₃	Ethyl nitrate	87.7	Nonazeotrope		574
3859	C₂H₆O	Ethyl alcohol	78.3	Nonazeotrope		574
		" "	78.32	78.03	12.6	806
3860	C₃H₆O₂	Propionic acid	141.3	Nonazeotrope		575
3861	C₃H₇ClO	1-Chloro-2-propanol	127.0	Nonazeotrope		554
3862	C₃H₇NO₃	Propyl nitrate	110.5	<109.6	>21	554
3863	C₃H₈O	Isopropyl alcohol	82.40	81.82	10.6	806
3864	C₃H₈O	Propyl alcohol	97.15	94.49	31.8	806
		" "	97.2	< 95.0	>23	554
3865	C₄H₈O₂	Dioxane	101.35	Nonazeotrope		554
3866	C₄H₉Br	1-Bromobutane	101.5	96.0	25	554
3867	C₄H₉Br	1-Bromo-2-methylpropane	91.4	89.5	10	554
3868	C₄H₉Cl	1-Chlorobutane	78.5	Nonazeotrope		554
3869	C₄H₁₀O	Butyl alcohol	117.8	107.7	55	554
		" "	117.75	107.94	58.6	806
3870	C₄H₁₀O	*sec*-Butyl alcohol	99.53	97.16	27.6	806
3871	C₄H₁₀O	*tert*-Butyl alcohol	82.41	82.22	4.5	806
3872	C₄H₁₀O	Isobutyl alcohol	107.89	102.68	40.8	806
		" "	108.0	102.5	40	554
3873	C₅H₁₀O	3-Pentanone	102.05	Nonazeotrope		575
3874	C₅H₁₀O₂	Propyl acetate	101.6	Nonazeotrope		575
3875	C₅H₁₁Br	1-Bromo-3-methylbutane	120.65	<108.5	>55	554
3876	C₅H₁₂O	Amyl alcohol	138.2	<137.8	>83	554
3877	C₅H₁₂O	*tert*-Amyl alcohol	102.35	<98.6	>30	554
3878	C₅H₁₂O	Isoamyl alcohol	131.9	112.0	78	554
3879	C₅H₁₂O₂	2-Propoxyethanol	151.35	Nonazeotrope		554
3880	C₆H₅NO₂	Nitrobenzene	210.9	Nonazeotrope	v-l	963
3881	C₆H₆	Benzene	25°C.	Nonazeotrope	v-l	845
		"	80.15	Nonazeotrope		554
3882	C₆H₁₂	1-Hexene	63.84	63.2	3.9	806
3883	C₆H₁₂	Methylcyclopentane	72.0	< 71.2	> 4	554
3884	C₆H₁₂O	4-Methyl-2-pentanone	116.05	<113.0		575
3885	C₆H₁₂O₂	Butyl acetate	126.0	Nonazeotrope		554
3886	C₆H₁₂O₂	Ethyl butyrate	121.5	<113.7	>73	554
3887	C₆H₁₂O₂	Ethyl isobutyrate	110.1	108.5	27	554
3888	C₆H₁₂O₂	Isobutyl acetate	117.4	112.5	60	554
3889	C₆H₁₄	Hexane	68.74	59.4	10.6	806
3890	C₆H₁₄S	Isopropyl sulfide	120.5	<110.9	>60	554
3891	C₇H₈	Toluene	110.75	106.2	25	162.554
3892	C₇H₁₄	Methylcyclohexane	101.15	90.8	30	554
3893	CₙH₂ₙ₊₂	Paraffins	107-110	82-104		162

	B-Component			Azeotropic Data		
No.	Formula	Name	B.P., °C	B.P., °C	Wt.%A	Ref.
A =	C₂H₅NO₂	Nitroethane (continued)	114.2			
3894	C₇H₁₆	n-Heptane	98.4	89.2	28	554
		"	98.43	89.5	29.8	806
3895	C₈H₁₀	m-Xylene	139.2	Nonazeotrope		554
3896	C₈H₁₈	2,5-Dimethylhexane	109.4	< 96.9	>62	554
3897	C₈H₁₈	Octane	125.66	90.5	55.6	806
A =	C₂H₅NO₃	Ethyl Nitrate	87.68			
3898	C₂H₆O	Ethyl alcohol	78.3	71.85	56	536
3899	C₃H₅Br	3-Bromopropene	70.5	Nonazeotrope		560
3900	C₃H₅I	3-Iodopropene	101.8	< 87.0		560
3901	C₃H₆Cl₂	2,2-Dichloropropane	70.4	Nonazeotrope		575
3902	C₃H₆O	Allyl alcohol	96.95	83.15	77.5	527
3903	C₃H₆O₃	Methyl carbonate	90.25	Nonazeotrope		527
3904	C₃H₇Br	1-Bromopropane	71.0	Nonazeotrope		527
3905	C₃H₇I	1-Iodopropane	102.4	87.4		560
3906	C₃H₇I	2-Iodopropane	89.45	83.2	52	527
3907	C₃H₈O	Isopropyl alcohol	82.35	77.0	53	560
3908	C₃H₈O	Propyl alcohol	97.25	82.55	70	536
3909	C₄H₄S	Thiophene	84.7	Nonazeotrope		527
3910	C₄H₈O	2-Butanone	79.6	Nonazeotrope		527
3911	C₄H₈O₂	Dioxane	101.35	Nonazeotrope		527
3912	C₄H₉Br	1-Bromobutane	101.6	Nonazeotrope		527
3913	C₄H₉Br	2-Bromobutane	91.2	< 85.5	<68	560
3914	C₄H₉Br	1-Bromo-2-methylpropane	91.4	85.0	65	570
3915	C₄H₉Br	2-Bromo-2-methylpropane	73.25	Nonazeotrope		527
3916	C₄H₉Cl	1-Chlorobutane	78.5	< 78	<20	527
3917	C₄H₉Cl	1-Chloro-2-methylpropane	68.85	Nonazeotrope		527
3918	C₄H₁₀O	Butyl alcohol	117.75	87.45	96	527
3919	C₄H₁₀O	sec-Butyl alcohol	99.5	84.8	78	527
3920	C₄H₁₀O	tert-Butyl alcohol	82.55	78.1	48	527
3921	C₄H₁₀O	Isobutyl alcohol	107.85	86.4	86	527
3922	C₄H₁₀S	Ethyl sulfide	92.1	85.0	58	560
3923	C₅H₁₀	Cyclopentane	49.3	Nonazeotrope		527
3924	C₅H₁₀O	Isovaleraldehyde	92.1	Nonazeotrope		527
3925	C₅H₁₁Cl	1-Chloro-3-methylbutane	99.4	87.55	92	527
3926	C₅H₁₂O	tert-Amyl alcohol	102.35	< 87.0	<95	527
3927	C₅H₁₂O	2-Pentanol	119.8	Nonazeotrope		527
3928	C₅H₁₂O₂	Diethoxymethane	87.95	85.85	49	569
3929	C₆H₅F	Fluorobenzene	84.9	< 82.5	<42	560
3930	C₆H₆	Benzene	80.15	80.03	12	560
3931	C₆H₈	1,3-Cyclohexadiene	80.4	76	<38	560
3932	C₆H₁₂	Cyclohexane	80.75	74.5	36	560
3933	C₆H₁₂	Methylcyclopentane	72.0	68.7	20	560
3934	C₆H₁₄	Hexane	68.8	66.25	24	570
3935	C₆H₁₄O	Propyl ether	90.1	< 87.0	>65	527
3936	C₆H₁₄O₂	Acetal	103.55	Nonazeotrope		557
3937	C₇H₈	Toluene	110.7	Nonazeotrope		527
3938	C₇H₁₄	Methylcyclohexane	101.15	83.85		571

TABLE I. *Binary Systems*   131

No.	Formula	Name	B.P., °C	B.P., °C	Wt.%A	Ref.
		B-Component		**Azeotropic Data**		

No.	Formula	Name	B.P., °C	B.P., °C	Wt.%A	Ref.
A =	$C_2H_5NO_3$	**Ethyl Nitrate** *(continued)*	**87.68**			
3939	C_7H_{16}	Heptane	98.4	82.8	63	527
3940	C_8H_{16}	1,3-Dimethylcyclohexane	120.7	Nonazeotrope		560
3941	C_8H_{18}	2,5-Dimethylhexane	109.2	86	84	560
A =	C_2H_6	**Ethane**	**— 93**			
3942	C_2H_6O	Ethyl alcohol	78.3	Nonazeotrope		563
3943	C_3H_6O	Acetone crit. press.		Nonazeotrope		442
3944	C_3H_8O	Isopropyl alcohol	82.45	Nonazeotrope		563
3945	C_3H_8O	Propyl alcohol	97.2	Nonazeotrope		834
3946	C_4H_{10}	Butane	0.6	Nonazeotrope		563
3947	$C_4H_{10}O$	Isobutyl alcohol	108	Nonazeotrope		563
3948	C_7F_{16}	Perfluoroheptane, crit. region		Nonazeotrope	v-l	430,431
A =	$C_2H_6Cl_2Si$	**Dichlorodimethylsilane**				
3948a	$C_4H_8Cl_2O$	Bis(2-chloroethyl) ether	178	Nonazeotrope		886c
		"	178	B.p. curve		886c
3949	C_7H_{16}	2-Methylhexane	90.1	Nonazeotrope		844
3950	C_7H_{16}	3-Methylhexane	91.96	Nonazeotrope		844
A =	C_3H_6O	**Ethyl Alcohol**	**78.3**			
3951	C_2H_6S	Methyl sulfide	37.4	Nonazeotrope		566
3951a	C_3H_3N	Acrylonitrile	77.3	70.8	41	498i
3952	C_3H_5Br	*trans*-1-Bromopropene	63.25	58.7	11	563
3953	C_3H_5Br	*cis*-1-Bromopropene	57.8	56.4	9	563
3954	C_3H_5Br	2-Bromopropene	48.35	46.2	6	563
3955	C_3H_5Br	3-Bromopropene	70.8	62.9		580
3956	C_3H_5Cl	*cis*-1-Chloropropene	32.8	32.1		533
3957	C_3H_5Cl	*trans*-1-Chloropropene	37.4	36.7	4	575
3958	C_3H_5Cl	2-Chloropropene	22.65	Nonazeotrope		573
3959	C_3H_5Cl	3-Chloropropene	45.7	44	5	573
3960	C_3H_5ClO	Epichlorohydrin	116.4	Nonazeotrope		556
3961	C_3H_5I	3-Iodopropene	102	75.4	42	532
3962	C_3H_5N	Propionitrile	97.1	77.5		563
		760 mm.		81	25.0	319,396
		760 mm.	97.1		27.5	319,396
		200 mm.			28.0	319,396
		100 mm.			35.5	319,396
		25 mm.			38.0	319,396
3963	$C_3H_6Cl_2$	1,2-Dichloropropane	96.2	74.7	52.74	533
3964	$C_3H_6Cl_2$	2,2-Dichloropropane	69.8	63.2	14.5	573
3965	C_3H_6O	Acetone	56.1	Nonazeotrope		371,960
				B.p. curve		
		"	56.4	Nonazeotrope	v-l	16,376
		" 680 mm.		Nonazeotrope	v-l	439f
3966	C_3H_6O	Propionaldehyde	48.7	Nonazeotrope		575
3967	C_3H_6OS	Methyl thioacetate	95.5	77.8		575
3968	$C_3H_6O_2$	Ethyl formate	54.1	54.05		536
3969	$C_3H_6O_2$	Methyl acetate	56.95	56.9	~3	536
3970	$C_3H_6O_3$	Methyl carbonate	90.35	73.5	~45	536

No.	Formula	B-Component Name	B.P., °C	Azeotropic Data B.P., °C	Wt.%A		Ref.
A =	**C₂H₆O**	**Ethyl Alcohol** *(continued)*	**78.3**				
3971	C₃H₇Br	1-Bromopropane	71	63.6	16.24		389
				B.p.	curve		
3972	C₃H₇Br	2-Bromopropane	59.8	55.5	11.5		527
3973	C₃H₇Cl	1-Chloropropane	46.65	44.95	6		555
3974	C₃H₇Cl	2-Chloropropane	36.25	35.6	2.8		573
3975	C₃H₇I	1-Iodopropane	102.4	75.4	44		573
3976	C₃H₇I	2-Iodopropane	89.35	70.2	25		555
3977	C₃H₇NO₂	1-Nitropropane	131.18	Nonazeotrope			806
3978	C₃H₇NO₂	2-Nitropropane	120.25	78.28	93.6		806
3979	C₃H₇NO₃	Propyl nitrate	110.5	75.0			560
3980	C₃H₈O	Isopropyl alcohol	82.3	Nonazeotrope		v-l	320,563
3981	C₃H₈O	Propyl alcohol	97.2	Nonazeotrope			813
3982	C₃H₈O₂	2-Methoxyethanol	124.5	Nonazeotrope			575
3983	C₃H₈O₂	Methylal	42.1	Nonazeotrope			556
3984	C₃H₈S	Propanethiol	67.3	< 63.5	<19		566
3985	C₃H₉BO₃	Methyl borate	68.7	63.0	~25		536
3986	C₄H₄N₂	Pyrazine	114-115	Nonazeotrope			751
3987	C₄H₄S	Thiophene	84	70.0	45		555
3988	C₄H₆	1,3-Butadiene	— 4.5	Nonazeotrope		v-l	111
3989	C₄H₆O	Crotonaldehyde	102.2	Nonazeotrope			241
3990	C₄H₆O₂	Allyl formate	80.0	71.5			567
3991	C₄H₆O₂	Biacetyl	87.5	73.9	47?		552
3993	C₄H₆O₂	Methyl acrylate	80	73.5	42.4		800
3994	C₄H₇Br	*trans*-1-Bromo-1-butene	94.70	72.3	35.71		580
3994	C₄H₇Br	*cis*-1-Bromo-1-butene	86.15	69.6	77.48		580
3995	C₄H₇Br	2-Bromo-1-butene	81.0	67.4	22.18		580
3996	C₄H₇Br	*cis*-2-Bromo-2-butene	93.9	72.3	33.7		580
3997	C₄H₇Br	*trans*-2-Bromo-2-butene	85.55	69.1	26.7		580
3998	C₄H₇Cl	*trans*-1-Chloro-1-butene	68	61.2	20.2		690
3999	C₄H₇Cl	*cis*-1-Chloro-1-butene	63.4	57	14.8		690
4000	C₄H₇Cl	2-Chloro-1-butene	58.4	53.6	11.5		690
4001	C₄H₇Cl	*trans*-2-Chloro-2-butene	66.6	60	18.4		690
4002	C₄H₇Cl	*cis*-2-Chloro-2-butene	62.4	56.8	15.4		690
4003	C₄H₇ClO₂	Ethyl chloroacetate	143.5	Nonazeotrope			121
4004	C₄H₇N	Isobutyronitrile	103.85	Nonazeotrope			575
4005	C₄H₈O	2-Butanone	79.6	75.7	>46		527,633
		"	79.6	74.0	39	v-l	376,31c
		"	79.6	74.0	39	v-l	376
		"	79.6	74.8	34		981
4006	C₄H₈O	Butyraldehyde	75.7	70.7	60.6		982
4007	C₄H₈O	Ethyl vinyl ether	35.5	Nonazeotrope			882
4008	C₄H₈O	Isobutyraldehyde	63.5	Nonazeotrope			575
4009	C₄H₈O	Tetrahydrofuran	66		10		224
4010	C₄H₈OS	Ethyl thioacetate	116.6	Nonazeotrope			575

TABLE I. *Binary Systems* 133

No.	Formula	B-Component Name	B.P., °C	Azeotropic Data B.P., °C	Wt.%A		Ref.
A =	C_2H_6O	**Ethyl Alcohol** *(continued)*	**78.3**				
4011	$C_4H_8O_2$	Dioxane	101.4	Nonazeotrope			201
			101.07	78.13	90.7	v-l	397
		" 200 mm.		46.4	68	v-l	348
		" 400 mm.		62.4	82		348
		" 600 mm.		72.19	88		348
		" 760 mm.		78.25	>98		348
		" 190 mm.				v-l	617a
		" 380 mm.				v-l	617a
		" 760 mm.	101			v-l	617a
		" 1140 mm.				v-l	617a
		" 1520 mm.				v-l	617a
		" 3040 mm.		Nonazeotrope		v-l	617a
4012	$C_4H_8O_2$	Ethyl acetate, 40°-60°C.		% Alc. increases with press.		v-l	674
		" 77.4 mm.			15.95		322
		" 760 mm.			30.97		322
		"	77.05	72.18	25.8		674
		" 25 mm.		− 1.39	12.81		
		" 300 mm.		47.83	23.22		650,834,
		" 760 mm.	77.05	71.81	30.98		1034
		" 1500 mm.		91.86	39.07		
		" 50°-80°C.				v-l	452c,31c
4013	$C_4H_8O_2$	Methyl propionate	79.7	72.0	33		536
4014	$C_4H_8O_2$	Propyl formate	80.8	71.75	~41		572
4015	C_4H_9Br	1-Bromobutane	100.3	75.0	43		573
4016	C_4H_9Br	2-Bromobutane	91.2	72.5	33		567
4017	C_4H_9Br	1-Bromo-2-methylpropane	89.2	71.4	41.0		389,555
4018	C_4H_9Br	2-Bromo-2-methylpropane	73.3	B.p. curve 63.8	15		563
4019	C_4H_9Cl	1-Chlorobutane	78.05	65.7	20.3		573
4020	C_4H_9Cl	2-Chlorobutane	68.25	61.2	15.8		567
4021	C_4H_9Cl	1-Chloro-2-methylpropane	68.9	61.45	16.3		563
4022	C_4H_9Cl	2-Chloro-2-methylpropane	51	~ 49	~ 6.5		563
4023	C_4H_9I	1-Iodobutane	130.4	< 78.15			567
4024	C_4H_9I	2-Iodobutane	120.0	77.2	70		567
4025	C_4H_9I	1-Iodo-2-methylpropane	120	77	70		834
		"	120.4	77.65	73		573
4026	$C_4H_{10}O$	Butyl alcohol	117.75	Nonazeotrope		v-l	376,452e
4027	$C_4H_{10}O$	sec-Butyl alcohol	99.4	Nonazeotrope		v-l	376
4028	$C_4H_{10}O$	tert-Butyl alcohol	82.55	Nonazeotrope			563
		"	82.9			v-l	937c
4029	$C_4H_{10}O$	Ethyl ether	34.5	Nonazeotrope B.p. curve			371,1033
4029a	$C_4H_{10}O$	Isobutyl alcohol	108	Nonazeotrope		v-l	937c
4030	$C_4H_{10}O$	Methyl propyl ether	38.95	Nonazeotrope			576
4031	$C_4H_{10}O_2$	Acetaldehyde dimethyl acetal	64.3	61.6	12		576
4032	$C_4H_{10}O_2$	2-Ethoxyethanol	133	Nonazeotrope		v-l	34

No.	B-Component Formula	B-Component Name	B.P., °C	Azeotropic Data B.P., °C	Wt.%A		Ref.
A =	C_2H_6O	**Ethyl Alcohol** (*continued*)	**78.3**				
4033	$C_4H_{10}O_2$	Ethoxymethoxymethane	65.90	63.95	13.3		1035
4034	$C_4H_{10}S$	Ethyl sulfide	92.2	72.6	56		555
4035	$C_4H_{11}ClSi$	Chloromethyltrimethyl-silane	97	72			907
4036	$C_4H_{11}N$	Butylamine	77.8	82.2	49		982
4037	$C_4H_{11}N$	Diethylamine	55.5	Nonazeotrope			981
4038	C_5H_5N	Pyridine	115.4	Nonazeotrope			553
4039	C_5H_6O	2-Methylfuran	63.8	< 60.5	<15		575
4040	C_5H_8	Isoprene	34.3	32.65	3		537
4041	C_5H_8	3-Methyl-1,2-butadiene	40.8	~39			563
4042	$C_5H_8O_2$	Ethyl acrylate		77.5	72.7		800
		" 100 mm.	44.9	32	54.4		982
		"		Nonazeotrope		v-l	591e
4043	C_5H_{10}	Cyclopentane	49.4	44.7	7.5		567
4044	C_5H_{10}	2-Methyl-1-butene	31.10	30.1	22 vol. %		826
4045	C_5H_{10}	3-Methyl-1-butene	22.5	21.9	~2		537
4046	C_5H_{10}	2-Methyl-2-butene	37.15	Nonazeotrope			563
		"	37.15	37.3	~4		256,537
4047	$C_5H_{10}O$	Allyl ethyl ether	63-65	60.5			580
4048	$C_5H_{10}O$	Isopropyl vinyl ether,					
		737 mm.	54.8	52.6			981
4049	$C_5H_{10}O$	2-Pentanone	102.35	78	93.3	v-l	376
		"	102	77.7	91.17		215
		"	102	Effect of pressure			83
4050	$C_5H_{10}O$	3-Methyl-2-butanone	95.4	Nonazeotrope			552
4051	$C_5H_{10}O$	Propyl vinyl ether	65.1	60	18.4		981
4052	$C_5H_{10}O_2$	Ethyl propionate	99.15	78.0	75		537
4053	$C_5H_{10}O_2$	Isobutyl formate	97.9	77.0	67		536
4054	$C_5H_{10}O_2$	Isopropyl acetate	91.0	76.8	53		536
		"		76.7	51		498c
4055	$C_5H_{10}O_2$	Methyl butyrate	102.65	78.0	~83		536
4056	$C_5H_{10}O_2$	Methyl isobutyrate	92.3	77.0			536
4057	$C_5H_{10}O_2$	Propyl acetate	101.6	78.18	~85		536
4058	$C_5H_{11}Br$	1-Bromo-3-methylbutane	118.2	77.3	72.0		388,573
				B.p. curve			
4059	$C_5H_{11}Cl$	1-Chloro-3-methylbutane	99.8	74.8	41		573
4060	$C_5H_{11}Cl$	1-Chloropentane	108.35	72.5			409
4061	C_5H_{12}	2-Methylbutane	27.95	26.75	3.5		537
4062	C_5H_{12}	Pentane	36.15	34.3	5		537
4063	$C_5H_{12}O$	Amyl alcohol	137.8	Nonazeotrope		v-l	376
4064	$C_5H_{12}O$	Butyl methyl ether	70.3	65.5	20		981
4065	$C_5H_{12}O$	Ethyl propyl ether	63.6	61.2	25		573
4066	$C_5H_{12}O$	Isoamyl alcohol	131.8	Nonazeotrope			563
4067	$C_5H_{12}O_2$	Diethoxymethane	87.5	74.2	42		324
4068	$C_5H_{12}O_2$	2,2-Dimethoxypropane	80	Min. b.p.			594

TABLE I.   *Binary Systems*   135

No.	Formula	Name	B.P., °C	B.P., °C	Wt.%A		Ref.
		B-Component		**Azeotropic Data**			
A =	**C_2H_6O**	**Ethyl Alcohol** *(continued)*	**78.3**				
4069	$C_5H_{14}OSi$	Ethoxytrimethylsilane	75-76		30		192
			75	66.4			839
4070	C_6H_5Cl	Chlorobenzene	132.0	Nonazeotrope			574
4071	C_6H_5F	Fluorobenzene	85.15	70.0	25		545
4072	$C_6H_5NO_2$	Nitrobenzene	210.75	Nonazeotrope			554
4073	C_6H_6	Benzene	80.1	68.24	32.4		1044
			198 mm.	34.8	25		855
			382 mm.	50	25		855
			570 mm.	60	25		855
		"	711 mm.	66	25		855
		"	310 mm.	45	26.2	v-l	91
			180 mm.	32.5	23.2	v-l	698
		"	400 mm.	51.2	28.1	v-l	698
		"	168.4 mm.	29.97	21.33	v-l	952
		"	233.5 mm.	38.37	23.72		952
		"	336.4 mm.	47.15	26.32		952
		"	584 mm.	61.06	30.35		952
		"	209 mm.	35.0	24.3		735
		"	760 mm.	67.9	31.7	v-l	517,1011
		at crit. pt.		Nonazeotrope		v-l	888
		"	760 mm.			v-l	310
		"		80.1	68.1	31.6	v-l	1050
		"	600 mm.	61.6	30.4	v-l	1050
		"	450 mm.	54.1	29.1	v-l	1050
		"	384 mm.	50	27.4	v-l	1050
		"	300 mm.	44.2	26.6	v-l	1050
		"	5570 mm.	132.9	56		735
		"	11,720 mm.	166.9	69.5		735
		"	19,160 mm.	191.1	81		735
		"		See also Refs. 109, 178, 297, 834, 843, 944, 960, 964, 1043, 1073			
4074	$C_6H_6O_2$	Resorcinol	281.4	Nonazeotrope			813
4075	C_6H_7N	Aniline	184.35			v-l	390
4076	C_6H_8	1,3-Cyclohexadiene	80.8	66.7	34		563
4077	C_6H_8	1,4-Cyclohexadiene	85.6	68.5			563
4078	C_6H_{10}	Biallyl	60.2	53.5	13		563
4079	C_6H_{10}	Cyclohexene	82.7	66.7	34		537
4080	C_6H_{10}	1,3-Hexadiene	72.9	Min. b.p.			221
4081	C_6H_{10}	2,4-Hexadiene	82	Min. b.p.			221
4082	C_6H_{10}	1-Hexyne	70.2	62.8	23.2		377
4083	C_6H_{10}	3-Hexyne	80.5	67.5	34.4		377
4084	C_6H_{10}	Methylcyclopentene	75.85	63.3	28		567
4085	C_6H_{10}	3-Methylcyclopentene	64.9	57.2	20 vol. %		826
4086	C_6H_{10}	3-Methyl-1,3-pentadiene	77	Min. b.p.			221
4087	C_6H_{12}	Cyclohexane, 296 mm.		41.2	25.5		924
		" 420 mm.		49.3	27.3		924

No.	Formula	B-Component Name	B.P., °C	Azeotropic Data B.P., °C	Wt.%A		Ref.
A =	C₂H₆O	**Ethyl Alcohol** *(continued)*	**78.3**				
		Cyclohexane, 643 mm.		60.8	29.8		924
		" 760 mm.		64.8	31.3		924
		" 760 mm.	80.8	64.9	40		413
		"	80.7	64.8	29.2	v-l	1049
		"	80.7	64.8	31.4	v-l	666
		" 430 mm.		50	27.7	v-l	666
		" 300 mm.		41.4	25.1	v-l	666
		"	80.75	64.9	30.5		563
4088	C₆H₁₂	2-Ethyl-1-butene	64.95	57.0	23 vol. %		826
4089	C₆H₁₂	1-Hexene	63.49	56.1	22 vol. %		826
		" 200 mm.		24.9	11.6		498m
		" 400 mm.		39.2	13.8		498m
		" 600 mm.		48.6	16.3		498m
		" 760 mm.	63.6	54.8	17.6		498m
4090	C₆H₁₂	*cis*-2-Hexene	68.8	59.5	22 vol. %		826
4091	C₆H₁₂	*cis*-3-Hexene	66.4	49.6	26 vol. %		826
4092	C₆H₁₂	Methylcyclopentane	72.0	60.05	22.7	v-l	886
		"	72.0	60.3	25		567
4093	C₆H₁₂	*cis*-3-Methyl-2-pentene	70.52	60.4	24 vol. %		826
4094	C₆H₁₂	*trans*-3-Methyl-2-pentene	67.6	58.8	20 vol. %		826
4095	C₆H₁₂	*trans*-4-Methyl-2-pentene	58.4	52.6	15 vol. %		826
4096	C₆H₁₂O	Butyl vinyl ether	94.2	73	48		982
4097	C₆H₁₂O	Isobutyl vinyl ether	83.4	69.2	33		982
4098	C₆H₁₂O	*trans*-2-Butenyl ethyl ether	100.45	77.5			580
4099	C₆H₁₂O	*cis*-2-Butenyl ethyl ether	100.3	76.2			580
4100	C₆H₁₂O	(1-Methylallyl) ethyl ether	76.65	69			580
4101	C₆H₁₂O	4-Methyl-2-pentanone	118	Nonazeotrope		v-1	654
4102	C₆H₁₂O₂	Ethyl Isobutyrate	110.1	Nonazeotrope			536
4103	C₆H₁₂O₂	Methyl isovalerate	116.3	Nonazeotrope			536
4104	C₆H₁₂O₃	Paraldehyde	124.5	Nonazeotrope			981
4105	C₆H₁₄	2,3-Dimethylbutane	58.0	51.5	12		567
4106	C₆H₁₄	*n*-Hexane	68.95	58.68	21.0		1044
		"	68.7	58.4	21.2	v-1	498
		" 674 mm.		55	23.5	v-1	1049
		" 670 mm.		55	20.5	v-l	498
		" 458 mm.		45	18.6	v-l	498
		" 301 mm.		35	16.1	v-l	498
			68.95	58	20.8	v-1	886
		" 1545 mm.			26.3 vol. %		719
4107	C₆H₁₄	2-Methylpentane	60.27	53.1	12 vol. %		517
4108	C₆H₁₄O	*tert*-Butyl ethyl ether	73	66.6	21		256
4109	C₆H₁₄O	Ethyl butyl ether	92.2	73.8	49.3		982
4110	C₆H₁₄O	Isopropyl ether	68.3	64	17.1		982
4111	C₆H₁₄O	Propyl ether	90.4	74.4	44		556
4112	C₆H₁₄O₂	Acetal	103.6	78.2	65.5		45,572
4113	C₆H₁₄O₂	1,2-Diethoxyethane	121.1	Nonazeotrope			981
4114	C₆H₁₄O₂	Ethoxypropoxymethane	113.7	Nonazeotrope			1035

TABLE I. *Binary Systems*   137

No.	Formula	B-Component Name	B.P., °C	B.P., °C	Wt.%A	Ref.
A =	**C₂H₆O**	**Ethyl Alcohol** *(continued)*	**78.3**			
4115	C₆H₁₄S	Isopropyl sulfide	120.5	Nonazeotrope		576
4116	C₆H₁₅BO₃	Triethylborate	120	76.6	68	637,990
4117	C₆H₁₅N	Triethylamine	89.7	76.9	51	982
		"	89.4	~75		563
4118	C₆H₁₆SiO	Ethoxymethyltrimethyl-silane	102	74		907
4119	C₆H₁₆O₂Si	Diethoxydimethylsilane	114	77	83	215
4120	C₇H₈	Toluene	110.7	76.7	68	50,834,943
		"		0.5	45.5	814
		"		25	62.5	814
		"		50	64.5	814
		"		75.5	65.5	814
		"	35° C., 55° C.		v-l	489
		"	110.6	77.0	69.2	662
		"	710 mm.	—	75.0	68.5	662
		"	610 mm.	—	71.0	67.1	662
		"	510 mm.	—	66.5	65.6	662
		"	412 mm.	—	61.5	63.8	662
		"	110.7	76.5	66.7 v-l	390,517
		"	327 mm.		76.2	322
		"	800 mm.		81.6	322
4121	C₇H₁₂	1,3-Heptadiene		Min. b.p.		221
4122	C₇H₁₂	2,4-Heptadiene		Min. b.p.		221
4123	C₇H₁₂	1-Heptyne	99.5	74.2	54.6	377
4124	C₇H₁₂	5-Methyl-1-hexyne	90.8	71.0	39.8	377
4125	C₇H₁₄	1,1-Dimethylcyclopentane	87.84		36	928
			87.85	68.0	37 vol. %	826
4126	C₇H₁₄	*cis*-1,2-Dimethyl-cyclopentane	99.53	72.1	46 vol. %	826
		"	99.53		~47	928
4127	C₇H₁₄	*trans*-1,2-Dimethyl-cyclopentane			~39	928
		"	91.87	69.6	39 vol. %	826
4128	C₇H₁₄	*cis*-1,3-Dimethyl-cyclopentane	91.73	69.5	38 vol. %	826
4129	C₇H₁₄	*trans*-1,3-Dimethyl-cyclopentane	90.77	69.1	37 vol. %	826
		"	90.77		~37	928
4130	C₇H₁₄	2,3-Dimethyl-1-pentene	84.2	67.1	35 vol. %	826
4131	C₇H₁₄	Ethylcyclopentane	103.47	73.1	48 vol. %	826
		"	103.45		~48	928
4132	C₇H₁₄	Methylcyclohexane	100.8	72.1	47	50
		"	100.95	71.95		572
				v-l, at 35.55°C.		490
4133	C₇H₁₄	1,1,2,2-Tetramethyl-cyclopropane	75.9	62.6	30 vol. %	826
4134	C₇H₁₆	2,2-Dimethylpentane	79.20	63.9	25 vol. %	826
		"	79.1		~26	928

No.	Formula	B-Component Name	B.P., °C	B.P., °C	Wt.%A		Ref.
A =	**C₂H₆O**	**Ethyl Alcohol** *(continued)*	**78.3**				
4135	C₇H₁₆	2,3-Dimethylpentane	89.79		~36		928
		"	89.78	68.6	34	vol. %	826
4136	C₇H₁₆	2,4-Dimethylpentane	80.50	64.6	29	vol. %	826
		"	80.8		29		928
4137	C₇H₁₆	3,3-Dimethylpentane	86.0		32		928
		"	86.07	67.1	38	vol. %	826
4138	C₇H₁₆	3-Ethylpentane	93.47	70	38	vol. %	826
		"	93.5		35		928
4139	C₇H₁₆	Heptane	98.45	70.9	49		537
		"	98.4	72	48	v-l	
							390,697,982
		" 180 mm.		37.5	43	v-l	439
		" 400 mm.		54.5	43	v-l	439
		" 750 mm.		71.0	45	v-l	439
		" 393 p.s.i.a		194.0	78.1	v-l	697
		" 220 p.s.i.a		169.0	67.0	v-l	697
		" 139 p.s.i.a.		147.7	62.5	v-l	697
		" 110 p.s.i.a.		135.5	60.6	v-l	697
		" 753 mm.		70	47	v-l	779c
		" 1155 mm.		80	48	v-l	779c
		" 1675 mm.		90	48	v-l	779c
4140	C₇H₁₆	2-Methylhexane	90.05	68.7	36	vol. %	826
		"	90.0		~36		928
4141	C₇H₁₆	3-Methylhexane	91.8		~36		928
		"	91.85	69.3	36	vol. %	826
4142	C₇H₁₆O	*tert*-Amyl ethyl ether	101-2	66.6	21		256
4143	C₈H₈	Styrene	145.8	Nonazeotrope			545
4144	C₈H₁₀	Ethylbenzene	136.15	Nonazeotrope			537
		"	—	Nonazeotrope		v-l	308,587
4145	C₈H₁₀	*m*-Xylene	**139.0**	Nonazeotrope			537
4146	C₈H₁₀	*o*-Xylene	143.6	Nonazeotrope			545
4147	C₈H₁₀	*p*-Xylene	138.3	Nonazeotrope		v-l	308,308c
4148	C₈H₁₆	1,1-Dimethylcyclo- hexane	119.54	76.2	65	vol. %	826
		"			~36		928
4149	C₈H₁₆	1,3-Dimethylcyclo- hexane	120.7	175.8	70		575
4150	C₈H₁₆	*cis*-1,4-Dimethylcyclo- hexane			~70		928
		"	124.32	76.9	70	vol. %	826
4151	C₈H₁₆	*trans*-1,3-Dimethylcyclo- hexane	124.45	76.9	70	vol. %	826
4152	C₈H₁₆	*trans*-1,4-Dimethylcyclo- hexane	119.35	76.2	64	vol. %	826
		"			~64		928
4153	C₈H₁₆	1-Ethyl-1-methyl- cyclopentane	121.52	76.5	66	vol. %	826
4154	C₈H₁₆	1,*cis*-2,*trans*-3-Trimethyl- cyclopentane	117.5	75.9	62	vol. %	826

TABLE I. *Binary Systems*   139

| No. | Formula | B-Component | B.P., °C | Azeotropic Data | | Ref. |
		Name		B.P., °C	Wt.%A	
A =	C₂H₆O	**Ethyl Alcohol** *(continued)*	**78.3**			
4155	C₈H₁₆	*cis-trans-cis*-1,2,4-Tri-methylcyclopentane			~52	928
4156	C₈H₁₆	1,*trans*-2,*cis*-4-Trimethyl-cyclopentane	109.29	74.3	52 vol. %	826
4157	C₈H₁₆	2,4,4-Trimethyl-2-pentene	104.91	73.9	50 vol. %	826
4158	C₈H₁₈	2,2-Dimethylhexane	106.84	73.6	46 vol. %	826
		"	106.84		36	928
4159	C₈H₁₈	2,3-Dimethylhexane	115.8		55	928
		"	115.61	75.5	57 vol. %	826
4160	C₈H₁₈	2,5-Dimethylhexane	109.2	73.6	59	545
4161	C₈H₁₈	3,4-Dimethylhexane	117.9		60	928
		"	117.73	75.8	60 vol. %	826
4162	C₈H₁₈	2-Methylheptane	117.65	75.8	59 vol. %	826
		"	117.2		59	928
4163	C₈H₁₈	3-Methylheptane	119.0		61	928
		"	118.93	76.0	61 vol. %	826
4164	C₈H₁₈	4-Methylheptane	117.71	75.8	61 vol. %	826
		"	118		61	928
4165	C₈H₁₈	Octane	125.6	77	78	537
		" 189.8 mm.		45	66.5	72c
		" 304.6 mm.		55	67.8	72c
		" 473.3 mm.		65	69.2	72c
		" 712.1 mm.		75	70.6	72c
4166	C₈H₁₈	2,2,3-Trimethylpentane	109.8		53	928
		"	109.84	74.3	53 vol. %	826
4167	C₈H₁₈	2,2,4-Trimethylpentane	99.24	71.8	40 vol. %	826
		" 96.1 mm.	—	25	30.4 v-l	488
		" 318.8 mm.	—	50	36.7 v-l	488
		"	—	0	24.8	488
		"	99.3	<72.4	<53	575
4168	C₈H₁₈	2,3,3-Trimethylpentane	113.6		57	928
		"	114.76	75.3	56 vol. %	826
4169	C₈H₁₈	2,3,4-Trimethylpentane	113.47	75.1	57 vol. %	826
		"	113.4		57	928
4170	C₈H₁₈O	Butyl ether	142.1	Nonazeotrope		981
4171	C₈H₁₈O	Isobutyl ether	122.1	Nonazeotrope		556
4172	C₈H₁₈O₂	2-Ethyl-1,3-hexanediol	243.1	Nonazeotrope		981
4173	C₉H₁₂	Cumene	152.8	Nonazeotrope		575
4174	C₉H₁₂	Mesitylene	164.6	Nonazeotrope		537
4175	C₉H₁₂	Propylbenzene	159.3	Nonazeotrope		575
4176	C₁₀H₁₄	Cymene	176.7	Nonazeotrope		537
4177	C₁₀H₁₆	Camphene	159.6	Nonazeotrope		537
4178	C₁₀H₁₆	*d*-Limonene	177.8	Nonazeotrope		537
4179	C₁₀H₁₆	α-Pinene	155.8	Nonazeotrope		528
		"	155.8	Min.b.p.		563

No.	Formula	B-Component Name	B.P., °C	Azeotropic Data B.P., °C	Wt.%A	Ref.
A =	C₂H₆O	**Ethyl Alcohol** *(continued)*	**78.3**			
4180	C₁₀H₁₆	α-Terpinene	173.4	Nonazeotrope		575
4181	C₁₀H₁₆	Thymene	179.7	Nonazeotrope		537
4182	C₁₀H₂₂	2,7-Dimethyloctane	160.2	Nonazeotrope		537
A =	C₂H₆O	**Methyl Ether**	**— 23.65**			
4183	C₃H₉N	Trimethylamine	3.5	Nonazeotrope		378
A =	C₂H₆OS	**Dimethylsulfoxide**				
4183a	C₄H₁₀O	Butyl alcohol	120.2	Nonazeotrope	v-l	145c
	"	"	150.3	Nonazeotrope	v-l	145c
4184	C₆H₆	Benzene, 25°-70°		V.p. curve		449
A =	C₂H₆O₂	**Glycol**	**197.4**			
4185	C₂H₇NO	2-Aminoethanol	170.8	Nonazeotrope		527
4186	C₃H₄O₃	Ethylene carbonate,				
		10 mm.		88	13.9	747
		" 25 mm.		107	7.5	747
		" 50 mm.		122	2.6	747
		" 72 mm.		**163**	**0**	**747**
4187	C₃H₅Cl₃	1,2,3-Trichloropropane	156.85	150.8	13	573
4188	C₃H₅I	3-Iodopropene	101.8	<101.5	<1.5	575
4189	C₃H₆Br₂	1,2-Dibromopropane	140.5	139.0	6	567
4190	C₃H₆Br₂	1,3-Dibromopropane	166.9	160.2	10.2	527
4191	C₃H₆Cl₂O	1,3-Dichloro-2-propanol	175.8	Nonazeotrope		541
4192	C₃H₆Cl₂O	2,3-Dichloro-1-propanol	182.5	Nonazeotrope		575
4193	C₃H₇ClO	1-Chloro-2-propanol	127.5	Nonazeotrope		575
4194	C₃H₇NO	Propionamide	222.1	Nonazeotrope		527
4195	C₃H₇NO₂	Ethyl carbamate	185.25	Nonazeotrope		527
4195a	C₃H₈O	Propyl alcohol				
		750 mm.	97	Nonazeotrope	v-l	882c
4196	C₃H₈O₃	Glycerol	290.5	Nonazeotrope		575
4197	C₄H₅N	Pyrrol	130.0	Nonazeotrope		527
4198	C₄H₅NS	Allyl isothiocyanate	152.05	<151.8		575
4199	C₄H₆O₄	Methyl oxalate	164.2	~163.5	~15	530
4200	C₄H₇BrO₂	Ethyl bromoacetate	158.8	157.3	12	527
4201	C₄H₇ClO₂	Ethyl chloroacetate	143.55	Nonazeotrope		575
4202	C₄H₈Br₂O	Bis(2-bromoethyl) ether,				
		760 mm.		180-185	~50	215
		50 mm.		105-115	~50	215
4203	C₄H₈Cl₂O	Bis(2-chloroethyl) ether	178	170.5	12.5	121
		"	178.65	171.05	21	556
		" 50 mm.	96	92.7		982
		"	178.6	164	17.8	183
4204	C₄H₈Cl₂S	Bis(2-chloroethyl) sulfide	216.8	186.0?		575
4204a	C₄H₈O	Tetrahydrofuran	66	Nonazeotrope	v-l	859c
4205	C₄H₈OS	Ethyl thioacetate	116.8	Nonazeotrope		575
4206	C₄H₈O₂	Dioxane	101.4	Nonazeotrope		201
4207	C₄H₈O₂	2-Vinyloxethanol	143		13	263

TABLE I. *Binary Systems* 141

No.	Formula	B-Component Name	B.P., °C	Azeotropic Data B.P., °C	Wt.%A	Ref.
A =	C₂H₆O₂	**Glycol** *(continued)*	197.4			
4208	C₄H₈O₃	Ethylene glycol monoacetate		Nonazeotrope		413
		" 150 mm.		Nonazeotrope		413
		"	190.9	184.75	25	569
4209	C₄H₉Br	1-Bromobutane	101.5	101.3	1.7	575
4210	C₄H₉Br	1-Bromo-2-methylpropane	91.4	< 91.35	< 0.8	575
4211	C₄H₉I	1-Iodobutane	130.4	128.5	5	567
4212	C₄H₉I	1-Iodo-2-methylpropane	120.8	119.5	3.5	575
4213	C₄H₁₀O	Butyl alcohol	117.75	Nonazeotrope		535
4214	C₅H₄O₂	2-Furaldehyde	161.45	Nonazeotrope		527
4215	C₅H₅N	Pyridine	115	Nonazeotrope		791
4216	C₅H₉ClO₂	Propyl chloroacetate	163.5	162	20	575
4217	C₅H₁₀O₃	Ethyl carbonate	125.9	Nonazeotrope		537
4218	C₅H₁₀O₃	2-Methoxyethyl acetate	144.6	Nonazeotrope		575
4219	C₅H₁₁Br	1-Bromo-3-methylbutane	120.65	119.45	5.5	527
4220	C₅H₁₁I	1-Iodo-3-methylbutane	147.65	143.0	7	567
4221	C₅H₁₂O	Isoamyl alcohol	131.35	Nonazeotrope		527
4222	C₅H₁₂O₂	2-Propoxyethanol	151.35	Nonazeotrope		526
4223	C₅H₁₂O₃	2-(2-Methoxyethoxy) ethanol	194.2	192	30	129,527, 808,982
		2-(2-Methoxyethoxy)- ethanol 50 mm.		114	4.0 ⎫	129,527,
		" 200 mm.		149	12.0 ⎭	808,982
4224	C₆H₃Cl₃	1,3,5-Trichlorobenzene	208.4	181.0		567
4225	C₆H₄BrCl	*p*-Bromochlorobenzene	196.4	173.8	28	567
4226	C₆H₄Br₂	*p*-Dibromobenzene	220.25	183.9	32.5	574
4227	C₆H₄ClNO₂	*m*-Chloronitrobenzene	235.5	192.5	53	554
4228	C₆H₄ClNO₂	*o*-Chloronitrobenzene	246.0	193.5	68	554
4229	C₆H₄ClNO₂	*p*-Chloronitrobenzene	239.1	192.85	57.8	554
4230	C₆H₄Cl₂	*o*-Dichlorobenzene	179.5	165.8	20	567
4231	C₆H₄Cl₂	*p*-Dichlorobenzene	174.35	163	~23	573
		"	174.35	162.7	18	574
4232	C₆H₅Br	Bromobenzene	156.15	150.2	12.5	530
4233	C₆H₅Cl	Chlorobenzene	132	130.05	5.6	526
4234	C₆H₅ClO	*o*-Chlorophenol	175.8	Nonazeotrope		575
4235	C₆H₅ClO	*p*-Chlorophenol	219.75	Nonazeotrope		535
4236	C₆H₅I	Iodobenzene	188.55	171.5		573
4237	C₆H₅NO₂	Nitrobenzene	210.75	185.9	59	554
4238	C₆H₅NO₃	*o*-Nitrophenol	217.2	189.35	49	527
4239	C₆H₆	Benzene	80.2	Nonazeotrope		537
4240	C₆H₆O	Phenol	182.2	Nonazeotrope		542,368, 374
		"	181.5	199	78	563
4241	C₆H₆O₂	Pyrocatechol	245.9	Nonazeotrope		564

No.	Formula	Name	B.P., °C	B.P., °C	Wt.%A		Ref.
A =	**C₂H₆O₂**	**Glycol** *(continued)*	**197.4**				
4242	C₆H₇N	Aniline	184.35	180.55	24		551
		" 37.1 mm.		95	8.75	v-l	181
		" 104.7 mm.		120	12.7	v-l	181
		" 257.9 mm.		145	16.8	v-l	181
4243	C₆H₇N	2-Picoline	128.8	Nonazeotrope			791
4244	C₆H₇N	3-Picoline	143.5	Nonazeotrope			790,791
4245	C₆H₈N₂	o-Phenylenediamine	258.6	Nonazeotrope			527
4246	C₆H₈O₄	Methyl maleate	204.05	189.6	42		570
4247	C₆H₉N	N-Ethylpyrrol	130.4	Nonazeotrope			575
4248	C₆H₁₀	Cyclohexene	82.7	Nonazeotrope			540
4249	C₆H₁₀O₂	2-5-Hexadione	191.3	<180.5	< 45		552
4250	C₆H₁₀O₄	Ethylidene diacetate	168.5	167.45	8.2		527
4251	C₆H₁₀O₄	Ethyl oxalate	185.65	176.5	25		575
		"	185.0	Reacts			563
4252	C₆H₁₀O₄	Glycol diacetate	186.3	<179.5	< 24		575
4253	C₆H₁₀O₄	Methyl succinate	195	Reacts			563
4254	C₆H₁₁ClO₂	Butyl chloroacetate	181.9	176.0	30		526
4255	C₆H₁₂	Cyclohexane	80.75	Nonazeotrope			537
4256	C₆H₁₂	Methylcyclohexane	72.0	Nonazeotrope			575
4257	C₆H₁₂O	Cyclohexanol	160.65	Nonazeotrope			530
4258	C₆H₁₂O₂	Butyl acetate	126.0	Nonazeotrope			575
4259	C₆H₁₂O₂	Ethyl butyrate	121.5	Nonazeotrope			575
4260	C₆H₁₂O₂	Isoamyl formate	123.8	Nonazeotrope			536
4261	C₆H₁₂O₃	2-Ethoxyethyl acetate	156.8	Nonazeotrope			526
4262	C₆H₁₂O₃	Paraldehyde	124.35	Nonazeotrope			556
4263	C₆H₁₃Br	1-Bromohexane	156.5	150.5	14		567
4264	C₆H₁₄	2,3-Dimethylbutane	58.0	Nonazeotrope			575
4265	C₆H₁₄	Hexane	68.8	Nonazeotrope			575
4266	C₆H₁₄O	Hexyl alcohol	157.8	Nonazeotrope			573
4267	C₆H₁₄O₂	Acetal	103.55	Nonazeotrope			535
4268	C₆H₁₄O₂	2-Butoxyethanol	171.25	Nonazeotrope			526
4269	C₆H₁₄O₂	Pinacol	174.35	Nonazeotrope			530
4270	C₆H₁₄O₃	2-(2-Ethoxyethoxy)ethanol	202.8	192	45.5		808,982
		" 100 mm.	137.3	134	33		982
		" 36 mm		108.5	26.6		707
		" 10 mm.		85.4	20.5		313
		"		194.0	46.0		313
4271	C₇H₅Cl₃	α,α,α-Trichlorotoluene	220.9	Reacts			535
4272	C₇H₅N	Benzonitrile	191.3	186.5			563
4273	C₇H₆Cl₂	α,α-Dichlorotoluene	205.1	Nonazeotrope			542
4274	C₇H₆O	Benzaldehyde	179.2	<173.5	> 15		575
4275	C₇H₇Br	m-Bromotoluene	184.3	168.3	23		527
4276	C₇H₇Br	o-Bromotoluene	181.75	166.8	25		528
4277	C₇H₇Br	p-Bromotoluene	~ 184.5	169.2	30		573
4278	C₇H₇Cl	α-Chlorotoluene	179.3	~ 167.0	~ 30		530

TABLE I. *Binary Systems* 143

No.	Formula	Name	B.P., °C	B.P., °C	Wt.%A		Ref.
		B-Component		**Azeotropic Data**			
A =	**C₂H₆O₂**	**Glycol** *(continued)*	**197.4**				
4279	C₇H₇Cl	*o*-Chlorotoluene	159.2	152.5	13		567
4280	C₇H₇Cl	*p*-Chlorotoluene	162.4	155.0			573
4281	C₇H₇I	*p*-Iodotoluene	214.5	181.5	30		567
4282	C₇H₇NO₂	*m*-Nitrotoluene	230.8	192.5	57?		554
4283	C₇H₇NO₂	*o*-Nitrotoluene	221.75	188.55	48.5		554
4284	C₇H₇NO₂	*p*-Nitrotoluene	238.9	192.4	63.5		554
4285	C₇H₈	Toluene	110.75	110.20	6.5		573
		"	110.6	110.1	2.3		982
4286	C₇H₈O	Anisole	153.85	150.45	10.5		530
4287	C₇H₈O	Benzyl alcohol	205.25	193.35	53.5		549
		"	205.2	193.1	56		183
4288	C₇H₈O	*m*-Cresol	202.1	195.2	60		541
		"	202.4		61	v-l	730
4289	C₇H₈O	*o*-Cresol	191.1	189.6	27		574
		"	191	189.52	26		505
4290	C₇H₈O	*p*-Cresol	202.0		59.5	v-l	730
		"	201.6	195.2	53.5		528
4291	C₇H₈O₂	Guaiacol	205.1	190.4	46		556
4292	C₇H₈O₂	*m*-Methoxyphenol	243.8	195.5	~ 80		575
4293	C₇H₉N	2,6-Lutidine	142	Nonazeotrope			791
4294	C₇H₉N	*N*-Methylaniline,					
		" 31.8 mm.		95	22.9	v-l	181
		" 95.3 mm.		120	26.6	v-l	181
		" 244 mm.		145	30.0	v-l	181
		"	196.25	181.6	40.2		551
4295	C₇H₉N	*m*-Toluidine	200.3	188.55	42		551
4296	C₇H₉N	*o*-Toluidine	200.35	186.45	42.5		551
4297	C₇H₉N	*p*-Toluidine	200.55	187.0	27		551
4298	C₇H₉NO	*o*-Anisidine	219.0	<193.5	< 59		551
4299	C₇H₁₂O₄	Ethyl malonate	198.9	Reacts			563
4300	C₇H₁₃ClO₂	Isoamyl chloroacetate	195.0	<187.5	> 38		575
4301	C₇H₁₄	Methylcyclohexane	101.1	100.8	~ 4		537
4302	C₇H₁₄O₂	Amyl acetate	148.7	147.6	6		564
4303	C₇H₁₄O₂	Butyl propionate	146.8	<146.0	< 7		575
4304	C₇H₁₄O₂	Ethyl isovalerate	134.7	Nonazeotrope			575
4305	C₇H₁₄O₂	Ethyl valerate	145.45	144.7	3		564
4306	C₇H₁₄O₂	Isoamyl acetate	142.1	141.95	~ 3		536
4307	C₇H₁₄O₂	Isobutyl propionate	137.5	Nonazeotrope			575
4308	C₇H₁₄O₂	Methyl caproate	149.8	148.0	7		575
4309	C₇H₁₄O₂	Propyl butyrate	142.8	142.7	~ 3		536
4310	C₇H₁₄O₂	Propyl isobutyrate	134.0	Nonazeotrope			575
4311	C₇H₁₄O₃	1,3-Butanediol methyl ether acetate	171.75	<171.0	< 12		575
4312	C₇H₁₆	Heptane	98.45	97.9	3		563
4313	C₇H₁₆O	Heptyl alcohol	176.15	174.1	20		564

		B-Component		Azeotropic Data		
No.	Formula	Name	B.P., °C	B.P., °C	Wt.%A	Ref.
A =	C₂H₆O₂	**Glycol** *(continued)*	**197.4**			
4314	C₈H₇N	Indole	253.5	Nonazeotrope		575
4315	C₈H₈	Styrene	145.8	139.5	16.5	537
4316	C₈H₈O	Acetophenone	202.0	185.65	52	552
		" 5–6 mm.		63.0	19.2	581c
4317	C₈H₈O₂	Benzyl formate	202.3	Reacts		535
4318	C₈H₈O₂	Methyl benzoate	199.45	182.2	36.5	530
4319	C₈H₈O₂	Phenyl acetate	195.7	182.9	34	530
4320	C₈H₈O₃	Methyl salicylate	222.35	188.8	48	574
4321	C₈H₁₀	Ethylbenzene	136.15	133.0	13.5	537
4322	C₈H₁₀	*m*-Xylene	139.0	135.6	~ 15	574
		"	139.1	135.1	6.55	856
4323	C₈H₁₀	*o*-Xylene	144.4	135.7	6.9	982
		"	143.6	139.6	16	537
4324	C₈H₁₀	*p*-Xylene	138.3	136.95	14.5	537
		"	138.4	134.5	6.42	856
4325	C₈H₁₀O	Benzyl methyl ether	167.8	159.8	18	567
4325a	C₈H₁₀O	2,6-Dimethylphenol				
		" 25 mm.		102	20	738c
		" 50 mm.		116	24	738c
		" 100 mm.		132	28	738c
		" 200 mm.		149	32	738c
		" 735 mm.		189	40	738c
4326	C₈H₁₀O	*p*-Methylanisole	177.05	166.6	22.8	541
4327	C₈H₁₀O	*p*-Ethylphenol	218.8	Nonazeotrope		556
4328	C₈H₁₀O	Phenethyl alcohol	219.4	194.4	69	549
4329	C₈H₁₀O	Phenetole	170.45	161.45	19	545
4330	C₈H₁₀O	3,4-Xylenol	226.8	197.2	89	564
4331	C₈H₁₀O₂	*o*-Ethoxyphenol	216.5	192.6		545
4332	C₈H₁₀O₂	Veratrol	205.5	178.5	35	574
4333	C₈H₁₁N	Dimethylaniline	194.05	178.55	33.5	551
		" 39.3 mm.		95	17.6	v-l	181
		" 115 mm.		120	21.8	v-l	181
		" 293 mm.		145	26.5	v-l	181
4334	C₈H₁₁N	*s*-Collidine	171.3	170.5	9.7	505
		"	171	170.7	8.8	791
4335	C₈H₁₁N	2,4-Xylidine	214.0	188.6	47	551
4336	C₈H₁₁N	3,4-Xylidine	225.5	<189.0	<91.6	551
4337	C₈H₁₁N	Ethylaniline	205.5	183.7	43	551
4338	C₈H₁₁NO	*o*-Phenetidine	232.5	194.8	67.8	551
4339	C₈H₁₁NO	*p*-Phenetidine	249.9	197.35	97	551
4340	C₈H₁₂O₄	Ethyl fumarate	217.85	189.35	48.5	569
4341	C₈H₁₂O₄	Ethyl maleate	223.3	193.1	55	570
4342	C₈H₁₄O	Methyl heptenone	173.2	168.1	23	552
4343	C₈H₁₄O₄	Ethyl succinate	217.25	Reacts		535
4344	C₈H₁₆	1,3-Dimethylcyclohexane	120.7	119.2	9	575
4345	C₈H₁₆O	2-Octanone	172.85	168.0	20	552
4346	C₈H₁₆O₂	Butyl butyrate	166.4	160.6	16	564

TABLE I. *Binary Systems*   145

	B-Component			Azeotropic Data		
No.	Formula	Name	B.P., °C	B.P., °C	Wt.%A	Ref.
A =	C₂H₆O₂	**Glycol** *(continued)*	**197.4**			
4347	C₈H₁₆O₂	Isoamyl propionate	160.3	155.5	12	536
4348	C₈H₁₆O₂	Isobutyl butyrate	155.7	153.7	10	564
4349	C₈H₁₆O₂	Isobutyl isobutyrate	148.6	147.5	6	564
4350	C₈H₁₆O₂	Propyl isovalerate	155.7	~ 152	10	536
4351	C₈H₁₆O₄	2-(2-Ethoxyethoxy)ethyl acetate	218.5	195.0		575
4352	C₈H₁₈	2,5-Dimethylhexane	109.4	108.65	7.5	575
4353	C₈H₁₈	Octane	125.75	123.5	11.5	567
4354	C₈H₁₈O	Butyl ether	142.1	140.0	10	545
		"	142.1	139.5	6.4	981,982
4355	C₈H₁₈O	Isobutyl ether	122.1	121.9	7	576
4356	C₈H₁₈O	*n*-Octyl alcohol	195.2	184.35	36.5	549
4357	C₈H₁₈O	*sec*-Octyl alcohol	180.4	175.55	21	549
4358	C₈H₁₈O₂	2-(2-Butoxyethoxy)ethanol	230.4	196.2	72.5	129,808
4359	C₈H₁₈O₃	Bis(2-ethoxyethyl)ether	186	178.0	26.1	129
4360	C₈H₁₉NO	2-Diisopropylamino-ethanol, 10 mm.	79	74	10	982
		" 50 mm.	111	104	15	982
		" 100 mm.	127	121	18	982
4361	C₉H₇N	Isoquinoline	240.3	Nonazeotrope		575
4362	C₉H₇N	Quinoline	237.3	196.35	79.5	553
4363	C₉H₈	Indene	183.0	168.4	26	541
4364	C₉H₈O	Cinnamaldehyde	253.5	Nonazeotrope		575
4365	C₉H₉N	*β*-Methylindole	**266.5**	Nonazeotrope		575
4366	C₉H₁₀O	*p*-Methylacetophenone	226.35	192.2	60	552
4367	C₉H₁₀O	Propiophenone	217.7	190.2	57	552
4368	C₉H₁₀O₂	Benzyl acetate	214.9	186.5	45	536
4369	C₉H₁₀O₂	Ethyl benzoate	212.5	186.1	46	570
4370	C₉H₁₀O₃	Ethyl salicylate	234.0	190.7	51.5	542
4371	C₉H₁₂	Cumene	152.8	147.0	18	567
4372	C₉H₁₂	Mesitylene	164.6	156	13	573
4373	C₉H₁₂	Propylbenzene	158.8	152	19	526
4374	C₉H₁₂	Pseudocumene	168.2	<157.7	83.2	541
4375	C₉H₁₂O	**Benzyl ethyl ether**	**185.0**	169.0	22	545
4375a	C₉H₁₂O	α,α-Dimethylbenzyl alcohol 5–6 mm.		74.5	17.1	581c
4376	C₉H₁₂O	3-Phenylpropanol	235.6	195.5	75	550
4377	C₉H₁₂O	Phenyl propyl ether	190.2	171.0	26	538
4378	C₉H₁₃N	*N,N*-Dimethyl-*o*-toluidine	185.3	169.3	23	551
4379	C₉H₁₃N	*N,N*-Dimethyl-*p*-toluidine	210.2	182.0	47	551
4380	C₉H₁₄O	Phorone	197.8	184.5	50	552
		Phorone	197.8	184.5	42	181
4381	C₉H₁₈O	2,6-Dimethyl-4-heptanone	168.0	164.2	35	552
4382	C₉H₁₈O₂	Butyl isovalerate	177.6	169.0	23	564
4383	C₉H₁₈O₂	Ethyl enanthate	188.7	174.0	30	575
4384	C₉H₁₈O₂	Isoamyl butyrate	178.5	167.9	24.5	536
4385	C₉H₁₈O₂	Isoamyl isobutyrate	168.5	161.5	20	536

No.	Formula	B-Component Name	B.P., °C	Azeotropic Data B.P., °C	Wt.%A	Ref.
A =	**C₂H₆O₂**	**Glycol** *(continued)*	**197.4**			
4386	C₉H₁₈O₂	Isobutyl isovalerate	171.4	163.7	21.7	541
4387	C₉H₁₈O₂	Methyl caprylate	192.9	175.5	31	563
4388	C₉H₁₈O₃	Isobutyl carbonate	190.3	<180.5	28	567
4389	C₁₀H₇Br	1-Bromonaphthalene	281.8	194.95	71.2	541
4390	C₁₀H₇Cl	1-Chloronaphthalene	262.7	192.9	65.2	541
4391	C₁₀H₈	Naphthalene	218.05	183.9	51	528
		"	217.9	183.6	46	413
4392	C₁₀H₁₀O₂	Isosafrol	252.1	192.8	64	574
4393	C₁₀H₁₀O₂	Methyl cinnamate	261.9	196.2	85	574
4394	C₁₀H₁₀O₂	Safrol	235.9	190.05	55	574
4395	C₁₀H₁₀O₄	Methyl phthalate	283.7	Nonazeotrope		564
4396	C₁₀H₁₂O	Anethole	235.7	189.35	56	527
4397	C₁₀H₁₂O	Estragol	215.6	182.3	40	545
4398	C₁₀H₁₂O₂	Ethyl α-toluate	228.7	190.0	54	536
4399	C₁₀H₁₂O₂	Eugenol	255.0	196.8	90	556
4400	C₁₀H₁₂O₂	Isoeugenol	268.8	Nonazeotrope		575
4401	C₁₀H₁₂O₂	Propyl benzoate	230.85	190.35	55	570
4402	C₁₀H₁₄	Butylbenzene	183.1	166.2	27	567
4403	C₁₀H₁₄	Cymene	176.7	163.2	25.5	537
4404	C₁₀H₁₄O	Carvone	231.0	192.5	60.8	552
4405	C₁₀H₁₄O	Thymol	232.9	195.5	62	564
4406	C₁₀H₁₄O₂	m-Diethoxybenzene	235	192.5	53	538
4407	C₁₀H₁₅N	Diethylaniline	217.05	183.4	33	551
4408	C₁₆H₁₆	Camphene	159.5	152.5	20	528
4409	C₁₀H₁₆	d-Limonene	177.8	163.5	26	537
4410	C₁₀H₁₆	Nopinene	163.8	155.0	19	526
4411	C₁₀H₁₆	α-Pinene	155.8	149.5	18.5	540
4412	C₁₀H₁₆	α-Terpinene	173.4	161.0	23.5	567
4413	C₁₀H₁₆	γ-Terpinene	183	166.5	26	575
4414	C₁₀H₁₆	Terpinolene	184.6	167.4	28.5	567
4415	C₁₀H₁₆	Thymene	179.7	164.5	27.5	573
4416	C₁₀H₁₆O	Camphor	209.1	186.15	40	552
4417	C₁₀H₁₆O	Pulegone	223.8	191.2	58	552
4418	C₁₀H₁₈	m-Menthene-8	170.8	159.5	~ 21	575
4419	C₁₀H₁₈O	Borneol	215.0	189.25	54.2	549
4420	C₁₀H₁₈O	Cineol	176.4	164.75	~ 15	528
4421	C₁₀H₁₈O	Citronellal	207.8	~188.5	~ 53	574
4422	C₁₀H₁₈O	Geraniol	229.6	194.65	67.5	549
4423	C₁₀H₁₈O	Linalool	198.6	182.2	40	549
4424	C₁₀H₁₈O	Menthone	209.5	<190.0	< 62	552
4425	C₁₀H₁₈O	α-Terpineol	218.85	189.55	56	549
4426	C₁₀H₁₈O	β-Terpineol	210.5	188.4	50	527
4427	C₁₀H₁₈O₄	Propyl succinate	250.5	Nonazeotrope		575
4428	C₁₀H₂₀O	Citronellol	224.4	193.5	63	549
4429	C₁₀H₂₀O	Menthol	216.3	188.55	51.5	549
4430	C₁₀H₂₀OS	2-Hexylthioethyl vinyl ether		Min. b.p.		953

TABLE I.   *Binary Systems*   147

No.	Formula	B-Component Name	B.P., °C	Azeotropic Data B.P., °C	Wt.%A		Ref.
A =	$C_2H_6O_2$	**Glycol** *(continued)*	**197.4**				
4431	$C_{10}H_{20}O_2$	Ethyl caprylate	208.35	182.5	41		575
4432	$C_{10}H_{20}O_2$	Isoamyl isovalerate	192.7	114.85	27.2		541
4433	$C_{10}H_{20}O_2$	Methyl pelargonate	213.8	186.0	45		567
4434	$C_{10}H_{22}$	Decane	173.3	161.0	23		567
4435	$C_{10}H_{22}$	2,7-Dimethyloctane	160.1	<153.0	<21		575
4436	$C_{10}H_{22}O$	n-Decyl alcohol	232.8	193.0	67		545
4437	$C_{10}H_{22}O$	Amyl ether	187.5	168.8	26		556
4438	$C_{10}H_{22}O$	Isoamyl ether	172.6	162.8	19		573
4439	$C_{10}H_{22}O_3$	Isoamyl carbonate	232.2	188.45	49		566
4440	$C_{10}H_{22}O_4$	Tripropylene glycol methyl ether	243	192	82	v-l	215
		"		138.5	77.2		215
		"		111.5	75.1		215
4441	$C_{11}H_{10}$	1-Methylnaphthalene	245.1	190.25	60		574
4442	$C_{11}H_{10}$	2-Methylnaphthalene	241.15	189.1	57.2		527
4443	$C_{11}H_{12}O_2$	Ethyl cinnamate	272.0	197.0	72		567
4444	$C_{11}H_{14}O_2$	1-Allyl 3,4-dimethoxy-benzene	255.2	195.1	68.5		574
4445	$C_{11}H_{14}O_2$	Butyl benzoate	251.2	193.2	68		535
4446	$C_{11}H_{14}O_2$	**1,2-Dimethoxy-4-propenyl-**benzene	270.5	196.5	80		545
4447	$C_{11}H_{14}O_2$	Isobutyl benzoate	242.15	192.0	63		574
4448	$C_{11}H_{16}O$	Methyl thymyl ether	216.5	183.0	40		575
4449	$C_{11}H_{20}O$	Methyl isobornyl ether	192.2	191	<25		563
4450	$C_{11}H_{20}O$	Methyl terpineol ether	216.2	184.5	40		538
4451	$C_{11}H_{22}O_2$	Ethyl pelargonate	227	190.8			575
4452	$C_{11}H_{22}O_3$	Isoamyl carbonate	232.2	188.45	46		568
4453	$C_{12}H_{10}$	Acenaphthene	277.9	194.65	74.2		541
4454	$C_{12}H_{10}$	Biphenyl	256.1	192.25	66.5		568
4455	$C_{12}H_{10}O$	Phenyl ether	259.3	193.05	60		542
		" 50 mm.	161.0	120.4	62.3		982
		"	259.3	192.3	64.5		181
4456	$C_{12}H_{16}O_2$	Isoamyl benzoate	262.05	193.95	66.2		574
4457	$C_{12}H_{18}$	1,3,5-Triethylbenzene	215.4	183	49		573
4458	$C_{12}H_{20}O_2$	Bornyl acetate	227.6	190.0	53		536
4459	$C_{12}H_{22}O$	Bornyl ethyl ether	204.9	177.0	34		567
4460	$C_{12}H_{22}O$	Ethyl isobornyl ether	203.8	176.5	33		556
4461	$C_{12}H_{26}$	Dodecane, 748 mm.	216	179			424
		" 200 mm.		142			424
		" 150 mm.		135			424
		" 100 mm.		125			424
		" 50 mm.		110			424
4462	$C_{12}H_{26}O$	Hexyl ether, 50 mm.	137.0	112.8	35.6		982
4463	$C_{13}H_{10}$	Fluorene	296.4	196.0	82		564
4464	$C_{13}H_{10}O_2$	Phenyl benzoate	315	Nonazeotrope			575
4465	$C_{13}H_{12}$	Diphenylmethane	265.6	193.3	68.5		574

No.	Formula	B-Component Name	B.P., °C	Azeotropic Data B.P., °C	Wt.%A	Ref.
A =	**C₂H₆O₂**	**Glycol** *(continued)*	**197.4**			
4466	C₁₃H₁₂O	Benzyl phenyl ether	286.5	195.5	87	567
4467	C₁₃H₂₈	Tridecane	234.0	188.0	55	526
4468	C₁₄H₁₀	Anthracene	340	197	98.3	887
4469	C₁₄H₁₂	Stilbene	306.4	196.8	87	564
4470	C₁₄H₁₂O₂	Benzyl benzoate	324	Nonazeotrope		575
4471	C₁₄H₁₄	1,2-Diphenylethane	284	195.2	77	537
4472	C₁₄H₁₄O	Benzyl ether	297	<196.5	<96	575
4473	C₁₄H₃₀	Tetradecane, 748 mm.	252.5	187.5		424
		" 200 mm.		150.5		424
		" 133 mm.		142.5		424
		" 118 mm.		118		424
4474	C₁₆H₃₄O	2-Ethylhexyl ether,				
		" 10 mm.	135	87		982
A =	**C₂H₆S**	**Ethanethiol**	**35.8**			
4475	C₂H₆S	Methyl sulfide	37.4	< 34.8	< 62	575
4476	C₃H₆O	Acetone	56.25	Nonazeotrope		563
4477	C₃H₇Cl	2-Chloropropane	36.25	36.15	~ 45	563
4478	C₃H₇NO₂	Propyl nitrite	47.75	Nonazeotrope		566
4479	C₃H₈O₂	Methylal	42.3	34.5	>80	566
4480	C₄H₁₀O	Ethyl ether	34.6	34.0	35	566
4481	C₅H₈	Isoprene	34.1	Reacts		563
4482	C₅H₁₀	Cyclopentane	49.263	34.95	89	202
4483	C₅H₁₀	2-Methyl-2-butene	37.15	32.95	~ 60	563
4484	C₅H₁₀	3-Methyl-1-butene	20.6	Nonazeotrope		575
4485	C₅H₁₂	2-Methylbutane	27.854	25.72	29	202
		"	27.95	Nonazeotrope		563
4486	C₅H₁₂	Pentane	36.074	30.46	51	202,563
4487	C₆H₁₄	2,2-Dimethylbutane	49.743	34.41	83	202
A =	**C₂H₆S**	**Methyl Sulfide**	**37.3**			
4488	C₃H₆O	Acetone	56.25	Nonazeotrope		563
4489	C₃H₇Cl	1-Chloropropane	46.6	Nonazeotrope		563
4490	C₃H₇Cl	2-Chloropropane	36.25	36		563
4491	C₃H₇NO₂	Isopropyl nitrite	40.1	< 36.6	> 19	566
4492	C₃H₇NO₂	Propyl nitrite	47.75	Nonazeotrope		550
4493	C₃H₈O	Isopropyl alcohol	82.4	Nonazeotrope		566
4494	C₃H₈O₂	Methylal	42.25	35.7		555
4495	C₄H₈O	2-Butanone	79.6	Nonazeotrope		575
4496	C₄H₉ClO	Chloroethyl ethyl ether	98.5	Nonazeotrope		566
4497	C₄H₁₀O	tert-Butyl alcohol	82.45	Nonazeotrope		566
4498	C₄H₁₀O	Ethyl ether	34.6	34.0	20	566
4499	C₄H₁₀O	Methyl propyl ether	38.95	< 37.0	> 65	575
4500	C₅H₈	Isoprene	34.3	32.5	35	566
4501	C₅H₁₀	Cyclopentane	49.35	37.09	87.5	205
4502	C₅H₁₀	2-Methyl-2-butene	38.60	34.83	53.6	205,555
4503	C₅H₁₀	2-Methyl-1-butene	31.25	30.64	17.0	205

TABLE I.   *Binary Systems*   149

No.	Formula	B-Component Name	B.P., °C	Azeotropic Data B.P., °C	Wt.%A	Ref.
A =	**C$_2$H$_6$O$_2$**	**Methyl Sulfide** *(continued)*	**37.3**			
4504	C$_5$H$_{10}$	3-Methyl-1-butene	20.6	Nonazeotrope		566
4505	C$_5$H$_{12}$	2-Methylbutane	27	27.3	15	531
		2-Methylbutane	27.90	26.62	25.0	205
4506	C$_5$H$_{12}$	Pentane	36.15	31.80	46.6	205
		"	36.15	~ 33.5	~ 45	531
4507	C$_6$H$_{14}$	2,3-Dimethylbutane	58.0	Nonazeotrope		566
4508	C$_6$H$_{14}$	2,2-Dimethylbutane	49.70	36.50	79.8	205
A =	**C$_2$H$_6$SO$_4$**	**Methyl Sulfate**	**189.1**			
4509	C$_5$H$_{10}$O$_2$	Isovaleric acid	176.5	<175.0	< 40	575
4510	C$_5$H$_{10}$O$_2$	Valeric acid	186.35	<182	< 60	575
4511	C$_6$H$_4$Cl$_2$	*p*-Dichlorobenzene	174.6	Nonazeotrope		547
4512	C$_6$H$_5$Br	Bromobenzene	156.1	Nanozeotrope		547
4513	C$_6$H$_5$I	Iodobenzene	188.45	<184	> 50	547
4514	C$_6$H$_6$O	Phenol	181.5	Reacts		563
4515	C$_6$H$_{10}$O$_4$	Ethyl oxalate	185	Nonazeotrope		563
4516	C$_6$H$_{13}$Br	1-Bromohexane	156.5	Nanozeotrope		575
4517	C$_7$H$_7$Br	*m*-Bromotoluene	184.3	<181.5	< 27	575
4518	C$_7$H$_7$Br	*o*-Bromotoluene	181.5	<179.5	< 28	575
4519	C$_7$H$_7$Br	*p*-Bromotoluene	185	181.5		563
4520	C$_7$H$_7$Cl	α-Chlorotoluene	179.35	Nonazeotrope		563
4521	C$_7$H$_7$Cl	*p*-Chlorotoluene	162.4	Nonazeotrope		547
4522	C$_7$H$_8$O	*m*-Cresol	202.2	Reacts		542
4523	C$_8$H$_{10}$O	Phenetole	170.45	Nonazeotrope		557
4524	C$_9$H$_{12}$O	Benzyl ethyl ether	185.0	<182.8	< 47	557
4525	C$_9$H$_{18}$O$_2$	Ethyl enanthate	188.7	185.5	38	575
4526	C$_9$H$_{18}$O$_2$	Isoamyl butyrate	181.05	179.5	18	575
4527	C$_{10}$H$_{16}$	*d*-Limonene	177.8	~173		563
4528	C$_{10}$H$_{20}$O$_2$	Isoamyl isovalerate	192.7	185.8	63	549
4529	C$_{11}$H$_{20}$O	Isobornyl methyl ether	192.2	<185.5	< 70	557
A =	**C$_2$H$_6$S$_2$**	**Methyl Disulfide**	**109.44**			
4530	C$_5$H$_6$S	2-Methylthiophene	111.92	Nonazeotrope		205
4531	C$_5$H$_{12}$S	Ethyl isopropyl sulfide	107.22	106.37		205
4532	C$_7$H$_8$	Toluene	110.85	108.93		205
4533	C$_7$H$_{14}$	Methylcyclohexane	101.05	98.92	28.6	205
4534	C$_7$H$_{16}$	Heptane	98.40	96.44	26.3	205
4535	C$_8$H$_{16}$	*trans*-1,3-Dimethyl-cyclohexane	120.30	107.22	73.3	205
4536	C$_8$H$_{18}$	2,3-Dimethylhexane	109.15	102.84	48.2	205
4537	C$_8$H$_{18}$	2-Methylheptane	117.70	106.22	69.5	205
A =	**C$_2$H$_7$N**	**Dimethylamine**	**6.8**			
4538	C$_3$H$_9$N	Trimethylamine	3.5	3	26	818
		" 107 p.s.i.g.		73	72	818
		" 370 p.s.i.g.		Nonazeotrope		30,31;818
		"	3.5	Nonazeotrope		353
4539	C$_4$H$_6$	1,3-Butadiene	— 4.6	Nonazeotrope		305
4540	C$_4$H$_8$	1-Butene	— 6.3	Nonazeotrope		305

No.	Formula	B-Component Name	B.P., °C	B.P., °C	Azeotropic Data Wt.%A	Ref.
A =	**C₂H₇S**	**Dimethylamine** *(continued)*	**6.8**			
4541	C₄H₈	*cis*-2-Butene	3.7	Nonazeotrope		305
4542	C₄H₈	*trans*-2-Butene	0.4	Nonazeotrope		305
4543	C₄H₈	2-Methylpropene	— 6.9	Nonazeotrope		305
4544	C₄H₁₀	Butane	0.6	<0.2	< 12	575
		Butane	— 0.5	Nonazeotrope		305
4545	C₄H₁₀	2-Methylpropane	— 11.7	Nonazeotrope		305
4546	C₄H₁₁NO	2-(Dimethylamino)-ethanol	134.6	Nonazeotrope		981
4547	C₅H₁₀	3-Methyl-1-butene	20.6	Nonazeotrope		575
A =	**C₂H₇N**	**Ethylamine**	**16.55**			
4548	C₃H₆O	Acetone	56.15	Nonazeotrope		575
4549	C₃H₆O	Propylene oxide	34.1	Nonazeotrope		575
4550	C₄H₄O	Furan	31.7	Nonazeotrope		551
4551	C₄H₁₀O	Ethyl ether	34.6	Nonazeotrope		551
4552	C₄H₁₀O	Methyl propyl ether	38.95	Nonazeotrope		551
4553	C₄H₁₁N	Diethylamine	55.5	Nonazeotrope		981
4554	C₄H₁₁NO₂	2,2'-Iminodiethanol	268	Nonazeotrope		981
4555	C₅H₁₀	3-Methyl-1-butene	20.6	<15.4	> 54	551
4556	C₅H₁₂	2-Methylbutane	27.95	Nonazeotrope		551
A =	**C₂H₇NO**	**2-Aminoethanol**	**171.0**			
4557	C₃H₆O	Acetone	56.1	Nonazeotrope		981
4558	C₃H₇NO	Propionamide	222.2	Nonazeotrope		527
4559	C₃H₇NO₂	Ethyl carbamate	185.25	Reacts		527
4560	C₄H₈Cl₂O	Bis(2-chlorethyl) ether	178.65	Reacts		527
4561	C₄H₁₀O₂	2-Ethoxyethanol	135.3	Nonazeotrope		551
4562	C₄H₁₁NO₂	2,2'-Iminodiethanol, 10 mm.	150	Nonazeotrope		981
4563	C₅H₈O	Cyclopentanone	130.65	Nonazeotrope		551
4564	C₅H₁₂O₂	2-Propoxyethanol	151.35	Nonazeotrope		551
4565	C₅H₁₂O₃	2-(2-Methoxyethoxy)-ethanol	192.95	Nonazeotrope		551
4566	C₆H₄Cl₂	*o*-Dichlorobenzene	179.5	157.3	40	551
4567	C₆H₄Cl₂	*p*-Dichlorobenzene	174.4	154.6	35	569
4568	C₆H₅Br	Bromobenzene	156.1	145.0	22	551
4569	C₆H₅Cl	Chlorobenzene	131.75	128.55	13.5	551
			132	124		906
4570	C₆H₅I	Iodobenzene	188.45	161.0	45	551
4571	C₆H₅NO₂	Nitrobenzene	210.75	Nonazeotrope		575
4572	C₆H₆	Benzene	80.15	Nonazeotrope		551
4573	C₆H₆O	Phenol	182.2	Nonazeotrope		551
4574	C₆H₇N	Aniline	184.35	170.3	90	551
4575	C₆H₁₀O	Cyclohexanone	155.7	Nonazeotrope		551
4576	C₆H₁₀S	Allyl sulfide	139	137.2	8	555
4577	C₆H₁₁NO₂	Nitrocyclohexane	205.3	Nonazeotrope		575
4578	C₆H₁₂	Cyclohexane	80.75	Nonazeotrope		551
4579	C₆H₁₂	Methylcyclopentane	72.0	Nonazeotrope		575
4580	C₆H₁₄	Hexane	68.8	Nonazeotrope		551

TABLE I. *Binary Systems* 151

No.	Formula	Name	B.P., °C	B.P., °C	Wt.%A	Ref.
		B-Component		**Azeotropic Data**		

No.	Formula	Name	B.P., °C	B.P., °C	Wt.%A	Ref.
A =	C_2H_7NO	**2-Aminoethanol** *(continued)*	**171.0**			
4581	$C_6H_{14}O_2$	2-Butoxyethanol	171.15	166.95	43	551
4582	$C_6H_{14}S$	Propyl sulfide	141.5	<139.7	<13	566
4583	C_7H_7Br	*m*-Bromotoluene	184.3	159.3	44	551
4584	C_7H_7Br	*o*-Bromotoluene	181.5	157.8	42	551
4585	C_7H_7Cl	*o*-Chlorotoluene	159.2	146.5	26	551
4586	C_7H_7Cl	*p*-Chlorotoluene	162.4	148.2	28	551
4587	$C_7H_7NO_2$	*o*-Nitrotoluene	221.75	Nonazeotrope		575
4588	C_7H_8O	Anisole	153.85	145.75	25.5	551
4589	C_7H_8O	*o*-Cresol	191.1	Nonazeotrope		551
4590	C_7H_8O	*p*-Cresol	201.7	Nonazeotrope		551
4591	C_7H_9N	Methylaniline	196.25	167.5	70	551
4592	C_7H_9N	*o*-Toluidine	200.35	Nonazeotrope		551
4593	C_7H_{14}	Methylcyclohexane	101.15	<100.5	<10	551
4594	$C_7H_{14}O$	4-Heptanone	143.55	Nonazeotrope		551
4595	C_7H_{16}	Heptane	98.4	<98.0		575
4596	C_8H_8O	Acetophenone	202.0	Nonazeotrope		551
4597	C_8H_{10}	Ethylbenzene	136.15	131.0	15	551
4598	C_8H_{10}	*m*-Xylene	139.2	133.0	18	551
4599	C_8H_{10}	*o*-Xylene	144.3	<138.0	20	551
4600	$C_8H_{10}O$	Benzyl methyl ether	167.8	150.5	28	551
4601	$C_8H_{10}O$	*p*-Methylanisole	177.05	154.5	37	551
4602	$C_8H_{10}O$	Phenetole	170.45	151.0	30	551
4603	$C_8H_{11}N$	Dimethylaniline	194.15	163.5	55	551
4604	$C_8H_{11}N$	2,4-Xylidine	214.0	Nonazeotrope		551
4605	$C_8H_{11}N$	Ethylaniline	206.05	<170.0		575
4606	C_8H_{18}	*n*-Octane	125.75	<123.0	<16	551
4607	$C_8H_{18}O$	Butyl ether	142.4	136.5	16	551
4608	$C_8H_{18}O$	Isobutyl ether	122.3	Nonazeotrope		551
4609	$C_8H_{18}S$	Butyl sulfide	185.0	<164.5	<53	566
4610	$C_8H_{18}S$	Isobutyl sulfide	172.	156.0	33	555
4611	C_9H_8	Indene	187.4		Min. b.p.	276
4612	C_9H_{12}	Cumene	152.8	142.5		575
4613	C_9H_{12}	Mesitylene	164.6	148.5	30	551
4614	C_9H_{12}	Propylbenzene	159.3	<147.0	<30	551
4615	$C_9H_{12}O$	Benzyl ethyl ether	185.0	159.8	45	551
4616	$C_9H_{12}O$	Phenyl propyl ether	190.5	162.5	55	551
4617	$C_9H_{13}N$	*N,N*-Dimethyl-*o*-toluidine	185.3	161.0	50	551
4618	$C_9H_{13}N$	*N,N*-Dimethyl-*p*-toluidine	210.2	<169.0	>75	551
4619	$C_{10}H_8$	Naphthalene	218.0	Nonazeotrope		551
4620	$C_{10}H_{14}$	Butylbenzene	183.1	<158.5	<48	551
4621	$C_{10}H_{14}$	Cymene	176.7	154.7	37	551
4622	$C_{10}H_{15}N$	Diethylaniline	217.05	<169.0	>82	551
4623	$C_{10}H_{16}$	Camphene	159.6	144.0	28	551
4624	$C_{10}H_{16}$	*α*-Pinene	155.8	142.0	25	551
4625	$C_{10}H_{16}$	*α*-Terpinene	173.4	<154.0	<36	551
4626	$C_{10}H_{16}$	Dipentene	177.7	153.0	37	551

No.	Formula	Name	B.P., °C	B.P., °C	Wt.%A	Ref.
		B-Component		**Azeotropic Data**		

No.	Formula	Name	B.P., °C	B.P., °C	Wt.%A		Ref.
A =	**C₂H₇NO**	**2-Aminoethanol** *(continued)*	**171.0**				
4627	C₁₀H₁₈O	Cineol	176.35	153.4	36		551
4628	C₁₀H₂₂O	Amyl ether	187.5	<160.0	<50		551
4629	C₁₀H₂₂	Isoamyl ether	173.2	149.5	30.5		551
4630	C₁₁H₁₀	1-Methylnaphthalene	244.6	Nonazeotrope			551
4631	C₁₁H₁₀O	2-Methylnaphthalene	241.15	Nonazeotrope			527
4632	C₁₁H₂₀O	Isobornyl methyl ether	192.4	<165.0	<62		551
4633	C₁₂H₁₀	Acenaphthene	277.9	Nonazeotrope			575
4634	C₁₃H₁₂	Diphenylmethane	265.4	Nonazeotrope			551
A =	**C₂H₇PO₃**	**Dimethylphosphite**					
4635	C₄H₇NO	α-Hydroxyiso-butyronitrile		72.5	46.3		24
A =	**C₂H₈N₂**	**Ethylenediamine**	**116.5**				
4636	C₃H₈O₂	2-Methoxyethanol	124.5	130.0	31-32		129
4637	C₄H₈O₂	*p*-Dioxane	101.3	Nonazeotrope			981
4638	C₄H₁₀O	Butyl alcohol	117.7	124.7	35.7		982
4639	C₄H₁₀O	Isobutyl alcohol	107.9	120.5	50		982
4640	C₆H₆	Benzene	80.1	Nonazeotrope			981
4641	C₇H₈	Toluene	110.7	104	30.8		215
		"		103	30		981
4642	C₈H₆	Phenylacetylene	142	min. b.pt.			246
4643	C₈H₈	Styrene	144.7	min. b.pt.			246
4644	C₈H₁₀	Ethylbenzene	136.2	min. b.pt.			246
		"	136	Min. b.p.			341
4645	C₈H₁₀	*m*-Xylene	139	Min. b.p.			341
4646	C₈H₁₀	*o*-Xylene	143.6	Min. b.p.			341
4647	C₈H₁₀	*p*-Xylene	138.4	Min. b.p.			341
4648	C₈H₁₀	Xylenes	140	min. b.pt.			246
4649	CₙH₂ₙ₊₂	Paraffins		Min. b.p.			341
A =	**C₃F₇I**	**Heptafluoro-1-iodopropane**					
4649a	C₄H₁₀O	Ethyl ether	34.6	32–35			637c
A =	**C₃HF₅O₂**	**Pentafluoropropionic Acid**					
4650	C₆F₁₄	Perfluorohexane, 25°C.		Nonazeotrope		v-l	695
A =	**C₃H₂ClF₃O₂**	**3-Chloro-2,2,3-trifluoropropionic Acid**					
4651	C₃H₇NO	*N,N,*-Dimethylform-amide, 20 mm.		115-120			250
A =	**C₃H₂F₄O₂**	**2,2,3,3-Tetrafluoropropionic Acid**					
4652	C₃H₇NO	*N,N,*-Dimethylform-amide, 20 mm.		40	67		250
A =	**C₃H₂F₆O**	**1,1,1,3,3,3-Hexafluoro-2-propanol**					
4652a	C₆H₆	Benzene	167.7 mm.	25.0	82.5	v-l	674f
		"	344.5 mm.	40.0	89.2	v-l	674f
		"	656.7 mm.	55.0	95.8	v-l	674f
		"	1100 mm.	Nonazeotrope			674f

TABLE I. *Binary Systems* 153

	B-Component			Azeotropic Data		
No.	Formula	Name	B.P., °C	B.P., °C	Wt.%A	Ref.

A = C₃H₃ClF₃NO 3-Chloro-2,2,3-trifluoropropionamide

4653	C₃H₇NO	N,N,-Dimethylform-				
		amide, 20 mm.		101-106		250

A = C₃H₃Cl₃O Methyl Trichloroacetate 152.8

4654	C₃H₈O	Propyl alcohol	97.2	Nonazeotrope		575
4655	C₄H₈O₂	Butyric acid	164.0	Nonazeotrope		575
4656	C₄H₈O₂	Isobutyric acid	154.6	151.0		562
4657	C₅H₁₀O₃	Ethyl lactate	155	Azeotrope doubtful		563
4658	C₆H₁₂O	Cyclohexanol	160.8	<151.0	>72	575
4659	C₇H₈O	Anisole	153.85	149	>60	563
4660	C₇H₁₄O₂	Propyl butyrate	143	Azeotrope doubtful		563

A = C₃H₃F₄NO 2,2,3,3-Tetrafluoropropionamide

4661	C₃H₇NO	N,N,-Dimethylform-				
		amide, 20 mm.		91-4	66	250
		"	153	187		250

A = C₃H₃N Acrylonitrile 77.3

4662	C₃H₄O	Acrolein	200 mm.		Nonazeotrope	v-l	904
			400 mm.		"	v-l	904
			600 mm.		"	v-l	904
			760 mm.		"	v-l	904
		"	200–760 mm.		Nonazeotrope	v-l	905f
		"		52.4	Nonazeotrope	v-l	904,905c
4663	C₃H₅N	Propionitrile		97.4	Nonazeotrope		981
4664	C₃H₈O	Isopropyl alcohol		82.55	71.7	56	215
4665	C₃H₉ClSi	Chlorotrimethylsilane		57.5	57	7	841,843
4666	C₄H₅Cl	2-Chloro-1,3-					
		butadiene	150 mm.		15.1	7.4 v-l	420
4666a	C₄H₆O₂	Vinyl acetate		72	69.2	35.8 v-l	335c
4666b	C₅H₁₂	Pentane	60°–120°C.			v-l	1049e
4667	C₆H₆	Benzene		80.2	73.3	47	215

A = C₃H₄ Propadiene — 32

4668	C₃H₄	Propyne		— 23.2	Nonazeotrope		981
4669	C₃H₆	Propene		— 47.7	Nonazeotrope		981
		"	0°–20.1°C.		Nonazeotrope	v-l	354c
4670	C₃H₈	Propane		— 42.1	— 42	11.6 vol. %	981
		"	4.9 kg./cm²		0	18.5 v-l	354c
		"	8.7 kg./cm²		20.1	24.2 v-l	354c
4671	C₄H₆	1,3-Butadiene		— 4.5	Nonazeotrope		981

A = C₃H₄ Propyne — 23.2

4672	C₃H₆	Propene		— 47.7	Nonazeotrope		981
4673	C₃H₈	Propane		— 42.1	— 42	11.7 vol. %	981
		"	322.5 p.s.i.g.	62.1	60.1	14.3	1006
4674	C₄H₆	1,3-Butadiene		— 4.5	Nonazeotrope		981

No.	B-Component Formula	Name	B.P., °C	Azeotropic Data B.P., °C	Wt.%A	Ref.
A =	C₃H₄Br₂	*cis*-1,2-Dibromopropene	135.2			
4675	C₃H₈O	Propyl alcohol	97.2	97.05	3.45	563
A =	C₃H₄Br₂	*trans*-1,2-Dibromopropene	125.95			
4676	C₃H₈O	Propyl alcohol	97.2	95.75	41.95	563
A =	C₃H₄Cl₂	1,3-Dichloropropene				
4677	C₃H₅Cl	3-Chloropropene	45.7	Nonazeotrope	v-l	1024
A =	C₃H₄Cl₄	1,1,2,2-Tetrachloropropane	153			
4678	C₆H₁₀O	Cyclohexanone	156	Max. b.p.		262
4679	C₇H₈O	Anisole	155	Max. b.p.		262
4680	C₇H₁₄O	Heptaldehyde	155	Max. b.p.		262
4681	C₇H₁₄O	2-Heptanone	150	Max. b.p.		262
A =	C₃H₄Cl₄	1,1,2,3-Tetrachloropropane	180			
4682	C₇H₆O	Benzaldehyde	179	Max. b.p.		262
4683	C₉H₁₈O	2,6-Dimethyl-4-heptanone	165	Max. b.p.		262
A =	C₃H₄Cl₄	Tetrachloropropane				
4684	C₅H₈Cl₄	Tetrachloropentane, 12-150 mm.		Nonazeotrope	v-l	718
A =	C₃H₄O	Acrolein	52.45			
4685	C₃H₆	Propene	— 48	Nonazeotrope		64
4686	C₃H₆O	Acetone 200 mm.	22.5	Nonazeotrope	v-l	469
4687	C₃H₆O	Propionaldehyde	48.7	Nonazeotrope		575
4688	C₄H₈O	2-Butanone	79.6	Nonazeotrope	v-l	998
4689	C₄H₈O	Isobutyraldehyde	63.5	Nonazeotrope		575
4690	C₅H₁₂	Pentane	36.15	Nonazeotrope		575
4691	C₆H₁₄	Hexane	68.8	Nonazeotrope		575
A =	C₃H₄O	2-Propyn-1-ol	115			
4692	C₆H₆	Benzene	80.1	78	9	264
		"	80.1		v-l	885
A =	C₃H₄O₂	Acrylic Acid	140.5			
4693	C₃H₆O₂	Propionic acid	140.7	140.3?		563
4694	C₅H₈O₂	Ethyl acrylate	99.3	Nonazeotrope		981
4695	CₙHₘ	Hydrocarbon	138-140	133	68.2	804
A =	C₃H₄O₃	Pyruvic Acid	166.8			
4696	C₃H₆O₂	Propionic acid	141.3	Nonazeotrope		552
4697	C₄H₈O₂	Butyric acid	164.0	162.4	34	569
4698	C₅H₁₀O₃	2-Methoxy ethyl acetate	144.6	Nonazeotrope		552
4699	C₆H₅Br	Bromobenzene	156.1	147.0	34	527
4700	C₆H₅Cl	Chlorobenzene	131.75	128.6	15	552
4701	C₆H₆	Benzene	80.15	Nonazeotrope		552
4702	C₆H₁₂O₃	2-Ethoxyethyl acetate	156.8	Nonazeotrope		552
4703	C₇H₇Cl	*o*-Chlorotoluene	159.2	149.5	37	552
4704	C₇H₇Cl	*p*-Chlorotoluene	162.4	151.5	40	552
4705	C₇H₈	Toluene	110.75	110.05	7.5	552
4706	C₇H₈O	Anisole	153.85	148.5	28	552
4707	C₈H₁₀	Ethylbenzene	136.15	130.5	22	552

TABLE I. *Binary Systems* 155

No.	Formula	B-Component Name	B.P., °C	Azeotropic Data B.P., °C	Wt.%A		Ref.
A =	C₃H₄O₂	**Pyruvic Acid** *(continued)*	**166.8**				
4708	C₈H₁₀	*m*-Xylene	139.2	132.85	24		552
4709	C₈H₁₀	*o*-Xylene	144.3	137.0	28		552
4710	C₈H₁₈O	Butyl ether	142.4	138.0	15		552
4711	C₉H₁₂	Cumene	152.8	143.0	33		552
4712	C₉H₁₂	Mesitylene	164.6	151.2	40		552
4713	C₉H₁₂	Propylbenzene	159.3	147.6	37		552
A =	C₃H₄N₂	**Pyrazole**	**187.5**				
4714	C₄H₈Cl₂O	Bis(2-chloroethyl) ether	178.65	Nonazeotrope			575
4715	C₆H₆O	Phenol	182.2	Nonazeotrope			575
4716	C₇H₈O	*o*-Cresol	191.1	>194.8	> 26		575
4717	C₈H₁₀O	*p*-Methylanisole	177.05	Nonazeotrope			575
4718	C₈H₁₈S	Butyl sulfide	185.0	<181.2	> 28		575
4719	C₉H₁₂O	Benzyl ethyl ether	185.0	<184.2	> 20		575
4720	C₁₀H₂₂O	Isoamyl ether	173.2	Nonazeotrope			575
A =	C₃H₅Br	**3-Bromopropene**	**70.5**				
4721	C₃H₆O	Acetone	56.15	56.05	8		552
4722	C₃H₆O	Allyl alcohol	96.85	< 69.2	92		575
4723	C₃H₆O₂	Ethyl formate	54.15	Nonazeotrope			547
4724	C₃H₆O₂	Methyl acetate	57.0	Nonazeotrope			547
4725	C₃H₇Br	1-Bromopropane	71.0	Nonazeotrope			575
4726	C₃H₈O	Isopropyl alcohol	82.45	66.5	80		573
4727	C₃H₈O	Propyl alcohol	97.2	69.0	90		573
4728	C₃H₉BO₃	Methyl borate	68.7	67.5			542
4729	C₄H₈O	2-Butanone	79.6	Nonazeotrope			557
4730	C₄H₈O₂	Ethyl acetate	77.15	Nonazeotrope			547
4731	C₄H₈O₂	Propyl formate	80.85	Nonazeotrope			575
4732	C₄H₉Cl	1-Chloro-2-methylpropane	68.85	68.75	15		549
4733	C₄H₉NO₂	Isobutyl nitrite	67.1	66.9	12		550
4734	C₄H₁₀O	Butyl alcohol	117.8	Nonazeotrope			575
4735	C₄H₁₀O	*tert*-Butyl alcohol	82.45	< 68.5	< 90		575
4736	C₄H₁₀O	Isobutyl alcohol	108.0	Nonazeotrope			575
4737	C₄H₁₀O₂	Acetaldehyde dimethyl acetal	64.3	Nonazeotrope			559
4738	C₄H₁₀O₂	Ethoxymethoxymethane	65.9	Nonazeotrope			559
4739	C₆H₆	Benzene	80.15	Nonazeotrope			575
4739a	C₆H₁₂	1-Hexene	200 mm.	27.2	—	v-l	498f
		"	400 mm.	43.8	26.0	v-l	498f
		"	600 mm.	55.6	28.1	v-l	498f
		"	760 mm.	62.8	29.0	v-l	498f
4740	C₆H₁₄	Hexane	68.8	66.9	45		562
4740a	C₇H₁₄	1-Heptene	400 mm.	51.7	99.3	v-l	498f
		"	600 mm.	63.5	99.4	v-l	498f
		"	760 mm.	70.3	99.7	v-l	498f
A =	C₃H₅BrO	**Epibromohydrin**	**138.5**				
4741	C₃H₆O₂	Propionic acid	141.3	<138.0	< 88		575
		"	141.3	Nonazeotrope			556
4742	C₃H₈O	Propyl alcohol	97.2	Nonazeotrope			575

No.	Formula	Name	B.P., °C	B.P., °C	Wt.%A	Ref.

| | | **B-Component** | | **Azeotropic Data** | | |

No.	Formula	Name	B.P., °C	B.P., °C	Wt.%A	Ref.
A =	C_3H_5BrO	**Epibromohydrin** (continued)	**138.5**			
4743	$C_4H_8O_2$	Isobutyric acid	154.6	Nonazeotrope		556
4744	C_4H_9Br	1-Bromobutane	101.5	Nonazeotrope		575
4745	$C_4H_{10}O$	Butyl alcohol	117.8	117.0	20	556
4746	$C_4H_{10}O$	Isobutyl alcohol	108.0	Nonazeotrope		575
4747	$C_5H_{10}O_3$	2-Methoxyethyl acetate	144.6	<137.5		575
4748	$C_5H_{11}I$	1-Iodo-3-methylbutane	147.65	<136.0	>75	575
4749	$C_5H_{12}O$	Isoamyl alcohol	131.9	129.5	40	575
4750	$C_6H_{10}O$	Mesityl oxide	129.45	Nonazeotrope		552
4751	$C_6H_{10}S$	Allyl sulfide	139.35	133.3	60	566
4752	$C_7H_{14}O$	4-Heptanone	143.55	Nonazeotrope		575
4753	C_8H_{10}	Ethylbenzene	136.15	133.3	40	575
4754	C_8H_{10}	m-Xylene	139.2	134.5	55	575
4755	C_9H_{12}	Cumene	152.8	Nonazeotrope		575
A =	$C_3H_5BrO_2$	**α-Bromopropionic Acid**	**205.8**			
4756	$C_6H_4Br_2$	o-Dibromobenzene	181.5	179.0	12	575
4757	$C_6H_4Cl_2$	p-Dichlorobenzene	174.4	<173.5	> 7	575
4758	C_6H_5I	Iodobenzene	188.45	<184.8		575
4759	$C_6H_5NO_2$	Nitrobenzene	210.75	203.3	60	554
4760	C_7H_7Br	α-Bromotoluene	198.5	~ 195		563
4761	$C_7H_8O_2$	Guaiacol	205.05	<204.2	>45	575
4762	$C_8H_8O_2$	Methyl benzoate	199.4	Nonazeotrope		575
4763	C_9H_{12}	Mesitylene	164.6	Nonazeotrope		
4764	$C_{10}H_8$	Naphthalene	218.0	<202.5	>73	562
4765	$C_{10}H_{14}$	Cymene	176.7	<176.4	> 4	575
4766	$C_{11}H_{10}$	2-Methylnaphthalene	241.15	Nonazeotrope		575
A =	$C_3H_5Br_3$	**1,2,3-Tribromopropane**	**220**			
4767	$C_6H_5NO_2$	Nitrobenzene	210.85	Nonazeotrope		563
5768	C_6H_6O	Phenol	182.2	Nonazeotrope		575
4769	$C_7H_6O_2$	Benzoic acid	250.8	<220.5	>94	575
4770	$C_7H_7NO_2$	o-Nitrotoluene	~222.3	Nonazeotrope		563
4771	C_7H_8O	p-Cresol	201.8	Nonazeotrope		563
4772	$C_7H_{14}O_2$	Enanthic acid	222	<218	>62	575
4773	$C_8H_{14}O_4$	Ethyl succinate	216.5	Azeotrope doubtful		563
4774	$C_9H_{10}O$	Propiophenone	217.7	223	~ 70	573
4775	$C_9H_{10}O_2$	Benzyl acetate	215.0	Nonazeotrope		575
4776	$C_9H_{10}O_2$	Ethyl benzoate	213	Nonazeotrope		563
4777	$C_9H_{10}O_3$	Ethyl salicylate	234.0	Nonazeotrope		548
4778	$C_9H_{18}O_2$	Pelargonic acid	254.0	Nonazeotrope		575
4779	$C_{10}H_8$	Naphthalene	218.05	Nonazeotrope		535
4780	$C_{10}H_{12}O_2$	Propyl benzoate	230.85	Nonazeotrope		547
4781	$C_{10}H_{14}O$	Thymol	232.9	Nonazeotrope		538
4782	$C_{10}H_{15}N$	Diethylaniline	216.5	<215	>15	563
4783	$C_{10}H_{16}O$	Pulegone	~224	226.5	~ 55	573
4784	$C_{10}H_{18}O$	Borneol	211.8	Nonazeotrope		563
4785	$C_{11}H_{20}O$	Terpineol methyl ether	216	Nonazeotrope		563
4786	$C_{11}H_{22}O_3$	Isoamyl carbonate	232.2	Nonazeotrope		547

TABLE I.   *Binary Systems*   **157**

No.	Formula	B-Component Name	B.P., °C	Azeotropic Data B.P., °C	Wt.%A		Ref.
A =	**C₃H₅Cl**	**2-Chloropropene**	**22.65**				
4787	C₃H₅Cl	3-Chloropropene	45.7	Nonazeotrope,		v-1	1024
4788	C₃H₇NO₂	Isopropyl nitrite	40.1	Nonazeotrope			550
4789	C₄H₄O	Furan	31.7	Nonazeotrope			559
4790	C₄H₁₀O	Ethyl ether	34.6	Nonazeotrope			559
4791	C₅H₁₀	3-Methyl-1-butene	20.6	<18.5	>45		562
4792	C₅H₁₂	2-Methylbutane	27.95	19.0	64		562
4793	C₅H₁₂	Pentane	36.15	<22.4	>72		575
A =	**C₃H₅Cl**	**3-Chloropropene**	**45.15**				
4794	C₃H₆O	Acetone	56.15	44.6	90		552
4795	C₃H₆O₂	Ethyl formate	54.15	45.0	90.0		547
4796	C₃H₆O₂	Methyl acetate	56.95	Nonazeotrope			575
4797	C₃H₇Cl	1-Chloropropane	46.6	Nonazeotrope			549
4798	C₃H₇Cl	2-Chloropropane	34.9	Nonazeotrope		v-1	226
4799	C₃H₇NO₂	Isopropyl nitrite	40.1	Nonazeotrope			550
4800	C₃H₇NO₂	Propyl nitrite	47.75	44.8	80		550
4801	C₃H₈O	Isopropyl alcohol	82.4	45.1	98		575
4802	C₃H₈O₂	Methylal	42.3	41.4	20		559
4803	C₄H₁₀O	*tert*-Butyl alcohol	82.45	Nonazeotrope			575
4804	C₄H₁₀O	Ethyl ether	34.6	Nonazeotrope			559
4805	C₅H₁₀	Cyclopentane	49.3	44.3	63		575
4806	C₅H₁₂	Pentane	36.15	<35.5	>28		575
4807	C₆H₁₄	2,3-Dimethylbutane	58.0	Nonazeotrope			575
A =	**C₃H₅ClO**	**2-Chloro-2-propen-1-ol**					
4808	C₅H₇ClO	2-Chloroallyl vinyl ether		10			1008
A =	**C₃H₅ClO**	**1-Chloro-2-propanone**	**119.7**				
4809	C₃H₈O	Isopropyl alcohol	82.4	Nonazeotrope			552
4810	C₄H₁₀O	Butyl alcohol	117.8	112.5	57		552
4811	C₄H₁₀O	*sec*-Butyl alcohol	99.5	Nonazeotrope			552
4812	C₄H₁₀O	Isobutyl alcohol	108.0	106.0	37		552
4813	C₅H₁₀O	Cyclopentanol	140.85	Nonazeotrope			552
4814	C₅H₁₀O₂	Butyl formate	106.7	Nonazeotrope			548
4815	C₅H₁₀O₂	Methyl butyrate	102.65	Nonazeotrope			552
4816	C₅H₁₀O₂	Propyl acetate	101.6	Nonazeotrope			552
4817	C₅H₁₂O	Amyl alcohol	138.2	Nonazeotrope			552
4818	C₅H₁₂O	*tert*-Amyl alcohol	102.35	Nonazeotrope			552
4819	C₅H₁₂O	Isoamyl alcohol	131.9	<119.0	>83		552
4820	C₅H₁₂O	2-Pentanol	119.8	<116.0	<68		552
4821	C₆H₁₀S	Allyl sulfide	139.35	Nonazeotrope			566
4822	C₆H₁₂O₂	Ethyl butyrate	121.5	117.5	53		552
4823	C₆H₁₂O₂	Ethyl isobutyrate	110.1	Nonazeotrope			552
4824	C₆H₁₂O₂	Isobutyl acetate	117.4	116.9	30		552
4825	C₆H₁₄S	Propyl sulfide	141.5	Nonazeotrope			566
4826	C₆H₁₄S	Isopropyl sulfide	120.5	<116.0			566
4827	C₆H₁₅BO₃	Ethyl borate	118.6	109.4	36		552
4828	C₇H₈	Toluene	110.75	109.2	28.5		552

No.	Formula	Name	B.P., °C	B.P., °C	Wt.%A	Ref.
		B-Component		**Azeotropic Data**		
A =	**C₃H₅ClO**	**1-Chloro-2-propanone**	**119.7**			
		(continued)				
4829	C₇H₁₄	Methylcyclohexane	101.15	<100.5		552
4830	C₇H₁₄O₂	Ethyl isovalerate	134.7	Nonazeotrope		552
4831	C₇H₁₄O₂	Isoamyl acetate	142.1	Nonazeotrope		552
4832	C₇H₁₄O₂	Isopropyl isobutyrate	120.8	117.2	50	552
4833	C₇H₁₄O₂	Propyl butyrate	143.7	Nonazeotrope		552
4834	C₇H₁₄O₂	Propyl isobutyrate	134.0	Nonazeotrope		552
4835	C₈H₁₀	Ethylbenzene	136.15	Nonazeotrope		552
4836	C₈H₁₆	1,3-Dimethylcyclohexane	120.7	<114.0		552
4837	C₈H₁₈	2,5-Dimethylhexane	109.3	<107.5	<35	552
4838	C₈H₁₈	Octane	125.75	<115.5	65	552
A =	**C₃H₅ClO**	**Epichlorohydrin**	**116.45**			
4839	C₃H₅Cl₃	1,2,3-Trichloropropane	156.85	Nonazeotrope	v-1	984
4840	C₃H₆Br₂	1,2-Dibromopropane	140.5	Nonazeotrope		556
4841	C₃H₆O	Allyl alcohol	96.95	95.8	22	556,875
4842	C₃H₆O₂	Propionic acid	141.3	Nonazeotrope		575
4843	C₃H₇I	1-Iodopropane	102.4	<100.5	<28	575
4844	C₃H₈O	Isopropyl alcohol	82.45	Nonazeotrope		556
4845	C₃H₈O	Propyl alcohol	97.2	96.0	23	563,980
4846	C₄H₅N	Pyrrole	130.5	Reacts		563
4847	C₄H₈S	Tetrahydrothiophene	118.8	<112.5	<70	566
4848	C₄H₉Br	1-Bromobutane	101.6	100.0		548
4849	C₄H₉Cl	1-Chlorobutane	78.5	Nonazeotrope		575
4850	C₄H₉I	1-Iodobutane	130.4	<115	<92	548
4851	C₄H₉I	1-Iodo-2-methylpropane	120.8	111.0	~47	548
4852	C₄H₁₀O	Butyl alcohol	116.9	112.0	57	556
4853	C₄H₁₀O	*sec*-Butyl alcohol	99.5	98.0	25	556
4854	C₄H₁₀O	*tert*-Butyl alcohol	82.45	Nonazeotrope		556
4855	C₄H₁₀O	Isobutyl alcohol	108.0	105.0	39.5	563
4856	C₅H₅N	Pyridine	115.5	Reacts		563
4857	C₅H₈O	Cyclopentanone	130.65	Nonazeotrope		575
4858	C₅H₁₀O	3-Pentanone	102.05	Nonazeotrope		552
4859	C₅H₁₀O₂	Methyl butyrate	102.65	Nonazeotrope		575
4860	C₅H₁₀O₂	Propyl acetate	101.6	Nonazeotrope		575
4861	C₅H₁₀O₃	Ethyl carbonate	126.0	Azeotrope doubtful		563
4862	C₅H₁₀O₃	2-Methoxyethyl acetate	144.6	Nonazeotrope		575
4863	C₅H₁₁Br	1-Bromo-3-methylbutane	120.65	111.2	63	556
4864	C₅H₁₂O	Amyl alcohol	138.2	<116.2	<95	575
4865	C₅H₁₂O	*tert*-Amyl alcohol	102.0	100.7	30	556
4866	C₅H₁₂O	Isoamyl alcohol	131.8	115.35	81	563
4867	C₅H₁₂O	3-Methyl-2-butanol	112.9	109.5	48	556
4868	C₅H₁₂O	2-Pentanol	119.8	113.0	60	556
4869	C₅H₁₂O	3-Pentanol	116.0	111.5	54	556
4870	C₆H₅Cl	Chlorobenzene	131.75	<116.2		575
		"	131.8	Azeotrope doubtful		563
4871	C₆H₆	Benzene	80.15	Nonazeotrope		575
4872	C₆H₁₀O	Mesityl oxide	129.45	Nonazeotrope		552
4873	C₆H₁₂	Cyclohexane	80.75	Nonazeotrope		575

TABLE I. *Binary Systems*   159

No.	Formula	B-Component Name	B.P., °C	Azeotropic Data B.P., °C	Wt.%A	Ref.
A =	C₃H₅ClO	Epichlorohydrin *(continued)*	116.45			
4874	C₆H₁₂O	Cyclohexanol	160.65	Nonazeotrope		556
4875	C₆H₁₂O	3-Hexanone	123.3	Nonazeotrope		575
4876	C₆H₁₂O	4-Methyl-2-pentanone	116.05	<115.5	> 32	575
4877	C₆H₁₂O	Pinacolone	106.2	Nonazeotrope		575
4878	C₆H₁₂O₂	Butyl acetate	125.0	Nonazeotrope		548
4879	C₆H₁₂O₂	Ethyl butyrate	121.5	115.75	75	556
4880	C₆H₁₂O₂	Ethyl isobutyrate	110.1	109.8	~ 10	575
		"	110.1	Azeotrope doubtful		563
4881	C₆H₁₂O₂	Isoamyl formate	123.6	~116.2		563
4882	C₆H₁₂O₂	Isobutyl acetate	117.2	<115.3	> 50	548
4883	C₆H₁₂O₂	Methyl isovalerate	116.3	115	45	563
4884	C₆H₁₂O₂	Propyl propionate	123.0	<116.3	> 88	575
4885	C₆H₁₄O	Hexyl alcohol	157.85	Nonazeotrope		556
4886	C₆H₁₄S	Isopropyl sulfide	120.5	111.5	67	566
4887	C₇H₈	Toluene	110.75	108.4	29	548
		"	110.6	108.3	26	980
4888	C₇H₁₄	Methylcyclohexane	101.15	<100.8	> 5	575
4889	C₇H₁₄O	2-Methylcyclohexanol	168.5	Nonazeotrope		575
4890	C₇H₁₆	Heptane	98.4	< 98.1	> 4	575
4891	C₈H₁₀	Ethylbenzene	136.15	Nonazeotrope		548
4892	C₈H₁₆	1,3-Dimethylcyclohexane	120.7	113.6	65	575
4893	C₈H₁₈	2,5-Dimethylhexane	109.3	~107.0	25	548
4894	C₈H₁₈	Octane	125.8	114.5	~ 80	548
			125.8	<116	>90	563
4895	C₈H₁₈O	Isobutyl ether	122.2	Nonazeotrope		548
A =	C₃H₅ClO₂	Methyl Chloroacetate	129.95			
4896	C₃H₆Br₂	1,2-Dibromopropane	140.5	Nonazeotrope		575
4897	C₃H₆O	Allyl alcohol	96.85	Nonazeotrope		575
4898	C₃H₆O₂	Propionic acid	141.3	Nonazeotrope		527
4899	C₃H₈O	Isopropyl alcohol	82.45	Nonazeotrope		573
4900	C₃H₈O	Propyl alcohol	97.2	Nonazeotrope		573
4901	C₃H₈O₂	2-Methoxyethanol	124.5	122.5	65	575
4902	C₄H₈O₃	Methyl lactate	143.8	Nonazeotrope		575
4903	C₄H₉I	1-Iodobutane	130.4	125.5	42	562
4904	C₄H₉I	1-Iodo-2-methylpropane	120.8	<119.5	< 22	562
4905	C₄H₁₀O	Butyl alcohol	117.5	116.3	26	573
4906	C₄H₁₀O	*sec*-Butyl alcohol	99.5	Nonazeotrope		575
4907	C₄H₁₀O	Isobutyl alcohol	107.85	107.55	12	530
4908	C₄H₁₀O₂	2-Ethoxyethanol	135.3	128.6	77	526
4909	C₅H₈O	Cyclopentanone	130.65	<129.6		552
4910	C₅H₁₀O	Cyclopentanol	140.85	127.5	77	567
4911	C₅H₁₀O₃	Ethyl carbonate	125.9	Nonazeotrope		572
4912	C₅H₁₀O₃	2-Methoxyethyl acetate	144.6	Nonazeotrope		556
4913	C₅H₁₁Br	1-Bromo-3-methylbutane	120.65	Nonazeotrope		527
4914	C₅H₁₁I	1-Iodo-3-methylbutane	147.65	Nonazeotrope		575
4915	C₅H₁₂O	Amyl alcohol	138.2	126.8	70	567

No.		B-Component		Azeotropic Data		
	Formula	Name	B.P., °C	B.P., °C	Wt.%A	Ref.
A =	C₃H₅ClO₂	**Methyl Chloroacetate** *(continued)*	**129.95**			
4916	C₅H₁₂O	*tert*-Amyl alcohol	102.35	Nonazeotrope		575
4917	C₅H₁₂O	Isoamyl alcohol	131.3	124.9	60.5	527
4918	C₅H₁₂O	2-Pentanol	119.8	117.0	40	567
4919	C₅H₁₂O	3-Pentanol	116.0	114.0	32	567
4920	C₆H₅Cl	Chlorobenzene	132.0	126	~ 60	532
4921	C₆H₁₀O	Mesityl oxide	129.45	128.8	42	552
4922	C₆H₁₀S	Allyl sulfide	139.35	Nonazeotrope		566
4923	C₆H₁₂O	Cyclohexanol	160.8	Nonazeotrope		575
4924	C₆H₁₂O	3-Hexanone	123.3	Nonazeotrope		552
4925	C₆H₁₂O	4-Methyl-2-pentanone	116.05	Nonazeotrope		552
4926	C₆H₁₂O₂	Butyl acetate	125.0	Nonazeotrope		548
4927	C₆H₁₂O₂	Ethyl butyrate	121.5	Nonazeotrope		575
4928	C₆H₁₂O₂	Isoamyl formate	123.8	Nonazeotrope		532
4929	C₆H₁₂O₂	Propyl propionate	122.1	Nonazeotrope		532
4930	C₆H₁₂O₃	Paraldehyde	124	Azeotrope doubtful		563
4931	C₆H₁₄O	Hexyl alcohol	157.85	Nonazeotrope		575
4932	C₆H₁₄S	Propyl sulfide	141.5	Nonazeotrope		575
4933	C₇H₈	Toluene	110.7	Nonazeotrope		563
4934	C₇H₁₄	Methylcyclohexane	101.15	Nonazeotrope		575
4935	C₇H₁₄O	4-Heptanone	143.55	Nonazeotrope		552
4936	C₇H₁₄O₂	Ethyl isovalerate	134.7	Nonazeotrope		532
4937	C₇H₁₄O₂	Isobutyl propionate	136.9	Nonazeotrope		532
4938	C₇H₁₄O₂	Propyl isobutyrate	134.0	Nonazeotrope		548
4939	C₈H₈	Styrene	145.8	Nonazeotrope		538
4940	C₈H₁₀	Ethylbenzene	136.15	127.2	62.5	572
4941	C₈H₁₀	*m*-Xylene	139.2	128.25	90	575
		"	139.0	Nonazeotrope		563
4942	C₈H₁₀	*p*-Xylene	138.45	128.3	85	562
4943	C₈H₁₆	1,3-Dimethylcyclohexane	120.7	118.5	15	562
4944	C₈H₁₈	Octane	125.8	123.5	~40	563
4945	C₈H₁₈O	Butyl ether	142.4	Nonazeotrope		575
4946	C₈H₁₈O	Isobutyl ether	122.3	<121.9	<18	575
4947	C₁₀H₁₆	α-Pinene	155.8	Nonazeotrope		575
A =	C₃H₅Cl₃	**1,1,3-Trichloropropane**	**148**			
4948	C₇H₁₄O	2-Heptanone	150	Max. b.p.		262
4949	C₇H₁₄O₂	Amyl acetate	148	Max. b.p.		262
A =	C₃H₅Cl₃	**1,2,2-Trichloropropane**	**122**			
4950	C₅H₅N	Pyridine	115	Nonazeotrope		262
4951	C₅H₈O	Cyclopentanone	129	Nonazeotrope		262
4952	C₅H₁₀O₃	Ethyl carbonate	126	Max. b.p.		262
4953	C₆H₁₂O₂	Butyl acetate	125	126.4	38 v-1	262
4954	C₇H₁₄O	2,4-Dimethyl-3-pentanone	124	Max. b.p.		292
4955	C₇H₁₄O₂	Isopropyl butyrate	128	Nonazeotrope		262
A =	C₃H₅Cl₃	**1,2,3-Trichloropropane**	**158**			
4956	C₃H₆O₂	Propionic acid	140.7	~ 140.5	30?	563

TABLE I.   *Binary Systems*   161

No.	Formula	Name	B.P., °C	B.P., °C	Wt.%A		Ref.
		B-Component		**Azeotropic Data**			
A =	**C₃H₅Cl₃**	**1,2,3-Trichloropropane**	**158**				
		(continued)					
4957	C₃H₇NO₂	Ethyl carbamate	185.25	155.0	90		564
4958	C₄H₅Cl₃O	α·α,β-Trichlorobutyralde-					
		hyde	164	Nonazeotrope			563
4959	C₄H₆O₄	Methyl oxalate	164.2	154.0	72		538
4960	C₄H₇ClO₂	Ethyl chloroacetate	143.5	Nonazeotrope			538
4961	C₄H₈O₂	Butyric acid	162.45	153.0	75		541
4962	C₄H₈O₂	Isobutyric acid	154.35	149.2	62		542
4963	C₄H₈O₃	Methyl lactate	143.8	Nonazeotrope			573
4964	C₅H₄O₂	2-Furaldehyde	161.45	Nonazeotrope			572
4965	C₅H₁₀O₂	Isovaleric acid	176.5	155.0	93?		575
4966	C₅H₁₀O₃	Ethyl lactate	153.9	~153.5	~15		572
4967	C₅H₁₁NO₃	Isoamyl nitrate	149.75	<149.5	>12		560
4968	C₅H₁₂O	Isoamyl alcohol	131.3	Nonazeotrope			527
4969	C₆H₅Br	Bromobenzene	156.1	155.6	30		549
4970	C₆H₆O	Phenol	182.2	Nonazeotrope			530
4971	C₆H₁₀O	Cyclohexanone	155.7	160.0	61		552
4972	C₆H₁₀O₃	Ethyl acetoacetate	180.4	Nonazeotrope			535
4973	C₆H₁₂	1-Hexene	63.5	Nonazeotrope		v-1	933
4974	C₆H₁₂O	Cyclohexanol	160.7	154.9	67		572
4975	C₆H₁₄	Hexane	68.74	Nonazeotrope		v-1	933
4976	C₆H₁₄O	Hexyl alcohol	157.85	152.8	60		567
4977	C₇H₇Cl	o-Chlorotoluene	159.2	Nonazeotrope			575
4978	C₇H₈O	Anisole	153.85	Nonazeotrope			530
		"	155	Max. b.p.			262
4979	C₇H₈O	o-Cresol	191.1	Nonazeotrope			575
4980	C₇H₁₄O	Heptaldehyde	155	Max. b.p.			262
4981	C₈H₁₀O	Benzyl methyl ether	167.8	Nonazeotrope			559
4982	C₈H₁₀O	Phenetole	170.45	Nonazeotrope			548
4983	C₈H₁₆O₂	Isobutyl isobutyrate	147.3	Nonazeotrope			547
4984	C₈H₁₈O	Butyl ether	142.4	Nonazeotrope			559
4985	C₈H₁₈O	sec-Octyl alcohol	179.0	Nonazeotrope			573
4986	C₈H₂₀SiO₄	Ethyl silicate	165	Nonazeotrope?			563
4987	C₉H₁₂	Pseudocumene	168.2	Nonazeotrope			575
4988	C₉H₁₈O₂	Isoamyl isobutyrate	170.0	Nonazeotrope			547
4989	C₉H₁₈O₂	Isobutyl isovalerate	171.35	Nonazeotrope			547
4990	C₁₀H₁₄	Cymene	176.7	Nonazeotrope			575
4991	C₁₀H₁₆	Camphene	159.6	~152.9	~65		529
4992	C₁₀H₁₆	d-Limonene	177.8	Nonazeotrope			535
4993	C₁₀H₁₆	α-Pinene	155.8	150.0	~85		572
4994	C₁₀H₂₂	2,7-Dimethyloctane	160.25	~155.5	~70		573
A =	**C₃H₅I**	**3-Iodopropene**	**102**				
4995	C₃H₆O	Allyl alcohol	96.95	89.4	72		532
4996	C₃H₆O₂	Propionic acid	141.3	Nonazeotrope			575
4997	C₃H₆O₃	Methyl carbonate	90.25	< 90.0			575

No.	Formula	B-Component Name	B.P., °C	Azeotropic Data B.P., °C	Wt.%A	Ref.
A =	C₃H₅I	3-Iodopropene *(continued)*	102			
4998	C₃H₇I	1-Iodopropane	102.4	Nonazeotrope		575
4999	C₃H₈O	Isopropyl alcohol	82.45	~ 79	~ 58	563
5000	C₃H₈O	Propyl alcohol	97.2	90.0	71	563
5001	C₃H₈O₂	2-Methoxyethanol	124.5	100.5	~ 95	575
5002	C₄H₇N	Isobutyronitrile	103.85	< 93.2	< 68	562
5003	C₄H₈O₂	Dioxane	101.35	98.5	56	527
5004	C₄H₁₀O	Butyl alcohol	117.8	98.7	87	567
5005	C₄H₁₀O	Isobutyl alcohol	108	96	~ 83	563
5006	C₅H₁₀O	2-Pentanone	102.35	100.7	66	552
5007	C₅H₁₀O	3-Pentanone	102.05	100.5	65	552
5008	C₅H₁₀O₂	Butyl formate	106.7	100.0	> 75	547
5009	C₅H₁₀O₂	Ethyl propionate	99.1	98.0	35	547
5010	C₅H₁₀O₂	Isobutyl formate	98.3	95.8	38	563
5011	C₅H₁₀O₂	Isopropyl acetate	89.5	Nonazeotrope		575
5012	C₅H₁₀O₂	Methyl butyrate	102.75	101.0	65	573
5013	C₅H₁₀O₂	Methyl isobutyrate	92.5	Nonazeotrope		547
5014	C₅H₁₀O₂	Propyl acetate	101.6	99.5	56	538
5015	C₅H₁₁Cl	1-Chloro-3-methylbutane	99.4	Nonazeotrope		549
5016	C₅H₁₁NO₂	Isoamyl nitrite	97.15	96.0		550
5017	C₅H₁₂O	*tert*-Amyl alcohol	102.35	< 97.2	< 75	567
5018	C₅H₁₂O	Isoamyl alcohol	131.9	Nonazeotrope		527
5019	C₆H₁₂O₂	Ethyl isobutyrate	110.1	Nonazeotrope?		563
5020	C₆H₁₄O₂	Acetal	103.55	100.0	67	559
5021	C₇H₈	Toluene	110.7	Nonazeotrope		563
5022	C₇H₁₄	Methylcyclohexane	101.8	99	~ 70	563
5023	C₇H₁₆	Heptane	98.45	97.0	48	538
A =	C₃H₅N	**Propionitrile**	97.2			
5024	C₃H₇I	2-Iodopropane	89.45	81.2	30	562
5025	C₃H₈O	Isopropyl alcohol	82.4	81.5	12	565
5026	C₃H₈O	Propyl alcohol	97.2	90.5	50	565
5027	C₃H₉ClSi	Chlorotrimethylsilane	57.7	Nonazeotrope		841
5028	C₄H₈O₂	Ethyl acetate	77.1	Nonazeotrope		565
5029	C₄H₈O₂	Methyl propionate	79.85	Nonazeotrope		565
5030	C₄H₈O₂	Propyl formate	80.85	Nonazeotrope		565
5031	C₄H₉Br	1-Bromo-2-methylpropane	91.4	85.0	35	562
5032	C₄H₁₀O	Butyl alcohol	117.75	Nonazeotrope		565
5033	C₄H₁₀O	Isobutyl alcohol	108.0	95.5	76	565
5034	C₅H₁₀O₂	Ethyl propionate	99.1	< 94.5	> 40	565
5035	C₅H₁₀O₂	Methyl butyrate	102.65	< 96.0	> 54	575
5036	C₅H₁₀O₂	Propyl acetate	101.55	95.4	55	565
5037	C₅H₁₂O	*tert*-Amyl alcohol	102.35	< 94.5	> 57	565
5038	C₆H₁₂O₂	Ethyl isobutyrate	110.1	Nonazeotrope		565
5039	C₆H₁₂O₂	Methyl isovalerate	116.5	Nonazeotrope		565
5040	C₆H₁₄	Hexane	68.8	63.5	9	565
5041	C₆H₁₄O	Isopropyl ether	68.3	< 67.5	> 4	575
5042	C₆H₁₄O	Propyl ether	90.1	< 83.5	> 18	562
5043	C₇H₈	Toluene	110.75	Min. b.p.		515

TABLE I.   *Binary Systems*   163

No.	Formula	B-Component Name	B.P., °C	Azeotropic Data B.P., °C	Wt.%A		Ref.
A =	**C₃H₅N**	**Propionitrile** *(continued)*	**97.2**				
5044	C₇H₁₄	Methylcyclohexane	101.15	< 85.0	> 45		562
5045	C₇H₁₆	Heptane	98.4	< 80.5			565
5046	C₈H₁₀	Ethylbenzene	136.15	Nonazeotrope			575
A =	**C₃H₅N₃O₉**	**Nitroglycerin**					
5047	C₃H₆O	Acetone	56.15	Nonazeotrope		v-l	633
A =	**C₃H₆**	**Propene**	**— 48**				
5048	C₃H₈	Propane, 10°-190°F.		Nonazeotrope		v-l	361
		" 320 p.s.i.a.		Nonazeotrope		v-l	618
		" 0°-20.1°C.		Nonazeotrope		v-l	354c
A =	**C₃H₆Br₂**	**1,2-Dibromopropane**	**140.5**				
5049	C₃H₆O₂	Propionic acid	141.3	134.5	67		527
5050	C₃H₇NO	Propionamide	222.2	Nonazeotrope			575
5051	C₃H₈O	Propyl alcohol	97.2	Nonazeotrope			575
5052	C₃H₈O₂	2-Methoxyethanol	124.5	124.0			526
5053	C₄H₅N	Pyrrole	130	Nonazeotrope			555
5054	C₄H₇BrO₂	Ethyl bromoacetate	158.8	Nonazeotrope			575
5055	C₄H₈O₂	Butyric acid	164.0	138.5	92		562
5056	C₄H₈O₂	Isobutyric acid	154.6	137.0	85		562
5057	C₄H₁₀O	Butyl alcohol	117.8	<117.1	39		575
5058	C₄H₁₀O	Isobutyl alcohol	108.0	Nonazeotrope			575
5059	C₄H₁₀O₂	2-Ethoxyethanol	135.3	132.5	50		555
5060	C₅H₄O₂	2-Furaldehyde	161.45	Nonazeotrope			556
5061	C₅H₁₀O₂	Isovaleric acid	176.5	Nonazeotrope			575
5062	C₅H₁₀O₂	Valeric acid	186.35	Nonazeotrope			575
5063	C₅H₁₁NO₃	Isoamyl nitrate	149.5	Nonazeotrope			547
5064	C₅H₁₂O	Isoamyl alcohol	131.9	<128.5	> 52		527
5065	C₆H₆O	Phenol	182.2	Nonazeotrope			575
5066	C₆H₁₀O	Mesityl oxide	129.45	Nonazeotrope			552
5067	C₆H₁₀S	Allyl sulfide	~138.7	Nonazeotrope			563
5068	C₆H₁₂O	Cyclohexanol	160.65	Nonazeotrope			563
5069	C₆H₁₂O₃	2-Ethoxyethyl acetate	156.8	Nonazeotrope			575
5070	C₆H₁₄O	Hexyl alcohol	157.85	Nonazeotrope			575
5071	C₇H₈O	Anisole	153.85	Nonazeotrope			559
5072	C₇H₁₄O	4-Heptanone	143.55	Nonazeotrope			552
5073	C₇H₁₄O	5-Methyl-2-hexanone	144.2	Nonazeotrope			552
5074	C₇H₁₄O₂	Amyl acetate	148.8	Nonazeotrope			575
5075	C₇H₁₄O₂	Butyl propionate	146.5	Nonazeotrope			547
5076	C₇H₁₄O₂	Ethyl isovalerate	134.7	Nonazeotrope			575
5077	C₇H₁₄O₂	Ethyl valerate	145.45	Nonazeotrope			575
5078	C₇H₁₄O₂	Isoamyl acetate	142.1	<140.2	>91		575
5079	C₇H₁₄O₂	Isobutyl propionate	136.9	Nonazeotrope			547
5080	C₇H₁₄O₂	Propyl butyrate	134.7	Nonazeotrope			547
5081	C₇H₁₄O₂	Propyl isobutyrate	134.0	Nonazeotrope			547
5082	C₈H₁₀	Ethylbenzene	136.15	135.95	5		563
5083	C₈H₁₀	*m*-Xylene	139.0	138	30		563
5084	C₈H₁₀	*o*-Xylene	142.6	139.2	~70		563

| No. | Formula | B-Component | | Azeotropic Data | | |
		Name	B.P., °C	B.P., °C	Wt.%A	Ref.
A =	$C_3H_6Br_2$	**1,2-Dibromopropane**	**140.5**			
		(continued)				
5085	C_8H_{10}	*p*-Xylene	138.2	137.5	~22	563
5086	$C_8H_{18}O$	Butyl ether	142.4	146.0	40	559
5087	C_9H_{12}	Cumene	152.8	Nonazeotrope		575
5088	$C_{10}H_{16}$	α-Pinene	155.8	Nonazeotrope		575
A =	$C_3H_6Br_2$	**1,3-Dibromopropane**	**166.9**			
5089	$C_3H_6O_2$	Propionic acid	141.3	Nonazeotrope		527
5090	C_3H_7NO	Propionamide	222.2	Nonazeotrope		527
5091	$C_3H_7NO_2$	Ethyl carbamate	185.25	164.05	87.8	527
5092	$C_4H_7BrO_2$	Ethyl bromoacetate	158.8	Nonazeotrope		527
5093	$C_4H_7Cl_3O$	Ethyl 1,1,2-trichloroethyl ether	172.5	Nonazeotrope		575
5094	$C_4H_8Cl_2O$	Bis(2-chloroethyl) ether	178.65	Nonazeotrope		527
5095	$C_4H_8O_2$	Butyric acid	163.5	158.4	70	527
5096	$C_4H_8O_2$	Isobutyric acid	154.6	151.5	40	562
5097	$C_5H_4O_2$	2-Furaldehyde	161.45	159.45	54	527
5098	$C_5H_6O_2$	Furfuryl alcohol	169.35	164.0	74	575
5099	$C_5H_8O_4$	Methyl malonate	181.5	Nonazeotrope		547
5100	$C_5H_{10}O_2$	Isovaleric acid	176.5	163.35	84.5	527
5101	$C_5H_{10}O_2$	Valeric acid	186.35	166.0	92	527
5102	$C_5H_{11}NO_3$	Isoamyl nitrate	149.5	Nonazeotrope		547
5103	$C_5H_{12}O_2$	2-Propoxyethanol	151.35	Nonazeotrope		556
5104	C_6H_6O	Phenol	182.2	Nonazeotrope		575
5105	C_6H_7N	Aniline	184.35	Nonazeotrope		575
5106	$C_6H_{10}O$	Cyclohexanone	155.7	Nonazeotrope		552
5107	$C_6H_{10}O_4$	Ethyl oxalate	185.65	Nonazeotrope		527
5108	$C_6H_{12}O$	Cyclohexanol	160.65	158.5		563
5109	$C_6H_{12}O_2$	Caproic acid	205.15	Nonazeotrope		575
5110	$C_6H_{12}O_2$	Isocaproic acid	199.5	Nonazeotrope		575
5111	$C_6H_{12}O_3$	2-Ethoxyethyl acetate	156.8	Nonazeotrope		556
5112	$C_6H_{14}O_2$	2-Butoxyethanol	171.15	164.55	77	527
5113	C_7H_5N	Benzonitrile	191.1	Nonazeotrope		565
5114	C_7H_6O	Benzaldehyde	179.2	Nonazeotrope		575
5115	C_7H_7Cl	*p*-Chlorotoluene	162.4	Nonazeotrope		575
5116	C_7H_8O	Benzyl alcohol	205.25	Nonazeotrope		575
5117	C_7H_8O	*o*-Cresol	191.1	Nonazeotrope		575
5118	$C_7H_{14}O_3$	1,3-Butanediol methyl ether acetate	171.75	Nonazeotrope		575
5119	$C_8H_{10}O$	Benzyl methyl ether	167.8	>170	>45	559
5120	$C_8H_{10}O$	*p*-Methylanisole	177.05	Nonazeotrope		559
5121	$C_8H_{10}O$	Phenetole	170.45	Nonazeotrope		559
5122	$C_8H_{16}O$	2-Octanone	172.85	Nonazeotrope		552
5123	$C_8H_{16}O_2$	Hexyl acetate	171.5	Nonazeotrope		547
5124	$C_8H_{16}O_2$	Isoamyl propionate	160.4	Nonazeotrope		547
5125	$C_8H_{18}O$	Octyl alcohol	195.2	Nonazeotrope		575
5126	$C_8H_{18}O$	*sec*-Octyl alcohol	180.4	Nonazeotrope		575
5127	C_9H_8	Indene	182.6	Nonazeotrope		575
5128	C_9H_{12}	Mesitylene	164.6	Nonazeotrope		575

TABLE I.   *Binary Systems*   165

No.	Formula	B-Component Name	B.P., °C	Azeotropic Data B.P., °C	Wt.%A	Ref.
A =	C₃H₆Br₂	**1,3-Dibromopropane**	**166.9**			
		(continued)				
5129	C₉H₁₈O₂	Isobutyl isovalerate	171.35	Nonazeotrope		547
5130	C₁₀H₁₄	Cymene	176.7	Nonazeotrope		575
5131	C₁₀H₁₆	Dipentene	177.7	Nonazeotrope		575
5132	C₁₀H₁₆	α-Pinene	155.8	Nonazeotrope		575
A =	**C₃H₆Br₂O**	**2,3-Dibromo-1-propanol**	**219.5**			
5133	C₄H₁₀O₃	Diethylene glycol	245.5	Nonazeotrope		575
5134	C₆H₄Cl₂	o-Dichlorobenzene	179.5	Nonazeotrope		575
5135	C₆H₅I	Iodobenzene	188.45	Nonazeotrope		575
5136	C₆H₁₄O₃	Dipropylene glycol	229.2	Nonazeotrope		575
5137	C₇H₇Br	o-Bromotoluene	181.5	Nonazeotrope		575
5138	C₇H₇I	p-Iodotoluene	214.5	<209.0	< 40	575
5139	C₈H₁₀O₂	2-Phenoxyethanol	245.2	Nonazeotrope		575
5140	C₈H₁₀O₂	Veratrole	206.8	Nonazeotrope		575
5141	C₈H₁₈O	Octyl alcohol	195.2	Nonazeotrope		575
5142	C₈H₁₈O₃	2-(2-Butoxyethoxy)ethanol	231.2	Nonazeotrope		575
5143	C₉H₈	Indene	182.6	Nonazeotrope		575
5144	C₉H₁₀O	p-Methylacetophenone	226.35	228.2		575
5145	C₉H₁₀O	Propiophenone	217.7	<222.0		575
5146	C₁₀H₁₄	Butylbenzene	183.1	Nonazeotrope		575
5147	C₁₀H₁₄	Cymene	176.7	Nonazeotrope		575
5148	C₁₀H₁₆	Dipentene	177.7	<176.5	> 12	575
5149	C₁₀H₂₀O	Menthol	216.3	<216.2	< 22	575
5150	C₁₁H₂₀O	Methyl α-terpineol ether	216.2	Nonazeotrope		575
5151	C₁₁H₂₀O	Isobornyl methyl ether	192.4	Nonazeotrope		575
5152	C₁₂H₂₂O	Bornyl ethyl ether	204.9	Nonazeotrope		575
A =	**C₃H₆Cl₂**	**1,1-Dichloropropane**	**90**			
5153	C₅H₁₀O₂	Isopropyl acetate	90	Max. b.p.		262
5154	C₆H₁₅N	Triethylamine	89	Max. b.p.		262
A =	**C₃H₆Cl₂**	**1,2-Dichloropropane**	**97**			
5155	C₃H₆O	Propylene oxide,				
		20 p.s.i.g.	60	Nonazeotrope		981
5156	C₃H₈O	Isopropyl alcohol	82.4		50 v-l	282
		At 30°C.			75 v-l	282
5157	C₃H₈O₂	1,2-Propanediol	187.3	Nonazeotrope		981
5158	C₄H₈O₂	Butyric acid	162.4	Nonazeotrope		678
5159	C₄H₈O₂	Dioxane	101	Max. b.p.		262
5160	C₅H₁₀O	2-Pentanone	102	Max. b.p		262
5161	C₅H₁₀O₂	Ethyl propionate	99	Max. b.p.		262
5162	C₆H₁₂	Cyclohexane	80	80.4	16	276
A =	**C₃H₆Cl₂**	**1,3-Dichloropropane**	**129.8**			
5163	C₅H₁₀O₃	2-Methoxy ethyl acetate	144.6	Nonazeotrope		575
5164	C₆H₆O	Phenol	182.2	Nonazeotrope		575
5165	C₆H₁₂O₃	2-Ethoxyethyl acetate	156.8	Nonazeotrope		575
A =	**C₃H₆Cl₂**	**2,2ˉDichloropropane**	**70.4**			
5166	C₃H₆O	Allyl alcohol	96.85	< 70.0		575
5167	C₃H₆O₂	Ethyl formate	54.15	Nonazeotrope		547

No.	Formula	B-Component Name	B.P., °C	Azeotropic Data B.P., °C	Wt.%A	Ref.
A =	C₃H₆Cl₂	**2,2-Dichloropropane** *(continued)*	**70.4**			
5168	C₃H₆O₂	Methyl acetate	57.0	Nonazeotrope		547
5169	C₃H₈O	Isopropyl alcohol	82.4	66.8	83	567
5170	C₃H₈O	Propyl alcohol	97.2	< 70.1	> 89	575
5171	C₄H₈O₂	Ethyl acetate	77.15	Nonazeotrope		547
5172	C₄H₈O₂	Methyl propionate	79.85	Nonazeotrope		547
5173	C₄H₈O₂	Propyl formate	80.85	Nonazeotrope		547
5174	C₄H₉NO₂	Butyl nitrite	78.2	Nonazeotrope		550
5175	C₄H₉NO₂	Isobutyl nitrite	67.1	Nonazeotrope		550
5176	C₄H₁₀O	Butyl alcohol	117.8	Nonazeotrope		575
5177	C₄H₁₀O	Isobutyl alcohol	108.8	Nonazeotrope		575
5178	C₆H₆	Benzene	80.15	Nonazeotrope		575
5179	C₆H₁₂	Cyclohexane	80.75	Nonazeotrope		575
5180	C₆H₁₂	Methylcyclopentane	72.0	< 69.5	< 70	575
5181	C₆H₁₄	Hexane	68.8	< 68.0	> 40	575
5182	C₆H₁₄O	Isopropyl ether	68.3	74.0	60	559
A =	C₃H₆Cl₂O	**1,3-Dichloro-2-propanol**	**175.8**			
5183	C₃H₇NO₂	Ethyl carbamate	185.25	Nonazeotrope		575
5186	C₄H₆O₄	Methyl oxalate	164.45	Nonazeotrope		575
5185	C₅H₄O₂	2-Furaldehyde	161.45	Nonazeotrope		527
5186	C₅H₁₁I	1-Iodo-3-methylbutane	147.65	<147.4	> 4	575
5187	C₅H₁₂O₂	2-Propoxyethanol	151.35	Nonazeotrope		526
5188	C₆H₃Cl₃	1,3,5-Trichlorobenzene	208.4	Nonazeotrope		575
5189	C₆H₄BrCl	*p*-Bromochlorobenzene	196.4	Nonazeotrope		575
5190	C₆H₄Cl₂	*o*-Dichlorobenzene	179.5	170.5	60	567
5191	C₆H₄Cl₂	*p*-Dichlorobenzene	174.35	168.2	45	530
5192	C₆H₅Br	Bromobenzene	156.1	155.5	~ 9	572
5193	C₆H₅I	Iodobenzene	188.55	173	~ 70	573
5194	C₆H₆O	Phenol	181.5	Nonazeotrope		563
5195	C₆H₁₀O	Cyclohexanone	155.7	Nonazeotrope		552
5196	C₆H₁₀O₄	Ethyl oxalate	185.65	Nonazeotrope		575
5197	C₆H₁₂O	Cyclohexanol	160.7	Nonazeotrope		530
5198	C₆H₁₂O₃	Propyl lactate	171.7	~170		563
5199	C₆H₁₃Br	1-Bromohexane	156.5	154.5	15	567
5200	C₆H₁₄O	Hexyl alcohol	157.85	Nonazeotrope		575
5201	C₆H₁₄O₂	2-Butoxyethanol	171.25	Nonazeotrope		526
5202	C₆H₁₄O₂	Pinacol	174.35	<173.6	< 45	575
5203	C₇H₆O	Benzaldehyde	179.2	<174	> 85	563
5204	C₇H₇Br	*m*-Bromotoluene	184.3	171.8	36	564
5205	C₇H₇Br	*o*-Bromotoluene	181.45	170.45	61	573
5206	C₇H₇Br	*p*-Bromotoluene	185.0	172.8	~ 68	534
5207	C₇H₇Cl	α-Chlorotoluene	179.3	168.9	57	573
5208	C₇H₇Cl	*o*-Chlorotoluene	159.3	158.0	15	538
5209	C₇H₇Cl	*p*-Chlorotoluene	162.4	160.0	22	538
5210	C₇H₈	Toluene	110.75	Nonazeotrope		575
5211	C₇H₈O	Anisole	153.85	Nonazeotrope		556
5212	C₇H₈O	*o*-Cresol	191.1	Nonazeotrope		575
5213	C₇H₁₄	Methylcyclohexane	101.15	Nonazeotrope		575

TABLE I.   *Binary Systems*   167

No.	B-Component			Azeotropic Data			
	Formula	Name	B.P., °C	B.P., °C	Wt.%A		Ref.
A =	$C_3H_6Cl_2O$	1,3-Dichloro-2-propanol	175.8				
	(continued)						
5214	C_7H_{16}	Heptane	98.4	Nonazeotrope			527
5215	$C_7H_{16}O$	Heptyl alcohol	176.15	174.2	47		575
5216	C_8H_8	Styrene	145.8	~143.5	~15		532
5217	C_8H_8O	Acetophenone	202.0	Nonazeotrope			552
5218	C_8H_{10}	Ethylbenzene	136.15	Nonazeotrope			575
5219	C_8H_{10}	*m*-Xylene	139.0	Nonazeotrope			563
5220	C_8H_{10}	*p*-Xylene	138.45	Nonazeotrope			575
5221	$C_8H_{10}O$	Benzyl methyl ether	167.8	<167.0			575
5222	$C_8H_{10}O$	*p*-Methylanisole	177.05	173.1	59		556
5223	$C_8H_{10}O$	Phenetole	170.45	168.8	37		556
5224	$C_8H_{14}O$	Methyl heptenone	173.2	179.0	65?		552
5225	$C_8H_{16}O$	2-Octanone	172.85	179.0	67?		552
5226	$C_8H_{16}O_2$	Isoamyl propionate	160.7	Nonazeotrope			575
5227	$C_8H_{18}O$	Octyl alcohol	195.2	Nonazeotrope			575
5228	$C_8H_{18}O$	*sec*-Octyl alcohol	180.4	175.35	85		564
5229	C_9H_8	Indene	183.0	173.5	66.5		541
5230	C_9H_{12}	Cumene	152.8	<152.5			575
5231	C_9H_{12}	Mesitylene	164.6	161.5	32		567
5232	C_9H_{12}	Propylbenzene	159.3	157.5	20		575
5233	C_9H_{12}	Pseudocumene	168.2	164.4	37		541
5234	$C_9H_{18}O$	2,6-Dimethyl-4-heptanone	168.0	177.5	>85		552
5235	$C_9H_{18}O_2$	Isoamyl butyrate	181.05	Nonazeotrope			575
5236	$C_9H_{18}O_2$	Isoamyl isobutyrate	169.8	Nonazeotrope			575
5237	$C_9H_{18}O_2$	Isobutyl isovalerate	171.35	Nonazeotrope			541
5238	$C_9H_{18}O_3$	Isobutyl carbonate	190.3	Nonazeotrope			575
5239	$C_{10}H_8$	Naphthalene	218.0	Nonazeotrope			575
5240	$C_{10}H_{14}$	Butylbenzene	183.1	172.0	65		567
5241	$C_{10}H_{14}$	Cymene	~176.7	165.5	55		532
5242	$C_{10}H_{16}$	Camphene	159.5	152.8	~38		572
5243	$C_{10}H_{16}$	*d*-Limonene	177.8	165.75	57		563
5244	$C_{10}H_{16}$	Nopinene	163.8	156.5	43		567
5245	$C_{10}H_{16}$	α-Terpinene	173.4	<165.0	<56		575
5246	$C_{10}H_{16}$	α-Phellandrene	171.5	163	43		563
5247	$C_{10}H_{16}$	α-Pinene	155.8	150.4	36.5		528
5248	$C_{10}H_{16}$	γ-Terpinene	181.5	166.8	62		538
5249	$C_{10}H_{16}$	Terpinolene	185	168	70		563
5250	$C_{10}H_{16}$	Thymene	179.7	166.5	60		532
5251	$C_{10}H_{18}O$	Linalool	198.6	Nonazeotrope			575
5252	$C_{10}H_{20}O_2$	Isoamyl isovalerate	192.7	Nonazeotrope			575
5253	$C_{10}H_{22}$	2,7-Dimethyloctane	160.2	155	~38		532
5254	$C_{10}H_{22}O$	Isoamyl ether	173.4	165.9	48		541
5255	$C_{11}H_{20}O$	Isobornyl methyl ether	192.4	Nonazeotrope			575
5256	$C_{12}H_{18}$	1,3,5-Triethylbenzene	215.5	Nonazeotrope			575
A =	$C_3H_6Cl_2O$	2,3-Dichloro-1-propanol	182.5				
5257	$C_3H_7NO_2$	Ethyl carbamate	185.25	>186.5	>20		575

No.	Formula	B-Component Name	B.P., °C	Azeotropic Data B.P., °C	Wt.%A	Ref.
A =	C₃H₆Cl₂O	2,3-Dichloro-1-propanol *(continued)*	182.5			
5258	C₄H₆O₄	Methyl oxalate	164.45	Nonazeotrope		575
5259	C₅H₁₂O₃	2-(2-Methoxyethoxy)-ethanol	192.95	Nonazeotrope		575 575
5260	C₆H₄Cl₂	o-Dichlorobenzene	179.5	174.2	40	567
5261	C₆H₄Cl₂	p-Dichlorobenzene	174.4	170.8	30	575
5262	C₆H₅Br	Bromobenzene	156.1	Nonazeotrope		575
5263	C₆H₅I	Iodobenzene	188.45	177.2	57	567
5264	C₆H₆O	Phenol	181.5	Azeotrope doubtful		563
5265	C₆H₇N	Aniline	184.35	~181		563
5266	C₆H₁₀O	Cyclohexanone	155.7	Nonazeotrope		552
5267	C₆H₁₀O₄	Ethyl oxalate	185.65	Nonazeotrope		575
5268	C₆H₁₂O	Cyclohexanol	160.8	Nonazeotrope		575
5269	C₆H₁₄O	Hexyl alcohol	157.85	Nonazeotrope		575
5270	C₆H₁₄O₂	2-Butoxyethanol	171.15	Nonazeotrope		575
5271	C₇H₇Br	m-Bromotoluene	184.3	175.8	50	567
5272	C₇H₇Br	o-Bromotoluene	181.75	171.6	45	563
5273	C₇H₇Br	p-Bromotoluene	185	176.2	52	575
5274	C₇H₇Cl	α-Chlorotoluene	179.35	171	40	563
5275	C₇H₇Cl	o-Chlorotoluene	159.2	Nonazeotrope		575
5276	C₇H₈	Toluene	110.75	Nonazeotrope		575
5277	C₇H₈O	Benzyl alcohol	205.25	Nonazeotrope		575
5278	C₇H₈O	o-Cresol	191.1	Nonazeotrope		542
5279	C₇H₁₄	Methylcyclohexane	101.15	Nonazeotrope		575
5280	C₇H₁₄O	2-Methylcyclohexanol	168.5	Nonazeotrope		575
5281	C₇H₁₆	Heptane	98.4	Nonazeotrope		575
5282	C₈H₈O	Acetophenone	202.0	Nonazeotrope		552
5283	C₈H₁₀	m-Xylene	139.0	Nonazeotrope		532
5284	C₈H₁₀	o-Xylene	144.3	Nonazeotrope		575
5285	C₈H₁₀O	Benzyl methyl ether	167.8	Nonazeotrope		575
5286	C₈H₁₀O	p-Methylanisole	177.05	175.5	32	575
5287	C₈H₁₆O	2-Octanone	172.85	184.0		552
5288	C₈H₁₈O	Octyl alcohol	195.2	Nonazeotrope		575
5289	C₈H₁₈O	sec-Octyl alcohol	178.7	~175		563
5290	C₉H₈	Indene	182.4	172.5	~57	532
5291	C₉H₁₂	Mesitylene	164.6	163	18	575
5292	C₉H₁₈O₂	Isoamyl butyrate	181.05	180.9?		575
5293	C₉H₁₈O₂	Isobutyl isovalerate	171.2	Nonazeotrope		575
5294	C₁₀H₈	Naphthalene	218.1	Nonazeotrope		563
5295	C₁₀H₁₄	Cymene	176.7	172.5	42	567
5296	C₁₀H₁₆	Camphene	159.6	156.0	25	532
5297	C₁₀H₁₆	d-Limonene	177.8	169.3	40	563
5298	C₁₀H₁₆	Nopinene	163.8	158.0	37	567
5299	C₁₀H₁₆	α-Pinene	155.8	153	20	532
5300	C₁₀H₁₆	α-Terpinene	173.4	<167.5	>40	575
5301	C₁₀H₁₆	γ-Terpinene	183	<173.5	<60	575
5302	C₁₀H₁₆	Terpinolene	~185	~174		563

TABLE I. *Binary Systems*   169

No.	Formula	B-Component Name	B.P., °C	B.P., °C	Wt.%A		Ref.
A =	C₃H₆Cl₂O	2,3-Dichloro-1-propanol	182.5				
		(continued)					
5303	C₁₀H₁₆	Thymene	179.7	170.8	50		532
5304	C₁₀H₁₈O	Cineole	176.35	Nonazeotrope			
				(reacts)			575
5305	C₁₀H₂₀O₂	Isoamyl isovalerate	192.7	Nonazeotrope			575
5306	C₁₂H₁₈	1,3,5-Triethylbenzene	215.5	Nonazeotrope			575
A =	**C₃H₆O**	**Acetone**	**56.15**				
5307	C₃H₆O	Allyl alcohol	96.85	Nonazeotrope			552
5308	C₃H₆O	Propionaldehyde	48.7	Nonazeotrope			552
5309	C₃H₆O₂	Ethyl formate	54.15	Nonazeotrope			552
5310	C₃H₆O₂	Methyl acetate	57	55	50		552,834
		"		55.8	48.3		497
		"	57.1	55.7	49.2	v-l	682
5311	C₃H₇Br	1-Bromopropane	71.0	56.13	98		552
		"	71.0	Nonazeotrope			389
5312	C₃H₇Br	2-Bromopropane	59.4	54.12	42		552
5313	C₃H₇Cl	1-Chloropropane	46.65	45.8	15		555
5314	C₃H₇Cl	2-Chloropropane	34.9	Nonazeotrope			552
5315	C₃H₇I	2-Iodopropane	89.45	Nonazeotrope			552
5316	C₃H₇NO₂	Isopropyl nitrite	40.1	Nonazeotrope			552
5317	C₃H₇NO₂	Propyl nitrite	47.75	Nonazeotrope			552
5318	C₃H₈	Propane crit press.		Nonazeotrope		v-l	442
5319	C₃H₈O	Isopropyl alcohol	82.3	Nonazeotrope		v-l	151,152,
		55°C.		Nonazeotrope		v-l	290c
		" 710 mm.		Nonazeotrope		v-l	36k
		"	82.3	Nonazeotrope		v-l	552,290c
5320	C₃H₈O	*n*-Propyl alcohol	97.2	Nonazeotrope			552
5321	C₃H₈O₂	Methylal	42.3	Nonazeotrope			552
5322	C₃H₈S	Propanethiol	67.5	54.5	~67		563
5323	C₃H₉BO₃	Methyl borate	68.7	55.45	82.5		552
5324	C₃H₉N	Propylamine	49.7	< 48.0	>20		551
5325	C₄H₄O₂	Diketene		Nonazeotrope			981
5326	C₄H₄S	Thiophene	84.7	Nonazeotrope			552
5327	C₄H₆O₂	Vinyl acetate	72.7	Nonazeotrope			981
		"	72	Nonazeotrope		v-l	862
5328	C₄H₆O₃	Acetic anhydride	139.6	Nonazeotrope		v-l	428
5329	C₄H₇Cl	1-Chloro-2-methyl-					
		propene	68	55.6	81		703
5330	C₄H₈O	2-Butanone,					
		15-500 p.s.i.a.		Nonazeotrope		v-l	726
		"	79.6	Nonazeotrope			813
		"	79.5	Nonazeotrope		v-l	31f,190b
5331	C₄H₈O	Butyraldehyde	75.2	Nonazeotrope			552
5332	C₄H₈O	Isobutyraldehyde	63.5	Nonazeotrope			552
5333	C₄H₈O₂	Dioxane	101.4	Nonazeotrope			201
5334	C₄H₈O₂	Ethyl acetate	77.1	Nonazeotrope			834
		"	77.1	Nonazeotrope		v-l	931

		B-Component		Azeotropic Data			
No.	Formula	Name	B.P., °C	B.P., °C	Wt.%A		Ref.
A =	C₃H₆O	**Acetone** *(continued)*	**56.15**				
5335	C₄H₈O₂	Isopropyl formate	68.8	Nonazeotrope			552
5336	C₄H₉Br	2-Bromo-2-methylpropane	73.25	Nonazeotrope			552
5337	C₄H₉Ci	1-Chlorobutane	78.5	Nonazeotrope			552
5338	C₄H₉Ci	2-Chlorobutane	68.25	55.75	80		552
5339	C₄H₉Cl	1-Chloro-2-methylpropane	68.85	55.75	75		552
5340	C₄H₉Cl	2-Chloro-2-methylpropane	50.8	49.2	25		552
5341	C₄H₉NO₂	Butyl nitrite	78.2	Nonazeotrope			552
5342	C₄H₉NO₂	Isobutyl nitrite	67.1	Nonazeotrope			552
5343	C₄H₁₀	Butane crit. press		Nonazeotrope		v-l	442
5344	C₄H₁₀O	Butyl alcohol	117.7	Nonazeotrope		v-l	96
		At 25°C.		Nonazeotrope		v-l	282
5345	C₄H₁₀O	*tert*-Butyl alcohol	82.45	Nonazeotrope			552
5346	C₄H₁₀O	Ethyl ether	34.6	Nonazeotrope			371,563,834
5347	C₄H₁₀O	Isobutyl alcohol	108.0	Nonazeotrope			552
5348	C₄H₁₀O	Methyl propyl ether	38.9	Nonazeotrope			563
5349	C₄H₁₀S	Ethyl sulfide	92.1	Nonazeotrope			566
5350	C₄H₁₁N	Butylamine	77.8	Nonazeotrope			551
5351	C₄H₁₁N	Diethylamine	55.5	51.39	38.21		551,633
5352	C₄H₁₁N	Isobutylamine	68.0	< 56.0	<96	·	551
5353	C₅H₅N	Pyridine	115	Nonazeotrope		v-l	428
5354	C₅H₆	Cyclopentadiene	41.0	Min. b.p.			258
5355	C₅H₈	Isoprene	34.3	30.5	20		258,552
		"	34.1	33.8	5.3	v-l	716
5356	C₅H₈	3-Methyl-1,2-butadiene	40.8	35.3	27		552
5357	C₅H₈	Piperylene	42.5	Min. b.p.			258
		"	42.3	41.4	18.1	v-l	716
5358	C₅H₈O₂	Isopropenyl acetate	96.5	Nonazeotrope			413
5359	C₅H₁₀	Cyclopentane	49.3	41.0	36		552
5360	C₅H₁₀	2-Methyl-1-butene	31.05	Min. b.p.			258
		"	31.1	30.1	11.9	v-l	716
5361	C₅H₁₀	2-Methyl-2-butene	38.5	35.6	21.1	v-l	716
		"	37.1	32.5	22		258,552
5362	C₅H₁₀	3-Methyl-1-butene	20.6	19.7	7		258,552
5363	C₅H₁₀	1-Pentene	30.1	Min. b.p.			258
		"	29.97	28.9	19 vol. %		826
5364	C₅H₁₀	2-Pentene	36.4	Min. b.p.			258
5365	C₅H₁₀O₂	Isopropyl acetate	88.7	Nonazeotrope			981
5366	C₅H₁₀O₂	Propyl acetate	101.6	Nonazeotrope		v-l	931
5367	C₅H₁₂	2-Methylbutane	27.9	25.6	14.4	v-l	716
		"	27.95	25.7	12		552
5368	C₅H₁₂	Pentane	36.15	31.9	21		552
		"	36.15	32.5	20		366
		" <100 mm.		Nonazeotrope			366
		" —35°–25°C.				v-l	779
		" crit. pres.		194.5	17	v-l	442
		"	36.1	31.86	21.9	v-l	590
		" crit. press.		194.5	17	v-l	442,148c

TABLE I. *Binary Systems*   171

No.	Formula	B-Component Name	B.P., °C	Azeotropic Data B.P., °C	Wt.%A		Ref.
A =	**C₃H₆O**	**Acetone** *(continued)*	**56.15**				
5369	C₅H₁₂O	*tert*-Amyl alcohol	102.35	Nonazeotrope			552
5370	C₅H₁₂O	Ethyl propyl ether	63.6	< 56.1	<95		552
5371	C₅H₁₂O₂	Diethoxymethane	87.95	Nonazeotrope			552
							552
5372	C₆H₅Cl	Chlorobenzene	131.6	Nonazeotrope		v-l	722
5373	C₆H₅F	Fluorobenzene	84.9	Nonazeotrope			552
5374	C₆H₆	Benzene	80.1	Nonazeotrope		v-l	261,722, 802,903
		" 45°C.		Nonazeotrope		v-l	92
		" 80.1		Nonazeotrope		v-l	126
		" 2 atm. to crit. press.		Nonazeotrope		v-l	122c
5375	C₆H₆O	Phenol	181.5	Nonazeotrope		v-l	668c,563
5376	C₆H₈	1,3-Cyclohexadiene	80.4	< 55	<85		552
5377	C₆H₁₀	Biallyl	60.1	47.1	45		552
5378	C₆H₁₂	Cyclohexane	80.75	53.0	67		273,552
		"	80.75	53.0	67.5	v-l	503
		"	80.75		67		735
		"			68	v-l	621
		"		55	69	v-l	621
		"		45	69.3	v-l	621
		"		35	66.7	v-l	621
5379	C₆H₁₂	1-Hexene	63.6	50.1	51.4	v-l	716
		" 35°C.			52	v-l	496g
5380	C₆H₁₂	Methylcyclopentane	72.0	50.3	57		516,552
		"	72.0		57		735
5381	C₆H₁₂	1,1,2-Trimethyl-cyclopropane	52.6	42.3	32 vol. %		826
5382	C₆H₁₂O	4-Methyl-2-pentanone	115.9	Nonazeotrope		v-l	438
5383	C₆H₁₂O₂	Butyl acetate	126.2	Nonazeotrope		v-l	186
5384	C₆H₁₄	2,3-Dimethylbutane	58.0	46.3	42		552
		"	58.0		46.5		735
		"	58	44.7	37.4	v-l	638c
		"		45.5	~40	v-l	1026c
5385	C₆H₁₄	Hexane	68.95	49.8	59		735,981
		"	68.7	49.6	53		716
		" crit. press.		220	47	v-l	442
		" —35°–25°C.				v-l	779
		" 900 mm.		55	55.2	v-l	498
		"	68.7	49.7	54.5	v-l	498
		" 648 mm.		45.0	54.4	v-l	498
		" 447 mm.		35.0	53.9	v-l	498
							498
		"		51.8	63.4		497
		"	68.8	49.7	53.5		552
5385a	C₆H₁₄O	Hexyl alcohol		Nonazeotrope		v-l	779e
5386	C₆H₁₄O	Isopropyl ether	69.0	54.2	61		261

		B-Component		Azeotropic Data		
No.	Formula	Name	B.P., °C	B.P., °C	Wt.%A	Ref.
A =	C_3H_6O	**Acetone** *(continued)*	**56.15**			
5387	$C_6H_{14}O$	Propyl ether	90.1	Nonazeotrope		552
5388	$C_6H_{14}O_2$	Acetal	104.5	Nonazeotrope		563
5389	$C_6H_{15}N$	Triethylamine	89.35	Nonazeotrope		551
5390	$C_7H_6O_2$	Benzoic acid	249.5	Nonazeotrope		575
5391	C_7H_8	Toluene	110.75	Nonazeotrope		552
5392	C_7H_{14}	Methylcyclohexane	101.15	Nonazeotrope		552
5393	C_7H_{16}	Heptane	98.4	55.85	89.5	552
		"	98.45	Nonazeotrope		359
		"	98.4		89.5	735
		"	98.4	b.p. curve		716
		" crit. press.		232	80 v-l	442
5394	C_8H_{18}	2,5-Dimethylhexane	109.4	Nonazeotrope		552
5395	C_8H_{18}	Octane crit. press.		Nonazeotrope		442
5396	$C_{10}H_{22}$	Decane crit. pres.		Nonazeotrope		442
5397	$C_{13}H_{28}$	Tridecane crit. press.		Nonazeotrope		442
A =	C_3H_6O	**Allyl Alcohol**	**96.95**			
5397a	C_3H_6O	Propylene oxide	35	Nonazeotrope	v-l	190b
5398	$C_3H_6O_3$	Methyl carbonate	90.5	86.4	23	527
5399	C_3H_7Br	1-Bromopropane	71.0	69.5	9	532
5400	C_3H_7Br	2-Bromopropane	59.4	Nonazeotrope		527
5401	C_3H_8O	Isopropyl alcohol	82.3	Nonazeotrope	v-l	468
5402	C_3H_8O	Propyl alcohol	97.2	96.73	74	569
5403	$C_4H_7ClO_2$	Ethyl chloroacetate	143.55	Nonazeotrope		575
5404	C_4H_8O	2-Butanone	79.6	Nonazeotrope		552
5405	C_4H_8OS	Ethyl thioacetate	116.5	< 96.5		575
5406	$C_4H_8O_2$	Dioxane	101.35	Nonazeotrope		527
5407	$C_4H_8O_2$	Ethyl acetate	77.05	Nonazeotrope		563
5408	$C_4H_8O_2$	Methyl propionate	79.7	Nonazeotrope	·	527
5409	$C_4H_8O_2$	Propyl formate	80.8	80.5		536
5410	C_4H_9Br	1-Bromobutane	101.5	89.5	30	527
5411	C_4H_9Br	1-Bromo-2-methylpropane	91.6	83.9	18	875
5412	C_4H_9Cl	1-Chlorobutane	78.5	74.5	15	567
5413	C_4H_9Cl	1-Chloro-2-methylpropane	68.85	67	~ 7	875
5414	C_4H_9I	1-Iodobutane	130.4	< 96.4	< 74	575
5414a	C_4H_9I	1-Iodo-2-methylpropane	120.8	93.8		571c
5415	$C_4H_{10}O$	*sec*-Butyl alcohol	99.6	Nonazeotrope		563
		"	99.5	Nonazeotrope	v-l	998
		" 100 mm.		Nonazeotrope	v-l	998
5416	$C_4H_{10}S$	Ethyl sulfide	92.1	85.1	45	527
5417	C_5H_5N	Pyridine	115.4	Nonazeotrope		553
5418	C_5H_8O	Allyl vinyl ether			10	1008
		"	67.4	66.6	5	981
5419	$C_5H_8O_2$	Allyl acetate		95.1	63	3
		"	104	Min. b.p		4
5420	$C_5H_{10}O$	3-Methyl-2-butanone	95.4	93.5	36	552
5421	$C_5H_{10}O$	2-Pentanone	102.35	96.0	70	552
5422	$C_5H_{10}O$	3-Pentanone	102.05	95.95	72	552

TABLE I.   *Binary Systems*   173

No.	Formula	B-Component Name	B.P., °C	Azeotropic Data B.P., °C	Wt.%A	Ref.
A =	C$_3$H$_6$O	**Allyl Alcohol** *(continued)*	**96.95**			
5423	C$_5$H$_{10}$O$_2$	Ethyl propionate	99.1	~ 93.2	~ 54	875
5424	C$_5$H$_{10}$O$_2$	Isobutyl formate	98.3	93	~ 52	875
5425	C$_5$H$_{10}$O$_2$	Methyl butyrate	102.75	< 94.7	< 51	527
5426	C$_5$H$_{10}$O$_2$	Methyl isobutyrate	92.5	89.8	28	527
5427	C$_5$H$_{10}$O$_2$	Propyl acetate	101.6	94.6	52	527
5427a	C$_5$H$_{11}$Br	1-Bromo-3-methylbutane	120.65	94.4		571c
5428	C$_5$H$_{11}$Cl	1-Chloro-3-methylbutane	99.4	88.3	29	527
5429	C$_5$H$_{12}$O$_2$	Diethoxymethane	87.95	< 87.0	>11	527
5430	C$_6$H$_5$Cl	Chlorobenzene	131.8	96.5	82.5	527
5431	C$_6$H$_6$	Benzene	80.2	76.75	17.36	834,875,1005
5432	C$_6$H$_8$	1,3-Cyclohexadiene	80.8	75.9	~21	875
5433	C$_6$H$_{10}$	Cyclohexene	82.75	76.3	~21.7	875
5434	C$_6$H$_{10}$O	Allyl ether	94.84	89.8	30.0	875
5435	C$_6$H$_{12}$	Cyclohexane	80.75	74	~20	875
		"	80.8	74.0	58	413
5436	C$_6$H$_{12}$	Methylcyclopentane	72.0	67.8	<10	567
5437	C$_6$H$_{12}$O$_2$	Ethyl isobutyrate	110.1	~96.2		536
5438	C$_6$H$_{12}$O$_2$	Isobutyl acetate	117.4	Nonazeotrope		527
5439	C$_6$H$_{14}$	2,3-Dimethylbutane	58.0	< 56.7		575
5440	C$_6$H$_{14}$	Hexane	68.95	65.5	4.5	875
		"	68.8		12.5	480
5441	C$_6$H$_{14}$O	Isopropyl ether	68.3	Nonazeotrope		981
5442	C$_6$H$_{14}$O	Propyl ether	90.1	85.7	30	527
5443	C$_7$H$_8$	Toluene	110.6	91-92	50	834,875
5444	C$_7$H$_{14}$	Methylcyclohexane	101.1	85.0	42	537
5445	C$_7$H$_{16}$	Heptane	98.45	84.5	~37	537
5446	C$_8$H$_{10}$	m-Xylene	139.0	Nonazeotrope		537
5447	C$_8$H$_{18}$	2,5-Dimethylhexane	109.4	89.3	50	567
5448	C$_8$H$_{18}$	Octane	125.75	93.4	68	567
5449	C$_9$H$_{12}$	Cumene	152.8	Nonazeotrope		575
5450	C$_{10}$H$_{16}$	Camphene	159.6	Nonazeotrope		575
5451	C$_{10}$H$_{16}$	d-Limonene	177.8	Nonazeotrope		537
5452	C$_{10}$H$_{16}$	α-Pinene	155.8	Nonazeotrope		537
A =	**C$_3$H$_6$O**	**Propionaldehyde**	**48.7**			
5453	C$_3$H$_6$O	Propylene oxide, 30 p.s.i.g.	69	Nonazeotrope		981
5454	C$_3$H$_6$O$_2$	Methyl acetate	56.95	Nonazeotrope		575
5455	C$_3$H$_7$Cl	1-Chloropropane	46.65	< 46.4		575
5456	C$_3$H$_7$NO$_2$	Isopropyl nitrite	40.0	Nonazeotrope		548
5457	C$_3$H$_7$NO$_2$	Propyl nitrite	57.75	< 47.3	>18	548
5458	C$_4$H$_8$O	Cyclopropyl methyl ether	44.73	43		878
5459	C$_5$H$_6$O	2-Methylfuran	63.7	Nonazeotrope		769
A =	**C$_3$H$_6$O**	**Propylene Oxide**	**35**			
5460	C$_4$H$_{10}$O	Ethyl ether	34.5	32.6	49.6	707
5461	C$_5$H$_8$	Cyclopentene	43.6	Min. b.p.		1014
5462	C$_5$H$_8$	Isoprene	34.5	31.6	60	1014

No.	Formula	B-Component Name	B.P., °C	B.P., °C	Wt.%A	Ref.
A =	C₃H₆O₂	**Propylene Oxide**	**35**			
		(continued)				
5463	C₅H₁₀	Cyclopentane	49.4	Min.	b.p.	1014
5464	C₅H₁₀	2-Methyl-1-butene	32	27	47	1014
5465	C₅H₁₀	Pentenes		Min.	b.p.	1014
5466	C₅H₁₀	2-Pentene	35.8	30˙	54	1014
5467	C₅H₁₂	2-Methylbutane	27.95	Nonazeotrope		558
5468	C₅H₁₂	Pentanes		Min.	b.p. azeotrope	1014
5469	C₅H₁₂	Pentane	36	27.5	57	1014
5470	C₆H₁₂	Cyclohexane	80.75	Min.	b.p.	1014
5471	C₆H₁₂	Hexenes		Min.	b.p.	1014
5472	C₆H₁₄	Hexanes		Min.	b.p., azeotrope	1014
A =	C₃H₆OS	**Methyl Thioacetate**	**95.5**			
5473	C₃H₈O	Isopropyl alcohol	82.4	< 81.5		575
5474	C₃H₈O	Propyl alcohol	97.2	< 91.5		575
5475	C₄H₉ClO	Chloroethyl ethyl ether	98.5	< 95.2	>85	575
5476	C₄H₁₀S	Ethyl sulfide	92.1	< 91.0	>28	575
5477	C₄H₁₀S	2-Methyl-1-propanethiol	87.8	< 87.2	<12	575
A =	C₃H₆O₂	**1,3-Dioxolane**	**75.6**			
5478	C₆H₆	Benzene	80.2	74	85	515
5479	C₇H₈	Toluene	110.6	Nonazeotrope		981·
5480	C₇H₁₆	Heptane	98.4	72.3	81	981
5481	C₈H₁₈O	Butyl ether	142.1	Nonazeotrope		981
5482	C₉H₂₀	Nonane	150.8	Nonazeotrope		981
A =	C₃H₆O₂	**Ethyl Formate**	**54.1**			
5483	C₃H₆O₂	Methyl acetate	56.25	Nonazeotrope		532
5484	C₃H₇Br	1-Bromopropane	71.0	Nonazeotrope		538
5485	C₃H₇Br	2-Bromopropane	59.35	53	69	527
		"	59.35	53.0	59.4	579,720
5486	C₃H₇Cl	1-Chloropropane	46.65	46.25	15	555
5487	C₃H₇Cl	2-Chloropropane	34.15	Nonazeotrope		547
5488	C₃H₇NO₂	Isopropyl nitrite	40.1	Nonazeotrope		549
5489	C₃H₇NO₂	Propyl nitrite	47.75	47.4	12	527
5490	C₃H₈O	Isopropyl alcohol	82.35	Nonazeotrope		536
5491	C₃H₈O₂	Methylal	42.25	Nonazeotrope		557
5492	C₃H₈S	1-Propanethiol	67.5	~52		561
5493	C₄H₈O	Isobutyraldehyde	63.5	Nonazeotrope		575
5494	C₄H₉Cl	1-Chloro-2-methylpropane	68.9	Nonazeotrope		563
5495	C₄H₉Cl	2-Chloro-2-methylpropane	51.6	48.5	35	547
5496	C₄H₁₀O	Ethyl ether	34.6	Nonazeotrope		557
5497	C₄H₁₀O₂	Acetaldehyde dimethyl acetal	64.3	Nonazeotrope		557
5498	C₅H₈	Isoprene	34.2	< 32.5	<24	562
5499	C₅H₁₀	Cyclopentane	49.4	< 42.0	<45.0	562
5500	C₅H₁₀	2-Methyl-2-butene	37.15	Nonazeotrope		563
5501	C₅H₁₂	2-Methylbutane	27.95	26.5	18	531
5502	C₅H₁₂	Pentane	36.2	32.5	30	546
5503	C₅H₁₂O	Ethyl propyl ether	63.6	Nonazeotrope		557
5504	C₆H₆	Benzene	80.2	Nonazeotrope		563
5505	C₆H₁₀	Biallyl	60.2	~45.2	~58	573

TABLE I.   *Binary Systems*   175

No.	Formula	Name	B.P., °C	B.P., °C	Wt.%A		Ref.
		B-Component			Azeotropic Data		
A =	C₃H₆O₂	**Ethyl Formate** *(continued)*	54.1				
5506	C₆H₁₂	Methylcyclopentane	72.0	51.2	75		562
5507	C₆H₁₄	2,3-Dimethylbutane	58.0	45.0	52		562
5508	C₆H₁₄	n-Hexane	68.95	49.0	~67		573
5509	C₉H₁₀O₂	Ethyl benzoate	213	Vapor pressure curve			563
5510	C₉H₁₂	Pseudocumene	169	Vapor pressure data			563
A =	C₃H₆O₂	**Glycidol**					
5510a	C₉H₁₂	Cumene	14 mm.	38			54c
		"	50 mm.	63	21		54c
A =	C₃H₆O₂	**Methyl Acetate**	56.95				
5511	C₃H₇Br	1-Bromopropane	71.0	Nonazeotrope			538
5512	C₃H₇Br	2-Bromopropane	59.35	56	68		527
		"	59.35	55.6	14.5		579
5513	C₃H₇Cl	1-Chloropropane	46.65	Nonazeotrope			555
5514	C₃H₇NO₂	Isopropyl nitrite	40.1	Nonazeotrope			550
5515	C₃H₇NO₂	Propyl nitrite	47.75	Nonazeotrope			549
5516	C₃H₈O	Isopropyl alcohol	82.35	Nonazeotrope		v-l	536,680
5517	C₃H₈O₂	Methylal	42.25	Nonazeotrope			557
5518	C₃H₉BO₃	Methyl borate	68.7	Nonazeotrope			575
5519	C₄H₈O	2-Butanone	79.6	Nonazeotrope			527
		"	79.5	Nonazeotrope		v-l	31f
5520	C₄H₈O	Butyraldehyde	75.5	Nonazeotrope			548
5521	C₄H₈O	Isobutyraldehyde	63.5	Nonazeotrope			575
5521a	C₄H₈O₂	Ethyl acetate	77	Nonazeotrope		v-l	507k
5522	C₄H₈O₂	Isopropyl formate	68.8	Nonazeotrope			575
5523	C₄H₉Cl	2-Chlorobutane	68.25	Nonazeotrope			575
5524	C₄H₉Cl	1-Chloro-2-methylpropane	68.9	Nonazeotrope			563
5525	C₄H₉Cl	2-Chloro-2-methylpropane	51.6	Nonazeotrope			538
5526	C₄H₉NO₂	Isobutyl nitrite	67.1	Nonazeotrope			550
5527	C₄H₁₀O	Ethyl ether	34.6	Nonazeotrope			557
5528	C₄H₁₀O₂	Ethoxymethoxymethane	65.9	Nonazeotrope			557
5529	C₄H₁₀O₂	Acetaldehyde dimethyl acetal	64.3	Nonazeotrope			557
5530	C₄H₁₁N	Diethylamine	56	~53			563
5531	C₅H₆O	2-Methylfuran	63.7	Nonazeotrope			557,769
5532	C₅H₈	Cyclopentene	44.4	41.7	27.7		354
5533	C₅H₁₀	Cyclopentane	49.3	43.2	37.9		354
5534	C₅H₁₀	2-Methyl-2-butene	37.2	< 36.9	<12		575
5535	C₅H₁₀	1-Pentene	30.1	30.0	3.3		354
5536	C₅H₁₂	Pentane	36.08	34.05	22		354
		"	36.15	Nonazeotrope			537
5537	C₆H₆	Benzene	80.2	Nonazeotrope		v-l	563,681,406a
		"	80.1	56.7	99.7		354
5538	C₆H₈	1,3-Cyclohexadiene	80.25	56.7	98.0		354
5539	C₆H₁₀	Biallyl	60.0	51	60		537
5540	C₆H₁₀	Cyclohexene	83	Nonazeotrope			546
		"	83.1	56.5	90.2		354

No.	Formula	Name	B.P., °C	B.P., °C	Wt.%A		Ref.
		B-Component		**Azeotropic Data**			
A =	$C_3H_6O_2$	**Methyl Acetate** *(continued)*	**56.95**				
5541	C_6H_{12}	Cyclohexane	80.6	54.9	83.0		354
		"	80.8	Nonazeotrope			546
		"	80.7	55.5	78	v-l	681
5542	C_6H_{12}	2,3-Dimethyl-1-butene	55.62	48.95	42.8		354
5543	C_6H_{12}	2,3-Dimethyl-2-butene	73.38	55.1	71.8		354
5544	C_6H_{12}	3,3-Dimethyl-1-butene	41.4	39.9	8.8		354
5545	C_6H_{12}	2-Ethyl-1-butene	64.8	52.8	60.1		354
5546	C_6H_{12}	1-Hexene	63.58	52.5	63.6		354
5547	C_6H_{12}	*cis*-2-Hexene	68.55	53.7	69.8		354
5548	C_6H_{12}	3-Methyl-2-pentene	70.64	54.45	73.7		354
5549	C_6H_{12}	4-Methyl-1-pentene	54.0	48.3	36.7		354
5550	C_6H_{12}	*trans*-4-Methyl-2-pentene	58.45	50.0	51.3		354
5551	C_6H_{12}	Methylcyclopentane	71.8	53.0	68.0		354
5552	C_6H_{14}	2,2-Dimethylbutane	49.65	43.7	38.2		354
5553	C_6H_{14}	2,3-Dimethylbutane	58.05	48.0	48.25		354
		"	58.0	51.2	50		575
5554	C_6H_{14}	*n*-Hexane	68.95	< 56.65	<90		575
		"		51.8	63.4		497
		"	68.85	51.75	60.7		354
5555	C_6H_{14}	2-Methylpentane	60.2	49.25	51.6		354
5556	C_6H_{14}	3-Methylpentane	63.25	50.05	57.4		354
5557	C_7H_{16}	2,4-Dimethylpentane	80.7	54.7	72.4		354
5558	C_7H_{16}	Heptane	98.45	56.65	96.45		354
5559	C_7H_{16}	2-Methylhexane	90.0	56.0	88.6		354
5560	C_7H_{16}	3-Methylhexane	91.85	56.3	84.9		354
5561	C_7H_{16}	2,2,3-Trimethylbutane	80.9	55.1	74.2		354
A =	$C_3H_6O_2$	**Propionic Acid**	**140.9**				
5562	C_3H_7I	1-Iodopropane	102.4	Nonazeotrope			542
5563	$C_4H_6Cl_2O_2$	Ethyl dichloroacetate	158.1	Nonazeotrope			575
5564	$C_4H_6O_3$	Acetic anhydride	138	Nonazeotrope		v-l	718
5565	$C_4H_6O_3$	Methyl pyruvate	137.5	<137.2	>75		552
5566	$C_4H_7BrO_2$	Ethyl bromoacetate	158.8	Nonazeotrope			527
5567	$C_4H_7ClO_2$	Ethyl chloroacetate	143.55	<140.35	<61		527
5568	C_4H_8O	2-Butanone	79.6	Nonazeotrope		v-l	722
5569	$C_4H_8O_2$	Butyric acid	163.5	Nonazeotrope		v-l	26
5570	$C_4H_8O_2$	Dioxane	101.35	Nonazeotrope			527
5571	C_4H_9Br	1-Bromobutane	101.5	Nonazeotrope			527
5572	C_4H_9I	1-Iodobutane	130.4	126.8	15		527
5573	C_4H_9I	1-Iodo-2-methylpropane	120.4	119.3	7		527
5574	$C_4H_9NO_3$	Isobutyl nitrate	123.5	122.0	9		559
5575	$C_4H_{10}S$	Ethyl sulfide	92.1	Nonazeotrope			566
5576	$C_5H_4O_2$	2-Furaldehyde	161.5	Nonazeotrope			563
5577	C_5H_5N	Pyridine	115.5	148-150	74		419
		"	115.5	148.6	67.2		1068
5578	C_5H_8O	Cyclopentanone	130.65	Nonazeotrope			552

TABLE I.   *Binary Systems*   177

No.	Formula	B-Component Name	B.P., °C	B.P., °C	Wt.%A	Ref.
				Azeotropic Data		
A =	$C_3H_6O_2$	**Propionic Acid** *(continued)*	**140.9**			
5579	$C_5H_8O_2$	2,4-Pentanedione	138	144	~70	563
5580	$C_5H_8O_3$	Ethyl pyruvate	155.5	Nonazeotrope		552
5581	$C_5H_9ClO_2$	Propyl chloroacetate	163.5	Nonazeotrope		575
5582	$C_5H_{10}O_3$	Ethyl carbonate	126.5	Nonazeotrope		575
5583	$C_5H_{10}O_3$	2-Methoxy ethyl acetate	144.6	146.85	36	568
5584	$C_5H_{11}Br$	1-Bromo-3-methylbutane	120.65	119.45	7.5	527
5585	$C_5H_{11}Cl$	1-Chloro-3-methylbutane	99.4	Nonazeotrope		527
5586	$C_5H_{11}I$	1-Iodo-3-methylbutane	147.65	136.5	42	527
5587	$C_5H_{11}NO$	N,N-Dimethylpropion- amide	175.5	179.3	23.6	832
5588	$C_5H_{11}NO_3$	Isoamyl nitrate	~149.6	138.8	59	527
5589	$C_6H_4Cl_2$	o-Dichlorobenzene	179.5	Nonazeotrope		527
5590	$C_6H_4Cl_2$	p-Dichlorobenzene	174.6	Nonazeotrope		527
5591	C_6H_5Br	Bromobenzene	156.1	140.15		571
5592	C_6H_5Cl	Chlorobenzene	132.0	128.9	18	542
5593	C_6H_6	Benzene	80.15	Nonazeotrope		527
5594	C_6H_7N	2-Picoline	131	~ 164		563
5595	C_6H_7N	3-Picoline	144	155-163		1032
		At 212 mm.		122	48.5	174,175
5596	C_6H_7N	4-Picoline	145.3	155-163		810
		At 212 mm.		124	48.1	174,175
5597	$C_6H_{10}O$	Cyclohexanone	155.7	Nonazeotrope		552
5598	$C_6H_{10}O$	Mesityl oxide	129.45	Nonazeotrope		552
5599	$C_6H_{10}S$	Allyl sulfide	139.35	134.6	40	555
5600	C_6H_{12}	Cyclohexane	80.75	Nonazeotrope		527
5601	$C_6H_{12}O$	2-Hexanone	127.2	Nonazeotrope		552
5602	$C_6H_{12}O$	3-Hexanone	123.3	Nonazeotrope		552
5603	$C_6H_{12}O$	4-Methyl-2-pentanone	116.05	Nonazeotrope		527
5604	$C_6H_{12}O_2$	Butyl acetate	126.0	Nonazeotrope		527
5605	$C_6H_{12}O_2$	Isoamyl formate	123.6	Nonazeotrope		563
5606	$C_6H_{12}O_2$	Propyl propionate	123.0	Nonazeotrope		527
5607	$C_6H_{12}O_3$	2-Ethoxyethyl acetate	156.8	Nonazeotrope		526
5608	$C_6H_{13}Br$	1-Bromohexane	156.5	139.0	60	575
5609	C_6H_{14}	Hexane	68.85	Nonazeotrope		507
5610	$C_6H_{14}O$	Propyl ether	90.1	Nonazeotrope		527
5611	$C_6H_{14}S$	Propyl sulfide	141.5	136.5	45	566
5612	C_7H_6O	Benzaldehyde	179.2	Nonazeotrope		575
5613	C_7H_7Br	m-Bromotoluene	184.3	Nonazeotrope		527
5614	C_7H_7Br	o-Bromotoluene	181.5	Nonazeotrope		527
5615	C_7H_7Cl	α-Chlorotoluene	179.3	Nonazeotrope		527
5616	C_7H_7Cl	o-Chlorotoluene	159.3	139.4	67	538
5617	C_7H_7Cl	p-Chlorotoluene	162.4	139.8	~ 75	538
5618	C_7H_8	Toluene	110.75	110.45	3	527
5619	C_7H_8O	Anisole	153.85	141.17	87	527
5620	C_7H_9N	2,6-Lutidine	144	155-163		810
		At 212 mm.		119	48.8	174,175

No.	Formula	B-Component Name	B.P., °C	Azeotropic Data B.P., °C	Wt.%A	Ref.
A =	C₃H₆O₂	**Propionic Acid** (*continued*)	**140.9**			
5621	C₇H₁₄	Methylcyclohexane	101.15	Nonazeotrope		575
5622	C₇H₁₄O₂	Ethyl isovalerate	134.7	Nonazeotrope		575
5623	C₇H₁₄O₂	Ethyl valerate	145.45	Nonazeotrope		575
5624	C₇H₁₄O₂	Isoamyl acetate	142.1	Nonazeotrope		527
5625	C₇H₁₄O₂	Methyl caproate	149.7	Nonazeotrope		575
5626	C₇H₁₄O₂	Propyl butyrate	142.8	Nonazeotrope		541
5627	C₇H₁₆	Heptane	98.15	97.82	2.0	507
5628	CₙHₓ	Hydrocarbons	138-140	134	67	804
5629	C₈H₈	Styrene	145.8	135.0	~ 47	545
5630	C₈H₁₀	Ethylbenzene	136.15	131.1	28	563
		At 60 mm.	60.5	58.5	10	53
5631	C₈H₁₀	*m*-Xylene	139.0	132.65	35.5?	563
5632	C₈H₁₀	*o*-Xylene	143.6	135.4	43	527
5633	C₈H₁₀	*p*-Xylene	138.2	132.5	34	527
5634	C₈H₁₀O	Benzyl methyl ether	167.8	Nonazeotrope		527
5635	C₈H₁₀O	*p*-Methylanisole	177.05	Nonazeotrope		575
5636	C₈H₁₀O	Phenetole	170.45	Nonazeotrope		527
5637	C₈H₁₆	1,3-Dimethylcyclohexane	120.7	118.2	18	562
5638	C₈H₁₆O₂	Amyl propionate		Nonazeotrope		804
5639	C₈H₁₆O₂	Isobutyl isobutyrate	148.6	Nonazeotrope		575
5640	C₈H₁₆O₂	Propyl isovalerate	155.7	Nonazeotrope		527
5641	C₈H₁₈	2,5-Dimethylhexane	109.4	108.0	8	563
5642	C₈H₁₈	Octane	125.75	121.5	< 30	563
			125.12	120.89	24	507
		"	125.12		24.2 v-l	427
		" Satd. with Na propionate			6 v-l	427
5643	C₈H₁₈O	Butyl ether	142.4	136.0	45	527
5644	C₈H₁₈O	Isobutyl ether	122.3	< 121.5	< 6	575
5645	C₉H₇N	Quinoline	237.3	Nonazeotrope		553
5646	C₉H₈	Indene	182.6	Nonazeotrope		527
5647	C₉H₁₂	Cumene	152.8	139.0	65	527
5648	C₉H₁₂	Mesitylene	164.0	139.3	77	563
5649	C₉H₁₂	Propylbenzene	158	139.5	75	527
5650	C₉H₁₂	Pseudocumene	168.2	Nonazeotrope		541
5651	C₉H₂₀	Nonane	150.67	134.27	54	507
5652	C₁₀H₁₄	Butylbenzene	183.1	Nonazeotrope		527
5653	C₁₀H₁₄	Cymene	175.5	Nonazeotrope		527
5654	C₁₀H₁₆	Camphene	159.6	138	65	527
5655	C₁₀H₁₆	*d*-Limonene	177	Nonazeotrope		563
5656	C₁₀H₁₆	Nopinene	164	~ 139.0	~ 24	563
5657	C₁₀H₁₆	α-Phellandrene	171.5	Nonazeotrope		563
5658	C₁₀H₁₆	α-Pinene	155.8	136.4	58.5	542
5659	C₁₀H₁₆	α-Terpinene	173.4	141.2	97	575
5660	C₁₀H₁₆	Terpinolene	184.6	Nonazeotrope		575
5661	C₁₀H₁₆	Thymene	179.7	139	~ 88	563
5662	C₁₀H₁₈O	Cineol	176.35	Nonazeotrope		527

TABLE I.   *Binary Systems*   179

No.	Formula	B-Component Name	B.P., °C	Azeotropic Data B.P., °C	Wt.%A	Ref.
A =	$C_3H_6O_2$	**Propionic Acid** *(continued)*	**140.9**			
5663	$C_{10}H_{20}O_2$	Isoamyl isovalerate	193.5	Nonazeotrope		542
		Decane	173.3	< 140.5	< 95	562
		"	174.06	139.76	80.5	507
5665	$C_{10}H_{22}$	2.7-Dimethyloctane	160.25	138.3	70	563
5666	$C_{10}H_{22}O$	Isoamyl ether	173.2	Nonazeotrope		527
5667	$C_{11}H_{24}$	Undecane	193.85	Nonazeotrope		507
A =	$C_3H_6O_3$	**Ethylene Glycol Monoformate**	**180**			
5668	C_nH_m	Hydrocarbons		Min. b.pt.		644
A =	$C_3H_6O_3$	**Methyl Carbonate**	**90.35**			
5669	C_3H_7Br	1-Bromopropane	71.0	Nonazeotrope		547
5670	C_3H_7I	1-Iodopropane	102.4	89.5	90	563
5671	C_3H_7I	2-Iodopropane	89.35	86.0	< 45	547
5672	C_3H_8O	Isopropyl alcohol	82.45	78.75	44	572
5673	C_3H_8O	Propyl alcohol	97.2	87	75	532
5674	C_4H_8O	2-Butanone	79.6	Nonazeotrope		527
5675	$C_4H_8O_2$	Dioxane	101.35	Nonazeotrope		557
5676	$C_4H_8O_2$	Ethyl acetate	77.1	Nonazeotrope		575
5677	C_4H_9Br	1-Bromobutane	101.6	Nonazeotrope		547
5678	C_4H_9Br	2-Bromobutane	91.2	< 88.5	< 54	575
5679	C_4H_9Br	1-Bromo-2-methylpropane	91.6	87.5	< 50	547
5680	C_4H_9Br	2-Bromo-2-methylpropane	73.25	Nonazeotrope		575
5681	C_4H_9Cl	1-Chlorobutane	78.5	Nonazeotrope		547
5682	$C_4H_{10}O$	Butyl alcohol	117.75	Nonazeotrope		527
5683	$C_4H_{10}O$	*sec*-Butyl alcohol	99.5	89.0	85	563
5684	$C_4H_{10}O$	*tert*-Butyl alcohol	82.45	80.65	33	570
5685	$C_4H_{10}O$	Isobutyl alcohol	108.0	90.05	92	527
5686	$C_4H_{10}S$	Butanethiol	97.5	88.2	70	568
5687	$C_4H_{10}S$	Ethyl sulfide	92.1	86.8	53	568
5688	$C_5H_{10}O$	3-Methyl-2-butanone	95.4	Nonazeotrope		552
5689	$C_5H_{10}O_2$	Isobutyl formate	98.2	Nonazeotrope		575
5690	$C_5H_{10}O_2$	Isopropyl acetate	91.0	Nonazeotrope		531
5691	$C_5H_{10}O_2$	Methyl isobutyrate	92.5	Nonazeotrope		549
5692	$C_5H_{11}Cl$	1-Chloro-3-methylbutane	99.4	< 90		575
5693	$C_5H_{12}O$	2-Pentanol	119.8	Nonazeotrope		575
5694	$C_5H_{12}O_2$	Diethoxymethane	87.95	86.0	40	527
5695	C_6H_6	Benzene	80.2	80.17	1	572
5696	$C_6H_{10}O$	Mesityl oxide	129.45	126.45	94	552
5697	C_6H_{12}	Cyclohexane	80.75	~ 75		563
5698	C_6H_{12}	Methylcyclopentane	72.0	< 69.5	> 12	575
5699	C_6H_{14}	*n*-Hexane	68.95	< 67.0	< 20	575
5700	$C_6H_{14}O$	Propyl ether	90.55	< 87.5	< 58	557
5701	$C_6H_{14}O_2$	Acetal	103.55	Nonazeotrope		557
5702	C_7H_8	Toluene	110.75	Nonazeotrope		575
5703	C_7H_{14}	Methylcyclohexane	101.15	< 85.0	< 75	563
5704	C_7H_{16}	*n*-Heptane	98.4	82.35	6 1	569
		"	98.45	~ 88.5	~ 70	563
5705	C_8H_8	Styrene	145	Min. b.p.		342

No.	Formula	B-Component Name	B.P., °C	Azeotropic Data B.P., °C Wt.%A		Ref.
A =	C₃H₆O₃	**Methyl Carbonate** *(continued)*	90.35			
5706	C₈H₁₀	Ethylbenzene	136	Min. b.p.		342
5707	C₈H₁₀	Xylenes	140	Min. b.p.		342
5708	C₈H₁₆	1,3-Dimethylcyclohexane	120.7	Nonazeotrope		575
5709	C₈H₁₈	2,5-Dimethylhexane	109.4	87.0	80	562
5710	C₈H₁₈	Octane	125.75	Nonazeotrope		575
A =	C₃H₆O₃	**Methyl Glycolate**	151.2			
5711	C₈H₁₀	Ethylbenzene	136.15	Min. b.p.		163
5712	C₈H₁₀	*m*-Xylene	139	Min. b.p.		163
A =	C₃H₆O₃	*s*-**Trioxane**	114.5			
5713	C₆H₆	Benzene	80.1		v-l	866
5714	C₆H₁₂	Cyclohexane	80.75	Min. b.p.		163
5715	C₆H₁₂	Naphthenes	~ 80	Min. b.p.		513
5716	C₆H₁₄	Hexanes	~ 70	Min. b.p.		513
5717	C₇H₁₄	Naphthenes	~ 100	Min. b.p.		513
5718	C₇H₁₆	Heptanes	~ 100	Min. b.p.		513
5719	C₈H₁₀	Xylene	140	Min. b.p.		515
5720	C₈H₁₆	Naphthenes	~ 120	Min. b.p.		513
5721	C₈H₁₈	Octanes	~ 120	Min. b.p.		513
5722	C₉H₂₀	Nonanes	~ 130	Min. b.p.		513
A =	C₃H₇Br	**1-Bromopropane**	71.0			
5723	C₃H₈O	Isopropyl alcohol	82.45	66.75	79.5	573
		"	82.45	65.2	84	563
5724	C₃H₈O	Propyl alcohol	97.6	69.75	90	389
5725	C₃H₈S	1-Propanethiol	67.5	Nonaeotrope		563
5726	C₃H₉BO₃	Methyl borate	68.75	~ 67.8	~ 55	531
5727	C₄H₄S	Thiophene	84.7	Nonazeotrope		527
5728	C₄H₈O	Butyraldehyde	75.2	Nonazeotrope		575
5729	C₄H₈O	Isobutyraldehyde	63.5	Nonazeotrope		575
5730	C₄H₈O	2-Butanone	79.6	Nonazeotrope		527
5731	C₄H₈O₂	Ethyl acetate	77.05	70	~ 80	563
		"	77.05	Nonazeotrope		532
5732	C₄H₈O₂	Isopropyl formate	68.8	66.0	< 45	547
5733	C₄H₈O₂	Methyl propionate	79.7	Azeotrope doubtful		563
5734	C₄H₈O₂	Propyl formate	80.85	Nonazeotrope		538
5735	C₄H₉Cl	1-Chloro-2-methylpropane	68.85	Nonazeotrope		549
		"	68.85	68.8	5	532
5736	C₄H₉NO₂	Butyl nitrite	78.2	Nonazeotrope		550
5737	C₄H₉NO₂	Isobutyl nitrite	67.1	67.05	5	550
5738	C₄H₁₀O	Butyl alcohol	117.75	Nonazeotrope		527
5739	C₄H₁₀O	*tert*-Butyl alcohol	82.45	68.0	88	567
5740	C₄H₁₀O	Isobutyl alcohol	108.5	Nonazeotrope, b.p. curve		389
5741	C₄H₁₀O₂	Acetaldehyde dimethyl acetal	64.3	Nonazeotrope		559
5742	C₄H₁₀O₂	Ethoxymethoxymethane	65.9	Nonazeotrope		559
5743	C₅H₁₀O₂	Isopropyl acetate	89.5	Nonazeotrope		575

TABLE I.   *Binary Systems*   181

No.	B-Component		B.P., °C	Azeotropic Data		
	Formula	Name		B.P., °C	Wt.%A	Ref.
A =	**C₃H₇Br**	**1-Bromopropane** *(continued)*	**71.0**			
5744	C₅H₁₂O	*tert*-Amyl alcohol	102.0	Nonazeotrope		535
5745	C₅H₁₂O	Ethyl propyl ether	63.85	Nonazeotrope		559
5746	C₅H₁₂O	Isoamyl alcohol	129.3	Nonazeotrope,		
				b.p. curve		389
5747	C₆H₆	Benzene	80.2	Nonazeotrope		563
5748	C₆H₁₂	Cyclohexane	80.75	Nonazeotrope		563
5748a	C₆H₁₂	1-Hexene 200 mm.	27.9	29.5		v-l	498f
		" 400 mm.	44.8	45.6	12.2	v-l	498f
		" 600 mm.	56.3	57.0	13.1	v-l	498f
		" 760 mm.	63.5	63.6	13.1	v-l	498f
5749	C₆H₁₂	Methylcyclopentane	72.0	68.8	58	562
5750	C₆H₁₄	2,3-Dimethylbutane	58.0	Nonazeotrope		575
5751	C₆H₁₄	Hexane	68.85	67.2	50	538
A =	**C₃H₇Br**	**2-Bromopropane**	**59.4**			
5752	C₃H₇NO₂	Propyl nitrite	47.45	Nonazeotrope		550
5753	C₃H₈O	Isopropyl alcohol	82.45	57.7	93	527
5754	C₃H₈O	Propyl alcohol	97.2	Nonazeotrope		527
		"	97.2	58.4	96	573
5755	C₃H₈S	Propanethiol	67.3	Nonazeotrope		575
5756	C₃H₉BO₃	Methyl borate	68.7	Nonazeotrope		547
5757	C₄H₈O	2-Butanone	79.6	Nonazeotrope		527
5758	C₄H₈O	Butyraldehyde	75.2	Nonazeotrope		575
5759	C₄H₈O	Isobutyraldehyde	63.5	Nonazeotrope		575
5760	C₄H₈O₂	Ethyl acetate	77.1	Nonazeotrope		527
5761	C₄H₉NO₂	Isobutyl nitrite	67.1	Nonazeotrope		550
5762	C₄H₁₀O	*sec*-Butyl alcohol	99.5	Nonazeotrope		527
5763	C₄H₁₀O	*tert*-Butyl alcohol	82.45	59.0	94.8	527
5764	C₄H₁₀O₂	Acetaldehyde dimethyl				
		acetal	64.3	Nonazeotrope		559
5765	C₅H₁₂O	Ethyl propyl ether	63.6	Nonazeotrope		548
5766	C₆H₆	Benzene	80.15	Nonazeotrope		527
5767	C₆H₁₂	Cyclohexane	80.75	Nonazeotrope		575
5768	C₆H₁₂	Methylcyclopentane	72.0	Nonazeotrope		527
5769	C₆H₁₄	2,3-Dimethylbutane	58.0	55.8	50	527
5770	C₆H₁₄	Hexane	68.8	59.3	98.5	527
		"	68.85	Nonazeotrope		538
5771	C₆H₁₄O	Isopropyl ether	68.3	Nonazeotrope		527
A =	**C₃H₇Cl**	**1-Chloropropane**	**46.65**			
5772	C₃H₇NO₂	Isopropyl nitrite	40.1	Nonazeotrope		550
5773	C₃H₇NO₂	Propyl nitrite	47.75	45.6	62	555
5774	C₃H₈O	Isopropyl alcohol	82.4	46.4	97.2	555
5775	C₃H₈O	Propyl alcohol	97.2	Nonazeotrope		555
5776	C₃H₈O₂	Methylal	42.15	Nonazeotrope		555
5777	C₃H₈S	Propanethiol	67.3	Nonazeotrope		575
5778	C₃H₉BO₃	Methyl borate	68.7	Nonazeotrope		575
5779	C₄H₈O₂	Ethyl acetate	77.05	Nonazeotrope		555
5780	C₄H₉NO₂	Isobutyl nitrite	67.1	Nonazeotrope		555
5781	C₄H₁₀O	*tert*-Butyl alcohol	82.55	Nonazeotrope		555

		B-Component		Azeotropic Data		
No.	Formula	Name	B.P., °C	B.P., °C	Wt.%A	Ref.
A =	**C₃H₇Cl**	**1-Chloropropane** *(continued)*	**46.65**			
5782	C₄H₁₀O	Ethyl ether	34.5	Nonazeotrope		555
5783	C₄H₁₀O	Methyl propyl ether	38.9	Nonazeotrope		559
5784	C₅H₈	Isoprene	34.3	Nonazeotrope		575
5785	C₅H₁₀	Cyclopentane	49.3	< 44.5	<64	575
5786	C₅H₁₂	Pentane	36	< 34.8	<32	555
5787	C₆H₁₄	Hexane	68.8	Nonazeotrope		575
A =	**C₃H₇Cl**	**2-Chloropropane**	**34.9**			
5788	C₃H₇NO₂	Isopropyl nitrite	40.1	Nonazeotrope		550
5789	C₃H₇NO₂	Propyl nitrite	47.75	Nonazeotrope		550
5790	C₃H₈O	Isopropyl alcohol	82.5	Nonazeotrope		215
5791	C₃H₈O₂	Methylal	42.3	Nonazeotrope		548
5792	C₅H₁₀	Cyclopentane	49.3	< 44.5	<64	562
		"		Nonazeotrope		575
5793	C₅H₁₀	2-Methyl-1-butene	33.1	32.8	58	562
5794	C₅H₁₀	2-Methyl-2-butene	37.1	34	61	563
5795	C₅H₁₂	2-Methylbutane	27.95	~ 24		563
5796	C₅H₁₂	Pentane	36.15	~ 32	~ 52	563
A =	**C₃H₇ClO**	**1-Chloro-2-propanol**	**127.0**			
5797	C₄H₉I	1-Iodobutane	130.4	120.0	45	567
5798	C₄H₉I	1-Iodo-2-methylpropane	120.8	115.0	25	567
5799	C₄H₁₀O	Butyl alcohol	117.8	Nonazeotrope		575
5800	C₄H₁₀O	Isobutyl alcohol	108.0	Nonazeotrope		575
5801	C₄H₁₀O₂	2-Ethoxyethanol	135.3	Nonazeotrope		575
5802	C₅H₁₁Br	1-Bromo-3-methylbutane	120.65	115.3	~ 30	575
5803	C₅H₁₂O	Isoamyl alcohol	131.9	< 127.3	>81	575
5804	C₅H₁₂O₂	2-Propoxyethanol	151.35	Nonazeotrope		575
5805	C₆H₅Br	Bromobenzene	156.1	Nonazeotrope		575
5806	C₆H₅Cl	Chlorobenzene	131.75	122.2	55	567
5807	C₆H₆	Benzene	80.15	Nonazeotrope		575
5808	C₆H₁₂	Cyclohexane	80.75	Nonazeotrope		575
5809	C₆H₁₂O	3-Hexanone	123.3	Nonazeotrope		552
5810	C₆H₁₂O	4-Methyl-2-pentanone	116.05	Nonazeotrope		527
5811	C₆H₁₂O₂	Butyl acetate	126.0	125.5	~ 25	575
5812	C₆H₁₂O₂	Ethyl isobutyrate	110.1	Nonazeotrope		575
5813	C₆H₁₂O₂	Isoamyl formate	123.8	123.0	~ 30	575
5814	C₆H₁₂O₂	Isobutyl acetate	117.4	Nonazeotrope		575
5815	C₇H₈	Toluene	110.75	109.0	15	567
5816	C₇H₈O	Anisole	153.85	Nonazeotrope		575
5817	C₇H₁₆	Heptane	98.4	96.5	17	575
5818	C₈H₁₀	*m*-Xylene	139.2	124.5	75	567
5819	C₈H₁₀	*o*-Xylene	144.3	125.5	85	575
5820	C₈H₁₈	2,5-Dimethylhexane	109.4	105.0	30	567
5821	C₈H₁₈O	Isobutyl ether	122.3	< 118.0	>35	575
A =	**C₃H₇ClO**	**2-Chloro-1-propanol**	**133.7**			
5822	C₄H₉I	1-Iodobutane	130.4	123.5	30	567
5823	C₅H₁₁Br	1-Bromo-3-methylbutane	120.65	118.0	15	575

TABLE I.   *Binary Systems*   183

No.	Formula	Name	B.P., °C	B.P., °C	Wt.%A	Ref.
		B-Component		**Azeotropic Data**		
A =	**C₃H₇ClO**	**2-Chloro-1-propanol**	**133.7**			
		(continued)				
5824	C₆H₅Br	Bromobenzene	156.1	Nonazeotrope		575
5825	C₆H₅Cl	Chlorobenzene	131.75	126.0	36	567
5826	C₆H₆O	Phenol	182.2	Nonazeotrope		575
5827	C₆H₁₂	Cyclohexane	80.75	Nonazeotrope		575
5828	C₆H₁₂O₂	Isoamyl formate	123.8	< 123.7	> 5	575
5829	C₆H₁₂O₂	Isobutyl acetate	117.2	Nonazeotrope		575
5830	C₇H₁₄O₂	Ethyl isovalerate	134.7	133.5?	60?	575
5831	C₇H₁₄O₂	Isoamyl acetate	142.1	Nonazeotrope		575
5832	C₈H₁₀	*m*-Xylene	139.2	129.0	53	567
5833	C₈H₁₀	*o*-Xylene	144.3	130.5	70	567
5834	C₈H₁₆	1,3-Dimethylcyclohexane	120.7	115.0	35	567
5835	C₈H₁₈O	Butyl ether	142.4	130.5	70	575
5836	C₈H₁₈O	Isobutyl ether	122.3	120.0	25	575
A =	**C₃H₇ClO**	**Propylene Chlorohydrin**	**73/100**			
5837	C₆H₁₂Cl₂O	Bis(chloroisopropyl) ether, 100 mm.	121.9	Nonazeotrope		981
A =	**C₃H₇ClO₂**	**Chloromethylal**	**95**			
5838	C₄H₈O	2-Butanone	79.6	Nonazeotrope		575
5839	C₅H₁₀O	3-Pentanone	95.0	Nonazeotrope		575
5840	C₆H₆	Benzene	80.15	Nonazeotrope		575
5841	C₆H₁₂	Cyclohexane	80.75	Nonazeotrope		575
5842	C₇H₁₆	Heptane	98.4	93.0	62	562
A =	**C₃H₇ClO₂**	**1-Chloro-2,3-propanediol**	**213**			
5843	C₆H₅NO₂	Nitrobenzene	210.85	~ 208		563
5844	C₇H₇Cl	α-Chlorotoluene	179.35	179.2?		563
5845	C₇H₈O	Benzyl alcohol	205.5	204.5		563
5846	C₇H₈O	*p*-Cresol	201.8	Nonazeotrope		563
5847	C₁₀H₁₆O	Camphor	208.9	Nonazeotrope		563
A =	**C₃H₇I**	**1-Iodopropane**	**102.4**			
5848	C₃H₈O	Isopropyl alcohol	82.45	79.8	58	573
5849	C₃H₈O	Propyl alcohol	97.2	90.2	70	563
5850	C₃H₈O₂	2-Methoxyethanol	124.5	101.0		575
5851	C₄H₆O	Crotonaldehyde	102.15	< 99.7		563
5852	C₄H₈O₂	Dioxane	101.35	98.75	60	559
5853	C₄H₁₀O	Butyl alcohol	117.75	99.5	86.5	535
5854	C₄H₁₀O	Isobutyl alcohol	108	96	~ 82	563
5855	C₄H₁₀O₂	2-Ethoxyethanol	135.3	Nonazeotrope		556
5856	C₅H₁₀O	3-Methyl-2-butanone	95.4	Nonazeotrope		552
5857	C₅H₁₀O	2-Pentanone	102.35	100.8	65	552
5858	C₅H₁₀O	3-Pentanone	102.05	100.8	62	552
5859	C₅H₁₀O₂	Isopropyl acetate	89.5	Nonazeotrope		575
5860	C₅H₁₀O₂	Methyl butyrate	102.65	101.0	56	527
5861	C₅H₁₀O₂	Methyl isobutyrate	92.5	Nonazeotrope		527
5862	C₅H₁₀O₂	Propyl acetate	101.6	99.0	>46	527
5863	C₅H₁₁NO₂	Isoamyl nitrite	97.15	< 96.7		550
5864	C₅H₁₂O	3-Pentanol	116.0	97.2	70	567

No.	Formula	B-Component Name	B.P., °C	Azeotropic Data B.P., °C	Wt.%A	Ref.
A =	C₃H₇I	**1-Iodopropane** *(continued)*	**102.4**			
5865	C₅H₁₂O	*tert*-Amyl alcohol	102.35	100.5	89	567
5866	C₆H₁₂O₂	Ethyl isobutyrate	110.1	Nonazeotrope		547
5867	C₆H₁₄O₂	Acetal	103.55	101.0	60	559
5868	C₇H₈	Toluene	110.7	Nonazeotrope		563
5869	C₇H₁₄	Methylcyclohexane	101.1	99.4	~ 60	573
5870	C₇H₁₆	Heptane	98.4	< 97.5	>40	562
A =	C₃H₇I	**2-Iodopropane**	**89.35**			
5871	C₃H₈O	Isopropyl alcohol	82.45	76.0	68	573
5872	C₃H₈O	Propyl alcohol	97.2	82.95	83	527
5873	C₄H₈O	2-Butanone	79.6	Nonazeotrope		527
5874	C₄H₈O₂	Dioxane	101.35	Nonazeotrope		559
5875	C₄H₈O₂	Methyl propionate	79.84	Nonazeotrope		547
5876	C₄H₈O₂	Propyl formate	80.85	Nonazeotrope		547
		"	80.85	< 80.2	>16	575
5877	C₄H₁₀O	Butyl alcohol	117.8	88.6	94	567
5878	C₄H₁₀O	*sec*-Butyl alcohol	99.5	85.4	83	567
5879	C₄H₁₀O	*tert*-Butyl alcohol	82.45	< 77.75	<69	567
5880	C₄H₁₀O	Isobutyl alcohol	107.85	86.8	88	527
5881	C₅H₁₀O₂	Isobutyl formate	98.2	Nonazeotrope		547
		"	98.2	< 88.5	>82	575
5882	C₅H₁₀O₂	Isopropyl acetate	90.8	87.0	60	547
5883	C₅H₁₀O₂	Methyl isobutyrate	92.5	< 88.8	>80	575
5884	C₅H₁₁NO₂	Isoamyl nitrite	97.15	Nonazeotrope		550
5885	C₅H₁₂O	*tert*-Amyl alcohol	102.35	88.6	92	567
5886	C₅H₁₂O	3-Methyl-2-butanol	112.6	Nonazeotrope		575
5887	C₅H₁₂O₂	Diethoxymethane	87.95	86.15	37	527
5888	C₆H₆	Benzene	80.2	Nonazeotrope		538
5889	C₆H₁₂	Cyclohexane	80.75	Nonazeotrope		575
5890	C₆H₁₄O	Propyl ether	90.55	~ 89.0	~ 65	548
5891	C₇H₁₄	Methylcyclohexane	100.8	88	65	929
A =	C₃H₇NO	**Acetoxime**	**135.8**			
5892	C₆H₁₀S	Allyl sulfide	138.7	134		563
A =	C₃H₇NO	**N,N-Dimethylformamide**	**153**			
5893	C₆H₆	Benzene	80.1	Nonazeotrope		163
5893a	C₇H₈	Toluene	110.7	Nonazeotrope	v-l	2c
5893b	C₈H₁₀	*m*-Xylene	139	135.45	19.1	v-l	2c
5893c	C₈H₁₀	*o*-Xylene	143.6	138.85	26.4	v-l	2c
5894	C₈H₁₀	*p*-Xylene	138.4	135.1	18.7	224
		"	138.4	134.10	19.2	v-l	2c
A =	C₃H₇NO	**Propionamide**	**222.2**			
5895	C₃H₇NO₂	Ethyl carbamate	185.25	Nonazeotrope		575
5896	C₄H₅NS	Allyl isothiocyanate	152.0	Nonazeotrope		575
5897	C₄H₆O₄	Methyl oxalate	164.45	Nonazeotrope		575
5898	C₄H₇BrO₂	Ethyl bromoacetate	158.8	Nonazeotrope		527
5899	C₄H₈Cl₂O	Bis(2-chloroethyl) ether	178.65	Nonazeotrope		527

TABLE I. *Binary Systems* 185

No.	Formula	Name	B.P., °C	B.P., °C	Wt.%A	Ref.
		B-Component		Azeotropic Data		
A =	C₃H₇NO	**Propionamide** *(continued)*	**222.2**			
5900	C₄H₈O₃	Glycol monoacetate	190.9	Nonazeotrope		575
5901	C₄H₉I	1-Iodobutane	130.4	Nonazeotrope		527
5902	C₄H₁₀O₃	Diethylene glycol	245.5	Nonazeotrope		526
5903	C₅H₆O₂	Furfuryl alcohol	169.35	Nonazeotrope		575
5904	C₅H₈O₃	Levulinic acid	252	Nonazeotrope		527
5905	C₅H₉ClO₂	Propyl chloroacetate	163.5	Nonazeotrope		575
5906	C₅H₁₁Br	1-Bromo-3-methylbutane	120.65	Nonazeotrope		527
5907	C₅H₁₁I	1-Iodo-3-methylbutane	147.65	Nonazeotrope		527
5908	C₅H₁₂O	Isoamyl alcohol	131.3	Nonazeotrope		527
5909	C₅H₁₂O₂	2-Propoxyethanol	151.35	Nonazeotrope		526
5910	C₅H₁₂O₃	2-(2-Methoxyethoxy)-ethanol	192.95	Nonazeotrope		526
5911	C₆H₄BrCl	*p*-Bromochlorobenzene	196.4	189.5	16	575
5912	C₆H₄Br₂	*p*-Dibromobenzene	220.25	204.9	22	527
5913	C₆H₄ClNO₂	*m*-Chloronitrobenzene	235.5	216.5	>48	554
5914	C₆H₄ClNO₂	*o*-Chloronitrobenzene	246.0	< 220.6	>54	554
5915	C₆H₄ClNO₂	*p*-Chloronitrobenzene	239.1	217.5	49.8	527
5916	C₆H₄Cl₂	*o*-Dichlorobenzene	179.5	177.0	9	564
5917	C₆H₄Cl₂	*p*-Dichlorobenzene	174.4	172.55	8	527
5918	C₆H₅Br	Bromobenzene	156.1	Nonazeotrope		527
5919	C₆H₅BrO	*o*-Bromophenol	194.8	Nonazeotrope		575
5920	C₆H₅Cl	Chlorobenzene	132.0	Nonazeotrope		527
5921	C₆H₅ClO	*p*-Chlorophenol	219.75	228.0	33	562
5922	C₆H₅I	Iodobenzene	188.45	183.5	10	569
5923	C₆H₅NO₂	Nitrobenzene	210.75	205.4	24	527
5924	C₆H₅NO₃	*o*-Nitrophenol	217.25	211.15	24.8	542
5925	C₆H₆O	Phenol	182.2	Nonazeotrope		542
5926	C₆H₆O₂	Pyrocatechol	245.9	Nonazeotrope		527
5927	C₆H₆O₂	Resorcinol	281.4	Nonazeotrope		544
5928	C₆H₇N	Aniline	184.35	Nonazeotrope		527
5929	C₆H₈N₂	*o*-Phenylenediamine	258.6	Nonazeotrope		551
5930	C₆H₈O₄	Methyl fumarate	193.25	Nonazeotrope		527
5931	C₆H₈O₄	Methyl maleate	204.05	Nonazeotrope		527
5932	C₆H₁₀O₄	Ethylidene diacetate	168.5	Nonazeotrope		527
5933	C₆H₁₀O₄	Ethyl oxalate	185.65	Nonazeotrope		527
5934	C₆H₁₀O₄	Glycol diacetate	186.3	Nonazeotrope		575
5935	C₆H₁₁ClO₂	Butyl chloroacetate	181.9	Nonazeotrope		575
5936	C₆H₁₁NO₂	Nitrocyclohexane	205.3	< 203.0	>11	575
5937	C₆H₁₂O	Cyclohexanol	160.7	Nonazeotrope		527
5938	C₆H₁₂O₂	Caproic acid	205.15	Nonazeotrope		575
5939	C₆H₁₂O₃	2-Ethoxyethyl acetate	156.8	Nonazeotrope		526
5940	C₆H₁₂O₃	Propyl lactate	171.7	Nonazeotrope		575
5941	C₆H₁₃Br	1-Bromohexane	156.5	Nonazeotrope		575
5942	C₆H₁₄O	Hexyl alcohol	157.85	Nonazeotrope		527
5943	C₆H₁₄O₂	2-Butoxyethanol	171.15	Nonazeotrope		527

No.	B-Component			Azeotropic Data		
	Formula	Name	B.P., °C	B.P., °C	Wt.%A	Ref.
A =	C₃H₇NO	**Propionamide** *(continued)*	**222.2**			
5944	C₆H₁₄O₂	Pinacol	174.35	Nonazeotrope		575
5945	C₆H₁₄S	Propyl sulfide	141.5	Nonazeotrope		566
5946	C₇H₅Cl₃	α,α,α-Trichlorotoluene	220.9	Reacts		535
5947	C₇H₆O	Benzaldehyde	179.2	Nonazeotrope		575
5948	C₇H₆O₂	Benzoic acid	250.5	Nonazeotrope		542
5949	C₇H₇Br	o-Bromotoluene	181.5	178.2		527
5950	C₇H₇Br	m-Bromotoluene	184.3	180.4		527
5951	C₇H₇Br	p-Bromotoluene	185.0	181.0	10	562
5952	C₇H₇BrO	o-Bromoanisole	217.7	208.0	27	562
5953	C₇H₇Cl	o-Chlorotoluene	159.2	Nonazeotrope		527
5954	C₇H₇ClO	o-Chloroanisole	195.7	194.0	10	575
5955	C₇H₇ClO	p-Chloroanisole	197.8	< 196.5	~ 12	575
5956	C₇H₇I	p-Iodotoluene	214.5	201.5	20	562
5957	C₇H₇NO₂	m-Nitrotoluene	230.8	214.5	44	554
5958	C₇H₇NO₂	o-Nitrotoluene	221.75	210.2	30	527
5959	C₇H₇NO₂	p-Nitrotoluene	238.9	217.5	50	527
5960	C₇H₈	Toluene	110.75	Nonazeotrope		527
5961	C₇H₈O	Benzyl alcohol	205.1	Nonazeotrope		527
5962	C₇H₈O	m-Cresol	202.2	Nonazeotrope		542
5963	C₇H₈O	o-Cresol	191.1	Nonazeotrope		544
5964	C₇H₈O	p-Cresol	201.7	Nonazeotrope		544
5965	C₇H₈O₂	Guaiacol	205.05	Nonazeotrope		527
5966	C₇H₈O₂	m-Methoxyphenol	244	Nonazeotrope		535
5967	C₇H₉N	Methylaniline	196.25	Nonazeotrope		551
5968	C₇H₉N	m-Toluidine	203.1	Nonazeotrope		527
5969	C₇H₉N	o-Toluidine	200.35	200.25	2.5	527
5970	C₇H₉N	p-Toluidine	200.55	Nonazeotrope		551
5971	C₇H₁₄O₃	1,3-Butanediol methyl ether acetate	171.75	Nonazeotrope		575
5972	C₇H₁₄O₃	Isobutyl lactate	182.15	Nonazeotrope		575
5973	C₇H₁₆O₄	2-[2-(2-Methoxyethoxy)-ethoxy]-ethanol	245.25	Nonazeotrope		575
5974	C₈H₈O	Acetophenone	202.0	200.35	15	552
5975	C₈H₈O₂	Methyl benzoate	199.4	196.95		571
5976	C₈H₈O₂	Phenyl acetate	195.7	Nonazeotrope		535
5977	C₈H₈O₃	Methyl salicylate	222.35	210.55	34	530
5978	C₈H₁₀	m-Xylene	139.2	Nonazeotrope		527
		"	139.0	138.5		531
5979	C₈H₁₀	o-Xylene	144.3	144.0	2	575
5980	C₈H₁₀O	Phenethyl alcohol	219.4	217.8	31	529
5981	C₈H₁₀O	Phenetole	170.5	Nonazeotrope		535
5982	C₈H₁₀O	3,4-Xylenol	226.8	221.1	96	564
		"	226.8	Nonazeotrope		575
5983	C₈H₁₀O₂	o-Ethoxyphenol	216.5	Nonazeotrope		575
5984	C₈H₁₁N	Dimethylaniline	194.15	190.5	15.5	551
5985	C₈H₁₁N	2,4-Xylidine	214.0	< 212.0	<27	551

TABLE I.   *Binary Systems*   187

No.	Formula	Name	B.P., °C	B.P., °C	Wt.%A	Ref.
		B-Component		**Azeotropic Data**		
A =	**C₃H₇NO**	**Propionamide** *(continued)*	**222.2**			
5986	C₈H₁₁N	3,4-Xylidine	225.5	217.2	28	575
5987	C₈H₁₁N	Ethylaniline	205.5	< 204.0	>12	551
5988	C₈H₁₁NO	*o*-Phenetidine	232.5	< 222.0		551
5989	C₈H₁₂O₄	Ethyl fumarate	217.85	< 211.0		562
5990	C₈H₁₂O₄	Ethyl maleate	223.3	214.0	38	570
5991	C₈H₁₄O	Methyl heptenone	173.2	Nonazeotrope		552
5992	C₈H₁₆O	2-Octanone	172.85	Nonazeotrope		552
5993	C₈H₁₆O₂	Butyl butyrate	166.4	Nonazeotrope		527
5994	C₈H₁₆O₂	Isoamyl propionate	160.7	Nonazeotrope		527
5995	C₈H₁₆O₂	Isobutyl butyrate	156.9	Nonazeotrope		527
5996	C₈H₁₆O₂	Propyl isovalerate	155.7	Nonazeotrope		527
5997	C₈H₁₈O	Octyl alcohol	195.2	Nonazeotrope		527
5998	C₈H₁₈O	*sec*-Octyl alcohol	179.0	Nonazeotrope		527
5999	C₉H₇N	Quinoline	237.3	Nonazeotrope		553
6000	C₉H₈	Indene	182.6	179.5	12	562
6001	C₉H₈O	Cinnamaldehyde	253.5	Nonazeotrope		575
6002	C₉H₁₀O	Cinnamyl alcohol	257.0	Nonazeotrope		575
6003	C₉H₁₀O	*p*-Methylacetophenone	226.35	214.0	40	552
6004	C₉H₁₀O	Propiophenone	217.7	207.0	28	552
6005	C₉H₁₀O₂	Benzyl acetate	~214.9	208.8	29	574
6006	C₉H₁₀O₂	Ethyl benzoate	212.6	205.0	25	574
6007	C₉H₁₀O₂	Methyl-*α*-toluate	215.3	206.5	28	562
6008	C₉H₁₀O₃	Ethyl salicylate	233.7	214.0	~ 50	536
6009	C₉H₁₂	Cumene	152.8	151.8	4	575
6010	C₉H₁₂	Mesitylene	164.6	162.3	10	562
6011	C₉H₁₂	Propyl benzene	159.3	157.7		527
6012	C₉H₁₂O	Benzyl ethyl ether	185.0	182.5	8	575
6013	C₉H₁₂O	3-Phenylpropanol	235.6	Nonazeotrope		527
6014	C₉H₁₃N	*N,N*-Dimethyl-*m*-toluidine	203.1	Nonazeotrope		564
6015	C₉H₁₃N	*N,N*-Dimethyl-*o*-toluidine	185.3	182.5		551
6016	C₉H₁₃N	*N,N*-Dimethyl-*p*-toluidine	210.5	Nonazeotrope		564
		"	210.5	199.0	20	551
6017	C₉H₁₄O	Phorone	197.8	Nonazeotrope		552
6018	C₉H₁₈O	2,6-Dimethyl-4-heptanone	168.0	Nonazeotrope		552
6019	C₉H₁₈O₂	Isoamyl butyrate	181.05	180.6	3?	575
		"	178.5	Nonazeotrope		536
6020	C₉H₁₈O₂	Isoamyl isobutyrate	169.8	Nonazeotrope		564
6021	C₉H₁₈O₂	Isobutyl isovalerate	171.2	Nonazeotrope		527
6022	C₉H₁₈O₃	Isobutyl carbonate	190.3	< 186.5	> 8	575
6023	C₁₀H₇Br	1-Bromonaphthalene	281.2	222.0?	95?	575
		"	281.8	Nonazeotrope		564
6024	C₁₀H₇Cl	1-Chloronaphthalene	262.7	218.6	39	527

No.	Formula	Name	B.P., °C	B.P., °C	Wt.%A	Ref.
		B-Component		**Azeotropic Data**		
A =	**C₃H₇NO**	**Propionamide** *(continued)*	**222.2**			
6025	$C_{10}H_8$	Naphthalene	218.05	204.65	31.5	527
6026	$C_{10}H_9N$	1-Naphthylamine	300.8	Nonazeotrope		551
6027	$C_{10}H_9N$	Quinaldine	246.5	Nonazeotrope		575
6028	$C_{10}H_{10}O_2$	Isosafrol	252.1	~ 218.5		535
6029	$C_{10}H_{10}O_2$	Methyl cinnamate	261.95	Nonazeotrope		527
6030	$C_{10}H_{10}O_2$	Safrol	235.9	~ 213.2	35	535
6031	$C_{10}H_{10}O_4$	Methyl phthalate	283.2	Nonazeotrope		527
6032	$C_{10}H_{12}O$	Anethole	235.7	212.0	39	562
6033	$C_{10}H_{12}O_2$	Ethyl α-toluate	228.1	220.0	60	562
6034	$C_{10}H_{12}O_2$	Eugenol	255.0	Nonazeotrope		527
6035	$C_{10}H_{12}O_2$	Propyl benzoate	230.85	213.0	45	564
6036	$C_{10}H_{14}$	Cymene	176.7	172.8	15	562
6037	$C_{10}H_{14}O$	Carvacrol	237.85	Nonazeotrope		575
6038	$C_{10}H_{14}O$	Carvone	231.0	214.5	48	552
6038	$C_{10}H_{14}O$	Thymol	232.8	Nonazeotrope		535
6040	$C_{10}H_{14}O_2$	*m*-Diethoxybenzene	235.4	< 213.5		575
6041	$C_{10}H_{15}N$	Diethylaniline	217.05	203.15	23	551
6042	$C_{10}H_{16}$	Camphene	159.6	156.5	13	527
6043	$C_{10}H_{16}$	Dipentene	177.7	171.8	15	527
6044	$C_{10}H_{16}$	*d*-Limonene	177.8	172	20	573
6045	$C_{10}H_{16}$	Nopinene	163.8	161.0	10	562
6046	$C_{10}H_{16}$	α-Pinene	155.8	154.0	5	575
6047	$C_{10}H_{16}$	α-Terpinene	173.4	169.8	13	562
6048	$C_{10}H_{16}O$	Camphor	209.1	203.5	17	552
6049	$C_{10}H_{16}O$	Pulegone	223.8	212.0	38	552
6050	$C_{10}H_{18}O$	Borneol	213.4	209.2	22	527
6051	$C_{10}H_{18}O$	Cineol	176.35	173.8	8	527
6052	$C_{10}H_{18}O$	Citronellal	208.0	203? (reacts)		575
6053	$C_{10}H_{18}O$	Geraniol	229.6	217.0	54	527
6054	$C_{10}H_{18}O$	Linalool	198.6	Nonazeotrope		527
6055	$C_{10}H_{18}O$	α-Terpineol	217.8	209.6	25	527
6056	$C_{10}H_{20}O$	Citronellol	224.5	~ 211.5	~ 40	535
6057	$C_{10}H_{20}O$	Menthol	216.4	208.5	25	564
6058	$C_{10}H_{20}O_2$	Ethyl caprylate	208.35	200.2	22	562
6059	$C_{10}H_{20}O_2$	Methyl pelargonate	213.8	204.0	18	527
6060	$C_{10}H_{20}O_2$	Isoamyl isovalerate	192.7	188.45	12.2	541
6061	$C_{10}H_{22}$	Decane, 50 mm.		88	3	95
		" 100 mm.		106	5	95
		" 200 mm.		126	7.5	95
		" 760 mm.	173.3	168	11.8	95
6062	$C_{10}H_{22}O$	Amyl ether	187.5	181.0	12	562
6063	$C_{10}H_{22}O$	Decyl alcohol	~ 232.9	215.9	70	531
6064	$C_{10}H_{22}O$	Isoamyl ether	173.2	170.5	7	575
6065	$C_{10}H_{22}S$	Isoamyl sulfide	214.8	204.0	20	566
6066	$C_{11}H_{10}$	1-Methylnaphthalene	245.1	213.8	52	527
6067	$C_{11}H_{10}$	2-Methylnaphthalene	241.15	213.0	50	527
6068	$C_{11}H_{12}O_2$	Ethyl cinnamate	272.0	Nonazeotrope		527

TABLE I.   *Binary Systems*   189

No.	Formula	B-Component Name	B.P., °C	Azeotropic Data B.P., °C	Wt.%A	Ref.
A =	C₃H₇NO	**Propionamide** *(continued)*	**222.2**			
6069	$C_{11}H_{14}O_2$	1-Allyl-3,4-di-methoxybenzene	255.2	220.0	60	574
6070	$C_{11}H_{14}O_2$	Butyl benzoate	249.0	218.0	64	562
6071	$C_{11}H_{14}O_2$	1,2-Dimethoxy-4-propenylbenzene	270.5	Nonazeotrope		535
6072	$C_{11}H_{14}O_2$	Isobutyl benzoate	242.15	215.5	60	527
6073	$C_{11}H_{20}O$	Methyl isobornyl ether	192.4	187.5	13	562
6074	$C_{11}H_{20}O$	Methyl α-terpineol ether	216.2	203.5	27	562
6075	$C_{11}H_{22}O_2$	Ethyl pelargonate	227	211.0	40	575
6076	$C_{11}H_{22}O_3$	Isoamyl carbonate	232.2	208.5	35	564
6077	$C_{11}H_{24}$	Undecane, 50 mm.		105	15	95
		" 100 mm.		123	16	95
		" 200 mm.		142	17.3	95
		" 760 mm.	194.5	183	21	95
6078	$C_{12}H_{10}$	Acenaphthene	277.9	220.8	75	527
6079	$C_{12}H_{10}$	Biphenyl	256.1	216.0	55	575
6080	$C_{12}H_{10}O$	Phenyl ether	259.3	219.0	~ 62	574
6081	$C_{12}H_{16}O_2$	Isoamyl benzoate	262.05	219	67	**527**
6082	$C_{12}H_{16}O_3$	Isoamyl salicylate	277.5	Nonazeotrope		575
6083	$CC_{12}H_{20}O_2$	Bornyl acetate	227.6	209.5	38	570
6084	$C_{12}H_{22}O$	Ethyl isobornyl ether	203.8	196.0	20	562
6085	$C_{12}H_{26}$	Dodecane, 50 mm.		115	26	95
		Dodecane 100 mm.		132	26	95
		" 200 mm.		152	26	95
		" 760 mm.	216	193	31.6	**95**
6086	$C_{13}H_{10}$	Fluorene	295	221.5	90	575
6087	$C_{13}H_{12}$	Diphenylmethane	265.6	218.2	60	527
6088	$C_{14}H_{12}$	Stilbene	306.5	Nonazeotrope		575
6089	$C_{14}H_{14}$	1,2-Diphenylethane	284.5	221.0	80	575
6090	$C_{14}H_{14}O$	Benzyl ether	297	Nonazeotrope		575
A =	C₃H₇NO₂	**Ethyl Carbamate**	**185.25**			
6091	$C_3H_8O_2$	1,2-Propanediol	187.8	< 183.5		575
6092	$C_4H_6O_4$	Methyl oxalate	164.45	Nonazeotrope		**527**
6093	$C_4H_7Cl_3O$	Ethyl 1,1,2-trichloroethyl ether	173	169.5		575
6094	$C_4H_8Cl_2O$	Bis(2-chloroethyl) ether	178.65	171.5	25	562
6095	C_4H_9I	1-Iodobutane	130.4	Nonazeotrope		564
6096	C_4H_9I	1-Iodo-2-methylpropane	120.8	Nonazeotrope		564
6097	$C_5H_4O_2$	2-Furaldehyde	161.45	Nonazeotrope		527
6098	$C_5H_6O_2$	Furfuryl alcohol	169.35	Nonazeotrope		575
6099	$C_5H_8O_4$	Methyl malonate	181.4	< 178.65	<35	575
6100	$C_5H_{10}O_3$	2-Methoxyethyl acetate	144.6	Nonazeotrope		556
6101	$C_5H_{11}Br$	1-Bromo-3-methyl-butane	120.3	Nonazeotrope		564
6102	$C_5H_{11}I$	1-Iodo-3-methylbutane	147.65	146.5	2	564

No.	Formula	B-Component Name	B.P., °C	Azeotropic Data B.P., °C	Wt.%A	Ref.
A =	$C_3H_7NO_2$	**Ethyl Carbamate** (*continued*)	**185.25**			
6103	$C_5H_{11}NO_3$	Isoamyl nitrate	149.75	149.1	7	560
6104	$C_5H_{12}O$	Isoamyl alcohol	131.9	Nonazeotrope		527
6105	$C_5H_{12}O_2$	2-Propoxyethanol	151.35	Nonazeotrope		527
6106	$C_5H_{12}O_3$	2-(2-Methoxyethoxy)-ethanol	192.95	Nonazeotrope		527
6107	$C_6H_4Br_2$	p-Dibromobenzene	220.25	183.6	64	563
6108	$C_6H_4Cl_2$	o-Dichlorobenzene	179.5	170.0	27	564
6109	$C_6H_4Cl_2$	p-Dichlorobenzene	174.35	167.0	24.2	555
6110	C_6H_5Br	Bromobenzene	156.1	153.95	9.8	564
6111	C_6H_5Cl	Chlorobenzene	131.75	Nonazeotrope		527
6112	C_6H_5I	Iodobenzene	188.45	174.5	33	564
6113	$C_6H_5NO_2$	Nitrobenzene	210.75	184.95	88	554
6114	$C_6H_5NO_3$	o-Nitrophenol	217.2	Nonazeotrope		575
6115	C_6H_6O	Phenol	182.2	190.75	53.5	564
6116	$C_6H_8O_4$	Methyl fumarate	193.25	184.2	79	527
6117	$C_6H_8O_4$	Methyl maleate	204.05	Nonazeotrope		527
6118	$C_6H_{10}O$	Cyclohexanone	155.75	Nonazeotrope		564
6119	$C_6H_{10}O_4$	Ethylidene diacetate	168.5	Nonazeotrope		527
6120	$C_6H_{10}O_4$	Ethyl oxalate	185.65	181.0	38	564
6121	$C_6H_{10}O_4$	Methyl succinate	195.5	184.3	80	527
6122	$C_6H_{10}S$	Allyl sulfide	139.35	Nonazeotrope		566
6123	$C_6H_{12}O$	Cyclohexanol	160.8	Nonazeotrope		527
6124	$C_6H_{12}O_3$	2-Ethoxyethyl acetate	156.8	Nonazeotrope		556
6125	$C_6H_{13}Br$	1-Bromohexane	156.5	154.0	10	564
6126	$C_6H_{13}ClO_2$	Chloroacetal	157.4	156.8	10	575
6127	$C_6H_{14}O$	n-Hexanol	157.85	Nonazeotrope		527
6128	$C_6H_{14}O_2$	2-Butoxyethanol	171.15	Nonazeotrope		527
6129	$C_6H_{14}O_2$	Pinacol	174.35	173.5		575
6130	$C_6H_{14}O_3$	2-(2-Ethoxyethoxy)ethanol	201.9	Nonazeotrope		575
6131	$C_6H_{14}S$	Isopropyl sulfide	120.5	Nonazeotrope		566
6132	C_7H_5N	Benzonitrile	191.1	182.1	57	569
6133	C_7H_7Br	m-Bromotoluene	184.3	171.9	30.5	527
6134	C_7H_7Br	o-Bromotoluene	181.5	170.5	28	527
6135	C_7H_7Br	p-Bromotoluene	185.0	172.3	32	527
6136	C_7H_7Cl	o-Chlorotoluene	159.2	156.4	13	564
6137	C_7H_7Cl	p-Chlorotoluene	162.4	158.7	15	564
6138	C_7H_7ClO	m-Chloroanisole	193.3	179.5	20	562
6139	C_7H_7ClO	o-Chloroanisole	195.7	180.0	18	562
6140	C_7H_7I	p-Iodotoluene	214.5	183.2	58	564
6141	$C_7H_7NO_2$	m-Nitrotoluene	230.8	Nonazeotrope		554
6142	$C_7H_7NO_2$	o-Nitrotoluene	221.75	Nonazeotrope		554
6143	$C_7H_7NO_2$	p-Nitrotoluene	238.9	Nonazeotrope		554
6144	C_7H_8	Toluene	110.75	Nonazeotrope		527
6145	C_7H_8O	Anisole	153.85	153.5	5	564
6146	C_7H_8O	Benzyl alcohol	205.25	Nonazeotrope		527

TABLE I. *Binary Systems* 191

No.	Formula	B-Component Name	B.P., °C	Azeotropic Data B.P., °C	Wt.%A	Ref.
A =	C₃H₇NO₂	**Ethyl Carbamate** *(continued)*	**185.25**			
6147	C₇H₈O	*m*-Cresol	202.2	202.6	8	564
6148	C₇H₈O	*o*-Cresol	191.1	193.45	30	564
6149	C₇H₈O	*p*-Cresol	201.7	202.2	10	564
6150	C₇H₁₂O₄	Ethyl malonate	199.2	185.15	95	564
6151	C₇H₁₄O	2-Methylcyclo- hexanol	168.5	Nonazeotrope		527
6152	C₇H₁₄O₂	Amyl acetate	148.8	Nonazeotrope		527
6153	C₇H₁₄O₃	1,3-Butanediol methyl ether acetate	171.75	Nonazeotrope		575
6154	C₇H₁₆O	*n*-Heptyl alcohol	176.15	175.1	28.5	564
6155	C₈H₈	Styrene	145.8	Nonazeotrope		527
6156	C₈H₈O	Acetophenone	202.0	184.85	86	552
6157	C₈H₈O₂	Benzyl formate	203.0	182.5	62	564
6158	C₈H₈O₂	Methyl benzoate	199.4	183.8	67	564
6159	C₈H₈O₂	Phenyl acetate	195.7	180.0	52	564
6160	C₈H₁₀	Ethylbenzene	136.15	Nonazeotrope		527
6161	C₈H₁₀	*m*-Xylene	139.2	Nonazeotrope		527
6162	C₈H₁₀O	Benzyl methyl ether	167.8	163.5	18	527
6163	C₈H₁₀O	*p*-Ethylphenol	218.8	Nonazeotrope		575
6164	C₈H₁₀O	*m*-Methylanisole	177.2	171.5	26	564
6165	C₈H₁₀O	*p*-Methylanisole	177.05	171.3	25	564
6166	C₈H₁₀O	Phenethyl alcohol	219.4	Nonazeotrope		575
6167	C₈H₁₀O	Phenetole	170.45	166.2	22	564
6168	C₈H₁₀O	2,4-Xylenol	210.5	Nonazeotrope		575
6169	C₈H₁₀O	3,4-Xylenol	226.8	Nonazeotrope		564
6170	C₈H₁₀O₂	Veratrole	206.8	182.0	67	564
6171	C₈H₁₂O₄	Ethyl fumarate	217.85	Nonazeotrope		527
6172	C₈H₁₂O₄	Ethyl maleate	223.3	Nonazeotrope		527
6173	C₈H₁₄O	Methylheptenone	173.2	171.5	30	552
6174	C₈H₁₄O₄	Ethyl succinate	217.25	Nonazeotrope		527
6175	C₈H₁₆O	2-Octanone	172.85	171.5	28	552
6176	C₈H₁₆O₂	Butyl butyrate	166.4	164.8	15	564
6177	C₈H₁₆O₂	Ethyl caproate	167.7	165.0	16	564
6178	C₈H₁₆O₂	Isoamyl propionate	160.7	< 159.5	> 7	527
6179	C₈H₁₆O₂	Isobutyl butyrate	156.9	< 156.3	> 6.5	575
6180	C₈H₁₆O₂	Isobutyl isobutyrate	148.6	Nonazeotrope		527
6181	C₈H₁₈O	Butyl ether	142.4	< 141.5	< 5	562
6182	C₈H₁₈O	Octyl alcohol	195.2	183.5	72.5	564
6183	C₈H₁₈O	*sec*-Octyl alcohol	180.4	177.0	37	564
6184	C₈H₁₈S	Butyl sulfide	185.0	< 175.5	<44	566
6185	C₈H₁₈S	Isobutyl sulfide	172.0	166.5	23	555
6186	C₉H₈	Indene	182.6	172.65	35	527
6187	C₉H₁₀O	*p*-Methylaceto- phenone	226.35	Nonazeotrope		552
6188	C₉H₁₀O	Propiophenone	217.7	Nonazeotrope		552
6189	C₉H₁₀O₂	Ethyl benzoate	212.5	Nonazeotrope		527
6190	C₉H₁₂	Cumene	152.8	151.5	6	575

No.	Formula	B-Component Name	B.P., °C	Azeotropic Data B.P., °C	Wt.%A	Ref.
A =	C₃H₇NO₂	**Ethyl Carbamate** *(continued)*	185.25			
6191	C₉H₁₂	Mesitylene	164.6	159.0	22	564
6192	C₉H₁₂	Propylbenzene	159.3	157.0	15	527
6193	C₉H₁₂	Pseudocumene	168.2	161.4	25	564
6194	C₉H₁₂O	Benzyl ethyl ether	185.2	175.0	34	564
6195	C₉H₁₂O	Phenyl propyl ether	190.5	176.2	45	562
6196	C₉H₁₄O	Phorone	197.8	< 184.5	<82	552
6197	C₉H₁₈O₂	Butyl isovalerate	177.6	171.3	28	527
6198	C₉H₁₈O₂	Ethyl enanthate	188.7	< 178.0	<48	562
6199	C₉H₁₈O₂	Isoamyl butyrate	181.05	173.7	33	564
6200	C₉H₁₈O₂	Isoamyl isobutyrate	169.8	166.5	21	562
6201	C₉H₁₈O₂	Isobutyl isovalerate	171.2	167.65	20	570
6202	C₉H₁₈O₂	**Methyl caprylate**	192.9	178.5	48	564
6203	C₉H₁₈O₃	Isobutyl carbonate	190.3	176.5	42	564
6204	C₁₀H₈	Naphthalene	218.0	184.05	77	564
6205	C₁₀H₁₂O₂	Ethyl α-toluate	228.75	Nonazeotrope		527
6206	C₁₀H₁₂O₂	Propyl benzoate	230.85	Nonazeotrope		527
6207	C₁₀H₁₄	Butylbenzene	183.1	172.0	37	527
6208	C₁₀H₁₄	Cymene	176.7	169.0	31	564
6209	C₁₀H₁₆	Camphene	159.6	157.0	15	564
6210	C₁₀H₁₆	Limonene	177.6	168.0	32	564
6211	C₁₀H₁₆	α-Terpinene	173.5	166.0	28	564
6212	C₁₀H₁₆	γ-Terpinene	183.0	171.5	38	564
6213	C₁₀H₁₆O	Camphor	209.1	184.85	84	552
6214	C₁₀H₁₆O	Fenchone	193.6	< 182.0	<75	552
6215	C₁₀H₁₈O	Borneol	215.0	Nonazeotrope		527
6216	C₁₀H₁₈O	Cineol	176.35	168.4	28	564
6217	C₁₀H₁₈O	Linalool	198.6	< 185.0		575
6218	C₁₀H₁₈O	α-Terpineol	218.85	Nonazeotrope		527
6219	C₁₀H₁₈O	β-Terpineol	210.5	Nonazeotrope		575
6220	C₁₀H₂₀O	Menthol	216.3	Nonazeotrope		527
6221	C₁₀H₂₀O₂	Ethyl caprylate	208.35	< 184.0	72	527
6222	C₁₀H₂₀O₂	Isoamyl isovalerate	192.7	177.75	46	564
6223	C₁₀H₂₀O₂	Methyl pelargonate	213.8	184.3	85	575
6224	C₁₀H₂₂	2,7-Dimethyloctane	160.1	< 157.5	<19	527
6225	C₁₀H₂₂O	Amyl ether	187.4	171.0	37	564
6226	C₁₀H₂₂O	Isoamyl ether	173.35	163.15	27	564
6227	C₁₁H₁₀	1-Methylnaphthalene	244.6	Nonazeotrope		527
6228	C₁₁H₁₀	2-Methylnaphthalene	241.15	Nonazeotrope		527
6229	C₁₁H₂₀O	Isobornyl methyl ether	192.4	177.0	45	564
6230	C₁₁H₂₀O	Methyl α-terpineol ether	216.2	184.9	96	575
6231	C₁₁H₂₂O₃	Isoamyl carbonate	232.2	Nonazeotrope		527
6232	C₁₂H₂₀O₂	Bornyl acetate	227.6	Nonazeotrope		527
6233	C₁₂H₂₂O	Ethyl isobornyl ether	203.8	181.2	82	562
A =	C₃H₇NO₂	**Isopropyl Nitrite**	40.1			
6234	C₃H₇NO₂	Propyl nitrite	47.75	Nonazeotrope		550
6235	C₃H₇NO₃	Propyl nitrate	110.5	Nonazeotrope		560
6236	C₃H₈O₂	Methylal	42.3	39.75	80	550

TABLE I.   *Binary Systems*   193

No.	Formula	B-Component Name	B.P., °C	Azeotropic Data B.P., °C	Wt.%A		Ref.
A =	**C₃H₇NO₂**	**Isopropyl Nitrite** *(continued)*	**40.1**				
6237	C₄H₄O	Furan	31.7	Nonazeotrope			550
6238	C₄H₉Cl	2-Chloro-2-methylpropane	50.8	Nonazeotrope			550
6239	C₄H₁₀O	Ethyl ether	34.6	Nonazeotrope			537
6240	C₄H₁₀O	Methyl propyl ether	38.85	< 37.5	33		550
6241	C₅H₈	Isoprene	34.3	33.5	28		550
6242	C₅H₁₀	Cyclopentane	49.3	< 39.9	<92		550
6243	C₅H₁₀	2-Methyl-2-butene	37.1	35.5	38		550
6244	C₅H₁₀	3-Methyl-1-butene	20.6	Nonazeotrope			550
6245	C₅H₁₂	2-Methylbutane	27.95	27.65	7.5		527
6246	C₅H₁₂	Pentane	36.15	34.5	35		550
6247	C₆H₁₀	Biallyl	60.1	Nonazeotrope			550
6248	C₆H₁₄	2,3-Dimethylbutane	58.0	Nonazeotrope			550
6249	C₆H₁₄	Hexane	68.8	Nonazeotrope			550
A =	**C₃H₇NO₂**	**1-Nitropropane**	**131**				
6250	C₃H₈O	Isopropyl alcohol	82.40	Nonazeotrope			806
6251	C₃H₈O	Propyl alcohol	97.15	96.95	8.8		806
6252	C₄H₁₀O	Butyl alcohol	117.73	115.30	32.2		806
6253	C₄H₁₀O	sec-Butyl alcohol	99.53	99.40	4.1		806
6254	C₄H₁₀O	tert-Butyl alcohol	82.41	Nonazeotrope			806
6255	C₄H₁₀O	Isobutyl alcohol	107.89	105.28	15.2		806
6256	C₅H₁₂O	n-Amyl alcohol	138.06	127.4	63.9		866
6257	C₆H₅Cl	Chlorobenzene, 130 mm.		75	44	v-l	510
		" 598 mm.		120	44	v-l	510
6258	C₆H₆	Benzene 250 mm.		Nonazeotrope		v-l	845
		" 25°C.		Nonazeotrope			806
6259	C₆H₁₂	Cyclohexane	80.74	Azeotropic			806
6260	C₆H₁₄	Hexane 25°C		Nonazeotrope		v-l	845
		"	68.72	68.7	0.4		806
6261	C₇H₈	Toluene	110.62	110.6	0.3		806
6262	C₇H₁₄	Methyl cyclohexane	100.93	100.0	15.0		806
6263	C₇H₁₆	Heptane	98.43	96.6	13.5		806
6264	C₈H₆	Phenylacetylene	142	Azeotropic			247
6265	C₈H₈	Styrene	145	Azeotropic			247
		"	68/60 mm.	Nonazeotrope			53
6266	C₈H₁₀	Ethylbenzene, 60 mm.	60.5	56.4	61		53
		"	136	127.5	59		52
		"	136.19	129.0	56.0		806
6267	C₈H₁₈	Octane	125.66	115.8	34.2		806
6268	C₈H₁₈	2,2,4-Trimethylpentane	99.24	99.0	13		806
6269	C₉H₂₀	Nonane	150.80	126.6	61.6		806
A =	**C₃H₇NO₂**	**2-Nitropropane**	**120**				
6270	C₃H₈O	Isopropyl alcohol	82.40	82.24	4.2		806
6271	C₃H₈O	Propyl alcohol	97.15	95.97	24.9		806
6272	C₄H₈O	2-Butanone	79.50	Nonazeotrope			806
6273	C₄H₈O	Tetrahydrofuran	66	Nonazeotrope			806

No.	Formula	B-Component Name	B.P., °C	B.P., °C	Wt.%A	Ref.
A =	$C_3H_7NO_2$	**2-Nitropropane** *(continued)*	**120**			
6274	$C_4H_8O_2$	Ethyl acetate	77.11	Nonazeotrope		806
6275	$C_4H_{10}O$	Butyl alcohol	117.73	111.61	52.4	806
6276	$C_4H_{10}O$	*sec*-Butyl alcohol	99.53	98.70	18.0	806
6277	$C_4H_{10}O$	*tert*-Butyl alcohol	82.41	Nonazeotrope		806
6278	$C_4H_{10}O$	Isobutyl alcohol	107.89	105.28	33.1	806
6279	$C_4H_{10}O_2$	2-Ethoxyethanol	134.8	119.4	85	806
		" 100 mm.		62.0	92.2	806
6280	$C_5H_{12}O$	Amyl alcohol	138.06	119.5	85.2	806
6281	C_6H_6	Benzene	80.1	Nonazeotrope v-l		806,845
6282	$C_6H_{10}O$	Cyclohexanone	155.65	Nonazeotrope		806
6283	C_6H_{12}	Cyclohexane	80.74	81.2	9.8	806
		" 500 mm.		66.8	8.6	806
		" 300 mm.		52.4	8.4	806
6284	C_6H_{14}	Hexane 25°C		Nonazeotrope v-l		845
		"	68.74	68.7	3	806
6285	C_7H_8	Toluene	110.63	109.8	18.3	806
		"	110.8	110		162
6286	C_nH_{2n+2}	Paraffins	107-110	96-108		162
6287	C_7H_{14}	Methyl cyclohexane	100.93	96.5	22.7	806
		" 300 mm.		67.4	21.6	806
6288	$C_7H_{14}O_2$	4-Methoxy-4-methyl-2-pentanone	165	Nonazeotrope		806
6289	C_7H_{16}	Heptane	98.43	94.5	20.5	806
		" 500 mm.		83.5	19.7	806
		" 300 mm.		66.3	20.0	806
		" 100 mm.		38.5	19.2	806
6290	C_8H_{10}	Ethylbenzene	136.19	120.2	91.6	806
6291	C_8H_{18}	Octane	125.66	110.7	46.9	806
		" 500 mm.		97.5	46.2	806
		" 300 mm.		81.7	46.0	806
		" 100 mm.		54.2	45.9	806
6292	C_8H_{18}	2,2,4-Trimethylpentane	99.24	95.4	20.9	806
		" 300 mm.		66.9	18.7	806
6293	C_9H_{20}	Nonane	150.79	118.4	75.4	806
		" 500 mm.		104.7	75.1	806
		" 300 mm.		88.3	75.0	806
		" 100 mm.		60.5	74.5	806
A =	$C_3H_7NO_2$	**Propyl Nitrite**	**47.75**			
6294	C_3H_8O	Propyl alcohol	97.25	Nonazeotrope		539
6295	$C_3H_8O_2$	Methylal	42.3	Nonazeotrope		550
6296	C_4H_9Cl	2-Chloro-2-methyl-propane	50.8	47.5 >79		550 / 550
6297	$C_4H_{10}O$	Ethyl ether	34.6	Nonazeotrope		550
6298	C_5H_{10}	Cyclopentane	49.3	45.5	54	550
6299	C_5H_{12}	2-Methylbutane	27.95	Nonazeotrope		550
6300	C_5H_{12}	Pentane	36.15	35.8	9	550

TABLE I.   *Binary Systems*   195

No.	Formula	Name	B.P., °C	B.P., °C	Wt.%A	Ref.
		B-Component		Azeotropic Data		
A =	$C_3H_7NO_2$	**Propyl Nitrite** *(continued)*	**47.75**			
6301	$C_5H_{12}O$	Ethyl propyl ether	63.85	Nonazeotrope		550
6302	C_6H_{14}	2,3-Dimethylbutane	58.0	Nonazeotrope		550
6303	C_6H_{14}	Hexane	68.8	Nonazeotrope		550
A =	$C_3H_7NO_3$	**Propyl Nitrate**	**110.5**			
6304	C_3H_8O	Isopropyl alcohol	82.42	< 81.5		560
6305	C_3H_8O	Propyl alcohol	97.2	93.7	30	560
6306	$C_{33}H_8O_2$	2-Methoxyethanol	124.5	108.0	80	560
6307	$C_4H_8O_2$	Dioxane	101.35	Nonazeotrope		557
6308	C_4H_8S	Tetrahydrothiophene	118.8	109.0	73	560
6309	C_4H_9Br	1-Bromobutane	101.5	< 101.0		560
6310	C_4H_9Br	1-Bromo-2-methyl-propane	91.4	Nonazeotrope		547
6311	C_4H_9I	2-Iodobutane	120.0	< 109.5	<85	560
6312	C_4H_9I	1-Iodo-2-methylpropane	120.8	< 109.5	<89	560
6313	$C_4H_{10}O$	Butyl alcohol	117.8	106.5	68	560
6314	$C_4H_{10}O$	Isobutyl alcohol	108.0	< 103.5	>47	560
6315	$C_4H_{10}O_2$	2-Ethoxyethanol	135.3	Nonazeotrope		556
6316	$C_5H_{12}O$	*tert*-Amyl alcohol	102.35	< 100.1	<23	560
6317	$C_5H_{12}O$	Isoamyl alcohol	131.9	< 110.0		560
6318	$C_5H_{12}O$	2-Pentanol	119.8	< 108.0	<90	560
6319	C_6H_6	Benzene	80.15	Nonazeotrope		560
6320	$C_6H_{12}O_2$	Ethyl isobutyrate	110.1	109.7		563
6321	$C_6H_{14}O_2$	Acetal	103.55	Nonazeotrope		560
6322	$C_6H_{14}O_2$	Ethoxypropoxymethane	113.7	< 110.0		557
6323	C_7H_8	Toluene	110.75	< 109.0	>47	560
6324	C_7H_{14}	Methylcyclohexane	101.15	97.0	25	560
6325	C_7H_{16}	Heptane	98.4	95.0	25	560
6326	C_8H_{18}	2,5-Dimethylhexane	109.4	101.2	45	560
A =	C_3H_8	**Propane**	**— 42.1**			
6327	C_4H_{10}	2-Methylpropane —7 to 121°		Nonazeotrope	v-l	384
6328	C_7F_{16}	Perfluoroheptane crit. press.		Nonazeotrope	v-l	431
A =	C_3H_8O	**Isopropyl Alcohol**	**82.45**			
6329	$C_3H_8O_2$	Methylal	42.3	Nonazeotrope		556
6330	C_4H_4S	Thiophene	84.7	< 76.0	<43	566
6331	$C_4H_6O_2$	Biacetyl	88	77.3	~ 60	640
		"	87.5	< 79	<60	552
6332	$C_4H_6O_2$	Methyl acrylate	80	76.0	46.5	799,800
6333	$C_4H_6O_2$	Vinyl acetate	72.7	70.8	22.4	982
6334	C_4H_7N	Isobutyronitrile	103.85	Nonazeotrope		575
6335	C_4H_8O	2-Butanone	79.6	77.9	32	29,527
		"	79.6	~ 78	~ 66	v-l	680
		"	79.6	77.7	34	597c,680
6335a	C_4H_8O	Tetrahydrofuran	66	Nonazeotrope	v-l	879f
6336	C_4H_8OS	Ethyl thioacetate	116.6	Nonazeotrope		575

No.	Formula	B-Component Name	B.P., °C	B.P., °C	Azeotropic Data Wt.%A		Ref.
A =	C₃H₈O	**Isopropyl Alcohol** *(continued)*	**82.45**				
6337	C₄H₈O₂	Dioxane	101.35	Nonazeotrope			527
		"				v-l	153
6338	C₄H₈O₂	Ethyl acetate	77.05	75.9	25	v-l	674
		" 40°-60°C.				v-l	674
		"	77.1	74	26		572,834
6339	C₄H₈O₂	Methyl propionate	79.8	76.35	38		572
		"	79.6	77	28		982
6340	C₄H₈O₂	Propyl formate	80.8	75.85	~ 36		572
6341	C₄H₈S	Tetrahydrothiophene	118.8	Nonazeotrope			566
6341a	C₄H₉Br	1-Bromobutane	101.5	79.6			571c
6342	C₄H₉Br	2-Bromobutane	91.2	77.5	34		567
6343	C₄H₉Br	1-Bromo-2-methylpropane	90.95	77.5	33		555
6344	C₄H₉Br	2-Bromo-2-methylpropane	73.3	67	<20		563
6345	C₄H₉Cl	1-Chlorobutane	78.05	70.8	23		573
6346	C₄H₉Cl	2-Chlorobutane	68.25	64.0	18		567
6347	C₄H₉Cl	1-Chloro-2-methylpropane	68.85	64.8	17		573
		"	68.9	63.8	19		982
6348	C₄H₉I	1-Iodobutane	130.4	Nonazeotrope			575
6349	C₄H₉I	1-Iodo-2-methylpropane	120	81-82	70		532,834
6350	C₄H₁₀O	*tert*-Butyl alcohol	82.45	Nonazeotrope			549
6351	C₄H₁₀O	Ethyl ether	34.6	Nonazeotrope			981
6352	C₄H₁₀S	Ethyl sulfide	92.2	78.0	~ 52		531
6353	C₄H₁₁N	Butylamine	77.8	84.7	60		982
6354	C₅H₁₀	Cyclopentane	49.4	< 47.3			567
6355	C₅H₁₀	2-Methyl-2-butene	37.15	Nonazeotrope			563
6356	C₅H₁₀	3-Methyl-1-butene	22.5	Nonazeotrope			540
6357	C₅H₁₀O	Isopropyl vinyl ether	55.7		16.5		1008
6358	C₅H₁₀O	3-Methyl-2-butanone	95.4	Nonazeotrope			552
6359	C₅H₁₀O	2-Pentanone	102.35	Nonazeotrope		v-l	37
6360	C₅H₁₀O	3-Pentanone	102.05	Nonazeotrope			552
6361	C₅H₁₀O₂	Butyl formate	106.8	Nonazeotrope			575
6362	C₅H₁₀O₂	Isobutyl formate	97.9	Nonazeotrope			532
6363	C₅H₁₀O₂	Isopropyl acetate	91	80.1	52.3		164,536
6364	C₅H₁₀O₂	Methyl butyrate	102.65	Nonazeotrope			536
6365	C₅H₁₀O₂	Methyl isobutyrate	92.5	81.4	65		575
6366	C₅H₁₀O₂	Propyl acetate	101.6	Nonazeotrope			537
6367	C₅H₁₁Br	1-Bromo-3-methylbutane	120.65	Nonazeotrope			527
		1-Bromo-3-methylbutane	120.3	82.2	~ 82		535
6368	C₅H₁₁Cl	1-Chloro-3-methylbutane	99.8	79.2	43		573
6369	C₅H₁₂	2-Methylbutane	27.95	Nonazeotrope			537
		"	27.95	27.8	5		538
6370	C₅H₁₂	Pentane	36.15	35.5	6		538
6371	C₅H₁₂O	Ethyl propyl ether	63.6	62.0	10		545
6372	C₅H₁₂O₂	Diethoxymethane	87.95	79.6	52		556
6373	C₆H₅Cl	Chlorobenzene	132.0	Nonazeotrope			532

TABLE I.   *Binary Systems*   197

No.	Formula	B-Component Name	B.P., °C	Azeotropic Data B.P., °C	Wt.%A		Ref.
A =	C₃H₈O	**Isopropyl Alcohol** *(continued)*	**82.45**				
6374	C₆H₅F	Fluorobenzene	85.15	74.5	30		545
6375	C₆H₆	Benzene	80.2	71.92	33.3		834,1042, 1044
		"	155 mm.	31.8	20.6		924
		"	243 mm.	41.8	23.6		924
		"	509 mm.	60.4	29.9		924
		"	607 mm.	65.3	31.4		924
		"	760 mm.	71.74	33.7		924
		"	196 mm.	37.2	22.4		735
		"	512 mm.	60.3	30		735
		"	4920 mm.	134.7	62		735
		"	10,180 mm.	166.3	79		735
		"	15,380 mm..	186.1	91		735
		"	500 mm.	60.1	29.3	v-l	684
		"	146 mm.	30	16	v-l	979c
		"	280 mm.	45	19	v-l	979c
		"	504 mm.	60	24.8	v-l	979c
6376	C₆H₈	1,3-Cyclohexadiene	80.8	70.4	36		563
6377	C₆H₈	1,4-Cyclohexadiene	85.6	72.3			563
6378	C₆H₁₀	Biallyl	60.0	55.8	11		537
6379	C₆H₁₀	Cyclohexene	82.7	70.5	27		537
6380	C₆H₁₀	1,3-Hexadiene	72.9	Min. b.p.			221
6381	C₆H₁₀	2,4-Hexadiene	82	Min. b.p.			221
6382	C₆H₁₀	3-Methyl-1,3-pentadiene	77	Min. b.p.			221
6383	C₆H₁₀O	Mesityl oxide	128.3	Nonazeotrope			981
6384	C₆H₁₂	Cyclohexane, 129 mm.		26.3	18.3		924
		"	270 mm.	42.5	23.3		924
		"	434 mm.	54.1	27.1		924
		"	549 mm.	60.2	29.2		924
		"	760 mm.	69.4	32		924
		"	80.84	69.6	32.7	v-l	682
		"	80.75	68.6	33		563
		"	**80.7**	68.80	33.0	v-l	683,1049
		"	500 mm.	57.8	28.7	v-l	684
		"		69.6	32	v-l	597c
6384a	C₆H₁₂	1-Hexene	200 mm.	27.2	8.9		498m
		"	400 mm.	41.9	10.9		498m
		"	600 mm.	52.1	12.6		498m
		"	760 mm.	63.6	59.2	14.1	498m
6385	C₆H₁₂	Methylcyclopentane	72.0	63.3	25		568
6386	C₆H₁₂O	4-Methyl-2-pentanone	115.9	Nonazeotrope			37
6387	C₆H₁₂O	Pinacolone	106.2	Nonazeotrope			552
6388	C₆H₁₂O₂	Methyl isovalerate	116.5	Nonazeotrope			575
6389	C₆H₁₄	2,3-Dimethylbutane	58.0	53.8	9		567
6390	C₆H₁₄	Hexane	68.85	62.7	23		538
6391	C₆H₁₄O	Isopropyl ether	69.0	66.2	16.3		215
		"				v-l	1042g
		"	67.5	65.6	14.2	v-l	991c

No.	Formula	B-Component Name	B.P., °C	Azeotropic Data B.P., °C	Wt.%A		Ref.
A =	C₃H₈O	**Isopropyl Alcohol** *(continued)*	**82.45**				
6392	C₆H₁₄O	Propyl ether	90.55	78.2	52		573
6393	C₆H₁₄O₂	Acetal	103.55	Nonazeotrope			556
		Acetal	103.55	81.3	~ 63		573
6394	C₆H₁₄S	Isopropyl sulfide	120.5	Nonazeotrope			566
6395	C₆H₁₅N	Diisopropylamine	84.1	79.7	40		982
6396	C₆H₁₅N	Hexylamine	132.7	Nonazeotrope			981
6397	C₇H₈	Toluene	110.7		52		950
		"	110.6	80.6	69		982
		"	110.6	81.5	77.3		662
		" 712 mm.		79.0	76.3		662
		" 611 mm.		75.5	73.1		662
		" 513 mm.		71.0	71.8		662
		" 411 mm.		65.6	69.1		662
		"	110.7	80.6	58		50
		"		20	47.7		814
		"		40	58.8		814
		"		60	67.4		814
		"		78	73.1		814
6398	C₇H₁₄	Methylcyclohexane	100.8	77.6	53		50,537
		"	100.98	77.7	53	v-l	682
		" 500 mm.		66.5	34.2	v-l	684
6399	C₇H₁₆	Heptane 684 mm.		72.2	54.3	v-l	993
		"	78.45	76.4	50.5		527
6400	C₇H₁₆O	Butyl isopropyl ether	103	79	71.91		73
6401	C₈H₈	Styrene	145.8	Nonazeotrope			545
6402	C₈H₁₀	Ethylbenzene	136.15	Nonazeotrope			537
6403	C₈H₁₀	o-Xylene	144.3	Nonazeotrope			575
6404	C₈H₁₀	m-Xylene	139.0	Nonazeotrope			537
6405	C₈H₁₀	p-Xylene	138.2	Nonazeotrope			541
6406	C₈H₁₄	Diisobutylene	102.3	77.8	54.5		982
6407	C₈H₁₆	*trans*-1,2-Dimethyl- cyclohexane	123.42	81.4	79 vol. %		826
		"			~ 79		928
6408	C₈H₁₆	1,3-Dimethylcyclohexane	120.7	81.0	78		575
6409	C₈H₁₆	*cis*-1-Ethyl-2-methyl- cyclopentane	128.05	82.2	83 vol. %		826
6410	C₈H₁₆	*trans*-1-Ethyl-2- methylcyclopentane	121.2	81.6	76 vol. %		826
6411	C₈H₁₆	*trans*-1-Ethyl-3- methylcyclopentane	120.8	81.4	75 vol. %		826
6412	C₈H₁₆	1,1,2-Trimethyl- cyclopentane	113.73	80.4	66 vol. %		826
		"			~ 67		928
6413	C₈H₁₆	1,1,3-Trimethyl- cyclopentane	104.9		~ 54		928
		"	104.89	78.5	53 vol. %		826

TABLE I.   *Binary Systems*   199

No.	Formula	Name	B.P., °C	B.P., °C	Wt.%A		Ref.
		B-Component		**Azeotropic Data**			
A =	**C₃H₈O**	**Isopropyl Alcohol** *(continued)*	**82.45**				
6414	C₈H₁₆	1,*cis*-2,*trans*-3-Trimethylcyclopentane	117.5	81.1	71 vol. %		826
6415	C₈H₁₆	1,*cis*-2,*trans*-4-Trimethylcyclopentane	116.73	80.9	71 vol. %		826
6416	C₈H₁₆	*cis-cis-trans*-1,2,4-Trimethylcyclopentane		~ 70			928
6417	C₈H₁₈	2,5-Dimethylhexane	109.2	79.0	62		535
6418	C₈H₁₈	Octane	124.75	81.6	84		575
6419	C₈H₁₈	2,2,4-Trimethylpentane	99.3	76.8	54		575
		"	99.3	77.3	48.5	v-l	104
6420	C₈H₁₈O	Isobutyl ether	122.1	Nonazeotrope			556
6421	C₉H₈	Indene	182.6	Nonazeotrope			575
6422	C₉H₁₂	Cumene	152.8	Nonazeotrope			575
6423	C₉H₁₂	Mesitylene	164.6	Nonazeotrope			540
6424	C₉H₁₂	Propylbenzene	159.3	Nonazeotrope			575
6425	C₉H₂₁BO₃	Isopropylborate	140.8	82	94.6		251
6426	C₁₀H₁₄	Butylbenzene	183.1	Nonazeotrope			575
6427	C₁₀H₁₄	Cymene	176.7	Nonazeotrope			575
6428	C₁₀H₁₆	Camphene	159.6	Nonazeotrope			540
6429	C₁₀H₁₆	*d*-Limonene	177.8	Nonazeotrope			537
6430	C₁₀H₁₆	α-Pinene	155.8	Nonazeotrope			537
6431	C₁₀H₁₆	α-Terpinene	173.4	Nonazeotrope			575
6432	C₁₀H₁₆	Thymene	179.7	Nonazeotrope			537
6433	C₁₀H₂₂	2,7-Dimethyloctane	160.2	Nonazeotrope			537
6434	TiC₁₂H₂₈O₄	Titanium isopropoxide 740 mm.		230	4		61
A =	**C₃H₈O**	**Propyl Alcohol**	**97.2**				
6435	C₃H₈O₂	Methylal	42.3	Nonazeotrope			575
6436	C₃H₈S	1-Propanethiol, 766 mm.	67.8	66.4	8.65		487
6437	C₄H₆O	Crotonaldehyde	102.15	< 97?			563
6438	C₄H₆O₂	Biacetyl	87.5	85.0	25		552
6439	C₄H₆O₂	Methyl acrylate	80	70.9	5.4		800
6440	C₄H₇ClO₂	Ethyl chloroacetate	143.55	Nonazeotrope			575
6441	C₄H₇N	Butyronitrile	118.5	Azeotrope doubtful			563
6442	C₄H₇N	Isobutyronitrile	103.85	95	70		567
6443	C₄H₇N	Pyrroline	90.9	< 89.0			575
6444	C₄H₈Cl₂O	1,2-Dichloroethyl ethyl ether	145.5	Nonazeotrope			575
6445	C₄H₈O	2-Butanone	79.6	Nonazeotrope			29
6446	C₄H₈OS	Ethyl thioacetate	116.6	Nonazeotrope			575
6447	C₄H₈O₂	Dioxane	101.35	95.3	55		527
6448	C₄H₈O₂	Ethyl acetate	77.05	Nonazeotrope			834
		" 40°-60°C.		Nonazeotrope		v-l	674
6449	C₄H₈O₂	Methyl propionate	79.85	Nonazeotrope			532
6450	C₄H₈O₂	Propyl formate	80.9	80.6	9.8		359
		"	80.8	80.65	< 3		572

No.	Formula	B-Component Name	B.P., °C	B.P., °C	Wt.%A	Ref.
				Azeotropic Data		
A =	C₃H₈O	**Propyl Alcohol** (*continued*)	**97.2**			
6451	C₄H₈S	Tetrahydrothiophene	118.8	96.5	90	555
6452	C₄₄H₉Br	1-Bromobutane	100.3	89.5	29	573
6453	C₄H₉Br	2-Bromobutane	91.2	85.3	20.5	567
6454	C₄₄H₉Br	1-Bromo-2-methyl-propane	89.2	86.1 B.p. curve	19.25	389,555
6455	C₄H₉Br	2-Bromo-2-methyl-propane	73.3	72.3		563
6456	C₄H₉Cl	1-Chlorobutane	78.05	74.8	~ 18	573
6457	C₄H₉Cl	2-Chlorobutane	68.25	67.2	> 9	567
6458	C₄H₉Cl	1-Chloro-2-methyl-propane	68.85	67.7	22	572
6459	C₄H₉Cl	2-Chloro-2-methyl-propane	68.25	67.2	> 9	575
6460	C₄H₉I	1-Iodobutane	130.4	96.2	66	567
6461	C₄H₉I	1-Iodo-2-methylpropane	120	93	45	573,834
6462	C₄H₉I	2-Iodobutane	120.0	94.5	53	575
6463	C₄H₁₀O	*sec*-Butyl alcohol	99.5	Nonazeotrope		549
6464	C₄H₁₀O	Ethyl ether	34.6	Nonazeotrope		556
6465	C₄H₁₀O	Isobutyl alcohol	108.0	Nonazeotrope		834
6466	C₄H₁₀O₂	Acetaldehyde dimethyl acetal	64.3	Nonazeotrope		575
				Nonazeotrope		575
6467	C₄H₁₀S	Butanethiol	97.5	< 92.0	<41	575
6468	C₄H₁₀S	Ethyl sulfide	92.2	85.5	28	555
6469	C₅H₅N	Pyridine	115.4	Nonazeotrope		553
6470	C₅H₇N	N-Methylpyrrol	112.8	Nonazeotrope		575
6471	C₅H₉ClO₂	Propyl chloroacetate	162.3	Nonazeotrope		121
6472	C₅H₁₀O	3-Methyl-2-butanone	95.4	93.5	35	552
6473	C₅H₁₀O	2-Pentanone	102.35	96.0	68	552
6474	C₅H₁₀O	3-Pentanone	102.05	96.0	63	552
		"	101.8	94.9	57	981
6475	C₅H₁₀O₂	Butyl formate	106.8	95.5	64	567
6476	C₅H₁₀O₂	Ethyl propionate	99.1	93.4	51	563
6477	C₅H₁₀O₂	Isobutyl formate	97.9	93.2	40	532
6478	C₅H₁₀O₂	Methyl butyrate	102.65	94.4	47	572
6479	C₅H₁₀O₂	Methyl isobutyrate	92.3	89.5	~ 26	532
6480	•C₅H₁₀O₂	Propyl acetate	101.6	94.2	40	359,572
		"	101.6	94.7	49 v-l	752
		" 200 mm.		59.96	31.4 v-l	890
		" 400 mm.		77.06	39.2 v-l	890
		" 600 mm.		88.04	44.8 v-l	890
		" 760 mm.		94.7	48.9 v-l	890
6481	C₅H₁₁Br	1-Bromo-3-methyl-butane	118.2	94.0 B.p. curve	70.7	446,555
6482	C₅H₁₁Cl	1-Chloro-3-methyl-butane	99.8	89.4	31	553

TABLE I. *Binary Systems* 201

No.	Formula	Name	B.P., °C	B.P., °C	Wt.%A		Ref.
		B-Component		Azeotropic Data			
A =	C_3H_8O	**Propyl Alcohol** *(continued)*	**97.2**				
6483	$C_5H_{11}I$	1-Iodo-3-methylbuatne	146.5	Nonazeotrope			446
6484	C_5H_{12}	Pentane	36.15	Nonazeotrope			537
6485	$C_5H_{12}O$	Ethyl propyl ether	63.85	Nonazeotrope			575
6486	$C_5H_{12}O$	Isoamyl alcohol	131.9	Nonazeotrope			575
6487	$C_5H_{12}O_2$	Diethoxymethane	88.0	86.15	11		1035
6488	C_6H_5Br	Bromobenzene	156.1	Nonazeotrope			575
6489	C_6H_5Cl	Chlorobenzene		96.9	63		573
		"	132	96.5	80		981
6490	C_6H_5F	Fluorobenzene	85.15	80.2	18		545
6491	C_6H_6	Benzene	80.2	77.12	16.9		1044,1051
		"		76-77	16.5		834
		"		0	4.5		775
		"		35.5	12		775
		"		76.5	21		775
		"	10.5 atm.	160	45		775
		"	239 mm.	45	10.5	v-l	93
		"	44.7-309.7 p.s.i.g.	Effect of press.		v-l	785
		"	123 mm.	28.0	8.0		924,1011
		"	289 mm.	49.8	11.6		924,1011
		"	423 mm.	59.9	13.6		924,1011
		"	610 mm.	70.1	15.7		924,1011
		"	760 mm.	77.10	17.1		924,1011
		"	342 mm.	53.7	12.3		735
		"	573 mm.	68.6	15.3		735
		"	2420 mm.	117.6	27.5		735
		"	5020 mm.	147.5	37		735
		"	10,050 mm.	183.8	50.5		735
		"	18,200 mm.	218.3	66.1		735
		"		Azeotropic at crit pt.		v-l	888
		"	80.1	77.1	17.9	v-l	661
		"	711 mm.	75	17.4	v-l	302c
6492	C_6H_8	1,3-Cyclohexadiene	80.4	75.8	20		537
6493	C_6H_{10}	Cyclohexene	82.75	76.6	21.6		563
6494	C_6H_{10}	Methylcyclopentene	75.85	< 71.7	<13		567
6495	C_6H_{12}	Cyclohexane	80.75	74.3	20		563
		"	161 mm.	33.8	9.9		924
		"	250 mm.	44.3	11.8		924
		"	429 mm.	58.0	15.0		924
		"	560 mm.	65.4	16.5		924
		"	760 mm.	74.69	18.5		924
		"	4-15 atm.	Effect of press.			786
		"	80.7	75.0	16.8	v-l	661
		"	80.7	74.7	19.82	v-l	100
		"	652.6 mm.	70	18.82	v-l	100
		"	453.3 mm.	59.81	16.6	v-l	100
		"	311.6 mm.	49.86	14.3	v-l	100

No.	Formula	Name	B.P., °C	B.P., °C	Wt.%A		Ref.
		B-Component		**Azeotropic Data**			
A =	**C₃H₈O**	**Propyl Alcohol** *(continued)*	**97.2**				
6495a	C₆H₁₂	1-Hexene	63.6	Nonazeotrope			498m
6496	C₆H₁₂	Methylcyclopentane	72.0	68.5	7		567
6497	C₆H₁₂O	2-Methylpentanal	118.3	95	86		981
6498	C₆H₁₂O	Pinacolone	106.2	Nonazeotrope			548
6499	C₆H₁₂O₂	Ethyl butyrate	120.0	Nonazeotrope			536
6500	C₆H₁₂O₂	Ethyl isobutyrate	110.1	96.8			536
6501	C₆H₁₂O₂	Isobutyl acetate	117.2	Nonazeotrope			532
6502	C₆H₁₂O₂	Methyl isovalerate	116.3	Nonazeotrope			536
6503	C₆H₁₂O₂	Propyl propionate	123.0	Nonazeotrope			575
6504	C₆H₁₂O₃	Paraldehyde	123.9	Nonazeotrope			576
6505	C₆H₁₄	2,3-Dimethylbutane	58.0	< 56.8	< 6		567
6506	C₆H₁₄	Hexane	68.95	65.65	4		563
6507	C₆H₁₄O	Propyl ether	90.7	85.8	32.2		545,760
6508	C₆H₁₄O₂	Acetal	103.55	92.4	37		572
6509	C₆H₁₄O₂	Ethoxypropoxymethane	113.7	Nonazeotrope			1035
6510	C₆H₁₆OSi	(Trimethylsiloxy)					
		propane, 735 mm.	100.3	87.5			520
6511	C₆H₁₈OSi₂	Hexamethyldisiloxane	99.85	85.15	27.4	v-l	458
6512	C₇H₈	Toluene	110.7		50.5		950
		"	110.6	92.6	49		981
		"	110.7	92.6	51.5	v-l	595
		"	110.6	92.5	51.2		662
		" 710 mm.		90.5	50.3		662
		" 612 mm.		86.0	48.3		662
		" 513 mm.		82.0	46.6		662
		" 413 mm.		75.5	44.2		662
		"	110.7	92.6	43		51,537, 834,1051
				0.5	19.5		814
				25	29.2		814
				50	38.9		814
				71.1	45.5		814
				91.1	50.5		814
6513	C₇H₁₄	Methylcyclohexane	100.8	86.3	35		50
		"	100.8	87.0	34.8	v-l	778f
6514	C₇H₁₆	*n*-Heptane	98.45	87.5	36		573
		"	98.4	84.6	34.7		351
		" 552 mm.		75	33.8	v-l	577c
		" 643 mm.		75	32	v-l	302c
6515	C₇H₁₆O₂	Dipropoxymethane	137.2	Nonazeotrope			324
6516	C₈H₈	Styrene	145.8	97.0	8		545
		" 50 mm.		38.5	84	v-l	617
				% PrOH increases with press.			617
6517	C₈H₁₀	Ethylbenzene, 50 mm.		% PrOH increases with press.			617

TABLE I. *Binary Systems* 203

No.	Formula	B-Component Name	B.P., °C	Azeotropic Data B.P., °C	Wt.%A		Ref.
A =	C₃H₈O	**Propyl Alcohol** *(continued)*	**97.2**				
		"		96.85	91.5	v-l	587
		" 60 mm.	60.5	41	68		53
		"	136	Nonazeotrope			541
6518	C₈H₁₀	*m*-Xylene	139.2	97.08	94		527
6519	C₈H₁₀	*o*-Xylene	143.6	Nonazeotrope			545
6520	C₈H₁₀	*p*-Xylene	138.4	96.88	92.2	v-l	308c
		"	138.45	97.0			575
		"	138.4		92		950
6521	C₈H₁₆	1,3-Dimethylcyclohexane	120.5	< 94	<70		563
6522	C₈H₁₈	2,5-Dimethylhexane	109.2	89.5	47		545
6523	C₈H₁₈	Octane	125.6	93.9	70		537
6524	C₈H₁₈	2,2,4-Trimethylpentane	99.3	< 85.3	<41		575
6525	C₈H₁₈O	Isobutyl ether	122.3	Nonazeotrope			556
		"	122.1	96.8			576
6526	C₈H₁₈O₂	1,1-Dipropoxyethane	147.7	Nonazeotrope			45
6527	C₉H₈	Indene	182.6	Nonazeotrope			575
6528	C₉H₁₂	Cumene	152.8	Nonazeotrope			575
6529	C₉H₁₂	Mesitylene	164.6	Nonazeotrope			537
6530	C₉H₁₂	Propylbenzene	158.9	Nonazeotrope			541
6531	C₁₀H₁₄	Cymene	176.7	Nonazeotrope			540
6532	C₁₀H₁₄	Butylbenzene	183.1	Nonazeotrope			575
6533	C₁₀H₁₆	Camphene	159.6	Nonazeotrope			540
6534	C₁₀H₁₆	*d*-Limonene	177.8	Nonazeotrope			537
6535	C₁₀H₁₆	α-Pinene	155.8	97.1	98-99?		563
6536	C₁₀H₁₆	α-Terpinene	173.4	Nonazeotrope			575
A =	C₃H₈OS	**2-(Methylthio)ethanol**					
6537	C₅H₁₀OS	2-Methylthioethyl vinyl ether, 22 mm.		75	20		953
A =	C₃H₈O₂	**2-Methoxyethanol**	**124.5**				
6538	C₄H₄N₂	Pyrazine	117.2	Nonazeotrope			575
6539	C₄H₅N	Pyrrol	130.0	Nonazeotrope			527
6540	C₄H₇ClO₂	Ethyl chloroacetate	143.45	Nonazeotrope			575
6541	C₄H₈O₂	Dioxane	101.35	Nonazeotrope			527
6542	C₄H₈O₃	Methyl lactate	143.8	Nonazeotrope			575
6543	C₄H₉I	1-Iodobutane	130.4	115.5			526
6544	C₄H₉I	1-Iodo-2-methylpropane	120.8	110.5	25		526
6545	C₄H₉NO₃	Isobutyl nitrate	123.5	< 115.0	<44		560
6546	C₄H₁₀O	Butyl alcohol	117.8	Nonazeotrope			526
6547	C₄H₁₀O	*sec*-Butyl alcohol	99.5	Nonazeotrope			575
6548	C₄H₁₀O	Isobutyl alcohol	108.0	Nonazeotrope			575
6549	C₅H₄O₂	2-Furaldehyde	161.45	Nonazeotrope			575
6550	C₅H₅N	Pyridine	115.4	Nonazeotrope			553
6551	C₅H₇N	1-Methylpyrrol	112.8	Nonazeotrope			575
6552	C₅H₉N	Isovaleronitrile	130.5	< 130.0			575
6553	C₅H₉N	Valeronitrile	141.3	Nonazeotrope			556
6554	C₅H₁₀O	Cyclopentanol	140.85	Nonazeotrope			575

No.	Formula	B-Component Name	B.P., °C	Azeotropic Data B.P., °C	Wt.%A		Ref.
A =	$C_3H_8O_2$	**2-Methoxyethanol** *(continued)*	**124.5**				
6555	$C_5H_{10}O_3$	2-Methoxyethyl acetate	144.6	Nonazeotrope			556
6556	$C_5H_{11}Br$	1-Bromo-3-methylbutane	120.65	111.5	20		526
6557	$C_5H_{11}Cl$	1-Chloro-3-methylbutane	99.4	Nonazeotrope			526
6558	$C_5H_{11}I$	1-Iodo-3-methylbutane	147.65	Nonazeotrope			526
6559	$C_5H_{11}NO_3$	Isoamyl nitrate	149.75	Nonazeotrope			556
6560	$C_5H_{12}O$	Amyl alcohol	138.2	Nonazeotrope			575
6561	$C_5H_{12}O$	*tert*-Amyl alcohol	102.15	Nonazeotrope			575
6562	$C_5H_{12}O$	Isoamyl alcohol	131.9	Nonazeotrope			527
6563	$C_5H_{12}O$	2-Pentanol	119.8	119.7	4		526
6564	$C_5H_{12}O$	3-Pentanol	116.0	Nonazeotrope			575
6565	$C_5H_{12}O_3$	2-(2-Methoxyethoxy)-ethanol	192.95	Nonazeotrope			884
6566	C_6H_5Cl	Chlorobenzene	131	119.45	47.5		527
6567	C_6H_6	Benzene		Nonazeotrope			556
		"	80.1	Nonazeotrope		v-l	961
6568	C_6H_6O	Phenol	181.2	Nonazeotrope			575
6569	C_6H_7N	2-Picoline	130.7	Nonazeotrope			575
6570	$C_6H_{10}O$	Mesityl oxide	129.45	122.5	59		552
6571	$C_6H_{10}S$	Allyl sulfide	139	122.5	75		555
6572	C_6H_{12}	Cyclohexane	80.75	< 79.8	8		575
		"	80.7	Nonazeotrope			981
		"		77.5	15	v-l	961
6573	$C_6H_{12}O$	2-Hexanone	127.2	< 121.5	<56		552
6574	$C_6H_{12}O$	3-Hexanone	123.3	< 119.5	<43		552
6575	$C_6H_{12}O$	4-Methyl-2-petanone	116.05	114.2	25		527
6576	$C_6H_{12}O_2$	Butyl acetate	126.0	119.45	48		556
6577	$C_6H_{12}O_2$	Ethyl butyrate	121.5	117.8	32		556
6578	$C_6H_{12}O_2$	Ethyl isobutyrate	110.1	Nonazeotrope			526
6579	$C_6H_{12}O_2$	Isoamyl formate	123.8	119.25	40		556
6580	$C_6H_{12}O_2$	Isobutyl acetate	117.2	115.5	16		556
6581	$C_6H_{12}O_2$	Methyl isovalerate	116.5	115.0	15		526
6582	$C_6H_{12}O_2$	Propyl propionate	123.0	118.5	38		526
6583	$C_6H_{12}O_3$	2-Ethoxy ethyl acetate	156.8	Nonazeotrope			556
6584	$C_6H_{12}O_3$	Paraldehyde	124.35	118.6	35		556
6585	$C_6H_{14}O$	Propyl ether	90.1	Nonazeotrope			556
6586	C_7H_8	Toluene	110.75	106.1	25.5		527
6587	C_7H_8O	Anisole	153.85	Nonazeotrope			556
6588	C_7H_{14}	*trans*-2-Heptene	98.0	92.9	19 vol. %		826
6589	C_7H_{14}	Methylcyclohexane	101.15	94.2	25		556
6590	$C_7H_{14}O_2$	Isoamyl acetate	142.1	Nonazeotrope			526
6591	$C_7H_{14}O_2$	Isobutyl propionate	137.5	Nonazeotrope			526
6592	C_7H_{16}	Heptane	98.4	92.5	23		527
6593	$C_7H_{16}O_2$	1-*tert*-Butoxy-2-methoxyethane		119	45		215
6594	$C_7H_{16}O_4$	2-[2-(2-Methoxyethoxy)-ethoxy]ethanol	245.25	Nonazeotrope			575

TABLE I. *Binary Systems*   205

No.	Formula	B-Component Name	B.P., °C	Azeotropic Data B.P., °C	Wt.%A		Ref.
A =	$C_3H_8O_2$	**2-Methoxyethanol** *(continued)*	**124.5**				
6595	C_8H_8	Styrene	145.8	121.0	62		567
		" 742 mm.		120.8	72		311
		" 100 mm.		69.5	62		311
		" 60 mm.		57.5	60		311
		" 40 mm.		49.6	59		311
		" 60 mm.				v-l	311c
		" 57 mm.	67.9	54.8	62 vol. %		826
		" 62 mm.		56.8	50.1	v-l	421
6596	C_8H_{10}	Ethylbenzene 62 mm.		51.9	34.3	v-l	421
		"	136	117	51.2		65
		" 62 mm.		51	39		65
		" 60 mm.	60.5	51	43		53,556
6597	C_8H_{10}	*m*-Xylene	139.2	119.5	58		527
		" 748 mm.		119.0	56		311
		" 100 mm.		65.5	47		311
		" 60 mm.		53.5	45	v-l	311,311c
"		" 40 mm.		45.5	44		311
6598	C_8H_{10}	*o*-Xylene 741 mm.		119.5	64		311
		" 100 mm.		67.5	56		311
		" 60 mm.		54.5	54	v-l	311,311c
		o-Xylene 40 mm.		47.5	52		311
		"	144.3	121.0	63		526
6599	C_8H_{10}	*m,p*-Xylene	139	120			514
6600	C_8H_{10}	Xylenes	140	Min. b.p.			65
6601	C_8H_{10}	*p*-Xylene 750 mm.		119.5	55		311
		" 100 mm.		65.5	46		311
		" 60 mm.		53.6	44	v-l	311,311c
		" 40 mm.		45.5	42		311
		"	138.35	119.3	54 vol. %		826
6602	C_8H_{12}	4-Vinylcyclohexene, 57 mm.		44.4	30 vol. %		826
6603	C_8H_{16}	*cis*-1,3-Dimethyl-cyclohexane	120.9	105.6	36 vol. %		826
6604	C_8H_{16}	*trans*-1-Ethyl-2-methylcyclopentane	121.2	106.3	32 vol. %		826
6605	C_8H_{16}	*trans*-1-Ethyl-3-methylcyclopentane	120.8	106.0	35 vol. %		826
6606	C_8H_{16}	1,1,3-Trimethyl-cyclopentane	104.89	96.7	20 vol. %		826
		"	104.9		~ 20		928
6607	C_8H_{16}	1,*cis*-2,*cis*-3-Tri-methylcyclopentane	123.0	107.4	35 vol. %		826
6608	C_8H_{16}	1,*trans*-2,*cis*-3-Tri-methylcyclopentane	110.2	100.2	20 vol. %		826
6609	C_8H_{16}	2,4,4-Trimethyl-1-pentene	101.44	95.5	20 vol. %		
6610	$C_8H_{16}O_2$	Propyl isovalerate	155.7	Nonazeotrope			526

No.	Formula	B-Component Name	B.P., °C	B.P., °C	Azeotropic Data Wt.%A	Ref.
A =	C₃H₈O₂	**2-Methoxyethanol** *(continued)*	**124.5**			
6611	C₈H₁₈	2,5-Dimethylhexane	109.4	100.0	33	556
6612	C₈H₁₈	2,4-Dimethylhexane	109.4		~ 25	928
		"	109.43	99.3	26 vol. %	826
6613	C₈H₁₈	2,2,3-Trimethylpentane	109.84	99.7	25 vol. %	826
		"	109.8		~ 24	928
6614	C₈H₁₈	Octane	125.75	110.0	48	556
6615	C₈H₁₈O	Butyl ether	142.4	122.0	68	526
6616	C₈H₁₈O	Isobutyl ether	122.3	115.0	48	556
6617	C₉H₁₂	Cumene	152.8	122.4	73.5	527
6618	C₉H₁₂	Mesitylene	164.6	< 124.3		575
		"	164.6	Nonazeotrope		526
6619	C₉H₁₂	Propylbenzene	159.3	< 124.0	>82	575
		"	159.3	Nonazeotrope		556
6620	C₉H₁₈	1,1,3-Trimethyl-cyclohexane	136.6	113.1	41 vol. %	826
6621	C₉H₂₀	2,2,3,4-Tetramethyl-pentane	133.02	111.4	39 vol. %	826
		"			~ 42	928
6622	C₉H₂₀	2,3,4-Trimethylhexane	139.0	113.5	39 vol. %	826
6623	C₉H₂₀	2,3,5-Trimethylhexane	131.34	110.6	40 vol. %	826
6624	C₁₀H₁₄	Cymene	176.7	Nonazeotrope		575
6625	C₁₀H₁₆	Camphene	159.6	121.0	70	526
6626	C₁₀H₁₆	Nopinene	163.8	121.8	5	567
6627	C₁₀H₁₆	α-Pinene	155.8	120.2	66	526
6628	C₁₀H₂₂	Decane	173.3	<123.5	<92	575
6629	C₁₀H₂₂	2,7-Dimethyloctane	160.1	121.0	70	526
A =	**C₃H₈O₂**	**Methylal**	**42.3**			
6630	C₃H₈S	Propanethiol	67.3	Nonazeotrope		575
6631	C₃H₉N	Propylamine	49.7	Nonazeotrope		551
6632	C₄H₉Cl	2-Chloro-2-methylpropane	50.8	Nonazeotrope		559
6633	C₄H₁₀O	Methyl propyl ether	38.9	Nonazeotrope		561
6634	C₄H₁₁N	Diethylamine	55.9	Nonazeotrope		551
6635	C₅H₈	3-Methyl-1,2-pentadiene	40.8	38.0	45	558
6636	C₅H₈	Isoprene	34.3	32.8	30	558
		"		Nonazeotrope		581c
6637	C₅H₁₀	Cyclopentane	49.3	40.0	62	558
6638	C₅H₁₀	3-Methyl-1-butene	21.5	Nonazeotrope		558
6639	C₅H₁₀	2-Methyl-2-butene	37.15	35.2	32	558
6640	C₅H₁₀	1-Pentene	30.1	29.8	26 vol.	836
6641	C₅H₁₀	2-Pentene	36.5	34.9	29 vol.	836
6642	C₅H₁₂	2-Methylbutane	27.9	24.1	30 vol.	836
		"	27.95	27.0	23	558
6643	C₅H₁₂	Pentane	36.08	31.5	28 vol.	836
		"	36.15	33.6	35	558
6643a	C₅H₁₂O	*tert*-Butyl methyl ether	55	Nonazeotrope		581c
6644	C₆H₆	Benzene	80.15	Nonazeotrope		575
6645	C₆H₁₀	Biallyl	60.1	41.8	85	558

TABLE I.   *Binary Systems*   207

No.	Formula	Name	B.P., °C	B.P., °C	Wt.%A	Ref.
		B-Component		**Azeotropic Data**		
A =	**C₃H₈O₂**	**Methylal** *(continued)*	**42.3**			
6646	C₆H₁₄	2,3-Dimethylbutane	58.0	41.5	80	558
6647	C₆H₁₄	Hexane	68.85	Nonazeotrope		558
A =	**C₃H₈O₂**	**1,2-Propanediol**	**187.8**			
6648	C₄H₅N	Pyrrol	130.0	Nonazeotrope		575
6649	C₄H₅NS	Allyl isothiocyanate	152.05	< 151.5		575
6650	C₄H₈Br₂O	Bis(2-bromoethyl) ether		176-180		215
6650a	C₄H₁₀O₂	1-Methoxy-2-propanol	120.8	Nonazeotrope		313c
6651	C₅H₁₀O₂	3-Vinyloxypropanol		Min. b.p.		263
6652	C₆H₅ClO	p-Chlorophenol	219.75	Nonazeotrope		575
6653	C₆H₅NO₃	o-Nitrophenol	217.2	< 186.0	>62	575
6654	C₆H₆	Benzene	80.1	Nonazeotrope		981
6655	C₆H₇N	Aniline	184.35	179.5	43	551
6655a	C₆H₁₂O₂	Butyl acetate				
		730 mm.		Nonazeotrope	v-l	684b
6656	C₆H₁₂O₃	2-Ethoxyethyl acetate	156.8	Nonazeotrope		575
6657	C₆H₁₄O₃	Dipropylene glycol,				
		10 mm.		Nonazeotrope	v-l	183
6658	C₇H₈	Toluene	110.6	110.5	1.5	982
6659	C₇H₈O	p-Cresol	201.8	Azeotrope doubtful		563
6660	C₇H₈O₂	m-Methoxyphenol	243.8	242.2	~ 7	575
6661	C₇H₉N	Methylaniline	196.25	< 181.0	>46	551
6662	C₇H₁₄O₃	1,3-Butanediol methyl ether				
		acetate	171.75	<170		575
6662a	C₇H₁₆O₂	1-Butoxy-2-propanol	170	Nonazeotrope		313c
6662b	C₇H₁₆O₃	Dipropylene glycol methyl ether		183.7	40.38	313c
		" 10 mm.			2.57	313c
6663	C₈H₆O	Coumarone	173	Azeo. distillation		330
6664	C₈H₈O	Acetophenone	202.0	< 183.5		552
6665	C₈H₁₀	o-Xylene	144.4	135.8	10	982
6665a	C₈H₁₀O	2,6-Dimethylphenol		Min. b.p.		738c
6666	C₈H₁₁N	Dimethylaniline	194.05	< 177.0	>45	551
6667	C₈H₁₆O	2-Octanone	172.85	<169.5		552
6668	C₈H₁₈O	Butyl ether	142.1	136		981,982
6669	C₉H₈	Indene	182.4	Min. b.p.		276
6670	C₉H₁₃N	N,N-Dimethyl-o-toluidine	185.3	< 174.0	37	551
6671	C₉H₁₃N	N,N-Dimethyl-p-toluidine	210.2	178.0	60	575
6672	C₁₀H₈	Naphthalene	218.1	Azeo. distillation		330
6673	C₁₀H₁₆O	Camphor	209.1	< 185.0		552
6674	C₁₀H₁₈O	Menthone	209.5	< 185.0	<85	552
6674a	C₁₀H₂₂O₃	Dipropylene glycol butyl ether		186.5	93.30	313c
		" 10 mm.			79.08	313c
6675	C₁₀H₂₂O₄	Tripropylene glycol				
		methyl ether,				
		50 mm.		Nonazeotrope		215
6676	C₁₂H₂₆	Dodecane	216	175	67	183
		" 743 mm.	216	175		424
		" 200 mm.		137		424

No.	Formula	B-Component Name	B.P., °C	B.P., °C	Wt.%A	Ref.
				Azeotropic Data		
A =	C₃H₈O₂	**1,2-Propanediol** *(continued)*		**187.8**		
		" 150 mm.		130		424
		" 100 mm.	145	120.5	60	424
		" 50 mm.		105.7		424
6677	C₁₄H₃₀	Tetradecane, 748 mm.	252.5	179	70	424
		" 200 mm.		142.5		424
		" 150 mm.		135		424
		" 100 mm.		126		424
		" 50 mm.		111		424
		"	252.5	179	76	183
6678	C₁₆H₃₄O	Bis(2-ethylhexyl) ether, 10 mm.	135	84		982
A =	C₃H₈O₂	**1,3-Propanediol**		**214**		
6679	C₅H₁₀O₂	3-Vinyloxy-1-propanol			10-15	216
6679a	C₈H₁₀O	2,6-Dimethylphenol		Min. b.p.		738c
A =	C₃H₈O₃	**Glycerol**		**290.5**		
6680	C₄H₁₀O₃	Diethylene glycol	245.5	Nonazeotrope		526
6681	C₆H₄Br₂	*p*-Dibromobenzene	220.25	217.1	10	574
6682	C₆H₄ClNO₂	*m*-Chloronitrobenzene	235.5	232.2	10	554
6683	C₆H₄ClNO₂	*o*-Chloronitrobenzene	246.0	242.1	15?	554
6684	C₆H₄ClNO₂	*p*-Chloronitrobenzene	239.1	235.6	13	554
6685	C₆H₅NO₂	Nitrobenzene	210.75	Nonazeotrope		530
6686	C₆H₆O₂	Pyrocatechol	245.9	Nonazeotrope		542
6687	C₆H₆O₂	Resorcinol	281.4	Nonazeotrope		542
6688	C₆H₈O₄	Methyl maleate	204.05	Nonazeotrope		575
6689	C₆H₁₀O₄	Ethyl oxalate	185.65	Nonazeotrope		575
6690	C₆H₁₀O₄	Glycol diacetate	186.3	Nonazeotrope		575
6691	C₆H₁₄O₄	Triethylene glycol	288.7	285.1	37	527
6692	C₇H₇NO₂	*m*-Nitrotoluene	230.8	228.8	13	554
6693	C₇H₇NO₂	*o*-Nitrotoluene	221.75	220.7	8	554
6694	C₇H₇NO₂	*p*-Nitrotoluene	238.9	235.6	17	554
6695	C₇H₈	Toluene	110.75	Nonazeotrope		537
6696	C₇H₈O	*o*-Cresol	191.1	Nonazeotrope		542
6697	C₇H₈O	*p*-Cresol	201.7	Nonazeotrope		544
6698	C₇H₈O₂	Guaiacol	205.05	Nonazeotrope		556
6699	C₈H₈	Styrene	145.8	Nonazeotrope		540
6700	C₈H₈O₂	Benzyl formate	202.3	Nonazeotrope		537
6701	C₈H₈O₂	Methyl benzoate	199.45	Nonazeotrope		537
6702	C₈H₈O₂	Phenyl acetate	195.7	Nonazeotrope		575
6703	C₈H₈O₃	Methyl salicylate	222.35	221.4	7.5	537
6704	C₈H₁₀	*m*-Xylene	139.0	Nonazeotrope		527
6705	C₈H₁₀	*o*-Xylene	143.6	Nonazeotrope		537
6706	C₈H₁₀O	Phenethyl alcohol	219.4	Nonazeotrope		549
6707	C₈H₁₀O	3,4-Xylenol	226.8	Nonazeotrope		575
6708	C₈H₁₀O₂	*m*-Dimethoxybenzene	214.7	212.5	7	576
6709	C₈H₁₀O₂	*o*-Ethoxyphenol	216.5	Nonazeotrope		575
6710	C₈H₁₂O₄	Ethyl fumarate	217.85	Nonazeotrope		575

TABLE I.   *Binary Systems*   209

No.	Formula	B-Component Name	B.P., °C	Azeotropic Data B.P., °C	Wt.%A	Ref.
A =	C₃H₈O₃	**Glycerol** *(continued)*	**290.5**			
6711	C₈H₁₂O₄	Ethyl maleate	223.3	Nonazeotrope		575
6712	C₈H₁₈	2,5-Dimethylhexane	109.4	Nonazeotrope		575
6713	C₈H₁₈O₃	2-(2-Butoxyethoxy)ethanol	231.2	Nonazeotrope		575
6714	C₉H₇N	Quinoline	237.3	Nonazeotrope		553
6715	C₉H₈	Indene	182.6	182.4	2	575
6716	C₉H₁₀O	*p*-Methylacetophenone	226.35	Nonazeotrope		552
6717	C₉H₁₀O₂	Benzyl acetate	214.9	Nonazeotrope		536
6718	C₉H₁₀O₂	Ethyl benzoate	212.6	Nonazeotrope		536
6719	C₉H₁₀O₃	Ethyl salicylate	233.7	230.5	10.3	537
6720	C₉H₁₂	Mesitylene	164.6	Nonazeotrope		537
6721	C₉H₁₂	Propylbenzene	158.8	Nonazeotrope		540
6722	C₉H₁₂O	3-Phenylpropanol	235.6	Nonazeotrope		549
6723	C₉H₁₂O	Phenyl propyl ether	190.5	190.0	< 8	575
6724	C₉H₁₈O₃	Isobutyl carbonate	190.3	Nonazeotrope		575
6725	C₁₀H₇Br	1-Bromonaphthalene	281.0	272.5		575
6726	C₁₀H₇Cl	1-Chloronaphthalene	262.7	256.0	17	575
6727	C₁₀H₈	Naphthalene	218.05	215.2	10	530
6728	C₁₀H₁₀O₂	Isosafrole	252.0	243.8	~ 16	538
6729	C₁₀H₁₀O₂	Methyl cinnamate	261.9	Reacts		535
6730	C₁₀H₁₀O₂	Safrol	235.9	231.3	14.5	530
6731	C₁₀H₁₀O₄	Methyl phthalate	283.2	271.5	31	567
6732	C₁₀H₁₂O	Anethol	235.7	230.8	14	556
6733	C₁₀H₁₂O	Estragol	215.6	213.5	7.5	545
6734	C₁₀H₁₂O₂	Ethyl α-toluate	228.75	228.6	7	530
6735	C₁₀H₁₂O₂	Eugenol	254.5	251.3	14	556
6736	C₁₀H₁₂O₂	Isoeugenol	268.8	263.5	25	575
6737	C₁₀H₁₂O₂	Propyl benzoate	230.85	228.8	8	536
6738	C₁₀H₁₄	Butylbenzene	183.1	< 182.9		575
6739	C₁₀H₁₄	Cymene	176.7	Nonazeotrope		575
6740	C₁₀H₁₄O	Carvacrol	237.85	Nonazeotrope		575
6741	C₁₀H₁₄O	Carvone	231.0	230.85	3	552
6742	C₁₀H₁₄O	Thymol	232.8	Nonazeotrope		530
6743	C₁₀H₁₄O₂	*m*-Diethoxybenzene	235.4	231.0	13	576
6744	C₁₀H₁₆	Camphene	159.6	Nonazeotrope		537
6745	C₁₀H₁₆	*d*-Limonene	177.8	177.7	~ 1	537
6746	C₁₀H₁₆	α-Pinene	155.6	Nonazeotrope		537
6747	C₁₀H₁₆	Nopinene	163.8	Nonazeotrope		575
6748	C₁₀H₁₆	α-Terpinene	173.4	Nonazeotrope		575
6749	C₁₀H₁₆	Terpinolene	184.6	184.2		575
6750	C₁₀H₁₆	Thymene	179.7	179.6	1	541
6751	C₁₀H₁₈O	α-Terpineol	218.85	Nonazeotrope		575
6752	C₁₀H₂₀O	Menthol	216.3	Nonazeotrope		575
6753	C₁₀H₂₀O₂	Ethyl caprylate	208.35	Nonazeotrope		575
6754	C₁₀H₂₀O₂	Isoamyl isovalerate	192.7	Nonazeotrope		575
6755	C₁₀H₂₀O₂	Methyl pelargonate	213.8	Nonazeotrope		575
6756	C₁₀H₂₂	Decane	173.3	Nonazeotrope		575
6757	C₁₀H₂₂	2,7-Dimethyloctane	160.1	Nonazeotrope		575
6758	C₁₁H₁₀	1-Methylnaphthalene	244.9	237.25	~ 18	537

No.	Formula	B-Component Name	B.P., °C	Azeotropic Data B.P., °C	Wt.%A	Ref.
A =	C₃H₈O₃	**Glycerol** *(continued)*	290.5			
6759	C₁₁H₁₀	2-Methylnaphthalene	241.15	233.7	16.5	569
6760	C₁₁H₁₂O₂	Ethyl cinnamate	271.5	Reacts		536
6761	C₁₁H₁₄O₂	1-Allyl-3,4-dimethoxy-benzene	255.0	248.3	18	538
6762	C₁₁H₁₄O₂	Butyl benzoate	249.8	243	17	536
6763	C₁₁H₁₄O₂	1,2-Dimethoxy-4-propenylbenzene	270.5	258.4	25	574
6764	C₁₁H₁₄O₂	Ethyl β-phenyl propionate	248.1	242.0	15	567
6765	C₁₁H₁₄O₂	Isobutyl benzoate	241.9	~ 237.4	14	536
6766	C₁₁H₂₀O	Isobornyl methyl ether	192.4	< 192.0	7.5	575
6767	C₁₁H₂₀O	Terpineol methyl ether	216.2	214.0	8	545
6768	C₁₂H₁₀	Acenaphthene	277.9	259.1	29	543
6769	C₁₂H₁₀	Biphenyl	254.9	246.1	25	564
6770	C₁₂H₁₀O	Phenyl ether	259.3	247.9	22	530
6771	C₁₂H₁₆O₂	Isoamyl benzoate	262.05	251.6	22	536
6772	C₁₂H₁₆O₃	Isoamyl salicylate	279	267		545
6773	C₁₂H₁₈	1,3,5-Triethylbenzene	215.5	212.9	8	538
6774	C₁₂H₂₀O₂	Bornyl acetate	227.7	226.0	10	530
6775	C₁₂H₂₂O	Bornyl ethyl ether	204.9	203.5	~ 5	575
6776	C₁₃H₁₀O₂	Phenyl benzoate	315	279	~ 55	536
6777	C₁₃H₁₂	Diphenylmethane	265.6	250.8	27	530
6778	C₁₃H₁₂O	Benzyl phenyl ether	286.5	264.5	30	567
6779	C₁₄H₁₂O₂	Benzyl benzoate	324	282.5		536
6780	C₁₄H₁₄	1,2-Diphenylethane	284	261.3	32	537
6781	C₁₄H₁₄O	Benzyl ether	297.0	269.5	36	567
A =	C₃H₈S	**Ethyl Methyl Sulfide**	66.61			
6782	C₆H₁₂	Cyclohexane	80.35	Nonazeotrope		205
6783	C₆H₁₂	1-Hexene	63.50	62.71	29.4	205
6784	C₆H₁₂	Methylcyclopentane	71.85	65.59	64.1	205
6785	C₆H₁₄	2,3-Dimethylbutane	58.10	57.41	18.7	205
6786	C₆H₁₄	Hexane	68.75	63.94	56.6	205
6787	C₇H₁₆	2,2-Dimethylpentane	79.20	66.37	88.2	205
A =	C₃H₈S	**1-Propanethiol**	67.3			
6788	C₄H₄S	Thiophene	84.7	Nonazeotrope		575
6789	C₄H₈O	2-Butanone	79.6	~ 55.5	~ 75	563
6790	C₅H₈	3-Methyl-1,2-butadiene	40.8	Reacts		563
6791	C₅H₁₀	Cyclopentane	49.4	Nonazeotrope		566
6792	C₅H₁₀	2-Methyl-2-butene	37.15	Nonazeotrope		563
6793	C₅H₁₂	Pentane	36.07	Nonazeotrope		202
6794	C₅H₁₂O	Ethyl propyl ether	63.85	< 63.5	> 9	575
6795	C₆H₆	Benzene	80.103	Nonazeotrope		202
6796	C₆H₁₀	Biallyl	60.2	Reacts		575
6797	C₆H₁₂	Cyclohexane	80.738	67.77	97.6	202
6798	C₆H₁₂	Methylcyclopentane	71.812	60.45	64.2 v-l	202
6799	C₆H₁₄	2,2-Dimethylbutane	49.743	Nonazeotrope		202
6800	C₆H₁₄	2,3-Dimethylbutane	57.990	57.54	16.3	202
6801	C₆H₁₄	Hexane	68.742	64.35	52.6 v-l	202,566

TABLE I. *Binary Systems* 211

No.	Formula	B-Component Name	B.P., °C	Azeotropic Data B.P., °C	Wt.%A		Ref.
A =	**C₃H₈S**	**1-Propanethiol** *(continued)*	**67.3**				
6802	C₆H₁₄	2-Methylpentane	60.274	59.20	23.9	v-l	202
6803	C₆H₁₄	3-Methylpentane	63.284	61.26	34.2		202
6804	C₆H₁₄O	Isopropyl ether	68.3	66.0	65		562
6805	C₇H₁₆	2,2-Dimethylpentane	79.205	67.20	81.3		202
6806	C₇H₁₆	2,4-Dimethylpentane	80.51	67.48	85.1		202
6807	C₇H₁₆	2,2,3-Trimethylbutane	80.871	67.57	87.4		202
A =	**C₃H₈S**	**2-Propanethiol**	**52.60**				
6808	C₅H₁₀	Cyclopentane	49.263	47.75	35.3		202
6809	C₅H₁₂	Pentane	34.074	Nonazeotrope			202
6810	C₆H₁₄	2,2-Dimethylbutane	49.743	47.41	37.7		202
6811	C₆H₁₄	2,3-Dimethylbutane	57.990	51.24	67.5		202
6812	C₆H₁₄	Hexane	68.742	Nonazeotrope			202
6813	C₆H₁₄	2-Methylpentane	60.274	51.70	75.9		202
6814	C₆H₁₄	3-Methylpentane	63.284	52.40	87.0		202
A =	**C₃H₉BO₃**	**Methyl Borate**	**68.7**				
6815	C₄H₈O	2-Butanone	79.6	68.0	85		552
6816	C₄H₈O	Butyraldehyde	75.5	Nonazeotrope			548
6817	C₄H₈O	Tetrahydrofuran	65	Nonazeotrope		v-l	318
6818	C₄H₈O₂	Ethyl acetate	77.1	Nonazeotrope			549
6819	C₄H₈O₂	Isopropyl formate	68.8	< 67.0	<58		549
6820	C₄H₉Br	2-Bromo-2-methylpropane	73.3	Nonazeotrope			538
6821	C₄H₉Cl	1-Chlorobutane	78.5	Nonazeotrope			575
6822	C₄H₉Cl	2-Chlorobutane	68.25	66.9	45		562
6823	C₄H₉Cl	1-Chloro-2-methylpropane	68.85	67.3	54		531
6824	C₄H₉Cl	2-Chloro-2-methylpropane	50.8	Nonazeotrope			575
6825	C₄H₉NO₂	Butyl nitrite	78.2	Nonazeotrope			549
6826	C₄H₉NO₂	Isobutyl nitrite	67.1	< 66.9			549
6827	C₄H₁₀O	*tert*-Butyl alcohol	82.45	< 66.0	>75		575
6828	C₅H₁₂	Pentane	36.2	Nonazeotrope			546
6829	C₆H₅F	Fluorobenzene	84.9	Nonazeotrope			575
6830	C₆H₆	Benzene	80.2	Nonazeotrope			538
6831	C₆H₈	1,3-Cyclohexadiene	80.4	Nonazeotrope			546
6832	C₆H₁₂	Cyclohexane	80.8	Nonazeotrope			546
6833	C₆H₁₂	Methylcyclopentane	72.0	67.5	58		562,929
6834	C₆H₁₄	2,3-Dimethylbutane	58.0	Nonazeotrope			575
6835	C₆H₁₄	*n*-Hexane	63.95	~ 66.3	50		531
A =	**C₃H₉ClSi**	**Chlorotrimethysilane**	**57.7**				
6835a	C₄H₈Cl₂O	Bis(2-chloroethyl) ether	178.6	Nonazeotrope			886c
6836	C₆H₁₄	2-Methylpentane	60.4	56.4	65		844
6837	C₆H₁₄	3-Methylpentane	63.3	57.3	70		844
A =	**C₃H₉N**	**Isopropylamine**	**32.4**				
6837a	C₅H₈	Diolefins		Min. b.p.			258c
6837b	C₅H₁₀	Amylenes		Min. b.p.			258c
6837c	C₅H₁₂	Pentanes		Min. b.p.			258c
6838	C₆H₁₄	Hexane	68.7	Nonazeotrope			981

No.	Formula	B-Component Name	B.P., °C	B.P., °C	Wt.%A	Ref.
A =	C_3H_9N	**Propylamine**	**49.7**			
6839	C_4H_8O	2-Butanone	79.6	Nonazeotrope		527
6840	$C_4H_{10}O$	Ethyl ether	34.6	Nonazeotrope		551
6841	C_5H_{10}	Cyclopentane	49.3	47.0	52	551
6842	C_5H_{10}	2-Methyl-2-butene	37.15	~ 32	~ 32	563
6843	C_5H_{12}	2-Methylbutane	27.95	Nonazeotrope		551
6844	C_6H_{14}	2,3-Dimethylbutane	58.0	Nonazeotrope		551
A =	C_3H_9N	**Trimethylamine**	**3.5**			
6845	C_4H_4	1-Buten-3-yne	5.0	Nonazeotrope		84
6846	C_4H_6	1,3-Butadiene	— 4.6	Nonazeotrope		84
6847	C_4H_8	1-Butene	— 6	Nonazeotrope		84,378
6848	C_4H_8	cis-2-Butene	1.0	Nonazeotrope		84
6849	C_4H_8	trans-2-Butene	3.5	Nonazeotrope		84
6850	C_4H_8	2-Methylpropene	— 6	Nonazeotrope		84,378
6851	C_4H_{10}	Butane	0	Nonazeotrope		84,378
6852	C_4H_{10}	2-Methylpropane	— 10	Nonazeotrope		84,378
A =	C_3H_9NO	**1-Amino-2-propanol**	**159.9**			
6853	C_6H_5Cl	Chlorobenzene	131	128.30	13	741
6854	$C_6H_{15}NO_2$	1,1'-Iminodi-2-propanol, 100 mm.	185	Nonazeotrope		981
6855	C_7H_8	Toluene	110.7	110	5	741
6856	C_7H_{16}	Heptane	98.4	96.6	6	981
A =	$C_3H_{10}N_2$	**1,2-Propanediamine**	**120.9**			
6857	$C_4H_{10}O$	Butyl alcohol	117.7	126.5	49	982
6858	$C_4H_{10}O$	Isobutyl alcohol	107.9	123	65	982
6859	C_7H_8	Toluene	110.6	105	32	981
A =	$C_3H_{10}OSi$	**Trimethylsilanol**	**99**			
6860	$C_6H_{18}OSi_2$	Hexamethyldisiloxane	100	90	33-35	839
A =	$C_4Cl_3F_7$	**2,2,3-Trichloro-heptafluorobutane**	**97.4**			
6861	$C_5Cl_2F_6$	1,2-Dichlorohexafluorocyclopentene	90.6	Nonazeotrope	v-l	1041
6862	C_7H_{16}	Heptane	98.53	92.3	76 v-l	1041
6863	$C_8F_{16}O$	Perfluorocyclic oxide	102.6	96.35	67 v-l	1041
A =	$C_4HF_7O_2$	**Perfluorobutyric Acid**	**122.0**			
6864	C_8H_{10}	Ethylbenzene	136.15	115.4	80	163
6865	C_8H_{10}	m-Xylene	139	117.5	83	163
6866	C_8H_{10}	p-Xylene	138.4	117.6	82	163
A =	$C_4H_2O_3$	**Maleic Anhydride**				
6867	C_8H_{10}	m-Xylene, 150 mm.		Nonazeotrope		981
6867a	C_8H_{10}	o-Xylene 40 mm.			v-l	29e
		" 60 mm.		Nonazeotrope	v-l	438c
6868	$C_{16}H_{22}O_4$	Dibutyl phthalate, 50 mm.	238	Nonazeotrope		981
6869	C_nH_m	Hydrocarbons		Min. b.pt.		643

TABLE I.   *Binary Systems*   213

No.	Formula	B-Component Name	B.P., °C	B.P., °C	Azeotropic Data Wt.%A	Ref.
**A = **	**C₄H₄**	**Vinylacetylene**				
6870	C₄H₅Cl	2-Chloro-1,3-butadiene, 740 mm.		Nonazeotrope	v-l	419
6871	C₄H₈	2-Butene	3.5	Min. b.p.		102
**A = **	**C₄H₄N₂**	**Pyrazine**	**117.2**			
6872	C₆H₁₄S	Isopropyl sulfide	120.5	116.0	> 5	575
**A = **	**C₄H₄N₂**	**Pyridazine**	**207.2**			
6873	C₆H₅NO₃	o-Nitrophenol	217.2	Nonazeotrope		575
6874	C₆H₆O	Phenol	182.2	209.0	88	575
6875	C₇H₇ClO	m-Chloroanisole	193.3	Nonazeotrope		575
6876	C₇H₇ClO	p-Chloroanisole	197.8	Nonazeotrope		575
6877	C₇H₈O	m-Cresol	202.2	211.8	68	575
6878	C₇H₈O	p-Cresol	201.7	211.5	70	575
6879	C₇H₈O₂	Guaiacol	205.05	203.5		575
6880	C₈H₁₀O	p-Ethylphenol	218.8	220.5	15	575
6881	C₈H₁₀O	2,4-Xylenol	210.5	215.5	25	575
6882	C₁₂H₂₂O	Ethyl isobornyl ether	203.8	<203.5	>13	575
**A = **	**C₄H₄O**	**Furan**	**31.7**			
6883	C₅H₁₀	3-Methyl-1-butene	20.6	Nonazeotrope		558
6884	C₅H₁₂	2-Methylbutane	27.95	< 27.0	> 8	558
**A = **	**C₄H₄O₂**	**Diketene**				
6885	C₇H₈	Toluene, 60 mm.		41	10	227
**A = **	**C₄H₄S**	**Thiophene**	**84.7**			
6886	C₄H₇N	Pyrroline	90.9	Nonazeotrope		575
6887	C₄H₈O	2-Butanone	79.6	Nonazeotrope		527
6888	C₄H₈O	Butyraldehyde	75.2	Nonazeotrope		566
6889	C₄H₈O₂	Ethyl acetate	77.1	Nonazeotrope		527
		"		< 73	>20	563
6890	C₄H₈O₂	Methyl propionate	79.85	Nonazeotrope		527
6891	C₄H₈O₂	Propyl formate	80.85	Nonazeotrope		527
6892	C₄H₉Cl	1-Chlorobutane	78.5	Nonazeotrope		527
6893	C₄H₉ClO	2-Chloroethyl ethyl ether	98.5	Nonazeotrope		566
6894	C₄H₉NO₂	Butyl nitrite	78.2	Nonazeotrope		527
6895	C₄H₉NO₂	Isobutyl nitrite	67.1	Nonazeotrope		550
6896	C₄H₁₀S	2-Butanethiol	85.15	82.27		205
6897	C₄H₁₀S	Isopropyl methyl sulfide	84.76	83.42		205
6898	C₅H₁₀O	3-Methyl-2-butanone	95.4	Nonazeotrope		205,575
6899	C₅H₁₁NO₂	Isoamyl nitrite	97.15	Nonazeotrope		527
6900	C₅H₁₂O₂	Diethoxymethane	87.95	< 83.9		566
6901	C₆H₆	Benzene	80.2	Nonazeotrope		527
		"		"	v-l	605
6902	C₆H₁₀	Cyclohexene	82.75	< 82.5	>15	561
6903	C₆H₁₂	Cyclohexane	80.85	77.90	41.2	205
6904	C₆H₁₂	Methylcyclopentane	71.85	71.47	14.0	205
		"		72.0	Nonazeotrope	566

		B-Component			Azeotropic Data		
No.	Formula	Name	B.P., °C	B.P., °C	Wt.%A		Ref.
A =	C₄H₄O₂	**Thiophene** *(continued)*	**84.7**				
6905	C₆H₁₄	Hexane	68.95	Nonazeotrope			527
		"	68.75	68.46	11.2		205
6906	C₆H₁₄O	Isopropyl ether	68.3	Nonazeotrope			566
6907	C₇H₁₄	*trans*-1,3-Dimethyl-					
		cyclopentane	90.80	82.00	67.7		205
6908	C₇H₁₆	2,3-Dimethylpentane	89.90	80.90	64		205
6909	C₇H₁₆	2,4-Dimethylpentane	80.55	76.58	42.7		205
6910	C₇H₁₆	Heptane	98.40	83.09	83.2		205
A =	C₄H₅Cl	**2-Chloro-1,3-butadiene**					
6911	C₄H₆Cl₂	1,3-Dichloro-2-					
		butene, 100 mm.		Nonazeotrope		v-l	419
		" 340 mm.		Nonazeotrope		v-l	419
6912	C₄H₆O	1-Butene-3-					
		one, 100 mm.		Nonazeotrope		v-l	419
		" 340 mm.		Nonazeotrope		v-l	419
A =	C₄H₅ClO₂	α-**Chlorocrotonic Acid**	**212.5**				
6913	C₆H₅NO₂	Nitrobenzene	210.75	< 208.0	>30		554
6914	C₇H₇NO₂	*o*-Nitrotoluene	221.75	< 211.2	>72		554
A =	C₄H₅Cl₃	**2,3,4-Trichloro-1-butene**					
6914a	C₄H₆Cl₂	*trans*-1,3-Dichloro-2-					
		butene 40 mm.		Nonazeotrope		v-l	158f
A =	C₄H₅Cl₃O₂	**Ethyl Trichloroacetate**	**167.2**				
6915	C₄H₈O₂	Butyric acid	164.0	< 163.5			575
6916	C₄H₈O₂	Isobutyric acid	154.6	Nonazeotrope			575
6917	C₅H₁₀O₂	Valeric acid	186.35	Nonazeotrope			575
6918	C₇H₆O	Benzaldehyde	179.2	Nonazeotrope			563
6919	C₇H₁₄O	2-Methylcyclohexanol	168.5	< 165.5	>62		575
6920	C₈H₁₈O	Octyl alcohol	195.2	Nonazeotrope			575
A =	C₄H₅N	**Pyrrol**	**129.2**				
6921	C₄H₉I	1-Iodobutane	130.4	< 123.2	32		553
6922	C₄H₁₀O	Butyl alcohol	117.8	Nonazeotrope			527
6923	C₄H₁₀O	Isobutyl alcohol	108.0	Nonazeotrope			575
6924	C₄H₁₀O₂	2-Ethoxyethanol	135.3	Nonazeotrope			575
6925	C₄H₁₀S	Butanethiol	97.8	Nonazeotrope			575
6926	C₄H₁₀S	Ethyl sulfide	92.1	Nonazeotrope			553
6927	C₅H₁₀O	Cyclopentanol	140.85	Nonazeotrope			553
6928	C₅H₁₀O₃	Ethyl carbonate	126.5	131.6	49		553
6929	C₅H₁₁Br	1-Bromo-3-methylbutane	120.65	< 116.4	>10		553
6930	C₅H₁₂O	Amyl alcohol	138.2	Nonazeotrope			527
6931	C₅H₁₂O	*tert*-Amyl alcohol	102.35	Nonazeotrope			575
6932	C₅H₁₂O	Isoamyl alcohol	131.9	< 129.4	>21		553
6933	C₅H₁₂O	2-Pentanol	119.8	Nonazeotrope			527
6934	C₅H₁₂O₂	2-Propoxyethanol	151.35	Nonazeotrope			527
6935	C₆H₅Br	Bromobenzene	156.1	Nonazeotrope			553
6936	C₆H₅Cl	Chlorobenzene	131.75	124.5	43		527

TABLE I. *Binary Systems* 215

No.	Formula	B-Component Name	B.P., °C	B.P., °C	Azeotropic Data Wt.%A	Ref.
A =	C₄H₅N	**Pyrrol** *(continued)*	129.2			
6937	C₆H₇N	3-Picoline	143.8	145-148		242
6938	C₆H₇N	4-Picoline	144.8	145-148		242
6939	C₆H₁₀O	Mesityl oxide	130.5	~ 128		561
6940	C₆H₁₀S	Allyl sulfide	139.35	127.0	70	575
6941	C₆H₁₂O₂	Isoamyl formate	123.8	~ 130.0	~ 60	548
6942	C₆H₁₂O₂	Isobutyl acetate	117.4	Nonazeotrope		527
6943	C₆H₁₄S	Isopropyl sulfide	120.5	117.5	20	553
6944	C₆H₁₄S	Propyl sulfide	140.8	127.5	65	553
6945	C₇H₇Cl	*o*-Chlorotoluene	159.2	Nonazeotrope		553
6946	C₇H₈	Toluene	110.75	Nonazeotrope		553
6947	C₇H₁₄O₂	Isoamyl acetate	142.1	Nonazeotrope		527
6948	C₇H₁₄O₂	Propyl isobutyrate	134.0	>134.8	>25	548
6949	C₈H₁₀	Xylenes	140	Min. b.p.		515
6950	C₈H₁₈	*n*-Octane	125.75	< 124.3	<36	553
6951	C₈H₁₈O	Isobutyl ether	122.3	< 121.5	>12	562
A =	C₄H₅NS	**Allyl Isothiocyanate**	152.0			
6952	C₄H₈Cl₂O	1,2-Dichloroethyl ethyl ether	145.5	Nonazeotrope		575
6953	C₅H₄O₂	2-Furaldehyde	161.45	Nonazeotrope		575
6954	C₅H₁₀O	Cyclopentanol	140.85	Nonazeotrope		575
6955	C₆H₁₂O	Cyclohexanol	160.8	Nonazeotrope		575
6956	C₆H₁₃ClO₂	Chloroacetal	157	Nonazeotrope		575
6957	C₆H₁₄O	Hexyl alcohol	157.85	< 151.8		575
6958	C₆H₁₄O₂	Pinacol	174.35	Nonazeotrope		575
6959	C₆H₁₄S	Propyl sulfide	141.5	< 141.1	<19	575
6960	C₇H₈O	Anisole	153.85	151.5	68	575
6961	C₈H₁₀O	Benzyl methyl ether	167.8	Nonazeotrope		575
6962	C₈H₁₈S	Isobutyl sulfide	172.0	Nonazeotrope		575
A =	C₄H₆	**1,3-Butadiene**	— 4.5			
6962a	C₄H₆	1-Butyne	7	Nonazeotrope	v-l	149c
6963	C₄H₈	1-Butene	— 5	Nonazeotrope		561
6964	C₄H₈	2-Butene	1.5-3	—5.53	76.5 v-l	111
		"	1	Nonazeotrope	v-l	149c
6964a	C₄H₈O₂	Ethyl acetate	77	Nonazeotrope	v-l	1030c
6965	C₄H₁₀	Butane	— 0.5	Min. b.p.		102
6966	C₄H₁₀O	Ethyl ether	34.5	Nonazeotrope	v-l	111,1030c
6967	C₅H₇Cl	Chloroprene	59.4	Nonazeotrope	v-l	428
6967a	C₆H₆	Benzene	80.1	Nonazeotrope	v-l	1030c
A =	C₄H₆	**1-Butyne**	9			
6968	C₄H₈	*cis*-2-Butene	1	Min. b.p.	9.5	102
6969	C₄H₈	*trans*-2-Butene	3.5		25.5	102
A =	C₄H₆Cl₂	**1,3-Dichloro-2-butene**				
6970	C₄H₆O	1-Butene-3-one				
			150 mm.	Nonazeotrope	v-l	420
A =	C₄H₆Cl₂O₂	**Ethyl Dichloroacetate**	158.1			
6971	C₄H₈O₂	Butyric acid	164.0	157.0		562

No.	Formula	B-Component Name	B.P., °C	Azeotropic Data B.P., °C	Wt.%A	Ref.
A =	C₄H₆Cl₂O₂	**Ethyl Dichloroacetate** *(continued)*	158.1			
6972	C₄H₈O₂	Isobutyric acid	154.6	< 153.8		562
6973	C₄H₁₀O	Butyl alcohol	117.8	Nonazeotrope		575
6974	C₅H₄O₂	2-Furaldehyde	161.5	Nonazeotrope		563
6975	C₅H₁₀O₃	Ethyl lactate	154.1	Nonazeotrope		575
6976	C₅H₁₁NO₃	Isoamyl nitrate	149.75	Nonazeotrope		560
6977	C₅H₁₂O₂	2-Propoxyethanol	151.35	Nonazeotrope		575
6978	C₆H₁₄O	Hexyl alcohol	157.85	< 156.0	58	575
6979	C₆H₁₄O₂	2-Butoxyethanol	171.15	Nonazeotrope		575
6980	C₇H₁₆O	Heptyl alcohol	176.15	Nonazeotrope		575
6981	C₈H₁₆O₂	Butyl butyrate	166.4	Nonazeotrope		575
6982	C₈H₁₆O₂	Isoamyl propionate	160.7	Nonazeotrope		575
A =	**C₄H₆O**	**Crotonaldehyde**	102.15			
6983	C₄H₉Br	1-Bromo-2-methylpropane	91.6	Nonazeotrope		563
6984	C₅H₁₀O	3-Methyl-2-butanone	95.4	Nonazeotrope		552
6985	C₅H₁₀O	2-Pentanone	102.35	101.2		552
6986	C₅H₁₀O	3-Pentanone	102.05	< 101.4		552
6987	C₅H₁₀O₂	Ethyl propionate	99.1	98.0	25	575
6988	C₅H₁₀O₂	Methyl butyrate	102.75	< 101		563
6989	C₆H₆	Benzene	80.2	Nonazeotrope		563
6990	C₆H₁₂	Cyclohexane	80.75	Nonazeotrope		575
6991	C₇H₈	Toluene	110.65	Min. b.p.		932
			110.65	Nonazeotrope		575
6992	C₇H₁₄	Methylcyclohexane	101.15	< 99.5		575
6993	CₙH₂ₙ₊₂	Paraffins	109.5-110.5	102.8		241,932
A =	**C₄H₆O₂**	**Allyl Formate**	80.0			
6994	C₄H₉Cl	1-Chlorobutane	78.5	< 76.0	>40	575
6995	C₄H₉NO₂	Butyl nitrite	78.2	< 77.0	>30	549
6996	C₆H₆	Benzene	80.15	79.2	>45	575
6997	C₆H₁₄	Hexane	68.8	< 64.5	>26	575
A =	**C₄H₆O₂**	**Biacetyl**	87.5			
6998	C₄H₈O	2-Butanone	79.6	Nonazeotrope		981
6999	C₅H₁₂O	Isoamyl alcohol	131.9	Nonazeotrope		552
7000	C₆H₆	Benzene	80	79.3	~ 55	640
7001	C₆H₁₅N	Dipropylamine	109.2	Nonazeotrope		575
A =	**C₄H₆O₂**	**2-Butyne-1,4-diol**				
7002	C₅H₉NO	N-Methyl-2-pyrrolidone	251	Nonazeotrope	v-l	339
A =	**C₄H₆O₂**	**Dioxene**	94.7			
7003	C₄H₁₀O₃	Diethylene glycol	245.1	Nonazeotrope	v-l	215
A =	**C₄H₆O₂**	**Methyl Acrylate**	80			
7004	C₄H₁₀O	Butyl alcohol	117	Nonazeotrope		800
7005	C₄H₁₀O	Isobutyl alcohol	108	Nonazeotrope		800
7006	C₅H₈O₂	Ethyl acrylate, 103 mm.	43	Nonazeotrope		800
A =	**C₄H₆O₂**	**Methacrylic Acid**	160.5			
7007	C₅H₈O₂	Methyl methacrylate	99.5	Nonazeotrope		1032

TABLE I.   *Binary Systems*   217

No.	Formula	B-Component Name	B.P., °C	Azeotropic Data B.P., °C	Wt.%A		Ref.
A = C₄H₆O₂		**Vinyl Acetate**	**72.7**				

No.	Formula	Name	B.P., °C	B.P., °C	Wt.%A		Ref.
A =	**$C_4H_6O_2$**	**Vinyl Acetate**	**72.7**				
7008	$C_4H_{10}O$	Butyl alcohol	117.7	Nonazeotrope			981
7009	C_6H_{12}	Cyclohexane	80.7	67.4	61.3		981
7010	C_7H_{16}	2,4-Dimethylpentane	80.5	67.2	56.9	v-l	941
1011	C_7H_{16}	Heptane	98.4	72	83.5		981
1012	$C_8H_{18}O$	Butyl ether	142.1	Nonazeotrope			981
A =	**$C_4H_6O_3$**	**Acetic Anhydride**	**138**				
7013	C_5H_5N	Pyridine	115	Nonazeotrope		v-l	694,428
7014	$C_5H_8O_2$	Isopropenyl acetate	97.4	Nonazeotrope			981
7015	$C_5H_8O_4$	Methylene diacetate	164	Nonazeotrope		v-l	428
		" 100 mm.		Nonazeotrope		v-l	428
		" low pressure	92			v-l	954c
7016	$C_5H_{10}O_2$	Isopropyl acetate	88.7	Nonazeotrope			981
7016a	C_6H_6	Benzene	80.1	Nonazeotrope		v-l	659c
7017	C_6H_{12}	Cyclohexane	80.7	80.1	7.8	v-l	428
7018	$C_6H_{14}O$	Isopropyl ether	68.3	Nonazeotrope			981
7019	C_7H_{14}	Methylcyclohexane	101	99	~ 18		277
7020	C_7H_{16}	*n*-Heptane	98.4	Azeotropic			277
7021	C_8H_{16}	Ethylcyclohexane	131	118	~ 37		277
7022	C_8H_{18}	*n*-Octane	125.8	Azeotropic			277
7023	C_9H_{20}	*n*-Nonane	150	Azeotropic			277
7024	$C_{10}H_{22}$	*n*-Decane	173	Azeotropic			277
7025	$C_{11}H_{24}$	*n*-Undecane	194.5	Azeotropic			277
A =	**$C_4H_6O_3$**	**Methyl Pyruvate**	**137.5**				
7026	$C_4H_8O_2$	Isobutyric acid	154.6	Nonazeotrope			552
7027	C_4H_9I	1-Iodobutane	130.4	< 127.0			552
7028	$C_5H_8O_2$	2,4-Pentanedione	137.7	< 136.2			552
7029	$C_5H_{10}O_2$	Methyl butyrate	102.65	Nonazeotrope			575
7030	$C_5H_{10}O_2$	Propyl acetate	101.6	Nonazeotrope			552
7031	$C_6H_{11}I$	1-Iodo-3-methylbutane	147.65	< 136.0			552
7032	C_6H_5Br	Bromobenzene	156.1	Nonazeotrope			552
7033	C_6H_5Cl	Chlorobenzene	131.75	129.0	30		552
7034	$C_6H_{10}O$	Mesityl oxide	129.45	Nonazeotrope			552
7035	$C_6H_{10}S$	Allyl sulfide	139.35	< 134.4	>53		566
7036	$C_6H_{12}O$	2-Hexanone	127.2	Nonazeotrope			552
7037	$C_6H_{12}O_2$	Isobutyl acetate	117.4	Nonazeotrope			552
7038	$C_6H_{12}O_2$	Methyl isovalerate	116.5	Nonazeotrope			575
7039	$C_6H_{14}O$	Propyl ether	90.1	Nonazeotrope			552
7040	$C_6H_{14}S$	Isopropyl sulfide	120.5	Nonazeotrope			566
7041	C_7H_8O	Anisole	153.85	Nonazeotrope			552
7042	$C_7H_{14}O_2$	Ethyl isovalerate	134.7	< 132.0			552
7043	$C_7H_{14}O_2$	Isoamyl acetate	142.1	135.0	65		552
7044	C_8H_{10}	*m*-Xylene	139.2	130.0	50		552
7045	C_8H_{16}	1,3-Dimethylcyclohexane	120.7	< 117.0			552
7046	$C_8H_{16}O_2$	Isoamyl propionate	160.7	Nonazeotrope			575
7047	$C_8H_{18}O$	Butyl ether	142.4	130.2			552
7048	$C_8H_{18}O$	Isobutyl ether	122.3	< 121.5			552

No.	Formula	B-Component Name	B.P., °C	Azeotropic Data B.P., °C	Wt.%A	Ref.
A =	**C₄H₆O₃**	**Methyl Pyruvate** *(continued)*	**137.5**			
7049	C₁₀H₁₆	Camphene	159.6	< 135.2		552
7050	C₁₀H₁₆	α-Pinene	155.8	< 134.5		552
A =	**C₄H₆O₄**	**Ethylene Glycol Diformate**	**174**			
7051	CₙHₘ	Hydrocarbons		Min. b.p.		644
A =	**C₄H₆O₄**	**Methyl Oxalate**	**163.3**			
7052	C₄H₇ClO₂	Ethyl chloroacetate	143.5	Nonazeotrope		563
7053	C₄H₈O₂	Butyric acid	164.0	< 160.8	>54	562
7054	C₄H₈O₂	Isobutyric acid	154.6	< 154.2	<18	562
7055	C₄H₈O₃	Glycol monoacetate	190.9	Nonazeotrope		575
7056	C₅H₄O₂	2-Furaldehyde	161.45	Nonazeotrope		572
7057	C₅H₁₀O₃	Ethyl lactate	154.1	Nonazeotrope		575
7058	C₅H₁₀O₃	2-Methoxyethyl acetate	144.6	Nonazeotrope		575
7059	C₅H₁₁I	1-Iodo-3-methylbutane	147.6	Nonazeotrope		547
7060	C₆H₄Cl₂	o-Dichlorobenzene	179.5	< 163.8	<89	575
7061	C₆H₄Cl₂	p-Dichlorobenzene	174.35	162.05	65	530
7062	C₆H₅Br	Bromobenzene	156.1	153.05	28	563
7063	C₆H₆O	Phenol	182.2	182.35	~ 8	573
7064	C₆H₁₀O	Cyclohexanone	155.7	Nonazeotrope		552
7065	C₆H₁₂O	Cyclohexanol	160.65	155.6	41	563
7066	C₆H₁₃Br	1-Bromohexane	156.5	< 154.0	<30	575
7067	C₆H₁₃ClO₂	Chloroacetal	157.4	Nonazeotrope		531
7068	C₆H₁₄O	Hexyl alcohol	157.85	< 155.5		567
7069	C₆H₁₄O₂	2-Butoxyethanol	171.25	Nonazeotrope		575
7070	C₆H₁₄O₂	Pinacol	174.35	163.15	81	530
7071	C₇H₇Br	m-Bromotoluene	184.3	Nonazeotrope		527
7072	C₇H₇Br	o-Bromotoluene	181.5	164.1	98	538
7073	C₇H₇Br	p-Bromotoluene	185.0	Nonazeotrope		538
7074	C₇H₇Cl	α-Chlorotoluene	179.3	Nonazeotrope		530
7075	C₇H₇Cl	o-Chlorotoluene	159.2	154.8	35	570
7076	C₇H₇Cl	p-Chlorotoluene	162.4	156.6	30	538
7077	C₇H₈	Toluene	110.75	Nonazeotrope		575
7078	C₇H₈O	Anisole	153.85	153.65	~ 15	557
7079	C₇H₁₄O	2-Methylcyclohexanol	168.5	< 161.2		575
7080	C₇H₁₆O	Heptyl alcohol	176.15	< 163.8		575
7081	C₈H₈	Styrene	145.7	< 142.5	~ 12	563
7082	C₈H₁₀	Ethylbenzene	136.15	Nonazeotrope		575
7083	C₈H₁₀	m-Xylene	139.2	< 138.8		575
7084	C₈H₁₀	o-Xylene	144.3	<143.0		562
7085	C₈H₁₀O	Benzyl methyl ether	167.8	< 161.9	<60	557
7086	C₈H₁₀O	Phenetole	170.45	161.35		571
7087	C₈H₁₆	1,3-Dimethylcyclohexane	120.7	Nonazeotrope		575
7088	C₈H₁₆O₂	Butyl butyrate	166.4	160.5	58	549
7089	C₈H₁₆O₂	Ethyl caproate	167.7	161.0	60	549
7090	C₈H₁₆O₂	Hexyl acetate	171.5	< 162.5	<76	575
7091	C₈H₁₆O₂	Isoamyl propionate	160.7	157.5	38	549
7092	C₈H₁₆O₂	Isobutyl butyrate	156.9	< 155.5	>23	549

TABLE I. *Binary Systems*   219

No.	Formula	B-Component Name	B.P., °C	Azeotropic Data B.P., °C	Wt.% A	Ref.
A =	**C₄H₆O₄**	**Methyl Oxalate** *(continued)*	**163.3**			
7093	C₈H₁₆O₂	Propyl isovalerate	155.7	< 154.5	>20	549
7094	C₈H̄₁₈O	*sec*-Octyl alcohol	179.0	~ 163.8	86?	530
7095	C₈H₂₀SiO₄	Ethyl silicate	165	162.5		563
7096	C₉H₈	Indene	182.6	163.6	83	564
7097	C₉H₁₂	Cumene	152.8	148.5		562
7098	C₉H₁₂	Mesitylene	164.0	154.8	49.8	563·
7099	C₉H₁₂	Propylbenzene	158	~ 152	~ 38	563
7100	C₉H₁₂	Pseudocumene	169	~ 157	~ 65	563
7101	C₉H₁₈O₂	Isoamyl isobutyrate	169.8	161.0	65	549
7102	C₁₀H₈	Naphthalene	218.0	Nonazeotrope		575
7103	C₁₀H₁₄	Butylbenzene	183.2	< 163.5		546
7104	C₁₀H₁₄	Cymene	175.3	~ 161	~ 80	563
7105	C₁₀H₁₆	Camphene	159.6	146.65	42	570
7106	C₁₀H₁₆	*d*-Limonene	177.8	156.7	~ 75	563
7107	C₁₀H₁₆	Nopinene	163.8	147.1	51	563
7108	C₁₀H₁₆	α-Phellandrene	171.5	153	~ 68	563
7109	C₁₀H₁₆	α-Pinene	155.8	144.1	39	563
7110	C₁₀H₁₆	Terpinene	180.5	~ 159.5	~ 88	563
7111	C₁₀H₁₆	α-Terpinene	173.3	159.5	82	562
7112	C₁₀H₁₆	γ-Terpinene	183	159.5	82	575
7113	C₁₀H₁₆	Terpinolene	185.2	160.0	<90	546
	"	"	185	Azeotrope doubtful		563
7114	C₁₀H₁₆	Terpinylene	175	~ 155	<80	563
7115	C₁₀H₁₆	Thymene	179	150	54	563
7116	C₁₀H₁₈	*p*-Menthen	170.8	154.0	70	546
7117	C₁₀H₁₈O	Cineol	176.35	158.85	55	557
7118	C₁₀H₂₀O₂	Isoamyl isovalerate	193.5	162.2	70	549
7119	C₁₀H₂₂	2,7-Dimethyloctane	160.2	147.0	45	546
7120	C₁₀H₂₂O	Isoamyl ether	173.2	154.8	54	557
	"	"	173.4	162.2	~ 80	548
7121	C₁₂H₁₈	1,3,5-Triethylbenzene	215.5	Nonazeotrope		575
7122	C₁₂H₂₂O₄	Isoamyl oxalate	268	Azeotrope doubtful		563
A =	**C₄H₇BrO₂**	**Ethyl Bromoacetate**	**158.8**			
7123	C₄H₈O₂	Butyric acid	164.0	157.4	84	527
7124	C₄H₈O₂	Isobutyric acid	154.6	153.0	40	527
7125	C₄H₈O₃	Methyl lactate	143.8	Nonazeotrope		527
7126	C₄H₁₀O	*n*-Butyl alcohol	117.8	Nonazeotrope		527
7127	C₄H₁₀O	Isobutyl alcohol	108.0	Nonazeotrope		575
7128	C₄H₁₀O₂	2-Ethoxyethanol	135.3	Nonazeotrope		575
7129	C₅H₁₀O₂	Isovaleric acid	176.5	Nonazeotrope		527
7130	C₅H₁₀O₃	Ethyl lactate	154.1	Nonazeotrope		527
	"	"	155	152.5		563
7131	C₅H₁₀O₃	2-Methoxyethyl acetate	144.6	Nonazeotrope		527
7132	C₅H₁₁I	1-Iodo-3-methylbutane	147.65	< 147.5	<10	527
7133	C₅H₁₁NO₃	Isoamyl nitrate	149.75	Nonazeotrope		560
7134	C₅H₁₂O	Isoamyl alcohol	131.9	Nonazeotrope		527

No.	Formula	B-Component Name	B.P., °C	Azeotropic Data B.P., °C	Wt.%A	Ref.
A =	C₄H₇BrO₂	**Ethyl Bromoacetate**	**158.8**			
		(continued)				
7135	C₅H₁₂O₂	2-Propoxyethanol	151.35	151.25	5	527
7136	C₆H₄Cl₂	*p*-Dichlorobenzene	174.4	Nonazeotrope		527
7137	C₆H₅Br	Bromobenzene	156.1	155.3	28	527
7138	C₆H₁₀O₄	Ethylidene diacetate	168.5	Nonazeotrope		527
7139	C₆H₁₂O	Cyclohexanol	160.8	155.5	65	527
		"	160.65	~ 156		563
7140	C₆H₁₂O₃	2-Ethoxyethyl acetate	156.8	Nonazeotrope		527
7141	C₆H₁₂O₃	Isopropyl lactate	166.8	Nonazeotrope		575
7142	C₆H₁₂O₃	Propyl lactate	171.7	Nonazeotrope		527
7143	C₆H₁₃Br	1-Bromohexane	156.5	< 155.0	<39	563
7144	C₆H₁₄O	*n*-Hexanol	157.85	154.0	55	527
7145	C₆H₁₄O₂	2-Butoxyethanol	171.15	Nonazeotrope		527
7146	C₆H₁₄S	Propyl sulfide	141.5	Nonazeotrope		566
7147	C₇H₇Cl	*o*-Chlorotoluene	159.3	156.2	52	527
7148	C₇H₇Cl	*p*-Chlorotoluene	162.4	< 158.5	<90	575
7149	C₇H₈O	Anisole	153.85	153.8		527
7150	C₇H₁₄O	2-Methylcyclohexanol	168.5	157.5	85	575
7151	C₇H₁₄O₃	Methyl-1,3-butanediol acetate	171.35	Nonazeotrope		527
7152	C₇H₁₆O	*n*-Heptyl alcohol	176.15	Nonazeotrope		527
7153	C₈H₈	Styrene	145.8	Nonazeotrope		575
7154	C₈H₁₀O	Benzyl methyl ether	167.8	Nonazeotrope		575
7155	C₈H₁₀O	*p*-Methylanisole	177.05	Nonazeotrope		575
7156	C₈H₁₀O	Phenetole	170.45	Nonazeotrope		575
7157	C₈H₁₆O	2-Octanone	172.85	Nonazeotrope		552
7158	C₈H₁₆O₂	Isobutyl isobutyrate	147.3	Nonazeotrope		532
7159	C₈H₁₈O	Butyl ether	142.4	Nonazeotrope		575
7160	C₈H₁₈O	*sec*-Octanol	180.4	Nonazeotrope		527
7161	C₈H₁₈S	Isobutyl sulfide	172.0	Nonazeotrope		566
7162	C₉H₁₂	Mesitylene	164.6	< 158.4	<88	575
7163	C₉H₁₂	Propylbenzene	159.3	155.8	50	562
7164	C₉H₁₈O₂	Isoamyl butyrate	181.05	Nonazeotrope		575
7165	C₉H₁₈O₂	Isobutyl isovalerate	171.2	Nonazeotrope		527
7166	C₁₀H₁₄	Cymene	176.7	Nonazeotrope		575
7167	C₁₀H₁₆	Camphene	~ 158	~ 154		563
7168	C₁₀H₁₆	Dipentene	177.7	Nonazeotrope		575
7169	C₁₀H₁₆	Nopinene	163.8	156.5	78	562
7170	C₁₀H₁₆	α-Pinene	155.8	152.5	~ 46	563
7171	C₁₀H₁₆	α-Terpinene	173.4	Nonazeotrope		575
7172	C₁₀H₁₈O	Cineol	176.35	Nonazeotrope		575
7173	C₁₀H₂₂O	Isoamyl ether	173.2	Nonazeotrope		575
A =	C₄H₇ClO	**2-Chloroethyl Vinyl Ether**	**108**			
7174	C₄H₈O₂	Dioxane	101	Nonazeotrope		180,798
7175	C₅H₁₂O	Isoamyl alcohol	131.8	109	99	982
		" 50 mm.	67	39	99	982

TABLE I.   *Binary Systems*   221

No.	Formula	B-Component Name	B.P., °C	Azeotropic Data B.P., °C	Wt.%A	Ref.
A =	$C_4H_7ClO_2$	**Ethyl Chloroacetate**	**143.55**			
7176	$C_4H_8O_2$	Butyric acid	164	Nonazeotrope		527
7177	$C_4H_8O_2$	Isobutyric acid	154.6	Nonazeotrope		575
7178	$C_4H_8O_3$	Methyl lactate	144.8	140.4	~ 52	563
7179	C_4H_9I	1-Iodobutane	130.4	< 130.0	<10	575
7180	$C_4H_{10}O$	Butyl alcohol	117.75	Nonazeotrope		535
7181	$C_4H_{10}O$	Isobutyl alcohol	108.0	Nonazeotrope		575
7182	$C_4H_{10}O_2$	2-Ethoxyethanol	135.3	134.8	32	556
7183	C_5H_8O	Cyclopentanone	130.65	Nonazeotrope		552
7184	$C_5H_{10}O$	Cyclopentanol	140.85	137.6	50	567
7185	$C_5H_{10}O_3$	Ethyl lactate	154.1	Nonazeotrope		575
7186	$C_5H_{10}O_3$	2-Methoxyethyl acetate	144.6	144.95	38	556,569
7187	$C_5H_{11}I$	1-Iodo-3-methylbutane	147.65	140.2	49	572
7188	$C_5H_{11}NO_3$	Isoamyl nitrate	149.75	Nonazeotrope		560
7189	$C_5H_{12}O$	Isoamyl alcohol	131.3	131	23	527
7190	$C_5H_{12}O$	2-Pentanol	119.8	Nonazeotrope		575
7191	$C_5H_{12}O_2$	2-Propoxyethanol	151.35	Nonazeotrope		526
7192	C_6H_5Br	Bromobenzene	156.1	Nonazeotrope		575
7193	C_6H_5Cl	Chlorobenzene	131.75	Nonazeotrope		575
7194	$C_6H_{10}O$	Cyclohexanone	155.7	Nonazeotrope		552
7195	$C_6H_{10}O$	Mesityl oxide	129.45	Nonazeotrope		527,552
7196	$C_6H_{10}S$	Allyl sulfide	139.35	138.5	22	566
7197	$C_6H_{12}O$	Cyclohexanol	160.8	Nonazeotrope		575
7198	$C_6H_{12}O$	2-Hexanone	127.2	Nonazeotrope		552
7199	$C_6H_{12}O_3$	2-Ethoxyethyl acetate	156.8	Nonazeotrope		556
7200	$C_6H_{13}Br$	1-Bromohexane	156.5	Nonazeotrope		575
7201	$C_6H_{14}O$	Hexyl alcohol	157.8	142	~ 75	535
7202	$C_6H_{14}S$	Propyl sulfide	141.5	< 140.3	<44	566
7203	C_7H_7Cl	o-Chlorotoluene	159.3	Nonazeotrope		532
7204	C_7H_8O	Anisole	153.85	Nonazeotrope		575
7205	$C_7H_{14}O$	4-Heptanone	143.55	142.75	47	552
7206	$C_7H_{14}O$	2-Methylcyclohexanol	168.5	Nonazeotrope		575
7207	$C_7H_{14}O_2$	Butyl propionate	146.5	Nonazeotrope		548
7208	$C_7H_{14}O_2$	Ethyl isovalerate	134.7	Nonazeotrope		575
7209	$C_7H_{14}O_2$	Ethyl valerate	145.45	< 143.4		575
7210	$C_7H_{14}O_2$	Isoamyl acetate	142.1	141.7	40	572
7211	$C_7H_{14}O_2$	Isobutyl propionate	136.9	N nazeotrope		532
7212	$C_7H_{14}O_2$	Propyl butyrate	142.8	141.7	47	530
7213	C_8H_8	Styrene	145.7	140.2	~ 60	563
7214	C_8H_{10}	Ethylbenzene	136.15	135.3	18	562
7215	C_8H_{10}	m-Xylene	139.0	137.45	32	527
7216	C_8H_{10}	o-Xylene	144.3	140.2	58	562
7217	C_8H_{10}	p-Xylene	138.2	137.0	~ 28	563
7218	$C_8H_{16}O_2$	Isobutyl isobutyrate	147.3	Nonazeotrope		532
7219	C_8H_{18}	Octane	125.75	Nonazeotrope		575
7220	$C_8H_{18}O$	Butyl ether	142.4	139.8	45	562
7221	C_9H_{12}	Propylbenzene	159.3	Nonazeotrope		575

		B-Component		Azeotropic Data		
No.	Formula	Name	B.P., °C	B.P., °C	Wt.%A	Ref.
A =	C₄H₇ClO₂	**Ethyl Chloroacetate**	**143.55**			
		(continued)				
7222	C₁₀H₁₆	Camphene	159.6	Nonazeotrope		575
7223	C₁₀H₁₆	α-Pinene	155.8	< 142.8	>88	575
A =	C₄H₇Cl₃O	**1,1,1-Trichloro-*tert*-butyl alcohol**				
7224	C₄H₁₁PO₃	Diethyl phosphite				
		1.5 mm.		55	53.3	24
A =	C₄H₇Cl₃O	**Ethyl 1,1,2-Trichloroethyl ether**	**173.0**			
7225	C₅H₄O₂	2-Furaldehyde	161.45	Nonazeotrope		575
7226	C₅H₈O₃	Methyl acetoacetate	169.5	Nonazeotrope		575
7227	C₆H₄Cl₂	*o*-Dichlorobenzene	179.5	Nonazeotrope		575
7228	C₆H₄Cl₂	*p*-Dichlorobenzene	174.4	171.3	75	556
7229	C₆H₅Br	Bromobenzene	156.1	Nonazeotrope		556
7230	C₆H₁₀O₃	Ethyl acetoacetate	180.4	Nonazeotrope		575
7231	C₇H₇Br	*o*-Bromotoluene	181.5	Nonazeotrope		556
7232	C₇H₇Cl	*p*-Chlorotoluene	162.4	Nonazeotrope		575
7233	C₈H₁₈S	Butyl sulfide	185.0	Nonazeotrope		575
7234	C₈H₁₈S	Isobutyl sulfide	172.0	< 171.3	<55	575
7235	C₉H₁₂	Mesitylene	164.6	Nonazeotrope		575
7236	C₉H₁₈O	2,6-Dimethyl-4-heptanone	168.0	Nonazeotrope		575·
7237	C₁₀H₁₆	Nopinene	163.8	Nonazeotrope		575
7238	C₁₀H₁₆	α-Terpinene	173.4	172.0	58	562
A =	C₄H₇N	**Butyronitrile**	**117.9**			
7239	C₄H₉I	1-Iodo-2-methylpropane	120.8	108.5	46	562
7240	C₄H₁₀O	Butyl alcohol	117.8	113.0	50	567
7241	C₄H₁₀O	Isobutyl alcohol	108	< 106.8	>10	567
		"	108	< 105	>25	563
7242	C₅H₁₁Br	1-Bromo-3-methylbutane	120.65	109.8	50	562
7243	C₆H₁₂	Cyclohexane	80.75	< 79.0	> 5	575
7244	C₇H₈	Toluene	110.75	107.0	27	562
7245	C₇H₁₄	Methylcyclohexane	101.15	90.5	20	562
A =	C₄H₇N	**Isobutyronitrile**	**103.85**			
7246	C₅H₁₁Cl	1-Chloro-3-methylbutane	99.4	91.0	35	562
7247	C₅H₁₂O	*tert*-Amyl alcohol	102.35	< 99.5	>42	567
7248	C₆H₆	Benzene	80.15	Nonazeotrope		575
7249	C₆H₁₂	Cyclohexane	80.75	< 74.5	>13	562
7250	C₇H₁₄	Methylcyclohexane	101.15	85.5	40	562
7251	C₇H₁₆	Heptane	98.4	80.5	38	562
A =	C₄H₇N	**Pyrroline**	**90.9**			
7252	C₆H₁₄O	Propyl ether	90.1	< 88.5	<43	575
A =	C₄H₇NO	**α-Hydroxy Isobutyronitrile**				
7253	C₄H₁₁PO₃	Diethyl phosphite				
		92 mm.		129.5	40.0	24
7254	C₆H₁₅PO₃	Dipropyl phosphite				
		8 mm.		96	20.1	24
7255	C₈H₁₉PO₃	Diisobutyl phosphite				
		6.5 mm.		100	12.7	24

TABLE I. *Binary Systems*   223

No.	Formula	Name	B.P., °C	B.P., °C	Wt.%A	Ref.
		B-Component			**Azeotropic Data**	
A =	C_4H_8	**1-Butene**	— 5			
7255a	C_4H_8	2-Butene	1	Nonazeotrope	v-l	149c
7256	C_4H_8	2-Methylpropene	— 6	Nonazeotrope		575
7256a	C_4H_{10}	Butane	0	Nonazeotrope	v-l	149c
7256b	C_4H_{10}	2-Methylpropane	—10	Nonazeotrope	v-l	149c
A =	C_4H_8	**2-Butene**	1–3			
7256c	C_4H_8	2-Methylpropene	—6	Nonazeotrope	v-l	149c
7256d	C_4H_{10}	Butane	0	Nonazeotrope	v-l	149c
7256e	C_4H_{10}	2-Methylpropane	—10	Nonazeotrope	v-l	149c
A =	C_4H_8	**2-Methylpropene**	—6			
7256f	C_4H_{10}	2-Methylpropane				
			10.56 atm.	Nonazeotrope	v-l	386f
A =	$C_4H_8Cl_2O$	**Bis(2-chloroethyl) Ether**	178.65			
7257	$C_4H_8O_2$	Butyric acid	164.0	Nonazeotrope		575
7258	$C_4H_8O_3$	Glycol monoacetate	190.9	Nonazeotrope		527
7259	$C_4H_{10}O$	Butyl alcohol	117.8	Nonazeotrope		527
7260	$C_4H_{10}O_3$	Diethylene glycol	245.5	174.6	92	183
7261	$C_5H_4O_2$	2-Furaldehyde	161.45	Nonazeotrope		527
7262	$C_5H_8O_3$	Methyl acetoacetate	169.5	Nonazeotrope		552
7263	$C_5H_{10}O$	Cyclopentanol	140.85	Nonazeotrope		527
7264	$C_5H_{12}O$	Isoamyl alcohol	131.9	Nonazeotrope		527
7265	$C_5H_{12}O_2$	2-Propoxyethanol	151.35	Nonazeotrope		527
7266	$C_6H_4Cl_2$	o-Dichlorobenzene	179.5	176.5	60	527
7267	$C_6H_4Cl_2$	p-Dichlorobenzene	174.4	173.45	28	527
7268	C_6H_5Br	Bromobenzene	156.1	Nonazeotrope		527
7269	C_6H_5BrO	o-Bromophenol	195.0	Nonazeotrope		575
7270	C_6H_5ClO	o-Chlorophenol	176.8	< 176.5	>14	575
7271	C_6H_5I	Iodobenzene	188.45	Nonazeotrope		527
7272	C_6H_6O	Phenol	182.2	< 176.2	>60	562
7273	$C_6H_{10}O_3$	Ethyl acetoacetate	180.4	Nonazeotrope		527
7274	$C_6H_{10}O_4$	Ethyl oxalate	185.65	Nonazeotrope		527
7275	$C_6H_{12}O_3$	2-Ethoxyethyl acetate	156.8	Nonazeotrope		527
7276	$C_6H_{12}O_3$	Propyl lactate	171.7	Nonazeotrope		575
7277	$C_6H_{13}Br$	Bromohexane	156.5	Nonazeotrope		527
7278	$C_6H_{14}O$	Hexyl alcohol	157.85	< 157.5	<22	527
7279	$C_6H_{14}O_2$	2-Butoxyethanol	171.15	170.85	25	527
7280	C_7H_6O	Benzaldehyde	179.2	Nonazeotrope		527
7281	C_7H_7Br	o-Bromotoluene	181.45	< 177.9	>63	527
7282	C_7H_7Cl	p-Chlorotoluene	162.4	Nonazeotrope		527
7283	$C_7H_{14}O$	2-Methylcyclohexanol	168.5	< 167.5	<40	575
7284	$C_7H_{14}O_3$	1,3-Butanediol methyl ether acetate	171.75	Nonazeotrope		575
7285	$C_7H_{16}O$	n-Heptyl alcohol	176.15	173.5	50	527
7286	$C_7H_{16}O$	3-Heptanol	156.4	141.2	28	981
7287	$C_8H_{16}O$	2-Octanone	172.85	Nonazeotrope		552
7288	$C_8H_{16}O_2$	Butyl butyrate	166.4	Nonazeotrope		527
7289	$C_8H_{18}O$	2-Ethyl-1-hexanol,				
			50 mm. 109	96	90	982

No.	Formula	Name	B.P., °C	B.P., °C	Wt.%A	Ref.
		B-Component		**Azeotropic Data**		
A =	$C_4H_8Cl_2O$	**Bis(2-chloroethyl) Ether**	**178.65**			
		(*continued*)				
7290	$C_8H_{18}O$	*n*-Octyl alcohol	195.2	Nonazeotrope		527
7291	$C_8H_{18}O$	*sec*-Octyl alcohol	180.4	< 177.2	<62	527
7292	$C_8H_{18}S$	Butyl sulfide	185	178.4	88	556
7293	C_9H_{12}	Mesitylene	164.6	Nonazeotrope		575
7294	$C_9H_{18}O_2$	Butyl isovalerate	177.6	177.0	80	575
7295	$C_9H_{18}O_2$	Isoamyl isobutyrate	169.8	Nonazeotrope		575
7296	$C_9H_{18}O_2$	Isobutyl isovalerate	171.2	Nonazeotrope		527
7297	$C_{10}H_{14}$	Butylbenzene	183.1	< 178.0		527
7298	$C_{10}H_{14}$	Cymene	176.7	< 176.4	>11	575
7299	$C_{10}H_{16}$	Dipentene	177.7	< 176.5		527
7300	$C_{10}H_{16}$	Terpinolene	184.6	Nonazeotrope		575
7301	$C_{10}H_{18}O$	Cineol	176.35	173.35	43	527
7302	$C_{10}H_{22}O$	Amyl ether	187.5	< 176.5		527
7303	$C_{10}H_{22}O$	Isoamyl ether	173.2	169.35	39	556
A =	$C_4H_8Cl_2O$	**1,2-Dichloroethyl Ethyl Ether**	**145.5**			
7304	$C_4H_{10}O$	Butyl alcohol	117.8	< 117.0	> 0.6	575
7305	$C_4H_{10}O$	*sec*-Butyl alcohol	99.5	Nonazeotrope		575
7306	$C_5H_4O_2$	2-Furaldehyde	161.45	Nonazeotrope		575
7307	$C_5H_{10}O$	Cyclopentanol	140.85	< 136.5	<50	575
7308	$C_5H_{10}O_3$	Ethyl lactate	154.1	Nonazeotrope		575
7309	$C_5H_{10}O_3$	2-Methoxyethyl acetate	144.6	< 143.0	>38	575
7310	$C_5H_{12}O$	*tert*-Amyl alcohol	102.35	Nonazeotrope		575
7311	$C_5H_{12}O$	Isoamyl alcohol	131.9	129.2	30	575
7312	$C_5H_{12}O_2$	2-Propoxyethanol	151.35	144.3	70	575
7313	C_6H_5Br	Bromobenzene	156.1	Nonazeotrope		556
7314	C_6H_5Cl	Chlorobenzene	131.75	Nonazeotrope		556
7315	$C_6H_{10}O$	Cyclohexanone	155.7	Nonazeotrope		575
7316	$C_6H_{12}O_3$	2-Ethoxyethyl acetate	156.8	Nonazeotrope		575
7317	$C_6H_{13}Br$	1-Bromohexane	156.5	Nonazeotrope		575
7318	$C_6H_{14}S$	Isopropyl sulfide	120.5	Nonazeotrope		566
7319	$C_6H_{14}S$	Propyl sulfide	141.5	< 141.0	>23	566
7320	C_7H_8O	Anisole	153.85	Nonazeotrope		575
7321	$C_7H_{14}O$	4-Heptanone	143.55	< 143.4 Nonazeotrope		575
7322	$C_7H_{14}O_2$	Butyl propionate	146.8	145.3	70	575
7323	$C_7H_{14}O_2$	Ethyl isovalerate	134.7	Nonazeotrope		575
7324	$C_7H_{14}O_2$	Isobutyl propionate	137.5	Nonazeotrope		575
7325	$C_7H_{14}O_2$	Propyl butyrate	143.7	< 143.55	>10	575
7326	$C_7H_{16}O$	Heptyl alcohol	176.15	Nonazeotrope		575
7327	C_8H_8	Styrene	145.8	144.0	53	575
7328	C_8H_{10}	Ethylbenzene	136.15	Nonazeotrope		575
7329	$C_8H_{16}O_2$	Propyl isovalerate	155.7	Nonazeotrope		575
7330	$C_8H_{18}O$	Butyl ether	142.4	138.0	72	562
7331	$C_8H_{18}O$	*sec*-Octyl alcohol	180.4	Nonazeotrope		575
7332	$C_{10}H_{16}$	α-Pinene	155.8	Nonazeotrope		575
A =	$C_4H_8Cl_2S$	**Bis(2-chloroethyl) Sulfide**	**216.8**			
7333	$C_6H_5NO_3$	*o*-Nitrophenol	217.2	<215.5	>48	575

TABLE I.   *Binary Systems*   225

No.	Formula	Name	B.P., °C	B.P., °C	Wt.%A		Ref.
		B-Component		**Azeotropic Data**			
A =	**C₄H₈Cl₂S**	**Bis(2-chloroethyl) Sulfide**	**216.8**				
		(continued)					
7334	C₇H₇ClO	o-Chloroanisole	195.7	Nonazeotrope			575
7335	C₇H₈O	Benzyl alcohol	205.25	195.5			575
7336	C₈H₁₀O	p-Ethylphenol	218.8	220.8	42		575
7337	C₈H₁₀O	2,4-Xylenol	210.5	>218.5	>75		575
7338	C₈H₁₀O	3,4-Xylenol	226.8	227.5	10		575
7339	C₈H₁₀O₂	o-Ethoxyphenol	216.5	<215.2	>42		575
7340	C₉H₁₂O	Mesitol	220.5	223.0	28		575
7341	C₁₀H₁₈O	β-Terpineol	210.5	Nonazeotrope			575
A =	**C₄H₈O**	**2-Butanone**	**79.6**				
7342	C₄H₈O	Butyraldehyde	75.2	Nonazeotrope			527
7343	C₄H₈O	Isobutyraldehyde	63.5	Nonazeotrope			527
7344	C₄H₈O₂	Dioxane	101.35	Nonazeotrope			552
7345	C₄H₈O₂	Ethyl acetate	77.1	77.0	18		527
		"	77.1		~ 26	v-l	680
		"	77.1	77.05	11.8	v-l	331,31c
7346	C₄H₈O₂	Isopropyl formate	68.8	Nonazeotrope			552,897
7347	C₄H₈O₂	Methyl propionate	79.85	79.0	60		552
7348	C₄H₈O₂	Propyl formate	80.85	Nonazeotrope			527
7349	C₄H₉Br	2-Bromobutane	91.2	Nonazeotrope			552
7350	C₄H₉Br	1-Bromo-2-methylpropane	91.4	Nonazeotrope			527
7351	C₄H₉Br	2-Bromo-2-methylpropane	73.25	Nonazeotrope			527
7352	C₄H₉Cl	1-Chlorobutane	78.5	77.0	38		527
7353	C₄H₉Cl	2-Chlorobutane	68.25	Nonazeotrope			552
7354	C₄H₉Cl	1-Chloro-2-methylpropane	68.85	Nonazeotrope			527,897
7355	C₄H₉NO₂	Butyl nitrite	78.2	76.7	30		527
7356	C₄H₉NO₂	Isobutyl nitrite	67.1	Nonazeotrope			527
7357	C₄H₁₀O	n-Butyl alcohol	117.8	Nonazeotrope			527
		"		"		v-l	27
7358	C₄H₁₀O	sec-Butyl alcohol	99.5	Nonazeotrope		v-l	19
		" 374 mm.		Nonazeotrope		v-l	19
7359	C₄H₁₀O	tert-Butyl alcohol	82.45	78.7	69		29,527
7360	C₄H₁₀O	Isobutyl alcohol	108.0	Nonazeotrope			527
7361	C₄H₁₀S	Ethyl sulfide	92.1	<79.4			566
7362	C₄H₁₁N	Butylamine	77.8	74.0	35		551
7363	C₄H₁₁N	Diethylamine	55.9	Nonazeotrope			527
7364	C₅H₆O	2-Methylfuran	63.7	Nonazeotrope			769
		"		Nonazeotrope		v-l	897
7365	C₅H₁₀O	Isovaleraldehyde	92.1	Nonazeotrope			552
7366	C₅H₁₀O₂	Isopropyl acetate	89.5	Nonazeotrope			552
7367	C₅H₁₂O₂	Diethoxymethane	87.95	Nonazeotrope			552
7368	C₆H₅F	Fluorobenzene	84.9	79.3	75		552
7369	C₆H₆	Benzene	80.1	78.33	44	v-l	917, 1051
		"	80.1	78.1	47	v-l	212,598c
		" 14.7 p.s.i.a.		78.2	45	v-l	940
		" 66.7 p.s.i a.		133.0	67.6	v-l	940

No.	Formula	B-Component Name	B.P., °C	B.P., °C	Wt.%A		Ref.
A =	C₄H₈O₂	2-Butanone *(continued)*	79.6				
		Benzene 118.0 p.s.i.a.		160.7	90.0	v-l	940
		" 125.0 p.s.i.a.		Nonazeotrope		v-l	940
		" 44.7 p.s.i.a.		116.5	56	v-l	784c
		" 64.7 p.s.i.a.		132.6	66	v-l	784c
		" 84.7 p.s.i.a.		145.0	76	v-l	784c
		" 112.7 p.s.i.a.		160.0	90	v-l	784c
7370	C₆H₆O	Phenol, 200-760 mm.		Nonazeotrope		v-l	116
7371	C₆H₈	1,3-Cyclohexadiene	80.8	~73	~40		563
7372	C₆H₈O	2,5-Dimethylfuran	93.3	Nonazeotrope			981
7373	C₆H₁₀	Cyclohexene	82.75	73.0	47		552
7374	C₆H₁₂	Cyclohexane	80.75	71.8	40		527
		"		71.8	44	v-l	597c
		" 14.7 p.s.i.a.		71.0	52.5		940
		" 66.7 p.s.i.a.		128.7	61.0		940
		" 118.0 p.s.i.a.		156.4	64.0		940
		" 125.0 p.s.i.a.		182.5	69.0		940
		"	80.85	71.6	45.5		503
		"	80.85	71.5	44		211
		"	80.85		42		735
7374a	C₆H₁₂	1-Hexene 710 mm.		60	17.6	v-l	362c
7375	C₆H₁₄	2,3-Dimethylbutane	58		15.1		735
		"	58.0	56.0	15		527
7376	C₆H₁₄	Hexane 661 mm.		60	31	v-l	362c
			68.8	64.3	29.5		527
		"		64.2	28.6	v-l	331
		"	68.95		29.6		735
		"	69	64	29.4		623c
7377	C₆H₁₄O	Propyl ether	90.1	Nonazeotrope			552
7378	C₆H₁₄O₂	Acetal	104.5	Nonazeotrope			563
7379	C₆H₁₅N	Dipropylamine	109.2	Nonazeotrope			527
7380	C₆H₁₅N	Triethylamine	89.35	<79.0	>75		551
7381	C₇F₁₆	Perfluoroheptane	81.6	62-63			257
7382	C₇H₈	Toluene	110.75	Nonazeotrope		v-l	917
7383	C₇H₁₄	Methylcyclohexane	101.15	77.7	80		527
		"	101.15		80		735
7384	C₇H₁₆	Heptane	98.4		73		735
		"	98.5	77	70	v-l	27
		" 293 mm.		50	70	v-l	27
		"	98.4	77	73	v-l	917
7385	C₈H₁₈	2,5-Dimethylhexane	109.4	109.0	95		576,735
A =	C₄H₈O	1-Butene-3-ol					
7386	C₄H₁₀O₂	2,3-Butanediol		Nonazeotrope		v-l	731
A =	C₄H₈O	Butyraldehyde	74.8				
7387	C₄H₈O	Isobutyraldehyde	64	Nonazeotrope			266
		"	64	Nonazeotrope		v-l	864
7388	C₄H₈O₂	Butyric acid	163.3	Nonazeotrope			981

TABLE I. *Binary Systems* 227

No.	Formula	B-Component Name	B.P., °C	Azeotropic Data B.P., °C	Wt.%A	Ref.
A =	**C₄H₈O**	**Butyraldehyde** *(continued)*	**74.8**			
7389	C₄H₈O₂	Propyl formate	80.85	Nonazeotrope		575
7390	C₄H₉Cl	1-Chloro-2-methylpropane	68.85	Nonazeotrope		575
7391	C₄H₁₀O	Butyl alcohol	117.7	Nonazeotrope	v-l	863
7392	C₄H₁₀O	Isobutyl alcohol	107.8	Nonazeotrope	v-l	863
	" "		107.9	Nonazeotrope		981
7393	C₄H₁₀O₂	1,1-Dimethoxyethane	64.5	Nonazeotrope		981
7394	C₆H₆	Benzene	80.1	Nonazeotrope		340
7395	C₆H₁₀O	Mesityl oxide	128.3	Nonazeotrope		981
7396	C₆H₁₄	Hexane	68.7	60	26	981
7397	C₆H₁₄O₂	1,1-Dimethoxybutane	114	Nonazeotrope		981
7398	C₇H₈	Toluene	110.6	Nonazeotrope	v-l	863
7399	C₇H₁₆	Paraffins	75-80	~ 61		340
A =	**C₄H₈O**	**Isobutyraldehyde**	**63.5**			
7400	C₄H₉Cl	2-Chloro-2-methylpropane	50.8	Nonazeotrope		575
7401	C₄H₁₀O	Butyl alcohol	117.7	Nonazeotrope	v-l	863
7402	C₄H₁₀O	Isobutyl alcohol	107.8	Nonazeotrope	v-l	665,863
7403	C₆H₆	Benzene	81	Nonazeotrope?		340
7404	C₇H₈	Toluene	110.6	Nonazeotrope	v-l	863
7405	C₇H₁₆	Paraffins	75-80	~ 50		340
A =	**C₄H₈O**	**2-Methyl-2-propen-1-ol**	**113.8**			
7406	C₈H₁₄O	2-Methylallyl ether	134.6	114.1	81.3	899
A =	**C₄H₈O**	**Tetrahydrofuran**	**65**			
7406a	C₅H₁₀O₂	Tetrahydrofurfuryl alcohol		Nonazeotrope	v-l	879c
7407	C₆H₁₄	Hexane	68.9	63	53.5	233
	"		68.7	63.0	46.5	224
A =	**C₄H₈OS**	**Ethyl Thioacetate**	**116.6**			
7408	C₄H₁₀O	Butyl alcohol	117.8	113.5		575
7409	C₄H₁₀O	*sec*-Butyl alcohol	99.5	Nonazeotrope		575
7410	C₄H₁₀O	Isobutyl alcohol	108.0	< 107.2		575
7411	C₅H₁₂O	Amyl alcohol	138.2	Nonazeotrope		575
7412	C₅H₁₂O	*tert*-Amyl alcohol	102.35	Nonazeotrope		575
7413	C₅H₁₂O	3-Pentanol	116.0	< 114.0		575
7414	C₆H₁₀S	Allyl sulfide	139.35	Nonazeotrope		575
7415	C₈H₁₈O	Isobutyl ether	122.3	Nonazeotrope		575
A =	**C₄H₈O₂**	**Butyric Acid**	**164.0**			
7416	C₄H₉I	Iodobutane	130.4	129.8	2.5	562
7417	C₄H₉I	1-Iodo-2-methylpropane	120.8	Nonazeotrope		527
7418	C₄H₁₀O	Ethyl ether	34.6	Nonazeotrope		563
7419	C₅H₄O₂	2-Furaldehyde	161.45	159.4	42.5	574
7420	C₅H₅N	Pyridine	115.5	163.2	92	1068
7421	C₅H₈O₃	Ethyl pyruvate	155.5	Nonazeotrope		552
7422	C₅H₉ClO₂	Propyl chloroacetate	162.5	160.5	40	562
7423	C₅H₁₀O₃	2-Methoxyethyl acetate	144.6	Nonazeotrope		526
7424	C₅H₁₁Br	1-Bromo-3-methylbutane	120.65	Nonazeotrope		527
7425	C₅H₁₁I	1-Iodo-3-methylbutane	147.6	144.4	13	527

		B-Component		Azeotropic Data		
No.	Formula	Name	B.P., °C	B.P., °C	Wt.%A	Ref.
A =	C₄H₈O₂	**Butyric Acid** *(continued)*	**164.0**			
7426	C₅H₁₁NO₃	Isoamyl nitrate	149.75	147.85	12	570
7427	C₅H₁₂	2-Methylbutane	27.95	Nonazeotrope		563
7428	C₆H₄BrCl	*p*-Bromochlorobenzene	196.4	Nonazeotrope		527
7429	C₆H₄Cl₂	*o*-Dichlorobenzene	179.5	163.0	65	527
7430	C₆H₄Cl₂	*p*-Dichlorobenzene	174.4	162.0	57	527
7431	C₆H₅Br	Bromobenzene	156	147-148	19	563,834
7432	C₆H₅Cl	Chlorobenzene	132.0	131.75	2.8	527
7433	C₆H₅ClO	*o*-Chlorophenol	175.5	Nonazeotrope		563
7434	C₆H₅I	Iodobenzene	188.55	161.6		538
7435	C₆H₆O	Phenol	181.5	Nonazeotrope		527
7436	C₆H₁₀	Cyclohexene	82.75	Nonazeotrope		678
7437	C₆H₁₀O	Cyclohexanone	156.7	164.5		563
7438	C₆H₁₀O₃	Ethyl acetoacetate	180.4	Nonazeotrope		527
7439	C₆H₁₀O₄	Ethylidene diacetate	168.5	Nonazeotrope		527
7440	C₆H₁₀S	Allyl sulfide	139.35	Nonazeotrope		566
7441	C₆H₁₁BrO₂	Ethyl α-bromoisobutyrate	178	161.5		575
7442	C₆H₁₂	Cyclohexane	80.75	Nonazeotrope		678
7443	C₆H₁₂O₂	Isoamyl formate	123.3	Nonazeotrope		575
7444	C₆H₁₂O₃	2-Ethoxyethyl acetate	156.8	164.3?	18	526
7445	C₆H₁₃Br	1-Bromohexane	156.5	151.5	25	527
7446	C₆H₁₃NO	*N,N*-Dimethylbutyr- amide, 100 mm.	124.5	130	32.6	832
7447	C₆H₁₄S	Propyl sulfide	141.5	Nonazeotrope		566
7448	C₇H₆Cl₂	α,α-Dichlorotoluene	205.2	Nonazeotrope		542
7449	C₇H₆O	Benzaldehyde	179.2	Nonazeotrope		542
7450	C₇H₇Br	α-Bromotoluene	198.5	Nonazeotrope		575
7451	C₇H₇Br	*m*-Bromotoluene	184.3	163.62	79.5	527
7452	C₇H₇Br	*o*-Bromotoluene	181.5	163	72	527
7453	C₇H₇Br	*p*-Bromotoluene	185.0	161.5	75	541
7454	C₇H₇Cl	α-Chlorotoluene	179.3	160.8	65	542
		"	179.35	161.5	93	563
7455	C₇H₇Cl	*o*-Chlorotoluene	159.3	154.5	27	527
7456	C₇H₇Cl	*p*-Chlorotoluene	162.4	156.8	32	527
7457	C₇H₈	Toluene	110.7	Nonazeotrope		527
7458	C₇H₈O	Anisole	153.85	152.85	12	527
7459	C₇H₁₄	Methylcyclohexane	101.8	Nonazeotrope		527,678
7460	C₇H₁₄O	5-Methyl-2-hexanone	144.2	Nonazeotrope		527
7641	C₇H₁₄O₃	Methyl-1,3-butanediol acetate	171.75	172.0?	5?	575
7462	C₇H₁₆	*n*-Heptane	98.4	Nonazeotrope		527
7463	C₈H₈	Styrene	145.8	143.5	15	575
7464	C₈H₉Cl	*o,m,p*-Chloroethylben- zene, 10 mm.	67.5	63.3	34	51
7465	C₈H₁₀	Ethylbenzene	136.15	135.8	4	575
7466	C₈H₁₀	*m*-Xylene	139.0	138.5	6	527
7467	C₈H₁₀	*o*-Xylene	144.3	143.0	10	527
7468	C₈H₁₀	*p*-Xylene	138.45	137.8	5.5	527

TABLE I. *Binary Systems* 229

No.	Formula	B-Component Name	B.P., °C	B.P., °C	Wt.%A	Ref.
					Azeotropic Data	
A =	C₄H₈O₂	**Butyric Acid** *(continued)*	**164.0**			
7469	C₈H₁₀O	Benzyl methyl ether	167.8	160.0	55	562
7470	C₈H₁₀O	Phenetole	170.5	162.35	65	556
7471	C₈H₁₄O	Methylheptenone	173.2	Nonazeotrope		527
7472	C₈H₁₆	1,3-Dimethylcyclohexane	120.7	Nonazeotrope		575
7473	C₈H₁₆O	2-Octanone	172.85	Nonazeotrope		527
7474	C₈H₁₆O₂	Butyl butyrate	166.4	Nonazeotrope		527
7475	C₈H₁₆O₂	Isoamyl propionate	160.7	Nonazeotrope		527
7476	C₈H₁₆O₂	Isobutyl butyrate	156.8	Nonazeotrope		527
7477	C₈H₁₆O₂	Isobutyl isobutyrate	148.6	Nonazeotrope		575
7478	C₈H₁₈	Octane	125.75	< 124.5	<15	562
7479	C₈H₁₈O	Isobutyl ether	122.3	Nonazeotrope		527
7480	C₈H₁₈O	Butyl ether	141.0	Nonazeotrope		543
7481	C₈H₁₈S	Butyl sulfide	185.0	Nonazeotrope		566
7482	C₈H₁₈S	Isobutyl sulfide	172.0	< 162.5	<78	566
7483	C₉H₈	Indene	182.6	163.65	84	527
7484	C₉H₁₂	Cumene	152.8	149.5	20	527
7485	C₉H₁₂	Mesitylene	164.6	158.0	38	527
7486	C₉H₁₂	Propylbenzene	158.9	154.5	28	527
7487	C₉H₁₂	Pseudocumene	169	159.5	45	527
7488	C₉H₁₂O	Benzyl ethyl ether	185.0	Nonazeotrope		527
7489	C₉H₁₂O	Phenyl propyl ether	190.5	Nonazeotrope		575
7490	C₉H₁₈O	2,6-Dimethyl-4-heptanone	168.0	Nonazeotrope		527
7491	C₉H₁₈O₂	Isoamyl butyrate	178.5	Nonazeotrope		527
7492	C₉H₁₈O₂	Isoamyl isobutyrate	170.0	Nonazeotrope		545
7493	C₉H₁₈O₂	Isobutyl isovalerate	172.2	Nonazeotrope		527
7494	C₁₀H₈	Naphthalene	218.1	Nonazeotrope		527
7495	C₁₀H₁₄	Butylbenzene	183.1	162.5	75	562
7496	C₁₀H₁₄	Cymene	176.7	161.0	60	527
7497	C₁₀H₁₆	Camphene	159.6	152.3	2.8	527
7498	C₁₀H₁₆	*d*-Limonene	177.8	160.75	55	563
7499	C₁₀H₁₆	Nopinene	164	156	38	527
7500	C₁₀H₁₆	α-Phellandrene	~ 171.5	160	~ 47	563
7501	C₁₀H₁₆	α-Pinene	155.8	150.2	28	527
7502	C₁₀H₁₆	α-Terpinene	173.4	160.65	46	575
7503	C₁₀H₁₆	γ-Terpinene	183	161.5	70	563
7504	C₁₀H₁₆	Terpinolene	184.6	162.5	72	575
7505	C₁₀H₁₆	Terpinylene	~ 175	160.5	40	563
7506	C₁₀H₁₆	Thymene	179.7	160.5	68	541
7507	C₁₀H₁₇Cl	Bornyl chloride	207.5	Nonazeotrope		575
7508	C₁₀H₁₈O	Cineol	176.35	Nonazeotrope		527
7509	C₁₀H₂₂	2,7-Dimethyloctane	160.2	152.5	33	527
7510	C₁₀H₂₂O	Amyl ether	187.5	Nonazeotrope		527
7511	C₁₀H₂₂O	Isoamyl ether	173.2	161.8	54	556
7512	C₁₁H₂₄	Undecane	194.5	162.4	84.5	1068
7513	C₁₂H₁₈	1,3,5-Triethylbenzene	215.5	Nonazeotrope		527
A =	C₄H₈O₂	**Dioxane**	**101.35**			
7514	C₄H₈O₂	Ethyl acetate		Nonazeotrope	v-l	249

No.	Formula	B-Component Name	B.P., °C	Azeotropic Data B.P., °C	Wt.%A		Ref.
A =	C₄H₈O₂	Dioxane *(continued)*	101.35				
		" 200–700 mm.				v-l	249c
		" 20°C.				v-l	916
		" 30°C.				v-l	916
7515	C₄H₈O₂	Isobutyric acid	154.6	Nonazeotrope			527
7516	C₄H₉Br	1-Bromobutane	101.5	98.0	47		527
7517	C₄H₉Br	1-Bromo-2methylpropane	91.4	Nonazeotrope			527
7518	C₄H₉Cl	1-Chloro-2-methylpropane	68.8	Nonazeotrope			215
7519	C₄H₁₀O	Butyl alcohol	117.8	Nonazeotrope			527
		"	117.75	Nonazeotrope		v-l	639
7520	C₄H₁₀O	*sec*-Butyl alcohol	99.5	< 98.8	<60		527
7521	C₄H₁₀O	*tert*-Butyl alcohol	82.45	Nonazeotrope			527
7522	C₄H₁₀O	Isobutyl alcohol	108	101.3			612
		"	108.0	Nonazeotrope			527,1038
7523	C₄H₁₁N	Diethylamine		Nonazeotrope		v-l	249
		200–700 mm.				v-l	249c
7524	C₅H₅N	Pyridine	115.4	Nonazeotrope			527
7525	C₅H₁₀O	3-Methyl-2-butanone	95.4	Nonazeotrope			575
7526	C₅H₁₀O₂	Isobutyl formate	98.2	Nonazeotrope			527
7527	C₅H₁₀O₂	Isopropyl acetate	89.5	Nonazeotrope			557
7528	C₅H₁₀O₂	Methyl butyrate	102.65	< 100.9			557
7529	C₅H₁₀O₂	Propyl acetate	101.6	< 100.8			557
7530	C₅H₁₁Br	1-Bromo-3-methylbutane	120.65	Nonazeotrope			557
7531	C₅H₁₁Cl	1-Chloro-3-methylbutane	99.4	97.5	36		527
7532	C₅H₁₁N	Piperidine	106.4	Nonazeotrope			575
7533	C₅H₁₁NO₂	Isoamyl nitrite	97.15	Nonazeotrope			527
7534	C₅H₁₂O	*tert*-Amyl alcohol	102.35	100.65	80		575
7535	C₅H₁₂O	2-Pentanol	119.8	Nonazeotrope			527
7536	C₅H₁₂O	3-Pentanol	116.0	Nonazeotrope			575
7537	C₆H₆	Benzene	80.15	Nonazeotrope			527
		"	80.2	82.4	12		201
		" 25°C.		Nonazeotrope		v-l	957
		" 200-760 mm.		Nonazeotrope		v-l	348
7538	C₆H₁₀	Cyclohexene	82.75	< 81.8	>20		558
7539	C₆H₁₀O	Cyclohexanone	156.7	Nonazeotrope			201
7540	C₆H₁₂	Cyclohexane	80.75	79.5	24.6		201
7541	C₆H₁₂	1-Hexene	63.5	Nonazeotrope		v-l	933
7542	C₆H₁₂	Methylcyclopentane	72.0	< 71.5	> 5		527
7543	C₆H₁₂O	Cyclohexanol	160.65	Nonazeotrope			201
7544	C₆H₁₂O	Pinacolone	106.2	Nonazeotrope			575
7545	C₆H₁₂O₂	Ethyl isobutyrate	110.1	Nonazeotrope			557
7546	C₆H₁₂O₂	Isobutyl acetate	117.4	Nonazeotrope			527
7547	C₆H₁₄	Hexane	68.74	Nonazeotrope		v-l	933
7548	C₆H₁₄O	Isopropyl ether	68.3	Nonazeotrope			981
7549	C₆H₁₅BO₃	Ethyl borate	118.6	100.7	92		557
7550	C₇H₈	Toluene	110.75	Nonazeotrope			527,1038
		"	110.7	101.8	80		201,241
		" 200-760 mm.		Nonazeotrope		v-l	348

TABLE I. *Binary Systems*   231

No.	Formula	B-Component Name	B.P., °C	Azeotropic Data B.P., °C	Wt.%A		Ref.
A =	$C_4H_8O_2$	**Dioxane** *(continued)*	**101.35**				
7551	C_7H_{14}	Methylcyclohexane	101.15	93.7	>45		527
7552	C_7H_{16}	Heptane	98.4	91.85	44		527
7553	C_8H_{18}	2,5-Dimethylhexane	109.4	97.0	65		527
7554	C_8H_{18}	*n*-Octane	125.75	< 100.5			527
7555	C_nH_{2n+2}	Paraffins	109.5-110.5	96.6-98.9			241
A =	$C_4H_8O_2$	**m-Dioxane**	**105**				
7556	C_7H_8	Toluene	110.7		85		528
A =	$C_4H_8O_2$	**Ethyl Acetate**	**77.1**				
7557	$C_4H_8O_2$	Isopropyl formate	68.8	Nonazeotrope			575
7558	$C_4H_8O_2$	Methyl propionate	79.85	Nonazeotrope			532
7559	$C_4H_8O_2$	Propyl formate	80.85	Nonazeotrope			575
7560	C_4H_9Br	1-Bromo-2-methylpropane	91.4	Nonazeotrope			547
7561	C_4H_9Br	2-Bromo-2-methylpropane	73.5	71.5	30		563
7562	C_4H_9Cl	1-Chlorobutane	78.05	76.0	<35		547
7563	C_4H_9Cl	2-Chlorobutane	68.25	Nonazeotrope			575
7564	C_4H_9Cl	1-Chloro-2-methylpropane	68.9	Nonazeotrope			563
7565	$C_4H_9NO_2$	Butyl nitrite	78.2	76.3	71		527
7566	$C_4H_9NO_2$	Isobutyl nitrite	67.1	Nonazeotrope			550
7567	$C_4H_{10}O$	Butyl alcohol	117.7	Nonazeotrope			589
		" 725 mm.		Nonazeotrope		v-l	607
7568	$C_4H_{10}O$	*sec*-Butyl alcohol	99.5	Nonazeotrope			575
7569	$C_4H_{10}O$	*tert*-Butyl alcohol	82.45	76.0	73		570
7570	$C_4H_{10}O$	Isobutyl alcohol	108.0	Nonazeotrope			575
		"	108.0	B.p. curve			563
		" 100-760 mm.				v-l	924
7571	$C_4H_{10}O_2$	2-Ethoxyethanol	135.1	Nonazeotrope		v-l	657
7572	$C_4H_{10}S$	Ethyl sulfide	92.2	Nonazeotrope			532
7573	$C_4H_{11}N$	Diethylamine	55.9	Nonazeotrope		v-l	249
		200–700 mm.				v-l	249c
7574	$C_5H_4O_2$	2-Furaldehyde	161.45	Nonazeotrope			1030
7574a	$C_5H_{10}O$	2-Methyl-3-buten-2-ol	97	Nonazeotrope			581c
7575	$C_5H_{10}O$	Isovaleraldehyde	92.3	Nonazeotrope			548
7576	$C_5H_{10}O_2$	Ethyl propionate	99.12	Nonazeotrope (b.p. curve)			1045
7577	$C_5H_{12}O_2$	Diethoxymethane	87.95	Nonazeotrope			527
7578	C_6H_5Cl	Chlorobenzene	131.8	Nonazeotrope			563
7579	C_6H_5F	Fluorobenzene	84.9	Nonazeotrope			575
7580	C_6H_6	Benzene	80.2	Nonazeotrope			834,943
		"	80.1	Nonazeotrope		v-l	146,137
							147
7581	C_6H_8	1,3-Cyclohexadiene	80.8	73.5			563
7582	C_6H_{10}	Cyclohexene	82.75	75.5	<85		563
7582a	$C_6H_{10}O_3$	Ethyl acetoacetate					
		100 mm.		Nonazeotrope		v-l	491e
7583	C_6H_{12}	Cyclohexane	80.75	72.8	54		563
		"	80.75	71.6	56	v-l	146,147

	B-Component			Azeotropic Data		
No.	Formula	Name	B.P., °C	B.P., °C	Wt.%A	Ref.
A =	C₄H₈O₂	**Ethyl Acetate** *(continued)*	**77.1**			
		Cyclohexane 233 mm.		38.7	50.1	924
		" 301 mm.		45.1	51	924
		" 415 mm.		53.6	52.3	924
		" 581 mm.		63.0	54.1	924
		" 756 mm.		71.1	55.3	924
7584	C₆H₁₂	Methylcyclopentane	72.0	67.2	38	575
7585	C₆H₁₂O₂	Butyl acetate	126.1	Nonazeotrope		981
7586	C₆H₁₂O₂	Ethyl butyrate	119.9	Nonazeotrope		563
7587	C₆H₁₄	2,3-Dimethylbutane	58.0	< 57.2	10	575
7588	C₆H₁₄	*n*-Hexane	68.8	65.1		571
		"	68.7	65.15	37.9 v-l	331
7589	C₆H₁₄O	Propyl ether	90.55	Nonazeotrope		557
7590	C₆H₁₄O₂	2-Butoxyethanol	171.1	Nonazeotrope		981
7591	C₇H₈	Toluene	110.6	Nonazeotrope	v-l	43
		"	110.7	Nonazeotrope		589
7592	C₇H₁₄	Methylcyclohexane	101.1	Nonazeotrope		537
7594	C₇H₁₆	Heptane	98.4	< 76.9	<94	575
		"	98.45	Nonazeotrope		537
7594	C₈H₁₀	*p*-Xylene	138.4	Nonazeotrope	v-l	43
A =	C₄H₈O₂	**Isobutyric Acid**	**154.6**			
7595	C₄H₉I	Iodobutane	130.4	128.8	7	562
7596	C₄H₁₀O	*sec*-Butyl alcohol	99.5		v-l	644
7597	C₄H₁₀O	Ethyl ether	34.6	Vapor pressure data		563
7598	C₅H₄O₂	2-Furaldehyde	161.45	153.8		575
7599	C₅H₈O	Cyclopentanone	130.65	Nonazeotrope		575
7600	C₅H₈O₃	Ethyl pyruvate	155.5	153.0	60	552
7601	C₅H₈O₃	Methyl acetoacetate	169.5	Nonazeotrope		552
7602	C₅H₁₀O₃	2-Methoxyethyl acetate	144.6	Nonazeotrope		575
		"	144.6	159.5	62	526
7603	C₅H₁₁Br	1-Bromo-3-methylbutane	120.65	120.2	3	575
7604	C₅H₁₁I	1-Iodo-3-methylbutane	147.65	143.8	22	541
7605	C₅H₁₁NO₃	Isoamyl nitrate	149.75	146.25	30	570
7606	C₆H₄Cl₂	*p*-Dichlorobenzene	174.5	153.0	~ 75	538
7607	C₆H₅Br	Bromobenzene	156.15	148.6	35	563
7608	C₆H₅Cl	Chlorobenzene	132.0	131.2	8	541
7609	C₆H₅I	Iodobenzene	188.55	154.2		542
7610	C₆H₆O	Phenol	182.2	Nonazeotrope		575
7611	C₆H₁₀O	Cyclohexanone	155.7	152.5		575
7612	C₆H₁₂O₃	2-Ethoxyethyl acetate	156.8	159.2	38	575
7613	C₆H₁₂O₃	Paraldehyde	123.2	Nonazeotrope		541
7614	C₆H₁₃Br	1-Bromohexane	156.5	148.0	35	562
7615	C₆H₁₃ClO₂	Chloroacetal	156.8	~ 153		563
7616	C₇H₆Cl₂	α,α-Dichlorotoluene	205.2	Nonazeotrope		575
7617	C₇H₆O	Benzaldehyde	179.2	Nonazeotrope		563
7618	C₇H₇Br	α-Bromotoluene	198.5	Nonazeotrope		575
7619	C₇H₇Br	*o*-Bromotoluene	181.5	153.9	85	541

TABLE I. *Binary Systems* 233

No.	Formula	Name	B.P., °C	B.P., °C	Wt.%A	Ref.
		B-Component		**Azeotropic Data**		
A =	$C_4H_8O_2$	**Isobutyric Acid** *(continued)*	**154.6**			
7620	C_7H_7Cl	α-Chlorotoluene	179.3	153.5	80	541
7621	C_7H_7Cl	o-Chlorotoluene	159.3	< 150.0	42	538
7622	C_7H_7Cl	p-Chlorotoluene	162.4	151.5	47	538
7623	C_7H_8	Toluene	110.75	Nonazeotrope		542
7624	C_7H_8O	Anisole	153.85	149	42	556
7625	$C_7H_{14}O$	4-Heptanone	143.55	Nonazeotrope		552
7626	$C_7H_{14}O$	5-Methyl-2-hexanone	144.2	Nonazeotrope		552
7627	$C_7H_{14}O_2$	Isoamyl acetate	142.1	Nonazeotrope		575
7628	$C_7H_{14}O_2$	Propyl butyrate	143.7	Nonazeotrope		575
7629	$C_7H_{14}O_2$	Methyl,-1,3-butanediol acetate	171.75	Nonazeotrope		575
7630	C_8H_8	Styrene	145.8	142.0	27	562
7631	C_8H_{10}	Ethylbenzene	136.15	134.3	12	541
		"	136.15	133.0	8.8	227
		" 30 mm.		48.0	0.8	227
7632	C_8H_{10}	Mixed xylenes		133.0	10.0	227
		" 56 mm.		62	1.0	227
7633	C_8H_{10}	m-Xylene	139.0	136.9	15	527
7634	C_8H_{10}	o-Xylene	144.3	141.0	22	562
7635	C_8H_{10}	p-Xylene	138.4	136.4	13	541
7636	$C_8H_{10}O$	Benzyl methyl ether	170.5	Nonazeotrope		563
7637	$C_8H_{10}O$	p-Methylanisole	177.05	Nonazeotrope		575
7638	$C_8H_{10}O$	Phenetole	170.45	Nonazeotrope		542
7639	C_8H_{16}	1,3-Dimethylcyclohexane	120.7	< 120.2	<10	575
7640	$C_8H_{16}O_2$	Isoamyl propionate	160.7	Nonazeotrope		575
7641	$C_8H_{16}O_2$	Ethyl caproate	167.7	Nonazeotrope		575
7642	$C_8H_{16}O_2$	Isobutyl butyrate	156.9	Nonazeotrope		575
7643	$C_8H_{16}O_2$	Isobutyl isobutyrate	148.6	Nonazeotrope		575
7644	$C_8H_{16}O_2$	Propyl isovalerate	155.7	Nonazeotrope		575
7645	C_8H_{18}	Octane	125.75	< 124.0	<18	575
7646	$C_8H_{18}O$	Butyl ether	142.4	< 140.5	<22	562
7647	$C_8H_{18}O$	Isobutyl ether	122	Nonazeotrope		556
7648	C_9H_8	Indene	182.4	Nonazeotrope		543
7649	C_9H_{12}	Cumene	152.8	146.8	35	562
7650	C_9H_{12}	Mesitylene	164.6	151.8	~ 57	541
		"	164.0	148.5	~ 48	563
7651	C_9H_{12}	Propylbenzene	158.9	149.3	49	541
7652	C_9H_{12}	Pseudocumene	168.2	152.3	63	541
7653	$C_9H_{12}O$	Benzyl ethyl ether	185.0	Nonazeotrope		575
7654	$C_9H_{18}O$	2,6-Dimethyl-4-heptanone	168.0	Nonazeotrope		552
7655	$C_{10}H_{14}$	Butylbenzene	183.1	Nonazeotrope		575
7656	$C_{10}H_{14}$	Cymene	176.7	153.4	80	562
7657	$C_{10}H_{16}$	Camphene	159.6	148.1	45	541
7658	$C_{10}H_{16}$	d-Limonene	177.8	152.5	78	542
7659	$C_{10}H_{16}$	Nopinene	163.8	149.2	52	562

No.	Formula	B-Component Name	B.P., °C	Azeotropic Data B.P., °C	Wt.%A	Ref.
A =	**C₄H₈O₂**	**Isobutyric Acid** *(continued)*	**154.6**			
7660	C₁₀H₁₆	α-Pinene	155.8	146.7	35	563
7661	C₁₀H₁₆	α-Phellandrene	171.5	150	~ 72	563
7662	C₁₀H₁₆	α-Terpinene	173.4	152.0	70	562
7663	C₁₀H₁₆	Thymene	179.7	~ 154.0		541
7664	C₁₀H₁₈	Cineol	176.35	Nonazeotrope		575
7665	C₁₀H₂₂	Decane	173.3	< 151.2	<72	562
7666	C₁₀H₂₂	2,7-Dimethyloctane	160.2	148.55	48	542
7667	C₁₀H₂₂O	Isoamyl ether	173.2	154.2	93	575
A =	**C₄H₈O₂**	**Isopropyl Formate**	**68.8**			
7668	C₄H₉Cl	1-Chloro-2-methylpropane	68.85	65	48	547
7669	C₄H₉NO₂	Isobutyl nitrite	67.1	65.5	40	549
7670	C₅H₁₀	Cyclopentane	49.4	< 47.0	18	562
7671	C₆H₆	Benzene	68.8	Nonazeotrope		575
7672	C₆H₁₂	Methylcyclopentane	72.0	< 61.5	55	562
7673	C₆H₁₄	Hexane	68.8	57.0	48	562
A =	**C₄H₈O₂**	**Methyl Propionate**	**79.85**			
7674	C₄H₈O₂	Propyl formate	80.85	Nonazeotrope		531
7675	C₄H₉Br	2-Bromobutane	91.2	Nonazeotrope		575
7676	C₄H₉Br	1-Bromo-2-methylpropane	91.4	Nonazeotrope		547
7677	C₄H₉Br	2-Bromo-2-methylpropane	73.25	Nonazeotrope		547
7678	C₄H₉Cl	1-Chlorobutane	78.05	76.8	~ 38	538
7679	C₄H₉Cl	1-Chloro-2-methylpropane	68.9	Nonazeotrope		563
7680	C₄H₉NO₂	Butyl nitrite	78.2	77.7	12	549
7681	C₄H₁₀O	Butyl alcohol	117.8	Nonazeotrope		527
7682	C₄H₁₀O	*sec*-Butyl alcohol	99.5	Nonazeotrope		575
7683	C₄H₁₀O	*tert*-Butyl alcohol	82.55	77.6	~63	536
7684	C₄H₁₀S	Ethyl sulfide	92.2	Nonazeotrope		532
7685	C₅H₁₀O	Isovaleraldehyde	92.3	Nonazeotrope		548
7686	C₅H₁₂O₂	Diethoxymethane	87.95	Nonazeotrope		557
7687	C₆H₆	Benzene	80.2	79.45	52	572
7688	C₆H₁₀	Cyclohexene	82.75	~ 75.5		563
7689	C₆H₁₂	Cyclohexane	80.75	75	52	573
7690	C₆H₁₂	Methylcyclopentane	72.0	69.5	28	562
7691	C₆H₁₄	*n*-Hexane	68.95	67	~12	573
7692	C₆H₁₄O	Propyl ether	90.55	Nonazeotrope		557
7693	C₇H₁₄	Methylcyclohexane	101.1	Nonazeotrope		546
		"	101.1	79.3	88.5	1038
7694	C₇H₁₆	Heptane	98.4	<79.6	<92	575
		"	98.5	Nonazeotrope		546
A =	**C₄H₈O₂**	**Propyl Formate**	**80.85**			
7695	C₄H₉Br	2-Bromobutane	91.2	Nonazeotrope		575
7696	C₄H₉Br	1-Bromo-2-methylpropane	91.4	Nonazeotrope		547
7697	C₄H₉Br	2-Bromo-2-methylpropane	73.3	71.8	28	573
7698	C₄H₉Cl	1-Chlorobutane	78.5	76.1	38	570
7699	C₄H₉Cl	2-Chlorobutane	68.25	Nonazeotrope		575
7700	C₄H₉Cl	1-Chloro-2-methylpropane	68.85	Nonazeotrope		570
7701	C₄H₉NO₂	Butyl nitrite	78.2	76.8	35	549
7702	C₄H₁₀O	Butyl alcohol	117.75	Nonazeotrope		527

TABLE I. *Binary Systems*   235

No.	Formula	Name	B.P., °C	B.P., °C	Wt.%A	Ref.
		B-Component		**Azeotropic Data**		
A =	$C_4H_8O_2$	**Propyl Formate** *(continued)*	**80.85**			
7703	$C_4H_{10}O$	*sec*-Butyl alcohol	99.5	Nonazeotrope		575
7704	$C_4H_{10}O$	*tert*-Butyl alcohol	82.6	78.0	60	537
7705	$C_4H_{10}O$	Isobutyl alcohol	107.85	Nonazeotrope		536
7706	$C_4H_{10}S$	Ethyl sulfide	92.1	<80.2	<87	575
		"	92.2	Nonazeotrope		532
7707	$C_4H_{10}S$	Butanethiol	97.5	Nonazeotrope		566
7708	C_5H_{10}	Cyclopentane	49.3	Nonazeotrope		575
7709	$C_5H_{11}Cl$	1-Chloro-3-methylbutane	99.8	Nonazeotrope		547
7710	$C_5H_{12}O_2$	Diethoxymethane	87.95	Nonazeotrope		527
7711	C_6H_5F	Fluorobenzene	84.9	<79.5	<78	575
7712	C_6H_6	Benzene	80.2	78.5	47	572
7713	C_6H_{10}	Cyclohexene	82.75	<75.0	<53	575
7714	C_6H_{12}	Cyclohexane	80.75	75	48	573
7715	C_6H_{12}	Methylcyclopentane	72.0	<67.5	<35	575
7716	C_6H_{14}	2,3-Dimethylbutane	58.0	56.0	15	562
7717	C_6H_{14}	*n*-Hexane	68.95	63	~20	571
		"	68.95	63.6	29.5	946
7718	$C_6H_{14}O$	Propyl ether	90.55	Nonazeotrope		557
7719	$C_6H_{14}O_2$	Acetal	103.55	Nonazeotrope		557
7720	C_7H_8	Toluene	110.75	Nonazeotrope		575
7721	C_7H_{14}	Methylcyclohexane	101.15	< 80.2	<88	575
7722	C_7H_{16}	Heptane	98.5	78.2	71	527
7723	C_8H_{18}	2,5-Dimethylhexane	109.4	Nonazeotrope		575
A =	$C_4H_8O_3$	**Glycol Monoacetate**	**190.9**			
7724	$C_5H_4O_2$	2-Furaldehyde	161.45	Nonazeotrope		527
7725	$C_5H_{12}O_3$	2-(2-Methoxyethoxy)ethanol	192.95	<188.0	>65	575
7726	$C_6H_4Cl_2$	*o*-Dichlorobenzene	179.5	<179.3		575
7727	$C_6H_4Cl_2$	*p*-Dichlorobenzene	174.4	Nonazeotrope		527
7728	C_6H_5I	Iodobenzene	188.45	184.0		567
7729	$C_6H_5NO_2$	Nitrobenzene	210.75	Nonazeotrope		527
7730	C_6H_6O	Phenol	182.2	197.5	65	527
7731	$C_6H_8O_4$	Methyl fumarate	193.25	<189.0	<65	527
7732	$C_6H_{10}O_4$	Ethyl oxalate	185.65	Nonazeotrope		527
7733	$C_6H_{11}NO_2$	Nitrocyclohexane	205.4	Nonazeotrope		575
7734	$C_6H_{14}O_2$	2-Butoxyethanol	171.15	Nonazeotrope		527
7735	C_7H_7Br	*m*-Bromotoluene	184.3	182.0	32	567
7736	C_7H_7Cl	*p*-Chlorotoluene	162.4	Nonazeotrope		575
7737	C_7H_8O	Benzyl alcohol	205.25	Nonazeotrope		527
7738	C_7H_8O	*m*-Cresol	202.2	206.5	31	527
7739	C_7H_8O	*o*-Cresol	191.1	199.45	51	569
7740	C_7H_8O	*p*-Cresol	201.7	206.0	33	527
7741	$C_7H_8O_2$	Guaiacol	205.05	Nonazeotrope		527
7742	$C_7H_{13}ClO_2$	Isoamyl chloroacetate	195	189.3	50	575
7743	$C_7H_{14}O_3$	1,3-Butanediol methyl ether acetal	171.75	Nonazeotrope		575

No.	Formula	B-Component Name	B.P., °C	Azeotropic Data B.P., °C	Wt.%A	Ref.
A =	C₄H₈O₃	**Glycol Monoacetate** (*continued*)	**190.9**			
7744	C₇H₁₆O	Heptyl alcohol	176.15	Nonazeotrope		575
7745	C₈H₈O	Acetophenone	202.0	Nonazeotrope		527
7746	C₈H₈O₂	Methyl benzoate	199.4	Nonazeotrope		527
7647	C₈H₈O₂	Phenyl acetate	195.7	<190.0		**575**
7748	C₈H₁₀O	Benzyl methyl ether	167.8	<167.0		575
7749	C₈H₁₀O	Phenethyl alcohol	219.4	Nonazeotrope		527
7750	C₈H₁₀O	2,4-Xylenol	~210.5	<212.0	<18	575
7751	C₈H₁₂O₄	Ethyl fumarate	217.85	Nonazeotrope		527
7752	C₈H₁₈O	Octyl alcohol	195.2	189.5	71	527
7753	C₈H₁₈O	*sec*-Octyl alcohol	180.4	<180.3		527
7754	C₉H₈	Indene	182.6	180.0	20	575
7755	C₉H₁₀O₂	Ethyl benzoate	212.5	Nonazeotrope		527
7756	C₉H₁₂O	Benzyl ethyl ether	185.0	180.5	35	575
7757	C₉H₁₄O	Phorone	197.8	Nonazeotrope		552
7758	C₉H₁₈O₂	Isoamyl butyrate	181.05	180.2	21	527
7759	C₉H₁₈O₂	Isobutyl isovalerate	171.2	Nonazeotrope		575
7760	C₁₀H₈	Naphthalene	218.0	Nonazeotrope		527
7761	C₁₀H₁₄	Butylbenzene	183.1	< 181.5		527
7762	C₁₀H₁₈O	Borneol	215.0	Nonazeotrope		575
7763	C₁₀H₁₈O	Cineol	176.35	174.1	22	527
7764	C₁₀H₁₈O	Citronellal	208.0	Nonazeotrope		527
7765	C₁₀H₂₀O	Menthol	216.3	Nonazeotrope		527
7766	C₁₀H₂₀O₂	Isoamyl isovalerate	192.7	187.0	57	527
7767	C₁₀H₂₀O₂	Ethyl caprylate	208.35	Nonazeotrope		527
7768	C₁₀H₂₂O	Amyl ether	187.5	180.8	42	556
7769	C₁₀H₂₂O	Isoamyl ether	173.2	170.2	28	556
7770	C₁₁H₂₀O	Isobornyl methyl ether	192.4	185.0	60	575
7771	C₁₂H₁₈	1,3,5-Triethylbenzene	215.5	Nonazeotrope		575
A =	C₄H₈O₃	**Methyl Lactate**	**143.8**			
7772	C₄H₉I	1-Iodobutane	130.4	<128.5	>20	567
7773	C₄H₉I	1-Iodo-2-methylpropane	120.8	<120.0	> 6	575
7774	C₄H₁₀O	Butyl alcohol	117.8	Nonazeotrope		575
7775	C₄H₁₀O₂	2-Ethoxyethanol	135.3	Nonazeotrope		575
7776	C₅H₈O	Cyclopentanone	130.65	Nonazeotrope		552
7777	C₅H₁₀O	Cyclopentanol	140.85	<140.2	<81	575
7778	C₅H₁₀O₃	2-Methoxyethyl acetate	144.6	143.2	55	575
7779	C₅H₁₁I	1-Iodo-3-methylbutane	147.65	139.0	52	567
7780	C₅H₁₁NO₃	Isoamyl nitrate	149.75	141.4	168	527
7781	C₅H₁₂O	Amyl alcohol	138.2	<138.0		575
7782	C₅H₁₂O	Isoamyl alcohol	131.9	Nonazeotrope		527
7783	C₅H₁₂O	2-Pentanol	119.8	Nonazeotrope		575
7784	C₅H₁₂O₂	2-Propoxyethanol	151.35	Nonazeotrope		526
7785	C₆H₅Br	Bromobenzene	156.1	141.5	22	567
	"		156.1	Nonazeotrope		535
7786	C₆H₅Cl	Chlorobenzene	131.75	<130.8		575
7787	C₆H₆O	Phenol	182.2	Nonazeotrope		575

TABLE I. *Binary Systems* 237

No.	Formula	Name	B.P., °C	B.P., °C	Wt.%A	Ref.
		B-Component		**Azeotropic Data**		
A =	**C₄H₈O₃**	**Methyl Lactate** *(continued)*	**143.8**			
7788	C₆H₁₀O	Cyclohexanone	155.7	Nonazeotrope		552
7789	C₆H₁₀O	Mesityl oxide	129.45	Nonazeotrope		552
7790	C₆H₁₂O	Cyclohexanol	160.65	Nonazeotrope		563
7791	C₆H₁₂O₃	2-Ethoxyethyl acetate	156.8	Nonazeotrope		526
7792	C₆H₁₄O	Hexyl alcohol	157.85	Nonazeotrope		575
7793	C₆H₁₄O₂	2-Butoxyethanol	171.15	Nonazeotrope		575
7794	C₆H₁₄S	Propyl sulfide	141.5	<138.0	<40	566
7795	C₇H₈	Toluene	110.75	~ 110.4	~ 18	573
7796	C₇H₈O	Anisole	153.85	142.8	82	556
7797	C₇H₈O	o-Cresol	191.1	Nonazeotrope		575
7798	C₇H₁₄O	4-Heptanone	143.55	142.7	47	552
7799	C₇H₁₄O₂	Butyl propionate	146.5	~ 141.3	>55	548
7800	C₇H₁₄O₂	Isobutyl propionate	137.5	135.8	40	575
7801	C₇H₁₄O₂	Ethyl isovalerate	134.7	Nonazeotrope		532
7802	C₇H₁₄O₂	Ethyl valerate	145.45	140.0	58	527
7803	C₇H₁₄O₂	Isoamyl acetate	142.1	~ 138.5	44	529
7804	C₇H₁₄O₂	Methyl caproate	149.8	141.7	70	575
7805	C₇H₁₄O₂	Propyl butyrate	142.8	137.5	46	572
7806	C₇H₁₄O₂	Propyl isobutyrate	134.7	Nonazeotrope		532
7807	C₈H₈	Styrene	145.8	~ 134.5	~ 50	548
		26 mm.			~ 33 vol. %	342
7808	C₈H₁₀	Ethylbenzene	136.15	129.4	35	573
		26 mm.			~ 26 vol. %	342
7809	C₈H₁₀	m-Xylene	139.0	131.2	42.5	527
7810	C₈H₁₀	p-Xylene	138.2	130.8	40	573
7811	C₈H₁₀O	Benzyl methyl ether	167.8	Nonazeotrope		575
7812	C₈H₁₆O₂	Butyl butyrate	166.4	Nonazeotrope		575
7813	C₈H₁₆O₂	Isoamyl propionate	160.7	Nonazeotrope		575
7814	C₈H₁₆O₂	Isobutyl isobutyrate	147.3	141.5	70	527
7815	C₈H₁₆O₂	Propyl isovalerate	155.7	Nonazeotrope		548
7816	C₈H₁₈	2,5-Dimethylhexane	109.4	<108.5	<17	575
7817	C₈H₁₈	Octane	125.8	120.3	30	567
7818	C₈H₁₈O	Butyl ether	142.4	137.0	42	575
7819	C₉H₁₂	Cumene	152.8	137.8	62	567
7820	C₉H₁₂	Mesitylene	164.6	142.0	>85	548
7821	C₉H₁₂	Propylbenzene	158.9	140	~ 88	538
7822	C₉H₁₂	Pseudocumene	168.2	~ 143.0	<90	575
7823	C₁₀H₁₄	Cymene	176.7	Nonazeotrope		575
7824	C₁₀H₁₆	Camphene	159.6	140	85	573
7825	C₁₀H₁₆	d-Limonene	177.8	Nonazeotrope		535
7826	C₁₀H₁₆	Nopinene	163.8	138.5	70	567
7827	C₁₀H₁₆	α-Pinene	155.8	<144.2	>90	563
7828	C₁₀H₁₆	α-Terpinene	173.4	<142.5	<88	575
7829	C₁₀H₂₂	2,7-Dimethyloctane	160.1	137.8	68	567
A =	**C₄H₈O₃**	**Propylene Glycol Monoformate**				
7830	CₙHₘ	Hydrocarbons		min. b.pt.		618

| | B-Component | | | Azeotropic Data | | |
No.	Formula	Name	B.P., °C	B.P., °C	Wt.%A	Ref.
A =	**C₄H₈S**	**Tetrahydrothiophene**	**118.8**			
7831	C₅H₅N	Pyridine	115.4	113.5	45	553
7832	C₅H₇N	1-Methylpyrrol	112.8	111.5	18	575
7833	C₅H₁₀O	3-Pentanone	102.05	Nonazeotrope		575
7834	C₅H₁₀O₂	Isobutyl formate	98.2	Nonazeotrope		566
7835	C₅H₁₀O₂	Propyl acetate	101.6	Nonazeotrope		566
7836	C₅H₁₁Cl	1-Chloro-3-methylbutane	99.4	Nonazeotrope		566
7837	C₅H₁₂O₂	Diethoxymethane	87.95	Nonazeotrope		575
7838	C₆H₁₂O	Pinacolone	106.2	Nonazeotrope		575
7839	C₆H₁₄O	Propyl ether	90.1	Nonazeotrope		575
7840	C₆H₁₄O₂	Acetal	103.55	Nonazeotrope		575
7841	C₆H₁₄S	Isopropyl sulfide	120.5	<117.5	>60	575
		"	119.25	118.40		205
7842	C₇H₈	Toluene	110.75	Nonazeotrope		205,566
7843	C₇H₁₆	Heptane	98.4	Nonazeotrope		566
7844	C₈H₁₆	trans-1,3-Dimethyl-				
		cyclohexane	120.30	115.90	43.1	205
7845	C₈H₁₆	Ethylcyclohexane	131.85	120.46	80.7	205
7846	C₈H₁₈	2,5-Dimethylhexane	109.15	107.95	16.8	205
		"	109.4	<109.1	> 6	575
7847	C₈H₁₈	2-Methylheptane	117.70	113.96	38.2	205
7849	C₈H₁₈	Octane	125.70	117.79	60.3	205
A =	**C₄H₉Br**	**1-Bromobutane**	**101.5**			
7849	C₄H₉Cl	1-Chlorobutane	77.9	Nonazeotrope		
				Vapor pressure data		899
7850	C₄H₁₀O	Butyl alcohol	117.8	98.6	87	527
7851	C₄H₁₀O	sec-Butyl alcohol	99.5	93.0	70	567
7852	C₄H₁₀O	tert-Butyl alcohol	82.45	< 81.8	<37	575
7853	C₄H₁₀O	Isobutyl alcohol	107.85	95	79	573
7854	C₄H₁₀S	Ethyl sulfide	92.1	Nonazeotrope		566
7855	C₅H₁₀O	2-Pentanone	102.35	100.1	63	552
7856	C₅H₁₀O	3-Pentanone	102.05	100.0	63	552
7857	C₅H₁₀O₂	Butyl formate	106.7	100.0	75	547
7858	C₅H₁₀O₂	Ethyl propionate	99.1	< 98.8		575
7859	C₅H₁₀O₂	Isobutyl formate	98.2	95.5	>35	547
7860	C₅H₁₀O₂	Isopropyl acetate	89.5	Nonazeotrope		575
7861	C₅H₁₀O₂	Methyl butyrate	102.65	99.5	65	547
7862	C₅H₁₀O₂	Methyl isobutyrate	92.5	Nonazeotrope		547
7863	C₅H₁₀O₂	Propyl acetate	101.6	99.9	52	526
		"	101.6	100.0	55	547
7864	C₅H₁₁NO₂	Isoamyl nitrite	97.15	Nonazeotrope		550
7865	C₅H₁₂O	tert-Amyl alcohol	102.35	< 97.8	<74	575
7866	C₅H₁₂O	Isoamyl alcohol	131.9	Nonazeotrope		527
7867	C₅H₁₂O	3-Methyl-2-butanol	112.9	99.7	86	567
7868	C₅H₁₂O	3-Pentanol	116.0	<100.7	>86	575
7869	C₆H₅NO₂	Nitrobenzene	210.75	Nonazeotrope		554
7870	C₆H₁₀S	Allyl sulfide	139.35	Nonazeotrope		566
7871	C₆H₁₂O	4-Methyl-2-pentanone	116.05	Nonazeotrope		527

TABLE I. *Binary Systems*   239

No.	Formula	Name		B.P., °C	B.P., °C	Wt.%A		Ref.
		B-Component			**Azeotropic Data**			
A =	**C₄H₉Br**	**1-Bromobutane** *(continued)*		**101.5**				
7872	C₆H₁₂O	Pinacolone		106.2	101.1	86		552
		"		106.2	Nonazeotrope			548
7873	C₆H₁₂O₂	Ethyl isobutyrate		110.1	Nonazeotrope			547
7874	C₆H₁₄O	Propyl ether		90.1	Nonazeotrope			559
7875	C₆H₁₄S	Isopropyl sulfide		120.5	Nonazeotrope			566
7876	C₇H₈	Toluene		110.75	Nonazeotrope			575
7876a	C₇H₁₄	1-Heptene	200 mm.		52.1		v-l	498f
		"	400 mm.		73.6	20.0	v-l	498f
		"	600 mm.		86.2	17.6	v-l	498f
		"	760 mm.		93.6	16.4	v-l	498f
7877	C₇H₁₄	Methylcyclohexane		101.15	< 99.5	55		562
7878	C₇H₁₆	Heptane		98.45	96.7	50		538
		"	50°C.	Vapor pressure data		42.5		899
A =	**C₄H₉Br**	**2-Bromobutane**		**91.2**				
7879	C₄H₉Br	1-Bromo-2-methylpropane		91.4	Nonazeotrope			549
7880	C₄H₁₀O	Butyl alcohol		117.8	90.6	94		575
7881	C₄H₁₀O	*sec*-Butyl alcohol		99.5	87.2	81.9		405
7882	C₄H₁₀O	Isobutyl alcohol		108.0	88.6	86		567
7883	C₅H₁₀O	3-Pentanone		102.05	Nonazeotrope			552
7884	C₅H₁₀O₂	Methyl isobutyrate		92.5	90.5	70		562
7885	C₅H₁₁NO₂	Isoamyl nitrite		97.15	Nonazeotrope			550
7886	C₆H₁₂	Cyclohexane		80.75	Nonazeotrope			575
7887	C₇H₁₆	Heptane		98.4	< 91.0	>80		575
A =	**C₄H₉Br**	**1-Bromo-2-methylpropane**		**91.4**				
7888	C₄H₉ClO	Chloroethyl ethyl ether		98.5	Nonazeotrope			575
7889	C₄H₉NO₂	Butyl nitrite		78.2	Nonazeotrope			550
7890	C₄H₁₀O	Butyl alcohol		117.75	90.2	93		535
7891	C₄H₁₀O	*sec*-Butyl alcohol		99.5	87.0	80.5		567
7892	C₄H₁₀O	*tert*-Butyl alcohol		82.45	79.0	58		575
7893	C₄H₁₀O	Isobutyl alcohol		107.85	89.2	<84		527
				108	Nonazeotrope			
					B.p. curve			389
7894	C₄H₁₀S	Ethyl sulfide		92.1	< 90.2	<54		566
7895	C₅H₁₀O	3-Methyl-2-butanone		95.4	90.8	82		552
7896	C₅H₁₀O	2-Pentanone		102.35	Nonazeotrope			552
7897	C₅H₁₀O	3-Pentanone		102.05	Nonazeotrope			552
7898	C₅H₁₀O₂	Butyl formate		106.8	Nonazeotrope			575
7899	C₅H₁₀O₂	Ethyl propionate		99.15	Nonazeotrope			547
7900	C₅H₁₀O₂	Isobutyl formate		97.9	90.0	~ 70		538
7901	C₅H₁₀O₂	Isopropyl acetate		90.8	89.0	55		538
7902	C₅H₁₀O₂	Methyl butyrate		102.65	Nonazeotrope			575
7903	C₅H₁₀O₂	Methyl isobutyrate		92.3	90	61		573
7904	C₅H₁₀O₂	Propyl acetate		101.6	Nonazeotrope			547
7905	C₅H₁₁NO₂	Isoamyl nitrite		97.15	Nonazeotrope			550
7906	C₅H₁₂O	*tert*-Amyl alcohol		102.0	87.5	82		532
7907	C₅H₁₂O	Isoamyl alcohol		131.3	Nonazeotrope			527

No.	Formula	B-Component Name	B.P., °C	Azeotropic Data B.P., °C	Wt.%A	Ref.
A =	C₄H₉Br	**1-Bromo-2-methylpropane** (*continued*)	**91.4**			
7808	C₅H₁₂O	3-Methyl-2-butanol	112.6	Nonazeotrope		575
7909	C₅H₁₂O	2-Pentanol	119.8	Nonazeotrope		575
7910	C₅H₁₂O	3-Pentanol	116.0	Nonazeotrope		575
7911	C₆H₆	Benzene	80.2	Nonazeotrope		563
7912	C₆H₁₂	Cyclohexane	80.75	Nonazeotrope		575
7913	C₆H₁₄O₂	Acetal	103.55	Nonazeotrope		559
7914	C₇H₁₄	Methylcyclohexane	101.15	Nonazeotrope		575
7915	C₇H₁₆	Heptane	98.4	< 91.0	>80	527
7916	C₈H₁₈	2,5-Dimethylhexane	109.4	Nonazeotrope		575
A =	C₄H₉Br	**2-Bromo-2-methylpropane**	**73.25**			
7917	C₄H₉NO₂	Butyl nitrite	78.2	Nonazeotrope		550
7918	C₄H₉NO₂	Isobutyl nitrite	67.1	Nonazeotrope		550
7919	C₄H₁₀O	*tert*-Butyl alcohol	82.45	69.95		571c,962
7920	C₄H₁₀O	Isobutyl alcohol	108	Nonazeotrope		563
7921	C₄H₁₀O₂	Ethoxymethoxymethane	65.9	Nonazeotrope		559
7922	C₆H₆	Benzene	80.2	Nonazeotrope		563
7923	C₆H₁₂	Cyclohexane	80.75	Nonazeotrope		575
7924	C₆H₁₂	Methylcyclopentane	72.0	< 70.5	>48	562
7925	C₆H₁₄	Hexane	68.85	68.0	~ 38	538
7926	C₆H₁₄O	Isopropyl ether	68.3	Nonazeotrope		559
A =	C₄H₉Cl	**1-Chlorobutane**	**78.5**			
7927	C₄H₉ClO	2-Chloroethyl ethyl ether	98.5	Nonazeotrope		575
7928	C₄H₉NO₂	Butyl nitrite	78.2	77.0	48	550
7929	C₄H₉NO₂	Isobutyl nitrite	67.1	Nonazeotrope		550
7930	C₄H₁₀O	Butyl alcohol	117.75	Nonazeotrope		527
		"	117	77.7	98.1	215
		"	117.75	Nonazeotrope	v-l	497,982
7931	C₄H₁₀O	*sec*-Butyl alcohol	99.5	77.7	92	575
7932	C₄H₁₀O	*tert*-Butyl alcohol	82.45	72.8	80	567
7933	C₄H₁₀O	Isobutyl alcohol	107.85	77.65	96	573
7934	C₄H₁₀O₂	Acetaldehyde dimethyl acetal	64.3	Nonazeotrope		559
7935	C₄H₁₀S	Butanethiol	97.5	Nonazeotrope		575
7936	C₅H₁₀O	Isovaleraldehyde	92.1	Nonazeotrope		575
7937	C₅H₁₀O	3-Methyl-2-butanone	95.4	Nonazeotrope		552
7938	C₅H₁₀O₂	Isopropyl acetate	89.5	Nonazeotrope		575
7939	C₅H₁₁NO₂	Isoamyl nitrite	97.15	Nonazeotrope		550
7940	C₅H₁₂O	*tert*-Amyl alcohol	102.35	Nonazeotrope		575
7941	C₅H₁₂O	3-Pentanol	116.0	Nonazeotrope		575
7942	C₅H₁₂O₂	Diethoxymethane	87.95	Nonazeotrope		527
7943	C₆H₅NO₂	Nitrobenzene	210.75	Nonazeotrope		575
7944	C₆H₁₂	Cyclohexane	80.75	< 78.0	>64	575
7945	C₆H₁₄	Hexane	68.8	Nonazeotrope		575
7946	C₆H₁₄O	Isopropyl ether	68.3	Nonazeotrope		559
7947	C₆H₁₄O	Propyl ether	90.1	Nonazeotrope		559

TABLE I. *Binary Systems*   241

No.	Formula	Name	B.P., °C	B.P., °C	Wt.%A	Ref.
		B-Component		**Azeotropic Data**		
A =	**C_4H_9Cl**	**1-Chlorobutane** *(continued)*	**78.5**			
7948	C_7H_{16}	Heptane	98.4	Nonazeotrope		
				Vapor pressure data		899
7949	$C_8H_{18}O$	Butyl ether	141.97	Nonazeotrope		806
A =	**C_4H_9Cl**	**2-Chlorobutane**	**68.25**			
7950	C_4H_9Cl	1-Chloro-2-methylpropane	68.85	Nonazeotrope		575
7951	$C_4H_9NO_2$	Butyl nitrite	78.2	Nonazeotrope		550
7952	$C_4H_9NO_2$	Isobutyl nitrite	67.1	66.2	38	550
7953	$C_5H_{12}O$	Ethyl propyl ether	63.85	Nonazeotrope		559
7954	$C_6H_5NO_2$	Nitrobenzene	210.75	Nonazeotrope		554
7955	C_6H_6	Benzene	80.15	Nonazeotrope		575
7956	C_6H_{14}	Hexane	68.8	65.85	57	562
A =	**C_4H_9Cl**	**1-Chloro-2-methylpropane**	**68.85**			
7957	$C_4H_9NO_2$	Butyl nitrite	78.2	Nonazeotrope		550
7958	$C_4H_9NO_2$	Isobutyl nitrite	67.1	66.5	33	550
7959	$C_4H_{10}O$	Butyl alcohol	117.75	Nonazeotrope		527
7960	$C_4H_{10}O$	*sec*-Butyl alcohol	99.5	Nonazeotrope		575
7961	$C_4H_{10}O$	*tert*-Butyl alcohol	82.55	65.5	83	535
7962	$C_4H_{10}O$	Isobutyl alcohol	107.85	Nonazeotrope		532
7963	$C_4H_{10}O_2$	Acetaldehyde dimethyl acetal	64.3	Nonazeotrope		559
7964	C_5H_{10}	Cyclopentane	49.3	Nonazeotrope		575
7965	$C_5H_{12}O$	*tert*-Amyl alcohol	102.0	Nonazeotrope		535
7966	$C_5H_{12}O$	Ethyl propyl ether	63.6	Nonazeotrope		548
7967	C_6H_6	Benzene	80.2	Nonazeotrope		529
7968	C_6H_8	1,3-Cyclohexadiene	80.8	Nonazeotrope		563
7969	C_6H_{10}	Cyclohexene	82.75	Nonazeotrope		575
7970	C_6H_{12}	Cyclohexane	80.75	Nonazeotrope		563
7971	C_6H_{12}	Methylcyclopentane	72.0	67.8	63	562
7972	C_6H_{14}	2,3-Dimethylhexane	58.0	Nonazeotrope		575
7973	C_6H_{14}	Hexane	68.95	66.3	55	563
7974	$C_6H_{14}O$	Isopropyl ether	68.3	> 69.0		559
A =	**C_4H_9Cl**	**2-Chloro-2-methylpropane**	**50.8**			
7975	$C_4H_9NO_2$	Isobutyl nitrite	67.1	Nonazeotrope		550
7976	$C_4H_{10}O$	*tert*-Butyl alcohol	82.5	Nonazeotrope		962
7977	C_5H_{10}	Cyclopentane	49.3	47.5	50	562
7978	C_5H_{12}	Pentane	36.15	< 35.8	>16	562
7879	C_6H_{10}	Biallyl	60.2	Nonazeotrope		563
7980	C_6H_{12}	Methylcyclopentane	72.0	Nonazeotrope		575
7981	C_6H_{14}	2,3-Dimethylbutane	58.0	< 50.5	<40	575
7982	C_6H_{14}	Hexane	68.9	Nonazeotrope		563
A =	**C_4H_9ClO**	**Chloroethyl Ethyl Ether**	**98.5**			
7983	$C_4H_{10}S$	Ethyl sulfide	92.1	91.8	6	575
7984	C_5H_7N	1-Methylpyrrol	112.8	Nonazeotrope		575
7985	$C_5H_{10}O$	3-Methyl-2-butanone	95.4	Nonazeotrope		575
7986	$C_5H_{10}O$	3-Pentanone	102.05	Nonazeotrope		575
7987	C_6H_6	Benzene	80.15	Nonazeotrope		575
7988	C_6H_{12}	Cyclohexane	80.75	Nonazeotrope		575
7989	$C_6H_{14}O$	Propyl ether	90.1	Nonazeotrope		575

		B-Component		Azeotropic Data		
No.	Formula	Name	B.P., °C	B.P., °C	Wt.%A	Ref.
A =	C₄H₉ClO	**Chloroethyl Ethyl Ether**	**98.5**			
		(continued)				
7990	C₇H₁₄	Methylcyclohexane	101.15	< 97.5	>65	562
7991	C₇H₁₆	Heptane	98.4	96.0	48	562
A =	C₄H₉Cl₃Sn	**Butyltin Trichloride**	**113/17**			
7992	C₈H₁₈Cl₂Sn	Dibutyltin dichloride,				
		17 mm.	157		Nonazeotrope	981
7993	C₁₂H₂₇ClSn	Tributyltin chloride,				
		17 mm.	166		Nonazeotrope	981
A =	C₄H₉I	**1-Iodobutane**	**130.4**			
7994	C₄H₉NO₃	Isobutyl nitrate	123.5	< 121.7	>27	560
7995	C₄H₁₀O	Butyl alcohol	117.8	113.8	58.5	575
7996	C₄H₁₀O	*tert*-Butyl alcohol	82.45		Nonazeotrope	575
7997	C₄H₁₀O	Isobutyl alcohol	108.0	106.2	50	567
7998	C₄H₁₀O₂	2-Ethoxyethanol	135.3	123.0	70	526
7999	C₅H₄O₂	2-Furaldehyde	161.45		Nonazeotrope	527
8000	C₅H₅N	Pyridine	115.5		Nonazeotrope	548
8001	C₅H₈O	Cyclopentanone	130.65	129.0	60	552
8002	C₅H₉N	Isovaleronitrile	130.5	118.5	60	562
8003	C₅H₁₀O	Cyclopentanol	140.85	127.0	84	567
8004	C₅H₁₀O₂	Isovaleric acid	176.5		Nonazeotrope	527
8005	C₅H₁₀O₃	Ethyl carbonate	126.0	124.5	30	547
8006	C₅H₁₀O₃	2-Methoxyethyl acetate	144.6	< 129.5	<13	575
8007	C₅H₁₂O	Amyl alcohol	138.2	125.0	78	527
8008	C₅H₁₂O	Isoamyl alcohol	131.9	123.2	72	527
8009	C₅H₁₂O	2-Pentanol	119.8	117.0	54	567
8010	C₆H₆O	Phenol	182.2		Nonazeotrope	575
8011	C₆H₁₀O	Mesityl oxide	129.5	128.0	56	527
8012	C₆H₁₂O	3-Hexanone	123.3		Nonazeotrope	552
8013	C₆H₁₂O₂	Butyl acetate	126.0	124.8	25	562
8014	C₆H₁₂O₂	Ethyl butyrate	121.5		Nonazeotrope	575
8015	C₆H₁₂O₂	Isoamyl formate	123.8	122.0	26	562
8016	C₆H₁₂O₂	Isobutyl acetate	117.4		Nonazeotrope	575
8017	C₆H₁₂O₂	Propyl propionate	122.5		Nonazeotrope	547
8018	C₆H₁₂O₃	2-Ethoxyethyl acetate	156.8		Nonazeotrope	575
8019	C₆H₁₄O	Hexyl alcohol	157.85		Nonazeotrope	575
8020	C₇H₈	Toluene	110.75		Nonazeotrope	575
8021	C₇H₁₄O₂	Ethyl isovalerate	134.7	< 130.3		575
8022	C₇H₁₄O₂	Isobutyl propionate	136.9		Nonazeotrope	547
8023	C₈H₁₀	Ethylbenzene	136.15	< 130.0	>85	575
A =	C₄H₉I	**2-Iodobutane**	**120.0**			
8024	C₆H₁₂O₂	Ethyl isobutyrate	110.1		Nonazeotrope	575
8025	C₆H₁₂O₂	Isobutyl acetate	117.4	< 116.0	>30	575
8026	C₆H₁₂O₂	Methyl isovalerate	116.5	< 116.0	>28	575
8027	C₇H₈	Toluene	110.75		Nonazeotrope	575
A =	C₄H₉I	**1-Iodo-2-methylpropane**	**120.8**			
8028	C₄H₉NO₃	Isobutyl nitrate	123.5	< 117.5	>60	560
8029	C₄H₁₀O	Butyl alcohol	117.75	110.6	70	535

TABLE I. *Binary Systems*

		B-Component		Azeotropic Data		
No.	Formula	Name	B.P., °C	B.P., °C	Wt.%A	Ref.
A =	C₄H₉I	**1-Iodo-2-methylpropane**	**120.8**			
		(continued)				
8030	C₄H₁₀O	Isobutyl alcohol	108	101	>67	834
			107.85	104	64	573
8031	C₄H₁₀O₂	2-Ethoxyethanol	135.3	117.5		575
8032	C₅H₅N	Pyridine	115.5	~ 114.0	~ 35	548
8033	C₅H₁₀O₂	Butyl formate	106.8	Nonazeotrope		575
8034	C₅H₁₀O₂	Methyl butyrate	102.65	Nonazeotrope		575
8035	C₅H₁₀O₂	Propyl acetate	101.6	Nonazeotrope		575
8036	C₅H₁₀O₃	Ethyl carbonate	126.0	118.2	80	547
8037	C₅H₁₀O₃	2-Methoxyethyl acetate	144.6	Nonazeotrope		526
8038	C₅H₁₁Br	1-Bromo-3-methylbutane	120.2	~ 119.0		563
8039	C₅H₁₂O	Isoamyl alcohol	131.8	115	<80	834
			131.3	117.6	83	527
8040	C₅H₁₂O₂	2-Propoxyethanol	151.35	<130.0		575
8041	C₆H₁₂O₂	Butyl acetate	125.0	120.0		547
8042	C₆H₁₂O₂	Ethyl butyrate	120.0	119	64	573
8043	C₆H₁₂O₂	Ethyl isobutyrate	110.1	Nonazeotrope		563
8044	C₆H₁₂O₂	Isoamyl formate	123.6	117.5	70	563
8045	C₆H₁₂O₂	Isobutyl acetate	117.2	116.0	50	571
8046	C₆H₁₂O₂	Methyl isovalerate	116.5	Nonazeotrope		547
8047	C₆H₁₅BO₃	Ethyl borate	118.6	117.2	35	547
8048	C₇H₈	Toluene	110.7	Nonazeotrope		834
8049	C₇H₁₄O₂	Ethyl isovalerate	134.7	Nonazeotrope		575
8050	C₇H₁₄O₂	Isoamyl acetate	142.1	Nonazeotrope		389
8051	C₇H₁₄O₂	Isopropyl isobutyrate	120.8	119.5	53	547
8052	C₇H₁₄O₂	Propyl isobutyrate	134.0	Nonazeotrope		547
8053	C₈H₁₆	1,3-Dimethylcyclohexane	120.7	< 119.0	>60	562
A =	C₄H₉N	**Pyrrolidine**	**88**			
8054	C₆H₆	Benzene	80.1	Min. b.p.		515
A =	C₄H₉NO	**2-Butanone Oxime**				
8055	C₆H₆	Benzene 150-300 mm.		Nonazeotrope	v-l	774
8056	C₇H₁₆	Heptane 150-300 mm.		Nonazeotrope	v-l	774
A =	C₄H₉NO	**Morpholine**	**128.3**			
8057	C₈H₁₀	o-Xylene	143.6	Nonazeotrope		340
8058	C₈H₁₈O	Butyl ether	142.1	126.7	73	981
8059	C₉H₁₈O	2,6-Dimethyl-4-heptanone	169.4	128	98	981
A =	C₄H₉NO₂	**Butyl Nitrite**	**78.2**			
8060	C₄H₁₀S	Ethyl sulfide	92.1	Nonazeotrope		550
8061	C₅H₁₀O	3-Methyl-2-butanone	95.4	Nonazeotrope		552
8062	C₅H₁₀O₂	Isopropyl acetate	89.5	Nonazeotrope		550
8063	C₅H₁₂O₂	Diethoxymethane	87.95	Nonazeotrope		527
8064	C₆H₅F	Fluorobenzene	84.9	Nonazeotrope		550
8065	C₆H₆	Benzene	80.15	77.95	75	550
8066	C₆H₁₂	Cyclohexane	80.75	76.5	63	570
8067	C₆H₁₂	Methylcyclopentane	72.0	< 71.5	< 2.8	575
8068	C₆H₁₄	Hexane	68.8	68.5	18	550

No.	Formula	B-Component Name	B.P., °C	Azeotropic Data B.P., °C	Wt.%A		Ref.
A =	C$_4$H$_9$NO$_2$	**Butyl Nitrite** *(continued)*	**78.2**				
8069	C$_6$H$_{14}$O	Propyl ether	90.1	Nonazeotrope			550
8070	C$_7$H$_{14}$	Methylcyclohexane	101.15	Nonazeotrope			550
8071	C$_7$H$_{16}$	Heptane	98.4	Nonazeotrope			527
A =	C$_4$H$_9$NO$_2$	**Isobutyl Nitrite**	**67.1**				
8072	C$_4$H$_{10}$O$_2$	Acetaldehyde dimethyl acetal	64.3	Nonazeotrope			550
8073	C$_5$H$_{10}$	Cyclopentane	49.3	Nonazeotrope			550
8074	C$_5$H$_{12}$	Pentane	36.15	Nonazeotrope			550
8075	C$_5$H$_{12}$O	Ethyl propyl ether	63.85	< 63.7	5		550
8076	C$_6$H$_6$	Benzene	80.15	Nonazeotrope			550
8077	C$_6$H$_{12}$	Cyclohexane	80.75	Nonazeotrope			550
8078	C$_6$H$_{12}$	Methylcyclopentane	72.0	65.9	68		570
8079	C$_6$H$_{14}$	Hexane	68.8	65.0	54		527
A =	C$_4$H$_9$NO$_3$	**Isobutyl Nitrate**	**123.5**				
8080	C$_4$H$_{10}$O	Butyl alcohol	117.8	112.8	45		527
8081	C$_4$H$_{10}$O	Isobutyl alcohol	107.85	105.6	36		560
8082	C$_4$H$_{10}$O$_2$	2-Ethoxyethanol	135.3	121.0	82		560
8083	C$_5$H$_{10}$O	Cyclopentanol	140.85	< 122.2			560
8084	C$_5$H$_{10}$O$_3$	Ethyl carbonate	126.5	Nonazeotrope			549
8085	C$_5$H$_{10}$O$_3$	2-Methoxyethyl acetate	144.6	Nonazeotrope			560
8086	C$_5$H$_{11}$Br	1-Bromo-3-methylbutane	120.65	118.0	32		560
8087	C$_5$H$_{12}$O	Amyl alcohol	138.2	122.0			560
8088	C$_5$H$_{12}$O	Isoamyl alcohol	131.3	~ 120.0	~ 74		560
8089	C$_5$H$_{12}$O	2-Pentanol	119.8	< 115.3	< 48		560
8090	C$_5$H$_{12}$O$_2$	2-Propoxyethanol	151.35	Nonazeotrope			560
8091	C$_6$H$_5$Cl	Chlorobenzene	131.75	Nonazeotrope			560
8092	C$_6$H$_{12}$O$_2$	Isoamyl formate	123.8	< 122.0	>54		549
8093	C$_6$H$_{12}$O$_2$	Propyl propionate	123.0	<121.7	>41		549
8094	C$_6$H$_{12}$O$_3$	Paraldehyde	124.35	< 122.8			557
8095	C$_6$H$_{14}$S	Propyl sulfide	141.5	Nonazeotrope			560
8096	C$_7$H$_8$	Toluene	110.75	Nonazeotrope			560
8097	C$_8$H$_{10}$	Ethylbenzene	136.15	Nonazeotrope			560
8098	C$_8$H$_{16}$	1,3-Dimethylcyclohexane	120.7	< 114.5	<41		560
8099	C$_8$H$_{18}$O	Isobutyl ether	122.3	< 121.0			557
A =	C$_4$H$_9$NO$_3$	**2-Methyl-2-nitro-1-Propanol**					
8100	C$_6$H$_{11}$NO$_3$	2-Methyl-2-nitro-propyl vinyl ether, 10 mm.		71-81	8.6		1008
8100a	C$_4$H$_{10}$	2-Methylpropane	—10	Nonazeotrope		v-l	386f
A =	C$_4$H$_{10}$	**Butane**	**— 0.5**				
8101	C$_7$H$_{16}$	Perfluoroheptane crit. press.		147.8	47.8	v-l	431
A =	C$_4$H$_{10}$O	**Butyl Alcohol**	**117.75**				
8102	C$_4$H$_{10}$O	*sec*-Butyl alcohol	99.5	Nonazeotrope			981
		" 100–700 mm.		Nonazeotrope		v-l	773c
8102a	C$_4$H$_{10}$O	*tert*-Butyl alcohol 100–700 mm.		Nonazeotrope		v-l	773c

TABLE I.   *Binary Systems*   245

No.	Formula	B-Component Name	B.P., °C	Azeotropic Data B.P., °C	Wt.%A		Ref.
A =	C₄H₁₀O	**Butyl Alcohol** *(continued)*	**117.75**				
8103	C₄H₁₀O	Isobutyl alcohol, to crit. region		Nonazeotrope		v-l	213,981
		" 750 mm.	107	Nonazeotrope		v-l	982
8104	C₄H₁₀O	Ethyl ether, to crit. region	34.5	Nonazeotrope		v-l	213
8105	C₄H₁₀O₂	2-Ethoxyethanol	135.3	Nonazeotrope			526
8106	C₄H₁₀S	1-Butanethiol, 770 mm.	98	97.8	14.84		487
8107	C₄H₁₀S	Ethyl sulfide	92.1	Nonazeotrope			527
8108	C₄H₁₁N	Butylamine	77.1	Nonazeotrope		v-l	481e
8109	C₅H₅N	Pyridine	115.4	118.7	71		553
		"	115.5	118.6	69	v-l	393
		"		118.85	69.0		497
8110	C₅H₇N	*N*-Methylpyrrol	112.8	< 112.2			575
8111	C₅H₈O₂	Methyl methacrylate	99.8	Nonazeotrope			413
8112	C₅H₉ClO₂	Propyl chloroacetate	163.5	Nonazeotrope			575
8113	C₅H₉N	Valeronitrile	141.3	Nonazeotrope			565
8114	C₅H₁₀	2-Methyl-2-butene	37.75	Nonazeotrope			256
8115	C₅H₁₀O	3-Methyl-2-butanone	95.4	Nonazeotrope			552
8116	C₅H₁₀O	2-Pentanone	102.35	Nonazeotrope			527
8117	C₅H₁₀O	3-Pentanone	102.05	Nonazeotrope			527
8118	C₅H₁₀O₂	Butyl formate	106.6	105.8	23.6		359
8119	C₅H₁₀O₂	Ethyl propionate	99.1	Nonazeotrope			527
8120	C₅H₁₀O₂	Isobutyl formate	97.9	Nonazeotrope			527
8121	C₅H₁₀O₂	Isopropyl acetate	89.5	Nonazeotrope			575
8122	C₅H₁₀O₂	Methyl butyrate	102.75	Nonazeotrope			527
8123	C₅H₁₀O₂	Methyl isobutyrate	92.3	Nonazeotrope			527
8124	C₅H₁₀O₂	Propyl acetate	101.6	Nonazeotrope			527
8125	C₅H₁₀O₃	Ethyl carbonate	125.9	116.5	63		527
8126	C₅H₁₀O₃	2-Methoxyethyl acetate	144.6	Nonazeotrope			526
8127	C₅H₁₁Br	1-Bromo-3-methylbutane	120.3	110.65	31.5		555
8128	C₅H₁₁Cl	1-Chloro-3-methylbutane	99.4	97.0	12		567
8129	C₅H₁₁I	1-Iodo-3-methylbutane	147.65	117.3	~ 78		535
8130	C₅H₁₂O	2-Pentanol	119.8	Nonazeotrope			575
8131	C₅H₁₂O	3-Pentanol	116.0	Nonazeotrope			575
8132	C₆H₅Br	Bromobenzene	156.1	Nonazeotrope			527
8133	C₆H₅Cl	Chlorobenzene	132.0	115.3	56		574
8134	C₆H₅F	Fluorobenzene	84.9	Nonazeotrope			575
8135	C₆H₅NO₂	Nitrobenzene	210.75	Nonazeotrope			554
8136	C₆H₆	Benzene	80.2	Nonazeotrope			563
		" 45°		Nonazeotrope		v-l	93
		"	80.1	Nonazeotrope		v-l	1011, 1060
		" crit. pt.		min. b.p.		v-l	888
		" 685 mm.				v-l	994
		" 1445 mm.				v-l	994
		" 2205 mm.				v-l	994

No.	Formula	B-Component Name	B.P., °C	Azeotropic Data B.P., °C	Wt.%A		Ref.
A =	C₄H₁₀O	**Butyl Alcohol** *(continued)*	**117.75**				
8137	C₆H₆O	Phenol	182.2	Nonazeotrope			575
8138	C₆H₇N	Aniline	184.35	Nonazeotrope			551
8139	C₆H₇N	2-Picoline	130.7	Nonazeotrope			575
8140	C₆H₈	1,3-Cyclohexadiene	80.8	Nonazeotrope			563
8141	C₆H₉N	*N*-Ethylpyrrol	130.4	Nonazeotrope			575
8142	C₆H₁₀	Cyclohexene	82.7	82.0	5		537
8143	C₆H₁₀O	Mesityl oxide	129.45	Nonazeotrope			527,552
8144	C₆H₁₀S	Allyl sulfide	130.35	Nonazeotrope			527
8145	C₆H₁₁ClO₂	Butyl chloroacetate	181.9	Nonazeotrope			121
8146	C₆H₁₂	Cyclohexane	80.75	79.8	4		537
		"	80.8	79.8	9.5		413,982
		" 762 mm.		80	9.8	v-l	779c
		" 1024 mm.		90	11	v-l	779c
		" 1342 mm.		100	11.5	v-l	779c
		" 1720 mm.		110	13.5	v-l	779c
8147	C₆H₁₂	Methylcyclopentane	72.0	71.8	< 8		575
8148	C₆H₁₂O	Butyl vinyl ether	93.8	93.3	7.75		882
			93.8	Nonazeotrope?			254,882
		"	94.2	93.3	7.8		982
8149	C₆H₁₂O	Hexaldehyde	128.3	116.8	77.1		982
8150	C₆H₁₂O	2-Hexanone	127.2	116.5	81.8		918
		"	127.2	Nonazeotrope			527
8151	C₆H₁₂O	3-Hexanone	123.3	117.2	80		527
8152	C₆H₁₂O	4-Methyl-2-pentanone	116.05	114.35	30		552
8153	C₆H₁₂O₂	Butyl acetate	125.5	116.2	63.3	v-l	96,359, 527
		" 50 mm.			27.3		322
		"	126.2	117.6	67.2		322,982
8154	C₆H₁₂O₂	Ethyl butyrate	120.0	115.7	~ 64		536
8155	C₆H₁₂O₂	Ethyl isobutyrate	110.1	109.2	17		537
8156	C₆H₁₂O₂	Isoamyl formate	123.8	115.9	69		536
8157	C₆H₁₂O₂	Isobutyl acetate	117.2	114.5	50		536
8158	C₆H₁₂O₂	Methyl isovalerate	116.3	113.5	40		537
8159	C₆H₁₂O₂	Propyl propionate	123.0	117.5			575
8160	C₆H₁₂O₃	2-Ethoxyethyl acetate	156.8	Nonazeotrope			575
8161	C₆H₁₂O₃	Paraldehyde	123.9	115.75	52		527
8162	C₆H₁₃Br	1-Bromohexane	156.5	Nonazeotrope			575
8163	C₆H₁₄	Hexane	68.85	Nonazeotrope			541
		"	68.95	68.2	3.2		477,982
8164	C₆H₁₄O	Propyl ether	90.4	Nonazeotrope			527
8165	C₆H₁₄O₂	Acetal	103.55	101	13		573
8166	C₆H₁₄O₂	2-Butoxyethanol	171.1	Nonazeotrope		v-l	982
8167	C₆H₁₄S	Isopropyl sulfide	120.5	112.0	45		555
8168	C₆H₁₄S	Propyl sulfide	141.5	Nonazeotrope			566
8169	C₆H₁₅BO₃	Ethyl borate	118.6	113	52		536
8170	C₇H₈	Toluene	110.7	105.5	32		50,527,589
		"		0.5	5.6		814
				25	6.0		814

TABLE I. *Binary Systems*   247

No.	Formula	B-Component Name	B.P., °C	B.P., °C	Azeotropic Data Wt.%A		Ref.
A =	C₄H₁₀O	**Butyl Alcohol** *(continued)*	**117.75**				
		Toluene		50	7.1		814
		"		73	11.5		814
		"		103.1	28.1		814
		"	110.7	105.5	27.5	v-l	393
		" 200 mm.		66.8	17.7	v-l	348
		" 400 mm.		85.45	22.9	v-l	348
		" 470 mm.		90.0	21.2	v-l	859f
		" 600 mm.		97.7	26.5	v-l	348
		" 760 mm.		105.3	29.7	v-l	348
		" 760 mm.	110.7	105.5	27.8	v-l	859f
		" 409 mm.		86.0	22.6		662
		" 510 mm.		92.5	24.0		662
		" 610 mm.		98.0	25.8		662
		" 710 mm.		102.5	27.1		662
		"	110.6	105.1	28.0		662
		"	110.6	105.5	27.6	v-l	619
		"		105.7	28.6		497
8171	C₇H₈O	Anisole	153.85	Nonazeotrope			527
8172	C₇H₁₂O₂	Butyl acrylate,					
		" 100 mm.	69.77	69	75		981
		" 20 mm.		39	87.7		215
		" 150 mm.		77	92.2		215
		"	147	117	98.2		215
		"			85		999
8173	C₇H₁₄	1-Heptene, 729 mm.		90	13		757
8174	C₇H₁₄	Methylcyclohexane	100.8	95.3	20		50,571
		"	100.8	96.5	20.6	v-l	778f
		"	100.8	96.2	20	v-l	859f
		" 626 mm.		90	17.1		859f
8175	C₇H₁₄O₂	Butyl propionate	146.8	Nonazeotrope			575
8176	C₇H₁₄O₂	Ethyl isovalerate	134.7	Nonazeotrope			527
8177	C₇H₁₄O₂	Isoamyl acetate	142.1	Nonazeotrope			527
8178	C₇H₁₄O₂	Isobutyl propionate	137.5	Nonazeotrope			575
8179	C₇H₁₄O₂	Isopropyl isobutyrate	120.8	115.5	54		567
8180	C₇H₁₄O₂	Propyl butyrate	143.7	Nonazeotrope			575
8181	C₇H₁₄O₂	Propyl isobutyrate	133.9	Nonazeotrope			527
8182	C₇H₁₆	Heptane	98.45	93.95	18		527
		"	98.4	93.85	18		477,478
		"	98.4	~ 94	~ 16	v-l	393
		"	98.5	93.8	17.6	v-l	27
		" 153 mm.		50	10.3	v-l	27
		" 684 mm.		88.8	15.0	v-l	993
		" 1445 mm.	122	114.8	23.4	v-l	993
		" 2205 mm.	139	129.8	29.8	v-l	993

No.	Formula	B-Component Name	B.P., °C	B.P., °C	Wt.%A		Ref.
A =	C₄H₁₀O	**Butyl Alcohol** *(continued)*	**117.75**				
		" 2965 mm.	153	142.4	32.8	v-l	993
		" 3725 mm.	166	150.5	35.8	v-l	993
8183	C₇H₁₈SiO	(Trimethylsiloxy) butane		Azeotropic			520
		"	124.5	111.0	40-44		839,907
8184	C₈H₈	Styrene	145.8	~ 116.5	79		537
		60 mm.	68	57	59		53
8185	C₈H₁₀	Ethylbenzene	136.15	114.8	~ 67		537
		60 mm.	60.5	53	37		53
		" 50 mm.			36.3	v-l	239
		" 100 mm.		63.65	42.1	v-l	239
		" 300 mm.			51.0	v-l	239
		" 500 mm.			59.7	v-l	239
		" 760 mm.	136.15	115.85	65.1	v-l	239
		"		115.05	67.1	v-l	587
8186	C₈H₁₀	Xylene		20	29.6		814
		"		40	38.4		814
		"		60	47.5		814
		"		80	56.5		814
		"		115	73.0		814
8187	C₈H₁₀	*m*-Xylene	**139.**	116.5	71.5		527
8188	C₈H₁₀	*o*-Xylene	143.6	116.8	75		541
8189	C₈H₁₀	*p*-Xylene	138.3	116.0	69.5	v-l	308c,537
8189	C₈H₁₀	*p*-Xylene	138.3	115.7	68		537
8190	C₈H₁₄O₂	Butyl methacrylate			~ 80		999
8191	C₈H₁₆	1,3-Dimethylcyclohexane	120.7	108.5	43		567
8192	C₈H₁₆O₂	Butyl butyrate	166	Nonazeotrope		v-l	722
8193	C₈H₁₈	2,5-Dimethylhexane	109.4	101.9	28		567
8194	C₈H₁₈	Octane	125.75	110.2	50		567
		"	125.75	108.45	43.2		477,478
8195	C₈H₁₈O	Butyl ether	142.1	117.6	82.5		982
		Butyl ether	141.9	117.25	88		760
		"	142.4	Nonazeotrope			527
8196	C₈H₁₈O	Isobutyl ether	122.3	113.5	48		527
8197	C₈H₁₉N	Dibutylamine	159.6	Nonazeotrope		v-l	481e
8197	C₈H₁₉N	Dibutylamine	159.6	Nonazeotrope			981
8198	C₉H₈	Indene	182.6	Nonazeotrope			575
8199	C₉H₁₂	Cumene	152.8	Nonazeotrope			527
8200	C₉H₁₂	Mesitylene	164.6	Nonazeotrope			541
8201	C₉H₁₂	Propylbenzene	158.8	Nonazeotrope			537
8202	C₉H₁₂	Pseudocumene	168.2	Nonazeotrope			575
8203	C₉H₂₀	Nonane	150.7	115.9	71.5		477,478
8204	C₉H₂₀O₂	Diisobutoxymethane	163.8	Nonazeotrope			575
8205	C₉H₂₀O₂	Dibutoxymethane	181.8	Nonazeotrope			324
8206	C₁₀H₁₄	Butylbenzene	183.1	Nonazeotrope			575
8207	C₁₀H₁₄	Cymene	176.7	Nonazeotrope			537
8208	C₁₀H₁₆	Camphene	159.6	117.73?	98		574

TABLE I. *Binary Systems* 249

No.	Formula	Name	B.P., °C	B.P., °C	Wt.%A		Ref.
		B-Component		**Azeotropic Data**			
A =	C₄H₁₀O	**Butyl Alcohol** *(continued)*	117.75				
8209	C₁₀H₁₆	*d*-Limonene	177.8	Nonazeotrope			537
8210	C₁₀H₁₆	Nopinene	163.8	Nonazeotrope			527
8211	C₁₀H₁₆	α-Pinene	155.8	117.4	~ 88		537
8212	C₁₀H₁₆	Thymene	179.7	Nonazeotrope			541
8213	C₁₀H₂₂	Decane	173.3	Nonazeotrope			575
		" 386 mm.		100	52	v-l	577c
8214	C₁₀H₂₂	2,7-Dimethyloctane	160.2	Nonazeotrope			537
8215	C₁₀H₂₂O₂	1,1-Dibutoxyethane	187.8	Nonazeotrope		v-l	45,170
A =	C₄H₁₀O	*sec*-**Butyl Alcohol**	99.5				
8216	C₄H₁₀S	Ethyl sulfide	92.1	< 89.0	<32		566
8217	C₅H₅N	Pyridine	115.4	Nonazeotrope			575
8218	C₅H₁₀O	3-Pentanone	102.05	98.0	58		534
8219	C₅H₁₀O₂	Butyl formate	106.8	98.0	68		567
8220	C₅H₁₀O₂	Ethyl propionate	99.15	95.7	47		536
8221	C₅H₁₀O₂	Isobutyl formate	98.2	94.7	40		575
8222	C₅H₁₀O₂	Methyl butyrate	102.65	< 97.7	<59		575
8223	C₅H₁₀O₂	Methyl isobutyrate	92.5	< 92.0	<23		575
8224	C₅H₁₀O₂	Propyl acetate	101.55	~ 96.5	~ 52		563
8225	C₅H₁₁Cl	1-Chloro-3-methylbutane	99.4	91.5	29		569
8226	C₅H₁₁I	1-Iodo-3-methylbutane	147.65	Nonazeotrope			575
8227	C₅H₁₂	2-Methylbutane	27.95	Nonazeotrope			575
8228	C₅H₁₂	Pentane	36.15	Nonazeotrope			537
8229	C₅H₁₂O	*tert*-Amyl alcohol	102.35	Nonazeotrope			575
8230	C₆H₅Cl	Chlorobenzene	131.75	Nonazeotrope			575
8231	C₆H₅NO₂	Nitrobenzene	210.75	Nonazeotrope			554
8232	C₆H₆	Benzene	80.2	78.5	15.4,	v-l	537,734
8233	C₆H₁₀	Cyclohexene	82.7	78.7	21		537
8234	C₆H₁₂	Cyclohexane	80.75	76.0	18		541
8234a	C₆H₁₂	1-Hexene		Nonazeotrope		v-l	362c
8235	C₆H₁₂	Methylcyclopentane	72.0	69.7	11.5		567
8236	C₆H₁₂O	Pinacolone	106.2	99.1	84		552
8237	C₆H₁₂O₂	Butyl acetate	126.1	Nonazeotrope			981
8238	C₆H₁₂O₂	*sec*-Butyl acetate	112.2	Nonazeotrope			981
		"	112.2	99.6	86.3		164
8239	C₆H₁₂O₂	Isobutyl acetate	117.4	Nonazeotrope			575
8240	C₆H₁₂O₂	Methyl isovalerate	116.5	Nonazeotrope			575
8241	C₆H₁₄	2,3-Dimethylbutane	58.0	< 57.75	< 8		575
8242	C₆H₁₄	Hexane 597 mm.		60	8	v-l	362c
		"	68.9	67.2	8		537
8243	C₆H₁₄O	*tert*-Amyl methyl ether	86.7	86.0	7		256
8244	C₆H₁₄O	*tert*-Butyl ethyl ether	73	Nonazeotrope			256
8245	C₆H₁₄O	Propyl ether	90.4	87.0	22		576
8246	C₇H₈	Toluene	110.7	95.3	55		50,537
8247	C₇H₁₄	Methylcyclohexane	100.8	89.9	41		50
		"	101.5	89.7	38.2	v-l	1039
8248	C₇H₁₆	Heptane	98.4	88.1	36.7	v-l	1039
		"	98.45	89	38		537

No.	Formula	Name		B.P., °C	B.P., °C	Wt.%A		Ref.
		B-Component			**Azeotropic Data**			
A =	**C₄H₁₀O**	*sec*-**Butyl Alcohol** *(continued)*		**99.5**				
8249	C₇H₁₆O	*tert*-Amyl ethyl ether		101-2	94.5	39		256
8250	C₈H₈	Styrene,	60 mm.	68	45	96		53
8251	C₈H₁₀	Ethylbenzene		136.15	Nonazeotrope			575
		"	60 mm.	60.5	44	84		53
8252	C₈H₁₀	*m*-Xylene		139.2	Nonazeotrope			575
8253	C₈H₁₄	Diisobutylene		102.3	91	35		132
8254	C₈H₁₈	Iso-octane		99.3	88.0	33.8	v-l	1039
8255	C₈H₁₈	2,5-Dimethylhexane		109.4	93.0	54		567
A =	**C₄H₁₀O**	*tert*-**Butyl Alcohol**		**82.9**				
8256	C₄H₁₀O	Isobutyl alcohol		108	Nonazeotrope			215
8257	C₄H₁₀S	Ethyl sulfide		92.1	79.8	70		566
8258	C₄H₁₁PO₃	Diethyl phosphite			Nonazeotrope			24
8258a	C₅H₈	Isoprene		34	Nonazeotrope			581c
8259	C₅H₁₀	Cyclopentane		49.4	48.2	~ 7		575
8260	C₅H₁₀	2-Methyl-2-butene		37.15	Nonazeotrope			563
8260a	C₅H₁₀O	2-Methyl-3-buten-2-ol		97.0	Nonazeotrope			581c
8261	C₅H₁₀O	3-Pentanone		102.05	Nonazeotrope			552
8262	C₅H₁₀O₂	Isobutyl formate		97.9	Nonazeotrope			536
8263	C₅H₁₀O₂	Methyl isobutyrate		92.3	82.2			536
8264	C₅H₁₀O₂	Propyl acetate		101.6	Nonazeotrope			575
8265	C₅H₁₁Br	1-Bromo-3-methylbutane		120.65	Nonazeotrope			575
8266	C₅H₁₁Cl	1-Chloro-3-methylbutane		99.4	< 81.15	>59		567
8267	C₅H₁₂	2-Methylbutane		27.95	Nonazeotrope			537
8268	C₅H₁₂	*n*-Pentane		36.15	35.9	3		575
		"		36	Nonazeotrope			581c
8268a	C₅H₁₂O	*tert*-Butyl methyl ether		55.2	Nonazeotrope			581c
8269	C₆H₅F	Fluorobenzene		85.15	76.0	31		545
8270	C₆H₆	Benzene		80.2	73.95	36.6		1044
		"		80.1	72.6	35.4	v-l	792
8271	C₆H₈	1,3-Cyclohexadiene		80.8	73.4	38.5		563
8272	C₆H₁₀	Cyclohexene		82.7	73.2	40		537
8273	C₆H₁₀	Methylcyclopentene		75.85	69.5	30		567
8273a	C₆H₁₀O	Methyldihydropyran		118.5	Nonazeotrope			581c
8274	C₆H₁₂	Cyclohexane		80.75	71.3	37		541
		"		80.7	71.2	34.2	v-l	792
8275	C₆H₁₂	Methylcyclopentane		72.0	66.6	26		567
8276	C₆H₁₄	2,3-Dimethylbutane		58.0	55.3	13		567
8277	C₆H₁₄	Hexane		68.85	63.7	22		541
8277a	C₆H₁₄O	Isopropyl ether		68.3	67.3	7.9		581c
8278	C₆H₁₄O	Propyl ether		90.4	79.0	52		576
8279	C₇H₈	Toluene		110.7	Nonazeotrope			50
8280	C₇H₁₄	Methycyclohexane		100.8	78.8	66		50
8281	C₇H₁₆	Heptane		98.45	78	62		537
8282	C₈H₈	Styrene,	60 mm.	68	Nonazeotrope			53
8283	C₈H₁₀	Ethylbenzene,	60 mm.	60.5	28	95		53
8284	C₈H₁₀	*p*-Xylene		138.45	Nonazeotrope			575

TABLE I. *Binary Systems* 251

No.	Formula	Name	B.P., °C	B.P., °C	Wt.%A	Ref.
		B-Component		**Azeotropic Data**		
A =	**C₄H₁₀O**	*tert*-**Butyl Alcohol**	**82.9**			
		(continued)				
8285	C₈H₁₆	1,3-Dimethylcyclohexane	120.7	< 82.2	>90	575
8286	C₈H₁₈	2,5-Dimethylhexane	109.2	81.5	77	545
8287	C₁₀H₁₆	α-Pinene	155.8	Nonazeotrope		537
A =	**C₄H₁₀O**	**Ethyl Ether**	**34.6**			
8288	C₄H₁₀O	Methyl propyl ether	38.9	Nonazeotrope		563
8289	C₄H₁₁N	Diethylamine	55.9	Nonazeotrope		551
8289a	C₄H₁₂Ge	Tetramethylgermane	43.5	34		580c
8290	C₅H₈	Isoprene	34.3	33.2	48	558
8291	C₅H₈	3-Methyl-1,2-butadiene	40.8	Nonazeotrope		563
8292	C₅H₁₀	Cyclopentane	49.3	Nonazeotrope		558
8293	C₅H₁₀	2-Methyl-2-butene	37.1	34.2	85	558
8294	C₅H₁₀	3-Methyl-1-butene	20.6	Nonazeotrope		558
8295	C₅H₁₂	2-Methylbutane	27.95	Nonazeotrope		563
8296	C₅H₁₂	Pentane	36.15	33.4	68	558
		"		33.7	56	497
8297	C₆H₅NO₂	Nitrobenzene	210.75	Nonazeotrope		554
8293	C₆H₆	Benzene	80.2	Nonazeotrope		558
8294	C₆H₁₀	Biallyl	60.1	Nonazeotrope		558
8300	C₆H₁₄	2,3-Dimethylbutane	58.0	Nonazeotrope		558
8301	C₆H₁₄	Hexane	68.85	Nonazeotrope		558
8302	C₆H₁₄O	Hexyl alcohol	155.8	Nonazeotrope		215
8303	C₆H₁₅N	Triethylamine	89.35	Nonazeotrope		551
8304	C₇H₈	Toluene	110.75	Nonazeotrope		558
8305	C₈H₁₈O	Butyl ether, 600 mm.		Ideal system	v-l	737
A =	**C₄H₁₀O**	**Isobutyl Alcohol**	**108.0**			
8306	C₄H₁₀O₂	2-Ethoxyethanol	135.3	Nonazeotrope		575
8307	C₅H₅N	Pyridine	115.4	Nonazeotrope		553
8308	C₅H₇N	N-Methylpyrrol	112.8	<107.5		575
8309	C₅H₉ClO₂	Propyl chloroacetate	163.5	Nonazeotrope		575
8310	C₅H₁₀	Cyclopentane	49.4	Nonazeotrope		575
8311	C₅H₁₀O	Isovaleraldehyde	92.5	Nonazeotrope	v-l	665
8312	C₅H₁₀O	3-Methyl-2-butanone	95.4	Nonazeotrope		552
8313	C₅H₁₀O	2-Pentanone	102.35	101.8	19	552
8314	C₅H₁₀O	3-Pentanone	102.05	101.7	20	552
8315	C₅H₁₀O₂	Butyl formate	106.8	103.0	40	567
8316	C₅H₁₀O₂	Ethyl propionate	99.1	< 98.9	13	575
8317	C₅H₁₀O₂	Isobutyl formate	98.3	Nonazeotrope		1033
		"	98.3	97.8	16.9	498i
		"	98.4	97.8	20.6	359,536
8318	C₅H₁₀O₂	Isopropyl acetate	89.5	Nonazeotrope		575
8319	C₅H₁₀O₂	Methyl butyrate	102.65	101.3	25	536
8320	C₅H₁₀O₂	Methyl isobutyrate	92.3	Nonazeotrope		536
8321	C₅H₁₀O₂	Propyl acetate	101.6	101.0	17	572
8322	C₅H₁₀O₃	Ethyl carbonate	125.9	Nonazeotrope		536
8323	C₅H₁₀O₃	2-Methoxyethyl acetate	144.6	Nonazeotrope		575
8324	C₅H₁₁Br	1-Bromo-3-methylbutane	118.1	103.4	63.6	388,555
				B.p. curve		
8325	C₅H₁₁Cl	1-Chloro-3-methylbutane	99.8	94.5	22	573

		B-Component		Azeotropic Data		
No.	Formula	Name	B.P., °C	B.P., °C	Wt.%A	Ref.
A =	C₄H₁₀O	**Isobutyl Alcohol** *(continued)*	**108.0**			
8326	C₅H₁₁I	1-Iodo-3-methylbutane	146.5	Nonazeotrope,		
				b.p. curve		388
8327	C₅H₁₂	*n*-Pentane	36.15	Nonazeotrope		575
8328	C₅H₁₂O	Isoamyl alcohol	131.9	Nonazeotrope		575
8329	C₅H₁₂O₂	Diethoxymethane	87.95	Nonazeotrope		527
8330	C₆H₅Br	Bromobenzene	156.1	Nonazeotrope		532
8331	C₆H₅Cl	Chlorobenzene	132.0	107.1	63	532
8332	C₆H₅F	Fluorobenzene	84.9	84.0	9	575
8333	C₆H₆	Benzene	80.2	79.84	9.3	1044
		"	80.1	78.36	12	306
		" 111 mm.		28.4	2.7	924
		" 240 mm.		45.0	4.2	924
		" 525 mm.		67.4	6.4	924
		" 760 mm.	80.1	79.3	7.4	924
		" 206 mm.		43.0	4.2	735
		" 394 mm.		59.5	6.0	735
		" 759 mm.	80.1	79.4	7.9	735
		" 5420 mm.		159.9	21.0	735
		" 12,930 mm.		207.5	33	735
		"		Nonazeotrope		834
		"		79.8	9 v-l	686c
8334	C₆H₈	1,3-Cyclohexadiene	80.8	79.35	12	563
8335	C₆H₁₀	Cyclohexene	82.7	80.5	14.2	541
8336	C₆H₁₀S	Allyl sulfide	139.35	Nonazeotrope		566
8337	C₆H₁₁ClO₂	Isobutyl chloroacetate	174.4	Nonazeotrope		121
8338	C₆H₁₂	Cyclohexane	80.75	78.1	14	541
		"	80.75	78.3	14 v-l	686c
8339	C₆H₁₂	Methylcyclopentane	72.0	71.0	5	575
8340	C₆H₁₂O	2-Hexanone	127.2	Nonazeotrope		552
8341	C₆H₁₂O	3-Hexanone	123.3	Nonazeotrope		552
8342	C₆H₁₂O	Isobutyl vinyl ether	83.0	82.7	6.2	882
8343	C₆H₁₂O	4-Methyl-2-pentanone	116.05	107.85	91	552
8344	C₆H₁₂O	Pinacolone	106.2	< 105.5	<42	548
8345	C₆H₁₂O₂	Butyl acetate	126.0	Nonazeotrope		527
8346	C₆H₁₂O₂	Ethyl butyrate	120.6	Nonazeotrope,		
				b.p. curve		389
8347	C₆H₁₂O₂	Ethyl isobutyrate	110.1	105.5	52	563
8348	C₆H₁₂O₂	Isoamyl formate	123.8	Nonazeotrope		536
8349	C₆H₁₂O₂	Isobutyl acetate	117.2	107.4	55	359
		"	116.3	Nonazeotrope		
				b.p. curve		389,572
8350	C₆H₁₂O₂	Methyl isovalerate	116.3	~ 107.5	~ 90	563
8351	C₆H₁₂O₂	Propyl propionate	123.0	Nonazeotrope		575
8352	C₆H₁₂O₃	2-Ethoxyethyl acetate	156.8	Nonazeotrope		575
8353	C₆H₁₃Br	1-Bromohexane	156.5	Nonazeotrope		575
8354	C₆H₁₄	Hexane	68.9	68.3	2.5	537
8355	C₆H₁₄O	Ethyl isobutyl ether 743 mm.	79	78	18.43	73

TABLE I. *Binary Systems* 253

No.	B-Component			Azeotropic Data		
	Formula	Name	B.P., °C	B.P., °C	Wt.%A	Ref.
A =	**C₄H₁₀O**	**Isobutyl Alcohol** *(continued)*	**108.0**			
8356	C₆H₁₄O	Propyl ether	90.55	89.5	10	556
8357	C₆H₁₄O₂	Acetal	103.55	98.2	20	573
8358	C₆H₁₄S	Isopropyl sulfide	100.5	105.8	73	555
8359	C₆H₁₄S	Propyl sulfide	141.5	Nonazeotrope		566
8360	C₆H₁₅BO₃	Ethyl borate	118.6	Nonazeotrope		530
8361	C₇H₈	Toluene	110.7	101.2	45	50,834,1051
		"	110.6	101.2	44	1038
		" 758 mm.	110.6	100.5	43.3	662
		" 709 mm.		98.0	42.4	662
		" 609 mm.		94.0	40.8	662
		" 509 mm.		89 0	38.2	662
		" 409 mm.		82.5	35.7	662
8362	C₇H₁₄	Methylcyclohexane	100.8	92.6	32	50
8363	C₇H₁₄O₂	Ethyl isovalerate	134.7	Nonazeotrope		575
8364	C₇H₁₄O₂	Isoamyl acetate	142.1	Nonazeotrope, b p. curve		389
8365	C₇H₁₄O₂	Isobutyl propionate	137.5	Nonazeotrope		575
8366	C₇H₁₄O₂	Isopropyl isobutyrate	120.8	Nonazeotrope		575
8367	C₇H₁₄O₂	Propyl isobutyrate	134.0	Nonazeotrope		575
8368	C₇H₁₆	Heptane	98.45	90.8	27	537
8369	C₇H₁₆O₂	Dipropoxymethane	137.2	Nonazeotrope		575
8370	C₈H₈	Styrene	145.8	Nonazeotrope		537
		" 60 mm.	68	49	75	53
8371	C₈H₁₀	Ethylbenzene	136.15	107.2	80	541
		" 60 mm.	60.5	48	61	53
8372	C₈H₁₀	*m*-Xylene	139	107.78	85.5	527
		"		Nonazeotrope		834
		"	139.1	107.2	90.1 v-l	312
		" 40 mm.		42 5	56.5	312
8373	C₈H₁₀	*o*-Xylene	144.4	Nonazeotrope	v-l	312
		" 40 mm.		42	67.6	312
		"	143.6	Nonazeotrope		537
8374	C₈H₁₀	*p*-Xylene	138.2	~ 107.5	~ 83	541
		"	138.4	107.1	88.6 v-l	312
		" 40 mm.		43.0	53.3	312
8375	C₈H₁₆	1,3-Dimethylcyclohexane	120.7	102.2	56	575
8376	C₈H₁₈	2.5-Dimethylhexane	109.2	98.7	42	545
8377	C₈H₁₈	Octane	125.8	104		563
8378	C₈H₁₈	2,2,4-Trimethylpentane	99.3	92.0	27	575
8379	C₈H₁₈O	Butyl ether	142.4	Nonazeotrope		575
8380	C₈H₁₈O	Isobutyl ether	122.3	107.8?		563
8381	C₉H₁₂	Cumene	152.8	Nonazeotrope		575
8382	C₉H₁₂	Mesitylene	164.6	Nonazeotrope		575
8383	C₉H₁₂	Propylbenzene	158.8	Nonazeotrope		537
8384	C₉H₂₀O₂	Diisobutoxymethane	163.8	Nonazeotrope		324
8385	C₁₀H₁₄	Cymene	176.7	Nonazeotrope		575

No.	Formula	Name	B.P., °C	B.P., °C	Wt.%A		Ref.
		B-Component		**Azeotropic Data**			
A =	**C₄H₁₀O**	**Isobutyl Alcohol** *(continued)*	**108.0**				
8386	C₁₀H₁₆	Camphene	159.6	Nonazeotrope			537
8387	C₁₀H₁₆	*d*-Limonene	177.8	Nonazeotrope			541
8388	C₁₀H₁₆	Nopinene	163.8	Nonazeotrope			575
8389	C₁₀H₁₆	α-Pinene	155.8	107.95	>99		528
8390	C₁₀H₁₆	Thymene	179.7	Nonazeotrope			537
8391	C₁₀H₂₂	2,7-Dimethyloctane	160.1	Nonazeotrope			575
8392	C₁₀H₂₂O₂	Acetaldehyde diisobutyl acetal	171.3	Nonazeotrope			45
A =	**C₄H₁₀O**	**Methyl Propyl Ether**	**38.95**				
8393	C₄H₁₁N	Diethylamine	55.9	Nonazeotrope			551
8394	C₅H₈	Isoprene	34.3	Nonazeotrope			558
8395	C₅H₁₀	2-Methyl-2-butene	37.15	36.3	25		558
8396	C₅H₁₂	Pentane	36.2	35.3	22		558
A =	**C₄H₁₀O₂**	**Acetaldehyde Dimethyl Acetal**	**64.3**				
8797	C₄H₁₁N	Diethylamine	55.9	Nonazeotrope			551
8398	C₆H₆	Benzene	80.15	Nonazeotrope			558
8399	C₆H₁₂	Methylcyclopentane	72.0	64.0	83		558
8400	C₆H₁₄	Hexane	68.8	64.0	70		558
A =	**C₄H₁₀O₂**	***l*-2,3-Butanediol**	**183-184**				
8401	C₈H₁₄O₄	*meso*-2,3-Butanediol diacetate	190-193	177.6	60.5	v-l	731
		" 500 mm.		164.6	55.5	v-l	731
		" 350 mm.		153.0	49.9	v-l	731
		" 250 mm.		143.5	46.6	v-l	731
A =	**C₄H₁₀O₂**	**2,3-Butylene Glycol**	**182**				
8402	C₈H₁₀	Xylene		135			1037
8403	C₉H₁₂	Cumene	152.4	146.8			1037
A =	**C₄H₁₀O₂**	**1,4-Butanediol**	**230**				
8404	C₆H₁₂O₂	4-Vinyloxybutanol		Min. b.p.			263
A =	**C₄H₁₀O₂**	**1,2-Dimethoxyethane**	**85.2**				
8405	C₆H₁₄O	Isopropyl ether	68.3	Nonazeotrope			981
A =	**C₄H₁₀O₂**	**2⁻Ethoxyethanol**	**135.3**				
8406	C₅H₄O₂	2-Furaldehyde	161.45	Nonazeotrope			527
8407	C₅H₅N	Pyridine	115.4	Nonazeotrope			553
8408	C₅H₇N	2-Methylpyrrol	147.5	Nonazeotrope			575
8409	C₅H₈O	Cyclopentanone	130.65	< 130.2	<27		552
8410	C₅H₈O₂	Methyl methacrylate	99.8	Nonazeotrope			413
8411	C₅H₉N	Valeronitrile	141.3	< 135.0			575
8412	C₅H₁₀O	Cyclopentanol	140.85	Nonazeotrope			526
8413	C₅H₁₀O₂	Propyl acetate	101.6	Nonazeotrope		v-l	657
8414	C₅H₁₀O₃	Ethyl lactate	154.1	Nonazeotrope			575
8415	C₅H₁₀O₃	2-Methoxyethyl acetate	144.6	Nonazeotrope			556
8416	C₅H₁₁Br	1-Bromo-3-methylbutane	120.65	118.0	~ 8		575
8417	C₅H₁₁I	1-Iodo-3-methylbutane	147.65	132.0	60?		526
8418	C₅H₁₁NO₃	Isoamyl nitrate	149.75	133.7	72		527

TABLE I. *Binary Systems* 255

No.	Formula	B-Component Name	B.P., °C	Azeotropic Data B.P., °C	Wt.%A		Ref.
A =	$C_4H_{10}O_2$	**2-Ethoxyethanol** *(continued)*	**135.3**				
8419	$C_5H_{12}O$	Amyl alcohol	138.2	Nonazeotrope			526
8420	$C_5H_{12}O$	Isoamyl alcohol	131.9	Nonazeotrope			527
8421	$C_5H_{12}O$	2-Pentanol	119.8	Nonazeotrope			575
8422	C_6H_5Br	Bromobenzene	156.1	135.22	86		556
8423	C_6H_5Cl	Chlorobenzene	131.75	127.15	32		527
8424	C_6H_5I	Iodobenzene	188.45	Nonazeotrope			575
8425	C_6H_6	Benzene	80.15	Nonazeotrope			575
8426	C_6H_6O	Phenol	182.2	Nonazeotrope			556
8427	C_6H_{10}	Cyclohexene	82.75	Nonazeotrope			526
8428	$C_6H_{10}O$	Mesityl oxide	129.45	128.9	18		527
8429	$C_6H_{11}N$	Capronitrile	163.9	Nonazeotrope			575
8430	C_6H_{12}	Cyclohexane	80.75	Nonazeotrope			575
8431	C_6H_{12}	Hexene	63.5	Nonazeotrope		v-l	933
8432	$C_6H_{12}O$	3-Hexanone	123.3	Nonazeotrope			552
8433	$C_6H_{12}O$	4-Methyl-2-pentanone	116.05	Nonazeotrope			552
8434	$C_6H_{12}O_2$	Butyl acetate	124.8	125.8	35.7		129
		"	126.2	125.7	13	v-l	658
8435	$C_6H_{12}O_2$	Ethyl butyrate	121.5	Nonazeotrope			575
8436	$C_6H_{12}O_2$	Isoamyl formate	123.8	Nonazeotrope			556
8437	$C_6H_{12}O_2$	Isobutyl acetate	117.4	Nonazeotrope			575
8438	$C_6H_{12}O_2$	Methyl isovalerate	116.5	Nonazeotrope			575
8439	$C_6H_{12}O_2$	Propyl propionate	123.0	Nonazeotrope			526
8440	$C_6H_{12}O_3$	2-Ethoxyethyl acetate	156.8	Nonazeotrope			556
8441	$C_6H_{12}O_3$	Paraldehyde	124.35	123.8	14		1048
8442	C_6H_{14}	Hexane	68.74	Nonazeotrope		v-l	933
8443	$C_6H_{14}N_2$	2,5-Dimethyl piperazine	164	Nonazeotrope			981
8444	$C_6H_{14}O_2$	1,2-Diethoxyethane	123	121.0	3.1		129
8445	$C_6H_{14}O_3$	2-(2-Ethoxyethoxy) ethanol	202.8	Nonazeotrope			981
8446	$C_6H_{14}S$	Propyl sulfide	140.8	130.2	52		555
8447	$C_6H_{15}NO$	2-Diethylaminoethanol	162.2	Nonazeotrope			552
8448	C_7H_7Cl	o-Chlorotoluene	159.2	Nonazeotrope			526
8449	C_7H_7Cl	p-Chlorotoluene	162.4	Nonazeotrope			556
8450	C_7H_8	Toluene	110.75	110.15	10.8		556
8451	C_7H_8O	Anisole	153.85	135.25	94		556
8452	C_7H_{14}	Methylcyclohexane	101.15	98.6	15		526
8453	$C_7H_{14}O$	5-Methyl-2-hexanone	144.2	Nonazeotrope			552
8454	$C_7H_{14}O_2$	Amyl acetate	148.8	Nonazeotrope			575
8455	$C_7H_{14}O_2$	Ethyl isovalerate	134.7	130.5	42		526
8456	$C_7H_{14}O_2$	Isoamyl acetate	142.1	133.8	70		526
8457	$C_7H_{14}O_2$	Isobutyl propionate	137.5	131.5	35		567
8458	$C_7H_{14}O_2$	Methyl caproate	149.8	Nonazeotrope			575
8459	$C_7H_{14}O_2$	Propyl butyrate	143.7	133.5	72		556
8460	$C_7H_{14}O_3$	1,3-Butanediol methyl ether acetate	171.75	Nonazeotrope			575
8461	C_7H_{16}	Heptane	98.4	96.5	14		556

No.	Formula	B-Component Name	B.P., °C	B.P., °C	Wt.%A		Ref.
A =	$C_4H_{10}O_2$	**2-Ethoxyethanol** *(continued)*	**135.3**				
8462	C_8H_8	Styrene	145.8	130.0	55		575,750
		" 50 mm.		59.8	42.5	v-l	294
		" 60 mm.				v-l	311c
8463	C_8H_{10}	Ethylbenzene,					
		50 mm.		53.9	27.6	v-l	294
		" 57 mm.	60.62	50.0	42 vol. %		826
		" 735 mm.	134.9	126.2	43.3	v-l	455
		"	136.15	128	45	v-l	673
		"	136.15	127.8	48		526,750
8464	C_8H_{10}	*m*-Xylene 60°C.				v-l	311c
		" 735 mm.	137.9	127.7	48.9	v-l	455
		"	139.2	128.85	51		527
8465	C_8H_{10}	*o*-Xylene 60 mm.				v-l	311c
		" 735 mm.	143.1	129.6	57.2	v-l	455
		"	144.3	130.8	55		526
8466	C_8H_{10}	*p*-Xylene 60°C.				v-l	311c
		"	138.45	128.6	50		556
		" 735 mm.	137.4	127.3	47.9	v-l	455
8467	$C_8H_{10}O$	Benzyl methyl ether	167.8	Nonazeotrope			526
8468	$C_8H_{10}O$	*p*-Methylanisole	177.05	Nonazeotrope			575
8469	$C_8H_{10}O$	Phenetole	170.45	Nonazeotrope			575
8470	C_8H_{16}	*trans*-1,2-Dimethyl- cyclohexane	123.42	115.6	27 vol. %		826
8471	C_8H_{16}	1,3-Dimethylcyclohexane	120.7	114.0	30		575,928
8472	C_8H_{16}	Ethylcyclohexane	131.8		37		928
		"	131.78	120.2	33 vol. %		826
8473	C_8H_{16}	*cis*-2-Octene	125.6	117.9	28 vol. %		826
8474	$C_8H_{16}O_2$	Propyl isovalerate	155.7	Nonazeotrope			575
8475	C_8H_{18}	2,5-Dimethylhexane	109.4		~ 16		928
		"	109.4	105.0	22.5		526
		"	109.10	105.1	16 vol. %		826
8476	C_8H_{18}	3,3-Dimethylhexane	111.97	107.1	17 vol. %		826
		"	111.9		~ 17		928
8477	C_8H_{18}	3-Ethyl-3-methylpentane			~ 24		928
		"	118.26	111.7	23 vol. %		826
8478	C_8H_{18}	Octane	125.75	122.5	33.6	v-l	673
		"	125.75	116.0	38		569
		"	125.75		~ 28		928
8479	$C_8H_{18}O$	Butyl ether	141	127.0	50		129
8480	$C_8H_{18}O$	Isobutyl ether	122.3	119.0	33		526
8481	C_9H_8	Indene	182.8	Nonazeotrope			575
8482	C_9H_{12}	Cumene	152.8	133.2	67		527
8483	C_9H_{12}	*o*-Ethyltoluene			~ 92		928
		"	165.15	135.0	91 vol. %		826
8484	C_9H_{12}	Mesitylene, 735 mm.	163.4	133.7	85.7	v-l	455
		"	164.6	Nonazeotrope			575
8485	C_9H_{12}	Propylbenzene	159.3	134.6	80		526

TABLE I. *Binary Systems*   257

No.	Formula	Name	B.P., °C	B.P., °C	Wt.%A	Ref.
		B-Component		**Azeotropic Data**		
A ⇌	**C₄H₁₀O₂**	**2-Ethoxyethanol** *(continued)*	**135.3**			
				~ 77		928
8486	C₉H₁₂	Pseudocumene	168.2	Nonazeotrope		575
8487	C₉H₁₈	Butylcyclopentane	156.56	130.3	61 vol. %	826
8488	C₉H₁₈	Isobutylcyclopentane	147.6	127.4	49 vol. %	826
8489	C₉H₁₈	Isopropylcyclohexane	154.5	129.5	56 vol. %	826
8490	C₉H₁₈	1-Nonene	146.87	128.1	48 vol. %	826
8491	C₉H₁₈	Propylcyclohexane	156.72	130.2	59 vol. %	826
8492	C₉H₂₀	3,3-Diethylpentane	146.17	126.4	45 vol. %	826
		"		~ 45		928
8493	C₉H₂₀	*n*-Nonane	**150.7**	~ 51		928
		"	150.8	128.0	50 vol. %	826
8494	C₉H₂₀	2,2,3,3-Tetramethyl-pentane	140.27	124.1	40 vol. %	826
		"		~ 39		928
8495	C₉H₂₀	2,2,4,4-Tetramethyl-pentane		~ 24		928
		"	122.28	114.3	26 vol. %	826
8496	C₉H₂₀	2,3,3,4-Tetramethyl-pentane	141.55	124.6	41 vol. %	826
		"		~ 42		928
8497	C₉H₂₀	2,2,3-Trimethylhexane	133.60	120.8	34 vol. %	826
8498	C₉H₂₀	2,2,4-Trimethylhexane	126.54	116.8	26 vol. %	826
8499	C₉H₂₀	2,3,3-Trimethylhexane	137.68	122.8	41 vol. %	826
8500	C₉H₂₀	2,3,5-Trimethylhexane	131.34	119.5	32 vol. %	826
8501	C₉H₂₀	2,4,4-Trimethylhexane	130.65	119.1	34 vol. %	826, 928
8502	C₉H₂₀	3,3,4-Trimethylhexane	140.46	124.0	40 vol. %	826
8503	C₁₀H₁₄	Cymene	176.7	Nonazeotrope		556
8504	C₁₀H₁₆	Camphene	159.6	131.0	65	526
8505	C₁₀H₁₆	Nopinene	163.8	< 133.0		575
8506	C₁₀H₁₆	α-Pinene	155.8	< 131.0	57	575
8507	C₁₀H₁₆	α-Terpinene	173.4	< 135.0	<87	575
8508	C₁₀H₁₈O	Cineol	176.35	Nonazeotrope		575
8509	C₁₀H₂₀	*tert*-Butylcyclohexane	171.5	133.3	73 vol. %	826
8510	C₁₀H₂₂	2,7-Dimethyloctane	160.2	130.8	63	556
8511	C₁₀H₂₂O	Isoamyl ether	173.2	Nonazeotrope		575
A =	**C₄H₁₀O₂**	**1-Methoxy-2-propanol**	**118**			
8512	C₇H₈	Toluene	110.7	106.5	30	215
A =	**C₄H₁₀O₃**	**Diethylene Glycol**	**245.5**			
8513	C₅H₁₂O₃	2-(2-Methoxyethoxy)ethanol	193.6	Nonazeotrope		981
8514	C₆H₄Br₂	*p*-Dibromobenzene	220.25	212.85	13	527
8515	C₆H₄ClNO₂	*m*-Chloronitrobenzene	235.5	228.2	32	554
8516	C₆H₄ClNO₂	*o*-Chloronitrobenzene	246.0	233.5	41	554
8517	C₆H₄ClNO₂	*p*-Chloronitrobenzene	239.1	229.5	34	527
8518	C₆H₅NO₂	Nitrobenzene	210.75	210.0	10	527

No.	Formula	B-Component Name	B.P., °C	B.P., °C	Wt.%A	Ref.
A =	C₄H₁₀O₃	**Diethylene Glycol**	**245.5**			
		(continued)				
8519	C₆H₅NO₃	o-Nitrophenol	217.2	216.0	10.5	527
8520	C₆H₆	Benzene	80.1	Nonazeotrope		981
8521	C₆H₆O₂	Pyrocatechol	245.9	259.5	46	569
8522	C₆H₈O₄	Methyl fumarate	193.25	Nonazeotrope		526
8523	C₆H₈O₄	Methyl maleate	204.05	Nonazeotrope		526
8524	C₆H₁₂O₃	2-(2-Vinyloxyethoxy) ethanol		Min. b.p.		263
8525	C₆H₁₄O₃	2-(2-Ethoxyethoxy) ethanol	202.8	Nonazeotrope		981
8526	C₆H₁₄O₄	Triethylene glycol, 3 mm.		Nonazeotrope	v-l	183
8527	C₇H₇BrO	o-Bromoanisole	217.7	211.0	25	575
8528	C₇H₇NO₂	m-Nitrotoluene	230.8	224.2	25	554
8529	C₇H₇NO₂	o-Nitrotoluene	221.75	218.2	71.5	527
8530	C₇H₇NO₂	p-Nitrotoluene	238.9	228.75	35	527
8531	C₇H₈	Toluene 130°		Nonazeotrope	v-l	88
		" 100°			v-l	883
		" 50°			v-l	883
8532	C₇H₈O	Benzyl alcohol	205.25	Nonazeotrope		526
8533	C₇H₈O	m-Cresol	202.4	Nonazeotrope	v-l	730
8534	C₇H₈O	p-Cresol	202.0	Nonazeotrope	v-l	730
8535	C₇H₁₂O₄	Ethyl malonate	199.35	Reacts		526
8536	C₇H₁₆O₄	2-[2-(2-Methoxyethoxy) ethoxy]-ethanol	245.25	245.0	22	527
8537	C₈H₇N	Indole	253	Azeo. distillation		330
8538	C₈H₈O	Acetophenone	202.0	Nonazeotrope		552
8539	C₈H₈O₂	Anisaldehyde	249.5	< 244		575
8540	C₈H₈O₂	Benzyl formate	202.3	Nonazeotrope		526
8541	C₈H₈O₂	Methyl benzoate	199.4	Nonazeotrope		526
8542	C₈H₈O₂	Phenyl acetate	195.7	Nonazeotrope		575
8543	C₈H₈O₃	Methyl salicylate	222.95	220.55	16	527
8544	C₈H₉BrO	p-Bromophenetole	234.2	222.0	32	575
8545	C₈H₁₀	Ethylbenzene	136.15	Azeo. distillation		330
8546	C₈H₁₀	o-Xylene 150°			v-l	883
		" 125°			v-l	883
		" 100°			v-l	883
8547	C₈H₁₀	p-Xylene	138.2	Azeo. distillation		330
8548	C₈H₁₀O	Phenethyl alcohol	219.4	Nonazeotrope		526
8549	C₈H₁₀O	3,4-Xylenol	226.8	Nonazeotrope		556
8550	C₈H₁₀O₂	2-Phenoxyethanol	245.2	< 244.5		575
8551	C₈H₁₁NO	o-Phenetidine	232.5	< 225.0	<18	575
8552	C₈H₁₁NO	p-Phenetidine	249.9	< 232.0	>52	575
8553	C₈H₁₂O₄	Ethyl fumarate	217.85	217.1	10	527
8554	C₈H₁₂O₄	Ethyl maleate	223.3	222.65	10.0	527
8555	C₈H₁₄O₄	Ethyl succinate	217.25	Reacts		526
8556	C₈H₁₈O₃	2-(2-Butoxyethoxy) ethanol, 10 mm.	109	Nonazeotrope		981

TABLE I. *Binary Systems* 259

No.	Formula	B-Component Name	B.P., °C	Azeotropic Data B.P., °C	Wt.%A	Ref.
A =	C₄H₁₀O₃	**Diethylene Glycol**	**245.5**			
		(continued)				
8557	C₈H₁₈O₄	2-[2-(2-Ethoxyethoxy) ethoxy]ethanol,				
		2 mm.	98	87	43	982
		" 3 mm.		135	83.4	183
8558	C₉H₇N	Quinoline	237.3	233.6	29	527
8559	C₉H₁₀O₂	Benzyl acetate	215.0	214.85	7	527
8560	C₉H₁₀O₂	Ethyl benzoate	212.5	211.65	10	527
8561	C₉H₁₀O₃	Ethyl salicylate	233.8	225.15	30	569
8562	C₉H₁₂O	3-Phenylpropanol	235.6	Nonazeotrope		575
8563	C₉H₁₂O	Phenyl propyl ether	190.5	Nonazeotrope		526
8564	C₁₀H₇Br	1-Bromonaphthalene	281.2	240.8	59.5	527
8565	C₁₀H₇Cl	1-Chloronaphthalene	262.7	234.1	47	527
8566	C₁₀H₈	Naphthalene	218.0	212.6	22.0	527
		"	218.1	Azeo. distillation		330
8567	C₁₀H₈O	1-Naphthol	288.5	Nonazeotrope		556
8568	C₁₀H₉N	Quinaldine	246.5	< 241.0		575
8569	C₁₀H₁₀O₂	Isosafrol	252.0	233.5	46	526
8570	C₁₀H₁₀O₂	Methyl cinnamate	261.9	240.0	63	527
8571	C₁₀H₁₀O₂	Safrole	235.9	225.5	33	556
8572	C₁₀H₁₀O₄	Methyl phthalate	283.7	245.4	96.3	556
8573	C₁₀H₁₂O	Anethole	235.7	210.0	20	567
8574	C₁₀H₁₂O₂	Ethyl α-toluate	228.75	224.0	20	567
8575	C₁₀H₁₂O₂	Propyl benzoate	230.85	222.7	26	556
8576	C₁₀H₁₄	Butylbenzene	183.1	Nonazeotrope		556
8577	C₁₀H₁₄	Cymene	176.7	Nonazeotrope		575
8578	C₁₀H₁₄O	Carvacrol	237.85	236.0	27	526
8579	C₁₀H₁₄O	Thymol	232.9	232.25	13	527
8580	C₁₀H₁₆O	Camphor	209.1	Nonazeotrope		552
8581	C₁₀H₁₈O	α-Terpineol	218.85	217.45	13.5	527
8582	C₁₀H₂₀O	Citronellol	224.4	Nonazeotrope		575
8583	C₁₁H₁₀	1-Methylnaphthalene	244.6	227.0	45	526
8584	C₁₁H₁₀	2-Methylnaphthalene	241.15	225.45	39	556
8585	C₁₁H₁₂O₂	Ethyl cinnamate	272.0	244.5	85?	526
8586	C₁₁H₁₄OS	2-(Benzylmercapto) ethyl vinyl ether		Min. b.p.		953
8587	C₁₁H₁₄O₂	1-Allyl-3,4-dimethoxy-benzene	254.7	235.0	47	526
8588	C₁₁H₁₄O₂	Butyl benzoate	249.0	232.2	43	556
8589	C₁₁H₁₄O₂	1,2-Dimethoxy-4-propenyl-benzene	270.5	238.8	60	567
8590	C₁₁H₁₄O₂	Isobutyl benzoate	241.9	228.65	37	556
8591	C₁₁H₁₆O	Methyl thymyl ether	216.5	210.5	~ 19	575
8592	C₁₁H₂₀O	Isobornyl methyl ether	192.4	< 191.0	< 9	575
8593	C₁₁H₂₀O	Methyl α-Terpineol ether	216.2	210.5	20	567
8594	C₁₂H₉N	Carbazole, >10 mm.	294	Nonazeotrope		272
8595	C₁₂H₁₀	Acenaphthene	277.9	239.6	62	527
8596	C₁₂H₁₀	Biphenyl	256.1	232.65	48	527,330

		B-Component		Azeotropic Data		
No.	Formula	Name	B.P., °C	B.P., °C	Wt.%A	Ref.
A =	$C_4H_{10}O_3$	**Diethylene Glycol**	**245.5**			
		(continued)				
8597	$C_{12}H_{10}O$	Phenyl ether, 4 mm.	100		23	982
		"	259.0	234.4	49.5	527
8598	$C_{12}H_{14}O_4$	Ethyl phthalate	297.5	Nonazeotrope		526
8599	$C_{12}H_{16}O_2$	Isoamyl benzoate	262.0	236.55	52.5	527
8600	$C_{12}H_{16}O_3$	Isoamyl salicylate	277.5	Nonazeotrope		526
8601	$C_{12}H_{18}$	1,3,5-Triethylbenzene	215.5	210.0	22	526
8602	$C_{12}H_{20}O_2$	Bornyl acetate	227.6	223.0	18	567
8603	$C_{12}H_{22}O_4$	Isoamyl oxalate	268.0	Reacts		526
8604	$C_{12}H_{24}OS$	2-(2-Ethylhexylthio)				
		ethyl vinyl ether		Min. b.p.		953
8605	$C_{12}H_{26}O$	Hexyl ether, 50 mm.	137	129.9	15.5	982
8606	$C_{13}H_{10}$	Fluorene, 10-760 mm.	294	Min. b.p.		272
		"	295.0	243.0	80	526
8607	$C_{13}H_{12}$	Diphenylmethane	265.4	236.0	52	556
8608	$C_{13}H_{12}O$	Benzyl phenyl ether	286.5	241.5	80	575
8609	$C_{14}H_{10}$	Phenanthrene, 20 mm.		146	93	272
		" 100 mm.		180	96.2	272
		" 200 mm.		203	98.5	272
		" 300 mm.		217	99.5	272
		" 400 mm.		226	99.9	272
8610	$C_{14}H_{14}$	1,2-Diphenylethane	284.5	241.0	66	526
8611	$C_{14}H_{14}O$	Benzyl ether	297	< 243.8	>87	575
		" 5 mm.			40	982
8612	$C_{16}H_{34}O$	Bis(2-ethylhexyl) ether,				
		10 mm.	135	114		982
A =	$C_4H_{10}S$	**1-Butanethiol**	**97.8**			
8613	$C_4H_{10}S$	Ethyl sulfide	92.1	Nonazeotrope		575
8614	C_5H_5N	Pyridine	115.4	Nonazeotrope		575
8615	C_6H_6	Benzene	80.15	Nonazeotrope		566
8616	C_6H_{12}	Cyclohexane	80.75	Nonazeotrope		566
8617	C_7H_8	Toluene	110.623	Nonazeotrope		202
8618	C_7H_{14}	*cis*-1,2-Dimethylcyclo-				
		pentane	99.53	96.35	48.0	202
8619	C_7H_{14}	Ethylcyclopentane	103.45	97.76	72.15	202
8620	C_7H_{14}	*trans*-1,3-Dimethylcyclo-				
		pentane	90.77	90.54	12.7	202
8621	C_7H_{14}	Methylcyclohexane	100.934	97.00	58.2	202
8622	C_7H_{16}	2,3-Dimethylpentane	89.79	59.53	15.1	202
8623	C_7H_{16}	Heptane	98.428	95.45	49.4	202
8624	C_7H_{16}	2-Methylhexane	90.05	89.74	15.4	202
8625	C_7H_{16}	3-Methylhexane	91.95	91.20	22.8	202
8626	C_8H_{18}	2,2-Dimethylhexane	106.843	98.01	78.8	202
8627	C_8H_{18}	2,5-Dimethylhexane	109.106	98.22	88.0	202
8628	C_8H_{18}	3,3-Dimethylhexane	111.927	98.56	97.6	202
8629	C_8H_{18}	2,2,4-Trimethylpentane	99.237	95.50	50.3	202
A =	$C_4H_{10}S$	**2-Butanethiol**	**85.15**			
8630	$C_4H_{10}S$	Isopropyl methyl sulfide	84.76	Nonazeotrope		205
8631	C_6H_6	Benzene	80.103	Nonazeotrope		202

TABLE I. *Binary Systems* 261

No.	Formula	B-Component Name	B.P., °C	Azeotropic Data B.P., °C	Wt.%A	Ref.
A =	C₄H₁₀S	**2-Butanethiol** *(continued)*	85.15			
8632	C₆H₁₂	Cyclohexane	80.73 8	79.97	25.5	202
8633	C₆H₁₂	Methylcyclopentane	71.81 2	Nonazeotrope		202
8634	C₇H₁₄	1,1-Dimethylcyclo-pentane	87.84	83.90	64.1	202
8635	C₇H₁₄	*trans*-1,3-Dimethylcyclo-pentane	90.77	84.75	78.1	202
8636	C₇H₁₆	2,2-Dimethylpentane	79.20 5	78.60	23.1	202
8637	C₇H₁₆	2,3-Dimethylpentane	89.79	84.16	68.6	202
8638	C₇H₁₆	2,4-Dimethylpentane	80.51	79.55	28.1	202
8639	C₇H₁₆	Heptane	98.42 8	Nonazeotrope		202
8640	C₇H₁₆	2-Methylhexane	90.05	84.30	72.1	202
8641	C₇H₁₆	3-Methylhexane	91.95	84.70	80.8	202
A =	C₄H₁₀S	**Ethyl Sulfide**	92.1			
8642	C₄H₁₀S	2-Methyl-1-propanethiol	87.8	87.0	85	575
8643	C₅H₅N	Pyridine	115.4	Nonazeotrope		566
8644	C₅H₇N	1-Methyl pyrrol	112.8	Nonazeotrope		575
8645	C₅H₁₀O	Isovaleraldehyde	92.1	88.5	53	566
8646	C₅H₁₀O	3-Methyl-2-butanone	95.4	78.0	70	566
8647	C₅H₁₀O	3-Pentanone	102.05	Nonazeotrope		566
8648	C₅H₁₀O₂	Methyl butyrate	102.65	Nonazeotrope		548
8649	C₅H₁₀O₂	Methyl isobutyrate	92.5	< 91.7	>56	566
8650	C₅H₁₀O₂	Propyl acetate	101.6	Nonazeotrope		548
8651	C₅H₁₂O	Isoamyl alcohol	131.9	Nonazeotrope		527
8652	C₅H₁₂O₂	Diethoxymethane	87.95	85.9	35	566
8653	C₆H₆	Benzene	80.2	Nonazeotrope		205,531
8654	C₆H₁₂	Cyclohexane	80.75	Nonazeotrope		205,531
8655	C₆H₁₄O	Isoproyl ether	68.3	Nonazeotrope		566
8656	C₆H₁₄O	Propyl ether	90.1	< 89.5	>25	566
8657	C₆H₁₄O₂	Acetal	104.5	Nonazeotrope		563
8658	C₇H₁₄	*trans*-1,3-Dimethyl-cyclopentane	90.80	88.89	41.0	205
8658	C₇H₁₄	1,1-Dimethylcyclo-pentane	87.90	86.98	26.1	205
8660	C₇H₁₄	Methylcyclohexane	101.05	92.10	94.5	205
		"	101.1	Nonazeotrope		531
8661	C₇H₁₆	Heptane	98.4	< 91.8	>78	566
8662	C₇H₁₆	3-Methylhexane	91.60	89.19	48.3	205
8663	C₇H₁₆	2,3-Dimethylpentane	89.90	87.93	38.6	205
8664	C₇H₁₆	2,4-Dimethylpentane	80.55	80.53	2.26	205
8665	C₈H₁₈	2,2,4-Trimethylpentane	99.30	91.44	77.0	205
A =	C₄H₁₀S	**Isopropyl Methyl Sulfide**	84.76			
8666	C₆H₁₂	Cyclohexane	80.85	79.76	30	205
8667	C₆H₁₂	Methylcyclopentane	71.85	Nonazeotrope		205
8668	C₇H₁₄	*trans*-1,3-Dimethyl-cyclopentane	90.80	84.38	80.4	205

No.	Formula	B-Component Name	B.P., °C	Azeotropic Data B.P., °C	Wt.%A	Ref.
A =	C₄H₁₀S	**Isopropyl Methyl Sulfide** *(continued)*	**84.76**			
8669	C₇H₁₄	1,1-Dimethylcyclo-pentane	87.90	83.62	64.9	205
8670	C₇H₁₆	3-Methylhexane	91.60	84.38	82.4	205
8671	C₇H₁₆	2,3-Dimethylpentane	89.90	83.83	72.8	205
8672	C₇H₁₆	2,4-Dimethylpentane	80.55	79.39	29.7	205
8673	C₇H₁₆	2,2-Dimethylpentane	79.20	78.40	23.3	205
A =	C₄H₁₀S	**2-Methyl-1-propanethiol**	**88.72**			
8674	C₆H₆	Benzene	80.103	Nonazeotrope		202
8675	C₆H₈	1,3-Cyclohexadiene	80.8	Reacts		563
8676	C₆H₈	1,4-Cyclohexadiene	85.6	Reacts		563
8677	C₆H₁₀	Cyclohexene	82.75	Reacts		563
8678	C₆H₁₂	Cyclohexane	80.738	80.70	11.7	202
8679	C₆H₁₄	Hexane	68.8	Nonazeotrope		575
8680	C₇H₁₄	1,1-Dimethylcyclopentane	87.84	85.69	44.25	202
3681	C₇H₁₄	*cis*-1,2-Dimethylcyclopentane	99.53	88.52	98.6	202
8682	C₇H₁₄	*trans*-1,3-Dimethyl-cyclopentane	90.77	87.02	58.6	202
8683	C₇H₁₄	Ethylcyclopentane	103.46	Nonazeotrope		202
3684	C₇H₁₄	Methylcyclohexane	100.934	88.55	98.9	202'
3685	C₇H₁₆	2,2-Dimethylpentane	79.205	79.12	10.3	202
8686	C₇H₁₆	2,3-Dimethylpentane	89.79	86.28	54.1	202
8687	C₇H₁₆	2,4-Dimethylpentane	80.51	80.28	14.1	202
8688	C₇H₁₆	Heptane	98.428	88.50	91.3	202
8689	C₇H₁₆	3-Methylhexane	91.95	87.16	62.8	202
8690	C₇H₁₆	2,2,3-Trimethylbutane	80.871	80.60	16.4	202
8691	C₈H₁₈	2,2,4-Trimethylpentane	99.237	88.41	90.0	202
A =	C₄H₁₀S	**2-Methyl-2-propanethiol**	**64.35**			
8692	C₆H₁₂	Methylcyclopentane	71.812	63.37	95.3	202
8693	C₆H₁₄	2,3-Dimethylbutane	57.990	57.82	21.1	202
8694	C₆H₁₄	Hexane	68.742	63.78	75.8	202
8695	C₆H₁₄	2-Methylpentane	60.274	59.55	30.4	202
8696	C₆H₁₄	3-Methylpentane	63.284	61.51	46.5	202
A =	C₄H₁₀S	**Methyl Propyl Sulfide**	**95.47**			
8697	C₇H₁₄	Ethylcyclopentane	103.45	95.41	90.7	205
8698	C₇H₁₄	Methylcyclohexane	101.05	95.06	78.0	205
8699	C₇H₁₄	*trans*-1,3-Dimethyl-cyclopentane	90.80	90.11	24.3	205
8700	C₇H₁₄	1,1-Dimethylcyclo-pentane	87.90	87.66	9.7	205
8701	C₇H₁₆	3-Methylhexane	91.60	90.53	32.95	205
8702	C₇H₁₆	2,3-Dimethylpentane	89.90	89.10	22.75	205
8703	C₈H₁₈	2,2-Dimethylhexane	106.85	95.42	94.4	205
8704	C₈H₁₈	2,2,4-Trimethylpentane	99.30	94.00	62.2	205
A =	C₄H₁₀S₂	**Ethyl Disulfide**	**154.11**			
8705	C₉H₂₀	Nonane	150.65	148.62	41.2	205
8706	C₁₀H₂₂	3-Ethyl-3-methyl-heptane	163.00	153.02	80.2	205

TABLE I.   *Binary Systems*   263

No.	Formula	B-Component Name	B.P., °C	B.P., °C	Wt.%A	Ref.
A =	$C_4H_{11}N$	**Butylamine**	**77.8**			
8707	C_6H_{12}	Cyclohexane	80.75	76.5	60	551
8707a	C_6H_{12}	1-Hexene 60°C.		Nonazeotrope	v-l	406c
8708	C_6H_{12}	Methylcyclopentane	72.0	< 77.5		551
8708a	C_6H_{14}	Hexane 617 mm.		60	22	v-l 406c
A =	$C_4H_{11}N$	**Diethylamine**	**55.9**			
8709	C_5H_{10}	2-Methyl-2-butene	37.1	Nonazeotrope		551
8710	$C_5H_{10}O$	3-Methyl-2-butanone	95.4	Nonazeotrope		551
8711	C_5H_{12}	Pentane	36.15	Nonazeotrope		551
8712	$C_5H_{12}O$	Ethyl propyl ether	63.85	Nonazeotrope		551
8713	C_6H_{10}	Biallyl	60.1	< 55.5		551
8713a	C_6H_{12}	1-Hexene 60°C.		Nonazeotrope	v-l	406c
8714	C_6H_{12}	Methylcyclopentane	72.0	Nonazeotrope		551
8715	C_6H_{14}	2,3-Dimethylbutane	58.0	< 55.0	<62	551
8716	C_6H_{14}	Hexane	68.8	Nonazeotrope	v-l	406c,551
8716	C_6H_{14}	n-Hexane	68.8	Nonazeotrope		551
8717	$C_6H_{14}O$	Isopropyl ether	68.3	Nonazeotrope		981
8718	$C_6H_{15}NO$	2-(Diethylamino) ethanol	162.1	Nonazeotrope		981
A =	$C_4H_{11}N$	**Isobutylamine**	**68.0**			
8719	C_5H_{10}	Cyclopentane	49.3	Nonazeotrope		551
8720	$C_5H_{10}O$	3-Methyl-2-butanone	95.4	Nonazeotrope		551
8721	C_5H_{12}	n-Pentane	36.15	Nonazeotrope		551
8722	C_6H_6	Benzene	80.15	Nonazeotrope		551
8723	C_6H_{12}	Cyclohexane	80.75	Nonazeotrope		551
8724	C_6H_{12}	Methylcyclopentane	72.0	< 67.6	>59	575
8725	C_6H_{14}	n-Hexane	68.8	< 66.5	>52	551
A =	$C_4H_{11}NO$	**2-Amino-2-methyl-1-propanol**	**165.4**			
8726	C_8H_9Cl	o,m,p-Chloroethylbenzene,				
		10 mm.	67.5	59.0	46	51
A =	$C_4H_{11}NO$	**2-Dimethylaminoethanol**	**134.6**			
8727	C_8H_6	Phenylacetylene	142	min. b.pt.		244
8728	C_8H_8	Styrene	144.7	"		244
8729	C_8H_{10}	Ethylbenzene	136.2	"		244
8730	C_8H_{10}	o,m,p-Xylenes	140	"		244
A =	$C_4H_{11}NO_2$	**2,2'-Iminodiethanol**	**268.0**			
8731	$C_6H_{15}NO_3$	2,2'-Iminodiethanol				
		2,2',2''-Nitrilo-				
		triethanol 2 mm.	195	Nonazeotrope		981
8732	$C_{10}H_{10}O_2$	Isosafrole	252.0	<246.0		575
8733	$C_{10}H_{15}N$	N,N-Diethylaniline	217.05	Nonazeotrope		575
8734	$C_{11}H_{14}O_2$	1-Allyl-3,4-dimethoxy-				
		benzene	254.7	<247.0		575
8735	$C_{12}H_{10}O$	Phenyl ether	259.0	<250.0		575
A =	$C_4H_{11}PO_3$	**Diethylphosphite**				
8736	C_5H_9NO	α-Hydroxyvaleronitrile				
		4 mm.		90	33.8	24
8737	$C_5H_{12}O$	tert-Amyl alcohol	102.25	Nonazeotrope		24
8738	$C_6H_{14}O$	Hexyl alcohol 8 mm		63	57.5	24

| | | B-Component | | | Azeotropic Data | | |
No.	Formula	Name	B.P., °C	B.P., °C	Wt.%A		Ref.
A =	**C₄H₁₁PO₃**	**Diethylphosphite** *(continued)*					
8739	C₇H₁₁NO	α-Hydroxycyclohexane-nitrile	1.5 mm	92	15.1		24
A =	**C₄H₁₂SiO₄**	**Methyl Silicate**	**121.8**				
8740	C₆H₁₂O₃	Paraldehyde	124.35	<121.3			557
A =	**C₅Cl₂F₆**	**1,2-Dichlorohexa-fluorocyclopentene**	**90.6**				
8741	C₈F₁₆O	Perfluorocyclic oxide	102.6	90.4	80	v-l	1041
A =	**C₅F₁₀**	**Perfluorocyclopentane**					
8742	C₅F₁₂	Perfluoropentane, 9.6°-25° C.		Nonazeotrope		v-l	695
8743	C₆F₁₄	Perfluorohexane 15°-25° C.		Nonazeotrope		v-l	695
A =	**C₅H₄F₈O**	**2,2,3,3,4,4,5,5-Octafluoro-1-pentanol**					
8744	C₅H₁₂O	Active amyl alcohol	128.5	Nonazeotrope			958
8745	C₅H₁₂O	Isoamyl alcohol	132.0	Nonazeotrope			958
A =	**C₅H₄O₂**	**2-Furaldehyde**	**161.45**				
8746	C₅H₆O₂	Furfuryl alcohol	169	Nonazeotrope		v-l	220
		" 25 mm.		Nonazeotrope		v-l	1028
8747	C₅H₈O₃	Methyl acetoacetate	~ 169.5	Reacts			563
8748	C₅H₁₀O₂	Isovaleric acid	176.5	Nonazeotrope			542
8749	C₅H₁₀O₃	Ethyl lactate	154.1	Nonazeotrope			575
8750	C₅H₁₀O₃	2-Methoxyethyl acetate	144.6	Nonazeotrope			575
8751	C₅H₁₁Br	1-Bromo-3-methylbutane	120.65	Nonazeotrope			527
8752	C₅H₁₁I	1-Iodo-3-methylbutane	147.6	146.5	~ 15		548
8753	C₅H₁₂O₂	2-Propoxyethanol	151.35	151.1	14		527
8754	C₅H₁₂O₃	2-(2-Methoxyethoxy)ethanol	192.95	Nonazeotrope			575
8755	C₆H₄Cl₂	o-Dichlorobenzene	179.5	161.0	78		527
8766	C₆H₄Cl₂	p-Dichlorobenzene	174.35	160.3	63.5		527
8757	C₆H₅Br	Bromobenzene	156.1	153.3	23		556
8758	C₆H₅Cl	Chlorobenzene	132.0	Nonazeotrope			527
8759	C₆H₅I	Iodobenzene	188.45	Nonazeotrope			527
8760	C₆H₆	Benzene	80.1	Nonazeotrope		v-l	163,961
8761	C₆H₆O	Phenol	181.5	Nonazeotrope			563
8762	C₆H₁₀O	Cyclohexanone	155.6	Nonazeotrope			538
8763	C₆H₁₂	Cyclohexane	80.75	Nonazeotrope		v-l	163,961
8764	C₆H₁₂O	Cyclohexanol	160.7	156.5	5.5		527
8765	C₆H₁₂O₂	Methyl isovalerate	155.8	Nonazeotrope			563
8766	C₆H₁₂O₃	2-Ethoxyethyl acetate	156.8	Nonazeotrope			527
8767	C₆H₁₂O₃	Propyl lactate	171.7	Nonazeotrope			575
8768	C₆H₁₄O	Hexyl alcohol	157.85	154.1	44		564
8769	C₆H₁₄O₂	2-Butoxyethanol	171.15	161.2	88		527
8770	C₆H₁₄S	Propyl sulfide	141.5	Nonazeotrope			556

TABLE I.   *Binary Systems*   265

No.	Formula	Name	B.P., °C	B.P., °C	Wt.%A		Ref.
		B-Component		**Azeotropic Data**			
A =	C$_5$H$_4$O$_2$	**2-Furaldehyde** *(continued)*	**161.45**				
8771	C$_7$H$_7$Br	*o*-Bromotoluene	181.5	< 161.3	>80		527
	"	"	181.45	Nonazeotrope			532
8772	C$_7$H$_7$Cl	α-Chlorotoluene	179.3	Nonazeotrope			532
8773	C$_7$H$_7$Cl	*o*-Chlorotoluene	159.3	155.4	35		527
8774	C$_7$H$_7$Cl	*p*-Chlorotoluene	162.4	157.2	42		527
8775	C$_7$H$_7$ClO	*m*-Chloroanisole	193.3	Nonazeotrope			575
8776	C$_7$H$_8$	Toluene	110.75	Nonazeotrope			528
8777	C$_7$H$_8$O	Anisole	153.85	153.25	22		556
8778	C$_7$H$_{14}$	Methylcyclohexane	101.05	100.8	4.1	v-l	317
8779	C$_7$H$_{14}$O	2-Methylcyclohexanol	168.5	< 160.9	<94		527
8780	C$_7$H$_{14}$O$_3$	1,3-Butanediol methyl ether acetate	171.75	Nonazeotrope			527
8781	C$_7$H$_{16}$	Heptane	98.40	98.3	5.3	v-l	317
8782	C$_8$H$_8$	Styrene	145.8	< 145			527
8783	C$_8$H$_{10}$	Ethylbenzene	136.15	Nonazeotrope			527
8784	C$_8$H$_{10}$	*m*-Xylene	139.0	138.4	12		531
8785	C$_8$H$_{10}$	*o*-Xylene	143.6	140.5	13		545
8786	C$_8$H$_{10}$	*p*-Xylene	138.4	138.0	5		545
8787	C$_8$H$_{10}$O	Benzyl methyl ether	167.8	< 160.3	>85		575
8788	C$_8$H$_{10}$O	*p*-Methylanisole	177.05	161.35	89		564
8789	C$_8$H$_{10}$O	Phenetole	170.45	~ 161.0	~ 83		548
8790	C$_8$H$_{14}$O	Methylheptenone	173.2	Nonazeotrope			545
8791	C$_8$H$_{16}$	1,3-Dimethylcyclohexane	120.7	Nonazeotrope			556
8792	C$_8$H$_{16}$O	2-Octanone	172.9	Nonazeotrope			545
8793	C$_8$H$_{16}$O$_2$	Butyl butyrate	166.4	Nonazeotrope			548
8794	C$_6$H$_{16}$O$_2$	Isoamyl propionate	160.7	< 159.5	>52		575
8795	C$_8$H$_{16}$O$_2$	Isobutyl butyrate	156.8	Nonazeotrope			532
8796	C$_8$H$_{16}$O$_2$	Propyl isovalerate	155.7	Nonazeotrope			538
8797	C$_8$H$_{18}$	Octane	125.8	Nonazeotrope			527
8788	C$_8$H$_{18}$O	Butyl ether	142.4	< 138.5	>11		527
8789	C$_8$H$_{18}$O	Octyl alcohol	195.2	Nonazeotrope			527
8800	C$_8$H$_{18}$S	Butyl sulfide	185.0	Nonazeotrope			556
8801	C$_8$H$_{18}$S	Isobutyl sulfide	172.0	< 161.3			575
8802	C$_9$H$_8$	Indene	182.6	Nonazeotrope			527
8803	C$_9$H$_{12}$	Cumene	152.8	148.5	27		527
8804	C$_9$H$_{12}$	Mesitylene	164.6	155.2	60		556
8805	C$_9$H$_{12}$	Pseudocumene	168.2	157.0	67		527
8806	C$_9$H$_{12}$	Propylbenzene	159.2	151.4	42		527
8807	C$_9$H$_{12}$O	Benzyl ethyl ether	185.0	Nonazeotrope			527
8808	C$_9$H$_{18}$O$_2$	Butyl isovalerate	177.6	Nonazeotrope			575
8809	C$_9$H$_{18}$O$_2$	Isoamyl isobutyrate	169.8	Nonazeotrope			575
8810	C$_9$H$_{18}$O$_2$	Isobutyl isovalerate	168.7	Nonazeotrope			532
8811	C$_{10}$H$_8$	Naphthalene	218.0	Nonazeotrope			527
8812	C$_{10}$H$_{14}$	Butylbenzene	183.1	Nonazeotrope			575
	"	"	183.2	160.5	82		548
8813	C$_{10}$H$_{14}$	Cymene	176.7	157.8	68		531

No.	Formula	B-Component Name	B.P., °C	B.P., °C	Wt.%A	Ref.
A =	$C_5H_4O_2$	**2-Furaldehyde** *(continued)*	**161.45**			
8814	$C_{10}H_{16}$	Camphene	159.5	146.75	40	556
8815	$C_{10}H_{16}$	*d*-Limonene	177.8	155.95	35	529
8816	$C_{10}H_{16}$	α-Pinene	155.8	143.4	38	556
8817	$C_{10}H_{16}$	Nopinene	163.8	147.1	50	527
8818	$C_{10}H_{16}$	α-Terpinene	173.3	155.0	60	527
8819	$C_{10}H_{16}$	γ-**Terpinene**	183	< 160.0		575
8820	$C_{10}H_{16}$	Terpinolene	185.2	159.5	80	527
8821	$C_{10}H_{16}$	Thymene	179.7	158.5	72	531
8822	$C_{10}H_{16}$	Dipentene	177.7	155.95	65	527
8823	$C_{10}H_{18}O$	Cineol	176.35	157.25	59	569
8824	$C_{10}H_{18}O$	Linalool	198.6	Nonazeotrope		575
8825	$C_{10}H_{22}$	2,7-Dimethyloctane	160.25	< 147.0	<48	527
8826	$C_{10}H_{22}O$	Amyl ether	187.5	< 158.5	>83	527
8827	$C_{10}H_{22}O$	Isoamyl ether	173.4	153.9	55	556
8828	$C_{11}H_{20}O$	Isobornyl methyl ether	192.4	Nonazeotrope		575
8829	$C_{12}H_{18}$	1,3,5-Triethylbenzene	215.5	Nonazeotrope		527
A =	C_5H_5N	**Pyridine**	**115.4**			
8830	$C_5H_{10}O$	2-Pentanone	102.35	Nonazeotrope		575
8831	$C_5H_{10}O$	3-Pentanone	102.05	Nonazeotrope		553
8832	$C_5H_{10}O_2$	Butyl formate	106.8	Nonazeotrope		553
8833	$C_5H_{10}O_3$	Ethyl carbonate	126.0	Nonazeotrope		562
8834	$C_5H_{11}Br$	1-Bromo-3-methylbutane	120.3	< 114.5	>60	548
8835	$C_5H_{11}N$	Piperidine	105.8	106.1	8	913,914
		"	106	105.8	3.4	413
8836	$C_5H_{12}O$	Amyl alcohol	138.2	Nonazeotrope		553
8837	$C_5H_{12}O$	*tert*-Amyl alcohol	102.35	Nonazeotrope		553
8838	$C_5H_{12}O$	Isoamyl alcohol	131.9	Nonazeotrope		527
8839	$C_5H_{12}O$	3-Pentanol	116.0	117.4	45	533
8840	C_6H_5Cl	Chlorobenzene	132.0	Nonazeotrope		548
8841	C_6H_6	Benzene	80.15	Nonazeotrope		553
8842	C_6H_6O	Phenol	181.4	183.10	13.1	v-l	791,27c
8843	C_6H_7N	2-Picoline	130.7	Nonazeotrope		575
8844	$C_6H_{10}O$	Mesityl oxide	129.45	Nonazeotrope		527
8845	$C_6H_{10}S$	Allyl sulfide	139.35	Nonazeotrope		566
8846	C_6H_{12}	Cyclohexane	80.75	Nonazeotrope		553
		"	80.7	Nonazeotrope		v-l	428
8847	$C_6H_{12}O$	Pinacoline	106.2	Nonazeotrope		553
8848	$C_6H_{12}O$	3-Hexanone	123.3	Nonazeotrope		553
8849	$C_6H_{12}O$	4-Methyl-2-pentanone	116.05	114.9	60	527
8850	$C_6H_{12}O_2$	Butyl acetate	126.0	Nonazeotrope		553
8851	$C_6H_{12}O_2$	Ethyl butyrate	121.5	Nonazeotrope?		548
8852	$C_6H_{12}O_2$	Isoamyl formate	123.8	Nonazeotrope		553
8853	$C_6H_{12}O_2$	Isobutyl acetate	117.4	114.5		553
8854	$C_6H_{12}O_2$	Methyl isovalerate	116.5	< 115.0	>52	553
8855	$C_6H_{12}O_2$	Propyl propionate	123.0	Nonazeotrope		553
8856	$C_6H_{14}O$	Propyl ether	90.1	Nonazeotrope		553

TABLE I. *Binary Systems*   267

No.	Formula	B-Component Name	B.P., °C	Azeotropic Data B.P., °C	Wt.%A	Ref.
A =	C_5H_5N	**Pyridine** *(continued)*	**115.4**			
8857	$C_6H_{14}S$	Isopropyl sulfide	120.5	< 114.5	<72	553
8858	C_7H_8	Toluene	110.75	110.15	22	553
		"		110.1	20.3	497
		"	110.8	110.1	22.2	v-l 393,1069
8859	C_7H_{14}	Methylcyclohexane	100	Min. b.p.		553
8860	C_7H_{16}	*n*-Heptane	98.4	< 97.0	<14	553
		"	98.40	95.60	25.3	1052,1071
		"	98.40	95	13.3	v-l 393
8861	C_8H_{10}	Ethylbenzene	136.15	Nonazeotrope		1061
		"	136.15	Nonazeotrope		553
8862	C_8H_{10}	*m*-Xylene	139.2	Nonazeotrope		553
8863	C_8H_{10}	*o*-Xylene	143.6	Nonazeotrope		1070
8864	C_8H_{10}	*p*-Xylene	138.4	Nonazeotrope		307
8865	C_8H_{16}	1,3-Dimethylcyclohexane	120.7	< 111.0		553
8866	C_8H_{18}	2,5-Dimethylhexane	109.4	< 105.5	<40	553
8867	C_8H_{18}	*n*-Octane	125.75	< 112.8	<90	553
		"	125.75	109.5	56.1	1052
8868	C_8H_{18}	2,2,4-Trimethylpentane	99.3	95.75	23.4	553
8869	$C_8H_{18}O$	Isobutyl ether	122.3	Nonazeotrope		553
8870	C_9H_{20}	Nonane	150.8	115.1	90	307
		"	150.7	115.1	89.9	1052
8871	$C_{10}H_8N_2$	2,2'-Dipyridyl	274	Nonazeotrope		413
8872	$C_{10}H_{22}$	Decane	173.3	Nonazeotrope		1052
A =	C_5H_6	**Cyclopentadiene**	**41**			
8872a	C_5H_8	Isoprene	34.07	Nonazeotrope		581e
A =	$C_5H_6O_2$	**Furfuryl Alcohol**	**169.35**			
8873	$C_5H_{11}NO_3$	Isoamyl nitrate	149.75	< 149.6		560
8874	$C_6H_4Cl_2$	*p*-Dichlorobenzene	174.4	172.5	70	575
8875	C_6H_5Cl	Chlorobenzene	131.75	Nonazeotrope		575
8876	C_6H_6O	Phenol	182.2	187.0	30	567
8877	C_6H_7N	Aniline	184.35	Nonazeotrope		575
8878	$C_6H_{12}O$	Cyclohexanol	160.8	Nonazeotrope		575
8879	$C_6H_{12}O_3$	2-Ethoxyethyl acetate	156.8	Nonazeotrope		575
8880	$C_6H_{14}O$	Hexyl alcohol	157.85	Nonazeotrope		575
8881	$C_6H_{14}O_2$	2-Butoxyethanol	171.15	< 167.5	>60	575
8882	C_7H_6O	Benzaldehyde	179.2	Nonazeotrope		575
8883	C_7H_8O	Anisole	153.85	153.3	10	545
8885	$C_7H_{14}O$	2-Methylcyclohexanol	168.5	< 168.3		575
8885	$C_7H_{14}O_3$	1,3-Butanediol methyl ether acetate	171.75	168.5	82	575
8886	C_8H_9Cl	*o,m,p*-Chloroethyl-benzene, 10 mm.	67.5	60.5	32	51
8887	$C_8H_{10}O$	Phenetole	170.45	165.0	46	545
8888	$C_8H_{11}N$	Dimethylaniline	194.15	Nonazeotrope		575
8889	$C_8H_{16}O_2$	Butyl butyrate	166.4	164.0	30	575

No.	Formula	B-Component Name	B.P., °C	B.P., °C	Wt.%A	Ref.
A =	C₅H₆O₂	**Furfuryl Alcohol** *(continued)*	**169.35**			
8890	C₈H₁₆O₂	Isoamyl propionate	160.7	Nonazeotrope		575
8891	C₈H₁₆O₂	Propyl isovalerate	155.7	Nonazeotrope		575
8892	C₉H₇N	Quinoline	237.3	Nonazeotrope		553
8893	C₉H₁₃N	Dimethyl-o-toluidine	185.3	Nonazeotrope		575
8894	C₁₀H₂₂O	Isoamyl ether	173.4	165.7	50	545
A =	C₅H₆S	**2-Methylthiophene**	**111.92**			
8895	C₇H₁₆	Heptane	98.40	97.77	2.2	205
8896	C₈H₁₈	2-Methylheptane	117.70	109.97	67.8	205
8897	C₈H₁₈	2,5-Dimethylhexane	109.15	106.12	39.6	205
8898	C₈H₁₈	2,2-Dimethylhexane	106.85	104.62	33.2	205
A =	C₅H₆S	**3-Methylthiophene**	**114.96**			
8899	C₇H₁₄	Ethylcyclopentane	103.45	102.82	3.9	205
8900	C₇H₁₆	Heptane	98.40	Nonazeotrope		205
8901	C₈H₁₆	trans-1,3-Dimethyl-cyclohexane	120.3	113.17	66	205
8902	C₈H₁₆	1,1,2-Trimethylcyclo-pentane	113.75	110.47	43.2	205
8903	C₈H₁₈	Octane	**125.70**	114.15	82	205
8904	C₈H₁₈	2-Methylheptane	117.70	111.86	58.8	205
8905	C₈H₁₈	2,5-Dimethylhexane	109.15	107.12	31.7	205
A =	C₅H₇N	**2-Methylpyrrol**	**147.5**			
8906	C₅H₁₂O	Isoamyl alcohol	131.9	Nonazeotrope		575
A =	C₅H₈	**Cyclopentene**	**43.6**			
8907	C₅H₈	cis-Piperylene	43.6	43.2		178
		"			50 vol. %	1013
A =	C₅H₈	**Isoprene**	**34.2**			
8908	C₅H₈	3-Methyl-1,2-butadiene	40.8	Nonazeotrope		563
8909	C₅H₈	trans-Piperylene	42		9	1013
8910	C₅H₁₀	Cyclopentane	49.4	Nonazeotrope		561
8911	C₅H₁₀	2-Methyl-1-butene	32	Nonazeotrope		714
8912	C₅H₁₀	2-Methyl-2-butene		Nonazeotrope		714
		"	37.1	34.0	86	561
8913	C₅H₁₀	3-Methyl-1-butene	20.6	Nonazeotrope		561, 714
8914	C₅H₁₂	2-Methylbutane	27.95	< 27.7	> 8	561
		"	27.6	Nonazeotrope		714
8915	C₅H₁₂	Pentane, 758 mm.	36	33.6	72.5 v-l	714
		"	36.15	33.8	90	561
8915a	C₅H₁₂O	tert-Butyl methyl ether	55	Nonazeotrope		581c
8916	C₆F₁₅N	Perfluorotriethylamine		30.2	45.5	717
A =	C₅H₈	**3⁻Methyl-1,2-butadiene**	**40.8**			
8917	C₅H₁₀	2-Methyl-2-butene	37.15	Nonazeotrope		563
A =	C₅H₈	**Piperylene**	**42.5**			
8917a	C₅H₉NO	N-Methyl-2-pyrrolidinone 30°–40°C.		Nonazeotrope	v-l	347c

TABLE I. *Binary Systems* 269

No.	Formula	B-Component Name	B.P., °C	Azeotropic Data B.P., °C	Wt.%A		Ref.
A =	$C_5H_8Cl_4$	**Tetrachloropentane**					
8918	$C_7H_{12}Cl_4$	Tetrachloroheptane,					
		12-150 mm.		Nonazeotrope		v-l	718
A =	C_5H_8O	**Cyclopentanone**	130.65				
8919	$C_5H_{10}O_3$	Ethyl carbonate	126.5	Nonazeotrope			552
8920	$C_5H_{11}Br$	1-Bromo-3-methylbutane	120.65	Nonazeotrope			552
8921	$C_5H_{12}O$	Active amyl alcohol	128.5	Nonazeotrope			958
8922	$C_5H_{12}O$	Isoamyl alcohol	131.85	127.8	60	v-l	233,234
		"	131.85	129.4			919
		"	131.85	Nonazeotrope			6
		"	131.9	< 130.0	>58		527
8923	$C_5H_{12}O$	2-Methyl-1-butanol	128.9	127			233,234
		"	128.9	124.6			919
8924	$C_5H_{12}O$	2-Pentanol	119.8	Nonazeotrope			552
8925	C_6H_5Cl	Chlorobenzene	131.75	Nonazeotrope			552
8926	$C_6H_{12}O_2$	Butyl acetate	126.0	Nonazeotrope			552
8927	$C_6H_{12}O_2$	Isoamyl formate	123.8	Nonazeotrope			552
8928	C_7H_8	Toluene	110.75	Nonazeotrope			552
8929	$C_7H_{14}O_2$	Ethyl isovalerate	134.7	Nonazeotrope			552
8930	C_8H_{10}	Ethylbenzene	136.15	Nonazeotrope			552
8931	C_8H_{16}	1,3-Dimethylcyclohexane	120.7	118.0	20		552
A =	$C_5H_8O_2$	**Ethyl Acrylate**	**99.3**				
8932	$C_6H_{14}O$	Isopropyl ether	68.3	Nonazeotrope			981
A =	$C_5H_8O_2$	**Methyl Methacrylate**	**61.8/200 mm.**				
8933	$C_8H_{14}O_2$	Butyl methacrylate,					
		200 mm.	117.7	Nonazeotrope			413
8934	$C_8H_{14}O_3$	2-Ethoxyethyl					
		methacrylate,					
		200 mm.	134.3	Nonazeotrope			413
A =	$C_5H_8O_2$	**2,4-Pentanedione**	**140.2**				
8935	$C_5H_8O_2$	Isopropenyl acetate	96.5	Nonazeotrope			413
8936	$C_5H_{10}O$	Cyclopentanol	140.85	< 135.5	>68		552
8937	$C_5H_{12}O$	Isoamyl alcohol	131.9	< 130.0	>35		552
		" "	131.8	Nonazeotrope			563
8938	C_6H_5Br	Bromobenzene	156.15	154.7	~ 10		563
8939	C_6H_5Cl	Chlorobenzene	131.8	Nonazeotrope			563
8940	C_6H_5I	Iodobenzene	188.55	~ 169	>90		563
8941	C_7H_7Cl	α-Chlorotoluene	179.35	~ 167.5	80		563
8942	C_7H_8	Toluene	110.75	Nonazeotrope			548
8943	$C_7H_{14}O_2$	Isobutyl propionate	137.5	136.4	45		552
8944	$C_7H_{14}O_2$	Propyl isobutyrate	134.0	Nonazeotrope			552
8945	C_8H_{10}	Ethylbenzene	136.15	~ 135	~ 35		548
8946	$C_8H_{18}O$	Isobutyl ether	122.2	Nonazeotrope			548
A =	$C_5H_8O_3$	**Ethyl Pyruvate**	**155.5**				
8947	C_6H_5Br	Bromobenzene	156.1	149.5	48		552
8948	C_6H_5Cl	Chlorobenzene	131.75	Nonazeotrope			552
8949	$C_6H_{10}O$	Cyclohexanone	155.7	153.5			552

No.	Formula	B-Component Name	B.P., °C	B.P., °C	Azeotropic Data Wt.%A	Ref.
A =	**C₅H₈O₃**	**Ethyl Pyruvate** *(continued)*	**155.5**			
8950	C₆H₁₂O₂	Butyl acetate	126.0	Nonazeotrope		552
8951	C₆H₁₂O₂	Isoamyl formate	123.8	Nonazeotrope		575
8952	C₆H₁₄S	Propyl sulfide	141.5	Nonazeotrope		566
8953	C₇H₇Br	*o*-Bromotoluene	181.5	Nonazeotrope		552
8954	C₇H₇Cl	*o*-Chlorotoluene	159.2	151.5	52	552
8955	C₇H₇Cl	*p*-Chlorotoluene	162.4	153.2	58	552
8956	C₇H₈O	Anisole	153.85	148.0	50	**552**
8957	C₇H₁₄O	5-Methyl-2-hexanone	144.2	Nonazeotrope		552
8958	C₇H₁₄O₂	Butyl propionate	146.8	< 145.5	>23	552
8959	C₇H₁₄O₂	Ethyl isovalerate	134.7	Nonazeotrope		552
8960	C₇H₁₄O₂	Propyl isobutyrate	134.0	Nonazeotrope		575
8961	C₈H₁₀	*m*-Xylene	139.2	137.2	30	552
8962	C₈H₁₀O	Phenetole	170.45	Nonazeotrope		552
8963	C₈H₁₆O	2-Octanone	172.85	Nonazeotrope		552
8964	C₈H₁₆O₂	Isoamyl propionate	160.7	153.0	67	552
8965	C₈H₁₆O₂	Isobutyl isobutyrate	148.6	147.0	33	552
8966	C₈H₁₆O₂	Propyl isovalerate	155.7	< 151.8		552
8967	C₈H₁₈O	Butyl ether	142.4	140.4		552
8968	C₈H₁₈S	Isobutyl sulfide	172.0	Nonazeotrope		566
8969	C₉H₁₂	Cumene	152.8	146.2	45	552
8970	C₉H₁₂	Mesitylene	164.6	< 151.5		552
8971	C₉H₁₈O	2,6-Dimethyl-4-heptanone	168.0	Nonazeotrope		552
8972	C₁₀H₁₆	Camphene	159.6	< 148.0		552
8973	C₁₀H₁₆	α-Pinene	155.8	< 147.0		552
8974	C₁₀H₁₈O	Cineol	176.35	Nonazeotrope		552
A =	**C₅H₈O₃**	**Levulinic Acid**	**252**			
8975	C₆H₄ClNO₂	*p*-Chloronitrobenzene	239.1	Reacts		565
8976	C₆H₅NO₂	Nitrobenzene	210.75	Nonazeotrope		552
8977	C₇H₇NO₂	*m*-Nitrotoluene	230.8	229.5	15	552
8978	C₇H₇NO₂	*o*-Nitrotoluene	221.75	221.55	4	552
8979	C₇H₇NO₂	*p*-Nitrotoluene	238.9	236.4	22	552
8980	C₇H₁₄O₂	Enanthic acid	222.0	Nonazeotrope		552
8981	C₈H₈O₃	Methyl salicylate	222.95	222.75	6	552
8982	C₈H₁₀O	3,4-Xylenol	226.8	Nonazeotrope		552
8983	C₈H₁₂O₄	Ethyl maleate	223.3	Nonazeotrope		552
8984	C₈H₁₆O₂	Caprylic acid	238.5	Nonazeotrope		552
8985	C₉H₁₀O₃	Ethyl salicylate	233.8	230.5	18	527,552
8986	C₁₀H₈	Naphthalene	218.0	216.7	11	552
8987	C₁₀H₁₀O₂	Safrol	235.9	232.5	17	527
8988	C₁₀H₁₂O	Anethole	235.7	232.0	22	552
8989	C₁₀H₁₂O₂	Propyl benzoate	230.85	230.0	7	552
8990	C₁₀H₁₄O	Thymol	232.9	Nonazeotrope		552
8991	C₁₀H₁₄O	Carvacrol	237.85	Nonazeotrope		552
8992	C₁₁H₁₀	1-Methylnaphthalene	244.6	237.0	36	552
8993	C₁₁H₁₀	2-Methylnaphthalene	241.15	234.55	29	527

TABLE I. *Binary Systems* 271

No.	Formula	Name	B.P., °C	B.P., °C	Wt.%A	Ref.
		B-Component		**Azeotropic Data**		
A =	**C₅H₈O₃**	**Levulinic Acid** *(continued)*	**252**			
8994	$C_{11}H_{14}O_2$	Isobutyl benzoate	241.9	238.6	25	552
8995	$C_{11}H_{22}O_3$	Isoamyl carbonate	232.2	Nonazeotrope		552
8996	$C_{12}H_{18}$	1,3,5-Triethylbenzene	215.5	214.0	11	552
8997	$C_{12}H_{22}O$	Bornyl ethyl ether	204.9	Nonazeotrope		552
A =	**C₅H₈O₃**	**Methyl Acetoacetate**	**169.5**			
8998	$C_5H_{10}O_2$	Valeric acid	186.35	Nonazeotrope		552
8999	$C_6H_4Cl_2$	o-Dichlorobenzene	179.5	Nonazeotrope		575
9000	$C_6H_4Cl_2$	p-Dichlorobenzene	174.4	167.2	33	552
9001	C_6H_5Br	Bromobenzene	156.15	154.7	~ 10	563
9002	C_6H_5I	Iodobenzene	188.45	Nonazeotrope		575
9003	C_6H_6O	Phenol	181.5	Reacts		563
9004	$C_6H_{10}O$	Cyclohexanone	155.7	Nonazeotrope		552
9005	$C_6H_{10}O_4$	Ethyl oxalate	185.65	Nonazeotrope		575
9006	$C_6H_{12}O$	Cyclohexanol	160.65	Azeotrope doubtful		563
9007	C_7H_6O	Benzaldehyde	179.2	Reacts		563
9008	C_7H_7Cl	α-Chlorotoluene	179.35	~ 167.5	<80	563
9009	C_7H_7Cl	o-Chlorotoluene	159.2	< 158.2	>16	552
9010	C_7H_7Cl	p-Chlorotoluene	162.4	160.0	26	552
9011	C_7H_8O	Anisole	153.85	Nonazeotrope		552
9012	C_8H_8	Styrene	145.8	< 145.0	27	552
9013	C_8H_9Cl	o,m,p-Chloroethyl-benzene, 10 mm.	67.5	60.0	52	51
9014	C_8H_{10}	m-Xylene	139.2	Nonazeotrope		552
9015	$C_8H_{10}O$	Benzyl methyl ether	167.8	< 160.0	>47	552
9016	$C_8H_{10}O$	Phenetole	170.45	< 163.5	>55	552
9017	$C_8H_{14}O$	Methylheptenone	173.2	167.7		552
9018	$C_8H_{16}O$	2-Octanone	172.85	168.5		552
9019	$C_8H_{16}O_2$	Ethyl caproate	167.7	164.0	55	552
9020	$C_8H_{16}O_2$	Isoamyl propionate	160.7	< 159.5	>20	552
9021	$C_8H_{16}O_2$	Isobutyl butyrate	156.9	< 156.5	> 5	575
9022	$C_8H_{16}O_2$	Isobutyl isobutyrate	148.6	Nonazeotrope		552
9023	$C_8H_{18}O$	Butyl ether	142.4	Nonazeotrope		552
9024	$C_8H_{18}S$	Isobutyl sulfide	172.0	166.0	58	566
9025	C_9H_{12}	Mesitylene	164.6	159.5	43	552
9026	C_9H_{12}	Pseudocumene	169	~ 165		563
9027	$C_9H_{18}O$	2,6-Dimethyl-4-heptanone	168.0	< 166.8		552
9028	$C_9H_{18}O_2$	Isoamyl butyrate	181.05	< 168.5	>75	552
9029	$C_9H_{18}O_2$	Isobutyl isovalerate	171.2	165.0	60	552
9030	$C_{10}H_{14}$	Cymene	176.7	165.0	56	552
9031	$C_{10}H_{16}$	Camphene	159.6	152.8	40	552
9032	$C_{10}H_{16}$	Dipentene	177.7	162.3	61	552
9033	$C_{10}H_{16}$	d-Limonene	177.8	162.7	61	563
9034	$C_{10}H_{16}$	α-Phellandrene	171.5	~ 160		563
9035	$C_{10}H_{16}$	α-Pinene	155.8	150.0	36	552
9036	$C_{10}H_{16}$	Terpinene	180.5	< 165		563
9037	$C_{10}H_{18}$	Menthene	170.8	160	52	563

No.	Formula	B-Component Name	B.P., °C	Azeotropic Data B.P., °C	Wt.%A	Ref.
A =	$C_5H_8O_3$	**Methyl Acetoacetate** *(continued)*	**169.5**			
9038	$C_{10}H_{18}O$	Cineol	176.35	< 164.5	80	552
9039	$C_{10}H_{20}O_2$	Isoamyl isovalerate	192.7	Nonazeotrope		552
9040	$C_{10}H_{22}O$	Amyl ether	187.5	Nonazeotrope		552
9041	$C_{10}H_{22}O$	Isoamyl ether	173.2	160.5	60	552
A =	$C_5H_8O_4$	**Methyl Malonate**	**181.4**			
9042	$C_5H_{10}O_2$	Isovaleric acid	176.5	< 180.5	<45	527
9043	$C_5H_{10}O_2$	Valeric acid	186.35	< 180.5	<85	527
9044	$C_6H_4Cl_2$	o-Dichlorobenzene	179.5	173.0	46	562
9045	$C_6H_4Cl_2$	p-Dichlorobenzene	174.4	171.0	30	570
9046	C_6H_5Br	Bromobenzene	156.1	Min. b.p.		547
9047	C_6H_5I	Iodobenzene	188.55	178.0	30	547
9048	C_6H_6O	Phenol	181.5	Reacts		563
9049	C_6H_7N	Aniline	184.35	Reacts		563
9050	$C_6H_{10}O_4$	Ethylidene diacetate	168.5	Nonazeotrope		527
9051	$C_6H_{10}O_4$	Ethyl oxalate	185.65	Nonazeotrope		527
9052	$C_6H_{10}O_4$	Glycol diacetate	186.3	Nonazeotrope		549
9053	$C_6H_{11}BrO_2$	Ethyl α-bromoisobutyrate	178	< 176.5	<40	563
9054	$C_6H_{12}O_2$	Isocaproic acid	199.5	Nonazeotrope		575
9055	$C_6H_{12}O_3$	2-Ethoxyethyl acetate	156.8	Nonazeotrope		575
9056	$C_6H_{13}Br$	1-Bromohexane	156.5	Nonazeotrope		575
9057	C_7H_7Br	α-Bromotoluene	198.5	Nonazeotrope		575
9058	C_7H_7Br	m-Bromotoluene	184.3	176.0	62	527
9059	C_7H_7Br	o-Bromotoluene	181.4	174.45	44.5	529
9060	C_7H_7Br	p-Bromotoluene	185.0	176.5	55	538
9061	C_7H_7Cl	α-Chlorotoluene	179.35	~ 178		563
9062	C_7H_7Cl	o-Chlorotoluene	159.15	Nonazeotrope		547
9063	C_7H_7Cl	p-Chlorotoluene	162.4	Nonazeotrope		547
9064	C_7H_8O	Anisole	153.85	Nonazeotrope		557
9065	C_7H_8O	o-Cresol	190.8	Reacts		563
9066	C_7H_8O	p-Cresol	201.7	Nonazeotrope		575
9067	C_8H_8	Styrene	145.8	Nonazeotrope		575
9068	C_8H_8O	Acetophenone	202.0	201.0	39	552
9069	$C_8H_{10}O$	Benzyl methyl ether	167.8	Nonazeotrope		557
9070	$C_8H_{10}O$	p-Methylanisole	177.05	< 174.8	40?	557
9071	$C_8H_{10}O$	Phenetole	171.5	169.9	23	557
9072	$C_8H_{10}O_2$	Veratrole	206.8	Nonazeotrope		557
9083	$C_8H_{14}O$	Methylheptenone	173.2	Nonazeotrope		552
9074	$C_8H_{16}O_2$	Hexyl acetate	171.5	< 170.8	>12	575
9075	$C_8H_{18}O$	Octyl alcohol	195.15	Reacts		536
9076	$C_8H_{18}O$	sec-Octyl alcohol	178.5	Chem. action		563
9077	$C_8H_{18}S$	Butyl sulfide	185.0	176.2	50	566
9078	C_9H_8	Indene	182.6	< 176.2	50?	562
9079	C_9H_{12}	Mesitylene	164.6	162	>10	546
9080	C_9H_{12}	Propylbenzene	158.9	< 159		546
9081	C_9H_{12}	Pseudocumene	168.2	< 165.5	>20	546
9082	$C_9H_{12}O$	Benzyl ethyl ether	185.0	178.0	37	1051
9083	$C_9H_{18}O_2$	Butyl isovalerate	177.6	175.0	30	449

TABLE I. *Binary Systems* 273

No.	Formula	B-Component Name	B.P., °C	B.P., °C	Wt.%A	Ref.
A =	**C₅H₈O₄**	**Methyl Malonate** *(continued)*	**181.4**			
9084	C₉H₁₈O₂	Isoamyl butyrate	181.05	< 177.2	>39	449
9085	C₉H₁₈O₂	Isobutyl isovalerate	171.2	< 170.5	>17	449
9086	C₁₀H₈	Naphthalene	218.0	Nonazeotrope		575
9087	C₁₀H₁₄	Butylbenzene	183.2	173	52	546
9088	C₁₀H₁₄	Cymene	176.7	169.0	40	546
9089	C₁₀H₁₆	Camphene	159.6	154.6	26	529
9090	C₁₀H₁₆	*d*-Limonene	177.8	167.3	48	529
9091	C₁₀H₁₆	Nopinene	164	158	28	546
9092	C₁₀H₁₆	α-Pinene	155.8	151.5	~ 22	529
9093	C₁₀H₁₆	α-Terpinene	173.3	167	<45	546
9094	C₁₀H₁₆	Terpinene	181.5	164.5	51	538
9095	C₁₀H₁₆	Terpinolene	185.2	171.0	<62	546
9096	C₁₀H₁₆	Thymene	179.7	~ 169.0	50	537
9097	C₁₀H₁₈	Menthene	170.8	164	37	546
9098	C₁₀H₁₈O	Cineol	176.35	169.1	40.5	557
9099	C₁₀H₁₈O	Linalool	198.6	Reacts		536
9100	C₁₀H₂₀O₂	Isoamyl isovalerate	192.7	< 180.8	>75	549
9101	C₁₀H₂₂	2,7-Dimethyloctane	160.2	< 157	<30	546
9102	C₁₀H₂₂O	Amyl ether	187.5	< 175.0	<62	557
9103	C₁₀H₂₂O	Isoamyl ether	173.4	165.5	35	557
9104	C₁₁H₂₀O	Isobornyl methyl ether	192.4	< 177.5	<90	557
9105	C₁₂H₁₈	1,3,5-Triethylbenzene	215.5	Nonazeotrope		575
A =	**C₅H₈O₄**	**Propylene Glycol Diformate**				
9106	CₙHₘ	Hydrocarbons		Min. b.pt.		644
A =	**C₅H₉ClO₂**	**Propyl Chloroacetate**	**163.5**			
9107	C₅H₁₀O₂	Isovaleric acid	176.5	Nonazeotrope		575
9108	C₅H₁₂O	Isoamyl alcohol	131.9	Nonazeotrope		575
9109	C₆H₁₂O	Cyclohexanol	160.8	159.0	47	575
9110	C₆H₁₂O₃	2-Ethoxyethyl acetate	156.8	Nonazeotrope		575
9111	C₆H₁₄O	Hexyl alcohol	157.85	156.4	40	575
9112	C₇H₇Cl	*o*-Chlorotoluene	159.2	< 158.5	<35	575
9113	C₇H₇Cl	*p*-Chlorotoluene	162.4	160.2	49	562
9114	C₇H₈O	Anisole	153.85	Nonazeotrope		575
9115	C₇H₁₆O	Heptyl alcohol	176.15	Nonazeotrope		575
9116	C₈H₁₀O	Phenetole	170.45	Nonazeotrope		575
9117	C₈H₁₆O	2-Octanone	172.85	Nonazeotrope		575
9118	C₈H₁₆O₂	Butyl butyrate	166.4	Nonazeotrope		575
9119	C₈H₁₆O₂	Isoamyl propionate	160.7	< 160.5	>20	575
9120	C₈H₁₆O₂	Isobutyl butyrate	156.9	Nonazeotrope		575
9121	C₈H₁₈O	*sec*-Octyl alcohol	180.4	Nonazeotrope		575
9122	C₉H₁₂	Mesitylene	164.6	< 161.0	<72	575
9123	C₉H₁₂	Propylbenzene	159.3	157.0	40	562
9124	C₉H₁₈O₂	Isobutyl isovalerate	171.2	Nonazeotrope		575
9125	C₁₀H₁₄	Cymene	176.7	Nonazeotrope		575
9126	C₁₀H₁₆	Camphene	159.6	156.2	42	562
9127	C₁₀H₁₆	α-Pinene	155.8	154.0	25	562
9128	C₁₀H₂₂O	Isoamyl ether	173.2	Nonazeotrope		575

No.	Formula	B-Component Name	B.P., °C	B.P., °C	Azeotropic Data Wt.%A		Ref.
A =	C_5H_9N	**Isovaleronitrile**	**130.5**				
9129	$C_6H_{14}O$	Propyl ether	90.1	Nonazeotrope			575
9130	C_8H_{10}	Ethylbenzene	136.15	126.3	60		562
9131	$C_8H_{18}O$	Isobutyl ether	122.3	115.5	24		562
A =	C_5H_9N	**Valeronitrile**	**141.3**				
9132	$C_5H_{12}O$	Amyl alcohol	138.2	< 136.5	<42		565
9133	$C_5H_{12}O$	2-Pentanol	119.8	Nonazeotrope			565
9134	$C_6H_{12}O_2$	Butyl acetate	126.0	Nonazeotrope			565
9135	$C_6H_{12}O_2$	Ethyl butyrate	121.5	Nonazeotrope			565
9136	$C_6H_{12}O_2$	Isobutyl acetate	117.4	Nonazeotrope			565
9137	$C_6H_{14}O$	Propyl ether	90.1	Nonazeotrope			575
9138	$C_6H_{14}S$	Propyl sulfide	141.5	< 137.5			565
9139	C_7H_8	Toluene	110.75	Nonazeotrope			565
9140	$C_7H_{14}O_2$	Isobutyl propionate	137.5	136.0	27		565
9141	C_8H_{10}	m-Xylene	139.2	< 136.5			565
9142	$C_8H_{18}O$	Butyl ether	142.4	< 130.5	>42		562
9143	$C_8H_{18}O$	Isobutyl ether	122.3	119.0	10		575
9144	C_9H_{12}	Propylbenzene	159.3	Nonazeotrope			565
A =	C_5H_{10}	**Amylene**	**37**				
9145	$C_6H_5NO_2$	Nitrobenzene	210.75	Nonazeotrope			554
9146	C_6H_7N	Aniline	184.35	Nonazeotrope			563
A =	C_5H_{10}	**Cyclopentane**	**49.4**				
9147	C_5H_{10}	2-Methyl-2-butene	37.1	Nonazeotrope			561
9148	C_5H_{12}	Pentane	36.15	Nonazeotrope			561
9149	$C_5H_{12}O$	Ethyl propyl ether	63.85	Nonazeotrope			558
9150	C_6H_6	Benzene	80.1	Nonazeotrope		v-l	676
9151	C_6H_{10}	Biallyl	60.1	Nonazeotrope			561
9152	C_6H_{14}	2,2-Dimethylbutane	49.7	Nonazeotrope			788
		"	49.7	49.1	82.3		630
9153	C_6H_{14}	2,3-Dimethylbutane	58.0	Nonazeotrope			561
A =	C_5H_{10}	**2-Methyl-1-butene**	**31.1**				
9154	$C_6F_{15}N$	Perfluorotriethylamine		28.8	50.7		717
A =	C_5H_{10}	**2-Methyl-2-butene**	**37.15**				
9155	C_5H_{12}	2-Methylbutane	27.95	27.7?			563
9156	C_5H_{12}	Pentane	36.15	35.5	~ 43		563
		"	36.15	Nonazeotrope			714
9157	$C_6F_{15}N$	Perfluorotriethylamine		34.5	45.8		717
A =	C_5H_{10}	**3-Methyl-1-butene**	**22.5**				
9158	C_5H_{12}	2-Methylbutane	27.95	< 20.4	>86		561
		"		27.6	Nonazeotrope		714
A =	$C_5H_{10}O$	**Cyclopentanol**	**140.85**				
9159	$C_5H_{10}O_3$	Ethyl carbonate	126.5	125			567
9160	$C_5H_{10}O_3$	Ethyl lactate	154.1	Nonazeotrope			575
9161	$C_5H_{10}O_3$	2-Methoxyethyl acetate	144.6	139.0	75		575
9162	$C_5H_{11}Br$	1-Bromo-3-methylbutane	120.65	< 120.2	> 5		575
9163	$C_5H_{12}O_2$	2-Propoxyethanol	151.35	Nonazeotrope			526
9164	C_6H_5Cl	Chlorobenzene	131.75	< 128.5	>20		567

TABLE I.   *Binary Systems*   **275**

No.	Formula	B-Component Name	B.P., °C	Azeotropic Data B.P., °C	Wt.%A	Ref.
A =	**C₅H₁₀O**	**Cyclopentanol** *(continued)*	**140.85**			
9165	C₆H₆	Benzene	80.15	Nonazeotrope		575
9166	C₆H₆O	Phenol	182.2	Nonazeotrope		575
9167	C₆H₇N	Aniline	184.35	Nonazeotrope		551
9168	C₆H₇N	2-Picoline	130.7	Nonazeotrope		575
9169	C₆H₁₀O	Mesityl oxide	129.45	Nonazeotrope		552
9170	C₆H₁₀S	Allyl sulfide	139.35	< 135.5	>33	566
9171	C₆H₁₁N	Capronitrile	163.9	Nonazeotrope		575
9172	C₆H₁₂	Cyclohexane	80.75	Nonazeotrope		575
9173	C₆H₁₂O₂	Butyl acetate	126.0	Nonazeotrope		575
9174	C₆H₁₂O₂	Isoamyl formate	123.8	Nonazeotrope		575
9175	C₆H₁₂O₃	Paraldehyde	124.35	Nonazeotrope		575
9176	C₇H₈	Toluene	110.75	Nonazeotrope		575
9177	C₇H₁₄O₂	Ethyl isovalerate	134.7	< 134.5	~ 15	575
9178	C₇H₁₄O₂	Isoamyl acetate	142.1	< 139.4	>48	567
9179	C₇H₁₄O₂	Isobutyl propionate	137.5	136.5	28	567
9180	C₈H₁₀	*m*-Xylene	139.2	132.8	40	567
9181	C₈H₁₀	*p*-Xylene	138.45	132.2	38	575
9182	C₈H₁₆	1,3-Dimethylcyclohexane	120.7	119.0	15	567
9183	C₈H₁₈O	Butyl ether	142.4	< 136.7	>39	527
9184	C₈H₁₈O	Isobutyl ether	122.3	< 122.0	> 3	575
A =	**C₅H₁₀O**	**Isovaleraldehyde**	**92.1**			
9185	C₅H₁₀O	3-Pentanone	102.05	Nonazeotrope		552
9186	C₅H₁₀O₂	Methyl isobutyrate	92.5	< 92.2	>30	548
9187	C₆H₆	Benzene	80.15	Nonazeotrope		575
9188	C₆H₁₄	Hexane	68.8	Nonazeotrope		575
A =	**C₅H₁₀O**	**3-Methyl-2-butanone**	**95.4**			
9189	C₅H₁₀O₂	Ethyl propionate	99.1	Nonazeotrope		552
9190	C₅H₁₀O₂	Isopropyl acetate	90.8	Nonazeotrope		548
9191	C₅H₁₀O₂	Methyl isobutyrate	92.5	Nonazeotrope		552
9192	C₅H₁₁Cl	1-Chloro-3-methylbutane	99.4	95.0	65	552
9193	C₅H₁₁NO₂	Isoamyl nitrite	97.15	94.0	50	552
9194	C₅H₁₂O₂	Diethoxymethane	87.95	Nonazeotrope		575
9195	C₆H₆	Benzene	80.15	Nonazeotrope		552
9196	C₆H₁₂	Cyclohexane	80.75	78.5	15	552
		"	80.75		14.8	735
9197	C₆H₁₅N	Triethylamine	89.35	< 88.0		575
9198	C₇H₁₆	Heptane	98.4	89.5	48	552
		"	98.4		48	735
A =	**C₅H₁₀O**	**2-Methyl-3-buten-2-ol**	**97**			
9198a	C₅H₁₂O	*tert*-Butyl methyl ether	55	Nonazeotrope		581c
9198b	C₆H₁₀O	Methyldihydropyran	118.5	Nonazeotrope		581c
9198c	C₆H₁₂O₂	4,4-Dimethyl-1,3-dioxane	133.4	Nonazeotrope		581c
9198d	C₆H₁₄	Hexane	68.8	67.2	9.3	581c
9198e	C₆H₁₄O	Isopropyl ether	69	Nonazeotrope		581c
A =	**C₅H₁₀O**	**3-Methyl-2-buten-1-ol**	**140**			
9198f	C₆H₁₂O₂	4,4-Dimethyl-1,3-dioxane	133.4	Nonazeotrope		581c

No.	B-Component Formula	B-Component Name	B.P., °C	Azeotropic Data B.P., °C	Azeotropic Data Wt.%A	Ref.
A =	**C₅H₁₀O**	**3-Methyl-3-buten-1-ol**	**130**			
9198g	C₆H₁₂O₂	4,4-Dimethyl-1,3-dioxane	133.4	129.5	76	581c
A =	**C₅H₁₀O**	**2-Pentanone**	**102.25**			
9199	C₅H₁₀O	3-Pentanone	102.2	Nonazeotrope		563
9200	C₅H₁₀O₂	Butyl formate	106.8	Nonazeotrope		552
9201	C₅H₁₀O₂	Ethyl propionate	99.1	Nonazeotrope		552
9202	C₅H₁₀O₂	Isobutyl formate	98.2	Nonazeotrope		552
9203	C₅H₁₀O₂	Methyl butyrate	102.65	101.9	50	552
9204	C₅H₁₀O₂	Methyl isobutyrate	92.5	Nonazeotrope		552
9205	C₅H₁₀O₂	Propyl acetate	101.6	100.8	35	552
9206	C₅H₁₁NO₂	Isoamyl nitrite	97.15	96.5	20	552
9207	C₅H₁₂O	tert-Amyl alcohol	102.35	100.9	58	552
9208	C₆H₆	Benzene	80.15	Nonazeotrope		575
9209	C₆H₁₂	Cyclohexane	80.75	79.8	5	552,735
9210	C₆H₁₄	Hexane	68.8	Nonazeotrope		575
9211	C₇H₈	Toluene	110.7	Nonazeotrope		563
9212	C₇H₁₄	Methylcyclohexane	101.15	95.2	40	552
9213	C₇H₁₆	Heptane	98.4	93.2	34	552
A =	**C₅H₁₀O**	**3-Pentanone**	**102.05**			
9214	C₅H₁₀O₂	Butyl formate	106.8	Nonazeotrope		552
9215	C₅H₁₀O₂	Ethyl propionate	99.1	Nonazeotrope		552
9216	C₅H₁₀O₂	Isobutyl formate	98.2	Nonazeotrope		552
9217	C₅H₁₀O₂	Methyl butyrate	102.65	101.45	55	552
9218	C₅H₁₀O₂	Methyl isobutyrate	92.5	Nonazeotrope		552
9219	C₅H₁₀O₂	Propyl acetate	101.6	100.75	40	552
9220	C₅H₁₁Cl	1-Chloro-3-methylbutane	99.4	98.5	25	552
9221	C₅H₁₁N	Piperidine	106.4	Nonazeotrope		575
9222	C₅H₁₁NO₂	Isoamyl nitrite	97.15	96.45	21	552
9223	C₅H₁₂O	tert-Amyl alcohol	102.35	100.7	60	552
9224	C₅H₁₂O	Isoamyl alcohol	131.9	Nonazeotrope		527
9225	C₅H₁₂O	2-Pentanol	119.8	Nonazeotrope		552
9226	C₅H₁₂O	3-Pentanol	116.0	Nonazeotrope		552
9227	C₆H₆	Benzene	80.15	Nonazeotrope		552
9228	C₆H₁₂	Cyclohexane	80.8	Nonazeotrope		548
9229	C₆H₁₂	Methylcyclopentane	72.0	Nonazeotrope		552
9230	C₆H₁₂O₂	Ethyl isobutyrate	110.1	Nonazeotrope		552
9231	C₆H₁₄	Hexane	68.8	Nonazeotrope		552
9232	C₆H₁₄O	Propyl ether	90.1	Nonazeotrope		552
9233	C₆H₁₄O₂	Acetal	103.55	< 101.8	>75	552
9234	C₆H₁₄S	Isopropyl sulfide	120.5	Nonazeotrope		566
9235	C₆H₁₅N	Dipropylamine	109.2	< 101.0	>82	551
9236	C₇H₈	Toluene	110.75	Nonazeotrope		552
9237	C₇H₁₄	Methylcyclohexane	101.15	95.0	40	552,735
9238	C₇H₁₆	Heptane	98.45	93.0	35	527,735
9239	C₈H₁₆	1,3-Dimethylcyclohexane	120.7	100.5	83	552,735
9240	C₈H₁₈	2,5-Dimethylhexane	109.4	97.5	60	552,735
A =	**C₅H₁₀O₂**	**Butyl Formate**	**106.8**			
9241	C₅H₁₀O₂	Methyl butyrate	102.65	Nonazeotrope		575

TABLE I.   *Binary Systems*   277

No.	B-Component			Azeotropic Data			
	Formula	Name	B.P., °C	B.P., °C	Wt.%A		Ref.
A =	$C_5H_{10}O_2$	**Butyl Formate** (*continued*)	106.8				
9242	$C_5H_{10}O_2$	Propyl acetate	101.6	Nonazeotrope			575
9243	$C_5H_{11}Br$	1-Bromo-3-methylbutane	120.65	Nonazeotrope			575
9244	$C_5H_{12}O$	*tert*-Amyl alcohol	102.35	101.0	35		567
9245	$C_5H_{12}O$	Isoamyl alcohol	131.9	Nonazeotrope			527
9246	$C_5H_{12}O$	2-Pentanol	119.8	Nonazeotrope			575
9247	$C_5H_{12}O$	3-Pentanol	116.0	< 106.5	<98.5		575
9248	C_6H_6	Benzene	80.15	Nonazeotrope			575
9249	C_6H_{12}	Cyclohexane	80.75	Nonazeotrope			575
9250	$C_6H_{12}O$	Pinacolone	106.2	106.0	38		552
9251	$C_6H_{12}O_2$	Ethyl isobutyrate	110.1	Nonazeotrope			575
9252	$C_6H_{14}O$	Propyl ether	90.1	Nonazeotrope			557
9253	$C_6H_{14}O_2$	Acetal	103.55	Nonazeotrope			557
9254	C_7H_8	Toluene	110.75	< 106.4	>70		575
9255	C_7H_{14}	Methylcyclohexane	101.15	96.0	35		562
9256	C_7H_{16}	Heptane	98.45	90.7	40		538
9257	$C_8H_{18}O$	Isobutyl ether	122.3	Nonazeotrope			557
A =	$C_5H_{10}O_2$	**Ethyl Propionate**	99.15				
9258	$C_5H_{10}O_2$	Isobutyl formate	97.9	Nonazeotrope			531
9259	$C_5H_{10}O_2$	Methyl butyrate	102.65	Nonazeotrope			575
9260	$C_5H_{10}O_2$	Propyl acetate	101.55	Nonazeotrope			563
9261	$C_5H_{11}Cl$	1-Chloro-3-methylbutane	99.8	98.4	55		538
9262	$C_5H_{12}O$	*tert*-Amyl alcohol	102.0	98	62		536
9263	$C_5H_{12}O$	3-Pentanol	116.0	Nonazeotrope			575
9264	C_6H_6	Benzene	80.15	Nonazeotrope			575
9265	C_6H_{12}	Cyclohexane	80.8	Nonazeotrope			546
9266	$C_6H_{12}O$	Pinacolone	106.2	Nonazeotrope			552
9267	C_6H_{14}	Hexane	69.0	Nonazeotrope			546
9268	$C_6H_{14}O$	Propyl ether	90.1	Nonazeotrope			557
9269	$C_6H_{14}O_2$	Acetal	103.55	Nonazeotrope			557
9270	$C_6H_{14}O_2$	Ethoxypropoxymethane	113.7	Nonazeotrope			557
9271	C_7H_8	Toluene	110.7	Nonazeotrope			546
9272	C_7H_{14}	Methylcyclohexane	101.1	94.5	~ 53		573
9273	C_7H_{16}	*n*-Heptane	98.45	93.0	47		527
9274	C_8H_{18}	2,5-Dimethylhexane	109.4	< 97.5	<78		562
A =	$C_5H_{10}O_2$	**Isobutyl Formate**	98.2				
9275	$C_5H_{10}O_2$	Methyl isobutyrate	92.5	Nonazeotrope			575
9276	$C_5H_{10}O_2$	Propyl acetate	101.6	Nonazeotrope			575
9277	$C_5H_{11}Cl$	1-Chloro-3-methylbutane	99.8	94.5	50		538
9278	$C_5H_{11}NO_2$	Isoamyl nitrite	97.15	95.5	43		549
9279	$C_5H_{12}O$	*tert*-Amyl alcohol	102.35	< 97.0	<81		567
9280	$C_5H_{12}O_2$	Diethoxymethane	87.95	Nonazeotrope			557
9281	C_6H_6	Benzene	80.2	Nonazeotrope			573
9282	C_6H_{12}	Cyclohexane	80.8	80	<20		546
9283	$C_6H_{12}O$	Pinacolone	106.2	Nonazeotrope			552
9284	C_6H_{14}	Hexane	69.0	68.5	12		546
9285	$C_6H_{14}O_2$	Acetal	103.55	Nonazeotrope			557

No.	Formula	B-Component Name	B.P., °C	Azeotropic Data B.P., °C	Wt.%A	Ref.
A =	C₅H₁₀O₂	Isobutyl Formate *(continued)*	98.2			
9286	C₇H₈	Toluene	110.7	Azeotrope doubtful		563
9287	C₇H₁₄	Methylcyclohexane	100.95	92.4	~ 57	572
9288	C₇H₁₆	n-Heptane	98.45	< 90.5	<50	527
9289	C₈H₁₈	2,5-Dimethylhexane	109.4	93.5	63	562
A =	C₅H₁₀O₂	Isopropyl Acetate	91.0			
9290	C₅H₁₀O₂	Methyl isobutyrate	92.3	Nonazeotrope		531
9291	C₅H₁₁Cl	1-Chloro-3-methylbutane	99.8	Nonazeotrope		547
9292	C₅H₁₁NO₂	Isoamyl nitrite	97.15	Nonazeotrope		549
9293	C₅H₁₂O₂	Diethoxymethane	87.95	< 87.6	<42	575
9294	C₆H₆	Benzene	80.2	Nonazeotrope		538
9295	C₆H₈O	2,5-Dimethylfuran	93.3	Nonazeotrope		981
9296	C₆H₁₂	Cyclohexane	80.75	78.9	25	538
9297	C₆H₁₄	Hexane	68.8	< 68.5	< 9	575
		"	69.0	Nonazeotrope		546
9298	C₆H₁₄O	Propyl ether	90.55	88.5	50	557
9299	C₆H₁₄O₂	Acetal	103.55	Nonazeotrope		548
9300	C₇H₈	Toluene	110.75	Nonazeotrope		575
9301	C₇H₁₄	Methylcyclohexane	101.1	89	78	546
9302	C₇H₁₆	Heptane	98.45	87.5	67	538
9303	C₈H₁₈	2,5-Dimethylhexane	109.4	< 89.0	<95	575
A =	C₅H₁₀O₂	Isovaleric Acid	176.5			
9304	C₅H₁₁Br	1-Bromo-3-methylbutane	120.65	Nonazeotrope		527
9305	C₅H₁₁I	1-Iodo-3-methylbutane	147.65	147.0	3	562
9306	C₆H₃Cl₃	1,3,5-Trichlorobenzene	208.4	Nonazeotrope		542
9307	C₆H₄BrCl	p-Bromochlorobenzene	196.4	175.5	75	562
9308	C₆H₄Cl₂	o-Dichlorobenzene	179.5	171.2	42	527
9309	C₆H₄Cl₂	p-Dichlorobenzene	174.5	168.85	28	527
9310	C₆H₅Br	Bromobenzene	156.15	154.75	8	527
9311	C₆H₅Cl	Chlorobenzene	131.75	Nonazeotrope		527
9312	C₆H₅ClO	o-Chlorophenol	175.5	172		563
9313	C₆H₅I	Iodobenzene	188.55	174.0	~ 55	538
9314	C₆H₆O	Phenol	181.5	Nonazeotrope		527
9315	C₆H₈O₄	Methyl fumarate	193.25	Nonazeotrope		527
9316	C₆H₁₀O	Cyclohexanone	155.7	Nonazeotrope		527
9317	C₆H₁₀O₃	Ethyl acetoacetate	180.4	176.1	77	527
9318	C₆H₁₀O₄	Ethylidene diacetate	168.5	Nonazeotrope		527
9319	C₆H₁₀O₄	Ethyl oxalate	185.65	176.3	84	570
9320	C₆H₁₀O₄	Glycol diacetate	186.3	Nonazeotrope		527
9321	C₆H₁₀S	Allyl sulfide	139.35	Nonazeotrope		566
9322	C₆H₁₂O₃	2-Ethoxyethyl acetate	156.8	Nonazeotrope		527
9323	C₆H₁₃Br	1-Bromohexane	156.5	155.0	10	562
9324	C₆H₁₄S	Propyl sulfide	141.5	Nonazeotrope		575
9325	C₇H₆Cl₂	α,α-Dichlorotoluene	205.2	Nonazeotrope		527
9326	C₇H₆O	Benzaldehyde	179.2	174.5	~ 68	541
9327	C₇H₇Br	α-Bromotoluene	198.5	175.2	72	562
		"	198.5	Nonazeotrope		563
9328	C₇H₇Br	m-Bromotoluene	184.3	172.5	45	527

TABLE I. *Binary Systems*   279

No.	B-Component Formula	Name	B.P., °C	Azeotropic Data B.P., °C	Wt.%A	Ref.
A =	$C_5H_{10}O_2$	**Isovaleric Acid** *(continued)*	**176.5**			
9329	C_7H_7Br	*o*-Bromotoluene	181.75	172.1	39.5	527
9330	C_7H_7Br	*p*-Bromotoluene	185.2	173.0	48	555
9331	C_7H_7Cl	α-Chlorotoluene	179.35	171.2	38	527
9332	C_7H_7Cl	o-Chlorotoluene	159	157.5	12	527
9333	C_7H_7Cl	*p*-Chlorotoluene	161.3	160.0	15	527
9334	C_7H_8	Toluene	110.95	Nonazeotrope		527
9335	C_7H_8O	Anisole	153.85	Nonazeotrope		527
9336	C_7H_8O	*o*-Cresol	191.1	Nonazeotrope		575
9337	$C_7H_{13}ClO_2$	Isoamyl chloroacetate	190.5	Nonazeotrope		575
9338	$C_7H_{14}O_3$	1,3-Butanediol methyl ether acetate	171.75	178.0	66	527
9339	C_8H_8	Styrene	145.8	145.2	8	575
9340	$C_8H_8O_2$	Phenyl acetate	195.5	Nonazeotrope		563
9341	C_8H_{10}	Ethylbenzene	136.15	Nonazeotrope		527
9342	C_8H_{10}	*m*-Xylene	139.0	Nonazeotrope		527
9343	C_8H_{10}	*o*-Xylene	144.3	143.8	5	527
9344	C_8H_{10}	*p*-Xylene	138	Nonazeotrope		527
9346	$C_8H_{10}O$	Benzyl methyl ether	167.8	< 167.0	<22	575
9345	$C_8H_{10}O$	*p*-Methylanisole	177.05	172.0	45	562
9347	$C_8H_{10}O$	Phenetole	171.5	168.5	23	527
9348	$C_8H_{14}O$	Methylheptenone	173.2	Nonazeotrope		552
9349	$C_8H_{16}O$	2-Octanone	172.85	Nonazeotrope		527
9350	$C_8H_{16}O_2$	Isobutyl butyrate	157	Nonazeotrope		563
9351	$C_8H_{16}O_2$	Hexyl acetate	171.5	Nonazeotrope		575
9352	C_8H_{18}	Octane	125.75	Nonazeotrope		527
9353	$C_8H_{18}O$	Butyl ether	142.4	Nonazeotrope		527
9354	$C_8H_{18}S$	Butyl sulfide	185	175	73	555
9355	C_9H_8	Indene	183.0	173.0	60	527
9356	C_9H_{12}	Cumene	152.8	152.0	12	527
9357	C_9H_{12}	Mesitylene	164.6	162.5	19	527
9358	C_9H_{12}	Propylbenzene	159.3	157.5	14	562
9359	C_9H_{12}	Pseudocumene	168.2	165.7	23	541
9360	$C_9H_{12}O$	Phenyl propyl ether	190.2	Nonazeotrope		541
9361	$C_9H_{18}O$	2,6-Dimethyl-4-heptanone	168.0	Nonazeotrope		527
9362	$C_9H_{18}O_2$	Isoamyl butyrate	181.05	Nonazeotrope		575
		"	178.5	176.1	70	538
9363	$C_9H_{18}O_2$	Isobutyl isovalerate	171.2	Nonazeotrope		527
9364	$C_{10}H_8$	Naphthalene	218.05	Nonazeotrope		527
9365	$C_{10}H_{14}$	Butylbenzene	183.1	173.0	50	527
9366	$C_{10}H_{14}$	Cymene	175.3	170.8	38	527
9367	$C_{10}H_{16}$	Camphene	159.6	156.5	17	527
9368	$C_{10}H_{16}$	*d*-Limonene	177.8	168.9	41	563
9369	$C_{10}H_{16}$	Nopinene	163.8	160.5	22	527
9370	$C_{10}H_{16}$	α-Phellandrene	171.5	165	~ 35	563
9371	$C_{10}H_{16}$	α-Pinene	155.8	154.2	11	527
9372	$C_{10}H_{16}$	α-Terpinene	173.4	168.0	32	562

No.	Formula	B-Component Name	B.P., °C	Azeotropic Data B.P., °C	Wt.%A	Ref.
A =	$C_5H_{10}O_2$	**Isovaleric Acid** *(continued)*	**176.5**			
9373	$C_{10}H_{16}$	γ-Terpinene	183	172.5	47	562
9374	$C_{10}H_{16}$	Terpinene	180.5	170	~ 43	563
9375	$C_{10}H_{16}$	Terpinolene	184.6	171.5	52	562
9376	$C_{10}H_{16}$	Thymene	179.7	170.5	44	541
9377	$C_{10}H_{18}O$	Cineol	176.3	175.0	42.5	527
9378	$C_{10}H_{20}O_2$	Isoamyl isovalerate	192.7	Nonazeotrope		527
9379	$C_{10}H_{22}$	Decane	173.3	167.0	33	562
9380	$C_{10}H_{22}$	2,7-Dimethyloctane	160.25	158.0	20	527
9381	$C_{10}H_{22}O$	Amyl ether	187.5	< 175.0	<70	527
9382	$C_{10}H_{22}O$	Isoamyl ether	173.4	168.85	27	564
9383	$C_{11}H_{20}O_2$	Isobornyl methyl ether	192.4	Nonazeotrope		575
9384	$C_{12}H_{18}$	Triethylbenzene	215.5	Nonazeotrope		527
9385	$C_{13}H_{28}$	Tridecane	234.0	Nonazeotrope		575
A =	$C_5H_{10}O_2$	**Methyl Butyrate**	**102.65**			
9386	$C_5H_{10}O_2$	Propyl acetate	101.60	Nonazeotrope		549
9387	$C_5H_{11}Cl$	1-Chloro-3-methylbutane	99.4	Nonazeotrope		575
9388	$C_5H_{12}O$	*tert*-Amyl alcohol	102.0	~ 99	~ 57	563
9389	$C_5H_{12}O$	3-Pentanol	116.0	Nonazeotrope		575
9390	C_6H_6	Benzene	80.15	Nonazeotrope		575
9391	$C_6H_{12}O$	Pinacolone	106.2	Nonazeotrope		552
9392	$C_6H_{14}O$	Propyl ether	90.1	Nonazeotrope		557
9393	$C_6H_{14}O_2$	Acetal	103.55	102	~ 55	557
9394	C_7H_8	Toluene	110.7	Nonazeotrope		563
9395	C_7H_{14}	Methylcyclohexane	101.1	97.0	45	546
9396	C_7H_{16}	*n*-Heptane	98.45	95.1	35	527
9397	C_8H_{16}	1,3-Dimethylcyclohexane	120.7	Nonazeotrope		575
9398	C_8H_{18}	2,5-Dimethylhexane	109.2	100.0	<75	546
9399	C_8H_{18}	*n*-Octane	125.8	Nonazeotrope		546
A =	$C_5H_{10}O_2$	**Methyl Isobutyrate**	**92.5**			
9400	$C_5H_{11}Cl$	1-Chloro-3-methylbutane	99.8	Nonazeotrope		547
9401	$C_5H_{11}NO_2$	Isoamyl nitrite	97.15	Nonazeotrope		549
9402	$C_5H_{12}O$	*tert*-Amyl alcohol	102.35	Nonazeotrope		575
9403	$C_5H_{12}O_2$	Diethoxymethane	87.95	Nonazeotrope		527
9403	C_6H_6	Benzene	80.2	Nonazeotrope		573
9405	C_6H_{12}	Cyclohexane	80.75	~ 78.6	~ 12	573
9406	C_6H_{12}	Methylcyclopentane	72.0	Nonazeotrope		575
9407	C_6H_{14}	Hexane	69.0	Nonazeotrope		546
9408	$C_6H_{14}O$	Propyl ether	90.1	89.7	75	557
9409	$C_6H_{14}O_2$	Acetal	104.5	Nonazeotrope		557
9410	C_7H_8	Toluene	110.75	Nonazeotrope		575
9411	C_7H_{14}	Methylcyclohexane	101.1	91	75	546
9412	C_7H_{16}	*n*-Heptane	98.45	89.7	65	527
A =	$C_5H_{10}O_2$	**Propyl Acetate**	**101.6**			
9413	$C_5H_{11}Cl$	1-Chloro-3-methylbutane	99.8	98.5	40	547
9414	$C_5H_{11}NO_2$	Isoamyl nitrite	97.15	Nonazeotrope		550

TABLE I. *Binary Systems* 281

No.	Formula	B-Component Name	B.P., °C	Azeotropic Data B.P., °C	Wt.%A	Ref.
A =	**C₅H₁₀O₂**	**Propyl Acetate (continued)**	**101.6**			
9415	C₅H₁₂O	tert-Amyl alcohol	102.0	99.5	58	536
9416	C₅H₁₂O	3-Pentanol	116.0	Nonazeotrope		575
9417	C₆H₆	Benzene	80.2	Nonazeotrope		537
9418	C₆H₁₂	Cyclohexane	80.75	Nonazeotrope		537
9419	C₆H₁₂O	Pinacolone	106.2	Nonazeotrope		552
9420	C₆H₁₄	Hexane	69.0	Nonazeotrope		546
9421	C₆H₁₄O	Propyl ether	90.55	Nonazeotrope		557
9422	C₆H₁₄O₂	Acetal	103.55	101.25	68	557
9423	C₆H₁₄O₂	Ethoxypropoxymethane	113.7	Nonazeotrope		557
9424	C₆H₁₄S	Isopropyl sulfide	120.5	Nonazeotrope		566
9425	C₇H₈	Toluene	110.6	Nonazeotrope		572
9426	C₇H₁₄	Methylcyclohexane	101.15	95.45		571
9427	C₇H₁₆	n-Heptane	98.4	93.6		571
9428	C₈H₁₈	Octane	125.8	Nonazeotrope		546
A =	**C₅H₁₀O₂**	**Tetrahydrofurfuryl Alcohol**	**72.1/10 mm.**			
9429	C₈H₉Cl	o,m,p-Chloroethyl-benzene, 10 mm.	67.5	63.0	29.5	51
A =	**C₅H₁₀O₂**	**Valeric Acid**	**186.35**			
9430	C₅H₁₁I	1-Iodo-3-methylbutane	147.65	Nonazeotrope		527
9430a	C₅H₁₂O₂	3-Methyl-1,3-butanediol 3 mm.		Nonazeotrope		581c
9431	C₆H₄Cl₂	o-Dichlorobenzene	179.5	175.8	22	527
9432	C₆H₄Cl₂	p-Dichlorobenzene	174.6	171.8	14.7	564
9433	C₆H₅Br	Bromobenzene	156.1	155.65	3.5	564
9434	C₆H₅I	Iodobenzene	188.45	180.15	35	527
9435	C₆H₆O	Phenol	182.2	Nonazeotrope		527
9436	C₆H₁₀O₃	Ethyl acetoacetate	180.4	Nonazeotrope		552
9437	C₆H₁₀O₄	Ethyl oxalate	185.65	182.5	37	569,570
9438	C₆H₁₀O₄	Glycol diacetate	186.3	< 185.6	>38	527
9438a	C₆H₁₂O₂	4-Methyl-4-hydroxy-tetrahydropyran 3 mm.		Azeotropic		581c
9439	C₆H₁₃Br	1-Bromohexane	156.5	155.5	4.5	562
9440	C₆H₁₄S	Propyl sulfide	141.5	Nonazeotrope		566
9441	C₇H₆O	Benzaldehyde	189.2	178.5		527
9442	C₇H₆O₂	Salicylaldehyde	196.7	Nonazeotrope		575
9443	C₇H₇Br	α-Bromotoluene	198.5	183.0	53	562
9444	C₇H₇Br	m-Bromotoluene	184.3	178.55	25.5	527
9445	C₇H₇Br	o-Bromotoluene	181.5	176.8	23	527
9446	C₇H₇Br	p-Bromotoluene	185.0	179.2	32	562
9447	C₇H₇Cl	α-Chlorotoluene	179.3	175.0	25	527
9448	C₇H₇Cl	o-Chlorotoluene	159.2	158.5	5	527
9449	C₇H₇Cl	p-Chlorotoluene	162.4	161.2	6	527

No.	Formula	B-Component Name	B.P., °C	Azeotropic Data B.P., °C	Wt.%A		Ref.
A =	$C_5H_{10}O_2$	**Valeric Acid** *(continued)*	**186.35**				
9450	C_7H_7I	*p*-Iodotoluene	214.5	184.5	80		575
9451	C_7H_8O	Anisole	153.8	Nonazeotrope			527
9452	C_7H_8O	*o*-Cresol	191.1	Nonazeotrope			527
9453	C_7H_8O	*p*-Cresol	201.7	Nonazeotrope			527
9454	$C_7H_{13}ClO_2$	Isoamyl chloroacetate	190.5	< 185.8			575
9455	$C_7H_{14}O_3$	1,3-Butanediol methyl ether acetate	171.75	Nonazeotrope			527
9456	$C_7H_{15}NO$	*N,N*-Dimethylvaler-amide, 100 mm.	141	145.8	30.8		832
9456a	C_7H_{16}	Heptane 50°–100°C.		Nonazeotrope		v-l	591c
9457	C_8H_8O	Acetophenone	202.0	Nonazeotrope			552
9458	C_8H_{10}	*m*-Xylene	139.2	Nonazeotrope			527
9459	C_8H_{10}	*o*-Xylene	144.3	Nonazeotrope			575
9460	$C_8H_{10}O$	Benzyl methyl ether	167.8	Nonazeotrope			527
9461	$C_8H_{10}O$	*p*-Methylanisole	177.05	< 176.0	<22		562
9462	$C_8H_{10}O$	Phenetole	170.45	Nonazeotrope			527
9463	$C_8H_{16}O$	2-Octanone	172.85	Nonazeotrope			552
9464	C_9H_8	Indene	182.6	178.5	30		562
9465	C_9H_{12}	Mesitylene	164.6	164.0	10		527
9466	C_9H_{12}	Propylbenzene	159.3	158.4	7		575
9467	$C_9H_{12}O$	Benzyl ethyl ether	185.0	180.5	40		562
9468	$C_9H_{12}O$	Phenyl propyl ether	190.5	184.3	58		562
9469	$C_9H_{14}O$	Phorone	197.8	Nonazeotrope			552
9470	$C_9H_{18}O_2$	Butyl isovalerate	177.6	Nonazeotrope			527
9471	$C_9H_{18}O_2$	Isoamyl butyrate	181.05	Nonazeotrope			527
9472	$C_{10}H_8$	Naphthalene	218.0	186.0	96		575
9473	$C_{10}H_{14}$	Cymene	176.7	176.5	22		527
9474	$C_{10}H_{16}$	Camphene	159.6	158.5	8		527
9475	$C_{10}H_{16}$	Dipentene	177.7	173.4	27		527
9476	$C_{10}H_{16}$	Nopinene	163.8	162.2	10		527
9477	$C_{10}H_{16}$	α-Pinene	155.8	155.5	5?		575
9478	$C_{10}H_{16}$	α-Terpinene	173.4	171.0	20		562
9479	$C_{10}H_{16}$	γ-Terpinene	183	178.5	33		562
9480	$C_{10}H_{16}$	Terpinolene	184.6	178.0	35		562
9481	$C_{10}H_{18}O$	Cineol	176.35	176.3	3		527
9482	$C_{10}H_{18}O$	Citronellal	208.0	Nonazeotrope			575
9483	$C_{10}H_{20}O_2$	Isoamyl isovalerate	192.7	Nonazeotrope			527
9484	$C_{10}H_{22}O$	Amyl ether	187.5	181.5	45		527
9485	$C_{10}H_{22}O$	Isoamyl ether	173.2	171.8	12.5		556
A =	$C_5H_{10}O_3$	**Ethyl Carbonate**	**126.5**				
9486	$C_5H_{10}O_3$	2-Methoxyethyl acetate	144.6	Nonazeotrope			526
9487	$C_5H_{11}Br$	1-Bromo-3-methylbutane	120.65	< 119.8	<28		575
9488	$C_5H_{11}I$	1-Iodo-3-methylbutane	147.65	Nonazeotrope			575
9489	$C_5H_{11}I$	2-Iodo-2-methylbutane	127.5	123.4	~ 50		563
9490	$C_5H_{12}O$	Amyl alcohol	138.2	< 125.5	<96		575
9491	$C_5H_{12}O$	Isoamyl alcohol	131.8	125.3	73.5		527

TABLE I. *Binary Systems*   283

No.	Formula	B-Component Name	B.P., °C	B.P., °C	Azeotropic Data Wt.%A	Ref.
A =	**C₅H₁₀O₃**	**Ethyl Carbonate** *(continued)*	**126.5**			
9492	C₆H₅Cl	Chlorobenzene	131.75	Nonazeotrope		575
9493	C₆H₁₀O	Mesityl oxide	129.4	126.45	6	527
9494	C₆H₁₀S	Allyl sulfide	139.35	126.0	90	566
9495	C₆H₁₂O	2-Hexanone	127.2	125.7	65	552
9496	C₆H₁₂O	3-Hexanone	123.3	Nonazeotrope		552
9497	C₆H₁₂O₂	Isoamyl formate	123.8	Nonazeotrope		549
9498	C₆H₁₂O₃	2-Ethoxyethyl acetate	156.8	Nonazeotrope		575
9499	C₆H₁₂O₃	Paraldehyde	124	Nonazeotrope		557
9500	C₇H₈	Toluene	110.7	Nonazeotrope		563
9501	C₇H₁₄O₂	Ethyl isovalerate	134.7	Nonazeotrope		575
9502	C₇H₁₄O₂	Propyl isobutyrate	134.0	Nonazeotrope		575
9503	C₇H₁₆O₂	Dipropoxymethane	137.2	Nonazeotrope		557
9504	C₈H₁₀	Ethylbenzene	136.15	Nonazeotrope		573
9505	C₈H₁₀	*m*-Xylene	139.0	Nonazeotrope		527
9506	C₈H₁₀	*p*-Xylene	138.45	Nonazeotrope		575
9507	C₈H₁₆	1,3-Dimethylcyclohexane	120.7	< 115.0	<42	575
9508	C₈H₁₈O	Isobutyl ether	122.3	< 120.8	<65	557
A =	**C₅H₁₀O₃**	**Butylene Glycol Monoformate**				
9509	CₙHₘ	Hydrocarbons		Min. b.p.		618
A =	**C₅H₁₀O₃**	**Ethyl Lactate**	**154.1**			
9510	C₅H₁₀O₃	2-Methoxyethyl acetate	144.6	Nonazeotrope		526
9511	C₅H₁₁I	1-Iodo-3-methylbutane	147.6	~ 146.0	< 25	548
9512	C₅H₁₁NO₃	Isoamyl nitrate	149.75	146.7	33	562
9513	C₅H₁₂O	Isoamyl alcohol	131.9	Nonazeotrope		527
9514	C₅H₁₂O₂	2-Propoxyethanol	151.35	151.33	5	526
9515	C₆H₄Cl₂	*p*-Dichlorobenzene	174.5	Nonazeotrope		538
9516	C₆H₅Br	Bromobenzene	156.1	149.7	53	572
9517	C₆H₅Cl	Chlorobenzene	132.0	Nonazeotrope		548
9518	C₆H₆O	Phenol	182.2	Nonazeotrope		542
9519	C₆H₁₀O	Cyclohexanone	155.7	153.7	66	552
9520	C₆H₁₂O	Cyclohexanol	160.7	153.75	~95	572
9521	C₆H₁₂O₃	2-Ethoxyethyl acetate	156.8	Nonazeotrope		575
9522	C₆H₁₃ClO₂	Chloroacetal	157.4	~ 152.5	73	572
9523	C₆H₁₄O	Hexyl alcohol	157.95	153.6	82	541
9524	C₆H₁₄O₂	2-Butoxyethanol	171.15	Nonazeotrope		575
9525	C₇H₇Cl	*o*-Chlorotoluene	159.15	152.0	~65	548
9526	C₇H₇Cl	*p*-Chlorotoluene	162.4	~ 153.0		548
9527	C₇H₈	Toluene	110.75	Nonazeotrope		575
9528	C₇H₈O	Anisole	153.85	150.1	55.5	556
9529	C₇H₈O	*o*-Cresol	191.1	Nonazeotrope		575
9530	C₇H₁₄	Methylcyclohexane	101.45	Nonazeotrope		575
9531	C₇H₁₄O	2-Methylcyclohexanol	168.5	Nonazeotrope		575
9532	C₇H₁₄O	5-Methyl-2-hexanone	144.2	Nonazeotrope		552
9533	C₇H₁₄O₂	Methyl caproate	151.0	< 150.0	<32	548
9534	C₇H₁₆	Heptane	98.4	Nonazeotrope		527

No.	Formula	B-Component Name	B.P., °C	Azeotropic Data B.P., °C	Wt.%A	Ref.
A =	$C_5H_{10}O_3$	**Ethyl Lactate** *(continued)*	**154.1**			
9535	C_8H_8	Styrene	145.8	140.5	25	548
		" 32 mm.			16 vol. %	342
9536	C_8H_{10}	Ethylbenzene			~ 16 vol. %	342
9537	C_8H_{10}	m-Xylene	139.0	137.4	19.5	527
9538	C_8H_{10}	o-Xylene	144.3	140.2	30	567
9539	C_8H_{10}	p-Xylene	138.45	136.6	17	575
9540	$C_8H_{10}O$	Phenetole	170.45	Nonazeotrope		556
9541	C_8H_{16}	1,3-Dimethylcyclohexane	120.7	Nonazeotrope		575
9542	$C_8H_{16}O_2$	Butyl butyrate	166.4	Nonazeotrope		548
9543	$C_8H_{16}O_2$	Isoamyl propionate	160.7	152.8	78	575
9544	$C_8H_{16}O_2$	Isobutyl butyrate	156.9	151.5	62	527
9545	$C_8H_{16}O_2$	Isobutyl isobutyrate	148.6	146.5	30	527
9546	$C_8H_{16}O_2$	Propyl isovalerate	155.7	150	~60	532,563
9547	$C_8H_{18}O$	Butyl ether	142.4	<141.5		575
9548	C_9H_8	Indene	182.6	Nonazeotrope		575
9549	C_9H_{12}	Cumene	152.8	143.5	48	569
9550	C_9H_{12}	Mesitylene	164.9	150.05	73	530
9551	C_9H_{12}	Propylbenzene	159.2	147	58	548
9552	C_9H_{12}	Pseudocumene	168.2	152.4	73	541
9553	$C_9H_{18}O$	2,6-Dimethyl-4-heptanone	168.0	Nonazeotrope		552
9554	$C_{10}H_{14}$	Butylbenzene	183.1	Nonazeotrope		575
9555	$C_{10}H_{14}$	Cymene	176.7	Nonazeotrope		538
9556	$C_{10}H_{16}$	Camphene	159.5	144.95	55	528
9557	$C_{10}H_{16}$	d-Limonene	177.8	Nonazeotrope		573
9558	$C_{10}H_{16}$	Nopinene	163.8	147.3	62	567
9559	$C_{10}H_{16}$	α-Pinene	155.8	143.1	49.8	528
		"	155.8	<152.0	<82	575
9560	$C_{10}H_{16}$	Terpinolene	181.6	Nonazeotrope		575
9561	$C_{10}H_{16}$	Thymene	179.7	Nonazeotrope		573
9562	$C_{10}H_{22}$	2,7-Dimethyloctane	160.2	146.0	60	573
A =	$C_5H_{10}O_3$	**2-Methoxyethyl Acetate**	**144.6**			
9563	$C_5H_{11}Br$	1-Bromo-3-methylbutane	120.65	Nonazeotrope		575
9564	$C_5H_{11}I$	1-Iodo-3-methylbutane	147.65	<141.5	<65	575
9565	$C_5H_{11}NO_3$	Isoamyl nitrate	149.75	144.4	87	559
9666	$C_5H_{12}O$	Amyl alcohol	138.2	<137.0		575
9567	$C_5H_{12}O$	tert-Amyl alcohol	102.35	Nonazeotrope		575
9568	$C_5H_{12}O$	Isoamyl alcohol	131.9	Nonazeotrope		527
9569	$C_5H_{12}O$	2-Pentanol	119.8	Nonazeotrope		575
9570	$C_5H_{12}O_2$	2-Propoxyethanol	151.35	Nonazeotrope		556
9571	C_6H_5Br	Bromobenzene	156.1	Nonazeotrope		556
9572	C_6H_5Cl	Chlorobenzene	131.75	Nonazeotrope		556
9573	C_6H_6O	Phenol	182.2	183.6	18	556
9574	$C_6H_{10}O_4$	Ethylidene diacetate	168.5	Nonazeotrope		527
9575	$C_6H_{12}O_2$	Butyl acetate	126.0	Nonazeotrope		526
9576	$C_6H_{12}O_2$	Ethyl butyrate	121.5	Nonazeotrope		575
9577	$C_6H_{12}O_2$	Isoamyl formate	123.8	Nonazeotrope		575

TABLE I. *Binary Systems*   285

No.	B-Component			Azeotropic Data			
	Formula	Name	B.P., °C	B.P., °C	Wt.%A	Ref.	
A =	C₅H₁₀O₃	**2-Methoxyethyl Acetate**	**144.6**				
		(continued)					
9578	C₆H₁₂O₂	Isobutyl acetate	117.4	Nonazeotrope		575	
9579	C₆H₁₃Br	1-Bromohexane	156.5	<144.2	<92	575	
9580	C₆H₁₄O	Hexyl alcohol	157.85	Nonazeotrope		526	
9581	C₇H₇Cl	o-Chlorotoluene	159.2	Nonazeotrope		575	
9582	C₇H₈	Toluene	110.75	Nonazeotrope		556	
9583	C₇H₈O	Anisole	153.85	Nonazeotrope		556	
9584	C₇H₈O	m-Cresol	202.2	Nonazeotrope		575	
9585	C₇H₈O	o-Cresol	191.1	Nonazeotrope		556	
9586	C₇H₈O	p-Cresol	201.7	Nonazeotrope		526	
9587	C₇H₁₄O	5-Methyl-2-hexanone	144.2	<144.0	>35	575	
9588	C₇H₁₄O₂	Amyl acetate	148.8	<144.45	<92	575	
9589	C₇H₁₄O₂	Ethyl isovalerate	134.7	Nonazeotrope		575	
9590	C₇H₁₄O₂	Ethyl valerate	145.45	143.8	70	526	
9591	C₇H₁₄O₂	Isoamyl acetate	142.1	141.5	20	526	
9592	C₇H₁₄O₂	Propyl butyrate	143.7	<143.2	<68	575	
9593	C₇H₁₆O₃	Ethyl orthoformate	145.75	143.45	51	527,556	
9594	C₈H₈	Styrene	145.8	143.0	61	562	
9595	C₈H₁₀	Ethylbenzene	136.15	135.5	15	526	
9596	C₈H₁₀	m-Xylene	139.2	137.7	28	514,527	
9597	C₈H₁₀	o-Xylene	144.3	141.5	50	526	
9598	C₈H₁₀	p-Xylene	138.45	137.2	26	514,526	
9599	C₈H₁₆	1,3-Dimethylcyclohexane	120.7	Nonazeotrope		575	
9600	C₈H₁₆O₂	Ethyl caproate	167.7	Nonazeotrope		575	
9601	C₈H₁₆O₂	Isoamyl propionate	160.7	Nonazeotrope		575	
9602	C₈H₁₆O₂	Isobutyl butyrate	156.9	Nonazeotrope		575	
9603	C₈H₁₆O₂	Isobutyl isobutyrate	147.3	Nonazeotrope		575	
9604	C₈H₁₆O₂	Propyl isovalerate	155.7	Nonazeotrope		575	
9605	C₈H₁₈	Octane	125.75	<125.2	<11	575	
9606	C₈H₁₈O	Butyl ether	142.4	138.0	30	526	
9607	C₈H₁₈O	Isobutyl ether	122.3	Nonazeotrope		526	
9608	C₉H₁₂	Cumene	152.8	144.3	94	527	
9609	C₉H₁₂	Mesitylene	164.6	Nonazeotrope		575	
9610	C₉H₁₂	Propylbenzene	159.3	Nonazeotrope		526	
9611	C₉H₁₂	Pseudocumene	168.2	Nonazeotrope		575	
9612	C₉H₁₈O₂	Isoamyl isobutyrate	169.8	Nonazeotrope		575	
9613	C₁₀H₁₆	Camphene	159.6	143.3	82	526	
9614	C₁₀H₁₆	Nopinene	163.8	143.5	83	575	
9615	C₁₀H₁₆	α-Terpinene	173.4	Nonazeotrope		575	
9616	C₁₀H₁₆	α-Pinene	155.8	142.0	80	526	
9617	C₁₀H₂₂	2,7-Dimethyloctane	160.1	142.5	80	562	
A =	C₅H₁₁Br	**1-Bromo-3-methylbutane**	**120.65**				
9618	C₅H₁₂O	Amyl alcohol	138.2	118.2	85	567	
9619	C₅H₁₂O	tert-Amyl alcohol	102.35	Nonazeotrope		527	
9620	C₅H₁₂O	Isoamyl alcohol	129.0	116.15	87.3	388,527	
9621	C₅H₁₂O	2-Pentanol	119.8	<115.0	<74	527	
9622	C₅H₁₂O₂	2-Propoxyethanol	151.35	Nonazeotrope		527	

No.	Formula	B-Component Name	B.P., °C	Azeotropic Data B.P., °C	Wt.%A	Ret.
A =	$C_5H_{11}Br$	1-Bromo-3-methylbutane	120.65			
		(continued)				
9623	$C_6H_{10}O$	Mesityl oxide	129.45	Nonazeotrope		527
9624	$C_6H_{10}S$	Allyl sulfide	139.35	Nonazeotrope		566
9625	$C_6H_{12}O$	3-Hexanone	123.3	119.8	45	552
9626	$C_6H_{12}O$	4-Methyl-2-pentanone	116.05	115.6	30	527
9627	$C_6H_{12}O_2$	Butyl acetate	125.0	Nonazeotrope		527
9628	$C_6H_{12}O_2$	Ethyl butyrate	121.5	119.8	65	388,527
9629	$C_6H_{12}O_2$	Ethyl isobutyrate	110.1	Nonazeotrope		547
9630	$C_6H_{12}O_2$	Isoamyl formate	123.8	120.0	76	527
9631	$C_6H_{12}O_2$	Isobutyl acetate	117.4	117.2	<28	527
9632	$C_6H_{12}O_2$	Methyl isovalerate	116.5	Nonazeotrope		547
9633	$C_6H_{12}O_2$	Propyl propionate	123.0	120.2	75	562
9634	$C_6H_{12}O_3$	Paraldehyde	124	118.5	~24	563
9635	$C_6H_{14}S$	Isopropyl sulfide	120.5	<118.9	<48	566
9636	$C_6H_{15}BO_3$	Ethyl borate	118.6	117.7	38	527
9637	C_7H_8	Toluene	109.5	Nonazeotrope		567
9638	$C_7H_{14}O_2$	Ethyl isovalerate	134.7	Nonazeotrope		527
9639	$C_7H_{14}O_2$	Isoamyl acetate	137.5	Nonazeotrope		388
9640	$C_7H_{14}O_2$	Isopropyl isobutyrate	120.8	119.5	60	547
9641	C_8H_{16}	1,3-Dimethylcyclohexane	120.7	<118.9	<60	527
9642	C_8H_{18}	*n*-Octane	125.75	<120.2	<90	527
A =	$C_5H_{11}Br$	1-Bromopentane	130.0			
9643	C_7H_8	Toluene	110.7	Nonazeotrope		813
A =	$C_5H_{11}Cl$	1-Chloro-3-methylbutane	99.4			
9644	$C_5H_{11}NO_2$	Isoamyl nitrite	97.15	< 96.9	<20	550
9645	$C_5H_{12}O$	*tert*-Amyl alcohol	102.25	95.85	73.5	545
9646	$C_5H_{12}O$	Isoamyl alcohol	131.9	Nonazeotrope		527
9647	$C_5H_{12}O_2$	Diethoxymethane	87.95	Nonazeotrope		527
9648	C_6H_6	Benzene	80.15	Nonazeotrope		575
9649	$C_6H_{10}S$	Allyl sulfide	139.35	Nonazeotrope		575
9650	$C_6H_{12}O$	Pinacolone	106.2	Nonazeotrope		548
9651	$C_6H_{12}O_2$	Ethyl isobutyrate	110.1	Nonazeotrope		547
9652	$C_6H_{14}O$	Propyl ether	90.1	Nonazeotrope		559
9653	$C_6H_{14}O_2$	Acetal	103.55	Nonazeotrope		559
9654	C_7H_{14}	Methylcyclohexane	101.15	98.0	64	562
9655	C_7H_{16}	Heptane	98.4	96.5	52	527
9656	C_8H_{18}	2,5-Dimethylhexane	109.4	Nonazeotrope		575
A =	$C_5H_{11}I$	1-Iodo-3-methylbutane	147.65			
9657	$C_5H_{11}NO_3$	Isoamyl nitrate	149.75	<144.5	>57	560
9658	$C_5H_{12}O$	Isoamyl alcohol	128.9	127.3	48	388,527
9659	$C_5H_{12}O_2$	2-Propoxyethanol	151.35	143.0		526
9660	C_6H_5ClO	*o*-Chlorophenol	176.8	Nonazeotrope		575
9661	C_6H_6O	Phenol	182.2	Nonazeotrope		542
9662	$C_6H_{10}O$	Cyclohexanone	155.7	Nonazeotrope		552
9663	$C_6H_{10}O$	Mesityl oxide	129.45	Nonazeotrope		527
9664	$C_6H_{12}O$	Cyclohexanol	160.65	147.0	~90	573
9665	$C_6H_{12}O_3$	2-Ethoxyethyl acetate	156.8	<147.4		575

TABLE I. *Binary Systems* 287

No.		B-Component			Azeotropic Data		
	Formula	Name	B.P., °C	B.P., °C	Wt.%A		Ref.
A =	C₅H₁₁I	**1-Iodo-3-methylbutane**	**147.65**				
		(continued)					
9666	C₆H₁₄O	Hexyl alcohol	157.85	145.2	87		567
9667	C₆H₁₄O₂	2-Butoxyethanol	171.15	Nonazeotrope			575
9668	C₆H₁₄O₂	Pinacol	174.35	145.5	~90		575
9669	C₇H₈O	Anisole	153.85	Nonazeotrope			573
9670	C₇H₁₄O	4-Heptanone	143.55	143.0	35		552
9671	C₇H₁₄O	2-Methylcyclohexanol	168.5	Nonazeotrope			575
9672	C₇H₁₄O₂	Isoamyl acetate	142.1	141.7	~18		528
		"	137.5	Nonazeotrope			388
9673	C₇H₁₄O₂	Amyl acetate	148.8	145.9	60		562
9674	C₇H₁₄O₂	Ethyl isovalerate	134.7	Nonazeotrope			575
9675	C₇H₁₄O₂	Ethyl valerate	145.45	<145.1	<30		575
9676	C₇H₁₄O₂	Isobutyl propionate	136.9	Nonazeotrope			546
9677	C₇H₁₄O₂	Propyl butyrate	143.7	Nonazeotrope			547
9678	C₇H₁₄O₂	Methyl caproate	149.8	<147.5	<70		575
9679	C₇H₁₄O₂	Propyl isobutyrate	134.0	Nonazeotrope			547
9680	C₇H₁₄O₃	1,3-Butanediol methyl ether					
		acetate	171.75	Nonazeotrope			575
9681	C₇H₁₆O	Heptyl alcohol	176.15	Nonazeotrope			575
9682	C₈H₈	Styrene	145.8	<145.0			575
9683	C₈H₁₀	*m*-Xylene	139.0	Nonazeotrope			538
9684	C₈H₁₆O₂	Isobutyl butyrate	156.8	Nonazeotrope			547
9685	C₈H₁₆O₂	Isobutyl isobutyrate	147.3	146.5	58		538
9686	C₈H₁₈O	Butyl ether	142.4	Nonazeotrope			559
9687	C₁₀H₁₆	Camphene	159.6	Nonazeotrope			575
9688	C₁₀H₁₆	Nopinene	163.8	Nonazeotrope			575
9689	C₁₀H₁₆	α-Pinene	155.8	<147.4	>80		575
A =	C₅H₁₁N	**Piperidine**	**106.4**				
9690	C₆H₁₄O	Propyl ether	90.1	Nonazeotrope			575
9691	C₇H₈	Toluene	110.7			Min. b.p.	516
9692	C₇H₁₄	Methylcyclohexane	100			Min. b.p.	240
9693	C₇H₁₆	Heptane	98.4	< 97.5	>9		575
A =	C₅H₁₁NO₂	**Ethyl-*N*-Ethylaminoformate**					
9694	C₆H₄Cl₂	*p*-Dichlorobenzene	174.4	167.0	24.2		555
9695	C₈H₁₈S	Isobutyl sulfide	172	166.5	23		555
A =	C₅H₁₁NO₂	**Isoamyl Nitrite**	**97.15**				
9696	C₅H₁₂O₂	Diethoxymethane	87.95	Nonazeotrope			527
9697	C₆H₆	Benzene	80.15	Nonazeotrope			550
9698	C₆H₁₂	Cyclohexane	80.75	Nonazeotrope			550
9699	C₆H₁₂	Methylcyclopentane	72.0	Nonazeotrope			550
9700	C₆H₁₂O	Pinacolone	106.2	Nonazeotrope			552
9701	C₆H₁₄	Hexane	68.8	Nonazeotrope			550
9702	C₆H₁₄O	Propyl ether	90.1	Nonazeotrope			550
9703	C₆H₁₄O₂	Acetal	103.55	Nonazeotrope			550
9704	C₇H₈	Toluene	110.75	Nonazeotrope			550
9705	C₇H₁₄	Methylcyclohexane	101.15	95.5	79		550

No.	Formula	Name	B.P., °C	B.P., °C	Wt.%A		Ref.
		B-Component		**Azeotropic Data**			
A =	**$C_5H_{11}NO_2$**	**Isoamyl Nitrite** *(continued)*	**97.15**				
9706	C_7H_{16}	Heptane	98.4	94.8	52		550
9707	C_8H_{16}	1,3-Dimethylcyclohexane	120.7	Nonazeotrope			550
9708	C_8H_{18}	2,5-Dimethylhexane	109.4	Nonazeotrope			550
9709	C_8H_{18}	Octane	125.75	Nonazeotrope			550
A =	**$C_5H_{11}NO_3$**	**Isoamyl Nitrate**	**149.75**				
9710	$C_5H_{12}O_2$	2-Propoxyethanol	151.35	<143.5	>57		560
9711	$C_5H_{12}O_3$	2-(2-Methoxyethoxy) ethanol	192.95	Nonazeotrope			560
9712	C_6H_5Br	Bromobenzene	156.1	Nonazeotrope			560
9713	$C_6H_{12}O$	Cyclohexanol	160.8	<148			560
9714	$C_6H_{12}O_3$	2-Ethoxyethyl acetate	156.8	Nonazeotrope			560
9715	$C_6H_{13}Br$	1-Bromohexane	156.5	<148.5	<80		560
9716	$C_6H_{14}O$	Hexyl alcohol	157.85	<148.0	>11		560
9717	$C_6H_{14}O_2$	2-Butoxyethanol	171.15	Nonazeotrope			556
9718	C_7H_7Cl	*o*-Chlorotoluene	159.2	Nonazeotrope			560
9719	C_7H_7Cl	*p*-Chlorotoluene	162.4	Nonazeotrope			547
9720	C_7H_8O	Anisole	153.85	Nonazeotrope			557
9721	$C_7H_{14}O$	2-Methylcyclohexanol	168.5	Nonazeotrope			560
9722	$C_7H_{14}O_2$	Methyl caproate	149.8	148.5	55		549
9723	$C_7H_{14}O_3$	1,3-Butanediol methyl ether acetate	171.75	Nonazeotrope			527
9724	C_8H_8	Styrene	145.8	<145.6	<38		560
9725	C_8H_{10}	*m*-Xylene	139.0	Nonazeotrope			527
9726	$C_8H_{16}O_2$	Isoamyl propionate	160.7	Nonazeotrope			560
9727	$C_8H_{16}O_2$	Isobutyl butyrate	156.9	Nonazeotrope			549
9728	$C_8H_{16}O_2$	Isobutyl isobutyrate	148.6	<147.5	<40		549
9729	C_9H_{12}	Mesitylene	164.6	Nonazeotrope			560
9730	C_9H_{12}	Propylbenzene	158.9	Nonazeotrope			546
9731	$C_9H_{20}O_2$	Diisobutoxymethane	163.8	Nonazeotrope			557
9732	$C_{10}H_{16}$	Camphene	159.6	149.0	72		560
9733	$C_{10}H_{16}$	Nopinene	163.8	149.2	80		560
9734	$C_{10}H_{16}$	*α*-Pinene	155.8	147.75	65		560
9735	$C_{10}H_{22}$	2,7-Dimethyloctane	160.1	<148.6	<83		560
A =	**C_5H_{12}**	**2-Methylbutane**	**27.95**				
9736	C_5H_{12}	Pentane	36.15	Nonazeotrope			563
9737	$C_6F_{15}N$	Perfluorotriethylamine		26.5	64.0		717
9738	$C_6H_5NO_2$	Nitrobenzene	210.75	Nonazeotrope			553
A =	**C_5H_{12}**	**Pentane**	**36.15**				
9739	$C_5H_{12}O$	*tert*-Amyl alcohol	102.35	Nonazeotrope			575
9740	$C_6H_5NO_2$	Nitrobenzene	210.75	Nonazeotrope			554
9741	C_6H_6	Benzene	80.2	Nonazeotrope		v-l	163,675
9742	C_6H_{12}	Cyclohexane	80.75	Nonazeotrope		v-l	677
9743	C_6H_{12}	Methylcyclopentane	72.0	Nonazeotrope		v-l	163,677
9744	C_7F_{16}	Perfluoroheptane, crit. region	82.5	176.3	23.1	v-l	430,431

TABLE I.   *Binary Systems*   289

No.	Formula	B-Component Name	B.P., °C	B.P., °C	Wt.%A		Ref.
				Azeotropic Data			
A =	**C₅H₁₂**	**Pentane** *(continued)*	**36.15**				
9745	C₇H₁₄	Methylcyclohexane	101.15	Nonazeotrope		v-l	163,677
A =	**C₅H₁₂N₂O**	**Tetramethylurea**	**176.5**				
9746	C₆H₆O	Phenol	181.4		48.2		596
	"	"		max. b.p.			596
A =	**C₅H₁₂O**	**Amyl Alcohol**	**138.2**				
9746a	C₅H₁₂O	Isoamyl alcohol					
			746 mm.	Nonazeotrope		v-l	213f
9747	C₆H₅Cl	Chlorobenzene	131.75	126.2	25		567
9748	C₆H₆	Benzene	80.15	Nonazeotrope			575
	"	"	80.2	Nonazeotrope		v-l	1011
9749	C₆H₆O	Phenol	182.2	Nonazeotrope			575
9750	C₆H₁₀O	Methylcyclopentanone	138	Min. b.p.			233,234, 919
9751	C₆H₁₀S	Allyl sulfide	139.35	'<134.5	>42		566
9752	C₆H₁₂	Cyclohexane	80.75	Nonazeotrope			575
9753	C₆H₁₂O₂	Amyl formate	132	131.4	43		359
9754	C₆H₁₂O₂	Butyl acetate	126.0	Nonazeotrope			527
9755	C₆H₁₂O₂	Ethyl butyrate	121.5	Nonazeotrope			575
9756	C₆H₁₂O₂	Isoamyl formate	123.8	Nonazeotrope			575
9757	C₆H₁₂O₃	Paraldehyde	123.9	Nonazeotrope			576
9758	C₆H₁₄	Hexane	69.0	Nonazeotrope			813
9759	C₆H₁₁N₂	2,5-Dimethylpiperazine	164	Nonazeotrope			981
9760	C₇H₈	Toluene	110.75	Nonazeotrope			575
9761	C₇H₁₄	Methylcyclohexane	101.15	<101.0			575
9762	C₇H₁₄O	4-Heptanone	143.55	Nonazeotrope			548
9763	C₇H₁₄O₂	Amyl acetate	148.8	Nonazeotrope			359
9764	C₇H₁₄O₂	Propyl isobutyrate	134.0	<133.5	>19		575
9765	C₈H₁₀	Ethylbenzene	136.15	129.8	40		567
	"	"	60 mm.	60.5	57.5	20	53
	"	"		129.57	39.5	v-l	587
9766	C₈H₁₀	*p*-Xylene	138.45	130.90	41.9	v-l	308c,567
9767	C₈H₁₆	1,3-Dimethylcyclohexane	120.7	118.2	20		567
9768	C₈H₁₈	C₈ paraffins	120-130	Min. b.p.			919
9769	C₈H₁₈	Octane	125.75	121.8			919
9770	C₈H₁₈O	Butyl ether	142.1	134.5	50		556
9771	C₈H₁₈O	Isobutyl ether	122.2	121.2	10		576
9772	C₁₀H₂₂O	Amyl ether	188	Nonazeotrope			760
9773	C₁₁H₂₄O₂	Diamyloxymethane	221.6	Nonazeotrope			324
9774	C₁₂H₂₆O₂	Acetaldehyde diamyl acetal	225.3	Nonazeotrope			45
A =	**C₅H₁₂O**	**Active Amyl Alcohol**	**128.5**				
9775	C₅H₁₂O	Isoamyl alcohol	131	Nonazeotrope		v-l	711
9776	C₆H₅Cl	Chlorobenzene	132	124.4	43		958
9777	C₆H₅FO	*o*-Fluorophenol		Nonazeotrope			958
9778	C₆H₇N	2-Picoline	129	132.8	49		958
9779	C₆H₁₀O	Mesityl oxide	129.5	Nonazeotrope			958
9780	C₇H₇F	*o*-Fluorotoluene	114	112.0	16		958

No.	Formula	Name	B.P., °C	B.P., °C	Wt.%A	Ref.
		B-Component		**Azeotropic Data**		
A =	**C₅H₁₂O**	**Active Amyl Alcohol** (*continued*)	**128.5**			
9781	C₇H₈	Toluene	111	109.9	12	958
9782	C₇H₉N	2,6-Lutidine	144	Nonazeotrope		958
9783	C₇H₁₄O	2,4-Dimethyl-3-pentanone	125	124.1	21	958
9784	C₇H₁₅N	1,2-Dimethylpiperidine	128	130.3		958
9785	C₇H₁₅N	2,6-Dimethylpiperidine	128	130.7	54	958
9786	C₈H₁₀	Ethylbenzene	136	125.0	53	958
9787	C₈H₁₈	n-Octane	126.0	117.0	34	958
9788	C₉H₂₀	2,2,5-Trimethylhexane	124	115.5	29	958
A =	**C₅H₁₂O**	**tert-Amyl Alcohol**	**102.35**			
9789	C₅H₁₂O₂	Diethoxymethane	87.95	Nonazeotrope		527
9790	C₆H₅Cl	Chlorobenzene	131.75	Nonazeotrope		575
9791	C₆H₆	Benzene	80.2	~ 80.0	~ 15	537
		"	80.2	80.0	15	163
		" 715 mm.			4.95 v-l	773
9792	C₆H₈	1,3-Cyclohexadiene	80.4	79.7	~15	541
9793	C₆H₁₀	Cyclohexene	82.7	80.8	17	537
9794	C₆H₁₂	Cyclohexane	80.75	78.5	16	537
9795	C₆H₁₂	Methylcyclopentane	72.0	71.5	5	575
9796	C₆H₁₂O₂	Methyl isovalerate	116.5	Nonazeotrope		575
9797	C₆H₁₄	2,3-Dimethylbutane	58.0	Nonazeotrope		575
9798	C₆H₁₄	Hexane	68.9	68.3	4	537
9799	C₆H₁₄O	Propyl ether	90.4	88.8	20	545
9800	C₇H₈	Toluene	110.7	100.5	56	50,537
		"	110.7	100.5	56	163
		" 715 mm.			32.5 v-l	773
9801	C₇H₁₄	Methylcyclohexane	100.8	92.0	40	50,571
9802	C₇H₁₆	Heptane	98.45	92.2	26.5	545,571
9803	C₈H₁₀	Ethylbenzene	136.15	Nonazeotrope		537
		60 mm.	60.5	45	83	53
9804	C₈H₁₀	m-Xylene	139.0	Nonazeotrope		540
9805	C₈H₁₀	o-Xylene	144.3	Nonazeotrope		575
9806	C₈H₁₆	1,3-Dimethylcyclohexane	120.7	100.1	68	567
9807	C₈H₁₈	2,5-Dimethylhexane	109.4	97.0	50	567
9808	C₈H₁₈	Octane	125.75	101.1	75	567
9809	C₈H₁₈O	Isobutyl ether	122.1	Min. b.p.		576
9810	C₁₀H₁₆	α-Pinene	155.8	Nonazeotrope		537
A =	**C₅H₁₂O**	**Ethyl Propyl Ether**	**63.6**			
9811	C₆H₁₀	Biallyl	60.1	< 60.0	> 5	558
9812	C₆H₁₂	Methylcyclopentane	72.0	Nonazeotrope		558
9813	C₆H₁₄	2,3-Dimethylbutane	58.0	Nonazeotrope		558
9814	C₆H₁₄	Hexane	68.85	Nonazeotrope		558
9815	C₆H₁₅N	Triethylamine	89.35	Nonazeotrope		551
A =	**C₅H₁₂O**	**Isoamyl Alcohol**	**131.9**			
9816	C₅H₁₂O₂	2-Propoxyethanol	151.35	Nonazeotrope		527
9817	C₅H₁₂S	3-Methyl-1-butanethiol	116	115.6	22.89	487

TABLE I. *Binary Systems*   **291**

No.	Formula	Name	B.P., °C	B.P., °C	Wt.%A	Ref.
		B-Component		**Azeotropic Data**		
A =	**C₅H₁₂O**	**Isoamyl Alcohol** *(continued)*	**131.9**			
9818	C₆H₅Br	Bromobenzene	156.15	131.65	85	527
9819	C₆H₅Cl	Chlorobenzene	131.8	124.35	34	527
	"		132	123.9	38	958
9820	C₆H₅FO	o-Fluorophenol		Nonazeotrope		958
9821	C₆H₆	Benzene	80.2	Nonazeotrope		1044
9822	C₆H₆O	Phenol	181.5	Nonazeotrope		527
9823	C₆H₆S	Benzenethiol	169.5	Nonazeotrope		575
9824	C₆H₇N	Aniline	184.35	Nonazeotrope		527
9825	C₆H₇N	2-Picoline	130.7	>132.5		575
	"		129	132.8	61	958
9826	C₆H₇N	3-Picoline	143.4	Nonazeotrope		575
9827	C₆H₈	1,3-Cyclohexadiene	80.4	Nonazeotrope		575
9828	C₆H₉N	N-Ethylpyrrol	130.4	<129.0		575
9829	C₆H₁₀	Cyclohexene	82.7	Nonazeotrope		537
9830	C₆H₁₀O	Mesityl oxide	129.45	129.15	24	552
	" "		129.5	Nonazeotrope		958
9831	C₆H₁₀S	Allyl sulfide	139.35	<131.5		527
9832	C₆H₁₁ClO₂	Butyl choroacetate	181.9	Nonazeotrope		575
9833	C₆H₁₁N	Capronitrile	163.9	Nonazeotrope		575
9834	C₆H₁₂	Cyclohexane	80.75	Nonazeotrope		537
9835	C₆H₁₂	Methylcyclopentane	72.0	Nonazeotrope		575
9836	C₆H₁₂O	4-Methyl-2-pentanone	116.05	Nonazeotrope		527
9837	C₆H₁₂O₂	Butyl acetate	126.0	125.85	17.5	527
9838	C₆H₁₂O₂	Ethyl butyrate	120.6	Nonazeotrope		388
9839	C₆H₁₂O₂	Isoamyl formate	124.2	123.6	25.5	359,527
9840	C₆H₁₂O₂	Isobutyl acetate	117.4	Nonazeotrope		527
9841	C₆H₁₂O₂	Propyl propionate	122.1	Nonazeotrope		532
9842	C₆H₁₂O₃	2-Ethoxyethyl acetate	156.8	Nonazeotrope		527
9843	C₆H₁₂O₃	Paraldehyde	124	123.5	22	556
9844	C₆H₁₄	2,3-Dimethylbutane	58	Nonazeotrope		575
9845	C₆H₁₄	Hexane	68.95	Nonazeotrope		563
9846	C₆H₁₄S	Propyl sulfide	141.5	<130.5	<79	566
9847	C₆H₁₅BO₃	Ethyl borate	118.6	Nonazeotrope		532
9847a	C₇H₇Cl	α-Chlorotoluene				
		20 mm.		Nonazeotrope	v-l	494f
	"		179	Nonazeotrope	v-l	494f
9848	C₇H₇Cl	o-Chlorotoluene	159.2	Nonazeotrope		527
9849	C₇H₇Cl	p-Chlorotoluene	162.4	Nonazeotrope		527
9850	C₇Hₙ	C₇ hydrocarbons	95-120	Min b.p.		919
9851	C₇H₇F	o-Fluorotoluene	114	112.1	14	958
9852	C₇H₈	Toluene	111	109.7	10	958,1069
	"		110.7		v-l	793
	"	690 mm.		107	v-l	793
	"	482 mm.		95	v-l	793
	"	295 mm.		80	v-l	793

No.	Formula	B-Component Name	B.P., °C	Azeotropic Data B.P., °C	Wt.%A		Ref.
A =	C₅H₁₂O	**Isoamyl Alcohol** *(continued)*	**131.9**				
		% alcohol decreases with decreasing press.					793
		"	110.7	Nonazeotrope			50,527, 834
9853	C₇H₈O	Anisole	153.85	Nonazeotrope			1048
9854	C₇H₉N	2,6-Lutidine, 70 mm.		Max. b.p.			958
		"	144	Nonazeotrope			958
9855	C₇H₁₃ClO₂	Isoamyl chloroacetate	195.2	Nonazeotrope			121
9856	C₇H₁₄	Methylcyclohexane	100.8	98.2	13		50,527
9857	C₇H₁₄O	2,4-Dimethyl-3-pentanone	125	124.5	8		958
9858	C₇H₁₄O	4-Heptanone	143.55	Nonazeotrope			527
9899	C₇H₁₄O	Isoamyl vinyl ether	112.6	112.1	12		882
9860	C₇H₁₄O	5-Methyl-2-hexanone	144.2	Nonazeotrope			527,552
9861	C₇H₁₄O₂	Ethyl isovalerate	134.7	130.5	58		527
9862	C₇H₁₄O₂	Ethyl valerate	145.45	Nonazeotrope			527
9863	C₇H₁₄O₂	Isoamyl acetate 20 mm.	52.8	51.4	40.8	v-l	496e
		"	142.2	130	89.7	v-l	496e
		"		130.8	98.7		498c
		"	137.5	129.1	97.4		567
		"	142	Nonazeotrope			359
9864	C₇H₁₄O₂	Isobutyl propionate	136.9	131.2	72		527
9865	C₇H₁₄O₂	Propyl butyrate	143	Nonazeotrope			527
9866	C₇H₁₄O₂	Propyl isobutyrate	134.0	130.2	53		527
9867	C₇H₁₅N	1,2-Dimethyl piperidine	128	132.5	81		958
9868	C₇H₁₅N	2,6-Dimethyl piperidine	128	132.6	76		958
9869	C₇H₁₆	Heptane	98.45	97.7	7		527
9870	C₈H₈	Styrene, 60 mm.	68	64.8	43		53
		"	145.8	128.5	63		537
9871	C₈H₁₀	Ethylbenzene,	136.15	125.9	49		539
		" 60 mm.	60.5	58.5	26		53
		"	136	125.7	49		958
9872	C₈H₁₀	*m*-Xylene	139	125-126	52		563,834
9873	C₈H₁₀	*o*-Xylene	142.6	127	>52		527,834
9874	C₈H₁₀	*p*-Xylene	138.2	125-126	52		541,834
9875	C₈H₁₀O	Benzyl methyl ether	167.8	Nonazeotrope			575
9876	C₈H₁₀O	Phenetole	170.45	Nonazeotrope			575
9877	C₈H₁₆	1,3-Dimethylcyclohexane	120.7	116.6	27		567
9878	C₈H₁₆	6-Methyl-1-heptene, 751 mm.		109	18		757
9879	C₈H₁₆O₂	Isoamyl propionate	160.7	Nonazeotrope			575
9880	C₈H₁₈	2,5-Dimethylhexane	109.4	107.6	15		567
9881	C₈H₁₈	Octane	125.8	120.0	35		545
		"	126	117.0	30		958
9882	C₈H₁₈	2,2,4-Trimethylpentane	99.3	99.0	5		575

TABLE I.   *Binary Systems*   293

No.	Formula	B-Component Name	B.P., °C	Azeotropic Data B.P., °C	Wt.%A	Ref.
A =	**$C_5H_{12}O$**	**Isoamyl Alcohol** *(continued)*	**131.9**			
9883	$C_8H_{18}O$	Butyl ether	142.1	129.8	65	527
9884	$C_8H_{18}O$	Isobutyl ether	122.1	119.8	22	527
9885	C_9H_8	Indene	181.7	Nonazeotrope		537
9886	C_9H_{12}	Cumene	152.8	131.6	94	537
9887	C_9H_{12}	Mesitylene	164.0	Nonazeotrope		563
9888	C_9H_{12}	Propylbenzene	159.3	Nonazeotrope		527
9889	C_9H_{12}	Pseudocumene	169	Nonazeotrope		563
9890	C_9H_{20}	2,2,5-Trimethylhexane	124	116.0	26	958
9891	$C_{10}H_{14}$	Butylbenzene	183.1	Nonazeotrope		575
9892	$C_{10}H_{14}$	Cymene	175.3	Nonazeotrope		563
9893	$C_{10}H_{16}$	Camphene	159.6	130.9	24	527
9894	$C_{10}H_{16}$	*d*-Limonene	177.8	Nonazeotrope		563
9895	$C_{10}H_{16}$	α-Phellandrene	171.5	Nonazeotrope		563
9896	$C_{10}H_{16}$	α-Pinene	155.8	130.7	74	527
9897	$C_{10}H_{16}$	Terpinolene	184.6	Nonazeotrope		575
9898	$C_{10}H_{16}$	Thymene	179.7	Nonazeotrope		537
9899	$C_{10}H_{22}$	2,7-Dimethyloctane	160.2	129.7	~85	537
9900	$C_{10}H_{22}O$	Isoamyl ether	171	Nonazeotrope		1033
9900a	$C_{12}H_{18}O$	Amyl benzyl ether				
		20 mm.		Nonazeotrope	v-l	494i
		"		Nonazeotrope	v-l	494i
9901	$C_{12}H_{26}O_2$	Acetaldehyde diisoamyl				
		acetal	213.6	Nonazeotrope		45
A =	**$C_5H_{12}O$**	**2-Methyl-1-butanol**	**128.9**			
9902	C_7H_n	C_7 hydrocarbons	95-120	Min. b.p.		919
9903	C_7H_8	Toluene	110.7	Min. b.p.		919
9904	C_8H_8	Styrene, 60 mm.	68	60	52	53
9905	C_8H_{10}	Ethylbenzene, 60 mm.	60.5	56	33	53
A =	**$C_5H_{12}O$**	**3-Methyl-2-butanol**	**112.9**			
9906	C_6H_6	Benzene	80.15	Nonazeotrope		575
9907	C_6H_{10}	Cyclohexene	82.75	< 82.5	> 3.5	575
9908	C_6H_{14}	Hexane	68.8	Nonazeotrope		575
9909	C_7H_8	Toluene	110.75	<105.8	>38	575
9910	C_7H_{14}	Methylcyclohexane	101.15	97.0	25	567
9911	C_7H_{16}	Heptane	98.4	95.0	23	567
9912	C_8H_{10}	Ethylbenzene, 60 mm.	60.5	51	62	53
9913	C_8H_{18}	2,5-Dimethylhexane	109.4	<103.5	>32	575
A =	**$C_5H_{12}O$**	**2-Pentanol**	**119.8**			
9914	C_6H_5Cl	Chlorobenzene	131.75	<118.2	>55	567
9915	C_6H_6	Benzene	80.15	Nonazeotrope		575
9916	C_6H_7N	2-Picoline	130.7	Nonazeotrope		575
9917	C_6H_{10}	Cyclohexene	82.75	Nonazeotrope		575
9918	$C_6H_{10}O$	Mesityl oxide	129.45	Nonazeotrope		527
9919	C_6H_{12}	Cyclohexane	80.75	Nonazeotrope		575
9920	C_6H_{12}	Methylcyclopentane	72.0	Nonazeotrope		575

No.	Formula	B-Component Name	B.P., °C	B.P., °C	Wt.%A	Ref.
				Azeotropic Data		
A =	$C_5H_{12}O$	**2-Pentanol** *(continued)*	**119.8**			
9921	$C_6H_{12}O$	2-Hexanone	127.2	Nonazeotrope		552
9922	$C_6H_{12}O_2$	Butyl acetate	126.0	Quasi-azeotrope		527
9923	$C_6H_{12}O_2$	Ethyl butyrate	121.5	<118.5	>47	567
9924	$C_6H_{12}O_2$	Ethyl isobutyrate	110.1	Nonazeotrope		575
9925	$C_6H_{12}O_2$	Isobutyl acetate	117.4	116.5	32	567
9926	$C_6H_{12}O_2$	Methyl isovalerate	116.5	<115.8	>20	575
9927	$C_6H_{12}O_3$	Paraldehyde	124.35	118.5	52	575
9928	C_6H_{14}	Hexane	68.8	Nonazeotrope		575
9929	$C_6H_{14}O$	*tert*-Amyl methyl ether	86-7	Nonazeotrope		256
9930	$C_6H_{14}O$	*tert*-Butyl ethyl ether	73	Nonazeotrope		256
9931	$C_6H_{14}S$	Propyl sulfide	141.5	Nonazeotrope		566
9932	C_7H_8	Toluene	110.75	107.0	28	567
9933	C_7H_{14}	Methylcyclohexane	101.15	98.6	18	567
9934	C_7H_{16}	Heptane	98.4	96.0	15	567
9935	C_8H_8	Styrene, 60 mm.	68	60	69	53
9936	C_8H_{10}	Ethylbenzene	136.15	118.0	67	483
		" 60 mm.	60.5	54	50	53
9937	C_8H_{10}	*m*-Xylene	139.2	118.3	70	575
9938	C_8H_{16}	1,3-Dimethylcyclohexane	120.7	<113.0	>38	567
9939	C_8H_{18}	Octane	125.75	<114.8	<56	567
9940	$C_8H_{18}O$	Isobutyl ether	122.1	115.0	41	576
A =	$C_5H_{12}O$	**3-Pentanol**	**116.0**			
9941	C_6H_6	Benzene	80.2	Nonazeotrope		537
9942	C_6H_{12}	Cyclohexane	80.8	80.0	3	540
9943	$C_6H_{12}O$	4-Methyl-2-pentanone	116.05	<115.0	>35	552
9944	C_6H_{14}	Hexane	68.95	Nonazeotrope		537
9945	$C_6H_{14}O$	Propyl ether	90.4	Nonazeotrope		576
9946	C_7H_8	Toluene	110.75	~ 106	~ 35	537
9947	C_7H_{14}	Methylcyclohexane	101.1	97.4	23	537
9948	C_7H_{16}	Heptane	98.4	96.0	20	567
9949	C_8H_{10}	Ethylbenzene, 60 mm.	60.5	51	50	53
9950	$C_8H_{18}O$	Isobutyl ether	122.1	112		576
A =	$C_5H_{12}O_2$	**Diethoxymethane**	**87.95**			
9951	C_6H_6	Benzene	80.15	Nonazeotrope		558
9952	C_6H_{12}	Cyclohexane	80.75	80.1	17	527
9953	C_6H_{14}	*n*-Hexane	68.8	Nonazeotrope		527
9954	$C_6H_{14}O$	Isopropyl ether	68.3	Nonazeotrope		575
9955	$C_6H_{15}N$	Triethylamine	89.35	< 86.8		551
9956	C_7H_{14}	Methylcyclohexane	101.15	Nonazeotrope		527
9957	C_7H_{16}	*n*-Heptane	98.4	87.8	96	558
A =	$C_5H_{12}O_2$	**3-Methyl-1,3-butanediol**				
9957a	$C_6H_{12}O_2$	Caproic acid 4 mm.		Azeotropic		581c
A =	$C_5H_{12}O_2$	**2-Propoxyethanol**	**151.35**			
9958	$C_6H_4Cl_2$	*p*-Dichlorobenzene	174.4	Nonazeotrope		556
9959	C_6H_5Br	Bromobenzene	156.1	148.2	48	556

TABLE I. *Binary Systems* 295

No.	Formula	Name	B.P., °C	B.P., °C	Wt.%A	Ref.
		B-Component		**Azeotropic Data**		
A =	**C₅H₁₂O₂**	**2-Propoxyethanol** *(continued)*	**151.35**			
9960	C₆H₅Cl	Chlorobenzene	131.75	Nonazeotrope		526
9961	C₆H₅I	Iodobenzene	188.45	Nonazeotrope		526
9962	C₆H₆O	Phenol	182.2	182.65	14	556
9963	C₆H₇N	Aniline	184.35	Nonazeotrope		551
9964	C₆H₁₀O₄	Ethylidene diacetate	168.5	Nonazeotrope		527
9965	C₆H₁₀S	Allyl sulfide	139.35	<137.5	<20	566
9966	C₆H₁₁N	Capronitrile	163.9	Nonazeotrope		575
9967	C₆H₁₂O	Cyclohexanol	160.8	Nonazeotrope		526
9968	C₆H₁₂O₂	Butyl acetate	126.0	Nonazeotrope		575
9969	C₆H₁₂O₃	2-Ethoxyethyl acetate	156.8	151.25	87.5	556
9970	C₆H₁₂O₃	Paraldehyde	124.35	Nonazeotrope		556
9971	C₆H₁₂O₃	Propyl lactate	171.7	Nonazeotrope		575
9972	C₆H₁₄O	Hexyl alcohol	157.85	Nonazeotrope		526
9973	C₆H₁₄O₂	Pinacol	174.35	Nonazeotrope		575
9974	C₆H₁₅NO	2-Diethylaminoethanol	162.2	Nonazeotrope		551
9975	C₇H₆O	Benzaldehyde	179.2	Nonazeotrope		526
9976	C₇H₇Cl	o-Chlorotoluene	159.2	149.5	60	526
9977	C₇H₇Cl	p-Chlorotoluene	162.4	149.7	70	556
9978	C₇H₈	Toluene	110.75	Nonazeotrope		526
9979	C₇H₈O	Anisole	153.85	148.15	58	527
9980	C₇H₈O	o-Cresol	191.1	Nonazeotrope		556
9981	C₇H₉N	Benzylamine	185.0	Nonazeotrope		551
9982	C₇H₉N	N-Methylaniline	196.25	Nonazeotrope		526
9983	C₇H₁₄O	4-Heptanone	143.55	Nonazeotrope		552
9984	C₇H₁₄O	5-Methyl-2-hexanone	144.2	Nonazeotrope		552
9985	C₇H₁₄O₂	Butyl propionate	146.8	<145.0	~20	575
9986	C₇H₁₄O₂	Ethyl isovalerate	134.7	Nonazeotrope		526
9987	C₇H₁₄O₂	Ethyl valerate	145.75	144.0	22	556
9978	C₇H₁₄O₂	Isobutyl propionate	137.5	Nonazeotrope		575
9989	C₇H₁₄O₃	1,3-Butanediol methyl ether acetate	171.75	Nonazeotrope		527
9990	C₈H₈	Styrene	145.8	140.5	37	567
9991	C₈H₁₀	Ethylbenzene	136.15	134.5	20	556
9992	C₈H₁₀	m-Xylene	139.2	136.95	25.5	527
9993	C₈H₁₀	o-Xylene	144.3	140.5	35	526
9994	C₈H₁₀	p-Xylene	138.45	136.3	24	526
9995	C₈H₁₀O	p-Methylanisole	177.05	Nonazeotrope		526
9996	C₈H₁₀O	Phenetole	170.45	Nonazeotrope		556
9997	C₈H₁₁N	N-Dimethylaniline	194.15	Nonazeotrope		575
9998	C₈H₁₆	1,3-Dimethylcyclohexane	120.7	119.0	15	575
9999	C₈H₁₆O₂	Butyl butyrate	166.4	Nonazeotrope		526
10000	C₈H₁₆O₂	Isobutyl butyrate	156.9	149.0	62	526
10001	C₈H₁₆O₂	Propyl isovalerate	155.7	147.5	65	567
10002	C₈H₁₈	Octane	125.75	122.8	18	526
10003	C₈H₁₈O	Butyl ether	142.4	138.5	37	526

No.	Formula	Name	B.P., °C	B.P., °C	Wt.%A	Ref.

		B-Component		**Azeotropic Data**		
A =	**C₅H₁₂O₂**	**2-Propoxyethanol** *(continued)*	**151.35**			
10004	C₈H₁₈O	Isobutyl ether	122.3	<122.0		575
10005	C₉H₈	Indene	182.6	Nonazeotrope		575
10006	C₉H₁₂	Cumene	152.8	147.0	50	526
10007	C₉H₁₂	Mesitylene	164.6	149.4	68	526
10008	C₉H₁₂	Propylbenzene	159.3	147.8	60	556
10009	C₉H₁₂	Pseudocumene	168.2	150.2	82	575
10010	C₉H₁₃N	N,N-Dimethyl-o-toluidine	185.3	Nonazeotrope		551
10011	C₉H₁₈O	2,6-Dimethyl-4-heptanone	168.0	Nonazeotrope		552
10012	C₁₀H₁₄	Butylbenzene	183.1	Nonazeotrope		575
10013	C₁₀H₁₄	Cymene	176.7	Nonazeotrope		556
10014	C₁₀H₁₆	Camphene	159.6	144	52	526
10015	C₁₀H₁₆	Dipentene	177.7	148.5	68	567
10016	C₁₀H₁₆	α-Pinene	155.8	142.0	48	567
10017	C₁₀H₁₆	α-Terpinene	173.4	148.0	65	567
10018	C₁₀H₁₆	Terpinolene	184.6	<150.8		575
10019	C₁₀H₁₈O	Cineole	176.35	Nonazeotrope		556
10020	C₁₀H₂₂	2,7-Dimethyloctane	160.1	143.7	52	527
10021	C₁₀H₂₂O	Amyl ether	187.5	Nonazeotrope		575
10022	C₁₀H₂₂O	Isoamyl ether	173.2	150.1	77	556
A =	**C₅H₁₂O₃**	**2-(2-Methoxyethoxy)ethanol**	**192.95**			
10023	C₆H₅NO₂	Nitrobenzene	210.75	Nonazeotrope		554
10024	C₆H	Phenol	182.2	199.65	61	556
10025	C₆H₇N	Aniline	184.35	Nonazeotrope		551
10026	C₆H₈O₄	Methyl fumarate	193.25	185.5	44	526
10027	C₆H₁₀O₄	Ethylidene diacetate	168.5	Nonazeotrope		575
10028	C₆H₁₀O₄	Glycol diacetate	186.0	181.5	30	567
10029	C₆H₁₁NO₂	Nitrocyclohexane	205.3	<192.7		554
10030	C₇H₅N	Benzonitrile	191.1	<190.5		575
10031	C₇H₇NO₂	o-Nitrotoluene	221.75	Nonazeotrope		554
10032	C₇H₈O	Benzyl alcohol	205.25	<192.5		575
10033	C₇H₈O	o-Cresol	191.1	201.5	52	527
10034	C₇H₈O	p-Cresol	201.7	208.0	30	526
10035	C₇H₈O₂	Guaiacol	205.05	Nonazeotrope		526
10036	C₇H₉N	Benzylamine	185.0	Nonazeotrope		575
10037	C₇H₉N	Methylaniline	196.25	190.0	60	551
10038	C₇H₁₃ClO₂	Isoamyl chloroacetate	190.5	187.0	55	575
10039	C₇H₁₄O₃	Isobutyl lactate	182.15	Nonazeotrope		575
10040	C₇H₁₆O₄	2-[2-(2-Methoxyethoxy)-ethoxy]ethanol	245.25	Nonazeotrope		575
10041	C₈H₆Cl₂	ar-Dichlorostyrene,				
		" 15 mm.		86-90		155
		" 29 mm.		100-101		155
10042	C₈H₈O	Acetophenone	202.0	191.9	80	527
10043	C₈H₈O₂	Methyl benzoate	199.4	188.8	50	526
10044	C₈H₈O₂	Phenyl acetate	195.7	188.6	45	526

TABLE I.   *Binary Systems*   297

No.	Formula	Name	B.P., °C	B.P., °C	Wt.%A	Ref.
		B-Component			**Azeotropic Data**	
A =	$C_5H_{12}O_3$	**2-(2-Methoxyethoxy)ethanol**	**192.95**			
		(continued)				
10045	$C_8H_8O_3$	Methyl salicylate	222.95	Nonazeotrope		575
10046	$C_8H_{10}O$	Benzyl methyl ether	167.8	Nonazeotrope		575
10047	$C_8H_{10}O$	Phenethyl alcohol	219.4	Nonazeotrope		575
10048	$C_8H_{11}N$	Dimethylaniline	194.15	184.85	49	551
10049	$C_8H_{16}O_2$	Isoamyl propionate	160.7	Nonazeotrope		575
10050	$C_8H_{16}O_3$	Isoamyl lactate	202.4	Nonazeotrope		575
10051	C_9H_7N	Quinoline	237.3	Nonazeotrope		553
10052	C_9H_8	Indene	182.3	177.5	30	276,567
10053	$C_9H_{10}O_2$	Benzyl acetate	215.0	Nonazeotrope		575
10054	C_9H_{12}	*m*-Ethyltoluene	161.3		~ 8	631
		"	161.31	160.9	13 vol. %	826
10055	C_9H_{12}	*o*-Ethyltoluene	165.15	164.3	16 vol. %	826
		"	165.1		~ 16	928
10056	C_9H_{12}	*p*-Ethyltoluene	162.0		~ 9	928
		"	161.99	161.4	9 vol. %	826
10057	C_9H_{12}	Mesitylene	164.72	163.8	12 vol. %	826
		"	164.6	162.5	13	575,928
10058	C_9H_{12}	Pseudocumene	168.2		~ 15	928
		"	169.35	167.9	21 vol. %	826
10059	C_9H_{12}	1,2,3-Trimethyl-				
		benzene	176.08	173.4	26 vol. %	826
		"	176.1		~ 26	928
10060	$C_9H_{12}O$	Benzyl ethyl ether	185.0	< 183.2		575
10061	$C_9H_{13}N$	*N,N*-Dimethyl-*o*-toluidine	185.3	< 183.0		551
10062	$C_9H_{13}N$	Dimethyl-*p*-toluidine	210.2	Nonazeotrope		575
10063	$C_9H_{14}O$	Phorone	197.8	190.5	<75	552
10064	$C_9H_{18}O_2$	Isoamyl butyrate	181.05	176.55	22	527
10065	$C_9H_{18}O_2$	Isobutyl isovalerate	171.2	< 170.5		575
10066	$C_{10}H_8$	Naphthalene	218.0	192.2	89	556
10067	$C_{10}H_{14}$	Butylbenzene	183.1	178.5	33	526
		"	183.27	177.9	32 vol. %	826
10068	$C_{10}H_{14}$	*sec*-Butylbenzene	173.30	170.7	16 vol. %	826
		"	173.1		~ 17	928
10069	$C_{10}H_{14}$	*tert*-Butylbenzene	168.5		~ 14	928
		"	169.11	167.6	13 vol. %	826
10070	$C_{10}H_{14}$	Isobutylbenzene	172.76	170.3	24 vol. %	826
10071	$C_{10}H_{14}$	*m*-Diethylbenzene	181.13	176.3	29 vol. %	826
10072	$C_{10}H_{14}$	*p*-Diethylbenzene	183.78	177.9	31 vol. %	826
10073	$C_{10}H_{14}$	5-Ethyl-*m*-xylene	183.75	177.9	30 vol. %	826
10074	$C_{10}H_{14}$	*p*-Cymene	177.10	173.3	22 vol. %	826
		"	176.7	172.0	27	575
10075	$C_{10}H_{14}$	1,2,3,5-Tetramethyl-				
		benzene	197.93	185.9	48 vol. %	826
10076	$C_{10}H_{15}N$	Diethylaniline	217.05	Nonazeotrope		551
10077	$C_{10}H_{16}$	Dipentene	177.7	168.5	33	575
10078	$C_{10}H_{16}$	Nopinene	163.8	159.0	~ 22	575

No.	Formula	B-Component Name	B.P., °C	B.P., °C	Wt.%A	Ref.
A =	$C_5H_{12}O_3$	2-(2-Methoxyethoxy)ethanol	192.95			
		(continued)				
10079	$C_{10}H_{16}$	α-Terpinene	173.4	166.0	30	575
10080	$C_{10}H_{16}O$	Camphor	209.1	Nonazeotrope		552
10081	$C_{10}H_{18}O$	Borneol	215.0	Nonazeotrope		526
10082	$C_{10}H_{18}O$	Cineole	176.35	173.0	22	556
10083	$C_{10}H_{18}O$	Citronellal	208.0	Nonazeotrope		575
10084	$C_{10}H_{18}O$	Geraniol	229.6	Nonazeotrope		575
10085	$C_{10}H_{18}O$	α-Terpineol	218.85	Nonazeotrope		575
10086	$C_{10}H_{20}O$	Menthol	216.3	Nonazeotrope		575
10087	$C_{10}H_{20}O_2$	Isoamyl isovalerate	192.7	< 185.0	<45	567
10088	$C_{10}H_{22}O$	Amyl ether	187.5	179.5	46	526
10089	$C_{10}H_{22}O$	Decyl alcohol	232.8	Nonazeotrope		575
10090	$C_{10}H_{22}O$	Isoamyl ether	173.2	168.85	23	527
10091	$C_{11}H_{10}$	2-Methylnaphthalene	241.15	Nonazeotrope		527
		"	241.1	Nonazeotrope		270
10092	$C_{11}H_{16}$	*tert*-Amylbenzene	198.1	182.8	40 vol. %	826
10093	$C_{11}H_{20}O$	Isobornyl methyl ether	192.4	187.5	50	567
10094	$C_{11}H_{22}$	*tert*-Amylcyclohexane	198.1	180.6	40 vol. %	826
10095	$C_{11}H_{24}$	*n*-Undecane	195.88	178.7	40 vol. %	826
10096	$C_{12}H_{18}$	1,3,5-Triethylbenzene	215.5	190.0	65	567
10097	$C_{12}H_{26}$	*n*-Dodecane, 217 mm.	169.79	144.2	52 vol. %	826
10098	$C_{12}H_{26}$	2,2,4,4,6-Pentamethyl-heptane	185.6	173.6	30 vol. %	826
10099	$C_{12}H_{26}$	2,2,4,6,6-Pentamethyl-heptane	177.9	168.9	23 vol. %	826
10100	$C_{13}H_{26}$	1-Tridecene	232.78	191.6	70 vol. %	826
A =	$C_5H_{12}S$	3-Methyl-1-butanethiol	~ 120			
10101	$C_6H_{10}O$	1-Hexene-5-one	129	Reacts		563
A =	$C_5H_{14}OSi$	Ethoxytrimethylsilane	75-76			
10102	C_6H_6	Benzene	80.2	Min. b.p.		192
A =	$C_6F_{12}O$	Perfluorocyclic Ether				
10103	C_6F_{14}	Perfluorohexane, 25°		Nonazeotrope	v-l	695
A =	C_6F_{14}	Perfluorohexane				
10104	C_6H_{14}	Hexane, 325 mm.		25	83.4	v-l	219
		" 479 mm.		35	83.7	v-l	219
		" 689 mm.		45	80.0	v-l	219
10105	$C_{12}F_{27}N$	Tris(perfluorobutyl) amine, 25°		Nonazeotrope	v-l	695
A =	$C_6H_{15}N$	Perfluorotriethylamine				
10106	C_6H_6	Benzene	80.1	56.8	87.2	717
10107	C_6H_{12}	Cyclohexane	80.7	56.2	85.4	717
10108	C_6H_{14}	Hexane	68.7	54.5	80.4	717
A =	$C_6H_3Cl_3$	1,2,4-Trichlorobenzene				
10109	$C_9H_6N_2O_2$	2,4-Tolylene diisocyanate, 40 mm.		Nonazeotrope	v-l	327

TABLE I. *Binary Systems*   299

No.	Formula	B-Component Name	B.P., °C	Azeotropic Data B.P., °C	Wt.%A	Ref.
A =	**C$_6$H$_3$Cl$_3$**	**1,3,5-Trichlorobenzene**	**208.4**			
10110	C$_6$H$_5$NO$_2$	Nitrobenzene	210.75	~ 207.0		545
10111	C$_6$H$_6$O	Phenol	181.5	181.3	5	563
		"	182.2	Nonazeotrope		544
10112	C$_6$H$_6$O$_2$	Pyrocatechol	245.9	Nonazeotrope		544
10113	C$_6$H$_7$N	Aniline	184.35	Nonazeotrope		575
10114	C$_6$H$_{10}$O$_3$	Ethyl acetoacetate	180.4	Nonazeotrope		545
10115	C$_6$H$_{12}$O$_2$	Caproic acid	205.2	204.0	57	543
10116	C$_7$H$_6$O$_2$	Benzoic acid	250.8	Nonazeotrope		575
10117	C$_7$H$_7$NO$_2$	o-Nitrotoluene	221.75	Nonazeotrope		554
10118	C$_7$H$_8$O	Benzyl alcohol	202.25	202.5		575
10119	C$_7$H$_8$O	m-Cresol	202.2	200.5	40	542
10120	C$_7$H$_8$O	o-Cresol	190.8	Nonazeotrope		563
10121	C$_7$H$_8$O	p-Cresol	201.7	200.2	40	542
10122	C$_7$H$_9$N	Methylaniline	196.25	Nonazeotrope		551
10123	C$_7$H$_9$N	m-Toluidine	203.1	< 202.5	>25	575
10124	C$_7$H$_9$N	p-Toluidine	200.3	~ 199		563
10125	C$_7$H$_{12}$O$_4$	Ethyl malonate	198.9	Nonazeotrope		563
10126	C$_8$H$_8$O	Acetophenone	202	Nonazeotrope		563
10127	C$_8$H$_8$O$_2$	Methyl benzoate	199.55	Nonazeotrope		563
10128	C$_8$H$_8$O$_3$	Methyl salicylate	222.95	Nonazeotrope		548
10129	C$_8$H$_{10}$O	Phenethyl alcohol	219.4	< 207.5		575
10130	C$_8$H$_{11}$N	Dimethylaniline	194.15	Nonazeotrope		551
10131	C$_8$H$_{11}$N	Ethylaniline	206.5	203	65	563
10132	C$_8$H$_{18}$O	Octyl alcohol	195.2	Nonazeotrope		575
10133	C$_8$H$_{18}$O$_3$	2-(2-Butoxyethoxy)ethanol	231.2	Nonazeotrope		575
10134	C$_9$H$_{10}$O$_2$	Benzyl acetate	215.6	Nonazeotrope?		563
10135	C$_9$H$_{10}$O$_2$	Ethyl benzoate	213	Nonazeotrope		563
10136	C$_9$H$_{13}$N	N,N-Dimethyl-o-toluidine	185.3	Nonazeotrope		575
10137	C$_{10}$H$_8$	Naphthalene	218.0	Nonazeotrope		575
10138	C$_{10}$H$_{14}$O	Thymol	232.9	Nonazeotrope		544
10139	C$_{10}$H$_{16}$O	Camphor	209.1	211.5	52	551
10140	C$_{10}$H$_{18}$O	Borneol	215.0	Nonazeotrope		575
10141	C$_{10}$H$_{18}$O	Menthone	~ 207	~ 209.5		563
10142	C$_{10}$H$_{20}$O	Menthol	216.3	Nonazeotrope		575
10143	C$_{11}$H$_{24}$O$_2$	Diisoamyloxymethane	210.8	213.0	35	559
A =	**C$_6$H$_4$BrCl**	**p-Bromochlorobenzene**	**196.4**			
10144	C$_6$H$_6$O	Phenol	182.2	181.0	38	562
10145	C$_6$H$_7$N	Aniline	184.35	Nonazeotrope		551
10146	C$_6$H$_{10}$O$_3$	Ethyl acetoacetate	180.4	Nonazeotrope		575
10147	C$_6$H$_{10}$O$_4$	Methyl succinate	195.5	< 191.3	>46	575
10148	C$_6$H$_{12}$O$_2$	Caproic acid	205.15	193.0	80	562
10149	C$_6$H$_{14}$O$_2$	2-Butoxyethanol	171.15	Nonazeotrope		575
10150	C$_7$H$_5$N	Benzonitrile	191.1	< 190.5	<30	575
10151	C$_7$H$_8$O	Benzyl alcohol	205.25	194.0		575
10152	C$_7$H$_8$O	o-Cresol	191.1	189.0	47	562
10153	C$_7$H$_8$O	p-Cresol	201.7	194.5	75	562

No.	Formula	B-Component Name	B.P., °C	B.P., °C	Azeotropic Data Wt.%A	Ref.
A =	C₆H₄BrCl	*p*-Bromochlorobenzene *(continued)*	196.4			
10154	C₇H₉N	*o*-Toluidine	200.35	194.6		575
10155	C₇H₉N	*p*-Toluidine	200.55	< 195.2	>68	575
10156	C₇H₁₂O₄	Ethyl malonate	199.35	< 193.5	>40	575
10157	C₈H₈O	Acetophenone	202.0	Nonazeotrope		552
10158	C₈H₈O₂	Methyl benzoate	199.4	Nonazeotrope		575
10159	C₈H₁₁N	Dimethylaniline	194.15	Nonazeotrope		575
10160	C₈H₁₆O₃	Isoamyl lactate	202.4	Nonazeotrope		575
10161	C₈H₁₈O	*sec*-Octyl alcohol	180.4	Nonazeotrope		575
10162	C₉H₁₂O	Benzyl ethyl ether	185.0	Nonazeotrope		559
10163	C₉H₁₂O	Phenyl propyl ether	190.5	Nonazeotrope		575
10164	C₉H₁₃N	*N,N*-Dimethyl-*o*-toluidine	185.3	Nonazeotrope		551
10165	C₉H₁₃N	*N,N*-Dimethyl-*p*-toluidine	210.2	Nonazeotrope		551
10166	C₉H₁₄O	Phorone	197.8	Nonazeotrope		552
10167	C₁₀H₂₀O₂	Isoamyl isovalerate	192.7	Nonazeotrope		575
A =	C₆H₄Br₂	*p*-Dibromobenzene	220.25			
10168	C₆H₄ClNO₂	*m*-Chloronitrobenzene	235.5	Nonazeotrope		554
10169	C₆H₄ClNO₂	*p*-Chloronitrobenzene	239.1	Nonazeotrope		554
10170	C₆H₅ClO	*p*-Chlorophenol	219.75	215.05	65	574
10171	C₆H₅NO₂	Nitrobenzene	210.75	210.45	22.5	554
10172	C₆H₅NO₃	*o*-Nitrophenol	217.2	215.15	48	564
10173	C₆H₆O	Phenol	182.2	Nonazeotrope		535
10174	C₆H₆O₂	Pyrocatechol	245.9	218.15	90	538
10175	C₆H₆O₂	Resorcinol	281.4	Nonazeotrope		542
10176	C₆H₁₂O₂	Caproic acid	205.15	203.4	42	564
10177	C₇H₅Cl₃	*α,α,α*-Trichlorotoluene	220.9	219.6	72	549
10178	C₇H₆O₂	Benzoic acid	250.5	219.5	96.2	538
10179	C₇H₇BrO	*o*-Bromoanisole	217.7	< 217.4	<12	575
10180	C₇H₇NO₂	*m*-Nitrotoluene	230.8	Nonazeotrope		554
10181	C₇H₇NO₂	*o*-Nitrotoluene	221.75	218.0	73	554
10182	C₇H₇NO₂	*p*-Nitrotoluene	238.9	Nonazeotrope		554
10183	C₇H₈O	Benzyl alcohol	205.2	204.2	34.5	574
10184	C₇H₈O	*m*-Cresol	202.1	201.9	7	541
10185	C₇H₈O	*o*-Cresol	191.1	Nonazeotrope		538
10186	C₇H₈O	*p*-Cresol	201.7	Nonazeotrope		542
10187	C₇H₈O₂	Guaiacol	205.05	Nonazeotrope		556
10188	C₇H₈O₂	*m*-Methoxyphenol	244	Nonazeotrope		535
10189	C₇H₉N	*m*-Toluidine	203.1	Nonazeotrope		551
10190	C₇H₉N	*o*-Toluidine	200.35	Nonazeotrope		551
10191	C₇H₉N	*p*-Toluidine	200.55	Nonazeotrope		551
10192	C₇H₉NO	*o*-Anisidine	219.0	217.5		575
10193	C₇H₁₄O₂	Enanthic acid	220.0	215.5	70	562
10194	C₇H₁₆O₄	2-[2-(2-Methoxyethoxy) ethoxy]ethanol	245.25	Nonazeotrope		575
10195	C₈H₈O₂	*α*-Toluic acid	266.5	Nonazeotrope		575
10196	C₈H₈O₃	Methyl salicylate	222.35	219.4	75	574
10197	C₈H₉BrO	*p*-Bromophenetole	234.5	Nonazeotrope		575

TABLE I. *Binary Systems*   301

No.	Formula	Name	B.P., °C	B.P., °C	Wt.%A	Ref.
		B-Component		**Azeotropic Data**		
A =	**C₆H₄Br₂**	*p*-**Dibromobenzene**	**220.25**			
		(continued)				
10198	C₈H₁₀O	3,4-Xylenol	226.8	218.65	75	550
10199	C₈H₁₀O	*p*-Ethylphenol	218.8	216.0	50	562
10200	C₈H₁₀O	Phenethyl alcohol	219.4	215.0	67.5	574
10201	C₈H₁₀O	2,4-Xylenol	210.5	209.8	10	575
10202	C₈H₁₀O₂	*m*-Dimethoxybenzene	214.7	Nonazeotrope		535
10203	C₈H₁₀O₂	*o*-Ethoxyphenol	216.5	214.0	32	575
10204	C₈H₁₀O₂	2-Phenoxyethanol	245.2	Nonazeotrope		575
10205	C₈H₁₁N	Ethylaniline	205.5	Nonazeotrope		551
10206	C₈H₁₁N	3,4-Xylidine	225.5	< 219.9	<89	575
10207	C₈H₁₁NO	*o*-Phenetidine	232.5	Nonazeotrope		548
10208	C₈H₁₂O₄	Ethyl fumarate	217.85	< 216.5	<47	575
10209	C₈H₁₄O₄	Ethyl succinate	217.25	< 215.0	>25	547
10210	C₈H₁₄O₄	Propyl oxalate	214	< 213	<32	575
10211	C₈H₁₆O₂	Caprylic acid	237.5	218.8	~ 90	541
10212	C₈H₁₈O	Octyl alcohol	195.2	Nonazeotrope		575
10213	C₈H₁₈O₃	2-(2-Butoxyethoxy)ethanol	231.2	Nonazeotrope		575
10214	C₉H₇N	Quinoline	237.3	Nonazeotrope		553
10215	C₉H₁₀O	*p*-Methylacetophenone	226.35	220.15	95	552
10216	C₉H₁₀O	Propiophenone	217.7	Nonazeotrope		552
10217	C₉H₁₀O₂	Benzyl acetate	214.9	Nonazeotrope		538
10218	C₉H₁₀O₂	Ethyl benzoate	212.6	Nonazeotrope		535
10219	C₉H₁₀O₃	Ethyl salicylate	234.0	Nonazeotrope		548
10220	C₉H₁₂O	3-Phenylpropanol	220.25	< 219.9	>85	575
10221	C₉H₁₃N	*N,N*-Dimethyl-*p*-toluidine	210.2	Nonazeotrope		551
10222	C₉H₁₈O₂	Pelargonic acid	254.0	Nonazeotrope		575
10223	C₁₀H₈	Naphthalene	218.05	Nonazeotrope		574
10224	C₁₀H₁₂O	Estragole	215.6	Nonazeotrope		535
10225	C₁₀H₁₂O₂	Ethyl α-toluate	228.75	Nonazeotrope		547
10226	C₁₀H₁₂O₂	Propyl benzoate	230.85	Nonazeotrope		575
10227	C₁₀H₁₄O	Carvacrol	237.85	Nonazeotrope		575
10228	C₁₀H₁₄O	Carvone	231.0	Nonazeotrope		552
10229	C₁₀H₁₄O	Thymol	232.9	Nonazeotrope		542
10230	C₁₀H₁₄O₂	*m*-Diethoxybenzene	235.0	Nonazeotrope		559
10231	C₁₀H₁₅N	Diethylaniline	217.05	Nonazeotrope		551
10232	C₁₀H₁₆O	Pulegone	223.8	Nonazeotrope		552
10233	C₁₀H₁₈O	Borneol	213.4	213.3	~ 18	535
10234	C₁₀H₁₈O	Geraniol	229.6	220.2	97	535
10235	C₁₀H₁₈O	α-Terpineol	217.8	Reacts		535
10236	C₁₀H₂₀O	Citronellol	224.5	Nonazeotrope		535
		"	224.5	218.5		533
10237	C₁₀H₂₀O	Menthol	216.4	215.4	43	574
10238	C₁₀H₂₀O₂	Methyl pelargonate	213.8	Nonazeotrope		575
10239	C₁₀H₂₂O	Decyl alcohol	~ 232.9	220.2	98	535
10240	C₁₁H₁₆O	Methyl thymol ether	216.5	Nonazeotrope		559
10241	C₁₁H₂₀O	Terpineol methyl ether	216.3	Nonazeotrope		548

| | B-Component | | | Azeotropic Data | | |
No.	Formula	Name	B.P., °C	B.P., °C	Wt.%A	Ref.
A =	**C₆H₄Br₂**	***p*-Dibromobenzene**	**220.25**			
		(continued)				
10242	C₁₁H₂₂O₃	Isoamyl carbonate	232.2	Nonazeotrope		547
10243	C₁₂H₁₈	1,3,5-Triethylbenzene	215.5	Nonazeotrope		575
10244	C₁₂H₂₀O₂	Bornyl acetate	227.6	Nonazeotrope		538
A =	**C₆H₄ClNO₂**	***m*-Chloronitrobenzene**	**235.5**			
10245	C₆H₆O₂	Pyrocatechol	245.9	Nonazeotrope		554
10246	C₆H₁₄O₃	Dipropylene glycol	229.2	< 227.0		554
10247	C₇H₅Cl₃	α,α,α-Trichlorotoluene	220.8	Nonazeotrope		554
10248	C₇H₇NO₂	*m*-Nitrotoluene	230.8	Nonazeotrope		554
10249	C₇H₇NO₂	*p*-Nitrotoluene	238.9	Nonazeotrope		575
10250	C₇H₉NO	*o*-Anisidine	219.0	Nonazeotrope		575
10251	C₇H₁₄O₂	Enanthic acid	222.0	< 221.5		554
10252	C₈H₈O₃	Methyl salicylate	222.95	Nonazeotrope		554
10253	C₈H₁₀O	*p*-Ethylphenol	220.0	Nonazeotrope		554
10254	C₈H₁₀O	3,4-Xylenol	226.8	Nonazeotrope		554
10255	C₈H₁₁NO	*o*-Phenetidine	232.5	Nonazeotrope		551
10256	C₈H₁₁NO	*p*-Phenetidine	249.9	Nonazeotrope		551
10257	C₉H₇N	Quinoline	237.3	Nonazeotrope		554
10258	C₉H₁₀O	*p*-Methylacetophenone	226.35	Nonazeotrope		575
10259	C₉H₁₀O	Cinnamyl alcohol	257.0	Nonazeotrope		554
10260	C₉H₁₀O₃	Ethyl salicylate	233.8	Nonazeotrope		554
10261	C₁₀H₈	Naphthalene	218.0	Nonazeotrope		554
10262	C₁₀H₁₂O₂	Propyl benzoate	230.85	Nonazeotrope		554
10263	**C₁₀H₁₄O**	Carvacrol	237.85	< 235.4		554
10264	C₁₀H₁₄O	Carvone	231.0	Nonazeotrope		575
10265	C₁₀H₁₄O	Thymol	232.9	Nonazeotrope		554
10266	C₁₀H₂₂S	Isoamyl sulfide	214.3	Nonazeotrope		575
10267	C₁₁H₁₀	1-Methylnaphthalene	244.6	Nonazeotrope		554
10268	C₁₁H₂₂O₃	Isoamyl carbonate	232.2	< 231.8		554
10269	C₁₂H₂₀O₂	Bornyl acetate	227.6	Nonazeotrope		557
A =	**C₆H₄ClNO₂**	***o*-Chloronitrobenzene**	**246.0**			
10270	C₆H₆O₂	Pyrocatechol	245.9	243.5		554
10271	C₆H₆O₂	Resorcinol	281.4	Nonazeotrope		554
10272	C₆H₁₄O₄	Triethylene glycol	288.7	Nonazeotrope		554
10273	C₇H₆O₂	Benzoic acid	250.8	243.0	67	554
10274	C₇H₇NO₂	*p*-Nitrotoluene	238.9	Nonazeotrope		544
10275	C₇H₈O	*m*-Cresol	202.2	Nonazeotrope		554
10276	C₇H₁₄O₂	Enanthic acid	222.0	Nonazeotrope		554
10277	C₈H₁₁NO	*o*-Phenetidine	232.5	Nonazeotrope		551
10278	C₈H₁₁NO	*p*-Phenetidine	249.9	Nonazeotrope		551
10279	C₉H₇N	Quinoline	237.3	Nonazeotrope		553
10280	C₁₀H₇Cl	1-Chloronaphthalene	262.7	Nonazeotrope		554
10281	C₁₀H₁₀O₂	Isosafrole	252.0	Nonazeotrope		554
10282	C₁₀H₁₀O₂	Safrole	235.9	Nonazeotrope		554
10283	C₁₀H₁₄O	Carvacrol	237.85	Nonazeotrope		554
10284	C₁₀H₁₄O	Thymol	232.9	Nonazeotrope		554

TABLE I.   *Binary Systems*   303

No.	Formula	Name	B.P., °C	B.P., °C	Wt.%A	Ref.
		B-Component		**Azeotropic Data**		
A =	C$_6$H$_4$ClNO$_2$	*o*-Chloronitrobenzene	**246.0**			
		(continued)				
10285	C$_{11}$H$_{10}$	2-Methylnaphthalene	241.15	Nonazeotrope		554
10286	C$_{11}$H$_{14}$O$_2$	Butyl benzoate	249.5	Nonazeotrope		554
10287	C$_{11}$H$_{14}$O$_2$	Isobutyl benzoate	241.9	Nonazeotrope		554
10288	C$_{12}$H$_{16}$O$_3$	Isoamyl salicylate	277.5	Nonazeotrope		554
A =	C$_6$H$_4$ClNO$_2$	*p*-Chloronitrobenzene	**239.1**			
10289	C$_6$H$_6$O$_2$	Pyrocat‿chol	247.9	238.6	82.5	554
10290	C$_6$H$_6$O$_2$	Resorcinol	281.4	Nonazeotrope		554
10291	C$_6$H$_{14}$O$_3$	Dipropylene glycol	229.2	< 228.3	<89	554
10292	C$_7$H$_6$O$_2$	Benzoic acid	250.8	237.75	84	554
10293	C$_7$H$_7$NO$_2$	*p*-Nitrotoluene	238.9	238.85	33	554
10294	C$_7$H$_8$O	Benzyl alcohol	205.25	Nonazeotrope		
10295	C$_7$H$_{16}$O$_4$	2-[2-(2-Methoxyethoxy)				
		ethoxy]ethanol	245.25	< 234.0		554
10296	C$_8$H$_8$O$_2$	Anisaldehyde	249.5	Nonazeotrope		575
10297	C$_8$H$_{10}$O	3,4-Xylenol	226.8	Nonazeotrope		554
10298	C$_8$H$_{11}$NO	*o*-Phenetidine	232.5	Nonazeotrope		551
10299	C$_8$H$_{11}$NO	*p*-Phenetidine	249.9	Nonazeotrope		551
10300	C$_8$H$_{16}$O$_2$	Caprylic acid	238.5	< 235.5		551.
10301	C$_9$H$_7$N	Quinoline	237.3	Nonazeotrope		553
10302	C$_9$H$_8$O	Cinnamyl aldehyde	253.5	Nonazeotrope		554
10303	C$_9$H$_{10}$O	Cinnamyl alcohol	257.0	Nonazeotrope		554
10304	C$_9$H$_{10}$O$_3$	Ethyl salicylate	233.8	Nonazeotrope		543
10305	C$_{10}$H$_9$N	Quinaldine	246.5	Nonazeotrope		554
10306	C$_{10}$H$_{10}$O$_2$	Safrole	235.9	Nonazeotrope		554
10307	C$_{10}$H$_{12}$O$_2$	Ethyl α-toluate	228.75	Nonazeotrope		554
10308	C$_{10}$H$_{12}$O$_2$	Propyl benzoate	230.85	Nonazeotrope		553
10309	C$_{10}$H$_{14}$O	Carvacrol	237.85	237.4		554
10310	C$_{10}$H$_{14}$O	Thymol	232.9	Nonazeotrope		575
10311	C$_{10}$H$_{14}$O	Carvone	231.0	Nonazeotrope		554
10312	C$_{11}$H$_{10}$	1-Methylnaphthalene	244.6	Nonazeotrope		527
10313	C$_{11}$H$_{10}$	2-Methylnaphthalene	241.15	Nonazeotrope		554
10314	C$_{11}$H$_{14}$O$_2$	Butyl benzoate	249.5	Nonazeotrope		554
10315	C$_{11}$H$_{14}$O$_2$	Isobutyl benzoate	241.9	Nonazeotrope		554
10316	C$_{11}$H$_{22}$O$_3$	Isoamyl carbonate	232.2	232.1	5?	554
10317	C$_{12}$H$_{10}$	Biphenyl	256.1	Nonazeotrope		554
10318	C$_{12}$H$_{20}$O$_2$	Bornyl acetate	227.6	Nonazeotrope		554
A =	C$_6$H$_4$Cl$_2$	*o*-Dichlorobenzene	**179.5**			
10319	C$_6$H$_5$Br	Bromobenzene	156.1	Nonazeotrope		575
10320	C$_6$H$_5$ClO	*o*-Chlorophenol	176.8	173.6	52	562
10321	C$_6$H$_5$NO$_2$	Nitrobenzene	210.75	Nonazeotrope		554
10322	C$_6$H$_6$O	Phenol	182.2	173.7	65	562
10323	C$_6$H$_7$N	Aniline	184.35	177.4	70	551
10324	C$_6$H$_8$O$_4$	Methyl fumarate	193.25	Nonazeotrope		527
10325	C$_6$H$_{10}$O$_3$	Ethyl acetoacetate	180.4	175.5	58	552
10326	C$_6$H$_{10}$O$_4$	Ethylidene diacetate	168.5	Nonazeotrope		527
10327	C$_6$H$_{10}$O$_4$	Ethyl oxalate	185.65	< 178.2	<82	575

No.	Formula	B-Component Name	B.P., °C	Azeotropic Data B.P., °C	Wt.%A	Ref.
A =	C₆H₄Cl₂	o-Dichlorobenzene (continued)	179.5			
10328	C₆H₁₂O	Cyclohexanol	160.8	Nonazeotrope		575
10329	C₆H₁₂O₂	Caproic acid	205.15	179.0	92	564
10330	C₆H₁₂O₂	Isocaproic acid	199.5	178.5	94	575
10331	C₆H₁₄O	Hexyl alcohol	157.85	Nonazeotrope		575
10332	C₆H₁₄O₂	2-Butoxyethanol	171.15	170.0	27	556
10333	C₇H₆O	Benzaldehyde	179.2	< 178.5	>48	575
10334	C₇H₈O	Benzyl alcohol	205.25	Nonazeotrope		575
10335	C₇H₈O	m-Cresol	202.2	Nonazeotrope		575
10336	C₇H₈O	o-Cresol	191.1	179.1	85	575
10337	C₇H₈O	p-Cresol	201.7	Nonazeotrope		575
10338	C₇H₉N	Methylaniline	196.25	Nonazeotrope		551
10339	C₇H₁₄O₂	Isoamyl acetate	142.1	Nonazeotrope		575
10340	C₇H₁₄O₃	1,3-Butanediol methyl ether acetate	171.75	Nonazeotrope		575
10341	C₇H₁₆O	Heptyl alcohol	176.15	173.5	45	567
10342	C₈H₈O₂	Phenyl acetate	195.7	Nonazeotrope		575
10343	C₈H₁₀O	p-Methylanisole	177.05	179.6	~ 5	559
10344	C₈H₁₀O	Phenetole	170.45	Nonazeotrope		559
10345	C₈H₁₁N	Dimethylaniline	194.15	Nonazeotrope		551
10346	C₈H₁₂N₂O₂	Hexamethylene diisocyanate 40 mm.	163.2	Nonazeotrope		413
10347	C₈H₁₆O₂	Butyl butyrate	166.4	Nonazeotrope		575
10348	C₈H₁₈O	Octyl alcohol	195.2	Nonazeotrope		575
10349	C₈H₁₈O	sec-Octyl alcohol	180.4	177.7	58	567
10350	C₈H₂₀SiO₄	Ethyl silicate	168.8	Nonazeotrope		575
10351	C₉H₆O₂N₂	2,4-Tolylene diisocyanate, 15 mm.	128.7	Nonazeotrope		413
10352	C₉H₈	Indene	182.6	> 183.0		575
10353	C₉H₁₂	Pseudocumene	168.2	Nonazeotrope		575
10354	C₉H₁₂O	Benzyl ethyl ether	185.0	Nonazeotrope		559
10355	C₉H₁₃N	N,N-Dimethyl-o-toluidine	185.3	Nonazeotrope		551
10356	C₉H₁₈O₂	Isobutyl isovalerate	171.2	Nonazeotrope		575
10357	C₁₀H₁₆	Dipentene	177.7	177.5	> 20	575
10358	C₁₀H₁₆	Nopinene	163.8	Nonazeotrope		575
10359	C₁₀H₁₆	α-Terpinene	173.4	Nonazeotrope		575
10360	C₁₀H₁₈O	Cineole	176.35	Nonazeotrope		559
10361	C₁₀H₁₈O	Linalool	198.6	Nonazeotrope		575
10362	C₁₀H₁₉N	Bornylamine	199.8	Nonazeotrope		575
10363	C₁₀H₂₀O₂	Isoamyl isovalerate	192.7	Nonazeotrope		575
10364	C₁₀H₂₂O	Amyl ether	187.5	Nonazeotrope		559
10365	C₁₀H₂₂O	Isoamyl ether	173.2	Nonazeotrope		559
10366	C₁₅H₁₀O₂N₂	Di-p-isocyanatodiphenylmethane, 5 mm.	192.0	Nonazeotrope		413

TABLE I.   *Binary Systems*   305

No.	B-Component			Azeotropic Data		
	Formula	Name	B.P., °C	B.P., °C	Wt.%A	Ref.
A =	**C₆H₄Cl₂**	**p-Dichlorobenzene**	**174.4**			
10367	C₆H₅BrO	o-Bromophenol	195.0	Nonazeotrope		575
10368	C₆H₅ClO	o-Chlorophenol	176.8	171.0	65	562
10369	C₆H₆O	Phenol	182.2	171.05	74.8	555
10370	C₆H₆S	Benzenethiol	169.5	< 168.2	<29	575
10371	C₆H₇N	Aniline	184.35	173.95	88	551
10372	C₆H₁₀O₃	Ethyl acetoacetate	180.4	172.65	71	552
10373	C₆H₁₀O₄	Ethyl oxalate	185.65	174.25?	~ 5	535
10374	C₆H₁₂O	Cyclohexanol	160.8	160.2		571
10375	C₆H₁₂O₂	Caproic acid	205.2	Nonazeotrope		541
10376	C₆H₁₂O₂	Isocaproic acid	199.5	174.2	98	575
10377	C₆H₁₂O₃	2-Ethoxyethyl acetate	156.8	Nonazeotrope		526
10378	C₆H₁₂O₃	Propyl lactate	171.7	< 170.0	<38	567
10379	C₆H₁₄O	Hexyl alcohol	157.85	157.65		571
10380	C₆H₁₄O₂	2-Butoxyethanol	171.2	168.3	48	527
10381	C₆H₁₄O₂	Pinacol	174.35	< 167.0	<70	567
10382	C₇H₅N	Benzonitrile	191.1	Nonazeotrope		565
10383	C₇H₆O	Benzaldehyde	179.2	174.1	83	536
10384	C₇H₈O	Benzyl alcohol	205.2	Nonazeotrope		535
10385	C₇H₈O	o-Cresol	191.1	Nonazeotrope		536
10386	C₇H₈O	p-Cresol	201.7	Nonazeotrope		542
10387	C₇H₉N	Methylaniline	196.25	Nonazeotrope		551
10388	C₇H₁₄O	2-Methylcyclohexanol	168.5	167.3	43	567
10389	C₇H₁₄O₃	Isobutyl lactate	182.15	Nonazeotrope		538
10390	C₇H₁₄O₃	1,3-Butanediol methyl ether acetate	171.75	Nonazeotrope		527,556
10391	C₇H₁₆O	Heptyl alcohol	176.15	171.2	65	567
10392	C₈H₁₀O	Benzyl methyl ether	167.8	Nonazeotrope		559
10393	C₈H₁₀O	p-Methylanisole	177.65	Nonazeotrope		559
		"	177.05	177.07	~ 6	541
10394	**C₈H₁₀O**	Phenetole	170.45	Nonazeotrope		538
10395	C₈H₁₁N	Dimethylaniline	194.15	Nonazeotrope		551
10396	C₈H₁₄O	Methylheptenone	173.2	Nonazeotrope		552
10397	C₈H₁₆O	2-Octanone	172.85	Nonazeotrope		552
10398	C₈H₁₆O₂	Butyl butyrate	166.4	Nonazeotrope		547
10399	C₈H₁₆O₂	Ethyl caproate	167.7	Nonazeotrope		575
10400	C₈H₁₆O₂	Hexyl acetate	171.5	171.4		547
10401	C₈H₁₆O₂	Isoamyl propionate	164.4	Nonazeotrope		547
10402	C₈H₁₈O	n-Octyl alcohol	195.15	Nonazeotrope		530
10403	C₈H₁₈O	sec-Octyl alcohol	180.4	173.85	78	564
10404	C₈H₁₈O₃	Bis(2-ethoxyethyl) ether	186.0	Nonazeotrope		575
10405	C₈H₁₈S	Butyl sulfide	185.0	Nonazeotrope		575
10406	C₈H₁₈S	Isobutyl sulfide	172.0	< 171.0	<42	566
10407	C₈H₂₀SiO₄	Ethyl silicate	168.8	Nonazeotrope		564
10408	C₉H₈	Indene	183.0	Nonazeotrope		541
10409	C₉H₁₂	Cumene	152.8	Nonazeotrope		575
10410	C₉H₁₂	Mesitylene	164.6	Nonazeotrope		575
10411	C₉H₁₂	Pseudocumene	168.2	Nonazeotrope		541

No.	Formula	Name	B.P., °C	B.P., °C	Wt.%A		Ref.
		B-Component		**Azeotropic Data**			
A =	C₆H₄Cl₂	*p*-**Dichlorobenzene**	**174.4**				
		(continued)					
10412	C₉H₁₂O	Benzyl ethyl ether	185.0	Nonazeotrope			559
10413	C₉H₁₃N	*N,N*-Dimethyl-*o*-toluidine	185.3	Nonazeotrope			551
10414	C₉H₁₈O	2,6-Dimethyl-4-heptanone	168.0	Nonazeotrope			552
10415	C₉H₁₈O₂	Isoamyl butyrate	178.5	Nonazeotrope			547
10416	C₉H₁₈O₂	Isoamyl isobutyrate	170.0	Nonazeotrope			545
10417	C₉H₁₈O₂	Isobutyl isovalerate	171.4	Nonazeotrope			538
10418	C₉H₁₈O₃	Isobutyl carbonate	190.3	Nonazeotrope			547
10419	C₁₀H₁₄	Butylbenzene	183.1	Nonazeotrope			575
10420	C₁₀H₁₄	Cymene	176.7	Nonazeotrope			575
10421	C₁₀H₁₆	Camphene	159.6	Nonazeotrope			538
10422	C₁₀H₁₆	*d*-Limonene	177.8	174.2	86		530
10423	C₁₀H₁₆	Nopinene	163.8	Nonazeotrope			575
10424	C₁₀H₁₆	α-Pinene	155.8	Nonazeotrope			538
10425	C₁₀H₁₆	α-Terpinene	173.4	173.15	50		527
10426	C₁₀H₁₆	γ-Terpinene	183	Nonazeotrope			575
10427	C₁₀H₁₆	Terpinene	181.5	Nonazeotrope			538
10428	C₁₀H₁₆	Terpinolene	184.6	Nonazeotrope			575
10429	C₁₀H₁₆	Thymene	179.7	Nonazeotrope			535
10430	C₁₀H₁₈O	Cineole	176.4	174.1	~ 80		559
10431	C₁₀H₂₂O	Amyl ether	187.5	Nonazeotrope			559
10432	C₁₀H₂₂O	Isoamyl ether	172.6	172.1	36.5		555
A =	C₆H₅Br	**Bromobenzene**	**156.1**				
10433	C₆H₅Cl	Chlorobenzene	132	Nonazeotrope			563
10434	C₆H₅ClO	*o*-Chlorophenol	176.8	Nonazeotrope			575
10435	C₆H₅NO₂	Nitrobenzene	210.75	Nonazeotrope			554
10436	C₆H₆O	Phenol	182.2	Nonazeotrope			542
10437	C₆H₇N	Aniline	184.35	Nonazeotrope			551
10438	C₆H₁₀O	Cyclohexanone	155.7	Nonazeotrope			552
10439	C₆H₁₀O₃	Ethyl acetoacetate	156.1	Nonazeotrope			552
10440	C₆H₁₀O₄	Ethylidene diacetate	168.5	155.95	92.5		527
10441	C₆H₁₀O₄	Ethyl oxalate	185.65	Nonazeotrope			527
10442	C₆H₁₀S	Allyl sulfide	139.35	Nonazeotrope			566
10443	C₆H₁₁ClO₂	Isobutyl chloroacetate	174.5	Nonazeotrope			575
10444	C₆H₁₂O	Cyclohexanol	160.65	153.6	66.5		563
		" 250 mm.	127.0	113.6	85.5	v-l	930
		" 500 mm.	144.4	136.8	81.5	v-l	930
		" 730 mm.	158.6	150.6	74.8	v-l	930
10445	C₆H₁₂O₃	2-Ethoxyethyl acetate	156.8	155.45	63		556
10446	C₆H₁₃ClO₂	Chloroacetal	156.8	~ 156			563
10447	C₆H₁₄O	Hexyl alcohol	157.95	151.6	66		538
10448	C₆H₁₄O₂	2-Butoxyethanol	171.15	155.85	93.5		556
10449	C₆H₁₄O₂	Pinacol	174.3	153.2	~ 85		532
		"	171.5	152	~ 86		563
10450	C₆H₁₄S	Propyl sulfide	141.5	Nonazeotrope			575
10451	C₇H₇Cl	*o*-Chlorotoluene	159.2	Nonazeotrope			549
10452	C₇H₈	Toluene	110.7	Nonazeotrope			563
10453	C₇H₈O	Anisole	153.85	Nonazeotrope			563

TABLE I. *Binary Systems* 307

No.	Formula	B-Component Name	B.P., °C	Azeotropic Data B.P., °C	Wt.%A	Ref.
A =	**C₆H₅Br**	**Bromobenzene** *(continued)*	**156.1**			
10454	C₇H₈O	o-Cresol	190.8	Nonazeotrope		563
10455	C₇H₁₄O₂	Ethyl valerate	145.45	Nonazeotrope		575
10456	C₇H₁₄O₂	Isoamyl acetate	142.1	Nonazeotrope		547
10457	C₇H₁₄O₂	Methyl caproate	151.0	Nonazeotrope		547
10458	C₇H₁₄O₂	Propyl butyrate	143.7	Nonazeotrope		547
10459	C₇H₁₄O₃	1,3-Butanediol methyl ether acetate	171.75	Nonazeotrope		575
10460	C₇H₁₆O	Heptyl alcohol	176.15	Nonazeotrope		575
10461	C₇H₁₆O₃	Ethyl orthoformate	145.75	Nonazeotrope		559
10462	C₈H₈	Styrene	145.8	Nonazeotrope		535
10463	C₈H₁₀	Ethylbenzene	136.15	Nonazeotrope		563
10464	C₈H₁₀	m-Xylene	139	Nonazeotrope		527,563
10465	C₈H₁₀O	Benzyl methyl ether	167.8	Nonazeotrope		559
10466	C₈H₁₆O₂	Butyl butyrate	166.4	Nonazeotrope		547
10467	C₈H₁₆O₂	Isoamyl propionate	~ 160.3	~ 155.2	~ 73	563
10468	C₈H₁₆O₂	Isobutyl butyrate	156.8	155.2		545
10469	C₈H₁₆O₂	Isobutyl isobutyrate	147.3	Nonazeotrope		573
10470	C₈H₁₆O₂	Propyl isovalerate	155.7	154.5	57	573
10471	C₈H₁₈O	Butyl ether	142.4	Nonazeotrope		559
10472	C₈H₁₈O	sec-Octyl alcohol	178.7	Nonazeotrope		563
10473	C₈H₂₀SiO₄	Ethyl silicate	168.8	Nonazeotrope		575
10474	C₉H₁₂	Cumene	152.8	Nonazeotrope		575
10475	C₉H₁₂	Mesitylene	164.0	Nonazeotrope		563
10476	C₉H₁₂	Propylbenzene	159.3	Nonazeotrope		575
10477	C₉H₁₈O₂	Isobutyl isovalerate	171.35	Nonazeotrope		541
10478	C₁₀H₁₆	Camphene	159.5	155.0	~ 56	528
10479	C₁₀H₁₆	Nopinene	163.8	< 155.9	>72	575
10480	C₁₀H₁₆	α-Pinene	155.8	153.4	50	563
10481	C₁₀H₂₂	2,7-Dimethyloctane	160.25	155.9	~ 87	563
A =	**C₆H₅BrO**	**o-Bromophenol**	**195.0**			
10482	C₆H₅NO₃	o-Nitrophenol	217.2	Nonazeotrope		575
10483	C₇H₇Br	p-Bromotoluene	185.0	183.8	20	575
10484	C₇H₇ClO	p-Chloroanisole	197.8	Nonazeotrope		575
10485	**C₇H₈O**	**o-Cresol**	**191.1**	**189.8**	**25**	**575**
10486	C₇H₈O	p-Cresol	201.7	194.0	20	575
10487	C₈H₈O	Acetophenone	202.0	212.5	52	575
10488	C₈H₈O₂	Methyl benzoate	199.4	206.2	42	562
10489	C₈H₈O₂	Phenyl acetate	195.7	205.0	50	562
10490	C₈H₁₆O	2-Octanone	172.85	198.5		575
10491	C₈H₁₆O₂	Butyl butyrate	166.4	Nonazeotrope		575
10492	C₈H₁₈O	Octyl alcohol	195.2	204.0	50	575
10493	C₈H₁₈O	sec-Octyl alcohol	180.8	Nonazeotrope		575
10494	C₈H₁₈S	Butyl sulfide	185.0	Nonazeotrope		575
10495	C₉H₈	Indene	182.6	Nonazeotrope		575
10496	C₉H₁₀O₂	Ethyl benzoate	212.5	214.2	15?	575
10497	C₉H₁₂O	Benzyl ethyl ether	185.0	Nonazeotrope		575
10498	C₉H₁₈O₂	Isoamyl butyrate	181.05	197.5	72	575

	B-Component			Azeotropic Data		
No.	Formula	Name	B.P., °C	B.P., °C	Wt.%A	Ref.
A =	C_6H_5BrO	o-Bromophenol (continued)	195.0			
10499	$C_{10}H_{12}O_2$	Ethyl α-toluate	228.75	Nonazeotrope		575
10500	$C_{10}H_{12}O_2$	Propyl benzoate	230.85	Nonazeotrope		575
10501	$C_{10}H_{14}$	Butylbenzene	183.1	Nonazeotrope		575
10502	$C_{10}H_{16}O$	Camphor	209.1	216.5	40	575
10503	$C_{10}H_{20}O_2$	Isoamyl isovalerate	192.7	203.0	54	83
10504	$C_{10}H_{22}O$	Isoamyl ether	173.2	Nonazeotrope		575
10505	$C_{10}H_{22}S$	Isoamyl sulfide	214.8	Nonazeotrope		575
10506	$C_{11}H_{20}O$	Isobornyl methyl ether	192.4	< 192.2	<25	575
10507	$C_{12}H_{20}O_2$	Bornyl acetate	227.6	Nonazeotrope		575
A =	C_6H_5Cl	Chlorobenzene	131.75			
10508	$C_6H_5NO_2$	Nitrobenzene	210.75	Nonazeotrope		554
10509	C_6H_6	Benzene	80.2	Nonazeotrope		563
10510	C_6H_6O	Phenol	181.5	Nonazeotrope		563
10511	C_6H_7N	Aniline, 95-380 mm.		Nonazeotrope	v-l	176
	"		184.35	Nonazeotrope		551
10512	$C_6H_{10}O$	Cyclohexanone	155.7	Nonazeotrope		552
10513	$C_6H_{10}O$	Mesityl oxide	129.45	Nonazeotrope		527
10514	$C_6H_{10}S$	Allyl sulfide	139.35	Nonazeotrope		566
10515	C_6H_{12}	Cyclohexane	80.75	Nonazeotrope		575
10516	$C_6H_{12}O_2$	Butyl acetate	124.8	Nonazeotrope		527
10517	$C_6H_{12}O_2$	Ethyl butyrate	121.5	Nonazeotrope		575
10518	$C_6H_{12}O_3$	Paraldehyde	124	Nonazeotrope		563
10519	C_6H_{14}	n-Hexane	68.95	Nonazeotrope		163
10520	$C_6H_{14}O$	Hexyl alcohol	157.85	Nonazeotrope		575
10521	$C_6H_{14}O_2$	2-Butoxyethanol	171.25	Nonazeotrope		526
10522	$C_6H_{14}O_2$	Pinacol	174.35	Nonazeotrope		575
10523	$C_6H_{14}S$	Propyl sulfide	141.5	Nonazeotrope		575
10524	C_7H_8	Toluene	110.7	Nonazeotrope		563
10525	C_7H_{14}	Methylcyclohexane	101.15	Nonazeotrope		575
10526	$C_7H_{14}O$	4-Heptanone	143.55	Nonazeotrope		552
10527	$C_7H_{14}O_2$	Ethyl isovalerate	134.7	Nonazeotrope		547
10528	$C_7H_{14}O_2$	Isoamyl acetate	~ 138.8	Nonazeotrope		563
10529	$C_7H_{14}O_2$	Isobutyl propionate	136.9	Nonazeotrope		547
10530	$C_7H_{14}O_2$	Propyl butyrate	143	Nonazeotrope		563
10531	C_7H_{16}	Heptane	98.4	Nonazeotrope		527
10532	C_8H_{10}	Ethylbenzene	136.15	Nonazeotrope		563
10533	C_8H_{10}	m-Xylene	139.0	Nonazeotrope		527
10534	C_8H_{10}	p-Xylene	138.2	Nonazeotrope		563
10535	C_8H_{16}	1,3-Dimethylcyclohexane	120.7	Nonazeotrope		575
10536	C_8H_{18}	Octane	125.8	Nonazeotrope		563
10537	$C_8H_{18}O$	Butyl ether	142.2	Nonazeotrope		548
10538	$C_8H_{18}O$	Isobutyl ether	122.3	Nonazeotrope		559
10539	$C_9H_6N_2O_2$	2,4-Tolylene diiso-cyanate, 40 mm.		Nonazeotrope	v-l	327

TABLE I.   *Binary Systems*   309

No.	B-Component				Azeotropic Data		
	Formula	Name	B.P., °C	B.P., °C	Wt.%A	Ref.
A =	**C$_6$H$_5$ClO**	**o-Chlorophenol**	**176.8**			
10540	C$_6$H$_5$I	Iodobenzene	188.45	< 176.0	<78	575
10541	C$_6$H$_6$O	Phenol	182.2	174.5	75	562
	"	"	181.9		94.5	215
10542	C$_6$H$_7$N	Aniline	184.35	Nonazeotrope		563
10543	C$_6$H$_7$N	3-Picoline	144	178-184		810
10544	C$_6$H$_7$N	4-Picoline	145	178-184		810
10545	C$_6$H$_{12}$O	Cyclohexanol	160.8	Nonazeotrope		575
10546	C$_6$H$_{13}$ClO$_2$	Chloroacetal	157.4	Nonazeotrope		575
10547	C$_7$H$_7$Br	α-Bromotoluene	~ 198.5	Reacts		563
10548	C$_7$H$_7$Br	o-Bromotoluene	181.75	171.5	~ 68	563
10549	C$_7$H$_7$Br	p-Bromotoluene	185.0	< 175.5	>64	562
10550	**C$_7$H$_7$Cl**	α-Chlorotoluene	179.35	Reacts		563
10551	C$_7$H$_7$ClO	o-Chloroanisole	195.7	Nonazeotrope		575
10552	C$_7$H$_8$O	o-Cresol	191.1	Nonazeotrope		575
10553	C$_7$H$_9$N	2,6-Lutidine	144	178-184		810
10554	C$_7$H$_{14}$O$_2$	Isoamyl acetate	142.1	Nonazeotrope		575
10555	C$_8$H$_8$O	Acetophenone	202.0	> 204.5		575
10556	C$_8$H$_8$O$_2$	Benzyl formate	203.0	Nonazeotrope		575
10557	C$_8$H$_8$O$_2$	Phenyl acetate	195.7	197.0	12	575
10558	**C$_8$H$_{10}$O**	Phenetole	170.45	Nonazeotrope		575
10559	C$_8$H$_{16}$O	2-Octanone	173	177	~ 75	563
10560	C$_8$H$_{18}$O	Octyl alcohol	195.2	Nonazeotrope		575
10561	C$_8$H$_{18}$O	sec-Octyl alcohol	180.4	183.5	25	575
10562	C$_8$H$_{18}$S	Butyl sulfide	185.0	175.0	82	566
10563	C$_8$H$_{18}$S	Isobutyl sulfide	172.0	169.5		566
10564	C$_9$H$_8$	Indene	182.4	Min. b.p.		276
10565	C$_9$H$_{10}$O$_2$	Ethyl benzoate	212.5	Nonazeotrope		575
10566	C$_9$H$_{12}$	Mesitylene	164.6	Nonazeotrope		575
10567	C$_9$H$_{12}$	Propylbenzene	159.3	Nonazeotrope		575
10568	C$_9$H$_{18}$O$_2$	Isoamyl butyrate	181.05	188.0	38	562
10569	C$_9$H$_{18}$O$_2$	Isobutyl isovalerate	171.2	182.8	57	562
10570	C$_{10}$H$_{14}$	Cymene	175.3	173.5	~ 50	563
10571	C$_{10}$H$_{16}$	d-Limonene	177.8	< 175		563
10572	C$_{10}$H$_{16}$	α-Pinene	155.8	< 155.2	> 5	575
10573	C$_{10}$H$_{16}$	α-Terpinene	173.4	< 169.5	>28	575
10574	C$_{10}$H$_{22}$O	Isoamyl ether	173.2	171.0	30	575
A =	**C$_6$H$_5$ClO**	**p-Chlorophenol**	**219.75**			
10575	C$_6$H$_5$NO$_2$	Nitrobenzene	210.75	219.9	8	554
10576	C$_6$H$_5$NO$_3$	o-Nitrophenol	217.2	< 217.05	> 7	575
10577	C$_6$H$_8$O$_4$	Methyl fumarate	193.25	> 221.0	<92	575
10578	C$_6$H$_8$O$_4$	Methyl maleate	204.05	223.0	68	562
10579	C$_6$H$_{10}$O$_4$	Ethyl oxalate	185.65	> 221.5	>88	575
10580	C$_6$H$_{10}$O$_4$	Methyl succinate	195.5	222.5	<90	548
10581	C$_7$H$_5$Cl$_3$	α,α,α-Trichlorotoluene	220.9	Reacts		535
10582	C$_7$H$_6$Cl$_2$	α,α-Dichlorotoluene	205.1	Reacts		563
10583	C$_7$H$_7$BrO	o-Bromoanisole	217.7	Nonazeotrope		575
10584	C$_7$H$_7$I	p-Iodotoluene	214.5	212.0	22	562

No.	Formula	B-Component Name	B.P., °C	B.P., °C	Wt.%A	Ref.
A =	C$_6$H$_5$ClO	p-Chlorophenol (continued)	219.75			
10585	C$_7$H$_7$NO$_2$	m-Nitrotoluene	230.8	Nonazeotrope		554
10586	C$_7$H$_7$NO$_2$	o-Nitrotoluene	221.75	223.2	43	554
10587	C$_7$H$_8$O	Benzyl alcohol	205.2	Nonazeotrope		575
10588	C$_7$H$_8$O	p-Cresol	201.7	Nonazeotrope		575
10589	C$_7$H$_8$O$_2$	m-Methoxyphenol	243.8	Nonazeotrope		575
10590	C$_7$H$_8$O$_2$	Guaiacol	205.05	Nonazeotrope		535
10591	C$_8$H$_8$O	Acetophenone	202.0	224.5	85	575
10592	C$_8$H$_8$O$_2$	Benzyl formate	202.3	221.4	75	548
10593	C$_8$H$_8$O$_2$	Methyl benzoate	199.45	220.75	79	536
10594	C$_8$H$_8$O$_2$	Phenyl acetate	195.7	220.2	~ 90	548
10595	C$_8$H$_9$BrO	p-Bromophenetole	234.2	Nonazeotrope		575
10596	C$_8$H$_{10}$O	Phenethyl alcohol	219.4	227.7	52.5	574
10597	C$_8$H$_{10}$O	2,4-Xylenol	210.5	< 210.0		575
10598	C$_8$H$_{10}$O	3,4-Xylenol	226.8	219.0	89	575
10599	C$_8$H$_{10}$O$_2$	Veratrol	206.8	Nonazeotrope		575
10600	C$_8$H$_{10}$O$_2$	m-Dimethoxybenzene	214.7	Nonazeotrope		575
10601	C$_8$H$_{10}$O$_2$	o-Ethoxyphenol	216.5	222.0	70	575
10602	C$_8$H$_{12}$O$_4$	Ethyl fumarate	217.85	> 230.5	<54	575
10603	C$_8$H$_{12}$O$_4$	Ethyl maleate	223.3	232.5	53	562
10604	C$_8$H$_{14}$O$_4$	Ethyl succinate	217.25	~ 231.8		529
10605	C$_8$H$_{18}$O	Octyl alcohol	195.15	Nonazeotrope		535
10606	C$_9$H$_{10}$O	p-Methylacetophenone	226.35	235.4	52	552
10607	C$_9$H$_{10}$O	Propiophenone	217.7	230.2		552
10608	C$_9$H$_{10}$O$_2$	Benzyl acetate	214.9	226.5	~ 55	529
10609	C$_9$H$_{10}$O$_2$	Ethyl benzoate	212.6	224.9	60	574
10610	C$_9$H$_{12}$O	Mesitol	220.5	217.2	58	562
10611	C$_9$H$_{12}$O	3-Phenylpropanol	235.6	Nonazeotrope		538
10612	C$_9$H$_{18}$O$_2$	Ethyl enanthate	188.7	Nonazeotrope		575
10613	C$_9$H$_{18}$O$_3$	Isobutyl carbonate	190.3	> 220.5		575
10614	C$_{10}$H$_8$	Naphthalene	218.05	216.3	36.5	574
10615	C$_{10}$H$_{10}$O$_2$	Methyl cinnamate	261.9	Nonazeotrope		575
10616	C$_{10}$H$_{10}$O$_2$	Safrole	235.9	Nonazeotrope		575
10617	C$_{10}$H$_{12}$O	Anethole	235.7	Nonazeotrope		575
10618	C$_{10}$H$_{12}$O$_2$	Ethyl α-toluate	228.75	233.0	27	535
10619	C$_{10}$H$_{12}$O$_2$	Propyl benzoate	230.85	234.5	25	548
10620	C$_{10}$H$_{14}$O	Carvone	231.0	238.2	<45	552
10621	C$_{10}$H$_{14}$O	Thymol	232.9	Nonazeotrope		575
10622	C$_{10}$H$_{14}$O$_2$	m-Diethoxybenzene	235.4	Nonazeotrope		575
10623	C$_{10}$H$_{16}$O	Camphor	209.1	227.5	>75	552
10624	C$_{10}$H$_{17}$Cl	Bornyl chloride	~ 210	~ 206.5		563
10625	C$_{10}$H$_{18}$O	Borneol	213.2	222.5	52.5	529
10626	C$_{10}$H$_{18}$O	Geraniol	229.7	~ 230.7	~ 10	538
10627	C$_{10}$H$_{18}$O	Linalool	198.6	Nonazeotrope		535
10628	C$_{10}$H$_{18}$O	α-Terpineol	217.4	225.7	49.8	529
10629	C$_{10}$H$_{18}$O	β-Terpineol	210.5	Nonazeotrope		575
10630	C$_{10}$H$_{20}$O	Citronellol	224	~ 227.5	~ 30	535
10631	C$_{10}$H$_{20}$O	Menthol	216.4	223.5	57.5	529

TABLE I.   *Binary Systems*   311

No.	Formula	Name	B.P., °C	B.P., °C	Wt.%A	Ref.
		B-Component		**Azeotropic Data**		

No.	Formula	Name	B.P., °C	B.P., °C	Wt.%A		Ref.
A =	**C₆H₅ClO**	**p-Chlorophenol** (continued)	**219.75**				
10632	$C_{10}H_{20}O_2$	Ethyl caprylate	208.35	223.2	65		562
10633	$C_{10}H_{20}O_2$	Isoamyl isovalerate	192.7	Nonazeotrope?			548
10634	$C_{10}H_{22}S$	Isoamyl sulfide	214.8	212.5	28		566
10635	$C_{11}H_{10}$	2-Methylnaphthalene	241.15	Nonazeotrope			575
10636	$C_{11}H_{14}O_2$	Butyl benzoate	249.5	Nonazeotrope			548
10637	$C_{11}H_{14}O_2$	Isobutyl benzoate	241.9	242.7	7		548
10638	$C_{11}H_{20}O$	Methyl α-terpineol ether	216.2	< 215.9	<15		575
10639	$C_{11}H_{22}O_3$	Isoamyl carbonate	232.2	235.3	22		548
10640	$C_{12}H_{10}$	Biphenyl	256.1	Nonazeotrope			575
10641	$C_{12}H_{18}$	1,3,5-Triethylbenzene	215.4	214.7	18		548
10642	$C_{12}H_{20}O_2$	Bornyl acetate	227.7	232.7	28		529
10643	$C_{13}H_{28}$	Tridecane	234.0	Nonazeotrope			575
A =	**C₆H₅Cl₃Si**	**Phenyltrichlorosilane**	**201.0**				
10644	$C_6H_6Cl_2Si$	Phenyl dichlorosilane					
		" 40 mm.		Nonazeotrope		v-l	229
		" 100 mm.		Nonazeotrope		v-l	229
			182.9	Nonazeotrope		v-l	229
10645	$C_7H_8Cl_2Si$	Methylphenyldichloro-silane	203.6	Nonazeotrope		v-l	228
A =	**C₆H₅F**	**Fluorobenzene**	**85.2**				
10646	C_6H_5I	Iodobenzene	188.55	Vapor pressure data			563
10647	C_6H_6	Benzene	80.15	Nonazeotrope			575
		"	80.1	Ideal system		v-l	40
10648	C_6H_{12}	Cyclohexane	80.75	Nonazeotrope			556
10649	C_6H_{12}	Methylcyclopentane	72.0	Nonazeotrope			575
10650	C_6H_{14}	Hexane	68.8	Nonazeotrope			575
A =	**C₆H₅I**	**Iodobenzene**	**188.55**				
10651	$C_6H_5NO_2$	Nitrobenzene	210.75	Nonazeotrope			554
10652	C_6H_6O	Phenol	181.5	177.7	53		563
10653	C_6H_7N	Aniline	184.35	181.6	>40		551
10654	$C_6H_8O_4$	Methyl fumarate	193.25	186.2	70		527
10655	$C_6H_{10}O_3$	Ethyl acetoacetate	180.4	178.0	52		552
10656	$C_6H_{10}O_4$	Ethyl oxalate	185.65	181.0	48		538
10657	$C_6H_{10}O_4$	Glycol diacetate	186.3	< 183.5	>42		562
10658	$C_6H_{10}O_4$	Methyl succinate	195	~ 186.5			563
10659	$C_6H_{11}ClO_2$	Butyl chloroacetate	181.8	< 181.2	>82		575
10660	$C_6H_{11}ClO_2$	Isobutyl chloroacetate	174.5	Nonazeotrope			575
10661	$C_6H_{12}O$	Cyclohexanol	160.65	Nonazeotrope			573
10662	$C_6H_{12}O_2$	Caproic acid	205.15	186.8	88		564
10663	$C_6H_{12}O_2$	Isocaproic acid	199.5	185.5	85		562
10664	$C_6H_{14}O$	Hexyl alcohol	157.85	Nonazeotrope			575
10665	$C_6H_{14}O_2$	2-Butoxyethanol	171.17	< 170.8			575
10666	C_7H_5N	Benzonitrile	191.1	< 187.0			565
10667	C_7H_6O	Benzaldehyde	179.2	Nonazeotrope			575
10668	C_7H_7Br	m-Bromotoluene	184.3	Nonazeotrope			549

No.	Formula	B-Component Name	B.P., °C	Azeotropic Data B.P., °C	Wt.%A	Ref.
A =	C₆H₅I	**Iodobenzene** *(continued)*	**188.55**			
10669	C₇H₈O	Benzyl alcohol	205.2	187.75	88	535
10670	C₇H₈O	*o*-Cresol	190.8	185	~ 32 ~ 53	563
10671	C₇H₈O	*p*-Cresol	201.7	188.1	90	542
10672	C₇H₉N	Methylaniline	196.25	Nonazeotrope		551
10673	C₇H₉N	*m*-Toluidine	203.1	Nonazeotrope		551
10674	C₇H₉N	*o*-Toluidine	200.35	Nonazeotrope		551
10675	C₇H₉N	*p*-Toluidine	200.55	Nonazeotrope		551
10676	C₇H₁₂O₄	Ethyl malonate	199.2	< 188	>80	547
10677	C₇H₁₄O₃	Isobutyl lactate	182.15	‾180.5	30	567
10678	C₈H₈O₂	Methyl benzoate	199.45	Nonazeotrope		547
10679	C₈H₈O₂	Phenyl acetate	195.7	< 188.3		575
10680	C₈H₁₀O	*p*-Methylanisole	177.05	Nonazeotrope		559
10681	C₈H₁₀O	Phenetole	170.45	Nonazeotrope		559
10682	C₈H₁₁N	*N,N*-Dimethylaniline	194.05	186.7	75	535
10683	C₈H₁₁N	Ethylaniline	205.5	Nonazeotrope		551
10684	C₈H₁₆O₃	Isoamyl lactate	202.4	Nonazeotrope		575
10685	C₈H₁₈O	*n*-Octyl alcohol	195.15	187.5		531
10686	C₈H₁₈O	*sec*-Octyl alcohol	179.0	178.4		531
10687	C₉H₈	Indene	182.6	Nonazeotrope		575
10688	C₉H₁₂O	Benzyl ethyl ether	185.0	Nonazeotrope		559
10689	C₉H₁₃N	*N,N*-Dimethyl-*o*-toluidine	185.3	Nonazeotrope		551
10690	C₉H₁₄O	Phorone	197.8	Nonazeotrope		552
10691	C₉H₁₈O₂	Butyl isovalerate	177.6	Nonazeotrope		547
10692	C₉H₁₈O₂	Isoamyl butyrate	178.5	Nonazeotrope		538
10693	C₉H₁₈O₃	Isobutyl carbonate	190.3	185.5	~ 65	563
10694	C₁₀H₁₄	Butylbenzene	183.1	Nonazeotrope		575
10695	C₁₀H₁₄	Cymene	176.7	Nonazeotrope		575
10696	C₁₀H₁₆	Dipentene	177.7	Nonazeotrope		575
10697	C₁₀H₁₆	α-Terpinene	173.4	Nonazeotrope		575
10698	C₁₀H₁₆	Terpinene	181.5	Nonazeotrope		538
10699	C₁₀H₁₆O	Fenchone	193	Nonazeotrope		563
10700	C₁₀H₁₈O	Linalool	198.6	Nonazeotrope		532
10701	C₁₀H₂₀O₂	Isoamyl isovalerate	192.7	< 188.3	>87	575
A =	C₆H₅NO₂	**Nitrobenzene**	**210.75**			
10702	C₆H₅NO₃	*o*-Nitrophenol	217.2	Nonazeotrope		554
10703	C₆H₆	Benzene	80.15	Nonazeotrope		554
		" 25°C.		Nonazeotrope	v.p.	845
10704	C₆H₇N	Aniline	184.35	Nonazeotrope	v-l	551,835
10705	C₆H₈O₄	Methyl maleate	204.05	203.9	7	527
10706	C₆H₁₂	Cyclohexane	80.75	Nonazeotrope		163
10707	C₆H₁₂O₂	Caproic acid	205.15	<202.5	<35	554
10708	C₆H₁₄	*n*-Hexane	68.8	Nonazeotrope		554
10709	C₆H₁₄O	*n*-Hexanol	157.85	Nonazeotrope		554
10710	C₆H₁₄O₂	2-Butoxyethanol	171.15	Nonazeotrope		554
10711	C₆H₁₄O₂	Pinacol	174.35	Nonazeotrope		576
10712	C₇H₅Cl₃	α,α,α-Trichlorotoluene	220.8	Nonazeotrope		554

TABLE I. *Binary Systems*   313

No.	Formula	Name	B.P., °C	B.P., °C	Wt.%A	Ref.
		B-Component		**Azeotropic Data**		
$A =$	$C_6H_5NO_2$	**Nitrobenzene** *(continued)*	**210.75**			
10713	$C_7H_6Cl_2$	α,α-Dichlorotoluene	205.2	Nonazeotrope		554
10714	C_7H_6O	Benzaldehyde	179.2	Nonazeotrope		554
10715	C_7H_7Br	o-Bromotoluene	181.75	Nonazeotrope		563
10716	C_7H_7Cl	α-Chlorotoluene	179.35	Nonazeotrope		563
10717	C_7H_7I	p-Iodotoluene	214.5	< 208.8		554
10718	C_7H_8	Toluene	110.7	Nonazeotrope		554
10719	C_7H_8O	Benzyl alcohol	205.25	204.2	38	554
10720	C_7H_8O	m-Cresol	202.2	Nonazeotrope		554
10721	C_7H_8O	o-Cresol	191.1	Nonazeotrope		554
10722	C_7H_8O	p-Cresol	201.7	Nonazeotrope		554
10723	$C_7H_8O_2$	Guaiacol	205.05	Nonazeotrope		554
10724	C_7H_9N	Benzylamine	185.0	Nonazeotrope		551
10725	C_7H_9N	Methylaniline	196.25	Nonazeotrope		551
10726	C_7H_9N	m-Toluidine	203.1	Nonazeotrope		551
10727	C_7H_9N	o-Toluidine	200.35	Nonazeotrope		551
10728	C_7H_9N	p-Toluidine	200.55	Nonazeotrope		551
10729	C_7H_9NO	o-Anisidine	219.0	Nonazeotrope		575
10730	$C_7H_{12}O_4$	Ethyl malonate	199.35	Nonazeotrope		554
10731	$C_7H_{14}O_2$	**Enanthic acid**	**222.0**	< 209.5	<88	554
10732	$C_7H_{16}O_4$	2-[2-(2-Methoxyethoxy) ethoxy]ethanol	245.25	Nonazeotrope		554
10733	C_8H_8O	Acetophenone	202.0	Nonazeotrope		552
10734	$C_8H_8O_2$	**Benzyl formate**	**203.0**	Nonazeotrope		554
10735	$C_8H_8O_2$	Methyl benzoate	199.4	Nonazeotrope		575
10736	$C_8H_8O_2$	Phenyl acetate	215.3	Nonazeotrope		554
10737	$C_8H_8O_3$	Methyl salicylate	222.95	Nonazeotrope		554
10738	$C_8H_{10}O$	p-Ethylphenol	220.0	Nonazeotrope		554
10739	$C_8H_{10}O$	Phenethyl alcohol	219.4	210.6	92	554
10740	$C_8H_{10}O$	3,4-Xylenol	226.8	Nonazeotrope		554
10741	$C_8H_{10}O_2$	m-Dimethoxybenzene	214.7	207.5	>62	554
10742	$C_8H_{10}O_2$	o-Ethoxyphenol	216.5	Nonazeotrope		554
10743	$C_8H_{10}O_2$	Veratrol	206.8	< 203.8		554
10744	$C_8H_{11}N$	Dimethylaniline	194.15	Nonazeotrope		551
10745	$C_8H_{11}N$	Ethylaniline	205.5	Nonazeotrope		551
10746	$C_8H_{11}N$	2,4-Xylidine	214.0	Nonazeotrope		551
10747	$C_8H_{11}N$	3,4-Xylidine	225.5	Nonazeotrope		551
10748	$C_8H_{12}O_4$	Ethyl fumarate	217.85	Nonazeotrope		554
10749	$C_8H_{12}O_4$	Ethyl maleate	223.3	Nonazeotrope		554
10750	$C_8H_{14}O_4$	Ethyl succinate	217.25	< 210.6		554
10751	$C_8H_{14}O_4$	Propyl oxalate	214.2	210.0		554
10752	$C_8H_{16}O_2$	Caprylic acid	238.5	Nonazeotrope		554
10753	$C_8H_{16}O_3$	Isoamyl lactate	202.4	Nonazeotrope		554
10754	$C_8H_{18}O$	n-Octyl alcohol	195.2	Nonazeotrope		554
10755	$C_9H_{10}O$	Cinnamyl alcohol	257.0	Nonazeotrope		554
10756	$C_9H_{10}O$	Propiophenone	217.7	Nonazeotrope		552
10757	$C_9H_{10}O_2$	Benzyl acetate	215.0	Nonazeotrope		554
10758	$C_9H_{10}O_2$	Ethyl benzoate	212.5	210.6	81	554

No.	B-Component		B.P., °C	Azeotropic Data		
	Formula	Name		B.P., °C	Wt.%A	Ref.
A =	**C₆H₅NO₂**	**Nitrobenzene** *(continued)*	**210.75**			
10759	C₉H₁₂O	3-Phenylpropanol	235.6	Nonazeotrope		554
10760	C₉H₁₃N	N,N-Dimethyl-o-toluidine	185.3	Nonazeotrope		575
10761	C₉H₁₃N	N,N-Dimethyl-p-toluidine	210.2	< 210		551
10762	C₉H₁₄O	Phorone	197.8	Nonazeotrope		552
10763	C₁₀H₈	Naphthalene	218.0	Nonazeotrope		554
10764	C₁₀H₁₄O	Thymol	232.9	Nonazeotrope		554
10765	C₁₀H₁₅N	Diethylaniline	217.05	210.72	97	551
10766	C₁₀H₁₆O	Camphor	208.9	208.4	35	563
10767	C₁₀H₁₆O	Fenchone	193.6	Nonazeotrope		575
10768	C₁₀H₁₆O	Pulegone	223.8	Nonazeotrope		552
10769	C₁₀H₁₇Cl	Bornyl chloride	207.5	205.0		554
10770	C₁₀H₁₈O	Borneol	215.0	207.8	58	554
10771	C₁₀H₁₈O	Citronellal	208.0	207.0	22	554
10772	C₁₀H₁₈O	Geraniol	229.6	Nonazeotrope		554
10773	C₁₀H₁₈O	Linalool	198.6	Nonazeotrope		554
10774	C₁₀H₁₈O	Menthone	206.5	Nonazeotrope		563
10775	C₁₀H₁₈O	α-Terpineol	218.85	209.7	78	554
10776	C₁₀H₁₈O	β-Terpineol	210.5	204.8	50	554
10777	C₁₀H₂₀O	Citronellol	224.5	Min. b.p.		574
10778	C₁₀H₂₀O	Menthol	216.3	208.35	67.3	554
10779	C₁₀H₂₂O	n-Decyl alcohol	232.8	Nonazeotrope		551
10780	C₁₀H₂₂S	Isoamyl sulfide	214.8	209.5	<93	554
10781	C₁₁H₁₆O	Methyl thymol ether	216.5	< 209.2	<82	554
10782	C₁₁H₂₀O	Methyl α-terpineol ether	216.2	208.6	75?	554
10783	C₁₁H₂₄O₂	Diisoamyloxymethane	210.8	206.5	>42	554
10784	C₁₂H₁₀O	1-and 2-Acetyl-naphthalene,				
		100 mm.	228.3	Nonazeotrope		413
10785	C₁₂H₁₈	1,3,5-Triethylbenzene	215.5	Nonazeotrope		554
10786	C₁₂H₂₂O	Ethyl bornyl ether	204.9	203.0	30	554
10787	C₁₂H₂₂O	Ethyl isobornyl ether	203.8	202.5?	25?	554
A =	**C₆H₅NO₃**	**o-Nitrophenol**	**217.25**			
10788	C₆H₆O₂	Pyrocatechol	245.9	Nonazeotrope		542
10789	C₆H₈O₄	Methyl maleate	204.05	Nonazeotrope		527
10790	C₆H₁₄O₂	Pinacol	174.35	Nonazeotrope		575
10791	C₆H₁₄O₃	Dipropylene glycol	229.2	215.0?		575
10792	C₇H₇BrO	o-Bromoanisole	217.7	Nonazeotrope		575
10793	C₇H₇ClO	p-Chloroanisole	197.8	Nonazeotrope		575
10794	C₇H₇I	p-Iodotoluene	214.5	212.0	18	575
10795	C₇H₇NO₂	o-Nitrotoluene	221.75	Nonazeotrope		554
10796	C₇H₈O	Benzyl alcohol	205.25	Nonazeotrope		575
10797	C₇H₈O	m-Cresol	202.2	Nonazeotrope		542
10798	C₇H₈O	o-Cresol	191.1	Nonazeotrope		575
10799	C₇H₈O	p-Cresol	201.7	Nonazeotrope		544
10800	C₇H₉NO	o-Anisidine	219.0	Nonazeotrope		575
10801	C₇H₁₄O	2-Methylcyclohexanol	168.5	Nonazeotrope		575

TABLE I.   *Binary Systems*   315

		B-Component			Azeotropic Data		
No.	Formula	Name	B.P., °C	B.P., °C	Wt.%A		Ref.
A =	**C₆H₅NO₃**	*o*-**Nitrophenol** *(continued)*	**217.25**				
10802	C₇H₁₆O	Heptyl alcohol	176.16	Nonazeotrope			575
10803	C₈H₈O	Acetophenone	202.0	Nonazeotrope			552
10804	C₈H₈O₂	Benzyl formate	202.3	Nonazeotrope			548
10805	C₈H₈O₂	Methyl benzoate	199.4	Nonazeotrope			575
10806	C₈H₈O₃	Methyl salicylate	222.95	Nonazeotrope			575
10807	C₈H₉BrO	*p*-Bromophenetole	234.2	Nonazeotrope			575
10808	C₈H₁₀O	Phenethyl alcohol	219.4	214.0	59		567
10809	C₈H₁₀O	3,4-Xylenol	226.8	Nonazeotrope			575
10810	C₈H₁₀O₂	*m*-Dimethoxybenzene	214.7	Nonazeotrope			575
10811	C₈H₁₀O₂	Veratrole	206.8	Nonazeotrope			575
10812	C₈H₁₁NO	*o*-Phenetidine	232.5	Nonazeotrope			575
10813	C₈H₁₂O₄	Ethyl fumarate	217.85	Nonazeotrope			526
10814	C₈H₁₂O₄	Ethyl maleate	223.3	Nonazeotrope			575
10815	C₈H₁₄O₄	Ethyl succinate	217.25	<216.9	<54		575
10816	C₈H₁₈O	Octyl alcohol	195.2	Nonazeotrope			575
10817	C₈H₁₈O	*sec*-Octyl alcohol	180.4	Nonazeotrope			575
10818	C₈H₁₈S	Butyl sulfide	185.0	Nonazeotrope			575
10819	C₉H₇N	Quinoline	237.3	Nonazeotrope			575
10820	C₉H₁₀O	Cinnamyl alcohol	257.0	Nonazeotrope			575
10821	C₉H₁₀O	*p*-Methylacetophenone	226.35	Nonazeotrope			552
10822	C₉H₁₀O₂	Benzyl acetate	215.0	Nonazeotrope			575
10823	C₉H₁₀O₂	Ethyl benzoate	212.6	Nonazeotrope			542
10824	C₉H₁₂O	3-Phenylpropanol	235.6	Nonazeotrope			575
10825	C₉H₁₄O	Phorone	197.8	Nonazeotrope			552
10826	C₁₀H₈	Naphthalene	218.05	215.75	60		542
10827	C₁₀H₁₂O₂	Ethyl α-toluate	228.75	Nonazeotrope			548
10828	C₁₀H₁₂O₂	Propyl benzoate	230.85	Nonazeotrope			548
10829	C₁₀H₁₄O	Carvacrol	237.85	Nonazeotrope			575
10830	C₁₀H₁₄O	Thymol	232.9	Nonazeotrope			542
10831	C₁₀H₁₄O₂	*m*-Diethoxybenzene	235.4	Nonazeotrope			575
10832	C₁₀H₁₆O	Camphor	209.1	Nonazeotrope			552
10833	C₁₀H₁₆O	Fenchone	193.6	Nonazeotrope			575
10834	C₁₀H₁₈O	Borneol	213.4	211.9	~ 40		542
10835	C₁₀H₁₈O	Menthone	209.5	Nonazeotrope			575
10836	C₁₀H₁₈O	α-Terpineol	218.85	213.9	58		567
10837	C₁₀H₁₈O	β-Terpineol	210.5	209.0	22		567
10838	C₁₀H₂₀O	Citronellol	224.4	214.5	78		575
10839	C₁₀H₂₀O	Menthol	216.4	212.2	46		564
10840	C₁₀H₂₀O₂	Ethyl caprylate	208.35	Nonazeotrope			575
10841	C₁₀H₂₀O₂	Methyl pelargenate	213.8	Nonazeotrope			575
10842	C₁₀H₂₂O	Decyl alcohol	232.8	216.5	90		575
10843	C₁₀H₂₂S	Isoamyl sulfide	214.8	212.5	30		566
10844	C₁₁H₁₀	1-Methylnaphthalene	244.6	Nonazeotrope			575
10845	C₁₁H₁₀	2-Methylnaphthalene	241.15	Nonazeotrope			575
10846	C₁₁H₂₀O	Methyl α-terpineol ether	216.2	215.9	28		575

No.	Formula	Name	B.P., °C	B.P., °C	Wt.%A		Ref.
		B-Component		**Azeotropic Data**			
A =	**C₆H₅NO₃**	**o-Nitrophenol** *(continued)*	**217.25**				
10847	C₁₂H₁₈	1,3,5-Triethylbenzene	215.4	~ 214.3	<45		548
10848	C₁₂H₂₀O₂	Bornyl acetate	227.7	Nonazeotrope			548
10849	C₁₃H₂₈	Tridecane	234.0	<215.0	<94		575
A =	**C₆H₆**	**Benzene**	**80.15**				
10850	C₆H₇N	Aniline	184.35	Nonazeotrope			551
10851	C₆H₈	1,3-Cyclohexadiene	80.4	< 79.9			561
10852	C₆H₈	1,4-Cyclohexadiene	85.6	Nonazeotrope			562
10853	C₆H₁₀	Cyclohexene	82.1	78.9	64.7	v-l	365,563
10854	C₆H₁₂	Cyclohexane, 206 mm.		40	48	v-l	848
		" 600 mm.		70	48	v-l	848
		"		80.6	77.7	51.8	631
		"		80.60	77.4	49.7 v-l	805
		" 1204 mm.			53.65		74
		" 93 mm.			46.70		74,561
		"		80.7	77.7	53.2 v-l	666
		"		80.7	77.4	52.1 v-l	888
		" 304 mm.		50	48.2	v-l	666
		" 300 mm.		49.6	48.2	v-l	666
		" 45°55°C				v-l	604
		"		80.75	77.6	51.2 v-l	147,517, 1010
		"		80.75	77.4	52.5	212,306
		" 128 mm.		28.4	47.6		924
		" 155 mm.		33.1	48.0		924
		" 287 mm.		48.3	49.3		924
		" 307 mm.		50.4	49.4		924
		" 495 mm.		63.7	50.8		924
		" 602 mm.		69.8	51.3		924
		" 760 mm.		77.56	51.9		924
		" 14.7 p.s.i.a.		77.4	50.2	v-l	940
		" 66.7 p.s.i.a.		137.1	61.5	v-l	940
		" 118.0 p.s.i.a.		165.8	67.0	v-l	940
		" 186.8 p.s.i.a.		193.0	71.5	v-l	940
		" 66.7 p.s.i.a.			59.7	v-l	784
		" 116.5 p.s.i.a.			64.9	v-l	784
		" 165.9 p.s.i.a.			67.6	v-l	784
		" 217.0 p.s.i.a.			71	v-l	784
		" 268.7 p.s.i.a.			74	v-l	784
		"		77.6	51.5	v-l	686c,807c
10854a	C₆H₁₂	1-Hexene 25°C.				v-l	209c
10855	C₆H₁₂	Methylcyclopentane	71.85	71.7	16	v-l	676
		"	71.8	min. b.p.		v-l	936
		" 524 mm.		60	13.1	v-l	214
		"	71.8	71.5	9.4	v-l	347
		5 p.s.i.g.			9		709
		Methylcyclopentane 150 p.s.i.g.			14		516,561, 709

TABLE I.   *Binary Systems*   317

No.	Formula	B-Component Name	B.P., °C	Azeotropic Data B.P., °C	Wt.%A		Ref.
A =	**C₆H₆**	**Benzene** *(continued)*	**80.15**				
10856	C₆H₁₂O	Cyclohexanol	160.65	Nonazeotrope			563
10857	C₆H₁₂O	4-Methyl-2-pentanone, 450-760 mm.		Nonazeotrope		v-l	189
10858	C₆H₁₂O	Pinacolone	106.2	Nonazeotrope			552
10859	C₆H₁₂O₂	Butyl acetate	126	Nonazeotrope		v-l	682
10860	C₆H₁₂O₂	Ethyl isobutyrate	110.1	Nonazeotrope			575
10861	C₆H₁₄	Hexane	68.7	Nonazeotrope		v-l	966
		"	69.0	68.5	4.7		418,561, 631,1045
		" 575 mm.		60	7.3	v-l	55
		"	68.76	68.74	0.2	v-l	97
		"	68.7	68.3	4.5		316,336
		" 400 mm.		49.6	6.2	v-l	336
		" 380 mm.		48	9.2		316
		" 300 mm.		41.8	7.3	v-l	336
		" 210 mm.		32	10.0		316
		" 200 mm.		31.6	7.6	v-l	336
		" 100 mm.		17	13.1		316
		" 50 mm.		7.5	13.8		316
			68.95	Nonazeotrope		v-l	675
		" 4-18 atm.		Nonazeotrope		v-l	784
		"	68.7	Nonazeotrope		v-l	807c,966
10862	C₆H₁₄O	Hexyl alcohol	155	Nonazeotrope			215
10863	C₆H₁₄O	Isopropyl ether	68.3	Nonazeotrope			558
10864	C₆H₁₄O	4-Methyl-2-pentanol	131.8	Nonazeotrope		v-l	781
10865	C₆H₁₄O	Propyl ether	90.55	Nonazeotrope			538
10866	C₆H₁₄O₂	Acetal	104.5	Nonazeotrope			563
10867	C₆H₁₅N	Triethylamine	89.35	Nonazeotrope			551
10868	C₆H₁₅NO	2-(Diethylamino)ethanol	162.2	Nonazeotrope			575
10869	C₇F₁₄	Perfluoromethylcyclo- hexane	73-78	59			160
10870	C₇F₁₆	Perfluoroheptane	83	61			160
10871	C₇H₈	Toluene	110.68	Nonazeotrope		v-l	449c,1045
10872	C₇H₁₄	Methylcyclohexane	101.05	Nonazeotrope		v-l	676
10873	C₇H₁₆	2,2-Dimethylpentane	79.1	75.85	46.3		60
10874	C₇H₁₆	2,3-Dimethylpentane	89.8	79.2	79.5		631
		"	89.79	79.4	78.8	v-l	509
10875	C₇H₁₆	2,4-Dimethylpentane	81	> 75	48.4		160
		"	80.8	75.2	48.3	v-l	60,631, 825
10876	C₇H₁₆	Heptane	98.4	80.1	99.3		631
		"	98.45	Nonazeotrope			527
		" 180-450 mm.		Nonazeotrope		v-l	698
		"	98.4	Nonazeotrope		v-l	675
			98.4	Nonazeotrope		v-l	98

| | | B-Component | | | Azeotropic Data | | |
No.	Formula	Name	B.P., °C	B.P., °C	Wt.%A		Ref.
A =	**C₆H₆**	**Benzene** *(continued)*	**80.15**				
10877	C₇H₁₆	Heptane <160 mm. 2,2,3-Trimethylbutane,		Azeotropic			98
		736 mm.	79.9	75.6	50.5	v-l	365
		"		76.6	49.7		631
10878	C₈F₁₈O	Perfluorobutyl ether	100	68			160
10878a	C₈H₁₀	Ethylbenzene	136.18	Nonazeotrope		v-l	449c
10879	C₈H₁₈	Octane	125.75	Nonazeotrope		v-l	163,239
10880	C₈H₁₈	2,2,4-Trimethylpentane,					
		" 35°-75° C.		Nonazeotrope		v-l	1012
			99.2	80.1	97.7		631
10881	C₈H₁₈O₃	2-(2-Butoxyethoxy) ethanol	230.6	Nonazeotrope			981
10882	C₉H₁₀O₂	Ethyl benzoate	213	Vapor pressure data			563
10882a	C₉H₁₂	Propylbenzene	159.24	Nonazeotrope		v-l	449c
A =	**C₆H₆O**	**Phenol**	**182.2**				
10883	C₆H₇N	Aniline	184.35	186.2	42		551
		"	183.91	185.84	41.9		911
		" 600 mm.	175.38	177.42	42.3		911
		" 500 mm.	168.84	170.98	43.0		911
		" 400 mm.	161.11	163.40	43.4		911
10884	C₆H₇N	2-Picoline	128.8	185.17	78.5		791
		"	129.20	185.5	75.4		27c
10885	C₆H₇N	3-Picoline	143.5	188.93	70.2		790
		"	143.0	185.5	76	v-l	729
		" 600 mm.	135.3	178.0	74	v-l	729
		" 400 mm.	121.0	166.3	71	v-l	729
		" 200 mm.	99.9	146.2	32	v-l	729,809
		"	143.2	187.2	71.4	v-l	27c,790
10886	C₆H₇N	4-Picoline	144.8	190	67.5	v-l	729
		" 600 mm.	136.0	181.2	66	v-l	729
		" 400 mm.	122.6	167.5	65	v-l	729
		" 200 mm.	101.5	147.0	64.5	v-l	729,809
		"	144.90	190.7	67.3	v-l	27c,729
10887	C₆H₈O₄	Methyl fumarate	193.25	194.85	23		527
10888	C₆H₈O₄	Methyl maleate	204.05	Nonazeotrope			527
10889	C₆H₁₀O	Cyclohexanone	184.5	72			274
				Composition independent of			
				pressure			274
		"	155.7	Nonazeotrope			274
		" 50 mm.	73		71.5		215
		"	155.6	Azeotropic			177
		" 50 mm.		Max. b.p.	75.8		981
		" 200 mm.		143.0	75.2	v-l	898c
		" 90 mm.		122.5	77.2	v-l	898c
10890	C₆H₁₀O₃	Ethyl acetoacetate	180.7	188?	Reacts		563
10891	C₆H₁₀O₄	Ethylidene diacetate	168.5	>182.5	<18		527
10892	C₆H₁₀O₄	Ethylene diacetate	189.86	195.53	39.2		563, 721

TABLE I. *Binary Systems* 319

No.	Formula	Name	B.P., °C	B.P., °C	Wt.%A		Ref.
		B-Component		**Azeotropic Data**			
A =	**C_6H_6O**	**Phenol** *(continued)*	**182.2**				
10893	$C_6H_{10}O_4$	Ethyl oxalate	185.65	189.5	41		572
10894	$C_6H_{10}O_4$	Methyl succinate	195	~ 197			563
10895	$C_6H_{12}O$	Cyclohexanol	160.7	183.0	87		574
		"		Nonazeotrope		v-l	5,274
		" 60 mm.		111	70		215
		" 70 mm.		111	73		177
		" 90 mm.		120	70	v-l	177
		" 200 mm.		140	71		215
		Cyclohexanol	160.65	180	87		215
10896	$C_6H_{12}O_2$	Butyl acetate					
		200-760 mm.		Nonazeotrope		v-l	472
		"		Azeotropic at low pressure			473
10897	$C_6H_{12}O_2$	Isocaproic acid	199.5	Nonazeotrope			575
10898	$C_6H_{12}O_3$	2-Ethoxyethyl acetate	156.8	184.9	72		556
10899	$C_6H_{12}O_3$	Ethyl α-hydroxy isobutyrate	150	Nonazeotrope			575
10900	$C_6H_{12}O_3$	Isopropyl lactate	167.5	184.8	73		542
10901	$C_6H_{12}O_3$	Propyl lactate	171.7	~ 185	~ 78		563
10902	$C_6H_{13}Br$	1-Bromohexane	156.5	Nonazeotrope			575
10903	$C_6H_{14}O$	*n*-Hexyl alcohol	157.8	Nonazeotrope			536
10904	$C_6H_{14}O_2$	2-Butoxyethanol	171.25	186.35	63		556
10905	$C_6H_{14}O_2$	Pinacol	174.35	185.5	71		573
10906	$C_6H_{14}O_3$	2-(2-Ethoxyethoxy)ethanol	201.9	208.0	36		567
10907	$C_6H_{15}N$	Triethylamine 15°C.		Nonazeotrope		v-l	627
10908	C_7H_5N	Benzontrile	191.1	192.0	80		565
10909	$C_7H_6Cl_2$	α,α-Dichlorotoluene	205.1	Reacts			563
10910	C_7H_6O	Benzaldehyde	179.2	185.6	51		563
10911	C_7H_7Br	α-Bromotoluene	198.5	Reacts			563
10912	C_7H_7Br	*m*-Bromotoluene	183.8	175.7	43		527
10913	C_7H_7Br	*o*-Bromotoluene	181.75	174.35	40		563
10914	C_7H_7Br	*p*-Bromotoluene	185	176.2	44		555
10915	C_7H_7Cl	α-Chlorotoluene	179.35	Reacts			563
10916	C_7H_7Cl	*o*-Chlorotoluene	159.2	159.0	3		575
10917	C_7H_7Cl	*p*-Chlorotoluene	162.4	161.5	~12		538
10918	C_7H_7ClO	*o*-Chloroanisole	195.7	Nonazeotrope			575
10919	C_7H_7I	*p*-Iodotoluene	215.0	Nonazeotrope			542
10920	C_7H_8	Toluene	110.75	Nonazeotrope			575
10921	C_7H_8O	Anisole	153.85	Nonazeotrope			544
10922	C_7H_8O	Benzyl alcohol	205.15	Nonazeotrope			573
10923	C_7H_8O	*m*-Cresol	202.2	Nonazeotrope			952
10924	C_7H_8O	*o*-Cresol	191.1	Nonazeotrope			952
10925	C_7H_8O	*p*-Cresol	201.7	Nonazeotrope			952
10926	C_7H_9N	Benzylamine	185.0	196.8	45		551
10926a	C_7H_9N	2,4-Lutidine 400 mm.		171.66	57.1		263g
		" 500 mm.		178.63	57		263g
		" 600 mm.		184.76	56.8		263g
		"	159.0	193.4	57	v-l	27c

No.	Formula	B-Component Name	B.P., °C	Azeotropic Data B.P., °C	Wt.%A		Ref.
A =	C$_6$H$_6$O	**Phenol** *(continued)*	**182.2**				
10927	C$_7$H$_9$N	2,6-Lutidine	143.3	185.5	72.5	v-l	826
		" 600 mm.	134.5	178.5	71	v-l	826
		" 400 mm.	121.0	163.5	67	v-l	826
		" 200 mm.	100.8	143.5	64.5	v-l	826,809
		"	142	185.81	72.1		791
10928	C$_7$H$_9$N	Methylaniline	196.25	Nonazeotrope			551
10929	C$_7$H$_9$N	*m*-Toluidine	203.1	Nonazeotrope			551
10930	C$_7$H$_9$N	*o*-Toluidine	200.35	Nonazeotrope			551
10931	C$_7$H$_9$N	*p*-Toluidine	200.55	Nonazeotrope			551
10932	C$_7$H$_{12}$O$_4$	Ethyl malonate	198.6	Reacts			563
10933	C$_7$H$_{14}$O	2-Methylcyclohexanol	168.5	183.1	80		575
10934	C$_7$H$_{14}$O$_3$	Isobutyl lactate	182.15	189.05	~46		563
10935	C$_7$H$_{14}$O$_3$	1,3-Butanediol methyl ether acetate	171.75	187.0	55		527
10936	C$_7$H$_{16}$	Heptane	98.4	Nonazeotrope			575
10937	C$_7$H$_{16}$O	Heptyl alcohol	176.15	185.0	72		569
10938	C$_8$H$_8$	Styrene	145.8	Nonazeotrope			575
10939	C$_8$H$_8$O	Acetophenone	202.0	202.0	7.8		552
		" 300 mm.		168.6	17.2	v-l	293
		" 100 mm.		135.7	25	v-l	293
10940	C$_8$H$_8$O$_2$	Benzyl formate	202.4	Nonazeotrope			542
10941	C$_8$H$_8$O$_2$	Methyl benzoate	199.55	Nonazeotrope			563
10942	C$_8$H$_8$O$_2$	Phenyl acetate	195.7	196.6	~12		573
		" "	195.14	195.89	8.9		721
10943	C$_8$H$_{10}$	Ethylbenzene	136.15	Nonazeotrope			575
10944	C$_8$H$_{10}$	*m*-Xylene	139.0	Nonazeotrope			527
10945	C$_8$H$_{10}$	*o*-Xylene	142.6	Nonazeotrope			563
10946	C$_8$H$_{10}$O	Benzyl methyl ether	167.8	Nonazeotrope			575
10947	C$_8$H$_{10}$O	*p*-Methylanisole	177.05	177.02	~ 3		541
10948	C$_8$H$_{10}$O	Phenetole	170.45	Nonazeotrope			542,556
10949	C$_8$H$_{10}$O$_2$	Veratrol	206.8	Nonazeotrope			575
10950	C$_8$H$_{11}$N	2,4,6-Collidine	171	195.23	52.3		791
		"	171.20	194.8	46.7	v-l	27c
10951	C$_8$H$_{11}$N	Dimethylaniline	194.15	Nonazeotrope			551
10952	C$_8$H$_{11}$N	Ethylaniline	205.5	Nonazeotrope			551
10953	C$_8$H$_{14}$O	Methylheptenone	173.2	184.6	67		552
10954	C$_8$H$_{16}$O	2-Octanone	172.85	184.5	68		552
10955	C$_8$H$_{16}$O$_2$	Butyl butyrate	166.4	Nonazeotrope			575
10956	C$_8$H$_{16}$O$_2$	Ethyl caproate	167.85	Nonazeotrope			542
10957	C$_8$H$_{16}$O$_2$	Isoamyl propionate	160.3	Nonazeotrope			531
10958	C$_8$H$_{16}$O$_3$	Isoamyl lactate	202.4	~203.5	12		542
10959	C$_8$H$_{18}$	Octane	125.75	Nonazeotrope			575
10960	C$_8$H$_{18}$O	Butyl ether	142.4	Nonazeotrope			556
10961	C$_8$H$_{18}$O	2-Ethyl-1-hexanol, 25 mm.		95.6	95		124,982
10962	C$_8$H$_{18}$O	*n*-Octyl alcohol	195.15	195.4	13		573
10963	C$_8$H$_{18}$O	*sec*-Octyl alcohol	179.0	184.5	50		535

TABLE I. *Binary Systems* 321

No.	Formula	Name	B.P., °C	B.P., °C	Wt.%A	Ref.
		B-Component		**Azeotropic Data**		
A =	**C₆H₆O**	**Phenol** *(continued)*	**182.2**			
10964	C₈H₁₈S	Butyl sulfide	172.0	<170.5	<28	566
10965	C₈H₁₈S	Isobutyl sulfide	172	<170.5	<28	555
10966	C₈H₂₀SiO₄	Ethyl silicate	165	Nonazeotrope		563
10967	C₉H₈	Indene	182.2	173.2	4⁵	573
		"	183.0	177.8	47	541
10968	C₉H₁₀	α-Methylstyrene		162	7	874
10969	C₉H₁₀O	Propiophenone	217.7	Nonazeotrope		552
10970	C₉H₁₂	Cumene	152.8	149	2	215
		"		152.8	Nonazeotrope	575
10971	C₉H₁₂	Mesitylene	164.6	163.5	21	575
10972	C₉H₁₂	Propylbenzene	158.9	158.0	~ 4	542
		"	158.9	158.5	14	946
10973	C₉H₁₂	Pseudocumene	168.2	166.0	25	542
10974	C₉H₁₂O	Benzyl ethyl ether	185.0	<181.9	<93	575
10975	C₉H₁₂O	Phenyl propyl ether	190.2	Nonazeotrope		542
10976	C₉H₁₃N	N,N-Dimethyl-o-toluidine	185.35	180.6	69.5	551
10977	C₉H₁₄O	Phorone	197.8	198.8	18	552
10978	C₉H₁₄SiO	(Trimethylsiloxy)benzene	181.9	175.5	39.5	520
10979	C₉H₁₈O	2,6-Dimethyl-4-heptanone	168.0	183.4	80	552
10980	C₉H₁₈O₂	Butyl isovalerate	177.6	184.0	70	562
10981	C₉H₁₈O₂	Ethyl enanthate	188.7	190.0	12	562
10982	C₉H₁₈O₂	Isoamyl butyrate	178.5	185.0	~58	573
10983	C₉H₁₈O₂	Isoamyl isobutyrate	169.8	Nonazeotrope		575
10984	C₉H₁₈O₂	Isobutyl isovalerate	168.7	182.8	92	573
10985	C₉H₁₈O₂	Isobutyl valerate	171.2	Nonazeotrope		564
		"	171.35	Nonazeotrope		542
10986	C₉H₁₈O₃	Isobutyl carbonate	190.3	192.5	26	563
10987	C₁₀H₈	Naphthalene	218.1	Nonazeotrope		563
10988	C₁₀H₁₀O₂	Safrole	235.9	Nonazeotrope		575
10989	C₁₀H₁₄	Butylbenzene	183.1	175.0	46	562
10990	C₁₀H₁₄	sec-Butylbenzene	173.3	166.5	31.8 v-l	973
10991	C₁₀H₁₄	Cymene	176.7	~170.5	37	542
10992	C₁₀H₁₆	Camphene	159.6	156.1	22	530
10993	C₁₀H₁₆	d-Limonene	177.8	169.0	40.5	563
10994	C₁₀H₁₆	Nopinene	163.8	~ 159	~ 25	563
10995	C₁₀H₁₆	α-Phellandrene	171.5	165	35	563
10996	C₁₀H₁₆	α-Pinene	155.8	152.75	19	563
10997	C₁₀H₁₆	α-Terpinene	173.4	166.7	36	562
10998	C₁₀H₁₆	Terpinene	181.5	171.5	45	542
10999	C₁₀H₁₆	Terpinolene	185	173	~ 62	563
		"	184.6	172.8	46	562
11000	C₁₀H₁₆	Thymene	179.7	172.25	40	530
11001	C₁₀H₁₆O	Camphor	209.1	Nonazeotrope		552
11002	C₁₀H₁₆O	Carvenone	234.5	Max b.p.		563
11003	C₁₀H₁₆O	Fenchone	193.6	196.2	25	552

No.	B-Component Formula	Name	B.P., °C	B.P., °C	Wt.%A		Ref.
A =	**C₆H₆O**	**Phenol** *(continued)*	**182.2**				
11004	C₁₀H₁₈	Menthene	170.5	~ 164	~ 33		563
11005	C₁₀H₁₈O	Borneol	211.8	Nonazeotrope			563
11006	C₁₀H₁₈O	Cineole	176.4	182.85	72		528
11007	C₁₀H₁₈O	1,4-Cineole, 100 mm.	105-106	119.3-120	88.7		426
11008	C₁₀H₁₈O	1,8-Cineole, 100 mm.	107.9	121-121.2	67		426
11009	C₁₀H₁₈O	Linalool	198.6	Nonazeotrope			535
11010	C₁₀H₁₈O	Menthone	~206	Nonazeotrope			563
11011	C₁₀H₁₈O	α-Terpineol	218.85	Nonazeotrope			575
11012	C₁₀H₁₈O	β-Terpineol	210.5	Nonazeotrope			575
11013	C₁₀H₂₀O	Citronellol	224.4	Nonazeotrope			575
11014	C₁₀H₂₀O	Menthol	212	Nonazeotrope			563
11015	C₁₀H₂₀O₂	Isoamyl isovalerate	193.5	Nonazeotrope			564
11016	C₁₀H₂₂	Decane	173.3	168.0	35		562
11017	C₁₀H₂₂	2,7-Dimethyloctane	160.25	159.5	6		544
11018	C₁₀H₂₂O	Amyl ether	187.5	180.2	78		562
11019	C₁₀H₂₂O	Isoamyl ether	173.2	172.2	15		556
11020	C₁₀H₂₂S	Isoamyl sulfide	214.8	Nonazeotrope			566
11021	C₁₁H₂₀O	Isobornyl methyl ether	192.4	Nonazeotrope			556
11021a	C₁₁H₂₄	Undecane 350 mm.		142.20	41	v-l	263g
		" 550 mm.		158.90	44.5	v-l	263g
		"		172.46	45.90	v-l 263c,263e	
11022	C₁₂H₁₈	1,3,5-Triethylbenzene	216	Nonazeotrope			563
11023	C₁₃H₂₈	Tridecane	235.42	180.56	83.1		911
		" 600 mm.	225.60	172.24	82.3		911
		" 500 mm.	218.20	165.87	82.1		911
		" 450 mm.	214.13	162.20	81.8		911
A =	**C₆H₆O₂**	**Pyrocatechol**	**245.9**				
11024	C₆H₆O₂	Resorcinol	281.4	Nonazeotrope			575
11025	C₆H₁₄O₃	Dipropylene glycol	229.2	253.0	~ 88		575
11026	C₇H₆Cl₂	α,α-Dichlorotoluene	205.2	Reacts			542
11027	C₇H₆O₂	Benzoic acid	250.5	245.85	98		538
11028	C₇H₇BrO	o-Bromoanisole	217.7	Nonazeotrope			575
11029	C₇H₇I	p-Iodotoluene	215.0	214.0	7		542
11030	C₇H₇NO₂	m-Nitrotoluene	230.8	Nonazeotrope			554
11031	C₇H₇NO₂	o-Nitrotoluene	221.75	Nonazeotrope			554
11032	C₇H₇NO₂	p-Nitrotoluene	238.9	238.7	11		554
11033	C₇H₈O₂	m-Methoxyphenol	243.8	241.5			542
11034	C₈H₇N	Indole	253.5	255.0	15		575
11035	C₈H₈O₂	Anisaldehyde	249.5	253	25		556
11036	C₈H₈O₂	α-Toluic acid	266.5	Nonazeotrope			575
11037	C₈H₉BrO	p-Bromophenetole	234.2	231.5	20		575
11038	C₈H₁₀O	3,4-Xylenol	226.8	Nonazeotrope			549
11039	C₈H₁₁NO	o-Phenetidine	232.5	246.0	92		551
11040	C₈H₁₁NO	p-Phenetidine	249.9	253.8	34		551
11041	C₈H₁₂O₄	Ethyl maleate	223.3	Nonazeotrope			575
11042	C₈H₁₆O₂	Caprylic acid	238.5	Nonazeotrope			575

TABLE I.   *Binary Systems*   323

No.	Formula	B-Component Name	B.P., °C	Azeotropic Data B.P., °C	Wt.%A	Ref.
A =	$C_9H_6O_2$	**Pyrocatechol** *(continued)*	**245.9**			
11043	C_9H_7N	Quinoline	237.4	257.9	61	564
11044	C_9H_8O	Cinnamaldehyde	253.5	Nonazeotrope		545
11045	$C_9H_{10}O$	Cinnamyl alcohol	257.0	Nonazeotrope		575
11046	$C_9H_{10}O$	p-Methylacetophenone	226.35	246.3	87.5	552
11047	$C_9H_{10}O_3$	Ethyl salicylate	234.0	Nonazeotrope		538
11048	$C_9H_{12}O$	3-Phenylpropanol	235.6	Nonazeotrope		575
11049	$C_9H_{18}O_2$	Pelargonic acid	254.0	Nonazeotrope		575
11050	$C_{10}H_7Br$	1-Bromonaphthalene	281.8	245.5	~ 80	542
11051	$C_{10}H_7Cl$	1-Chloronaphthalene	262.7	241.0	59	542
11052	$C_{10}H_8$	Naphthalene	218.05	217.45	11.5	538
11053	$C_{10}H_9N$	Quinaldine	246.5	252.5	48	575
11054	$C_{10}H_{10}O_2$	Isosafrole	252.0	243.0	70	544
11055	$C_{10}H_{10}O_2$	Methyl cinnamate	261.9	Nonazeotrope		542
11056	$C_{10}H_{10}O_2$	Safrole	235.9	233.55	23	536
11057	$C_{10}H_{12}O$	Anethole	235.7	233.0	25	562
11058	$C_{10}H_{12}O_2$	Ethyl α-toluate	228.75	Nonazeotrope		573
11059	$C_{10}H_{12}O_2$	Eugenol	254.8	245.85	98.5	538
11060	$C_{10}H_{12}O_2$	Propyl benzoate	230.9	Nonazeotrope		538
11061	$C_{10}H_{14}O$	Carvacrol	237.85	236.7	30	575
11062	$C_{10}H_{14}O$	Carvone	231.0	248.3	71	552
11063	$C_{10}H_{14}O$	Thymol	232.9	232.2	17	549
11064	$C_{10}H_{14}O_2$	m-Diethoxybenzene	235.4	<233.5	< 29	575
11065	$C_{10}H_{16}$	Terpinolene	184.6	Nonazeotrope		575
11066	$C_{10}H_{16}O$	Pulegone	223.8	246.5	90	552
11067	$C_{10}H_{18}O$	Geraniol	229.7	Nonazeotrope		542
11068	$C_{10}H_{20}O_2$	Capric acid	268.8	Nonazeotrope		575
11069	$C_{10}H_{22}O$	Decyl alcohol	232.9	Nonazeotrope		573
11070	$C_{11}H_{10}$	1-Methylnaphthalene	244.9	235.1	40	536
11071	$C_{11}H_{10}$	2-Methylnaphthalene	241.15	233.25	37	527
11072	$C_{11}H_{12}O_2$	Ethyl cinnamate	272.0	Nonazeotrope		575
11073	$C_{11}H_{14}O_2$	1-Allyl-3,4-dimethoxy-benzene	255.0	Nonazeotrope		538
11074	$C_{11}H_{14}O_2$	Butyl benzoate	249.8	Nonazeotrope		542
11075	$C_{11}H_{14}O_2$	1,2-Dimethoxy-4-propenylbenzene	270.5	Nonazeotrope		542
11076	$C_{11}H_{14}O_2$	Isobutyl benzoate	241.9	Nonazeotrope		538
11077	$C_{11}H_{16}O$	Methyl thymyl ether	216.5	Nonazeotrope		575
11078	$C_{11}H_{20}O$	α-Terpineol methyl ether	216.2	Nonazeotrope		542
11079	$C_{11}H_{22}O_3$	Isoamyl carbonate	232.2	Nonazeotrope		542
11080	$C_{12}H_{10}$	Acenaphthene	277.9	245.25	84	542
11081	$C_{12}H_{10}$	Biphenyl	255.9	239.85	56.5	542
11082	$C_{12}H_{10}O$	Phenyl ether	259.3	242.0	59.3	538
11083	$C_{12}H_{16}O_2$	Isoamyl benzoate	262.0	Nonazeotrope		542
11084	$C_{12}H_{18}$	1,3,5-Triethylbenzene	215.5	214.7		542
		" "	215.5	214.7	8.9	946
11085	$C_{12}H_{20}O_2$	Bornyl acetate	227.7	Nonazeotrope		542

No.	Formula	B-Component Name	B.P., °C	Azeotropic Data B.P., °C	Wt.%A	Ref.
A =	C₆H₆O₂	**Pyrocatechol** *(continued)*	**245.9**			
11086	C₁₂H₂₂O₄	Isoamyl oxalate	268.0	Nonazeotrope		544
11087	C₁₃H₁₀	Fluorene	295	Nonazeotrope		575
11088	C₁₃H₁₂	Diphenyl methane	265.6	243.05	65	536
11089	C₁₃H₁₂O	Benzyl phenyl ether	286.5	Nonazeotrope		575
11090	C₁₃H₂₆	Tridecane	234.0	229.7	30	542
11091	C₁₄H₁₄	1,2-Diphenylethane	284.9	Nonazeotrope		542
A =	C₆H₆O₂	**Resorcinol**	**281.4**			
11092	C₆H₆O₃	Pyrogallol	309	Nonazeotrope		575
11093	C₇H₇NO₂	*p*-Nitrotoluene	238.9	Nonazeotrope		575
11094	C₈H₈O₂	α-Toluic acid	266.5	Nonazeotrope		541
11095	C₈H₁₁NO	*o*-Phenetidine	232.5	Nonazeotrope		551
11096	C₈H₁₁NO	*p*-Phenetidine	249.9	Nonazeotrope		544
11097	C₉H₁₀O	Cinnamyl alcohol	257.0	Nonazeotrope		575
11098	C₉H₁₂O	3-Phenylpropanol	235.6	Nonazeotrope		544
11099	C₁₀H₇Br	1-Bromonaphthalene	281.8	266.3	45	542
11100	C₁₀H₇Cl	1-Chloronaphthalene	262.7	255.8	26	542
11101	C₁₀H₈	Naphthalene	218.05	Nonazeotrope		538
11102	C₁₀H₈O	1-Naphthol	288.0	280.2	70	575
11103	C₁₀H₈O	2-Naphthol	295	280.8	85	575
11104	C₁₀H₁₀O₂	Isosafrole	252.0	Nonazeotrope		542
11105	C₁₀H₁₀O₂	Methyl cinnamate	261.9	Nonazeotrope		538
11106	C₁₀H₁₀O₄	Methyl phthalate	283.7	287.5	38	544
11107	C₁₀H₁₂O₂	Eugenol	254.8	Nonazeotrope		542
11108	C₁₀H₁₂O₂	Isoeugenol	268.5	Nonazeotrope		542
11109	C₁₀H₁₈O₄	Propyl succinate	250.5	Nonazeotrope		575
11110	C₁₀H₂₀O	Citronellol	224.4	Nonazeotrope		575
11111	C₁₀H₂₀O₂	Capric acid	268.8	Nonazeotrope		575
11112	C₁₁H₁₀	1-Methylnaphthalene	244.6	243.1	14.5	538
11113	C₁₁H₁₀	2-Methylnaphthalene	241.15	240.05	10.5	527
11114	C₁₁H₁₂O₂	Ethyl cinnamate	271.5	Nonazeotrope		542
11115	C₁₁H₁₄O₂	1-Allyl-3,4-dimethoxy-benzene	255.0	Nonazeotrope		542
11116	C₁₁H₁₄O₂	1,2-Dimethoxy-4-propenyl-benzene	270.5	Nonazeotrope		544
11117	C₁₁H₁₆O	*p-tert*-Amylphenol	266.5	265.8	15	575
11118	C₁₂H₁₀	Acenaphthene	277.9	266.2	41	542
11119	C₁₂H₁₀	Biphenyl	255.9	252.15	21	542
11120	C₁₂H₁₀O	Phenyl ether	259.3	255.65	23	556
11121	C₁₂H₁₆O₂	Isoamyl benzoate	262.0	Nonazeotrope		538
11122	C₁₂H₁₈	1,3,5-Triethylbenzene	215.5	Nonazeotrope		542
11123	C₁₂H₂₂O₄	Isoamyl oxalate	268.0	282.5	85	544
11124	C₁₃H₁₀	Fluorene	295.0	274.0	48	562
11125	C₁₃H₁₂	Diphenylmethane	265.6	258.95	26	536
11126	C₁₃H₁₂O	Benzyl phenyl ether	286.5	<275.0	<83	562
11127	C₁₃H₂₈	Tridecane	234.0	233.25	12	542
11128	C₁₄H₁₂	Stilbene	306.5	277.5	56	562

TABLE I. *Binary Systems*   325

	B-Component			Azeotropic Data		
No.	Formula	Name	B.P., °C	B.P., °C	Wt.%A	Ref.
A =	$C_6H_6O_2$	**Resorcinol** *(continued)*	**281.4**			
11129	$C_{14}H_{14}$	1,2-Diphenylethane	284.9	269.7	47	542
A =	$C_6H_6O_3$	**Pyrogallol**	**309**			
11130	$C_{10}H_8O$	2-Naphthol	295.0	293.5	78	575
11131	$C_{11}H_{10}$	2-Methylnaphthalene	241.15	<240.6	< 6	575
11132	$C_{12}H_{10}$	Acenaphthene	277.9	272.8	20	562
11133	$C_{12}H_{10}$	Biphenyl	256.1	253.5	10	562
11134	$C_{13}H_{12}$	Diphenylmethane	265.4	<263.5	>11	562
11135	$C_{13}H_{12}O$	Benzyl phenyl ether	286.5	<283.5	<20	575
A =	C_6H_6S	**Benzenethiol**	**169.5**			
11136	C_7H_7Cl	*p*-Chlorotoluene	162.4	<161.5	79	566
11137	$C_8H_{16}O_2$	Isobutyl butyrate	157	~ 155	~ 15?	563
11138	$C_{10}H_{16}$	Camphene	~158	Reacts		563
11139	$C_{10}H_{16}$	α-Phellandrene	171.5	Reacts		563
11140	$C_{10}H_{16}$	α-Pinene	155.8	Reacts		563
11141	$C_{10}H_{18}$	Menthene	170.8	Reacts		563
A =	C_6H_7N	**Aniline**	**184.35**			
11142	C_nH_{2n-6}	Aromatic hydrocarbons	160-175	Min. b.p.		196
11143	C_nH_{2n+2}	Paraffins	160-175	Min. b.p.		196
11144	$C_6H_{10}O$	Cyclohexanone	155.7	Nonazeotrope		551
11145	$C_6H_{10}O_4$	Ethyl oxalate	185.0	~ 181.5	~ 40	563
11146	$C_6H_{11}NO$	Caprolactam 5 mm.		Nonazeotrope	v-l	191
		" 10 mm.		Nonazeotrope	v-l	191
11147	$C_6H_{11}NO_2$	Nitrocyclohexane	205.4	Nonazeotrope		551
11148	C_6H_{12}	Cyclohexane	80.75	Nonazeotrope		551
11149	$C_6H_{12}O$	Cyclohexanol	160.8	Nonazeotrope	v-l	704,705
		"		Nonazeotrope		551
11150	$C_6H_{13}N$	Cyclohexylamine	134	Nonazeotrope		413,704
		"	134		v-l	705
11151	C_6H_{14}	Hexane, 556-731 mm.		Nonazeotrope	v-l	1053
		"	68.8	Nonazeotrope		551
11152	$C_6H_{14}O$	*n*-Hexyl alcohol	157.85	Nonazeotrope		551
11153	$C_6H_{14}O_2$	2-Butoxyethanol	171.15	Nonazeotrope		551
11154	$C_6H_{14}O_2$	Pinacol	**174.35**	172.0	45	551
11155	$C_6H_{14}O_3$	2-(2-Ethoxyethoxy)ethanol	201.9	Nonazeotrope		575
11156	$C_6H_{15}NO$	2-Diethylaminoethanol	162.2	Nonazeotrope		551
11157	C_7H_5N	Benzonitrile	191.1	Nonazeotrope		565
11158	C_7H_6O	Benzaldehyde	179.2	Reacts		563
11159	C_7H_7Br	α-Bromotoluene	198.5	Reacts		563
11160	C_7H_7Br	*m*-Bromotoluene	184.3	179.9	39	551
11161	C_7H_7Br	*o*-Bromotoluene	181.5	178.45	35	551
11162	C_7H_7Br	*p*-Bromotoluene	185.0	180.2	44	551
11163	C_7H_7Cl	α-Chlorotoluene	179.35	Reacts		563
11164	C_7H_7Cl	*o*-Chlorotoluene	159.2	Nonazeotrope		551
11165	C_7H_7Cl	*p*-Chlorotoluene	162.4	Nonazeotrope		551

		B-Component			Azeotropic Data		
No.	Formula	Name	B.P., °C	B.P., °C	Wt.%A		Ref.
A =	C₆H₇N	**Aniline** *(continued)*	**184.35**				
11166	C₇H₇NO₂	o-Nitrotoluene	221.75	Nonazeotrope			551
11167	C₇H₈	Toluene	110.75	Nonazeotrope			551
		"	110.7			v-l	390
11168	C₇H₈O	Anisole	153.85	Nonazeotrope			551
11169	C₇H₈O	Benzyl alcohol	205.25	Nonazeotrope			551
11170	C₇H₈O	m-Cresol	202.2	Nonazeotrope			551
11171	C₇H₈O	o-Cresol	191.1	191.25	8		551
11172	C₇H₈O	p-Cresol	201.7	Nonazeotrope			551
11173	C₇H₈O₂	Guaiacol	205.05	Nonazeotrope			551
11174	C₇H₉N	Benzylamine	185.0	185.55	44		568
11175	C₇H₉N	N-Methylaniline,					
		95°-145°C.		Nonazeotrope		v-l	181
11176	C₇H₁₄	Methylcyclohexane	101.15	Nonazeotrope			551
11177	C₇H₁₄O	2-Methylcyclohexanol	168.5	168			576
11178	C₇H₁₄O₃	Isobutyl lactate	182.15	~ 180			563
11179	C₇H₁₆	Heptane	98.4	Nonazeotrope			527
		"	98.4			v-l	390
11180	C₇H₁₆O	n-Heptyl alcohol	176.15	175.4	22		551
11181	C₈H₈	Styrene	145.8	Nonazeotrope			551
11182	C₈H₈O	Acetophenone	202.0	Nonazeotrope			551
11183	C₈H₁₀	Ethylbenzene	136.15	Nonazeotrope			551
11184	C₈H₁₀	m-Xylene	139.2	Nonazeotrope			527
11185	C₈H₁₀	o-Xylene	144.3	Nonazeotrope			551
11186	C₈H₁₀	p-Xylene	138.45	Nonazeotrope			551
11187	C₈H₁₀O	Benzyl methyl ether	167.8	Nonazeotrope			551
11188	C₈H₁₀O	p-Methylanisole	177.05	Nonazeotrope			551
11189	C₈H₁₀O	Phenetole	170.45	Nonazeotrope			551
11190	C₈H₁₀O₂	o-Ethoxyphenol	216.5	Nonazeotrope			551
11191	C₈H₁₀O₂	Veratrole	206.8	Nonazeotrope			551
11192	C₈H₁₁N	N,N-Dimethylaniline,					
		36.7 mm.		95	74.5	v-l	181
		" 101.4 mm.		120	76.1	v-l	181
		" 243.1 mm.		145	77.5	v-l	181
11193	C₈H₁₄O	Methylheptenone	173.2	Reacts			535
11194	C₈H₁₆	1,3-Dimethylcyclohexane	120.7	Nonazeotrope			551
11195	C₈H₁₆O	2-Octanone	~ 173	Nonazeotrope			563
11196	C₈H₁₈	Isooctane, 86-741 mm.		Nonazeotrope		v-l	1053
11197	C₈H₁₈	Octane 400 mm.		103.80	0.3		911
		" <400 mm.		Min. b.p.			911
		" >400 mm.		Nonazeotrope			911
		"	125.75	Nonazeotrope			551
11198	C₈H₁₈O	Butyl ether	142.4	Nonazeotrope			551
11199	C₈H₁₈O	Isobutyl ether	122.3	Nonazeotrope			551
11200	C₈H₁₈O	n-Octyl alcohol	195.2	183.95	83		551
11201	C₈H₁₈O	sec-Octyl alcohol	180.4	179.0	36		551

TABLE I.   *Binary Systems*   327

No.	Formula	Name	B.P., °C	B.P., °C	Wt.%A		Ref.
		B-Component		**Azeotropic Data**			
A =	**C₆H₇N**	**Aniline** *(continued)*	**184.35**				
11202	C₉H₈	Indene	182.6	179.75	41.5		551
11203	C₉H₁₂	Cumene	152.8	Nonazeotrope			551
11204	C₉H₁₂	Mesitylene	164.6	Nonazeotrope			196,551
11205	C₉H₁₂	Propylbenzene	159.3	Nonazeotrope			551
11206	C₉H₁₂	Pseudocumene	168.2	<167.8	<13		551
11207	C₉H₁₂O	Benzyl ethyl ether	185.0	179.8	51		551
11208	C₉H₁₂O	Phenyl propyl ether	190.5	<183.5	<82		551
11209	C₉H₁₃N	Dimethyl-*o*-toluidine	185.3	180.55	51.5		549
11210	C₉H₂₀	Nonane	150.7	149.20	13.5		506
		"		150.4	148.94	13.2	911
		"	600 mm.	140.29	13.9		911
		"	500 mm.	133.38	14.6		911
		"	400 mm.	126.31	14.9		911
11211	C₁₀H₈	Naphthalene	218.0	Nonazeotrope			551
11212	C₁₀H₁₄	Butylbenzene	183.1	177.8	46		551
11213	C₁₀H₁₄	Cymene	176.7	173.5	27		551
		"	50 mm.		21.3	v-l	239
		"	100 mm.	106.3	23	v-l	239
		"	300 mm.			v-l	239
		"	500 mm.			v-l	239
		"	760 mm.	172.80	31.3	v-l	239
11214	C₁₀H₁₆	Camphene	159.6	157.5	13		551
11215	C₁₀H₁₆	Dipentene	177.7	171.3	39		551
11216	C₁₀H₁₆	*d*-Limonene	177.8	171.35	38.8		563
11217	C₁₀H₁₆	Nopinene	163.8	161.8	23		551
11218	C₁₀H₁₆	α-Phellandrene	171.5	167	~30		563
11219	C₁₀H₁₆	α-Pinene	155.8	155.25	15		551
11220	C₁₀H₁₆	α-Terpinene	173.4	169.5	32		551
11221	C₁₀H₁₆	γ-Terpinene	181.5	174	~42		538
11222	C₁₀H₁₆	Terpinolene	184.6	175.8	52		551
11223	C₁₀H₁₆	Thymene	179.7	173.5	41		532
11224	C₁₀H₁₆O	Camphor	209.1	Nonazeotrope			551
11225	C₁₀H₁₆O	Fenchone	193	Nonazeotrope			563
11226	C₁₀H₁₈	*d*-Menthene	170.8	<167.5	<34		551
11227	C₁₀H₁₈O	Cineole	176.35	174.65	30		551
11228	C₁₀H₁₈O	Linalool	198.6	Nonazeotrope			551
11229	C₁₀H₁₈O	β-Terpineol	210.75	Nonazeotrope			551
11230	C₁₀H₂₂	*n*-Decane	173.3	<169.5	<36		551
		"		174.6	167.28	36	506
11231	C₁₀H₂₂	2,7-Dimethyloctane	160.1	<159.5	<22		551
11232	C₁₀H₂₂O	Amyl ether	187.5	177.5	55		551
11233	C₁₀H₂₂O	Isoamyl ether	173.2	169.35	28		551
11234	C₁₁H₁₀	2-Methylnaphthalene	241.15	Nonazeotrope			527,551
11235	C₁₁H₂₀O	Isobornyl methyl ether	192.4	<183.8	<80		551
		"		192.2	Nonazeotrope		563

No.	Formula	Name	B.P., °C	B.P., °C	Wt.%A		Ref.
		B-Component		**Azeotropic Data**			
A =	**C₆H₇N**	**Aniline** *(continued)*	**184.35**				
11236	C₁₁H₂₄	Undecane	195.5	175.90	57.8		506
		"	194.5	175.31	57.5		911
		" 600 mm.		166.03	56.1		911
		" 500 mm.		159.34	55.6		911
		" 450 mm.		155.34	55.6		911
11237	C₁₂H₁₈	1,3,5-Triethylbenzene	215.5	Nonazeotrope			550
11238	C₁₂H₂₂O	Ethyl isobornyl ether	203.8	Nonazeotrope			575
11239	C₁₂H₂₆	Dodecane	216.5	180.37	71.5		506
11240	C₁₃H₂₈	Tridecane	234.6	183.07	87.0		506
		"	235.42	182.93	86.2		911
		" 600 mm.	225.60	174.20	85.4		911
		" 500 mm.	218.20	167.42	85.4		911
		" 400 mm.	209.60	159.43	84.2		911
11241	C₁₄H₃₀	Tetradecane	252.5	183.90	95.2		506
A =	**C₆H₇N**	**Picolines**					
11242	C₈H₈	Styrene	145	Min. b.p.			243
11243	C₈H₁₀	Ethylbenzene	136	Min. b.p.			243
11244	C₈H₁₀	Xylenes	140	Min. b.p.			243
A =	**C₆H₇N**	**2-Picoline**	**130.7**				
11245	C₆H₁₀S	Allyl sulfide	139.35	<130.2	<95		575
11246	C₆H₁₂O₃	Paraldehyde	124.5	Nonazeotrope			981
11247	C₆H₁₄S	Propyl sulfide	141.5	129.8	90		575
11248	C₈H₁₈	Octane	125.75	121.12	42		1067
11249	C₈H₁₈	2,2,4-Trimethylpentane	99.3	Nonazeotrope			575
11250	C₉H₂₀	Nonane	150.7	129.2	84.1		1067
A =	**C₆H₇N**	**3-Picoline**	**144**				
11251	C₆H₁₀S	Allyl sulfide	139.35	135.5	30		575
11252	C₇H₈	Toluene	110.7	Nonazeotrope			175
11253	C₇H₉N	2,6-Lutidine	144.06	143.5	27.3	v-l	99
11254	C₈H₁₈	2,3,4-Trimethylpentane	113.4	Nonazeotrope			175
A =	**C₆H₇N**	**4-Picoline**	**145.3**				
11255	C₇H₈	Toluene	110.7	Nonazeotrope			175
11256	C₈H₁₈	2,3,4-Trimethylpentane	113.4	Nonazeotrope			175
A =	**C₆H₈ClN**	**1,3-Cyclohexadiene**	**80.8**				
11257	C₆H₁₀	Cyclohexene	82.75	Nonazeotrope			563
11258	C₆H₁₂	Cyclohexane	80.75	79.0	45		561
A =	**C₆H₈ClN**	**Aniline Hydrochloride**					
11259	C₁₂H₁₁N	Diphenylamine,					
		100 mm			45.8		425
		" 250 mm.			48		425
		" 350 mm.	265	215	50		425
		" 740 mm		233	65		425
		" 2500 mm.		270			425

TABLE I. *Binary Systems* 329

No.	Formula	Name	B.P., °C	B.P., °C	Wt.%A	Ref.
		B-Component		**Azeotropic Data**		
A =	$C_6H_8N_2$	**2-Amino-3-methylpyridine**	**221**			
11260	$C_{11}H_{10}$	1-Methylnaphthalene,				
		" 20 mm.		115	68.2	270,271
		" 50 mm.		136	75.2	270,271
		" 150 mm.		166	89.7	270,271
		" 290 mm.		187	96.4	270,271
		" 400 mm.		198	98.7	270,271
		" 760 mm.	244.8	Nonazeotrope		270,271
11261	$C_{11}H_{10}$	2-Methylnaphthalene,				
		" 16 mm.		109	57.5	270,271
		" 50 mm.		137	69.5	270,271
		" 150 mm.		165	76.8	270,271
		" 400 mm.		196	92	270,271
		" 550 mm.		209	96	270,271
		" 760 mm.	241.1	Nonazeotrope		270,271
A =	$C_6H_8N_2$	**o-Phenylenediamine**	**258.6**			
11262	$C_7H_7NO_2$	m-Nitrotoluene	230.8	Nonazeotrope		551
11263	$C_7H_7NO_2$	p-Nitrotoluene	238.9	Nonazeotrope		527
11264	$C_7H_8O_2$	m-Methoxyphenol	243.8	Nonazeotrope		551
11265	$C_8H_{10}O$	Phenethyl alcohol	219.4	Nonazeotrope		527
11266	$C_9H_{12}O$	3-Phenylpropanol	235.6	Nonazeotrope		527
11267	$C_{10}H_8O$	1-Naphthol	288.0	Nonazeotrope		527
11268	$C_{10}H_{10}O_2$	Isosafrole	252.0	249.2	30	527
11269	$C_{10}H_{10}O_2$	Safrole	235.9	Nonazeotrope		527
11270	$C_{10}H_{12}O$	Anethole	235.7	Nonazeotrope		527
11271	$C_{10}H_{12}O_2$	Eugenol	254.8	Nonazeotrope		551
11272	$C_{10}H_{12}O_2$	Isoeugenol	268.8	Nonazeotrope		575
11273	$C_{10}H_{20}O$	Menthol	216.3	Nonazeotrope		527
11274	$C_{11}H_{10}$	1-Methylnaphthalene	244.6	<243.0	<17	527
11275	$C_{11}H_{14}O_2$	1-Allyl-3,4-dimethylbenzene	254.7	250.5	38	527
11276	$C_{11}H_{14}O_2$	1,2-Dimethoxy-4-propenylbenzene	270.5	Nonazeotrope		551
11277	$C_{12}H_{10}$	Acenaphthene	277.9	<258.0		527
11278	$C_{12}H_{10}$	Biphenyl	256.1	249.7	37	527
11279	$C_{12}H_{10}O$	Phenyl ether	259.0	251.2	46	527
11280	$C_{13}H_{12}$	Diphenylmethane	265.4	254.0	70	527
11281	$C_{14}H_{14}$	1,2-Diphenylethane	284.5	Nonazeotrope		527
A =	$C_6H_8O_4$	**Methyl Fumarate**	**193.25**			
11282	$C_6H_8O_4$	Methyl maleate	204.05	Nonazeotrope		527
11283	$C_6H_{10}O_4$	Ethyl oxalate	185.65	Nonazeotrope		527
11284	$C_6H_{10}O_4$	Glycol diacetate	186.3	Nonazeotrope		549
11285	$C_6H_{12}O_2$	Caproic acid	205.15	Nonazeotrope		575
11286	$C_6H_{14}O_2$	2-Butoxyethanol	171.15	Nonazeotrope		527
11287	C_7H_7Br	α-Bromotoluene	198.5	<192.3		575
11288	C_7H_7Br	m-Bromotoluene	184.3	183.65	16	527
11289	C_7H_7Br	o-Bromotoluene	181.5	Nonazeotrope		527
11290	C_7H_7Cl	α-Chlorotoluene	179.3	Nonazeotrope		527

No.		B-Component			Azeotropic Data	
	Formula	Name	B.P., °C	B.P., °C	Wt.%A	Ref.
A =	**C₈H₈O₄**	**Methyl Fumarate** (*continued*)	**193.25**			
11291	C₇H₈O	*m*-Cresol	202.2	204.3	72	526
11292	C₇H₈O	*o*-Cresol	191.1	197.8	60	570
11293	C₇H₈O	*p*-Cresol	201.7	204.0	29	527
11294	C₇H₁₂O₄	Ethyl malonate	199.35	Nonazeotrope		527
11295	C₈H₈O₂	Methyl benzoate	199.4	Nonazeotrope		527,549
11296	C₈H₁₀O	*p*-Methylanisole	177.05	Nonazeotrope		527
11297	C₈H₁₈O	*n*-Octyl alcohol	195.2	<190.1	<72	527
11298	C₉H₈	Indene	182.6	Nonazeotrope		575
11299	C₉H₁₂O	Benzyl ethyl ether	185.0	183.5	32	527
11300	C₉H₁₈O₂	Methyl caprylate	192.9	189.4	46	527
11301	C₁₀H₈	Naphthalene	218.0	Nonazeotrope		575
11302	C₁₀H₁₄	Butylbenzene	183.1	Nonazeotrope		575
11303	C₁₀H₁₆	Dipentene	177.7	172.5	70	562
11304	C₁₀H₁₆	α-Pinene	155.8	Nonazeotrope		575
11305	C₁₀H₁₆	α-Terpinene	173.4	170.5	75	562
11306	C₁₀H₁₈O	Borneol	215	Nonazeotrope		575
11307	C₁₀H₁₈O	Cineole	176.35	175.75	15	557
11308	C₁₀H₁₈O	Citronellal	208.0	Nonazeotrope		575
11309	C₁₀H₂₀O₂	Ethyl-caprylate	208.35	Nonazeotrope		575
11310	C₁₀H₂₀O₂	Isoamyl isovalerate	192.7	189.3	95	569
		"	192.7	189.3	43	549
11311	C₁₀H₂₂O	Isoamyl ether	173.2	172.35	16	527
11312	C₁₁H₂₀O	Isobornyl methyl ether	192.4	185.5	48	527
11313	C₁₂H₂₂O	Bornyl ethyl ether	204.9	191.2	80	557
11314	C₁₂H₂₂O	Ethyl isobornyl ether	203.8	<191.5	<81	557
A =	**C₆H₈O₄**	**Methyl Maleate**	**204.05**			
11315	C₆H₁₀O₄	Methyl succinate	195.5	Nonazeotrope		527
11316	C₆H₁₂O₂	Caproic acid	205.15	201.5	63	562
11317	C₆H₁₂O₂	Isocaproic acid	199.5	198.3	40	562
11318	C₇H₇Br	α-Bromotoluene	198.5	197.7	12	575
11319	C₇H₈O	Benzyl alcohol	205.25	Reacts		527
11320	C₇H₈O	*m*-Cresol	202.2	208.75	55	527
11321	C₇H₈O	*o*-Cresol	191.1	204.65	78	527
11322	C₇H₈O	*p*-Cresol	201.7	208.6	56	569,570
11323	C₇H₈O₂	Guaiacol	205.05	205.15	20	527
11324	C₇H₁₂O₄	Ethyl malonate	199.35	Nonazeotrope		527
11325	C₈H₈O	Acetophenone	202.0	201.0	39	569
11326	C₈H₈O₂	Methyl benzoate	199.4	198.95	25	527
11327	C₈H₈O₂	Phenyl acetate	195.7	Nonazeotrope		527
11328	C₈H₁₀O	3,4-Xylenol	226.8	Nonazeotrope		527
11329	C₈H₁₀O₂	*m*-Dimethoxybenzene	214.7	<202.8	>55	557
11330	C₈H₁₀O₂	*o*-Ethoxyphenol	216.5	Nonazeotrope		575
11331	C₈H₁₀O₂	Veratrole	206.8	<200.9		557
11332	C₈H₁₄O₄	Propyl oxalate	214	Nonazeotrope		575
11333	C₈H₁₆O₃	Isoamyl lactate	202.4	200.0	45	575
11334	C₈H₁₈O	Octyl alcohol	195.2	193.55	32	550

TABLE I. *Binary Systems* 331

No.	Formula	Name	B.P., °C	B.P., °C	Wt.%A		Ref.
		B-Component		Azeotropic Data			
A =	C$_6$H$_8$O$_4$	**Methyl Maleate** *(continued)*	**204.05**				
11335	C$_9$H$_{10}$O$_2$	Ethyl benzoate	212.5	Nonazeotrope			527
11336	C$_{10}$H$_8$	Naphthalene	218.0	203.7	87		527
11337	C$_{10}$H$_{12}$O	Estragol	215.6	Nonazeotrope			557
11338	C$_{10}$H$_{18}$O	Borneol	215.0	202.95	78		527
11339	C$_{10}$H$_{18}$O	Citronellal	208.0	Nonazeotrope			575
11340	C$_{10}$H$_{18}$O	Geraniol	229.6	Nonazeotrope			575
11341	C$_{10}$H$_{18}$O	Linalool	198.6	<197.2	<40		575
11342	C$_{10}$H$_{18}$O	α-Terpineol	218.85	<203.8			575
11343	C$_{10}$H$_{20}$O$_2$	Isoamyl isovalerate	192.7	190.65	25		527
11344	C$_{10}$H$_{22}$S	Isoamyl sulfide	214.8	203.0	82		566
11345	C$_{11}$H$_{10}$	2-Methylnaphthalene	241.15	Nonazeotrope			527
11346	C$_{11}$H$_{16}$O	Methyl thymyl ether	216.5	Nonazeotrope			557
11347	C$_{12}$H$_{18}$	Triethylbenzene	215.5	<202.8	>72		527
11348	C$_{12}$H$_{22}$O	Ethyl isobornyl ether	203.8	<197.8			557
A =	C$_6$H$_{10}$	**Biallyl**	**60.1**				
11349	C$_6$H$_{14}$	2,3-Dimethylbutane	58.0	< 57.5	42		561
A =	C$_6$H$_{10}$	**Cyclohexene**	**82.75**				
11350	C$_6$H$_{12}$	Cyclohexane	80.0	Nonazeotrope		v-l	365
		"	80.75	< 80.6	>10		561
11351	C$_6$H$_{14}$	Hexane	68.95	Nonazeotrope			563
11352	C$_6$H$_{14}$O	Propyl ether	90.55	Nonazeotrope			548
11353	C$_6$H$_{14}$O$_2$	Acetal	103.55	Nonazeotrope			558
11354	C$_7$H$_{14}$	Methylcyclohexane	101.15	Nonazeotrope			575
11355	C$_7$H$_{16}$	Heptane	98.4	Nonazeotrope			575
A =	C$_6$H$_{10}$	**Hexyne**	**70.2**				
11355a	C$_6$H$_{12}$	Hexene	63.6	Nonazeotrope		v-l	498f
11355b	C$_6$H$_{14}$	Hexane 400 mm.		47.8	31.7	v-l	498f
		" 600 mm.		59.8	34.2	v-l	498f
		" 760 mm.		67.2	34.0	v-l	498f
A =	C$_6$H$_{10}$	**Methylcyclopentene**	**75.85**				
11356	C$_6$H$_{14}$	Hexane	68.8	< 68.6	> 7		561
A =	C$_6$H$_{10}$O	**Cyclohexanone**	**155.7**				
11357	C$_6$H$_{12}$O	Cyclohexanol	160.8	Nonazeotrope			274,552
		" 100 mm.		Nonazeotrope		v-l	177
		" 300 mm.		119.6	87.7	v-l	248
		" 100 mm.		91	82.7	v-l	248
		" 30 mm.		Nonazeotrope		v-l	248
11358	C$_6$H$_{12}$O$_3$	Propyl lactate	171.7	Nonazeotrope			552
11359	C$_6$H$_{13}$ClO$_2$	Chloroacetal	157.4	155.3			552
11360	C$_6$H$_{14}$O	Hexyl alcohol	157.85	155.65	94		552
11361	C$_6$H$_{14}$S	Propyl sulfide	141.5	Nonazeotrope			575
11362	C$_7$H$_7$Cl	o-Chlorotoluene	159.2	Nonazeotrope			552
11363	C$_7$H$_7$Cl	p-Chlorotoluene	162.4	Nonazeotrope			552
11364	C$_7$H$_8$O	Anisole	153.85	Nonazeotrope			552
11365	C$_7$H$_{14}$O	2-Methylcyclohexanol	168.5	Nonazeotrope			552

No.	Formula	B-Component Name	B.P., °C	Azeotropic Data B.P., °C	Wt.%A	Ref.
A =	C$_6$H$_{10}$O	**Cyclohexanone** (*continued*)	**155.7**			
11366	C$_7$H$_{14}$O$_2$	Methyl caproate	149.7	Nonazeotrope		552
11367	C$_8$H$_{10}$	o-Xylene	144.3	Nonazeotrope		552
11368	C$_8$H$_{16}$O$_2$	Butyl butyrate	166.4	Nonazeotrope		552
11369	C$_8$H$_{16}$O$_2$	Isoamyl propionate	160.7	Nonazeotrope		552
11370	C$_8$H$_{16}$O$_2$	Isobutyl butyrate	156.9	155.3	60	552
11371	C$_8$H$_{16}$O$_2$	Propyl isovalerate	155.7	155.2	45	552
11372	C$_9$H$_{12}$	Cumene	152.8	152.0	65	552
11373	C$_9$H$_{12}$	Mesitylene	164.6	Nonazeotrope		552
11374	C$_9$H$_{12}$	Pseudocumene	168.2	Nonazeotrope		575
11375	C$_{10}$H$_{16}$	Camphene	159.6	150.55	57.5	552
11376	C$_{10}$H$_{16}$	Nopinene	163.8	152.2	65	552
11377	C$_{10}$H$_{16}$	α-Pinene	155.8	149.8	40	552
11378	C$_{10}$H$_{16}$	α-Terpinene	173.4	Nonazeotrope		552
11379	C$_{10}$H$_{22}$	2,7-Dimethyloctane	160.1	151.5	55	552
A =	C$_6$H$_{10}$O	**Mesityl Oxide**	**130.5**			
11380	C$_6$H$_{10}$S	Allyl sulfide	139.35	Nonazeotrope		566
11381	C$_6$H$_{12}$O	4-Methyl-2-pentanone	116.2	Nonazeotrope		981
11382	C$_6$H$_{12}$O$_2$	Butyl acetate	126.0	Nonazeotrope		527
11383	C$_6$H$_{12}$O$_2$	4-Hydroxy-4-methyl-2-pentanone	169.2	Nonazeotrope		981
11384	C$_6$H$_{12}$O$_2$	Isoamyl formate	123.8	Nonazeotrope		527
11385	C$_6$H$_{12}$O$_2$	Propyl propionate	123.0	Nonazeotrope		527
11386	C$_6$H$_{12}$O$_3$	Paraldehyde	124.35	Nonazeotrope		552
11387	C$_7$H$_8$	Toluene	110.75	Nonazeotrope		527
11388	C$_7$H$_{12}$O	3-Hepten-2-one	162.9	Nonazeotrope		981
11389	C$_7$H$_{14}$	Methylcyclohexane	101.15	Nonazeotrope		552
11390	C$_7$H$_{14}$O$_2$	Ethyl isovalerate	134.7	Nonazeotrope		527
11391	C$_7$H$_{14}$O$_2$	Isobutyl propionate	134.0	Nonazeotrope		552
11392	C$_7$H$_{14}$O$_2$	Propyl isobutyrate	133.9	Nonazeotrope		531
11393	C$_8$H$_{10}$	Ethylbenzene	136.15	Nonazeotrope		527
11394	C$_8$H$_{10}$	m-Xylene	139.2	Nonazeotrope		527
11395	C$_8$H$_{16}$	1,3-Dimethylcyclohexane	120.7	118.0	25	552
11396	C$_8$H$_{16}$O$_2$	4-Methyl-2-pentyl acetate	146.1	Nonazeotrope		981
11397	C$_8$H$_{16}$O$_2$	Propyl isovalerate	134.7	Nonazeotrope		552
11398	C$_8$H$_{18}$	Octane	125.75	121.0	35	527
11399	C$_8$H$_{18}$O	Butyl ether	142.4	Nonazeotrope		552
11400	C$_8$H$_{18}$O	Isobutyl ether	122.3	Nonazeotrope		552
11401	C$_8$H$_{19}$N	Diisobutylamine	138.5	<128.5	>25	575
11402	C$_9$H$_{18}$O	2,6-Dimethyl-4-heptanone	169.4	Nonazeotrope		981
A =	C$_6$H$_{10}$O	**Methyldihydropyran**				
11402a	C$_6$H$_{12}$O$_2$	4,4-Dimethyl-1,3-dioxane		Nonazeotrope		581c
		"	133.3	Nonazeotrope	v-l	581g
11402b	C$_6$H$_{14}$O	Isopropyl ether	69	Nonazeotrope		581c

TABLE I.   *Binary Systems*   333

| No. | B-Component | | | Azeotropic Data | | |
	Formula	Name	B.P., °C	B.P., °C	Wt.%A	Ref.
A =	$C_6H_{10}O$	**2,5-Hexanedione**	**192.2**			
11403	C_7H_8O	*m*-Cresol	202.4		36.3 v-l	730
11404	C_7H_8O	*p*-Cresol	202.0		32.2 v-l	730
11405	C_8H_9Cl	*o,m,p*-Chloroethylbenzene,				
		10 mm.	67.5	66.0	24	51
11406	$C_8H_{18}O$	Octyl alcohol	195.2	190.0	65	860
11407	$C_8H_{18}O$	*sec*-Octyl alcohol	180.4	<179.0	>18	575
A =	$C_6H_{10}O_3$	**Ethyl Acetoacetate**	**180.4**			
11408	$C_6H_{10}O_4$	Ethyl oxalate	185.65	Nonazeotrope		527
11409	$C_6H_{12}O_2$	Isocaproic acid	199.5	Nonazeotrope		552
11410	$C_7H_6Cl_2$	α,α-Dichlorotoluene	205.1	Nonazeotrope		563
11411	C_7H_6O	Benzoic acid	179.2	Reacts		563
11412	C_7H_7Br	α-Bromotoluene	198.5	Azeotrope doubtful		563
11413	C_7H_7Br	*m*-Bromotoluene	184.3	176.5	55	527
11414	C_7H_7Br	*o*-Bromotoluene	181.5	174.7	51	552
11415	C_7H_7Br	*p*-Bromotoluene	185	176.5	55	552
11416	C_7H_7Cl	α-Chlorotoluene	179.3	175	35	552
11417	C_7H_7Cl	*o*-Chlorotoluene	159.2	Nonazeotrope		552
11418	C_7H_7Cl	*p*-Chlorotoluene	162.4	Nonazeotrope		552
11419	C_7H_8O	Anisole	153.85	Nonazeotrope		552
11420	C_7H_8O	*o*-Cresol	190.8	Reacts		563
11421	C_8H_8	Styrene	145.8	Nonazeotrope		552
11422	C_8H_8O	Acetophenone	202.0	Nonazeotrope		552
11423	$C_8H_8O_2$	Phenyl acetate	195.7	Nonazeotrope		552
11424	C_8H_{10}	*m*-Xylene	139	Nonazeotrope		564
11425	C_8H_{10}	*o*-Xylene	144.3	Nonazeotrope		552
11426	C_8H_{10}	*p*-Xylene	138.4	Nonazeotrope		564
11427	$C_8H_{10}O$	*p*-Methylanisole	177.05	175.7		571
11428	$C_8H_{10}O$	Phenetole	170.45	169.8	24	552
11429	$C_8H_{10}O_2$	Veratrole	206.8	Nonazeotrope		552
11430	$C_8H_{14}O$	Methylheptenone	173.2	173.0	30?	552
11431	$C_8H_{16}O_2$	Butyl butyrate	166.4	Nonazeotrope		552
11432	$C_8H_{16}O_2$	Isoamyl propionate	160.7	Nonazeotrope		552
11433	$C_8H_{18}O$	Butyl ether	142.4	Nonazeotrope		552
11434	$C_8H_{18}O$	*sec*-Octyl alcohol	179.0	Nonazeotrope		572
11435	$C_8H_{18}S$	Butyl sulfide	185.0	<178.5	<78	566
11436	$C_8H_{18}S$	Isobutyl sulfide	172.0	171.0	10	567
11437	C_9H_8	Indene	182.6	177.15	68	552
11438	C_9H_{12}	Mesitylene	164.6	162.5	32	552
11439	C_9H_{12}	Propylbenzene	159.3	158.3	24	552
11440	C_9H_{12}	Pseudocumene	168.2	165.2	37	552
11441	$C_9H_{12}O$	Benzyl ethyl ether	185.0	175.5	>75	552
11442	$C_9H_{18}O$	2,6-Dimethyl-4-heptanone	168.0	Nonazeotrope		552
11443	$C_9H_{18}O_2$	Isoamyl butyrate	181.05	174.5	60	564
11444	$C_9H_{18}O_2$	Isoamyl isobutyrate	169.8	169.0	20	552
11445	$C_9H_{18}O_2$	Isobutyl isovalerate	171.2	170.2	25	552

No.	Formula	Name	B.P., °C	B.P., °C	Wt.%A	Ref.
		B-Component		Azeotropic Data		
A =	$C_6H_{10}O_3$	**Ethyl Acetoacetate**	**180.4**			
		(continued)				
11446	$C_9H_{18}O_2$	Methyl caprylate	192.9	180.0	80	575
11447	$C_{10}H_8$	Naphthalene	218.0	Nonazeotrope		552
11448	$C_{10}H_{14}$	Butylbenzene	183.1	174.0	52	552
11449	$C_{10}H_{14}$	Cymene	176.7	170.5	41	552
11450	$C_{10}H_{16}$	Camphene	159.6	156.15	30	552
11451	$C_{10}H_{16}$	Dipentene	177.7	169.05	43	552
11452	$C_{10}H_{16}$	d-Limonene	177.8	169.05	43	563
11453	$C_{10}H_{16}$	Nopinene	163.8	159.3	<35	552
11454	$C_{10}H_{16}$	α-Phellandrene	171.5	165	~ 40	563
11455	$C_{10}H_{16}$	α-Pinene	155.8	153.35	22	552
11456	$C_{10}H_{16}$	α-Terpinene	173.4	166.6	40	552
11457	$C_{10}H_{16}$	Terpinene	181.5	171.0	50	545
11458	$C_{10}H_{16}$	Terpinolene	184.6	172.2	55	552
11459	$C_{10}H_{16}O$	Fenchone	193.6	Nonazeotrope		552
11460	$C_{10}H_{18}$	m-Menthene-8	170.8	164.9		552
11461	$C_{10}H_{18}O$	Cineol	176.35	168.75	43	552
11462	$C_{10}H_{20}O_2$	Isoamyl isovalerate	192.7	179.5	77	552
11463	$C_{10}H_{22}$	2,7-Dimethyloctane	160.1	156.0	24	552.
11464	$C_{10}H_{22}O$	Amyl ether	187.5	174.5	70	552
11465	$C_{10}H_{22}O$	Isoamyl ether	173.2	167.4	40	552
11466	$C_{11}H_{20}O$	Isobornyl methyl ether	192.4	<179.0		552
11467	$C_{12}H_{18}$	1,3,5-Triethylbenzene	216	Nonazeotrope		563
11468	$C_{12}H_{22}O$	Ethyl isobornyl ether	203.8	Nonazeotrope		552
A =	$C_6H_{10}O_4$	**Butylene Glycol Diformate**				
11469	C_nH_m	Hydrocarbon		min. b.p.		618
A =	$C_6H_{10}O_4$	**Ethylidene Diacetate**	**168.5**			
11470	$C_6H_{12}O_3$	2-Ethoxyethyl acetate	156.8	Nonazeotrope		527
11471	$C_6H_{14}O$	Hexyl alcohol	157.85	<157.3		575
11472	$C_6H_{14}O_2$	2-Butoxyethanol	171.15	166.7	64	527
11473	$C_6H_{14}O_2$	Pinacol	174.35	<167.0		575
11474	C_7H_7Br	m-Bromotoluene	184.3	Nonazeotrope		527
11475	C_7H_7Br	o-Bromotoluene	181.5	Nonazeotrope		575
11476	C_7H_7Cl	α-Chlorotoluene	179.3	Nonazeotrope		575
11477	C_7H_7Cl	p-Chlorotoluene	162.4	<161.0	>70	527
11478	C_7H_8O	Anisole	153.85	Nonazeotrope		527
11479	C_7H_8O	o-Cresol	191.1	Nonazeotrope		527
11480	$C_7H_{14}O$	2-Methylcyclohexanol	168.5	<165.8	<57	575
11481	$C_7H_{14}O_3$	1,3-Butanediol methyl ether acetate	171.75	Nonazeotrope		527
11482	C_8H_{10}	m-Xylene	139.2	Nonazeotrope		527
11483	C_8H_{10}	o-Xylene	144.3	Nonazeotrope		527
11484	$C_8H_{10}O$	Benzyl methyl ether	167.8	164.0	48	527
11485	$C_8H_{10}O$	p-Methylanisole	177.05	<168.3	>62	557
11486	$C_8H_{10}O$	Phenetole	170.45	164.5	56	527
11487	$C_8H_{14}O$	Methylheptenone	173.2	Nonazeotrope		552
11488	$C_8H_{16}O$	2-Octanone	172.85	Nonazeotrope		552

TABLE I. *Binary Systems*   335

No.	B-Component				Azeotropic Data		
	Formula	Name	B.P., °C	B.P., °C	Wt.%A	Ref.
A =	$C_6H_{10}O_4$	**Ethylidene Diacetate** *(continued)*	**168.5**			
11489	$C_8H_{16}O_2$	Butyl butyrate	166.4	163.5	37	527
11490	$C_8H_{16}O_2$	Ethyl caproate	167.7	164.0	45	549
11491	$C_8H_{16}O_2$	Hexyl acetate	171.5	<166.5	<67	527
11492	$C_8H_{16}O_2$	Isoamyl propionate	160.7	159.3	23	549
11493	$C_8H_{18}O$	sec-Octyl alcohol	180.4	168.3	93.5	527
11494	$C_9H_{18}O_2$	Butyl isovalerate	177.6	167.5		575
11495	$C_9H_{18}O_2$	Isoamyl isobutyrate	169.8	165.0	60	527
11496	$C_9H_{18}O_2$	Isobutyl isovalerate	171.2	165.5	65	527
11497	$C_{10}H_{14}$	Cymene	176.7	165.5	>62	567
11498	$C_{10}H_{16}$	Camphene	159.6	<157.0	>32	527
11499	$C_{10}H_{16}$	α-Pinene	155.8	<154.0	>25	527
11500	$C_{10}H_{18}O$	Cineole	176.35	164.95	66	527
11501	$C_{10}H_{22}O$	Isoamyl ether	173.2	161.5	57	527
A =	$C_6H_{10}O_4$	**Ethyl Oxalate**	**185.65**			
11502	$C_6H_{10}O_4$	Methyl succinate	195.5	Nonazeotrope		527
11503	$C_6H_{12}O_2$	Isocaproic acid	199.7	Nonazeotrope		527
11504	$C_6H_{12}O_3$	2-Ethoxyethyl acetate	156.8	Nonazeotrope		526
11505	$C_6H_{13}Br$	1-Bromohexane	156.5	Nonazeotrope		575
11506	$C_6H_{14}O_2$	2-Butoxyethanol	171.25	Reacts		526
11507	C_7H_5N	Benzonitrile	191.1	Nonazeotrope		565
11508	C_7H_6O	Benzaldehyde	179.2	Nonazeotrope		563
11509	C_7H_7Br	m-Bromotoluene	184.3	179.0	46	527
11510	C_7H_7Br	o-Bromotoluene	181.75	177.40	38	527
11511	C_7H_7Br	p-Bromotoluene	185	<180.2	<49	528
11512	C_7H_7Cl	α-Chlorotoluene	179.35	Nonazeotrope		563
11513	C_7H_7Cl	o-Chlorotoluene	159.2	Nonazeotrope		527
11514	C_7H_7Cl	p-Chlorotoluene	162.4	Nonazeotrope		527
11515	C_7H_8O	Anisole	153.85	Nonazeotrope		527
11516	C_7H_8O	m-Cresol	202.2	202.3	~ 3	542
11517	C_7H_8O	o-Cresol	191.1	194.1	36	542
11518	C_7H_8O	p-Cresol	201.7	202.0	6.5	542
11519	C_7H_9N	Methylaniline	196.1	Reacts		563
11520	$C_7H_{13}ClO_2$	Isoamyl chloroacetate	190.5	181.5	~ 65	564
11521	$C_7H_{14}O$	2-Methylcyclohexanol	168.5	Nonazeotrope		575
11522	$C_7H_{14}O_3$	1,3-Butanediol methyl ether acetate	171.75	Nonazeotrope		527
11523	$C_7H_{16}O$	Heptyl alcohol	176.15	175.5		527
11524	C_8H_8	Styrene	145.8	Nonazeotrope		575
11525	$C_8H_8O_2$	Phenyl acetate	195.7	Nonazeotrope		527
11526	$C_8H_{10}O$	Benzyl methyl ether	167.8	Nonazeotrope		527,557
11527	$C_8H_{10}O$	p-Methylanisole	177.05	< 176.3		557
11528	$C_8H_{10}O$	Phenetole	171.5	Nonazeotrope		527
11529	$C_8H_{10}O_2$	Veratrol	205.5	Nonazeotrope		557
11530	$C_8H_{16}O_2$	Hexyl acetate	171.5	Nonazeotrope		575
11531	$C_8H_{18}O$	Octyl alcohol	195.15	Reacts		535
11532	$C_8H_{18}O$	sec-Octyl alcohol	180.4	178.85	33	567
11533	C_9H_8	Indene	182.6	< 181.0	<43	575

No.	Formula	Name	B.P., °C	B.P., °C	Wt.%A		Ref.
		B-Component		**Azeotropic Data**			
A =	**C₆H₁₀O₄**	**Ethyl Oxalate** *(continued)*	**185.65**				
11534	C₉H₁₂	Cumene	152.8	Nonazeotrope			575
11535	C₉H₁₂	Mesitylene	164.6	Nonazeotrope			546
11536	C₉H₁₂	Pseudocumene	168.2	167.95	~ 6		541
11537	C₉H₁₂O	Benzyl ethyl ether	185.0	< 181.8	<50		527
11538	C₉H₁₈O₂	Butyl isovalerate	177.6	176.3	25		527
11539	C₉H₁₈O₂	Ethyl enanthate	188.7	183.0	60		549
11540	C₉H₁₈O₂	Isoamyl butyrate	181.05	179.45	32.5		568
11541	C₉H₁₈O₂	Isobutyl isovalerate	185.65	Nonazeotrope			527
11542	C₉H₁₈O₂	Methyl caprylate	192.9	184.2	70		549
11543	C₁₀H₈	Naphthalene	218.0	Nonazeotrope			575
11544	C₁₀H₁₄	Butylbenzene	183.1	< 180.0	<44		562
11545	C₁₀H₁₄	Cymene	175.3	~ 173	~ 15		563
11546	C₁₀H₁₆	Camphene	159.6	158.5	16		574
11547	C₁₀H₁₆	Dipentene	177.7	172.2	40		575
11548	C₁₀H₁₆	d-Limonene	177.8	172.2	41		563
11549	C₁₀H₁₆	Nopinene	163.8	161.5	27		546
11550	C₁₀H₁₆	α-Pinene	155.8	154.8	20		537
11551	C₁₀H₁₆	α-Terpinene	173.3	170.5	30		546
11552	C₁₀H₁₆	γ-Terpinene	181.5	173.5	45		538
11553	C₁₀H₁₆	Terpinolene	185	173	~ 50		563
11554	C₁₀H₁₆	Thymene	179.7	~ 176.0	40.5		537
11555	C₁₀H₁₈	m-Menthene-8	170.8	168.0	28		575
11556	C₁₀H₁₈O	Cineole	176.35	173.5	28		557
11557	C₁₀H₁₈O	Linalool	198.6	Nonazeotrope			575
		"	198.6	185.6	~ 97		574
11558	C₁₀H₂₀O₂	Isoamyl isovalerate	192.7	184.1	69		527
11559	C₁₀H₂₂	2,7-Dimethyloctane	160.1	188.5	28		562
11560	C₁₀H₂₂O	Amyl ether	187.5	177.7	54		527
11561	C₁₀H₂₂O	Isoamyl ether	173.2	170.15	29		527
		"	172.6	Nonazeotrope			535
11562	C₁₁H₂₀O	Methyl isobornyl ether	192.2	181.15	88?		557
11563	C₁₂H₁₈	1,3,5-Triethylbenzene	215.5	Nonazeotrope			575
11564	C₁₂H₂₂O	Ethyl isobornyl ether	203.8	Nonazeotrope			557
A =	**C₆H₁₀O₄**	**Glycol Diacetate**	**186.3**				
11565	C₆H₁₂O₃	2-Ethoxyethyl acetate	156.8	Nonazeotrope			575
11566	C₆H₁₄O₂	2-Butoxyethanol	171.15	Nonazeotrope			575
11567	C₇H₇Br	o-Bromotoluene	181.5	< 179.8	<32		575
11568	C₇H₇Br	p-Bromotoluene	185.0	< 182.0	<45		575
11569	C₇H₈O	m-Cresol	202.4		24	v-l	730
11570	C₇H₈O	o-Cresol	191.1	194.5	35		562
11571	C₇H₈O	p-Cresol	202.0		23	v-l	730
11572	C₇H₁₄O₃	1,3-Butanediol methyl ether acetate	171.75	Nonazeotrope			527
11573	C₈H₈O₂	Phenyl acetate	195.7	Nonazeotrope			575
11574	C₈H₁₀O	Benzyl methyl ether	167.8	Nonazeotrope			557
11575	C₈H₁₀O	Phenetole	170.45	Nonazeotrope			557

TABLE I. *Binary Systems* 337

No.		B-Component			Azeotropic Data		
	Formula	Name	B.P., °C	B.P., °C	Wt.%A		Ref.
A =	**C₆H₁₀O₄**	**Glycol Diacetate** *(continued)*	**186.3**				
11576	C₈H₁₀O₂	Veratrole	206.8	Nonazeotrope			557
11577	C₈H₁₈O	Octyl alcohol	195.2	< 186.0			575
11578	C₈H₁₈O	*sec*-Octyl alcohol	180.4	179.2			567
11579	C₉H₁₂	Mesitylene	164.6	Nonazeotrope			575
11580	C₉H₁₂	Propylbenzene	159.3	Nonazeotrope			575
11581	C₉H₁₂O	Benzyl ethyl ether	185.0	< 181.2			557
11582	C₉H₁₈O₂	Butyl isovalerate	177.6	< 177.0	>15		549
11583	C₉H₁₈O₂	Isoamyl butyrate	181.05	179.0	38		549
11584	C₁₀H₁₄	Butylbenzene	183.1	< 181.2	<42		575
11585	C₁₀H₁₆	Dipentene	177.7	< 173.5	<37		575
11586	C₁₀H₂₀O₂	Isoamyl isovalerate	192.7	184.6	75		549
11587	C₁₀H₂₂O	Amyl ether	187.5	< 179.0	<60		557
11588	C₁₀H₂₂O	Isoamyl ether	173.2	170.1			557
11589	C₁₁H₂₀O	Isobornyl methyl ether	192.4	< 183.5	<82		557
A =	**C₆H₁₀O₄**	**Methyl Succinate**	**195.5**				
11590	C₆H₁₂O₂	Caproic acid	205.15	Nonazeotrope			575
11591	C₆H₁₂O₂	Isocaproic acid	199.5	< 194.2	<80		562
11592	C₇H₅N	Benzonitrile	191.1	Nonazeotrope			565
11593	C₇H₆Cl₂	α,α-Dichlorotoluene	205.2	Nonazeotrope			547
11594	C₇H₇Br	α-Bromotoluene	198.5	< 192.5	>55		575
11595	C₇H₇Br	*m*-Bromotoluene	184.3	182.6	<21		575
11596	C₇H₇Br	*o*-Bromotoluene	181.5	< 181.0	<10		575
11597	C₇H₇Br	*p*-Bromotoluene	185.0	180.0			547
11598	C₇H₈O	*o*-Cresol	190.8	198.8	~ 60		563
11599	C₇H₈O	*p*-Cresol	201.8	204.7			563
11600	C₇H₁₂O₄	Ethyl malonate	199.35	Nonazeotrope			575
11601	C₈H₈O	Acetophenone	202.0	Nonazeotrope			552
11602	C₈H₈O₂	Methyl benzoate	199.4	Nonazeotrope			575
11603	C₈H₈O₂	Phenyl acetate	195.5	Nonazeotrope			572
11604	C₈H₁₀O	*p*-Methylanisole	177.05	Nonazeotrope			557
11605	C₈H₁₈O	*n*-Octyl alcohol	195.15	192.5	50		572
11606	C₉H₈	Indene	182.6	Nonazeotrope			575
11607	C₉H₁₂	Mesitylene	164.6	Nonazeotrope			546
11608	C₉H₁₂	Propylbenzene	159.3	Nonazeotrope			575
11609	C₉H₁₂	Pseudocumene	168.2	Nonazeotrope			575
11610	C₉H₁₄O	Phorone	197.8	Nonazeotrope			552
11611	C₉H₁₈O₃	Isobutyl carbonate	190.3	Nonazeotrope			549
11612	C₁₀H₈	Naphthalene	218.1	Nonazeotrope			546
11613	C₁₀H₁₆	Camphene	159.6	~ 159.0	10		546
11614	C₁₀H₁₆	*d*-Limonene	177.8	175.5	26		529
11615	C₁₀H₁₆	α-Pinene	155.8	155.5	<10		546
11616	C₁₀H₁₆	α-Terpinene	173.4	172.5	19		562
11617	C₁₀H₁₆	γ-Terpinene	181.5	178.0	32		538
11618	C₁₀H₁₆	Terpinolene	185	~ 178	~ 28		563
11619	C₁₀H₁₆	Thymene	179.7	178.2	~ 32		530
11620	C₁₀H₁₇Cl	Bornyl chloride	207.5	< 195.2			575
11621	C₁₀H₁₈O	Cineole	176.35	< 176.0	<95		557

No.	Formula	Name	B.P., °C	B.P., °C	Wt.%A	Ref.
		B-Component		**Azeotropic Data**		
A =	**C₆H₁₀O₄**	**Methyl Succinate** *(continued)*	**195.5**			
11622	$C_{10}H_{18}O$	Linalool	198.6	Reacts		536
11623	$C_{10}H_{20}O$	Menthol	212	Nonazeotrope?		563
11624	$C_{10}H_{20}O_2$	Ethyl caprylate	208.35	Nonazeotrope		549
11625	$C_{10}H_{20}O_2$	Isoamyl isovalerate	192.7	191.0	30	549
11626	$C_{10}H_{22}O$	Isoamyl ether	173.2	< 172.5		557
11627	$C_{11}H_{20}O$	Isobornyl methyl ether	192.4	186.4		557
11628	$C_{12}H_{18}$	1,3,5-Triethylbenzene	216	Nonazeotrope		546
11629	$C_{12}H_{22}O$	Ethyl isobornyl ether	203.8	193.0	75	557
A =	**C₆H₁₀S**	**Allyl Sulfide**	**139.35**			
11630	$C_6H_{12}O$	Cyclohexanol	160.8	Nonazeotrope		566
11631	$C_6H_{12}O$	2-Hexanone	127.2	Nonazeotrope		566
11632	$C_6H_{12}O$	4-Methyl-2-pentanone	116.05	Nonazeotrope		566
11633	$C_6H_{12}O_2$	Butyl acetate	126.0	Nonazeotrope		566
11634	$C_6H_{12}O_2$	Ethyl butyrate	121.5	Nonazeotrope		566
		"	119.9	~ 117.5	~ 15	563
11635	$C_6H_{12}O_2$	Isoamyl formate	123.6	~ 120	~ 20	563
11636	$C_6H_{14}O$	Hexyl alcohol	157.85	Nonazeotrope		566
11637	$C_7H_{14}O$	4-Heptanone	143.55	138.2	75	566
11638	$C_7H_{16}O_2$	Dipropoxymethane	137.2	< 135.5	>68	562
11639	C_8H_{10}	Ethylbenzene	136.15	< 136.0	>11	566
11640	C_8H_{10}	*m*-Xylene	139.2	< 138.3	>52	566
11641	C_8H_{16}	1,3-Dimethylcyclohexane	120.7	Nonazeotrope		566
11642	$C_8H_{18}O$	Butyl ether	142.4	< 139.0	70	566
A =	**C₆H₁₁BrO₂**	**Ethyl α-bromoisobutyrate**	**178**			
11643	C_7H_6O	Benzaldehyde	179.2	Azeotrope doubtful		563
11644	C_7H_7Cl	α-Chlorotoluene	179.3	~ 173.5	~ 60	532
11645	$C_8H_{18}O$	*sec*-Octyl alcohol	178.7	~ 175		563
11646	$C_{10}H_{16}$	*d*-Limonene	177.8	174	~ 55	563
A =	**C₆H₁₁ClO₂**	**Butyl Chloroacetate**	**181.8**			
11647	C_7H_7Br	*o*-Bromotoluene	181.5	179.5	45	562
11648	$C_8H_{10}O$	Benzyl methyl ether	167.8	Nonazeotrope		575
11649	$C_8H_{10}O$	Phenetole	170.45	Nonazeotrope		575
11650	$C_8H_{18}O$	Octyl alcohol	195.2	Nonazeotrope		575
11651	$C_{10}H_{14}$	Butylbenzene	183.1	< 179.5	<70	562
11652	$C_{10}H_{14}$	Cymene	176.7	175.4	25	575
11653	$C_{10}H_{16}$	Dipentene	177.7	175.0	32	562
11654	$C_{10}H_{22}O$	Isoamyl ether	173.2	Nonazeotrope		575
A =	**C₆H₁₁ClO₂**	**Isobutyl Chloroacetate**	**174.5**			
11655	C_7H_7Cl	*p*-Chlorotoluene	162.4	Nonazeotrope		575
11656	$C_8H_{10}O$	Phenetole	170.45	170.0	12	575
11657	C_9H_{12}	Propylbenzene	159.3	Nonazeotrope		575
11658	$C_{10}H_{14}$	Cymene	176.7	172.2	65	562
11659	$C_{10}H_{16}$	Camphene	159.6	Nonazeotrope		575
11660	$C_{10}H_{16}$	α-Pinene	155.8	Nonazeotrope		575
11661	$C_{10}H_{18}O$	Cineole	176.35	173.2	70	562

TABLE I. *Binary Systems* 339

No.	Formula	Name	B.P., °C	B.P., °C	Wt.%A	Ref.
		B-Component		**Azeotropic Data**		
A =	**C₆H₁₁ClO₂**	**Isobutyl Chloroacetate**	**174.5**			
		(*continued*)				
11662	C₁₀H₁₈O	Linalool	198.6	Nonazeotrope		575
11663	C₁₀H₂₂O	Amyl ether	187.5	Nonazeotrope		575
11664	C₁₀H₂₂O	Isoamyl ether	173.2	172.0	38	562
A =	**C₆H₁₁N**	**Capronitrile**	**163.9**			
11665	C₆H₁₂O	Hexyl alcohol	157.85	158.0	36	567
11666	C₆H₁₄O	Cyclohexanol	160.8	< 156.6	>19	567
11667	C₆H₁₄O₂	2-Butoxyethanol	171.15	Nonazeotrope		575
11668	C₈H₁₀	*m*-Xylene	139.2	Nonazeotrope		575
11669	C₉H₁₂	Cumene	152.8	150.8	18	575
11670	C₁₀H₁₆	Camphene	159.6	143.0	35	562
11671	C₁₀H₁₆	α-Pinene	155.8	142.0	30	562
A =	**C₆H₁₁NO₂**	**Nitrocyclohexane**	**205.3**			
11672	C₆H₁₅NO	2-(Diethylamino)ethanol	162.2	Nonazeotrope		575
11673	C₇H₆O	Benzaldehyde	179.2	Nonazeotrope		554
11674	C₇H₉N	Methylaniline	196.25	Nonazeotrope		551
11675	C₇H₉N	*m*-Toluidine	203.1	< 203.0	> 4	551
11676	C₇H₉N	*o*-Toluidine	200.35	Nonazeotrope		551
11677	C₇H₁₄O₃	Isobutyl lactate	182.15	Nonazeotrope		575
11678	C₈H₁₁N	Dimethylaniline	194.15	Nonazeotrope		551
11679	C₈H₁₁N	Ethylaniline	205.5	< 204.8		551
11680	C₈H₁₆O₃	Isoamyl lactate	202.4	< 201.0	>28	575
11681	C₈H₁₈S	Butyl sulfide	185.0	Nonazeotrope		575
11682	C₉H₁₃N	*N,N*-Dimethyl-*o*-toluidine	185.3	Nonazeotrope		551
A =	**C₆H₁₂**	**Cyclohexane**	**80.75**			
11683	C₆H₁₂	Methylcyclopentane				
			60°C	Nonazeotrope	v-l	55
		"	71.8	Nonazeotrope	v-l	936
11684	C₆H₁₂O	Cyclohexanol	35-55°	Nonazeotrope	v-l	149
		"	161.1	Nonazeotrope		981
11685	C₆H₁₂O	4-Methyl-2-pentanone,				
			450-760 mm.	Nonazeotrope	v-l	189
		"	116.05	Nonazeotrope		527
11686	C₆H₁₂O	Pinacolone	106.2	Nonazeotrope		552
11687	C₆H₁₂O₂	Butyl acetate	126		v-l	682
11688	C₆H₁₂O₂	Ethyl isobutyrate	110.1	Nonazeotrope		575
11689	C₆H₁₃N	Cyclohexylamine	134	Nonazeotrope	v-l	704,718c
11690	C₆H₁₄	Hexane	68.95	Nonazeotrope	v-l	677,807c
11691	C₆H₁₄O	4-Methyl-2-pentanol	131.8	Nonazeotrope		781
11692	C₆H₁₄O	Propyl ether	90.55	Nonazeotrope		548
11693	C₆H₁₄O₂	Acetal	103.55	Nonazeotrope		538
11694	C₇H₈	Toluene	110.7	Nonazeotrope	v-l	200,676
		"	110.7	Nonazeotrope		563,812c
11695	C₇H₁₄	Methylcyclohexane	100.80	Nonazeotrope	v-l	805

No.	Formula	B-Component Name	B.P., °C	Azeotropic Data B.P., °C	Wt.%A		Ref.
A =	C₆H₁₂	Cyclohexane *(continued)*	**80.75**				
11696	C₇H₁₆	2,4-Dimethylpentane	80.5	80.2	48.6		630
11697	C₇H₁₆	Heptane	98.4			v-l	163,677
11698	C₇H₁₆	2,2,3-Trimethylbutane	80.8	80.0	46.6		630
		" 744 mm.	80.1	79.45	47.8	v-l	365
		"	80.75	Nonazeotrope			575
11699	C₈H₁₈	2,5-Dimethylhexane	109.4	Nonazeotrope			575
11699a	C₈H₁₈	2,2,4-Trimethylpentane		Nonazeotrope		v-l	43e
A =	C₆H₁₂	**1-Hexene**	**63.5**				
11700	C₆H₁₄	Hexane	68.74	Nonazeotrope		v-l	362c,933
11700a	C₆H₁₅N	Diisopropylamine		Nonazeotrope		v-l	406c
11700b	C₆H₁₅N	Dipropylamine		Nonazeotrope		v-l	406c
11700c	C₆H₁₅N	Hexylamine		Nonazeotrope		v-l	406c
11700d	C₆H₁₅N	Triethylamine		Nonazeotrope		v-l	406c
A =	C₆H₁₂	**Methylcyclopentane**	**71.95**				
11701	C₆H₁₄	Hexane	68.8	< 67.9	>25		418,561
		" 60°C				v-l	55
		" 200-760 mm.	68.95	Nonazeotrope		v-l	230,677
11702	C₆H₁₄O	Isopropyl ether	68.3	< 68.0	<20		558
11703	C₆H₁₅N	Triethylamine	89.35	Nonazeotrope			551
11704	C₇H₈	Toluene	110.7	Nonazeotrope		v-l	676
11704a	C₆H₁₂O₂	Butyl acetate					
		725 mm.		Nonazeotrope		v-l	684c
A =	C₆H₁₂O	**Cyclohexanol**	**160.8**				
11705	C₆H₁₂O₃	Isopropyl lactate	166.8	< 160.7			575
11706	C₆H₁₂O₃	Paraldehyde	124.35	Nonazeotrope			575
11707	C₆H₁₂O₃	Propyl lactate	171.7	Nonazeotrope			575
11708	C₆H₁₃Br	1-Bromohexane	156.5	< 153.7	<34		575
11709	C₆H₁₃ClO₂	Chloroacetal	156.8	155.6	15		563
11710	C₆H₁₃N	Cyclohexylamine	134			v-l	705,718c
11711	C₆H₁₄O	Hexyl alcohol	157.95	Nonazeotrope			538
11712	C₆H₁₄O₂	2-Butoxyethanol	171.15	Nonazeotrope			575
11713	C₇H₆O	Benzaldehyde	179.2	Nonazeotrope			575
11714	C₇H₇Br	*m*-Bromotoluene	184.3	Nonazeotrope			527
11715	C₇H₇Br	*o*-Bromotoluene	181.45	160.6?	~ 98		530
11716	C₇H₇Br	*p*-Bromotoluene	185.0	Nonazeotrope			575
11717	C₇H₇Cl	α-Chlorotoluene	179.35	Nonazeotrope			563
11718	C₇H₇Cl	*o*-Chlorotoluene	159.3	155.5	38		573
11719	C₇H₇Cl	*p*-Chlorotoluene	162.4	156.5	55		531
11720	C₇H₈	Toluene	110.75	Nonazeotrope			541
11721	C₇H₈O	Anisole	153.85	152.45	30		529
11722	C₇H₈O	*o*-Cresol	191.1	Nonazeotrope			542
11723	C₇H₁₂O₂	Cyclohexyl formate,					
		50 mm.		79.4	50		413
11724	C₇H₁₄	Methylcyclohexane	101.1	Nonazeotrope			537
11725	C₇H₁₄O₂	Methyl caproate	149.8	Nonazeotrope			575
11726	C₇H₁₄O₃	1,3-Butanediol methyl ether					
		acetate	144.6	Nonazeotrope			575
11727	C₇H₁₆	Heptane	98.45	Nonazeotrope			541
11728	C₈H₈	Styrene	145.8	144			537

TABLE I.   *Binary Systems*   341

No.	Formula	B-Component Name	B.P., °C	B.P., °C	Wt.%A	Ref.
A =	C₆H₁₂O	**Cyclohexanol** *(continued)*	**160.8**			
11729	C₈H₁₀	*m*-Xylene	139.0	138.9	5	563
11730	C₈H₁₀	*o*-Xylene	143.6	143.0	14	563
11731	C₈H₁₀	*p*-Xylene	138.2	Nonazeotrope		541
11732	C₈H₁₀O	Benzyl methyl ether	167.8	159.0	62	556,878
11733	C₈H₁₀O	Phenetole	170.35	159.2	~ 72	529
11734	C₈H₁₀O	*p*-Methylanisole	177.05	160.5	92	564
11735	C₈H₁₁N	Dimethylaniline	194.05	Nonazeotrope		551
11736	C₈H₁₄O	Cyclohexyl vinyl ether, 45 mm.		71-80	21	1008
11737	C₈H₁₄O	Methylheptenone	173.2	Nonazeotrope		552
11738	C₈H₁₆	1,3-Dimethylcyclohexane	120.7	Nonazeotrope		575
11739	C₈H₁₆O	2-Octanone	172.85	Nonazeotrope		552
11740	C₈H₁₆O₂	Butyl butyrate	166.4	< 160.5		575
11741	C₈H₁₆O₂	Isoamyl propionate	~ 160.3	157.7	~ 63	**563**
11742	C₈H₁₆O₂	Isobutyl butyrate	156.8	156	~ 20	537
11743	C₈H₁₆O₂	Isobutyl isobutyrate	147.3	Nonazeotrope		537
11744	C₈H₁₆O₂	Propyl isovalerate	155.7	155.1	17	575
11745	C₈H₁₈O	Butyl ether	142.1	Nonazeotrope		576
11746	C₈H₁₈O	Isobutyl ether	122.3	Nonazeotrope		575
11747	C₈H₁₈S	Butyl sulfide	185.0	Nonazeotrope		566
11748	C₉H₈	Indene	181.7	160	75	537
11749	C₉H₁₂	Cumene	152.8	150.0	28	567
11750	C₉H₁₂	Mesitylene	164.0	156.3	~ 50	563
11751	C₉H₁₂	Propylbenzene	158.8	153.8	40	537
11752	C₉H₁₂	Pseudocumene	169	158	~ 60	563
11753	C₉H₁₂O	Benzyl ethyl ether	185.0	Nonazeotrope		575
11754	C₉H₁₃N	*N,N*-Dimethyl-*o*-toluidine	185.3	Nonazeotrope		551
11755	C₉H₁₈O	2,6-Dimethyl-4-heptanone	168.0	Nonazeotrope		552
11756	C₉H₁₈O₂	Isobutyl isovalerate	168.7	Nonazeotrope		536
11757	C₁₀H₈	Naphthalene	218.05	Nonazeotrope		540
11758	C₁₀H₁₄	Butylbenzene	183.1	Nonazeotrope		575
11759	C₁₀H₁₄	Cymene	176.7	159.5	72	537
11760	C₁₀H₁₆	Camphene	159.5	151.9	41	528
11761	C₁₀H₁₆	*d*-Limonene	177.8	159.25	73.5	541
11762	C₁₀H₁₆	α-Phellandrene	171.5	158	65	563
11763	C₁₀H₁₆	α-Pinene	155.8	149.9	35.5	563
11764	C₁₀H₁₆	α-Terpinene	173.4	158.3	65	567
11765	C₁₀H₁₆	γ-Terpinene	183	160.3	83	575
11766	C₁₀H₁₆	Terpinene	181	159.8		563
11767	C₁₀H₁₆	Terpinolene	184.6	160.5	87	575
11768	C₁₀H₁₆	Thymene	179.7	159.8	78	573
11769	C₁₀H₁₈	*d*-Menthene	170.8	~ 157.5	~ 62	563
11770	C₁₀H₁₈O	Cineole	176.35	160.55	92	574
11771	C₁₀H₂₂	2,7-Dimethyloctane	160.2	153.0	~ 62	537
11772	C₁₀H₂₂O	Amyl ether	187.5	Nonazeotrope		556
11773	C₁₀H₂₂O	Isoamyl ether	172.6	158.8	78	574

No.	\multicolumn{3}{c}{B-Component}			Azeotropic Data		

No.	Formula	Name	B.P., °C	B.P., °C	Wt.%A	Ref.
A =	$C_6H_{12}O$	2-Hexanone	**127.2**			
11774	$C_6H_{12}O_2$	Butyl acetate	126.0	125.4	32	527
11775	$C_6H_{12}O_2$	Isoamyl formate	123.8	Nonazeotrope		552
11776	$C_7H_{14}O_2$	Propyl isobutyrate	134.0	Nonazeotrope		552
A =	$C_6H_{12}O$	3-Hexanone	**123.3**			
11777	$C_6H_{12}O_2$	Butyl acetate	126.0	123.1		552
11778	$C_6H_{12}O_2$	Ethyl butyrate	121.5	Nonazeotrope		552
11779	$C_6H_{12}O_2$	Isoamyl formate	123.8	123.0	50	552
11780	$C_6H_{12}O_2$	Isobutyl acetate	117.4	Nonazeotrope		552
11781	$C_6H_{12}O_2$	Methyl isovalerate	116.5	Nonazeotrope		552
11782	$C_6H_{12}O_2$	Propyl propionate	123.0	122.5	40	552
11783	$C_6H_{14}S$	Isopropyl sulfide	120.5	119.0	32	555
11784	$C_6H_{15}BO_3$	Ethyl borate	118.6	116.7	28	552
11785	$C_6H_{15}N$	Dipropylamine	109.2	Nonazeotrope		551
11786	C_7H_8	Toluene	110.75	Nonazeotrope		552
11787	C_7H_{16}	n-Heptane	98.4	Nonazeotrope		527
11788	C_8H_{10}	Ethylbenzene	136.15	Nonazeotrope		575
11789	C_8H_{10}	m-Xylene	139.2	Nonazeotrope		527
11790	C_8H_{16}	1,3-Dimethylcyclohexane	120.7	116.0	37	552
11791	$C_8H_{19}N$	Diisobutylamine	138.5	Nonazeotrope		575
A =	$C_6H_{12}O$	4-Methyl-2-pentanone	**116.05**			
11792	$C_6H_{12}O_2$	Ethyl butyrate	121.5	Nonazeotrope		527
11793	$C_6H_{12}O_2$	Ethyl isobutyrate	110.1	Nonazeotrope		527
11794	$C_6H_{12}O_2$	Isobutyl acetate	117.4	115.6		552
11795	$C_6H_{12}O_2$	Isopropyl propionate	110.5	Nonazeotrope		552
11796	$C_6H_{12}O_2$	Methyl isovalerate	116.5	115.6	55	527
11797	$C_6H_{14}S$	Isopropyl sulfide	120.5	114.9	72	555
11798	$C_6H_{15}N$	Dipropylamine	109.2	<105.5	<32	551
11799	C_7H_8	Toluene	110.75	110.7	3	527
11800	C_7H_{14}	Methylcyclohexane	101.15	<100.1	<20	527
		" 410 mm.		80	~14	v-l	859i
		" 760 mm.	100.8	99.9	20.8	v-l	859i
11801	C_7H_{16}	Heptane	98.4	97.5	13	552
11802	C_8H_{10}	Ethylbenzene	136.15	Nonazeotrope		527
11803	C_8H_{16}	1,3-Dimethylcyclohexane	120.7	112.0	53	552
11804	$C_8H_{16}O_2$	4-Methyl-2-pentyl acetate	146.1	Nonazeotrope		981
11805	C_8H_{18}	n-Octane	125.75	113.4	65	527
11806	$C_8H_{18}O$	Isobutyl ether	122.3	Nonazeotrope		575
11807	$C_8H_{19}N$	Diisobutylamine	138.5	Nonazeotrope		575
11808	$C_9H_{18}O$	2,6-Dimethyl-4-heptanone	169.4	Nonazeotrope		981
A =	$C_6H_{12}O$	Pinacolone	**106.2**			
11809	$C_6H_{12}O_2$	Ethyl isobutyrate	110.1	Nonazeotrope		552
11810	$C_6H_{12}O_2$	Isopropyl propionate	110.5	Nonazeotrope		552
11811	C_6H_{14}	Hexane	68.8	Nonazeotrope		552
11812	$C_6H_{14}O_2$	Acetal	103.55	Nonazeotrope		575
11813	$C_6H_{15}N$	Dipropylamine	109.2	<104.5		575

TABLE I.   *Binary Systems*   343

No.	Formula	B-Component Name	B.P., °C	B.P., °C	Azeotropic Data Wt.%A		Ref.
A =	C₆H₁₂O	**Pinacolone** *(continued)*	**106.2**				
11814	C₇H₈	Toluene	110.75	106.0	85		552
11815	C₇H₁₄	Methylcyclohexane	101.15	97.0	32		552
11816	C₇H₁₆	Heptane	98.4	95.5	28		552
11817	C₈H₁₆	1,3-Dimethylcyclohexane	120.7	104.0	75		**552**
A =	C₆H₁₂O₂	**Butyl Acetate**	**126.0**				
11818	C₆H₁₂O₂	*sec*-Butyl acetate	112.2	Nonazeotrope			981
11819	C₆H₁₂O₂	Isoamyl formate	123.8	Nonazeotrope			575
11820	C₆H₁₂O₂	Propyl propionate	123.0	Nonazeotrope			575
11821	C₆H₁₂O₃	2-Ethoxyethyl acetate	156.8	Nonazeotrope			575
11822	C₆H₁₂O₃	Paraldehyde	124.35	124.25	9		**527**
11823	C₆H₁₄O₂	2-Butoxyethanol	171.1	Nonazeotrope			981
11824	C₆H₁₄S	Propyl sulfide	141.5	Nonazeotrope			575
11825	C₇H₈	Toluene	110.75	Nonazeotrope			527
11826	C₇H₁₄O₂	Propyl isobutyrate	134.0	Nonazeotrope			575
11826a	C₇H₁₆	Heptane 74.7°C.		Nonazeotrope		v-l	848e
		" 100°C.		Nonazeotrope		v-l	848e
11827	C₇H₁₆O₂	Dipropoxymethane	137.2	Nonazeotrope			557
11828	C₈H₁₀	Ethylbenzene	136.1	Nonazeotrope			**527**
11829	C₈H₁₀	*m*-Xylene	139.0	Nonazeotrope			527
11830	C₈H₁₀	*p*-Xylene	138.45	Nonazeotrope			527
11831	C₈H₁₆	1,3-Dimethylcyclohexane	120.7	<118.0	<37		575
11832	C₈H₁₈	Octane	125.8	119	52		538
11833	C₈H₁₈O	Butyl ether	142.1	125.9	95		981
11834	C₈H₁₈O	Isobutyl ether	122.3	Nonazeotrope			557
11835	C₁₀H₁₆	Camphene	159.6	Nonazeotrope			575
11836	C₁₀H₁₆	Nopinene	163.8	Nonazeotrope			527
11837	C₁₀H₁₆	α-Pinene	155.8	Nonazeotrope			575
11837a	C₆H₁₂O₂	4-Methyl-4-hydroxy-tet-rahydropyran 3 mm.		Azeotropic			581c
11837b	C₆H₁₄O₃	3-Methyl-1,3,5-pentanetriol 3 mm.		Nonazeotrope			581c
A =	C₆H₁₂O₂	**Caproic Acid**	**205.3**				
11838	C₇H₆Cl₂	α,α-Dichlorotoluene	205.2	199.0	36		542
11839	C₇H₇Br	α-Bromotoluene	198.5	~ 196.5	77		563
11840	C₇H₇Br	*o*-Bromotoluene	181.5	180.8	6		541
11841	C₇H₇Br	*p*-Bromotoluene	185.0	184.0	8		541
11842	C₇H₇BrO	*o*-Bromoanisole	217.7	Nonazeotrope			575
11843	C₇H₇Cl	α-Chlorotoluene	179.3	179.0	~ 3		541
11844	C₇H₇Cl	*m*-Chlorotoluene	162.3	Nonazeotrope			564
11845	C₇H₇Cl	*o*-Chlorotoluene	159.2	Nonazeotrope			564
11846	C₇H₇Cl	*p*-Chlorotoluene	162.4	Nonazeotrope			564
11847	C₇H₇I	*p*-Iodotoluene	214.5	202.2	50		562
11848	C₇H₇NO₂	*o*-Nitrotoluene	221.85	~ 205.0	~ 96		542
11849	C₇H₇NO₂	*p*-Nitrotoluene	238.9	Nonazeotrope			554
11850	C₇H₈O	*m*-Cresol	202.2	201.9	13		564
11851	C₇H₈O	*o*-Cresol	190.8	Nonazeotrope			563

No.	Formula	Name	B.P., °C	B.P., °C	Wt.%A		Ref.
		B-Component		**Azeotropic Data**			
A =	$C_6H_{12}O_2$	**Caproic Acid** *(continued)*	**205.3**				
11852	C_7H_8O	*p*-Cresol	201.7	201.5	11		564
		"	201.8	Nonazeotrope			563
11853	$C_7H_8O_2$	Guaiacol	205.05	200.8	42		556
11854	$C_7H_{12}O_4$	Ethyl malonate	199.35	198.5	12		562
11855	$C_7H_{13}ClO_2$	Isoamyl chloroacetate	190.5	Nonazeotrope			575
11856	C_8H_8O	Acetophenone	202.0	200.5	32		552
11857	$C_8H_8O_2$	Benzyl formate	203.0	<202.2	20		575
11858	$C_8H_8O_2$	Phenyl acetate	195.7	Nonazeotrope			541
11859	$C_8H_{10}O$	*p*-Ethylphenol	218.8	Nonazeotrope			575
11860	$C_8H_{10}O$	3,4-Xylenol	226.8	Nonazeotrope			564
11861	$C_8H_{10}O_2$	*m*-Dimethoxybenzene	216.2	Nonazeotrope			543
11862	$C_8H_{10}O_2$	*o*-Ethoxyphenol	216.5	Nonazeotrope			575
11863	$C_8H_{10}O_2$	Veratrole	206.5	~ 202.5	~ 42		537
11864	$C_8H_{14}O_4$	Propyl oxalate	214	Nonazeotrope			575
11865	$C_8H_{16}O_2$	Octanoic acid, 20-100 mm.		Nonazeotrope		v-l	822
11866	$C_8H_{16}O_4$	2-(2-Ethoxyethoxy)ethyl acetate	218.5	Nonazeotrope			575
11867	$C_8H_{17}NO$	*N,N*-Dimethylhexanamide, 100 mm.		Max. b.p.			832
11868	C_9H_8	Indene	182.6	Nonazeotrope			575
11869	$C_9H_{10}O$	Propiophenone	217.7	Nonazeotrope			552
11870	$C_9H_{10}O_2$	Benzyl acetate	215.0	Nonazeotrope			575
11871	$C_9H_{10}O_2$	Ethyl benzoate	212.5	Nonazeotrope			575
11872	C_9H_{12}	Mesitylene	164.6	Nonazeotrope			564
11873	C_9H_{12}	Pseudocumene	168.2	Nonazeotrope			543
11874	$C_9H_{12}O$	Benzyl ethyl ether	185.0	Nonazeotrope			575
11875	$C_9H_{12}O$	Phenyl propyl ether	190.5	Nonazeotrope			575
11876	$C_9H_{14}O$	Phorone	197.8	Nonazeotrope			552
11877	$C_9H_{18}O_2$	Methyl caprylate	192.9	Nonazeotrope			575
11878	$C_9H_{18}O_3$	Isobutyl carbonate	190.3	Nonazeotrope			575
11879	$C_{10}H_7Cl$	1-Chloronaphthalene	262.7	Nonazeotrope			564
11880	$C_{10}H_8$	Naphthalene	218.05	203.75	71		564
11881	$C_{10}H_{14}$	Cymene	176.7	Nonazeotrope			543
11882	$C_{10}H_{16}$	*d*-Limonene	177.8	177.0	~ 5		541
11883	$C_{10}H_{16}$	Nopinene	163.8	Nonazeotrope			575
11884	$C_{10}H_{16}$	Terpinolene	185	Azeotrope doubtful			563
11885	$C_{10}H_{16}$	Thymene	179.7	179.0	~ 3		541
11886	$C_{10}H_{16}O$	Camphor	209.1	204.0			552
11887	$C_{10}H_{17}Cl$	Bornyl chloride	207.5	200.0	38		562
11888	$C_{10}H_{18}O$	Citronellal	207.8	~ 203.5			541
11889	$C_{10}H_{20}O_2$	Isoamyl isovalerate	192.7	Nonazeotrope			575
11890	$C_{10}H_{20}O_2$	Methyl pelargonate	213.8	<204.5	<95		566
11891	$C_{10}H_{22}S$	Isoamyl sulfide	214.8	Nonazeotrope			575
11892	$C_{11}H_{10}$	1-Methylnaphthalene	244.6	Nonazeotrope			575
11893	$C_{11}H_{10}$	2-Methylnaphthalene	241.15	Nonazeotrope			527

TABLE I.   *Binary Systems*   345

No.	Formula	B-Component Name	B.P., °C	Azeotropic Data B.P., °C	Wt.%A	Ref.
A =	**C₆H₁₂O₂**	**Caproic Acid** *(continued)*	**205.3**			
11894	C₁₁H₂₀O	Isobornyl methyl ether	192.4	Nonazeotrope		575
11895	C₁₂H₁₈	1,3,5-Triethylbenzene	215.5	202.0	63	564
11896	C₁₂H₂₂O	Ethyl isobornyl ether	203.5	<201.5	>30	563
A =	**C₆H₁₂O₂**	**4,4-Dimethyl-1,3-dioxane**	**133.4**			
11896a	C₇H₁₂O₂	4-Methyl-4-vinyl-1,3-dioxane		Nonazeotrope		581c
A =	**C₆H₁₂O₂**	**Ethyl Butyrate**	**121.5**			
11897	C₆H₁₂O₂	Isoamyl formate	123.8	Nonazeotrope		575
11898	C₆H₁₂O₂	Isobutyl acetate	117.4	Nonazeotrope		575
11899	C₆H₁₂O₂	Methyl isovalerate	116.5	Nonazeotrope		575
11900	C₆H₁₂O₂	Propyl propionate	123.0	Nonazeotrope		575
11901	C₆H₁₂O₃	Paraldehyde	124	Nonazeotrope		557
11902	C₆H₁₄S	Isopropyl sulfide	120.5	<120.0	<42	566
11903	C₆H₁₅BO₃	Ethyl borate	118.6	117.6	35	549
11904	C₇H₈	Toluene	110.7	Nonazeotrope		563
11905	C₇H₁₄	Methylcyclohexane	101.1	Nonazeotrope		546
11906	C₇H₁₄O₂	Isoamyl acetate	137.5	Nonazeotrope		388
11907	C₇H₁₆	Heptane	98.5	Nonazeotrope		527
11908	C₈H₁₀	Ethylbenzene	136.15	Nonazeotrope		575
11909	C₈H₁₆	1,3-Dimethylcyclohexane	120.7	116.7	<50	562
11910	C₈H₁₈	Octane	125.8	Nonazeotrope		563
		"	125.8	118.0	>60	546
11911	C₈H₁₈O	Isobutyl ether	122.3	120.5	20	557
A =	**C₆H₁₂O₂**	**Ethyl Isobutyrate**	**110.1**			
11912	C₆H₁₂O₂	Methyl isovalerate	116.5	Nonazeotrope		575
11913	C₆H₁₄O₂	Acetal	103.55	Nonazeotrope		557
11914	C₆H₁₄O₂	Ethoxypropoxymethane	113.7	Nonazeotrope		557
11915	C₆H₁₄S	Isoproyl sulfide	120.5	Nonazeotrope		566
11916	C₇H₈	Toluene	110.75	109.8		573
11917	C₇H₁₄	Methylcyclohexane	101.1	100.1	<20	546
11918	C₇H₁₆	Heptane	98.5	97.0	17	527
11919	C₈H₁₆	1,3-Dimethylcyclohexane	120.7	<109.5	<88	575
11920	C₈H₁₈	Octane	125.75	<109.8	<96	575
11921	C₈H₁₈O	Isobutyl ether	122.2	Nonazeotrope		557
A =	**C₆H₁₂O₂**	**4-Hydroxy-methyl-2-pentanone**	**166**			
11922	C₈H₉Cl	*o,m,p*-Chloroethylbenzene, 10 mm.	67.5	59.0	58	51
11923	C₉H₁₂	*x*-Ethyltoluene, 20 mm.	< 80	25		1040
A =	**C₆H₁₂O₂**	**Isoamyl Formate**	**123.8**			
11924	C₆H₁₂O₂	Propyl propionate	122.5	Nonazeotrope		545
11925	C₆H₁₂O₃	Paraldehyde	124.1	123.0	56	557
11926	C₇H₈	Toluene	110.7	Nonazeotrope		563
11927	C₈H₁₀	Ethylbenzene	136.15	Nonazeotrope		573
11928	C₈H₁₈	Octane	125.8	<116.5	~ 55	563
11929	C₈H₁₈O	Isobutyl ether	122.3	121.5	65	557

No.	Formula	B-Component Name	B.P., °C	Azeotropic Data B.P., °C	Wt.%A	Ref.
A =	$C_8H_{16}O_2$	**Isobutyl Acetate**	**117.2**			
11930	$C_6H_{12}O_2$	Methyl isovalerate	116.5	Nonazeotrope		549
11931	$C_6H_{12}O_3$	Paraldehyde	124	Nonazeotrope		557
11932	$C_6H_{14}O_2$	Acetal	103.55	Nonazeotrope		557
11934	$C_6H_{14}S$	Isopropyl sulfide	120.5	115.2	57	566
11934	$C_6H_{15}BO_3$	Ethyl borate	118.6	117.0	63	549
11935	C_7H_8	Toluene	110.6	Nonazeotrope		572
11936	C_7H_{16}	Heptane	98.5	Nonazeotrope		546
11937	C_8H_{10}	Ethylbenzene	136.15	Nonazeotrope		575
11938	C_8H_{16}	1,3-Dimethylcyclohexane	120.7	<114.0	<62	562
11939	C_8H_{18}	Octane	125.8	114.5	>70	546
11940	$C_8H_{18}O$	Isobutyl ether	122.3	Nonazeotrope		557
A =	$C_6H_{12}O_2$	**Isocaproic Acid**	**199.5**			
11941	C_7H_6O	Benzaldehyde	179.2	Nonazeotrope		575
11942	$C_7H_6O_2$	Salicylaldehyde	196.7	<196.4		575
11943	C_7H_7Br	α-Bromotoluene	198.5	193.0	32	562
11944	C_7H_7Br	m-Bromotoluene	184.3	183.0	10	562
11945	C_7H_7Br	o-Bromotoluene	181.5	180.5	9	575
11946	C_7H_7Br	p-Bromotoluene	185.0	183.0	12	575
11947	C_7H_7Cl	α-Chlorotoluene	179.3	178.0	8	575
11948	C_7H_7Cl	o-Chlorotoluene	159.2	Nonazeotrope		575
11949	C_7H_7Cl	p-Chlorotoluene	162.4	Nonazeotrope		575
11950	C_7H_8O	o-Cresol	191.1	Nonazeotrope		575
11951	C_7H_8O	p-Cresol	201.7	199.1	80?	575
11952	$C_7H_8O_2$	Guaiacol	205.05	<198.5	>80	575
11953	$C_7H_{12}O_4$	Ethyl malonate	199.35	196.5	42	562
11954	C_8H_8O	Acetophenone	202.0	<199.2		575
11955	$C_8H_8O_2$	Benzyl formate	203.0	198.8	62	575
11956	$C_8H_{10}O$	Benzyl methyl ether	167.8	Nonazeotrope		575
11957	$C_8H_{10}O$	p-Methylanisole	177.05	Nonazeotrope		575
11958	$C_8H_{10}O$	Phenetole	170.45	Nonazeotrope		575
11959	C_9H_8	Indene	182.6	Nonazeotrope		575
11960	C_9H_{12}	Mesitylene	164.6	Nonazeotrope		575
11961	$C_9H_{12}O$	Benzyl ethyl ether	185.0	Nonazeotrope		575
11962	$C_9H_{12}O$	Phenyl propyl ether	190.5	190.0	10	575
11963	$C_{10}H_8$	Naphthalene	218.0	199.0	75	562
11964	$C_{10}H_{14}$	Cymene	176.7	<176.2	> 3	575
11965	$C_{10}H_{16}$	Dipentene	177.7	176.5	10	562
11966	$C_{10}H_{16}$	Limonene	177.7	176.5	10	575
11967	$C_{10}H_{16}O$	Camphor	209.1	Nonazeotrope		575
11968	$C_{10}H_{18}O$	Cineole	176.35	Nonazeotrope		575
11969	$C_{10}H_{20}O_2$	Isoamyl isovalerate	192.7	Nonazeotrope		575
11970	$C_{10}H_{22}O$	Amyl ether	187.5	<186.5	> 8	575
11971	$C_{10}H_{22}O$	Isoamyl ether	172.6	Nonazeotrope		564
11972	$C_{10}H_{22}S$	Isoamyl sulfide	214.8	Nonazeotrope		566

TABLE I.   *Binary Systems*   347

No.	Formula	B-Component Name	B.P., °C	Azeotropic Data B.P., °C	Wt.%A	Ref.
A =	$C_6H_{12}O_2$	**Methyl Isovalerate**	116.5			
11973	$C_6H_{12}O_3$	Paraldehyde	124.35	Nonazeotrope		557
11974	$C_6H_{14}O_2$	Acetal	103.55	Nonazeotrope		557
11975	C_7H_8	Toluene	110.75	Nonazeotrope		573
11976	C_7H_{14}	Methylcyclohexane	101.15	Nonazeotrope		575
11977	C_7H_{16}	Heptane	98.5	Nonazeotrope		527
11978	C_8H_{16}	1,3-Dimethylcyclohexane	120.7	<115.0	<75	575
11979	C_8H_{18}	Octane	125.75	<115.5	<88	575
11980	$C_8H_{18}O$	Isobutyl ether	122	Nonazeotrope		557
A =	$C_6H_{12}O_2$	**Propyl Propionate**	122.5			
11981	C_7H_8	Toluene	110.75	Nonazeotrope		538
11982	C_8H_{10}	Ethylbenzene	136.15	Nonazeotrope		538
11983	C_8H_{18}	Octane	125.8	118.2	60	538
A =	$C_6H_{12}SO_2$	**2,4-Dimethylsulfolane**				
11984	$C_{15}H_{18}$	Amyl naphthalene,				
		20 mm.		151	75	667
11985	$C_{16}H_{34}$	Hexadecane, 20 mm.		142	75	667
A =	$C_6H_{12}O_3$	**2-Ethoxyethyl Acetate**	156.8			
11986	$C_6H_{12}O_3$	Isopropyl lactate	166.8	Nonazeotrope		575
11987	$C_6H_{13}Br$	1-Bromohexane	156.5	<155.0	<49	575
11988	$C_6H_{14}O$	Hexyl alcohol	157.85	<156.0	<63	575
11989	$C_6H_{14}O_2$	2-Butoxyethanol	171.15	Nonazeotrope		556
11990	C_7H_6O	Benzaldehyde	179.2	Nonazeotrope		526
11991	C_7H_7Cl	o-Chlorotoluene	159.2	156.6	90	526
11992	C_7H_7Cl	p-Chlorotoluene	162.4	Nonazeotrope		575
11993	C_7H_8	Toluene	110.6	Nonazeotrope		981
11994	C_7H_8O	Anisole	153.85	Nonazeotrope		556
11995	C_7H_8O	m-Cresol	202.2	Nonazeotrope		526
11996	C_7H_8O	o-Cresol	191.1	191.5	10	556
11997	C_7H_8O	p-Cresol	201.7	Nonazeotrope		526
11998	$C_7H_{14}O$	2-Heptanone	143.55	Nonazeotrope		575
11999	$C_7H_{14}O$	2-Methylcyclohexanol	168.5	Nonazeotrope		575
12000	$C_7H_{14}O$	5-Methyl-2-hexanone	144.2	Nonazeotrope		575
12001	$C_7H_{14}O_2$	Ethyl isovalerate	134.7	Nonazeotrope		575
12002	$C_7H_{14}O_2$	Isoamyl acetate	142.1	Nonazeotrope		575
12003	$C_7H_{14}O_2$	Propyl butyrate	143.7	Nonazeotrope		575
12004	$C_7H_{16}O$	Heptyl alcohol	176.15	Nonazeotrope		575
12005	$C_7H_{16}O_3$	Ethyl orthoformate	145.75	Nonazeotrope		556
12006	C_8H_8	Styrene	145.8	Nonazeotrope		575
12007	C_8H_{10}	Ethylbenzene	136.15	Nonazeotrope		526
12008	C_8H_{10}	m-Xylene	139.2	Nonazeotrope		527
12009	C_8H_{10}	o-Xylene	144.3	Nonazeotrope		526
12010	C_8H_{10}	p-Xylene	138.45	Nonazeotrope		526
12011	$C_8H_{10}O$	Benzyl methyl ether	167.8	Nonazeotrope		575
12012	$C_8H_{10}O$	p-Methylanisole	177.05	Nonazeotrope		575
12013	$C_8H_{10}O$	Phenetole	170.45	Nonazeotrope		526

No.		B-Component			Azeotropic Data		
	Formula	Name	B.P., °C	B.P., °C	Wt.%A		Ref.
A =	$C_6H_{12}O_3$	**2-Ethoxyethyl Acetate**	**156.8**				
		(continued)					
12014	$C_8H_{16}O$	2-Octanone	172.85	Nonazeotrope			575
12015	$C_8H_{16}O_2$	Ethyl caproate	167.7	Nonazeotrope			575
12016	$C_8H_{16}O_2$	Hexyl acetate	171.5	Nonazeotrope			575
12017	$C_8H_{16}O_2$	Isoamyl propionate	160.7	156.5	90		526
12018	$C_8H_{16}O_2$	Isobutyl butyrate	156.9	156.0	52		526
12019	$C_8H_{16}O_2$	Isobutyl isobutyrate	148.6	Nonazeotrope			575
12020	$C_8H_{16}O_2$	Propyl isovalerate	155.7	<155.0	<35		575
12021	$C_8H_{18}O$	Butyl ether	142.4	141.7	88		526
12022	$C_8H_{18}O$	sec-Octyl alcohol	180.4	Nonazeotrope			526
12023	C_9H_{12}	Cumene	152.8	152.0	15		575
12024	C_9H_{12}	Mesitylene	164.6	Nonazeotrope			575
12025	C_9H_{12}	Propylbenzene	159.3	<156.0	>70		575
12026	$C_9H_{18}O$	2,6-Dimethyl-4-heptanone	168.0	Nonazeotrope			575
12027	$C_9H_{18}O_2$	Isoamyl butyrate	181.05	Nonazeotrope			575
12028	$C_9H_{18}O_2$	Isobutyl isovalerate	171.3	Nonazeotrope			526
12029	$C_{10}H_{14}$	Butylbenzene	183.1	Nonazeotrope			526
12030	$C_{10}H_{14}$	Cymene	176.7	Nonazeotrope			575
12031	$C_{10}H_{16}$	Camphene	159.6	153.2	68		526
12032	$C_{10}H_{16}$	Nopinene	163.8	154.0	80		562
12033	$C_{10}H_{16}$	α-Pinene	155.8	151.0	50		526
12034	$C_{10}H_{16}$	α-Terpinene	173.4	<156.5	<93		575
12035	$C_{10}H_{16}$	Terpinolene	184.6	Nonazeotrope			575
12036	$C_{10}H_{18}O$	Cineole	176.35	Nonazeotrope			556
12037	$C_{10}H_{22}$	2,7-Dimethyloctane	160.1	153.0	75		562
12038	$C_{10}H_{22}O$	Isoamyl ether	173.2	156.45	94		527,556
A =	$C_6H_{12}O_3$	**Isopropyl Lactate**	**166.8**				
12039	$C_6H_{14}O$	Hexyl alcohol	157.85	Nonazeotrope			575
12040	C_7H_8O	o-Cresol	191.1	Nonazeotrope			575
12041	C_7H_8O	p-Cresol	201.7	Nonazeotrope			575
12042	$C_7H_{14}O$	2-Methylcyclohexanol	168.5	165.5	67		575
12043	$C_7H_{16}O$	Heptyl alcohol	176.15	Nonazeotrope			575
12044	C_9H_{12}	Mesitylene	164.6	159.5	60		567
12045	$C_{10}H_{16}$	Camphene	159.6	154.2	30		567
12046	$C_{10}H_{16}$	Nopinene	163.8	157.5	38		567
12047	$C_{10}H_{16}$	α-Pinene	155.8	152.5	22		575
A =	$C_6H_{12}O_3$	**Paraldehyde**	**124.35**				
12048	$C_6H_{14}O$	Hexyl alcohol	157.85	Nonazeotrope			575
12049	$C_6H_{15}BO_3$	Ethyl borate	118.6	Nonazeotrope			548
12050	C_7H_8	Toluene	110.7	Nonazeotrope			563
12051	$C_7H_{14}O_2$	Ethyl isovalerate	134.7	Nonazeotrope			557
12052	$C_7H_{14}O_2$	Isobutyl propionate	137.5	Nonazeotrope			557
12053	$C_7H_{14}O_2$	Propyl isobutyrate	134.0	Nonazeotrope			557
12054	C_8H_{10}	Ethylbenzene	136.15	Nonazeotrope			563
12055	C_8H_{10}	m-Xylene	139.0	Nonazeotrope			315,563
12056	C_8H_{10}	p-Xylene	138.4	Nonazeotrope			315,546

TABLE I. *Binary Systems* 349

No.	Formula	B-Component Name	B.P., °C	Azeotropic Data B.P., °C	Wt.%A	Ref.
A =	**$C_6H_{12}O_3$**	**Propyl Lactate**	**171.7**			
12057	$C_6H_{13}ClO_2$	Chloroacetal	157.4	Nonazeotrope		575
12058	$C_6H_{14}O$	Hexyl alcohol	157.85	Nonazeotrope		575
12059	$C_6H_{14}O_2$	2-Butoxyethanol	171.25	> 170.75	>55	575
12060	$C_6H_{14}O_2$	Pinacol	171.5	~ 168	~ 37	563
12061	C_7H_6O	**Benzaldehyde**	179.2	Nonazeotrope		563
12062	C_7H_7Br	o-Bromotoluene	181.5	171.0	~ 15	575
12063	C_7H_7Cl	α-Chlorotoluene	179.35	171.2	~ 78	563
12064	C_7H_7Cl	o-Chlorotoluene	159.2	< 159.0		575
12065	C_7H_7Cl	p-Chlorotoluene	162.4	160.5	18	567
12066	C_7H_8O	Anisole	153.85	Nonazeotrope		556
12067	C_7H_8O	m-Cresol	202.2	Nonazeotrope		575
12068	C_7H_8O	o-Cresol	191.1	Nonazeotrope		544
12069	C_7H_8O	p-Cresol	201.7	Nonazeotrope		544
12070	$C_7H_{14}O$	2-Methylcyclohexanol	168.5	< 167.8	<34	575
12071	$C_7H_{16}O$	Heptyl alcohol	176.15	< 171.55	<90	575
12072	C_8H_8	Styrene	145.8	Nonazeotrope		575
12073	$C_8H_{10}O$	Benzyl methyl ether	167.8	165.5	25	575
12074	$C_8H_{10}O$	p-Methylanisole	177.05	< 171.0	>82	575
12075	$C_8H_{10}O$	Phenetole	171.5	167.1	50	556
12076	$C_8H_{16}O$	2-Octanone	172.85	< 171.4	<75	552
12077	$C_8H_{18}O$	sec-Octyl alcohol	179.8	Nonazeotrope		575
12078	$C_8H_{18}S$	Isobutyl sulfide	172.0	169.0	48	566
12079	C_9H_{12}	Cumene	152.8	Nonazeotrope		575
12080	C_9H_{12}	Mesitylene	164.6	160.5	28	538
12081	C_9H_{12}	Pseudocumene	108.2	103.5	38	567
12082	$C_9H_{12}O$	Benzyl ethyl ether	185.0	Nonazeotrope		575
12083	$C_9H_{18}O_2$	Isoamyl isobutyrate	169.8	167.5	40	575
12084	$C_9H_{18}O_2$	Isobutyl isovalerate	171.2	< 169.0	<52	527
12085	$C_{10}H_{14}$	Cymene	176.7	~ 167.0	60	538
12086	$C_{10}H_{16}$	Camphene	159.6	~ 156.2	17	538
12087	$C_{10}H_{16}$	d-Limonene	177.8	166.35	63	563
12088	$C_{10}H_{16}$	Nopinene	163.8	159.0	33	567
12089	$C_{10}H_{16}$	α-Phellandrene	171.5	~ 162.5	~ 50	563
12090	$C_{10}H_{16}$	α-Pinene	155.8	< 154.5		575
12091	$C_{10}H_{16}$	α-Terpinene	173.3	~ 164.0	50	548
12092	$C_{10}H_{18}O$	Cineole	176.3	~ 169	~ 73	563
12093	$C_{10}H_{22}O$	Isoamyl ether	173.2	167.5	53	556
A =	**$C_6H_{13}Br$**	**1-Bromohexane**	**156.5**			
12094	C_6H_{14}	Hexyl alcohol	157.85	150.5	60	575
12095	$C_6H_{14}O_2$	2-Butoxyethanol	171.15	< 156.0		575
12096	C_7H_8O	o-Cresol	191.1	Nonazeotrope		575
12097	$C_7H_{14}O_2$	Isoamyl acetate	142.1	Nonazeotrope		575
12098	$C_7H_{16}O$	Heptyl alcohol	176.15	Nonazeotrope		575
12099	$C_8H_{10}O$	Benzyl methyl ether	167.8	Nonazeotrope		559
12100	$C_8H_{16}O_2$	Isobutyl isobutyrate	148.6	Nonazeotrope		575
12101	$C_8H_{16}O_2$	Propyl isovalerate	155.7	< 155.2	>28	575

	B-Component			Azeotropic Data		
No.	Formula	Name	B.P., °C	B.P., °C	Wt.%A	Ref.
A =	**C₆H₁₃Br**	**1-Bromohexane** *(continued)*	**156.5**			
12102	C₈H₁₈O	Butyl ether	142.4	Nonazeotrope		559
12103	C₈H₁₈O	*sec*-Octyl alcohol	180.4	Nonazeotrope		575
12104	C₉H₁₈O₂	Isobutyl isovalerate	171.2	Nonazeotrope		575
A =	**C₆H₁₃ClO₂**	**Chloroacetal**	**157.4**			
12105	C₆H₁₄O	Hexyl alcohol	157.85	< 154.5	<58	575
12106	C₆H₁₄O₂	Pinacol	171.5	155.5	<90	563
12107	C₇H₈O	Anisole	153.85	Nonazeotrope		548
12108	C₇H₁₄O	4-Heptanone	143.55	Nonazeotrope		575
12109	C₈H₈	Styrene	145.8	Nonazeotrope		548
12110	C₈H₁₀	*m*-Xylene	139.2	Nonazeotrope		575
12111	C₈H₁₀	*o*-Xylene	143.6	Nonazeotrope		548
12112	C₈H₁₀O	Phenetole	170.45	Nonazeotrope		575
12113	C₈H₁₆O	2-Octanone	172.85	Nonazeotrope		552
12114	C₈H₁₆O₂	Isoamyl propionate	160.3	Nonazeotrope		531
12115	C₈H₁₆O₂	Propyl isovalerate	155.8	154.7	~ 43	563
12116	C₉H₁₂	Cumene	152.8	< 152.0	<10	575
12117	C₉H₁₂	Propylbenzene	159.2	< 156.0	<75	548
12118	C₉H₁₈O	2,6-Dimethyl-4-heptanone	168.0	Nonazeotrope		575
12119	C₉H₁₈O₂	Isobutyl isovalerate	171.35	Nonazeotrope		545
12120	C₁₀H₁₄	Cymene	176.7	Nonazeotrope		531
12121	C₁₀H₁₆	Camphene	159.5	~ 155.2	56	556
12122	C₁₀H₁₆	*d*-Limonene	177.8	Nonazeotrope		531
12123	C₁₀H₁₆	Nopinene	163.8	156.2	23	556
12124	C₁₀H₁₆	α-Pinene	155.8	153.0	43	556
12125	C₁₀H₁₆	α-Terpinene	173.3	Nonazeotrope		548
12126	C₁₀H₂₂	2,7-Dimethyloctane	160.2	155.5	35	556
12127	C₁₀H₂₂O	Isoamyl ether	173.4	Nonazeotrope		548
A =	**C₆H₁₄**	**Hexane**	**68.8**			
12127a	C₆H₁₄O	Hexyl alcohol	157	Nonazeotrope	v-l	779g
12128	C₆H₁₄O	Isopropyl ether	68.3	67.5	47	558
12128a	C₆H₁₅N	Diisopropylamine 60°C.		Nonazeotrope	v-l	406c
12128b	C₆H₁₅N	Dipropylamine 60°C.		Nonazeotrope	v-l	406c
12128c	C₆H₁₅N	Hexylamine 60°C.		Nonazeotrope	v-l	406c
12129	C₆H₁₅N	Triethylamine	89.35	Nonazeotrope		551
	"	60°C.		Nonazeotrope	v-l	406c
12130	C₇F₁₆	Perfluoroheptane, crit. region		Azeotropic	v-l	430,431
12131	C₇H₈	Toluene	110.7	Nonazeotrope		163
	"	150-760 mm.		Nonazeotrope	v-l	675
12132	C₇H₁₄	Methylcyclohexane	101.15	Nonazeotrope	v-l	163,677
12133	C₇H₁₆	Heptane	98.45	Vapor pressure data		899
12134	C₈H₁₈	Octane	125.8	Nonazeotrope (b.p. curve)		1045
A =	**C₆H₁₄O**	**2-Ethylbutanol**	**147.0**			
12135	C₈H₉Cl	*o,m,p*-Chloroethylbenzene, 10 mm.	67.5	54.9	74	51

TABLE I. *Binary Systems* 351

No.	Formula	B-Component Name	B.P., °C	B.P., °C	Wt.%A		Ref.
				Azeotropic Data			
A =	C$_6$H$_{14}$O	2-Ethylbutanol *(continued)*	147.0				
12136	C$_8$H$_{16}$O	2-Ethylhexaldehyde	163.6	Nonazeotrope			981
12137	C$_8$H$_{17}$Cl	3-(Chloromethyl)					
		heptane, 50 mm.	89	77	61		982
		" 100 mm.	106.9	92	68		982
12138	C$_9$H$_{16}$O$_2$	2-Ethylbutyl acrylate			>80		999
A =	**C$_6$H$_{14}$O**	**Hexyl Alcohol**	**157.8**				
12139	C$_6$H$_{14}$O	Isopropyl ether	69.0	Nonazeotrope			215
12140	C$_6$H$_{14}$O$_2$	2-Butoxyethanol	171.25	Nonazeotrope			526
12141	C$_7$H$_6$O	Benzaldehyde	179.2	Nonazeotrope^l			575
12142	C$_7$H$_7$Br	o-Bromotoluene	181.5	Nonazeotrope			575
12143	C$_7$H$_7$Cl	α-Chlorotoluene	179.3	Nonazeotrope			575
12144	C$_7$H$_7$Cl	o-Chlorotoluene	159.2	153.5	44		567
12145	C$_7$H$_7$Cl	p-Chlorotoluene	166.4	< 154.0	<54		567
12146	C$_7$H$_8$	Toluene	110.75	Nonazeotrope			537
12147	C$_7$H$_8$O	Anisole	153.85	151.0	36.5		538
12148	C$_7$H$_8$O	o-Cresol	191.1	Nonazeotrope			575
12149	C$_7$H$_{14}$	Methylcyclohexane	101.1	Nonazeotrope			540
12150	C$_7$H$_{14}$O	5-Methyl-2-hexanone	144.2	Nonazeotrope			552
12151	C$_7$H$_{14}$O$_2$	Propyl butyrate	142.8	Nonazeotrope			536
12152	C$_7$H$_{14}$O$_3$	1,3-Butanediol methyl ether					
		acetate	171.75	Nonazeotrope			575
12153	C$_7$H$_{16}$	Heptane	98.45	Nonazeotrope		v-l	541,779g
12154	C$_8$H$_8$	Styrene	145.8	144	23		541
12155	C$_8$H$_9$Cl	o,m,p-Chloroethylben-					
		zene, 10 mm.	67.5	62.0	43		51
12156	C$_8$H$_{10}$	Ethylbenzene	136.15	Nonazeotrope			537
		"		136.10	7.7	v-l	587
12157	C$_8$H$_{10}$	m-Xylene	139.0	138.3	15		537
12158	C$_8$H$_{10}$	o-Xylene	143.6	142.3	~ 18		537
12159	C$_8$H$_{10}$	p-Xylene	138.2	~ 137.7	13		541
				137.86	10.2	v-l	308c
12160	C$_8$H$_{10}$O	Benzyl methyl ether	167.8	156.7	73		575
12161	C$_8$H$_{10}$O	p-Methylanisole	177.05	Nonazeotrope			575
12162	C$_8$H$_{10}$O	Phenetole	170.45	**157.65**	**81**		**538**
12163	C$_8$H$_{11}$N	Dimethylaniline	194.05	Nonazeotrope			551
12164	C$_8$H$_{16}$O	2-Octanone	172.85	Nonazeotrope			552
12165	C$_8$H$_{16}$O$_2$	Butyl butyrate	166.4	Nonazeotrope			575
12166	C$_8$H$_{16}$O$_2$	2-Ethylbutyl acetate	162.3	154.4	72.5		982
12167	C$_8$H$_{16}$O$_2$	Isoamyl propionate	160.7	156.7	60		567
12168	C$_8$H$_{16}$O$_2$	Isobutyl butyrate	156.8	~ 155.0	40		536
12169	C$_8$H$_{16}$O$_2$	Isobutyl isobutyrate	147.3	Nonazeotrope			536
12170	C$_8$H$_{16}$O$_2$	Propyl isovalerate	155.7	~ 154.2	33		536
12171	C$_8$H$_{18}$	Octane	125.75	Nonazeotrope			575
12172	C$_8$H$_{18}$O	Isobutyl ether	122.3	Nonazeotrope			556
12173	C$_8$H$_{18}$O	Octyl alcohol	194	Nonazeotrope		v-l	825
12174	C$_8$H$_{18}$S	Butyl sulfide	185.0	Nonazeotrope			566

No.	Formula	Name	B.P., °C	B.P., °C	Wt.%A	Ref.
		B-Component		**Azeotropic Data**		
A =	C$_6$H$_{14}$O	**Hexyl Alcohol** *(continued)*	**157.8**			
12175	C$_8$H$_{18}$S	Isobutyl sulfide	172.0	Nonazeotrope		566
12176	C$_9$H$_8$	Indene	182.6	Nonazeotrope		575
12177	C$_9$H$_{12}$	Cumene	152.8	149.5	35	567
12178	C$_9$H$_{12}$	Mesitylene	164.6	153.5	55	537
12179	C$_9$H$_{12}$	Pseudocumene	168.2	156.3	68	541
12180	C$_9$H$_{12}$	Propylbenzene	158.8	152.5	45	540
12181	C$_9$H$_{12}$O	Benzyl ethyl ether	**185.0**	Nonazeotrope		575
12182	C$_9$H$_{13}$N	*N,N*-Dimethyl-*o*-toluidine	185.3	Nonazeotrope		551
12183	C$_9$H$_{16}$O$_2$	Hexyl acrylate			90	999
12184	C$_9$H$_{18}$O	2,6-Dimethyl-4-heptanone	168.0	Nonazeotrope		552
12185	C$_9$H$_{18}$O$_2$	Isobutyl isovalerate	171.2	Nonazeotrope		575
12186	C$_{10}$H$_{14}$	Butylbenzene	183.1	Nonazeotrope		575
12187	C$_{10}$H$_{16}$	Camphene	159.6	~ 150.8	~ 48	573
12188	C$_{10}$H$_{16}$	*d*-Limonene	**177.8**	155.5	~ 79	537
12189	C$_{10}$H$_{16}$	Nopinene	163.8	153.0	52	567
12190	C$_{10}$H$_{16}$	α-Pinene	155.8	150.8	40	537
12191	C$_{10}$H$_{16}$	α-Terpinene	173.4	156.5	72	567
12192	C$_{10}$H$_{18}$O	Cineole	176.35	Nonazeotrope		556
12193	C$_{10}$H$_{22}$	2,7-Dimethyloctane	160.2	152.5	47	537
12194	C$_{10}$H$_{22}$O	Isoamyl ether	173.4	Nonazeotrope		576
		"	173.4	157	89	556
A =C$_6$H$_{14}$O		**Isopropyl Ether**	**68.3**			
12195	C$_6$H$_{14}$O$_2$	1,2-Diethoxyethane	121.1	Nonazeotrope		981
12196	C$_7$H$_{16}$	*n*-Heptane				
		685-2280 mm.		Nonazeotrope	v-l	995
A =	C$_6$H$_{14}$O	**4-Methyl-2-pentanol**	**131.8**			
12197	C$_8$H$_{16}$O$_2$	4-Methyl-2-pentyl acetate	146.1	Nonazeotrope		981
A =	C$_6$H$_{14}$O	**Propyl Ether**	**90.1**			
12198	C$_6$H$_{15}$N	Dipropylamine	109.2	Nonazeotrope		551
12199	C$_6$H$_{15}$N	Triethylamine	89.35	< 88.5		551
12200	C$_7$H$_8$	Toluene	110.75	Nonazeotrope		558
12201	C$_7$H$_{14}$	Methylcyclohexane	101.1	Nonazeotrope		573
12202	C$_7$H$_{16}$	Heptane	98.45	Nonazeotrope		527
12203	C$_8$H$_{18}$	2,5-Dimethylhexane	109.4	Nonazeotrope		558
A =	C$_6$H$_{14}$OS	**2-Butylthioethanol**				
12204	C$_8$H$_{16}$OS	2-Butylthioethyl vinyl ether		Min. b.p.		953
A =	C$_6$H$_{14}$O$_2$	**Acetal**	**103.55**			
12205	C$_6$H$_{14}$S	Isopropyl sulfide	120.5	Nonazeotrope		566
12206	C$_6$H$_{15}$N	Dipropylamine	109.2	Nonazeotrope		551
12207	C$_6$H$_{15}$N	Triethylamine	89.35	Nonazeotrope		551
12208	C$_7$H$_8$	Toluene	110.75	Nonazeotrope		573
12209	C$_7$H$_{14}$	Methylcyclohexane	101.15	99.65	40	558
12210	C$_7$H$_{16}$	*n*-Heptane	98.45	97.75	28	558

TABLE I.   *Binary Systems*   353

No.	Formula	B-Component Name	B.P., °C	Azeotropic Data B.P., °C	Wt.%A	Ref.
A =	**C₆H₁₄O₂**	**Acetal** *(continued)*	**103.55**			
12211	C₈H₁₈	2,5-Dimethylhexane	109.3	103.0	75	548
12212	C₈H₁₈	Octane	125.75	Nonazeotrope		558
A =	**C₆H₁₄O₂**	**2-Butoxyethanol**	**171.15**			
12213	C₆H₁₅NO	2-Diethylaminoethanol	162.2	Nonazeotrope		551
12214	C₇H₅N	Benzonitrile	191.1	Nonazeotrope		556
12215	C₇H₆O	Benzaldehyde	179.2	170.95	91	556
12216	C₇H₇Br	o-Bromotoluene	181.5	169.7	65	526
12217	C₇H₇Cl	o-Chlorotoluene	159.2	158.0	12	526
12218	C₇H₇Cl	p-Chlorotoluene	162.4	160.5	20	556
12219	C₇H₇ClO	o-Chloroanisole	195.7	Nonazeotrope		575
12220	C₇H₈O	Anisole	153.85	Nonazeotrope		556
12221	C₇H₈O	m-Cresol	202.2	Nonazeotrope		526
12222	C₇H₈O	o-Cresol	191.1	191.55	15	556
12223	C₇H₈O	p-Cresol	201.7	Nonazeotrope		556
12224	C₇H₉N	Benzylamine	185.0	Nonazeotrope		551
12225	C₇H₉N	Methylaniline	196.25	Nonazeotrope		551
12226	C₇H₁₃ClO₂	Isoamyl chloroacetate	190.5	Nonazeotrope		575
12227	C₇H₁₄O	2-Methylcyclohexanol	168.5	Nonazeotrope		575
12228	C₇H₁₄O₃	1,3-Butanediol methyl ether acetate	171.75	170.1	53	556
12229	C₇H₁₄O₃	Isobutyl lactate	182.15	Nonazeotrope		575
12230	C₇H₁₆O	Heptyl alcohol	176.15	Nonazeotrope		526
12231	C₈H₈O₂	Methyl benzoate	199.4	Nonazeotrope		556
12232	C₈H₈O₂	Phenyl acetate	195.7	Nonazeotrope		575
12233	C₈H₉Cl	o,m,p-Chloroethylbenzene, 10 mm.	67.5	62.5	37	51
12234	C₈H₁₀	m-Xylene	139.2	Nonazeotrope		575
12235	C₈H₁₀	o-Xylene	144.3	Nonazeotrope		556
12236	C₈H₁₀O	Benzyl methyl ether	167.8	165.0	43	526
12237	C₈H₁₀O	p-Methylanisole	177.05	169.3	62	526
12238	C₈H₁₀O	Phenetole	170.45	167.1	52	556
12239	C₈H₁₁N	Dimethylaniline	194.15	Nonazeotrope		551
12240	C₈H₁₆O₂	Butyl butyrate	166.4	164.7	20	526
12241	C₈H₁₆O₂	Ethyl caproate	167.7	166.0	25	567
12242	C₈H₁₆O₂	Hexyl acetate	171.5	167.7	45	567
12243	C₈H₁₆O₂	Isoamyl propionate	160.7	Nonazeotrope		575
12244	C₈H₁₆O₂	Isobutyl butyrate	156.9	Nonazeotrope		575
12245	C₈H₁₆O₂	Propyl isovalerate	155.7	Nonazeotrope		575
12246	C₈H₁₆O₄	2-(2-Ethoxyethoxy) ethyl acetate	218.5	Nonazeotrope		575
12247	C₈H₁₈O	Butyl ether	142.4	Nonazeotrope		575
12248	C₈H₁₈O	sec-Octyl alcohol	180.4	Nonazeotrope		526
12249	C₈H₁₈S	Isobutyl sulfide	172	163.8	42	555
12250	C₉H₇N	Quinoline	237.3	Nonazeotrope		553
12251	C₉H₁₂	Cumene	152.4	151.7	10.3	456
12252	C₉H₁₂	Mesitylene	164.6	162.0	32	556
12253	C₉H₁₂	Propylbenzene	159.3	158.0		526

	B-Component			Azeotropic Data		
No.	Formula	Name	B.P., °C	B.P., °C	Wt.%A	Ref.
A =	C₆H₁₄O₂	**2-Butoxyethanol** *(continued)*	**171.15**			
12254	C₉H₁₂	Pseudocumene	168.2	164.5	38	575
12255	C₉H₁₃N	*N,N*-Dimethyl-*o*-toluidine	185.3	170.95	88	551
12256	C₉H₁₆	*cis*-Hexahydroindan	167.7	159.9	38 vol. %	826
12257	C₉H₁₈O₂	Isoamyl butyrate	181.05	170.85	86	556
12258	C₉H₁₈O₂	Isoamyl isobutyrate	169.8	166.5	36	567
12259	C₉H₁₈O₂	Isobutyl isovalerate	171.2	167.75	43	527
12260	C₁₀H₁₄	Butylbenzene	183.0	170.2	80	526
		"	183.4	169.6	73.4	456
12261	C₁₀H₁₄	*sec*-Butylbenzene	173.3	166.0	47.9	456
12262	C₁₀H₁₄	*tert*-Butylbenzene	169.1	164.4	39.1	456
12263	C₁₀H₁₄	*p*-Cymene	177.2	167.4	56.6	456
		"	176.7	168.0	60	556
12264	C₁₀H₁₆	Camphene	159.6	154.5	30	526
12265	C₁₀H₁₆	Dipentene	177.7	164.0	53	567
12266	C₁₀H₁₆	Nopinene	163.8	158.0	37	526
12267	C₁₀H₁₆	α-Pinene	155.8	151.5	25	567
12268	C₁₀H₁₆	α-Terpinene	173.4	164.0	50	526
12269	C₁₀H₁₈O	Cineole	176.35	168.9	58.5	527
12270	C₁₀H₁₈O	Citronellal	207.8	Nonazeotrope		526
12271	C₁₀H₁₈O	Linalool	198.6	Nonazeotrope		575
12272	C₁₀H₂₀	*n*-Butylcyclohexane	180.95	165.6	56 vol. %	826
12273	C₁₀H₂₀	*sec*-Butylcyclohexane	179.3	165.1	53 vol. %	826
12274	C₁₀H₂₀	Isobutylcyclohexane	171.3	161.5	40 vol. %	826
12275	C₁₀H₂₀	*cis*-1-Methyl-4-isopropylcyclohexane	172.7	162.0	45 vol. %	826
12276	C₁₀H₂₀	*trans*-1-Methyl-4-isopropyl-cyclohexane	170.5	160.9	41 vol. %	826
12277	C₁₀H₂₂	3,3,5-Trimethylheptane	155.5	151.6	23 vol. %	826
12278	C₁₀H₂₂O	Amyl ether	187.5	169.0	67	556
12279	C₁₀H₂₂O	Isoamyl ether	173.2	164.95	54	569
12280	C₁₀H₂₂O₂	Acetaldehyde dibutyl acetal	188.8	170.6	42	129
12281	C₁₁H₂₀O	Isobornyl methyl ether	192.4	Nonazeotrope		556
12282	C₁₂H₁₈	Triethylbenzene	215.5	Nonazeotrope		526
A =	C₆H₁₄O₂	**Hexylene Glycol**				
12283	C₈H₁₀	Ethylben-zene, 400 mm.		Nonazeotrope	v-l	771
12284	C₈H₁₆	Ethylcyclohexane, 400 mm.		Nonazeotrope	v-l	771
A =	C₆H₁₄O₂	**Pinacol**	**174.35**			
12285	C₇H₇Cl	*o*-Chlorotoluene	159.2	< 157.0		575
12286	C₇H₇Cl	*p*-Chlorotoluene	162.4	158.0	>13	567
12287	C₇H₈	Toluene	110.7	Nonazeotrope		540
12288	C₇H₈O	Anisole	174.35	153.5		545
12289	C₇H₈O	*m*-Cresol	202.2	Nonazeotrope		544
12290	C₇H₈O	*o*-Cresol	191.1	191.5	8	575

TABLE I. *Binary Systems*   355

No.	Formula	B-Component Name	B.P., °C	B.P., °C	Wt.%A	Ref.
				Azeotropic Data		
A =	**$C_6H_{14}O_2$**	**Pinacol** *(continued)*	**174.35**			
12291	C_7H_8O	*p*-Cresol	201.7	Nonazeotrope		575
12292	C_7H_{14}	Methylcyclohexane	101.1	Nonazeotrope		537
12293	C_7H_{16}	Heptane	98.45	Nonazeotrope		537
12294	C_8H_{10}	*m*-Xylene	139.0	Nonazeotrope		537
12295	$C_8H_{10}O$	Benzyl methyl ether	167.8	163.5?	28?	575
12296	$C_8H_{10}O$	*p*-Methylanisole	177.05	168.7	44	576
12297	$C_8H_{10}O$	Phenetole	170.4	165.2	33	572
12298	$C_8H_{11}N$	Dimethylaniline	194.05	< 169.5	>60	551
12299	$C_8H_{14}O$	Methylheptenone	173.2	171.7	40	552
12300	$C_8H_{16}O$	2-Octanone	172.85	171.5	35	552
12301	C_8H_{18}	2,5-Dimethylhexane	109.4	Nonazeotrope		575
12302	C_8H_{18}	Octane	125.75	Nonazeotrope		575
12303	C_9H_{12}	Mesitylene	164.6	160.2	35	572
12304	C_9H_{12}	Propylbenzene	159.3	156.3	28	567
12305	C_9H_{12}	Pseudocumene	168.2	162.9	38	567
12306	$C_9H_{12}O$	Benzyl ethyl ether	185.0	< 171.5	>62	575
12307	$C_9H_{18}O_2$	Isobutyl isovalerate	171.2	< 169.8	>10	575
12308	$C_9H_{18}O_2$	Isoamyl butyrate	181.05	< 173.9		575
12309	$C_{10}H_8$	Naphthalene	218.05	Nonazeotrope		537
12310	$C_{10}H_{14}$	Cymene	176.7	167.7	50	567
12311	$C_{10}H_{16}$	Camphene	159.6	155.5	~ 28	537
12312	$C_{10}H_{16}$	Dipentene	177.7	166.7	~ 50	575
12313	$C_{10}H_{16}$	*d*-Limonene	177.8	171	~ 45	537
12314	$C_{10}H_{16}$	α-Pinene	155.8	152.5		537
12315	$C_{10}H_{18}O$	Cineole	176.35	168.5	45	567
12316	$C_{10}H_{22}$	2,7-Dimethyloctane	160.25	~ 144?		563
12317	$C_{10}H_{22}O$	Isoamyl ether	173.4	167.2	40	576
A =	**$C_6H_{14}O_3$**	**Dipropylene Glycol**	**229.2**			
12318	C_7H_7BrO	*o*-Bromoanisole	217.7	212.0	30	575
12319	$C_7H_7NO_2$	*o*-Nitrotoluene	221.75	216.9	>21	554
12320	$C_7H_7NO_2$	*p*-Nitrotoluene	238.9	225.0	62?	554
12321	C_7H_8O	Benzyl alcohol	205.25	Nonazeotrope		575
12322	C_7H_8O	*p*-Cresol	201.7	Nonazeotrope		575
12323	$C_8H_8O_2$	Anisaldehyde	249.5	Nonazeotrope		575
12324	$C_8H_8O_3$	Methyl salicylate	222.95	213.0	35	575
12325	C_8H_9BrO	*p*-Bromophenetole	234.2	221.0	45	575
12326	C_9H_7N	Quinoline	237.3	< 228.0	<72	575
12327	$C_9H_{10}O_3$	Ethyl salicylate	233.8	218.2	55	575
12328	$C_{10}H_8$	Naphthalene 100 mm.		142.9	12.4	v-l	600
12329	$C_{10}H_9N$	Quinaldine	246.5	Nonazeotrope		575
12330	$C_{10}H_{10}O_2$	Isosafrole	252.0	225.5	60	567
12331	$C_{10}H_{10}O_2$	Safrole	235.9	222.0	50	567
12332	$C_{10}H_{12}O$	Anethole	235.7	221.5	48	567
12333	$C_{10}H_{18}O$	Cineole	176.35	Nonazeotrope		575
12334	$C_{10}H_{22}O$	Isoamyl ether	173.2	Nonazeotrope		575
12335	$C_{11}H_{10}$	2-Methylnaphthalene	241.1	Nonazeotrope		270

No.	Formula	Name	B.P., °C	B.P., °C	Wt.%A		Ref.
		B-Component		**Azeotropic Data**			
A =	**C₆H₁₄O₃**	**Dipropylene Glycol**	**229.2**				
12336	C₁₁H₁₄O₂	1-Allyl-3,4-dimethoxy-benzene	254.7	226.5	65		575
12337	C₁₁H₁₆O	Methyl thymyl ether	216.5	211.0	30		575
12338	C₁₁H₂₀O	Methyl α-terpineol ether	216.2	< 211.5	>24		575
12339	C₁₂H₁₀O	Phenyl ether	259.0	< 228.0	<77		575
12340	C₁₂H₁₆O₃	Isoamyl salicylate	277.5	Nonazeotrope			575
12341	C₁₃H₁₂O	Benzyl phenyl ether	286.5	Nonazeotrope			575
12342	C₁₄H₁₄O	Benzyl ether	297.0	Nonazeotrope			575
A =	**C₆H₁₄O₃**	**2-(2-Ethoxyethoxy)ethanol**	**195.0**				
12343	C₇H₈O	m-Cresol	202.4		36.8	v-l	730
12344	C₇H₈O	o-Cresol	191.1	205.5	70		575
12345	C₇H₈O	p-Cresol	202.0		38	v-l	730
		"	202.0	209.0	50		567
12346	C₇H₁₆O₄	2-[2-(2-Methoxyethoxy)-ethoxy]ethanol	245.25	Nonazeotrope			575
12347	C₈H₈O₃	Methyl salicylate	222.95	Nonazeotrope			575
12348	C₈H₁₀O	3,4-Xylenol	226.8	Nonazeotrope			575
12349	C₈H₁₁N	Dimethylaniline	194.15	<193.0	>10		575
12350	C₈H₁₆O₃	Isoamyl lactate	202.4	<201.0	>38		575
12351	C₈H₁₈O₃	Bis(2-ethoxyethyl) ether	188.4	Nonazeotrope			981
		" 10 mm.	72	Nonazeotrope			981
12352	C₉H₇N	Quinoline	237.3	Nonazeotrope			575
12353	C₉H₁₃N	Dimethyl-o-toluidine	185.3	Nonazeotrope			575
12354	C₉H₁₃N	Dimethyl-p-toluidine	210.2	199.5			575
12355	C₁₀H₈	Naphthalene	218.0	200.5			575
12356	C₁₀H₁₂O	Estragole	215.6	201.0	87		575
12357	C₁₀H₁₄	Butylbenzene	183.1	181.3	18		575
12358	C₁₀H₁₅N	Diethylaniline	217.05	<200.5	>85		575
12359	C₁₀H₁₆	Dipentene	177.7	173.0	23		575
12360	C₁₀H₁₈O	Cineole	176.35	<175.5			575
12361	C₁₀H₂₂O	Amyl ether	187.5	<183.0			575
12362	C₁₁H₁₀	2-Methylnaphthalene	241.15	Nonazeotrope			270,575
12363	C₁₁H₂₀O	Isobornyl methyl ether	192.4	190.5	25		567
12364	C₁₂H₂₂O	Ethyl isobornyl ether	203.8	198.5	55		567
A =	**C₆H₁₄O₄**	**Triethylene Glycol**	**288.7**				
12365	C₉H₁₀O₃	Ethyl salicylate	233.8	Nonazeotrope			575
12366	C₁₀H₇Br	1-Bromonaphthalene	281.2	273.4	33		527
12367	C₁₀H₇Cl	1-Chloronaphthalene	262.7	261.5	5		527
12368	C₁₀H₁₀O₂	Isosafrole	252.0	Nonazeotrope			556
12369	C₁₀H₁₀O₂	Methyl cinnamate	261.9	Nonazeotrope			526
12370	C₁₀H₁₀O₂	Safrole	235.9	Nonazeotrope			526
12371	C₁₀H₁₀O₄	Methyl phthalate	283.2	277.0	33		526
12372	C₁₀H₁₂O	Anethole	235.7	Nonazeotrope			575
12373	C₁₁H₁₀	1-Methylnaphthalene	244.6	Nonazeotrope			526

TABLE I. *Binary Systems*   357

No.	Formula	Name	B.P., °C	B.P., °C	Wt.%A	Ref.
		B-Component		**Azeotropic Data**		
A =	**C₆H₁₄O₄**	**Triethylene Glycol**	**288.7**			
		(*continued*)				
12374	C₁₁H₁₀	2-Methylnaphthalene	241.15	Nonazeotrope		527
12375	C₁₁H₁₂O₂	Ethyl cinnamate	272.0	<271.5	> 7	575
12376	C₁₁H₁₄O₂	1-Allyl-3,4-dimethoxy-				
		benzene	254.7	Nonazeotrope		526
12377	C₁₁H₁₄O₂	Butyl benzoate	249.0	Nonazeotrope		526
12378	C₁₂H₉N	Carbazole		Nonazeotrope		272
		" Low press.		Min. b.p.		272
12379	C₁₂H₁₀	Acenaphthene	277.9	271.5	35	527
12380	C₁₂H₁₀	Biphenyl	256.1	255.3	10	556
12381	C₁₂H₁₀O	Phenyl ether	259.0	258.7	3	556
		" " 4 mm.	102	Nonazeotrope		981
12382	C₁₂H₁₄O₄	Ethyl phthalate	298.5	<285.5	>58	575
12383	C₁₂H₁₆O₂	Isoamyl benzoate	262.0	261.4	14	527
12384	C₁₂H₁₆O₃	Isoamyl salicylate	277.5	269.0	30	575
12385	C₁₂H₂₂O₄	Isoamyl oxalate	268.0	Reacts		526
12386	C₁₃H₁₀	Fluorene	294	Nonazeotrope		272
		" High press.		Min. b.p.		272
12387	C₁₃H₁₀O₂	Phenyl benzoate	315	286.0	80	526
12388	C₁₃H₁₂	Diphenylmethane	265.4	263.0	20	556
12389	C₁₃H₁₂O	Benzyl phenyl ether	286.5	280.0	40	526
12390	C₁₄H₁₀	Phenanthrene,				
		Low press.		Min. b.p.		272
		Glycol decreases with				
		decreasing pressure				
12391	C₁₄H₁₂	Stilbene	306.5	284.5	60	526
12392	C₁₄H₁₄	1,2-Diphenylethane	284.5	275.5	42	526
12393	C₁₄H₁₄O	Benzyl ether, 5 mm.	145.5		28	982
A =	**C₆H₁₄S**	**Isopropyl Sulfide**	**120.5**			
12394	40°C.	Toluene	110.75	Nonazeotrope		575
12395	70°C.	Methylcyclohexane	101.15	Nonazeotrope		575
12396	⁻₇¹I₁₆	Heptane	98.4	Nonazeotrope		575
12397	C₈H₁₈O	Isobutyl ether	122.3	<119.8	>64	566
A =	**C₆H₁₄S**	**Propyl Sulfide**	**141.5**			
12398	C₇H₁₄O	5-Methyl-2-hexanone	144.2	<140.7	>65	566
12399	C₇H₁₄O₂	Ethyl isovalerate	134.7	~ 134.0	~ 10	532
12400	C₈H₁₀	*m*-Xylene	139.0	~ 137.5		531
12401	C₈H₁₀	*p*-Xylene	138.45	<138.2	> 7	575
12402	C₈H₁₈O	Butyl ether	142.4	140.3	62	562
A =	**C₆H₁₅BO₃**	**Ethyl Borate**	**118.6**			
12403	C₇H₈	Toluene	110.75	Nonazeotrope		530
12404	C₇H₁₄	Methylcyclohexane	101.1	Nonazeotrope		546
12405	C₇H₁₆	Heptane	98.5	Nonazeotrope		546
12406	C₈H₁₈O	Isobutyl ether	122.3	<116.8		557
A =	**C₆H₁₅N**	**Dipropylamine**	**109.2**			
12407	C₇H₈	Toluene	110.75	<108.5	>53	551

No.	B-Component			Azeotropic Data			
	Formula	Name	B.P., °C	B.P., °C	Wt.%A		Ref.
A =	**C₆H₁₅N**	**Dipropylamine** *(continued)*	**109.2**				
12408	C₇H₁₆	*n*-Heptane	98.4	Nonazeotrope			527
12409	C₈H₁₆	1,3-Dimethylcyclohexane	120.7	Nonazeotrope			551
12410	C₈H₁₈	2,4-Dimethylhexane	109.4	<108.0	<54		551
12411	C₈H₁₈O	Isobutyl ether	122.3	Nonazeotrope			551
A =	**C₆H₁₅N**	**Isohexylamine**	**123.5**				
12412	C₇H₈	Toluene	110.75	Nonazeotrope			575
12413	C₈H₁₆	1,3-Dimethylcyclohexane	120.7	<120.0			575
12414	C₈H₁₈O	Isobutyl ether	122.3	<121.8			575
A =	**C₆H₁₅N**	**Triethylamine**	**89.35**				
12415	C₇H₁₄	Methylcyclohexane	101.15	Nonazeotrope			551
12416	C₇H₁₆	*n*-Heptane	98.4	Nonazeotrope			551
A =	**C₆H₁₅NO**	**2-(Diethylamino)ethanol**	**162.2**				
12417	C₇H₈	Toluene	110.75	Nonazeotrope			575
12418	C₇H₈O	Anisole	153.85	<148.0	>19		551
12419	C₇H₈O	*o*-Cresol	191.1	Nonazeotrope			551
12420	C₇H₉N	Methylaniline	196.25	Nonazeotrope			551
12421	C₇H₁₆	Heptane	98.4	Nonazeotrope			575
12422	C₈H₉Cl	*o,m,p*-Chloroethylbenzene, 10 mm.	67.5	57.0	91		51
12423	C₈H₁₀	*m*-Xylene	139.2	<136.0	> 8		575
12424	C₈H₁₁N	Dimethylaniline	194.15	<160.5	>58		551
12425	C₈H₁₈O	Isobutyl ether	122.3	Nonazeotrope			551
12426	C₁₀H₁₈	Camphene	159.6	<146.5			575
12427	C₁₀H₁₈O	Cineole	176.35	<158.0			575
12428	C₁₀H₂₂O	Isoamyl ether	173.2	<156.5	>58		551
A =	**C₆H₁₅NO₂**	**1,1'-Iminodi-2-propanol**	**133/10**				
12429	C₉H₂₁NO₃	1,1',1''-Nitrilotri-2-propanol, 10 mm.	177	Nonazeotrope			981
A =	**C₇F₁₆**	**Perfluoroheptane**	**82.5**				
12430	C₇H₁₆	Heptane, crit. region		198.5	95.3	v-l	430,431
12431	C₈F₁₆O	Perfluorocyclic oxide	102.6	Nonazeotrope		v-l	1041
12432	C₈H₁₈	Octane, crit. region		201.7	99.7	v-l	430,431
12433	C₉H₂₀	Nonane, crit. region		Nonazeotrope		v-l	430,431
A =	**C₇H₅Cl₃**	**α,α,α-Trichlorotoluene**	**220.8**				
12434	C₇H₇NO₂	*m*-Nitrotoluene	230.8	Nonazeotrope			554
12435	C₇H₇NO₂	*o*-Nitrotoluene	221.75	219.45	75.5		554
12436	C₇H₇NO₂	*p*-Nitrotoluene	238.9	Nonazeotrope			554
12437	C₇H₈O	Benzyl alcohol	205.2	Reacts			535
12438	C₇H₈O₂	Guaiacol	205.05	Reacts			535
12439	C₇H₉N	*o*-Toluidine	200.3	Nonazeotrope			538
12440	C₈H₈O₂	Benzyl formate	202.3	Nonazeotrope			547
12441	C₈H₈O₃	Methyl salicylate	222.35	220.75	~ 97		538
12442	C₈H₉BrO	*p*-Bromophenetole	234.5	Nonazeotrope			575
12443	C₈H₁₀O₂	*m*-Dimethoxybenzene	214.7	Nonazeotrope			535
12444	C₉H₁₀O	*p*-Methylacetophenone	226.35	Nonazeotrope			552

TABLE I. *Binary Systems* 359

No.	Formula	Name	B.P., °C	B.P., °C	Wt.%A	Ref.
		B-Component		**Azeotropic Data**		
A =	C₇H₅Cl₃	*a,a,a*-**Trichlorotoluene** (*continued*)	**220.8**			
12445	C₉H₁₀O	Propiophenone	217.7	Nonazeotrope		552
12446	C₉H₁₀O₂	Benzyl acetate	214.9	Nonazeotrope		535
12447	C₉H₁₀O₂	Ethyl benzoate	212.6	Nonazeotrope		574
12448	C₉H₁₀O₃	Ethyl salicylate	234.0	Nonazeotrope		538
12449	C₁₀H₇Cl	1-Chloronaphthalene	262.7	Nonazeotrope		545
12450	C₁₀H₈	Naphthalene	218.05	Nonazeotrope		574
12451	C₁₀H₁₂O	Estragole	215.6	Nonazeotrope		535
12452	C₁₀H₁₂O₂	Ethyl *a*-toluate	228.75	Nonazeotrope		538
12453	C₁₀H₁₂O₂	Propyl benzoate	230.85	Nonazeotrope		538
12454	C₁₀H₁₄O	Carvone	231.0	Nonazeotrope		551
12455	C₁₀H₁₄O	Thymol	232.9	Reacts		542
12456	C₁₀H₁₅N	Diethylaniline	217.05	Nonazeotrope		538
12457	C₁₁H₂₂O₃	Isoamyl carbonate	232.2	Nonazeotrope		547
12458	C₁₂H₁₈	1,3,5-Triethylbenzene	215.5	Nonazeotrope		575
12459	C₁₂H₂₀O₂	Bornyl acetate	~227.7	Nonazeotrope		535
A =	C₇H₅N	**Benzonitrile**	**191.1**			
12460	C₇H₇Br	*m*-Bromotoluene	184.3	183.8	11.5	569
12461	C₇H₇Br	*o*-Bromotoluene	181.5	181.4		565
12462	C₇H₇Br	*p*-Bromotoluene	185.0	184.3	15	565
		"	185	~ 181		563
12463	C₇H₇Cl	*p*-Chlorotoluene	162.4	Nonazeotrope		565
12464	C₇H₈O	Benzyl alcohol	205.25	Nonazeotrope		565
12465	C₇H₈O	*m*-Cresol	202.2	202.5	11	527
12466	C₇H₈O	*o*-Cresol	191.1	195.95	49	569
12467	C₇H₈O	*p*-Cresol	201.7	202.1	14	527
12468	C₇H₉N	Methylaniline	196.25	Nonazeotrope		575
12469	C₇H₉N	*o*-Toluidine	200.35	Nonazeotrope		575
12470	C₇H₁₂O₄	Ethyl malonate	199.35	Nonazeotrope		575
12471	C₇H₁₆O	Heptyl alcohol	176.15	Nonazeotrope		565
12472	C₈H₈O₂	Phenyl acetate	195.7	<189.5	>51	575
12473	C₈H₁₀O	Benzyl methyl ether	167.8	Nonazeotrope		565
12474	C₈H₁₀O	Phenetole	170.45	Nonazeotrope		565
12475	C₈H₁₁N	Dimethylaniline	194.15	Nonazeotrope		575
12476	C₈H₁₁N	Ethylaniline	205.5	Nonazeotrope		575
12477	C₈H₁₈O	*n*-Octyl alcohol	195.2	<189.2	<70	527
12478	C₈H₁₈O	*sec*-Octyl alcohol	180.4	180.05	11	569
12479	C₈H₁₈S	Butyl sulfide	185.0	<184.5	<12	575
12480	C₈H₁₈S	Isobutyl sulfide	172.0	Nonazeotrope		566
12481	C₉H₁₂O	Benzyl ethyl ether	185.0	182.5	27	565
12482	C₉H₁₈O₂	Butyl isovalerate	177.6	Nonazeotrope		565
12483	C₉H₁₈O₂	Isoamyl butyrate	181.05	180.85	8	527
12484	C₉H₁₈O₂	Isoamyl isobutyrate	169.8	Nonazeotrope		565
12485	C₉H₁₈O₂	Isobutyl isovalerate	171.2	Nonazeotrope		565
12486	C₁₀H₁₈O	Cineole	176.35	175.6	14	527
12487	C₁₀H₂₀O₂	Isoamyl isovalerate	192.7	<189.0	>42	527
12488	C₁₀H₂₂O	Amyl ether	187.5	180.5	42	527

No.	Formula	B-Component Name	B.P., °C	B.P., °C	Azeotropic Data Wt.%A	Ref.
A =	**C₇H₅N**	**Benzonitrile** *(continued)*	**191.1**			
12489	C₁₀H₂₂O	Isoamyl ether	173.2	171.4	16	527
12490	C₁₁H₂₀O	Isobornyl methyl ether	192.4	<186.0		565
A =	**C₇H₅NO**	**Phenyl Isocyanate**	**162.8**			
12491	C₈H₁₈S	Butyl sulfide	185.0	Nonazeotrope		575
12492	C₈H₁₈S	Isobutyl sulfide	172.0	Nonazeotrope		575
A =	**C₇H₆Cl₂**	**α,α-Dichlorotoluene**	**205.2**			
12497	C₇H₆O₂	Benzoic acid	250.8	Nonazeotrope		575
12494	C₇H₇NO₂	o-Nitrotoluene	221.75	Nonazeotrope		575
12495	C₇H₈O	Benzyl alcohol	205.5	182?		563
12496	C₇H₈O	m-Cresol	202.8	Reacts		563
12497	C₇H₈O	o-Cresol	190.8	Reacts		563
12498	C₇H₈O	p-Cresol	201.8	Reacts		563
12499	C₇H₉N	Methylaniline	196.1	Reacts		563
12500	C₇H₉N	o-Toluidine	200.3	Nonazeotrope		538
12501	C₇H₉N	p-Toludine	200.3	Reacts		563
12502	C₇H₁₂O₄	Ethyl malonate	198.9	Nonazeotrope		563
12503	C₈H₈O	Acetophenone	202	Nonazeotrope		563
12504	C₈H₈O₂	Benzyl formate	202.3	Nonazeotrope		538
12505	C₈H₈O₂	Methyl benzoate	199.55	Nonazeotrope		563
12506	C₈H₈O₂	Phenyl acetate	195.5	Nonazeotrope		563
12507	C₈H₁₁N	Dimethylaniline	194.15	Nonazeotrope		551
12508	C₈H₁₁N	Ethylaniline	206.3	Reacts		563
12509	C₈H₁₄O₄	Ethyl succinate	217.25	Nonazeotrope		547
12510	C₈H₁₄O₄	Propyl oxalate	212	Nonazeotrope		547
12511	C₈H₁₆O₃	Isoamyl lactate	202.4	201.3	45	563
12512	C₈H₁₈O	n-Octyl alcohol	195.15	194.5	~ 10	531
12513	C₉H₁₀O₂	Benzyl acetate·	214.9	Nonazeotrope		538
12514	C₉H₁₀O₂	Ethyl benzoate	213	Nonazeotrope		563
12515	C₁₀H₈	Naphthalene	218.05	Nonazeotrope		563
12516	C₁₀H₁₆O	Camphor	209.1	209.7	25	551
12617	C₁₀H₁₈O	Borneol	213.4	205.0	~ 85	538
12518	C₁₀H₁₈O	Citronellal	~207.8	Nonazeotrope		538
12519	C₁₀H₁₈O	Menthone	207	Azeotrope doubtful		563
		Menthone	209.5	Nonazeotrope		575
12520	C₁₀H₂₀O	Menthol	216.3	Nonazeotrope		575
12521	C₁₀H₂₀O₂	Isoamyl isovalerate	192.7	Nonazeotrope		575
12522	C₁₂H₁₈	1,3,5-Triethylbenzene	215.5	Nonazeotrope		538
A =	**C₇H₆O**	**Benzaldehyde**	**179.2**			
12523	C₇H₇Br	m-Bromotoluene	184.3	<179.0	<92	575
12524	C₇H₇Br	o-Bromotoluene	181.5	178.5		545
12525	C₇H₇Br	p-Bromotoluene	185.0	Nonazeotrope		545
12526	C₇H₇Cl	α-Chlorotoluene	179.35	177.9	50	563
12527	C₇H₇Cl	o-Chlorotoluene	159.15	Nonazeotrope		548
12528	C₇H₇Cl	p-Chlorotoluene	162.4	Nonazeotrope		545
12529	C₇H₇ClO	o-Chloroanisole	195.7	Nonazeotrope		575
12530	C₇H₈O	Anisole	153.85	Nonazeotrope		575

TABLE I.   *Binary Systems*   361

	B-Component			Azeotropic Data		
No.	Formula	Name	B.P., °C	B.P., °C	Wt.%A	Ref.
A =	C₇H₆O	**Benzaldehyde** *(continued)*	**179.2**			
12531	C₇H₈O	*m*-Cresol	202.2	Nonazeotrope		575
12532	C₇H₈O	*o*-Cresol	191.1	192.0	23	538
12533	C₇H₈O	*p*-Cresol	201.7	Nonazeotrope		545
12534	C₇H₁₄O₃	Isobutyl lactate	182.15	<178.8	<92	575
12535	C₇H₁₆O	Heptyl alcohol	176.15	<174.5	<45	575
12536	C₈H₈	Styrene	145.8	Nonazeotrope		575
12537	C₈H₉Cl	*o,m,p*-Chloroethylbenzene, 10 mm.	67.5	63.5	57	51
12538	C₈H₁₀	*o*-Xylene	144.3	Nonazeotrope		575
12539	C₈H₁₀O	Benzyl methyl ether	167.8	<167.0		575
12540	C₈H₁₀O	*p*-Ethylphenol	218.8	Nonazeotrope		575
12541	C₈H₁₀O	*p*-Methylanisole	177.05	<175.5		575
12542	C₈H₁₀O	Phenetole	170.45	<169.8	<12	575
12543	C₈H₁₁N	Dimethylaniline	194.05	Reacts		563
12544	C₈H₁₄O	Methylheptenone	173.2	Nonazeotrope		552
12545	C₈H₁₆O	2-Octanone	172.85	Nonazeotrope		552
12546	C₈H₁₆O₂	Butyl butyrate	166.4	Nonazeotrope		575
12547	C₈H₁₆O₂	Ethyl caproate	167.7	Nonazeotrope		575
12548	C₈H₁₆O₂	Hexyl acetate	171.5	<171.3		575
12549	C₈H₁₈O	Butyl ether	142.6	Nonazeotrope		575
12550	C₈H₁₈O	Octyl alcohol	195.2	Nonazeotrope		575
12551	C₈H₁₈O	*sec*-Octyl alcohol	178.7	174	~ 25	563
12552	C₉H₁₂	Cumene	152.8	Nonazeotrope		575
12553	C₉H₁₂O	Benzyl ethyl ether	185.0	<177.5	<92	575
12554	C₉H₁₄O	Phorone	197.8	Nonazeotrope		552
12555	C₉H₁₈O	2,6-Dimethyl-4 heptanone	168.0	Nonazeotrope		552
12556	C₉H₁₈O₂	Isoamyl butyrate	178.5	~ 176.3	38	536
12557	C₉H₁₈O₂	Isoamyl isobutyrate	169.8	Nonazeotrope		575
12558	C₉H₁₈O₂	Isobutyl isovalerate	171.2	170.85	10	575
		" "	171.35	Nonazeotrope		545
12559	C₉H₁₈O₃	Isobutyl carbonate	190.3	Nonazeotrope		575
12560	C₁₀H₈	Naphthalene	218.0	Nonazeotrope		575
12561	C₁₀H₁₄	Butylbenzene	183.1	<176.5	<65	575
12562	C₁₀H₁₄	Cymene	175.3	171	28	563
12563	C₁₀H₁₆	Camphene	159.6	158.45	15.5	548
12564	C₁₀H₁₆	*d*-Limonene	177.8	171.2	43	563
12565	C₁₀H₁₆	Nopinene	163.8	<162.0	<25	548
12566	C₁₀H₁₆	α-Phellandrene	171.5	170		563
12567	C₁₀H₁₆	α-Pinene	155.8	Nonazeotrope		575
		"	155.8	~ 155.0	~ 10	548
12568	C₁₀H₁₆	α-Terpinene	173.4	<170.0	<38	575
12569	C₁₀H₁₆	γ-Terpinene	179.9	~ 173.0	~ 48	548
12570	C₁₀H₁₆	Terpinolene	185	<176.5	>70	563
12571	C₁₀H₁₆O	Fenchone	193.6	Nonazeotrope		552
12572	C₁₀H₁₈O	Cineole	176.35	172.05	36	556
		"	176.3	Nonazeotrope		563

No.	Formula	Name	B.P., °C	B.P., °C	Wt.%A	Ref.
		B-Component		**Azeotropic Data**		
A =	**C₇H₆O**	**Benzaldehyde** *(continued)*	**179.2**			
12573	$C_{10}H_{22}$	2,7-Dimethyloctane	160.2	<159.5		575
12574	$C_{10}H_{22}O$	Amyl ether	187.5	175.2		556
12575	$C_{10}H_{22}O$	Isoamyl ether	173.2	168.6	37.5	527
12576	$C_{11}H_{20}O$	Isobornyl methyl ether	192.4	178.0	92?	575
12577	$C_{12}H_{18}$	1,3,5-Triethylbenzene	215.5	Nonazeotrope		575
12578	$C_{12}H_{22}O$	Bornyl ethyl ether	204.9	Nonazeotrope		575
12579	$C_{12}H_{22}O$	Ethyl isobornyl ether	203.8	Nonazeotrope		575
A =	**C₇H₆O₂**	**Benzoic Acid**	**250.8**			
12580	$C_7H_7NO_2$	*m*-Nitrotoluene	230.8	Nonazeotrope		554
12581	$C_7H_7NO_2$	*o*-Nitrotoluene	221.75	Nonazeotrope		554
12582	$C_7H_7NO_2$	*p*-Nitrotoluene	238.9	237.4	11	554
12583	$C_7H_8O_2$	*m*-Methoxyphenol	243.8	Nonazeotrope		575
12584	$C_8H_8O_2$	Anisaldehyde	249.5	Nonazeotrope		538
12585	$C_8H_{10}O$	3,4-Xylenol	226.8	Nonazeotrope		575
12586	$C_8H_{11}NO$	*p*-Phenetidine	249.9	Nonazeotrope		541
12587	C_9H_8O	Cinnamaldehyde	253.5	~ 250.2	~ 90	538
		"	253.5	Nonazeotrope		575
12588	$C_9H_{10}O$	*p*-Methylacetophenone	226.35	Nonazeotrope		552
12589	$C_9H_{10}O_3$	Ethyl salicylate	234.0	233.85	6	538
12590	$C_{10}H_7Br$	1-Bromonaphthalene	281.8	249.9	~ 95	541
		"	281.8	Nonazeotrope		564
12591	$C_{10}H_7Cl$	1-Chloronaphthalene	262.7	247.8	57	541
12592	$C_{10}H_8$	Naphthalene	218.05	217.7	5	538
12593	$C_{10}H_{10}O_2$	Isosafrole	252.0	246.5	53.5	556
12594	$C_{10}H_{10}O_2$	Methyl cinnamate	261.9	Nonazeotrope		541
12595	$C_{10}H_{10}O_2$	Safrole	235.9	Nonazeotrope		537
		"	235.9	234.75	12.5	556
12596	$C_{10}H_{12}O$	Anethole	235.7	234.6	12	562
12597	$C_{10}H_{12}O_2$	Eugenol	254.8	Nonazeotrope		575
		"	254.8	250.4	96.5	538
12598	$C_{10}H_{12}O_2$	Propyl benzoate	230.85	Nonazeotrope		542
12599	$C_{10}H_{14}O$	Carvone	231.0	Nonazeotrope		552
12600	$C_{10}H_{14}O$	Carvacrol	237.85	<237.75		575
12601	$C_{10}H_{14}O$	Thymol	232.9	Nonazeotrope		575
		"	232.9	232.85?	1.5?	538
12602	$C_{10}H_{14}O_2$	*m*-Diethoxybenzene	235.0	Nonazeotrope		541
12603	$C_{10}H_{18}O_4$	Propyl succinate	250.5	248.0	43	575
12604	$C_{10}H_{20}O_4$	2-(2-Butoxyethoxy)-ethyl acetate	245.3	251.8	70	562
12605	$C_{11}H_{10}$	1-Methylnaphthalene	244.6	239.6	27	538
12606	$C_{11}H_{10}$	2-Methylnaphthalene	241.15	237.25	25	527
12607	$C_{11}H_{14}O_2$	1-Allyl-3,4-dimethoxy-benzene	254.7	Nonazeotrope		537
		"	255.0	250.3	89	538
12608	$C_{11}H_{14}O_2$	Butyl benzoate	249.0	245.5	35	562

TABLE I.   *Binary Systems*   363

No.	Formula	Name	B.P., °C	B.P., °C	Wt.%A	Ref.
		B-Component			**Azeotropic Data**	
A =	**$C_7H_6O_2$**	**Benzoic Acid** *(continued)*	**250.8**			
12609	$C_{11}H_{14}O_2$	1,2-Dimethoxy-4-propenylbenzene	270.5	Nonazeotrope		541
12610	$C_{11}H_{14}O_2$	Isobutyl benzoate	241.9	241.15	~ 12	538
12611	$C_{11}H_{16}O$	Methyl thymol ether	216.5	Nonazeotrope		575
12612	$C_{11}H_{20}O$	Methyl α-terpineol ether	216.2	Nonazeotrope		575
12613	$C_{11}H_{22}O_3$	Isoamyl carbonate	232.2	Nonazeotrope		575
12614	$C_{12}H_{10}$	Acenaphthene	277.9	~ 250.0		541
12615	$C_{12}H_{10}$	Biphenyl	277.9	246.05	50.5	541
12616	$C_{12}H_{10}O$	Phenyl ether	257	Nonazeotrope		562
		" "	259.3	247.3	59	556
		" " 100 mm.	181	176.5	27	215
12617	$C_{12}H_{16}O_3$	Isoamyl salicylate	277.5	Nonazeotrope		575
12618	$C_{12}H_{18}$	1,3,5-Triethylbenzene	215.5	Nonazeotrope		543
12619	$C_{12}H_{22}O_4$	Isoamyl oxalate	268.0	Nonazeotrope		541
12620	$C_{13}H_{10}$	Fluorene	295	Nonazeotrope		575
12621	$C_{13}H_{12}$	Diphenylmethane	265.6	248.95	82	538
12622	$C_{13}H_{12}O$	Benzyl phenyl ether	286.5	Nonazeotrope		575
12623	$C_{14}H_{12}$	Stilbene	306.5	Nonazeotrope		575
12624	$C_{14}H_{14}$	1,2-Diphenylethane	284	Nonazeotrope		543
A =	**C_7H_7Br**	**α-Bromotoluene**	**198.5**			
12625	C_7H_8O	o-Cresol	190.8	Reacts		563
12626	C_7H_8O	p-Cresol	201.8	Reacts		563
12627	C_7H_9N	Methylaniline	196.1	Reacts		563
12628	C_7H_9N	p-Toluidine	200.3	Reacts		563
12629	$C_7H_{12}O_4$	Ethyl malonate	198.9	197.3	58	563
12630	C_8H_8O	Acetophenone	202	Nonazeotrope		563
12631	$C_8H_8O_2$	Benzyl formate	203.0	<198.0		575
12632	$C_8H_8O_2$	Methyl benzoate	199.45	Nonazeotrope		575
		" "	199.55	~ 197.5	~ 59	563
12633	$C_8H_8O_2$	Phenyl acetate	195.5	194.5	~ 43	563
12634	$C_8H_{11}N$	Dimethylaniline	194.05	Reacts		563
12635	$C_8H_{14}O_4$	Propyl oxalate	214.5	Nonazeotrope		575
12636	$C_8H_{16}O_3$	Isoamyl lactate	202.4	197.6	~ 73	563
12637	$C_8H_{18}O$	Octyl alcohol	195.2	193.5	68	575
12638	$C_8H_{18}O$	sec-Octyl alcohol	180.4	Nonazeotrope		575
12639	$C_9H_{12}O$	Phenyl propyl ether	190.5	Nonazeotrope		575
12640	$C_9H_{18}O_2$	Isoamyl butyrate	181.05	Nonazeotrope		575
12641	$C_9H_{18}O_3$	Isobutyl carbonate	190.3	Nonazeotrope		547
12642	$C_{10}H_{16}O$	Fenchone	193	Nonazeotrope		563
12643	$C_{10}H_{18}O$	Citronellal	208.0	Nonazeotrope (reacts)		575
12644	$C_{10}H_{18}O$	Menthone	~207	Nonazeotrope		563
12645	$C_{10}H_{20}O_2$	Ethyl caprylate	208.35	Nonazeotrope		575
12646	$C_{10}H_{20}O_2$	Isoamyl isovalerate	192.7	Nonazeotrope		547
12647	$C_{12}H_{22}O$	Bornyl ethyl ether	204.9	Nonazeotrope		559
A =	**C_7H_7Br**	**m-Bromotoluene**	**184.3**			
12648	C_7H_8O	Benzyl alcohol	205.25	<184.15		575
12649	C_7H_8O	o-Cresol	191.1	183.05	78	527

		B-Component		Azeotropic Data		
No.	Formula	Name	B.P., °C	B.P., °C	Wt.%A	Ref.
A =	C₇H₇Br	*m*-Bromotoluene	184.3			
		(continued)				
12650	C₇H₉N	Methylaniline	196.25	Nonazeotrope		527
12651	C₇H₉N	*o*-Toluidine	200.35	Nonazeotrope		527
12652	C₇H₁₄O₂	Enanthic acid	221.3	Nonazeotrope		527
12653	C₇H₁₄O₃	Isobutyl lactate	182.15	180.4	40	567
12654	C₈H₁₁N	Dimethylaniline	194.15	Nonazeotrope		527
12655	C₈H₁₆O	2-Octanone	172.85	Nonazeotrope		527
12656	C₈H₁₆O₃	Isoamyl lactate	202.4	Nonazeotrope		575
12657	C₈H₁₈O	Octyl alcohol	195.2	184.05	91	527
12658	C₈H₁₈O	*sec*-Octyl alcohol	180.4	178.9	43	527
12659	C₉H₁₂O	Phenyl propyl ether	190.5	Nonazeotrope		559
12660	C₉H₁₃N	*N,N*-Dimethyl-*o*-toluidine	185.3	184.25	87	564
		" "	185.3	Nonazeotrope		551
12661	C₉H₁₈O₂	Isobutyl isovalerate	171.2	Nonazeotrope		575
12662	C₉H₁₈O₃	Isobutyl carbonate	190.3	182.8	75	562
12663	C₁₀H₁₄	Butylbenzene	183.1	Nonazeotrope		575
12664	C₁₀H₁₆	Dipentene	177.7	Nonazeotrope		527
12665	C₁₀H₁₆	α-Terpinene	173.4	Nonazeotrope		527
12666	C₁₀H₁₈O	Cineole	176.35	Nonazeotrope		527
A =	C₇H₇Br	*o*-Bromotoluene	181.75			
12667	C₇H₇Cl	α-Chlorotoluene	179.35	Nonazeotrope		563
12668	C₇H₈O	Benzyl alcohol	205.15	181.25	93?	531
12669	C₇H₈O	*m*-Cresol	202.2	Nonazeotrope		544
12670	C₇H₈O	*o*-Cresol	191.1	180.3	81	542
12671	C₇H₈O	*p*-Cresol	201.8	Nonazeotrope		542
12672	C₇H₉N	*m*-Toluidine	200.55	Nonazeotrope		551
12673	C₇H₉N	*o*-Toluidine	200.35	Nonazeotrope		551
12674	C₇H₉N	*p*-Toluidine	200.55	Nonazeotrope		551
12675	C₇H₁₄O₃	1,3-Butanediol methyl ether acetate	171.75	Nonazeotrope		575
12676	C₇H₁₄O₃	Isobutyl lactate	182.15	180	56	563
12677	C₇H₁₆O	Heptyl alcohol	176.15	174.0	33	567
12678	C₈H₁₀O	*p*-Methylanisole	177.05	Nonazeotrope		548
12679	C₈H₁₀O	Phenetole	170.35	Nonazeotrope		573
12680	C₈H₁₁N	Dimethylaniline	194.15	Nonazeotrope		551
12681	C₈H₁₄O	Methylheptenone	173.2	Nonazeotrope		552
12682	C₈H₁₆O	2-Octanone	172.85	Nonazeotrope		552
12683	C₈H₁₈O	Octyl alcohol	195.15	181.0		529
12684	C₈H₁₈O	*sec*-Octyl alcohol	179.0	177.0	48	572
12685	C₈H₂₀SiO₄	Ethyl silicate	168.8	Nonazeotrope		575
12686	C₉H₈	Indene	182.3	<180.5		563
12687	C₉H₁₃N	*N,N*-Dimethyl-*o*-toluidine	185.3	Nonazeotrope		551
12688	C₉H₁₈O₂	Butyl isovalerate	177.6	Nonazeotrope		547
12689	C₉H₁₈O₂	Ethyl enanthate	188.7	Nonazeotrope		575
12690	C₉H₁₈O₂	Isoamyl butyrate	178.5	Nonazeotrope		547
12691	C₉H₁₈O₂	Isoamyl isobutyrate	170.0	Nonazeotrope		547
12692	C₉H₁₈O₂	Isobutyl isovalerate	168.7	Nonazeotrope		563

TABLE I. *Binary Systems* 365

No.	Formula	Name	B.P., °C	B.P., °C	Wt.%A	Ref.
		B-Component		Azeotropic Data		
A =	**C₇H₇Br**	*o*-**Bromotoluene** *(continued)*	**181.75**			
12693	C₉H₁₈O₃	Isobutyl carbonate	190.3	180.5	~ 90	563
		"	190.3	Nonazeotrope		547
12694	C₁₀H₁₄	Cymene	176.7	Nonazeotrope		538
12695	C₁₀H₁₆	*d*-Limonene	177.8	Nonazeotrope		535
		"	177.8	177.3	~ 17	563
12696	C₁₀H₁₆	α-Terpinene	173.4	Nonazeotrope		575
12697	C₁₀H₁₆	γ-Terpinene	181.5	181.0		538
12698	C₁₀H₁₆	Terpinolene	184.6	Nonazeotrope		575
12699	C₁₀H₁₆	Thymene	179.7	179.55	~ 15	573
12700	C₁₀H₁₈O	Cineole	176.4	Nonazeotrope		528
12701	C₁₀H₁₈O	Linalool	198.6	Nonazeotrope		529
12702	C₁₀H₁₉N	Bornylamine	199.8	Nonazeotrope		575
12703	C₁₀H₂₀O₂	Isoamyl isovalerate	192.7	Nonazeotrope		547
12704	C₁₀H₂₂O	Isoamyl ether	173.5	Nonazeotrope		548
A =	**C₇H₇Br**	*p*-**Bromotoluene**	**185.0**			
12705	C₇H₈O	Benzyl alcohol	205.2	~ 184.5	~ 92	535
12706	C₇H₈O	*m*-Cresol	202.2	184.8	~ 95	542
12707	C₇H₈O	*o*-Cresol	191.1	182.7	72	538
12708	C₇H₈O	*p*-Cresol	201.7	184.8	~ 93	542
12709	C₇H₉N	*o*-Toluidine	200.35	Nonazeotrope		551
12710	C₇H₉N	*p*-Toluidine	200.55	Nonazeotrope		551
12711	C₇H₁₂O₄	Ethyl malonate	199.2	Nonazeotrope		547
12712	C₇H₁₄O₃	Isobutyl lactate	182.15	180.2	38	567
12713	C₈H₈O₂	Phenyl acetate	195.7	Nonazeotrope		547
12714	C₈H₁₀O	2,4-Xylenol	210.5	Nonazeotrope		575
12715	C₈H₁₁N	Dimethylaniline	194.15	Nonazeotrope		551
		"	194.05	184.2	85	535
12716	C₈H₁₈O	Octyl alcohol	195.2	184.6	90	575
12717	C₉H₁₄O	Phorone	197.8	Nonazeotrope		552
12718	C₉H₁₈O₂	Butyl isovalerate	177.6	Nonazeotrope		575
12719	C₉H₁₈O₂	Isoamyl butyrate	178.5	Nonazeotrope		547
12720	C₉H₁₈O₃	Isobutyl carbonate	190.3	182.9	~ 35	563
12721	C₁₀H₁₄	Butylbenzene	183.1	Nonazeotrope		575
12722	C₁₀H₁₄	Cymene	176.7	Nonazeotrope		575
12723	C₁₀H₁₆	*d*-Limonene	177.8	Nonazeotrope		535
12724	C₁₀H₁₆	α-Terpinene	173.4	Nonazeotrope		575
12725	C₁₀H₁₆	γ-Terpinene	183	182.8	15	575
12726	C₁₀H₁₆	Terpinolene	185	~ 183		563
12727	C₁₀H₁₆	Thymene	179.7	Nonazeotrope		535
12728	C₁₀H₁₆O	Fenchone	193.6	Nonazeotrope		575
12729	C₁₀H₁₈O	Cineole	176.35	Nonazeotrope		573
12730	C₁₀H₁₈O	Linalool	198.6	Nonazeotrope		532
12731	C₁₀H₂₀O₂	Isoamyl isovalerate	192.7	Nonazeotrope		547
A =	**C₇H₇BrO**	*o*-**Bromoanisole**	**217.7**			
12732	C₇H₇I	*p*-Iodotoluene	214.5	<214.3	<10	575

No.	Formula	Name	B.P., °C	B.P., °C	Wt.%A		Ref.
		B-Component		**Azeotropic Data**			
A =	**C₇H₇BrO**	***o*-Bromoanisole**	**217.7**				
		(*continued*)					
12733	C₇H₈O	*m*-Cresol	202.2	Nonazeotrope			575
12734	C₈H₁₀O	3,4-Xylenol	226.8	Nonazeotrope			575
12735	C₉H₇N	Quinoline	237.3	Nonazeotrope			575
12736	C₁₀H₈	Naphthalene	218.0	<216.5	>55		562
12737	C₁₀H₁₂O	Estragole	215.6	Nonazeotrope			548
12738	C₁₀H₁₈O	Citronellal	208.0	Nonazeotrope			575
12739	C₁₁H₂₀O	Terpineol methyl ether	216.2	~ 215.0	>15		548
A =	**C₇H₇BrO**	***p*-Bromoanisole**	**217.7**				
12740	C₉H₁₀O	*p*-Methylacetophenone	226.25	Nonazeotrope			575
12741	C₉H₁₀O	Propiophenone	217.7	<217.4	>54		575
12742	C₁₀H₁₆O	Pulegone	223.8	Nonazeotrope			575
A =	**C₇H₇Cl**	**α-Chlorotoluene**	**179.3**				
12743	C₇H₈O	Benzyl alcohol	205.15	Nonazeotrope			530
		"		205.2	Nonazeotrope	v-l	494
		"	100 mm.	Nonazeotrope		v-l	494
		"	15 mm.	Nonazeotrope		v-l	494
12744	C₇H₈O	*o*-Cresol	190.8	Reacts			563
12745	C₇H₁₄O	2-Methylcyclohexanol	168.5	<168.2	<34		575
12746	C₇H₁₄O₂	Enanthic acid	222.0	204.0	88		575
12747	C₇H₁₄O₃	Isobutyl lactate	182.15	178.0	~ 70		563
12748	C₇H₁₆O	Heptyl alcohol	176.15	<173.5	<51		575
12749	C₈H₈O	Acetophenone	202.0	Nonazeotrope			552
12750	C₈H₁₀O	Phenetole	170.35	Nonazeotrope			530
12751	C₈H₁₁N	Dimethylaniline	194.05	Reacts			563
12752	C₈H₁₄O	Methylheptenone	173.2	Nonazeotrope			552
12753	C₈H₁₆O	2-Octanone	172.85	Nonazeotrope			552
12754	C₈H₁₆O₂	Butyl butyrate	166.4	Nonazeotrope			547
12755	C₈H₁₆O₂	Ethyl caproate	167.7	Nonazeotrope			575
12756	C₈H₁₆O₂	Hexyl acetate	171.5	Nonazeotrope			575
12757	C₈H₁₆O₂	Isoamyl propionate	160.4	Nonazeotrope			547
12758	C₈H₁₈O	*n*-Octyl alcohol	195.15	Nonazeotrope			530
12759	C₈H₁₈O	*sec*-Octyl alcohol	179.0	165.7			531
12760	C₉H₁₀O₂	Benzyl acetate		Nonazeotrope		v-l	495
		" "	100 mm.	Nonazeotrope		v-l	495
		" "	15 mm.	Nonazeotrope		v-l	495
12761	C₉H₁₂	Mesitylene	164.6	Nonazeotrope			575
12762	C₉H₁₂	Propylbenzene	159.3	Nonazeotrope			575
12763	C₉H₁₂	Pseudocumene	169	Nonazeotrope			563
12764	C₉H₁₈O₂	Butyl isovalerate	177.6	Nonazeotrope			547
12765	C₉H₁₈O₂	Isoamyl butyrate	178.5	~ 178.2	30?		530
12766	C₉H₁₈O₂	Isoamyl isobutyrate	170.0	Nonazeotrope			547
12767	C₉H₁₈O₂	Isobutyl isovalerate	171.35	Nonazeotrope			538
12768	C₉H₁₈O₃	Isobutyl carbonate	190.3	Nonazeotrope			547
12769	C₁₀H₁₄	Cymene	175.3	174	< 20		563
12770	C₁₀H₁₆	Camphene	159.6	Nonazeotrope			575

TABLE I. *Binary Systems* 367

No.	Formula	B-Component Name	B.P., °C	Azeotropic Data B.P., °C	Wt.%A	Ref.
A =	C_7H_7Cl	α-Chlorotoluene *(continued)*	**179.3**			
12771	$C_{10}H_{16}$	*d*-Limonene	177.8	174.8	46	563
12772	$C_{10}H_{16}$	Nopinene	163.8	Nonazeotrope		575
12773	$C_{10}H_{16}$	α-Phellandrene	171.5	170?		563
12774	$C_{10}H_{16}$	α-Terpinene	173.4	173.0		562
12775	$C_{10}H_{16}$	γ-Terpinene	181.5	176.9	~ 70	538
12776	$C_{10}H_{16}$	Terpinolene	185	~ 177.5		563
12777	$C_{10}H_{16}$	Thymene	179.7	177.2	~ 52	531
12778	$C_{10}H_{18}$	*m*-Menthene-8	170.8	<170.0	<15	562
12779	$C_{10}H_{18}O$	Cineol	176.3	175.5	~ 19	563
12780	$C_{10}H_{18}O$	Linalool	198.6	Nonazeotrope		532
12781	$C_{10}H_{22}O$	Isoamyl ether	172.6	Nonazeotrope		573
A =	C_7H_7Cl	*o*-Chlorotoluene	**159.15**			
12782	C_7H_8O	Anisole	153.85	Nonazeotrope		548
12783	C_7H_8O	*o*-Cresol	191.1	Nonazeotrope		575
12784	$C_7H_{14}O$	2-Methylcyclohexanol	168.5	158.4		575
12785	$C_7H_{14}O_2$	Methyl caproate	151.0	Nonazeotrope		547
12786	$C_8H_{10}O$	Benzyl methyl ether	167.8	Nonazeotrope		559
12787	$C_8H_{10}O$	Phenetole	170.45	Nonazeotrope		559
12788	$C_8H_{14}O$	Methylheptenone	173.2	Nonazeotrope		552
12789	$C_8H_{16}O_2$	Butyl butyrate	166.4	Nonazeotrope		547
12790	$C_8H_{16}O_2$	Isoamyl propionate	160.3	158.0	>65	538
12791	$C_8H_{16}O_2$	Isobutyl butyrate	157	155.5	<50	563
		"		156.8	Nonazeotrope	547
12792	$C_8H_{16}O_2$	Isobutyl isobutyrate	147.3	Nonazeotrope		547
12793	$C_8H_{16}O_2$	Propyl isovalerate	155.7	Nonazeotrope		547
12794	$C_8H_{18}O$	*sec*-Octyl alcohol	180.4	Nonazeotrope		575
12795	$C_8H_{20}SiO_4$	Ethyl silicate	168.8	Nonazeotrope		575
12796	C_9H_{12}	Cumene	152.8	Nonazeotrope		575
12797	C_9H_{12}	Mesitylene	164.6	Nonazeotrope		538
12798	C_9H_{12}	Pseudocumene	168.2	Nonazeotrope		575
12799	$C_9H_{18}O$	2,6-Dimethyl-4-heptanone	168.0	Nonazeotrope		552
12800	$C_9H_{18}O_2$	Isobutyl isovalerate	171.35	Nonazeotrope		538
12801	$C_{10}H_{16}$	Camphene	159.6	~ 158.0		538
12802	$C_{10}H_{16}$	Nopinene	163.8	<158.5	>63	562
12803	$C_{10}H_{16}$	α-Pinene	155.8	154.5		562
12804	$C_{10}H_{16}$	α-Terpinene	173.4	Nonazeotrope		575
A =	C_7H_7Cl	*p*-Chlorotoluene	**161.3**			
12805	C_7H_8O	Anisole	153.85	Nonazeotrope		563
12806	C_7H_8O	*o*-Cresol	191.1	Nonazeotrope		575
12807	$C_7H_{14}O$	2-Methylcyclohexanol	168.5	161.1	75	567
12808	$C_7H_{14}O_3$	1,3-Butanediol methyl ether acetate	171.75	Nonazeotrope		575
12809	$C_7H_{16}O$	Heptyl alcohol	176.15	161.9	~ 92	575
12810	$C_8H_{10}O$	Phenetole	170.35	Nonazeotrope		573
12811	$C_8H_{14}O$	Methylheptenone	173.2	Nonazeotrope		552
12812	$C_8H_{16}O$	2-Octanone	172.85	Nonazeotrope		552

No.	B-Component		B.P., °C	Azeotropic Data		
	Formula	Name	B.P., °C	B.P., °C	Wt.%A	Ref.
A =	**C₇H₇Cl**	**p-Chlorotoluene**	**161.3**			
		(continued)				
12813	C₈H₁₆O₂	Butyl butyrate	166.4	Nonazeotrope		547
12814	C₈H₁₆O₂	Ethyl caproate	167.9	Nonazeotrope		547
12815	C₈H₁₆O₂	Isoamyl propionate	160.3	159.5		547
12816	C₈H₁₈O	sec-Octyl alcohol	179.0	Nonazeotrope		573
12817	C₈H₁₈S	Isobutyl sulfide	172.0	Nonazeotrope		566
12818	C₈H₂₀SiO₄	Ethyl silicate	168.8	Nonazeotrope		575
12819	C₉H₁₂	Cumene	152.8	Nonazeotrope		575
12820	C₉H₁₂	Mesitylene	164.0	160.5	~ 72	563
12821	C₉H₁₂	Pseudocumene	168.2	Nonazeotrope		538
12822	C₉H₁₃N	N,N-Dimethyl-o-toluidine	185.3	Nonazeotrope		551
12823	C₉H₁₈O	2,6-Dimethyl-4-heptanone	168.0	Nonazeotrope		552
12824	C₉H₁₈O₂	Isoamyl butyrate	181.05	Nonazeotrope		575
12825	C₉H₁₈O₂	Isobutyl isovalerate	171.2	Nonazeotrope		575
12826	C₁₀H₁₆	Cymene	176.7	Nonazeotrope		575
12827	C₁₀H₁₆	Camphene	159.6	~ 158.0		535
12828	C₁₀H₁₆	Dipentene	177.7	Nonazeotrope		575
12829	C₁₀H₁₆	Nopinene	163.8	160.2		563
12830	C₁₀H₁₆	α-Pinene	155.8	<155.5	<20	575
12831	C₁₀H₁₆	α-Terpinene	173.4	Nonazeotrope		575
12832	C₁₀H₁₈	m-Menthene-8	170.8	Nonazeotrope		575
12833	C₁₀H₁₈O	Cineole	176.35	Nonazeotrope		559
12834	C₁₀H₂₂	2,7-Dimethyloctane	160.25	158.5	~ 50	563
A =	**C₇H₇ClO**	**m-Chloroanisole**	**193.3**			
12835	C₇H₁₄O₃	1,3-Butanediol methyl ether				
		acetate	171.75	Nonazeotrope		575
12836	C₈H₁₈S	Butyl sulfide	185.0	Nonazeotrope		575
A =	**C₇H₇ClO**	**o-Chloroanisole**	**195.7**			
12837	C₇H₈O	o-Cresol	191.1	<189.8	>20	562
A =	**C₇H₇ClO**	**p-Chloroanisole**	**193.3**			
12838	C₈H₈O	Acetophenone	202.0	Nonazeotrope		575
12839	C₈H₁₈S	Butyl sulfide	185.0	Nonazeotrope		575
12840	C₉H₁₂O	Benzyl ethyl ether	185.0	Nonazeotrope		575
12841	C₉H₁₄O	Phorone	197.8	<197.4		575
12842	C₁₀H₈	Naphthalene	218.0	Nonazeotrope		575
12843	C₁₀H₁₆O	Camphor	209.1	Nonazeotrope		575
12844	C₁₀H₁₆O	Fenchone	193.6	Nonazeotrope		575
12845	C₁₂H₁₈	1,3,5-Triethylbenzene	215.5	Nonazeotrope		575
A =	**C₇H₇I**	**p-Iodotoluene**	**214.5**			
12846	C₇H₇NO₂	o-Nitrotoluene	221.75	Nonazeotrope		554
12847	C₇H₈O	Benzyl alcohol	205.15	~ 203.0	25?	529
12848	C₇H₈O	m-Cresol	202.2	201.6	25	543
12849	C₇H₈O	o-Cresol	190.8	Nonazeotrope		563
12850	C₇H₈O	p-Cresol	201.7	201.0	23	542
12851	C₇H₉N	m-Toluidine	203.1	Nonazeotrope		551
12852	C₇H₉N	o-Toluidine	200.35	Nonazeotrope		551
12853	C₇H₉N	p-Toluidine	200.55	Nonazeotrope		551

TABLE I. *Binary Systems* 369

No.	Formula	B-Component Name	B.P., °C	Azeotropic Data B.P., °C	Wt.%A	Ref.
A =	**C₇H₇I**	**p-Iodotoluene** *(continued*	**214.5**			
12854	C₇H₉NO	o-Anisidine	219.0	213.0	70?	575
12855	C₇H₁₂O₄	Ethyl malonate	199.35	<198.8	> 8	575
12856	C₇H₁₄O₂	Enanthic-acid	222.0	211.5	83	562
12857	C₈H₈O₂	Methyl benzoate	199.4	Nonazeotrope		575
12858	C₈H₈O₃	Methyl salicylate	222.95	Nonazeotrope		575
12859	C₈H₁₀O	p-Ethylphenol	218.8	212.0	72	562
12860	C₈H₁₀O	Phenethyl alcohol	219.4	<211.5		575
12861	C₈H₁₀O	2,4-Xylenol	210.5	207.5	38	575
12862	C₈H₁₀O	3,4-Xylenol	226.8	214.0	85	575
12863	C₈H₁₀O₂	m-Dimethoxybenzene	214.7	Nonazeotrope		535
12864	C₈H₁₀O₂	Veratrole	205.5	Nonazeotrope		535
12865	C₈H₁₁N	Dimethylaniline	194.15	Nonazeotrope		551
12866	C₈H₁₁N	2,4-Xylidine	214.0	<212.5		575
12867	C₈H₁₄O₄	Propyl oxalate	214	<209.2	>53	575
12868	C₈H₁₆O₃	Isoamyl lactate	202.4	Nonazeotrope		575
12869	C₈H₁₈O	Octyl alcohol	195.2	Nonazeotrope		575
12870	C₈H₁₈O₃	2-(2-Butoxyethoxy)-ethanol	231.2	Nonazeotrope		575
12871	C₉H₇N	Quinoline	237.3	Nonazeotrope		553
12872	C₉H₁₀O	Propiophenone	217.7	Nonazeotrope		552
12873	C₉H₁₀O₂	Ethyl benzoate	212.5	<212.3	>14	575
12874	C₁₀H₂₀O	Menthol	216.3	<213.0		575
12875	C₁₀H₂₀O₂	Ethyl caprylate	208.35	Nonazeotrope		575
12876	C₁₀H₂₂S	Isoamyl sulfide	214.8	<213.3	>42	562
A =	**C₇H₇NO₂**	**m-Nitrotoluene**	**230.8**			
12877	C₇H₈O	Benzyl alcohol	205.25	Nonazeotrope		554
12878	C₇H₉NO	o-Anisidine	219.0	Nonazeotrope		575
12879	C₇H₁₄O₂	Enanthic acid	222.0	220.0	30	554
12880	C₇H₁₆O₄	2-[2-(2-Methoxyethoxy)-ethoxy]-ethanol	245.25	226.4 Nonazeotrope	77	554
12881	C₈H₈O₃	Methyl salicylate	222.95	Nonazeotrope		554
12882	C₈H₁₀O	3,4-Xylenol	226.8	Nonazeotrope		554
12883	C₈H₁₀O	p-Ethylphenol	220.0	Nonazeotrope		554
12884	C₈H₁₁NO	o-Phenetidine	232.5	233.0	30	551
12885	C₈H₁₁NO	p-Phenetidine	249.9	Nonazeotrope		551
12886	C₈H₁₂O₄	Ethyl fumarate	217.85	Nonazeotrope		554
12887	C₈H₁₂O₄	Ethyl maleate	223.3	Nonazeotrope		554
12888	C₈H₁₂O₄	Ethyl succinate	217.25	Nonazeotrope		554
12889	C₈H₁₆O₂	Caprylic acid	238.5	<229.8	<80	554
12890	C₈H₁₈O₃	2-(2-Butoxyethoxy)-ethanol	231.2	<229.0	<70	554
12891	C₉H₇N	Quinoline	237.6	Nonazeotrope		554
12892	C₈H₁₀O	Cinnamyl alcohol	257.0	Nonazeotrope		554
12893	C₈H₁₀O	p-Methylacetophenone	226.35	Nonazeotrope		575
12894	C₉H₁₀O	Propiophenone	217.7	Nonazeotrope		575
12895	C₉H₁₀O₂	Benzyl acetate	215.0	Nonazeotrope		554

		B-Component		Azeotropic Data		
No.	Formula	Name	B.P., °C	B.P., °C	Wt.%A	Ref.
A =	**$C_7H_7NO_2$**	***m*-Nitrotoluene**	**230.8**			
		(continued)				
12896	$C_9H_{10}O_3$	Ethyl salicylate	253.8	Nonazeotrope		554
12897	$C_9H_{12}O$	3-Phenylpropanol	235.6	229.5	68	554
12898	$C_9H_{18}O_2$	Pelargonic acid	254.0	Nonazeotrope		575
12899	$C_{10}H_8$	Naphthalene	218.0	Nonazeotrope		554
12900	$C_{10}H_{10}O_2$	Safrol	232	227	55	563
12901	$C_{10}H_{12}O_2$	Propyl benzoate	230.85	230.0	48	554
12902	$C_{10}H_{14}O$	Carvacrol	237.85	Nonazeotrope		554
12903	$C_{10}H_{14}O$	Carvone	231.0	230.5		545
12904	$C_{10}H_{14}O$	Thymol	232.9	Nonazeotrope		554
12905	$C_{10}H_{15}N$	Diethylaniline	217.05	Nonazeotrope		551
12906	$C_{10}H_{16}O$	Pulegone	223.8	Nonazeotrope		575
12907	$C_{10}H_{18}O$	Borneol	213.4	Nonazeotrope		576
12908	$C_{10}H_{18}O$	Geraniol	229.6	227.3	49	554
12909	$C_{10}H_{18}O$	α-Terpineol	218.85	218.65	8	554
12910	$C_{10}H_{20}O$	Citronellol	224.4	223.2	>26	554
12911	$C_{10}H_{20}O$	Menthol	216.3	<216.2		554
12912	$C_{10}H_{22}O$	*n*-Decanol	232.8	228.2	60	554
12913	$C_{11}H_{10}$	1-Methylnaphthalene	244.6	Nonazeotrope		554
12914	$C_{11}H_{10}$	2-Methylnaphthalene	241.15	Nonazeotrope		554
12915	$C_{11}H_{14}O_2$	Ethyl β-phenylpropionate	248.1	Nonazeotrope		554
12916	$C_{11}H_{14}O_2$	Isobutyl benzoate	241.9	Nonazeotrope		554
12917	$C_{11}H_{17}N$	Isoamylaniline	256.0	Nonazeotrope		551
12918	$C_{11}H_{22}O_3$	Isoamyl carbonate	232.2	<230.2	>56	554
12919	$C_{12}H_{20}O_2$	Bornyl acetate	227.6	<226.5	>28	554
A =	**$C_7H_7NO_2$**	***o*-Nitrotoluene**	**221.75**			
12920	C_7H_8O	Benzyl alcohol	202.25	Nonazeotrope		554
			205.2	204.75	9	536
12921	C_7H_8O	*m*-Cresol	202.2	Nonazeotrope		554
12922	C_7H_9N	Methylaniline	196.25	Nonazeotrope		551
12923	C_7H_9N	*o*-Toluidine	200.35	Nonazeotrope		551
12924	C_7H_9N	*p*-Toluidine	200.55	Nonazeotrope		551
12925	$C_7H_{14}O_2$	Enanthic acid	222.0	<218.0	<60	554
12926	$C_7H_{16}O_4$	2-[2-(2-Methoxyethoxy)-				
		ethoxy]-ethanol	245.25	<220.8	88	554
12927	$C_8H_8O_2$	Phenyl acetate	228.75	Nonazeotrope		554
12928	$C_8H_8O_3$	Methyl salicylate	222.95	221.65	86	554
12929	$C_8H_{10}O$	2-Phenethyl alcohol	219.4	217.6	43	554
12930	$C_8H_{10}O$	2,4-Xylenol	210.5	Nonazeotrope		575
12931	$C_8H_{10}O$	3,4-Xylenol	226.8	Nonazeotrope		554
12932	$C_8H_{10}O_2$	*m*-Dimethoxybenzene	214.7	Nonazeotrope		537
12933	$C_8H_{10}O_2$	*o*-Ethoxyphenol	216.5	Nonazeotrope		554
12934	$C_8H_{10}O_2$	Veratrol	206.5	Nonazeotrope		537
12935	$C_8H_{11}N$	Dimethylaniline	194.15	Nonazeotrope		551
12936	$C_8H_{11}N$	Ethylaniline	205.5	Nonazeotrope		551
12937	$C_8H_{11}N$	2,4-Xylidine	214.0	Nonazeotrope		551

TABLE I. *Binary Systems* 371

No.	Formula	B-Component Name	B.P., °C	Azeotropic Data B.P., °C	Wt.%A	Ref.
A =	**C₇H₇NO₂**	**o-Nitrotoluene**	**221.75**			
		(continued)				
12938	C₈H₁₁N	3,4-Xylidine	225.5	Nonazeotrope		551
12939	C₈H₁₁NO	o-Phenetidine	232.5	Nonazeotrope		551
12940	C₈H₁₂O₄	Ethyl fumarate	217.85	Nonazeotrope		554
12941	C₈H₁₂O₄	Ethyl maleate	223.3	221.1	62	554
12942	C₈H₁₄O₄	Ethyl succinate	217.25	217.1		554
12943	C₈H₁₆O₂	Caprylic acid	237.5	Nonazeotrope		541
12944	C₈H₁₈O	n-Octyl alcohol	195.2	Nonazeotrope		554
12945	C₉H₇N	Quinoline	237.3	Nonazeotrope		554
12946	C₉H₁₀O	Cinnamyl alcohol	257.0	Nonazeotrope		554
12947	C₉H₁₀O	p-Methylacetophenone	226.35			552
12948	C₉H₁₀O	Propiophenone	217.7	Nonazeotrope		552
12949	C₉H₁₀O₂	Benzyl acetate	215.0	Nonazeotrope		554
12950	C₉H₁₀O₂	Ethyl benzoate	212.5	Nonazeotrope		554
12951	C₉H₁₀O₃	Ethyl salicylate	233.8	Nonazeotrope		554
12952	C₉H₁₂O	3-Phenylpropanol	235.6	Nonazeotrope		554
		"	235.6	235.3	92	545
12953	C₉H₁₃N	N,N-Dimethyl-p-toluidine	210.2	Nonazeotrope		551
12954	C₁₀H₈	Naphthalene	218.0	Nonazeotrope		554
12955	C₁₀H₁₀O₂	Safrole	235.9	Nonazeotrope		554
12956	C₁₀H₁₂O₂	Propyl benzoate	230.85	Nonazeotrope		554
12957	C₁₀H₁₄O	Carvacrol	237.85	Nonazeotrope		554
12958	C₁₀H₁₄O	Thymol	232.9	Nonazeotrope		554
12959	C₁₀H₁₅N	Diethylaniline	217.05	216.85	12	551
12960	C₁₀H₁₆O	Camphor	209.1	Nonazeotrope		552
12961	C₁₀H₁₆O	Pulegone	223.8	Nonazeotrope		552
12962	C₁₀H₁₇Cl	Bornyl chloride	207.5	Nonazeotrope		554
12963	C₁₀H₁₈O	Borneol	215.0	213.5	25	554
12964	C₁₀H₁₈O	Citronellal	208.0	Nonazeotrope		554
12965	C₁₀H₁₈O	Geraniol	229.6	220.7	81	554
12966	C₁₀H₁₈O	Linalool	198.6	Nonazeotrope		554
12967	C₁₀H₁₈O	α-Terpineol	218.85	217.1	38	554
12968	C₁₀H₁₈O	β-Terpineol	210.5	209.7	10	554
12969	C₁₀H₂₀O	Citronellol	224.4	219.8	62	554
12970	C₁₀H₂₀O	Menthol	216.3	214.65	34	554
12971	C₁₀H₂₀O₂	Methyl pelargonate	213.8	Nonazeotrope		554
12972	C₁₀H₂₂O	n-Decyl alcohol	232.8	221.0	85	554
12973	C₁₀H₂₂S	Isoamyl sulfide	214.8	Nonazeotrope		575
12974	C₁₁H₁₀	1-Methylnaphthalene	244.6	Nonazeotrope		554
12975	C₁₁H₁₀	2-Methylnaphthalene	241.15	Nonazeotrope		554
12976	C₁₁H₂₀O	Methyl α-terpineol ether	216.2	215.0	15?	554
12977	C₁₁H₂₂O₃	Isoamyl carbonate	232.2	Nonazeotrope		554
12978	C₁₂H₁₈	1,3,5-Triethylbenzene	215.5	Nonazeotrope		554
12979	C₁₂H₂₀O₂	Bornyl acetate	227.6	221.15	73	554
A =	**C₇H₇NO₂**	**p-Nitrotoluene**	**238.9**			
12980	C₇H₈O	Benzyl alcohol	202.25	Nonazeotrope		554

No.	Formula	B-Component Name	B.P., °C	B.P., °C	Azeotropic Data Wt.%A	Ref.
A =	C₇H₇NO₂	*p*-Nitrotoluene *(continued)*	238.9			
12981	C₇H₁₆O₄	2-[2-(2-Methoxyethoxy)-ethoxy)-]ethanol	245.25	231.2	61	554
12982	C₈H₈O₂	Anisaldehyde	249.5	Nonazeotrope		538
12983	C₈H₈O₂	α-Toluic acid	266.8	Nonazeotrope		554
12984	C₈H₁₀O	3,4-Xylenol	226.8	Nonazeotrope		554
12985	C₈H₁₀O	2-Phenylethanol	219.4	Nonazeotrope		554
12986	C₈H₁₁NO	*o*-Phenetidine	232.5	Nonazeotrope		551
12987	C₈H₁₁NO	*p*-Phenetidine	249.9	Nonazeotrope		551
12988	C₈H₁₆O₂	Caprylic acid	238.5	<235.0	<38	554
12989	C₉H₇N	Quinoline	237.3	237.2	8	553
12990	C₉H₈O	Cinnamyl aldehyde	253.5	Nonazeotrope		554
12991	C₉H₁₀O₃	Ethyl salicylate	233.8	Nonazeotrope		554
12992	C₉H₁₂O	3-Phenylpropanol	235.6	234.0	38	554
12993	C₉H₁₂O₂	2-Benzyloxyethanol	265.2	Nonazeotrope		554
12994	C₁₀H₇Cl	1-Chloronaphthalene	262.7	Nonazeotrope		554
12995	C₁₀H₁₀O₂	Isosafrole	252.1	Nonazeotrope		554
12996	C₁₀H₁₀O₂	Safrole	235.9	234.5	18	554
12997	C₁₀H₁₂O₂	Eugenol	254.8	Nonazeotrope		554
12998	C₁₀H₁₂O₂	Ethyl α-toluate	228.75	Nonazeotrope		554
12999	C₁₀H₁₂O₂	Propyl benzoate	230.85	Nonazeotrope		554
13000	C₁₀H₁₄O	Carvacrol	237.85	237.7	>25	554
13001	C₁₀H₁₄O	Carvone	231.0	Nonazeotrope		552
13002	C₁₀H₁₄O	Thymol	232.9	Nonazeotrope		554
13003	C₁₀H₁₈O	Borneol	215.0	Nonazeotrope		554
13004	C₁₀H₁₈O	Geraniol	229.6	228.8	25	554
13005	C₁₀H₁₈O	α-Terpineol	217.8	~ 217.6	5	536
		"		Nonazeotrope		545
13006	C₁₀H₁₈O	β-Terpineol	210.5	Nonazeotrope		554
13007	C₁₀H₂₀O	Menthol	216.3	Nonazeotrope		554
		"	216.4	216.3	3	536
13008	C₁₀H₂₂O	*n*-Decyl alcohol	232.8	231.5	33	554
13009	C₁₁H₁₀	1-Methylnaphthalene	244.6	Nonazeotrope		554
13010	C₁₁H₁₀	2-Methylnaphthalene	241.15	Nonazeotrope		527
13011	C₁₁H₁₄O₂	Butyl benzoate	249.5	Nonazeotrope		554
13012	C₁₁H₁₄O₂	Ethyl β-phenylpropionate	248.1	Nonazeotrope		554
13013	C₁₁H₁₄O₂	Isobutyl benzoate	241.9	238.6	70	554
13014	C₁₁H₁₇N	Isoamylaniline	256.0	Nonazeotrope		551
13015	C₁₂H₁₀	Biphenyl	256.1	Nonazeotrope		554
13016	C₁₂H₁₀O	Phenyl ether	259.0	Nonazeotrope		554
13017	C₁₂H₁₆O₃	Isoamyl salicylate	277.5	Nonazeotrope		554
13018	C₁₂H₂₀O₂	Bornyl acetate	227.6	227.45	10	554
A =	C₇H₈	**Toluene**	110.7			
13019	C₇H₈O	Benzyl alcohol	204.7	Nonazeotrope	v-l	935
13020	C₇H₈O	*p*-Cresol	201.7	Nonazeotrope		625
13021	C₇H₉N	2,6-Lutidine	144	Nonazeotrope		562
13022	C₇H₁₄	Ethylcyclopentane	103.5	103.0	7	516,632

TABLE I.   *Binary Systems*   373

No.	Formula	Name	B.P., °C	B.P., °C	Wt.%A	Ref.
		B-Component		**Azeotropic Data**		
A =	**C₇H₈**	**Toluene** *(continued)*	**110.7**			
13023	C₇H₁₄	3-Heptene		Nonazeotrope	v-l	1016
13024	C₇H₁₄	Methylcyclohexane	101.0	Nonazeotrope	v-l	236
		" 500 mm.	86.7	Nonazeotrope	v-l	236
		" 350 mm.	75.5	Nonazeotrope	v-l	236
		" 200 mm.	59.6	Nonazeotrope	v-l	236
		"	101.1	Nonazeotrope	v-l	317,772,
						893
		" 60°-100° C.		Evaporation data		851
13025	C₇H₁₄O	2-Methylcyclohexanol	168.5	Nonazeotrope		575
13026	C₇H₁₄O₂	Isopropyl isobutyrate	120.8	Nonazeotrope		575
13027	C₇H₁₆	Heptane 25°C.		Nonazeotrope	v-l	932c
		" 90°C		Nonazeotrope	v-l	1022
		"	98	Evaporation data	v-l	317
		"	98.4	Nonazeotrope	v-l	1016
13028	C₇H₁₆O	Heptyl alcohol	176.15	Nonazeotrope		575
13029	C₈H₁₀	Ethylbenzene 90°C	136.18	Nonazeotrope		
				(b.p. curve)		1045
13030	C₈H₁₀	*p*-Xylene 90°C		Nonazeotrope	v-l	1022
13031	C₈H₁₁N	2-Methyl-5-ethyl-				
		pyridine	178.3	Nonazeotrope		981
13032	C₈H₁₆	1,3-Dimethylcyclohexane	120.7	Nonazeotrope		575
13033	C₈H₁₆	*cis*-1,3-Dimethylcyclo-				
		hexane	120.1	110.6	96	632
13034	C₈H₁₆	1,1,3-Trimethylcyclopen-				
		tane	104.9	103.8	16	632
13035	C₈H₁₆	*cis-trans-cis*-1,2,3-Trime-				
		thylcyclopentane	110.4	108.0	39	632
13036	C₈H₁₆	*cis-trans-cis*-1,2,4-Trime-				
		thylcyclopentane	109.3	107.0	39	632
13037	C₈H₁₆	2,3,4-Trimethyl-2-pentene	116	110	82	632
13038	C₈H₁₈	2,5-Dimethylhexane	109.4	107.0	35	632
13039	C₈H₁₈	2-Methylheptane	117.6	110.3	82	632
13040	C₈H₁₈	Isooctane	99.3	Nonazeotrope	v-l	759
13041	C₈H₁₈	*n*-Octane	125.4	Nonazeotrope	v-l	86,203
13042	C₈H₁₈	2,3,4-Trimethylpentane	113.5	109.5	60	632
13043	C₈H₁₈	2,2,4-Trimethylpentane	99.3	Nonazeotrope	v-l	163,779c
13044	C₈H₁₈O	Isobutyl ether	122	Nonazeotrope		537
13045	C₈H₁₈O	*sec*-Octyl alcohol	180.4	Nonazeotrope		575
13046	C₁₀H₂₂O	Decyl alcohol (isomers)		Nonazeotrope		981
13047	C₁₂H₂₆O	2,6,8-Trimethyl-				
		4-nonanol	225.5	Nonazeotrope		981
A =	**C₇H₈SiCl₂**	**Methylphenyldichlorosilane**	**203.6**			
13048	C₈H₁₁SiCl	Dimethylphenyl-				
		chlorosilane	194.6	Nonazeotrope	v-l	228
A =	**C₇H₈O**	**Anisole**	**153.85**			
13049	C₇H₁₄	4-Heptanone	143.3	Nonazeotrope		545

No.	B-Component Formula	B-Component Name	B.P., °C	B.P., °C	Wt.%A	Ref.
				Azeotropic Data		
A =	**C₇H₈O**	**Anisole** *(continued)*	**153.85**			
13050	C₇H₁₄O	2-Methylcyclohexanol	168.5	Nonazeotrope		576
13051	C₇H₁₄O₂	Butyl propionate	146.5	Nonazeotrope		557
13052	C₇H₁₄O₂	Isoamyl acetate	142.1	Nonazeotrope		557
13053	C₇H₁₄O₂	Propyl butyrate	143.7	Nonazeotrope		557
13054	C₇H₁₄O₃	1,3-Butanediol methyl ether acetate	171.75	Nonazeotrope		527
13055	C₇H₁₆O	Heptyl alcohol	176.15	Nonazeotrope		556
13056	C₇H₁₆O₃	Ethyl orthoformate	145.75	Nonazeotrope		549
13057	C₈H₈	Styrene	145.8	Nonazeotrope		573
13058	C₈H₁₀	m-Xylene	139.2	Nonazeotrope		527
13059	C₈H₁₀	o-Xylene	143.6	Nonazeotrope		548
13060	C₈H₁₆O	2-Octanone	172.85	Nonazeotrope		552
13061	C₈H₁₆O₂	Butyl butyrate	166.4	Nonazeotrope		557
13062	C₈H₁₆O₂	Isoamyl propionate	160.4	Nonazeotrope		557
13063	C₈H₁₆O₂	Isobutyl butyrate	156.8	Nonazeotrope		557
		"	157	151	67	563
13064	C₈H₁₆O₂	Isobutyl isobutyrate	148.0	Nonazeotrope		527
13065	C₈H₁₆O₂	Propyl isovalerate	155.7	<153.6		557
13066	C₈H₁₈O	Butyl ether	142.4	Nonazeotrope		549
13067	C₈H₁₉N	Diisobutylamine	138.5	Nonazeotrope		551
13068	C₈H₂₀SiO₄	Ethyl silicate	168.8	Nonazeotrope		557
13069	C₉H₁₂	Cumene	152.8	<152.0	>30	558
13070	C₉H₁₂	Mesitylene	164.6	Nonazeotrope		558
13071	C₉H₁₈O₂	Isobutyl isovalerate	171.2	Nonazeotrope		557
13072	C₁₀H₁₆	Camphene	159.5	151.85	63	572
13073	C₁₀H₁₆	Nopinene	163.8	152.3	74	558
13074	C₁₀H₁₆	α-Pinene	155.8	150.45	56	563
13075	C₁₀H₁₆	α-Terpinene	173.4	Nonazeotrope		558
13076	C₁₀H₂₂	2,7-Dimethyloctane	160.1	153.2	66	558
		"	160.25	Nonazeotrope		563
A =	**C₇H₈O**	**Benzyl Alcohol**	**205.2**			
13077	C₇H₈O	m-Cresol	202.2	207.1	61	542
13078	C₇H₈O	o-Cresol	191.1	Nonazeotrope		535
		"	190.8	206		563
13079	C₇H₈O	p-Cresol	201.7	206.8	62	542
13080	C₇H₈O₂	Guaiacol	205.05	204.25	43	556
13081	C₇H₉N	Methylaniline	196.25	195.8	30	551
		"	196.1	Nonazeotrope		545
13082	C₇H₉N	m-Toluidine	203.1	Nonazeotrope		551
		"	203.2	203.1	47	548
13083	C₇H₉N	o-Toluidine	200.35	Nonazeotrope		551
13084	C₇H₉N	p-Toluidine	200.55	Nonazeotrope		551
13085	C₇H₉NO	o-Anisidine	219.0	Nonazeotrope		551
13086	C₇H₁₃ClO₂	Isoamyl chloroacetate	195.0	Nonazeotrope		575
13087	C₈H₈O	Acetophenone	202.0	Nonazeotrope		571
		"	202	~ 201		563

TABLE I. *Binary Systems* 375

No.	Formula	B-Component Name	B.P., °C	B.P., °C	Wt.%A		Ref.
				Azeotropic Data			
A =	C₇H₈O	**Benzyl Alcohol**	**205.2**				
		(continued)					
13088	C₈H₈O₂	Benzyl formate	~ 202.3	~ 202.0			535
		" "			52		999
13089	C₈H₈O₂	Methyl benzoate	199.2	Nonazeotrope			529
13090	C₈H₈O₂	Phenyl acetate	195.7	Nonazeotrope			535
13091	C₈H₈O₃	Methyl salicylate	205.2	Nonazeotrope			545
13092	C₈H₁₀O	p-Ethylphenol	218.8	Nonazeotrope			575
13093	C₈H₁₀O	3,4-Xylenol	226.8	Nonazeotrope			575
13094	C₈H₁₀O₂	m-Dimethoxybenzene	214.7	Min. b.p.?			576
13095	C₈H₁₀O₂	o-Ethoxyphenol	216.5	Nonazeotrope			575
13096	C₈H₁₀O₂	2-Phenoxyethanol	245.2	Nonazeotrope			575
13097	C₈H₁₀O₂	Veratrole	206.5	202.5	50		545
13098	C₈H₁₁N	Dimethylaniline	194.05	193.9	6.5		551
13099	C₈H₁₁N	Ethylaniline	205.5	202.8	50		551
13100	C₈H₁₁N	2,4-Xylidine	214.0	Nonazeotrope			551
13101	C₈H₁₁N	3,4-Xylidine	225.5	Nonazeotrope			551
13102	C₈H₁₁NO	o-Phenetidine	232.5	Nonazeotrope			551
13103	C₈H₁₆O₄	2-(2-Ethoxyethoxy)ethyl acetate	218.5	Nonazeotrope			575
13104	C₉H₇N	Quinoline	237.3	Nonazeotrope			553
13105	C₉H₁₀O	Benzyl vinyl ether, 25 mm.		103			1008
13106	C₉H₁₀O	Propiophenone	217.7	Nonazeotrope			552
13107	C₉H₁₀O₂	Benzyl acetate	214.9	Nonazeotrope			529
		" "		201.1	93	v-l	495
		" " 100 mm.		140.0	84.7	v-l	495
		" " 50 mm.		98.8	79	v-l	495
13108	C₉H₁₀O₂	Ethyl benzoate	213	Nonazeotrope			563
13109	C₉H₁₀O₂	Methyl α-toluate	215.3	Nonazeotrope			575
13110	C₉H₁₂	Mesitylene	164.6	Nonazeotrope			540
13111	C₉H₁₂O	Phenyl propyl ether	190.5	Nonazeotrope			575
13112	C₉H₁₃N	N,N-Dimethyl-o-toluidine	185.3	185.2	7		551
		"		Nonazeotrope			545
13113	C₉H₁₃N	N,N-Dimethyl-p-toluidine	210.2	202.8	58		551
13114	C₉H₁₈O₃	Isobutyl carbonate	190.3	Nonazeotrope			575
13115	C₁₀H₈	Naphthalene	218.05	204.1	60		541
13116	C₁₀H₁₀O₂	Safrole	235.9	Nonazeotrope			556
13117	C₁₀H₁₂O	Anethole	235.7	Nonazeotrope			575
13118	C₁₀H₁₄	Cymene	176.7	Nonazeotrope			537
13119	C₁₀H₁₄O	Thymol	232.9	Nonazeotrope			575
13120	C₁₀H₁₅N	Diethylaniline	217.05	204.2	72		551
13121	C₁₀H₁₆	Camphene	159.6	Nonazeotrope			537
13122	C₁₀H₁₆	d-Limonene	177.8	176.4	11		541
13123	C₁₀H₁₆	α-Pinene	155.8	Nonazeotrope			563
13124	C₁₀H₁₆	α-Terpinene	173.4	Nonazeotrope			575
13125	C₁₀H₁₆	Terpinene	180.5	179	13?		563

No.	Formula	B-Component Name	B.P., °C	Azeotropic Data B.P., °C	Wt.%A	Ref.
A =	C₇H₈O	**Benzyl Alcohol**	**205.2**			
		(continued)				
13126	C₁₀H₁₆	Terpinolene	184.6	182.5	15	575
13127	C₁₀H₁₆	Thymene	179.7	179.0	14	530
13128	C₁₀H₁₆O	Camphor	209.1	Nonazeotrope		552
		"	208.9	205.45?		563
13129	C₁₀H₁₈O	Borneol	215.0	205.07	85.8	549
13130	C₁₀H₁₈O	Citronellal	207.8	202.9	56	529
13131	C₁₀H₁₈O	Menthone	209.5	Nonazeotrope		552
		"	207	~ 204.8		563
13132	C₁₀H₁₈O	α-Terpineol	217.8	Nonazeotrope		532
13133	C₁₀H₂₀O	Menthol	216.4	Nonazeotrope		545
13134	C₁₀H₂₀O₂	Ethyl caprylate	208.35	<204.8	<82	575
13135	C₁₀H₂₀O₂	Isoamyl isovalerate	192.7	Nonazeotrope		575
13136	C₁₀H₂₂S	Isoamyl sulfide	214.8	Nonazeotrope		566
13137	C₁₁H₁₀	1-Methylnaphthalene	244.9	Nonazeotrope		537
13138	C₁₁H₁₀	2-Methylnaphthalene	241.15	Nonazeotrope		575
13139	C₁₁H₁₆O	Methyl thymyl ether	216.5	Nonazeotrope		575
13140	C₁₁H₂₀O	Isobornyl methyl ether	192.4	Nonazeotrope		556
		" "	192.2	Min. b.p.		576
13141	C₁₁H₂₀O	Terpineol methyl ether	216.2	Nonazeotrope		545
13142	C₁₁H₂₄O₂	Diisoamyloxymethane	207.5	198.7	~ 50	563
13143	C₁₂H₁₈	1,3,5-Triethylbenzene	215.5	203.2	57	537
13144	C₁₂H₂₂O	Ethyl isobornyl ether	203.5	201	39	556
13145	C₁₂H₂₂O	Bornyl ethyl ether	204.9	<203.0	<50	575
A =	**C₇H₈O**	***m*-Cresol**	**202.2**			
13146	C₇H₈O	*o*-Cresol	191.1	Nonazeotrope		813
13147	C₇H₈O	*p*-Cresol	200.9	Nonazeotrope		328
13148	C₇H₈O₂	Guaiacol	205.05	Nonazeotrope		542
13149	C₇H₉N	Benzylamine	185.0	<207.2	<94	551
13150	C₇H₉N	Methylaniline	196.25	Nonazeotrope		551
13151	C₇H₉N	*m*-Toluidine	203.1	205.5	53	551
13152	C₇H₉N	*o*-Toluidine	200.35	203.65	61.5	551
13153	C₇H₉N	*p*-Toluidine	200.55	204.3	62	551
13154	C₇H₁₄O₂	Enanthic acid	222.0	Nonazeotrope		575
13155	C₇H₁₄O₃	Isobutyl lactate	182.15	Max. b.p.		544
13156	C₇H₁₆O₃	2-Ethoxyethyl 2-methoxyethyl ether	194.2		63.6 v-l	730
13157	C₈H₈O	Acetophenone	202.0	208.45	47.2	552
13158	C₈H₈O₂	Benzyl formate	202.4	207.1	46	542
13159	C₈H₈O₂	Methyl benzoate	199.45	204.6	63	542
13160	C₈H₈O₂	Phenyl acetate	195.7	204.4	70	542
13161	C₈H₁₀O	Phenethyl alcohol	219.4	Nonazeotrope		542
13162	C₈H₁₀O	2,4-Xylenol	210.5	Nonazeotrope		575
13163	C₈H₁₀O₂	*m*-Dimethoxybenzene	214.7	Nonazeotrope		544
13164	C₈H₁₀O₂	Veratrole	206.5	Nonazeotrope		542
13165	C₈H₁₁N	2,4,6-Collidine	171.2	206.19	73	506

TABLE I. *Binary Systems* 377

No.	Formula	Name	B.P., °C	B.P., °C	Wt.%A	Ref.
		B-Component		Azeotropic Data		
A =	C₇H₈O	*m*-**Cresol** *(continued)*	**202.2**			
13166	C₈H₁₁N	Dimethylaniline	194.15	Nonazeotrope		551
13167	C₈H₁₁N	Ethylaniline	205.5	Nonazeotrope		.551
13168	C₈H₁₂O₄	Ethyl fumarate	217.85	Nonazeotrope		575
13169	C₈H₁₂O₄	Ethyl maleate	223.3	Nonazeotrope		575
13170	C₈H₁₆O₂	2-Ethylcaproic acid	227	Nonazeotrope	v-l	730
13171	C₈H₁₆O₃	Isoamyl lactate	202.4	207.6	50	563
13172	C₈H₁₈O	Octyl alcohol	195.15	203.3	62	542
13173	C₈H₁₈O	*sec*-Octyl alcohol	179.0	Nonazeotrope		542
13174	C₈H₁₈O₃	Bis(2-ethoxyethyl)ether	188.9		62	v-l	730
13175	C₉H₈	Indene	182.6	Nonazeotrope		542
13176	C₉H₁₀O	*p*-Methylacetophenone	226.35	Nonazeotrope		552
13177	C₉H₁₀O	Propiophenone	217.7	218.6	17	552
13178	C₉H₁₀O₂	Benzyl acetate	215.0	Nonazeotrope		575
		" "	214.9	215.5	12	542
13179	C₉H₁₀O₂	Ethyl benzoate	212.6	212.75	~ 9	542
		" "	212.4		26.6	314
13180	C₉H₁₂O	Phenyl propyl ether	190.5	Nonazeotrope		575
13181	C₉H₁₃N	*N,N*-Dimethyl-*o*-toluidine	185.35	Nonazeotrope		551
13182	C₉H₁₃N	*N,N*-Dimethyl-*p*-toluidine	210.2	Nonazeotrope		551
13183	C₉H₁₄O	Phorone	197.8	206.5	55	552
13184	C₉H₁₈O₂	Isoamyl butyrate	181.05	Nonazeotrope		575
13185	C₉H₁₈O₃	Isobutyl carbonate	190.3	Nonazeotrope		544
13186	C₁₀H₇Cl	1-Chloronaphthalene	262.7	Nonazeotrope		544
13187	C₁₀H₈	Naphthalene	218.05	202.08	2.8?	541
		"	217.9	Nonazeotrope		506
13188	C₁₀H₁₂O	Estragole	215.6	Nonazeotrope		542
13189	C₁₀H₁₂O₂	Ethyl α-toluate	228.75	Nonazeotrope		575
13190	C₁₀H₁₄	Butylbenzene	183.1	Nonazeotrope		575
13191	C₁₀H₁₄	Cymene	176.7	Nonazeotrope		544
13192	C₁₀H₁₄O	Carvone	231.0	Nonazeotrope		552
13193	C₁₀H₁₅N	Diethylaniline	217.05	Nonazeotrope		551
13194	C₁₀H₁₆	*d*-Limonene	177.9	Nonazeotrope		544
13195	C₁₀H₁₆	Nopinene	163.8	Nonazeotrope		575
13196	C₁₀H₁₆	α-Terpinene	173.4	Nonazeotrope		575
13197	C₁₀H₁₆	Thymene	179.7	Nonazeotrope		542
13198	C₁₀H₁₆O	Camphor	209.1	213.35	36.5	552
13199	C₁₀H₁₈O	Borneol	213.4	Nonazeotrope		542
13200	C₁₀H₁₈O	Citronellal	207.8	211.0	30	545
13201	C₁₀H₁₈O	Geraniol	229.6	Nonazeotrope		575
13202	C₁₀H₁₈O	Linalool	198.6	Reacts		542
13203	C₁₀H₂₀O	Menthol	216.4	Nonazeotrope		544
13204	C₁₀H₂₀O₂	Isoamyl isovalerate	192.7	Nonazeotrope		542
13205	C₁₀H₂₀O₂	Methyl pelargonate	213.8	Nonazeotrope		575
13206	C₁₀H₂₂O	Isoamyl ether	173.35	Nonazeotrope		564
13207	C₁₁H₁₀	2-Methylnaphthalene	241.1	Nonazeotrope	v-l	730
13208	C₁₁H₂₀O	α-Terpineol methyl ether	216.2	Nonazeotrope		542

No.	Formula	B-Component Name	B.P., °C	Azeotropic Data B.P., °C	Wt.%A	Ref.
A =	C₇H₈O	*m*-Cresol *(continued)*	202.2			
13209	C₁₂H₁₂	1-Ethylnaphthalene	254.2	Nonazeotrope	v-l	730
13210	C₁₂H₁₈	1,3,5-Triethylbenzene	215.5	Nonazeotrope		542
13211	C₁₂H₂₀O₂	Bornyl acetate	227.7	Nonazeotrope		544
13212	C₁₃H₁₄	2-Isopropylnaphthalene	266.5	Nonazeotrope	v-l	730
13213	C₁₄H₃₀O	Tetradecanol	260.0	Nonazeotrope	v-l	730
13214	C₁₅H₁₈	2-Amylnaphthalene	292.3	Nonazeotrope	v-l	730
13215	C₁₆H₂₀	Diisopropylnaphthalene	305	Nonazeotrope	v-l	730
A =	C₇H₈O	*o*-Cresol	191.1			
13216	C₇H₈O	*p*-Cresol	201.7	Nonazeotrope		545
13217	C₇H₉N	Benzylamine	185.0	201.45	67	551
13218	C₇H₉N	Methylaniline	196.25	Nonazeotrope		551
		"	196.1	196.7	~ 10	563
13219	C₇H₉N	*m*-Toluidine	203.1	Nonazeotrope		551
13220	C₇H₉N	*o*-Toluidine	200.35	Nonazeotrope		551
13221	C₇H₉N	*p*-Toluidine	200.55	Nonazeotrope		551
13222	C₇H₉NO	*o*-Anisidine	219.0	Nonazeotrope		575
13223	C₇H₁₂O₄	Ethyl malonate	198.9	Reacts		563
13224	C₇H₁₄O	2-Methylcyclohexanol	168.5	Nonazeotrope		575
13225	C₇H₁₄O₃	1,3-Butanediol methyl ether acetate	171.75	194.1	68	556
13226	C₇H₁₄O₃	Isobutyl lactate	182.15	193.3	69	542
13227	C₇H₁₆O	Heptyl alcohol	176.16	Nonazeotrope		575
13228	C₈H₈	Styrene	145.8	Nonazeotrope		575
13229	C₈H₈O	Acetophenone	202.0	203.75	26	552
13230	C₈H₈O₂	Benzyl formate	202.3	~ 203.0	~ 15	574
13231	C₈H₈O₂	Methyl benzoate	199.45	200.3	21	542
13232	C₈H₈O₂	Phenyl acetate	195.7	198.5	36	542
13233	C₈H₁₀	*o*-Xylene	144.3	Nonazeotrope		575
13234	C₈H₁₀O	*p*-Methylanisole	177.05	Nonazeotrope		542
13235	C₈H₁₀O	Phenetole	170.45	Nonazeotrope		556
13236	C₈H₁₀O₂	Veratrole	206.5	Nonazeotrope		544
13237	C₈H₁₁N	*o*-Collidine	171.30	197.20	63.0	505
13238	C₈H₁₁N	Dimethylaniline	194.15	Nonazeotrope		551
		"	194.05	195.6	<30	563
13239	C₈H₁₁N	Ethylaniline	205.5	Nonazeotrope		551
13240	C₈H₁₁N	2,4-Xylidine	214.0	Nonazeotrope		551
13241	C₈H₁₂O₄	Ethyl fumarate	217.85	Nonazeotrope		575
13242	C₈H₁₂O₄	Ethyl maleate	223.3	Nonazeotrope		575
13243	C₈H₁₄O	Methylheptenone	173.2	191.8	85	552
13244	C₈H₁₄O₄	Ethyl succinate	216.5	Nonazeotrope		563
13245	C₈H₁₆O	2-Octanone	172.85	192.05	76	527
13246	C₈H₁₆O₃	Isoamyl lactate	202.4	204.2	18	542
13247	C₈H₁₈O	Octyl alcohol	195.15	196.9	38	574
13248	C₈H₁₈O	*sec*-Octyl alcohol	179.0	191.4	~ 92	535
13249	C₈H₁₈S	Butyl sulfide	185.0	183.8	25	566

TABLE I.   *Binary Systems*   379

No.	Formula	Name	B.P., °C	B.P., °C	Wt.%A	Ref.
		B-Component		**Azeotropic Data**		

No.	Formula	Name	B.P., °C	B.P., °C	Wt.%A	Ref.
A =	C₇H₈O	*o*-**Cresol** *(continued)*	**191.1**			
13250	C₈H₁₈S	Isobutyl sulfide	172.0	Nonazeotrope		566
13251	C₈H₂₀SiO₄	Ethyl silicate	168.8	Nonazeotrope		575
13252	C₉H₈	Indene	183.0	182.9	9	541
13253	C₉H₁₀O	Propiophenone	217.7	Nonazeotrope		552
13254	C₉H₁₀O₂	Benzyl acetate	215.0	Nonazeotrope		575
13255	C₉H₁₀O₂	Ethyl benzoate	212.9	Nonazeotrope		563
13256	C₉H₁₂	Mesitylene	164.0	Nonazeotrope		563
13257	C₉H₁₂	Cumene	152.8	Nonazeotrope		575
13258	C₉H₁₂	Propylbenzene	159.3	Nonazeotrope		575
13259	C₉H₁₂	Pseudocumene	168.2	Nonazeotrope		575
13260	C₉H₁₂O	Benzyl ethyl ether	185.0	Nonazeotrope		575
13261	C₉H₁₃N	*N,N*-Dimethyl-*o*-toluidine	185.35	185.3	5	551
13262	C₉H₁₃N	*N,N*-Dimethyl-*p*-toluidine	210.2	Nonazeotrope		575
13263	C₉H₁₄O	Phorone	197.8	201.3	35	552
13264	C₉H₁₈O	2,6-Dimethyl-4-heptanone	168.0	Nonazeotrope		552
13265	C₉H₁₈O₂	Butyl isovalerate	177.6	Nonazeotrope		575
13266	C₉H₁₈O₂	Ethyl enanthate	188.7	193.7	60	562
13267	C₉H₁₈O₂	Isoamyl butyrate	178.5	191.6	~ 83	573
13268	C₉H₁₈O₂	Isobutyl isovalerate	168.7	Nonazeotrope		563
13269	C₉H₁₈O₂	Methyl caprylate	192.9	195.8	33	562
13270	C₉H₁₈O₃	Isobutyl carbonate	190.3	194.5	49	542
13271	C₁₀H₈	Naphthalene	218.05	Nonazeotrope		538
13272	C₁₀H₁₄	Cymene	175.3	~ 175		563
13273	C₁₀H₁₅N	Diethylaniline	217.05	Nonazeotrope		551
13274	C₁₀H₁₆	Camphene	159.6	Nonazeotrope		542
13275	C₁₀H₁₆	*d*-Limonene	177.8	175.35	25	563
13276	C₁₀H₁₆	Nopinene	163.8	Azeotrope doubtful (reacts)		563
13277	C₁₀H₁₆	α-Phellandrene	171.5	171?		563
13278	C₁₀H₁₆	α-Pinene	155.8	Nonazeotrope		563
13279	C₁₀H₁₆	α-Terpinene	173.4	172.0	16	562
13280	C₁₀H₁₆	Terpinene	181.5	177.8	28	542
13281	C₁₀H₁₆	Terpinolene	184.6	179.5	34	562
13282	C₁₀H₁₆	Thymene	179.7	176.6	73	573
13283	C₁₀H₁₆O	Camphor	209.1	209.85	15	552
13284	C₁₀H₁₆O	Fenchone	193.6	199.6	43	552
13285	C₁₀H₁₇Cl	Bornyl chloride	~ 210	Nonazeotrope		563
13286	C₁₀H₁₈O	Borneol	211.8	Nonazeotrope		563
13287	C₁₀H₁₈O	Cineole	176.35	Nonazeotrope		556
13288	C₁₀H₁₈O	Citronellal	208.0	Nonazeotrope		575
13289	C₁₀H₁₈O	Linalool	198.6	199.0	~ 20	535
		"	198.6	Nonazeotrope		538
13290	C₁₀H₁₈O	α-Terpineol	218.85	Nonazeotrope		575
13291	C₁₀H₁₈O	β-Terpineol	210.5	Nonazeotrope		575
13292	C₁₀H₂₀O	Menthol	216.4	Nonazeotrope		542
13293	C₁₀H₂₀O₂	Ethyl caprylate	208.35	Nonazeotrope		575

No.	Formula	Name	B.P., °C	B.P., °C	Wt.%A	Ref.
		B-Component			**Azeotropic Data**	
A =	**C_7H_8O**	**o-Cresol** *(continued)*	**191.1**			
13294	$C_{10}H_{20}O_2$	Isoamyl isovalerate	192.7	195.45	33	570
13295	$C_{10}H_{22}$	2,7-Dimethyloctane	160.1	Nonazeotrope		575
13296	$C_{10}H_{22}O$	Amyl ether	187.5	186.2		556
13297	$C_{10}H_{22}O$	Isoamyl ether	173.4	Nonazeotrope		542,556
13298	$C_{10}H_{22}S$	Isoamyl sulfide	214.8	Nonazeotrope		566
13299	$C_{11}H_{20}O$	Isobornyl methyl ether	192.4	189.7	68	562
13300	$C_{12}H_{18}$	1,3,5-Triethylbenzene	216	Nonazeotrope		542
13301	$C_{12}H_{22}O$	Ethyl isobornyl ether	204.9	Nonazeotrope		575
A =	**C_7H_8O**	**x-Cresol**	**202**			
13302	C_7H_9N	Pyridine bases	163	204.9	78	1065
13303	C_7H_9N	Pyridine bases	157	204.4	80	1065
13304	C_7H_9N	Pyridine bases	142-5	202.5	90	1065
A =	**C_7H_8O**	**m,p-Cresols**	**202**			
13305	$C_{10}H_8$	Naphthalene	218.1	202	71.8	624
13306	$C_{11}H_{10}$	2-Methylnaphthalene	241.15	Nonazeotrope		270
A =	**C_7H_8O**	**p-Cresol**	**201.6**			
13307	$C_7H_8O_2$	Guaiacol	205.1	Nonazeotrope		528
13308	C_7H_9N	Benzylamine	185.0	> 206.5	<95	551
13309	C_7H_9N	Methylaniline	196.25	Nonazeotrope		551
		"	196.1	~ 202.2	~ 93	563
13310	C_7H_9N	m-Toluidine	203.1	204.9	47	551
13311	C_7H_9N	o-Toluidine	200.35	203.5	57	551
13312	C_7H_9N	p-Toluidine	200.55	204.05	57	551
13313	C_7H_9NO	o-Anisidine	219.0	Nonazeotrope		575
13314	$C_7H_{12}O_4$	Ethyl malonate	198.9	Reacts		563
13315	$C_7H_{14}O$	2-Methylcyclohexanol	168.5	Nonazeotrope		575
13316	$C_7H_{14}O_2$	Enanthic acid	222.0	Nonazeotrope		575
13317	$C_7H_{14}O_3$	1,3-Butanediol methyl ether acetate	171.75	203.3	82	527
13318	$C_7H_{14}O_3$	Isobutyl lactate	182.15	Nonazeotrope		542
13319	$C_7H_{14}O_3$	2-Ethoxyethyl 2-methoxy- ethyl ether	194.2	64.7	v-l	730
13320	C_8H_8O	Acetophenone	202.0	208.4	46.5	552
13321	$C_8H_8O_2$	Benzyl formate	202.4	207.0	42	542
13322	$C_8H_8O_2$	Methyl benzoate	199.4	204.35	40	527
13323	$C_8H_8O_2$	Phenyl acetate	195.7	204.3	68	573
13324	$C_8H_{10}O$	Phenethyl alcohol	219.4	Nonazeotrope		542
13325	$C_8H_{10}O_2$	m-Dimethoxybenzene	214.7	Nonazeotrope		538
13326	$C_8H_{10}O_2$	o-Ethoxyphenol	216.5	Nonazeotrope		542
13327	$C_8H_{10}O_2$	Veratrole	206.5	Nonazeotrope		542
13328	$C_8H_{11}N$	Dimethylaniline	194.15	Nonazeotrope		551
13329	$C_8H_{11}N$	2-4-Xylidine	214.0	Nonazeotrope		551
13330	$C_8H_{11}N$	Ethylaniline	205.5	Nonazeotrope		551
		"		207.2	<20	562
13331	$C_8H_{12}O_4$	Ethyl fumarate	217.85	Nonazeotrope		526

TABLE I.   *Binary Systems*   381

No.	Formula	B-Component Name	B.P., °C	Azeotropic Data B.P., °C	Wt.%A		Ref.
A =	**C₇H₈O**	***p*-Cresol** *(continued)*	**201.6**				
13332	C₈H₁₂O₄	Ethyl maleate	223.3	Nonazeotrope			526
13333	C₈H₁₄O₄	Ethyl succinate	216.5	Reacts			563
13334	C₈H₁₆O₂	2-Ethylcaproic acid	227	Nonazeotrope		v-l	730
13335	C₈H₁₆O₃	Isoamyl lactate	202.4	207.25	48		563
13335a	C₈H₁₈	Octane	125	Nonazeotrope		v-l	956c
13336	C₈H₁₈O	Octyl alcohol	195.2	202.25	70		564
13337	C₈H₁₈O	*sec*-Octyl alcohol	178.5	Nonazeotrope			563
13338	C₈H₁₈O₃	Bis(2-ethoxyethyl) ether	188.9		63	v-l	730
13339	C₈H₁₈S	Butyl sulfide	185.0	Nonazeotrope			566
13340	C₉H₈	Indene	182.6	Nonazeotrope			542
13341	C₉H₁₀O	*p*-Methylacetophenone	226.35	Nonazeotrope			552
13342	C₉H₁₀O	Propiophenone	217.7	218.5	16.2		552
13343	C₉H₁₀O₂	Benzyl acetate	214.9	~ 215.2	10		542
		"	215.6	Nonazeotrope			563
13344	C₉H₁₀O₂	Ethyl benzoate	212.6	Nonazeotrope			542
		" "	212.4		24.5		314
13345	C₉H₁₂O	Phenyl propyl ether	190.5	Nonazeotrope			575
13346	C₉H₁₃N	*N,N*-Dimethyl-*o*-toluidine	185.35	Nonazeotrope			551
13347	C₉H₁₃N	*N,N*-Dimethyl-*p*-toluidine	210.2	Nonazeotrope			551
13348	C₉H₁₄O	Phorone	197.8	206.0	55		552
13349	C₉H₁₈O₂	Butyl isovalerate	177.6	Nonazeotrope			575
13350	C₉H₁₈O₂	Ethyl enanthate	188.7	Nonazeotrope			575
13351	C₉H₁₈O₂	Isoamyl butyrate	178.5	Nonazeotrope			542
13352	C₉H₁₈O₂	Methyl caprylate	192.9	Nonazeotrope			575
13353	C₉H₁₈O₃	Isobutyl carbonate	190.3	203.2	~ 80		563
13354	C₁₀H₈	Naphthalene	218.05	Nonazeotrope			542
13355	C₁₀H₁₂O₂	Ethyl α-toluate	228.75	**Nonazeotrope**			**575**
13356	C₁₀H₁₂O₂	Propyl benzoate	230.85	Nonazeotrope			575
13357	C₁₀H₁₄	Butylbenzene	183.1	**Nonazeotrope**			**575**
13358	C₁₀H₁₄	Cymene	176.7	Nonazeotrope			542
13359	C₁₀H₁₄O	Carvone	231.0	Nonazeotrope			552
13360	C₁₀H₁₅N	Diethylaniline	217.05	Nonazeotrope			551
13361	C₁₀H₁₆	*d*-Limonene	177.8	177.6	4		542
13362	C₁₀H₁₆	Nopinene	163.8	Nonazeotrope			575
13363	C₁₀H₁₆	α-Pinene	155.8	**Nonazeotrope**			**575**
13364	C₁₀H₁₆	α-Terpinene	173.4	**Nonazeotrope**			**575**
13365	C₁₀H₁₆	γ-Terpinene	183	181.8	13		575
13366	C₁₀H₁₆	Terpinene	180.5	~ 179			563
13367	C₁₀H₁₆	Terpinolene	184.6	183	16		575
13368	C₁₀H₁₆	Thymene	179.7	Nonazeotrope			564
13369	C₁₀H₁₆O	Camphor	209.1	213.5	30.5		552
13370	C₁₀H₁₆O	Fenchone	193.6	205.5	72		552
13371	C₁₀H₁₆O	Pulegone	223.8	224.2	97		**552**
13372	C₁₀H₁₇Cl	Bornyl chloride	~210	200.5	70		563
13373	C₁₀H₁₈O	Borneol	213.4	213.6	~ 10		535
13374	C₁₀H₁₈O	Cineole	176.35	Nonazeotrope			556

No.	B-Component		B.P., °C	Azeotropic Data			Ref.
	Formula	Name	B.P., °C	B.P., °C	Wt.%A		Ref.
A =	**C₇H₈O**	*p*-**Cresol** *(continued)*	**201.6**				
13375	C₁₀H₁₈O	Citronellal	207.8	210.5			545
13376	C₁₀H₁₈O	Geraniol	229.5	Nonazeotrope			575
13377	C₁₀H₁₈O	Linalool	198.6	204	~ 55		535
13378	C₁₀H₁₈O	Menthone	~206	211	~ 38		563
13379	C₁₀H₁₈O	α-Terpineol	218.0	Nonazeotrope			542
13380	C₁₀H₁₈O	β-Terpineol	210.5	Nonazeotrope			575
13381	C₁₀H₂₀O	Citronellol	224.4	Nonazeotrope			575
13382	C₁₀H₂₀O	Menthol	216.4	Nonazeotrope			535
		"	212	212			563
13383	C₁₀H₂₀O₂	Ethyl caprylate	208.35	209.5	25		575
13384	C₁₀H₂₀O₂	Isoamyl isovalerate	~ 193.5	~203.5	~74		573
		"	192.7	Nonazeotrope			542
13385	C₁₀H₂₂O	Amyl ether	187.5	Nonazeotrope			556
13386	C₁₀H₂₂O	Isoamyl ether	173.35	Nonazeotrope			564
13387	C₁₁H₁₀	2-Methylnaphthalene	241.1	Nonazeotrope		v-l	527,730
13388	C₁₁H₂₀O	Isobornyl methyl ether	192.4	Nonazeotrope			575
13389	C₁₁H₂₂O₃	Isoamyl carbonate	232.2	Nonazeotrope			575
13390	C₁₂H₁₂	1-Ethylnaphthalene	254.2	Nonazeotrope		v-l	730
13391	C₁₂H₁₈	1,3,5-Triethylbenzene	216	201.5	~96		563
13392	C₁₂H₂₀O₂	Bornyl acetate	227.7	Nonazeotrope			544
13393	C₁₃H₁₄	2-Isopropylnaphthalene	266.5	Nonazeotrope		v-l	730
13394	C₁₄H₃₀O	Tetradecanol	260.0	Nonazeotrope		v-l	730
13395	C₁₅H₁₈	2-Amylnaphthalene	292.3	Nonazeotrope		v-l	730
13396	C₁₆H₂₀	Diisopropylnaphthalene	305	Nonazeotrope		v-l	730
A =	**C₇H₈O₂**	**Guaiacol**	**205.05**				
13397	C₇H₉N	Methylaniline	196.25	Nonazeotrope			551
13398	C₇H₉N	*m*-Toluidine	203.1	Nonazeotrope			551
13399	C₇H₉N	*o*-Toluidine	200.35	Nonazeotrope			551
13400	C₇H₉N	*p*-Toluidine	200.55	Nonazeotrope			551
13401	C₇H₁₂O₄	Ethyl malonate	198.9	Nonazeotrope			563
13402	C₈H₈O	Acetophenone	202.0	205.25	67.5		552
13403	C₈H₈O₂	Benzyl formate	202.3	206.2	~90		574
13404	C₈H₈O₂	Methyl benzoate	199.45	Nonazeotrope			556
13405	C₈H₈O₂	Phenyl acetate	195.5	Nonazeotrope			563
13406	C₈H₈O₃	Methyl salicylate	222.95	Nonazeotrope			575
13407	C₈H₁₀O	Phenethyl alcohol	219.4	Nonazeotrope			535
13408	C₈H₁₀O	2,4-Xylenol	210.5	Nonazeotrope			575
13409	C₈H₁₀O₂	*m*-Dimethoxybenzene	214.7	Nonazeotrope			535
13410	C₈H₁₁N	Dimethylaniline	194.15	Nonazeotrope			551
13411	C₈H₁₁N	Ethylaniline	205.5	204.4	55		551
13412	C₈H₁₁N	2,4-Xylidine	214.0	Nonazeotrope			551
13413	C₈H₁₆O₃	Isoamyl lactate	202.4	Nonazeotrope			563
13414	C₈H₁₈O	Octyl alcohol	195.2	Nonazeotrope			575
13415	C₈H₁₈S	Butyl sulfide	185.0	Nonazeotrope			566
13416	C₉H₁₀O	Propiophenone	217.7	Nonazeotrope			575

TABLE I. *Binary Systems* 383

No.	B-Component Formula	Name	B.P., °C	Azeotropic Data B.P., °C	Wt.%A	Ref.
A =	**C₇H₈O₂**	**Guaiacol** *(continued)*	**205.05**			
13417	C₉H₁₀O₂	Benzyl acetate	215.0	Nonazeotrope		545
13418	C₉H₁₀O₂	Ethyl benzoate	212.6	Nonazeotrope		556
13419	C₉H₁₃N	*N,N*-Dimethyl-*o*-toluidine	185.35	Nonazeotrope		551
13420	C₉H₁₄O	Phorone	197.8	Nonazeotrope		575
13421	C₁₀H₈	Naphthalene	218.05	Nonazeotrope		535
13422	C₁₀H₁₂O	Estragole	215.6	Nonazeotrope		535
13423	C₁₀H₁₅N	Diethylaniline	217.05	Nonazeotrope		551
13424	C₁₀H₁₆O	Camphor	209.1	Nonazeotrope		552
13425	C₁₀H₁₆O	Fenchone	193.6	Nonazeotrope		575
13426	C₁₀H₁₈O	Borneol	211.8	Nonazeotrope		556
13427	C₁₀H₁₈O	Citronellal	207.8	204.55	86.5	556
13428	C₁₀H₁₈O	Geraniol	229.6	Nonazeotrope		575
13429	C₁₀H₁₈O	Linalool	198.6	Nonazeotrope		545
13430	C₁₀H₁₈O	α-Terpineol	217.8	Nonazeotrope		535
13431	C₁₀H₂₀O	Menthol	216.4	Nonazeotrope		535
13432	C₁₀H₂₀O₂	Ethyl caprylate	208.35	208.9	15	575
13433	C₁₀H₂₀O₂	Methyl pelargonate	213.8	Nonazeotrope		575
13434	C₁₀H₂₂S	Isoamyl sulfide	214.8	Nonazeotrope		566
13435	C₁₂H₁₈	1,3,5-Triethylbenzene	215.5	Nonazeotrope		535
13436	C₁₃H₂₈	Tridecane	234.0	Nonazeotrope		575
A =	**C₇H₈O₂**	**m-Methoxyphenol**	**243.8**			
13437	C₈H₇N	Indole	253.5	Nonazeotrope		575
13438	C₉H₇N	Quinoline	237.3	Nonazeotrope		575
13439	C₉H₈O	Cinnamaldehyde	253.7	Nonazeotrope		575
13440	C₉H₁₀O₃	Ethyl salicylate	233.8	Nonazeotrope		575
13441	C₉H₁₂O	3-Phenylpropanol	235.6	Nonazeotrope		575
13442	C₁₀H₈	Naphthalene	218.0	Nonazeotrope		575
13443	C₁₀H₁₀O₂	Isosafrole	252.1	Nonazeotrope		535
13444	C₁₀H₁₀O₂	Safrole	235.9	Nonazeotrope		575
13445	C₁₀H₁₄O	Carvacrol	237.85	Nonazeotrope		575
13446	C₁₀H₁₄O	Thymol	232.9	Nonazeotrope		544
13447	C₁₀H₁₄O₂	*m*-Diethoxybenzene	235.0	Nonazeotrope		535
13448	C₁₁H₁₀	1-Methylnaphthalene	245.1	243		535
13449	C₁₁H₁₀	2-Methylnaphthalene	241.15	240.2	25	575
13450	C₁₁H₁₄O₂	1-Allyl-3,4-dimethoxyben--zene	255.2	Nonazeotrope		535
13451	C₁₁H₁₄O₂	Isobutyl benzoate	242.15	245.5	~60	535
13452	C₁₁H₁₇N	Isoamylaniline	256.0	Nonazeotrope		551
13453	C₁₂H₁₀	Biphenyl	256.1	Nonazeotrope		575
A =	**C₇H₈S**	**α-Toluenethiol**	**194.8**			
13454	C₁₀H₁₆	Terpinolene	185	Reacts		563
A =	**C₇H₉N**	**Benzylamine**	**185.0**			
13455	C₈H₁₀O	Benzyl methyl ether	167.8	Nonazeotrope		551
13456	C₈H₁₀O	*p*-Methylanisole	177.05	Nonazeotrope		551
13457	C₈H₁₀O	Phenetole	170.45	Nonazeotrope		551
13458	C₉H₁₂O	Benzyl ethyl ether	185.0	< 181.5		575

No.	Formula	B-Component Name	B.P., °C	Azeotropic Data B.P., °C	Wt.%A	Ref.
A =	**C₇H₉N**	**Benzylamine** *(continued)*	**185.0**			
13459	C₁₀H₁₈O	Cineole	176.35	175.6	16.5	527
13460	C₁₀H₂₂O	Amyl ether	187.5	< 180.0	<67	551
13461	C₁₀H₂₂O	Isoamyl ether	173.2	170.4	23	551
13462	C₁₁H₂₀O	Isobornyl methyl ether	192.4	< 184.2		575
A =	**C₇H₉N**	**2,4-Lutidine**	**159.0**			
13462a	C₉H₂₀	Nonane	150	148.30	32.26	263c
13462b	C₁₁H₂₄	Undecane		Nonazeotrope		v-l 263c,263e
A =	**C₇N₉N**	**2,6-Lutidine**	**144**			
13463	C₈H₈	Styrene	145	Min. b.p.		243
13464	C₈H₁₀	Ethylbenzene	136	Min. b.p.		243
13465	C₈H₁₀	Xylenes	140	Min. b.p.		243
13466	C₈H₁₈	2,3,4-Trimethylpentane	113.4	Nonazeotrope		175
13467	C₁₀H₂₂	Decane	174.0	Nonazeotrope		v-l 1056
A =	**C₇H₉N**	**Methylaniline**	**196.25**			
13468	C₇H₉N	*o*-Toluidine	200.3	Nonazeotrope		549
13469	C₇H₁₆O	*n*-Heptyl alcohol	176.75	Nonazeotrope		551
13470	C₈H₈O	Acetophenone	202.25	Nonazeotrope		545
13471	C₈H₁₀O	*p*-Methylanisole	177.05	Nonazeotrope		551
13472	C₈H₁₀O₂	*o*-Ethoxyphenol	216.5	Nonazeotrope		551
13473	C₈H₁₁N	Dimethylaniline	194.15	Nonazeotrope		549
		" 20 mm.	85.4	Nonazeotrope		413
		" 95°-145°C.		Nonazeotrope		v-l 181
13474	C₈H₁₈O	*n*-Octyl alcohol	195.2	193.0	45	551
13475	C₈H₁₈O	*sec*-Octyl alcohol	180.4	Nonazeotrope		551
13476	C₉H₈	Indene	182.6	Nonazeotrope		551
13477	C₉H₁₂	Mesitylene	164.6	Nonazeotrope		551
13478	C₁₀H₈	Naphthalene	218.0	Nonazeotrope		551
13479	C₁₀H₁₄	Cymene	176.7	Nonazeotrope		551
13480	C₁₀H₁₆	Camphene	159.6	Nonazeotrope		551
13481	C₁₀H₁₆	Dipentene	177.7	< 177.2	< 11	551
13482	C₁₀H₁₆	*d*-Limonene	177.8	174.5	13	563
13483	C₁₀H₁₆	Nopinene	163.8	Nonazeotrope		551
13484	C₁₀H₁₆	α-Pinene	155.8	Nonazeotrope		551
13485	C₁₀H₁₆	α-Terpinene	173.4	Nonazeotrope		551
13486	C₁₀H₁₆	Terpinolene	185	180	~32	563
13487	C₁₀H₁₈O	Borneol	215.0	Nonazeotrope		551
13488	C₁₀H₁₈O	Cineole	176.35	Nonazeotrope		551
13489	C₁₀H₁₈O	Linalool	198.6	195.6	70	551
13490	C₁₀H₁₈O	Menthone	209.5	Nonazeotrope		575
13491	C₁₀H₁₈O	β-Terpineol	210.5	Nonazeotrope		551
13492	C₁₀H₂₀O	Menthol	216.3	Nonazeotrope		551
13493	C₁₀H₂₂O	Isoamyl ether	173.2	Nonazeotrope		551
13494	C₁₂H₁₈	1,3,5-Triethylbenzene	215.5	Nonazeotrope		551
13495	C₁₂H₂₂O	Ethyl isobornyl ether	203.8	Nonazeotrope		551

TABLE I.   *Binary Systems*   385

No.	B-Component			Azeotropic Data			
	Formula	Name	B.P., °C	B.P., °C	Wt.% A		Ref.
A =	**C₇H₉N**	**m-Toluidine**	**203.1**				
13496	C₈H₁₀O	p-Ethylphenol	218.8	Nonazeotrope			551
13497	C₈H₁₀O	3,4-Xylenol	226.8	Nonazeotrope			551
13498	C₈H₁₀O₂	o-Ethoxyphenol	216.5	Nonazeotrope			551
13499	C₈H₁₁N	Ethylaniline	205.5	202.95	89		564
13500	C₈H₁₈O	n-Octyl alcohol	195.2	Nonazeotrope			551
13501	C₈H₁₈O	sec-Octyl alcohol	180.4	Nonazeotrope			551
13502	C₁₀H₈	Naphthalene	218.0	Nonazeotrope			551
13503	C₁₀H₁₄	Butylbenzene	183.1	Nonazeotrope			551
13504	C₁₀H₁₆O	Camphor	209.1	Nonazeotrope			551
13505	C₁₀H₁₆O	Pulegone	223.8	Nonazeotrope			551
13506	C₁₀H₁₈O	Borneol	215.0	Nonazeotrope			551
13507	C₁₀H₁₈O	Menthone	209.5	Nonazeotrope			579
13508	C₁₀H₁₈O	α-Terpineol	218.85	Nonazeotrope			551
13509	C₁₀H₁₈O	β-Terpineol	210.5	Nonazeotrope			575
13510	C₁₀H₂₀O	Menthol	216.3	Nonazeotrope			551
13511	C₁₁H₁₆O	Methyl thymyl ether	216.5	Nonazeotrope			575
13512	C₁₁H₂₀O	Methyl a-terpineol ether	216.2	Nonazeotrope			551
13513	C₁₂H₂₂O	Ethyl isobornyl ether	203.8	< 201.0	<60		551
A =	**C₇H₉N**	**o-Toluidine**	**200.7**				
13514	C₇H₁₂O₄	Ethyl malonate	198.9	Reacts			563
13515	C₇H₁₆O	n-Heptyl alcohol	176.15	Nonazeotrope			551
13516	C₈H₈O	Acetophenone	202.0	203.65	32		551
13517	C₈H₁₀O	Phenethyl alcohol	219.4	Nonazeotrope			551
13518	C₈H₁₀O₂	m-Dimethoxybenzene	214.7	Nonazeotrope			537
13519	C₈H₁₈O	n-Octyl alcohol	195.2	194.7	23		564
13520	C₈H₁₈O	sec-Octyl alcohol	180.4	Nonazeotrope			551
13521	C₉H₈	Indene	182.6	Nonazeotrope			551
13522	C₉H₁₀O	Propiophenone	217.7	Nonazeotrope			551
13523	C₉H₁₃N	N-Dimethyl-o-toluidine	185.3	Nonazeotrope			549
13524	C₁₀H₈	Naphthalene	218.0	Nonazeotrope			551
13525	C₁₀H₁₄	Butylbenzene	183.1	Nonazeotrope			551
13526	C₁₀H₁₄	Cymene	176.7	Nonazeotrope			551
13527	C₁₀H₁₆O	Camphor	209.1	Nonazeotrope			551
13528	C₁₀H₁₈O	Borneol	215.0	Nonazeotrope			551
13529	C₁₀H₁₈O	Cineole	176.35	Nonazeotrope			575
13530	C₁₀H₁₈O	Linalool	198.6	198.3	30		551
13531	C₁₀H₁₈O	α-Terpineol	218.85	Nonazeotrope			575
13532	C₁₀H₁₈O	β-Terpineol	210.75	Nonazeotrope			551
13533	C₁₀H₂₀O	Menthol	216.3	Nonazeotrope			551
13534	C₁₀H₂₂	Decane	174.6	173.76	13.0		506
13535	C₁₁H₁₇N	Diethyl-o-toluidine, 20 mm.		98.8	48		474
13536	C₁₁H₂₀O	Isobornyl methyl ether	192.4	< 192.0			575
13537	C₁₁H₂₀O	Terpineol methyl ether	216.0	Nonazeotrope			537
13538	C₁₁H₂₄	Undecane	195.5	188.25	39.7		506
13539	C₁₂H₁₈	1,3,5-Triethylbenzene	215.5	Nonazeotrope			551

		B-Component		Azeotropic Data		
No.	Formula	Name	B.P., °C	B.P., °C	Wt.%A	Ref.
A =	**C₇H₉N**	*o*-**Toluidine** *(continued)*	**200.7**			
13540	C₁₂H₂₂O	Ethyl isobornyl ether	203.8	< 198.5		575
13541	C₁₂H₂₆	Dodecane	216.5	195.75	63.0	506
13542	C₁₃H₂₈	Tridecane	234.6	199.45	85.5	506
A =	**C₇H₉N**	*p*-**Toluidine**	**200.5**			
13543	C₈H₈O	Acetophenone	202.0	203.65	32	551
		"	202	~ 199		563
13544	C₈H₁₈O	*n*-Octyl alcohol	195.2	194.65	23	551
		" "	195.15	194.4	33	545
13545	C₈H₁₈O	*sec*-Octyl alcohol	180.4	Nonazeotrope		551
13546	C₉H₈	Indene	182.6	Nonazeotrope		551
13547	C₉H₁₀O	Propiophenone	217.7	Nonazeotrope		551
13548	C₁₀H₈	Naphthalene	218.0	Nonazeotrope		551
13549	C₁₀H₁₆	Terpinolene	184.6	< 183.5		551
13550	C₁₀H₁₆O	Camphor	209.1	Nonazeotrope		551
13551	C₁₀H₁₈O	Borneol	215.0	Nonazeotrope		551
13552	C₁₀H₁₈O	Menthone	~ 207	Nonazeotrope		563
13553	C₁₀H₂₀O	Menthol	216.3	Nonazeotrope		551
A =	**C₇H₉NO**	*o*-**Anisidine**	**219.0**			
13554	C₈H₈O	Acetophenone	202.0	Nonazeotrope		575
13555	C₈H₈O₃	Methyl salicylate	222.95	Nonazeotrope		575
13556	C₈H₁₀O	3,4-Xylenol	226.8	Nonazeotrope		575
13557	C₉H₁₀O	*p*-Methylacetophenone	226.35	Nonazeotrope		575
13558	C₉H₁₀O	Propiophenone	217.7	219.7	~ 65	575
13559	C₉H₁₀O₃	Ethyl salicylate	233.8	Nonazeotrope		575
13560	C₉H₁₄O	Phorone	197.8	Nonazeotrope		575
13561	C₁₀H₈	Naphthalene	218.0	217.0	50	575
13562	C₁₀H₁₄O	Thymol	232.9	Nonazeotrope		575
13563	C₁₀H₂₀O	Menthol	216.3	< 216.0		551
13564	C₁₁H₁₀	2-Methylnaphthalene	241.15	Nonazeotrope		575
13565	C₁₁H₂₀O	Methyl-α-terpineol ether	216.2	215.2	35	575
13566	C₁₂H₁₈	1,3,5-Triethylbenzene	215.5	214.5	35	575
A =	**C₇H₁₂**	**Heptyne**	**99.5**			
13566a	C₇H₁₄	Heptene	93.6	Nonazeotrope	v-l	498f
13566b	C₇H₁₆	Heptane 400 mm.		76.2	44.8 v-l	498f
		" 600 mm.		89.3	42.3 v-l	498f
		" 760 mm.	98.4	96.8	39.9 v-l	498f
A =	**C₇H₁₂O₄**	**Ethyl Malonate**	**198.6**			
13567	C₈H₈O	Acetophenone	202.0	Nonazeotrope		552
13568	C₈H₈O₂	Benzyl formate	203.0	< 198.2		549
13569	C₈H₈O₂	Methyl benzoate	199.55	198.2	~ 54	528
		" "	199.4	198.7	56	527
13570	C₈H₈O₂	Phenyl acetate	195.7	Nonazeotrope		529
13571	C₈H₁₀O₂	*m*-Dimethoxybenzene	214.7	Nonazeotrope		537
13572	C₈H₁₄O₄	Propyl oxalate	214	Nonazeotrope		575
13573	C₈H₁₆O	2-Octanone	172.85	Nonazeotrope		552
13574	C₈H₁₆O₃	Isoamyl lactate	202.4	Nonazeotrope		575

TABLE I.   *Binary Systems*   387

		B-Component		Azeotropic Data		
No.	Formula	Name	B.P., °C	B.P., °C	Wt.%A	Ref.
A =	**C₇H₁₂O₄**	**Ethyl Malonate**	**198.6**			
		(continued)				
13575	C₈H₁₈O	Octyl alcohol	195.15	Reacts		536
13576	C₉H₁₀O₂	Ethyl benzoate	212.5	Nonazeotrope		575
13577	C₉H₁₂	Mesitylene	164.6	Nonazeotrope		575
13578	C₉H₁₂	Propylbenzene	159.3	Nonazeotrope		575
13579	C₉H₁₂O	Benzyl ethyl ether	185.0	Nonazeotrope		557
13580	C₉H₁₄O	Phorone	197.8	< 197.65	<47	552
13581	C₉H₁₈O₂	Methyl caprylate	192.9	191.9	26	549
13582	C₉H₁₈O₃	Isobutyl carbonate	190.3	Nonazeotrope		549
13583	C₁₀H₈	Naphthalene	218.1	Nonazeotrope		563
13584	C₁₀H₁₄	Cymene	176.7	Nonazeotrope		575
13585	C₁₀H₁₆	Camphene	159.6	Nonazeotrope		537
13586	C₁₀H₁₆	Dipentene	177.7	Nonazeotrope		575
13587	C₁₀H₁₆	*d*-Limonene	177.8	177.5	10	537
13588	C₁₀H₁₆	Nopinene	163.8	Nonazeotrope		575
13589	C₁₀H₁₆	α-Pinene	155.8	Nonazeotrope		546
13590	C₁₀H₁₆	α-Terpinene	173.4	Nonazeotrope		575
13591	C₁₀H₁₆	Terpinene	181.5	178.0	22	538
13592	C₁₀H₁₆O	Camphor	209.1	Nonazeotrope		552
13593	C₁₀H₁₇Cl	Bornyl chloride	207.5	< 198.0	<82	575
13594	C₁₀H₁₈O	Linalool	199	~ 198	~ 60	563
13595	C₁₀H₂₀O₂	Isoamyl isovalerate	192.7	191.75	30	527
13596	C₁₁H₁₆O	Methyl thymol ether	216.5	Nonazeotrope		557
13597	C₁₁H₂₀O	Methyl α-terpineol ether	216.2	Nonazeotrope		557
13598	C₁₁H₂₄O₂	Diisoamyloxymethane	207.5	Azeotrope doubtful		563
		"		Nonazeotrope		557
13599	C₁₂H₁₈	1,3,5-Triethylbenzene	215.5	Nonazeotrope		575
13600	C₁₂H₂₂O	Bornyl ethyl ether	204.9	< 196.0	<71	557
13601	C₁₂H₂₂O	Ethyl isobornyl ether	203.8	< 196.2	<70	557
A =	**C₇H₁₃ClO₂**	**Isoamyl Chloroacetate**	**190.5**			
13602	C₇H₁₄O₃	Isobutyl lactate	182.15	Nonazeotrope		575
13603	C₈H₈O	Acetophenone	202.0	Nonazeotrope		575
13604	C₈H₈O₂	Methyl benzoate	199.55	Nonazeotrope		563
13605	C₈H₁₆O	2-Octanone	172.85	Nonazeotrope		575
13606	C₈H₁₈O	Octyl alcohol	195.2	< 193.5	<62	575
13607	C₈H₁₈O	*sec*-Octyl alcohol	180.4	Nonazeotrope		575
13608	C₁₀H₁₆	*d*-Limonene	177.8	Nonazeotrope		563
13609	C₁₀H₁₈O	Linalool	198.6	< 194.2	<82	575
A =	**C₇H₁₄**	**3-Heptene**	**94.8**			
13610	C₇H₁₆	Heptane	98.4	Nonazeotrope		575
		"	98.4	Nonazeotrope	v-l	1016
A =	**C₇H₁₄**	**Methylcyclohexane**	**100.8**			
13611	C₇H₁₆	*n*-Heptane	98.4	Nonazeotrope	v-l	86,317,893
		"	98.45	98.3	10	381,572
13612	C₈F₁₆O	Perfluorocyclic oxide	102.5	85	40 vol. %	609
13613	C₈H₁₈	2,5-Dimethylhexane	109.4	Nonazeotrope		575

	B-Component			Azeotropic Data		
No.	Formula	Name	B.P., °C	B.P., °C	Wt.%A	Ref.
A =	**C₇H₁₄**	**Methylcyclohexane**	**100.8**			
		(continued)				
13614	C₈H₁₈	2,2,4-Trimethylpentane,				
		741 mm.	98.2	Nonazeotrope	v-l	365
13615	C₈H₁₈O	Isobutyl ether	122.3	Nonazeotrope		558
A =	**C₇H₁₄O**	**4-Heptanone**	**143.55**			
13616	C₇H₁₄O₂	Butyl propionate	146.8	Nonazeotrope		552
13617	C₇H₁₄O₂	Ethyl *n*-valerate	145.15	Nonazeotrope		552
13618	C₇H₁₄O₂	Isoamyl acetate	142.1	141.7	25	552
13619	C₇H₁₄O₂	Isobutyl propionate	137.5	Nonazeotrope		552
13620	C₇H₁₄O₂	Propyl butyrate	143.7	143.0	47	552
13621	C₈H₁₀	Ethylbenzene	136.15	Nonazeotrope		552
13622	C₈H₁₀	*m*-Xylene	139.2	139.0	10	552
13623	C₈H₁₀	*o*-Xylene	144.3	142.4	42	552
13624	C₈H₁₉N	Diisobutylamine	138.5	< 137.0	<32	575
13625	C₉H₁₂	Cumene	152.8	Nonazeotrope		552
13626	C₉H₁₂	Propylbenzene	159.3	Nonazeotrope		552
13627	C₁₀H₁₆	Camphene	159.6	142.5	95	552
13628	C₁₀H₁₆	α-Pinene	155.8	142.0	80	543
A =	**C₇H₁₄O**	**2-Methylcyclohexanol**	**168.5**			
13629	C₇H₁₄O₃	Isobutyl lactate	182.15	Nonazeotrope		575
13630	C₈H₁₀	Ethylbenzene	136.15	Nonazeotrope		575
13631	C₈H₁₀	*m*-Xylene	139.2	Nonazeotrope		575
13632	C₈H₁₀O	Benzyl methyl ether	167.8	165.0	46	575
13633	C₈H₁₀O	*p*-Methylanisole	177.05	167.5	71	576
13634	C₈H₁₀O	Phenetole	170.45	165.7	50	576
13635	C₈H₁₁N	Dimethylaniline	194.05	Nonazeotrope		551
13636	C₈H₁₄O	Methylheptenone	173.2	Nonazeotrope		552
13637	C₈H₁₆O	2-Octanone	172.85	Nonazeotrope		552
13638	C₈H₁₆O₂	Isoamyl propionate	160.7	Nonazeotrope		575
13639	C₈H₁₆O₂	Isobutyl butyrate	156.9	Nonazeotrope		575
13640	C₉H₁₂	Cumene	152.8	151.7	12	575
13641	C₉H₁₂	Mesitylene	164.6	160.5	34	567
13642	C₉H₁₂	Pseudocumene	168.2	< 164.0	<48	575
13643	C₉H₁₂O	Benzyl ethyl ether	185.0	Nonazeotrope		575
13644	C₉H₁₃N	*N,N*-Dimethyl-*o*-toluidine	185.3	Nonazeotrope		551
13645	C₉H₁₈O	2,6-Dimethyl-4-heptanone	168.0	167.5	40	552
13646	C₉H₁₈O₂	Isobutyl isovalerate	171.2	167.5	62	575
13647	C₁₀H₁₄	Butylbenzene	183.1	< 168.0	>70	575
13648	C₁₀H₁₄	Cymene	176.7	< 166.5	<68	575
13649	C₁₀H₁₆	Camphene	159.6	155.5	25	567
13650	C₁₀H₁₆	Dipentene	177.7	165.3	60	567
13651	C₁₀H₁₆	α-Pinene	155.8	152.8	20	567
13652	C₁₀H₁₆	α-Terpinene	173.4	163.7	52	567
13653	C₁₀H₁₈O	Cineole	176.35	167.2	70	576
13654	C₁₀H₂₂	2,7-Dimethyloctane	160.1	155.8	27	567
13655	C₁₀H₂₂O	Amyl ether	187.5	Nonazeotrope		575
13656	C₁₀H₂₂O	Isoamyl ether	173.4	166.2	60	545

TABLE I.   *Binary Systems*   389

No.	Formula	B-Component Name	B.P., °C	Azeotropic Data B.P., °C	Wt.%A	Ref.
A =	**C₇H₁₄O**	**3-Methylcyclohexanol**	**172**			
13657	C₈H₁₀O	Phenetole, 770 mm.	170.5	167.2	46.5	512
		13 mm.		60	24	512
		2 mm.		28.8	18.7	512
A =	**C₇H₁₄O**	**5-Methyl-2-hexanone**	**144.2**			
13658	C₇H₁₄O₂	Butyl propionate	146.8	Nonazeotrope		552
13659	C₇H₁₄O₂	Isoamyl acetate	142.1	141.8	18	552
13660	C₇H₁₄O₂	Isobutyl propionate	137.5	Nonazeotrope		552
13661	C₇H₁₄O₂	Propyl butyrate	143.7	143.3	35	552
13662	C₈H₁₀	Ethylbenzene	136.15	Nonazeotrope		552
13663	C₈H₁₀	o-Xylene	144.3	143.0	42	552
13664	C₈H₁₆O₂	Isobutyl isobutyrate	148.6	Nonazeotrope		552
13665	C₈H₁₉N	Diisobutylamine	138.5	136.3	30	575
13666	C₉H₁₂	Cumene	152.8	Nonazeotrope		552
13667	C₁₀H₁₆	α-Pinene	155.8	142.0	75	552
A =	**C₇H₁₄O₂**	**Amyl Acetate**	**148.8**			
13668	C₇H₁₄O₂	Butyl propionate	146.8	Nonazeotrope		575
13669	C₇H₁₄O₂	Isoamyl acetate	142.1	Nonazeotrope		575
13670	C₇H₁₄O₂	Propyl butyrate	143.7	Nonazeotrope		575
13671	C₈H₁₀	o-Xylene	143.6	Nonazeotrope		546
13672	C₈H₁₆O₂	Isobutyl isobutyrate	148.6	< 148.5	>10	549
13673	C₈H₁₈O	Butyl ether	142.4	Nonazeotrope		557
13674	C₁₀H₁₆	α-Pinene	155.8	< 148.0	75	546
A =	**C₇H₁₄O₂**	**Butyl Propionate**	**146.8**			
13675	C₇H₁₄O₂	Isoamyl acetate	142.1	Nonazeotrope		575
13676	C₇H₁₄O₂	Propyl butyrate	143.7	Nonazeotrope		575
13677	C₈H₈	Styrene	146	145.5		546
13678	C₈H₁₀	o-Xylene	143.6	Nonazeotrope		546
13679	C₈H₁₈O	Butyl ether	142.4	Nonazeotrope		557
13680	C₉H₁₂	Cumene	152.8	Nonazeotrope		575
13681	C₁₀H₁₆	α-Pinene	155.8	< 145.8	>85	546
A =	**C₇H₁₄O₂**	**Enanthic Acid**	**222.0**			
13682	C₈H₈O	Acetophenone	202.0	Nonazeotrope		552
13683	C₈H₁₀O	3,4-Xylenol	226.8	Nonazeotrope		575
13684	C₈H₁₀O₂	o-Ethoxyphenol	216.5	< 215.2	>15	575
13685	C₈H₁₂O₄	Ethyl fumarate	217.85	216.4	22	562
13686	C₈H₁₂O₄	Ethyl maleate	223.3	220.0	50	562
13687	C₈H₁₄O₄	Ethyl succinate	217.25	216.0	20	562
13688	C₈H₁₄O₄	Propyl oxalate	214	< 213.8	> 7	575
13689	C₈H₁₆O₄	2-(2-Ethoxyethoxy)-ethyl acetate	218.5	224.5	58	562
13690	C₉H₈	Indene	182.6	Nonazeotrope		575
13691	C₉H₁₀O	p-Methylacetophenone	226.35	< 221.2	>70	552
13692	C₉H₁₀O	Propiophenone	217.7	216.5	20	552

| | B-Component | | | Azeotropic Data | | |
No.	Formula	Name	B.P., °C	B.P., °C	Wt.%A	Ref.
A =	**C₇H₁₄O₂**	**Enanthic Acid** *(continued)*	**222.0**			
13693	C₉H₁₀O₂	Benzyl acetate	215.0	Nonazeotrope		575
13694	C₉H₁₀O₂	Ethyl benzoate	212.5	Nonazeotrope		575
13695	C₉H₁₉NO	N,N-Dimethylheptan-amide		Max. b.p.		832
13696	C₁₀H₇Cl	1-Chloronaphthalene	262.7	Nonazeotrope		575
13697	C₁₀H₈	Naphthalene	218.0	214.2	30	562
13698	C₁₀H₁₀O₂	Safrole	235.9	< 221.7	>85	575
13699	C₁₀H₁₂O₂	Ethyl α-toluate	228.75	Nonazeotrope		575
13700	C₁₀H₁₄O	Carvacrol	237.85	Nonazeotrope		575
13701	C₁₀H₁₄O	Carvone	231.0	Nonazeotrope		552
13702	C₁₁H₁₆O	Methyl thymol ether	216.5	215.0	25	575
13703	C₁₁H₂₀O	Methyl α-terpineol ether	216.2	< 215.3	<30	575
13704	C₁₂H₁₀	Biphenyl	256.1	Nonazeotrope		575
13705	C₁₂H₁₈	1,3,5-Triethylbenzene	215.5	211.0	27	562
13706	C₁₂H₂₀O₂	Bornyl acetate	227.6	Nonazeotrope		575
13707	C₁₃H₂₈	Tridecane	234.0	< 219.2	>55	562
A =	**C₇H₁₄O₂**	**Ethyl Isovalerate**	**134.7**			
13708	C₇H₁₆O₃	Ethyl orthoformate	145.75	Nonazeotrope		557
13709	C₈H₈	Styrene	145.8	Nonazeotrope		575
13710	C₈H₁₀	Ethylbenzene	136.15	Nonazeotrope		531
13711	C₈H₁₀	m-Xylene	139.0	Nonazeotrope		527
13712	C₈H₁₀	p-Xylene	138.45	Nonazeotrope		575
13713	C₈H₁₈O	Butyl ether	142.4	Nonazeotrope		557
13714	C₈H₁₈O	Isobutyl ether	122.3	Nonazeotrope		557
A =	**C₇H₁₄O₂**	**Ethyl Valerate**	**145.45**			
13715	C₇H₁₄O₂	Isoamyl acetate	142.1	Nonazeotrope		575
13716	C₈H₈	Styrene	145.8	< 145.0	>48	575
13717	C₈H₁₀	m-Xylene	139.2	Nonazeotrope		527
13718	C₈H₁₀	o-Xylene	144.3	Nonazeotrope		575
13719	C₈H₁₈O	Butyl ether	142.4	Nonazeotrope		557
A =	**C₇H₁₄O₂**	**Isoamyl Acetate**	**142.1**			
13720	C₇H₁₄O₂	Isobutyl propionate	137.5	Nonazeotrope		575
13721	C₇H₁₄O₂	Propyl butyrate	142.8	Nonazeotrope		572
13722	C₇H₁₆O₃	Ethyl orthoformate	145.75	Nonazeotrope		557
13723	C₈H₁₀	Ethylbenzene	136.15	Nonazeotrope		572
13724	C₈H₁₀	m-Xylene	139.0	Nonazeotrope		527
		"	139.0	136	50	563
13725	C₈H₁₀	o-Xylene	143.6	Nonazeotrope		546
13726	C₈H₁₀	p-Xylene	138.3	Nonazeotrope		546
13727	C₈H₁₆	1,3-Dimethylcyclohexane	120.7	Nonazeotrope		575
13728	C₈H₁₈O	Butyl ether	142.2	< 141.2	<55	557
13729	C₉H₁₂	Cumene	152.8	Nonazeotrope		575
13730	C₁₀H₁₆	Camphene	158	Nonazeotrope		546
13731	C₁₀H₁₆	Nopinene	163.8	Nonazeotrope		575
13732	C₁₀H₁₆	α-Pinene	155.8	142.05	97.5	537

TABLE I. *Binary Systems* 391

No.	Formula	B-Component Name	B.P., °C	B.P., °C	Wt.%A	Ref.
A =	C₇H₁₄O₂	Isobutyl Propionate	136.9			
13733	C₈H₈	Styrene, 60 mm.	68	Nonazeotrope		53
13734	C₈H₁₀	Ethylbenzene	136.15	135.8	~ 30	573
	"	" 60 mm.	60.5	60	13	53
13735	C₈H₁₀	*m*-Xylene	139.0	Nonazeotrope		575
	"	"	139.0	134.5		563
13736	C₈H₁₀	*o*-Xylene	143.6	Nonazeotrope		546
13737	C₈H₁₀	*p*-Xylene	138.3	136.8	85	546
13738	C₈H₁₈O	Butyl ether	142.4	Nonazeotrope		557
A =	C₇H₁₄O₂	Isopropyl Isobutyrate	120.8			
13739	C₇H₁₆	Heptane	98.4	Nonazeotrope		575
A =	C₇H₁₄O₂	Methyl Caproate	149.6			
13740	C₈H₁₀	*m*-Xylene	139.0	Nonazeotrope		563
13741	C₈H₁₀	*o*-Xylene	144.3	Nonazeotrope		575
13742	C₈H₁₆O₂	Isobutyl isobutyrate	149.75	Nonazeotrope		575
13743	C₈H₁₆O₂	Propyl isovalerate	155.7	Nonazeotrope		575
13744	C₉H₁₂	Cumene	152.8	Nonazeotrope		575
13745	C₉H₁₈O₂	Methyl caprylate 20-100 mm.		Nonazeotrope	v-l	825
A =	C₇H₁₄O₂	Propyl Butyrate	143.7			
13746	C₈H₈	Styrene	146	Nonazeotrope		546
	"	"	145.8	<143.5	<68	575
13747	C₈H₁₀	*m*-Xylene	139.0	Nonazeotrope		527
	"	"	139.0	138.7		563
13748	C₈H₁₀	*o*-Xylene	143.6	143.2	55	546
13749	C₈H₁₈O	Butyl ether	142.4	<142.0	<45	557
13750	C₉H₁₂	Cumene	152.8	Nonzaeotrope		575
13751	C₁₀H₁₆	α-Pinene	155.8	<143.4	<88	575
	"	"	155.8	Nonazeotrope		563
A =	C₇H₁₄O₂	Propyl Isobutyrate	134.0			
13752	C₇H₁₆O₂	Dipropoxymethane	137.2	Nonazeotrope		557
13753	C₈H₈	Styrene	146	Nonazeotrope		546
13754	C₈H₁₀	Ethylbenzene	136.15	Nonazeotrope		575
13755	C₈H₁₀	*m*-Xylene	139.0	Nonazeotrope		527
13756	C₈H₁₀	*p*-Xylene	138.2	Nonazeotrope		573
13757	C₈H₁₈O	Butyl ether	142.2	Nonazeotrope		557
13758	C₁₀H₁₆	α-Pinene	155.8	Nonazeotrope		546
A =	C₇H₁₄O₃	1,3-Butanediol Methyl Ether Acetate	171.75			
13759	C₇H₁₆O	Heptyl alcohol	176.15	Nonazeotrope		575
13760	C₈H₁₀O	*p*-Methylanisole	177.05	Nonazeotrope		575
13761	C₈H₁₀O	Phenetole	170.45	170.0	22	527
13762	C₈H₁₆O	2-Octanone	172.85	171.3	35	575
13763	C₈H₁₆O₂	Ethyl caproate	167.7	167.4	~ 10	575
13764	C₈H₁₆O₂	Hexyl acetate	171.5	170.7	49	562
13765	C₈H₁₆O₂	Isoamyl propionate	160.7	Nonazeotrope		527

No.	Formula	B-Component Name	B.P., °C	B.P., °C	Azeotropic Data Wt.%A	Ref.
A =	$C_7H_{14}O_3$	**1,3-Butanediol Methyl Ether Acetate** *(continued)*	171.75			
13766	$C_8H_{16}O_2$	Isobutyl butyrate	156.9	Nonazeotrope		575
13767	$C_8H_{18}O$	Octyl alcohol	195.2	Nonazeotrope		575
13768	$C_8H_{18}O$	sec-Octyl alcohol	180.4	Nonazeotrope		575
13769	C_9H_{12}	Mesitylene	164.6	Nonazeotrope		575
13770	C_9H_{12}	Propylbenzene	159.3	Nonazeotrope		575
13771	$C_9H_{12}O$	Benzyl ethyl ether	185.0	Nonazeotrope		575
13772	$C_9H_{18}O$	2,6-Dimethyl-4-heptanone	168.0	Nonazeotrope		575
13773	$C_9H_{18}O_2$	Isoamyl butyrate	181.05	Nonazeotrope		527
13774	$C_9H_{18}O_2$	Isobutyl isovalerate	171.2	170.35	47	569
13775	$C_9H_{18}O_3$	Isobutyl carbonate	190.3	Nonazeotrope		575
13776	$C_{10}H_{14}$	Cymene	176.7	Nonazeotrope		575
13777	$C_{10}H_{16}$	Camphene	159.6	<159.45	> 5	575
13778	$C_{10}H_{16}$	Dipentene	177.7	169.6	78	562
13779	$C_{10}H_{16}$	Nopinene	163.8	162.0	20	562
13780	$C_{10}H_{16}$	α-Terpinene	173.4	168.9	65	562
13781	$C_{10}H_{18}O$	Cineole	176.35	170.9	64	527
13782	$C_{10}H_{20}O_2$	Isoamyl isovalerate	192.7	Nonazeotrope		527
13783	$C_{10}H_{22}O$	Amyl ether	187.5	Nonazeotrope		575
13784	$C_{10}H_{22}O$	Isoamyl ether	173.2	<170.0	>52	527
A =	$C_7H_{14}O_3$	**Isobutyl Lactate**	182.15			
13785	$C_8H_{10}O$	Benzyl methyl ether	167.8	Nonazeotrope		575
13786	$C_8H_{10}O$	Phenetole	171.5	Nonazeotrope		563
13787	$C_8H_{10}O$	2,4-Xylenol	210.5	Nonazeotrope		575
13788	$C_8H_{18}O$	Octyl alcohol	195.2	Nonazeotrope		575
13789	$C_8H_{18}O$	sec-Octyl alcohol	178.5	177.3		563
13790	$C_8H_{18}S$	Butyl sulfide	185.0	<181.3	<78	566
13791	C_9H_8	Indene	182.8	177	48	548
13792	C_9H_{12}	Mesitylene	164.6	Nonazeotrope		575
13793	$C_9H_{12}O$	Benzyl ethyl ether	185.0	181.0	75?	575
13794	$C_9H_{18}O_2$	Isoamyl butyrate	181.05	<178.5	>28	527
13795	$C_{10}H_{14}$	Cymene	175.3	171.5	~ 35	563
13796	$C_{10}H_{16}$	Camphene	159.6	Nonazeotrope		538
13797	$C_{10}H_{16}$	d-Limonene	177.8	172.5	40	563
13798	$C_{10}H_{16}$	Nopinene	163.8	Nonazeotrope		575
13799	$C_{10}H_{16}$	α-Pinene	155.8	Nonazeotrope		575
13800	$C_{10}H_{16}$	Terpinene	180.5	172.5	~ 46	563
13801	$C_{10}H_{16}$	Terpinolene	185	175	55	563
13802	$C_{10}H_{18}O$	Cineole	176.35	174.0	32	556
13803	$C_{10}H_{18}O$	Linalool	198.6	Nonazeotrope		575
13804	$C_{10}H_{22}O$	Isoamyl ether	173.2	<172.0	>13	575
A =	C_7H_{16}	**2,4-Dimethylpentane,**				
13805	C_7H_{16}	2,2,3-Trimethylbutane,				
		505 mm.	67.58	67.71	~ 50	122
		"		Nonazeotropic below 55° C.		122
		"		Nonazeotropic above 75° C.		122

TABLE I.   *Binary Systems*   393

No.	Formula	B-Component Name	B.P., °C	Azeotropic Data B.P., °C	Wt.%A		Ref.
A =	**C_7H_{16}**	**Heptane**	**98**				
13806	$C_8F_{18}O$	Perfluorobutyl ether	100	Min. b.p.			159
13807	C_8H_{10}	Ethylbenzene	136.15	Nonazeotrope			163
		" 100-760 mm.		Nonazeotrope		v-l	675
13808	C_8H_{10}	*p*-Xylene 90°C		Nonazeotrope		v-l	1022
13809	C_8H_{18}	2,2,4-Trimethylpentane	99.3	Nonazeotrope			163
A =	**$C_7H_{16}O$**	**2-Heptanol, 10 mm.**	**65.4**				
13810	C_8H_9Cl	*o,m,p*-Chloroethylbenzene, 10 mm.	67.5	61.4	43		51
A =	**$C_7H_{16}O$**	**Heptyl Alcohol**	**176.15**				
13811	C_8H_{10}	*m*-Xylene	139.2	Nonazeotrope			575
13812	$C_8H_{10}O$	Benzyl methyl ether	167.8	167.0	20		575
13813	$C_8H_{10}O$	*p*-Methylanisole	177.05	173.3	52		567
13814	$C_8H_{10}O$	Phenetole	170.45	169.0	28		545
13815	$C_8H_{11}N$	Dimethylaniline	194.05	Nonazeotrope			551
13816	$C_8H_{14}O$	Methylheptenone	173.2	Nonazeotrope			552
13817	$C_8H_{16}O_2$	Ethyl caproate	167.7	Nonazeotrope			575
13818	$C_8H_{18}O$	*sec*-Octyl alcohol	180.4	Nonazeotrope			575
13819	C_9H_{12}	Cumene	152.8	Nonazeotrope			575
13820	$C_9H_{13}N$	*N,N*-Dimethyl-*o*-toluidine	185.3	175.5	82		551
13821	$C_9H_{18}O_2$	Isobutyl isovalerate	171.2	<171.0	> 8		575
13822	$C_{10}H_{14}$	Cymene	176.0	172.5	47		567
13823	$C_{10}H_{15}N$	Diethylaniline	217.05	Nonazeotrope			551
13824	$C_{10}H_{16}$	Camphene	159.6	<159.3	>10		575
13825	$C_{10}H_{16}$	Dipentene	177.7	171.7	50		567
13826	$C_{10}H_{16}$	Nopinene	163.8	<162.6	>15		575
13827	$C_{10}H_{16}$	*α*-Terpinene	173.4	169.7	40		567
13828	$C_{10}H_{18}O$	Cineole	176.35	173.0	48		556
13829	$C_{10}H_{22}O$	Isoamyl ether	173.35	170.35	37		564
A =	**$C_7H_{16}O_3$**	**Dipropylene Glycol Methyl Ether**					
13830	$C_{11}H_{10}$	2-Methylnaphthalene	241.15	Nonazeotrope			270
A =	**$C_7H_{16}O_3$**	**Ethyl Orthoformate**	**145.75**				
13831	C_8H_8	Styrene	145.8	<145.0	<45		558
13832	C_8H_{10}	*m*-Xylene	139.2	Nonazeotrope			527
13833	$C_8H_{16}O_2$	Propyl isovalerate	155.7	Nonazeotrope			557
13834	C_9H_{12}	Cumene	152.8	Nonazeotrope			558
13835	C_9H_{12}	Propylbenzene	159.3	Nonazeotrope			558
13836	$C_{10}H_{16}$	Camphene	159.6	Nonazeotrope			558
13837	$C_{10}H_{16}$	Nopinene	163.8	Nonazeotrope			558
A =	**$C_7H_{16}O_4$**	**2-[2-(2-Methoxyethoxy)-ethoxy] Ethanol**	**245.25**				
13838	$C_8H_8O_2$	Methyl benzoate	199.4	Nonazeotrope			575
13839	$C_8H_8O_3$	Methyl salicylate	222.95	222.0	8		575
13840	$C_8H_{10}O_2$	2-Phenoxyethanol	245.2	<244.0	>55		575

No.	Formula	B-Component Name	B.P., °C	Azeotropic Data B.P., °C	Wt.%A	Ref.
A =	C₇H₁₆O₄	2-[2-(2-Methoxyethoxy)-ethoxy] Ethanol (continued)	245.25			
13841	C₈H₁₂O₄	Ethyl fumarate	217.85	Nonazeotrope		575
13842	C₈H₁₂O₄	Ethyl maleate	223.3	Nonazeotrope		575
13843	C₉H₇N	Quinoline	237.3	235.55	22	553
13844	C₉H₁₀O₂	Benzyl acetate	215.0	Nonazeotrope		575
13845	C₉H₁₀O₂	Ethyl benzoate	212.5	Nonazeotrope		575
13846	C₉H₁₀O₃	Ethyl salicylate	233.8	227.7	28	556
13847	C₉H₁₂O₂	2-Benzyloxyethanol	265.2	Nonazeotrope		575
13848	C₁₀H₇Br	1-Bromonaphthalene	281.2	Nonazeotrope		575
13849	C₁₀H₇Cl	1-Chloronaphthalene	262.7	Nonazeotrope		575
13850	C₁₀H₈	Naphthlene	218.0	214.8	20	556
13851	C₁₀H₈O	1-Naphthol	288.5	Nonazeotrope		556
13852	C₁₀H₉N	Quinaldine	246.5	<243.0		575
13853	C₁₀H₁₀O₂	Isosafrole	252.0	241.5	65	567
13854	C₁₀H₁₀O₂	Methyl cinnamate	261.9	242.3	70	567
13855	C₁₀H₁₀O₂	Safrole	235.9	233.5	31	567
13856	C₁₀H₁₀O₄	Methyl phthalate	283.2	Nonazeotrope		575
13857	C₁₀H₁₂O	Anethole	235.7	233.0	30	567
13858	C₁₀H₁₂O₂	Propyl benzoate	230.85	226.0	32	567
13859	C₁₁H₁₀	1-Methylnaphthalene	244.6	232.0	46	567
13860	C₁₁H₁₀	2-Methylnaphthalene	241.15	229.4	44	556
13861	C₁₁H₁₄O₂	Butyl benzoate	249.0	235.0	52	567
13862	C₁₁H₁₄O₂	Isobutyl benzoate	241.9	231.2	40	567
13863	C₁₁H₁₆O	Methyl thymyl ether	216.5	Nonazeotrope		575
13864	C₁₁H₂₀O	Methyl α-terpineol ether	216.2	Nonazeotrope		575
13865	C₁₂H₁₀	Acenaphthene	277.9	242.5	71	556
13866	C₁₂H₁₀	Biphenyl	256.1	236.0	50	567
13867	C₁₂H₁₀O	Phenyl ether	259.0	243.0	80	567
13868	C₁₂H₁₆O₂	Isoamyl benzoate	262.0	239.4	60	556
13869	C₁₂H₁₆O₃	Isoamyl salicylate	277.5	Nonazeotrope		575
13870	C₁₂H₁₈	1,3,5-Triethylbenzene	215.5	212.0	18	567
13871	C₁₃H₁₂	Diphenylmethane	265.4	239.0	56	567
13872	C₁₄H₁₄	1,2-Diphenylethane	284.5	243.8	80	575
A =	C₈F₁₆O	Perfluorocyclic Oxide	102.5			
13873	C₈H₁₆	Ethylcyclohexane	131.78	96.3	80 vol. %	609
13874	C₈H₁₈	2,2,4-Trimethylpentane	99.24	87.5	60 vol. %	609
13875	C₉H₂₀	2,3,4-Trimethylhexane	131.34	98.4	80 vol. %	609
A =	C₈H₅Cl₃	ar-Trichlorostyrene				
13876	C₉H₂₀O₃	2-(2-Isoamyloxyethoxy) ethanol, 6.7 mm.		101		155
A =	C₈H₇N	Indole	253.5			
13877	C₈H₉BrO	p-Bromophenetole	234.2	Nonazeotrope		575
13878	C₉H₁₀O	Cinnamyl alcohol	257.0	Nonazeotrope		575
13879	C₁₀H₁₀O₂	Safrole	235.9	Nonazeotrope		575
13880	C₁₀H₁₂O₂	Eugenol	254.8	<251.8	>35	575

TABLE I. *Binary Systems*

395

No.	B-Component			Azeotropic Data		
	Formula	Name	B.P., °C	B.P., °C Wt.%A		Ref.
A =	**C₈H₇N**	**Indole** *(continued)*	**253.5**			
13881	C₁₀H₁₂O₂	Isoeugenol	268.8	Nonazeotrope		575
13882	C₁₀H₁₄O	Carvacrol	237.85	254.5	88	575
13883	C₁₁H₁₄O₂	1-Allyl-3,4-dimethoxy-benzene	254.7	<251.8	>55	575
13884	C₁₁H₁₄O₂	1,2-Dimethyl-4-propenyl-benzene	270.5	Nonazeotrope		575
13885	C₁₁H₁₆O	*p-tert*-Amylphenol	266.5	268.0	12	575
13886	C₁₂H₁₀O	Phenyl ether	259.0	Nonaozeotrope		575
A =	**C₈H₇N**	**α-Toluonitrile**	**232**			
13887	C₁₀H₁₈O	Geraniol	229.5	~ 226		563
A =	**C₈H₈**	**Styrene**	**145.8**			
13888	C₈H₁₀	Ethylbenzene	136.15	Nonazeotrope		561
		" 10-100 mm.		Nonazeotrope	v-l	143,294
		" 30°-120°C.		Nonazeotrope	v-l	432
13889	C₈H₁₀	Xylene, 20 mm.	50	Nonazeotrope		981
13890	C₈H₁₀	*m*-Xylene	139.2	Nonazeotrope		561
13891	C₈H₁₀	*o*-Xylene	142.6	Nonazeotrope		563
13892	C₈H₁₆O₂	Isobutyl isobutyrate	148.6	<145.5	>60	575
13893	C₈H₁₆O₂	Propyl isovalerate	155.7	Nonazeotrope		575
13894	C₉H₂₀	Nonane	149.5	144.0	75	561
A =	**C₈H₈O**	**Acetophenone**	**202.0**			
13895	C₈H₈O₂	Benzyl formate	203.0	Nonazeotrope		552
13896	C₈H₈O₂	Methyl benzoate	199.4	Nonazeotrope		552
13897	C₈H₈O₂	Phenyl acetate	195.7	Nonazeotrope		552
13898	C₈H₁₀O	*p*-Ethylphenol	218.8	219.5	15	552
13899	C₈H₁₀O	2,4-Xylenol	210.5	213.0	30	575
13900	C₈H₁₀O	3,4-Xylenol	226.8	Nonazeotrope		552
13901	C₈H₁₀O₂	*o*-Ethoxyphenol	216.5	Nonazeotrope		552
13902	C₈H₁₀O₂	Veratrol	205.5	Nonazeotrope		574
13903	C₈H₁₁N	Dimethylaniline	194.15	Nonazeotrope		551
13904	C₈H₁₁N	Ethylaniline	205.5	Nonazeotrope		551
13905	C₈H₁₁N	2,4-Xylidine	214.0	Nonazeotrope		551
13906	C₈H₁₄O₄	Propyl oxalate	214.2	Nonazeotrope		552
13907	C₈H₁₆O₃	Isoamyl lactate	202.4	<201.7	48	552
13908	C₈H₁₆O₄	2-(2-Ethoxyethoxy)-ethyl acetate	218.5	Nonazeotrope		575
13909	C₈H₁₈O	Octyl alcohol	195.2	194.95	12.5	552
13910	C₈H₁₈O	*sec*-Octyl alcohol	180.4	Nonazeotrope		552
13911	C₉H₁₀O₂	Benzyl acetate	215.0	Nonazeotrope		552
13912	C₉H₁₀O₂	Ethyl benzoate	212.5	Nonazeotrope		552
13913	C₉H₁₃N	*N,N*-Dimethyl-*o*-toluidine	185.3	Nonazeotrope		551
13914	C₉H₁₃N	*N,N*-Dimethyl-*p*-toluidine	210.2	Nonazeotrope		575
13915	C₁₀H₈	Naphthalene	218.0	Nonazeotrope		552
13916	C₁₀H₁₄O	Thymol	232.9	Nonazeotrope		552

No.	Formula	Name	B.P., °C	B.P., °C	Wt.%A	Ref.
		B-Component		**Azeotropic Data**		
A =	**C₈H₈O**	**Acetophenone** *(continued)*	**202.0**			
13917	C₁₀H₁₅N	Diethylaniline	217.05	Nonazeotrope		551
13918	C₁₀H₁₈O	Borneol	215.0	Nonazeotrope		552
13919	C₁₀H₁₈O	Citronellal	208.0	Nonazeotrope		552
13920	C₁₀H₁₈O	Linalool	198.6	198.0	14	552
13921	C₁₀H₁₈O	β-Terpineol	210.5	Nonazeotrope		552
13922	C₁₀H₂₀O	Menthol	216.3	Nonazeotrope		552
13923	C₁₀H₂₀O₂	Methyl pelargonate	213.8	Nonazeotrope		552
13924	C₁₀H₂₂S	Isoamyl sulfide	214.8	Nonazeotrope		566
13925	C₁₁H₂₀O	Isobornyl methyl ether	192.4	Nonazeotrope		575
13926	C₁₂H₁₈	1,3,5-Triethylbenzene	215.5	Nonazeotrope		552
13927	C₁₂H₂₂O	Bornyl ethyl ether	204.9	Nonazeotrope		575
A =	**C₈H₈O₂**	**Anisaldehyde**	**249.5**			
13928	C₈H₉BrO	p-Bromophenetole	234.2	Nonazeotrope		575
13929	C₉H₁₀O	Cinnamyl alcohol	257.0	<248.0		575
13930	C₁₀H₇Cl	1-Chloronaphthalene	262.7	Nonazeotrope		575
13931	C₁₀H₁₀O₂	Isosafrole	252.0	248.6	60	556
13932	C₁₀H₁₀O₂	Methyl cinnamate	261.9	Nonazeotrope		548
13933	C₁₀H₁₀O₂	Safrole	235.9	Nonazeotrope		556
13934	C₁₀H₁₄O	Carvacrol	237.85	Nonazeotrope		575
13935	C₁₀H₁₄O	Thymol	232.9	Nonazeotrope		575
13936	C₁₀H₂₀O	Citronellol	224.5	Nonazeotrope		575
13937	C₁₁H₁₄O₂	1-Allyl-3,4-dimethoxy-benzene	255.0	Nonazeotrope		556
13938	C₁₁H₁₄O₂	Butyl benzoate	249.5	<248.8	~ 50	548
13939	C₁₁H₁₄O₂	1,2-Dimethoxy-4-propenyl-benzene	270.5	Nonazeotrope		575
13940	C₁₁H₁₄O₂	Isobutyl benzoate	241.9	Nonazeotrope		538
13941	C₁₂H₁₀O	Phenyl ether	259.3	Nonazeotrope		556
13942	C₁₂H₁₆O₂	Isoamyl benzoate	262.0	Nonazeotrope		548
A =	**C₈H₈O₂**	**Benzyl Formate**	**203.0**			
13943	C₈H₈O₂	Methyl benzoate	199.4	Nonazeotrope		549
13944	C₈H₈O₂	Phenyl acetate	195.7	Nonazeotrope		575
13945	C₈H₁₀O	3,4-Xylenol	226.8	Nonazeotrope		575
13946	C₈H₁₀O₂	m-Dimethoxybenzene	214.7	Nonazeotrope		537
13947	C₈H₁₈O	Octyl alcohol	195.2	Nonazeotrope		575
		"	195.15	195.0	3	536
13948	C₁₀H₈	Naphthalene	218.05	Nonazeotrope		537
13949	C₁₀H₁₆	d-Limonene	177.9	Nonazeotrope		546
13950	C₁₀H₁₆	γ-Terpinene	179.7	Nonazeotrope		546
13951	C₁₀H₁₆O	Camphor	209.1	Nonazeotrope		548
13952	C₁₀H₁₈O	Borneol	213.4	Nonazeotrope		535
13953	C₁₀H₁₈O	Citronellal	208.0	Nonazeotrope		575
13954	C₁₀H₁₈O	Linalool	198.6	197.5		535
13955	C₁₀H₁₈O	α-Terpineol	217.8	Nonazeotrope		536
13956	C₁₀H₂₀O	Menthol	216.4	Nonazeotrope		535
13957	C₁₀H₂₀O₂	Isoamyl isovalerate	192.7	Nonazeotrope		575

TABLE I. *Binary Systems* 397

No.	B-Component Formula	Name	B.P., °C	Azeotropic Data B.P., °C	Wt.%A	Ref.
A =	C$_8$H$_8$O$_2$	**Benzyl Formate** (continued)	**203.0**			
13958	C$_{11}$H$_{16}$O	Methyl thymol ether	216.5	Nonazeotrope		557
13959	C$_{12}$H$_{18}$	1,3,5-Triethylbenzene	216	Nonazeotrope		546
A =	**C$_8$H$_8$O$_2$**	**Methyl Benzoate**	**199.4**			
13960	C$_8$H$_{10}$O	2,4-Xylenol	210.5	Nonazeotrope		575
13961	C$_8$H$_{10}$O$_2$	Veratrol	205.5	Nonazeotrope		557
13962	C$_8$H$_{14}$O$_4$	Propyl oxalate	214	Nonazeotrope		575
13963	C$_8$H$_{16}$O$_3$	Isoamyl lactate	202.4	<198.8		575
	" "		202.4	Nonazeotrope		563
13964	C$_8$H$_{16}$O$_4$	2-(2-Ethoxyethoxy)-ethyl acetate	218.5	Nonazeotrope		575
13965	C$_8$H$_{18}$O	n-Octyl alcohol	195.2	194.4	35	570
13966	C$_9$H$_8$	Indene	182.6	Nonazeotrope		575
13967	C$_9$H$_{10}$O$_2$	Ethyl benzoate	212.5	Nonazeotrope		575
13968	C$_9$H$_{12}$O	Benzyl ethyl ether	185.0	Nonazeotrope		557
13969	C$_9$H$_{14}$O	Phorone	197.8	Nonazeotrope		552
13970	C$_9$H$_{18}$O$_3$	Isobutyl carbonate	190.3	Nonazeotrope		575
13971	C$_{10}$H$_8$	Naphthalene	218.0	Nonazeotrope		575
13972	C$_{10}$H$_{14}$	Cymene	176.7	Nonazeotrope		575
13973	C$_{10}$H$_{16}$	d-Limonene	177.8	Nonazeotrope		530
13974	C$_{10}$H$_{16}$	α-Terpinene	173.4	Nonazeotrope		575
13975	C$_{10}$H$_{16}$	γ-Terpinene	179.7	Nonazeotrope		546
13976	C$_{10}$H$_{16}$O	Camphor	209.1	Nonazeotrope		552
13977	C$_{10}$H$_{17}$Cl	Bornyl chloride	207.5	Nonazeotrope		575
13978	C$_{10}$H$_{18}$O	Borneol	213.4	Nonazeotrope		536
13979	C$_{10}$H$_{18}$O	Citronellal	~ 207.8	Nonazeotrope		529
13980	C$_{10}$H$_{18}$O	Linalool	198.7	197.8	~ 42	528
13981	C$_{10}$H$_{18}$O	β-Terpineol	210.5	Nonazeotrope		575
13982	C$_{10}$H$_{20}$O	Menthol	216.3	Nonazeotrope		575
13983	C$_{10}$H$_{20}$O$_2$	Ethyl caprylate	208.35	Nonazeotrope		575
13984	C$_{10}$H$_{20}$O$_2$	Isoamyl isovalerate	192.7	Nonazeotrope		549
13985	C$_{11}$H$_{16}$O	Methyl thymol ether	216.5	Nonazeotrope		557
13986	C$_{11}$H$_{24}$O$_2$	Diisoamyloxymethane	210.8	Nonazeotrope		557
13987	C$_{12}$H$_{18}$	1,3,5-Triethylbenzene	215.5	Nonazeotrope		575
A =	**C$_8$H$_8$O**	**Phenyl Acetate**	**195.7**			
13988	C$_8$H$_{10}$O	2,4-Xylenol	210.5	Nonazeotrope		575
13989	C$_8$H$_{10}$O$_2$	Veratrole	205.5	Nonazeotrope		557
13990	C$_8$H$_{18}$O	n-Octyl alcohol	195.15	192.4	53	572
13991	C$_9$H$_8$	Indene	182.6	Nonazeotrope		575
13992	C$_9$H$_{12}$	Pseudocumene	168.2	Nonazeotrope		575
13993	C$_9$H$_{12}$O	Benzyl ethyl ether	185.0	Nonazeotrope		557
13994	C$_9$H$_{14}$O	Phorone	198.2	Nonazeotrope		573
	"		197.8	<195.6	<90	552
13995	C$_9$H$_{18}$O$_3$	Isobutyl carbonate	190.3	Nonazeotrope		575
13996	C$_{10}$H$_8$	Naphthalene	218.05	Nonazeotrope		537
13997	C$_{10}$H$_{14}$	Cymene	176.7	Nonazeotrope		537
13998	C$_{10}$H$_{16}$	Camphene	158	Nonazeotrope		546

No.	Formula	B-Component Name	B.P., °C	B.P., °C	Wt.%A	Ref.
A =	C₈H₈O	**Phenyl Acetate**	**195.7**			
		(continued)				
13999	C₁₀H₁₆	d-Limonene	177.8	177.5	7	538
14000	C₁₀H₁₆	Nopinene	163.8	Nonazeotrope		575
14001	C₁₀H₁₆	α-Terpinene	173.4	Nonazeotrope		575
14002	C₁₀H₁₆	γ-Terpinene	181.5	180.3	15	538
14003	C₁₀H₁₆	Thymene	179.7	179.3	18	530
14004	C₁₀H₁₈O	Borneol	213.2	Nonazeotrope		530
14005	C₁₀H₁₈O	Cineole	176.35	Nonazeotrope		557
14006	C₁₀H₁₈O	Citronellal	208.0	Nonazeotrope		575
14007	C₁₀H₁₈O	Linalool	198.6	193.5	61	529
14008	C₁₀H₁₈O	β-Terpineol	210.5	Nonazeotrope		575
14009	C₁₀H₂₀O₂	Ethyl caprylate	208.35	Nonazeotrope		575
14010	C₁₀H₂₂O	Isoamyl ether	173.2	Nonazeotrope		557
14011	C₁₁H₂₀O	Isobornyl methyl ether	192.4	Nonazeotrope		557
14012	C₁₁H₂₄O₂	Diisoamyloxymethane	210.8	Nonazeotrope		557
14013	C₁₂H₁₈	1,3,5-Triethylbenzene	216	Nonazeotrope		536
14014	C₁₂H₂₂O	Bornyl ethyl ether	204.9	Nonazeotrope		557
14015	C₁₂H₂₂O	Ethyl isobornyl ether	203.8	Nonazeotrope		557
A =	C₈H₈O₂	**α-Toluic Acid**	**266.5**			
14016	C₉H₈O	Cinnamaldehyde	253.5	Nonazeotrope		541
14017	C₁₀H₇Br	1-Bromonaphthalene	281.8	264.0	53.5	541
14018	C₁₀H₇Cl	1-Chloronaphthalene	262.7	255.9	30	541
14019	C₁₀H₈	Naphthalene	218.05	Nonazeotrope		541
14020	C₁₀H₈O	1-Naphthol	288.5	Nonazeotrope		541
14021	C₁₀H₁₀O₂	Isosafrole	252.0	251.5	11	541
14022	C₁₀H₁₀O₂	Methyl cinnamate	261.9	261.8	3	541
14023	C₁₀H₁₀O₄	Methyl phthalate	283.7	Nonazeotrope		541
14024	C₁₀H₁₂O	Anethole	235.7	Nonazeotrope		556
14025	C₁₀H₁₂O₂	Eugenol	254.8	Nonazeotrope		575
14026	C₁₀H₁₂O₂	Isoeugenol	268.8	<266.2	>58	575
14027	C₁₀H₁₈O₄	Propyl succinate	250.5	Nonazeotrope		575
14028	C₁₁H₁₀	1-Methylnaphthalene	244.6	243.2	~ 12	541
14029	C₁₁H₁₀	2-Methylnaphthalene	241.15	239.95	12	527
14030	C₁₁H₁₂O₂	Ethyl cinnamate	271.5	Nonazeotrope		541
14031	C₁₁H₁₄O₂	1-Allyl-3,4-dimethoxy-				
		benzene	255.0	Nonazeotrope		541
14032	C₁₁H₁₄O₂	Butyl benzoate	249.8	Nonazeotrope		541
14033	C₁₁H₁₄O₂	1,2-Dimethoxy-4-				
		propenylbenzene	270.5	265.4	60	541
14034	C₁₁H₁₄O₂	Ethyl β-phenylpropionate	248.1	Nonazeotrope		575
14035	C₁₂H₁₀	Acenaphthene	277.9	262.2	71	541
14036	C₁₂H₁₀	Biphenyl	255.9	252.15	23.3	541
14037	C₁₂H₁₀O	Phenyl ether	259.3	255.05	27.8	556
14038	C₁₂H₁₆O₂	Isoamyl benzoate	262.0	259.85	26	541
14039	C₁₂H₁₈	1,3,5-Triethyl benzene	215.5	Nonazeotrope		575
14040	C₁₂H₂₂O₄	Isoamyl oxalate	268.0	262.35	50	541
14041	C₁₃H₁₀	Fluorene	295	265.8	90	575

TABLE I. *Binary Systems*   399

No.	Formula	Name	B.P., °C	B.P., °C	Wt.%A	Ref.
		B-Component		**Azeotropic Data**		
A =	**C₈H₈O₂**	**α-Toluic Acid** *(continued)*	**266.5**			
14042	C₁₃H₁₂	Diphenylmethane	265.4	258.7	35	541
14043	C₁₃H₁₂O	Benzyl phenyl ether	286.5	<266.0	>90	575
14044	C₁₄H₁₂	Stilbene	306.5	Nonazeotrope		575
14045	C₁₄H₁₄	1,2-Diphenylethane	284.5	264.3	~ 90	541
A =	**C₈H₈O₃**	**Methyl Salicylate**	**222.95**			
14046	C₈H₁₀O	*p*-Ethylphenol	218.8	Nonazeotrope		575
14047	C₈H₁₀O	Phenethyl alcohol	219.4	218.0	43	529
14048	C₈H₁₀O	3,4-Xylenol	226.8	Nonazeotrope		564
14049	C₈H₁₀O₂	*m*-Dimethoxybenzene	214.7	Nonazeotrope		575
14050	C₈H₁₀O₂	*o*-Ethoxyphenol	216.5	Nonazeotrope		575
14051	C₈H₁₀O₂	2-Phenoxyethanol	245.2	Nonazeotrope		575
14052	C₈H₁₁NO	*o*-Phenetidine	232.5	Nonazeotrope		552
14053	C₈H₁₂O₄	Ethyl fumarate	217.85	Nonazeotrope		575
14054	C₈H₁₂O₄	Ethyl maleate	223.3	221.95	60	569
14055	C₈H₁₈O₃	2-(2-Butoxyethoxy)-ethanol	231.2	220.7	78	575
14056	C₉H₇N	Quinoline	237.3	Nonazeotrope		553
14057	C₉H₁₀O	*p*-Methylacetophenone	226.35	Nonazeotrope		552
14058	C₉H₁₀O	Propiophenone	217.7	Nonazeotrope		552
14059	C₉H₁₀O₂	Benzyl acetate	215.0	Nonazeotrope		545
14060	C₉H₁₀O₂	Ethyl benzoate	212.6	Nonazeotrope		545
14061	C₉H₁₀O₂	Methyl α-toluate	215.3	Nonazeotrope		575
14062	C₉H₁₂O	3-Phenylpropanol	235.6	Nonazeotrope		545
14063	C₁₀H₈	Naphthalene	218.05	Nonazeotrope		528
14064	C₁₀H₁₀O₂	Safrole	234.5	Nonazeotrope		556
14065	C₁₀H₁₂O₂	Ethyl α-toluate	228.75	Nonazeotrope		529
14066	C₁₀H₁₂O₂	Propyl benzoate	230.85	Nonazeotrope		548
14067	C₁₀H₁₄O	Carvone	231.0	Nonazeotrope		552
14068	C₁₀H₁₄O	Thymol	232.9	Nonazeotrope		536
14069	C₁₀H₁₄O₂	*m*-Diethoxybenzene	235.4	Nonazeotrope		575
14070	C₁₀H₁₆O	Pulegone	223.8	Nonazeotrope		552
14071	C₁₀H₁₈O	Borneol	213.4	Nonazeotrope		536
14072	C₁₀H₁₈O	Geraniol	229.7	222.2	97	536
14073	C₁₀H₁₈O	Linalool	198.6	Nonazeotrope		575
14074	C₁₀H₁₈O	α-Terpineol	217.8	216.0	~ 37	528
14075	C₁₀H₂₀O	Citronellol	224.5	220.5		536
14076	C₁₀H₂₀O	Menthol	216.4	216.25	15	529
14077	C₁₀H₂₀O₂	Ethyl caprylate	208.35	Nonazeotrope		575
14078	C₁₀H₂₂O	Decyl alcohol	232.9	Nonazeotrope		536
14079	C₁₁H₁₀	1-Methylnaphthalene	244.6	Nonazeotrope		575
14080	C₁₁H₁₀	2-Methylnaphthalene	241.15	Nonazeotrope		270,575
14081	C₁₁H₂₀O	Methyl α-terpineol ether	216.2	Nonazeotrope		575
14082	C₁₁H₂₂O₃	Isoamyl carbonate	232.2	Nonazeotrope		548
14083	C₁₂H₁₈	1,3,5-Triethylbenzene	215.5	Nonazeotrope		538
14084	C₁₂H₂₀O₂	Bornyl acetate	227.7	222.3	10?	530

	B-Component			Azeotropic Data		
No.	Formula	Name	B.P., °C	B.P., °C	Wt.%A	Ref.
A =	C$_8$H$_8$O$_3$	**Methyl Salicylate**	**222.95**			
		(continued)				
14085	C$_{13}$H$_{28}$	Tridecane	234.0	Nonazeotrope		575
A =	**C$_8$H$_9$BrO**	***p*-Bromophenetole**	**234.2**			
14086	C$_8$H$_{10}$O	3,4-Xylenol	226.8	226.0	12	575
14087	C$_9$H$_7$N	Isoquinoline	240.8	Nonazeotrope		575
14088	C$_9$H$_8$O	Cinnamaldehyde	253.7	Nonazeotrope		575
14089	C$_9$H$_{10}$O	*p*-Methylacetophenone	226.25	Nonazeotrope		575
14090	C$_{10}$H$_8$	Naphthalene	218.0	Nonazeotrope		575
14091	C$_{10}$H$_{10}$O$_2$	Safrole	235.9	233.5	78	575
14092	C$_{10}$H$_{12}$O	Anethole	235.7	233.0	70	562
14093	C$_{10}$H$_{14}$N$_2$	Nicotine	247.5	Nonazeotrope		575
14094	C$_{10}$H$_{14}$O	Carvone	231.0	Nonazeotrope		575
14095	C$_{11}$H$_{10}$	2-Methylnaphthalene	241.15	Nonazeotrope		575
A =	**C$_8$H$_9$N**	**2-Methyl-5-Vinylpridine**				
14096	C$_8$H$_{11}$N	2-Methyl-5-ethylpyridine				
		20 mm.		Nonazeotrope	v-l	298
A	**C$_8$H$_{10}$**	**Ethylbenzene**	**136.15**			
14097	C$_8$H$_{10}$	*m*-Xylene	139.2	Nonazeotrope		561
14098	C$_8$H$_{10}$	*p*-Xylene	138.2	Nonazeotrope		563
14099	C$_8$H$_{16}$	Ethylcyclohexane	131.8	131.2	15	578
		" 400 mm.		Nonazeotrope	v-l	771
14100	C$_8$H$_{16}$	1-Octene	121.6	Nonazeotrope	v-l	1009
14101	C$_8$H$_{18}$	Octane	125.75	Nonazeotrope		163
		"	125.75	<125.6	<12	561
14102	C$_8$H$_{18}$O	Butyl ether	142.2	Nonazeotrope		563
14103	C$_8$H$_{18}$O	Isobutyl ether	122.3	Nonazeotrope		558
14104	C$_8$H$_{19}$N	Diisobutylamine	138.5	<135.5	<62	575
14105	C$_9$H$_{12}$	Cumene	152.4	Nonazeotrope		981
14106	C$_9$H$_{20}$	Nonane	150.7	Nonazeotrope		1061
14107	C$_9$H$_{20}$	2,2,5-Trimethylhexane	120.1	Nonazeotrope	v-l	1009
A =	**C$_8$H$_{10}$**	***m*-Xylene**	**139.2**			
14108	C$_8$H$_{10}$	*o*-Xylene	144.3	Nonazeotrope		575
14109	C$_8$H$_{10}$	*p*-Xylene	138.2	Nonazeotrope		563
14110	C$_8$H$_{16}$O$_2$	Isobutyl isobutyrate	147.3	Nonazeotrope		527
14111	C$_8$H$_{18}$O	Butyl ether	142.2	Nonazeotrope		548
14112	C$_8$H$_{18}$O	2-Ethyl-1-hexanol	184.8	Nonazeotrope		981
14113	C$_8$H$_{18}$O	Octyl alcohol	195.2	Nonazeotrope		575
14114	C$_8$H$_{18}$O	*sec*-Octyl alcohol	179.0	Nonazeotrope		537
14115	C$_8$H$_{19}$N	Diisobutylamine	138.5	<137.5	<49	551
14116	C$_9$H$_{18}$O	2-Ethylheptanal		139.0	96.1	163
A =	**C$_8$H$_{10}$**	***o*-Xylene**	**143.6**			
14117	C$_8$H$_{18}$O	Butyl ether	142.4	<142.0	<22	558
14118	C$_9$H$_{20}$	Nonane	150.7	144.25	81	1070

TABLE I.   *Binary Systems*   401

	B-Component			Azeotropic Data		
No.	Formula	Name	B.P., °C	B.P., °C	Wt.%A	Ref.
A =	C$_8$H$_{10}$	*p*-Xylene	138.4			
14119	C$_8$H$_{18}$	Octane	125.75	Nonazeotrope		163
14120	C$_9$H$_{20}$	*n*-Nonane	150.8	Nonazeotrope		307
A =	C$_8$H$_{10}$O	**Benzyl Methyl Ether**	167.8			
14121	C$_8$H$_{14}$O	Methylheptenone	173.2	Nonazeotrope		575
14122	C$_8$H$_{16}$O	2-Octanone	172.85	Nonazeotrope		575
14123	C$_8$H$_{16}$O$_2$	Butyl butyrate	166.4	166.0	30	557
14124	C$_8$H$_{16}$O$_2$	Isoamyl propionate	160.7	Nonazeotrope		557
14125	C$_8$H$_{18}$O	*sec*-Octyl alcohol	180.4	Nonazeotrope		575
14126	C$_8$H$_{20}$SiO$_4$	Ethyl silicate	168.8	<165.5		557
14127	C$_9$H$_{12}$	Mesitylene	164.6	<163.5	>15	558
14128	C$_9$H$_{18}$O$_2$	Isoamyl butyrate	181.05	Nonazeotrope		557
14129	C$_9$H$_{18}$O$_2$	Isobutyl isovalerate	171.2	Nonazeotrope		557
14130	C$_{10}$H$_{16}$	Camphene	159.6	158.0	<30	558
14131	C$_{10}$H$_{16}$	Nopinene	163.8	161.2	35	558
14132	C$_{10}$H$_{16}$	α-Terpinene	173.4	166.4	65	558
A =	C$_8$H$_{10}$O	***o*-Ethylphenol**	216.5			
14133	C$_{10}$H$_{18}$O	Citronellal	208.0	Nonazeotrope		575
A =	C$_8$H$_{10}$O	***p*-Ethylphenol**	218.8			
14134	C$_8$H$_{10}$O	Phenethyl alcohol	219.4	>220.5	>55	575
14135	C$_8$H$_{10}$O	2,4-Xylenol	210.5	Nonazeotrope		575
14136	C$_8$H$_{10}$O$_2$	Veratrole	206.8	Nonazeotrope		575
14137	C$_8$H$_{10}$O$_2$	2-Phenoxyethanol	245.2	Nonazeotrope		575
14138	C$_8$H$_{11}$N	Ethylaniline	205.5	Nonazeotrope		215
14139	C$_8$H$_{11}$NO	*o*-Phenetidine	232.5	Nonazeotrope		551
14140	C$_8$H$_{12}$O$_4$	Ethyl fumarate	217.85	223.0	48	562
14141	C$_8$H$_{12}$O$_4$	Ethyl maleate	223.3	226.3	38	575
14142	C$_8$H$_{18}$O	Octyl alcohol	195.2	Nonazeotrope		575
14143	C$_9$H$_7$N	Quinoline	237.3	<239.5	>11	575
14144	C$_9$H$_{10}$O	*p*-Methylacetophenone	226.35	229.5	30	552
14145	C$_9$H$_{10}$O	Propiophenone	217.7	224.5		552
14146	C$_9$H$_{10}$O$_2$	Benzyl acetate	215.0	221.0	60	562
14147	C$_9$H$_{10}$O$_2$	Ethyl benzoate	212.5	219.8	80	575
14148	C$_9$H$_{13}$N	*N,N*-Dimethyl-*p*-toluidine	210.2	Nonazeotrope		551
14149	C$_{10}$H$_8$	Naphthalene	218.0	215.0	45	562
14150	C$_{10}$H$_{12}$O$_2$	Propyl benzoate	230.85	Nonazeotrope		575
14151	C$_{10}$H$_{15}$N	Diethylaniline	217.05	214.0	60	551
14152	C$_{10}$H$_{18}$O	Citronellal	208.0	Nonazeotrope		575
14153	C$_{10}$H$_{18}$O	α-Terpineol	218.85	<219.7	>58	575
14154	C$_{10}$H$_{22}$O	Decyl alcohol	232.8	Nonazeotrope		575
14155	C$_{10}$H$_{22}$S	Isoamyl sulfide	214.8	<213.5	>23	566
14156	C$_{11}$H$_{16}$O	Methyl thymyl ether	216.5	<216.3	>20	575
14157	C$_{11}$H$_{20}$O	Methyl α-terpineol ether	216.2	<215.9	>14	575
14158	C$_{11}$H$_{22}$O$_3$	Isoamyl carbonate	232.2	Nonazeotrope		575
14159	C$_{12}$H$_{16}$O$_2$	Isoamyl benzoate	262.9	Nonazeotrope		575

No.	Formula	B-Component Name	B.P., °C	Azeotropic Data B.P., °C	Wt.%A	Ref.
A =	**$C_8H_{10}O$**	*p*-**Ethylphenol** *(continued)*	**218.8**			
14160	$C_{12}H_{18}$	1,3,5-Triethylbenzene	215.5	212.0	40	562
A =	**$C_8H_{10}O$**	*p*-**Methylanisole**	**177.05**			
14161	$C_8H_{11}N$	Dimethylaniline	194.15	Nonazeotrope		551
14162	$C_8H_{16}O$	2-Octanone	172.85	Nonazeotrope		575
14163	$C_8H_{16}O_2$	Butyl butyrate	166.4	Nonazeotrope		557
14164	$C_8H_{16}O_2$	Isoamyl propionate	160.7	Nonazeotrope		557
14165	$C_8H_{18}O$	Octyl alcohol	195.15	Nonazeotrope		576
14166	$C_8H_{18}O$	*sec*-Octyl alcohol	180.4	176.3	79	576
14167	$C_8H_{18}S$	Butyl sulfide	185.0	Nonazeotrope		575
14168	C_9H_8	Indene	183.0	Nonazeotrope		541
14169	C_9H_{12}	Pseudocumene	~ 168.2	Nonazeotrope		541
14170	$C_9H_{13}N$	Dimethyl-*o*-toluidine	185.35	Nonazeotrope		551
14171	$C_9H_{18}O_2$	Butyl isovalerate	177.6	176.4	58	557
14172	$C_9H_{18}O_2$	Isoamyl butyrate	181.05	Nonazeotrope		557
14173	$C_9H_{18}O_2$	Isoamyl isobutyrate	169.8	Nonazeotrope		557
14174	$C_9H_{18}O_2$	Isobutyl isovalerate	171.35	Nonazeotrope		557
14175	$C_{10}H_{14}$	Butylbenzene	183.2	Nonazeotrope		548
14176	$C_{10}H_{14}$	Cymene	176.7	Nonazeotrope?		548
14177	$C_{10}H_{16}$	α-Terpinene	173.4	Nonazeotrope		558
14178	$C_{10}H_{16}$	Terpinolene	184.6	Nonazeotrope		558
14179	$C_{10}H_{18}O$	Cineole	176.35	175.35	35	527
14180	$C_{10}H_{20}O_2$	Isoamyl isovalerate	192.7	Nonazeotrope		557
14181	$C_{10}H_{22}O$	Isoamyl ether	173.2	172.5	29.5	549
14182	$C_{10}H_{23}N$	Diisoamylamine	188.2	Nonazeotrope		551
A =	**$C_8H_{10}O$**	**Phenethyl Alcohol**	**219.4**			
14183	$C_8H_{10}O$	3,4-Xylenol	226.8	Nonazeotrope		575
14184	$C_8H_{10}O_2$	2-Phenoxyethanol	245.2	Nonazeotrope		575
14185	$C_8H_{11}N$	Dimethylaniline	194.05	Nonazeotrope		551
14186	$C_8H_{11}N$	Ethylaniline	205.5	Nonazeotrope		545
14187	$C_8H_{11}N$	2,4-Xylidine	214.0	Nonazeotrope		551
14188	$C_8H_{11}N$	3,4-Xylidine	225.5	Nonazeotrope		551
14189	$C_8H_{11}NO$	*o*-Phenetidine	232.5	Nonazeotrope		551
14190	$C_8H_{12}O_4$	Ethyl fumarate	217.85	Nonazeotrope		575
14191	$C_8H_{12}O_4$	Ethyl maleate	223.3	Nonazeotrope		575
14192	$C_8H_{16}O_3$	Isoamyl lactate	202.4	Nonazeotrope		575
14193	$C_8H_{18}O_3$	2-(2-Butoxyethoxy)-ethanol	231.2	<219.0	<92	575
14194	$C_9H_{10}O$	*p*-Methylacetophenone	226.35	Nonazeotrope		552
14195	$C_9H_{10}O$	Propiophenone	217.7	Nonazeotrope		552
14196	$C_9H_{10}O_2$	Benzyl acetate	214.9	Nonazeotrope		529
14197	$C_9H_{10}O_2$	Ethyl benzoate	212.6	Nonazeotrope		535
14198	$C_9H_{10}O_3$	Ethyl salicylate	233.7	Nonazeotrope		536
14199	$C_9H_{13}N$	*N,N*-Dimethyl-*p*-toluidine	210.2	208.5	30	551
14200	$C_{10}H_8$	Naphthalene	218.05	214.2	44	528
14201	$C_{10}H_{10}O_2$	Safrole	235.9	Nonazeotrope		575

TABLE I.   *Binary Systems*   403

No.	Formula	B-Component Name	B.P., °C	Azeotropic Data B.P., °C	Wt.%A	Ref.
A =	$C_8H_{10}O$	**Phenethyl Alcohol** (*continued*)	**219.4**			
14202	$C_{10}H_{12}O$	Anethole	235.7	Nonazeotrope		575
14203	$C_{10}H_{12}O_2$	Ethyl α-toluate	228.75	Nonazeotrope		535
14204	$C_{10}H_{12}O_2$	Propyl benzoate	230.85	Nonazeotrope		575
14205	$C_{10}H_{14}O$	Carvacrol	237.85	Nonazeotrope		575
14206	$C_{10}H_{14}O$	Carvone	231.0	Nonazeotrope		552
14207	$C_{10}H_{14}O$	Thymol	232.8	Nonazeotrope		530
14208	$C_{10}H_{15}N$	Diethylaniline	217.05	213.95	40	551
14209	$C_{10}H_{16}O$	Pulegone	223.8	Nonazeotrope		552
14210	$C_{10}H_{18}O$	Borneol	213.4	213.0	20	545
14211	$C_{10}H_{18}O$	Citronellal	208.0	Nonazeotrope		575
14212	$C_{10}H_{18}O$	α-Terpineol	218.85	217.85	33	549
14213	$C_{10}H_{20}O$	Menthol	216.3	215.05	30	549
14214	$C_{11}H_{10}$	1-Methylnaphthalene	244.9	Nonazeotrope		537
14215	$C_{11}H_{16}O$	Methyl thymyl ether	216.5	~ 215.0		575
14216	$C_{11}H_{17}N$	Isoamylaniline	256.0	Nonazeotrope		551
14217	$C_{11}H_{20}O$	α-Terpineol methyl ether	216.2	215.5		545
14218	$C_{12}H_{10}$	Biphenyl	254.9	Nonazeotrope		537
14219	$C_{12}H_{18}$	1,3,5-Triethylbenzene	215.5	212.5		537
14220	$C_{12}H_{20}O_2$	Bornyl acetate	227.6	Nonazeotrope		535
A =	$C_8H_{10}O$	**Phenetole**	**170.45**			
14221	$C_8H_{11}N$	Dimethylaniline	194.15	Nonazeotrope		551
14222	$C_8H_{14}O$	Methylheptenone	173.2	170.1	90?	552
14223	$C_8H_{16}O$	2-Octanone	172.85	170.0	92	552
14224	$C_8H_{16}O_2$	Butyl butyrate	166.4	Nonazeotrope		557
14225	$C_8H_{16}O_2$	Hexyl acetate	171.5	169.9	<75	557
14226	$C_8H_{16}O_2$	Isoamyl propionate	160.3	Nonazeotrope		557
14227	$C_8H_{18}O$	*sec*-Octyl alcohol	179.0	Nonazeotrope		556
14228	$C_8H_{18}S$	Butyl sulfide	185.0	Nonazeotrope		556
14229	$C_8H_{20}SiO_4$	Ethyl silicate	168.8	<166.0		557
14230	C_9H_8	Indene	182.8	Nonazeotrope		548
14231	C_9H_{12}	Cumene	168.2	168.15	<10	558
14232	C_9H_{12}	Mesitylene	164.6	Nonazeotrope		530
14233	C_9H_{12}	Propylbenzene	159.3	Nonazeotrope		558
14234	C_9H_{12}	Pseudocumene	168.2	168.15	<10	548
14235	$C_9H_{13}N$	Dimethyl-*o*-toluidine	185.35	Nonazeotrope		551
14236	$C_9H_{18}O_2$	Butyl isovalerate	177.6	Nonazeotrope		557
14237	$C_9H_{18}O_2$	Isoamyl butyrate	178.5	Nonazeotrope		557
14238	$C_9H_{18}O_2$	Isoamyl isobutyrate	169.8	169.2	40?	557
14239	$C_9H_{18}O_2$	Isobutyl isovalerate	171.4	170.1	65	557
14240	$C_9H_{18}O_3$	Isobutyl carbonate	190.3	Nonazeotrope		557
14241	$C_{10}H_{14}$	Cymene	176.7	Nonazeotrope		548
14242	$C_{10}H_{16}$	Camphene	159.6	Nonazeotrope		548
14243	$C_{10}H_{16}$	Dipentene	177.7	Nonazeotrope		558
14244	$C_{10}H_{16}$	*d*-Limonene	177.8	170.35	97?	548
14245	$C_{10}H_{16}$	Nopinene	163.8	Nonazeotrope		558
14246	$C_{10}H_{16}$	α-Pinene	155.8	Nonazeotrope		548

No.	Formula	B-Component Name	B.P., °C	Azeotropic Data B.P., °C	Wt.%A		Ref.
A =	$C_8H_{10}O$	**Phenetole** *(continued)*	**170.45**				
14247	$C_{10}H_{16}$	α-Terpinene	173.4	170.0	86		558
14248	$C_{10}H_{16}$	γ-Terpinene	179.9	Nonazeotrope			548
14249	$C_{10}H_{18}O$	Cineole	176.35	Nonazeotrope			545
14250	$C_{10}H_{20}O_2$	Isoamyl isovalerate	192.7	Nonazeotrope			557
14251	$C_{10}H_{22}O$	Isoamyl ether	173.2	169.2	65		549
14252	$C_{10}H_{23}N$	Diisoamylamine	188.2	Nonazeotrope			551
A =	$C_8H_{10}O$	**2,4-Xylenol**	**210.5**				
14253	$C_8H_{12}O_4$	Ethyl fumarate	217.85	219.65	32		575
14254	$C_8H_{12}O_4$	Ethyl maleate	223.3	223.7			575
14255	$C_8H_{16}O_3$	Isoamyl lactate	202.4	>212.2	<30		575
14256	C_9H_7N	Quinoline	237.3	239.0	8		575
		" 200 mm.	182.5	184.2	26.3	v-l	440
		" 150 mm.		175	27.1		440
14257	$C_9H_{10}O$	*p*-Methylacetophenone	226.35	227.0	85		575
14258	$C_9H_{10}O$	Propiophenone	217.7	221.0	65		575
14259	$C_9H_{10}O_2$	Benzyl acetate	215.0	216.8	36		575
14260	$C_9H_{10}O_2$	Ethyl benzoate	212.5	>214.5	>32		575
14261	$C_{10}H_{16}O$	Camphor	209.1	217.0	50		575
14262	$C_{10}H_{20}O_2$	Isoamyl isovalerate	192.7	Nonazeotrope			575
14263	$C_{10}H_{22}S$	Isoamyl sulfide	214.8	<209.5	<88		575
14264	$C_{12}H_{20}O_2$	Bornyl acetate	227.6	Nonazeotrope			575
A =	$C_8H_{10}O$	**3,4-Xylenol**	**226.8**				
14265	$C_8H_{10}O_2$	*o*-Ethoxyphenol	216.5	Nonazeotrope			575
14266	$C_8H_{10}O_2$	2-Phenoxyethanol	245.2	Nonazeotrope			575
14267	$C_8H_{11}N$	2,4-Xylidine	214.0	Nonazeotrope			551
14268	$C_8H_{11}N$	Ethylaniline	205.5	Nonazeotrope			551
14269	$C_8H_{11}NO$	*o*-Phenetidine	232.5	232.65	8		551
14270	$C_8H_{11}NO$	*p*-Phenetidine	249.9	Nonazeotrope			551
14271	$C_8H_{12}O_4$	Ethyl fumarate	217.85	228.2	65		4,526
14272	$C_8H_{12}O_4$	Ethyl maleate	223.3	230.0	55		527
14273	$C_8H_{16}O_2$	Caprylic acid	238.5	Nonazeotrope			575
14274	$C_8H_{16}O_3$	Isoamyl lactate	202.4	Nonazeotrope			575
14275	$C_8H_{18}O$	Octyl alcohol	195.2	Nonazeotrope			575
14276	C_9H_7N	Quinoline	237.3	241.95	35		568
		" 200 mm.	182.5	191.5	42.8	v-l	440
		" 116 mm.		175	44.1		440
14277	$C_9H_{10}O$	Cinnamyl alcohol	257.0	Nonazeotrope			575
14278	$C_9H_{10}O$	*p*-Methylacetophenone	226.35	231.35	51		568
14279	$C_9H_{10}O$	Propiophenone	217.7	228.5	67		552
14280	$C_9H_{10}O_2$	Benzyl acetate	215.0	Nonazeotrope			575
14281	$C_9H_{10}O_2$	Ethyl benzoate	212.5	Nonazeotrope			575
14282	$C_9H_{10}O_3$	Ethyl salicylate	233.8	Nonazeotrope			575
14283	$C_9H_{12}O$	Mesitol	220.5	Nonazeotrope			575
14284	$C_9H_{12}O$	3-Phenylpropanol	235.6	Nonazeotrope			575
14285	$C_9H_{13}N$	*N,N*-Dimethyl-*p*-toluidine	210.2	Nonazeotrope			551

TABLE I. *Binary Systems* 405

No.	Formula	Name	B.P., °C	B.P., °C	Wt.%A		Ref.
		B-Component		Azeotropic Data			

No.	Formula	Name	B.P., °C	B.P., °C	Wt.%A		Ref.
A =	**$C_8H_{10}O$**	**3,4-Xylenol** *(continued)*	**226.8**				
14286	$C_{10}H_8$	Naphthalene	218.0	217.6	16		564
14287	$C_{10}H_9N$	Quinaldine	246.5	>248.0	20		575
14288	$C_{10}H_{12}O_2$	Ethyl α-toluate	228.75	230.8	42		562
14289	$C_{10}H_{12}O_2$	Propyl benzoate	230.85	231.9	33		575
14290	$C_{10}H_{11}O$	Thymol	232.9	Nonazeotrope			549
14291	$C_{10}H_{15}N$	Diethylaniline	217.05	217.0	8		551
14292	$C_{10}H_{16}O$	Camphor	209.1	227.55	73		568
14293	$C_{10}H_{18}O$	Borneol	215.0	Nonazeotrope			575
14294	$C_{10}H_{18}O$	Citronellal	208.0	Nonazeotrope			575
14295	$C_{10}H_{18}O$	Linalool	198.6	Nonazeotrope			575
14296	$C_{10}H_{18}O$	α-Terpineol	218.85	Nonazeotrope			575
14297	$C_{10}H_{20}O$	Menthol	216.3	Nonazeotrope			575
14298	$C_{10}H_{20}O_2$	Ethyl caprylate	208.35	Nonazeotrope			575
14299	$C_{10}H_{22}S$	Isoamyl sulfide	214.8	Nonazeotrope			566
14300	$C_{11}H_{14}O_2$	Isobutyl benzoate	241.9	Nonazeotrope			575
14301	$C_{11}H_{16}O$	Methyl thymyl ether	216.5	Nonazeotrope			575
14302	$C_{11}H_{20}O$	Methyl α-terpineol ether	216.2	Nonazeotrope			575
14303	$C_{11}H_{22}O_3$	Isoamyl carbonate	232.2	Nonazeotrope			575
14304	$C_{12}H_{18}$	1,3,5-Triethylbenzene	215.5	Nonazeotrope			575
14305	$C_{12}H_{20}O_2$	Bornyl acetate	227.6	>229.8	>37		575
14306	$C_{13}H_{28}$	Tridecane	234.0	223.5	58		562
A =	**$C_8H_{10}O$**	**3,5-Xylenol**	**175.7/200**				
14307	C_9H_7N	Quinoline 200 mm.	182.5	186.35	33.7	v-l	440
		" 130 mm.		175	33.7		440
A =	**$C_8H_{10}O_2$**	**m-Dimethoxybenzene**	**214.7**				
14308	$C_8H_{11}N$	2,4-Xylidine	214.0	<211.8	<56		575
14309	$C_8H_{11}N$	3,4-Xylidine	225.5	Nonazeotrope			575
14310	$C_8H_{12}O_4$	Ethyl fumarate	217.85	211.2			557
14311	$C_8H_{12}O_4$	Ethyl maleate	223.3	<212.5	>82		575
14312	$C_9H_{10}O_2$	Benzyl acetate	215.0	<214.0	<60		575
14313	$C_9H_{10}O_2$	Ethyl benzoate	212.5	<212.35			557
14314	$C_{10}H_{18}O$	Borneol	213.4	213.0			576
14315	$C_{10}H_{18}O$	α-Terpineol	218.85	<214.0	>70		575
		"	218.0	Nonazeotrope			576
14316	$C_{10}H_{20}O$	Menthol	216.4	214.2			576
14317	$C_{10}H_{22}S$	Isoamyl sulfide	214.8	<213.5	>44		575
14318	$C_{11}H_{20}O$	Terpineol methyl ether	216.2	Nonazeotrope			537
14319	$C_{12}H_{20}O_2$	Bornyl acetate	227.6	Nonazeotrope			557
A =	**$C_8H_{10}O_2$**	**m-Ethoxyphenol**	**243.8**				
14320	$C_{11}H_{14}O_2$	Butyl benzoate	249.0	Nonazeotrope			575
A =	**$C_8H_{10}O_2$**	**o-Ethoxyphenol**	**216.5**				
14321	$C_8H_{10}O_2$	2-Phenoxyethanol	245.2	Nonazeotrope			575
14322	$C_8H_{11}N$	Dimethylaniline	194.15	Nonazeotrope			551
14323	$C_8H_{11}N$	Ethylaniline	205.5	Nonazeotrope			551
14324	$C_8H_{11}N$	2,4-Xylidine	214.0	Nonazeotrope			551

No.	Formula	B-Component Name	B.P., °C	Azeotropic Data B.P., °C	Wt.%A	Ref.
A =	**C₈H₁₀O₂**	*o*-**Ethoxyphenol** *(continued)*	**216.5**			
14325	C₈H₁₂O₄	Ethyl maleate	223.3	Nonazeotrope		575
14326	C₈H₁₄O₄	Ethyl succinate	216.5	Azeotropic		563
14327	C₈H₁₈O₃	2-(2-Butoxyethoxy)ethanol	231.2	Nonazeotrope		575
14328	C₉H₁₀O	Propiophenone	217.7	218.3		552
14329	C₉H₁₀O₂	Benzyl acetate	~214.9	218		535
14330	C₉H₁₀O₂	Ethyl benzoate	212.6	Nonazeotrope		548
14331	C₉H₁₂O	Mesitol	220.5	Nonazeotrope		575
14332	C₉H₁₃N	*N,N*-Dimethyl-*p*-toluidine	210.2	Nonazeotrope		551
14333	C₁₀H₈	Naphthalene	218.0	<215.5	>72	575
14334	C₁₀H₁₄O	Thymol	232.9	Nonazeotrope		542
14335	C₁₀H₁₅N	Diethylaniline	217.05	<216.2	>57	551
14336	C₁₀H₁₆O	Camphor	209.1	Nonazeotrope		575
14337	C₁₀H₁₆O	Pulegone	223.8	Nonazeotrope		552
14338	C₁₀H₁₈O	Borneol	211.8	Nonazeotrope		563
14339	C₁₀H₂₀O	Menthol	216.3	<216.0		575
14340	C₁₀H₂₀O₂	Ethyl caprylate	208.35	Nonazeotrope		575
14341	C₁₀H₂₂S	Isoamyl sulfide	214.8	<214.2		566
14342	C₁₂H₁₈	1,3,5-Triethylbenzene	215.5	<214.5	>30	575
14343	C₁₂H₂₀O₂	Bornyl acetate	227.6	Nonazeotrope		535
A =	**C₈H₁₀O₂**	**2-Phenoxyethanol**	**245.2**			
14344	C₉H₁₀O	Cinnamyl alcohol	257.0	Nonazeotrope		575
14345	C₉H₁₀O₃	Ethyl salicylate	233.8	Nonazeotrope		575
14346	C₉H₁₂O	3-Phenylpropanol	235.6	Nonazeotrope		575
14347	C₁₀H₇Cl	1-Chloronaphthalene	262.7	Nonazeotrope		575
14348	C₁₀H₈	Naphthalene	218.0	Nonazeotrope		575
14349	C₁₀H₁₀O₂	Isosafrole	252.0	<244.5	>68	575
14350	C₁₀H₁₈O	Geraniol	229.6	Nonazeotrope		575
14351	C₁₁H₁₀	1-Methylnaphthalene	244.6	<243.0	>43	575
14352	C₁₁H₁₀	2-Methylnaphthalene	241.15	239.5	30	575
14353	C₁₂H₁₀	Acenaphthene	277.9	Nonazeotrope		575
14354	C₁₂H₁₀O	Phenyl ether	259.0	Nonazeotrope		575
14355	C₁₃H₁₂	Diphenylmethane	265.4	Nonazeotrope		575
A =	**C₈H₁₀O₂**	**Veratrole**	**206.8**			
14356	C₈H₁₁N	Dimethylaniline	194.15	Nonazeotrope		551
14357	C₈H₁₁N	Ethylaniline	205.5	<203.0		575
14358	C₈H₁₂O₄	Ethyl fumarate	217.85	<205.9	>69	557
14359	C₉H₁₀O₂	Benzyl acetate	215.0	Nonazeotrope		557
14360	C₉H₁₀O₂	Ethyl benzoate	212.6	Nonazeotrope		557
14361	C₉H₁₃N	Dimethyl-o-toluidine	185.35	Nonazeotrope		551
14362	C₁₀H₈	Naphthalene	218.05	Nonazeotrope		573
14363	C₁₀H₁₅N	Diethylaniline	217.05	Nonazeotrope		551
14364	C₁₀H₁₈O	Borneol	213.4	Nonazeotrope		573
14365	C₁₀H₂₀O₂	Isoamyl isovalerate	192.7	Nonazeotrope		557
14366	C₁₂H₁₈	1,3,5-Triethylbenzene	215.5	Nonazeotrope		537

TABLE I.   *Binary Systems*   407

No.	B-Component			Azeotropic Data			
	Formula	Name	B.P., °C	B.P., °C	Wt.%A	Ref.

No.	Formula	Name	B.P., °C	B.P., °C	Wt.%A	Ref.
A =	$C_8H_{11}N$	**Dimethylaniline**	**194.05**			
14367	$C_8H_{18}O$	Octyl alcohol	195.2	191.75	49.5	551
14368	$C_8H_{18}O$	sec-Octyl alcohol	180.4	Nonazeotrope		551
	"	"	180.4	180.0		545
14369	C_9H_8	Indene	182.6	Nonazeotrope		551
14370	$C_9H_{10}O$	Prophiophenone	217.7	Nonazeotrope		551
14371	C_9H_{12}	Mesitylene	164.6	Nonazeotrope		551
14372	C_9H_{12}	Propylbenzene	159.3	Nonazeotrope		551
14373	$C_9H_{12}O$	Benzyl ethyl ether	185.0	Nonazeotrope		575
14374	$C_{10}H_8$	Napthalene	218.0	Nonazeotrope		551
14375	$C_{10}H_{14}$	Cymene	176.7	Nonazeotrope		551
14376	$C_{10}H_{16}$	Camphene	159.6	Nonazeotrope		551
14377	$C_{10}H_{16}$	Dipentene	177.7	Nonazeotrope		551
14378	$C_{10}H_{16}$	d-Limonene	177.8	Nonazeotrope		545
	"	"	177.8	174	27	563
14379	$C_{10}H_{16}$	Nopinene	163.8	Nonazeotrope		551
14380	$C_{10}H_{16}$	α-Pinene	155.8	Nonazeotrope		551
14381	$C_{10}H_{16}$	α-Terpinene	173.4	Nonazeotrope		551
14382	$C_{10}H_{16}$	Terpinolene	185	~ 179	~ 35	563
14383	$C_{10}H_{16}$	Thymene	179.7	Nonazeotrope		532
14384	$C_{10}H_{16}O$	Camphor	209.1	Nonazeotrope		551
14385	$C_{10}H_{16}O$	Fenchone	193	191	~ 35	563
14386	$C_{10}H_{18}O$	Borneol	215.0	Nonazeotrope		551
14387	$C_{10}H_{18}O$	Cineole	176.35	Nonazeotrope		551
14388	$C_{10}H_{18}O$	Linalool	198.6	193.9	85	551
14389	$C_{10}H_{18}O$	α-Terpineol	218.85	Nonazeotrope		551
14390	$C_{10}H_{20}O$	Citronellol	224.4	Nonazeotrope		551
14391	$C_{10}H_{20}O$	Menthol	216.3	Nonazeotrope		551
14392	$C_{10}H_{22}O$	Amyl ether	187.5	<187.0	<27	551
14393	$C_{10}H_{22}O$	Isoamyl ether	173.2	Nonazeotrope		551
14394	$C_{12}H_{22}O$	Bornyl ethyl ether	204.9	Nonazeotrope		551
A =	$C_8H_{11}N$	**Ethylaniline**	**205.5**			
14395	$C_8H_{18}O$	n-Octyl alcohol	195.2	194.9	15	551
14396	$C_8H_{18}O$	sec-Octyl alcohol	180.4	Nonazeotrope		551
14397	$C_9H_{10}O$	Propiophenone	217.7	Nonazeotrope		551
14398	$C_9H_{12}O$	Phenyl propyl ether	190.5	Nonazeotrope		575
14399	$C_{10}H_8$	Napthalene	218.0	Nonazeotrope		551
		"	218.1	205	~ 10	563
14400	$C_{10}H_{14}O$	Thymol	232.9	Nonazeotrope		551
14401	$C_{10}H_{16}$	Terpinolene	184.6	Nonazeotrope		551
14402	$C_{10}H_{16}O$	Camphor	209.1	Nonazeotrope		551
14403	$C_{10}H_{18}O$	Borneol	215.0	Nonazeotrope		551
14404	$C_{10}H_{18}O$	Geraniol	229.6	Nonazeotrope		575
14405	$C_{10}H_{18}O$	Linalool	198.6	Nonazeotrope		551
14406	$C_{10}H_{18}O$	Menthone	207	<205	~ 60	563
		"	209.5	Nonazeotrope		575

No.	Formula	B-Component Name	B.P., °C	Azeotropic Data B.P., °C	Wt.%A	Ref.
A =	**C₈H₁₁N**	**Ethylaniline** *(continued)*	**205.5**			
14407	C₁₀H₁₈O	α-Terpineol	218.85	Nonazeotrope		551
14408	C₁₀H₂₀O	Citronellol	224.4	Nonazeotrope		551
14409	C₁₀H₂₀O	Menthol	216.3	Nonazeotrope		551
14410	C₁₀H₂₂O	n-Decyl alcohol	232.8	Nonazeotrope		551
14411	C₁₁H₁₆O	Methyl thymyl ether	216.5	Nonazeotrope		575
14412	C₁₁H₂₀O	Isobornyl methyl ether	192.4	Nonazeotrope		551
14413	C₁₁H₂₀O	Methyl α-terpineol ether	216.3	Nonazeotrope		563
14414	C₁₁H₂₄O₂	Diisoamyloxymethane	207.3	204	58	563
14415	C₁₂H₂₂O	Bornyl ethyl ether	204.9	<203.0	<48	551
A =	**C₈H₁₁N**	**s-Collidine**	**170.0**			
14416	C₈H₁₈	2,2,4-Trimethylpentane	99.3	Nonazeotrope		575
14417	C₁₀H₈	Napthalene	217.9	Nonazeotrope		506
A =	**C₈H₁₁N**	**2,4-Xylidine**	**214.0**			
14418	C₈H₁₈O	n-Octyl alcohol	195.2	Nonazeotrope		551
14419	C₉H₁₀O	Propiophenone	217.7	Nonazeotrope		551
14420	C₁₀H₁₄O	Thymol	232.9	Nonazeotrope		551
14421	C₁₀H₁₆O	Camphor	209.1	Nonazeotrope		551
14422	C₁₀H₂₀O	Menthol	216.3	213.5	70	551
14423	C₁₁H₁₆O	Methyl thymyl ether	216.5	<212.5		575
14424	C₁₁H₂₄	Undecane	195.5	194.98	12	506
14425	C₁₂H₁₈	1,3,5-Triethylbenzene	215.5	<212.5	>51	551
14426	C₁₂H₂₂O	Ethyl bornyl ether	204.9	Nonazeotrope		575
14427	C₁₂H₂₆	Dodecane	216.5	209.80	37.0	506
14428	C₁₃H₂₈	Tridecane	234.6	215.28	71.0	506
14429	C₁₄H₃₀	Tetradecane	252	217.38	97.5	506
A =	**C₈H₁₁N**	**3,4-Xylidine**	**225.5**			
14430	C₉H₁₂O	3-Phenylpropanol	235.6	Nonazeotrope		551
14431	C₁₀H₈	Naphthalene	218.0	Nonazeotrope		551
14432	C₁₀H₁₄O	Carvacrol	237.85	Nonazeotrope		575
14433	C₁₀H₂₀O	Citronellol	224.4	223.5	40	551
14434	C₁₁H₁₀	2-Methylnaphthalene	241.15	Nonazeotrope		527,551
14435	C₁₂H₂₂O	Bornyl ethyl ether	204.9	Nonazeotrope		575
A =	**C₈H₁₁NO**	**o-Phenetidine**	**232.5**			
14436	C₈H₁₈O₃	2-(2-Butoxyethoxy)ethanol	231.2	226.0	52	575
14437	C₉H₁₀O	p-Methylacetophenone	226.35	Nonazeotrope		551
14438	C₉H₁₀O₃	Ethyl salicylate	233.8	232.2	82	551
14439	C₉H₁₂O	3-Phenylpropanol	235.6	Nonazeotrope		551
14440	C₁₀H₈	Naphthalene	218.0	Nonazeotrope		551
14441	C₁₀H₁₀O₂	Isosafrole	252.0	Nonazeotrope		551
14442	C₁₀H₁₀O₂	Safrole	235.9	232.38	86	551
14443	C₁₀H₁₂O	Anethole	235.7	232.25	75	551
14444	C₁₀H₁₄O	Carvacrol	237.85	238.0	13	551
14445	C₁₀H₁₄O	Carvone	231.0	>232.8	<74	551
14446	C₁₀H₁₄O	Thymol	232.9	234.3	45.1	551
14447	C₁₀H₁₆O	Carvenone	234.5	235.0	30	551

TABLE I.   *Binary Systems*   409

No.	Formula	B-Component Name	B.P., °C	Azeotropic Data B.P., °C	Wt.%A	Ref.
A =	**C₈H₁₁NO**	*o*-**Phenetidine** *(continued)*	**232.5**			
14448	C₁₀H₁₆O	Pulegone	223.8	Nonazeotrope		551
14449	C₁₀H₁₈O	α-Terpineol	218.85	Nonazeotrope		551
14450	C₁₀H₂₀O	Menthol	216.3	Nonazeotrope		551
14451	C₁₀H₂₂O	*n*-Decyl alcohol	232.8	232.0	>52	551
14452	C₁₁H₁₀	1-Methylnaphthalene	244.6	Nonazeotrope		551
		"	244.6	Nonazeotrope		548
14453	C₁₁H₁₀	2-Methylnaphthalene	241.15	Nonazeotrope		527,551
14454	C₁₂H₁₈	1,3,5-Triethylbenzene	215.5	Nonazeotrope		551
A =	**C₈H₁₁NO**	*p*-**Phenetidine**	**249.9**			
14455	C₉H₁₀O	Cinnamyl alcohol	257.0	Nonazeotrope		551
14456	C₉H₁₀O₃	Ethyl salicylate	233.8	Nonazeotrope		551
14457	C₉H₁₂O	3-Phenylpropanol	235.6	Nonazeotrope		551
14458	C₉H₁₂O₂	Benzyloxyethanol	265.2	Nonazeotrope		575
14459	C₁₀H₇Cl	1-Chloronaphthalene	262.7	249.7	90	551
14460	C₁₀H₁₀O₂	Isosafrole	252.0	248.8	64	551
14461	C₁₀H₁₀O₂	Safrole	235.9	Nonazeotrope		551
14462	C₁₀H₁₂O	Anethole	235.7	Nonazeotrope		551
14463	C₁₀H₁₄O	Carvacrol	237.85	Nonazeotrope		551
14464	C₁₀H₁₄O	Carvone	231.0	Nonazeotrope		551
14465	C₁₀H₁₄O	Thymol	232.9	Nonazeotrope		551
14466	C₁₀H₁₆O	Carvenone	234.5	Nonazeotrope		551
14467	C₁₁H₁₀	1-Methylnaphthalene	244.6	243.95	27	551
14468	C₁₁H₁₀	2-Methylnaphthalene	241.15	240.85	15	527
14469	C₁₁H₁₄O₂	1-Allyl-3,4-dimethoxybenzene	254.7	249.4	75	551
14470	C₁₁H₁₄O₂	1,2-Dimethoxy-4-propenylbenzene	270.5	Nonazeotrope		551
14471	C₁₂H₁₀	Biphenyl	256.1	249.5	90	551
14472	C₁₂H₁₀O	Phenyl ether	259.0	249.75	85	551
14473	C₁₂H₁₆O₃	Isoamyl salicylate	277.5	Nonazeotrope		551
14474	C₁₃H₁₂	Diphenylmethane	266.4	Nonazeotrope		551
A =	**C₈H₁₂O₄**	**Ethyl Fumarate**	**217.85**			
14475	C₈H₁₄O₄	Propyl oxalate	214	Nonazeotrope		575
14476	C₈H₁₆O₂	Caprylic acid	238.5	Nonazeotrope		575
14477	C₈H₁₆O₄	2-(2-Ethoxyethoxy)ethyl acetate	218.5	217.0	62	562
14478	C₉H₁₀O	*p*-Methylacetophenone	226.35	Nonazeotrope		552
14479	C₉H₁₀O	Propiophenone	217.7	216.8	53	552
14480	C₉H₁₀O₂	Benzyl acetate	215.0	Nonazeotrope		549
14481	C₉H₁₀O₂	Ethyl benzoate	212.5	Nonazeotrope		575
14482	C₁₀H₈	Naphthalene	218.0	216.7	58	527
14483	C₁₀H₁₀O₂	Safrole	235.9	Nonazeotrope		557
14484	C₁₀H₁₂O	Anethole	235.7	Nonazeotrope		557
14485	C₁₀H₁₂O₂	Ethyl α-toluate	228.75	Nonazeotrope		575
14486	C₁₀H₁₄O	Thymol	232.9	233.35	12.5	562

		B-Component		Azeotropic Data			
No.	Formula	Name	B.P., °C	B.P., °C	Wt.%A		Ref.
A =	**C₈H₁₂O₄**	**Ethyl Fumarate** *(continued)*	**217.85**				
14487	C₁₀H₁₄O₂	*m*-Diethoxybenzene	235.0	Nonazeotrope			557
14488	C₁₀H₁₆O	Camphor	209.1	Nonazeotrope			552
14489	C₁₀H₁₆O	Pulegone	223.8	Nonazeotrope			552
14490	C₁₀H₁₈O	Borneol	215.0	Nonazeotrope			526
14491	C₁₀H₁₈O	Citronellal	208.0	Nonazeotrope			575
14492	C₁₀H₁₈O	Geraniol	229.6	Nonazeotrope			575
14493	C₁₀H₁₈O	α-Terpineol	218.85	Nonazeotrope			575
14494	C₁₀H₂₀O	Menthol	216.3	216.0	30		526
14495	C₁₀H₂₀O₂	Methyl pelargonate	213.8	Nonazeotrope			549
14496	C₁₁H₁₀	2-Methylnaphthalene	241.15	Nonazeotrope			527
14497	C₁₁H₁₆O	Methyl thymol ether	216.5	<212.8			557
14498	C₁₁H₂₀O	Methyl α-terpineol ether	216.2	209.5	43		557
14499	C₁₁H₂₂O₂	Ethyl pelargonate	227	Nonazeotrope			575
14500	C₁₂H₁₈	1,3,5-Triethylbenzene	215.5	<215.0	<43		575
14501	C₁₂H₂₀O₂	Bornyl acetate	227.6	Nonazeotrope			549
A =	**C₈H₁₂O₄**	**Ethyl Maleate**	**223.3**				
14502	C₈H₁₄O₄	Propyl oxalate	214	Nonazeotrope			575
14503	C₈H₁₆O₂	Caprylic acid	238.5	Nonazeotrope			575
14504	C₉H₁₀O	*p*-Methylacetophenone	226.35	223.15	88		552
14505	C₉H₁₀O	Priopiophenone	217.7	Nonazeotrope			559
14506	C₉H₁₀O₂	Benzyl acetate	215.0	Nonazeotrope			545
14507	C₉H₁₀O₂	Ethyl benzoate	212.5	Nonazeotrope			572
14508	C₉H₁₀O₃	Ethyl salicylate	233.8	Nonazeotrope			580
14509	C₁₀H₈	Naphthalene	218.0	217.65	23		525
14510	C₁₀H₁₀O₂	Safrole	235.9	Nonazeotrope			556
14511	C₁₀H₁₂O	Anethole	235.7	Nonazeotrope			557
14512	C₁₀H₁₂O₂	Propyl benzoate	230.85	Nonazeotrope			575
14513	C₁₀H₁₄O	Carvacrol	237.85	238.7	12		575
14514	C₁₀H₁₄O	Carvone	231.0	Nonazeotrope			552
14515	C₁₀H₁₄O	Thymol	232.9	234.9	27		562
14516	C₁₀H₁₆O	Pulegone	223.8	221.8	53		552
14517	C₁₀H₁₈O	Citronellal	208.0	Nonazeotrope			575
14518	C₁₀H₁₈O	α-Terpineol	218.85	218.3	20		526
14519	C₁₀H₂₀O	Citronellol	224.4	<222.3	<50		575
14520	C₁₀H₂₂O	Decyl alcohol	232.8	Nonazeotrope			575
14521	C₁₁H₁₀	2-Methylnaphthalene	241.15	Nonazeotrope			527
14522	C₁₁H₁₆O	Methyl thymyl ether	216.5	<215.9	<12		575
14523	C₁₁H₂₀O	Methyl terpineol ether	216.2	<214.8	<18		575
14524	C₁₁H₂₂O₃	Isoamyl carbonate	232.2	Nonazeotrope			575
14525	C₁₂H₁₈	1,3,5-Triethylbenzene	215.5	Nonazeotrope			575
A	**C₈H₁₄O**	**Methylheptenone**	**173.2**				
14526	C₈H₁₈O	Octyl alcohol	195.2	Nonazeotrope			552
14527	C₈H₁₈O	*sec*-Octyl alcohol	180.4	Nonazeotrope			552
14528	C₈H₁₈S	Butyl sulfide	185.0	Nonazeotrope			566
14529	C₉H₈	Indene	182.6	Nonazeotrope			552
14530	C₉H₁₂	Mesitylene	164.6	Nonazeotrope			552

TABLE I. *Binary Systems* 411

No.	Formula	B-Component Name	B.P., °C	Azeotropic Data B.P., °C	Wt.%A	Ref.
A =	**C$_8$H$_{14}$O**	**Methylheptenone** *(continued)*	**173.2**			
14531	C$_9$H$_{12}$	Propylbenzene	159.3	Nonazeotrope		575
14532	C$_9$H$_{18}$O$_2$	Isoamyl butyrate	181.05	Nonazeotrope		552
14533	C$_9$H$_{18}$O$_2$	Isoamyl isobutyrate	169.8	Nonazeotrope		552
14534	C$_9$H$_{18}$O$_2$	Isobutyl isovalerate	171.2	Nonazeotrope		552
14535	C$_{10}$H$_{14}$	Butylbenzene	183.1	Nonazeotrope		552
14536	C$_{10}$H$_{14}$	Cymene	176.7	172.7	72	552
14537	C$_{10}$H$_{16}$	Camphene	159.6	157.5	12	552
		"	159.6	Nonazeotrope		545
14538	C$_{10}$H$_{16}$	Dipentene	177.7	170.9	52.5	552
14539	C$_{10}$H$_{16}$	*d*-Limonene	177.8	170.9	52.5	529
14540	C$_{10}$H$_{16}$	α-Pinene	155.8	Nonazeotrope		573
14541	C$_{10}$H$_{16}$	α-Terpinene	173.4	170.0	42	552
14542	C$_{10}$H$_{18}$O	Cineole	176.35	171.9	52	552
14543	C$_{10}$H$_{22}$	Decane	173.3	169.0	42	552
14544	C$_{10}$H$_{22}$O	Isoamyl ether	172.6	~ 171.5		574
A =	**C$_8$H$_{14}$O$_3$**	**Bis(2-Vinyloxyethyl) Ether**	**196.5/10 mm.**			
14545	C$_8$H$_{18}$O$_3$	Bis(2-ethoxyethyl) ether 10 mm.		Nonazeotrope	v-l	199
A =	**C$_8$H$_{14}$O$_4$**	**Ethyl Succinate**	**217.25**			
14546	C$_8$H$_{16}$O$_2$	Caprylic acid	238.5	Nonazeotrope		575
14547	C$_9$H$_8$	Indene	182.6	Nonazeotrope		575
14548	C$_9$H$_{10}$O	*p*-Methylacetophenone	226.35	Nonazeotrope		552
14549	C$_9$H$_{10}$O	Propiophenone	217.7	216.7	67	552
14550	C$_9$H$_{10}$O$_2$	Ethyl benzoate	212.4	Nonazeotrope		529
14551	C$_{10}$H$_8$	Naphthalene	218.05	216.3	61.5	529
14552	C$_{10}$H$_{10}$O$_2$	Safrole	235.9	Nonazeotrope		557
14553	C$_{10}$H$_{12}$O$_2$	Ethyl α-toluate	228.75	Nonazeotrope		549
14554	C$_{10}$H$_{14}$O	Thymol	232.9	>233.0		575
14555	C$_{10}$H$_{16}$	*d*-Limonene	177.8	Nonazeotrope		538
14556	C$_{10}$H$_{16}$	γ-Terpinene	179.9	Nonazeotrope		546
14557	C$_{10}$H$_{16}$	Thymene	179.7	Nonazeotrope		538
14558	C$_{10}$H$_{16}$O	Camphor	209.1	Nonazeotrope		552
14559	C$_{10}$H$_{16}$O	Pulegone	~223.8	Nonazeotrope		552
14560	C$_{10}$H$_{18}$O	Borneol	213.4	Nonazeotrope		535
14561	C$_{10}$H$_{18}$O	Geraniol	229.7	Reacts		535
14562	C$_{10}$H$_{20}$O	Menthol	216.4	215		535
14563	C$_{10}$H$_{20}$O$_2$	Methyl pelargonate	213.8	212.5		549
14564	C$_{11}$H$_{10}$	1-Methylnaphthalene	245.1	Nonazeotrope		546
14565	C$_{11}$H$_{10}$	2-Methylnaphthalene	241.15	Nonazeotrope		527
14566	C$_{11}$H$_{16}$O	Methyl thymyl ether	216.5	<213.5	>38	575
14567	C$_{11}$H$_{20}$O	Methyl α-terpineol ether	216.2	<212	18	557
14568	C$_{11}$H$_{22}$O$_2$	Ethyl pelargonate	227	Nonazeotrope		549
14569	C$_{11}$H$_{24}$O$_2$	Diisoamyloxymethane	210.8	<210.4		557
14570	C$_{12}$H$_{18}$	1,3,5-Triethylbenzene	215.5	<214.0	<46	562
14571	C$_{12}$H$_{20}$O$_2$	Bornyl acetate	227.6	Nonazeotrope		549

No.	Formula	Name	B.P., °C	B.P., °C	Wt.%A	Ref.
		B-Component		**Azeotropic Data**		
A =	**$C_8H_{14}O_4$**	**Propyl Oxalate**	**214**			
14572	$C_9H_{10}O_2$	Benzyl acetate	215.0	<212.5		549
14573	$C_{10}H_8$	Naphthalene	218.1	Nonazeotrope		546
14574	$C_{10}H_{14}O$	Thymol	232.9	Nonazeotrope		575
14575	$C_{10}H_{16}$	d-Limonene	177.9	Nonazeotrope		546
14576	$C_{10}H_{16}$	α-Pinene	155.8	Nonazeotrope		546
14577	$C_{10}H_{17}Cl$	Bornyl chloride	207.5	205.5	25	575
14578	$C_{12}H_{18}$	1,3,5-Triethylbenzene	216	<210	>70	546
A =	**$C_8H_{15}N$**	**Caprylonitrile**	**205.2**			
14579	$C_8H_{18}O$	Octyl alcohol	195.2	Nonazeotrope		575
A =	**C_8H_{16}**	**1,3-Dimethylcyclohexane**	**120.7**			
14580	C_8H_{18}	Octane	125.75	Nonazeotrope		561
14581	$C_8H_{18}O$	Isobutyl ether	122.3	120.0	72	558
A =	**C_8H_{16}**	**Ethylcyclohexane**				
14582	C_8H_{18}	Octane 50-760 mm.		Nonazeotrope	v-l	761
A =	**$C_8H_{16}O$**	**2-Octanone**	**172.85**			
14583	$C_8H_{16}O_2$	Butyl butyrate	166.4	Nonazeotrope		552
14584	$C_8H_{16}O_2$	Ethyl caproate	167.7	Nonazeotrope		552
14585	$C_8H_{16}O_2$	Hexyl acetate	171.5	171.4?		552
14586	$C_8H_{18}O$	sec-Octyl alcohol	180.4	Nonazeotrope		552
14587	$C_8H_{18}S$	Butyl sulfide	185.0	Nonazeotrope		575
14588	$C_8H_{18}S$	Isobutyl sulfide	172.0	169.8	50	566
14589	C_9H_8	Indene	182.6	Nonazeotrope		575
14592	C_9H_{12}	Mesitylene	164.6	Nonazeotrope		552
14593	$C_9H_{18}O_2$	Propylbenzene	159.3	Nonazeotrope		552
14590	C_9H_{12}	Pseudocumene	168.2	168.0		552
14591	C_9H_{12}	Isoamyl butyrate	181.05	Nonazeotrope		552
14594	$C_9H_{18}O_2$	Isobutyl isovalerate	171.2	Nonazeotrope		552
14595	$C_{10}H_{14}$	Butylbenzene	183.2	Nonazeotrope		548
14596	$C_{10}H_{14}$	Cymene	176.7	172.5	75	552
		"	175.3	Nonazeotrope		563
14597	$C_{10}H_{16}$	Camphene	159.6	158.0	13	552
14598	$C_{10}H_{16}$	Dipentene	177.7	170.0	55	552
14599	$C_{10}H_{16}$	d-Limonene	177.8	170	~ 57	573
14600	$C_{10}H_{16}$	α-Pinene	155.8	Nonazeotrope		552
14601	$C_{10}H_{16}$	α-Terpinene	173.4	169.0	42	552
14602	$C_{10}H_{16}$	γ-Terpinene	183	171.0	75	552
14603	$C_{10}H_{18}O$	Cineole	176.35	172.0	55	552
A =	**$C_8H_{16}O_2$**	**Butyl Butyrate**	**166.4**			
14604	$C_8H_{16}O_2$	Ethyl caproate	167.7	Nonazeotrope		575
14605	$C_8H_{16}O_2$	Isoamyl propionate	160.7	Nonazeotrope		575
14606	$C_8H_{16}O_2$	Isobutyl butyrate	156.9	Nonazeotrope		575
14607	$C_8H_{20}SiO_4$	Ethyl silicate	168.8	Nonazeotrope		549
14608	C_9H_8	Indene	182.6	Nonazeotrope		575

TABLE I.   *Binary Systems*   413

No.	Formula	B-Component Name	B.P., °C	Azeotropic Data B.P., °C	Wt.%A	Ref.
A =	**C₈H₁₆O₂**	**Butyl Butyrate** *(continued)*	**166.4**			
14609	C₉H₁₂	Mesitylene	164.6	Nonazeotrope		546
14610	C₉H₁₂	Propylbenzene	159.3	Nonazeotrope		575
14611	C₉H₁₈O	2,6-Dimethyl-4-heptanone	168.0	Nonazeotrope		552
14612	C₉H₁₈O₂	Isoamyl isobutyrate	169.8	Nonazeotrope		575
14613	C₉H₁₈O₂	Isobutyl isovalerate	171.2	Nonazeotrope		575
14614	C₁₀H₁₄	Cymene	176.7	Nonazeotrope		575
14615	C₁₀H₁₆	Camphene	159.6	158.0	30	562
14616	C₁₀H₁₆	*d*-Limonene	177.9	Nonazeotrope		546
14617	C₁₀H₁₆	Nopinene	163.8	160.5	40	562
14618	C₁₀H₁₆	α-Pinene	155.8	<155.0	<20	575
		"	155.8	Nonazeotrope		546
14619	C₁₀H₁₆	α-Terpinene	173.4	<165.0	<74	575
14620	C₁₀H₁₈O	Cineole	176.35	Nonazeotrope		557
14621	C₁₀H₂₂O	Isoamyl ether	173.4	Nonazeotrope		557
A =	**C₈H₁₆O₂**	**Caprylic Acid**	**238.5**			
14622	C₉H₁₀O	*p*-Methylacetophenone	226.35	Nonazeotrope		575
14623	C₁₀H₇Cl	1-Chloronaphthalene	262.7	Nonazeotrope		575
		"	262.7	237.0		543
14624	C₁₀H₈	Naphthalene	218.05	216.2	6	541
14625	C₁₀H₁₀O₂	Safrole	235.9	232.5	~ 45	541
14626	C₁₀H₁₂O	Anethole	235.7	<234.0	>35	562
14627	C₁₀H₁₂O₂	Ethyl α-toluate	228.75	Nonazeotrope		575
14628	C₁₀H₁₄O	Carvacrol	237.85	237.6	25	575
14629	C₁₀H₁₄O	Thymol	232.9	<232.8		575
14630	C₁₀H₂₁NO	N,N-Dimethyloctan-amide, 100 mm.	187	190	26.0	832
14631	C₁₁H₁₀	1-Methylnaphthalene	244.6	233.5	52	542
14632	C₁₁H₁₀	2-Methylnaphthalene	241.15	235.0	48	527
14633	C₁₁H₁₆O	Methyl thymol ether	216.5	Nonazeotrope		575
14634	C₁₁H₂₀O	Methyl α-terpineol ether	216.2	Nonazeotrope		543
14635	C₁₁H₂₂O₃	Isoamyl carbonate	232.2	<231.8	>10	575
14636	C₁₂H₁₈	1,3,5-Triethylbenzene	215.5	~ 214.3	4	541
14637	C₁₂H₂₀O₂	Bornyl acetate	227.6	Nonazeotrope		575
A =	**C₈H₁₆O₂**	**1-3-Dimethylbutyl Acetate**	**146.1**			
14638	C₉H₁₈O	2,6-Dimethyl-4-heptanone	169.4	Nonazeotrope		981
A =	**C₈H₁₆O₂**	**Ethyl Caproate**	**167.7**			
14639	C₈H₁₈S	Isobutyl sulfide	172.0	Nonazeotrope		575
14640	C₉H₁₂	Mesitylene	164.6	Nonazeotrope		575
14641	C₉H₁₂	Propylbenzene	158.9	Nonazeotrope		546
14642	C₉H₁₂	Pseudocumene	168.2	167.6		546
14643	C₉H₁₈O	2,6-Dimethyl-4-heptanone	168.0	167.5	60	552
14644	C₉H₁₈O₂	Isobutyl isovalerate	171.2	Nonazeotrope		575
14645	C₁₀H₁₆	Camphene	158	159	15	546
14646	C₁₀H₁₆	α-Pinene	155.8	Nonazeotrope		546
14647	C₁₀H₁₆	α-Terpinene	173.4	Nonazeotrope		575
14648	C₁₀H₁₈O	Cineole	176.35	Nonazeotrope		557

		B-Component		Azeotropic Data		
No.	Formula	Name	B.P., °C	B.P., °C	Wt.%A	Ref.
A =	**C₈H₁₆O₂**	**Ethyl Caproate** *(continued)*	**167.7**			
14649	C₁₀H₂₂O	Isoamyl ether	173.2	Nonazeotrope		557
A =	**C₈H₁₆O₂**	**2-Ethylhexanoic Acid**	**227**			
14650	C₁₁H₁₀	2-Methylnaphthalene	241.15		<50	270
A =	**C₈H₁₆O₂**	**Hexyl Acetate**	**171.5**			
14651	C₈H₁₈O	*sec*-Octyl alcohol	180.4	Nonazeotrope		575
14652	C₉H₁₈O₂	Isoamyl butyrate	181.05	Nonazeotrope		575
14653	C₁₀H₁₈O	Cineole	176.35	Nonazeotrope		557
14654	C₁₀H₂₂O	Isoamyl ether	173.4	~ 171.2	>80	557
A =	**C₈H₁₆O₂**	**Isoamyl Propionate**	**160.7**			
14655	C₈H₁₆O₂	Isobutyl butyrate	156.9	Nonazeotrope		575
14656	C₈H₁₆O₂	Propyl isovalerate	155.7	Nonazeotrope		575
14657	C₈H₁₈O	*sec*-Octyl alcohol	180	Nonazeotrope		536
14658	C₈H₁₈S	Isobutyl suflide	172.0	Nonazeotrope		575
14659	C₈H₂₀SiO₄	Ethyl silicate	168.8	Nonazeotrope		575
14660	C₉H₁₂	Mesitylene	164.0	Nonazeotrope		546
14661	C₉H₁₈O	2,6-Dimethyl-4-heptanone	168.0	Nonazeotrope		552
14662	C₉H₂₀O₂	Diisobutoxymethane	163.8	Nonazeotrope		557
14663	C₁₀H₁₄	Cymene	176.7	Nonazeotrope		575
14664	C₁₀H₁₆	Camphene	159.6	155.5	46	570
		"	~158	~ 155.5	<50	563
14665	C₁₀H₁₆	Nopinene	163.8	157.0	57	562
14666	C₁₀H₁₆	α-Pinene	155.8	154	~ 25	563
14667	C₁₀H₁₈O	Cineole	176.35	Nonazeotrope		557
14668	C₁₀H₂₂	2,7-Dimethyloctane	160.25	157	~49	563
14669	C₁₀H₂₂O	Isoamyl ether	173.2	Nonazeotrope		557
A =	**C₈H₁₆O₂**	**Isobutyl Butyrate**	**156.9**			
14670	C₈H₂₀SiO₄	Ethyl silicate	168.8	Nonazeotrope		575
14671	C₉H₁₂	Mesitylene	164.6	Nonazeotrope		573
14672	C₁₀H₁₆	Nopinene	163.8	<155.4	<75	575
14673	C₁₀H₁₆	α-Pinene	155.8	<153.0	<50	546
14674	C₁₀H₁₆	α-Terpinene	173.3	Nonazeotrope		546
A =	**C₈H₁₆O₂**	**Isobutyl Isobutyrate**	**148.6**			
14675	C₈H₁₈O	Butyl ether	142.4	Nonazeotrope		557
14676	C₉H₁₂	Cumene	152.8	Nonazeotrope		575
14677	C₁₀H₁₆	Camphene	158	153	63	546
14678	C₁₀H₁₆	α-Pinene	155.8	Nonazeotrope		546
A =	**C₈H₁₆O₂**	**Propyl Isovalerate**	**155.7**			
14679	C₈H₁₈O	Butyl ether	142.4	Nonazeotrope		557
14680	C₈H₂₀SiO₄	Ethyl silicate	168.8	Nonazeotrope		575
14681	C₉H₁₂	Cumene	152.8	Nonazeotrope		575
14682	C₉H₁₂	Mesitylene	164.6	Nonazeotrope		573
14683	C₉H₁₂	Propylbenzene	158.9	Nonazeotrope		546
14684	C₉H₂₀O₂	Diisobutoxymethane	163.8	Nonazeotrope		557
14685	C₁₀H₁₆	Camphene	159.6	145	65	545

TABLE I.   *Binary Systems*   415

No.	Formula	Name	B.P., °C	B.P., °C	Wt.%A		Ref.
		B-Component		**Azeotropic Data**			
A =	**C₈H₁₆O₂**	**Propyl Isovalerate** *(continued)*	**155.7**				
14686	C₁₀H₁₆	Nopinene	163.8	155.0	75		562
14687	C₁₀H₁₆	α-Pinene	155.8	144.0	53		545
14688	C₁₀H₁₆	α-Terpinene	173.4	Nonazeotrope			575
14689	C₁₀H₂₂	2,7-Dimethyloctane	160.25	152	57		573
A =	**C₈H₁₆O₃**	**Isoamyl Lactate**	**202.4**				
14690	C₈H₁₈O	Octyl alcohol	195.2	Nonazeotrope			575
14691	C₉H₁₄O	Phorone	197.8	Nonazeotrope			552
14692	C₁₀H₈	Naphthalene	218.05	Nonazeotrope			538
14693	C₁₀H₁₄O	Carvacrol	237.85	Nonazeotrope			575
14694	C₁₀H₁₄O	Thymol	232.9	Nonazeotrope			542
14695	C₁₀H₁₆O	Camphor	209.1	Nonazeotrope			552
14696	C₁₀H₁₇Cl	Bornyl chloride	207.2	201.8			575
14697	C₁₀H₁₈O	Citronellal	208.0	<202.2			575
14698	C₁₀H₁₈O	Linalool	198.6	<198.5			575
14699	C₁₀H₂₀O₂	Ethyl caprylate	208.35	<202.2			575
14700	C₁₀H₂₂S	Isoamyl sulfide	214.8	Nonazeotrope			565
14701	C₁₂H₁₈	1,3,5-Triethylbenzene	215.5	Nonazeotrope			576
A =	**C₈H₁₆O₄**	**2-(2-Ethoxyethoxy) Ethyl Acetate**	**218.5**				
14702	C₉H₁₈O	p-Methylacetophenone	226.35	Nonazeotrope			575
14703	C₉H₁₀O₂	Benzyl acetate	215.0	<214.8	>9		575
14704	C₉H₁₀O₂	Ethyl benzoate	212.5	212.3	8		575
14705	C₉H₁₂O	3-Phenylpropanol	235.6	Nonazeotrope			575
14706	C₁₀H₁₈O	Borneol	215	Nonazeotrope			575
14707	C₁₀H₁₈O	Geraniol	229.6	Nonazeotrope			575
14708	C₁₀H₁₈O	α-Terpineol	218.85	<218.0	<53		575
14709	C₁₀H₂₀O	Citronellol	224.4	Nonazeotrope			575
14710	C₁₀H₂₀O₂	Ethyl caprylate	208.35	Nonazeotrope			575
14711	C₁₀H₂₂O	Decyl alcohol	232.8	Nonazeotrope			575
14712	C₁₁H₁₀	2-Methyl naphthalene	241.15	Nonazeotrope			270
14713	C₁₂H₂₀O₂	Bornyl acetate	227.6	Nonazeotrope			575
A =	**C₈H₁₇Cl**	**3-(Chloromethyl) heptane**	**106.9/100 mm.**				
14714	C₈H₁₈O	2-Ethyl-1-hexanol, 100 mm.	124.8	106	98		982
A =	**C₈H₁₈**	**n-Octane**	**125.4**				
14715	C₈H₁₈	2,2,4-Trimethylpentane	99.2	Nonazeotrope		v-l	86
14716	C₈H₁₈O	Isobutyl ether	122.3	122.0	90		558
		"	122.2	Nonazeotrope?			548
A =	**C₈H₁₈Cl₂Sn**	**Dibutyltin Dichloride**	**157/17**				
14717	C₁₂H₂₇ClSn	Tributyltin chloride, 17 mm.	166	Nonazeotrope			981
A =	**C₈H₁₈O**	**Butyl Ether**	**142.1**				
14718	C₈H₁₈O	2-Ethyl-1-hexanol	184.8	Nonazeotrope			981
14719	C₉H₁₂	Cumene	152.8	Nonazeotrope			558

		B-Component		Azeotropic Data		
No.	Formula	Name	B.P., °C	B.P., °C	Wt.%A	Ref.
A =	**C$_8$H$_{18}$O**	**Butyl Ether** (continued)	**142.1**			
14720	C$_{10}$H$_{16}$	α-Pinene	155.8	Nonazeotrope		558
A =	**C$_8$H$_{18}$O**	**2-Ethyl-1-hexanol**	**184.8**			
14721	C$_9$H$_{20}$	Nonane	150.8	Nonazeotrope		981
14722	C$_{10}$H$_{20}$O$_2$	2-Ethylhexyl acetate	198.4	Nonazeotrope		981
14723	C$_{11}$H$_{20}$O$_2$	2-Ethylhexyl acrylate			>85	999
14724	C$_{11}$H$_{25}$N	(2-Ethylhexyl) propylamine, 50 mm.	147	Nonazeotrope		981
A =	**C$_8$H$_{18}$O**	**Isobutyl Ether**	**122.3**			
14725	C$_8$H$_{19}$N	Diisobutylamine	138.5	Nonazeotrope		551
A =	**C$_8$H$_{18}$O**	**Octyl Alcohol**	**195.15**			
14726	C$_9$H$_8$	Indene	182.6	182.4	12	527
14727	C$_9$H$_{10}$O$_2$	Ethyl benzoate	212.6	Nonazeotrope		536
14728	C$_9$H$_{12}$	Mesitylene	164.6	Nonazeotrope		575
14729	C$_9$H$_{12}$	Propylbenzene	159.3	Nonazeotrope		575
14730	C$_9$H$_{12}$O	Benzyl ethyl ether	185.0	Nonazeotrope		545
14731	C$_9$H$_{12}$O	Phenyl propyl ether	190.2	190.0		538
14732	C$_9$H$_{13}$N	N,N-Dimethyl-o-toluidine	185.3	184.8	20	551
14733	C$_9$H$_{14}$O	Phorone	197.8	<193.5	<80	552
	"		197.8	Nonazeotrope		548
14734	C$_9$H$_{18}$O$_2$	Ethyl enanthate	188.7	Nonazeotrope		575
14735	C$_9$H$_{18}$O$_3$	Isobutyl carbonate	190.3	~ 189.5	20	536
14736	C$_{10}$H$_8$	Naphthalene	218.05	Nonazeotrope		537
14737	C$_{10}$H$_{14}$	Cymene	176.7	Nonazeotrope		537
14738	C$_{10}$H$_{14}$O	Thymol	232.9	Nonazeotrope		575
14739	C$_{10}$H$_{15}$N	Diethylaniline	217.05	Nonazeotrope		531
14740	C$_{10}$H$_{16}$	Camphene	159.6	Nonazeotrope		575
14741	C$_{10}$H$_{16}$	d-Limonene	177.8	177.45	~ 8	529
14742	C$_{10}$H$_{16}$	Nopinene	163.8	Nonazeotrope		575
14743	C$_{10}$H$_{16}$	α-Pinene	155.8	Nonazeotrope		575
14744	C$_{10}$H$_{16}$	α-Terpinene	173.4	Nonazeotrope		575
14745	C$_{10}$H$_{16}$	γ-Terpinene	183	182.5	>10	575
14746	C$_{10}$H$_{16}$	Thymene	179.7	179.6	~ 7	530
14747	C$_{10}$H$_{16}$O	Camphor	209.1	Nonazeotrope		552
14748	C$_{10}$H$_{18}$O	Cineole	176.35	Nonazeotrope		556
14749	C$_{10}$H$_{18}$O	Citronellal	208.0	Nonazeotrope		575
14750	C$_{10}$H$_{18}$O	Linalool	198.7	Nonazeotrope		528
14751	C$_{10}$H$_{18}$O	Menthone	209.5	Nonazeotrope		552
14752	C$_{10}$H$_{20}$O	Octyl vinyl ether 5 mm	64	64	17	1008
14753	C$_{10}$H$_{20}$O$_2$	Ethyl caprylate	208.35	Nonazeotrope		575
14754	C$_{10}$H$_{20}$O$_2$	Isoamyl isovalerate	192.7	192.55	15	564
14755	C$_{10}$H$_{22}$O	Isoamyl ether	173.2	Nonazeotrope		556
14756	C$_{10}$H$_{22}$S	Isoamyl sulfide	214.8	Nonazeotrope		566
14757	C$_{11}$H$_{20}$O	Isobornyl methyl ether	192.2	191.9	30	556
A =	**C$_8$H$_{18}$O**	**sec-Octyl Alcohol**	**179.0**			
14758	C$_9$H$_8$	Indene	181.7	176	~ 60	537

TABLE I.   *Binary Systems*   417

		B-Component		Azeotropic Data		
No.	Formula	Name	B.P., °C	B.P., °C	Wt.%A	Ref.
A =	C₈H₁₈O	*sec*-**Octyl Alcohol** *(continued)*	**179.0**			
14759	C₉H₁₂	Cumene	152.8	Nonazeotrope		575
14760	C₉H₁₂	Mesitylene	164.6	Nonazeotrope		541
14761	C₉H₁₂	Propylbenzene	159.3	Nonazeotrope		575
14762	C₉H₁₂O	Benzyl ethyl ether	185.0	180.0		545
14763	C₉H₁₂O	Phenyl propyl ether	190.2	Nonazeotrope		576
14764	C₉H₁₃N	*N,N*-Dimethyl-*o*-toluidine	185.3	179.0	70	551
14765	C₉H₁₃N	*N,N*-Dimethyl-*p*-toluidine	210.2	Nonazeotrope		551
14766	C₉H₁₈O₂	Butyl isovalerate	177.6	177.4	11	575
14767	C₉H₁₈O₂	Isoamyl butyrate	181.05	180.3	72	564
14768	C₉H₁₈O₂	Isoamyl isobutyrate	169.8	Nonazeotrope		575
14769	C₉H₁₈O₂	Isobutyl isovalerate	168.7	Nonazeotrope		536
14770	C₉H₁₈O₃	Isobutyl carbonate	190.3	<180.0		575
		"	190.3	Nonazeotrope		536
14771	C₁₀H₁₄	Butylbenzene	183.1	178.2	50	567
14772	C₁₀H₁₄	Cymene	176.7	174	44	537
14773	C₁₀H₁₅N	Diethylaniline	217.05	Nonazeotrope		551
14774	C₁₀H₁₆	Camphene	159.6	159.55?		537
14775	C₁₀H₁₆	*d*-Limonene	177.8	174.5	~ 45	537
14776	C₁₀H₁₆	Nopinene	163.8	163.5	~ 5	575
14777	C₁₀H₁₆	α-Phellandrene	171.5	~ 170		563
14778	C₁₀H₁₆	α-Pinene	155.8	Nonazeotrope		563
14779	C₁₀H₁₆	α-Terpinene	173.4	171.8	27	567
14780	C₁₀H₁₆	Terpinene	180.5	~ 175.5		563
14781	C₁₀H₁₆	Terpinolene	184.6	179.0	57	567
14782	C₁₀H₁₆	Thymene	179.7	176	52	537
14783	C₁₀H₁₈O	Cineole	176.35	175.85	26.5	572
14784	C₁₀H₂₂	2,7-Dimethyloctane	160.1	Nonazeotrope		575
14785	C₁₀H₂₂O	Amyl ether	187.5	179.8	86	556
14786	C₁₀H₂₂O	Isoamyl ether	173.2	172.65	17	527
		"	173.4	Nonazeotrope		576
A =	C₈H₁₈OS	**2-Hexylthioethanol**				
14787	C₁₀H₂₀OS	2-Hexylthioethyl vinyl ether		Min. b.p.		953
A =	C₈H₁₈O₂	**2-Ethyl-1,3-hexanediol**	**243.1**			
14788	C₁₆H₃₄O	Bis(2-ethylhexyl)ether,				
			10 mm. 135	123	40	982
		"	269.8	241		982
A =	C₈H₁₈O₃	**Bis(2-ethoxyethyl) Ether**	**186.0**			
14789	C₉H₈	Indene	182.5	Nonazeotrope		558
14790	C₁₀H₁₆	Dipentene	177.7	Nonazeotrope		558
A =	C₈H₁₈O₃	**2-(2-Butoxyethoxy) Ethanol**	**231.2**			
14791	C₉H₇N	Quinoline	237.3	<229.5	>56	553
14792	C₉H₁₀O	Cinnamyl alcohol	257.0	Nonazeotrope		575
14793	C₉H₁₀O₃	Ethyl salicylate	233.8	225.2	54	575
14794	C₁₀H₇Cl	1-Chloronaphthalene	262.7	Nonazeotrope		575
14795	C₁₀H₈	Naphthalene, 100 mm	144.35	Nonazeotrope v-l		404

	B-Component			Azeotropic Data		
No.	Formula	Name	B.P., °C	B.P., °C	Wt.%A	Ref.
A =	**$C_8H_{18}O_3$**	**2-(2-Butoxyethoxy) Ethanol**	**231.2**			
		(*continued*)				
14796	$C_{10}H_9N$	Quinaldine	246.5	Nonazeotrope		575
14797	$C_{10}H_{18}O$	Geraniol	229.6	<228.5		575
14798	$C_{10}H_{22}O$	Decyl alcohol	232.8	<230.5	<85	575
14799	$C_{11}H_{10}$	1-Methylnaphthalene,				
		20 mm.			46.8	270
		" 100 mm.			64.3	270
		" 200 mm.			74	270
14800	$C_{11}H_{10}$	2-Methylnaphthalene				
		20 mm.			38	270
		" 100 mm.			53.5	270
		"	241.15		82	270
14801	$C_{12}H_{26}$	Dodecane, 100 mm.	146.2	142.6	34 v-l	404
14802	$C_{15}H_{30}$	1-Pentadecene,				
		217 mm.	183.7	185.16	87 vol. %	826
A =	**$C_8H_{18}S$**	**Butyl Sulfide**	**185.0**			
14803	C_9H_{12}	Mesitylene	164.6	Nonazeotrope		566
14804	$C_9H_{12}O$	Benzyl ethyl ether	185.0	<184.2	>53	566
14805	$C_9H_{18}O$	2,6-Dimethyl-4-heptanone	168.0	Nonazeotrope		566
14806	$C_9H_{18}O_2$	Butyl isovalerate	177.6	Nonazeotrope		566
14807	$C_9H_{18}O_2$	Isobutyl isovalerate	171.2	Nonazeotrope		566
14808	$C_{10}H_{14}$	Butylbenzene	183.1	182.0	40	566
14809	$C_{10}H_{18}O$	Cineole	176.35	Nonazeotrope		566
14810	$C_{10}H_{22}O$	Isoamyl ether	173.2	Nonazeotrope		566
A =	**$C_8H_{18}S$**	**Isobutyl Sulfide**	**172.0**			
14811	C_9H_{12}	Mesitylene	164.6	Nonazeotrope		566
14812	$C_9H_{18}O$	2,6-Dimethyl-4-heptanone	168.0	<167.2		566
14813	$C_{10}H_{16}$	Camphene	159.6	Nonazeotrope		566
14814	$C_{10}H_{16}$	α-Pinene	155.8	Nonazeotrope		566
14815	$C_{10}H_{22}O$	Isoamyl ether	172.6	171.0	62	555
A =	**$C_8H_{20}SiO_4$**	**Ethyl Silicate**	**168.8**			
14816	$C_9H_{18}O_2$	Butyl isovalerate	177.6	Nonazeotrope		575
14817	$C_9H_{18}O_2$	Isoamyl isobutyrate	169.8	168.2		575
14818	$C_9H_{18}O_2$	Isobutyl isovalerate	171.2	168.75	93	549
14819	$C_{10}H_{16}$	Camphene	~158	~150	~37	563
14820	$C_{10}H_{16}$	α-Pinene	155.8	<149	<35	563
14821	$C_{10}H_{22}O$	Isoamyl ether	173.2	<165.5		557
A =	**$C_9F_{21}N$**	**Tris(perfluoropropyl)amine**	**130**			
14822	C_9H_{12}	Cumene	152	116		160
A =	**$C_9H_6N_2O_2$**	**2,4-Tolylene Diisocyanate**				
14823	$C_9H_6N_2O_2$	2,6-Tolylene diisocyanate,				
		5-60 mm.		Nonazeotrope	v-l	166
A =	**C_9H_7N**	**Isoquinoline**	**243.2**			
14824	C_9H_7N	Quinoline	237.6	Nonazeotrope	v-l	610
14825	$C_{10}H_9N$	Quinaldine	247.7	Nonazeotrope	v-l	610
14826	$C_{11}H_{10}$	2-Methylnaphthalene	241.15	<50		270

TABLE I. *Binary Systems* 419

No.	Formula	B-Component Name	B.P., °C	B.P., °C	Azeotropic Data Wt.%A	Ref.
A =	**C₉H₇N**	**Quinoline**	**237.3**			
14827	C₉H₁₀O	p-Methylacetophenone	226.35	Nonazeotrope		575
14828	C₉H₁₀O₃	Ethyl salicylate	233.8	Nonazeotrope		553
14829	C₉H₁₂O	Mesitol	220.5	240.4	85	575
14830	C₉H₁₂O₂	2-Benzyloxyethanol	265.2	Nonazeotrope		553
14831	C₁₀H₈	Naphthalene	218	Nonazeotrope		553
14832	C₁₀H₁₀O₂	Isosafrole	252.0	Nonazeotrope		553
14833	C₁₀H₁₀O₂	Safrole	235.9	235.15	27	553
14834	C₁₀H₁₂O	Anethole	235.7	234.7	30	553
14835	C₁₀H₁₂O₂	Eugenol	254.8	Nonazeotrope		575
14836	C₁₀H₁₄O	Carvacrol	237.85	244.3	48	575
14837	C₁₀H₁₄O	Thymol	232.9	243.1	55	564
14838	C₁₀H₁₄O₂	m-Diethoxybenzene	235.4	235.0	22	575
14839	C₁₀H₁₈O	α-Terpineol	218.85	Nonazeotrope		575
14840	C₁₀H₂₀O	Menthol	216.3	Nonazeotrope		553
14841	C₁₁H₁₀	1-Methylnaphthalene	244.6	Nonazeotrope		553
14842	C₁₁H₁₀	2-Methylnaphthalene	241.15	237.25	93	553
		"		237.25	93	527
		"	241.15	>50		270
14843	C₁₁H₁₄O₂	1-Allyl-3,4-dimethylbenzene	254.7	Nonazeotrope		575
14844	C₁₁H₁₆O	Methyl thymyl ether	216.5	Nonazeotrope		575
14845	C₁₁H₁₆O	p-tert-Amylphenol	266.5	267.5	6	575
14846	C₁₁H₂₀O	Methyl α-terpineol ether	216.2	Nonazeotrope		575
14847	C₁₂H₁₀	Biphenyl	256.1	Nonazeotrope		553
14848	C₁₂H₁₆O₃	Isoamyl salicylate	277.5	Nonazeotrope		553
A =	**C₉H₈**	**Indene**	**182.6**			
14849	C₉H₁₂O	Benzyl ethyl ether	185.0	Nonazeotrope		558
14850	C₉H₁₃N	N,N-Dimethyl-o-toluidine	185.3	Nonazeotrope		551
14851	C₉H₁₄O	Phorone	197.8	Nonazeotrope		548
14852	C₉H₁₈O₂	Ethyl enanthate	188.7	Nonazeotrope		575
14853	C₉H₁₈O₂	Isoamyl butyrate	178.5	178.0		546
14854	C₉H₁₈O₂	Isobutyl isovalerate	171.35	Nonazeotrope		541
14855	C₉H₁₈O₃	Isobutyl carbonate	190.3	Nonazeotrope		575
14856	C₁₀H₁₄	Cymene	176.7	Nonazeotrope		561
14857	C₁₀H₁₆	Dipentene	177.7	Nonazeotrope		575
14858	C₁₀H₁₆	Limonene	177.7	Nonazeotrope		561
14859	C₁₀H₁₈O	Borneol	215	Nonazeotrope		575
14860	C₁₀H₁₈O	Cineole	176.35	Nonazeotrope		558
14861	C₁₀H₁₈O	Linalool	198.6	Nonazeotrope		575
14862	C₁₀H₁₈O	β-Terpineol	210.5	Nonazeotrope		575
14863	C₁₀H₂₀O	Menthol	216.3	Nonazeotrope		575
14864	C₁₀H₂₀O₂	Isoamyl isovalerate	192.7	Nonazeotrope		546
14865	C₁₀H₂₂O	Amyl ether	187.5	Nonazeotrope		558
14866	C₁₀H₂₂O	Isoamyl ether	173.2	Nonazeotrope		558
14867	C₁₀H₂₃N	Diisoamylamine	188.2	Nonazeotrope		551
A =	**C₉H₈O**	**Cinnamaldehyde**	**253.5**			
14868	C₉H₁₀O	Cinnamyl alcohol	257.0	< 252.3		575

No.	Formula	B-Component Name	B.P., °C	Azeotropic Data B.P., °C	Wt.%A	Ref.
A =	**C₉H₈O**	**Cinnamaldehyde** *(continued)*	**253.5**			
14869	C₉H₁₂O	3-Phenylpropanol	235.6	Nonazeotrope		575
14870	C₁₀H₇Cl	1-Chloronaphthalene	262.7	Nonazeotrope		545
14871	C₁₀H₈	Naphthalene	218.0	Nonazeotrope		575
14872	C₁₀H₁₀O₂	Isosafrole	252.0	251.3	23	556
14873	C₁₀H₁₀O₂	Methyl cinnamate	261.9	Nonazeotrope		545
14874	C₁₀H₁₀O₂	Safrole	235.9	Nonazeotrope		548
14875	C₁₀H₁₂O	Anethole	235.7	Nonazeotrope		575
14876	C₁₀H₁₂O₂	Isoeugenol	268.8	Nonazeotrope		575
14877	C₁₀H₁₄O	Carvacrol	237.85	Nonazeotrope		575
14878	C₁₀H₁₄O	Thymol	232.9	Nonazeotrope		575
14879	C₁₁H₁₀	1-Methylnaphthalene	244.6	Nonazeotrope		575
	"		244.6	~ 244.4	~ 5	538
14880	C₁₁H₁₀	2-Methylnaphthalene	241.15	Nonazeotrope		527
14881	C₁₁H₁₄O₂	1-Allyl-3,4-dimethoxy-benzene	255.0	253.0	80?	538
14882	C₁₁H₁₄O₂	Butyl benzoate	249.5	Nonazeotrope		548
14883	C₁₁H₁₄O₂	1,2-Dimethoxy-4-propenylbenzene	270.5	Nonazeotrope		575
14884	C₁₁H₁₄O₂	Isobutyl benzoate	241.9	Nonazeotrope		548
14885	C₁₂H₁₀	Acenaphthene	277.9	Nonazeotrope		575
14886	C₁₂H₁₀	Biphenyl	255.0	~ 250.0	~ 40	548
14887	C₁₂H₁₀O	Phenyl ether	259.0	253.0	65	556
14888	C₁₂H₁₆O₂	Isoamyl benzoate	262.0	Nonazeotrope		548
14889	C₁₂H₁₈	1,3,5-Triethylbenzene	215.5	Nonazeotrope		575
14890	C₁₃H₁₂	Diphenylmethane	265.4	Nonazeotrope		548
A =	**C₉H₉N**	**2-Methylindole**	**268**			
14891	C₁₁H₁₆O	*p-tert*-Amylphenol	266.5	272.0	56	575
A =	**C₉H₁₀**	**Vinyltoluene**				
14892	C₉H₁₂	Ethyltoluene 60 mm.		Nonazeotrope	v-l	302
	"	" 30 mm.		Nonazeotrope	v-l	302
	"	" 15 mm.		Nonazeotrope	v-l	302
A =	**C₉H₁₀O**	**Cinnamyl Alcohol**	**257.0**			
14893	C₉H₁₂O₂	2-Benzyloxyethanol	265.2	Nonazeotrope		575
14894	C₁₀H₈	Naphthalene	218.0	Nonazeotrope		575
14895	C₁₀H₁₀O₂	Isosafrole	252.0	< 251.6		575
14896	C₁₀H₁₀O₂	Methyl cinnamate	261.9	Nonazeotrope		575
14897	C₁₀H₁₀O₂	Safrole	235.9	Nonazeotrope		575
14898	C₁₀H₁₂O₂	Isoeugenol	268.8	Nonazeotrope		575
14899	C₁₀H₁₄O	Carvone	231.0	Nonazeotrope		552
14900	C₁₀H₁₄O	Thymol	232.9	Nonazeotrope		575
14901	C₁₀H₁₅N	Diethylaniline	217.05	Nonazeotrope		551
14902	C₁₀H₂₀O₄	2-(2-Butoxyethoxy)ethyl acetate	245.3	Nonazeotrope		575

TABLE I.   *Binary Systems*   421

No.	Formula	B-Component Name	B.P., °C	Azeotropic Data B.P., °C	Wt.%A	Ref.
A =	**C₉H₁₀O**	**Cinnamyl Alcohol** *(continued)*	**257.0**			
14903	C₁₁H₁₀	1-Methylnaphthalene	244.6	< 244.3	>12	575
14904	C₁₁H₁₂O₂	Ethyl cinnamate	272.0	Nonazeotrope		575
14905	C₁₁H₁₄O₂	Butyl benzoate	249.0	Nonazeotrope		575
14906	C₁₁H₁₄O₂	Ethyl β-phenylpropionate	248.1	Nonazeotrope		575
14907	C₁₁H₁₄O₂	Isobutyl benzoate	241.9	Nonazeotrope		575
14908	C₁₁H₂₂O₃	Isoamyl carbonate	232.2	Nonazeotrope		575
14909	C₁₂H₁₀	Acenaphthene	277.9	Nonazeotrope		575
14910	C₁₂H₁₀	Biphenyl	256.1	253.0	~ 45	575
14911	C₁₂H₁₀O	Phenyl ether	259.0	< 256.0		575
14912	C₁₂H₁₆O₂	Isoamyl benzoate	262.0	Nonazeotrope		575
14913	C₁₂H₂₂O₄	Isoamyl oxalate	268.0	< 256.7		575
14914	C₁₃H₁₂	Diphenylmethane	265.4	< 256.2	>62	575
14915	C₁₃H₂₈	Tridecane	234.0	Nonazeotrope		575
A =	**C₉H₁₀O**	**p-Methylacetophenone**	**226.35**			
14916	C₉H₁₀O₃	Ethyl salicylate	233.8	Nonazeotrope		552
14917	C₉H₁₂O	3-Phenylpropanol	235.6	Nonazeotrope		552
14918	C₁₀H₈	Naphthalene	218.0	Nonazeotrope		552
14919	C₁₀H₁₀O₂	Safrole	235.9	Nonazeotrope		552
14920	C₁₀H₁₂O	Anethole	235.7	Nonazeotrope		552
14921	C₁₀H₁₂O₂	Ethyl α-toluate	228.75	226.2	75	552
14922	C₁₀H₁₂O₂	Propyl benzoate	230.85	Nonazeotrope		552
14923	C₁₀H₁₄O	Thymol	232.9	234.9	32	552
14924	C₁₀H₁₄O₂	m-Diethoxybenzene	235	Nonazeotrope		537
14925	C₁₀H₁₅N	Diethylaniline	217.05	Nonazeotrope		551
14926	C₁₀H₁₈O	Geraniol	229.6	226.25	95	552
14927	C₁₀H₁₈	α-Terpineol	218.85	Nonazeotrope		552
14928	C₁₀H₂₀O	Citronellol	224.4	223.7	32	552
14929	C₁₀H₂₀O	Menthol	216.4	Nonazeotrope		535
14930	C₁₀H₂₂O	Decyl alcohol	232.8	Nonazeotrope		552
14931	C₁₁H₁₀	2-Methylnaphthalene	241.15	Nonazeotrope		527
14932	C₁₁H₂₀O	Methyl terpenyl ether	216.2	Nonazeotrope		552
14933	C₁₁H₂₂O₃	Isoamyl carbonate	232.2	Nonazeotrope		552
14934	C₁₂H₂₀O₂	Bornyl acetate	227.6	225.8	60	552
A =	**C₉H₁₀O**	**Propiophenone**	**217.7**			
14935	C₉H₁₀O₂	Benzyl acetate	215.0	Nonazeotrope		552
14936	C₉H₁₀O₂	Ethyl benzoate	212.5	Nonazeotrope		552
14937	C₉H₁₃N	N,N-Dimethyl-p-toluidine	210.2	Nonazeotrope		575
14938	C₁₀H₈	Naphthalene	218.0	Nonazeotrope		552
14939	C₁₀H₁₄O	Carvacrol	237.85	Nonazeotrope		575
14940	C₁₀H₁₄O	Thymol	232.9	> 233.2	>13	552
14941	C₁₀H₁₅N	Diethylaniline	217.05	< 216.6	<47	551
14942	C₁₀H₁₈O	Borneol	215.0	Nonazeotrope		552
14943	C₁₀H₂₀O₂	Methyl pelargonate	213.8	Nonazeotrope		552
14944	C₁₂H₁₈	1,3,5-Triethylbenzene	215.5	215.4	25	552
14945	C₂₀H₂₀O₂	Bornyl acetate	227.6	Nonazeotrope		552

		B-Component		Azeotropic Data		
No.	Formula	Name	B.P., °C	B.P., °C	Wt.%A	Ref.
A =	$C_9H_{10}O_2$	**Benzyl Acetate**	**214.9**			
14946	$C_9H_{10}O_2$	Ethyl benzoate	212.4	212.35	2	529
	"	"	212.5	Nonazeotrope		549
14947	$C_{10}H_8$	Naphthalene	218.05	214.65	~ 72	529
14948	$C_{10}H_{14}O$	Thymol	232.8	Nonazeotrope		531
14949	$C_{10}H_{16}$	γ-Terpinene	179.7	Nonazeotrope		546
14950	$C_{10}H_{16}O$	Camphor	209.1	Nonazeotrope		552
14951	$C_{10}H_{16}O$	Pulegone	223.8	Nonazeotrope		552
14952	$C_{10}H_{17}Cl$	Bornyl chloride	207.5	Nonazeotrope		575
14953	$C_{10}H_{18}O$	Borneol	213.2	212.8	~ 36	529
14954	$C_{10}H_{18}O$	Citronellal	208.0	Nonazeotrope		575
14955	$C_{10}H_{18}O$	α-Terpineol	217.8	214.5	~ 65	529
14956	$C_{10}H_{18}O$	β-Terpineol	210.5	210.2	22	575
14957	$C_{10}H_{20}O$	Citronellol	224.4	Nonazeotrope		575
14958	$C_{10}H_{20}O$	Menthol	216.4	~ 213.5	73.5	529
14959	$C_{11}H_{10}$	2-Methylnaphthalene	241.15	Nonazeotrope		527
14960	$C_{11}H_{20}O$	Methyl α-terpineol ether	216.2	214.7	72	557
14961	$C_{11}H_{24}O_2$	Diisoamyloxymethane	207.5	Nonazeotrope		557
14962	$C_{12}H_{18}$	1,3,5-Triethylbenzene	216	214.5	50	546
A =	$C_9H_{10}O_2$	**Ethyl Benzoate**	**212.4**			
14963	$C_9H_{10}O_2$	Methyl α-toluate	215.3	Nonazeotrope		549
14964	$C_{10}H_8$	Naphthalene	218.05	Nonazeotrope		563
14965	$C_{10}H_{12}O$	Estragol	215.6	Nonazeotrope		557
14966	$C_{10}H_{14}O$	Thymol	232.8	Nonazeotrope		531
14967	$C_{10}H_{15}N$	Diethylaniline	216.1	Reacts		563
14968	$C_{10}H_{16}O$	Camphor	209.1	Nonazeotrope		552
14969	$C_{10}H_{17}Cl$	Bornyl chloride	207.5	Nonazeotrope		575
	"	"	~210	~ 209.5		563
14970	$C_{10}H_{18}O$	Borneol	213.2	212.2	90	529
14971	$C_{10}H_{18}O$	Citronellal	~207.8	Nonazeotrope		532
14972	$C_{10}H_{18}O$	Linalool	198.6	Nonazeotrope		535
14973	$C_{10}H_{18}O$	α-Terpineol	218.85	Nonazeotrope		575
	"	"	~217.8	212.55	~ 98	536
14974	$C_{10}H_{18}O$	β-Terpineol	210.5	< 209.8	<48	575
14975	$C_{10}H_{20}O$	Menthol	216.4	212.3	95	529
14976	$C_{11}H_{20}O$	Terpineol methyl ether	216	< 212.3	<78	557
14977	$C_{11}H_{24}O_2$	Diisoamyloxymethane	210.8	< 210.6	15?	557
14978	$C_{12}H_{18}$	1,3,5-Triethylbenzene	216.0	Nonazeotrope		546
A =	$C_9H_{10}O_2$	**Methyl α-Toluate**	**215.3**			
14979	$C_{10}H_{16}O$	Pulegone	223.8	Nonazeotrope		552
14980	$C_{10}H_{18}O$	Borneol	215.0	< 214.3	<52	575
14981	$C_{10}H_{18}O$	α-Terpineol	218.85	< 215.0	>75	575
14982	$C_{10}H_{20}O$	Menthol	216.3	< 214.5	>63	575
A =	$C_9H_{10}O_3$	**Ethyl Salicylate**	**233.8**			
14983	$C_{10}H_{10}O_2$	Isosafrole	252.0	Nonazeotrope		575
14984	$C_{10}H_{10}O_2$	Safrole	235.9	233.65	88	536
	"	"	235.9	Nonazeotrope		556

TABLE I.   *Binary Systems*   423

No.	Formula	Name	B.P., °C	B.P., °C	Wt.%A		Ref.
		B-Component		**Azeotropic Data**			
A =	$C_9H_{10}O_3$	**Ethyl Salicylate** *(continued)*	233.8				
14985	$C_{10}H_{12}O_2$	Ethyl α-toluate	234.0	Nonazeotrope			538
14986	$C_{10}H_{12}O_2$	Propyl benzoate	230.85	Nonazeotrope			548
14987	$C_{10}H_{14}O$	Carvacrol	237.85	Nonazeotrope			575
14988	$C_{10}H_{14}O$	Carvone	231.0	Nonazeotrope			552
14989	$C_{10}H_{14}O$	Thymol	232.9	235	~ 65		536
14990	$C_{10}H_{16}O$	Pulegone	223.8	Nonazeotrope			552
14991	$C_{10}H_{18}O$	Borneol	213.4	Nonazeotrope			545
14992	$C_{10}H_{18}O$	Geraniol	229.7	228.5	40		536
14993	$C_{10}H_{18}O$	α-Ternipeol	~217.8	Nonazeotrope			536
14994	$C_{10}H_{20}O$	Citronellol	224.5	Nonazeotrope			545
14995	$C_{10}H_{20}O$	Menthol	216.4	Nonazeotrope			536
14996	$C_{10}H_{22}O$	Decyl alcohol	232.9	230.5	48		536
14997	$C_{11}H_{10}$	1-Methylnaphthalene	244.9	Nonazeotrope			536
14998	$C_{11}H_{10}$	2-Methylnaphthalene	241.15	Nonazeotrope			527
14999	$C_{11}H_{14}O_2$	Ethyl β-phenylpropionate	248.1	Nonazeotrope			575
15000	$C_{11}H_{14}O_2$	Isobutyl benzoate	241.9	Nonazeotrope			538
15001	$C_{11}H_{22}O_2$	Ethyl pelargonate	227	Nonazeotrope			575
15002	$C_{11}H_{22}O_3$	Isoamyl carbonate	232.2	< 232.0	<28		575
15003	$C_{12}H_{10}$	Biphenyl	256.1	Nonazeotrope			575
15004	$C_{12}H_{10}O$	Phenyl ether, 5 mm.		Nonazeotrope		v-l	323
		" 50 mm.		Nonazeotrope		v-l	323
		" 180 mm.		Nonazeotrope		v-l	323
15005	$C_{12}H_{20}O_2$	Bornyl acetate	227.6	Nonazeotrope			538
A =	C_9H_{12}	**Cumene**	152.8				
15006	C_9H_{20}	Nonane	149.5	148.0	23		561
15007	$C_{10}H_{16}$	α-Pinene	155.8	151.8	80		561
15008	$C_{12}F_{27}N$	Tris(perfluorobutyl) amine	177	138			160
A =	C_9H_{12}	**Mesitylene**	164.6				
15009	C_9H_{12}	Propylbenzene	159.3	Nonazeotrope			561
15010	C_9H_{12}	Pseudocumene	169.0	Nonazeotrope			266,563
15011	$C_9H_{13}N$	N,N-Dimethyl-o-toluidine	185.3	Nonazeotrope			551
15012	$C_9H_{18}O_2$	Isoamyl butyrate	181.05	Nonazeotrope			575
15013	$C_9H_{18}O_2$	Isoamyl isobutyrate	169.8	Nonazeotrope			575
15014	$C_9H_{18}O_2$	Isobutyl isovalerate	168.7	163			563
		"	168.7	Nonazeotrope			546
15015	$C_9H_{18}O_3$	Isobutyl carbonate	190.3	Nonazeotrope			575
15016	$C_{10}H_{14}$	Cymene	176.7	Nonazeotrope			561
15017	$C_{10}H_{16}$	Camphene	159.6	Nonazeotrope			561
15018	$C_{10}H_{16}$	Nopinene	163.8	162.7	40		561
15019	$C_{10}H_{16}$	α-Pinene	155.8	Nonazeotrope			561
15020	$C_{10}H_{16}$	α-Terpinene	173.4	Nonazeotrope			561
15021	$C_{10}H_{18}O$	Linalool	198.6	Nonazeotrope			540
15022	$C_{10}H_{22}$	2,7-Dimethyloctane	160.1	158.6	28		561
15023	$C_{10}H_{22}O$	Isoamyl ether	173.4	Nonazeotrope			548

		B-Component		Azeotropic Data		
No.	Formula	Name	B.P., °C	B.P., °C	Wt.%A	Ref.
A =	**C_9H_{12}**	**Propylbenzene**	**159**			
15024	$C_9H_{13}N$	N,N-Dimethyl-o-toluidine	185.3	Nonazeotrope		575
15025	$C_{10}H_{16}$	Camphene	159.6	158.0	47	561
15026	$C_{10}H_{16}$	Nopinene	163.8	< 159.0	>85	561
15027	$C_{10}H_{16}$	α-Pinene	155.8	155.0	17	561
A =	**C_9H_{12}**	**Pseudocumene**	**168.2**			
15028	$C_9H_{13}N$	N,N-Dimethyl-o-toluidine	185.3	Nonazeotrope		551
15029	$C_9H_{18}O_2$	Isobutyl isovalerate	171.35	Nonazeotrope		541
		"	168.7	< 166.5	~ 49	563
15030	$C_{10}H_{14}$	Cymene	176.7	Nonazeotrope		561
15031	$C_{10}H_{18}$	Menthene	170.8	167.5	>85	561
15032	$C_{10}H_{18}O$	Cineole	176.35	Nonazeotrope		558
15033	$C_{10}H_{22}$	Decane	173.3	166.5	75	561
15034	$C_{10}H_{22}O$	Isoamyl ether	173.2	Nonazeotrope		558
A =	**$C_9H_{12}O$**	**Benzyl Ethyl Ether**	**185.0**			
15035	$C_9H_{18}O_2$	Butyl isovalerate	177.6	Nonazeotrope		557
15036	$C_9H_{18}O_2$	Isoamyl butyrate	181.05	Nonazeotrope		557
15037	$C_{10}H_{14}$	Cymene	176.7	Nonazeotrope		558
15038	$C_{10}H_{16}$	Dipentene	177.7	Nonazeotrope		558
15039	$C_{10}H_{16}O$	Fenchone	193.6	Nonazeotrope		575
15040	$C_{10}H_{18}O$	Citronellal	208.0	Nonazeotrope		575
15041	$C_{10}H_{18}O$	Linalool	198.6	Nonazeotrope		575
15042	$C_{10}H_{20}O_2$	Isoamyl isovalerate	192.7	Nonazeotrope		557
A =	**$C_9H_{12}O$**	**Mesitol**	**230.5**			
15043	$C_{10}H_8$	Naphthalene	218.0	215.5	37	562
15044	$C_{12}H_{18}$	1,3,5-Triethylbenzene	215.5	213.0	30	562
A =	**$C_9H_{12}O$**	**3-Phenylpropanol**	**235.6**			
15045	$C_9H_{12}O_2$	2-Benzyloxyethanol	265.2	Nonazeotrope		575
15046	$C_{10}H_8$	Naphthalene	218.05	217.8	~ 20	537
15047	$C_{10}H_{10}O_2$	Isosafrole	252.0	Nonazeotrope		575
15048	$C_{10}H_{10}O_2$	Safrole	235.9	233.8	47	545
15049	$C_{10}H_{12}O$	Anethole	235.7	234.0	48	567
15050	$C_{10}H_{12}O_2$	Ethyl α-toluate	228.75	Nonazeotrope		536
15051	$C_{10}H_{12}O_2$	Eugenol	254.8	Nonazeotrope		575
15052	$C_{10}H_{12}O_2$	Propyl benzoate	230.85	Nonazeotrope		536
15053	$C_{10}H_{14}O$	Carvacrol	237.85	> 238.5	<42	575
15054	$C_{10}H_{14}O$	Carvone	231.0	Nonazeotrope		552
15055	$C_{10}H_{14}O$	Thymol	232.9	237.5	~ 62	542
15056	$C_{10}H_{14}O_2$	m-Diethoxybenzene	235.4	< 234.8	>43	575
15057	$C_{10}H_{15}N$	Diethylaniline	217.05	216.9	7	551
		"	217.05	Nonazeotrope		548
15058	$C_{10}H_{18}O$	Geraniol	229.7	Nonazeotrope		545
15059	$C_{10}H_{20}O$	Citronellol	224.4	Nonazeotrope		549
15060	$C_{10}H_{20}O_4$	2-(2-Butoxyethoxy)ethyl acetate	245.3	Nonazeotrope		575

TABLE I.   *Binary Systems*   425

No.	Formula	Name	B.P., °C	B.P., °C	Wt.%A	Ref.
		B-Component		**Azeotropic Data**		
A =	**C₉H₁₂O**	**3-Phenylpropanol** *(continued)*	**235.6**			
15061	C₁₀H₂₂O	Decyl alcohol	232.9	232.0		545
15062	C₁₁H₁₀	1-Methylnaphthalene	244.6	234	~ 60	541
15063	C₁₁H₁₀	2-Methylnaphthalene	241.15	233.7		575
15064	C₁₁H₁₄O₂	Ethyl β-phenylpropionate	248.1	Nonazeotrope		575
15065	C₁₁H₁₄O₂	Isobutyl benzoate	241.9	Nonazeotrope		535
15066	C₁₁H₁₆O	Methyl thymyl ether	216.5	Nonazeotrope		575
15067	C₁₁H₁₇N	Isoamylaniline	256.0	Nonazeotrope		551
15068	C₁₁H₂₂O₃	Isoamyl carbonate	232.2	< 231.8	> 5	575
15069	C₁₂H₁₀	Biphenyl	254.9	235.4		537
15070	C₁₂H₁₆O₃	Isoamyl salicylate	277.5	Nonazeotrope		575
15071	C₁₂H₂₀O₂	Bornyl acetate	227.6	Nonazeotrope		575
15072	C₁₃H₁₂	Diphenylmethane	265.6	Nonazeotrope		537
A =	**C₉H₁₂O**	**Phenyl Propyl Ether**	**190.5**			
15073	C₉H₁₃N	Dimethyl-*o*-toluidine	185.35	Nonazeotrope		575
15074	C₁₀H₁₈O	Linalool	198.6	Nonazeotrope		545
A =	**C₉H₁₂OS**	**2-Benzylthioethanol**				
15075	C₁₁H₁₄OS	2-Benzylthioethyl vinyl ether		Min. b.p.		953
A =	**C₉H₁₂O₂**	**2-Benzyloxyethanol**	**265.2**			
15076	C₁₀H₇Cl	1-Chloronaphthalene	262.7	< 261.5		575
15077	C₁₀H₈	Naphthalene	218.0	Nonazeotrope		575
15078	C₁₀H₁₀O₂	Isosafrole	252.0	Nonazeotrope		575
15079	C₁₀H₁₀O₂	Methyl cinnamate	261.9	Nonazeotrope		575
15080	C₁₀H₁₀O₂	Safrole	235.9	Nonazeotrope		575
15081	C₁₀H₁₂O₂	Eugenol	254.8	Nonazeotrope		575
15082	C₁₁H₁₀	1-Methylnaphthalene	244.6	Nonazeotrope		575
15083	C₁₁H₁₀	2-Methylnaphthalene	241.15	Nonazeotrope		575
15084	C₁₁H₁₂O₂	Ethyl cinnamate	272.0	Nonazeotrope		575
15085	C₁₁H₁₄O₂	Butyl benzoate	249.0	Nonazeotrope		575
15086	C₁₁H₁₄O₂	Ethyl β-phenylpropionate	248.1	Nonazeotrope		575
15087	C₁₂H₁₀O	Phenyl ether	259.0	< 258.2	>15	575
15088	C₁₂H₁₆O₂	Isoamyl benzoate	262.0	261.0	~ 15	575
15089	C₁₂H₁₆O₃	Isoamyl salicylate	277.5	Nonazeotrope		575
15090	C₁₃H₁₂	Diphenylmethane	265.4	262.5	46	575
15091	C₁₄H₁₄	1,2-Diphenylethane	284.5	Nonazeotrope		575
A =	**C₉H₁₃N**	**N,N-Dimethyl-*o*-toluidine**	**185.3**			
15092	C₁₀H₈	Naphthalene	218.0	Nonazeotrope		551
15093	C₁₀H₁₄	Cymene	176.7	Nonazeotrope		551
15094	C₁₀H₁₆	Camphene	159.6	Nonazeotrope		551
15095	C₁₀H₁₆	α-Pinene	155.8	Nonazeotrope		551
15096	C₁₀H₁₆O	Camphor	209.1	Nonazeotrope		551
15097	C₁₀H₁₈O	Borneol	215.0	Nonazeotrope		551
15098	C₁₀H₁₈O	Cineole	176.35	Nonazeotrope		551
15099	C₁₀H₁₈O	Linalool	198.6	Nonazeotrope		551
15100	C₁₀H₁₈O	β-Terpineol	210.5	Nonazeotrope		551
15101	C₁₁H₂₀O	Isobornyl methyl ether	192.4	Nonazeotrope		551

		B-Component		Azeotropic Data		
No.	Formula	Name	B.P., °C	B.P., °C	Wt.%A	Ref.
A =	**C₉H₁₃N**	**N,N-Dimethyl-o-toluidine**	**185.3**			
		(continued)				
15102	C₁₂H₁₈	1,3,5-Triethylbenzene	215.5	Nonazeotrope		575
A =	**C₉H₁₃N**	**N,N-Dimethyl-p-toluidine**	**210.2**			
15103	C₁₀H₈	Naphthalene	218.0	Nonazeotrope		575
15104	C₁₀H₁₈O	Geraniol	229.6	Nonazeotrope		551
15105	C₁₀H₂₂O	n-Decyl alcohol	232.8	Nonazeotrope		551
15106	C₁₁H₂₀O	Methyl α-terpineol ether	216.2	Nonazeotrope		575
15107	C₁₂H₁₈	1,3,5-Triethylbenzene	215.5	Nonazeotrope		575
15108	C₁₂H₂₂O	Ethyl isobornyl ether	203.8	Nonazeotrope		575
A =	**C₉H₁₄O**	**Isophorone**	**215.2**			
15109	C₁₁H₁₀	2-Methylnaphthalene	241.15	Nonazeotrope		270
A =	**C₉H₁₄O**	**Phorone**	**197.8**			
15110	C₉H₁₈O₂	Methyl caprylate	192.9	Nonazeotrope		552
15111	C₉H₁₈O₃	Isobutyl carbonate	190.3	Nonazeotrope		552
15112	C₁₀H₁₄	Butylbenzene	183.1	Nonazeotrope		575
15113	C₁₀H₁₄O	Thymol	232.9	Nonazeotrope		575
15114	C₁₀H₁₅N	Diethylaniline	217.05	Nonazeotrope		575
15115	C₁₀H₂₀O₂	Isoamyl isovalerate	192.7	Nonazeotrope		552
15116	C₁₀H₂₂S	Isoamyl sulfide	214.8	Nonazeotrope		566
15117	C₁₁H₂₀O	Isobornyl methyl ether	192.4	Nonazeotrope		575
15118	C₁₂H₂₂O	Bornyl ethyl ether	204.9	Nonazeotrope		575
A =	**C₉H₁₈**	**Propylcyclohexane**	**156.72**			
15119	C₁₂F₂₇N	Perfluorotributylamine	178.4	145.4	55 vol. %	609
A =	**C₉H₁₈O**	**2,6-Dimethyl-4-heptanone**	**168.0**			
15120	C₉H₁₈O₂	Isoamyl isobutyrate	169.8	Nonazeotrope		552
15121	C₉H₁₈O₂	Isobutyl isovalerate	171.2	Nonazeotrope		552
A =	**C₉H₁₈O₂**	**Butyl Isovalerate**	**177.6**			
15122	C₉H₁₈O₂	Isoamyl butyrate	181.05	Nonazeotrope		575
15123	C₉H₁₈O₂	Isobutyl isovalerate	171.2	Nonazeotrope		575
15124	C₁₀H₁₆	Camphene	158	Nonazeotrope		546
15125	C₁₀H₁₆	d-Limonene	177.9	176	55	546
15126	C₁₀H₁₆	Nopinene	164	Nonazeotrope		546
15127	C₁₀H₁₆	α-Pinene	155.8	Nonazeotrope		575
15128	C₁₀H₁₈O	Cineole	176.35	< 176.2	<75	557
15129	C₁₀H₂₂O	Amyl ether	187.5	Nonazeotrope		557
15130	C₁₀H₂₂O	Isoamyl ether	173.2	Nonazeotrope		557
A =	**C₉H₁₈O₂**	**Ethyl Enanthate**	**188.7**			
15131	C₁₀H₁₄	Butylbenzene	183.1	Nonazeotrope		575
15132	C₁₀H₁₆	Dipentene	177.7	Nonazeotrope		575
15133	C₁₀H₁₆	α-Pinene	155.8	Nonazeotrope		575
A =	**C₉H₁₈O₂**	**Isoamyl Butyrate**	**181.05**			
15134	C₉H₁₈O₂	Isobutyl isovalerate	171.2	Nonazeotrope		575
15135	C₁₀H₁₄	Butylbenzene	183.2	Nonazeotrope		546
15136	C₁₀H₁₄	Cymene	176.7	Nonazeotrope		575
		"	175.3	< 173		563

TABLE I. *Binary Systems* 427

	B-Component			Azeotropic Data		
No.	Formula	Name	B.P., °C	B.P., °C	Wt.%A	Ref.
A =	**C₉H₁₈O₂**	**Isoamyl Butyrate** *(continued)*	**181.05**			
15137	C₁₀H₁₆	Camphene	158	Nonazeotrope		546
15138	C₁₀H₁₆	d-Limonene	177.8	~ 176.5	~ 45	528
15139	C₁₀H₁₆	Nopinene	163.8	Nonazeotrope		575
15140	C₁₀H₁₆	α-Terpinene	173.4	Nonazeotrope		575
15141	C₁₀H₁₆	γ-Terpinene	179.9	177.5	57	546
15142	C₁₀H₁₆	Terpinolene	185	~ 177		563
		"	185.2	Nonazeotrope		546
15143	C₁₀H₁₈O	Cineole	176.35	Nonazeotrope		557
		"	176.35	< 175.9	~ 25	572
15144	C₁₀H₁₈O	Linalool	198.6	Nonazeotrope		536
15145	C₁₀H₂₂O	Amyl ether	187.5	Nonazeotrope		557
15146	C₁₀H₂₂O	Isoamyl ether	173.2	Nonazeotrope		557
15147	C₁₁H₂₀O	Isobornyl methyl ether	192.4	Nonazeotrope		557
A =	**C₉H₁₈O₂**	**Isoamyl Isobutyrate**	**168.8**			
15148	C₉H₁₈O₂	Isobutyl isovalerate	168.7	168.4?		573
15149	C₁₀H₁₄	Cymene	176.7	Nonazeotrope		575
15150	C₁₀H₁₆	Camphene	159.6	< 159.5	<22	575
15151	C₁₀H₁₆	Dipentene	177.7	Nonazeotrope		575
15152	C₁₀H₁₆	α-Pinene	155.8	< 155.6	<16	575
15153	C₁₀H₁₈O	Cineole	176.35	Nonazeotrope		557
A =	**C₉H₁₈O₂**	**Isobutyl Isovalerate**	**171.2**			
15154	C₁₀H₁₄	Butylbenzene	183.1	Nonazeotrope		575
15155	C₁₀H₁₄	Cymene	176.7	Nonazeotrope		546
15156	C₁₀H₁₆	Camphene	159.6	Nonazeotrope		575
15157	C₁₀H₁₆	d-Limonene	177.9	Nonazeotrope		546
15158	C₁₀H₁₆	α-Pinene	155.8	Nonazeotrope		546
15159	C₁₀H₁₆	α-Terpinene	173.3	170.5	65	546
15160	C₁₀H₁₆	γ-Terpinene	183	Nonazeotrope		575
15161	C₁₀H₁₆	Terpinolene	185.2	Nonazeotrope		546
15162	C₁₀H₁₈	m-Menthene-8	170.8	< 170.5	<92	575
15163	C₁₀H₁₈O	Cineole	176.35	Nonazeotrope		557
15164	C₁₀H₂₂	2,7-Dimethyloctane	160.2	159	12	546
15165	C₁₀H₂₂O	Amyl ether	187.5	Nonazeotrope		557
15166	C₁₀H₂₂O	Isoamyl ether	173.2	170.95	90	557
A =	**C₉H₁₈O₂**	**Methyl Caprylate**	**192.9**			
15167	C₁₀H₁₄	Butylbenzene	183.1	Nonazeotrope		575
15168	C₁₀H₁₆	Dipentene	177.7	Nonazeotrope		575
15169	C₁₀H₂₀O₂	Isoamyl isovalerate	192.7	192.5	47	549
15170	C₁₁H₂₂O₂	Methyl caprate				
		20-100 mm.		Nonazeotrope	v-l	825
A =	**C₉H₁₈O₂**	**Pelargonic Acid**	**254.0**			
15171	C₁₀H₇Br	1-Bromonaphthalene	281.2	Nonazeotrope		575
15172	C₁₀H₇Cl	1-Chloronaphthalene	262.7	252.5	>50	575
15173	C₁₀H₁₈	Naphthalene	218.0	Nonazeotrope		575
15174	C₁₀H₁₀O₂	Isosafrole	252.0	249.5	35	556

		B-Component		Azeotropic Data		
No.	Formula	Name	B.P., °C	B.P., °C	Wt.%A	Ref.
A =	**C₉H₁₈O₂**	**Pelargonic Acid** *(continued)*	**254.0**			
15175	C₁₀H₁₀O₂	Safrole	235.9	Nonazeotrope		575
15176	C₁₀H₁₂O₂	Eugenol	254.8	250.5	52	575
15177	C₁₀H₁₄O	Thymol	232.9	Nonazeotrope		575
15178	C₁₀H₁₈O₄	Propyl succinate	250.5	< 249.8	20	575
15179	C₁₁H₁₀	1-Methylnaphthalene	244.6	243.0	18	562
15180	C₁₁H₁₀	2-Methylnaphthalene	241.15	240.2	10	527
15181	C₁₁H₁₄O₂	1,2-Dimethoxy-4-propenylbenzene	270.5	Nonazeotrope		575
15182	C₁₂H₁₀	Biphenyl	256.1	250	45	562
15183	C₁₂H₁₀O	Phenyl ether	259.0	250.5	55	556
15184	C₁₂H₁₈	1,3,5-Triethylbenzene	215.5	Nonazeotrope		575
15185	C₁₂H₂₂O₄	Isoamyl oxalate	268.0	Nonazeotrope		575
15186	C₁₃H₁₂	Diphenylmethane	265.4	252.7	75	563
A =	**C₉H₁₈O₃**	**Isobutyl Carbonate**	**190.3**			
15187	C₁₀H₁₆	Camphene	158	Nonazeotrope		546
15188	C₁₀H₁₆	Dipentene	177.7	< 174.5	<33	575
15189	C₁₀H₁₆	*d*-Limonene	177.9	Nonazeotrope		546
15190	C₁₀H₁₆	α-Pinene	155.8	Nonazeotrope		546
15191	C₁₀H₁₈O	Cineole	176.35	< 176.0	>18	557
		"	176.35	Nonazeotrope		568
15192	C₁₀H₁₈O	Linalool	198.6	< 189.8	<96	575
		"	198.6	Nonazeotrope		535
15193	C₁₀H₂₂O	Isoamyl ether	173.2	< 172.5		557
15194	C₁₁H₁₀	1-Methylnaphthalene	244.6	Nonazeotrope		538
A =	**C₉H₂₀**	**Nonane**	**151**			
15195	C₁₂F₂₇N	Perfluorotributylamine	177	Min. b.p.		159
A =	**C₉H₂₀O**	**2,6-Dimethyl-4-heptanol**	**104/52**			
15196	C₁₂H₂₄	2,6,8-Trimethylnonene, 8 mm.		56	18	981
		" 52 mm.		95	32	981
A =	**C₉H₂₀O₄**	**Tripropylene Glycol**				
15197	C₁₂H₉N	Carbazole		Nonazeotrope		272
		" Low press.		Min. b.p.		272
15198	C₁₃H₁₀	Fluorene, high press		Min. b.p.		272
		" Low press.		Nonazeotrope		272
15199	C₁₄H₁₀	Phenanthrene		Min. b.p.		272
		"		% glycol decreases with decreasing pressure		272
A =	**C₁₀H₇Br**	**1-Bromonaphthalene**	**281.2**			
15200	C₁₀H₈O	1-Naphthol	288	281		544
15201	C₁₀H₈O	2-Naphthol	295	Nonazeotrope		575
15202	C₁₀H₁₀O₂	Methyl cinnamate	261.9	Nonazeotrope		547
15203	C₁₀H₁₀O₄	Methyl phthalate	283.7	278.85	61	541
15204	C₁₁H₁₂O₂	Ethyl cinnamate	271.5	Nonazeotrope		541

TABLE I. *Binary Systems* 429

No.		B-Component			Azeotropic Data		
	Formula	Name	B.P., °C	B.P., °C	Wt.%A		Ref.
A =	**C₁₀H₇Br**	**1-Bromonaphthalene**	**281.2**				
		(continued)					
15205	C₁₁H₁₄O₂	1,2-Dimethoxy-4-					
		propenylbenzene	270.5	Nonazeotrope			559
15206	C₁₂H₁₀	Acenaphthene	277.9	Nonazeotrope			542
15207	C₁₂H₁₆O₂	Isoamyl benzoate	262.0	Nonazeotrope			547
15208	C₁₂H₁₆O₃	Isoamyl salicylate	277.5	Nonazeotrope			575
15209	C₁₂H₂₂O₄	Isoamyl oxalate	268.0	Nonazeotrope			542
15210	C₁₃H₁₀	Fluorene	295	Nonazeotrope			575
15211	C₁₃H₁₂	Diphenylmethane	265.4	Nonazeotrope			575
15212	C₁₃H₁₂O	Benzyl phenyl ether	286.5	Nonazeotrope			559
15213	C₁₄H₁₄	1,2-Diphenylethane	284.5	Nonazeotrope			545
A =	**C₁₀H₇Cl**	**1-Chloronaphthalene**	**262.7**				
15214	C₁₀H₈O	1-Naphthol	288	Nonazeotrope			542
15215	C₁₀H₈O	2-Naphthol	295	Nonazeotrope			542
15216	C₁₀H₁₀O₂	Isoafrole	252.0	Nonazeotrope			541
15217	C₁₀H₁₀O₂	Methyl cinnamate	261.9	260.7	55		542
15218	C₁₀H₁₀O₄	Methyl phthalate	283.7	Nonazeotrope			547
15219	C₁₀H₁₂O₂	Eugenol	254.8	Nonazeotrope			575
15220	C₁₀H₁₂O₂	Isoeugenol	268.8	< 262.4	<92		575
15221	C₁₀H₁₄O	Carvacrol	237.85	Nonazeotrope			575
15222	C₁₀H₁₄O	Thymol	232.9	Nonazeotrope			543
15223	C₁₀H₁₈O₄	Prophyl succinate	250.5	Nonazeotrope			547
15224	C₁₀H₂₀O₂	Capric acid	268.8	< 261.5	<88		575
15225	C₁₁H₁₀	1-Methylnaphthalene	244.6	Nonazeotrope			575
15226	C₁₁H₁₀	2-Methylnaphthalene	241.15	Nonazeotrope			527
15227	C₁₁H₁₂O₂	Ethyl cinnamate	271.5	Nonazeotrope			541
15228	C₁₁H₁₄O₂	1-Allyl-3,4-dimethoxy-					
		benzene	255.0	Nonazeotrope			541
15229	C₁₁H₁₄O₂	Butyl benzoate	249.5	Nonazeotrope			547
15230	C₁₁H₁₄O₂	1,2-Dimethoxy-4-					
		propenylbenzene	270.5	Nonazeotrope			559
15231	C₁₂H₁₀	Acenaphthene	277.9	Nonazeotrope			575
15232	C₁₂H₁₀	Biphenyl	254.8	Nonazeotrope			545
15233	C₁₂H₁₀O	Phenyl ether	259.3	258.92	~ 6		559
15234	C₁₂H₁₆O₂	Isomyl benzoate	262.0	261.65	23		542
15235	C₁₂H₁₆O₃	Isoamyl salicylate	277.5	Nonazeotrope			575
15236	C₁₂H₂₂O₄	Isoamyl oxalate	268.0	262.5	~ 92		542
15237	C₁₃H₁₂	Diphenylmethane	265.4	262.55	93		541
A =	**C₁₀H₈**	**Naphthalene**	**218.05**				
15238	C₁₀H₁₀O₂	Safrole	235.9	Nonazeotrope			548
15239	C₁₀H₁₂O	Anethole	235.7	Nonazeotrope			558
15240	C₁₀H₁₂O₂	Ethyl α-toluate	228.75	Nonazeotrope			529
15241	C₁₀H₁₂O₂	Propyl benzoate	231.2	Nonazeotrope			563
15242	C₁₀H₁₄O	Carvacrol	237.85	Nonazeotrope			575
15243	C₁₀H₁₄O	Carvone	231.0	Nonazeotrope			552
15244	C₁₀H₁₄O	Thymol	232.8	Nonazeotrope			530

		B-Component		Azeotropic Data		
No.	Formula	Name	B.P., °C	B.P., °C	Wt.%A	Ref.
A =	C₁₀H₈	**Naphthalene** *(continued)*	**218.05**			
15245	C₁₀H₁₅N	Diethylaniline	217.05	213		551
	"		216.5	Nonazeotrope		563
15246	C₁₀H₁₆O	Camphor	209.1	Nonazeotrope		552
15247	C₁₀H₁₆O	Citral	226	Nonazeotrope		563
15248	C₁₀H₁₆O	Pulegone	~224	Nonazeotrope		529
15249	C₁₀H₁₇Cl	Bornyl chloride	207.5	Nonazeotrope		575
15250	C₁₀H₁₈O	Borneol	213.4	213.0	35	574
15251	C₁₀H₁₈O	Geraniol	229.6	Nonazeotrope		541
	"		229.5	218.0?		563
15252	C₁₀H₁₈O	Linalool	198.6	Nonazeotrope		532
15253	C₁₀H₁₈O	α-Terpineol	217.8	212	~ 45	528
15254	C₁₀H₁₈O	β-Terpineol	210.5	Nonazeotrope		575
15255	C₁₀H₁₈O₄	Propyl succinate	250.5	Nonazeotrope		546
15256	C₁₀H₂₀O	Citronellol	224.5	217.8	70	537
15257	C₁₀H₂₀O	Menthol	216.4	215.15	25.5	529
15258	C₁₀H₂₀O₂	Capric acid	268.8	Nonazeotrope		575
15259	C₁₀H₂₀O₂	Ethyl caprylate	208.35	Nonazeotrope		575
15260	C₁₀H₂₀O₂	Isoamyl isovalerate	192.7	Nonazeotrope		575
15261	C₁₀H₂₂O	n-Decyl alcohol	232.9	Nonazeotrope		529
15262	C₁₀H₂₂S	Isoamyl sulfide	214.8	Nonazeotrope		575
15263	C₁₁H₁₄O₂	Isobutyl benzoate	241.9	Nonazeotrope		575
15264	C₁₁H₂₀O	Terpineol methyl ether	216	Nonazeotrope		563
15265	C₁₁H₂₂O₂	Ethyl pelargonate	227	Nonazeotrope		575
15266	C₁₁H₂₂O₃	Isoamyl carbonate	228.8	Nonazeotrope		531
15267	C₁₂H₁₈	1,3,5-Triethylbenzene	215.5	< 214.8	<20	561
15268	C₁₂H₂₀O₂	Bornyl acetate	227.7	Nonazeotrope		529
15269	C₁₂H₂₂O	Bornyl ethyl ether	204.9	Nonazeotrope		558
15270	C₁₂H₂₆	Dodecane　　100 mm.		140.2	59.2　v-l	404,600
15271	C₁₃H₂₈	Tridecane	234.0	Nonazeotrope		561
A =	C₁₀H₈O	**1-Naphthol**	**288.0**			
15272	C₁₀H₉N	1-Naphthylamine	300.8	Nonazeotrope		551
15273	C₁₀H₉N	2-Naphthylamine	306.1	Nonazeotrope		551
15274	C₁₀H₁₀O₂	Methyl cinnamate	261.9	Nonazeotrope		575
15275	C₁₁H₁₀	1-Methylnaphthalene	244.6	Nonazeotrope		542
15276	C₁₁H₁₀	2-Methylnaphthalene	241.15	Nonazeotrope		527
15277	C₁₁H₁₂O₂	Ethyl cinnamate	271.5	Nonazeotrope		542
15278	C₁₁H₁₄O₂	1,2-Dimethoxy-4-propenylbenzene	270.5	Nonazeotrope		542
15279	C₁₂H₁₀	Acenaphthene	277.9	Nonazeotrope		575
	"		277.9	274	20	544
15280	C₁₂H₁₀	Biphenyl	255.9	Nonazeotrope		542
15281	C₁₂H₁₀O	Phenyl ether	259.0	Nonazeotrope		556
15282	C₁₂H₁₁N	Diphenylamine	275	Azeotropic		563
15283	C₁₂H₁₆O₂	Isoamyl benzoate	262.0	Nonazeotrope		575
15284	C₁₂H₁₆O₃	Isoamyl salicylate	277.5	Nonazeotrope		575
15285	C₁₂H₂₂O₄	Isoamyl oxalate	268.0	Nonazeotrope		542

TABLE I. *Binary Systems* 431

No.	Formula	Name	B.P., °C	B.P., °C	Wt.%A	Ref.
		B-Component		**Azeotropic Data**		
A =	**C₁₀H₈O**	**1-Naphthol** *(continued)*	**288.0**			
15286	C₁₃H₁₀	Fluorene	295	Nonazeotrope		575
15287	C₁₃H₁₂	Diphenylmethane	265.4	Nonazeotrope		575
		"	265.6	265	10	544
15288	C₁₄H₁₂	1,2-Diphenylethylene	308.5	Nonazeotrope		575
A =	**C₁₀H₈O**	**2-Naphthol**	**295.0**			
15289	C₁₀H₁₀O₄	Methyl phthalate	283.2	> 296.0	>82	575
15290	C₁₁H₁₂O₂	Ethyl cinnamate	272.0	Nonazeotrope		575
15291	C₁₂H₁₀	Acenaphthene	277.9	Nonazeotrope		575
		"	277.9	277.0	10	544
15292	C₁₂H₁₀	Biphenyl	255.9	Nonazeotrope		524
15293	C₁₃H₁₂	Diphenylmethane	265.5	Nonazeotrope		524
15294	C₁₄H₁₂	Stilbene	308.5	Nonazeotrope		575
15295	C₁₄H₁₄	1,2-Diphenylethane	285.5	Nonazeotrope		575
		"	284	283.5		544
A =	**C₁₀H₉N**	**4-Methylquinoline**	**265.6**			
15296	C₁₀H₉N	Quinaldine	247.7	Nonazeotrope	v-l	610
15297	C₁₀H₉N	7-Methylquinoline	257.7	Nonazeotrope	v-l	610
A =	**C₁₀H₉N**	**1-Naphthylamine**	**300.8**			
15298	C₁₂H₁₀	Acenaphthene	277.9	Nonazeotrope		551
15299	C₁₃H₁₂O	Benzyl phenyl ether	286.5	Nonazeotrope		551
15300	C₁₄H₁₄	1,2-Diphenylethane	284.5	Nonazeotrope		551
15301	C₁₄H₁₄O	Benzyl ether	297	< 296		575
A =	**C₁₀H₉N**	**2-Naphthylamine**	**306.1**			
15302	C₁₃H₁₂O	Benzyl phenyl ether	286.5	Nonazeotrope		575
15303	C₁₄H₁₄O	Benzyl ether	297	Nonazeotrope		575
A =	**C₁₀H₉N**	**Quinaldine**	**246.5**			
15304	C₁₀H₁₀O₂	Safrole	235.9	Nonazeotrope		575
15305	C₁₀H₁₄O	Carvacrol	237.85	250.8	67	575
15306	C₁₀H₁₄O	Thymol	232.9	250.0	80	575
A =	**C₁₀H₁₀O₂**	**Isosafrol**	**252.1**			
15307	C₁₀H₁₀O₂	Methyl cinnamate	261.6	Nonazeotrope		531,557
15308	C₁₀H₁₂O₂	Eugenol	255.0	252.05?	~92	574
15309	C₁₀H₁₄O	Thymol	232.9	Nonazeotrope		542
15310	C₁₀H₁₈O₄	Propyl succinate	250.5	< 249.0	<70	557
15311	C₁₀H₂₀O₂	Capric acid	268.8	Nonazeotrope		575
15312	C₁₁H₁₀	1-Methylnaphthalene	244.6	Nonazeotrope		548
15313	C₁₁H₁₀	2-Methylnaphthalene	241.15	Nonazeotrope		527
15314	C₁₁H₁₄O₂	1-Allyl-3,4-dimethoxy-benzene	254.7	Nonazeotrope		549
15315	C₁₁H₁₄O₂	Butyl benzoate	249.5	Nonazeotrope		557
15316	C₁₁H₁₄O₂	Isobutyl benzoate	241.9	Nonazeotrope		557
15317	C₁₁H₁₇N	Isoamylaniline	256.0	< 250.0	>64	551
15318	C₁₂H₁₀	Biphenyl	255.0	Nonazeotrope		548
15319	C₁₂H₁₀O	Phenyl ether	259.0	Nonazeotrope		549
15320	C₁₂H₁₆O₂	Isoamyl benzoate	262.05	Nonazeotrope		557
15321	C₁₂H₂₂O₄	Isoamyl oxalate	268.0	Nonazeotrope		557
15322	C₁₃H₁₂	Diphenylmethane	265.6	Nonazeotrope		535
15323	C₁₅H₃₃BO₃	Isoamyl borate	255	< 250.8		557

		B-Component		Azeotropic Data		
No.	Formula	Name	B.P., °C	B.P., °C	Wt.%A	Ref.
A =	**C₁₀H₁₀O₂**	**Methyl Cinnamate**	**261.95**			
15324	C₁₀H₁₂O₂	Eugenol	255.0	Nonazeotrope		556
15325	C₁₀H₁₂O₂	Isoeugenol	268.8	Nonazeotrope		535
15326	C₁₀H₁₄O	Thymol	232.9	Nonazeotrope		575
15327	C₁₀H₂₀O₂	Capric acid	~268.8	Nonazeotrope		575
15328	C₁₀H₂₀O₄	2-(2-Butoxyethoxy)ethyl acetate	245.3	Nonazeotrope		575
15329	C₁₁H₁₀	1-Methylnaphthalene	245.1	Nonazeotrope		546
15330	C₁₁H₁₀	2-Methylnaphthalene	241.15	Nonazeotrope		527
15331	C₁₁H₁₄O₂	1-Allyl-3,4-dimethoxy-benzene	255.2	Nonazeotrope		557
15332	C₁₁H₁₄O₂	1,2-Dimethoxy-4-propenylbenzene	270.5	Nonazeotrope		557
15333	C₁₂H₁₀	Acenaphthene	277.9	Nonazeotrope		546
15334	C₁₂H₁₀	Biphenyl	255.9	Nonazeotrope		542
15335	C₁₂H₁₀O	Phenyl ether	259.3	258.8	17?	557
15336	C₁₂H₁₆O₂	Isoamyl benzoate	262.0	260.5	47.5	549
15337	C₁₂H₁₆O₃	Isoamyl salicylate	277.5	Nonazeotrope		575
15338	C₁₂H₂₂O₄	Isoamyl oxalate	268.0	Nonazeotrope		575
15339	C₁₃H₁₂	Diphenylmethane	265.6	261.55	~ 95	573
A =	**C₁₀H₁₀O₂**	**Safrole**	**235.9**			
15340	C₁₀H₁₂O	Anethole	235.7	234.65	60	527
15341	C₁₀H₁₂O₂	Ethyl α-toluate	228.75	Nonazeotrope		557
15342	C₁₀H₁₂O₂	Propyl benzoate	230.85	Nonazeotrope		557
		"	231.2	228	40	563
15343	C₁₀H₁₄N₂	Nicotine	247.5	Nonazeotrope		575
15344	C₁₀H₁₄O	Carvacrol	237.85	Nonazeotrope		556
15345	C₁₀H₁₄O	Carvone	231.0	Nonazeotrope		552
15346	C₁₀H₁₄O	Thymol	232.8	Nonazeotrope		529
15347	C₁₀H₁₅N	Diethylaniline	217.05	Nonazeotrope		551
15348	C₁₀H₁₆O	Menthenone	222.5	Nonazeotrope		564
15349	C₁₀H₁₆O	Pulegone	223.8	Nonazeotrope		575
15350	C₁₀H₁₈O	Borneol	215.0	Nonazeotrope		575
15351	C₁₀H₁₈O	Geraniol	235.9	Nonazeotrope		545
15352	C₁₀H₁₈O	α-Terpineol	218.85	Nonazeotrope		575
15353	C₁₀H₁₈O₄	Propyl succinate	250.5	Nonazeotrope		557
15354	C₁₀H₂₀O	Citronellol	224.4	Nonazeotrope		545
15355	C₁₀H₂₂O	Decyl alcohol	235.9	Nonazeotrope		545
15356	C₁₁H₁₀	1-Methylnaphthalene	244.9	Nonazeotrope		537
15357	C₁₁H₁₀	2-Methylnaphthalene	241.15	Nonazeotrope		527
15358	C₁₁H₁₄O₂	1-Allyl-3,4-dimethoxy-benzene	255.2	Nonazeotrope		535
15359	C₁₁H₁₄O₂	Butyl benzoate	249.0	Nonazeotrope		557
15360	C₁₁H₁₄O₂	Ethyl β-phenylpropionate	248.1	Nonazeotrope		557
15361	C₁₁H₁₄O₂	Isobutyl benzoate	241.9	Nonazeotrope		557
15362	C₁₁H₁₇N	Isoamylaniline	256.0	Nonazeotrope		551
15363	C₁₁H₂₂O₃	Isoamyl carbonate	232.2	< 231.8		557
		"	232.2	Nonazeotrope		548

TABLE I. *Binary Systems*

No.	Formula	Name	B.P., °C	B.P., °C	Wt.%A	Ref.
		B-Component		**Azeotropic Data**		
A =	**C₁₀H₁₀O₂**	**Safrole** *(continued)*	**235.9**			
15364	C₁₂H₂₀O₂	Bornyl acetate	227.6	Nonazeotrope		557
A =	**C₁₀H₁₀O₄**	**Methyl Phthalate**	**283.2**			
15365	C₁₁H₁₂O₂	Ethyl cinnamate	272.0	Nonazeotrope		549
15366	C₁₁H₁₄O₂	1,2-Dimethoxy-4-propenylbenzene	270.5	Nonazeotrope		557
15367	C₁₂H₁₀	Acenaphthene	277.9	276.35	33.5	542
15368	C₁₂H₁₀	Biphenyl	255.9	Nonazeotrope		546
15369	C₁₂H₁₀O	Phenyl ether	259.0	Nonazeotrope		557
15370	C₁₂H₁₆O₃	Isoamyl salicylate	277.5	Nonazeotrope		575
15371	C₁₃H₁₂	Diphenylmethane	265.6	Nonazeotrope		546
15372	C₁₃H₁₂O	Benzyl phenyl ether	286.5	< 282.5		557
15373	C₁₄H₁₄	1,2-Diphenylethane	284	280.5	53	546
15374	C₁₄H₁₄O	Benzyl ether	297	Nonazeotrope		557
A =	**C₁₀H₁₂O**	**Anethole**	**235.7**			
15375	C₁₀H₁₂O₂	Propyl benzoate	230.85	Nonazeotrope		557
15376	C₁₀H₁₄O	Carvacrol	237.85	Nonazeotrope		556
15377	C₁₀H₁₄O	Carvone	231.0	Nonazeotrope		552
15378	C₁₀H₁₄O	Thymol	232.9	Nonazeotrope		575
15379	C₁₀H₁₅N	Diethylaniline	217.05	Nonazeotrope		551
15380	C₁₀H₁₆O	Pulegone	223.8	Nonazeotrope		575
15381	C₁₀H₁₈O	α-Terpineol	218.85	Nonazeotrope		575
15382	C₁₀H₂₀O	Citronellol	224.4	Nonazeotrope		575
15383	C₁₀H₂₀O	Menthol	216.3	Nonazeotrope		575
15384	C₁₀H₂₂O	Decyl alcohol	232.8	< 232.6	<78	575
15385	C₁₁H₁₀	1-Methylnaphthalene	244.6	Nonazeotrope		558
15386	C₁₁H₁₀	2-Methylnaphthalene	241.15	Nonazeotrope		527
15387	C₁₁H₁₄O₂	Isobutyl benzoate	241.9	Nonazeotrope		557
A =	**C₁₀H₁₂O**	**Estragole**	**215.6**			
15388	C₁₀H₁₆O	Camphor	209.1	Nonazeotrope		575
A =	**C₁₀H₁₂O₂**	**Ethyl α-Toluate**	**288.75**			
15389	C₁₀H₁₂O₂	Propyl benzoate	230.9	228.7	97	529
		"	230.85	Nonazeotrope		549
15390	C₁₀H₁₄O	Carvacrol	237.85	238.3	20	575
15391	C₁₀H₁₄O	Carvone	231.0	228.6	93	552
15392	C₁₀H₁₄O	Thymol	232.8	235.75	37.5	529
15393	C₁₀H₁₄O₂	*m*-Diethoxybenzene	235.0	Nonazeotrope		557
15394	C₁₀H₁₆O	Carvenone	234.5	Nonazeotrope		552
15395	C₁₀H₁₆O	Pulegone	223.8	Nonazeotrope		552
15396	C₁₀H₁₈O	Geraniol	229.6	228.1	70	529
15397	C₁₀H₁₈O	α-Terpineol	217.8	Nonazeotrope		536
15398	C₁₀H₂₀O	Citronellol	224.5	Nonazeotrope		536
15399	C₁₀H₂₀O	Menthol	216.3	Nonazeotrope		575
15400	C₁₀H₂₀O₄	2-(2-Butoxyethoxy)ethyl acetate	245.3	Nonazeotrope		551
15401	C₁₀H₂₂O	Decyl alcohol	232.9	228.55	94	529

		B-Component		Azeotropic Data		
No.	Formula	Name	B.P., °C	B.P., °C	Wt.%A	Ref.
A =	**C₁₀H₁₂O₂**	**Ethyl α-Toluate** *(continued)*	**288.75**			
15402	C₁₁H₁₀	1-Methylnaphthalene	244.9	Nonazeotrope		537
15403	C₁₁H₁₂	2-Methylnaphthalene	241.15	Nonazeotrope		575
15404	C₁₁H₁₄O₂	Isobutyl benzoate	241.9	Nonazeotrope		575
15405	C₁₁H₁₆O	Methyl thymol ether	216.5	Nonazeotrope		557
15406	C₁₁H₂₂O₃	Isoamyl carbonate	228.5	227.9		573
15407	C₁₂H₂₀O₂	Bornyl acetate	227.6	226.6	44	549
A =	**C₁₀H₁₂O₂**	**Eugenol**	**254.8**			
15408	C₁₀H₁₄O	Carvone	231.0	Nonazeotrope		575
15409	C₁₀H₁₆O	Menthenone	222.5	Nonazeotrope		575
15410	C₁₀H₂₀O	Citronellol	224.4	Nonazeotrope		575
15411	C₁₁H₁₀	1-Methylnaphthalene	244.6	Nonazeotrope		556
15412	C₁₁H₁₀	2-Methylnaphthalene	241.15	Nonazeotrope		527
15413	C₁₁H₁₄O₂	1-Allyl-3,4-dimethoxy-benzene	255.2	255.3	~ 45	574
15414	C₁₁H₁₄O₂	Butyl benzoate	249.5	Nonazeotrope		548
15415	C₁₁H₁₄O₂	Isobutyl benzoate	242.15	Nonazeotrope		535
15416	C₁₁H₁₆O	*p-tert*-Amylphenol	266.5	Nonazeotrope		575
15417	C₁₁H₁₇N	Isoamylaniline	256.0	< 254.5		551
15418	C₁₂H₁₀	Biphenyl	255.0	253.5	50?	556
15419	C₁₂H₁₀O	Phenyl ether	259.3	254.9	~ 97	574
15420	C₁₂H₁₆O₂	Isoamyl benzoate	262.05	Nonazeotrope		556,574
15421	C₁₃H₁₂	Diphenylmethane	265.4	Nonazeotrope		556
A =	**C₁₀H₁₂O₂**	**Isoeugenol**	**268.8**			
15422	C₁₁H₁₀	2-Methylnaphthalene	241.15	Nonazeotrope		575
15423	C₁₁H₁₂O₂	Ethyl cinnamate	272.5	Nonazeotrope		548
15424	C₁₁H₁₄O₂	1,2-Dimethoxy-4-propenylbenzene	270.5	Nonazeotrope		535
15425	C₁₁H₁₆O	*p-tert*-Amylphenol	266.5	Nonazeotrope		575
15426	C₁₁H₁₇N	Isoamylaniline	256.0	Nonazeotrope		575
15427	C₁₂H₁₀	Acenaphthene	277.9	Nonazeotrope		556
15428	C₁₂H₁₀	Biphenyl	255.0	Nonazeotrope		556
15429	C₁₂H₁₀O	Phenyl ether	259.3	Nonazeotrope		571
15430	C₁₂H₁₆O₂	Isoamyl benzoate	262.05	Nonazeotrope		535
15431	C₁₂H₁₆O₃	Isoamyl salicylate	277.5	Nonazeotrope		575
15432	C₁₃H₁₂	Diphenylmethane	265.5	264.7	20?	556
15433	C₁₄H₁₄	1,2-Diphenylethane	284.5	Nonazeotrope		575
A =	**C₁₀H₁₂O₂**	**Propyl Benzoate**	**230.85**			
15434	C₁₀H₁₄O	Carvacrol	237.85	238.85	18	562
15435	C₁₀H₁₄O	Carvone	231.0	231.5?	50	552
15436	C₁₀H₁₄O	Thymol	232.8	235.5	45	529
15437	C₁₀H₁₆O	Carvenone	234.5	Nonazeotrope		552
15438	C₁₀H₁₆O	Citral	226	Nonazeotrope		563
15439	C₁₀H₁₆O	Pulegone	223.8	Nonazeotrope		552
15440	C₁₀H₁₈O	Geraniol	229.5	228.0	~ 45	563
15441	C₁₀H₁₈O	α-Terpineol	218.85	Nonazeotrope		575

TABLE I. *Binary Systems*   435

No.	Formula	Name	B.P., °C	B.P., °C	Wt.%A	Ref.
		B-Component		Azeotropic Data		
A =	**C₁₀H₁₂O₂**	**Propyl Benzoate** *(continued)*	**230.85**			
15442	C₁₀H₂₀O	Citronellol	224.5	Nonazeotrope		536
15443	C₁₀H₂₀O	Menthol	216.3	Nonazeotrope		575
15444	C₁₀H₂₀O₄	2-(2-Butoxyethoxy)ethyl acetate	245.3	Nonazeotrope		575
15445	C₁₀H₂₂O	*n*-Decyl alcohol	232.5	230.7	~ 75	528
15446	C₁₁H₁₀	1-Methylnaphthalene	244.9	Nonazeotrope		537
15447	C₁₁H₁₀	2-Methylnaphthalene	241.15	Nonazeotrope		527
15448	C₁₁H₂₂O₃	Isoamyl carbonate	232.2	< 230.8		549
A =	**C₁₀H₁₄**	**Butylbenzene**	**183.1**			
15449	C₁₀H₁₄	Cymene	176.7	Nonazeotrope		561
15450	C₁₀H₁₆	α-Terpinene	173.4	Nonazeotrope		561
15451	C₁₀H₁₆	Terpinolene	184.6	182.2	65	561
15452	C₁₀H₁₈O	Borneol	215	Nonazeotrope		575
15453	C₁₀H₁₈O	Cineole	176.35	Nonazeotrope		548
15454	C₁₀H₁₈O	Citronellal	208.0	Nonazeotrope		575
15455	C₁₀H₁₈O	Linalool	198.6	Nonazeotrope		575
15456	C₁₀H₂₀O	Menthol	216.3	Nonazeotrope		575
15457	C₁₀H₂₂O	Amyl ether	187.5	Nonazeotrope		558
15458	C₁₁H₂₀O	Isobornyl methyl ether	192.4	Nonazeotrope		558
A =	**C₁₀H₁₄**	**Cymene**	**176.7**			
15459	C₁₀H₁₆	Camphene	159.6	Nonazeotrope		561
15460	C₁₀H₁₆	Dipentene	177.7	175.8	60	561
15461	C₁₀H₁₆	*d*-Limonene	177.8	174.5	75	563
15462	C₁₀H₁₆	Nopinene	163.8	Nonazeotrope		561
15463	C₁₀H₁₆	α-Terpinene	173.4	173.0	20	561
15464	C₁₀H₁₈O	Cineole	176.35	176.2	45	558
15465	C₁₀H₁₈O	Linalool	198.6	Nonazeotrope		537
15466	C₁₀H₁₈O	α-Terpineol	218.85	Nonazeotrope		575
15467	C₁₀H₂₂O	Isoamyl ether	172.6	Nonazeotrope		537
15468	C₁₀H₂₃N	Diisoamylamine	188.2	Nonazeotrope		551
A =	**C₁₀H₁₄**	**Isobutylbenzene**	**172.76**			
15469	C₁₀H₁₄O₂	*m*-Diethoxybenzene	235.0	Nonazeotrope		557
A =	**C₁₀H₁₄N₂**	**Nicotine**	**247.5**			
15470	C₁₀H₁₄O	Thymol	232.9	> 250.2	>79	575
15471	C₁₀H₁₄O₂	*m*-Diethoxybenzene	235.4	Nonazeotrope		575
A =	**C₁₀H₁₄O**	**Carvacrol**	**237.85**			
15472	C₁₀H₁₄O	Carvone	231.0	242.2	>58	552
15473	C₁₀H₁₄O	Thymol	232.9	Nonazeotrope		575
15474	C₁₀H₁₅N	Diethylaniline	217.05	Nonazeotrope		551
15475	C₁₀H₁₆O	Carvenone	234.5	243.0	55	575
15476	C₁₀H₁₆O	Menthenone	222.5	239.5	75	575
15477	C₁₀H₁₆O	Pulegone	223.8	328.4		552
15478	C₁₀H₁₈O	Geraniol	229.6	> 238.2	>85	575
15479	C₁₀H₁₈O	Menthone	209.5	Nonazeotrope		575
15480	C₁₀H₁₈O	α-Terpineol	218.85	Nonazeotrope		575

No.	Formula	B-Component Name	B.P., °C	Azeotropic Data B.P., °C	Wt.%A	Ref.
A =	C₁₀H₁₄O	**Carvacrol** *(continued)*	**237.85**			
15481	C₁₀H₁₈O₄	Propyl succinate	250.5	251.5	25	575
15482	C₁₀H₂₂O	Decyl alcohol	232.8	Nonazeotrope		575
15483	C₁₁H₁₀	1-Methylnaphthalene	244.6	Nonazeotrope		575
15484	C₁₁H₁₄O₂	Butyl benzoate	249.0	Nonazeotrope		575
15485	C₁₁H₁₄O₂	Isobutyl benzoate	241.9	243.85	33	562
15486	C₁₁H₁₇N	Isoamylaniline	256.0	Nonazeotrope		551
15487	C₁₁H₂₂O₃	Isoamyl carbonate	232.2	> 239.0	>62	575
15488	C₁₂H₁₀	Biphenyl	256.1	Nonazeotrope		575
15489	C₁₂H₂₀O₂	Bornyl acetate	227.6	238.2	75	562
A =	C₁₀H₁₄O	**Carvone**	**230.95**			
15490	C₁₀H₁₄O	Thymol	232.9	238.65	48	552
15491	C₁₀H₁₄O₂	*m*-Diethoxybenzene	235	Nonazeotrope		537
15492	C₁₀H₁₅N	Diethylaniline	217.05	Nonazeotrope		551
15493	C₁₀H₁₈O	Borneol	215.0	Nonazeotrope		552
15494	C₁₀H₁₈O	Geraniol	229.6	229.2	40	552
15495	C₁₀H₂₀O	Citronellol	224.4	Nonazeotrope		552
15496	C₁₀H₂₀O	Menthol	216.3	Nonazeotrope		552
15497	C₁₀H₂₂O	*n*-Decyl alcohol	232.8	230.85	81	552
15498	C₁₁H₁₀	1-Methylnaphthalene	244.6	Nonazeotrope		552
15499	C₁₁H₁₀	2-Methylnaphthalene	241.15	Nonazeotrope		527
15500	C₁₁H₁₄O₂	Isobutyl benzoate	241.9	Nonazeotrope		552
15501	C₁₁H₁₆O	*p-tert*-Amylphenol	265	Nonazeotrope		575
15502	C₁₁H₁₇N	*N*-Isoamylaniline	256.0	Nonazeotrope		575
15503	C₁₁H₂₂O₃	Isoamyl carbonate	228.5	Nonazeotrope		573
		"	232.2	230.5	60	552
15504	C₁₂H₂₀O₂	Bornyl acetate	227.6	Nonazeotrope		552
A =	C₁₀H₁₄O	**Thymol**	**232.9**			
15505	C₁₀H₁₄O₂	*m*-Diethoxybenzene	235.0	Nonazeotrope		542
15506	C₁₀H₁₅N	Diethylaniline	217.05	Nonazeotrope		531
15507	C₁₀H₁₆O	Camphor	209.1	233.3	84	552
		"	209.1	Nonazeotrope		542
15508	C₁₀H₁₆O	Carvenone	234.5	241.0	50	575
15509	C₁₀H₁₆O	Pulegone	223.8	235.3	65	552
15510	C₁₀H₁₈O	Borneol	213.4	Nonazeotrope		542
15511	C₁₀H₁₈O	Geraniol	229.6	325.6	57.5	529
15512	C₁₀H₁₈O	Linalool	198.6	Nonazeotrope		575
15513	C₁₀H₁₈O	Menthone	209.5	233.2	92	575
15514	C₁₀H₁₈O	α-Terpineol	217.8	Nonazeotrope		529
15515	C₁₀H₁₈O₄	Propyl succinate	250.5	Nonazeotrope		575
15516	C₁₀H₂₀O	Citronellol	224	233.8	~ 85	573
15517	C₁₀H₂₀O	Menthol	216.4	Nonazeotrope		542
15518	C₁₀H₂₀O₂	Methyl pelargonate	213.8	Nonazeotrope		575
15519	C₁₀H₂₂O	*n*-Decyl alcohol	232.5	~ 234.5	~ 60	529
15520	C₁₁H₁₀	1-Methylnaphthalene	242	Nonazeotrope		573
15521	C₁₁H₁₀	2-Methylnaphthalene	241.15	Nonazeotrope		527

TABLE I.   *Binary Systems*   437

No.	Formula	B-Component Name	B.P., °C	Azeotropic Data B.P., °C	Wt.%A	Ref.
A =	$C_{10}H_{14}O$	**Thymol** *(continued)*	**232.9**			
15522	$C_{11}H_{14}O_2$	1-Allyl-3,4-dimethoxy-benzene	254.7	Nonazeotrope		575
15523	$C_{11}H_{14}O_2$	Butyl benzoate	249.8	Nonazeotrope		542
15524	$C_{11}H_{14}O_2$	Isobutyl benzoate	242.15	243.2	20	573
15525	$C_{11}H_{16}O$	Methyl thymyl ether	216.5	Nonazeotrope		575
15526	$C_{11}H_{20}O$	Methyl α-terpineol ether	216.2	Nonazeotrope		544
15527	$C_{11}H_{22}O_3$	Isoamyl carbonate	232.2	236.25	~ 48	542
15528	$C_{12}H_{10}$	Biphenyl	255.9	Nonazeotrope		542
15529	$C_{12}H_{10}O$	Phenyl ether	259.0	Nonazeotrope		575
15530	$C_{12}H_{18}$	1,3,5-Triethylbenzene	216	Nonazeotrope		544
15531	$C_{12}H_{22}O_2$	Bornyl acetate	227.7	235.6	60	529
A =	$C_{10}H_{14}O_2$	**m-Diethoxybenzene**	**235.4**			
15532	$C_{10}H_{15}N$	Diethylaniline	217.05	Nonazeotrope		575
15533	$C_{10}H_{18}O$	Geraniol	229.7	Nonazeotrope		576
15534	$C_{10}H_{20}O$	Citronellol	224.4	Nonazeotrope	.	575
15535	$C_{10}H_{22}O$	Decyl alcohol	232.8	232.2		576
15536	$C_{11}H_{22}O_3$	Isoamyl carbonate	232.2	< 231.0	>33	557
15537	$C_{12}H_{20}O_2$	Bornyl acetate	227.6	Nonazeotrope		557
A =	$C_{10}H_{15}N$	**Diethylaniline**	**217.05**			
15538	$C_{10}H_{16}O$	Camphor	209.1	Nonazeotrope		551
15539	$C_{10}H_{16}O$	Citral	226	Reacts		563
15540	$C_{10}H_{16}O$	Pulegone	223.8	Nonazeotrope		551
15541	$C_{10}H_{18}O$	Borneol	215.0	< 214.8	<20	551
		"	213.5	Nonazeotrope		542
15542	$C_{10}H_{18}O$	Geraniol	229.6	Nonazeotrope		551
15543	$C_{10}H_{18}O$	Linalool	198.6	Nonazeotrope		551
15544	$C_{10}H_{18}O$	α-Terpineol	218.85	215.5	56	551
15545	$C_{10}H_{18}O$	β-Terpineol	210.5	Nonazeotrope		551
15546	$C_{10}H_{20}O$	Citronellol	224.4	Nonazeotrope		551
15547	$C_{10}H_{20}O$	Menthol	216.3	215.3	43.5	551
15548	$C_{10}H_{22}O$	Decyl alcohol	232.8	Nonazeotrope		551
15549	$C_{11}H_{10}$	2-Methylnaphthalene	241.5	Nonazeotrope		527
15550	$C_{11}H_{16}O$	Methyl thymyl ether	216.5	< 216.0	<49	575
15551	$C_{11}H_{20}O$	Methyl α-terpinyl ether	216.2	< 215.0	<48	551
15552	$C_{11}H_{24}O_2$	Diisoamyloxymethane	210.8	Nonazeotrope		551
15553	$C_{12}H_{22}O$	Ethyl isobornyl ether	203.8	Nonazeotrope		551
A =	$C_{10}H_{16}$	**Camphene**	**159.6**			
15554	$C_{10}H_{16}$	Dipentene	177.7	Nonazeotrope		561
15555	$C_{10}H_{16}$	Nopinene	163.8	Nonazeotrope		561
15556	$C_{10}H_{16}$	α-Pinene	155.8	Nonazeotrope		561
15557	$C_{10}H_{18}O$	Linalool	198.6	Nonazeotrope		537
15558	$C_{10}H_{22}$	2,7-Dimethyloctane	160.25	158	62	561
15559	$C_{10}H_{23}N$	Diisoamylamine	188.2	Nonazeotrope		551
15560	$C_{12}H_{20}O_2$	Isobornyl acetate	225.8	Nonazeotrope		575
A =	$C_{10}H_{16}$	**Dipentene**	**177.7**			
15561	$C_{10}H_{16}$	α-Pinene	155.8	Nonazeotrope		561
15562	$C_{10}H_{16}$	α-Terpinene	173.4	Nonazeotrope		575

		B-Component		Azeotropic Data		
No.	Formula	Name	B.P., °C	B.P., °C	Wt.%A	Ref.
A =	C₁₀H₁₆	**Dipentene** *(continued)*	**177.7**			
15563	C₁₀H₂₂O	Amyl ether	187.5	Nonazeotrope		558
15564	C₁₀H₂₃N	Diisoamylamine	188.2	Nonazeotrope		551
15565	C₁₂H₂₀O₂	Isobornyl acetate	225.8	Nonazeotrope		575
A =	C₁₀H₁₆	*d*-**Limonene**	**177.8**			
15566	C₁₀H₁₆	Terpinene	180.5	Nonazeotrope		563
15567	C₁₀H₁₈O	Borneol	213.4	Nonazeotrope		537
15568	C₁₀H₁₈O	Cineole	176.35	Nonazeotrope		529
15569	C₁₀H₁₈O	Linalool	198.6	Nonazeotrope		537
15570	C₁₀H₂₀O	Menthol	216.4	Nonazeotrope		537
15571	C₁₀H₂₀O₂	Isoamyl isovalerate	~193.5	Nonazeotrope		573
15572	C₁₀H₂₂O	Isoamyl ether	172.7	Nonazeotrope		563
A =	C₁₀H₁₆	**Nopinene**	**163.8**			
15573	C₁₀H₁₆	α-Terpinene	173.4	Nonazeotrope		561
15574	C₁₀H₂₂O	Isoamyl ether	173.2	Nonazeotrope		558
A =	C₁₀H₁₆	α-**Phellandrene**	**171.5**			
15575	C₁₀H₁₈O	Cineole	176.3	Nonazeotrope		563
A =	C₁₀H₁₆	α-**Pinene**	**155.8**			
15576	C₁₀H₁₆	α-Terpinene	173.4	Nonazeotrope		561
15577	C₁₀H₁₈O	Borneol	155.8	Nonazeotrope		537
15578	C₁₀H₂₂	2,7-Dimethyloctane	160.1	< 155.5	<89	561
A =	C₁₀H₁₆	α-**Terpinene**	**173.4**			
15579	C₁₀H₁₈O	Cineole	176.35	Nonazeotrope		558
15580	C₁₀H₁₈O	Linalool	198.6	Nonazeotrope		575
15581	C₁₀H₂₂	Decane	173.3	< 171.5	<50	561
15582	C₁₀H₁₈O	Isoamyl ether	173.2	172.0	50	558
A =	C₁₀H₁₆	γ-**Terpinene**	**180.5**			
15583	C₁₀H₁₈O	Cineole	176.3	Nonazeotrope		563
15584	C₁₀H₂₀O₂	Isoamyl isovalerate	192.7	Nonazeotrope		546
15585	C₂₀H₂₂O	Isoamyl ether	173.4	Nonazeotrope		548
A =	C₁₀H₁₆	**Terpinolene**	**184.6**			
15586	C₁₀H₂₀O₂	Isoamyl isovalerate	192.7	Nonazeotrope		575
15587	C₁₁H₂₀O	Isobornyl methyl ether	192.4	Nonazeotrope		558
A =	C₁₀H₁₆	**Thymene**	**179.7**			
15588	C₁₀H₁₈O	Borneol	213.4	Nonazeotrope		537
15589	C₁₀H₁₈O	Cineole	176.35	Nonazeotrope		537
15590	C₁₀H₁₈O	Linalool	198.6	Nonazeotrope		574
15591	C₁₀H₁₈O	α-Terpineol	~217.8	Nonazeotrope		537
15592	C₁₀H₂₀O	Menthol	216.4	Nonazeotrope		540
15593	C₁₀H₂₀O₂	Isoamyl isovalerate	193.5	Nonazeotrope		573
A =	**C₁₀H₁₆O**	**Camphor**	**208.9**			
15594	C₁₀H₁₇Cl	Bornyl chloride	~210	Nonazeotrope		563
15595	C₁₀H₁₈O	Borneol	215.0	Nonazeotrope		552
15596	C₁₀H₁₈O	Citronellal	208.0	207.5		552

TABLE I.   *Binary Systems*   439

No.	Formula	Name	B.P., °C	B.P., °C	Wt.%A	Ref.
		B-Component		Azeotropic Data		
A =	**C₁₀H₁₆O**	**Camphor** *(continued)*	**208.9**			
15597	C₁₀H₁₈O	Linalool	198.6	Nonazoetrope		552
15598	C₁₀H₁₈O	Menthone	207	Nonazeotrope		563
15599	C₁₀H₂₀O	Menthol	216.3	Nonazeotrope		552
15600	C₁₀H₂₂S	Isoamyl sulfide	214.8	< 208.8		566
15601	C₁₁H₁₆O	Methyl thymyl ether	216.5	Nonazeotrope		575
15602	C₁₁H₂₀O	Methyl terpenyl ether	216.2	Nonazeotrope		552
15603	C₁₂H₁₈	1,3,5-Triethylbenzene	215.5	Nonazeotrope		552
A =	**C₁₀H₁₆O**	**Carvenone**	**234.5**			
15604	C₁₁H₁₄O₂	Isobutyl benzoate	241.9	Nonazeotrope		552
15605	C₁₁H₂₂O₃	Isoamyl carbonate	232.2	Nonazeotrope		552
A =	**C₁₀H₁₆O**	**Citral**	**226**			
15606	C₁₀H₁₈O	Geraniol	229	Nonazeotrope		563
15607	C₁₂H₁₈	1,3,5-Triethylbenzene	215.5	Nonazeotrope		575
A =	**C₁₀H₁₆O**	**Fenchone**	**193**			
15608	C₁₁H₂₀O	Methyl isobornyl ether	192.2	191		563
A =	**C₁₀H₁₆O**	**Pulegone**	**223.8**			
15609	C₁₀H₁₇Cl	Bornyl chloride	207.5	Nonazeotrope		552
15610	C₁₀H₁₈O	Borneol	215.0	Nonazeotrope		552
15611	C₁₀H₁₈O	α-Terpineol	218.85	Nonazeotrope		552
15612	C₁₀H₂₀O	Menthol	216.3	Nonazeotrope		552
15613	C₁₁H₁₆O	Methyl thymyl ether	216.5	Nonazeotrope		575
15614	C₁₁H₂₀O	Terpineol methyl ether	216.3	Nonazeotrope		563
15615	C₁₁H₂₂O₃	Isoamyl carbonate	232.2	Nonazeotrope		552
15616	C₁₂H₁₈	1,3,5-Triethylbenzene	215.5	Nonazeotrope		575
15617	C₁₂H₂₀O₂	Bornyl acetate	227.6	Nonazeotrope		552
A =	**C₁₀H₁₇Cl**	**Bornyl Chloride**	**207.5**			
15618	C₁₀H₂₂S	Isoamyl sulfide	214.8	Nonazeotrope		575
A =	**C₁₀H₁₈**	**Decahydronaphthalene**				
15619	C₁₀H₂₂	Decane, 10 mm.		Nonazeotrope	v-l	893
		" 20 mm.		Nonazeotrope	v-l	893
		" 50 mm.		Nonazeotrope	v-l	893
A =	**C₁₀H₁₈O**	**Borneol**	**211.8**			
15620	C₁₀H₁₈O	Menthone	207	Nonazeotrope		563
15621	C₁₀H₁₈O	α-Terpineol	218.0	Nonazeotrope		545
15622	C₁₀H₂₀O	Menthol	216.4	Nonazeotrope		545
15623	C₁₁H₁₆O	Methyl thymyl ether	216.5	< 214.0	<62	575
15624	C₁₁H₂₀O	Methyl α-terpineol ether	216.2	214.0	55	575
		"	216	Nonazeotrope		563
15625	C₁₂H₁₈	1,3,5-Triethylbenzene	215.5	212.2	62	545
15626	C₁₂H₂₂O	Ethyl isobornyl ether	204.9	Nonazeotrope		575
A =	**C₁₀H₁₈O**	**Cineole**	**176.35**			
15627	C₁₀H₁₈O	α-Terpineol	218.85	Nonazeotrope		575
15628	C₁₀H₂₀O₂	Isoamyl isovalerate	192.7	Nonazeotrope		557
15629	C₁₀H₂₂O	Isoamyl ether	173.2	Nonazeotrope		549

	B-Component			Azeotropic Data		
No.	Formula	Name	B.P., °C	B.P., °C	Wt.%A	Ref.
A =	**C₁₀H₁₈O**	**Cineole** *(continued)*	**176.35**			
15630	C₁₀H₂₃N	Diisoamylamine	188.2	Nonazeotrope		551
A =	**C₁₀H₁₈O**	**Citronellal**	**208.0**			
15631	C₁₀H₁₈O	α-Terpineol	218.85	Nonazeotrope		575
15632	C₁₀H₂₀O	Citronellol	224.4	Nonazeotrope		575
15633	C₁₀H₂₀O	Menthol	216.3	Nonazeotrope		575
15634	C₁₁H₂₀O	Isobornyl methyl ether	192.4	Nonazeotrope		575
A =	**C₁₀H₁₈O**	**Geraniol**	**229.6**			
15635	C₁₀H₁₈O	α-Terpineol	218.85	Nonazeotrope		575
15636	C₁₀H₂₂O	Decyl alcohol	232.9	Nonazeotrope		575
15637	C₁₁H₁₀	1-Methylnaphthalene	244.9	Nonazeotrope		537
15638	C₁₁H₁₆O	Methyl thymyl ether	216.5	Nonazeotrope		575
15639	C₁₁H₂₀O	Methyl α-terpineol ether	216.2	Nonazeotrope		575
15640	C₁₁H₂₂O₃	Isoamyl carbonate	232.2	< 229.2	>65	567
15641	C₁₂H₁₈	1,3,5-Triethylbenzene	215.5	Nonazeotrope		575
15642	C₁₂H₂₀O₂	Bornyl acetate	228	Nonazeotrope		528
A =	**C₁₀H₁₈O**	**Linalool**	**198.6**			
15643	C₁₀H₂₀O₂	Isoamyl isovalerate	192.7	< 192.4		575
15644	C₁₁H₂₀O	Isobornyl methyl ether	192.2	Nonazeotrope		576
15645	C₁₂H₁₈	1,3,5-Triethylbenzene	215.5	Nonazeotrope		537
A =	**C₁₀H₁₈O**	**Menthone**	**209.5**			
15646	C₁₀H₂₀O	Menthol, 5 mm.		Nonazeotrope	v-l	323
		" 50 mm.		Nonazeotrope	v-l	323
		" 180 mm.		Nonazeotrope	v-l	323
A =	**C₁₀H₁₈O**	**α-Terpineol**	**217.8**			
15647	C₁₀H₂₀O	Menthol	216.4	Nonazeotrope		529
15648	C₁₁H₁₀	1-Methylnaphthalene	244.9	Nonazeotrope		540
15649	C₁₁H₁₀	2-Methylnaphthalene	241.15	Nonazeotrope		527
15650	C₁₁H₁₆O	Methyl thymyl ether	216.5	< 215.5		575
15651	C₁₁H₂₀O	Methyl terpineol ether	216.2	Min. b.p.?		576
15652	C₁₂H₂₀O₂	Bornyl acetate	227.7	Nonazeotrope		529
A =	**C₁₀H₁₈O**	**β-Terpineol**	**210.5**			
15653	C₁₁H₂₀O	Isobornyl methyl ether	192.4	Nonazeotrope		575
15654	C₁₁H₂₀O	Methyl terpineol ether	216.2	< 210	>82	575
15655	C₁₂H₁₈	1,3,5-Triethylbenzene	215.5	210.0		575
A =	**C₁₀H₁₈O₄**	**Propyl Succinate**	**250.5**			
15656	C₁₁H₁₀	1-Methylnaphthalene	245.1	Nonazeotrope		546
15657	C₁₁H₁₀	2-Methylnaphthalene	241.15	Nonazeotrope		575
15658	C₁₁H₁₄O₂	1-Allyl-3,4-dimethoxy-benzene	254.7	Nonazeotrope		575
15659	C₁₁H₁₄O₂	Butyl benzoate	249.0	Nonazeotrope		549
15660	C₁₁H₁₄O₂	Isobutyl benzoate	241.9	Nonazeotrope		575
15661	C₁₂H₁₀	Biphenyl	256.1	Nonazeotrope		575
15662	C₁₂H₁₀O	Phenyl ether	259.0	< 250.0		557

TABLE I. *Binary Systems* 441

No.	Formula	Name	B.P., °C	B.P., °C	Wt.%A	Ref.
		B-Component		**Azeotropic Data**		
A =	**C₁₀H₂₀O**	**Citronellol**	**224.4**			
15663	C₁₁H₁₀	2-Methylnaphthalene	241.15	Nonazeotrope		575
15664	C₁₁H₂₀O	Methyl terpineol ether	216.2	Nonazeotrope		576
15665	C₁₁H₂₂O₃	Isoamyl carbonate	232.2	< 224.2		575
15666	C₁₂H₁₈	1,3,5-Triethylbenzene	215.5	< 215.3		575
15667	C₁₂H₂₀O₂	Bornyl acetate	227.6	Nonazeotrope		575
A =	**C₁₀H₂₀O**	**Menthol**	**216.3**			
15668	C₁₀H₂₀O₂	Ethyl caprylate	208.35	Nonazeotrope		575
15659	C₁₁H₁₀	1-Methylnaphthalene	244.9	Nonazeotrope		537
15670	C₁₁H₁₀	2-Methylnaphthalene	241.15	Nonazeotrope		575
15671	C₁₁H₂₀O	Terpineol methyl ether	216.2	215.3	50	545
15672	C₁₁H₂₂O₂	Ethyl pelargonate	227	Nonazeotrope		575
15673	C₁₂H₁₈	1,3,5-Triethylbenzene	215.5	214	~ 55	537
15674	C₁₂H₂₀O₂	Bornyl acetate	227.6	Nonazeotrope		535
A =	**C₁₀H₂₀O₂**	**Capric Acid**	**268.8**			
15675	C₁₁H₁₀	1-Methylnaphthalene	244.6	Nonazeotrope		575
15676	C₁₁H₁₀	2-Methylnaphthalene	241.15	Nonazeotrope		527
15677	C₁₂H₁₀O	Phenyl ether	259.0	< 258.0	>12	575
15678	C₁₂H₂₂O₄	Isoamyl oxalate	268.0	< 266.0	>35	575
15679	C₁₃H₁₂	Diphenylmethane	265.4	262.5	28	562
A =	**C₁₀H₂₀O₂**	**Ethyl Caprylate**	**208.35**			
15680	C₁₀H₂₂S	Isoamyl sulfide	214.8	Nonazeotrope		122
15681	C₁₁H₂₀O	Methyl α-terpineol ether	216.2	Nonazeotrope		557
15682	C₁₂H₁₈	1,3,5-Triethylbenzene	215.5	Nonazeotrope		575
A =	**C₁₀H₂₀O₂**	**Isoamyl Isovalerate**	**192.7**			
15683	C₁₀H₂₂O	Isoamyl ether	173.2	Nonazeotrope		557
15684	C₁₁H₂₀O	Isobornyl methyl ether	192.4	< 192	<55	557
15685	C₁₂H₁₈	1,3,5-Triethylbenzene	215.5	Nonazeotrope		575
15686	C₁₂H₂₂O	Bornyl ethyl ether	204.9	Nonazeotrope		557
15687	C₁₂H₂₂O	Ethyl isobornyl ether	203.8	Nonazeotrope		557
A =	**C₁₀H₂₀O₂**	**Methyl Pelargonate**	**213.8**			
15688	C₁₂H₁₈	1,3,5-Triethylbenzene	215.5	Nonazeotrope		575
A =	**C₁₀H₂₀O₄**	**2-(2-Butoxyethoxy) Ethyl Acetate**	**245.3**			
15689	C₁₁H₁₄O₂	Ethyl β-phenylpropionate	248.1	< 245.0	>82	575
15690	C₁₁H₁₄O₂	Isobutyl benzoate	241.9	< 241.7	>10	575
15691	C₁₁H₂₂O₃	Isoamyl carbonate	232.2	Nonazeotrope		575
15692	C₁₂H₂₀O₂	Bornyl acetate	227.6	Nonazeotrope		575
A =	**C₁₀H₂₂**	**3,3,5-Trimethylheptane**	**155.68**			
15693	C₁₂F₂₇N	Perfluorotributyl-amine	178.4	147.3	55 vol. %	609
A =	**C₁₀H₂₂O**	**Decyl Alcohol**	**~232.9**			
15694	C₁₁H₁₀	1-Methylnaphthalene	244.9	Nonazeotrope		537
15695	C₁₁H₁₀	2-Methylnaphthalene				
		" 1.5 mm.		68.85	5.2 v-l	955
		" 3.0 mm.		79.80	9.9 v-l	955

		B-Component		Azeotropic Data			
No.	Formula	Name	B.P., °C	B.P., °C	Wt.%A		Ref.
A =	**C₁₀H₂₂O**	**Decyl Alcohol** *(continued)*	**~232.9**				
15696	C₁₁H₁₄O₂	Isobutyl benzoate	241.9	Nonazeotrope			536
15697	C₁₁H₁₆O	Methyl thymyl ether	216.5	Nonazeotrope			575
15698	C₁₁H₂₀O	Methyl terpineol ether	216.0	Nonazeotrope			575
15699	C₁₁H₂₂O₃	Isoamyl carbonate	232.2	< 230.9	>36		567
15700	C₁₂H₁₀	Biphenyl	254.8	Nonazeotrope			540
15701	C₁₂H₁₈	1,3,5-Triethylbenzene	215.5	Nonazeotrope			537
15702	C₁₂H₂₀O₂	Bornyl acetate	228	Nonazeotrope			528
15703	C₁₂H₂₆O	Dodecyl alcohol, 20, 50, 100, 300 mm.		Ideal system		v-l	823
15704	C₁₃H₁₂	Diphenylmethane	265.6	Nonazeotrope			537
A =	**C₁₀H₂₂O**	**Isoamyl Ether**	**173.2**				
15705	C₁₀H₂₃N	Diisoamylamine	188.2	Nonazeotrope			551
A =	**C₁₀H₂₂OS**	**2-(2-Ethylhexylthio) ethanol**					
15706	C₁₂H₂₄OS	2-(2-Ethylhexylthio) ethyl vinyl ether		Min. b.p.			953
A =	**C₁₀H₂₂O₄**	**Tripropylene Glycol Methyl Ether**	**243**				
15707	C₁₁H₁₀	2-Methylnaphthalene	241.15		<50		270
A =	**C₁₀H₂₂S**	**Isoamyl Sulfide**	**214.8**				
15708	C₁₁H₂₀O	Methyl α-terpineol ether	216.2	213.8	70		566
15709	C₁₂H₁₈	1,3,5-Triethylbenzene	215.5	214.0	65		575
15710	C₁₂H₂₂O	Ethyl isobornyl ether	203.8	Nonazeotrope			566
A =	**C₁₁H₁₀**	**1-Methylnaphthalene**	**244.6**				
15711	C₁₁H₁₀	2-Methylnaphthalene	241.15	Nonazeotrope			561
		"				v-l	441
		" 55 mm.				v-l	441
15712	C₁₁H₁₄O₂	1-Allyl-3,4-dimethoxy-benzene	254.7	Nonazeotrope			548
15713	C₁₁H₁₄O₂	Butyl benzoate	249.5	Nonazeotrope			546
15714	C₁₁H₁₄O₂	Ethyl β-phenylpropionate	248.1	Nonazeotrope			575
15715	C₁₁H₁₄O₂	Isobutyl benzoate	242.15	Nonazeotrope			532
15716	C₁₁H₁₆O	p-tert-Amylphenol	266.5	Nonazeotrope			575
15717	C₁₁H₁₇N	Isoamylaniline	256.0	Nonazeotrope			575
15718	C₁₁H₂₂O₃	Isoamyl carbonate	232.2	Nonazeotrope			546
15719	C₁₁H₂₄O	5-Ethyl-2-nonanol, 19 mm.		121	41.4		270,271
		" 50 mm.		143	25.2		270,271
		" 150 mm.		173	5.25		270,271
		" 200 mm.		179.5	2		270.271
		5-Ethyl-2-nonanol 400 mm.		Nonazeotrope			270,271
15720	C₁₂H₁₀	Biphenyl	256.1	Nonazeotrope			561
15721	C₁₂H₁₀O	Phenyl ether	259.0	Nonazeotrope			558
15722	C₁₂H₁₆O₂	Isoamyl benzoate	262.0	Nonazeotrope			575
15723	C₁₂H₂₀O₂	Bornyl acetate	227.7	Nonazeotrope			535

TABLE I.   *Binary Systems*   443

No.	Formula	Name	B.P., °C	B.P., °C	Wt.%A	Ref.
		B-Component			Azeotropic Data	
A =	**C₁₁H₁₀**	**1-Methylnaphthalene**	**244.6**			
		(continued)				
15724	C₁₃H₁₂	Diphenylmethane	265.4	Nonazeotrope		561
A =	**C₁₁H₁₀**	**2-Methylnaphthalene**	**241.15**			
15725	C₁₁H₁₄O₂	Butyl benzoate	249.0	Nonazeotrope		527
15726	C₁₁H₁₄O₂	Ethyl β-phenylpropionate	248.1	Nonazeotrope		575
15727	C₁₁H₁₄O₂	Isobutyl benzoate	241.9	240.8	60	527
15728	C₁₁H₁₇N	Isoamylaniline	256.0	Nonazeotrope		575
15729	C₁₁H₂₂O₃	Isoamyl carbonate	232.2	Nonazeotrope		575
15730	C₁₁H₂₄O	5-Ethyl-2-nonanol,				
		20 mm.		120	49.8	270,271
	"	50 mm.		140.5	36.0	270,271
	"	90 mm.		157	24.5	270,271
	"	200 mm.		181.5	9.0	270,271
	"	300 mm.		193.5	3.5	270,271
	"	400 mm.		Nonazeotrope		270,271
15731	C₁₂H₂₀O₂	Bornyl acetate	227.6	Nonazeotrope		575
A =	**C₁₁H₁₂O₂**	**Ethyl Cinnamate**	**272.0**			
15732	C₁₁H₁₄O₂	1,2-Dimethyl-4-propenyl-				
		benzene	270.5	Nonazeotrope		557
	"	"	270.5	270.4	~ 7	**541**
15733	C₁₂H₁₀	Acenaphthene	277.9	Nonazeotrope		546
15734	C₁₂H₁₀	Biphenyl	256.1	Nonazeotrope		575
15735	C₁₂H₁₀O	Phenyl ether	259.3	Nonazeotrope		557
15736	C₁₂H₁₆O₂	Isoamyl benzoate	262.0	Nonazeotrope		545
15737	C₁₂H₁₆O₃	Isoamyl salicylate	277.5	Nonazeotrope		575
15738	C₁₂H₂₂O₄	Isoamyl oxalate	268.0	< 267.5	>21	549
15739	C₁₃H₁₂	Diphenylmethane	265.6	Nonazeotrope		546
15740	C₁₄H₁₄	1,2-Diphenylethane	284	Nonazeotrope		546
A =	**C₁₁H₁₄O₂**	**1-Allyl-3,4-**				
		dimethoxybenzene	**~249.8**			
15741	C₁₁H₁₄O₂	Butyl benzoate	254.7	Nonazeotrope		557
15742	C₁₁H₁₄O₂	Isobutyl benzoate	242.15	Nonazeotrope		557
15743	C₁₁H₁₇N	Isoamylaniline	256.0	250.5	58	551
15744	C₁₂H₁₀	Biphenyl	255.0	254.5	70	**558**
15745	C₁₂H₁₀O	Phenyl ether	259.0	Nonazeotrope		549
15746	C₁₂H₁₆O₂	Isoamyl benzoate	262.05	Nonazeotrope		557
15747	C₁₃H₁₂	Diphenylmethane	265.6	Nonazeotrope		535
A =	**C₁₁H₁₄O₂**	**Butyl Benzoate**	**249.8**			
15748	C₁₂H₁₀	Biphenyl	255.9	Nonazeotrope		546
15749	C₁₂H₁₀O	Phenyl ether	259.3	Nonazeotrope		537,557
15750	C₁₅H₃₃BO₃	Isoamyl borate	255	Nonazeotrope		575
A =	**C₁₁H₁₄O₂**	**1,2-Dimethoxy-4-propenyl-**				
		benzene	**270.5**			
15751	C₁₁H₁₇N	Isoamylaniline	256.0	Nonazeotrope		575
15752	C₁₂H₁₀	Acenaphthene	277.9	Nonazeotrope		548
15753	C₁₂H₁₀O	Phenyl ether	259.3	Nonazeotrope		535
15754	C₁₂H₁₆O₂	Isoamyl benzoate	262.05	Nonazeotrope		535,557

		B-Component		Azeotropic Data		
No.	Formula	Name	B.P., °C	B.P., °C	Wt.%A	Ref.
A =	$C_{11}H_{14}O_2$	**1,2-Dimethoxy-4-propenyl-benzene** *(continued)*	**270.5**			
15755	$C_{12}H_{16}O_3$	Isoamyl salicylate	277.5	Nonazeotrope		575
15756	$C_{12}H_{22}O_4$	Isoamyl oxalate	268.0	Nonazeotrope		557
	" "		268.0	267.95	4	541
15757	$C_{13}H_{12}$	Diphenylmethane	265.6	Nonazeotrope		535
A =	$C_{11}H_{14}O_2$	**Ethyl β-phenylpropionate**	**248.1**			
15758	$C_{11}H_{14}O_2$	Isobutyl benzoate	241.9	Nonazeotrope		575
15759	$C_{12}H_{10}$	Biphenyl	256.1	Nonazeotrope		575
15760	$C_{12}H_{10}O$	Phenyl ether	259.0	Nonazeotrope		557
A =	$C_{11}H_{16}O$	**p-tert-Amylphenol**	**266.5**			
15761	$C_{12}H_{10}$	Acenaphthene	277.9	Nonazeotrope		575
15762	$C_{12}H_{16}O_3$	Isoamyl salicylate	277.5	Nonazeotrope		575
15763	$C_{13}H_{10}$	Fluorene	295.	Nonazeotrope		575
15764	$C_{13}H_{12}$	Diphenylmethane	265.4	263.0	40	575
15765	$C_{14}H_{14}$	1,2-Diphenylethane	284.5	Nonazeotrope		575
A =	$C_{11}H_{16}O$	**Methyl Thymyl Ether**	**216.5**			
15766	$C_{12}H_{20}O_2$	Bornyl acetate	227.6	Nonazeotrope		557
A =	$C_{11}H_{17}N$	**Isoamylaniline**	**256.0**			
15767	$C_{12}H_{10}$	Biphenyl	256.1	< 255.0		575
15768	$C_{12}H_{10}O$	Phenyl ether	259.0	< 252.5		575
A =	$C_{11}H_{20}O$	**Methyl α-Terpineol Ether**	**216.2**			
15769	$C_{12}H_{18}$	1,3,5-Triethylbenzene	215.5	Nonazeotrope		558
15770	$C_{12}H_{20}O_2$	Bornyl acetate	227.6	Nonazeotrope		557
A =	$C_{11}H_{22}O_2$	**Methyl Caprate**				
15771	$C_{13}H_{26}O_2$	Methyl laurate				
		20-100 mm.		Nonazeotrope	v-l	825
A =	$C_{11}H_{22}O_3$	**Isoamyl Carbonate**	**232.2**			
15772	$C_{12}H_{16}O_2$	Isoamyl benzoate	241.9	Nonazeotrope		575
15773	$C_{12}H_{20}O_2$	Bornyl acetate	227.6	Nonazeotrope		542
A =	$C_{12}H_9N$	**Carbazole**	**355**			
15774	$C_{14}H_{30}$	Tetradecanol		Nonazeotrope		272
		Low press.		Min. b.p.		272
15775	$C_{17}H_{36}O$	Heptadecanol		Nonazeotrope		272
		Low press.		Min. b.p.		272
A =	$C_{12}H_{10}$	**Acenaphthene**	**277.9**			
15776	$C_{12}H_{14}O_4$	Ethyl phthalate	298.5	Nonazeotrope		575
15777	$C_{12}H_{16}O_2$	Isoamyl benzoate	262.0	Nonazeotrope		546
15778	$C_{12}H_{22}O_4$	Isoamyl oxalate	268.0	Nonazeotrope		542
15779	$C_{13}H_{12}$	Diphenylmethane	265.4	Nonazeotrope		561
15780	$C_{13}H_{12}O$	Benzyl phenyl ether	286.5	Nonazeotrope		558
15781	$C_{14}H_{14}$	1,2-Diphenylethane	284.5	Nonazeotrope		561
A =	$C_{12}H_{10}$	**Biphenyl**	**255.9**			
15782	$C_{12}H_{10}O$	Phenyl ether	259.3	Nonazeotrope		542
	" "		259	Nonazeotrope	v-l	470
15783	$C_{12}H_{14}O_4$	Ethyl phthalate	298.5	Nonazeotrope		575

TABLE I.   *Binary Systems*   445

	B-Component			Azeotropic Data		
No.	Formula	Name	B.P., °C	B.P., °C	Wt.%A	Ref.
A =	**C₁₂H₁₀**	**Biphenyl** *(continued)*	**255.9**			
15784	C₁₂H₁₆O₂	Isoamyl benzoate	262.0	Nonazeotrope		546
15785	C₁₂H₁₆O₃	Isoamyl salicylate	277.5	Nonazeotrope		575
15786	C₁₂H₂₂O₄	Isoamyl oxalate	268.0	Nonazeotrope		546
15787	C₁₃H₁₂	Diphenylmethane	265.4	Nonazeotrope		561
15788	C₂₄H₃₈O₄	Dioctyl phthalate,				
		10 mm.	248	Nonazeotrope		981
A =	**C₁₂H₁₀O**	**Phenyl Ether**	**259**			
15789	C₁₂H₁₄O₄	Ethyl phthalate	298.5	Nonazeotrope		557
15790	C₁₂H₁₆O₂	Isoamyl benzoate	262.05	258.9	90	557
15791	C₁₂H₁₆O₃	Isoamyl salicylate	277.5	Nonazeotrope		575
15792	C₁₂H₂₂O₄	Isoamyl oxalate	268.0	Nonazeotrope		541
15793	C₁₃H₁₂	Diphenylmethane	265.6	Nonazeotrope		529
15794	C₁₄H₁₄O	Benzyl ether	297	Nonazeotrope		576
A =	**C₁₂H₁₄O₂**	**Ethyl Phthalate**	**298.5**			
15795	C₁₃H₁₂	Diphenylmethane	265.4	Nonazeotrope		575
A =	**C₁₂H₁₆O₂**	**Isoamyl Benzoate**	**262.0**			
15796	C₁₂H₁₆O₃	Isoamyl salicylate	277.5	Nonazeotrope		575
15797	C₁₂H₂₂O₄	Isoamyl oxalate	268.0	Nonazeotrope		541
15798	C₁₃H₁₂	Diphenylmethane	265.6	Nonazeotrope		535
A =	**C₁₂H₁₆O₃**	**Isoamyl Salicylate**	**277.5**			
15799	C₁₃H₁₂	Diphenylmethane	265.4	Nonazeotrope		575
15800	C₁₃H₁₂O	Benzyl phenyl ether	286.5	Nonazeotrope		575
15801	C₁₄H₁₄	1,2-Diphenylethane	284.5	Nonazeotrope		575
A =	**C₁₂H₁₈**	**1,3,5-Triethylbenzene**	**215.5**			
15802	C₁₂H₂₀O₂	Bornyl acetate	227.2	Nonazeotrope		537
15803	C₁₂H₂₂O	Bornyl ethyl ether	204.9	Nonazeotrope		558
A =	**C₁₂H₂₂O₄**	**Isoamyl Oxalate**	**268.0**			
15804	C₁₃H₁₂	Diphenylmethane	265.4	265.25	14	545
15805	C₁₄H₁₄	1,2-Diphenylethane	284	Nonazeotrope		546
A =	**C₁₂H₂₄**	**2,6,8-Trimethylnonene**				
15806	C₁₂H₂₆O	2,6,8-Trimethyl-4-nonanol,				
		50 mm.	137	Nonazeotrope		981
		" 10 mm.	103	Nonazeotrope		981
A =	**C₁₂H₂₆**	**Dodecane**	**216**			
15807	C₁₆H₃₄	Hexadecane,				
		10-760 mm.		Nonazeotrope	v-l	447
A =	**C₁₃H₁₀**	**Fluorene**	**294**			
15808	C₁₄H₃₀O	Tetradecanol,				
		Low press.		Nonazeotrope		272
		High press.		Min. b.p.		272
15809	C₁₇H₃₆O	Heptadecanol,				
		Low press.		Nonazeotrope		272
		High press.		Min. b.p.		272
A =	**C₁₃H₁₀O₂**	**Phenyl Benzoate**	**315**			
15810	C₁₄H₁₂	Stilbene	306.5	Nonazeotrope		575

| | B-Component | | | Azeotropic Data | | |
No.	Formula	Name	B.P., °C	B.P., °C Wt.%A		Ref.
A =	**C₁₃H₁₀O₂**	**Phenyl Benzoate** *(continued)*				
15811	C₁₄H₁₄O	Benzyl ether	297	Nonazeotrope		557
A =	**C₁₃H₁₂O**	**Benzyl Phenyl Ether**	**286.5**			
15812	C₁₄H₁₄	1,2-Diphenylethane	284.5	Nonazeotrope		558
A =	**C₁₃H₂₆O₂**	**Methyl Laurate**				
15813	C₁₅H₃₀O₂	Methyl myristate				
			20-100 mm.	Nonazeotrope	v-l	825
A =	**C₁₄H₁₀**	**Phenanthrene**	**340**			
15814	C₁₄H₁₄	Bibenzyl	284	Nonazeotrope	v-l	645
15815	C₁₄H₃₀O	Tetradecanol		% Phenanthrene		
				increases with pressure;		
				min. b.p.		272
15816	C₁₆H₃₄	Hexadecane 100 mm.		Nonazeotrope	v-l	645
15817	C₁₇H₃₆O	Heptadecanol		% Phenanthrene		
				increases with pressure;		
				min. b.p.		272
A =	**C₁₄H₁₄**	**Bibenzyl**	**284**			
15818	C₁₆H₃₄	Hexadecane 100 mm.	209.5	200 76.3	v-l	645
A =	**C₁₅H₃₀O₂**	**Methyl Myristate**				
15819	C₁₇H₃₄O₂	Methyl palmitate				
			20-100 mm	Nonazeotrope	v-l	825
A =	**C₁₆H₃₂O₂**	**Palmitic Acid**				
15820	C₁₈H₃₆O₂	Stearic acid, 5 mm.			v-l	391
A =	**C₁₇H₃₄O₂**	**Methyl Palmitate**				
15821	C₁₉H₃₈O₂	Methyl stearate				
			20-100 mm.	Nonazeotrope	v-l	825
A =	**C₁₈H₃₄O₂**	**Oleic Acid**				
15822	C₁₈H₃₄O₃	Ricinoleic acid, 5 mm.			v-l	391
15823	C₁₈H₃₆O₂	Stearic acid, 5 mm.			v-l	391
15824	C₂₀H₂₀O₂	Abietic acid, 1-10 mm.		Nonazeotrope	v-l	454
A =	**C₁₉H₃₆O₂**	**Methyl Oleate**				
15825	C₁₉H₃₈O₂	Methyl stearate		Nonazeotrope	v-l	824

Table II.

No.	A-Component Formula	Name	B.P., °C.	B-Component Formula	Name	B.P., °C.
15826	A	Argon	—186	N_2	Nitrogen	—195
15827	BCl_3	Boron chloride	11.5	B_2H_6	Boron hydride	— 92.5
15827a	BeF_2	Beryllium fluoride		FLi	Lithium fluoride	1670
15828	BrH	Hydrobromic acid	— 67	H_2O	Water	100
		"			"	
15828a	CHN	Hydrocyanic acid	26	C_2H_3N	Acetonitrile	81.6
15828b	CHN	Hydrocyanic acid	26	C_3H_3N	Acrylonitrile	77.3
15828c	CO	Carbon monoxide	—192	CO_2	Carbon dioxide	— 79.1
15829	CO	Carbon monoxide	—192	H_2	Hydrogen	—252.7
15830	ClF_3	Chlorine trifluoride		FH	Hydrogen fluoride	19.4
15831	ClH	Hydrochloric acid	— 80	H_2O	Water	100
		"			"	
15831a	ClH	Hydrochloric acid	— 80	H_2O	Water	100
	"	"		"	"	
15832	ClH	Hydrochloric acid	— 80	H_2O	Water	100
15833	ClH	Hydrochloric acid	— 80	C_5H_5N	Pyridine	115.5
15834	$POCl_3$	Phosphorus oxychloride	107.2	$VOCl_3$	Vanadium oxychloride	127.2
15834a	Cl_4Si	Silicon tetrachloride	56.7	C_3H_9ClSi	Chlorotrimethylsilane	57.5
15834b	Cl_4Si	Silicon tetrachloride	56.7	C_3H_9ClSi	Chlorotrimethylsilane	57.5
15835	FH	Hydrofluoric acid	19.4	$FHSO_3$	Fluorosulfuric acid	
15836	FH	Hydrofluoric acid	19.4	F_6H_2Si	Fluosilicic acid	
15837	FH	Hydrofluoric acid	19.4	H_2O	Water	100
15838	FH	Hydrofluoric acid	19.5	H_2O	Water	100
15839	FH	Hydrofluoric acid	19.4	SO_2	Sulfur dioxide	— 10
		"			" "	
15839a	FH	Hydrofluoric acid	19	$C_2Cl_3F_3$	1,1,2-Trichlorotrifluoroethane	47
15840	F_4Si	Silicon tetrafluoride		C_2F_6	Hexafluoroethane	— 78
15841	HNO_3	Nitric acid	86	H_2O	Water	100
15842	HNO_3	Nitric acid		H_2O	Water	100
15842a	H_2O	Water	100	H_2S	Hydrogen sulfide	—63.5
15843	H_2O	Water	100	H_4N_2	Hydrazine	113.5
15844	H_2O	Water	100	SO_2	Sulfur dioxide	— 10

Ternary Systems

	C-Component			Azeotropic Data				
Formula	Name	B.P., °C.		B.P., °C.	Wt. % A	Wt. % B	Wt. % C	Ref.
O_2	Oxygen	—183	90°–120°K.		Nonazeotrope		v-1	685c
ClH	Hydrogen chloride	— 80			Nonazeotrope			638
F_4Th	Thorium tetrafluoride				Nonazeotrope		v-1	896c
C_6H_5Cl	Chlorobenzene	131.8		105	10.4	11.0	78.6	215
"			100 mm	56.4	12.2	12.3	75.5	215
C_3H_4O	Acrolein	52.45			Nonazeotrope		v-1	905f
C_3H_4O	Acrolein	52.45	400 mm		Nonazeotrope		v-1	905f
H_2	Hydrogen	—252.7	200 atm		Nonazeotrope		v-1	966c
N_2	Nitrogen	—195.8			20 atmos. to crit. pt.		v-1	6
F_6U	Uranium hexafluoride	56			Nonazeotrope		v-1	238,828
C_6H_5Cl	Chlorobenzene	131.8		96.9	5.3	20.2	74.5	762
"			100 mm	49.5	4.8	15.9	79.3	215
C_6H_5ClO	o-Chlorophenol		120 mm	62.6	14	47	39	215
"	"	175	750 mm	105.0	13	48	39	215
C_6H_6O	Phenol	182		107.33	15.8	64.8	19.4	762
$C_{10}H_8$	Naphthalene	218.1		189.6				942
$TiCl_4$	Titanium tetrachloride	136.4			Nonazeotrope		v-1	702
$C_4H_7ClO_2$	Ethyl chloroacetate	143.5			Nonazeotrope		v-1	886e
$C_4H_8Cl_2O$	Bis(2-chloroethyl) ether	178	60°C.		Nonazeotrope		v-1	886c
H_2O	Water	100					v-1	433
H_2O	Water	100		116.1	10	36	54	670
C_2H_6O	Ethyl alcohol	78.3		103	30	10	60	135
$C_4HF_7O_2$	Perfluorobutyric acid			108	12	28	60	659
CCl_2F_2	Dichlorodifluoromethane			— 36				48
"	"		44 p.s.i.g	4	3.5	12	84	48
C_3F_6O	Hexafluoroacetone			13	8.6	20.1	71.3	101c
C_2H_6	Ethane	— 88		—104	24.6	32.7	42.7	120
SO_3	Sulfur trioxide	47			Vapor pressure data			563
$CHCl_3$	Chloroform	61			92	3	5	738
H_3N	Ammonia	—33	600 mm		Nonazeotrope		v-1	326e
$C_2H_8N_2$	1,1-Dimethylhydrazine				Nonazeotrope		v-1	656,736
C_2H_4O	Acetaldehyde	20.2					v-1	777

	A-Component		B.P., °C.	B-Component		B.P., °C.
No.	Formula			Formula	Name	
15845	H_2O	Water	100	CCl_4	Carbon tetra-chloride	76.75
15846	H_2O	Water	100	CCl_4	Carbon tetra-chloride	76.75
15847	H_2O	Water	100	CCl_4	′ Carbon tetra-chloride	76.75
		"			"	
15848	H_2O	Water	100	CCl_4	Carbon tetra-chloride	76.75
15849	H_2O	Water	100	CCl_4	Carbon tetrachloride	76.75
		"			"	
15850	H_2O	Water	100	CCl_4	Carbon tetrachloride	76.75
15851	H_2O	Water	100	CCl_4	Carbon tetrachloride	76.75
15851a	H_2O	Water	100	CCl_4	Carbon tetrachloride	76.7
15852	H_2O	Water	100	CCl_4	Carbon tetrachloride	76.75
15853	H_2O	Water	100	CS_2	Carbon disulfide	46.25
15854	H_2O	Water	100	CS_2	Carbon disulfide	46.25
15855	H_2O	Water	100	CS_2	Carbon disulfide	46.25
15856	H_2O	Water	100	CS_2	Carbon disulfide	46.25
15857	H_2O	Water	100	CS_2	Carbon disulfide	46.25
15858	H_2O	Water	100	$CHBrCl_2$	Bromodichloro-methane	90.2
15859	H_2O	Water	100	$CHBrCl_2$	Bromodichloro-methane	90.2
15860	H_2O	Water	100	$CHBrCl_2$	Bromodichloro-methane	90.2
15861	H_2O	Water	100	$CHBrCl_2$	Bromodichloro-methane	90.2
15862	H_2O	Water	100	$CHCl_3$	Chloroform	61
15863	H_2O	Water	100	$CHCl_3$	Chloroform	61
15864	H_2O	Water	100	$CHCl_3$	Chloroform	61
		"			"	
15865	H_2O	Water	100	$CHCl_3$	Chloroform	61
15866	H_2O	Water	100	$CHCl_3$	Chloroform	61
		"			"	
		"			"	
15867	H_2O	Water	100	$CHCl_3$	Chloroform	61
15868	H_2O	Water	100	$CHCl_3$	Chloroform	61
15868a	H_2O	Water	100	$CHCl_3$	Chloroform	61
15869	H_2O	Water	100	CH_2Cl_2	Dichloromethane	41.5
15870	H_2O	Water	100	CH_2Cl_2	Dichloromethane	40
15871	H_2O	Water	100	CH_2O_2	Formic acid	100.8
15872	H_2O	Water	100	CH_2O_2	Formic acid	100.8
		"	"		"	"
15872a	H_2O	Water	100	CH_2O_2	Formic acid	100.8

TABLE II. *Ternary Systems* 451

C-Component			Azeotropic Data				
Formula	Name	B.P., °C.	B.P., °C.	Wt. % A	Wt. % B	Wt. % C	Ref.
C_2H_3N	Acetonitrile	81.6	60				763
C_2H_5ClO	2-Chloroethanol	128.8	min. b.p.				479
C_2H_6O	Ethyl alcohol	78.3	62	4.5	85.5	10	563
	" "		61.8	3.4	86.3	10.3	382
C_3H_6O	Acetone	57	Nonazeotrope				29
C_3H_6O	Allyl alcohol	96.95	65.15	5	84	11	563
	" "		65.4	4.13	90.43	5.44	358
C_3H_8O	Propyl alcohol	97.2	65.4	5	84	11	563
C_4H_8O	2-Butanone	79.6	65.7	3	74.8	22.2	29
$C_4H_{10}O$	*sec*-Butyl alcohol	99	65	4.05	91.0	4.95	623c
$C_4H_{10}O$	*tert*-Butyl alcohol	82.5	64.7	3.1	85.0	11.9	29
CH_4O	Methanol	64.7	Nonazeotrope				29
C_2H_3N	Acetonitrile	81.6	39				763
C_2H_6O	Ethyl alcohol	78.3	41.3	1.6	93.4	5.0	324
C_3H_6O	Acetone	56.4	38.042	0.81	75.21	23.98	947
$C_4H_8O_2$	Dioxane	101.4	Nonazeotrope				201
C_2H_6O	Ethyl alcohol	78.3	72.0	7.5	>70	<22.5	563
C_3H_6O	Allyl alcohol	96.95	76				563
C_3H_8O	Isopropyl alcohol	82.45	~ 74.5				563
$C_4H_{10}O$	Isobutyl alcohol	108	77.5				563
CH_2O_2	Formic acid	100.75	Nonazeotrope		v-l		171
CH_4O	Methanol	64.7	52.3	1.3	90.5	8.2	498i
C_2H_3N	Acetonitrile	81.6	Nonazeotrope				981
	"		Min. b.p.				763
$C_2H_4O_2$	Acetic acid	118.1	Nonazeotrope		v-l		171
C_2H_6O	Ethyl alcohol	78.3	78.0	3.9	91.2	4.9	982
	" "		55.3	2.3	94.2	3.5	497
	" "		55.4	3.5	92.5	4	1000
C_3H_6O	Acetone	56.4	60.4?	40	57.6	38.4	803
C_4H_8O	2-Butanone	79.6	Nonazeotrope				981
$C_5H_{10}O$	2-Methyl-3-buten-2-ol	97	Nonazeotrope				581c
C_2H_6O	Ethyl alcohol	78.3	Nonazeotrope				36
C_4H_8O	2-Butanone	79.6	Nonazeotrope		v-l		620
$C_2H_4Cl_2$	1,2-Dichloroethane	83.7	Nonazeotrope		v-l		112
$C_2H_4O_2$	Acetic acid	118.1	50 mm 40.4	26.2	39.1	65.3 v-l	507e
	" "	118.1	200 mm 69.8	23.4	45.7	30.9 v-l	507e
$C_3H_6O_2$	Propionic acid	140.7	107.2	18.6	71.9	9.5 v-l	507g

	A-Component		B.P., °C.	B-Component		B.P., °C.
No. Formula				Formula	Name	
15872b	H_2O	Water	100	CH_2O_2	Formic acid	100.8
15872c	H_2O	Water	100	CH_2O_2	Formic acid	100.8
15873	H_2O	Water	100	CH_2O_2	Formic acid	100.75
15873a	H_2O	Water	100	CH_2O_2	Formic acid	100.8
15873b	H_2O	Water	100	CH_2O_2	Formic acid	100.8
15874	H_2O	Water " "	100	CH_2O_2	Formic acid " " " "	100.8
15875	H_2O	Water	100		Formic acid	100.75
15876	H_2O	Water	100	CH_3NO_2	Nitromethane	101
15877	H_2O	Water	100	CH_3NO_2	Nitromethane	101.5
15878	H_2O	Water "	100	CH_3NO_2	Nitromethane " "	101
15879	H_2O	Water	100	CH_3NO_2	Nitromethane	101
15880	H_2O	Water	100	CH_3NO_2	Nitromethane	101.2
15881	H_2O	Water	100	CH_3NO_2	Nitromethane	101.2
15882	H_2O	Water	100	CH_3NO_2	Nitromethane	101.2
15883	H_2O	Water	100	CH_3NO_2	Nitromethane	101.2
15884	H_2O	Water	100	CH_3NO_2	Nitromethane	101.2
15885	H_2O	Water	100	CH_3NO_2	Nitromethane "	101.2
15886	H_2O	Water	100	CH_3NO_2	Nitromethane	101.2
15887	H_2O	Water "	100	CH_3NO_2	Nitromethane "	101.5
15888	H_2O	Water	100	CH_3NO_2	Nitromethane	101.2
15889	H_2O	Water "	100	CH_3NO_2	Nitromethane "	101.5
15890	H_2O	Water	100	CH_3NO_2	Nitromethane	101.2
15891	H_2O	Water "	100	CH_3NO_2	Nitromethane "	101.2
15892	H_2O	Water	100	CH_3NO_2	Nitromethane	101.5
15893	H_2O	Water	100	CH_3NO_2	Nitromethane	101.5
15894	H_2O	Water	100	CH_3NO_2	Nitromethane	101.5
15895	H_2O	Water	100	CH_4O	Methanol	64.3
15896	H_2O	Water	100	CH_4O	Methanol	64.7
15897	H_2O	Water	100	CH_4O	Methanol	64.7
15898	H_2O	Water	100	CH_4O	Methanol	64.7
15899	H_2O	Water "	100	CH_4O	Methanol "	64.7
15900	H_2O	Water	100	CH_4O	Methanol	64.7
15901	H_2O	Water	100	CH_4O	Methanol	64.7
15902	H_2O	Water	100	CH_4O	Methanol	64.7
15902a	H_2O	Water	100	CH_4O	Methanol	64.3

TABLE II. *Ternary Systems* 453

Formula	Name	B.P., °C.		B.P., °C.	Wt. % A	Wt. % B	Wt. % C	Ref.
		C-Component		Azeotropic Data				
C₄H₈O₂	Butyric acid	162.4		107.62	19.5	75.9	4.6	507i
C₄H₈O₂	Isobutyric acid	154		107.02	15.5	66.8	17.7	507i
C₅H₅N	Pyridine	115.5		Nonazeotrope			v-l	1058
C₅H₁₀O₂	Isovaleric acid	176.5		107.64	21.3	76.3	2.4	507i
C₅H₁₀O₂	Valeric acid	186		Nonazeotrope				507i
C₆H₆	Benzene	80.1	499 mm	60	3.43	6.76	89.83	979
	"		278.1 mm	45	3.16	6.34	90.49	979
	"		144.4 mm	30	4.07	6.27	89.65	979
C₈H₁₀	m-Xylene	139		97.5?	10.6	40.4	49.0	803
C₂Cl₄	Tetrachloroethylene	120.8		77.84				611
C₂H₅ClO	2-Chloroethanol	128.8		Nonazeotrope				479
C₃H₈O	Isopropyl alcohol	82.3			11.7	36.8	51.5	480
	" "			78	6	32	62	858
	" "			Liquid-vapor equilibrium				858
C₃H₈O	Propyl alcohol	97.2		82.3	17.5	55.9	26.6 v-l	285
C₅H₁₀O	3-Pentanone	102.2		82.4	18?	17?	65?	563
C₅H₁₂	Pentane	36.07		33.1	2.1	6.5	91.4	806
C₆H₁₂	1-Hexene	63.84		55.5	5.8	12.8	82.0	806
C₆H₁₄	Hexane	68.74		56.9	6.7	16.0	77.3	806
C₇H₁₄	1-Heptene	93.64		69.7	13.4	26.1	60.5	806
C₇H₁₆	Heptane	98.43		70.6	10.6	34.2	55.2	806
	"		748 mm	71.43	7.88	29.73	62.39	613
C₈H₁₆	1-Octene	121.28		77.4	14.2	41.8	44.0	806
C₈H₁₈	Octane	125.7	748 mm	77.35	12.40	44.25	43.35	613
	"			77.7	15.1	43.3	41.6	806
C₉H₁₈	1-Nonene	146.87		81.1	21.9	54.3	23.8	806
C₉H₂₀	Nonane	150.8	748 mm	80.72	17.4	58.3	24.3	613
	"			80.7	19.4	55.8	24.2	806
C₁₀H₂₀	1-Decene	170.57		82.5	29.2	57.8	13.0	806
C₁₀H₂₂	Decane	174.12		82.5	22.7	64.2	13.1	806
	"		748 mm	82.35	19.1	68.1	12.8	613
C₁₁H₂₄	Undecane	194.5	748 mm	82.82	20.6	73.3	6.1	613
C₁₂H₂₆	Dodecane	214.5	748 mm	83.13	21.5	75.3	3.2	613
C₁₃H₂₈	Tridecane	234	748 mm	83.21	22.8	75.4	1.8	613
C₂Cl₃F₃	1,1,2-Trichlorotrifluoroethane	47.5		39.4	0.6	3.0	96.4	982
C₂H₆O	Ethyl alcohol	78.3		Nonazeotrope			v-l	198
C₃H₅ClO₂	Methyl chloroacetate	131.4		67.85	5.26	81.20	13.54	121
C₃H₆O	Propionaldehyde	47.9		Nonazeotrope				981
C₃H₆O₂	Methyl acetate	57		Nonazeotrope				359
	" "	57.1		Nonazeotrope			v-l	179,36i
C₃H₈O	Isopropyl alcohol	82.3		Nonazeotrope				981
C₃H₈O₂	Methylal	42.3		Nonazeotrope				324,581c
C₄H₆O₂	Methyl acrylate	80.9		Nonazeotrope				981
C₄H₆O₂	Vinyl acetate			Nonazeotrope			v-l	830c

	A-Component		B.P., °C.		B-Component	B.P., °C.
No.	Formula			Formula	Name	
15903	H_2O	Water	100	CH_4O	Methanol	64.7
15904	H_2O	Water	100	CH_4O	Methanol	64.7
15905	H_2O	Water	100	CH_4O	Methanol	64.7
15905a	H_2O	Water	100	CH_4O	Methanol	64.3
15906	H_2O	Water	100	CH_4O	Methanol	64.7
15907	H_2O	Water	100	CH_4O	Methanol	64.7
		"				
15908	H_2O	Water	100	CH_4O	Methanol	64.7
15909	H_2O	Water	100	CH_4O	Methanol	64.7
15909a	H_2O	Water	100	CH_4O	Methanol	64.3
15909b	H_2O	Water	100	CH_4O	Methanol	64.3
15910	H_2O	Water	100	CH_4O	Methanol	64.7
15911	H_2O	Water	100	CH_4O	Methanol	64.7
15912	H_2O	Water	100	CH_4O	Methanol	64.7
15913	H_2O	Water	100	CH_4O	Methanol	64.7
15914	H_2O	Water	100	CH_4O	Methanol	64.7
15915	H_2O	Water	100	CH_4O	Methanol	64.7
15916	H_2O	Water	100	CH_4O	Methanol	64.7
15917	H_2O	Water	100	$C_2Cl_3F_3$	1,1,2-Trichlorotri-fluoroethane	47.5
15918	H_2O	Water	100	C_2Cl_4	Tetrachloro-ethylene	120.8
15919	H_2O	Water	100	C_2Cl_4	Tetrachloro-ethylene	120.8
		"			"	
15920	H_2O	Water	100	C_2HCl_3	Trichloro-ethylene	86.95
		"			"	
15921	H_2O	Water	100	C_2HCl_3	Trichloroethylene	86.95
		"			"	
15922	H_2O	Water	100	C_2HCl_3	Trichloroethylene	86.95
		"			"	
		"			"	
		"			"	
		"			"	
15923	H_2O	Water	100	C_2HCl_3	Trichloroethylene	86.95
		"			"	
15924	H_2O	Water	100	C_2HCl_3	Trichloroethylene	86.95
		"			"	
15925	H_2O	Water	100	C_2HCl_3	Trichloroethylene	86.2
		"			"	
15926	H_2O	Water	100	C_2HCl_3	Trichloroethylene	86.2

TABLE II. *Ternary Systems* 455

C-Component		B.P., °C.	B.P., °C.	Wt. % A	Wt. % B	Wt. % C	Ref.
Formula	Name						
$C_4H_8O_2$	Ethyl acetate	77.1		Nonazeotrope		v-l	7
$C_4H_8O_2$	Methyl propionate	79.85		Nonazeotrope			227
C_4H_{10}	2-Methylpropane	—11.7		Nonazeotrope			981
$C_4H_{10}O$	*tert*-Butyl alcohol	82.5		Nonazeotrope			581c
$C_4H_{10}O$	Isobutyl alcohol	108		Nonazeotrope			422
$C_4H_{10}O_2$	Acetaldehyde dimethylacetal	64.3		Nonazeotrope			45
$C_4H_{10}O_2$	Ethoxymethoxy-methane	65.90		Nonazeotrope			1035
C_5H_6O	2-Methylfuran	63.7	51.2				769
C_5H_8	Isoprene	34	30.2	0.6	5.4	94.0	581c
$C_5H_{12}O$	*tert*-Butyl methyl ether	55		Nonazeotrope			581c
C_6H_6	Benzene	80.2		Nonazeotrope			1044
C_6H_8	1,3-Cyclohexadiene	80.8		Nonazeotrope			563
C_6H_{10}	Biallyl	60.2		Nonazeotrope			563
C_6H_{10}	Cyclohexene	82.75		Nonazeotrope			563
C_6H_{12}	Cyclohexane	80.75		Nonazeotrope			563
C_6H_{14}	Hexane	68.95		Nonazeotrope			563
C_7H_8	Toluene	110.7		Nonazeotrope			563
C_2H_6O	Ethyl alcohol	78.3	42.6	0.6	3.9	95.5	982
C_2H_3N	Acetonitrile	81.6	72				763
C_3H_8O	Propyl alcohol	97.2	88				563
"	"		81.18	12.45	66.75	20.80	611
C_2H_3N	Acetonitrile	81.6	67	6.4	73.1	20.5	763
	"		Liquid-vapor equilibrium				763
C_2H_5ClO	2-Chloroethanol	128		Nonazeotrope			672
	"		70.8-71.5				253
C_2H_6O	Ethyl alcohol,	118 mm	25.1	3.4	85.1	11.5	584
	"	509 mm	52.5	5.2	79.6	15.2	584,803
	"	760 mm	67	5.5	78.4	16.1	165,584
	"	2060 mm	96	7.1	72.3	20.6	584
	"	5660 mm	131	8.3	70.5	21.2	584
C_3H_6O	Allyl alcohol	96.95	71.6	6.55	84.7	8.75	358,563
	" "		71.4	7.5	80	12.5	981
C_3H_8O	Isopropyl alcohol	82.45	~ 70				563
	" "		69.4	7	73	20	982
C_3H_8O	Propyl alcohol	97.3		7.1	84.8	8.1	480
	" "		71.55	7	81	12	563
$C_4H_{10}O$	Isobutyl alcohol	108	72.7				563

	A-Component		B.P., °C.		B-Component		B.P., °C.
No.	Formula			Formula	Name		
15927	H_2O	Water	100	$C_2H_2Cl_2$	cis-1,2-Dichloro-ethylene		60.25
15928	H_2O	Water	100	$C_2H_2Cl_2$	trans-1,2-Dichloro-ethylene		48.35
15929	H_2O	Water	100	$C_2H_2Cl_4$	1,1,2,2-Tetra-chloroethane		146.35
15930	H_2O	Water	100	C_2H_3N	Acetonitrile		81.6
15931	H_2O	Water	100	C_2H_3N	Acetonitrile		81.6
15932	H_2O	Water	100	C_2H_3N	Acetonitrile		81.6
15933	H_2O	Water	100	C_2H_3N	Acetonitrile		81.6
15934	H_2O	Water	100	C_2H_3N	Acetonitrile		81.6
15935	H_2O	Water	100	C_2H_3N	Acetonitrile		81.5
15936	H_2O	Water	100	C_2H_3N	Acetonitrile		81.5
15937	H_2O	Water	100	C_2H_3N	Acetonitrile		81.6
15938	H_2O	Water	100	C_2H_3N	Acetonitrile		81.5
15939	H_2O	Water	100	C_2H_3N	Acetonitrile		81.6
15940	H_2O	Water	100	C_2H_3N	Acetonitrile		81.6
15941	H_2O	Water	100	C_2H_3N	Acetonitrile		81.6
15942	H_2O	Water	100	C_2H_3N	Acetonitrile		81.6
15943	H_2O	Water	100	C_2H_3N	Acetonitrile		81.6
15944	H_2O	Water	100	$C_2H_4Cl_2$	1,2-Dichloroethane		83.7
		"					
15945	H_2O	Water	100	$C_2H_4Cl_2$	1,2-Dichloroethane		83.7
		"			"		
15946	H_2O	Water	100	$C_2H_4Cl_2$	1,2-Dichloroethane		83.5
15947	H_2O	Water	100	C_2H_4O	Acetaldehyde		20.2
		"			"		
15947a	H_2O	Water	100	C_2H_4O	Acetaldehyde		20.2
15948	H_2O	Water	100	C_2H_4O	Acetaldehyde		20.2
15948a	H_2O	Water	100	$C_2H_4O_2$	Acetic acid		118.1
15949	H_2O	Water	100	$C_2H_4O_2$	Acetic acid		118.1
15950	H_2O	Water	100	$C_2H_4O_2$	Acetic acid		118.1
15951	H_2O	Water	100	$C_2H_4O_2$	Acetic acid		118.5
15952	H_2O	Water	100	$C_2H_4O_2$	Acetic acid		118.1
15953	H_2O	Water	100	$C_2H_4O_2$	Acetic acid		118.1
15954	H_2O	Water	100	C_2H_5Br	Bromoethane		38.4
15955	H_2O	Water	100	C_2H_5ClO	2-Chloroethanol		128
15956	H_2O	Water	100	C_2H_5ClO	2-Chloroethanol		128.8
15957	H_2O	Water	100	C_2H_5ClO	2-Chloroethanol		128
		"			"		
15958	H_2O	Water	100	C_2H_5ClO	2-Chloroethanol		128.8
15959	H_2O	Water	100	C_2H_5ClO	2-Chloroethanol		128.8
15960	H_2O	Water	100	C_2H_5ClO	2-Chloroethanol		128.8
15961	H_2O	Water	100	C_2H_5I	Iodoethane		72.3
15962	H_2O	Water	100	$C_2H_5NO_2$	Nitroethane		114.07
15963	H_2O	Water	100	$C_2H_5NO_2$	Nitroethane		114.07

TABLE II. *Ternary Systems* 457

	C-Component			Azeotropic Data				
Formula	Name	B.P., °C.		B.P., °C.	Wt. % A	Wt. % B	Wt. % C	Ref.
C_2H_6O	Ethyl alcohol	78.3		53.8	2.85	90.5	6.65	148
C_2H_6O	Ethyl alcohol	78.3		44.4	1.1	94.5	4.4	148
C_2H_6O	Acetonitrile	81.6			Nonazeotrope			763
C_2H_6O	Ethyl alcohol	78.3		72.9	1	44	55	982
C_3H_3N	Acrylonitrile	72.1			Nonazeotrope v-l 63, 997c,953c,997f			
C_3H_6O	Acetone	56.4			Nonazeotrope		v-l	763
$C_4H_8O_2$	Ethyl acetate	77		70				763
$C_4H_{11}N$	Diethylamine	55.5			Nonazeotrope			981
C_5H_8	Isoprene			32.4	1.08	2.32	97.6	742
C_5H_{10}	β-amylene			34.6	1.3	7.8	90.9	742
$C_5H_{10}O_2$	Propyl acetate	101.6		74				763
C_5H_{12}	2-Methylbutane			24.7	0.76	5.70	93.54	742
C_6H_6	Benzene	80.2		66	8.2	23.3	68.5	763
$C_6H_{12}O_2$	Butyl acetate	124.8			Nonazeotrope			763
$C_6H_{14}O$	Isopropyl ether	68.3		59	5	13	82	981
$C_6H_{15}N$	Triethylamine	89.7		68.6	3.5	9.6	86.9	498i
C_7H_8	Toluene	110.7		73				763
C_2H_5ClO	2-Chloroethanol	128			Nonazeotrope			479,672
	"			69.6				253
C_2H_6O	Ethyl alcohol	78.3		66.7	5	78	17	563
	" "			67.8	7.2	77.1	15.7	982
C_3H_8O	Isopropyl alcohol	82.3		69.7	7.7	73.3	19.0	982
C_2H_6O	Ethyl alcohol	78.3			Nonazeotrope		v-l	451
	" "						v-l	374
$C_3H_6O_2$	Methyl acetate	57.1			Nonazeotrope		v-l	985c
$C_6H_{12}O_3$	Paraldehyde	124			Nonazeotrope		v-l	981
$C_3H_6O_2$	Methyl acetate	57.1			Nonazeotrope		v-l	36g
$C_3H_6O_2$	Propionic acid	141.1			Nonazeotrope		v-l	26
$C_5H_4O_2$	Furfural	161.7			Nonazeotrope		v-l	971
C_7H_8	Toluene	110.7			Nonazeotrope			563
C_8H_{10}	Ethylbenzene	136			Min. b.p.			65
C_8H_{10}	Xylene	140			Min. b.p.			65
C_2H_6O	Ethyl alcohol	78.3			Azeotropic ?			563
$C_4H_8Cl_2O$	Bis(2-chloro-ethyl) ether	178			Min. b.p.			672
$C_4H_8O_2$	Ethyl acetate	77.1			Nonazeotrope			479
C_6H_6	Benzene	80.1			Nonazeotrope			479,672
	"			67.0				253
C_6H_{12}	Cyclohexane	80.7			min. b.p.			
$C_6H_{14}O$	Isopropyl ether	68.5			Nonazeotrope			
C_7H_8	Toluene	110.6			min. b.p.			
C_2H_6O	Ethyl alcohol	78.3		61	~ 5	~ 86	~ 9	563
C_6H_{14}	1-Hexene	63.84		57.4	10.1	3.0	86.9	806
C_6H_{12}	Hexane	68.74		59.5	8.4	9.3	82.3	806

	A-Component		B.P., °C.	B-Component		B.P., °C.
No.	Formula			Formula	Name	
15964	H$_2$O	Water	100	C$_2$H$_5$NO$_2$	Nitroethane	114.07
15965	H$_2$O	Water	100	C$_2$H$_6$O	Ethyl alcohol	78.3
		"			" "	
15966	H$_2$O	Water	100	C$_2$H$_6$O	Ethyl alcohol	78.3
15967	H$_2$O	Water	100	C$_2$H$_6$O	Ethyl alcohol	78.3
15968	H$_2$O	Water	100	C$_2$H$_6$O	Ethyl alcohol	78.3
15969	H$_2$O	Water	100	C$_2$H$_6$O	Ethyl alcohol	78.3
15970	H$_2$O	Water	100	C$_2$H$_6$O	Ethyl alcohol	78.3
15971	H$_2$O	Water	100	C$_2$H$_6$O	Ethyl alcohol	78.3
15972	H$_2$O	Water	100	C$_2$H$_6$O	Ethyl alcohol	78.3
15973	H$_2$O	Water	100	C$_2$H$_6$O	Ethyl alcohol	78.3
15974	H$_2$O	Water	100	C$_2$H$_6$O	Ethyl alcohol	78.3
15975	H$_2$O	Water	100	C$_2$H$_6$O	Ethyl alcohol	78.3
15976	H$_2$O	Water	100	C$_2$H$_6$O	Ethyl alcohol	78.3
15977	H$_2$O	Water	100	C$_2$H$_6$O	Ethyl alcohol	78.3
15978	H$_2$O	Water	100	C$_2$H$_8$O	Ethyl alcohol	78.3
15979	H$_2$O	Water	100	C$_2$H$_6$O	Ethyl alcohol	78.3
15980	H$_2$O	Water	100	C$_2$H$_6$O	Ethyl alcohol	78.3
		"			" "	
		"			" "	
		"			" "	
15981	H$_2$O	Water	100	C$_2$H$_6$O	Ethyl alcohol	78.3
15982	H$_2$O	Water	100	C$_2$H$_6$O	Ethyl alcohol	78.3
15982a	H$_2$O	Water	100	C$_2$H$_6$O	Ethyl alcohol	78.3
15982b	H$_2$O	Water	100	C$_2$H$_6$O	Ethyl alcohol	78.3
15982c	H$_2$O	Water	100	C$_2$H$_6$O	Ethyl alcohol	78.3
15983	H$_2$O	Water	100	C$_2$H$_6$O	Ethyl alcohol	78.3
		"			" "	
15984	H$_2$O	Water	100	C$_2$H$_6$O	Ethyl alcohol	78.3
15985	H$_2$O	Water	100	C$_2$H$_6$O	Ethyl alcohol	78.3
15986	H$_2$O	Water	100	C$_2$H$_6$O	Ethyl alcohol	78.3
15987	H$_2$O	Water	100	C$_2$H$_6$O	Ethyl alcohol	78.3
15988	H$_2$O	Water	100	C$_2$H$_6$O	Ethyl alcohol	78.3
15989	H$_2$O	Water	100	C$_2$H$_6$O	Ethyl alcohol	78.3
15990	H$_2$O	Water	100	C$_2$H$_6$O	Ethyl alcohol	78.3
15991	H$_2$O	Water	100	C$_2$H$_6$O	Ethyl alcohol	78.3
15992	H$_2$O	Water	100	C$_2$H$_6$O	Ethyl alcohol	78.3

TABLE II. *Ternary Systems* 459

	C-Component			Azeotropic Data				
Formula	Name	B.P., °C.	B.P., °C.	Wt. % A	Wt. % B	Wt. % C		Ref.
C_7H_{16}	Heptane	98.43	75.1	11.5	24.5	64.0		806
C_3H_3N	Acrylonitrile	77.2	69.5	8.7	20.3	71.0		982
	" 100 mm		<30	6.6	9.0	84.4		982
C_3H_5Br	cis-1-Bromopropene	57.8	54?	3	6	91		563
C_3H_5Br	trans-1-Bromopropene	63.25	54.5	4	87.5	7.5		563
C_3H_5Br	2-Bromopropene	48.35	43.3?	1	4	95		563
C_3H_5I	3-Iodopropene	102	72					563
C_3H_6O	Acetone	56.1		Nonazeotrope				981
$C_3H_6O_2$	Ethyl formate	54.2		Nonazeotrope				981
C_3H_7Br	1-Bromopropane	71.0	60.0	3.5	9.6	86.9		498i
$C_3H_7NO_2$	2-Nitropropane	120.25	78.2	6.1	86.3	7.6		806
C_4H_6O	Crotonaldehyde	102.4	78.0	4.8	87.9	7.3		982
$C_4H_6O_2$	Biacetyl	88		Nonazeotrope ?				640
$C_4H_7ClO_2$	Ethyl chloroacetate	143.5	31.35	17.5	61.7	20.8		121
C_4H_8O	2-Butanone	79.6	73.2	11	14	75		982
C_4H_8O	Butyraldehyde	75.7	67.2	9	11	80		982
C_4H_8O	Ethyl vinyl ether	35.5		Nonazeotrope				981
$C_4H_8O_2$	Ethyl acetate	77.05		Nonazeotrope				29
	" 25 mm		− 1.40	4.0	4.0	92.0		650
	" 760 mm		70.23	9.0	8.4	82.6	v-l	650, 452g
	" 1446 mm		88.96	10.3	12.1	77.6		650
C_4H_9Br	1-Bromo-2-methylpropane	91.4	69.5	5.8	18.4	75.8		498i
C_4H_9Cl	1-Chloro-2-methylpropane	68.85	58.62	4.5	13	82.5		563
$C_4H_{10}O$	Butyl alcohol	117		Nonazeotrope			v-l	452b
$C_4H_{10}O$	tert-Butyl alcohol	82.5					v-l	937c
$C_4H_{10}O$	Isobutyl alcohol	108	50°–130°				v-l	748c, 937c
$C_4H_{10}O$	Ethyl ether	34.5		Nonazeotrope				1033
	" "			Nonazeotrope			v-l	452
$C_4H_{10}O_2$	2-Ethoxyethanol	133		Nonazeotrope				33c
$C_4H_{10}O_2$	Ethoxymethoxymethane	65.90		Nonazeotrope				1035
$C_4H_{11}N$	Butylamine	77.8	81.8	7.5	42.5	50.0		982
$C_5H_8O_2$	Ethyl acrylate	99.3	77.1	10.1	48.3	41.6		982
	" " 165 mm		44	8.6	36.3	55.1		982
$C_5H_{10}O$	1-Butenyl methyl ether		61.4	6.8	14.3	78.9		981
$C_5H_{10}O$	Propyl vinyl ether	65.1	57	5.1	21.2	73.7		982
$C_5H_{10}O_2$	Isopropyl acetate	88.7	74.8	9.8	19.4	70.8		982,498c
$C_5H_{12}O$	Butyl methyl ether	70.3	62	6.3	8.6	85.1		982
$C_5H_{12}O$	Isoamyl alcohol	132		Nonazeotrope			v-l	652

		A-Component	B.P., °C.		B-Component	B.P., °C.
No.	Formula			Formula	Name	
15993	H_2O	Water	100	C_2H_6O	Ethyl alcohol	78.3
15994	H_2O	Water	100	C_2H_6O	Ethyl alcohol	78.3
15995	H_2O	Water "	100	C_2H_6O	Ethyl alcohol " "	78.3
15996	H_2O	Water	100	C_2H_6O	Ethyl alcohol	78.3
15997	H_2O	Water	100	C_2H_6O	Ethyl alcohol	78.3
15998	H_2O	Water	100	C_2H_6O	Ethyl alcohol	78.3
15999	H_2O	Water	100	C_2H_6O	Ethyl alcohol	78.3
16000	H_2O	Water	100	C_2H_6O	Ethyl alcohol	78.3
16001	H_2O	Water	100	C_2H_6O	Ethyl alcohol	78.3
16002	H_2O	Water " "	100	C_2H_6O	Ethyl alcohol " " " "	78.3
16003	H_2O	Water	100	C_2H_6O	Ethyl alcohol	78.3
16004	H_2O	Water " "	100	C_2H_6O	Ethyl alcohol " "	78.3
16005	H_2O	Water "	100	C_2H_6O	Ethyl alcohol " "	78.3
16006	H_2O	Water	100	C_2H_6O	Ethyl alcohol	78.3
16007	H_2O	Water " "	100	C_2H_6O	Ethyl alcohol " " " "	78.3
16008	H_2O	Water	100	C_2H_6O	Ethyl alcohol	78.3
16009	H_2O	Water	100	C_2H_6O	Ethyl alcohol	78.3
16010	H_2O	Water	100	C_2H_6O	Ethyl alcohol	78.3
16011	H_2O	Water	100	C_2H_6O	Ethyl alcohol	78.3
16012	H_2O	Water "	100	C_2H_6O	Ethyl alcohol " "	78.3
16013	H_2O	Water	100	C_2H_6O	Ethyl alcohol	78.3
16014	H_2O	Water "	100	C_2H_6O	Ethyl alcohol " "	78.3
16015	H_2O	Water	100	C_2H_6O	Ethyl alcohol	78.3
16016	H_2O	Water "	100	C_2H_6O	Ethyl alcohol " "	78.3
16017	H_2O	Water	100	C_2H_6O	Ethyl alcohol	78.3
16018	H_2O	Water	100	C_2H_6O	Ethyl alcohol	78.3
16019	H_2O	Water	100	C_2H_6O	Ethyl alcohol	78.3
16020	H_2O	Water	100	$C_2H_6O_2$	Glycol	197.4
16021	H_2O	Water	100	C_2H_7N	Dimethylamine	7.4
16022	H_2O	Water	100	$C_2H_8N_2$	Ethylenediamine	116.9
16022a	H_2O	Water	100	C_3H_3N	Acrylonitrile	77.3
16023	H_2O	Water	100	C_3H_3N	Acrylonitrile	77.2
16023a	H_2O	Water	100	C_3H_3N	Acrylonitrile	77.3
16024	H_2O	Water	100	C_3H_4O	2-Propyn-l-ol	115

TABLE II. *Ternary Systems* 461

C-Component		B.P., °C.	Azeotropic Data				
Formula	Name		B.P., °C.	Wt. % A	Wt. % B	Wt. % C	Ref.
$C_5H_{12}O$	Diethoxymethane	87.5	73.2	12.8	18.4	69.5	324
C_6H_5Cl	Chlorobenzene	131.8	77.3	6.8	80.2	13.9	498i
C_6H_6	Benzene	80.2	64.86	7.4	18.5	74.1	1046
"			Effect of pressure, 1-19 atm.				436
C_6H_8	1,3-Cyclohexadiene	80.8	63.6	7	20	73	563
C_6H_8	1,4-Cyclohexadiene	85.6	~ 65.5				563
C_6H_{10}	Biallyl	60.2	~ 52				563
C_6H_{10}	Cyclohexene	82.75	64.05	7	20	73	563
C_6H_{10}	1-Hexyne	70.2	59.9				377
C_6H_{10}	3-Hexyne	80.5	64.4				377
C_6H_{12}	Cyclohexane	80.75	62.1				563
"			62.60	4.8	19.7	75.5	1059
"		80.7	62.1	7	17	76	982
$C_6H_{12}O$	Isobutyl vinyl ether	83.4	60	8	22	70	982
C_6H_{11}	Hexane	68.22	56.4	3	18.7	78.3	949
"		68.7	56.0	3	12	85	982
"			56.60				1044
$C_6H_{11}O$	Butyl ethyl ether	92.2	71.6	9.0	22.4	68.6	497
"			71.6	9.3	4.2	86.5	982
$C_6H_{14}O$	Ethyl isobutyl ether	79	66	6.5	15.8	77.7	981
$C_6H_{14}O$	Isopropyl ether	68.3	61.0	4.0	6.5	89.5	982
"	100 p.s.i.g.		128.5	9.1	14.2	76.7	982
"	50 p.s.i.g.		105.8	7.1	11.9	81	982
$C_6H_{14}O$	Isopropyl propyl ether		66	7.0	14.7	78.3	981
$C_6H_{14}O_2$	Acetal	103.6	77.8	11.4	27.6	61.0	45
$C_6H_{14}O_2$	Ethoxypropoxy-methane	113.7	Nonazeotrope				1035
$C_6H_{15}N$	Triethylamine	89.4	74.7	9	13	78	977
C_7H_8	Toluene	110.7	74.55				563
"		110.6	74.4	12	37	51	982
C_7H_{12}	1-Heptyne	99.5	71.0				377
C_7H_{14}	Methylcyclohexane	101.8	~ 70.5				563
"			69.59	6.8	32.4	60.8	949
$C_7H_{14}O_2$	Isoamyl acetate	90.8	69.0				377
C_7H_{16}	Heptane	98.45	~ 69.5				563
"			68.8	6.1	33.00	60.9	982
C_8H_8	Styrene	145.1	Nonazeotrope				981
$C_8H_{18}O$	Butyl ether	142.1	Nonazeotrope				981
$C_8H_{18}O_2$	2-Ethyl-1,3-hexanediol	243.1	Nonazeotrope				981
$C_4H_8O_2$	Dioxane	101.4	Nonazeotrope				201
$C_4H_{11}NO$	2-(Dimethyl-amino) ethanol	134.6	Nonazeotrope				981
C_6H_6	Benzene	80.1	Nonazeotrope				981
C_3H_4O	Acrolein	52.4	200-760 mm	Nonazeotrope v-l		905f	905p
C_3H_5N	Propionitrile	97.4	Nonazeotrope				981
$C_4H_6O_2$	Vinyl acetate					v-l	335c
$C_5H_8O_2$	3,3-Dimethoxy-propyne	111	88.95				264

	A-Component		B.P., °C.		B-Component	B.P., °C.
No.	Formula			Formula	Name	
16025	H₂O	Water	100	C₃H₄O	2-Propyn-1-ol	115
		"			"	
16026	H₂O	Water	100	C₃H₄O₂	Acrylic acid	141.2
16027	H₂O	Water	100	C₃H₅I	3-Iodopropene	102
16028	H₂O	Water	100	C₃H₅I	3-Iodopropene	102
16029	H₂O	Water	100	C₃H₆Cl₂	1,2-Dichloro-	
					propane	96.3
16030	H₂O	Water	100	C₃H₆O	Acetone	56.1
16031	H₂O	Water	100	C₃H₆O	Acetone	56.1
16032	H₂O	Water	100	C₃H₆O	Acetone	56.1
16033	H₂O	Water	100	C₃H₆O	Acetone	56.1
16034	H₂O	Water	100	C₃H₆O	Acetone	56.1
16035	H₂O	Water	100	C₃H₆O	Acetone	56.4
16036	H₂O	Water	100	C₃H₆O	Acetone	56.1
16037	H₂O	Water	100	C₃H₆O	Acetone	56.25
16038	H₂O	Water	100	C₃H₆O	Acetone	56.1
		"			"	
16039	H₂O	Water	100	C₃H₆O	Acetone	56.1
16040	H₂O	Water	100	C₃H₆O	Allyl alcohol	96.9
16041	H₂O	Water	100	C₃H₆O	Allyl alcohol	96.95
16042	H₂O	Water	100	C₃H₆O	Allyl alcohol	96.95
16043	H₂O	Water	100	C₃H₆O	Allyl alcohol	96.95
16044	H₂O	Water	100	C₃H₆O	Allyl alcohol	96.95
16045	H₂O	Water	100	C₃H₆O	Allyl alcohol	96.95
16046	H₂O	Water	100	C₃H₆O	Allyl alcohol	96.95
		"			" "	
16047	H₂O	Water	100	C₃H₆O	Allyl alcohol	96.6
		"			" "	
16048	H₂O	Water	100	C₃H₆O	Allyl alcohol	96.6
16049	H₂O	Water	100	C₃H₆O₂	Propionic acid	140.7
16050	H₂O	Water	100	C₃H₆O₃	Trioxane	114.5
16051	H₂O	Water	100	C₃H₆O₃	Trioxane	114.5
16052	H₂O	Water	100	C₃H₆O₃	Trioxane	114.5
16053	H₂O	Water	100	C₃H₆O₃	Trioxane	114.5
16054	H₂O	Water	100	C₃H₆O₃	Trioxane	114.5
16055	H₂O	Water	100	C₃H₆O₃	Trioxane	114.5
16056	H₂O	Water	100	C₃H₆O₃	Trioxane	114.5
16057	H₂O	Water	100	C₃H₇I	1-Iodopropane	102.4
16058	H₂O	Water	100	C₃H₇NO₂	1-Nitropropane	130.5
16059	H₂O	Water	100	C₃H₇NO₂	2-Nitropropane	120.25
16059a	H₂O	Water	100	C₃H₈O	Isopropyl alcohol	82.3
16060	H₂O	Water	100	C₃H₈O	Isopropyl alcohol	82.3
		"			"	
16061	H₂O	Water	100	C₃H₈O	Isopropyl alcohol	82.45

TABLE II. *Ternary Systems*   463

Formula	Name	B.P., °C.	B.P., °C.	Wt. % A	Wt. % B	Wt. % C	Ref.
			Azeotropic Data				
C_6H_6	Benzene	80.1	68.1	8.3	3.7	88.0	163
	"		69	9	4	87	264
$C_5H_8O_2$	Ethyl acrylate	99.3	Vapor-liquid equilibrium				885
C_3H_6O	Allyl alcohol	96.95	Nonazeotrope				981
C_3H_8O	Propyl alcohol	97.2	77.7				563
C_3H_7ClO	Propylene		78.15	8	72	20	563
	chlorohydrin	127.4	Nonazeotrope				981
C_3H_8O	Isopropyl alcohol	82.3	Nonazeotrope		v-l	152,	1042c
$C_4H_6O_2$	Vinyl acetate	72.7	Nonazeotrope				981
C_4H_8O	2-Butanone	79.6	Nonazeotrope				726
C_4H_8O	Butyraldehyde	74.8	Nonazeotrope				981
C_5H_6O	2-Methylfuran	63.7	55.6				769
C_5H_8	Isoprene	34.7	32.5	0.4	7.6	92.0	739
$C_5H_{10}O_2$	Isopropyl acetate	88.6	Nonazeotrope				981
C_6H_6O	Phenol	181.5	Nonazeotrope, vapor pressure curve				854,668c
$C_6H_{14}O$	Isopropyl ether	69	53.8	1.8	53.5	44.7	498i
	" 15 p.s.i.g.		75	3	49	48	304
C_nH_{2n+2}	Paraffin hydrocarbons						981
			61-71	1.4	42.1	56.5 vol. %	461
C_3H_8O	Propyl alcohol	97.8	Nonazeotrope		v-l		998
C_6H_6	Benzene	80.2	68.21	3.58	9.16	82.26	875,1005
C_6H_8	1,3-Cyclohexadiene	80.8	67.5				563
C_6H_{10}	Cyclohexene	82.75	67.95	8.5	11	80.5	563
$C_6H_{10}O$	Allyl ether	94.8	77.8	12.4	8.7	78.9	875
C_6H_{12}	Cyclohexane	80.75	66.18	8	11	81	563
C_6H_{14}	Hexane	68.95	59.7	5	5	90	563
	"			8.5	5.1	86.4	480
C_7H_8	Toluene	110.6	80.6	15.2	31.4	53.4	981
	"		80.2				563
$C_9H_{16}O_2$	2,2-Bis(allyl- oxy) propane		88	28	55	17	981
$C_4H_8O_2$	Methyl propionate	79.85	Nonazeotrope				227
C_6H_{12}	Naphthenes		Min. b.p.				513
C_6H_{11}	Hexanes		Min. b.p.				513
C_7H_{14}	Naphthenes		Min. b.p.				513
C_7H_{16}	Heptanes		Min. b.p.				513
C_8H_{16}	Naphthenes		Min. b.p.				513
C_8H_{18}	Octanes		Min. b.p.				513
C_9H_{20}	Nonanes		Min. b.p.				513
C_3H_8O	Propyl alcohol	97.2	78.25				563
C_8H_{10}	Ethylbenzene	136		28.8	32.2	39	52,806
$C_4H_{10}O_2$	2-Ethoxyethanol	134.8	Nonazeotrope				806
$C_3H_8O_3$	Glycerol	290	Nonazeotrope			v-l	991d
C_4H_8O	2-Butanone	79.6	73.4	11	1	88	982
	"		Nonazeotrope				29
C_4H_9Cl	1-Chloro-2- methylpropane	63.85	61				563

		A-Component	B.P., °C.		B-Component	B.P., °C.
No.	Formula			Formula	Name	
16062	H_2O	Water	100	C_3H_8O	Isopropyl alcohol	82.3
16063	H_2O	Water	100	C_3H_8O	Isopropyl alcohol	82.4
16064	H_2O	Water	100	C_3H_8O	Isopropyl alcohol	82.3
16065	H_2O	Water	100	C_3H_8O	Isopropyl alcohol	82.7
16066	H_2O	Water	100	C_3H_8O	Isopropyl alcohol	82.4
16067	H_2O	Water	100	C_3H_8O	Isopropyl alcohol	82.4
16068	H_2O	Water	100	C_3H_8O	Isopropyl alcohol	82.4
16069	H_2O	Water	100	C_3H_8O	Isopropyl alcohol	82.45
		"			" "	
		"			" "	
	H_2O	Water	100		Isopropyl alcohol	
16070	H_2O	Water	100	C_3H_8O	Isopropyl alcohol	82.45
16071	H_2O	Water	100	C_3H_8O	Isopropyl alcohol	82.45
16072	H_2O	Water	100	C_3H_8O	Isopropyl alcohol	82.45
16073	H_2O	Water	100	C_3H_8O	Isopropyl alcohol	82.3
16074	H_2O	Water	100	C_3H_8O	Isopropyl alcohol	82.45
16075	H_2O	Water	100	C_3H_8O	Isopropyl alcohol	82.3
16076	H_2O	Water	100	C_3H_8O	Isopropyl alcohol	82.45
16077	H_2O	Water	100	C_3H_8O	Isopropyl alcohol	82.3
		"			" "	
		"			" "	
		"			" "	
16078	H_2O	Water	100	C_3H_8O	Isopropyl alcohol	82.3
16079	H_2O	Water	100	C_3H_8O	Isopropyl alcohol	82.3
		"			" "	
16080	H_2O	Water	100	C_3H_8O	Isopropyl alcohol	82.3
16081	H_2O	Water	100	C_3H_8O	Propyl alcohol	97.3
16082	H_2O	Water	100	C_3H_8O	Propyl alcohol	97.2
16083	H_2O	Water	100	C_3H_8O	Propyl alcohol	97.16
16084	H_2O	Water	100	C_3H_8O	Propyl alcohol	97.2
16085	H_2O	Water	100	C_3H_8O	Propyl alcohol	97.2
16086	H_2O	Water	100	C_3H_8O	Propyl alcohol	97.2
		"			" "	
16087	H_2O	Water	100	C_3H_8O	Propyl alcohol	97.3
		"			" "	
		"			" "	
		"			" "	
		"			" "	
		"			" "	
16088	H_2O	Water	100	C_3H_8O	Propyl alcohol	97.2

TABLE II. *Ternary Systems* 465

	C-Component			Azeotropic Data				
Formula	Name	B.P., °C.		B.P., °C.	Wt. % A	Wt. % B	Wt. % C	Ref.
$C_4H_{11}N$	Butylamine	77.8		83	12.5	40.5	47	982
$C_5H_{10}O$	Allyl ethyl ether	67.6		Azeotropic				18
$C_5H_{10}O_2$	Isopropyl acetate	88.7		75.5	11	13	76	982
$C_5H_{12}O$	Butyl methyl ether	70.3		Azeotropic				18
$C_5H_{12}O$	Ethyl isopropyl ether	54		Azeotropic				18
$C_5H_{12}O$	Ethly propyl ether	64		Azeotropic				18
$C_5H_{12}O$	Isobutyl methyl ether	59		Azeotropic				18
C_6H_6	Benzene	80.2		66.51	7.5	18.7	73.8	1044
	"			66.3	7.5	19.0	73.5	1042
	"	80.1		65.7	8.2	19.8	72.0	982
	Benzene 20 p.s.i.g.			90	10	18	72	982
C_6H_8	1,3-Cyclohexadiene	80.8		65.7				563
C_6H_{10}	Cyclohexene	82.75		66.1	7.5	21.5	71	563
C_6H_{12}	Cyclohexane	80.75		64.3	7.5	18.5	74	v-l 563, 991a
$C_6H_{12}O$	4-Methyl-2-pentanone	116.2		Nonazeotrope				981
C_6H_{11}	Hexane	68.95		58.2				563
$C_6H_{11}O$	Butyl ethyl ether	92.2		73.4	10.4	21.9	67.7	982
$C_6H_{14}O$	Ethyl *tert*-butyl ether	68-69		Azeotropic				18
$C_6H_{14}O$	Isopropyl ether	69		61.9	4.55	4.45	91.0	v-l 991c
	"			61.8	5	4	91	982
	" 30 p.s.i.g.			95	6	9	85	982
	" 15 p.s.i.g.			81	6	7	87	982
$C_6H_{15}N$	Diisopropylamine	84.1		Nonazeotrope				981
C_7H_8	Toluene	110.6		76.3	13.1	38.2	48.7	v-l 982, 912c
	"			76.2				563
C_8H_{14}	Diisobutylene	102.3		72.3	9.3	31.6	59.1	982
C_3H_8S	1-Propanethiol	67.5	771 mm	60.8				487
C_4H_8O	2-Butanone	79.6		Nonazeotrope				29
$C_4H_8O_2$	Propyl formate	80.9		70.8	13	5	82	359,498c
C_4H_9Cl	1-Chloro-2-methylpropane	68.85		64.2				563
$C_5H_9ClO_2$	Propyl chloroacetate	162.3		88.6	25.25	58.27	16.48	121
$C_5H_{10}O$	3-Pentanone	102.2		~ 82.1	~ 20	~ 20	~ 60	563
	"			81.2	20	20	60	981
$C_6H_{10}O_2$	Propyl acetate	101.6		Nonazeotrope			v-l	752
	" 200 mm			50.23	13.3	4.7	82.0	v-l 892
	" 400 mm			66.07	15.0	6.5	78.5	v-l 892
	" 600 mm			76.26	16.0	8.5	75.5	v-l 892
	" 760 mm			82.45	17.0	10.0	73.0	v-l 892
	"			82.2	21	19.5	59.5	359
$C_5H_{12}O_2$	Diethoxymethane	88.0		Nonazeotrope				1035

	A-Component		B.P., °C.		B-Component		B.P., °C.
No.	Formula			Formula	Name		
16089	H_2O	Water	100	C_3H_8O	Propyl alcohol		97.2
		"			"	"	
		"			"	"	
		"			"	"	
		"			"	"	
16090	H_2O	Water	100	C_3H_8O	Propyl alcohol		97.2
16091	H_2O	Water	100	C_3H_8O	Propyl alcohol		97.2
16092	H_2O	Water	100	C_3H_8O	Propyl alcohol		97.2
16093	H_2O	Water	100	C_3H_8O	Propyl alcohol		97.3
16094	H_2O	Water	100	C_3H_8O	Propyl alcohol		97.3
16094a	H_2O	Water	100	C_3H_8O	Propyl alcohol		97.3
16095	H_2O	Water	100	C_3H_8O	Propyl alcohol		97.2
16096	H_2O	Water	100	C_3H_8O	Propyl alcohol		97.2
16097	H_2O	Water	100	C_3H_8O	Propyl alcohol		97.2
16098	H_2O	Water	100	C_3H_8O	Propyl alcohol		97.2
16099	H_2O	Water	100	C_3H_8O	Propyl alcohol		96.90
16100	H_2O	Water	100	C_3H_8O	Propyl alcohol		97.2
16101	H_2O	Water	100	$C_3H_8O_2$	2-Methoxy-ethanol		124.6
16102	H_2O	Water	100	$C_3H_8O_2$	2-Methoxy-ethanol		124.6
16103	H_2O	Water	100	$C_3H_8O_2$	2-Methoxy-ethanol		124.6
16104	H_2O	Water	100	$C_3H_8O_2$	2-Methoxy-ethanol		124
16105	H_2O	Water	100	$C_3H_8O_2$	2-Methoxy-ethanol		124
16105a	H_2O	Water	100	$C_3H_8O_2$	Dimethoxymethane		42.3
16106	H_2O	Water	100	$C_3H_8O_2$	1,2-Propanediol		187.8
16107	H_2O	Water	100	C_4H_6O	Crotonaldehyde		102.4
16108	H_2O	Water	100	C_4H_6O	Crotonaldehyde		102
16109	H_2O	Water	100	C_4H_6O	Crotonaldehyde		102
16110	H_2O	Water	100	C_4H_8O	2-Butanone		79.6
16111	H_2O	Water	100	C_4H_8O	2-Butanone		79.6
16112	H_2O	Water	100	C_4H_8O	2-Butanone		79.6
		"			"		
16113	H_2O	Water	100	C_4H_8O	2-Butanone		79.6
16114	H_2O	Water	100	C_4H_8O	2-Butanone		79.6
16115	H_2O	Water	100	C_4H_8O	2-Butanone		79.6
16116	H_2O	Water	100	C_4H_8O	2-Butanone		79.6
16117	H_2O	Water	100	C_4H_8O	2-Butanone		79.6
16118	H_2O	Water	100	C_4H_8O	2-Butanone		79.6

TABLE II. *Ternary Systems* 467

	C-Component			Azeotropic Data			
Formula	Name	B.P., °C.	B.P., °C.	Wt. % A	Wt. % B	Wt. % C	Ref.
C_6H_6	Benzene						
	740 mm		67	7.6	10.1	82.3	584
	" 2830 mm		107	9.5	13.1	77.4	584
	" 4900 mm		127	10.3	14.2	75.5	584
	" 5930 mm		135	12.3	15.0	72.7	584
	"		68.5	8.6	9.0	82.4	981
C_6H_8	1,3-Cyclohexadiene	80.8	67.75	9	12	79	563
C_6H_{10}	Cyclohexene	82.75	63.2	9	11.5	79.5	563
C_6H_{12}	Cyclohexane	80.75	66.55	8.5	10	81.5	563
$C_6H_{12}O$	2-Hexanone	127.2	87	27	63	10	343
$C_6H_{12}O$	2-Methyl-pentanal	118.3	86	28	58	14	981
$C_6H_{12}O_2$	Propyl propionate	122	86.2	20.7	46.2	33.1	668g
C_6H_{14}	Hexane	68.95	59.95				563
$C_6H_{14}O$	Propyl ether	113.7	74.8	11.7	20.2	68.1	760
$C_6H_{14}O_2$	Ethoxypropoxy-methane	113.7	83.8	17.6	22.9	59.5	1035
C_7H_8	Toluene	110.7	80.05				563
$C_7H_{16}O_2$	Dipropoxymethane	137.2	86.4	8	44.8	47.2	324
$C_8H_{18}O_2$	Acetaldehyde dipropylacetal	147.7	87.6	27.4	51.6	21.0	45
C_6H_6	Benzene	80.1		Nonazeotrope			981
C_6H_{12}	Cyclohexane	80.7		Nonazeotrope			981
C_7H_8	Toluene	110.6		Nonazeotrope			981
C_8H_{10}	Ethylbenzene	136	90	25.4	7.4	67.2	65
C_8H_{10}	Xylene	140		Min. b.p.			65
C_5H_8	Isoprene	34		Nonazeotrope			581c
C_7H_8	Toluene	110.6		Nonazeotrope			981
$C_6H_{10}O$	2-Ethylcroton-aldehyde	135.3		Nonazeotrope			981
C_7H_8	Toluene	110.7	85.3				932
C_nH_{2n+2}	Paraffins		80-85				932
$C_4H_{10}O$	sec-Butyl alcohol	99.5	200-760 mm	Nonazeotrope		v-l	15
$C_4H_{10}O$	tert-Butyl alcohol	82.4		Nonazeotrope			29
C_6H_6	Benzene	80.12	68.9	8.9	17.5	73.6	876
	"		68.2	8.8	26.1	65.1	981
C_6H_{12}	Cyclohexane	80.7	63.6	5	35	60	981
C_6H_{12}	1-Hexene	82		Min. b.p.			66
C_6H_{12}	2-Hexene			Min. b.p.			66
C_6H_{12}	3-Hexene			Min. b.p.			66
C_6H_{12}	2-Methyl-1-pentene			Min. b.p.			66
C_6H_{12}	2-Methyl-2-pentene			Min. b.p.			66

No.	Formula	(A-Component)	B.P., °C.	Formula	Name (B-Component)	B.P., °C.
16119	H_2O	Water	100	C_4H_8O	2-Butanone	79.6
16120	H_2O	Water	100	C_4H_8O	2-Butanone	79.6
		"			"	
16121	H_2O	Water	100	C_4H_8O	2-Butanone	
16122	H_2O	Water	100	C_4H_8O	2-Butanone	79.6
16123	H_2O	Water	100	C_4H_8O	Butyraldehyde	74.8
16124	H_2O	Water	100	C_4H_8O	Butyraldehyde	74.8
16125	H_2O	Water	100	C_4H_8O	Butyraldehyde	74.8
16126	H_2O	Water	100	C_4H_8O	Butyraldehyde	74.8
16127	H_2O	Water	100	C_4H_8O	Butyraldehyde	75.7
16128	H_2O	Water	100	C_4H_8O	Isobutyraldehyde	
16129	H_2O	Water	100	C_4H_8O	Isobutyraldehyde	63
16130	H_2O	Water	100	$C_4H_8O_2$	Isobutyric acid	154.5
16131	H_2O	Water	100	C_4H_9Cl	Chlorobutane	78.44
16132	H_2O	Water	100	C_4H_9Cl	Chlorobutane	78.44
16133	H_2O	Water	100	C_4H_9Cl	1-Chloro-2-methylpropane	68.85
16134	H_2O	Water	100	$C_4H_{10}O$	Butyl alcohol	117.7
16135	H_2O	Water	100	$C_4H_{10}O$	Butyl alcohol	117.7
16136	H_2O	Water	100		Butyl alcohol	117.7
16137	H_2O	Water	100	$C_4H_{10}O$	Butyl alcohol	117.8
16138	H_2O	Water	100	$C_4H_{10}O$	Butyl alcohol	117
16139	H_2O	Water	100	$C_4H_{10}O$	Butyl alcohol	116.9
16140	H_2O	Water	100	$C_4H_{10}O$	Butyl alcohol	117.4
16141	H_2O	Water	100	$C_4H_{10}O$	Butyl alcohol	117.7
		"			" "	
16142	H_2O	Water	100	$C_4H_{10}O$	Butyl alcohol	117.75
		"			" "	
16143	H_2O	Water	100	$C_4H_{10}O$	Butyl alcohol	117.75
16144	H_2O	Water	100	$C_4H_{10}O$	Butyl alcohol	117.75
		"				
16145	H_2O	Water	100	$C_4H_{10}O$	Butyl alcohol	117.75
16145a	H_2O	Water	100	$C_4H_{10}O$	Butyl alcohol	117.7
16146	H_2O	Water	100	$C_4H_{10}O$	Butyl alcohol	117.75
16147	H_2O	Water	100	$C_4H_{10}O$	Butyl alcohol	117.75
		"			" "	
		"			" "	
16148	H_2O	Water	100	$C_4H_{10}O$	Butyl alcohol	117.75
16149	H_2O	Water	100	$C_4H_{10}O$	Butyl alcohol	117.75
16150	H_2O	Water	100	$C_4H_{10}O$	Butyl alcohol	117
16151	H_2O	Water	100	$C_4H_{10}O$	Butyl alcohol	117

TABLE II. *Ternary Systems* 469

C-Component			Azeotropic Data				
Formula	Name	B.P., °C.	B.P., °C.	Wt. % A	Wt. % B	Wt. % C	Ref.
C_6H_{12}	3-Methyl-2-pentene		Min. b.p.				66
C_6H_{14}	Hexane	68.9	56	4.8	27.1	68.1	623c
"			55	1	22	77	981
C_6H_{14}	2-Methylpentane		Min. b.p.				66
C_6H_{14}	3-Methylpentane		Min. b.p.				66
$C_4H_{10}O$	Isobutyl alcohol		Nonazeotrope				981
$C_4H_{10}O_2$	1,1-Dimethoxyethane	64.5	Nonazeotrope				981
$C_6H_{12}O_2$	Butyl acetate	126.1	Nonazeotrope				981
C_6H_{14}	Hexane	68.7	55	4	21	75	981
C_7H_{16}	Heptanes		~ 57				340
$C_4H_{10}O$	Isobutyl alcohol	108	Nonazeotrope			v-l	665
C_7H_{16}	Heptanes		48				340
$C_4H_{10}O$	sec-Butyl alcohol	99.5	Nonazeotrope			v-l	626
$C_4H_{10}O$	Butyl alcohol	117.73	Nonazeotrope				806
$C_8H_{18}O$	Butyl ether	141.97	Nonazeotrope				806
$C_4H_{10}O$	tert-Butyl alcohol	82.55	62				563
$C_4H_{10}O$	sec-Butyl alcohol	99.5	Nonazeotrope				981
$C_4H_{10}S$	1-Butanethiol	97.5	78.6				487
$C_4H_{11}N$	Butylamine	77.8	Nonazeotrope				981
$C_5H_{10}O_2$	Butyl formate	106.6	83.6	21.3	10	68.7	359,498c
C_6H_6	Benzene	80.1	69	10.1	1.5	88.4 v-l	880
C_6H_{10}	Cyclohexene	82.75	70.22				563
$C_6H_{11}ClO_2$	Butyl chloroacetate	181.9	93.1	41.8	50.3	7.9	121
$C_6H_{12}O$	Butyl vinyl ether	94.2	77.3	11.2	1.9	85.9	497
"			77.4	10	2	88	982
$C_6H_{12}O_2$	Butyl acetate	126.1	90.7	29	8	63	982
"			89.4	37.3	27.4	35.3	77,333,359
C_6H_{14}	Hexane	68.95	61.5	19.2	2.9	77.9	477
$C_7H_{12}O_2$	Butyl acrylate	147	92	50	37.6	12.4	215
"	100 mm.		46	41	26	33	982
C_7H_{16}	Heptane	98.4	78.1	41.4	7.6	51	477
C_8H_{14}	Diisobutylene	101		12.6	9.5	77.9	1007a
C_8H_{18}	Octane	125.75	86.1	60	14.6	25.4	477
$C_8H_{18}O$	Butyl ether	142.1	90.6	29.9	34.6	35.5	982
"	100 mm.		45	31.2	24.6	44.2	982
"			91	29.3	42.9	27.7	760
$C_8H_{19}N$	Dibutylamine	159.6	Nonazeotrope				981
C_9H_{20}	Nonane	150.7	90	69.9	18.3	11.8	477
$C_9H_{20}O_2$	Dibutoxy-methane	181.8	Nonazeotrope				324
$C_{10}H_{22}O_2$	Acetaldehyde dibutyl acetal	188.8	Nonazeotrope				45

	A-Component			B-Component		
No.	Formula		B.P., °C.	Formula	Name	B.P., °C.
16152	H_2O	Water	100	$C_4H_{10}O$	sec-Butyl alcohol	99.4
16153	H_2O	Water	100	$C_4H_{10}O$	sec-Butyl alcohol	99.6
16154	H_2O	Water	100	$C_4H_{10}O$	sec-Butyl alcohol	99.6
16155	H_2O	Water	100	$C_4H_{10}O$	sec-Butyl alcohol	99.6
16156	H_2O	Water	100	$C_4H_{10}O$	sec-Butyl alcohol	99.6
16157	H_2O	Water	100	$C_4H_{10}O$	sec-Butyl alcohol	99.6
16158	H_2O	Water	100	$C_4H_{10}O$	sec-Butyl alcohol	99.6
		"			"	"
		"			"	"
		"			"	"
		"			"	"
		"			"	"
16159	H_2O	Water	100	$C_4H_{10}O$	sec-Butyl alcohol	99.6
		"			"	"
16160	H_2O	Water	100	$C_4H_{10}O$	sec-Butyl alcohol	99.4
16161	H_2O	Water	100	$C_4H_{10}O$	sec-Butyl alcohol	99.6
16162	H_2O	Water	100	$C_4H_{10}O$	sec-Butyl alcohol	99.6
16163	H_2O	Water	100	$C_4H_{10}O$	sec-Butyl alcohol	99.6
16163a	H_2O	Water	100	$C_4H_{10}O$	sec-Butyl alcohol	99.4
16164	H_2O	Water	100	$C_4H_{10}O$	sec-Butyl alcohol	99.4
16165	H_2O	Water	100	$C_4H_{10}O$	sec-Butyl alcohol	99.4
16166	H_2O	Water	100	$C_4H_{10}O$	sec-Butyl alcohol	99.4
		"			"	"
16167	H_2O	Water	100	$C_4H_{10}O$	sec-Butyl alcohol	99.4
16168	H_2O	Water	100	$C_4H_{10}O$	sec-Butyl alcohol	99.4
		"			"	"
16169	H_2O	Water	100	$C_4H_{10}O$	sec-Butyl alcohol	99.53
16169a	H_2O	Water	100	$C_4H_{10}O$	tert-Butyl alcohol	82.5
16169b	H_2O	Water	100	$C_4H_{10}O$	tert-Butyl alcohol	82.5
16169c	H_2O	Water	100	$C_4H_{10}O$	tert-Butyl alcohol	82.5
16169d	H_2O	Water	100	$C_4H_{10}O$	tert-Butyl alcohol	82.5
16170	H_2O	Water	100	$C_4H_{10}O$	tert-Butyl alcohol	82.55
16171	H_2O	Water	100	$C_4H_{10}O$	tert-Butyl alcohol	82.55
16172	H_2O	Water	100	$C_4H_{10}O$	tert-Butyl alcohol	82.55
16172a	H_2O	Water	100	$C_4H_{10}O$	tert-Butyl alcohol	82.5
16173	H_2O	Water	100	$C_4H_{10}O$	tert-Butyl alcohol	82.55
16174	H_2O	Water	100	$C_4H_{10}O$	tert-Butyl alcohol	82.55
16175	H_2O	Water	100	$C_4H_{10}O$	Isobutyl alcohol	108
16176	H_2O	Water	100	$C_4H_{10}O$	Isobutyl alcohol	108.0
16177	H_2O	Water	100	$C_4H_{10}O$	Isobutyl alcohol	108
16178	H_2O	Water	100	$C_4H_{10}O$	Isobutyl alcohol	108
16179	H_2O	Water	100	$C_4H_{10}O$	Isobutyl alcohol	108

TABLE II. *Ternary Systems*   471

	C-Component		Azeotropic Data				
Formula	Name	B.P., °C.	B.P., °C.	Wt. % A	Wt. % B	Wt. % C	Ref.
$C_4H_{10}O$	*tert*-Butyl alcohol	82.6		Nonazeotrope			981
$C_5H_{10}O$	Allyl ethyl ether	67.6		Nonazeotrope			18
$C_5H_{12}O$	Butyl methyl ether	70.3		Nonazeotrope			18
$C_5H_{12}O$	Ethyl isopropyl ether	54		Nonazeotrope			18
$C_5H_{12}O$	Ethyl propyl ether	64		Nonazeotrope			18
$C_5H_{12}O$	Isobutyl methyl ether	59		Nonazeotrope			18
C_6H_6	Benzene	80.2		Azeotrope doubtful			563
	" 200 mm.		38.2	7	5	88	194
	" 300 mm.		47.0	7	6	87	194
	" 400 mm.		53.8	7	6	87	194
	" 500 mm.		59.0	7	6	87	194
	" 665 mm.		65.5	8	6	86	194
C_6H_{12}	Cyclohexane	80.75	~ 67				563
	"		69.7	8.9	10.8	80.3	497
$C_6H_{12}O_2$	*sec*-Butyl acetate	112.2	85.5	20.2	27.4	52.4	981
C_6H_{14}	Hexane	68.95	61.1				563
$C_6H_{14}O$	Ethyl *tert*-butyl ether	68-69		Nonazeotrope			18
$C_6H_{14}O$	Isopropyl ether	69		Nonazeotrope			18
C_7H_8	Toluene	110.7	200-662 mm			v-l	257c
C_7H_{14}	Methylcyclohexane	101.1	77.1	11.9	21.9	66.4	1039
C_7H_{16}	Heptane	98.4	75.8	10.9	22.2	66.9	1039
C_8H_{14}	Diisobutylene	102.3	77.5	11	19	70	1039
	"		80.2				740
C_8H_{18}	Isoctane	99	76.3	9	19	72	498i
$C_8H_{18}O$	Butyl ether	142.1	86.6	24.7	56.1	19.2	981
	" "		86.5				255
$C_8H_{18}O$	*sec*-Butyl ether	121	83				255
C_5H_8	Isoprene	34		Nonazeotrope			581c
$C_5H_{10}O$	2-Methyl-3-buten-2-ol	97		Nonazeotrope			581c
C_5H_{12}	Pentane	36.15		Nonazeotrope			581c
$C_5H_{12}O$	*tert*-Butyl methyl ether	55		Nonazeotrope			581c
C_6H_6	Benzene	80.2	67.30	8.1	21.4	70.5	1044
C_6H_8	1,3-Cyclohexadiene	80.8	66.7				563
C_6H_{10}	Cyclohexene	82.75	67				563
$C_6H_{10}O$	Methyl dihydropyran	118.5		Nonazeotrope			581c
C_6H_{12}	Cyclohexane	80.75	65	8	21	71	563
C_6H_{14}	Hexane	68.95	58.9				563
$C_5H_{10}O$	Isovaleraldehyde			Nonazeotrope		v-l	665
$C_5H_{10}O$	3-Pentanone	102.2		Nonazeotrope			563
$C_5H_{10}O_2$	Isobutyl formate	98.3	79.3	12.3	1.15	86.55	498i
C_6H_6	Benzene	80.2		Nonazeotrope			1044
C_6H_8	1,3-Cyclohexadiene	80.8		Nonazeotrope			563

		A-Component	B.P., °C.		B-Component	B.P., °C.
No.	Formula			Formula	Name	
16180	H_2O	Water	100	$C_4H_{10}O$	Isobutyl alcohol	108
16181	H_2O	Water	100	$C_4H_{10}O$	Isobutyl alcohol	107.4
16182	H_2O	Water	100	$C_4H_{10}O$	Isobutyl alcohol	108
16183	H_2O	Water	100	$C_4H_{10}O$	Isobutyl alcohol	108
16184	H_2O	Water	100	$C_4H_{10}O$	Isobutyl alcohol	108
16185	H_2O	Water	100	$C_4H_{10}O$	Isobutyl alcohol	108
16186	H_2O	Water	100	$C_4H_{10}O$	Isobutyl alcohol	108
16187	H_2O	Water	100	$C_4H_{10}O$	Isobutyl alcohol	108
16188	H_2O	Water	100	$C_4H_{10}O$	Isobutyl alcohol	108
16189	H_2O	Water	100	$C_4H_{10}O$	Isobutyl alcohol	107.5
16190	H_2O	Water	100	$C_4H_{10}O$	Isobutyl alcohol	107.8
16191	H_2O	Water	100	$C_4H_{10}O_2$	2-Ethoxyethanol	135.6
16192	H_2O	Water	100	$C_4H_{10}O_2$	2-Ethoxyethanol	135.6
16193	H_2O	Water	100	$C_4H_{11}N$	Butylamine	77.8
16194	H_2O	Water	100	$C_4H_{11}N$	Diethylamine	55.5
16195	H_2O	Water	100	C_5H_5N	Pyridine	115.5
16196	H_2O	Water	100	C_5H_5N	Pyridine	115.5
16197	H_2O	Water	100	C_5H_5N	Pyridine	115.5
16198	H_2O	Water	100	C_5H_5N	Pyridine	115.5
16199	H_2O	Water	100	C_5H_5N	Pyridine	115.5
16200	H_2O	Water	100	C_5H_5N	Pyridine	115.5
16201	H_2O	Water	100	C_5H_5N	Pyridine	115.5
16202	H_2O	Water	100	C_5H_5N	Pyridine	115.5
16203	H_2O	Water	100	C_5H_5N	Pyridine	115.5
16204	H_2O	Water	100	C_5H_5N	Pyridine	115.5
16205	H_2O	Water "	100	C_5H_5N	Pyridine "	115.5
16206	H_2O	Water	100	C_5H_5N	Pyridine	115.5
16207	H_2O	Water	100	C_5H_5N	Pyridine	115.5
16208	H_2O	Water	100	C_5H_5N	Pyridine	115.5
16209	H_2O	Water	100	C_5H_5N	Pyridine	115.5
16210	H_2O	Water	100	C_5H_5N	Pyridine	115.5
16211	H_2O	Water	100	C_5H_5N	Pyridine	115.5
16212	H_2O	Water	100	C_5H_5N	Pyridine	115.5
16212a	H_2O	Water	100	C_5H_8	Isoprene	34
16213	H_2O	Water	100	$C_5H_8O_2$	Ethyl acrylate	99.3
16213a	H_2O	Water	100	$C_5H_{10}O$	2-Methyl-3-buten-2-ol	97

TABLE II. *Ternary Systems*   473

	C-Component			Azeotropic Data				
Formula	Name	B.P., °C.		B.P., °C.	Wt. % A	Wt. % B	Wt. % C	Ref.
C_6H_{10}	Cyclohexene	82.75		~ 69.5				563
$C_6H_{11}ClO_2$	Isobutyl chloro-acetate	174.4		90.2	33.64	53.1	13.26	121
C_6H_{12}	Cyclohexane	80.75		Nonazeotrope				563
$C_6H_{12}O_2$	Isobutyl acetate	117.2		86.8	30.4	23.1	46.5	333,359, 498c
C_6H_{14}	Hexane	68.95		Nonazeotrope				563
C_7H_8	Toluene	110.7		81.3	17.9	16.4	65.7 v-l	298c
C_8H_{10}	Ethylbenzene	136.15		~ 89.5				563
$C_8H_{18}O$	Butyl ether	141.9		89				760
$C_8H_{18}O$	Isobutyl ether	122		85.4				760
$C_9H_{20}O_2$	Diisobutoxy-methane	163.8		Nonazeotrope				324
$C_{10}H_{22}O_2$	Acetaldehyde diisobutyl acetal	171.3		Nonazeotrope				45
C_7H_8	Toluene	110.7		Nonazeotrope				981
$C_{10}H_{20}O$	2-Ethylhexyl vinyl ether	177.7		97.7	51	11	38	981
$C_8H_{19}N$	Dibutylamine	159.6		Nonazeotrope				981
$C_6H_{14}O$	Isopropyl ether	68.3		Nonazeotrope				981
C_6H_6	Benzene	80.1		Nonazeotrope				215
C_6H_7N	2-Picoline			Nonazeotrope				215
C_6H_8	1.3-Cyclohexadiene	80.8		Min. b.p.				913
C_6H_{10}	Cyclohexene	82.75		Min. b.p.				913
C_6H_{12}	Cyclohexane	80.75		Min. b.p.				913
C_7H_{10}	Methylcyclo-hexadiene			Min. b.p.				913
C_7H_{14}	1,1-Dimethyl-cyclopentane	87.8		Min. b.p.				913
C_7H_{14}	1.2-Dimethyl-cyclopentane			Min. b.p.				913
C_7H_{14}	1.3-Dimethyl-cyclopentane	90.8		Min. b.p.				913
C_7H_{14}	Methylcyclohexane	101.2		80.0		5		913
C_7H_{16}	*n*-Heptane	98.45		Min. b.p.				913
	"			78.6	14	13.5	70.5	968
C_7H_{16}	3-Methylhexane	91.8		Min. b.p.				913
C_8H_{14}	Diisobutylene	101		Min. b.p.				913
C_8H_{18}	Octane	125.75		86.7	22.5	25.5	52	968
C_9H_{20}	Nonane	150.7		90.5	40.5	37	32.5	968
$C_{10}H_{22}$	Decane	173.3		92.3	35.5	45.5	19	968
$C_{11}H_{24}$	Undecane	194.5		93.1	38.5	51	10.5	968
$C_{12}H_{26}$	Dodecane	216		93.5	40.5	54.5	5	968
$C_5H_{12}O$	*tert*-Butyl methyl ether 55			Nonazeotrope				581c
$C_6H_{14}O$	Isopropyl ether	68.3		Nonazeotrope				981
$C_5H_{12}O$	*tert*-Butyl methyl ether 55			Nonazeotrope				581c

	A-Component		B.P., °C.		B-Component	B.P., °C.
No.	Formula			Formula	Name	
16213b	H_2O	Water	100	$C_5H_{10}O$	2-Methyl-3-buten-2-ol	97
16213c	H_2O	Water	100	$C_5H_{10}O$	3-Methyl-3-buten-1-ol	130
16214	H_2O	Water	100	$C_5H_{10}O$	2-Pentanone	102.3
16215	H_2O	Water	100	$C_5H_{12}O$	Amyl alcohol	137.8
16216	H_2O	Water	100	$C_5H_{12}O$	Amyl alcohol	137.8
16217	H_2O	Water	100	$C_5H_{12}O$	Amyl alcohol	138
16218	H_2O	Water	100	$C_5H_{12}O$	Amyl alcohol	137.2
16219	H_2O	Water	100	$C_5H_{12}O$	Amyl alcohol	137.5
16220	H_2O	Water	100	$C_5H_{12}O$	tert-Amyl alcohol	102
16221	H_2O	Water	100	$C_5H_{12}O$	tert-Amyl alcohol	102
16222	H_2O	Water	100	$C_5H_{12}O$	tert-Amyl alcohol	102
16223	H_2O	Water	100	$C_5H_{12}O$	Isoamyl alcohol	132
16224	H_2O	Water	100	$C_5H_{12}O$	Isoamyl alcohol	131.3
16225	H_2O	Water	100	$C_5H_{12}O$	Isoamyl alcohol	131.5
16226	H_2O	Water	100	$C_5H_{12}O$	Isoamyl alcohol	131.3
16227	H_2O	Water	100	$C_5H_{12}O$	Isoamyl alcohol	131.5
16228	H_2O	Water	100	$C_5H_{12}O$	Isoamyl alcohol	132
16229	H_2O	Water	100	$C_5H_{12}O$	Isoamyl alcohol	131.6
16229a	H_2O	Water	100	C_6H_6	Benzene	80.1
16230	H_2O	Water	100	C_6H_6	Benzene	80.1
16231	H_2O	Water	100	C_6H_6O	Phenol	182
16232	H_2O	Water	100	C_6H_7N	2-Picoline	128.8
16233	H_2O	Water	100	$C_6H_{10}O$	Cyclohexanone	155.6
16233a	H_2O	Water	100	$C_6H_{10}O$	Cyclohexanone	155.6
16233b	H_2O	Water	100	C_6H_{12}	Cyclohexane	80.75
16233c	H_2O	Water	100	$C_6H_{12}O$	Cyclohexanol	160.65
16234	H_2O	Water	100	$C_6H_{12}O$	2-Methyl-2-penten-4-ol	
16235	H_2O	Water	100	$C_6H_{12}O_2$	Butyl acetate	126.1
16236	H_2O	Water	100	$C_6H_{12}O_2$	Butyl acetate	126.1
16237	H_2O	Water	100	$C_6H_{12}O_2$	sec-Butyl acetate	112.2
16238	H_2O	Water	100	$C_6H_{12}O_3$	2-Ethoxyethyl acetate	156.2
16239	H_2O	Water	100	$C_6H_{14}O$	Isopropyl ether	68.3
16240	H_2O	Water	100	C_7H_8	Toluene	110.6
16241	H_2O	Water	100	C_7H_8	Toluene	110.6
16242	H_2O	Water	100	C_7H_8	Toluene	110.6

TABLE II. *Ternary Systems* 475

Formula	Name	B.P., °C.	B.P., °C.	Wt. % A	Wt. % B	Wt. % C	Ref.
	C-Component			Azeotropic Data			
$C_6H_{14}O$	Isopropyl ether	68		Nonazeotrope			581c
$C_6H_{12}O_2$	4,4-Dimethyl-1,3-dioxane	133.4		Nonazeotrope			581c
C_6H_6	Benzene	80.1		Nonazeotrope			981
$C_6H_{12}O_2$	Amyl formate	132	91.4	37.5	21.5	41	359,498c
$C_7H_{14}O_2$	Amyl acetate	148.8	94.9	45.9	12.2	41.9	498i
$C_{10}H_{22}O$	Amyl ether	188	95.94				1033
$C_{11}H_{24}O_2$	Diamyloxymethane	221.6		Nonazeotrope			324
$C_{12}H_{26}O_2$	Acetaldehyde diamyl acetal	225.3		Nonazeotrope			45
C_6H_6	Benzene	80.2		Nonazeotrope			563
C_6H_{12}	Cyclohexane	80.75		Nonazeotrope			563
C_7H_8	Toluene	110.7	~82				563
$C_5H_{12}S$	3-Methyl-1-butane-thiol 765.4 mm	120	86.6				487
C_6H_6	Benzene	80.2		Nonazeotrope			1044
$C_6H_{12}O_2$	Isoamyl formate	124.2	89.8	32.4	19.6	48	359
$C_7H_{13}ClO_2$	Isoamyl chloro-acetate	195.2	95.4	46.2	47.3	6.5	121
$C_7H_{14}O_2$	Isoamyl acetate	142	93.6	44.8	31.2	24	359,411
$C_{10}H_{22}O$	Isoamyl ether	171	94.4				760
$C_{12}H_{26}O_2$	Acetaldehyde di-isoamyl acetal	213.6		Nonazeotrope			45
$C_6H_{13}N$	Cyclohexylamine	134		Nonazeotrope		v-l	718c
$C_6H_{11}O$	Hexyl alcohol	157.85		Nonazeotrope			480
C_8H_{10}	Xylene	137		Min. b.p.			90
$C_6H_{12}O_3$	Paraldehyde	124.5		Nonazeotrope			981
$C_6H_{12}O$	Cyclohexanol	160.65		Nonazeotrope		v-l	335
$C_6H_{13}N$	Cyclohexylamine					v-l	718c
$C_6H_{13}N$	Cyclohexylamine					v-l	718c
$C_6H_{13}N$	Cyclohexylamine					v-l	718c
$C_8H_{14}O$	2,4,6-Trimethyl-5,6-dihydro-1,2-pyran		90.7	27.0	9.7	63.3	849
$C_6H_{12}O_2$	sec-Butyl acetate	112.2		Nonazeotrope			981
$C_8H_{18}O$	Butyl ether	142.1		Nonazeotrope			981
$C_8H_{18}O$	Butyl ether	142.1		Nonazeotrope			981
C_7H_8	Toluene	110.6		Nonazeotrope			981
$C_6H_{15}N$	Triethylamine	89.7		Nonazeotrope			981
C_7H_8O	Benzyl alcohol	204.7		Nonazeotrope		v-l	935
$C_{10}H_{22}O$	Decyl alcohol (isomers)	217.3		Nonazeotrope			981
$C_{12}H_{26}O$	2,6,8-Trimethyl-4-nonanol	225.5		Nonazeotrope			981

		A-Component	B.P., °C.	B-Component		B.P., °C.
No.	Formula			Formula	Name	
16243	H_3N	Ammonia	— 33	C_2H_6O	Methyl ether	— 24
16244	H_3N	Ammonia	— 33	C_3H_9N	Trimethylamine	3.5
16245	H_3N	Ammonia	— 33	C_3H_9N	Trimethylamine	3.5
16246	H_3N	Ammonia	— 33	C_3H_9N	Trimethylamine	3.5
16247	H_3N	Ammonia	— 33	C_3H_9N	Trimethylamine	3.5
16247a	O_2S	Sulfur dioxide	—10	CH_4O	Methanol	64.7
16247b	O_2S	Sulfur dioxide	—10	CH_4O	Methanol	64.7
16248	CCl_4	Carbon tetrachloride	76.8	CH_4O	Methanol	64.7
16249	CCl_4	Carbon tetrachloride	76.75	CH_4O	Methanol	64.7
16250	CCl_4	Carbon tetrachloride	76.75	C_2HCl_3	Trichloroethylene	86.2
16251	CCl_4	Carbon tetrachloride	76.75	$C_2H_4Br_2$	1,2-Dibromoethane	131.5
16252	CCl_4	Carbon tetrachloride	76.75	C_2H_6O	Ethyl alcohol	78.3
16253	CCl_4	Carbon tetrachloride	76.75	C_2H_6O	Ethyl alcohol	78.3
16254	CCl_4	Carbon tetrachloride	76.75	C_2H_6O	Ethyl alcohol ”	78.3
16254a	CCl_4	Carbon tetrachloride	76.8	$C_3H_6O_2$	Methyl acetate	57.1
16255	CCl_4	Carbon tetrachloride	76.8	C_3H_8O	Propyl alcohol	97.8
16255a	CCl_4	Carbon tetrachloride	76.8	C_3H_8O	Isopropyl alcohol	82.5
16256	CCl_4	Carbon tetrachloride	76.8	C_3H_8O	Isopropyl alcohol	82.5
16257	CCl_4	Carbon tetrachloride	76.75	C_4H_8O	2-Butanone	79.6
16257a	CCl_4	Carbon tetrachloride	76.8	C_4H_8O	2-Butanone	79.6
16258	CCl_4	Carbon tetrachloride	76.75	C_4H_8O	2-Butanone	79.6
16259	CCl_4	Carbon tetrachloride	76.75	$C_4H_8O_2$	Ethyl acetate	77.05
16260	CCl_4	Carbon tetrachloride	76.8	$C_4H_{10}O$	Butyl alcohol	117
16261	CCl_4	Carbon tetrachloride	76.8	C_6H_6	Benzene	80.1
16262	CCl_4	Carbon tetrachloride	76.8	C_6H_6	Benzene	80.1
16263	CS_2	Carbon disulfide	46.25	CH_3I	Iodomethane	42.6
16264	CS_2	Carbon disulfide	46.25	CH_3I	Iodomethane	42.6

TABLE II. *Ternary Systems*   477

C-Component			Azeotropic Data					
Formula	Name	B.P., °C.	B.P., °C.	Wt. % A	Wt. % B	Wt. % C		Ref.
C_3H_9N	Trimethylamine	3.5		Nonazeotrope				378
C_4H_8	1-Butene	−6		Nonazeotrope				378
C_4H_8	2-Methylpropene	−6		Nonazeotrope				378
C_4H_{10}	Butane	0		Nonazeotrope				378
C_4H_{10}	2-Methylpropane	−10		Nonazeotrope				378
C_3H_6O	Acetone	56.1	20°–40°				v-l	44c
$C_3H_6O_2$	Methyl acetate	57.1	20°–40°				v-l	44c
C_6H_6	Benzene	80.1		Nonazeotrope			v-l	385
C_6H_{12}	Cyclohexane	80.75		Nonazeotrope				563
C_4H_8O	2-Butanone	79.6		Nonazeotrope			v-l	501
C_7H_8	Toluene	110.7		Nonazeotrope				563
C_4H_8O	2-Butanone	79.6		Nonazeotrope				563
$C_4H_8O_2$	Ethyl acetate	77.05		Nonazeotrope				563
C_6H_6	Benzene	80.2		Nonazeotrope				563
	"			Nonazeotrope			v-l	123,385
C_6H_6	Benzene	80.1					v-l	684a
C_6H_6	Benzene	80.1		Nonazeotrope			v-l	385
C_6H_{12}	Cyclohexane	80.75	68.55				v-l	1032c
C_6H_6	Benzene	80.1		Nonazeotrope			v-l	684
$C_4H_8O_2$	Methyl propionate	79.7		Nonazeotrope				563
C_6H_6	Benzene	80.1					v-l	598c
C_6H_{12}	Cyclohexane	80.75		Nonazeotrope				563
C_6H_{12}	Cyclohexane	80.75		Nonazeotrope				563
C_6H_6	Benzene	80.1		Nonazeotrope			v-l	385
C_6H_{12}	Cyclohexane			Nonazeotrope			v-l	598
$C_6H_{12}O_2$	Butyl acetate			Nonazeotrope			v-l	599
CH_4O	Methanol	64.7	35.95			<12		563
$C_2H_4O_2$	Methyl formate	31.9		Nonazeotrope				563

	A-Component		B.P., °C.	B-Component		B.P., °C.
No. Formula				Formula	Name	
16265	CS_2	Carbon disulfide	46.25	CH_3I	Iodomethane	42.5
16266	CS_2	Carbon disulfide	46.25	CH_4O	Methanol	64.7
16267	CS_2	Carbon disulfide	46.25	CH_4O	Methanol	64.7
16268	CS_2	Carbon disulfide	46.25	CH_4O	Methanol	64.7
16269	CS_2	Carbon disulfide	46.25	CH_4O	Methanol	64.7
16270	CS_2	Carbon disulfide	46.25	CH_4O	Methanol	64.7
16271	CS_2	Carbon disulfide	46.25	CH_4O	Methanol	64.7
16272	CS_2	Carbon disulfide	46.25	$C_2H_4O_2$	Methyl formate	31.9
16273	CS_2	Carbon disulfide	46.25	$C_2H_4O_2$	Methyl formate	31.9
16274	CS_2	Carbon disulfide	46.25	$C_2H_4O_2$	Methyl formate	31.9
16275	CS_2	Carbon disulfide	46.25	C_2H_6O	Ethyl alcohol	78.3
16276	CS_2	Carbon disulfide	46.25	C_2H_6O	Ethyl alcohol	78.3
16277	CS_2	Carbon disulfide	46.25	C_3H_6O	Acetone	56.25
16278	CS_2	Carbon disulfide	46.25	$C_3H_6O_2$	Ethyl formate	54.1
16279	CS_2	Carbon disulfide	46.25	C_3H_8O	Isopropyl alcohol	82.45
16280	CS_2	Carbon disulfide	46.25	$C_3H_8O_2$	Methylal	42.25
16281	$CHCl_3$	Chloroform	61	CH_2Cl_2	Dichloromethane	40
16282	$CHCl_3$	Chloroform	61	CH_2O_2	Formic acid	100.75
16283	$CHCl_3$	Chloroform "	61	CH_4O	Methanol "	64.7
16284	$CHCl_3$	Chloroform	61	CH_4O	Methanol	64.7
16285	$CHCl_3$	Chloroform	61	CH_4O	Methanol	64.7
16286	$CHCl_3$	Chloroform	61.2	CH_4O	Methanol	64.7
16286a	$CHCl_3$	Chloroform	61	CH_4O	Methanol	64.7
16287	$CHCl_3$	Chloroform	61	CH_4O	Methanol	64.7
16288	$CHCl_3$	Chloroform	61	C_2H_6O	Ethyl alcohol	78.3
16289	$CHCl_3$	Chloroform	61	C_2H_6O	Ethyl alcohol	78.3
16290	$CHCl_3$	Chloroform	61	C_2H_6O	Ethanol	78.3
16291	$CHCl_3$	Chloroform	61.2	C_2H_6O	Ethyl alcohol	78.3
16292	$CHCl_3$	Chloroform	61.2	C_2H_6O	Ethyl alcohol	78.3
16293	$CHCl_3$	Chloroform	61.2	C_2H_6O	Ethyl alcohol	78.3
16294	$CHCl_3$	Chloroform "	61.2	C_2H_6O	Ethyl alcohol " "	78.3
16295	$CHCl_3$	Chloroform	61	C_3H_6O	Acetone	56.4
16296	$CHCl_3$	Chloroform	61	C_3H_6O	Acetone	56.4
16297	$CHCl_3$	Chloroform " " "	61.2	C_3H_6O	Acetone " " "	56.5
16297a	$CHCl_3$	Chloroform	61	C_3H_6O	Acetone	56.1
16298	$CHCl_3$	Chloroform	61	C_3H_6O	Acetone	56.4
16299	$CHCl_3$	Chloroform	61	C_3H_6O	Acetone	56.4

TABLE II. *Ternary Systems* 479

	C-Component		Azeotropic Data					
Formula	Name	B.P., °C.	B.P., °C.	Wt. % A	Wt. % B	Wt. % C		Ref.
$C_3H_8O_2$	Methylal	42.25	37.2?					563
C_2H_5Br	Bromoethane	38.4	33.92	~40	~10	50		563
C_3H_6O	Acetone	56.25	Nonazeotrope					563
$C_3H_6O_2$	Methyl acetate	57.0	37					563
C_3H_7Cl	1-Chloropropane	46.6	37?					563
$C_3H_8O_2$	Methylal	42.25	35.55	55	7	38		563
C_5H_{10}	2-Methyl-2-butene	37.15	Nonazeotrope					563
C_2H_5Br	Bromoethane	38.4	24.7?	18?	60?	22?		563
C_5H_{10}	2-Methyl-2-butene	37.15	~24					563
C_5H_{12}	Pentane	36.15	21.5?					563
C_4H_8O	2-Butanone	79.6	Nonazeotrope					563
$C_4H_8O_2$	Ethyl acetate	77.05	Nonazeotrope					563
$C_3H_6O_2$	Methyl acetate	57.0	Nonazeotrope					563
C_3H_7Cl	1-Chloropropane	46.6	38.2?					563
$C_4H_8O_2$	Ethyl acetate	77.05	Nonazeotrope					563
C_5H_{10}	2-Methyl-2-butene	37.15	35.2?					563
C_3H_6O	Acetone	56.4	Nonazeotrope					261
$C_2H_4O_2$	Acetic acid	118.1	Nonazeotrope				v-l	171
C_3H_6O	Acetone	56.1	57.5	46.7	23.4	29.9	v-l	982
	"		57.5	47	23	30	v-l	261
$C_3H_6O_2$	Methyl acetate	57.1	56.4	52.5	21.6	25.9	v-l	406a
C_4H_8O	2-Butanone	79.6	Nonazeotrope					981
$C_4H_8O_2$	Ethyl acetate	77.1	Nonazeotrope				v-l	679
C_6H_{14}	2,3-Dimethylbutane	58	Nonazeotrope				v-l	461c
C_6H_{14}	Hexane	69.85	Nonazeotrope					563
C_3H_6O	Acetone	56.1	55.0	70.2	6.8	23	v-l	937
C_3H_6O	Acetone	56.1	63.2	65.3	10.4	24.3	v-l	664
C_6H_{14}	Hexane	68.95	~58.3					563
C_6H_{14}	Hexane	68.7	57.3	56.1	9.5	34.4	v-l	498
C_6H_{14}	Hexane	68.7	55				v-l	498
C_6H_{14}	Hexane		45				v-l	498
C_6H_{14}	Hexane		35				v-l	498
	"		60.6	69.2	4.5	26.3		497
C_4H_8O	2-Butanone	79.6	Nonazeotrope				v-l	185
C_6H_6	Benzene	80.2	Nonazeotrope				v-l	802
C_6H_{14}	Hexane	68.7	60.79	68.8	3.6	27.6	v-l	498
	"	631 mm	55	68.7	3.6	27.7	v-l	498
	"	444 mm	45	68.7	3.6	27.7	v-l	498
	"	300 mm	35	68.7	3.3	28.0	v-l	498
C_6H_{14}	2,3-Dimethylbutane	58	Nonazeotrope				v-l	317c
$C_6H_{12}O$	4-Methyl-2-pentanone	115.9	Nonazeotrope				v-l	438
$C_6H_{14}O$	Isopropyl ether	68.3	Nonazeotrope					981

A-Component			B.P., °C.	B-Component		B.P., °C.
No.	Formula			Formula	Name	
16300	CHCl$_3$	Chloroform	61	C$_3$H$_6$O	Acetone	56.4
16301	CHCl$_3$	Chloroform	61	C$_3$H$_6$O$_2$	Ethyl formate	54.1
16302	CHCl$_3$	Chloroform	61	C$_3$H$_6$O$_2$	Methyl acetate	57.1
16303	CHCl$_3$	Chloroform	61	C$_3$H$_7$Br	2-Bromopropane	59.4
16304	CHCl$_3$	Chloroform	61	C$_3$H$_8$O	Isopropyl alcohol	82.3
16305	CHCl$_3$	Chloroform	61.2	C$_4$H$_8$O	2-Butanone	79.6
16306	CH$_2$Cl$_2$	Dichloromethane	40.0	CH$_4$O	Methanol	64.7
16307	CH$_2$Cl$_2$	Dichloromethane	40.0	CH$_4$O	Methanol	64.7
16307a	CH$_3$Cl$_3$Si	Trichloromethylsilane	66.4	C$_2$H$_6$Cl$_2$Si	Dichlorodimethylsilane	
16307b	CH$_3$Cl$_3$Si	Trichloromethylsilane	66.4	C$_2$H$_6$Cl$_2$Si	Dichlorodimethylsilane	
16308	CH$_3$I	Iodomethane	42.6	CH$_4$O	Methanol	64.7
16309	CH$_3$I	Iodomethane	42.7	C$_2$H$_4$O$_2$	Methyl formate	31.9
16310	CH$_3$NO$_2$	Nitromethane	101.3	C$_2$Cl$_4$	Tetrachloro-ethylene	120.8
16311	CH$_3$NO$_2$	Nitromethane	101.3	C$_3$H$_8$O	Propyl alcohol	97.5
16312	CH$_3$NO$_2$	Nitromethane	101.3	C$_3$H$_8$O	Propyl alcohol	97.5
16313	CH$_3$NO$_2$	Nitromethane	101.3	C$_4$H$_8$O$_2$	Methyl propionate	79.8
16314	CH$_3$NO$_2$	Nitromethane	101.3	C$_5$H$_8$	Isoprene	34.1
16315	CH$_3$NO$_2$	Nitromethane	101.3	C$_5$H$_8$	Isoprene	34.1
16316	CH$_3$NO$_2$	Nitromethane	101.3	C$_5$H$_{10}$	2-Methyl-2-butene	38.5
16317	CH$_3$NO$_2$	Nitromethane	101.2	C$_5$H$_{10}$O	3-Pentanone	102.2
16318	CH$_3$NO$_2$	Nitromethane	101.2	C$_6$H$_6$	Benzene	80.1
16319	CH$_4$	Methane	—161.5	C$_2$H$_6$	Ethane	— 88.6
16320	CH$_4$O	Methanol	64.7	C$_2$H$_5$Br	Bromoethane	38.4
		,,			,,	
16321	CH$_4$O	Methanol	64.7	C$_2$H$_5$Br	Bromoethane	38.4
16322	CH$_4$O	Methanol	64.7	C$_2$H$_5$I	Iodoethane	72.3
16323	CH$_4$O	Methanol	64.7	C$_2$H$_5$I	Iodoethane	72.3
16324	CH$_4$O	Methanol	64.7	C$_2$H$_6$O	Ethyl alcohol	78.3
16324a	CH$_4$O	Methanol	64.7	C$_2$H$_6$O$_2$	Ethylene glycol	197
16325	CH$_4$O	Methanol	64.5	C$_3$H$_6$O	Acetone	56.1
		,,			,,	
16326	CH$_4$O	Methanol	64.7	C$_3$H$_6$O	Acetone	56.25
16327	CH$_4$O	Methanol	64.7	C$_3$H$_6$O	Acetone	56.4
		,,			,,	
		,,			,,	
16327a	CH$_4$O	Methanol	64.7	C$_3$H$_6$O	Acetone	56.1
16328	CH$_4$O	Methanol	64.7	C$_3$H$_6$O	Acetone	56.25
		,,			,,	
16329	CH$_4$O	Methanol	64.7	C$_3$H$_6$O$_2$	Methyl acetate	57

TABLE II. *Ternary Systems* 481

	C-Component			Azeotropic Data				
Formula	Name	B.P., °C.		B.P., °C.	Wt. % A	Wt. % B	Wt. % C	Ref.
C_7H_8	Toluene	110.7			Nonazeotrope		v-1	838
C_3H_7Br	2-Bromopropane	59.4		61.97	79	5.3	15.7	579
C_3H_7Br	2-Bromopropane	59.4			Nonazeotrope			579
C_4H_8O	Isopropyl formate	68.8			Nonazeotrope			579
C_4H_8O	2-Butanone	79.6			Nonazeotrope			981
C_6H_6	Benzene	80.1			Nonazeotrope		v-1	493
C_3H_6O	Acetone	56.4			Nonazeotrope			261
$C_8H_{18}O_3$	2 (2-Butoxy ethoxy) ethanol	230.6			Nonazeotrope			981
$C_4H_7ClO_2$	Ethyl chloroacetate	143.5			Nonazeotrope		v-1	886e
$C_4H_8Cl_2O$	Bis(2-chloroethyl) ether	179	60°C.		Nonazeotrope		v-1	886c
$C_3H_8O_2$	Methylal	42.25		38.5				563
C_5H_{12}	Pentane	36.15			Nonazeotrope			563
C_3H_8O	Propyl alcohol	97.5		86.68				611
$C_5H_{10}O$	3-Pentanone	102.2			Azeotropic			563
C_8H_{18}	*n*-Octane	125.75		85.52				611
C_7H_{14}	Methyl cyclohexane	101.1			Nonazeotrope			1038
C_5H_{10}	2-Methyl-2-butene	38.5			Nonazeotrope		v-1	716
C_5H_{12}	2-Methylbutane	27.9			Nonazeotrope		v-1	716
C_5H_{12}	2-Methylbutane	27.9			Nonazeotrope		v-1	716
$C_5H_{10}O_2$	Propyl acetate	101.55		99.0?				563
C_6H_{12}	Cyclohexane	80.75			Nonazeotrope		v-1	1010
C_3H_8	Propane 200° to 50°F.	—44			Nonazeotrope		v-1	766
C_5H_{10}	2-Methyl-2-butene	37.15		31.4	15	55	30	563
	"			32.0	7.1	48.5	44.4	497
C_3H_6O	2-Methyl-butane	27.95			Nonazeotrope			563
C_5H_{12}	Acetone	56.25			Nonazeotrope ?			563
$C_4H_8O_2$	Ethyl acetate	77.05			Nonazeotrope			563
C_3H_6O	Acetone	56.1			Nonazeotrope		v-1	16
C_4H_8O	Tetrahydrofuran	66			Nonazeotrope		v-1	859c
$C_3H_6O_2$	Methyl acetate	56.3		53.7	17.4	5.8	76.8	982,563
	" "	56.3			Nonazeotrope			303
C_4H_9Cl	1-Chloro-2-methylpropane	68.85		52.0				563
C_6H_{12}	Cyclohexane	80.75		51.1	16	43.5	40.5	276
	" 35°				8.9	52.4	38.7	v-1 622
	" 45°							v-1 622
	" 55°							v-1 622
C_6H_{14}	2,3-Dimethylbutane	58		44	12.4	20.2	67.4	v-1 1026c
C_6H_{14}	Hexane	68.95			Nonazeotrope			563
	"			47	14.6	30.8	59.6	283
C_6H_{12}	Cyclohexane	80.75		50.8	17.8	48.6	33.6	276

	A-Component		B.P., °C.	B-Component		B.P., °C.
No.	Formula			Formula	Name	
16330	CH$_4$O	Methanol "	64.5	C$_3$H$_6$O$_2$	Methyl acetate " "	56.3
16331	CH$_4$O	Methanol	64.7	C$_3$H$_8$O$_2$	Methylal	42.25
16332	CH$_4$O	Methanol	64.7	C$_3$H$_9$BO$_3$	Trimethyl borate	68.7
16333	CH$_4$O	Methanol	64.7	C$_4$H$_8$O$_2$	Ethyl acetate	77.05
16334	CH$_4$O	Methanol	64.7	C$_5$H$_8$	Isoprene	
16335	CH$_4$O	Methanol	64.7	C$_5$H$_8$	Isoprene	
16336	CH$_4$O	Methanol	64.7	C$_6$H$_6$	Benzene	80.2
16337	CH$_4$O	Methanol "	64.7	C$_6$H$_6$	Benzene "	80.2
16337a	CH$_4$O	Methanol	64.7	C$_6$H$_6$	Benzene	80.1
16338	CH$_4$O	Methanol	64.7	C$_6$H$_8$	1,3-Cyclohexa-diene	80.8
16338a	CH$_4$O	Methanol	64.7	C$_6$H$_{14}$	Hexane	68.9
16338b	CH$_4$O	Methanol	64.7	C$_7$H$_8$	Toluene	110.7
16339	CH$_4$O	Methylamine	— 6.7	C$_4$H$_6$	Butadiene	— 4.6
16340	C$_2$Cl$_4$	Tetrachloro-ethylene	120.8	C$_2$H$_4$O$_2$	Acetic acid	118.5
16341	C$_2$Cl$_4$	Tetrachloro-ethylene	120.8	C$_3$H$_5$ClO	Epichlorohydrin	116.45
16342	C$_2$Cl$_4$	Tetrachloro-ethylene	120.8	C$_3$H$_5$ClO	Epichlorohydrin	116.45
16343	C$_2$Cl$_4$	Tetrachloro-ethylene	120.8	C$_3$H$_5$ClO	Epichlorohydrin	116.45
16344	C$_2$Cl$_4$	Tetrachloro-ethylene	120.8	C$_3$H$_5$ClO	Epichlorohydrin	116.45
16345	C$_2$Cl$_4$	Tetrachloro-ethylene	120.8	C$_3$H$_5$ClO	Epichlorohydrin	116.45
16346	C$_2$Cl$_4$	Tetrachloro-ethylene	120.8	C$_5$H$_{12}$O$_3$	Ethyl carbonate	126.0
16347	C$_2$Cl$_4$	Tetrachloro-ethylene	120.8	C$_6$H$_{12}$O$_2$	Isoamyl formate	123.6
16348	C$_2$HCl$_3$	Trichloroethylene	87.2	C$_6$H$_6$	Benzene	80.1
16349	C$_2$H$_2$	Acetylene	— 84	C$_2$H$_4$	Ethylene	—104
16350	C$_2$H$_3$ClO$_2$	Chloroacetic acid	186.5	C$_7$H$_7$Br	o-Bromotoluene	181.75
16351	C$_2$H$_3$ClO$_2$	Chloroacetic acid	186.5	C$_7$H$_7$Cl	α-Chlorotoluene	179.35
16352	C$_2$H$_3$N	Acetonitrile	81.6	C$_2$H$_6$O	Ethyl alcohol	78.3
16353	C$_2$H$_3$N	Acetonitrile	81.6	C$_5$H$_8$	Isoprene	34.1
16354	C$_2$H$_3$N	Acetonitrile	81.6	C$_5$H$_8$	Isoprene	34.1
16355	C$_2$H$_3$N	Acetonitrile	81.6	C$_5$H$_{10}$	2-Methyl-2-butene	38.5
16356	C$_2$H$_4$Br$_2$	1,2-Dibromoethane	131.5	C$_2$H$_4$O$_2$	Acetic acid	118.5
16357	C$_2$H$_4$Br$_2$	1,2-Dibromoethane	131.5	C$_3$H$_6$O$_2$	Propionic acid	140.7
16358	C$_2$H$_4$Br$_2$	1,2-Dibromoethane	131.5	C$_5$H$_{12}$O	Isoamyl alcohol	131.8
16359	C$_2$H$_4$Br$_2$	1,2-Dibromoethane	131.5	C$_5$H$_{12}$O	Isoamyl alcohol	131.8
16360	C$_2$H$_4$Cl$_2$	1,2-Dichloroethane	83.45	C$_3$H$_6$O	Acetone	56.4

TABLE II. *Ternary Systems* 483

| C-Component | | | Azeotropic Data | | | | | |
Formula	Name	B.P., °C.	B.P., °C.	Wt. % A	Wt. % B	Wt. % C		Ref.
C_6H_{11}	Hexane	68.7	45	14	27	59		982
"			47.4	14.6	36.8	48.6		497
C_5H_{10}	2-Methyl-2-butene	37.15	Nonazeotrope					563
C_4H_8O	Tetrahydrofuran	65	Nonazeotrope				v-l	318
C_6H_{12}	Cyclohexane	80.75	Nonazeotrope					563
C_5H_{10}	2-Methyl-2-butene	38.5	Nonazeotrope				v-l	716
C_5H_{12}	2-Methyl-butane	27.9	Nonazeotrope				v-l	716
C_6H_{10}	Cyclohexene	82.75	Nonazeotrope					563
C_6H_{12}	Cyclohexane	80.75	Nonazeotrope					563
"		80.75	Nonazeotrope				v-l	663,1060
C_7H_8	Toluene	110.7					v-l	105c
C_6H_{12}	Cyclohexane	80.75	Nonazeotrope					563
C_7H_{14}	Methylcyclohexane	101.6					v-l	848c
C_7H_{14}	Methylcyclohexane	101.6					v-l	778c
C_4H_8	1-Butene	—6.1	Nonazeotrope				v-l	407
C_3H_5ClO	Epichlorohydrin	116.45	Nonazeotrope					563
C_3H_8O	Propyl alcohol	97.2	Nonazeotrope					563
C_4H_9I	1-Iodo-2-methyl-propane	120	Azeotrope ?					563
$C_4H_{10}O$	Isobutyl alcohol	108	Nonazeotrope					563
$C_5H_{12}O$	Isoamyl alcohol	131.8	Nonazeotrope					563
$C_6H_{12}O_2$	Ethyl butyrate	119.9	Nonazeotrope					563
$C_5H_{12}O$	Isoamyl alcohol	131.8	<116.0?					563
$C_6H_{12}O_3$	Paraldehyde	124	~ 117.6	45	25	30		563
C_6H_{12}	Cyclohexane	80.7	Nonazeotrope				v-l	780
C_2H_6	Ethane	—88	Nonazeotrope				v-l	387
$C_{10}H_{16}$	d-Limonene	177.8	Nonazeotrope					563
$C_{10}H_{16}$	d-Limonene	177.8	Nonazeotrope					563
$C_6H_{15}N$	Triethylamine	89.7	70.1	34	8	58		982
C_5H_{10}	2-Methyl-2-butene	38.5	Nonazeotrope				v-l	716
C_5H_{12}	2-Methylbutane	27.9	Nonazeotrope				v-l	716
C_5H_{12}	2-Methylbutane	27.9	Nonazeotrope				v-l	716
C_6H_5Cl	Chlorobenzene	131.8	Nonazeotrope					563
C_6H_5Cl	Chlorobenzene	131.8	127.5					563
C_6H_5Cl	Chlorobenzene	131.8	Nonazeotrope					563
C_8H_{10}	Ethylbenzene	136.15	Nonazeotrope					563
C_6H_6	Benzene	80.1	Nonazeotrope				v-l	125

	A-Component		B.P., °C.	B-Component		B.P., °C.
No.	Formula			Formula	Name	
16361	C₂H₄O	Acetaldehyde	20.2	C₂H₄O₂	Acetic acid	118.1
16362	C₂H₄O₂	Acetic acid	118.1	C₃H₅ClO	Epichlorohydrin	116.45
16362a	C₂H₄O₂	Acetic acid	118.1	C₃H₆O₂	Propionic acid	140.7
16363	C₂H₄O₂	Acetic acid	118.1	C₄H₆O₃	Acetic anhydride	139.6
16363a	C₂H₄O₂	Acetic acid	118.1	C₄H₆O₃	Acetic anhydride	139.6
16364	C₂H₄O₂	Acetic acid	118.1	C₄H₈O	2-Butanone	79.6
16365	C₂H₄O₂	Acetic acid	118.1	C₄H₈O	2-Butanone	79.6
16366	C₂H₄O₂	Acetic acid	118.1	C₄H₈O₂	Ethyl acetate	77.1
16367	C₂H₄O₂	Acetic acid "	118.1	C₅H₅N	Pyridine "	115.5
16368	C₂H₄O₂	Acetic acid	118.1	C₅H₅N	Pyridine	115.5
16369	C₂H₄O₂	Acetic acid	118.1	C₅H₅N	Pyridine	115.5
16370	C₂H₄O₂	Acetic acid	118.1	C₅H₅N	Pyridine	115
16371	C₂H₄O₂	Acetic acid	118.1	C₅H₅N	Pyridine	115.5
16372	C₂H₄O₂	Acetic acid "	118.1	C₅H₅N	Pyridine "	115.5
16373	C₂H₄O₂	Acetic acid	118.1	C₅H₅N	Pyridine	115.5
16374	C₂H₄O₂	Acetic acid	118.1	C₅H₅N	Pyridine	115.5
16374a	C₂H₄O₂ "	Acetic acid "	118.1 "	C₅H₁₂O "	Isoamyl alcohol "	132 "
16374b	C₂H₄O₂	Acetic acid	118.1	C₆H₆	Benzene	80.1
16375	C₂H₄O₂	Acetic acid	118.1	C₆H₇N	2-Picoline	134
16376	C₂H₄O₂	Acetic acid	118.1	C₆H₇N	2-Picoline	134
16377	C₂H₄O₂	Acetic acid	118.1	C₆H₇N	2-Picoline	134
16378	C₂H₄O₂	Acetic acid	118.1	C₆H₇N	2-Picoline	134
16379	C₂H₄O₂	Acetic acid	118.1	C₇H₉N	2,6-Lutidine	144
16380	C₂H₄O₂	Acetic acid "	118.1	C₇H₉N	2,6-Lutidine "	144
16381	C₂H₄O₂	Acetic acid	118.1	C₇H₉N	2,6-Lutidine	144
16382	C₂H₄O₂	Acetic acid	118.1	C₈H₁₀	Ethylbenzene	136.15
16383	C₂H₄O₂	Methyl formate	31.9	C₂H₅Br	Bromoethane	38.4
16384	C₂H₄O₂	Methyl formate	31.9	C₂H₅Br	Bromoethane	38.4
16385	C₂H₄O₂	Methyl formate " "	31.9	C₂H₅Br	Bromoethane "	38.4
16386	C₂H₄O₂	Methyl formate	31.9	C₂H₅Br	Bromoethane	38.4
16387	C₂H₄O₂	Methyl formate	31.9	C₂H₆S	Ethanethiol	24.3
16388	C₂H₄O₂	Methyl formate	31.9	C₄H₁₀O	Ethyl ether	34.6
16389	C₂H₄O₂	Methyl formate " "	31.9	C₄H₁₀O	Ethyl ether " "	34.6
16390	C₂H₄O₂	Methyl formate	31.9	C₅H₁₀	Cyclopentane	49.3
16391	C₂H₅I	Iodoethane	72.3	C₂H₆O	Ethyl alcohol	78.3
16392	C₂H₅NO₂	Nitroethane	114.2	C₄H₈O₂	p-Dioxane	101.3
16393	C₂H₆O	Ethyl alcohol	78.3	C₄H₈O	2-Butanone	79.6
16394	C₂H₆O	Ethyl alcohol	78.3	C₄H₈O	2-Butanone	79.6
16395	C₂H₆O	Ethyl alcohol	78.3	C₄H₈O	2-Butanone	79.6

TABLE II. *Ternary Systems*   485

Formula	Name	B.P., °C.	B.P., °C.	Wt. % A	Wt. % B	Wt. % C		Ref.
			C-Component			Azeotropic Data		
$C_4H_6O_2$	Vinyl acetate	72					v-l	831
C_7H_8	Toluene	110.7		Nonazeotrope				563
$C_4H_8O_2$	Butyric acid	162.4		Nonazeotrope			v-l	970c
C_5H_5N	Pyridine	115	134.4	23	55	22		428
$C_5H_8O_4$	Methylene diacetate	164		Nonazeotrope			v-l	954c
$C_4H_8O_2$	Ethyl acetate	77.1		Nonazeotrope			v-l	331
C_6H_{14}	Hexane	68.7		Nonazeotrope			v-l	331
C_6H_{14}	Hexane	68.7		Nonazeotrope			v-l	331
C_7H_{16}	Heptane	98.4	96.2	2	6.5	91.5		1071
	"		96.5	3.4	10.6	86.0		1052
C_8H_{10}	Ethylbenzene	136.15	129.08	13.5	25.2	61.3		309.1061
C_8H_{10}	o-Xylene	143.6	132.2	17.7	30.5	51.8		1070
C_8H_{10}	p-Xylene	138.4	129.22	10.2	22.5	67.3		307
C_8H_{18}	Octane	125.75	115.7	10.4	20.1	69.5		1052
C_9H_{20}	Nonane	150.7	128.0	20.7	29.4	49.9		1070
	"		128.0	20.9	29.3	49.8		307,1052
$C_{10}H_{22}$	Decane	173.3	134.1	31.4	38.2	30.4		1052
$C_{11}H_{24}$	Undecane	194.5	137.1	37.5	43.5	19.0		1068
$C_7H_{14}O_2$	Isoamyl acetate	142	20 mm		4.8	35.3	59.9 v-l	494c
	"	"	760 mm 132	15	54	31	v-l	494c
C_6H_{12}	Cyclohexane	80.75	77.2	7.6	34.4	58		40c
C_8H_{18}	Octane	125.75	121.3	3.6	24.8	71.6		1067
C_9H_{20}	Nonane	150.7	135.0	12.8	38.4	48.8		1067
$C_{10}H_{22}$	Decane	173.3	141.3	19.9	46.8	33.3		1067
$C_{11}H_{24}$	Undecane	194.5	143.4	30.5	55.2	14.3		1067
C_8H_{18}	Octane	125.75		Nonazeotrope				1062
$C_{10}H_{22}$	Decane	173.3	147.0	12.6	74.3	13.1		945,1062
	"						v-l	1055
$C_{11}H_{24}$	Undecane	194.5	162.0	75.0	13.8	11.3		1068
C_9H_{20}	Nonane	150.7		Nonazeotrope				309,1061
C_5H_8	Isoprene	34.1	<23					563
C_5H_{10}	2-Methyl-2-butene	37.15	24.1					563
C_5H_{12}	2-Methylbutane	27.95	16.95	~ 52	~ 5	~ 54		1045
	"			Nonazeotrope				497
C_5H_{12}	Pentane	36.15	21.7?					563
C_5H_{10}	2-Methyl-2-butene	37.15	24?					563
C_5H_{10}	2-Methyl-2-butene	37.15	24					563
C_5H_{12}	Pentane	36.15	20.4	40	8	52		563
	"			Nonazeotrope				497
C_6H_{14}	2,2-Dimethylbutane	49.7		Nonazeotrope				788
$C_4H_8O_2$	Ethyl acetate	77.05		Nonazeotrope				563
$C_4H_{10}O$	Isobutyl alcohol	108	102.87	31.7	17.7	50.6		612
$C_4H_8O_2$	Ethyl acetate	77.0		Nonazeotrope				296,31c
$C_4H_8O_2$	Methyl propionate	79.7		Nonazeotrope				563
C_6H_6	Benzene	80.2		Nonazeotrope				563

	A-Component		B.P., °C.		B-Component	B.P., °C.
No.	Formula			Formula	Name	
16396	C_2H_6O	Ethyl alcohol	78.3	C_4H_8O	2-Butanone	79.6
16397	C_2H_6O	Ethyl alcohol	78.3	$C_4H_8O_2$	Ethyl acetate	77
16398	C_2H_6O	Ethyl alcohol	78.3	$C_4H_8O_2$	Ethyl acetate	77.05
16399	C_2H_6O	Ethyl alcohol	78.3	C_4H_9Cl	1-Chloro-2-methylpropane	68.95
16400	C_2H_6O	Ethyl alcohol	78.3	$C_5H_{14}OSi$	Ethoxytri-methylsilane	75-76
16401	C_2H_6O	Ethyl alcohol "　　"	78.3	C_6H_6	Benzene "	80.2
		"　　"			"	
16402	C_2H_6O	Ethyl alcohol	78.3	C_6H_6	Benzene	80.1
16403	C_2H_6O	Ethyl alcohol "　　"	78.3	C_6H_6	Benzene "	80.1
16404	C_2H_6O	Ethyl alcohol	78.3	C_6H_6	Benzene	80.1
16405	C_2H_6O	Ethyl alcohol	78.3	C_6H_6	Benzene	80.1
		"　　"			"	
16406	C_2H_6O	Ethyl alcohol	78.3	C_6H_7N	Aniline	184.35
16407	C_2H_6O	Ethyl alcohol	78.3	C_6H_7N	Aniline	184.35
16408	C_2H_6O	Ethyl alcohol	78.3	C_7H_8	Toluene	110.7
16409	C_2H_6OS	Dimethyl sulfoxide		C_7H_8	Toluene	110.7
16409a	$C_2H_6O_2$	Ethylene glycol	197	$C_4H_{10}O$	Butyl alcohol	117.7
16410	$C_2H_6O_2$	Ethylene glycol	197.4	C_5H_5N	Pyridine	115
16411	$C_2H_6O_2$	Ethylene glycol	197.4	C_6H_6O	Phenol	181.4
16412	$C_2H_6O_2$	Ethylene glycol	197.4	C_6H_6O	Phenol	181.4
16413	$C_2H_6O_2$	Ethylene glycol	197.4	C_6H_6O	Phenol	181.4
16414	$C_2H_6O_2$	Ethylene glycol	197.4	C_6H_6O	Phenol	181.4
16415	$C_2H_6O_2$	Ethylene glycol	197.4	C_6H_7N	Aniline	184.35
16416	$C_2H_6O_2$	Ethylene glycol	196.7	C_7H_8O	o-Cresol	191
16417	C_3H_4	Propadiene	— 32	C_3H_4	Propyne	— 23.2
16418	C_3H_4	Propadiene	— 32	C_3H_4	Propyne	— 23.2
16419	C_3H_4	Propadiene	— 32	C_3H_4	Propyne	— 23.2
16420	C_3H_5ClO	Epichlorohydrin	116.45	C_3H_8O	Propyl alcohol	97.2
16421	C_3H_5ClO	Epichlorohydrin	116.45	C_4H_9I	1-Iodo-2-methyl-propane	120
16422	C_3H_5ClO	Epichlorohydrin	116.45	$C_4H_{10}O$	Isobutyl alcohol	108.0
16423	$C_3H_5Cl_3$	1,2,3-Trichloro-propane	156.85	C_6H_{12}	1-Hexene	63.5
16424	C_3H_5I	3-Iodopropene	102	C_3H_8O	Propyl alcohol	97.2
16425	C_3H_5I	3-Iodopropene	102	$C_5H_{10}O$	3-Pentanone	102.2
16426	C_3H_5I	3-Iodopropene	102	$C_5H_{10}O$	3-Pentanone	102.2
16427	$C_3H_6Cl_2O$	1,3-Dichloro-2-propanol	174.5	$C_4H_6O_4$	Methyl oxalate	163.3
16428	$C_3H_6Cl_2O$	1,3-Dichloro-2-propanol	174.5	$C_6H_{12}O_3$	Propyl lactate	171.7

TABLE II. *Ternary Systems* 487

C-Component			Azeotropic Data					
Formula	Name	B.P., °C.	B.P., °C.	Wt. % A	Wt. % B	Wt. % C		Ref.
C₆H₁₂	Cyclohexane	80.75		Nonazeotrope				563
C₄H₁₀O	Butyl alcohol	725 mm		Nonazeotrope			v-l	608
C₆H₁₂	Cyclohexane	80.75	64.3?					563
C₆H₁₁	Hexane	68.95		Azeotropic ?				563
C₆H₆	Benzene	80.2		Min. b.p.				192
C₆H₁₂	Cyclohexane	80.75		Nonazeotrope				563
	"		65.05	30.4	10.8	58.8		1060
	"		64.7	29.6	12.8	57.6	v-l	204, 666, 1032c
C₆H₁₂	Methylcyclopentane	72		Nonazeotrope				886
C₆H₁₁	Hexane	68.7		Nonazeotrope			v-l	1049
	"	68.95		Nonazeotrope				949
C₇H₁₄	Methylcyclohexane	100.88		Nonazeotrope				949
C₇H₁₆	Heptane	98.4		Nonazeotrope			v-l	698, 1002
	" 180 mm		32.38	22.4	74.2	3.4	v-l	708
C₇H₈	Toluene	110.7		Nonazeotrope				390
C₇H₁₆	Heptane	98.4		Nonazeotrope			v-l	390
C₇H₁₆	Heptane	98.4					v-l	390
C₇H₁₆	Heptane	98.4					v-l	1022
C₆H₁₂O₂	Butyl acetate	126	725 mm	Nonazeotrope			v-l	608c
C₆H₆O	Phenol	181.4		Nonazeotrope				791
C₆H₇N	2-Picoline	128.8	185.01	5.9	79.1	15.0		791
C₆H₇N	3-Picoline	143.5	186.41	15.9	67.7	16.4		790,791
C₇H₉N	2,6-Lutidine	142	185.04	8.7	74.6	16.7		791
C₈H₁₁N	2,4,6-Collidine	171	188.55	29.5	54.8	15.7		791
C₁₀H₁₆	d-Limonene	177.8	162.45					563
C₈H₁₁N	s-Collidine	171.30	189.65	33.6	62.4	4.0		505
C₃H₆	Propene	— 47.7		Nonazeotrope				981
C₃H₈	Propane	— 42.1		Nonazeotrope				981
C₄H₆	Butadiene	— 4.5		Nonazeotrope				981
C₇H₈	Toluene	110.7		Nonazeotrope				563
C₆H₁₂O₂	Ethyl butyrate	119.9		Nonazeotrope				563
C₇H₈	Toluene	110.7		Nonazeotrope				563
C₆H₁₄	Hexane	68.74		Nonazeotrope			v-l	933
C₅H₁₀O	3-Pentanone	102.2		Nonazeotrope				563
C₅H₁₀O₂	Propyl acetate	101.55		Azeotropic ?				563
C₇H₁₄	Methylcyclohexane	101.8		Azeotropic ?				563
C₁₀H₁₆	d-Limonene	177.8		Nonazeotrope				563
C₇H₇Cl	α-Chlorotoluene	179.35		Nonazeotrope				563

No.	Formula	A-Component	B.P., °C	Formula	B-Component Name	B.P., °C
16429	$C_3H_6Cl_2O$	1,3-Dichloro-2-propanol	174.5	$C_6H_{12}O_3$	Propyl lactate	171.7
16430	$C_3H_6Cl_2O$	1,3-Dichloro-2-propanol	174.5	C_7H_7Cl	α-Chlorotoluene	179.35
16431	$C_3H_6Cl_2O$	2,3-Dichloro-1-propanol	56.1	$C_8H_{18}O$	sec-Octyl alcohol	178.7
16431a	C_3H_6O	Acetone	56.1	$C_3H_6O_2$	Methyl acetate	57.1
16431b	C_3H_6O	Acetone	56.1	$C_3H_6O_2$	Methyl acetate	57.1
16432	C_3H_6O	Acetone "	56.1	$C_3H_6O_2$	Methyl acetate " "	56.3
16433	C_3H_6O	Acetone	56.25	C_3H_8O	Isopropyl alcohol	82.3
16433a	C_3H_6O	Acetone	56.1	C_3H_8O	Isopropyl alcohol	82.3
16433b	C_3H_6O	Acetone	56.1	C_3H_8O	Isopropyl alcohol	82.3
16433c	C_3H_6O	Acetone	56.1	C_4H_8O	2-Butanone	79.6
16434	C_3H_6O	Acetone	56.5	C_4H_9Cl	1-Chloro-2-methylpropane	68.85
16435	C_3H_6O	Acetone	56.5	C_5H_8	Isoprene	34.1
16436	C_3H_6O	Acetone	56.1	C_5H_8	Isoprene	34.1
16437	C_3H_6O	Acetone	56.1	C_6H_5Cl	Chlorobenzene	131.8
16438	C_3H_6O	Acetone	56.4	C_6H_6	Benzene	80.1
16438a	C_3H_6O	Acetone	56.1	C_6H_6	Benzene	80.1
16439	C_3H_6O	Acetone	56.4	C_7H_8	Toluene	110.4
16440	C_3H_6O	Allyl alcohol	97.1	$C_6H_{14}O$	Isopropyl ether	68.3
16441	$C_3H_6O_2$	Methyl acetate	57.1	C_6H_6	Benzene	80.1
16442	$C_3H_6O_2$	Propionic acid	140.7	C_5H_5N	Pyridine	115.5
16443	C_3H_7NO	Dimethylformamide "		C_7H_8	Toluene " "	110.7
16443a	C_3H_8O	Isopropyl alcohol	82.3	C_4H_8O	2-Butanone	79.6
16444	C_3H_8O	Isopropyl alcohol	82.45	$C_4H_8O_2$	Ethyl acetate	77.05
16445	C_3H_8O	Isopropyl alcohol	82.3	$C_5H_{10}O_2$	Isopropyl acetate	88.7
16446	C_3H_8O	Isopropyl alcohol " "	82.3	C_6H_6	Benzene "	80.1
16447	C_3H_8O	Propyl alcohol	97.2	$C_5H_{10}O$	3-Pentanone	102.2
16448	C_3H_8O	Propyl alcohol " " " "	97.2	C_6H_6	Benzene " " "	80.2
16448a	C_3H_8O	Propyl alcohol	97.2	C_6H_6	Benzene	80.1
16449	$C_3H_8O_2$	2-Methoxyethanol	124.5	C_6H_6	Benzene	80.1
16450	$C_3H_8O_2$	2-Methoxyethanol	124.5	C_8H_8	Styrene	144
16451	$C_3H_8O_2$	Methylal	42.25	C_5H_{10}	2-Methyl-2-butene	37.15

TABLE II. *Ternary Systems* 489

C-Component			Azeotropic Data				
Formula	Name	B.P., °C.	B.P., °C.	Wt. % A	Wt. % B	Wt. % C	Ref.
$C_{10}H_{16}$	*d*-Limonene	177.8	165.5?				563
$C_{10}H_{16}$	*d*-Limonene	177.8	165.5?				563
$C_{10}H_{16}$	*d*-Limonene	177.8	Nonazeotrope				563
C_4H_8O	2-Butanone	79.6	Nonazeotrope			v-l	31f
$C_4H_8O_2$	Ethyl acetate	77.0	Nonazeotrope			v-l	507k
C_6H_{14}	Hexane	68.7	49.7	51.1	5.6	43.3	497
	"	68.7	47	45	7	48	981
$C_5H_{10}O_2$	Isopropyl acetate	88.7	Nonazeotrope				981
C_6H_6	Benzene	80.1	Nonazeotrope			v-l	996f
C_7H_8	Toluene	110.7	Nonazeotrope			v-l	996f
$C_4H_8O_2$	Ethyl acetate	77.0	Nonazeotrope			v-l	31f
C_6H_{14}	Hexane	68.95	Nonazeotrope				563
C_5H_{10}	2-Methyl-2-butene	38.5	Nonazeotrope				716
C_5H_{12}	2-Methylbutane	27.9	Nonazeotrope				716
C_6H_6	Benzene	80.1	Nonazeotrope			v-l	290
C_6H_{12}	Cyclohexane	80.75	Nonazeotrope			v-l	502
C_7H_8	Toluene	110.7	Nonazeotrope			v-l	996f
C_7H_{14}	Methylcyclohexane	100.8	Liquid-vapor equilibrium				86
$C_9H_{16}O_2$	2,2-Bis(allyloxy) propane		Nonazeotrope				981
C_6H_{12}	Cyclohexane	80.7	Nonazeotrope				681
$C_{11}H_{24}$	Undecane	194.5	147.1	55.5	26.4	18.1	1068
C_7H_{16}	Heptane	90°C				v-l	1022
	"		~ 68.3				563
C_6H_{12}	Cyclohexane	80.7	68.9	23.3	16.7	60	597c
C_6H_{12}	Cyclohexane	80.75	68.3				563
$C_6H_{14}O$	Isopropyl ether	68.3	Nonazeotrope				981
C_6H_{12}	Cyclohexane	80.75	Azeotrope				614
	"		69.1	31.1	15.0	53.9	v-l	683
$C_5H_{10}O_2$	Propyl acetate	101.55	Nonazeotrope				563
C_6H_{12}	Cyclohexane	80.75	<74?				563
	"		73.81	15.5	30.4	54.2	v-l	661
	"		73.75	18	28	54		614,1060
	" 66.7-216.7 p.s.i.g.					v-l	939
C_7H_{16}	Heptane	98	Nonazeotrope			v-l	302e
C_6H_{12}	Cyclohexane	80.75	73	9	39.1	51.9	v-l	961
C_8H_{10}	Ethylbenzene	62 mm	Nonazeotrope			v-l	421
C_5H_{12}	Pentane	36.15	Nonazeotrope				563

	A-Component		B.P., °C.	B-Component		B.P., °C.
No.	Formula			Formula	Name	
16451a	C_4H_4	Vinyl acetylene	5.0	C_4H_6	1,3-Butadiene	—4.6
16451b	C_4H_4	Vinyl acetylene	5.0	C_4H_{10}	Butane	—0.5
16451c	C_4H_6	1,3-Butadiene	—4.6	C_4H_6	1-Butyne	7
16451d	C_4H_6	1,3-Butadiene	—4.6	C_4H_{10}	Butane	—0.5
16452	$C_4H_6O_4$	Methyl oxalate	163.3	$C_5H_4O_2$	2-Furaldehyde	161.5
16453	$C_4H_6O_4$	Methyl oxalate	163.3	C_6H_5Br	Bromobenzene	156.1
16454	$C_4H_6O_4$	Methyl oxalate	163.3	C_6H_5Br	Bromobenzene	156.1
16455	$C_4H_6O_4$	Methyl oxalate	163.3	$C_6H_{12}O$	Cyclohexanol	160.65
16456	$C_4H_6O_4$	Methyl oxalate	163.3	$C_6H_{12}O$	Cyclohexanol	160.65
16457	$C_4H_6O_4$	Methyl oxalate	163.3	C_9H_{12}	Mesitylene	164.0
16458	$C_4H_7BrO_2$	Ethyl bromo-acetate	158.2	C_6H_5Br	Bromobenzene	156.1
16459	$C_4H_7BrO_2$	Ethyl bromo-acetate	158.2	$C_6H_{12}O$	Cyclohexanol	160.65
16460	$C_4H_7BrO_2$	Ethyl bromo-acetate	158.2	C_7H_8O	Anisole	158.85
16461	$C_4H_7ClO_2$	Ethyl chloro-acetate	143.5	$C_4H_8O_3$	Methyl lactate	144.8
16462	$C_4H_7ClO_2$	Ethyl chloro-acetate	143.5	$C_7H_{14}O_2$	Propyl butyrate	143
16463	C_4H_8O	2-Butanone	79.6	$C_4H_8O_2$	Ethyl acetate	77.1
16464	C_4H_8O	2-Butanone	79.6	$C_4H_8O_2$	Propyl formate	80.8
16465	C_4H_8O	2-Butanone	79.6	$C_4H_{10}O$	Butyl alcohol	117.5
16466	C_4H_8O	2-Butanone	79.6	C_6H_6	Benzene	80.1
		,,			,,	
16467	C_4H_8O	2-Butanone	79.6	C_7H_8	Toluene	110.7
16468	$C_4H_8O_2$	Butyric acid	162.45	C_5H_5N	Pyridine	115.5
16469	$C_4H_8O_2$	Dioxane	101.1	$C_4H_{10}O$	Isobutyl alcohol	107
16470	$C_4H_8O_2$	Dioxane	101.1	C_6H_{12}	1-Hexene	63.5
16471	$C_4H_8O_2$	Ethyl acetate	77.1	$C_4H_{10}O$	Butyl alcohol	117.7
16472	$C_4H_8O_2$	Ethyl acetate	77.05	C_6H_6	Benzene	80.1
16473	$C_4H_8O_2$	Isobutyric acid	154.35	C_6H_5Br	Bromobenzene	156.1
16474	$C_4H_8O_2$	Isobutyric acid	154.35	C_7H_8O	Anisole	153.85
16475	$C_4H_8O_2S$	Sulfolane		C_7H_8O	Toluene	110.7
16476	C_4H_9Cl	Chlorobutane	78.44	$C_4H_{10}O$	Butyl alcohol	117.73
16477	$C_4H_9Cl_3Sn$	Butyltin trichloride	113/17	$C_8H_{18}Cl_2Sn$	Dibutyltin dichloride	157/17
16478	$C_4H_{10}O$	Butyl alcohol	117.7	$C_4H_{11}N$	Butylamine	77.8
16479	$C_4H_{10}O$	Butyl alcohol	117.75	C_5H_5N	Pyridine	115.5
		,,			,,	
16480	$C_4H_{10}O$	Butyl alcohol	117.75	C_6H_6	Benzene	80.1
		,,			,,	
16481	$C_4H_{10}O$	Butyl alcohol	117.7	C_6H_6	Benzene	80.1
16481a	$C_4H_{10}O$	Butyl alcohol	117.7	$C_6H_{12}O$	Cyclohexanol	160.6
16481b	$C_4H_{10}O$	Butyl alcohol	117.7	$C_6H_{12}O_2$	Butyl acetate	126

TABLE II. *Ternary Systems* 491

C-Component			Azeotropic Data				
Formula	Name	B.P., °C.	B.P., °C.	Wt. % A	Wt. % B	Wt. % C	Ref.
C_4H_8	cis-2-Butene	1		Nonazeotrope		v-l	149c
C_4H_{10}	2-Methylpropane	—10		Nonazeotrope		v-l	149c
C_4H_{10}	Butane	—0.5		Nonazeotrope		v-l	149c
C_4H_{10}	2-Methylpropane	—10		Nonazeotrope		v-l	149c
$C_{10}H_{16}$	α-Pinene	155.8		Nonazeotrope			563
$C_6H_{12}O$	Cyclohexanol	160.65		Nonazeotrope			563
$C_{10}H_{16}$	α-Pinene	155.8		Nonazeotrope			563
C_9H_{12}	Mesitylene	174	<154.5				563
$C_{10}H_{16}$	d-Limonene	177.8		Reacts			563
$C_{10}H_{16}$	Nopinene	163.8		Nonazeotrope			563
$C_{10}H_{16}$	α-Pinene	155.8	<152.3?				563
$C_{10}H_{16}$	α-Pinene	155.8		Nonazeotrope			563
$C_{10}H_{16}$	α-Pinene	155.8	<150.4				563
C_8H_{10}	m-Xylene	139.0		Nonazeotrope			563
C_8H_{10}	m-Xylene	139.0		Nonazeotrope			563
C_6H_{14}	Hexane	68.7		Nonazeotrope		v-l	331
C_6H_6	Benzene	80.2		Nonazeotrope			563
C_7H_{16}	Heptane	98.5		Nonazeotrope		v-l	27
C_6H_{12}	Cyclohexane	80.75		Nonazeotrope		v-l	184, 212
	” 14.7-186.8 p.s.i.g.			Nonazeotrope		v-l	184, 940
C_7H_{16}	n-Heptane	98.45		Nonazeotrope		v-l	917
$C_{11}H_{24}$	Undecane	194.5		Nonazeotrope			1068
C_7H_8	Toluene	110.6	101.8	44.3	26.7	30.0	1038
C_6H_{14}	Hexane	68.74		Nonazeotrope		v-l	933
C_7H_8	Toluene	110.7		Nonazeotrope			589
C_6H_{12}	Cyclohexane	80.75		Nonazeotrope		v-l	146, 147
$C_{10}H_{16}$	α-Pinene	155.8	146.4				563
$C_{10}H_{16}$	α-Pinene	155.8	143.9				563
C_7H_{16}	Heptane	90°C				v-l	1022
$C_8H_{18}O$	Butyl ether	141.97		Nonazeotrope			806
$C_{12}H_{27}ClSn$	Tributyltin chloride	166/17		Nonazeotrope			981
$C_8H_{19}N$	Dibutylamine	159.6		Nonazeotrope			981
C_7H_8	Toluene	110.7	108.7	11.9	20.7	67.4	392
	”			Nonazeotrope			497
C_6H_{12}	Cyclohexane	80.75	77.42	4	48	48	1060
	”			Nonazeotrope		v-l	930c
C_7H_{16}	Heptane	98.4		Nonazeotrope		v-l	993
$C_6H_{12}O_2$	Butyl acetate	126 725 mm		Nonazeotrope		v-l	684d
C_7H_8	Toluene	110.7 725 mm		Nonazeotrope		v-l	608f

		A-Component	B.P., °C.		B-Component	B.P., °C.
No.	Formula			Formula	Name	
16482	$C_4H_{10}O$	Butyl alcohol	117.7	$C_6H_{12}O_2$	Butyl acetate	126.1
16482a	$C_4H_{10}O$	Butyl alcohol	117.7	$C_6H_{12}O_2$	Isobutyl acetate	117.2
16482b	$C_4H_{10}O$	tert-Butyl alcohol	82.5	C_6H_6	Benzene	80.1
16483	$C_4H_{10}O$	Ethyl ether	34.6	C_5H_{10}	2-Methyl-2-butene	37.15
16484	$C_4H_{10}O$	Isobutyl alcohol	107.0	C_6H_6	Benzene	80.1
16484a	$C_4H_{10}O$	Isobutyl alcohol	108	C_6H_6	Benzene	80.1
16485	$C_4H_{10}O_2$	2-Ethoxyethanol	135.1	C_6H_{12}	1-Hexene	63.5
16486	$C_4H_{10}O_2$	2-Ethoxyethanol	135.1	C_8H_8	Styrene	145
16487	$C_4H_{10}O_2$	" 2-Ethoxyethanol	135.1	C_8H_{10}	" Ethylbenzene	136.15
16488	$C_5H_4O_2$	2-Furaldehyde	161.7	C_6H_6	Benzene	80.1
16489	$C_5H_4O_2$	2-Furaldehyde	161.7	C_7H_8	Toluene	110.7
16490	$C_5H_4O_2$	2-Furaldehyde	161.7	C_7H_8	Toluene	110.7
16491	$C_5H_4O_2$	2-Furaldehyde	161.7	C_7H_{14}	Methylcyclohexane	101.1
16492	C_5H_5N	Pyridine	115.3	$C_5H_{11}N$	Piperidine	105.8
16493	C_5H_5N	Pyridine	115.5	$C_5H_{12}O$	Isoamyl alcohol	131
16494	C_5H_5N	Pyridine	115.5	C_8H_{10}	Ethylbenzene	136.15
16495	$C_5H_{10}O_2$	Isovaleric acid	176.5	C_7H_6O	Benzaldehyde	179.2
16496	$C_5H_{10}O_2$	Isovaleric acid	176.5	C_7H_6O	Benzaldehyde	179.2
16497	$C_5H_{10}O_2$	Isovaleric acid	176.5	C_7H_7Cl	α-Chlorotoluene	179.35
16498	$C_5H_{10}O_2$	Isovaleric acid	176.5	C_7H_7Cl	α-Chlorotoluene	179.35
16499	$C_5H_{10}O_2$	Isovaleric acid	176.5	C_7H_7Cl	α-Chlorotoluene	179.35
16500	C_6H_5Br	Bromobenzene	156.1	C_6H_6O	Phenol	181.5
16501	C_6H_5Br	Bromobenzene	156:1	$C_6H_{12}O$	Cyclohexanol	160.65
16502	C_6H_5Br	Bromobenzene	156.1	$C_6H_{12}O$	Cyclohexanol	160.65
16503	C_6H_5Br	Bromobenzene	156.1	$C_6H_{13}ClO_2$	Chloroacetal	156.8
16504	C_6H_5ClO	o-Chlorophenol	175.5	C_7H_7Br	o-Bromotoluene	181.75
16505	$C_6H_5NO_2$	Nitrobenzene	210.85	C_7H_8O	Benzyl alcohol	205.5
16506	C_6H_6	Benzene	80.1	C_6H_7N	Aniline	184.4
16507	C_6H_6	Benzene	80.1	C_6H_{12}	Cyclohexane	80.7
16508	C_6H_6	Benzene	80.1	C_6H_{12}	Cyclohexane	80.75
16508a	C_6H_6	Benzene	80.1	C_6H_{12}	Cyclohexane	80.75
16509	C_6H_6	Benzene	80.1	C_6H_{12}	Methylcyclopentane	
16510	C_6H_6	Benzene	80.1	C_7H_{16}	2,3-Dimethyl-pentane	89.8
16511	C_6H_6O	Phenol " "	181.4	C_6H_7N	Aniline " "	184.4
16512	C_6H_6O	Phenol	182	$C_6H_{10}O$	Cyclohexanone	155.6
16513	C_6H_6O	Phenol	182	$C_6H_{10}O_4$	Ethylene diacetate	186
16514	C_6H_6O	Phenol	181.5	C_7H_7Br	o-Bromotoluene	181.75
16514a	C_6H_6O	Phenol	181.4	C_7H_9N	2,4-Lutidine	159

TABLE II. *Ternary Systems*

	C-Component		Azeotropic Data					
Formula	Name	B.P., °C.	B.P., °C.	Wt. % A	Wt. % B	Wt. % C		Ref.
$C_8H_{18}O$	Butyl ether	142.1	Nonazeotrope					981
C_7H_8	Toluene	110.7					v-l	331c
C_6H_{12}	Cyclohexane	80.75					v-l	793c
C_5H_{12}	Pentane	36.15	Nonazeotrope					563
C_6H_{12}	Cyclohexane	80.75	76.73	8	42	50		425,1060
C_6H_{12}	Cyclohexane	80.75	77.2	43.2	47.0	9.8	v-l	686f
C_6H_{14}	Hexane	68.74	Nonazeotrope				v-l	933
C_8H_{10}	Ethylbenzene	136.15	Nonazeotrope				v-l	294
	" 5-62 mm		Nonazeotrope				v-l	421
C_8H_{18}	Octane	125.75	Nonazeotrope				v-l	673
C_6H_{12}	Cyclohexane	80.75	Nonazeotrope				v-l	961
C_7H_{14}	Methylcyclohexane	101.1	Nonazeotrope				v-l	317
C_7H_{16}	Heptane	98.4	Nonazeotrope				v-l	317
C_7H_{16}	Heptane	98.4	Nonazeotrope				v-l	317
C_8H_{14}	Diisobutylene	102.5	98.6					240
C_7H_8	Toluene	110.7	110.19	8.6	4.1	87.3		1069
C_9H_{20}	Nonane	150.7	Nonazeotrope					309,1061
C_7H_7Cl	α-Chlorotoluene	179.35	Nonazeotrope					563
$C_{10}H_{16}$	d-Limonene	177.8	168.7					563
$C_{10}H_{14}$	Cymene	175.3	167.8?					563
$C_{10}H_{16}$	d-Limonene	177.8	168.7?					563
$C_{10}H_{18}O$	Cineole	176.3	Azeotropic ?					563
C_7H_{16}	α-Pinene	155.8	152.6?					563
C_7H_{16}	Camphene	~158	>153.4?					563
$C_{10}H_{16}$	α-Pinene	155.8	Azeotropic ?					563
$C_{10}H_{16}$	α-Pinene	155.8	Nonazeotrope					563
$C_{10}H_{16}$	d-Limonene	177.8	Nonazeotrope					563
$C_{11}H_{24}O_2$	Diisoamyloxy-methane	207.5	197?					563
C_6H_{12}	Cyclohexane	80.7	Nonazeotrope				v-l	758
C_6H_{12}	Methylcyclopentane	71.8	Nonazeotrope				v-l	936
$C_6H_{12}O$	4-Methyl-2-pentanone	115.9	Nonazeotrope				v-l	188
C_6H_{14}	Hexane	68.8	Nonazeotrope				v-l	807c
C_6H_{14}	Hexane		Nonazeotrope				v-l	55
$C_{12}F_{27}N$	Perfluorotributylamine		Nonazeotrope				v-l	509
$C_{13}H_{28}$	Tridecane	234	184.45	33.5	48.5	18.0		911
	"	600 mm	175.90	33.9	48.1	18.0		911
	"	500 mm	169.37	33.7	47.0	19.3		911
	"	400 mm	161.71	33.5	46.7	19.8		911
$C_6H_{12}O$	Cyclohexanol	160.65 90 mm	Nonazeotrope				v-l	177
$C_8H_8O_2$	Phenyl acetate	195.7	194.45	26.4	34.4	39.2		721
$C_{10}H_{16}$	d-Limonene	177.8	Nonazeotrope					563
$C_{11}H_{24}$	Undecane	194.5	181.78	19.88	21.52	58.60		263i

	A-Component		B.P., °C.	B-Component		B.P., °C.
No.	Formula			Formula	Name	
16515	C_6H_7N	Aniline	184.35	$C_6H_{10}O_4$	Ethyl oxalate	185
16516	C_6H_7N	Aniline	184.35	$C_6H_{10}O_4$	Ethyl oxalate	185
16517	C_6H_7N	Aniline	184.35	$C_6H_{10}O_4$	Ethyl oxalate	185
16518	C_6H_7N	Aniline	184.35	$C_6H_{10}O_4$	Ethyl oxalate	185
16519	C_6H_7N	Aniline	184.35	$C_6H_{12}O$	Cyclohexanol	160.65
16520	C_6H_7N	Aniline	184.35	C_7H_7Br	o-Bromotoluene	181.75
16521	C_6H_7N	Aniline	184.35	C_7H_7Br	o-Bromotoluene	181.75
16522	C_6H_7N	Aniline	184.35	C_7H_8	Toluene	110
16523	C_6H_7N	Aniline	184.35	C_7H_8	Toluene	110.7
16524	C_6H_7N	Aniline	184.35	C_7H_8O	Benzyl alcohol	205.5
16525	C_6H_7N	Aniline	184.35	$C_8H_{18}O$	sec-Octyl alcohol	178.7
16526	$C_6H_{10}O$	Cyclohexanone	156.7	C_7H_8O	Anisole	153.85
16527	$C_6H_{10}O_3$	Ethyl aceto-acetate	180.7	C_7H_7Br	o-Bromotoluene	181.75
16528	$C_6H_{10}O_3$	Ethyl aceto-acetate	180.7	C_7H_7Cl	α-Chlorotoluene	179.35
16529	$C_6H_{10}O_3$	Ethyl aceto-acetate	180.7	C_9H_{12}	Mesitylene	164.0
16530	$C_6H_{10}O_4$	Ethyl oxalate	185	C_7H_7Br	o-Bromotoluene	181.75
16531	$C_6H_{12}O$	Cyclohexanol	160.65	C_7H_8O	Anisole	153.85
16532	$C_6H_{12}O_3$	Propyl lactate	171.7	C_7H_7Cl	α-Chlorotoluene	179.35
16533	$C_6H_{12}O_3$	Propyl lactate	171.7	$C_8H_{10}O$	Phenetole	171.5
16534	$C_6H_{14}O_2$	2-Butoxyethanol		C_7H_8	Toluene	110.7
16535	$C_6H_{14}O_2$	Hexylene glycol		C_8H_{10}	Ethylbenzene	136.15
16536	C_7H_6O	Benzaldehyde	179.2	C_7H_7Cl	α-Chlorotoluene	179.35
16537	C_7H_6O	Benzaldehyde	179.2	C_7H_7Cl	α-Chlorotoluene	179.35
16538	C_7H_7Cl	α-Chlorotoluene "	179.35	C_7H_8O	Benzyl alcohol " "	
16539	C_7H_7Cl	α-Chlorotoluene		$C_7H_{14}O_3$	Isobutyl lactate	182.15
16540	C_7H_7Cl	α-Chlorotoluene	179.35	$C_7H_{14}O_3$	Isobutyl lactate	182.15
16541	C_7H_7Cl	α-Chlorotoluene	179.35	$C_8H_{18}O$	sec-Octyl alcohol	178.7
16542	C_7H_8	Toluene	110.4	C_7H_{14}	Methylcyclohexane	100.8
16543	C_7H_8	Toluene	110.6	C_7H_{14}	3-Heptene	
16544	C_7H_8	Toluene	110.6	C_7H_{16}	Heptane	
16545	C_7H_8O	x-Cresol	202	C_7H_9N	Pyridine bases	143
16546	C_7H_8O	x-Cresol	202	C_7H_9N	Pyridine bases	157
16547	C_7H_8O	x-Cresol	202	C_7H_9N	Pyridine bases	163
16548	C_7H_8O	m-Cresol	202.8	$C_8H_{11}N$	2,4,6-Collidine	171
16549	$C_7H_{14}O_3$	Isobutyl lactate	182.15	$C_8H_{18}O$	sec-Octyl alcohol	178.7
16550	$C_7H_{14}O_3$	Isobutyl lactate	182.15	$C_8H_{18}O$	sec-Octyl alcohol	178.7
16551	C_8H_{10}	Ethylbenzene	136.1	C_9H_{12}	Isopropylbenzene	152.8

TABLE II. *Ternary Systems* 495

C-Component			Azeotropic Data				
Formula	Name	B.P., °C.	B.P., °C.	Wt. % A	Wt. % B	Wt. % C	Ref.
C_7H_7Br	o-Bromotoluene	181.75	Reacts				563
C_7H_7Br	p-Bromotoluene	185	Reacts				563
$C_{10}H_{16}$	d-Limonene	177.8	Reacts				563
$C_{10}H_{16}$	Terpinene	180.5	Reacts				563
$C_6H_{13}N$	Cyclohexylamine		Nonazeotrope			v-l	704
$C_8H_{18}O$	sec-Octyl alcohol	178.7	Nonazeotrope				563
$C_{10}H_{16}$	d-Limonene	177.8	Nonazeotrope				563
C_7H_{14}	Methylcyclohexane	101.1 80° to 100°C.	Evaporation data				851
C_7H_{16}	Heptane	98.4	Nonazeotrope			v-l	390
$C_{10}H_{16}$	d-Limonene	177.8	Nonazeotrope				563
$C_{10}H_{16}$	d-Limonene	177.8	Nonazeotrope				563
$C_{10}H_{16}$	a-Pinene	155.8	Nonazeotrope ?				563
$C_{10}H_{16}$	d-Limonene	177.8	Nonazeotrope				563
$C_{10}H_{16}$	d-Limonene	177.8	168.8?				563
$C_{10}H_{16}$	Nopinene	163.8	Nonazeotrope				563
$C_{10}H_{16}$	d-Limonene	177.8	Nonazeotrope				563
$C_{10}H_{16}$	α-Pinene	155.8	Nonazeotrope				563
$C_{10}H_{16}$	d-Limonene	177.8	Nonazeotrope				563
$C_{10}H_{18}$	Menthene	170.8	163.0	31	33	36	563
C_7H_{16}	Heptane	90°				v-l	1022
C_8H_{16}	Ethylcyclohexane	131.8 400 mm	Nonazeotrope			v-l	771
$C_{10}H_{16}$	d-Limonene	177.8	Nonazeotrope				563
$C_{10}H_{16}$	Terpinene	180.5	Nonazeotrope				563
$C_9H_{10}O_2$	Benzyl acetate	15 mm	Nonazeotrope			v-l	496
"	"	760 mm	Nonazeotrope			v-l	496
$C_{10}H_{16}$	d-Limonene	177.8	172.5				563
$C_{10}H_{16}$	Terpinene	180.5	Azeotrope doubtful				563
$C_{10}H_{16}$	d-Limonene	177.8	Azeotropic ?				563
C_7H_{16}	n-Heptane	98.4	Liquid-vapor equilibrium				84
C_7H_{16}	Heptane	98.4	Nonazeotrope			v-l	1016
C_8H_{10}	p-Xylene	90°C	Nonazeotrope			v-l	1022
$C_{10}H_8$	Naphthalene	218.1	202.81	81	9	10	1065
$C_{10}H_8$	Naphthalene	218.1	202.03	65.5	16.5	18	1065
$C_{10}H_8$	Naphthalene	218.1	202.39	62	17	21	1065
$C_{10}H_8$	Naphthalene	217.9	205.8	61.5	20.8	17.7 v-l	506
$C_{10}H_{16}$	d-Limonene	177.8	Reacts				563
$C_{10}H_{16}$	Terpinene	180.5	Nonazeotrope				563
$C_{10}H_{14}$	Butylbenzene	183.1	Nonazeotrope			v-l	586

Table III. Quaternary

		Formula	Name	B.P.,°C
	A	CHN	Hydrocyanic acid	26
	B	H_2O	Water	100
16551a	C	C_3H_3N	Acrylonitrile	77.3
	D	C_3H_4O	Acrolein	52.4
	A	CHN	Hydrocyanic acid	26
	B	C_2H_3N	Acetonitrile	81.6
16551b	C	C_3H_3N	Acrylonitrile	77.3
	D	C_3H_4O	Acrolein	52.4
	A	H_2O	Water	100
	B	CH_2O_2	Formic acid	100.8
16551c	C	$C_2H_4O_2$	Acetic acid	118.1
	D	$C_4H_8O_2$	Butyric acid	162.4
	A	H_2O	Water	100.0
	B	CH_3NO_2	Nitromethane	101.2
16552	C	C_2Cl_4	Tetrachloroethylene	120.8
	D	C_3H_8O	Propyl alcohol	97.2
	A	H_2O	Water	100.0
	B	CH_3NO_2	Nitromethane	101.2
16553	C	C_2Cl_4	Tetrachloroethylene	120.8
	D	C_8H_{18}	n-Octane	125.75
	A	H_2O	Water	100.0
	B	CH_3NO_2	Nitromethane	101.2
16554	C	C_3H_8O	Propyl alcohol	97.2
	D	C_8H_{18}	n-Octane	125.75
	A	H_2O	Water	100.0
	B	C_2Cl_4	Tetrachloroethylene	120.8
16555	C	C_3H_8O	Propyl alcohol	97.2
	D	C_8H_{18}	n-Octane	125.75
	A	H_2O	Water	100
	B	C_2H_3N	Acetonitrile	81.6
16556	C	C_2H_6O	Ethyl alcohol	78.3
	D	$C_6H_{15}N$	Triethylamine	89.7
	A	H_2O	Water	100
	B	C_2H_6O	Ethyl alcohol	78.3
16557	C	C_4H_8O	Crotonaldehyde	102.2
	D	$C_4H_8O_2$	Ethyl acetate	77.1

Azeotropic Data

B.P.,°C	Azeotropic Composition				E	Ref.
	A	B	C	D		
	Nonazeotrope				v-1	905f, 905m
	Nonazeotrope				v-1	905f, 905m
	Nonazeotrope				v-1	507c
76.88	7.38	20.65	59.45	12.52	—	611
77.06	9.86	34.40	32.60	23.14	—	611
76.34	9.98	41.00	12.42	36.60	—	611
80.98	—	—	—	—	—	611
	Nonazeotrope					981
70	8.7	11.1	0.1	80.1	—	227

		Formula	Name	B.P.,°C
16558	A	H_2O	Water	100
	B	C_2H_6O	Ethyl alcohol	78.3
	C	C_6H_6	Benzene	80.1
	D	C_6H_{12}	Cyclohexane	80.75
16559	A	H_2O	Water	100
	B	C_2H_6O	Ethyl alcohol	78.3
	C	C_6H_6	Benzene	80.1
	D	C_6H_{14}	Hexane	68.95
16560	A	H_2O	Water	100
	B	C_2H_6O	Ethyl alcohol	78.3
	C	C_6H_6	Benzene	80.1
	D	C_7H_{14}	Methylcyclohexane	100.88
16561	A	H_2O	Water	100
	B	C_2H_6O	Ethyl alcohol	78.3
	C	C_6H_6	Benzene	80.1
	D	C_7H_{16}	Heptane	98.4
16562	A	H_2O	Water	100
	B	C_2H_6O	Ethyl alcohol	78.3
	C	C_6H_6	Benzene	80.1
	D	C_8H_{18}	Iso-octane	
16563	A	H_2O	Water	100
	B	C_4H_9Cl	1-Chlorobutane	78.44
	C	$C_4H_{10}O$	Butyl alcohol	117.73
	D	$C_8H_{18}O$	Butyl ether	
16563a	A	$CHCl_3$	Chloroform	61.2
	B	CH_4O	Methanol	64.6
	C	$C_3H_6O_2$	Methyl acetate	56.9
	D	C_6H_6	Benzene	80.1
16564	A	$C_2H_4O_2$	Acetic acid	118.1
	B	C_5H_5N	Pyridine	115
	C	C_8H_{10}	Ethylbenzene	136.4
	D	C_9H_{20}	Nonane	150.8
16565	A	$C_2H_4O_2$	Acetic acid	118.1
	B	C_5H_5N	Pyridine	115
	C	C_8H_{10}	p-Xylene	138.4
	D	C_9H_{20}	Nonane	150.8

TABLE III. *Quaternary and Quinary Systems* 499

Azeotropic Data

B.P.,°C	A	Azeotropic Composition			E	Ref.
		B	C	D		
62.19	7.1	17.4	21.5	54.0	——	1054
62.14	6.1	19.2	20.4	54.3	——	942, 1059
	Nonazeotrope					949
	Nonazeotrope					949
64.79	6.8	18.7	62.4	12.1	——	948
64.69	6.7	17.7	61.4	14.1	——	948
	Nonazeotrope					806
	Nonazeotrope				v-1	406a
27.9	17	27	18	38	——	309
	Nonazeotrope					307

		Formula	Name	B.P., °C
16565a	A	C_3H_6O	Acetone	56.1
	B	C_3H_8O	Isopropyl alcohol	82.3
	C	C_6H_6	Benzene	80.1
	D	C_7H_8	Toluene	110.7
16565b	A	C_3H_6O	Acetone	56.1
	B	C_6H_6	Benzene	80.1
	C	C_6H_{12}	Cyclohexane	80.9
	D	C_7H_8	Toluene	110.7
16565c	A	C_3H_8O	Isopropyl alcohol	82.3
	B	C_4H_8O	2-Butanone	79.6
	C	C_6H_6	Benzene	80.1
	D	C_6H_{12}	Cyclohexane	80.9
16566	A	H_2O	Water	100
	B	CH_3NO_2	Nitromethane	101.2
	C	C_2Cl_4	Tetrachloroethylene	120.8
	D	C_3H_8O	Propyl alcohol	97.2
	E	C_8H_{18}	n-Octane	125.75
16566a	A	$CHCl_3$	Chloroform	61.2
	B	CH_4O	Methanol	64.6
	C	C_3H_2O	Acetone	56.15
	D	$C_3H_6O_2$	Methyl acetate	56.9
	E	C_6H_6	Benzene	80.1

TABLE III. *Quaternary and Quinary Systems*

Azeotropic Data

B.P.,°C	Azeotropic Composition						
	A	B	C	D		E	Ref.
	Nonazeotrope					v-1	996f
	Nonazeotrope					v-1	996c
	Nonazeotrope					v-1	597a
76.5	9.45	37.30	21.15	10.58	21.52		611
						v-1	406a

Formula Index

Formula   Name and System No.

ClH   Hydrochloric acid. B.p. −80
             37, 78, 94, 97-104, 15827, 15831-15833, 6b, 15831a
ClHO₄   Perchloric acid. B.p. 110
             105
Cl₂   Chlorine. B.p. −37.6
             79, 97, 106-108
Cl₂Cu   Cupric chloride.
             109, 110
Cl₂OSe   Selenyl chloride. B.p. 179
             4b, 110a
Cl₂O₂S   Sulfuryl chloride. B.p. 69.1
             111-115
Cl₂Pb   Lead chloride. B.p. 954
             4, 109, 116
Cl₂S₂   Sulfur monochloride. B.p. 138
             116a
Cl₂Zn   Zinc chloride. B.p. 732
             110, 116
Cl₃Fe   Ferric chloride. B.p. 315
             116b, 116c
Cl₃HSi   Trichlorosilane. B.p. 32
             117-124, 116b, 116d
Cl₃OP   Phosphorus oxychloride. B.p. 107.2
             111, 125-126, 15834, 116d, 126a, 126b
Cl₃OV   Vanadium oxychloride. B.p. 126.5
             125, 15834, 126c-127e
Cl₃P   Phosphorus trichloride. B.p. 76
             117, 128-134, 4c, 125a
Cl₃Sb   Antimony trichloride. B.p. 220
             5, 135, 136
Cl₄Ge   Germanium chloride.
             6, 98, 126c, 136a-136g
Cl₄Si   Silicon tetrachloride. B.p. 56.9
             128, 137-152, 116c, 136a, 152b, 15834a, 15834b
Cl₄Sn   Tin tetrachloride. B.p. 113.85
             153-165, 6a, 110a, 126e
Cl₄Ti   Titanium tetrachloride. B.p. 136
             126, 127, 137, 153, 166-170, 15834
Cl₄V   Vanadium tetrachloride. B.p. 153
             127a
Cl₅Ta   Tantalum pentachloride. B.p. 242
             4a
Cu   Copper. B.p. 2310
             171, 172
DH   Deuterium hydride.
             173, 174
D₂   Deuterium. B.p. −249.7
             173, 175
D₂O   Deuterium oxide.
             176
FH   Hydrofluoric acid. B.p. 19.54
             44, 46, 51, 69, 95, 106, 117-186, 15830, 15835-15839
FHO₃S   Fluorosulfuric acid.
             15835

Formula   Name and System No.

FLi   Lithium fluoride.
               40a, 15827a
FNa   Sodium fluoride.
               41
F₃Sb   Antimony fluoride. B.p. 319
               187
F₄Si   Silicon tetrafluoride.
               15840
F₄Th   Thorium tetrafluoride.
               15827a
F₅Nb   Niobium pentafluoride.
               187a
F₅Sb   Antimony pentafluoride. B.p. 142.7
               177, 187
F₆H₂Si   Fluosilicic acid.
               15836
F₆S   Sulfur hexafluoride.
               188
F₆U   Uranium hexafluoride. B.p. 56
               45, 47, 52, 96, 178, 15830, 187a, 188a, 188b
F₆W   Tungsten hexafluoride.
               189, 190, 188a
HI   Hydroiodic acid. B.p. −34
               191, 192
HNO₃   Nitric acid. B.p. 86
               193, 194, 15841, 15842
HO₄Re   Rhenic acid.
               194a
H₂   Hydrogen. B.p. −252.7
               80, 174, 175, 195, 15829, 194b, 195a, 15828c
H₂O   Water. B.p. 100
               9, 48, 53, 72, 81, 99, 105, 107, 176, 179, 191, 193, 196-954, 15828, 15831,
               15835-15838, 15841-16242, 16552-16563, 16566, 68a, 194a, 16551a, 15831a
H₂O₂   Hydrogen peroxide. B.p. 152.1
               196
H₂O₄S   Sulfuric acid.
               197
H₂S   Hydrogen sulfide. B.p. −63.5
               49, 82, 192, 198, 955, 956, 6c, 15842a
H₃N   Ammonia. B.p. −33
               10, 199, 957-980, 16243-16247, 956a, 15842a
H₄N₂   Hydrazine. B.p. 113.5
               200, 981, 15843, 956a
H₄Si   Silane. B.p. −111.86
               982
He   Helium. B.p. −268.9
               983, 82a, 194b, 982a
I₂   Iodine. B.p. 185.3
               54
I₄Sn   Tin iodide. B.p. 340
               64
Kr   Krypton. B.p. −152
               983a
MnS   Manganese sulfide.
               67

Formula   Name and System No.

NO   Nitric oxide. B.p. -153.6
                  984
NO_2 Nitrogen peroxide. B.p. 26
                  984
N_2 Nitrogen. B.p. -195
                  83, 985-987, 15826, 15829, 1a, 984a
N_2O Nitrous oxide. B.p. -90.7
                  84, 988, 984a
N_2O_5 Nitrogen pentoxide.
                  201
Ne   Neon. B.p. -245.9
                  195, 985, 982a
O_2 Oxygen. B.p. -183
                  2, 85, 986, 15826
O_2S Sulfur dioxide. B.p. -10
                  50, 86, 100, 108, 180, 202, 989-1007, 15839, 15844, 988a, 16247a, 16247b
O_3S Sulfur trioxide. B.p. 47
                  203, 15841, 988a
$O_{10}P_4$ Phosphorus pentoxide.
                  204
Pb   Lead. B.p. 1525
                  171, 1008
S   Sulfur. B.p. 444.6
                  1009, 1009a
Se   Selenium. B.p. 688
                  1009, 1010
Sn   Tin. B.p. 2275
                  172, 1008
Te   Tellurium
                  1010, 1009a
Xe   Xenon. B.p. -109
                  1010a
CCl_2F_2 Trichlorodifluoromethane. B.p. -29.8
                  181, 1011-1018, 15839
CCl_3F Trichlorofluoromethane. B.p. 24.9
                  1019-1021
CCl_3NO_2 Trichloronitromethane. B.p. 111.9
                  1022-1084
CCl_4 Carbon tetrachloride. B.p. 76.75
                  55, 112, 118, 138, 154, 166, 205, 1085-1168, 15845-15852, 16248-16262,
                  127b, 136b, 1168a-1168c, 15851a, 16254a
CF_4 Carbon tetrafluoride. B.p. -128
                  1168d, 1168e
CS_2 Carbon disulfide. B.p. 46.2
                  87, 167, 206, 1085, 1169-1278, 15853-15857, 16263-16280
$CHBrCl_2$ Bromodichloromethane. B.p. 90.1
                  1022, 1279-1337, 15858-15861
$CHBr_3$ Bromoform. B.p. 149.5
                  1338-1418
$CHClF_2$ Chlorodifluoromethane. B.p. -40.8
                  182, 1011, 1419-1424
$CHCl_2F$ Dichlorofluoromethane.
                  1425

Formula   Name and System No.

CHCl$_3$ Chloroform. B.p. 61
                     139, 194, 207, 1086, 1169, 1426-1502, 15842, 15862-15868, 16281-16305,
                     136c, 15868a, 16286a, 16297a, 16563a, 16566a
CHCl$_3$S   Perfluoromethylmercaptan.
                     116a
CHF$_3$ Fluoroform.
                     1503, 1168d
CH$_2$BrCl   Bromochloromethane. B.p. 69
                     1504
CH$_2$Br$_2$ Dibromomethane. B.p. 97.0
                     1023, 1505-1534
CH$_2$ClF   Chlorofluoromethane.
                     1534a
CH$_2$ClNO$_2$ Chloronitromethane. B.p. 122.5
                     1535-1541
CH$_2$Cl$_2$ Dichloromethane. B.p. 41.5
                     208, 1170, 1426, 1504, 1542-1575, 15869, 15870, 16281, 16306, 16307, 136d
CH$_2$I$_2$ Diiodomethane. B.p. 181
                     1576-1602
CH$_2$O   Formaldehyde. B.p. −21
                     209
CH$_2$O$_2$ Formic acid. B.p. 100.75
                     11, 210, 1024, 1087, 1171, 1279, 1338, 1427, 1603-1712, 15862, 15871-15875,
                     16282, 16551a
CH$_3$Br   Bromoethane. B.p. 3.65
                     1713-1720
CH$_3$Cl   Chloromethane. B.p. −23.7
                     88, 1012, 1721, 1722
CH$_3$Cl$_3$Si   Trichloromethylsilane. B.p. 66.4
                     140, 1723, 126a, 167a, 1087a, 1724a, 16307a, 16307b
CH$_3$I   Iodomethane. B.p. 42.55
                     1172, 1428, 1542, 1603, 1725-1747, 16263-16265, 16308, 16309
CH$_3$NO$_2$ Methyl nitrite. B.p. −16
                     1748-1752
CH$_3$NO$_2$ Nitromethane. B.p. 101
                     141, 211, 1025, 1088, 1173, 1280, 1429, 1604, 1753-1879, 15876-15894,
                     16310-16318, 16552-16554, 16566
CH$_3$NO$_3$ Methyl nitrate. B.p. 64.8
                     212, 1089, 1174, 1281, 1543, 1725, 1880-1909
CH$_4$ Methane. B.p. −164
                     77, 983, 987, 1910, 1911, 16319, 88a, 1910a
CH$_4$Cl$_2$Si   Dichloromethylsilane.
                     142, 1912
CH$_4$O   Methanol. B.p. 64.7
                     12, 73, 213, 1026, 1090, 1175, 1282, 1430, 1505, 1544, 1713, 1726, 1753,
                     1880, 1913-2127, 15853, 15863, 15985-15916, 16248, 16249, 16263,
                     16266-16271, 16283-16287, 16306-16308, 16320-16338, 988b, 16247a,
                     16338a, 16338b, 16563a, 16566a
CH$_4$S   Methanethiol. B.p. 6.8
                     989, 2128-2134
CH$_5$N   Methylamine. B.p. −6
                     214, 957, 2135-2146, 16339
CH$_6$N   Methylhydrazine. B.p. 88
                     215

Formula   Name and System No.

$C_2Br_2Cl_2$ 1,2-Dibromodichloroethylene. B.p. 172
                                2147, 2148

$C_2Br_2F_4$ 1,2-Dibromotetrafluoroethane.
                                56

C_2ClF_5 Chloropentafluoroethane. B.p. -38.5
                                1419, 2149

$C_2Cl_2F_4$ 1,2-Dichlorotetrafluoroethane.
                                1425, 2150, 188b, 1534a

$C_2Cl_3F_3$ 1,1,2-Trichlorotrifluoroethane. B.p. 47.6
                                57, 216, 1545, 1913, 2151-2153, 15895, 15917, 15839a

C_2Cl_3N Trichloroacetonitrile.
                                2154

C_2Cl_4 Tetrachloroethylene. B.p. 121.1
                                217, 1027, 1091, 1431, 1535, 1605, 1754, 1914, 2155-2228, 15876, 15918,
                                15919, 16310, 16340-16347, 16552, 16553, 16555, 16566

$C_2Cl_4F_2$ 1,1,1,2-Tetrachlorodifluoroethane. B.p. 91.6
                                58

$C_2Cl_4F_2$ 1,1,2,2-Tetrachlorodifluoroethane. B.p. 92.4
                                2151

C_2Cl_4O Trichloroacetyl chloride. B.p. 118
                                168

C_2Cl_6 Hexachloroethane. B.p. 184.8
                                113, 1176, 2229-2269

$C_2F_4N_2O_4$ 1,1,2,2-Tetrafluorodinitroethane.
                                1092

C_2F_6 Hexafluoroethane. B.p. -78
                                2270, 15840

$C_2HBrClF_3$ 2-Bromo-2-chloro-1,1,1-trifluoroethane.
                                2271

C_2HBrCl_2 *cis*-1-Bromo-1,2-dichloroethylene. B.p. 113.8
                                2272

C_2HBrCl_2 *trans*-1-Bromo-1,2-dichloroethylene.
                                2273

C_2HBrCl_2 1-Bromo-2,2-dichloroethylene. B.p. 107
                                2274

C_2HBr_2Cl 1,2-Dibromo-1-chloroethylene. B.p. 140
                                2275, 2276

C_2HBr_3O Bromal. B.p. 174
                                2277

C_2HClF_2 Chlorodifluoroethylene.
                                2278

C_2HClF_4 Chlorotetrafluoroethane. B.p. -10
                                2279

C_2HCl_3 Trichloroethylene. B.p. 86.2
                                218, 1093, 1283, 1606, 1755, 1915, 2280-2335, 15920-15926, 16250, 16348

$C_2HCl_3F_2$ 1,2,2-Trichloro-1,1-difluoroethane. B.p. 71.1
                                59

C_2HCl_3O Chloral. B.p. 97.75
                                219, 1284, 1756, 2155, 2336-2370

$C_2HCl_3O_2$ Trichloroacetic acid. B.p. 196
                                2229, 2371-2394

C_2HCl_5 Pentachloroethane. B.p. 162.0
                                220, 1607, 2371, 2395-2486

$C_2HF_3O_2$ Trifluoroacetic acid.
                                183, 221, 2486a

Formula   Name and System No.

Formula | Name and System No.

C$_2$H$_3$Cl$_3$O$_2$ Chloral hydrate. B.p. 97.5
1094, 1432, 2753-2756

C$_2$H$_3$F$_3$O   2,2,2-Trifluoroethanol.
2757

C$_2$H$_3$N   Acetonitrile. B.p. 81.6
13, 143, 226, 1095, 1433, 1546, 1757, 1925, 2154, 2280, 2758-2816, 15845, 15854, 15864, 15918, 15920, 15929-15943, 16352-16355, 16556, 1609a, 15828a, 16551b

C$_2$H$_3$NO   Hydroxyacetonitrile (Glycolonitrile).
2817

C$_2$H$_3$NO   Methylisocyanate. B.p. 37.9
1095a, 2817a-2817d, 1433a

C$_2$H$_3$NS   Methyl thiocyanate. B.p. 132.5
2818

C$_2$H$_4$ Ethylene. B.p. −103.9
90, 982, 990, 2487, 2819, 16349, 195a, 2639a

C$_2$H$_4$BrCl   1-Bromo-2-chloroethane. B.p. 106.7
1926, 2820-2831

C$_2$H$_4$Br$_2$ 1,1-Dibromoethane. B.p. 109.5
1927, 2832-2856, 136e

C$_2$H$_4$Br$_2$ 1,2-Dibromoethane. B.p. 131.5
1096, 1342, 1610, 1928, 2644, 2857-2931, 16251, 16356-16359, 136f

C$_2$H$_4$Cl$_2$ 1,1-Dichloroethane. B.p. 57.4
144, 1097, 1178, 1434, 1611, 1929, 2730, 2932-2959

C$_2$H$_4$Cl$_2$ 1,2-Dichloroethane. B.p. 83.45
70, 115, 145, 227, 1098, 1435, 1612, 1881, 1930, 2281, 2336, 2640, 2729, 2731, 2758, 2857, 2932, 2960-3010, 15871, 15944-15946, 16360, 2817a

C$_2$H$_4$Cl$_2$O   Bis(chloromethyl) ether. B.p. 105
228, 1179, 3011-3020

C$_2$H$_4$Cl$_2$O   2,2-Dichloroethanol. B.p. 146.2
1343, 2157, 2530, 3021-3045

C$_2$H$_4$F$_2$ 1,1-Difluoroethane. B.p. −24.7
1014, 2149

C$_2$H$_4$O   Acetaldehyde. B.p. 20.2
229, 1019, 1547, 1714, 1931, 3046-3066, 15844, 15947, 15948, 16361, 15947a

C$_2$H$_4$O   Ethylene oxide. B.p. 10
230, 1932, 2960, 3046, 3067-3082

C$_2$H$_4$OS   Thioacetic acid. B.p. 89.5
3083-3085

C$_2$H$_4$O$_2$ Acetic acid. B.p. 118.1
14, 231, 1028, 1099, 1180, 1285, 1344, 1436, 1506, 1576, 1613, 1758, 1933, 2158, 2282, 2397, 2495, 2531, 2732, 2820, 2832, 2858, 2961, 3086-3240, 15865, 15872, 15948-15953, 16282, 16340, 16356, 16361-16382, 16564, 16565, 2758a, 16551a

C$_2$H$_4$O$_2$ Methyl formate. B.p. 31.9
15, 74, 232, 1020, 1181, 1437, 1548, 1715, 1727, 1934, 2128, 2373, 2521, 2606, 2645, 3047, 3067, 3241-3286, 16264, 16272-16274, 16309, 16383-16390

C$_2$H$_4$S   Ethylene sulfide. B.p. 55.7
1759, 1935, 3241, 3287-3296

C$_2$H$_5$Br   Bromoethane. B.p. 38.4
233, 1100, 1182, 1438, 1549, 1614, 1882, 1936, 3048, 3242, 3287, 3297-3331, 15954, 16266, 16272, 16320, 16321, 16383-16386

Formula   Name and System No.

C₂H₅BrO   2-Bromoethanol. B.p. 150.2
                     234, 1286, 2159, 2283, 2532, 2859, 3332-3362
C₂H₅BrO   Bromomethyl methyl ether. B.p. 87.5
                     3363, 3364
C₂H₅Cl   Chloroethane. B.p. 12.4
                     91, 1183, 1439, 1615, 1937, 2129, 2607, 3049, 3243, 3365-3372, 136g
C₂H₅ClO   2-Chloroethanol. B.p. 127
                     16, 235, 1029, 1287, 1345, 1507, 1760, 2160, 2284, 2398, 2533, 2833,
                     2860, 2962, 3373-3489, 15846, 15877, 15921, 15955-15960
C₂H₅ClO   Chloromethyl methyl ether. B.p. 59.5
                     1101, 1184, 1440, 1550, 1616, 1938, 2933, 3244, 3288, 3297, 3490-3518
C₂H₅I   Iodoethane. B.p. 72.3
                     236, 1102, 1185, 1441, 1617, 1761, 1883, 1939, 2759, 3086, 3298, 3519-
                     3548, 15961, 16322, 16323, 16391
C₂H₅IO   2-Iodoethanol. B.p. 176.5
                     237, 2534, 3549-3562
C₂H₅NO   Acetamide. B.p. 221.2
                     238, 1346, 1940, 2161, 2231, 2399, 2535, 2733, 2834, 2861, 3087, 3373,
                     3563-3843
C₂H₅NO   N-Methylformamide.
                     239
C₂H₅NO₂   Ethyl nitrite. B.p. 17.4
                     75, 1186, 1716, 2130, 2608, 3245, 3299, 3365, 3844-3857
C₂H₅NO₂   Nitroethane. B.p. 114.2
                     240, 1103, 1187, 1288, 1618, 1941, 2337, 3088, 3374, 3858-3897, 15962-
                     15964, 16392
C₂H₅NO₃   Ethyl nitrate. B.p. 87.68
                     241, 1104, 1188, 1289, 1508, 1762, 1942, 2285, 2963, 3089, 3858, 3898-
                     3941
C₂H₆   Ethane. B.p. −88
                     38, 92, 102, 955, 988, 991, 1503, 1910, 1943, 2270, 2488, 2819, 3942-
                     3948, 15840, 16319, 16349
C₂H₆Cl₂Si   Dichlorodimethylsilane.
                     1723, 3949, 3950, 3948a, 16307a, 16307b
C₂H₆O   Ethyl alcohol. B.p. 78.3
                     17, 242, 1030, 1105, 1189, 1290, 1442, 1509, 1551, 1728, 1763, 1884,
                     1944, 2147, 2152, 2162, 2272-2275, 2286, 2338, 2490, 2494, 2499, 2500,
                     2504, 2513, 2609, 2734, 2757, 2760, 2821, 2835, 2862, 2934, 2964, 3050,
                     3090, 3246, 3300, 3366, 3490, 3519, 3859, 3898, 3942, 3951-4182, 15837,
                     15847, 15855, 15858, 15866, 15869, 15896, 15917, 15922, 15927, 15928,
                     15930, 15945, 15947, 15954, 15961, 15965-16019, 16252-16254, 16275,
                     16276, 16288-16294, 16324, 16352, 16391, 16393-16408, 16556-16562,
                     241a
C₂H₆O   Methyl ether. B.p. −23.65
                     18, 93, 103, 960, 992, 1015, 1721, 4183, 16243
C₂H₆OS   Dimethyl sulfoxide.
                     243, 4184, 16409, 4183a
C₂H₆O₂   Ethylene Glycol. B.p. 197.4
                     244, 1291, 1347, 1577, 1764, 1945, 2163, 2232, 2287, 2400, 2536, 2735,
                     2822, 2836, 2863, 3375, 3563, 4185-4474, 16020, 16410-16416, 16324a,
                     16409a
C₂H₆O₄S   Methyl sulfate. B.p. 189.1
                     245, 2233, 2401, 2646, 4509-4529

Formula   Name and System No.

C_2H_6S Ethanethiol. B.p. 36.2
                 1552, 1765, 1946, 3247, 3301, 3492, 4475-4487, 16387
C_2H_6S Methyl sulfide. B.p. 37.4
                 1190, 1619, 1766, 1947, 3091, 3248, 3302, 3491, 3844, 3951, 4775, 4488-
                 4508
$C_2H_6S_2$ Methyl disulfide. B.p. 109.44
                 4530-4537
C_2H_7N Dimethylamine. B.p. 7.3
                 246, 961, 2135, 4538-4547, 16021
C_2H_7N Ethylamine. B.p. 16.55
                 247, 962, 4548-4556
C_2H_7NO 2-Aminoethanol. B.p. 170.8
                 248, 1348, 1578, 1767, 3564, 4185, 4557-4634
$C_2H_7O_3P$ Dimethyl phosphite.
                 4635
$C_2H_8N_2$ 1,1-Dimethylhydrazine.
                 249, 981, 15843
$C_2H_8N_2$ Ethylenediamine. B.p. 116
                 250, 4636-4649, 16022
$C_3Cl_3F_5$ 1,2,2-Trichloropentafluoropropane. B.p. 72.5
                 61
C_3F_6 Hexafluoropropene.
                 1016, 1421, 2278
C_3F_6O Hexafluoroacetone.
                 15839a
C_3F_7I Heptafluoro-1-iodopropane.
                 4649a
C_3F_8 Perfluoropropane.
                 1422
$C_3HF_5O_2$ Pentafluoropropionic acid.
                 251, 4650
C_3HF_7 Heptafluoropropane.
                 1017
$C_3H_2ClF_3O_2$ 3-Chloro-2,2,3-trifluoropropionic acid.
                 4651
$C_3H_2F_4O_2$ 2,2,3,3-Tetrafluoropropionic acid.
                 4652
$C_3H_2F_6O$ 1,1,1,3,3,3-Hexafluoro-2-propanol.
                 251a, 4652a
$C_3H_3ClF_3NO$ 3-Chloro-2,2,3-trifluoropropionamide.
                 4653
$C_3H_3Cl_3O_2$ Methyl trichloroacetate. B.p. 152.8
                 4654-4660
$C_3H_3F_4NO$ 2,2,3,3-Tetrafluoropropionamide.
                 4661
C_3H_3N Acrylonitrile. B.p. 77.3
                 76, 146, 252, 1106, 1948, 2761, 4662-4667, 15931, 15965, 16023, 3951a,
                 15828b, 16022a, 16022b, 16551a, 16551b
C_3H_3NS Thiazole. B.p. 111.5
                 253
C_3H_4 Propadiene. B.p. -32
                 963, 4668-4671, 16417-16419
C_3H_4 Propyne. B.p. -23
                 964, 2489, 4668, 4672-4674, 16417-16419

Formula   Name and System No.

$C_3H_4Br_2$	*cis*-1,2-Dibromopropene. B.p. 135.2
	4675
$C_3H_4Br_2$	*trans*-1,2-Dibromopropene. B.p. 125.95
	4676
$C_3H_4Cl_2$	1,2-Dichloropropene. B.p. 76.8
	1949
$C_3H_4Cl_2$	1,3-Dichloropropene.
	4677
$C_3H_4Cl_4$	1,1,2,2-Tetrachloropropane. B.p. 153
	4678-4681
$C_3H_4Cl_4$	1,1,2,3-Tetrachloropropane. B.p. 180
	4682, 4683
$C_3H_4N_2$	Pyrazole. B.p. 187.5
	4714-4720
C_3H_4O	Acrolein. B.p. 52.45
	254, 1191, 1950, 4662, 4685-4691, 76a, 15828a, 15828b, 16022a, 16551a,
	16551b
C_3H_4O	2-Propyne-1-ol. B.p. 113
	255, 4692, 16024, 16025
$C_3H_4O_2$	Acrylic acid. B.p. 140.5
	256, 4693-4695, 16026
$C_3H_4O_3$	Ethylene carbonate.
	257, 4186
$C_3H_4O_3$	Pyruvic acid. B.p. 166.8
	4696-4713
C_3H_5Br	*cis*-1-Bromopropene. B.p. 57.8
	1952, 3953, 15966
C_3H_5Br	*trans*-1-Bromopropene. B.p. 63.25
	1951, 3952, 15967
C_3H_5Br	2-Bromopropene. B.p. 48.35
	1953, 3954, 15968
C_3H_5Br	3-Bromopropene. B.p. 70.5
	1192, 1620, 1768, 1885, 1954, 3092, 3493, 3520, 3899, 3955, 4721-4740,
	4740a
C_3H_5BrO	Epibromohydrin. B.p. 138.5
	2164, 2864, 3093, 4741-4755
$C_3H_5BrO_2$	α-Bromopropionic acid. B.p. 205.8
	2402, 4756-4766
$C_3H_5Br_3$	1,2,3-Tribromopropane. B.p. 220
	3565, 4767-4786
C_3H_5Cl	*cis*-1-Chloropropene. B.p. 32.8
	3956
C_3H_5Cl	*trans*-1-Chloropropene. B.p. 37.4
	3957
C_3H_5Cl	2-Chloropropene. B.p. 22.65
	1621, 1955, 3249, 3845, 3958, 4787-4793
C_3H_5Cl	3-Chloropropene. B.p. 45.15
	258, 1193, 1622, 1886, 1956, 3250, 3494, 3959, 4677, 4787, 4794-4807
C_3H_5Cl	Methylvinylchloride.
	259
C_3H_5ClO	1-Chloro-2-propanone. B.p. 121
	260, 1623, 2165, 4809-4838
C_3H_5ClO	2-Chloro-2-propen-1-ol.
	4808

Formula   Name and System No.

C_3H_5ClO α-Chloropropionaldehyde. B.p. 86
266   261
C_3H_5ClO Epichlorohydrin. B.p. 116.45
               155, 262, 1031, 1624, 1957, 2166, 2823, 3094, 3960, 4839-4895, 16340-
               16345, 16362, 16420-16422
$C_3H_5ClO_2$ Methyl chloroacetate. B.p. 131.4
               263, 1349, 1958, 2167, 2537, 2865, 3095, 3376, 4896-4947, 15897
$C_3H_5Cl_3$ 1,1,3-Trichloropropane. B.p. 148
               4948, 4949
$C_3H_5Cl_3$ 1,2,2-Trichloropropane. B.p. 122
               4950-4955
$C_3H_5Cl_3$ 1,2,3-Trichloropropane. B.p. 156.85
               2647, 3096, 3566, 4187, 4839, 4956-4994, 16423
C_3H_5F 2-Fluoropropene. B.p. −24
               965
C_3H_5I 3-Iodopropene. B.p. 102.0
               264, 1032, 1625, 1769, 1959, 2339, 3097, 3377, 3900, 3961, 4188, 4995-
               5023, 15969, 16027, 16028, 16424-16426
C_3H_5N Propionitrile. B.p. 97
               147, 265, 1960, 3962, 4663, 5024-5046, 16023
C_3H_5NO Hydracrylonitrile. B.p. 229.7
               266
$C_3H_5N_3O_9$ Nitroglycerin.
               5047
C_3H_6 Cyclopropane. B.p. −31.5
               966
C_3H_6 Propene. B.p. −48
               967, 993, 4669, 4672, 4685, 5048, 16417, 93a, 983a, 1010a, 1168e, 1910a
$C_3H_6Br_2$ 1,2-Dibromopropane. B.p. 141
               2866, 3098, 3332, 3378, 3549, 3567, 4189, 4840, 4896, 5049-5088
$C_3H_6Br_2$ 1,3-Dibromopropane. B.p. 166.9
               3021, 3099, 3379, 3568, 4190, 5089-5132
$C_3H_6Br_2O$ 2,3-Dibromo-1-propanol. B.p. 219.5
               5133-5152
$C_3H_6Cl_2$ 1,1-Dichloropropane. B.p. 90
               5153, 5154
$C_3H_6Cl_2$ 1,2-Dichloropropane. B.p. 97
               267, 1961, 2288, 2965, 3963, 5155-5162, 16029
$C_3H_6Cl_2$ 1,3-Dichloropropane. B.p. 129.8
               1033, 5163-5165
$C_3H_6Cl_2$ 2,2-Dichloropropane. B.p. 70.4
               1626, 1962, 3011, 3100, 3901, 3964, 5166-5182
$C_3H_6Cl_2O$ 1,3-Dichloro-2-propanol. B.p. 175.8
               1350, 2403, 2538, 3569, 4191, 5183-5256, 16427-16430
$C_3H_6Cl_2O$ 2,3-Dichloro-1-propanol. B.p. 183.8
               268, 4192, 5257-5306, 16431
C_3H_6O Acetone. B.p. 56.1
               269, 1108, 1194, 1292, 1443, 1553, 1627, 1729, 1770, 1963, 2153, 2168,
               2289, 2505 2514, 2641, 2736, 2762, 2935, 2966, 3051, 3251, 3289, 3303,
               3495, 3521, 3943, 3965, 4476, 4488, 4548, 4557, 4686, 4721, 4794, 5047,
               5307-5397, 15848, 15856, 15867, 15932, 15970, 16030-16039, 16267,
               16277, 16281, 16283, 16288, 16289, 16295-16300, 16306, 16322, 16324-
               16328, 16360, 16432-16439, 993a, 16247a, 16431a, 16431b, 16565a,
               16565b, 16566a

Formula   Name and System No.

C$_3$H$_6$O   Allyl alcohol. B.p. 96.9
                  270, 1034, 1109, 1195, 1293, 1351, 1444, 1510, 1771, 2169, 2290, 2867,
                  2936, 2967, 3522, 3902, 4722, 4841, 4897, 4995, 5166, 5307, 5398-5452,
                  15849, 15859, 15923, 16027, 16040-16048, 16440, 5397a

C$_3$H$_6$O   Propionaldehyde. B.p. 48.7
                  371, 1196, 1445, 1554, 1964, 2937, 3304, 3966, 4687, 5308, 5453-5459,
                  15898

C$_3$H$_6$O   Propylene oxide. B.p. 35
                  272, 1446, 1555, 3052, 3068, 3305, 4549, 5155, 5453, 5460-5472, 5397a

C$_3$H$_6$OS   Methyl thioacetate. B.p. 95.5
                  1965, 3967, 5473-5477

C$_3$H$_6$O$_2$ 1,3-Dioxolane. B.p. 75
                  273, 5478-5482

C$_3$H$_6$O$_2$ Ethyl formate. B.p. 54.1
                  19, 274, 1197, 1447, 1556, 1730, 1966, 2506, 2515, 2938, 3290, 3306,
                  3496, 3523, 3968, 4723, 4795, 5167, 5309, 5483-5510, 15971, 16278,
                  16301

C$_3$H$_6$O$_2$ Glycidol.
                  5510a

C$_3$H$_6$O$_2$ Methoxyacetaldehyde. B.p. 92
                  275

C$_3$H$_6$O$_2$ Methyl acetate. B.p. 57.1
                  20, 276, 1110, 1198, 1448, 1557, 1731, 1967, 2507, 2516, 2763, 2939,
                  3307, 3497, 3524, 3969, 4724, 4796, 5168, 5310, 5454, 5483, 5511-5561,
                  15899, 16268, 16277, 16284, 16302, 16325, 16329, 16330, 16432, 16441,
                  993b, 3052a, 3101a, 15947a, 15948a, 16247b, 16254a, 16431a, 16431b,
                  16563a, 16566a

C$_3$H$_6$O$_2$ Propionic acid. B.p. 140.7
                  21, 277, 1935, 1199, 1352, 1511, 1579, 1628, 1772, 2170, 2404, 2496, 2539,
                  2737, 2824, 2837, 2868, 3102, 3860, 4693, 4696, 4741, 4842, 4898, 4956,
                  4996, 5049, 5089, 5562-5667, 15949, 16049, 16357, 16442, 15872a, 16362a,
                  16551a

C$_3$H$_6$O$_3$ Ethylene glycol monoformate. B.p. 180
                  5668

C$_3$H$_6$O$_3$ Methyl carbonate. B.p. 90.25
                  278, 1111, 1200, 1294, 1512, 1968, 2291, 2340, 2968, 3103, 3525, 3903,
                  3970, 4997, 5398, 5669-5710

C$_3$H$_6$O$_3$ Methyl glycolate. B.p. 151
                  22, 5711, 5712

C$_3$H$_6$O$_3$ Trioxane. B.p. 114.5
                  279, 1558, 5713-5722, 16050-16056

C$_3$H$_7$Br   1-Bromopropane. B.p. 71.0
                  1112, 1201, 1449, 1629, 1773, 1887, 1969, 2764, 3104, 3498, 3526, 3904,
                  3971, 4725, 5311, 5399, 5484, 5511, 5669, 5723-5751, 15972, 279a

C$_3$H$_7$Br   2-Bromopropane. B.p. 59.4
                  1202, 1450, 1630, 1774, 1888, 1970, 2940, 3105, 3499, 3972, 5312, 5400,
                  5485, 5512, 5752-5771, 16301-16303

C$_3$H$_7$Cl   1-Chloropropane. B.p. 46.4
                  280, 1203, 1631, 1732, 1889, 1971, 3012, 3252, 3500, 3973, 4489, 4797,
                  5313, 5455, 5486, 5513, 5772-5787, 16269, 16278

C$_3$H$_7$Cl   2-Chloropropane. B.p. 34.9
                  281, 1204, 1632, 1972, 3053, 3253, 3308, 3846, 3974, 4477, 4490, 4798,
                  5314, 5487, 5788-5796

Formula	Name and System No.

C_3H_7ClO 1-Chloro-2-propanol. B.p. 127
        282, 1036, 1295, 1353, 1513, 1776, 2171, 2292, 2540, 2869, 2969, 3861, 4193, 5797-5821, 5837, 16029

C_3H_7ClO 2-Chloro-1-propanol. B.p. 133.7
        283, 1777, 2172, 2293, 2870, 5822-5836

$C_3H_7ClO_2$ Chloromethoxymethoxymethane. B.p. 95
        5838-5842

$C_3H_7ClO_2$ 1-Chloro-2,3-propanediol. B.p. 213
        5843-5847

C_3H_7I 1-Iodopropane. B.p. 102.4
        1037, 1633, 1778, 1973, 2341, 3022, 3106, 3380, 3570, 3905, 3975, 4843, 4998, 5562, 5670, 5848-5870, 16057

C_3H_7I 2-Iodopropane. B.p. 89.45
        1296, 1634, 1779, 1974, 2294, 2342, 3107, 3381, 3906, 3976, 5024, 5315, 5671, 5871-5891

C_3H_7N Allylamine. B.p. 52.9
        284

C_3H_7NO Acetoxime. B.p. 135.8
        5892

C_3H_7NO Dimethylformamide. B.p. 153
        285, 1635, 2765, 3254, 4651, 4652, 4653, 4661, 5893, 5894, 16443

C_3H_7NO Propionamide. B.p. 222.1
        286, 1354, 1975, 2173, 2234, 2405, 2541, 2871, 3382, 3571, 4194, 4558, 5050, 5090 5257, 5895-6090

$C_3H_7NO_2$ Ethyl carbamate. B.p. 185.25
        1355, 1580, 2174, 2406, 2542, 2872, 3572, 4195, 4559, 4957, 5091, 5183, 5257, 5895, 6091-6233

$C_3H_7NO_2$ Isopropyl nitrite. B.p. 40.1
        287, 1205, 1451, 1559, 1717, 1733, 2610, 3054, 3255, 3291, 3309, 3367, 4491, 4788, 4799, 5316, 5456, 5488, 5514, 5772, 5788, 6234-6249

$C_3H_7NO_2$ 1-Nitropropane. B.p. 130.5
        289, 1113, 1976, 2873, 3977, 6250-6269, 16058

$C_3H_7NO_2$ 2-Nitropropane. B.p. 120
        290, 1114, 1977, 3978, 6270-6293, 15973, 16059

$C_3H_7NO_2$ Propyl nitrite. B.p. 47.75
        288, 1206, 1452, 1560, 2941, 3256, 3310, 3501, 4478, 4492, 4800, 5317, 5457, 5489, 5515, 5752, 5773, 5789, 6234, 6294-6303

$C_3H_7NO_3$ Propyl nitrate. B.p. 110.5
        291, 1780, 2175, 2838, 3108, 3862, 3979, 6235, 6304-6326

C_3H_8 Propane. B.p. −44
        956, 968, 994, 1423, 1911, 1978, 2766, 4670, 4673, 5048, 5318, 6327, 6328, 16319, 16418

C_3H_8O Ethyl methyl ether. B.p. 10.8
        23, 1207

C_3H_8O Isopropyl alcohol. B.p. 82.3
        292, 1038, 1115, 1208, 1297, 1453, 1514, 1561, 1734, 1781, 1890, 2176, 2295, 2767, 2839, 2874, 2942, 2970, 3311, 3527, 3847, 3863, 3907, 3944, 3980, 4493, 4664, 4726, 4801, 4809, 4844, 4899, 4999, 5025, 5156, 5169, 5319, 5401, 5473, 5490, 5516, 5672, 5723, 5753, 5774, 5790, 5848, 5871, 6250, 6270, 6304, 6329-6434, 15860, 15878, 15900, 15924, 15946, 16030, 16060-16080, 16255, 16279, 16304, 16433, 1978a, 16057a, 16255a, 16325a, 16433a, 16433b, 16443a-16446a, 16565a, 16565c

Formula   Name and System No.

C$_4$H$_7$BrO$_2$ Ethyl bromoacetate. B.p. 158.2
                     1359, 2408, 2546, 3577, 4200, 5054, 5092, 5566, 5898, 7123-7173, 16458-
                     16460

C$_4$H$_7$Cl   cis-1-Chloro-1-butene. B.p. 63.4
                     3999

C$_4$H$_7$Cl   trans-1-Chloro-1-butene. B.p. 68
                     3998

C$_4$H$_7$Cl   2-Chloro-1-butene. B.p. 58.4
                     4000

C$_4$H$_7$Cl   cis-2-Chloro-2-butene. B.p. 62.4
                     4002

C$_4$H$_7$Cl   trans-2-Chloro-2-butene. B.p. 66.6
                     4001

C$_4$H$_7$Cl   1-Chloro-2-methyl-1-propene. B.p. 68.1
                     327, 5329

C$_4$H$_7$ClO   α-3-Chloro-2-butene-1-ol. B.p. 164
                     328

C$_4$H$_7$ClO   β-3-Chloro-2-butene-1-ol. B.p. 166
                     329

C$_4$H$_7$ClO   2-Chloroethyl vinyl ether. B.p. 108
                     330, 3384, 7174, 7175

C$_4$H$_7$ClO$_2$ 4-Chloromethyl-1,3-dioxolane. B.p. 66/40mm
                     331

C$_4$H$_7$ClO$_2$ Ethyl chloroacetate. B.p. 143.5
                     332, 1360, 2409, 2547, 2878, 3334, 3385, 3578, 4003, 4201, 4960, 5403,
                     5567, 6440, 6540, 7052, 7176-7223, 15976, 16461, 16462, 15834a, 16307a

C$_4$H$_7$Cl$_3$O   Ethyl 1,1,2-trichloroethyl ether. B.p. 172.5
                     1361, 5093, 6093, 7225-7238

C$_4$H$_7$Cl$_3$O   1,1,1-Trichloro-tert-butanol.
                     7224

C$_4$H$_7$N   Butyronitrile. B.p. 118
                     333, 6441, 7239-7245

C$_4$H$_7$N   Isobutyronitrile. B.p. 103
                     334, 4004, 5002, 6334, 6442, 7246-7251

C$_4$H$_7$N   Pyrroline. B.p. 90.9
                     1215, 1992, 6443, 6886, 7252

C$_4$H$_7$NO   2-Hydroxyisobutyronitrile.
                     335, 4635, 7253-7255

C$_4$H$_8$ 1-Butene. B.p. −6
                     975, 995, 1719, 1749, 2139, 2643, 3070, 3259, 4540, 6847, 6963, 16244,
                     16339, 7255a-7256b

C$_4$H$_8$ 2-Butene. B.p. 1–3.7
                     997, 998, 2140, 2141, 3071, 3072, 3260, 3261, 4541, 4542, 6848, 6849,
                     6871, 6964, 6968, 6969, 7255a, 7256c-7256e, 16451a

C$_4$H$_8$ 2-Methylpropene. B.p. −6
                     976, 996, 1750, 2131, 2142, 3073, 3262, 4543, 6850, 7256, 16245, 7256c,
                     7256f

C$_4$H$_8$Br$_2$O   Bis(2-bromoethyl) ether.
                     4202, 6650

C$_4$H$_8$Cl$_2$O   Bis(2-chloroethyl) ether. B.p. 178.65
                     336, 1362, 2976, 3386, 3580, 4203, 4560, 4714, 5094, 5899, 6094, 7257-
                     7303, 15955, 118a, 148a, 1724a, 3948a, 6835a, 15834b, 16307b

C$_4$H$_8$Cl$_2$O   1,2-Dichloroethyl ethyl ether. B.p. 145.5
                     1363, 2818, 3116, 3579, 6444, 6952, 7304-7332

Formula   Name and System No.

C₄H₈Cl₂O$\quad$1,3-Dichloro-2-methyl-2-propanol. B.p. 174
$\qquad$337

C₄H₈Cl₂S$\quad$Bis(2-chloroethyl) sulfide. B.p. 216.8
$\qquad$4204, 7333-7341

C₄H₈O$\quad$2-Butanone. B.p. 79.6
$\qquad$338, 1121, 1216, 1300, 1460, 1564, 1640, 1785, 1993, 2299, 2343, 2509,
$\qquad$2518, 2947, 2977, 3117, 3292, 3505, 3532, 3910, 4005, 4495, 4688, 4729,
$\qquad$5330, 5404, 5519, 5568, 5674, 5730, 5757, 5838, 5873, 6272, 6335, 6445,
$\qquad$6789, 6815, 6839, 6887, 6998, 7342-7385, 15851, 15868, 15870, 15977,
$\qquad$16032, 16060, 16082, 16110-16122, 16250, 16252, 16257-16258, 16275,
$\qquad$16285, 1629, 16304, 16305, 16364, 16365, 16393-16396, 16463-16467,
$\qquad$16431a, 16433c, 16443a, 16565c

C₄H₈O$\quad$1-Butene-3-ol. B.p. 96
$\qquad$339, 7386

C₄H₈O$\quad$Butyraldehyde. B.p. 76
$\qquad$340, 1217, 1461, 1994, 4006, 5331, 5520, 5728, 5758, 6816, 6888, 7342,
$\qquad$7387-7399, 15978, 16033, 16123-16127

C₄H₈O$\quad$Crotonyl alcohol. B.p. 119
$\qquad$341

C₄H₈O$\quad$Cyclopropyl methyl ether. B.p. 44.73
$\qquad$5458

C₄H₈O$\quad$Ethyl vinyl ether. B.p. 35.5
$\qquad$342, 3057, 4007, 15979

C₄H₈O$\quad$Isobutene oxide. B.p. 50
$\qquad$1463, 1565, 2948

C₄H₈O$\quad$Isobutyraldehyde. B.p. 63.5
$\qquad$343, 1122, 1218, 1462, 1995, 2949, 4008, 4687, 5332, 5493, 5521, 5729,
$\qquad$5759, 7343, 7387, 7400-7405, 16128, 16129

C₄H₈O$\quad$2-Methyl-2-propen-1-ol. B.p. 113.8
$\qquad$7406

C₄H₈O$\quad$Methyl propenyl ether. B.p. 46.3
$\qquad$344

C₄H₈O$\quad$Tetrahydrofuran. B.p. 66
$\qquad$345, 1464, 1996, 2510, 2519, 4009, 6273, 6817, 7407, 16332, 4204a,
$\qquad$6335a, 7406a, 16324a

C₄H₈OS$\quad$Ethyl thioacetate. B.p. 116.6
$\qquad$4010, 4205, 5405, 6336, 6446, 7408-7415

C₄H₈OS$\quad$2-(Methylthio)propionaldehyde.
$\qquad$346

C₄H₈OS$\quad$1,4-Oxathiane. B.p. 149.2
$\qquad$347

C₄H₈O₂$\quad$Butyric acid. B.p. 162.45
$\qquad$27, 348, 1123, 1219, 1364, 1581, 1641, 2180, 2236, 2300, 2410, 2548,
$\qquad$2738, 2842, 2879, 2978, 3118, 3263, 3314, 4655, 4697, 4961, 5055, 5095,
$\qquad$5158, 5569, 6915, 6971, 7053, 7123, 7176, 7257, 7388, 7416-7513, 16468,
$\qquad$15872b, 16362a

C₄H₈O₂$\quad$1,2-Dimethoxyethylene. B.p. 102
$\qquad$1997

C₄H₈O₂$\quad$*p*-Dioxane. B.p. 101.32
$\qquad$349, 1041, 1124, 1465, 1642, 1786, 1998, 2181, 2301, 2739, 2979, 3119,
$\qquad$3387, 3865, 3911, 4011, 4206, 4637, 5003, 5159, 5333, 5406, 5570, 5675,
$\qquad$5852, 5874, 6307, 6337, 6447, 6541, 7174, 7344, 7514-7555, 15857,
$\qquad$16020, 16392, 16469, 16470, 118b, 148b, 2769a

C₄H₈O₂$\quad$1,3-Dioxane. B.p. 104
$\qquad$350, 7556

Formula   Name and System No.

C$_4$H$_8$O$_2$ Ethoxyacetaldehyde. B.p. 105
                        351
C$_4$H$_8$O$_2$ Ethyl acetate. B.p. 77.05
                        28, 352, 1125, 1220, 1301, 1466, 1737, 1787, 1999, 2302, 2753, 2770,
                        2980, 3120, 3315, 3506, 3533, 4012, 4730, 5028, 5171, 5334, 5407, 5676,
                        5731, 6760, 5779, 6274, 6338, 6448, 6818, 6889, 7345, 7514, 7557-7594,
                        15903, 15933, 15956, 15980, 16253, 16259, 16276, 16279, 16286, 16323,
                        16333, 16364, 16366, 16391, 16393, 16397, 16398, 16444, 16463, 16471,
                        16472, 16557, 5521a, 6964a, 16431b, 16433c
C$_4$H$_8$O$_2$ 2-Hydroxybutyraldehyde.
                        353
C$_4$H$_8$O$_2$ Isobutyric acid. B.p. 154.35
                        354, 1042, 1221, 1365, 1582, 2182, 2411, 2549, 2880, 4656, 4743, 4962,
                        5056, 5096, 6916, 6972, 7026, 7054, 7124, 7177, 7515, 7595-7667, 16130,
                        16473, 16474, 15872c
C$_4$H$_8$O$_2$ Isopropyl formate. B.p. 68.8
                        355, 1126, 1222, 1467, 2000, 3507, 3534, 5335, 5522, 5732, 6819, 7346,
                        7557, 7668-7673, 16303
C$_4$H$_8$O$_2$ 3-Methoxypropionaldehyde.
                        356
C$_4$H$_8$O$_2$ 2-Methyl-1,3-dioxolane. B.p. 82.5
                        357
C$_4$H$_8$O$_2$ Methyl propionate. B.p. 79.85
                        358, 1127, 1223, 1302, 1468, 1788, 2001, 2303, 2771, 3535, 4013, 5029,
                        5172, 5408, 5733, 5875, 6339, 6449, 6890, 7347, 7558 7674-7694, 15904,
                        16049, 16257, 16313, 16394
C$_4$H$_8$O$_2$ Propyl formate. B.p. 80.9
                        359, 1128, 1224, 1303, 1469, 1789, 2002, 2304, 2344, 2772, 2981, 3121,
                        3536, 4014, 4731, 5030, 5173, 5409, 5734, 5876, 6340, 6450, 6891, 7348,
                        7389, 7559, 7674, 7695-7723, 16083, 16464, 1642a
C$_4$H$_8$O$_2$ 2-Vinyloxyethanol. B.p. 143
                        360, 4207
C$_4$H$_8$O$_2$S   Sulfolane.
                        16475
C$_4$H$_8$O$_3$ Ethylene glycol monoacetate. B.p. 190.9
                        3581, 4208, 5900, 7055, 7258, 7724-7771
C$_4$H$_8$O$_3$ Methyl lactate. B.p. 144.8
                        361, 1366, 2183, 2412, 2550, 2881, 4902, 4963, 6542, 7125, 7178, 7772-
                        7829, 16461
C$_4$H$_8$O$_3$ Propylene glycol monoformate.
                        7830
C$_4$H$_8$S   Thiophane. B.p. 118.8
                        1043, 1643, 2003, 3122, 3388, 4847, 6308, 6341, 6451, 7831-7848
C$_4$H$_9$Br   1-Bromobutane. B.p. 101.5
                        1044, 1516, 1644, 1790, 2004, 2345, 2773, 3023, 3123, 3335, 3389, 3866,
                        3912, 4015, 4209, 4744, 4848, 5410, 5571, 5677, 6309, 6452, 7516, 7849-
                        7878, 6341a
C$_4$H$_9$Br   2-Bromobutane. B.p. 91.2
                        1304, 2005, 3124, 3913, 4016, 5678, 6342, 6453, 7349, 7675, 7695, 7879-
                        7887
C$_4$H$_9$Br   1-Bromo-2-methylpropane. B.p. 91.4
                        1305, 1645, 1791, 2006, 2346, 2774, 3125, 3336, 3390, 3867, 3914, 4017,
                        4210, 5031, 5411, 5679, 6310, 6343, 6454, 6983, 7350, 7517, 7560, 7676,
                        7696, 7879, 7888-7916, 15981, 361a

Formula | Name and System No.

Formula   Name and System No.

$C_4H_9NO_3$ Isobutyl nitrate. B.p. 122.9
                 371, 2185, 3134, 5574, 6545. 7994, 8028, 8080-8099
$C_4H_9NO_3$ 2-Methyl-2-nitro-1-propanol.
                 8100
C_4H_{10} Butane. B.p. −0.5
                 184, 977, 999, 1720, 1751, 2132, 2143, 2150, 3058, 3074, 3265, 3369,
                 3850, 3946, 4545, 5343, 6851, 6965, 8101, 16246, 7256a, 7256d, 8100a,
                 16451b, 16451c, 16451d
C_4H_{10} 2-Methylpropane. B.p. −10
                 185, 978, 1000, 1722, 1752, 2133, 2144, 3075, 3266, 4544, 6327, 6852,
                 15905, 16247, 7256b, 7256e, 7256f, 8100a, 16451b, 16451d
$C_4H_{10}O$ Butyl alcohol. B.p. 117.75
                 29, 372, 1046, 1133, 1233, 1307, 1368, 1797, 2015, 2148, 2186, 2276,
                 2306, 2497, 2502, 2551, 2843, 2884, 2984, 3135, 3393, 3538, 3584, 3869,
                 3918, 4026, 4213, 4638, 4734, 4745, 4810, 4852, 4905, 5004, 5032, 5057,
                 5176, 5344, 5682, 5738, 5799, 5853, 5877, 6252, 6275, 6313, 6546, 6857,
                 6922, 6973, 7004, 7008, 7126, 7180, 7240, 7259, 7304, 7357, 7391, 7401,
                 7408, 7519, 7681, 7702, 7774, 7850, 7880, 7890, 7930, 7959, 7995, 8029,
                 8080, 8102-8215, 16131, 16134-16151, 16260, 16397, 16465, 16471, 16476,
                 16478-16482, 16569, 15982a, 16409a, 16481a, 16481b, 16482a
$C_4H_{10}O$ sec-Butyl alcohol. B.p. 99.4
                 373, 1047, 1134, 1308, 1472, 1798, 2187, 2307, 2885, 2985, 3316, 3870,
                 3919, 4027, 4811, 4853, 4906, 5415, 5683, 5762, 5878, 6253, 6276, 6463,
                 6547, 7305, 7358, 7409, 7520, 7568, 7596, 7682, 7703, 7851, 7881, 7891,
                 7931, 7960, 8102, 8216-8255, 16110, 16130, 16134, 16152-16169, 15851a
$C_4H_{10}O$ tert-Butyl alcohol. B.p. 82.5
                 374, 1048, 1135, 1235, 1309, 1473, 1799, 1897, 2188, 2308, 2951, 2986,
                 3871, 3920, 4028, 4497, 4735, 4803, 4854, 5345, 5684, 5739, 5763, 5781,
                 5879, 6254, 6277, 6350, 6827, 7359, 7521, 7569, 7683, 7704, 7852, 7892,
                 7919, 7932, 7961, 7976, 7996, 8256-8287, 15852, 16111, 16133, 16152,
                 16170-16174, 2015a, 3538a, 8102a, 15905a, 15982b, 16169a, 16482b
$C_4H_{10}O$ Ethyl ether. B.p. 34.5
                 30, 39, 186, 375, 1136, 1234, 1474, 1566, 1652, 1738, 1898, 2016, 2271,
                 2374, 2522, 2650, 2778, 2987, 3059, 3136, 3267, 3293, 3317, 3511, 3851,
                 4029, 4480, 4498, 4551, 4790, 4804, 5346, 5460, 5496, 5527, 5782, 6239,
                 6297, 6351, 6464, 6840, 6966, 7418, 7597, 8104, 8288-8305, 15983, 16388,
                 16389, 16483, 4649a
$C_4H_{10}O$ Isobutyl alcohol. B.p. 107.1
                 376, 1049, 1137, 1236, 1310, 1369, 1475, 1518, 1800, 2189, 2309,
                 2349, 2552, 2740, 2779, 2825, 2844, 2886, 2988, 3394, 3539, 3872, 3921,
                 3947, 4639, 4736, 4746, 4812, 4855, 4907, 5005, 5033, 5058, 5177, 5347,
                 5685, 5740, 5800, 5854, 5880, 6255, 6278, 6314, 6465, 6548, 6858, 6923,
                 7005, 7127, 7181, 7241, 7360, 7392, 7402, 7410, 7522, 7570, 7705, 7853,
                 7882, 7893, 7920, 7933, 7962, 7997, 8030, 8081, 8103, 1256, 8306-8392,
                 15861, 15906, 15926, 16123, 16128, 16175-16190, 16343, 16392, 16422,
                 16469, 16484, 4029a, 15982c, 16484a
$C_4H_{10}O$ Methyl propyl ether. B.p. 38.9
                 377, 1237, 1476, 1567, 1739, 2017, 3268, 3318, 3852, 4030, 4499, 4552,
                 5348, 5783, 6240, 6633, 8288, 8393-8396
$C_4H_{10}O_2$ Acetaldehyde dimethyl acetal. B.p. 64.3
                 380, 1238, 1311, 1478, 2018, 3540, 4031, 4737, 5497, 5529, 5741, 5764,
                 6466, 7393, 7934, 7963, 8072, 8397-8400, 15907, 16124
$C_4H_{10}O_2$ levo-2,3-Butanediol. B.p. 183
                 378, 8401

Formula   Name and System No.

C₄H₁₀O₂   *meso*-2,3-Butanediol. B.p. 183
           379, 7386, 8402, 8403
C₄H₁₀O₂   1,4-Butanediol. B.p. 230
           8404
C₄H₁₀O₂   1,2-Dimethoxyethane. B.p. 83
           381, 8405
C₄H₁₀O₂   2-Ethoxyethanol. B.p. 135.1
           382, 1370, 1583, 1801, 2190, 2413, 2553, 2887, 3337, 3395, 4032, 4561,
           4908, 5059, 5801, 5855, 6279, 6315, 6924, 7128, 7182, 7571, 7775, 7998,
           8031, 8082, 8105, 8306, 8406-8511, 15984, 16059, 16191, 16485-16487
C₄H₁₀O₂   Ethoxymethoxymethane. B.p. 65.91
           383, 1477, 1899, 2019, 4033, 4738, 5528, 5742, 7921, 15908, 15985
C₄H₁₀O₂   1-Methoxy-2-propanol. B.p. 118
           384, 8512, 6650a
C₄H₁₀O₂   2-Methoxy-1-propanol. B.p. 130
           385
C₄H₁₀O₃   Diethylene glycol. B.p. 245.5
           386, 3585, 5133, 5902, 6680, 7003, 7260, 8513-8612
C₄H₁₀S   Butanethiol. B.p. 97.5
           1050, 1519, 1802, 3396, 5686, 6467, 6925, 7707, 7935, 8106, 8613-8629,
           16135
C₄H₁₀S   2-Butanethiol. B.p. 85.15
           6896, 8630-8641
C₄H₁₀S   Ethyl sulfide. B.p. 92.2
           1138, 1312, 1479, 1653, 1803, 2020, 2310, 3015, 3137, 3397, 3922, 4034,
           5349, 5416, 5476, 5575, 5687, 6352, 6468, 6926, 7361, 7572, 7684, 7706,
           7854, 7894, 7983, 8060, 8107, 8216, 8257, 8613, 8642-8665
C₄H₁₀S   Isopropyl methyl sulfide. B.p. 84.76
           6897, 8630, 8666-8673
C₄H₁₀S   2-Methyl-1-propanethiol. B.p. 88
           2311, 2989, 5477, 8642, 8674-8691
C₄H₁₀S   2-Methyl-2-propanethiol. B.p. 64.35
           8692-8696
C₄H₁₀S   Methyl propyl sulfide. B.p. 95.47
           8697-8704
C₄H₁₀S₂   Ethyl disulfide. B.p. 154.11
           8705, 8706
C₄H₁₁ClSi Chloromethyltrimethylsilane. B.p. 97
           4035
C₄H₁₁N   Butylamine. B.p. 77.8
           387, 4036, 5350, 6353, 7362, 8108, 8707-8708e, 15986, 16062, 16136,
           16193, 16478, 2020a
C₄H₁₁N   Diethylamine. B.p. 55.9
           388, 1139, 1239, 2021, 2952, 3319, 4037, 4553, 5351, 5530, 6634, 7363,
           7523, 7573, 8289, 8393, 8397, 8709-8718, 15934, 16194
C₄H₁₁N   Isobutylamine. B.p. 68
           389, 2022, 5352, 8719-8725
C₄H₁₁NO   2-Amino-2-methyl-1-propanol. B.p. 165.4
           8726
C₄H₁₁NO   2-Dimethylaminoethanol. B.p. 134.6
           390, 4546, 8727-8730, 16021
C₄H₁₁NO   3-Methoxypropylamine. B.p. 116
           391

Formula   Name and System No. ·

$C_4H_{11}NO_2$ 2,2'-Iminodiethanol. B.p. 268
                   392, 3586, 4554, 4562, 8731-8735
$C_4H_{11}O_3P$ Diethyl phosphite.
                   2752, 2817, 7224, 7253, 8258, 8736-8739
$C_4H_{12}Ge$ Tetramethylgermane. B.p. 43.5
                   8289a
$C_4H_{12}N_2$ Tetramethylhydrazine.
                   393
$C_4H_{12}Si$ Tetramethylsilane. B.p. 26.64
                   1740
$C_4H_{12}OSi$ Methoxytrimethylsilane. B.p. 57
                   2023
$C_4H_{12}O_4Si$ Tetramethoxysilane. B.p. 121.8
                   8740
$C_5Cl_2F_6$ 1,2-Dichlorohexafluorocyclopentene. B.p. 90.6
                   6861, 8741
C_5F_{10} Perfluorocyclopentane.
                   189, 8742, 8743
$C_5F_{11}IO$ Heptafluoroisopropyl 2-iodo-tetrafluoroethyl ether. B.p. 86
                   279a
C_5F_{12} Perfluoropentane.
                   3, 188, 190, 8742
$C_5H_4F_8O$ 2,2,3,3,4,4,5,5-Octafluoro-1-pentanol.
                   8744, 8745
$C_5H_4O_2$ 2-Furaldehyde. B.p. 161.45
                   394, 1140, 1371, 1480, 2191, 2237, 2414, 2554, 2888, 3060, 3138, 3398,
                   3587, 4214, 4964, 5060, 5097, 5185, 5576, 6097, 6549, 6953, 6974, 7056,
                   7225, 7261, 7306, 7419, 7574, 7598, 7724, 7999, 8406, 8746-8829, 15950,
                   16452, 16488-16491
C_5H_5N Pyridine. B.p. 115.5
                   31, 156, 395, 1051, 1141, 1536, 1654, 1804, 2024, 2192, 2741, 2889, 3139,
                   3399, 4038, 4215, 4856, 4950, 5353, 5417, 5577, 6469, 6550, 7013, 7420,
                   7524, 7831, 8000, 8032, 8109, 8217, 8307, 8407, 8614, 8643, 8830-8872,
                   15833, 15873, 16195-16212, 16363, 16367-16374, 16410, 16442, 16468,
                   16479, 16492-16494, 16564, 16565, 1480a, 2779b
C_5H_6 Cyclopentadiene. B.p. 41.0
                   3269, 5354, 8872a
$C_5H_6N_2$ 2-Methylpyrazine. B.p. 133
                   396
C_5H_6O 2-Methylfuran. B.p. 63.7
                   397, 1240, 2025, 4039, 5459, 5531, 7364, 15909, 16034
$C_5H_6O_2$ Furfuryl alcohol. B.p. 169.35
                   398, 1584, 2193, 2555, 3338, 3400, 3588, 5098, 5903, 6098, 8746, 8873-
                   8894
C_5H_6S 2-Methylthiophene. B.p. 111.92
                   4530, 8895-8898
C_5H_6S 3-Methylthiophene. B.p. 114.96
                   8899-8905
C_5H_7Cl Chloroprene. B.p. 59.4
                   6967
C_5H_7ClO 2-Chloroallyl vinyl ether.
                   4808
C_5H_7N 3-Methyl-3-butenenitrile. B.p. 137
                   399

Formula   Name and System No.

C$_5$H$_7$N   1-Methylpyrrole. B.p. 112.8
                  3016, 6470, 6551, 7832, 7984, 8110, 8308, 8644
C$_5$H$_7$N   2-Methylpyrrole. B.p. 147.5
                  8408, 8906
C$_5$H$_7$NO   Furfurylamine. B.p. 144
                  400
C$_5$H$_8$ Cyclopentene. B.p. 43
                  2026, 3270, 5461, 5532, 8907
C$_5$H$_8$ Isoprene. B.p. 34.3
                  1241, 1805, 2027, 2145, 2611, 2780, 3271, 3320, 4040, 4481, 4500, 5355,
                  5462, 5498, 5784, 6241, 6636, 8290, 8394, 8908-8916, 15935, 16035,
                  16314, 16315, 16334, 16335, 16353, 16354, 16383, 16435, 16436, 400a,
                  8258a, 8872a, 15909a, 16105a, 16169a, 16212a
C$_5$H$_8$ 3-Methyl-1,2-butadiene. B.p. 40.8
                  2028, 3272, 3321, 4041, 5356, 6635, 6790, 8291, 8908, 8917
C$_5$H$_8$ Piperylene. B.p. 42.5
                  1806, 2029, 2030, 3273, 5357, 8907, 8909, 2780a, 8917a
C$_5$H$_8$Cl$_4$ Tetrachloropentane.
                  4684, 8918
C$_5$H$_8$O   Allyl vinyl ether. B.p. 67.4
                  401, 5418
C$_5$H$_8$O   Cyclopentanone. B.p. 130.65
                  402, 2194, 2748, 3140, 4563, 4857, 4909, 4951, 5578, 7183, 7599, 7776,
                  8001, 8409, 8919-8931
C$_5$H$_8$O   1-Methoxy-1,3-butadiene. B.p. 90.71
                  403, 2031
C$_5$H$_8$O   3-Methyl-3-butene-2-one. B.p. 98.5
                  404
C$_5$H$_8$O   2-Methyl-3-butyn-2-ol. B.p. 103
                  405
C$_5$H$_8$O   4-Pentenal. B.p. 106
                  406
C$_5$H$_8$O   3-Penten-2-one. B.p. 123.5
                  407
C$_5$H$_8$O$_2$ Allyl acetate. B.p. 105
                  408, 5419
C$_5$H$_8$O$_2$ 3,3-Dimethoxypropyne. B.p. 111
                  16024
C$_5$H$_8$O$_2$ Ethyl acrylate. B.p. 100
                  409, 2032, 4042, 4694, 7006, 8932, 15987, 16026, 16213
C$_5$H$_8$O$_2$ Isopropenyl acetate. B.p. 96.5
                  410, 5358, 7014, 8935
C$_5$H$_8$O$_2$ Methyl methacrylate. B.p. 99.5
                  411, 2033, 7007, 8111, 8410, 8933, 8934
C$_5$H$_8$O$_2$ 2,3-Pentanedione. B.p. 109
                  412
C$_5$H$_8$O$_2$ 2,4-Pentanedione. B.p. 138
                  413, 5579, 7028, 8935-8946
C$_5$H$_8$O$_2$ Δ-Valerolatone.
                  414
C$_5$H$_8$O$_2$ Vinyl propionate. B.p. 95.0
                  415
C$_5$H$_8$O$_3$ Ethyl pyruvate. B.p. 155.5
                  5580, 7421, 7600, 8947-8974

Formula   Name and System No.

C$_5$H$_8$O$_3$ Levulinic acid. B.p. 252
                  3589, 5904, 8975-8997

C$_5$H$_8$O$_3$ Methyl acetoacetate. B.p. 169.5
                  2415, 2556, 7226, 7262, 7601, 8747, 8998-9041

C$_5$H$_8$O$_4$ Methylene diacetate. B.p. 164
                  3140a, 7015, 16363a

C$_5$H$_8$O$_4$ Methyl malonate. B.p. 181.5
                  2238, 2416, 5099, 6099, 9042-9105

C$_5$H$_8$O$_4$ Propylene glycol diformate.
                  9106

C$_5$H$_9$ClO$_2$ Propyl chloroacetate. B.p. 162.3
                  416, 1372, 2417, 2557, 3590, 4216, 5581, 5905, 6471, 7422, 8112, 8309,
                  9107-9128, 16085

C$_5$H$_9$N   Isovaleronitrile. B.p. 130.5
                  2195, 6552, 8002, 9129-9131

C$_5$H$_9$N   Valeronitrile. B.p. 141.3
                  2890, 6553, 8113, 8411, 9132-9144

C$_5$H$_9$NO   α-Hydroxyvaleronitrile.
                  8736

C$_5$H$_9$NO   N-Methyl-2-pyrrolidone. B.p. 251
                  7002, 416a, 8917a

C$_5$H$_{10}$ Amylenes.
                  2146, 2782, 2783, 5465, 9145, 9146, 15936, 6837b

C$_5$H$_{10}$ Cyclopentane. B.p. 49.3
                  1242, 1568, 1655, 1741, 1810, 1900, 2034, 2781, 2953, 3141, 3274, 3322,
                  3512, 3853, 3923, 4043, 4482, 4501, 4805, 5359, 5463, 5499, 5533, 5785,
                  5792, 6242, 6298, 6354, 6637, 6791, 6808, 6841, 7670, 7708, 7964, 7977,
                  8073, 8259, 8292, 8310, 8719, 8910, 9147-9153, 16390

C$_5$H$_{10}$ 2-Methyl-1-butene. B.p. 32
                  1001, 1807, 2035, 3076, 3275, 4044, 4503, 5360, 5464, 5793, 8911, 9154

C$_5$H$_{10}$ 3-Methyl-1-butene. B.p. 21.2
                  1002, 1244, 1657, 1743, 1809, 2037, 2612, 3077, 3276, 3370, 3854, 4045,
                  4484, 4504, 4547, 4555, 4791, 5362, 6244, 6356, 6638, 6883, 8294, 8913,
                  9158

C$_5$H$_{10}$ 2-Methyl-2-butene. B.p. 37.7
                  1003, 1243, 1481, 1569, 1656, 1742, 1808, 2036, 2784, 3078, 3277, 3323,
                  3513, 3855, 4046, 4483, 4502, 5361, 5500, 5534, 5794, 6243, 6355, 6639,
                  6792, 6842, 8114, 8260, 8293, 8395, 8709, 8912, 8917, 9147, 9155-9157,
                  16271, 16273, 16280, 16314, 16316, 16320, 16331, 16334, 16353, 16355,
                  16384, 16387, 16388, 16435, 16451, 16483

C$_5$H$_{10}$ 1-Pentene. B.p. 30.2
                  1004, 2038, 2785, 3079, 3278, 5363, 5535, 6640

C$_5$H$_{10}$ 2-Pentene. B.p. 35.8
                  1005, 2039, 2040, 2786, 3080, 3279, 5364, 5466, 6641

C$_5$H$_{10}$Cl$_2$O$_2$ Bis(2-chloroethoxy)methane. B.p. 218.1
                  417

C$_5$H$_{10}$N$_2$ 3-Dimethylaminopropionitrile. B.p. 174.5
                  418

C$_5$H$_{10}$O   Allyl ethyl ether. B.p. 63
                  4047, 16063, 16153

C$_5$H$_{10}$O   1-Butenyl methyl ether. B.p. 72-76
                  419, 420, 15988

Formula	Name and System No.

C₅H₁₀O Cyclopentanol. B.p. 140.85
421, 1245, 1811, 2196, 3339, 3401, 4813, 4910, 6554, 6927, 6954, 7184, 7263, 7307, 7777, 8003, 8083, 8412, 8936, 9159-9184

C₅H₁₀O Isopropenyl methyl ether. B.p. 61.9
422

C₅H₁₀O Isopropyl vinyl ether. B.p. 55.7
423, 4048, 6357

C₅H₁₀O Isovaleraldehyde. B.p. 92.5
424, 1052, 1142, 2312, 2990, 3142, 3924, 7365, 7575, 7685, 7936, 8311, 8645, 9185-9188, 16175

C₅H₁₀O 3-Methyl-2-butanone. B.p. 95.4
426, 1143, 1246, 1313, 1520, 1658, 1812, 2041, 2313, 2991, 4050, 5420, 5688, 5856, 6358, 6472, 6898, 7525, 7895, 7937, 7985, 8061, 8115, 8312, 8646, 8710, 8720 9189-9198

C₅H₁₀O 3-Methyl-2-buten-1-ol. B.p. 140
424a, 9198f

C₅H₁₀O 3-Methyl-3-buten-1-ol. B.p. 130
424b, 9198g, 16213c

C₅H₁₀O 2-Methyl-3-buten-2-ol. B.p. 97
424c, 1481a, 1143a, 2991a, 7574a, 8260a, 9198a-9198g, 15868a, 16169b, 16213a

C₅H₁₀O 2-Pentanone. B.p. 102.35
427, 1247, 1314, 1521, 1659, 1813, 2826, 4049, 5006, 5160, 5421, 5857, 6359 6473, 6985, 7855, 7896, 8116, 8313, 8830, 9199-9213, 16214

C₅H₁₀O 3-Pentanone. B.p. 102
428, 1053, 1248, 1315, 1522, 1660, 1814, 2042, 2314, 2350, 2827, 2845, 3143, 3873, 4858, 5007, 5422, 5839, 5858, 6360, 6474, 6986, 7833, 7856, 7883, 7897, 7986, 8117, 8218, 8261, 8314, 8647, 8831, 9185, 9199, 9214-9240, 15880, 16086, 16176, 16311, 16317, 16424-16426, 16447

C₅H₁₀O Propyl vinyl ether. B.p. 65.1
429, 4051, 15989

C₅H₁₀O Tetrahydropyran. B.p. 88
430

C₅H₁₀O Valeraldehyde. B.p. 103.3
431, 432

C₅H₁₀OS 2-Methylthioethyl vinyl ether.
6537

C₅H₁₀O₂ Butyl formate. B.p. 106.6
433, 1316, 1815, 2043, 2351, 3144, 4814, 5008, 6361, 6475, 7857, 7898, 8033, 8118, 8219, 8315, 8832, 9200, 9214, 9241-9257, 16137, 1660a

C₅H₁₀O₂ 4,5-Dimethyl-1,3-dioxolane.
434

C₅H₁₀O₂ 3-Ethoxy-1,2-epoxypropane. B.p. 124
435

C₅H₁₀O₂ Ethyl propionate. B.p. 99.15
32, 436, 1054, 1249, 1317, 1816, 2044, 2315, 2352, 2491, 2787, 3145, 4052, 5009, 5034, 5161, 5423, 6476, 6987, 7576, 7858, 7899, 8119, 8220, 8316, 9189, 9201, 9215, 9258-9274

C₅H₁₀O₂ 3-Hydroxy-3-methyl-2-butanone. B.p. 141
437

C₅H₁₀O₂ Isobutyl formate. B.p. 98.3
438, 1144, 1250, 1318, 1661, 1817, 2045, 2316, 2353, 3146, 3402, 4053, 5010, 5424, 5689, 5881, 6362, 6477, 7526, 7834, 7859, 7900, 8120, 8221, 8262, 8317, 9202, 9216, 9258, 9275-9289, 16177

Formula	Name and System No.

$C_5H_{10}O_2$ Isopropyl acetate. B.p. 90.8
439, 1145, 1251, 1319, 1523, 1818, 2046, 2317, 2354, 2788, 2992, 3147, 4054, 5011, 5153, 5365, 5690, 5743, 5859, 5882, 6363, 7016, 7366, 7527, 7860, 7901, 7938, 8062, 8121, 8318, 9190, 9290-9303, 15990, 16036, 16064, 16433, 16445

$C_5H_{10}O_2$ Isovaleric acid. B.p. 176.5
440, 1252, 1373, 1585, 2197, 2239, 2277, 2418, 2558, 2651, 2891, 3591, 4509, 4965, 5061, 5100, 7129, 8004, 8748, 9042, 9107, 9304-9385, 16495-16499, 15873a

$C_5H_{10}O_2$ 3-Methoxybutyraldehyde. B.p. 131
441

$C_5H_{10}O_2$ Methyl butyrate. B.p. 102.65
442, 1055, 1320, 1819, 2047, 2355, 2846, 3148, 3403, 4055, 4815, 4859, 5012, 5035, 5425, 5860, 6364, 6478, 6988, 7029, 7528, 7861, 7902, 8034, 8122, 8222, 8319, 8648, 9203, 9217, 9241, 9259, 9386-9399

$C_5H_{10}O_2$ Methyl isobutyrate. B.p. 92.3
443, 1146, 1321, 1524, 1820, 2048, 2318, 2356, 3149, 3541, 4056, 5013, 5426, 5691, 5861, 5883, 6365, 6479, 7862, 7884, 7903, 8123, 8223, 8263, 8320, 8649, 9186, 9191, 9204, 9218, 9275, 9290, 9400-9412

$C_5H_{10}O_2$ Propyl acetate. B.p. 101.6
33, 444, 1056, 1253, 1322, 1525, 1821, 2049, 2357, 2492, 2742, 2754, 2789, 2847, 3150, 3404, 3874, 4057, 4816, 4860, 5014, 5036, 5366, 5427, 5862, 6366, 6480, 7030, 7529, 7835, 7863, 7904, 8035, 8124, 8224, 8264, 8321, 8413, 8650, 9205, 9219, 9242, 9260, 9276, 9386, 9413-9428, 15937, 16087, 16317, 16425, 16447

$C_5H_{10}O_2$ Tetrahydrofurfuryl alcohol.
445, 9429, 7406a

$C_5H_{10}O_2$ Valeric acid. B.p. 186.35
446, 447, 1374, 2240, 2419, 2559, 2652, 3151, 4510, 5062, 5101, 6917, 8998, 9043, 9430-9485, 1661a, 15873b

$C_5H_{10}O_2$ 1-Vinyloxy-2-propanol
448

$C_5H_{10}O_2$ 3-Vinyloxypropanol.
449, 6651, 6679

$C_5H_{10}O_3$ Butylene glycol monoformate.
9509.

$C_5H_{10}O_3$ β-Ethoxypropionic acid. B.p. 219.2
450

$C_5H_{10}O_3$ α-Methoxybutyric acid.
451

$C_5H_{10}O_3$ Ethyl carbonate. B.p. 126.5
452, 2198, 2560, 2892, 3340, 3405, 3592, 4217, 4861, 4911, 4952, 5582, 6928, 8005, 8036, 8084, 8125, 8322, 8833, 8919, 9159, 9486-9508, 16346

$C_5H_{10}O_3$ Ethyl lactate. B.p. 153.9
2420, 2561, 4657, 4966, 6975, 7057, 7130, 7185, 7308, 8414, 8749, 9160, 9510-9562

$C_5H_{10}O_3$ 2-Methoxyethyl acetate. B.p. 144.6
453, 2199, 2421, 2562, 2893, 3152, 3341, 3406, 4218, 4698, 4747, 4862, 4912, 5163, 5583, 6100, 6555, 7058, 7131, 7186, 7309, 7423, 7602, 7778, 8006, 8037, 8085, 8126, 8323, 8415, 8750, 9161, 9486, 9510, 9563-9617, 2049a

$C_5H_{10}O_3$ Methoxymethyl propionate.
454

Formula   Name and System No.

$C_5H_{10}O_3$ Methyl β-methoxypropionate. B.p. 84/100 mm
     455
$C_5H_{11}Br$ 1-Bromo-3-methylbutane. B.p. 120.65
     1057, 1537, 1662, 1822, 2050, 2200, 2358, 2790, 3153, 3342, 3407, 3593, 3875, 4058, 4219, 4863, 4913, 5584, 5802, 5823, 5906, 6101, 6367, 6481, 6556, 6929, 7242, 7424, 7530, 7603, 8038, 8086, 8127, 8265, 8324, 8416, 8751, 8834, 8920, 9162, 9243, 9304, 9487, 9563, 9618-9642, 5427a
$C_5H_{11}Br$ 1-Bromopentane. B.p. 130.0
     9643
$C_5H_{11}Cl$ 1-Chloro-3-methylbutane. B.p. 99.4
     1058, 1538, 1663, 1823, 2052, 2359, 3026, 3154, 3408, 3925, 4059, 5015, 5428, 5585, 5692, 6368, 6482, 6557, 7246, 7531, 7709, 7836, 8128, 8225, 8266, 8325, 9192, 9220, 9261, 9277, 9291, 9387, 9400, 9413, 9644-9656
$C_5H_{11}Cl$ 1-Chloropentane. B.p. 108.35
     456, 2051, 4060
$C_5H_{11}I$ 1-Iodo-3-methylbutane. B.p. 147.65
     1375, 1664, 2563, 2653, 3027, 3155, 3409, 3550, 3594, 4220, 4748, 4914, 5186, 5586, 5907, 6102, 6483, 6558, 7031, 7059, 7132, 7187, 7425, 7604, 7779, 8129, 8226, 8326, 8417, 8752, 9305, 9430, 9488, 9511, 9564, 9657-9689
$C_5H_{11}I$ 2-Iodo-2-methylbutane. B.p. 127.5
     9489
$C_5H_{11}N$ Piperidine. B.p. 105.8
     457, 2053, 7532, 8835, 9221, 9690-9693, 16492
$C_5H_{11}NO$ N,N-Dimethylpropionamide. B.p. 175.5
     5587
$C_5H_{11}NO$ 4-Methylmorpholine. B.p. 115.6
     458
$C_5H_{11}NO$ Tetrahydrofurfurylamine. B.p. 153
     459
$C_5H_{11}NO_2$ Ethyl N-ethylaminoformate.
     9694, 9695
$C_5H_{11}NO_2$ Isoamyl nitrite. B.p. 97.15
     460, 1147, 1254, 1323, 1526, 1824, 2319, 2848, 2993, 5016, 5863, 5884, 6899, 7533, 7864, 7885, 7905, 7939, 9193, 9206, 9222, 9278, 9292, 9401, 9414, 9644, 9696-9709
$C_5H_{11}NO_3$ Isoamyl nitrate. B.p. 149.75
     461, 1376, 2422, 2564, 4967, 5063, 5102, 5588, 6103, 6559, 6976, 7133, 7188, 7426, 7605, 7780, 8418, 8873, 9512, 9565, 9657, 9710-9735
C_5H_{12} 2-Methylbutane. B.p. 27.6
     979, 1006, 1021, 1255, 1570, 1665, 1744, 1825, 2054, 2134, 2613, 2791, 3061, 3081, 3280, 3324, 3371, 3856, 4061, 4485, 4505, 4556, 4792, 5367, 5467, 5501, 5795, 6245, 6299, 6369, 6642, 6843, 6884, 7427, 8227, 8267, 8295, 8914, 9155, 9158, 9736-9738, 15938, 16315, 16316, 16321, 16340, 16354, 16355, 16385, 16436
C_5H_{12} Pentane. B.p. 36.15
     462, 1007, 1256, 1482, 1571, 1666, 1745, 1826, 1901, 2055, 2614, 2792, 3062, 3082, 3156, 3281, 3294, 3294, 3325, 3372, 3857, 4062, 4486, 4506, 4697, 4793, 4806, 5368, 5469, 5502, 5536, 5786, 5796, 6246, 6300, 6370, 6484, 6643, 6793, 6809, 6828, 7978, 8074, 8228, 8268, 8296, 8327, 8396, 8711, 8721, 8915, 9148, 9156, 9736, 9739-9745, 15881, 16274, 16309, 16386, 16389, 16451, 16483, 4666b, 6837c, 16169c
$C_5H_{12}N_2$ 1-Methylpiperazine. B.p. 138
     463

Formula   Name and System No.

$C_5H_{12}N_2O$ Tetramethylurea. B.p. 176.5
                9746
$C_5H_{12}O$ Amyl alcohol. B.p. 137.8
                464, 1257, 2056, 2201, 2894, 3343, 3410, 3595, 3876, 4063, 4817, 4864,
                4915, 6256, 6280, 6560, 6930, 7411, 7781, 8007, 8087, 8419, 8836, 9132,
                9490, 9566, 9618, 9747-9774, 16215-16219, 1147a, 9746a
$C_5H_{12}O$ Active amyl alcohol. B.p. 128.5
                8744, 8921, 9775-9788
$C_5H_{12}O$ tert-Amyl alcohol. B.p. 102.25
                465, 1059, 1148, 1258, 1324, 1827, 2202, 2320, 2793, 2849, 2895, 2994,
                3542, 3877, 3926, 4818, 4865, 4916, 5017, 5037, 5369, 5744, 5865, 5885,
                6316, 6561, 6931, 7247, 7310, 7412, 7534, 7865, 7906, 7940, 7965, 8229,
                8737, 8837, 9207, 9223, 9244, 9262, 9279, 9388, 9402, 9415, 9415, 9567,
                9619, 9645, 9739, 9789-9810, 16220-16222
$C_5H_{12}O$ Butyl methyl ether. B.p. 71
                2057, 4064, 15991, 16065, 16154
$C_5H_{12}O$ tert-Butyl methyl ether. B.p. 55
                466, 2059, 6643a, 8268a, 8915a, 9198a, 15909b, 16169d, 16212a, 16213a
$C_5H_{12}O$ Ethyl isopropyl ether. B.p. 54
                16066, 16155
$C_5H_{12}O$ Ethyl propyl ether. B.p. 63.6
                467, 1259, 1483, 1667, 1902, 2058, 2954, 3514, 4065, 5370, 5503, 5745,
                5765, 6301, 6371, 6485, 6794, 7953, 7966, 8077, 8712, 9149, 9811-9815,
                16067, 16156
$C_5H_{12}O$ Isoamyl alcohol. B.p. 132.05
                468, 1060, 1149, 1260, 1377, 1527, 1586, 1828, 2203, 2321, 2423, 2565,
                2850, 2896, 2995, 3344, 3411, 3543, 3596, 3878, 4066, 4221, 4749, 4819,
                4866, 4917, 4968, 5018, 5064, 5746, 5803, 5908, 6104, 6317, 6486, 6562,
                6933, 6999, 7134, 7175, 7189, 7264, 7311, 7782, 7866, 7907, 8008, 8039,
                8088, 8328, 8420, 8651, 8745, 8838, 8906, 8922, 8937, 9109, 9224, 9245,
                9491, 9513, 9568, 9620, 9646, 9658, 9775, 9816-9901, 15992, 16223-
                16229, 16344, 16346, 16358, 16359, 16493, 3156a, 9746a, 16374a
$C_5H_{12}O$ Isobutyl methyl ether. B.p. 59
                16068, 16157
$C_5H_{12}O$ 2-Methyl-1-butanol. B.p. 128.9
                8923, 9902-9905
$C_5H_{12}O$ 3-Methyl-2-butanol. B.p. 112.9
                469, 1061, 1261, 1325, 1829, 2322, 2996, 4867, 5886, 7867, 7908, 9906-
                9913
$C_5H_{12}O$ 2-Pentanol. B.p. 119.3
                470, 1062, 1150, 1262, 1830, 2204, 2323, 2897, 3412, 3597, 3927, 4820,
                4868, 4918, 5693, 6318, 6563, 6932, 7190, 7535, 7783, 7909, 8009, 8089,
                8130, 8421, 8924, 9133, 9225, 9246, 9569, 9621, 9914-9940
$C_5H_{12}O$ 3-Pentanol. B.p. 115.4
                471, 1063, 1151, 1263, 1831, 2324, 2794, 2997, 3413, 4869, 4919, 5864,
                6564, 7413, 7536, 7868, 7910, 7941, 8131, 8839, 9226, 9247, 9263, 9389,
                9416, 9941-9950
$C_5H_{12}O_2$ Diethoxymethane. B.p. 87.5
                472, 1264, 1326, 1528, 2060, 2325, 2998, 3157, 3928, 4067, 5371, 5429,
                5694, 5887, 6372, 6487, 6900, 7367, 7577, 7686, 7710, 7837, 7942, 8063,
                8329, 8652, 9194, 9280, 9293, 9403, 9647, 9696, 9789, 9951-9957, 15993,
                16088
$C_5H_{12}O_2$ 2,2-Dimethoxypropane. B.p. 80
                2061, 4068

Formula　　　　　　　　　　　Name and System No.

$C_6H_4Br_2$　　　*p*-Dibromobenzene. B.p. 220.25
　　　　　　　　2615, 2656, 3601, 4226, 5912, 6107, 6681, 8514, 10168-10244
$C_6H_4ClNO_2$　　*m*-Chloronitrobenzene. B.p. 235.5
　　　　　　　　3602, 4227, 5913, 6682, 8515, 10168, 10245-10269
$C_6H_4ClNO_2$　　*o*-Chloronitrobenzene. B.p. 246.0
　　　　　　　　3603, 4228, 5914, 6683, 8516, 10270-10288
$C_6H_4ClNO_2$　　*p*-Chloronitrobenzene. B.p. 239.1
　　　　　　　　3604, 4229, 5915, 6684, 8517, 8975, 10169, 10289-10318
$C_6H_4Cl_2$　　*c*-Dichlorobenzene. B.p. 179.5
　　　　　　　　2616, 2657, 3416, 3605, 4230, 4566, 5134, 5190, 5260, 5589, 5916, 6108,
　　　　　　　　7060, 7227, 7260, 7429, 7726, 8755, 8999, 9044, 9308, 9431, 10319-10366
$C_6H_4Cl_2$　　*p*-Dichlorobenzene. B.p. 174.4
　　　　　　　　1588, 1668, 2376, 2617, 2658, 3028, 3158, 3417, 3606, 4231, 4511, 4567,
　　　　　　　　4757, 5191, 5261, 5590, 5917, 6109, 7061, 7136, 7228, 7267, 7430, 7606,
　　　　　　　　7727, 8756, 8874, 9000, 9045, 9309, 9432, 9515, 9694, 9958, 10367-10432
C_6H_5Br　　Bromobenzene. B.p. 156.1
　　　　　　　　1379, 1589, 1669, 2377, 2425, 2567, 2618, 2659, 2899, 3029, 3159, 3418,
　　　　　　　　3551, 3607, 4232, 4512, 4568, 4699, 4969, 5192, 5262, 5591, 5805, 5824,
　　　　　　　　5918, 6110, 6488, 6935, 7032, 7062, 7137, 7137, 7192, 7229, 7268, 7313,
　　　　　　　　7431, 7607, 7785, 8132, 8330, 8422, 8757, 8938, 8947, 9001, 9046, 9310,
　　　　　　　　9433, 9516, 9571, 9712, 9818, 9959, 10319, 10433-10481, 16453, 16454,
　　　　　　　　16458, 16473, 16500-16503
C_6H_5BrO　　*o*-Bromophenol. B.p. 194.8
　　　　　　　　3608, 5919, 7269, 10367, 10482-10507
C_6H_5Cl　　Chlorobenzene. B.p. 131.8
　　　　　　　　484, 1152, 1380, 1484, 1670, 1833, 2063, 2660, 2755, 2900, 3030, 3160,
　　　　　　　　3345, 3419, 3609, 4070, 4233, 4569, 4700, 4870, 4920, 5372, 5430, 5592,
　　　　　　　　5806, 5825, 5920, 6111, 6257, 6373, 6489, 6566, 6853, 6936, 7033, 7193,
　　　　　　　　7314, 7432, 7578, 7608, 7786, 8091, 8133, 8230, 8331, 8423, 8758, 8840,
　　　　　　　　8875, 8925, 8939, 8948, 9164, 9311, 9492, 9517, 9572, 9747, 9776, 9790,
　　　　　　　　9819, 9914, 9960, 10433, 10508-10539, 15828, 15831, 15994, 16356-
　　　　　　　　16358, 16437, 2794a, 2817b
C_6H_5ClO　　*o*-Chlorophenol. B.p. 176.8
　　　　　　　　1381, 2426, 2568, 3610, 4234, 7270, 7433, 9312, 9660, 10320, 10368,
　　　　　　　　10434, 10540-10574, 16504, 484a, 15831a
C_6H_5ClO　　*p*-Chlorophenol. B.p. 219.75
　　　　　　　　3611, 4235, 5921, 6652, 10170, 10575-10643
$C_6H_5Cl_3Si$　　Phenyltrichlorosilane. B.p. 201.0
　　　　　　　　10644, 10645
C_6H_5F　　Fluorobenzene. B.p. 84.9
　　　　　　　　1671, 1903, 2064, 3161, 3420, 3929, 4071, 5373, 6374, 6490, 6829, 7368,
　　　　　　　　7579, 7711, 8064, 8134, 8269, 8332, 10646-10650
C_6H_5FO　　*o*-Fluorophenol.
　　　　　　　　9777, 9820
C_6H_5I　　Iodobenzene. B.p. 188.55
　　　　　　　　2378, 2619, 2661, 3612, 4236, 4513, 4570, 4758, 5135, 5193, 5263, 5922,
　　　　　　　　6112, 7271, 7434, 7609, 7728, 8424, 8759, 8940, 9002, 9047, 9313, 9434,
　　　　　　　　9961, 10540, 10646, 10651-10701
$C_6H_5NO_2$　　Nitrobenzene. B.p. 210.85
　　　　　　　　485, 1153, 1382, 1485, 1746, 1834, 2065, 2379, 2427, 2523, 2569, 2620,
　　　　　　　　2662, 2901, 3162, 3613, 3880, 4072, 4237, 4571, 4767, 5843, 5923, 6113,
　　　　　　　　6685, 6913, 7729, 7869, 7943, 7954, 8135, 8231, 8297, 8518, 8976, 9145,
　　　　　　　　9738, 9740, 10023, 10110, 10171, 10321, 10435, 10508, 10575, 10651,
　　　　　　　　10702-10787, 16505

Formula   Name and System No.

$C_6H_5NO_3$ o-Nitrophenol. B.p. 217.25
                 3614, 4238, 5924, 6114, 6653, 6873, 7333, 8519, 10172, 10482, 10576,
                 10702, 10788-10849

C_6H_6 Benzene. B.p. 80.1
                 119, 157, 486, 1064, 1154, 1265, 1327, 1486, 1529, 1672, 1835, 1904,
                 2066, 2326, 2360, 2511, 2795, 2902, 2999, 3063, 3083, 3163, 3326, 3363,
                 3421, 3515, 3544, 3881, 3930, 4073, 4184, 4239, 4572, 4640, 4667, 4692,
                 4701, 4739, 4871, 5178, 5374, 5431, 5478, 5504, 5537, 5593, 5695, 5713,
                 5747, 5766, 5807, 5840, 5888, 5893, 6258, 6281, 6319, 6375, 6491, 6567,
                 6644, 6654, 6795, 6830, 6901, 6989, 6996, 7000, 7248, 7369, 7394, 7403,
                 7537, 7580, 7671, 7687, 7712, 7911, 7922, 7955, 7967, 7987, 8054, 8055,
                 8065, 8076, 8136, 8232, 8270, 8298, 8333, 8398, 8425, 8520, 8615, 8631,
                 8653, 8674, 8722, 8760, 8841, 9150, 9165, 9187, 9195, 9208, 9227, 9247,
                 9264, 9281, 9294, 9390, 9404, 9417, 9648, 9697, 9741, 9748, 9791, 9821,
                 9906, 9915, 9941, 9951, 10102, 10106, 10509, 10647, 10793, 10850-10882,
                 15874, 15910, 15939, 15957, 15995, 16022, 16025, 16041, 16069, 16089,
                 16101, 16112, 16138, 16158, 16170, 16178, 16195, 16214, 16220, 16224,
                 16230, 16248, 16254-16256, 16260-16262, 16296, 16305, 16318, 16336,
                 16337, 16348, 16360, 16395, 16400-16405, 16437, 16438, 16441, 16446,
                 16448, 16449, 16464, 16466, 16472, 16480, 16481, 16484, 16483, 16506-
                 16510, 16558-16562, 4652a, 6967a, 7016a, 10882a, 16229a, 16254a,
                 16337a, 16374b, 16433a, 16438a, 16446a, 16482b, 16484a, 16563a,
                 16565a,, 16565b, 16565c, 16566a

$C_6H_6Cl_2Si$ Phenyldichlorosilane. B.p. 182.9
                 10644

C_6H_6O Phenol. B.p. 182
                 487, 1155, 1383, 2241, 2428, 2524, 2570, 2663, 2903, 3422, 3552, 3615,
                 4240, 4514, 4573, 4715, 4768, 4970, 5065, 5104, 5164, 5194, 5264, 5375,
                 5826, 5925, 6115, 6568, 6874, 7063, 7272, 7370, 7435, 7610, 7730, 7787,
                 8010, 8137, 8426, 8761, 8842, 8876, 9003, 9048, 9166, 9314, 9435, 9518,
                 9573, 9661, 9746, 9749, 9822, 9962, 10024, 10111, 10144, 10173, 10322,
                 10369, 10436, 10510, 10541, 10652, 10883-11023, 15832, 16037, 16231,
                 16410-16414, 16500, 16511-16514, 16514a

$C_6H_6O_2$ Pyrocatechol. B.p. 245.9
                 3616, 4241, 5926, 6686, 8521, 10112, 10174, 10245, 10270, 10289, 10788,
                 11024-11091

$C_6H_6O_2$ Resorcinol. B.p. 281.4
                 3617, 4074, 5927, 6687, 10175, 10271, 10290, 11024, 11092-11129

$C_6H_6O_3$ Pyrogallol. B.p. 309
                 11092, 11130-11135

C_6H_6S Benzenethiol. B.p. 169.5
                 3423, 9823, 10370, 11136-11141

C_6H_7N Aniline. B.p. 184.35
                 104, 488, 1384, 1673, 2242, 2429, 2571, 2904, 3164, 3618, 4075, 4242,
                 4574, 5105, 5265, 5928, 6655, 8138, 8877, 9049, 9146, 9167, 9824, 9963,
                 10025, 10113, 10145, 10323, 10371, 10437, 10511, 10542, 10653, 10704,
                 10850, 10883, 11142-11241, 16406, 16407, 16415, 16506, 16511, 16515-
                 16525

C_6H_7N 2-Picoline. B.p. 129
                 489, 1674, 2796, 3165, 4243, 5594, 6569, 8139, 8843, 9168, 9778, 9825,
                 9916, 10884, 11242, 11250, 16196, 16232, 16375-16378, 16411

C_6H_7N 3-Picoline. B.p. 144
                 490, 1675, 3166, 4244, 5595, 6937, 9826, 10543, 10885, 11251-11254,
                 16412

Formula   Name and System No.

C_6H_7N 4-Picoline. B.p. 145.3
             491, 1676, 3167, 5596, 6938, 10544, 10886, 11255, 11256

C_6H_8 1,3-Cyclohexadiene. B.p. 80.8
             493, 1156, 1266, 1487, 1677, 2067, 3168, 3931, 4076, 5376, 5432, 5538,
             6376, 6492, 6831, 7371, 7581, 7968, 8140, 8271, 8334, 8675, 9792, 9827,
             10851, 11257, 11258, 15911, 15996, 16042, 16070, 16090, 16171, 16179,
             16197, 16338

C_6H_8 1,4-Cyclohexadiene. B.p. 85.6
             492, 1157, 2068, 3169, 4077, 6377, 8676, 10852, 15997

C_6H_8ClN Aniline hydrochloride.
             11259

$C_6H_8N_2$ 2-Amino-3-methylpyridine. B.p. 221
             11260, 11261

$C_6H_8N_2$ 2,5-Dimethylpyrazine. B.p. 154
             494

$C_6H_8N_2$ 2-Ethylpyrazine.
             495

$C_6H_8N_2$ o-Phenylenediamine. B.p. 258.6
             3619, 4245, 5929, 11262-11281

$C_6H_8N_2$ Phenylhydrazine. B.p. 243
             496

C_6H_8O 2,5-Dimethylfuran. B.p. 93.3
             497, 2069, 7372, 9295

C_6H_8O 2,4-Hexadienal. B.p. 171
             498

$C_6H_8O_2$ 1,3-Butadienyl acetate. B.p. 138.5
             499

$C_6H_8O_2$ Vinyl crotonate. B.p. 132.7
             500

$C_6H_8O_4$ Methyl fumarate. B.p. 193.25
             501, 2664, 5930, 6116, 7731, 8522, 9315, 10026, 10324, 10577, 10654,
             10887, 11282-11314

$C_6H_8O_4$ Methyl maleate. B.p. 204.05
             2665, 3620, 4246, 5931, 6117, 6688, 8523, 10578, 10705, 10789, 10888,
             11282, 11315-11348

C_6H_9N 1-Ethylpyrrol. B.p. 130.4
             4247, 8141, 9828

$C_6H_9N_3$ 3,3'-Iminodipropionitrile.
             502

C_6H_{10} Biallyl. B.p. 60.2
             1488, 1678, 1837, 2070, 2955, 3282, 3327, 3516, 4078, 5377, 5505, 5539,
             6247, 6378, 6645, 6796, 7979, 8299, 8713, 9151, 9811, 11349, 15912,
             15998

C_6H_{10} Cyclohexene. B.p. 82.75
             503, 1065, 1158, 1267, 1328, 1489, 1679, 1836, 2071, 2327, 2905, 3000,
             3170, 3424, 4079, 4248, 5433, 5540, 6379, 6493, 6902, 7373, 7436, 7538,
             7582, 7688, 7713, 7969, 8142, 8233, 8272, 8335, 8427, 8677, 9793, 9829,
             9907, 9917, 10853, 11257, 11350-11355, 15913, 15999, 16043, 16071,
             16091, 16139, 16172, 16180, 16198, 16336

C_6H_{10} 2,3-Dimethyl-1,3-butadiene. B.p. 68.9
             2072

C_6H_{10} 2-Ethyl-1,3-butadiene. B.p. 66.9
             504

Formula Name and System No.

C_6H_{10}	1,3-Hexadiene. B.p. 72.9
	2073, 4080, 6380
C_6H_{10}	2,4-Hexadiene. B.p. 82
	2074, 4081, 6381
C_6H_{10}	1-Hexyne. B.p. 70.2
	4082, 16000, 11355a, 11355b
C_6H_{10}	3-Hexyne. B.p. 80.2
	4083, 16001
C_6H_{10}	Methylcyclopentene. B.p. 75.85
	1268, 2075, 4084, 4085, 6494, 8273, 11356
C_6H_{10}	3-Methyl-1,3-pentadiene. B.p. 77
	2076, 4086, 6382
C_6H_{10}	4-Methyl-1,3-pentadiene
	505
$C_6H_{10}O$	Allyl ether. B.p. 94.84
	5434, 16044
$C_6H_{10}O$	Cyclohexanone. B.p. 155.6
	506, 1385, 2430, 2572, 3346, 3621, 4575, 4678, 4971, 5106, 5195, 5266,
	5597, 6118, 6282, 7064, 7194, 7315, 7437, 7539, 7611, 7788, 8762, 8949,
	9004, 9316, 9519, 9662, 10438, 10512, 10889, 11144, 11357-11379, 16233,
	16512, 16526, 16233a
$C_6H_{10}O$	2-Ethylcrotonaldehyde. B.p. 135.3
	507, 16107
$C_6H_{10}O$	2-Hexenal. B.p. 149
	508
$C_6H_{10}O$	1-Hexene-5-one. B.p. 129
	509, 10101
$C_6H_{10}O$	Mesityl oxide. B.p. 129.5
	510, 1386, 2206, 2573, 2906, 3171, 3347, 3425, 4750, 4872, 4921, 5066,
	5598, 5696, 6383, 6570, 6939, 7034, 7195, 7395, 7789, 8011, 8143, 8428,
	8844, 9169, 9493, 9623, 9663, 9779, 9830, 9918, 10513, 11380-11402
$C_6H_{10}O$	Methylcyclopentanone. B.p. 138
	9750
$C_6H_{10}O$	Methyldihydropyran. B.p. 118.5
	510a, 2076a, 8273a, 9198b, 11402a, 11402b, 16172a
$C_6H_{10}O$	2-Methyl-2-pentenal. B.p. 138.2
	511
$C_6H_{10}O_2$	Crotonyl acetate. B.p. 129
	512
$C_6H_{10}O_2$	Ethyl crotonate. B.p. 137.8
	513
$C_6H_{10}O_2$	2,5-Hexanedione. B.p. 191.3
	4249, 11403-11407
$C_6H_{10}O_2$	Isopropyl acrylate.
	2077
$C_6H_{10}O_2$	Propyl acrylate.
	2078
$C_6H_{10}O_2$	Vinyl butyrate. B.p. 116.7
	514
$C_6H_{10}O_2$	4-Vinyl-1,3-dioxane. B.p. 144.9
	513a
$C_6H_{10}O_2$	Vinyl isobutyrate. B.p. 105.4
	515

Formula   Name and System No.

$C_6H_{10}O_3$ Ethyl acetoacetate. B.p. 180.4
                     2243, 2431, 4972, 7230, 7273, 7438, 7582a, 9317, 9436, 10114, 10146,
                     10325, 10372, 10439, 10655, 10890, 11408-11468, 16527-16529

$C_6H_{10}O_4$ Butylene glycol diformate.
                     11469

$C_6H_{10}O_4$ Ethylidene diacetate. B.p. 168.5
                     1387, 1590, 2244, 3426, 3622, 4250, 5932, 6119, 7138, 7439, 9050, 9318
                     9574, 9964, 10027, 10326, 10440, 10891, 11470-11501

$C_6H_{10}O_4$ Ethyl oxalate. B.p. 185.65
                     2245, 2432, 2667, 3623, 4251, 4515, 5107, 5196, 5267, 5933, 6120, 6689,
                     7274, 7732, 9005, 9051, 9319, 9437, 10327, 10373, 10441, 10579, 10656,
                     10893, 11145, 11283, 11408, 11502-11564, 16515-16518, 16530

$C_6H_{10}O_4$ Glycol diacetate. B.p. 186.3
                     516, 3624, 4252, 5934, 6690, 9052, 9320, 9438, 10028, 10657, 10892,
                     11284, 11565-11589, 16513

$C_6H_{10}O_4$ Methyl succinate. B.p. 195.5
                     2246, 2666, 4253, 6121, 10147, 10580, 10658, 10894, 11315, 11502,
                     11590-11629

$C_6H_{10}S$ Allyl sulfide. B.p. 139.35
                     1388, 1680, 1838, 2207, 2574, 3172, 3348, 3427, 3625, 4576, 4751, 4821,
                     4922, 5067, 5599, 5892, 6122, 6571, 6940, 7035, 7196, 7414, 7440, 7870,
                     8144, 8336, 8845, 9170, 9321, 9494, 9624, 9649, 9751, 9831, 9965, 10442,
                     10514, 11245, 11251, 11380, 11630-11642

$C_6H_{11}BrO_2$ Ethyl α-bromoisobutyrate. B.p. 178
                     2433, 7441, 9053, 11643-11646

$C_6H_{11}ClO_2$ Butyl chloroacetate. B.p. 181.9
                     517, 4254, 5935, 8145, 9832, 10659, 11647, 11654, 16140

$C_6H_{11}ClO_2$ Isobutyl chloroacetate. B.p. 174.4
                     518, 2434, 8337, 10443, 10660, 11655-11644, 16181

$C_6H_{11}N$ Capronitrile. B.p. 163.9
                     8429, 9171, 9833, 9966, 11665-11671

$C_6H_{11}N$ Diallylamine. B.p. 110.4
                     519

$C_6H_{11}NO$ Caprolactam.
                     520, 11146

$C_6H_{11}NO_2$ Nitrocyclohexane. B.p. 205.3
                     3626, 4577, 5936, 7733, 10029, 11147, 11672-11682

$C_6H_{11}NO_3$ 2-Methyl-2-nitropropyl vinyl ether.
                     521, 8100

C_6H_{12} Cyclohexane. B.p. 80.75
                     129, 158, 522, 1066, 1159, 1269, 1329, 1490, 1681, 1839, 1905, 2079,
                     2328, 2361, 2756, 2797, 2907, 3001, 3084, 3173, 3428, 3545, 3932, 4087,
                     4255, 4578, 4873, 5162, 5179, 5378, 5435, 5470, 5541, 5600, 5697, 5714,
                     5748, 5767, 5808, 5827, 5841, 5889, 6259, 6283, 6384, 6495, 6572, 6782,
                     6797, 6832, 6903, 6990, 7009, 7017, 7243, 7249, 7374, 7442, 7540, 7583,
                     7689, 7714, 7886, 7912, 7923, 7944, 7970, 7988, 8066, 8077, 8146, 8234,
                     8274, 8338, 8430, 8616, 8632, 8654, 8666, 8678, 8707, 8723, 8763, 8846,
                     9172, 9196, 1209, 9228, 9249, 9265, 9282, 9296, 9405, 9418, 9698, 9742,
                     9752, 9794, 9834, 9919, 9942, 9952, 10107, 10515, 10648, 10706, 10854,
                     11148, 11258, 11350, 11683-11699, 15914, 15958, 16002, 16045, 16050,
                     16072, 16092, 16102, 16113, 16159, 16173, 16182, 16199, 16221, 16249,
                     16258, 16259, 16261, 16318, 16327, 16329, 16333, 16337, 16338, 16348,
                     16396, 16398, 16401, 16438, 16441, 16444, 16446, 16448, 16449, 16466,
                     16472, 16480, 16484, 16488, 16506-16508, 16558, 16233b, 16255a, 16374b,
                     16443a, 16482b, 16484a, 16508a, 16565b, 16565c

Formula   Name and System No.

C$_6$H$_{12}$ 2-Ethyl-1-butene. B.p. 64.95
                       4088, 5545
C$_6$H$_{12}$ 2,3-Dimethyl-1-butene. B.p. 55.62
                       5542
C$_6$H$_{12}$ 2,3-Dimethyl-2-butene. B.p. 73.38
                       5543
C$_6$H$_{12}$ 3,3-Dimethyl-1-butene. B.p. 41.4
                       5544
C$_6$H$_{12}$ Hexene. B.p. 63.6
                       523, 1840, 2081, 3882, 4089, 4973, 5379, 5471, 5546, 6783, 7541, 8431,
                       11700, 15882, 15962, 16114, 16423, 16470, 16485, 1159a, 2079a, 2207a,
                       2328a 2574a, 2730a, 2742a, 4739a, 5748a, 6384a, 6495a, 7374a, 8234a,
                       8707a, 8713a, 10854a, 11355a, 11700a-11700d
C$_6$H$_{12}$ *cis*-2-Hexene. B.p. 68.8
                       4090, 5547, 16115
C$_6$H$_{12}$ *cis*-3-Hexene. B.p. 66.4
                       2080, 4091, 16116
C$_6$H$_{12}$ 2-Methyl-1-pentene.
                       16117
C$_6$H$_{12}$ Methylcyclopentane. B.p. 72.0
                       1160, 1270, 1330, 1491, 1572, 1682, 1841, 1906, 2082, 2329, 2362, 2798,
                       2956, 3002, 3085, 3174, 3283, 3328, 3429, 3883, 3933, 4092, 4256, 4579,
                       5180, 5380, 5436, 5506, 5551, 5698, 5749, 5768, 6385, 6496, 6784, 6798,
                       6833, 6904, 7542, 7584, 7672, 7690, 7715, 7924, 7971, 7980, 8067, 8078,
                       8147, 8235, 8275, 8339, 8399, 8633, 8667, 8692, 8708, 8714, 1724, 9229,
                       9406, 9699, 9743, 9795, 9812, 9835, 9920, 10649, 10855, 11683, 11701-
                       11704, 16397, 16507, 16509
C$_6$H$_{12}$ 2-Methyl-2-pentene.
                       16118
C$_6$H$_{12}$ *cis*-3-Methyl-2-pentene. B.p. 70.2
                       4093, 5548, 16119
C$_6$H$_{12}$ *trans*-3-Methyl-2-pentene. B.p. 67.6
                       4094
C$_6$H$_{12}$ 4-Methyl-1-pentene. B.p. 54.0
                       5549
C$_6$H$_{12}$ 4-Methyl-2-pentene. B.p. 56.7
                       524, 4095, 5550
C$_6$H$_{12}$ 1,1,2-Trimethylcyclopropane. B.p. 52.6
                       5381
C$_6$H$_{12}$Cl$_2$O   Bis(chloroisopropyl) ether. B.p. 187.0
                       525, 5837
C$_6$H$_{12}$Cl$_2$O$_2$ 1,2-Bis(2-chloroethoxy)ethane. B.p. 240.9
                       526
C$_6$H$_{12}$O   *trans*-2-Butenyl ethyl ether. B.p. 100.45
                       4098
C$_6$H$_{12}$O   *cis*-2-Butenyl ethyl ether. B.p. 100.3
                       4099
C$_6$H$_{12}$O   Butyl vinyl ether. B.p. 93.8
                       527, 2083, 4096, 8148, 16141
C$_6$H$_{12}$O   Cyclohexanol. B.p. 160.65
                       528, 1067, 1389, 1842, 2247, 2435, 2575, 3627, 4257, 4658, 4874, 4923,
                       4974, 5068, 5108, 5197, 5268, 5937, 6123, 6955, 7065, 7139, 7197, 7543,
                       7790, 8764, 8878, 9006, 9109, 9520, 9664, 9713, 9967, 10328, 10374,
                       10444, 10545, 10661, 10856, 10895, 11149, 11357, 11630, 11665, 11684,
                       11705-11773, 16233, 16453, 16455, 16456, 16459, 16501, 16502, 16512,
                       16519, 16531, 11704a, 16233c, 16481a

Formula	Name and System No.

$C_6H_{12}O$ 2,2-Dimethyltetrahydrofuran. B.p. 90
       529

$C_6H_{12}O$ 2,5-Dimethyltetrahydrofuran. B.p. 90
       530

$C_6H_{12}O$ 2-Ethylbutyraldehyde. B.p. 116.7
       531

$C_6H_{12}O$ Ethyl methallyl ether. B.p. 76.65
       4100

$C_6H_{12}O$ Hexaldehyde. B.p. 128.3
       532, 8149

$C_6H_{12}O$ 2-Hexanone. B.p. 127
       533, 2749, 3430, 5601, 6573, 7036, 7198, 8150, 8340, 9495, 9921, 11631, 11774-11776, 16093, 16118

$C_6H_{12}O$ 3-Hexanone. B.p. 124
       534, 1068, 2208, 3431, 4875, 4924, 5602, 5809, 6574, 8012, 8151, 8341, 8432, 8848, 9496, 9625, 11777-11791

$C_6H_{12}O$ Isobutyl vinyl ether. B.p. 83.0
       535, 4097, 8342, 16003

$C_6H_{12}O$ 2-Methylpentanal. B.p. 118.3
       536, 6497, 16094

$C_6H_{12}O$ 4-Methyl-2-pentanone. B.p. 115.9
       537, 1272, 1331, 1492, 1843, 2084, 2209, 2851, 3175, 3432, 3884, 4101, 4876, 4925, 5382, 5603, 5810, 6386, 6575, 7871, 8152, 8343, 8433, 8849, 9626, 9836, 9943, 10857, 11381, 11632, 11685, 11792-11808, 16073, 16298, 16508

$C_6H_{12}O$ 2-Methyl-2-pentene-4-ol.
       538, 16234

$C_6H_{12}O$ 3,3-Dimethyl-2-butanone (Pinacolone). B.p. 106
       539, 1271, 1332, 1683, 1844, 3176, 3433, 4877, 6387, 6498, 7544, 7838, 7872, 8236, 8344, 8847, 9250, 9266, 9283, 9391, 9419, 9650, 9700, 10858, 11686, 11809-11817

$C_6H_{12}OS$ 2-Ethylthioethyl vinyl ether. B.p. 169.7
       540

$C_6H_{12}O_2$ Amyl formate. B.p. 132
       541, 9753, 16215

$C_6H_{12}O_2$ Butyl acetate. B.p. 126.2
       542, 1161, 1493, 2210, 2908, 3177, 3349, 3434, 3628, 3885, 4258, 4878, 4926, 4953, 5383, 5604, 5811, 6576, 7585, 8013, 8041, 8153, 8237, 8345, 8434, 8850, 8926, 8950, 9134, 9173, 9575, 9627, 9754, 9837, 9922, 9968, 10516, 10859, 10896, 11382, 11633, 11687, 11774, 11777, 11818-11837, 15940, 16125, 16142, 16235, 16236, 16262, 16482, 6655a, 11704a, 11826a, 16409a, 16481a, 16481b

$C_6H_{12}O_2$ *sec*-Butyl acetate. B.p. 112.4
       543, 8238, 11818, 16160, 16235, 16237

$C_6H_{12}O_2$ 4,4-Dimethyl-1,3-dioxane. B.p. 133.4
       543a, 9198c, 9198f, 9198g, 11402a, 11896a, 16213c

$C_6H_{12}O_2$ Ethyl butyrate. B.p. 119.9
       159, 544, 1845, 2211, 2743, 2909, 3178, 3886, 4259, 4822, 4879, 4927, 6499, 6577, 7586, 8014, 8042, 8154, 8346, 8435, 8851, 9135, 9576, 9628, 9755, 9838, 9923, 10517, 11634, 11778, 11792, 11897-11911, 16345, 16421

$C_6H_{12}O_2$ 2-Ethylbutyric acid. B.p. 194.2
       545

Formula   Name and System No.

$C_6H_{12}O_2$ Ethyl isobutyrate. B.p. 110.1
546, 1069, 1846, 2212, 2363, 2493, 2744, 2828, 3179, 3887, 4102, 4823, 4880, 5019, 5038, 5437, 5812, 5866, 6320, 6500, 6578, 7545, 7873, 8024, 8043, 8155, 8347, 9230, 9251, 9629, 9651, 9924, 10860, 11688, 11793, 11809, 11912-11921

$C_6H_{12}O_2$ 2-Ethyl-2-methyl-1,3-dioxolane. B.p. 117.6
547

$C_6H_{12}O_2$ Caproic acid. B.p. 205.15
548, 2380, 2621, 2668, 3629, 5109, 5938, 10115, 10148, 10176, 10329, 10375, 10662, 10707, 11285, 11316, 11590, 11838-11896, 9957a, 11837a

$C_6H_{12}O_2$ 4-Hydroxy-4-methyl-2-pentanone. B.p. 166
549, 11383, 11922, 11923

$C_6H_{12}O_2$ Isoamyl formate. B.p. 124.2
550, 2213, 2910, 3180, 3435, 3630, 4260, 4881, 4928, 5605, 5813, 5828, 6579, 6941, 7443, 8015, 8044, 8092, 8156, 8348, 8436, 8852, 8927, 8951, 9174, 9497, 9577, 9630, 9756, 9839, 11384, 11635, 11775, 11779, 11819, 11897, 11924-11929, 16225, 16347

$C_6H_{12}O_2$ Isobutyl acetate. B.p. 117.2
551, 1070, 1847, 2214, 2576, 3436, 3888, 4824, 4882, 5438, 5814, 5829, 6501, 6580, 6942, 7037, 7546, 8016, 8025, 8045, 8157, 8239, 8349, 8437, 8853, 9136, 9578, 9631, 9840, 9925, 11780, 11794, 11898, 11930-11940, 16183, 16482a

$C_6H_{12}O_2$ Isocaproic acid. B.p. 199.5
2436, 5110, 9054, 10330, 10376, 10663, 10897, 11317, 11409, 11503, 11591, 11941-1972

$C_6H_{12}O_2$ Isopropyl propionate. B.p. 110.3
552, 11795, 11810, 11820

$C_6H_{12}O_2$ 4-Methyl-4-hydroxytetrahydropyran. B.p. 188
552a, 9438a, 11837a

$C_6H_{12}O_2$ Methyl isovalerate. B.p. 116.3
553, 1071, 1848, 2215, 2829, 2852, 3437, 4103, 4883, 5039, 6388, 6502, 6581, 7038, 8026, 8046, 8158, 8240, 8350, 8438, 8765, 8854, 9632, 9796, 9926, 11781, 11796, 11899, 11912, 11930, 11973-11980

$C_6H_{12}O_2$ 2-Methylpentanoic acid. B.p. 196.4
554

$C_6H_{12}O_2$ Propyl propionate. B.p. 122.1
555, 2216, 2911, 3181, 3438, 3631, 4884, 4929, 5606, 6503, 6582, 8017, 8093, 8159, 8351, 8439, 8855, 9633, 9841, 11385, 11782, 11820, 11900, 11924, 11981-11983, 16094a

$C_6H_{12}O_2$ Tetrahydropyran-2-methanol. B.p. 187.2
557

$C_6H_{12}O_2$ 4-Vinyloxy-1-butanol
556, 8404

$C_6H_{12}O_2S$ 2,4-Dimethylsulfolane.
11984, 11985

$C_6H_{12}O_3$ 2,2-Dimethoxy-3-butanone. B.p. 145
558

$C_6H_{12}O_3$ 2-Ethoxyethyl acetate. B.p. 156.8
559, 1591, 2437, 2577, 3632, 4261, 4702, 5069, 5111, 5165, 5607, 5939, 6124, 6583, 6656, 7140, 7199, 7275, 7316, 7444, 7612, 7791, 8018, 8160, 8352, 8440, 8766, 8879, 9055, 9110, 9322, 9498, 9521, 9665, 9714, 9842, 9969, 10377, 10445, 10898, 11470, 11504, 11565, 11821, 11986-12038, 16238

Formula	Name and System No.

$C_6H_{12}O_3$ Ethyl α-hydroxyisobutyrate. B.p. 150
  10899

$C_6H_{12}O_3$ Isopropyl lactate. B.p. 166.9
  7141, 10900, 11705, 11986, 12039-12047

$C_6H_{12}O_3$ Methyl 3-ethoxypropionate.
  560

$C_6H_{12}O_3$ Paraldehyde. B.p. 124
  561, 2217, 2912, 3064, 3439, 4104, 4262, 4930, 6504, 6584, 7613, 8094,
  8161, 8441, 8740, 9175, 9499, 9634, 9757, 9843, 9927, 9970, 10518,
  11246, 11386 11706, 11822, 11901, 11925, 11931, 11973, 12048-12056,
  15948, 16232, 16347

$C_6H_{12}O_3$ Propyl lactate. B.p. 171.7
  2438, 3633, 5198, 5940, 7142, 7276, 8767, 9971, 10378, 10901, 11358,
  11707, 12057-12093, 16428, 16429, 16532, 16533

$C_6H_{12}O_3$ Trioxane. B.p. 114.5
  562

$C_6H_{12}O_3$ 2(2-Vinyloxyethoxy)ethanol.
  563, 8524

$C_6H_{13}Br$ 1-Bromohexane. B.p. 156.5
  3182, 3440, 3634, 4263, 4516, 5199, 5608, 5941, 6125, 7066, 7143, 7200,
  7277, 7317, 7445, 7614, 8162, 8353, 9057, 9323, 9439, 9579, 9715, 10902,
  11505, 11708, 11987, 12094-12104

$C_6H_{13}Cl$ 1-Chlorohexane. B.p. 134.5
  564

$C_6H_{13}ClO_2$ Chloroacetal. B.p. 157.4
  6126, 6956, 7067, 7615, 9522, 10446, 10546, 11359, 11709, 12057, 12105-
  12127, 16503

$C_6H_{13}N$ Cyclohexylamine. B.p. 134.5
  565, 11150, 11689, 11710, 16519, 16229a, 16233a, 16233b, 16233c

$C_6H_{13}N$ Hexamethylenimine. B.p. 138
  566

$C_6H_{13}NO$ *N,N*-Dimethylbutyramide.
  7446

$C_6H_{13}NO$ 2,6-Dimethylmorpholine. B.p. 146.6
  567

$C_6H_{13}NO$ 4-Ethylmorpholine. B.p. 138.3
  568

$C_6H_{13}NO_2$ 4-Morpholinoethanol. B.p. 225.5
  569

C_6H_{14} 2,2-Dimethylbutane. B.p. 49.7
  1573, 2085, 3284, 4487, 4508, 5552, 6799, 6810, 9152, 16390, 1161a

C_6H_{14} 2,3-Dimethylbutane. B.p. 58.0
  1273, 1494, 1574, 1684, 1849, 1907, 2086, 2799, 2957, 3183, 3285, 3295,
  3329, 3517, 4105, 4264, 4507, 4807, 5384, 5439, 5507, 5553, 5750, 5769,
  6248, 6302, 6389, 6505, 6646, 6785, 6800, 6811, 6834, 6844, 7375, 7587,
  7716, 7972, 7981, 8241, 8276, 8300, 8693, 8715, 9153, 9797, 9813, 9844,
  11349

C_6H_{14} Hexane. B.p. 68.95
  130, 570, 1162, 1274, 1333, 1495, 1575, 1685, 1747, 1850, 1908, 2087,
  2330, 2364, 2577a, 2730b, 2744a, 2800 2913, 2958, 3003, 3184, 3286,
  3296, 3330, 3364, 3441, 3518, 3546, 3889, 3934, 4106, 4265, 4580, 4691,
  4740, 4975, 5040, 5181, 5385, 5440, 5472, 5508, 5554, 5609, 5699, 5716,
  5151, 5770, 5787, 6249, 6260, 6284, 6303, 6390, 6506, 6647, 6786, 6801,
  6812, 6835, 6838, 6905, 6997, 7376, 7396, 7407, 7547, 7588, 7673, 7691,

Formula   Name and System No.

C₆H₁₄   Hexane (*continued*)
                  7717, 7925, 7945, 7956, 7973, 7982, 8068, 8079, 8163, 8242, 8277, 8301,
                  8354, 8400, 8442, 8679, 8694, 8716, 8725, 9188, 9198d, 9210, 9231, 9267,
                  9284 9297, 9407, 9420, 9701, 9758, 9798, 9814, 9845, 9908, 9928, 9944,
                  9953, 10104, 10108, 10519, 10650, 10708, 10861, 11151, 11351, 11356,
                  11690, 11700, 11701, 11811, 12128-12134, 15883, 15915, 15963, 16004,
                  16046, 16051, 16074, 16095, 16120, 16126, 16143, 16161, 16174, 16184,
                  16290-16294, 16297, 16328, 16330, 16365, 16366, 16399, 16403, 16423,
                  16432, 16434, 16463, 16470, 16485, 16509, 16559, 2217a, 8708a, 11355b,
                  12127a, 16338a, 16508a
C₆H₁₄   2-Methylpentane. B.p. 60.4
                  150, 2088, 4107, 5555, 6802, 6813, 6836, 8695, 16121, 1162a
C₆H₁₄   3-Methylpentane. B.p. 63.3
                  149, 2089, 5556, 6803, 6814, 6837, 8696, 16122, 1162b
C₆H₁₄N₂   2,5-Dimethylpiperazine. B.p. 164
                  571, 8443, 9759
C₆H₁₄N₂O   4-(2-Aminoethyl)morpholine. B.p. 204.7
                  572
C₆H₁₄N₂O   1-Piperazineethanol. B.p. 246.3
                  573
C₆H₁₄O   Amyl methyl ether. B.p. 100
                  34
C₆H₁₄O   *tert*-Amyl methyl ether. B.p. 86
                  574, 2090, 8243, 9929
C₆H₁₄O   Butyl ethyl ether. B.p. 92.2
                  575, 2091, 4109, 16005, 16075, 1495a
C₆H₁₄O   *tert*-Butyl ethyl ether. B.p. 73
                  576, 4108, 8244, 9930, 16076, 16162
C₆H₁₄O   2-Ethyl-1-butanol. B.p. 148.9
                  577, 12135-12138
C₆H₁₄O   Ethyl isobutyl ether. B.p. 79
                  8355, 16006
C₆H₁₄O   Hexyl alcohol. B.p. 157.85
                  578, 1072, 1390, 1851, 2439, 2578, 2914, 3635, 4266, 4885, 4931, 4976,
                  5070, 5200, 5269, 5385a, 5942, 6127, 6957, 6978, 7068, 7144, 7201, 7278,
                  7792, 8019, 8302, 8738, 8768, 8880, 9111, 9523, 9580, 9666, 9716, 9972,
                  10331, 10379, 10447, 10520, 10664, 10709, 10862, 10903, 11152, 11360,
                  11471, 11636, 11666, 11711, 11988, 12039, 12048, 12058, 12094, 12105,
                  12139-12194, 16230, 12127a
C₆H₁₄O   Isopropyl ether. B.p. 69
                  35, 120, 151, 579, 1163, 1496, 2331, 2512, 2520, 2959, 3004, 3065, 3185,
                  3442, 3547, 4110, 5041, 5182, 5386, 5441, 5771, 6391, 6804, 6906, 7018,
                  7548, 7926, 7946, 7974, 8405, 8655 8717, 8932, 9198, 9954, 10863,
                  11702, 12128, 12139, 12195, 12196, 15941, 15959, 16007, 16038, 16077,
                  16163, 16194, 16213, 16239, 16299, 16440, 16445, 2091a, 8277a, 11402b,
                  16213b
C₆H₁₄O   Isopropyl propyl ether. B.p. 66
                  16008
C₆H₁₄O   2-Methyl-1-pentanol. B.p. 148
                  580
C₆H₁₄O   4-Methyl-2-pentanol. B.p. 131.8
                  581, 10864, 11691, 12197

Formula   Name and System No.

$C_6H_{14}O$ Propyl ether. B.p. 90.7
                                 121, 582, 1164, 1334, 1530, 2092, 2332, 3005, 3017, 3186, 3443, 3935,
                                 4111, 5042, 5387, 5442, 5610, 5700, 5890, 6392, 6507, 6585, 7039, 7252,
                                 7377, 7589, 7692, 7718, 7839, 7874, 7947, 7989, 8069, 8164, 8245, 8278,
                                 8356, 8656, 8856, 9129, 9137, 9232, 9252, 9268, 9298, 9392, 9408, 9421,
                                 9652, 9690, 9702, 9799, 9945, 10865, 11352, 11692, 12198-12203, 16096,
                                 151a, 1496a

$C_6H_{14}OS$ 2-Butylthioethanol.
                                 12204

$C_6H_{14}O_2$ Acetaldehyde diethyl acetal. B.p. 103.6
                                 583, 1073, 1165, 1335, 1531, 1852, 2093, 2333, 2830, 2853, 3187, 3936,
                                 4112, 4267, 5020, 5388, 5701, 5867, 6321, 6393, 6508, 7378, 7719, 7840,
                                 7913, 8165, 8357, 8657, 9233, 9253, 9269, 9285, 9299, 9393, 9409, 9422,
                                 9653, 9703, 10866, 11353, 11693, 11812, 11913, 11932, 11974, 12205-
                                 12212, 16009

$C_6H_{14}O_2$ 2-Butoxyethanol. B.p. 171.2
                                 584, 1391, 1592, 2218, 2440, 2579, 2915, 3350, 3444, 3636, 4268, 4581,
                                 5112, 5201, 5270, 5943, 6128, 6979, 7069, 7145, 7279, 7590, 7734, 7793,
                                 8166, 8769, 8881, 9524, 9667, 9717, 10149, 10332, 10380, 10448, 10521,
                                 10665, 10710, 10904, 11153, 11286, 11472, 11506, 11566, 11667, 11712,
                                 11823, 11989, 12059, 12095, 12140, 12213-12282, 16534

$C_6H_{14}O_2$ 1,2-Diethoxyethane. B.p. 123
                                 585, 4113, 8444, 12195

$C_6H_{14}O_2$ Butyraldehyde dimethyl acetal. B.p. 114
                                 2094, 7397

$C_6H_{14}O_2$ 2,2-Dimethoxybutane. B.p. 106
                                 2095

$C_6H_{14}O_2$ 1,3-Dimethoxybutane. B.p. 120.5
                                 586

$C_6H_{14}O_2$ Isobutyraldehyde dimethyl acetal. B.p. 104.7
                                 587

$C_6H_{14}O_2$ Ethoxypropoxymethane. B.p. 113.7
                                 588, 4114, 6322, 6509, 9270, 9423, 11914, 16010, 16097

$C_6H_{14}O_2$ Hexylene glycol.
                                 12289, 12284, 16535

$C_6H_{14}O_2$ 2-Methyl-1,5-pentanediol. B.p. 242.4
                                 589

$C_6H_{14}O_2$ 3-Methyl-1,5-pentanediol. B.p. 248.4
                                 590

$C_6H_{14}O_2$ Pinacol. B.p. 174.35
                                 591, 2441, 3637, 4269, 5202, 5944, 6129, 6958, 7070, 9668, 9973, 10381,
                                 10449, 10522, 10711, 10790, 10905, 11154, 11473, 12060, 12106, 12285-
                                 12317

$C_6H_{14}O_3$ Bis(2-methoxyethyl)ether. B.p. 162
                                 592

$C_6H_{14}O_3$ Dipropylene glycol. B.p. 229.2
                                 5136, 6657, 10246, 10291, 10791, 11025, 12318-12342

$C_6H_{14}O_3$ 2(2-Ethoxyethoxy)ethanol. B.p. 201.9
                                 593, 4270, 6130, 8445, 8525, 10906, 11155, 12343-12364

$C_6H_{14}O_3$ 3-Methyl-1,3,5-pentanetriol. B.p. 295
                                 593a, 11837b

$C_6H_{14}O_4$ Triethylene glycol. B.p. 288.7
                                 594, 6691, 8526, 10272, 12365-12393

Formula   Name and System No.

C₆H₁₄S   Isopropyl sulfide. B.p. 120.5
                1074, 1539, 1686, 1853, 2096, 2750, 3018, 3188, 3351, 3445, 3890, 4115,
                4826, 4886, 6131, 6394, 6872, 6943, 7040, 7318, 7841, 7875, 8167, 8358,
                8857, 9234, 9424, 9635, 11783, 11797, 11902, 11915, 11933, 12205,
                12394-12397

C₆H₁₄S   Propyl sulfide. B.p. 141.5
                1392, 1687, 1854, 2580, 3189, 3446, 4582, 4932, 5611, 5945, 6944, 6959,
                7146, 7202, 7319, 7447, 7794, 8095, 8168, 8359, 8446, 8770, 8952, 9138,
                9324, 9440, 9846, 9931, 10450, 10523, 11247, 11361, 11824, 12398-
                12402

C₆H₁₅B   Triethylborane.
                40

C₆H₁₅BO₃   Ethyl borate. B.p. 118.6
                2219, 4116, 4827, 7549, 8047, 8169, 8360, 9636, 9847, 11784, 11903,
                11934, 12049, 12403-12406

C₆H₁₅N   Diisopropylamine. B.p. 83.86
                595, 6395, 11700a, 12128a

C₆H₁₅N   3,3-Dimethyl-1-butylamine. B.p. 112.8
                597

C₆H₁₅N   1,3-Dimethylbutylamine. B.p. 108.5
                596

C₆H₁₅N   Dipropylamine. B.p. 109.2
                598, 7001, 7379, 9235, 11785, 11798, 11813, 12198, 12206, 12407-12411,
                11700b, 12128b

C₆H₁₅N   Ethylbutylamine. B.p. 111.2
                599

C₆H₁₅N   Ethyl-sec-butylamine.
                599a

C₆H₁₅N   Hexylamine. B.p. 132.7
                600, 6396, 11700c, 12128c

C₆H₁₅N   Isohexylamine. B.p. 123.5
                12412-12414

C₆H₁₅N   Triethylamine. B.p. 89.4
                601, 1275, 1497, 2097, 3190, 4117, 5154, 5389, 7380, 8303, 9197, 9815,
                9955, 10867, 10907, 11700d, 11703, 12129, 12199, 12207, 12415, 12416,
                15942, 16011, 16239, 16352, 16556, 2800a

C₆H₁₅NO   2-Butylaminoethanol. B.p. 199.31
                602

C₆H₁₅NO   2-Diethylaminoethanol. B.p. 162
                603, 1855, 3638, 8447, 8718, 9974, 10868, 11156, 11672, 12213, 12417-
                12428

C₆H₁₅NO   3-Isopropoxypropylamine. B.p. 147
                605

C₆H₁₅NO   1-Isopropylamino-2-propanol. B.p. 164.5
                604

C₆H₁₅NO₂   1,1′-Iminodi-2-propanol.
                6854, 12429

C₆H₁₅NO₃   2,2′,2″-Nitrilotriethanol.
                8731

C₆H₁₅N₃   4(2-Aminoethyl)piperazine. B.p. 222.0
                606

C₆H₁₅O₃P   Dipropyl phosphate.
                7254

Formula   Name and System No.

$C_6H_{16}N_2$ N,N-Diethylethylenediamine. B.p. 144.9
                  607
$C_6H_{16}N_2$ N,N,N',N'-Tetramethylethylenediamine. B.p. 120
                  608
$C_6H_{16}OSi$ Ethoxymethyltrimethylsilane. B.p. 102
                  4118
$C_6H_{16}OSi$ Trimethylpropoxysilane. B.p. 100.3
                  6510
$C_6H_{16}O_2Si$ Diethoxydimethylsilane. B.p. 114
                  4119
$C_5H_{18}OSi$ Hexamethyldisiloxane. B.p. 100
                  6511, 6860
C_7F_{14} Perfluoromethylcyclohexane. B.p. 73
                  10869
C_7F_{16} Perfluoroheptane. B.p. 81.6
                  3948, 6328, 7381, 8101, 9744, 10870, 12130, 12430-12433
$C_7H_5Cl_3$ α,α,α-Trichlorotoluene. B.p. 220.9
                  3639, 4271, 5946, 10177, 10247, 10581, 10712, 12434-12459
$C_7H_5F_3$ α,α,α-Trifluorotoluene. B.p. 103.9
                  62
C_7H_5N Benzonitrile. B.p. 191.3
                  4272, 5113, 6132, 10030, 10150, 10382, 10666, 10908, 11157, 11507,
                  11592, 12214, 12460-12490
C_7H_5NO Phenylisocyanate. B.p. 162.8
                  12491, 12492
$C_7H_6Cl_2$ α,α-Dichlorotoluene. B.p. 205.2
                  2669, 3640, 4273, 7448, 7616, 9325, 10582, 10713, 10909, 11026, 11410,
                  11593, 11838, 12493-12522
C_7H_6O Benzaldehyde. B.p. 179.2
                  2248, 2442, 2670, 3641, 4274, 4682, 5114, 5203, 5612, 5947, 6918, 7280,
                  7449, 7617, 8882, 9007, 9326, 9441, 9975, 10333, 10383, 10667, 10714,
                  10910, 11158, 11411, 11508, 11643, 11673, 11713, 11941, 11990, 12061,
                  12061, 12141, 12215, 12523-12579, 16495, 16496, 16536, 16537
$C_7H_6O_2$ Benzoic acid. B.p. 250.5
                  3642, 4769, 5390, 5948, 10116, 10178, 10273, 10292, 11027, 12493,
                  12580-12624
$C_7H_6O_2$ Salicyladehyde. B.p. 196.7
                  9442, 11942
C_7H_7Br α-Bromotoluene. B.p. 198.5
                  2671, 4760, 7450, 7618, 9057, 9327, 9443, 10547, 10911, 11159, 11287,
                  11318, 11412, 11594, 11839, 11943, 12625-12647
C_7H_7Br m-Bromotoluene. B.p. 184.5
                  2622, 2672, 3447, 3643, 4275, 4517, 4583, 5204, 5271, 5613, 5950, 6133,
                  7071, 7451, 7735, 9058, 9328, 9444, 10668, 10912, 11160, 11288, 11413,
                  11474, 11509, 11595, 11714, 11944, 12460, 12523, 12648-12666
C_7H_7Br o-Bromotoluene. B.p. 181.45
                  2381, 2525, 2623, 2673, 3448, 3644, 4276, 4518, 4584, 5137, 5205, 5272,
                  5614, 5949, 6134, 7072, 7231, 7281, 7452, 7619, 8771, 8953, 9059, 9329,
                  9445, 10548, 10715, 10913, 11161, 11289, 11414, 11475, 11510, 11567,
                  11596, 11647, 11715, 11840, 11945, 12062, 12142, 12216, 12461, 12524,
                  12667-12704, 16350, 16504, 16514, 16515, 16520, 16521, 16527, 16530

Formula	Name and System No.

C₇H₇Br *p*-Bromotoluene. B.p. 185

$$\text{C}_7\text{H}_7\text{Br}$$ *p*-Bromotoluene. B.p. 185
2249, 2674, 3449, 3645, 4277, 4519, 5206, 5273, 5951, 6135, 7073, 7435, 9060, 9330, 9446, 10483, 10549, 10914, 11162, 11415, 11511, 11568, 11597, 11716, 11841, 11946, 12462, 12525, 12705-12731, 16516

$$\text{C}_7\text{H}_7\text{BrO}$$ *o*-Bromoanisole. B.p. 217.7
3646, 5952, 8527, 10179, 10583, 10792, 11028, 11842, 12318, 12732-12739

$$\text{C}_7\text{H}_7\text{BrO}$$ *p*-Bromoanisole. B.p. 217.7
12740-12742

$$\text{C}_7\text{H}_7\text{Cl}$$ *α*-Chlorotoluene. B.p. 179.3
2382, 2675, 3191, 3647, 4278, 4520, 5207, 5274, 5615, 5844, 7074, 7454, 7620, 8772, 8941, 9008, 9061, 9331, 9447, 10550, 10716, 10915, 11163, 11290, 11416, 11476, 11512, 11644, 11717, 11843, 11947, 12063, 12143, 12526, 12667, 12743-12781, 16351, 16428, 16430, 16495, 16497-16499, 16528, 16532, 16536-16541, 9847a

$$\text{C}_7\text{H}_7\text{Cl}$$ *m*-Chlorotoluene. B.p. 162.3
11844

$$\text{C}_7\text{H}_7\text{Cl}$$ *o*-Chlorotoluene. B.p. 159.2
1593, 1688, 2383, 2443, 2676, 3192, 3450, 3553, 3648, 4279, 4585, 4703, 4977, 5208, 5275, 5616, 5953, 6136, 6945, 7075, 7147, 7203, 7455, 7621, 8448, 8773, 8954, 9009, 9062, 9112, 9332, 9448, 9525, 9581, 9718, 9848, 9976, 10451, 10916, 11164, 11362, 11417, 11513, 11718, 11845, 11948, 11991, 12064, 12144, 12217, 12285, 12527, 12782-12804

$$\text{C}_7\text{H}_7\text{Cl}$$ *p*-Chlorotoluene. B.p. 163.5
609, 1689, 2444, 2677, 3193, 3451, 3649, 4280, 4521, 4586, 4704, 5115, 5209, 5617, 6137, 7076, 7148, 7232, 7282, 7456, 7622, 7736, 8449, 8774, 8955, 9010, 9063, 9113, 9333, 9449, 9526, 9719, 9849, 9977, 10917, 11136, 1165, 11363, 11418, 11477, 11514, 11655, 11719, 11846, 11949, 11992, 12065, 12145, 12218, 12286, 12463, 12528, 12805-12834

$$\text{C}_7\text{H}_7\text{ClO}$$ *m*-Chloroanisole. B.p. 193.3
6138, 6875, 8775, 12835, 12836

$$\text{C}_7\text{H}_7\text{ClO}$$ *o*-Chloroanisole. B.p. 195.7
3650, 5954, 6139, 7334, 10551, 10918, 12219, 12529, 12837

$$\text{C}_7\text{H}_7\text{ClO}$$ *p*-Chloroanisole. B.p. 197.8
3651, 5955, 6876, 10484, 10793, 12838-12845

$$\text{C}_7\text{H}_7\text{F}$$ *o*-Fluorotoluene. B.p. 114
9780, 9851

$$\text{C}_7\text{H}_7\text{I}$$ *p*-Iodotoluene. B.p. 214.5
2384, 2624, 2678, 3652, 4281, 5138, 5956, 6140, 9450, 10584, 10717, 10794, 10919, 11029, 11847, 12732, 12846-12876

$$\text{C}_7\text{H}_7\text{NO}_2$$ *m*-Nitrotoluene. B.p. 230.8
3653, 4282, 5957, 6141, 6692, 8528, 8977, 10180, 10248, 10585, 11030, 11262, 12434, 12580, 12877-12919

$$\text{C}_7\text{H}_7\text{NO}_2$$ *o*-Nitrotoluene. B.p. 221.75
3654, 4283, 4587, 4770, 5958, 6142, 6693, 6914, 8529, 8978, 10031, 10117, 10181, 10586, 10795, 11031, 11166, 11848, 12319, 12435, 12494, 12581, 12846, 12920-12979

$$\text{C}_7\text{H}_7\text{NO}_2$$ *p*-Nitrotoluene. B.p. 238.9
3655, 4284, 5959, 6143, 6694, 8530, 8979, 10182, 10249, 10274, 10293, 11032, 11093, 11263, 11849, 12320, 12436, 12582, 12980-13018

$$\text{C}_7\text{H}_8$$ Toluene. B.p. 110.7
160, 610, 1075, 1166, 1276, 1393, 1498, 1532, 1540, 1690, 1856, 2098, 2220, 2365, 2581, 2801, 2854, 2916, 3006, 3019, 3031, 3066, 3194, 3452, 3554, 3656, 3891, 3937, 4120, 4285, 4532, 4641, 4705, 4828, 4887, 4933,

Formula	Name and System No.

C₇H₈   Toluene (*continued*)

5021, 5043, 5210, 5276, 5391, 5443, 5479, 5618, 5702, 5815, 5868, 5960, 6144, 6261, 6285, 6323, 6397, 6512, 6586, 6658, 6695, 6855, 6859, 6885, 6946, 6991, 7077, 7244, 7382, 7398, 7404, 7457, 7550, 7556, 7591, 7623, 7720, 7795, 7842, 7876, 8020, 8027, 8048, 8096, 8170, 8246, 8279, 8304, 8361, 8450, 8512, 8531, 8617, 8776, 8858, 8928, 8942, 9139, 9176, 9211, 9236, 9254, 9271, 9286, 9300, 9334, 9394, 9410, 9425, 9500, 9527, 9582, 9637, 9643, 9691, 9704, 9760, 9781, 9800, 9852, 9903, 9909, 9932, 9946, 9978, 10452, 10524, 10718, 10871, 10920, 11167, 11252, 11255, 11387, 11694, 11704, 11720, 11786, 11799, 11814, 11825, 11904, 11916, 11926, 11935, 11975, 11981, 11993, 12050, 12131, 12146, 12200, 12208, 12287, 12394, 12403, 12407, 12412, 12417, 13019-13047, 15916, 15943, 15951, 15960, 16012, 16047, 16079, 16098, 16103, 16108, 16185, 16191, 16222, 16238, 16240-16242, 16251, 16300, 16362, 16406, 16408, 16409, 16420, 16422, 16439, 16443, 16467, 16469, 16471, 16475, 16479, 16489, 16490, 16493, 16523, 16534, 16542-16544, 21817c, 5893a, 16163a, 16337a, 16338b, 16433b, 16438a, 16481b, 16482a, 16565a, 16565b

C₇H₈Cl₂Si   Methylphenyldichlorosilane. B.p. 203.6

10645, 13048

C₇H₈O   Anisole. B.p. 153.8

611, 1394, 2445, 2582, 3032, 3195, 3453, 3657, 4286, 4588, 4659, 4679, 4706, 4978, 5071, 5211, 5619, 5816, 6145, 6587, 6960, 7041, 7078, 7149, 7204, 7320, 7458, 7624, 7796, 8171, 8451, 8777, 8883, 8956, 9011, 9064, 9114, 9335, 9451, 9528, 9583, 9669, 9720, 9853, 9979, 10453, 10921, 11168, 11364, 11419, 11478, 11515, 11721, 11994, 12066, 12107, 12147, 12220, 12288, 12418, 12530, 12782, 12805, 13049-13076, 16460, 16474, 16526, 16531, 2098a

C₇H₈O   Benzyl alcohol. B.p. 205.2

612, 2250, 3658, 4287, 5116, 5277, 5845, 5961, 6146, 7335, 7737, 8532, 10032, 10118, 10151, 10183, 10294, 10334, 10384, 10587, 10669, 10719, 10796, 10922, 11169, 11319, 12321, 12437, 12464, 12495, 12648, 12668, 12705, 12743, 12847, 12877, 12920, 12980, 13019, 13077-13145, 16240, 16505, 16524, 16538

C₇H₈O   *m*-Cresol. B.p. 202.2

2251, 2385, 2526, 2679, 3659, 4288, 4522, 5962, 6147, 6877, 7738, 8533, 9584, 10119, 10184, 10275, 10335, 10720, 10797, 10923, 11170, 11291, 11320, 11403, 11516, 11569, 11850, 11995, 12067, 12221, 12289, 12343, 12465, 12496, 12531, 12669, 12706, 12733, 12848, 12921, 13077, 13146-13215, 13302-13306, 16548

C₇H₈O   *o*-Cresol. B.p. 191.1

1395, 2252, 2386, 2446, 2527, 2680, 3660, 4289, 4589, 4716, 4979, 5117, 5212, 5278, 5963, 6148, 6696, 7739, 7797, 9065, 9336, 9452, 9529, 9585, 9980, 10033, 10120, 10152, 10185, 10336, 10385, 10454, 10485, 10552, 10670, 10721, 10798, 10924, 11171, 11292, 11321, 11420, 11479, 11517, 11598, 11722, 11950, 11996, 12040, 12068, 12096, 12148, 12222, 12290, 12344, 12419, 12466, 12497, 12532, 12625, 12649, 12670, 12707, 12744, 12783, 12806, 12837, 12849, 13078, 13146, 13216-13301, 16416, 16545-16547

C₇H₈O   *p*-Cresol. B.p. 201.7

2253, 2387, 2625, 2681, 3661, 4290, 4590, 4771, 5964, 6149, 6659, 6697, 6878, 7740, 8534, 9066, 9453, 9586, 10034, 10121, 10153, 10186, 10337, 10386, 10486, 10588, 10671, 10722, 10799, 10925, 11172, 11293, 11322, 11404, 11518, 11571, 11599, 11852, 11951, 11997, 12041, 12069, 12223, 12291, 12322, 12345, 12467, 12498, 12533, 12626, 12671, 12708, 12850, 13020, 13079, 13147, 13216, 13307-13396

Formula	Name and System No.

$C_7H_8O_2$ Guaiacol. B.p. 205.05
613, 2388, 2528, 2626, 3662, 4291, 4761, 5965, 6698, 6879, 7741, 10035, 10187, 10590, 10723, 11173, 11323, 11853, 11952, 12438, 13080, 13148, 13307, 13397-13436

$C_7H_8O_2$ m-Methoxyphenol. B.p. 243
614, 3663, 4292, 5966, 6660, 10188, 10589, 11033, 11264, 12583, 13437-13453

C_7H_8S α-Toluenethiol. B.p. 194.8
13454

$C_7H_9ClO_4$ 2-Chloroallylidene diacetate. B.p. 212.2
615

C_7H_9N Benzylamine. B.p. 185.0
9981, 10036, 10724, 10926, 11174, 12224, 13149, 13217, 13308, 13455-13462

C_7H_9N 2,4-Lutidine. B.p. 159.0
10926a, 13462a, 13462b, 16514a

C_7H_9N 2,6-Lutidine. B.p. 144
616, 1691, 3196, 4293, 5620, 9782, 9854, 10553, 10927, 11253, 13021, 13463-13467, 16379-16381, 16413

C_7H_9N Methylaniline. B.p. 196.25
3664, 4294, 4591, 5967, 6661, 9982, 10037, 10122, 10338, 10387, 10672, 10725, 10928, 11175, 11519, 11674, 12225, 12420, 12468, 12499, 12627, 12650, 12922, 13081, 13150, 13218, 13309, 13397, 13468-13495

C_7H_9N Tetrahydrobenzonitrile. B.p. 195.1
619

C_7H_9N m-Toluidine. B.p. 203.1
3665, 4295, 5968, 10123, 10189, 10673, 10726, 10929, 11675, 12672, 12851, 13082, 13151, 13219, 13310, 13398, 13496-13513

C_7H_9N o-Toluidine. B.p. 200.35
617, 3666, 4296, 4592, 5969, 10154, 10190, 10674, 10727, 10930, 11676, 12439, 12469, 12500, 12651, 12673, 12709, 12852, 12923, 13083, 13152, 13220, 13311, 13399, 13468, 13514-13542

C_7H_9N p-Toluidine. B.p. 200.55
618, 3667, 4297, 5970, 10124, 10155, 10191, 10675, 10728, 10931, 12501, 12628, 12674, 12710, 12853, 12924, 13084, 13153, 13221, 13312, 13400, 13543-13553

C_7H_9NO o-Anisidine, B.p. 219.0
4298, 10192, 10250, 10729, 10800, 12854, 12878, 13085, 13222, 13313, 13554-13566

C_7H_{10} Methylcyclohexadiene.
16200

$C_7H_{10}O$ 1,2,3,6-Tetrahydrobenzaldehyde. B.p. 164.2
620

$C_7H_{10}O_4$ Allylidene diacetate.
621

$C_7H_{11}NO$ α-Hydroxycyclohexanenitrile.
8739

C_7H_{12} 2,4-Dimethyl-1,3-pentadiene.
622

C_7H_{12} 1,3-Heptadiene.
4121

C_7H_{12} 2,4-Heptadiene.
4122

Formula   Name and System No.

C_7H_{12} 1-Heptyne. B.p. 99.5
                   4123, 16013, 13566a, 13566b
C_7H_{12} 5-Methyl-1-hexyne. B.p. 90.8
                   4124
$C_7H_{12}Cl_4$ Tetrachloroheptane.
                   8918
$C_7H_{12}O$ 3-Heptene-2-one. B.p. 162.9
                   623, 11388
$C_7H_{12}O$ Methylcyclohexanone. B.p. 165.0
                   3197
$C_7H_{12}O_2$ Butyl acrylate. B.p. 147
                   624, 8172, 16144
$C_7H_{12}O_2$ Cyclohexyl formate.
                   11723
$C_7H_{12}O_2$ 2-Ethoxy-3,4-dihydro-1,2-pyran. B.p. 142.9
                   625
$C_7H_{12}O_2$ 4-Methyl-4-vinyl-1,3-dioxane. B.p. 151
                   625a, 11896a
$C_7H_{12}O_4$ Ethyl malonate. B.p. 198.9
                   65, 2254, 4299, 6150, 8535, 10125, 10156, 10676, 10730, 10932, 11294,
                   11324, 11600, 11854, 11953, 12470, 12502, 12629, 12711, 12855, 13223,
                   13314, 13401, 13514, 13567-13601
$C_7H_{12}O_4$ Pimelic acid.
                   626
$C_7H_{13}ClO_2$ Isoamyl chloroacetate. B.p. 195.2
                   627, 2683, 3668, 4300, 7742, 9337, 9454, 9855, 10038, 11520, 11855,
                   12226, 13086, 13602-13609, 16226
C_7H_{14} 1,1-Dimethylcyclopentane. B.p. 87.84
                   4125, 8634, 8659, 8680, 8700, 16201
C_7H_{14} cis-1,2-Dimethylcyclopentane. B.p. 99.53
                   4126, 8618, 8681, 16202
C_7H_{14} trans-1,2-Dimethylcyclopentane. B.p. 91.87
                   4127
C_7H_{14} cis-1,3-Dimethylcyclopentane. B.p. 91.72
                   4128, 16203
C_7H_{14} trans-1,3-Dimethylcyclopentane. B.p. 90.77
                   2099, 4129, 6907, 8620, 8635, 8658, 8668, 8682, 8699
C_7H_{14} 2,3-Dimethyl-1-pentene. B.p. 84.2
                   4130
C_7H_{14} Ethylcyclopentane. B.p. 103.45
                   4131, 8619, 8683, 8899, 13022
C_7H_{14} 1-Heptene. B.p. 93.64
                   629, 1857, 8173, 15884, 16543, 4740a, 13566a
C_7H_{14} trans-2-Heptene. B.p. 98.0
                   6588
C_7H_{14} 3-Heptene. B.p. 94.8
                   13023, 13610, 16543
C_7H_{14} Methylcyclohexane. B.p. 101.65
                   161, 628, 1076, 1277, 1336, 1499, 1533, 1692, 1853, 2100, 2334, 2366,
                   2745, 2802, 2831, 3007, 3198, 3454, 3669, 3892, 3938, 4132, 4301,
                   4533, 4593, 4829, 4888, 4934, 5022, 5044, 5213, 5279, 5392, 5444, 5621,
                   5703, 5869, 5891, 6262, 6287, 6324, 6398, 6513, 6589, 6992, 7019, 7245,
                   7250, 7383, 7459, 7551, 7592, 7693, 7721, 7877, 7914, 7990, 8070, 8174,
                   8247, 8280, 8362, 8452, 8621, 8660, 8684, 8698, 8778, 8859, 9212, 9237,

Formula   Name and System No.

C$_7$H$_{14}$ Methylcyclohexane (*continued*)
                      9255, 9287, 9301, 9395, 9411, 9426, 9530, 9654, 9692, 9705, 9745, 9761,
                      9801, 9856, 9910, 9933, 9947, 9956, 10525, 10872, 11176, 11354, 11389,
                      11695, 11724, 11800, 11815, 11905, 11917, 11976, 12132, 12149, 12201,
                      12209, 12292, 12395, 12404, 12415, 13024, 13611-13615, 16014, 16052,
                      16164, 16204, 16313, 16404, 16426, 16439, 16489, 16491, 16522, 16542,
                      16560, 1166a, 16338a, 16338b
C$_7$H$_{14}$ 1,1,2,2-Tetramethylcyclopropane. B.p. 75.9
                      4133
C$_7$H$_{14}$O   Butyl isopropenyl ether. B.p. 114.8
                      630
C$_7$H$_{14}$O   2,4-Dimethyl-3-pentanone. B.p. 124
                      4954, 9783, 9857
C$_7$H$_{14}$O   Heptaldehyde. B.p. 155
                      2447, 2583, 4680, 4980
C$_7$H$_{14}$O   2-Heptanone. B.p. 149
                      631, 3199, 4681, 4948, 11998
C$_7$H$_{14}$O   3-Heptanone. B.p. 147.6
                      632
C$_7$H$_{14}$O   4-Heptanone. B.p. 143
                      633, 1396, 2584, 2917, 3352, 3455, 3670, 4594, 4752, 4935, 5072, 7205,
                      7321, 7625, 7798, 9670, 9762, 9858, 9983, 10526, 11637, 12108, 13049,
                      13616-13628
C$_7$H$_{14}$O   Isoamyl vinyl ether. B.p. 112.6
                      9859
C$_7$H$_{14}$O   2-Methylcyclohexanol. B.p. 168.5
                      634, 1077, 3671, 4889, 5280, 6151, 6919, 7079, 7150, 7206, 7283, 8779,
                      8884, 9531, 9671, 9721, 10388, 10801, 10933, 11177, 11365, 11480,
                      11521, 11999, 12042, 12070, 12227, 12745, 12784, 12807, 13025, 13050,
                      13224, 13315, 13629-13656
C$_7$H$_{14}$O   3-Methylcyclohexanol. B.p. 168.5
                      1397, 13657
C$_7$H$_{14}$O   5-Methyl-2-hexanone. B.p. 144.2
                      635, 3456, 3672, 5073, 7460, 7626, 8453, 8957, 9532, 9587, 9860, 9984,
                      12000, 12150, 12398, 13658-13667
C$_7$H$_{14}$O$_2$ Amyl acetate. B.p. 148.8
                      636, 637, 1398, 2448, 2585, 3200, 3457, 3673, 4302, 4949, 5074, 6152,
                      8454, 9588, 9673, 9763, 13668-13674, 16216
C$_7$H$_{14}$O$_2$ *sec*-Amyl acetate. B.p. 133.5
                      638
C$_7$H$_{14}$O$_2$ Butyl propionate. B.p. 146.8
                      639, 1594, 2586, 3353, 3674, 4303, 5075, 7207, 7322, 7799, 8175, 8958,
                      9985, 13051, 13616, 13658, 13668, 13675-13681
C$_7$H$_{14}$O$_2$ Enanthic acid. B.p. 222.0
                      640, 2627, 3675, 4772, 8980, 10193, 10251, 10276, 10731, 12652, 12746,
                      12856, 12879, 12925, 13154, 13316, 13682-13707
C$_7$H$_{14}$O$_2$ Ethyl isovalerate. B.p. 134.7
                      641, 1399, 2221, 2587, 2918, 3201, 3676, 4304, 4830, 4936, 5076, 5622,
                      5830, 7042, 7208, 7323, 7801, 8021, 8049, 8176, 8363, 8455, 8929,
                      8959, 9177, 9501, 9589, 9638, 9674, 9861, 9986, 10527, 11390, 12001,
                      12051, 12399, 13708-13714
C$_7$H$_{14}$O$_2$ Ethyl valerate. B.p. 145.45
                      642, 1400, 2449, 3677, 4305, 5077, 5623, 7209, 7802, 9590, 9675, 9862,
                      9987, 10455, 13617, 13715-13719

Formula	Name and System No.

$C_7H_{14}O_2$ Isoamyl acetate. B.p. 142.1
643, 1401, 2450, 2588, 2919, 3202, 3354, 3458, 3678, 4306, 4831, 5078, 5624, 5831, 6590, 6947, 7043, 7210, 7627, 7803, 8050, 8177, 8364, 8456, 9178, 9591, 9639, 9672, 9863, 10339, 10456, 10528, 10554, 11906, 12002, 12097, 13052, 13618, 13659, 13669, 13675, 13715, 13720-13732, 16015, 16227, 16374a

$C_7H_{14}O_2$ Isobutyl propionate. B.p. 136.9
644, 1402, 2222, 2589, 2920, 3459, 3679, 4307, 4937, 5079, 6591, 7211, 7324, 7800, 8022, 8178, 8365, 8457, 8943 9140, 9179, 9676, 9864, 9988, 10529, 11391, 12052, 13619, 13660, 13720, 13733-13738

$C_7H_{14}O_2$ Isopropyl butyrate. B.p. 128
4955

$C_7H_{14}O_2$ Isopropyl isobutyrate. B.p. 120.8
645, 2223, 4832, 8051, 8179, 8366, 9640, 13026, 13739

$C_7H_{14}O_2$ 4-Methoxy-4-methyl-2-pentanone. B.p. 165
6288

$C_7H_{14}O_2$ Methyl caproate. B.p. 149.8
646, 2451, 2590, 4308, 5625, 7804, 8458, 9533, 9678, 9722, 10457, 11366, 11725, 12785, 13740-13745

$C_7H_{14}O_2$ Propyl butyrate. B.p. 142.8
647, 2452, 2591, 3460, 3680, 4309, 4660, 4833, 5080, 5626, 7212, 7325, 7628, 7805, 8180, 8459, 9592, 9677, 9865, 10458, 10530, 12003, 12151, 13053, 13620, 13661, 13670, 13676, 13721, 13746-13751, 16462

$C_7H_{14}O_2$ Propyl isobutyrate. B.p. 133.9
648, 2224, 2921, 3203, 3461, 4310, 4834, 4938, 5081, 6948, 7806, 8052, 8181, 8367, 8944, 8960, 9502, 9679, 9764, 9866, 11392, 11776, 11826, 12053, 13752-13758

$C_7H_{14}O_3$ 1,3-Butanediol methyl ether acetate. B.p. 171.75
649, 2453, 2592, 3681, 4311, 5118, 5971, 6153, 6662, 7151, 7284, 7461, 7629, 7743, 8460, 8780, 8885, 9338, 9455, 9680, 9723, 9989, 10340, 10390, 10459, 10935, 11481, 11522, 11572, 11726, 12152, 12228, 12675, 12808, 12835, 13054, 13225, 13317, 13759-13784

$C_7H_{14}O_3$ 2,2-Dimethoxy-3-pentanone. B.p. 162.5
650

$C_7H_{14}O_3$ Ethyl 3-ethoxypropionate. B.p. 170.1
651

$C_7H_{14}O_3$ 4-Hydroxyethyl-4-methyl-1,3-dioxane
651a

$C_7H_{14}O_3$ 4-Hydroxy-3-methylol-4-methyltetrahydropyran
651b

$C_7H_{14}O_3$ Isobutyl lactate. B.p. 182.15
3682, 5972, 10039, 10389, 10677, 10934, 11178, 11677, 12229, 12534, 12653, 12676, 12712, 12747, 13155, 13226, 13318, 13602, 13629, 13785-13804, 16539, 16540, 16549, 16550

$C_7H_{14}O_3$ 3-Methoxybutyl acetate. B.p. 171.3
652

$C_7H_{14}O_4$ 2(2-Methoxyethoxy)ethyl acetate. B.p. 208.9
653

$C_7H_{15}N$ 1,2-Dimethylpiperidine. B.p. 128
9784, 9867

$C_7H_{15}N$ 2,6-Dimethylpiperidine. B.p. 128
9785, 9868

$C_7H_{15}NO$ *N,N*-Dimethylvaleramide.
9456

Formula   Name and System No.

$C_7H_{16}O_2$ 1-Butoxy-2-propanol. B.p. 170.1
                  660, 6662a
$C_7H_{16}O_2$ Diisopropoxymethane.
                  661
$C_7H_{16}O_2$ Dipropoxymethane. B.p. 137.2
                  662, 6515, 8369, 9503, 11638, 11827, 13752, 16099
$C_7H_{16}O_2$ 2-Ethyl-1,5-pentanediol. B.p. 253.3
                  663
$C_7H_{16}O_3$ 1(2-Ethoxyethoxy)-2-propanol. B.p. 198.1
                  664
$C_7H_{16}O_3$ Dipropylene glycol methyl ether.
                  13830, 6662b
$C_7H_{16}O_3$ 2-Ethoxyethyl 2-methoxyethyl ether. B.p. 194.2
                  665, 13156, 13319
$C_7H_{16}O_3$ Ethyl orthoformate. B.p. 145.75
                  1404, 2455, 2593, 3463, 3685, 9593, 10461, 12005, 13056, 13708, 13722,
                  13831-13837
$C_7H_{16}O_3$ 2(2-Propoxyethoxy)ethanol. B.p. 215.8
                  666
$C_7H_{16}O_4$ 2(2(2-Methoxyethoxy)ethoxy)ethanol. B.p. 245.25
                  3686, 5973, 6594, 8536, 10040, 10194, 10295, 10732, 12346, 12880,
                  12926, 12981, 13838-13872
$C_7H_{17}NO$ 1-Diethylamino-2-propanol. B.p. 159.5
                  667
$C_7H_{18}N_2$ 3-Diethylaminopropylamine. B.p. 169.4
                  668
$C_7H_{18}OSi$ Butoxytrimethylsilane. B.p. 124.5
                  8183
$C_8F_{18}O$ Perfluorocyclic oxide. B.p. 102.6
                  6863, 8741, 12431, 13612, 13873-13875
$C_8F_{18}O$ Perfluorobutyl ether. B.p. 100
                  10878, 13806
$C_8H_5Cl_3$ *ar*-Trichlorostyrene.
                  13876
C_8H_6 Phenylacetylene. B.p. 142
                  3464, 4642, 6264, 8727
$C_8H_6Cl_2$ *ar*-Dichlorostyrene.
                  10041
C_8H_6O Coumarone. B.p. 173
                  6663
C_8H_7N Indole. B.p. 253.5
                  3687, 4314, 8537, 11034, 13437, 13877-13886
C_8H_7N α-Toluonitrile. B.p. 232
                  13887
C_8H_8 Styrene. B.p. 145
                  669, 1694, 1860, 2105, 2456, 2694, 2684, 2923, 3033, 3205, 3465, 3688,
                  4143, 4315, 4643, 4939, 5216, 5629, 5705, 6155, 6265, 6401, 6516, 6595,
                  6699, 7081, 7153, 7213, 7327, 7463, 7630, 7807, 8184, 8250, 8282, 8370,
                  8462, 8728, 8282, 9012, 9067, 9339, 9535, 9594, 9682, 9724, 9870, 9904,
                  9935, 9990, 10462, 10938, 11181, 11242, 11421, 11524, 11728, 12006,
                  12072, 12109, 12154, 12536, 13057, 13228, 13463, 13677, 13709, 13716,
                  13733, 13746, 13753, 13831, 13888-13894, 16017, 16450, 16486
$C_8H_8Cl_2O$ 2(2,4-Dichlorophenoxy)ethanol.
                  670

Formula	Name and System No.

C_8H_8O Acetophenone. B.p. 202.05

671, 2389, 2628, 3689, 4316, 4596, 5217, 5282, 5974, 6156, 6664, 7745, 8538, 9068, 9457, 10042, 10126, 10157, 10487, 10555, 10591, 10733, 10803, 10939, 11182, 11325, 11422, 11601, 11856, 11954, 12503, 12630, 12749, 12838, 13157, 13229, 13320, 13402, 13470, 13516, 13543, 13554, 13567, 13603, 13682, 13895-13927

C_8H_8O Epoxyethylbenzene. B.p. 194.2

672

$C_8H_8O_2$ Anisaldehyde. B.p. 249.5

8539, 10296, 11035, 12323, 12584, 12982, 13928-13942

$C_8H_8O_2$ Benzyl formate. B.p. 202.3

673, 3690, 4317, 6157, 6700, 8540, 10556, 10592, 10734, 10804, 10940, 11857, 11955, 12440, 12504, 12631, 13088, 13158, 13230, 13321, 13403, 13568, 13895, 13943-13959

$C_8H_8O_2$ Methyl benzoate. B.p. 199.45

674, 2629, 2685, 3691, 4318, 4762, 5975, 6158, 6701, 7746, 8541, 10043, 10127, 10158, 10488, 10593, 10678, 10735, 10805, 10941, 11295, 11326, 11602, 12231, 12505, 12632, 12857, 13089, 13159, 13231, 13322, 13404, 13569, 13604, 13838, 13896, 13943, 13960-13987

$C_8H_8O_2$ Phenyl acetate. B.p. 195.7

675, 2255, 2686, 3692, 4319, 5976, 6159, 6702, 7747, 8542, 9340, 10044, 10342, 10489, 10557, 10594, 10679, 10736, 10942, 11327, 11423, 11525, 11573, 11603, 11858, 12232, 12472, 12506, 12633, 12713, 12927, 13090, 13160, 13232, 13323, 13405, 13570, 13897, 13944, 13988-14015, 16513

$C_8H_8O_2$ α-Toluic acid. B.p. 266.5

3693, 10195, 11036, 11094, 12983, 14016-14045

$C_8H_8O_3$ Methyl salicylate. B.p. 222.3

3694, 4320, 5977, 6703, 8543, 8981, 10045, 10128, 10196, 10252, 10737, 10806, 12324, 12347, 12441, 12858, 12881, 12928, 13091, 13406, 13555, 13839, 14046-14085

C_8H_9BrO p-Bromophenetole. B.p. 234.2

3695, 8544, 10197, 10595, 10807, 11037, 12325, 12442, 13877, 13928, 14086-14095

C_8H_9Cl o,m,p-Chloroethylbenzene.

7464, 8726, 8886, 9013, 9429, 11405, 11922, 12135, 12155, 12233, 12422, 12537, 13810

C_8H_9N 2-Methyl-5-vinylpyridine.

14096

C_8H_{10} Ethylbenzene. B.p. 136.15

162, 676, 1080, 1405, 1695, 1861, 2106, 2225, 2595, 2687, 2751, 2804, 2924, 3206, 3356, 3466, 3696, 4144, 4321, 4597, 4644, 4707, 4753, 4835, 4891, 4940, 5046, 5082, 5218, 5630, 5706, 5711, 6160, 6266, 6290, 6402, 6517, 6596, 6864, 7082, 7214, 7328, 7465, 7631, 7808, 8023, 8097, 8185, 8251, 8283, 8371, 8463, 8545, 8729, 8783, 8861, 8930, 8945, 9130, 9341, 9504, 9536, 9595, 9765, 9786, 9803, 9871, 9905, 9912, 9936, 9949, 9991, 10463, 10532, 10943, 11183, 11243, 11393, 11639, 11788, 11802, 11828, 11908, 11927, 11937, 11982, 12007, 12054, 12156, 12283, 13029, 13464, 13621, 13630, 13662, 13710, 13723, 13754, 13807, 13888, 14097-14107, 15952, 16058, 16104, 16186, 16359, 16368, 16382, 16450, 16486, 16487, 16494, 16535, 16551, 16564, 1167a

C_8H_{10} m-Xylene. B.p. 139

677, 1081, 1406, 1596, 1696, 1862, 2107, 2596, 2689, 2805, 2925, 3034, 3207, 3357, 3468, 3697, 3895, 4145, 4322, 4598, 4645, 4708, 4754, 4941, 5083, 5219, 5283, 5446, 5631, 5707, 5712, 5719, 5818, 5832, 5978, 6161,

Formula   Name and System No.

C_8H_{10} *m*-Xylene (*continued*)
                6404, 6518, 6597, 6704, 6865, 6867, 7044, 7083, 7215, 7466, 7633, 7809,
                8187, 8252, 8372, 8402, 8464, 8730, 8784, 8862, 8961, 9014, 9141, 9180,
                9342, 9458, 9505, 9537, 9596, 1683, 9725, 9804, 9872, 9937, 9992, 10464,
                10533, 10944, 11184, 11244, 11394, 11424, 11482, 11640, 11668, 11729,
                11789, 11829, 12008, 12055, 12110, 12157, 12234, 12294, 12400, 12423,
                13058, 13465, 13622, 13631, 13711, 13717, 13724, 13735, 13740, 13747,
                13755, 13811, 13832, 13889, 13890, 14097, 14108-14116, 15875, 15953,
                16105, 16231, 16461, 16462, 1167b, 5893b

C_8H_{10} *o*-Xylene. B.p. 143.6
                1697, 1863, 2108, 2457, 2597, 2688, 3035, 3208, 3469, 3555, 3698, 4146,
                4323, 4599, 4646, 4709, 5084, 5284, 5632, 5819, 5833, 5979, 6403, 6519,
                6598, 6665, 6705, 7084, 7216, 7467, 7634, 8057, 8188, 8373, 8465, 8546,
                8785, 8863, 9343, 9459, 9538, 9597, 9805, 9873, 9993, 10945, 11185,
                11367, 11425, 11483, 11730, 12009, 12111, 12158, 12235, 12401, 12538,
                13059, 13233, 13623, 13663, 13671, 13678, 13718, 13725, 13736, 13741,
                13748, 13891, 14108, 14117, 14118, 16369, 1167c, 5893c, 6867a

C_8H_{10} *p*-Xylene. B.p. 138.4
                1698, 2109, 2690, 2926, 3209, 3358, 3470, 3699, 4147, 4324, 4647, 4942,
                5085, 5220, 5633, 5894, 6405, 6520, 6599, 6601, 6866, 6949, 7217, 7468,
                7594, 7635, 7810, 8189, 8284, 8374, 8466, 8547, 8786, 8864, 9181, 9344,
                9506, 9539, 9598, 9874, 9994, 10534, 11186, 11426, 11731, 11830,
                12010, 12056, 12159, 12401, 13030, 13712, 13726, 13737, 13756, 13808,
                14098, 14109, 14119, 14120, 16370, 16544, 16566, 1167d

$C_8H_{10}O$ Benzyl methyl ether. B.p. 167.8
                1407, 2458, 3211, 3471, 3556, 3700, 4325, 4600, 4981, 5119, 5221, 5285,
                5634, 6162, 6961, 7085, 7154, 7469, 7636, 7748, 7811, 8467, 8787, 9015,
                9069, 9345, 9460, 9875, 10046, 10392, 10465, 10946, 11187, 11484,
                11526, 11574, 11648, 11732, 11956, 12011, 12073, 12099, 12160, 12236,
                12295, 12473, 12539, 12786, 13455, 13632, 13785, 13812, 14121-14132

$C_8H_{10}O$ *p*-Ethylphenol. B.p. 218.8
                4327, 6163, 6880, 7336, 10199, 10253, 10738, 11859, 12540, 12859,
                12883, 13092, 13496, 13898, 14046, 14134-14160

$C_8H_{10}O$ *m*-Methylanisole. B.p. 177.2
                6164

$C_8H_{10}O$ *p*-Methylanisole. B.p. 177.05
                2459, 3701, 4326, 4601, 4717, 5120, 5222, 5286, 5635, 6165, 7155, 7637,
                8468, 8788, 9070, 9346, 9461, 9995, 10343, 10393, 10680, 10947, 11188,
                11296, 11427, 11485, 11527, 11604, 11734, 11957, 12012, 12074, 12161,
                12237, 12296, 12541, 12678, 13234, 13456, 13471, 13633, 13760, 13813,
                14161-14182

$C_8H_{10}O$ α-Methylbenzyl alcohol. B.p. 203.4
                678

$C_8H_{10}O$ Phenethyl alcohol. B.p. 219.5
                3702, 4328, 5980, 6166, 6706, 7749, 8548, 10047, 10129, 10200, 10596,
                10739, 10808, 11265, 12860, 12929, 12985, 13161, 13324, 13407, 13517,
                14047, 14134, 14183-14220

$C_8H_{10}O$ Phenetole. B.p. 170.4
                679, 2460, 2691, 3036, 3472, 3557, 3703, 4329, 4523, 4602, 4982, 5121,
                5223, 5636, 5981, 6167, 7086, 7156, 7470, 7638, 8469, 8789, 8887, 8962,
                9016, 9071, 9116, 9347, 9462, 9540, 9876, 9996, 10344, 10394, 10558,
                10681, 10948, 11189, 11428, 11486, 11528, 11575, 11649, 11656, 11733,
                11958, 12013, 12075, 12112, 12162, 12238, 12297, 12474, 12542, 12679,
                12750, 12787, 12810, 13235, 13457, 13634, 13657, 13761, 13786, 13814,
                14221-14252, 16533

Formula	Name and System No.

$C_8H_{10}O$ 2,4-Dimethylphenol. B.p. 210.5
3704, 6168, 6881, 7337, 7750, 10201, 10597, 12714, 12861, 12930, 13162, 13408, 13787, 13899, 13960, 13988, 14135, 14253-14264

$C_8H_{10}O$ 2,6-Dimethylphenol.
4325a, 6665a, 6679a

$C_8H_{10}O$ 3,4-Xylenol. B.p. 226.8
3705, 4330, 5982, 6169, 6707, 7338, 8549, 8982, 10198, 10254, 10297, 10598, 10740, 10809, 11038, 11328, 11860, 12348, 12585, 12734, 12862, 12882, 12931 12984, 13093, 13497, 13556, 13683, 13900, 13945, 14048, 14086, 14183, 14265-14306

$C_8H_{10}O$ 3,5-Xylenol.
14307

$C_8H_{10}O_2$ m-Dimethoxybenzene. B.p. 214.7
3706, 6708, 10202, 10600, 10741, 10810, 11329, 11861, 12443, 12863, 12932, 13094, 13163, 13325, 13409, 13518, 13571, 13946, 14049, 14308-14319

$C_8H_{10}O_2$ m-Ethoxyphenol. B.p. 243.8
14320

$C_8H_{10}O_2$ o-Ethoxyphenol. B.p. 216.5
2630, 3707, 4331, 5983, 6709, 7339, 10203, 10601, 10742, 11190, 11330, 11862, 12933, 13095, 13326, 13472, 13498, 13684, 13901, 14050, 14265, 14321-14343

$C_8H_{10}O_2$ 2-Phenoxyethanol. B.p. 245.2
5139, 8550, 10204, 13096, 13840, 14051, 14137, 14184, 14266, 14321, 14344-14355

$C_8H_{10}O_2$ Veratrole. B.p. 205.5
680, 3708, 4332, 5140, 6170, 9072, 10599, 10743, 10811, 10949, 11191, 11331, 11429, 11529, 11576, 11863, 12864, 12934, 13097, 13164, 13236, 13327, 13902, 13961, 13989, 14136, 14356-14366

$C_8H_{11}ClSi$ Dimethylphenylchlorosilane. B.p. 194.6
13048

$C_8H_{11}N$ 2,4,6-Collidine. B.p. 171
681, 4334, 10950, 13165, 13237, 14416, 14417, 16414, 16416, 16548

$C_8H_{11}N$ N,N-Dimethylaniline. B.p. 194.05
1699, 3212, 3709, 4333, 4603, 5984, 6666, 8888, 9997, 10048, 10130, 10159, 10345, 10395, 10682, 10744, 10951, 11192, 11678, 11735, 12163, 12239, 12298, 13349, 12424, 12475, 12507, 12543, 12634, 12654, 12680, 12715, 12751, 12865, 12935, 13098, 13166, 13238, 13328, 13410, 13473, 13635, 13815, 13903, 14161, 14185, 14221, 14322, 14356, 14367-14394

$C_8H_{11}N$ Ethylaniline. B.p. 205.5
682, 3712, 4337, 4605, 5987, 10131, 10205, 10683, 10745, 10952, 11679, 12476, 12508, 12936, 13099, 13167, 13239, 13330, 13411, 13499, 13904, 14138, 14186, 14268, 14323, 14357, 14395-14415

$C_8H_{11}N$ α-Methylbenzylamine. B.p. 188.6
683

$C_8H_{11}N$ 5-Ethyl-2-methylpyridine. B.p. 178.3
684, 13031, 14096

$C_8H_{11}N$ ar-Methyl-1,2,3,6-tetrahydrobenzonitrile. B.p. 205.4
685

$C_8H_{11}N$ 2,4-Xylidine. B.p. 214.0
3710, 4335, 4604, 5985, 10746, 12866, 12937, 13100, 13240, 13329, 13412, 13905, 14187, 14267, 14308, 14324, 14418-14429

Formula   Name and System No.

$C_8H_{11}N$ 3,4-Xylidine. B.p. 225.5
                     3711, 4336, 5986, 10206, 10747, 12938, 13101, 14188, 14309, 14430-
                     14435

$C_8H_{11}NO$ o-Phenetidine. B.p. 232.5
                     3713, 4338, 5988, 8551, 10207, 10255, 10277, 10298, 10812, 11039,
                     11095, 12884, 12939, 12986, 13102, 14052, 14139, 14189, 14269, 14436-
                     14454

$C_8H_{11}NO$ p-Phenetidine. B.p. 249.9
                     3714, 4339, 8552, 10256, 10278, 10299, 11040, 11096, 12586, 12885,
                     12987, 14270, 14455-14474

C_8H_{12} 4-Vinylcyclohexene.
                     6602, 2805c

C_8H_{12} 1,3,7-Octatriene. B.p. 125
                     2805b

C_8H_{12} 1,3-trans-6-cis-octatriene. B.p. 132
                     2805a

$C_8H_{12}N_2O_2$ Hexamethylenediisocyanate.
                     10346

$C_8H_{12}O$ 2-Methyl-1,2,3,6-tetrahydrobenzaldehyde. B.p. 176.4
                     686

$C_8H_{12}O_2$ 3,4-Dihydro-2,5-dimethyl-2-H-pyran-2-carboxaldehyde. B.p. 176.9
                     687

$C_8H_{12}O_4$ Ethyl fumarate. B.p. 217.85
                     688, 3715, 4340, 5989, 6171, 6710, 7751, 8553, 10208, 10602, 10748,
                     10813, 12886, 12940, 13168, 13241, 13331, 13685, 13841, 14053, 14140,
                     14190, 14253, 14271, 14310, 14358, 14475-14501

$C_8H_{12}O_4$ Ethyl maleate. B.p. 223.3
                     689, 2692, 3716, 4341, 5990, 6172, 6711, 8554, 8983, 10603, 10749,
                     10814, 11041, 12887, 12941, 13169, 13242, 13332, 13686, 13842, 14054,
                     14141, 14254, 14272, 14311, 14325, 14502-14525

C_8H_{14} Diisobutylene. B.p. 101
                     690, 6406, 8253, 16080, 16166, 16207, 16492

$C_8H_{14}O$ Bicyclo(2.2.1)heptane-2-methanol. B.p. 203.9
                     691

$C_8H_{14}O$ Cyclohexyl vinyl ether.
                     11736

$C_8H_{14}O$ Diisobutylene oxide.
                     692

$C_8H_{14}O$ 2-Ethyl-2-hexenal. B.p. 176
                     693, 2110

$C_8H_{14}O$ 2-Methallyl ether. B.p. 134.6
                     694, 7406

$C_8H_{14}O$ Methylheptenone. B.p. 173.2
                     2461, 3717, 4342, 5224, 5991, 6173, 7471, 8790, 9017, 9073, 9348, 10396,
                     10953, 11193, 11430, 11487, 11737, 12299, 12544, 12681, 12752, 12788,
                     12811, 13243, 13636, 13816, 14121, 14222, 14526-14544

$C_8H_{14}O$ 2-Octenal.
                     695

$C_8H_{14}O$ 2,4,6-Trimethyl-5,6-dihydro-1,2-pyran.
                     16234

$C_8H_{14}O_2$ Butyl methacrylate.
                     8190, 8933

$C_8H_{14}O_2$ Acetaldehyde diallyl acetal. B.p. 150.9
                     696

Formula   Name and System No.

$C_8H_{14}O_2$ Cyclohexyl acetate. B.p. 177.0
                  3213
$C_8H_{14}O_2$ 2-Ethyl-3-hexenoic acid. B.p. 231.8
                  697
$C_8H_{14}O_2$ Vinyl 2-methylvalerate. B.p. 148.8
                  698
$C_8H_{14}O_3$ Bis(2-vinyloxyethyl) ether. B.p. 198.7/10 mm
                  699, 14545
$C_8H_{14}O_3$ 2-Ethoxyethyl methacrylate.
                  8934
$C_8H_{14}O_3$ Butyl acetoacetate. B.p. 213.9
                  700 ·
$C_8H_{14}O_4$ meso-2,3-Butanediol diacetate. B.p. 190
                  3214, 8401
$C_8H_{14}O_4$ Ethyl succinate. B.p. 217.25
                  701, 4343, 4773, 6174, 8555, 10209, 10604, 10750, 10815, 12509, 12888,
                  12942, 13244, 13333, 13687, 14326, 14546-14571
$C_8H_{14}O_4$ Propyl oxalate. B.p. 214
                  10210, 10751, 11332, 11864, 12510, 12635, 12867, 13572, 13688, 13906,
                  13962, 14475, 14502, 14572-14578
$C_8H_{15}N$ 2(Aminomethyl)bicyclo(2.2.1)heptane. B.p. 185.9
                  702
$C_8H_{15}N$ Caprylonitrile. B.p. 205.2
                  14579
$C_8H_{15}N$ Dimethallylamine. B.p. 149.0
                  703
C_8H_{16}.   1,1-Dimethylcyclohexane. B.p. 119.54
                  4148
C_8H_{16} trans-1,2-Dimethylcyclohexane. B.p. 123.02
                  6407, 8470
C_8H_{16} 1,3-Dimethylcyclohexane. B.p. 120.7
                  163, 1082, 1700, 1864, 2111, 2226, 2368, 2598, 2927, 3215, 3359, 3473,
                  3940, 4149, 4151, 4344, 4535, 4836, 4892, 4943, 5637, 5708, 5834, 6408,
                  6521, 6603, 7045, 7087, 7472, 7639, 7844, 8053, 8098, 8191, 8285, 8375,
                  8471, 8791, 8865, 8901, 8931, 9182, 9239, 9397, 9507, 9541, 9599, 9641,
                  9707, 9767, 9806, 9877, 9938, 9998, 10535, 11194, 11395, 11641, 11738,
                  11790, 11803, 11817, 11831, 11909, 11919, 11938, 11978, 12409, 12413,
                  13032, 13033, 13727, 14580-14581
C_8H_{16} trans-1,4-Dimethylcyclohexane. B.p. 119.35
                  4152
C_8H_{16} cis-1,4-Dimethylcyclohexane.
                  4150
C_8H_{16} 1-Ethyl-1-methylcyclopentane. B.p. 121.52
                  4153
C_8H_{16} Ethylcyclohexane. B.p. 131.8
                  3216, 7021, 7845, 8472, 12284, 13873, 14099, 14582, 16535
C_8H_{16} cis-1-Ethyl-2-methylcyclopentane. B.p. 128.05
                  6409
C_8H_{16} trans-1-Ethyl-2-methylcyclopentane. B.p. 121.2
                  6410, 6604
C_8H_{16} trans-1-Ethyl-3-methylcyclopentane. B.p. 120.8
                  6411, 6605
C_8H_{16} 6-Methyl-1-heptene.
                  9878

Formula   Name and System No.

C_8H_{16} 1-Octene. B.p. 121.6
                     704, 1865, 2806, 14100, 15886, 1167e

C_8H_{16} 2-Octene. B.p. 125.2
                     2807, 8473

C_8H_{16} 1,1,2-Trimethylcyclopentane. B.p. 113.73
                     6412, 8902

C_8H_{16} 1,1,3-Trimethylcyclopentane. B.p. 104.9
                     6413, 6606, 13034

C_8H_{16} cis,cis,trans-1,2,4-Trimethylcyclopentane.
                     6416

C_8H_{16} 1-trans-2-cis-4-Trimethylcyclopentane. B.p. 109.29
                     4156

C_8H_{16} 1-cis-2-trans-4-Trimethylcyclopentane. B.p. 116.73
                     6415

C_8H_{16} cis,trans,cis-1,2,3-Trimethylcyclopentane. B.p. 110.4
                     8595, 13035

C_8H_{16} 1-trans-2-cis-3-Trimethylcyclopentane. B.p. 110.2
                     6608

C_8H_{16} 1-cis-2-trans-3-Trimethylcyclopentane. B.p. 117.5
                     4154, 6414

C_8H_{16} cis,trans,cis-1,2,4-Trimethylcyclopentane. B.p. 105.3
                     4155, 13036

C_8H_{16} 1-cis-2-cis-3-Trimethylcyclopentane. B.p. 123.0
                     6607

C_8H_{16} 2,3,4-Trimethyl-2-pentene. B.p. 116
                     13037

C_8H_{16} 2,4,4-Trimethyl-1-pentene. B.p. 101.44
                     6609

C_8H_{16} 2,4,4-Trimethyl-2-pentene. B.p. 104.91
                     4157

$C_8H_{16}O$ Allyl isoamyl ether. B.p. 120
                     705

$C_8H_{16}O$ 2-Ethylhexaldehyde. B.p. 163.6
                     706, 12136

$C_8H_{16}O$ 2-Octanone. B.p. 172.85
                     2462, 3718, 4345, 5122, 5225, 5287, 5992, 6175, 6667, 7157, 7287, 7473,
                     8792 8963, 9018, 9117, 9349, 9463, 10397, 10490, 10559, 10954, 11195,
                     11488, 11739, 12014, 12076, 12113, 12164, 12300, 12545, 12655, 12682,
                     12753, 12812, 13060, 13245, 13573, 13605, 13637, 13762, 14122, 14162,
                     14223, 14583-14603

$C_8H_{16}O$ 2,2,5,5-Tetramethyltetrahydrofuran. B.p. 115
                     707

$C_8H_{16}O$ 2,4,4-Trimethyl-1,2-epoxypentane. B.p. 140.9
                     708

$C_8H_{16}O$ 2,4,4-Trimethyl-2,3-epoxypentane. B.p. 127.3
                     709

$C_8H_{16}OS$ 2-Butylthioethyl vinyl ether. B.p. 210.5
                     710, 12204

$C_8H_{16}O_2$ Amyl propionate.
                     5638

$C_8H_{16}O_2$ 2-Butoxyethyl vinyl ether.
                     711

Formula   Name and System No.

$C_8H_{16}O_2$ Butyl butyrate. B.p. 166.4
712, 1408, 1597, 2498, 2599, 2693, 3719, 4346, 5993, 6176, 6981, 7088, 7288, 7474, 7812, 8192, 8793, 8889, 9118, 9542, 9999, 10347, 10398, 10466, 10491, 10955, 11368, 11431, 11489, 11740, 12165, 12240, 12546, 12754, 12789, 12813, 13061, 14123, 14163, 14224, 14583, 14604-14621

$C_8H_{16}O_2$ Caprylic acid. B.p. 238.5
3720, 8984, 10211, 10300, 10752, 11042, 11865, 12889, 12943, 12988, 14273, 14476, 14503, 14546, 14622-14637

$C_8H_{16}O_2$ 1,3-Dimethylbutyl acetate. B.p. 146.1
14638

$C_8H_{16}O_2$ 2,3-Epoxy-2-ethylhexanol.
713

$C_8H_{16}O_2$ 2-Ethylbutyl acetate. B.p. 162.3
714, 12166

$C_8H_{16}O_2$ Ethyl caproate. B.p. 166.8
715, 2463, 2694, 6177, 7089, 7641, 9019, 9600, 10399, 10956, 11490, 12015, 12241, 12547, 12755, 12814, 13763, 13817, 14584, 14604, 14639-14649

$C_8H_{16}O_2$ 2-Ethylhexanoic acid. B.p. 227
716, 13170, 13334, 14650

$C_8H_{16}O_2$ Hexyl acetate. B.p. 171.5
717, 2464, 2695, 3721, 5123, 7090, 9074, 9351, 10400, 11491, 11530, 12016, 12242, 12548, 12756, 13764, 14225, 14585, 14651-14654

$C_8H_{16}O_2$ Isoamyl propionate. B.p. 160.3
718, 1409, 1598, 2465, 2600, 3722, 4347, 5124, 5226, 5994, 6178, 6982, 7046, 7091, 7475, 7640, 7813, 8794, 8890, 8964, 9020, 9119, 9543, 9601, 9726, 9879, 10049, 10401, 10467, 10957, 11369, 11432, 11492, 11741, 12017, 12114, 12167, 12243, 12757, 12790, 12815, 13062, 13638, 13765, 14124, 14164, 14226, 14605, 14655-14669

$C_8H_{16}O_2$ Isobutyl butyrate. B.p. 156.8
619, 1410, 2466, 2601, 3474, 3723, 4348, 5995, 6179, 7092, 7476, 7642, 8795, 9021, 9120, 9350, 9544, 9602, 9684, 9727, 10000, 10468, 11137, 11370, 11742, 12018, 12168, 12244, 12791, 13063, 13639, 13766

$C_8H_{16}O_2$ Isobutyl isobutyrate. B.p. 147.3
720, 1411, 2602, 3724, 4349, 4983, 5639, 6180, 7158, 7218, 7477, 7643, 7814, 8965, 9022, 9545, 9603, 9685, 9728, 10469, 11743, 12019, 12100, 12169, 12792, 13064, 13664, 13672, 13742, 13892, 14110, 14675-14678

$C_8H_{16}O_2$ Isooctanoic acid. B.p. 220
721

$C_8H_{16}O_2$ Methylisoamyl acetate.
3217

$C_8H_{16}O_2$ 4-Methyl-2-pentyl acetate. B.p. 146.1
722, 11396, 11804, 12197

$C_8H_{16}O_2$ Propyl isovalerate. B.p. 155.8
723, 2467, 3725, 4350, 5640, 5996, 6610, 7093, 7329, 7644, 7815, 8474, 8796, 8891, 8966, 9546, 9604, 10001, 10470, 11371, 11397, 11744, 12020, 12101, 12115, 12170, 12245, 12793, 13065, 13743, 13833, 13893, 14656, 14679-14689

$C_8H_{16}O_3$ 2-Butoxyethyl acetate. B.p. 192.2
724

$C_8H_{16}O_3$ 2,2-Diethoxy-3-butanone. B.p. 163.5
725

$C_8H_{16}O_3$ 2,5-Diethoxytetrahydrofuran. B.p. 173
726

Formula   Name and System No.

$C_8H_{16}O_3$ 2-Ethoxyethyl 2-vinyloxyethyl ether. B.p. 194.0
                   727

$C_8H_{16}O_3$ Isoamyl lactate. B.p. 202.4
                   3726, 10050, 10160, 10684, 10753, 10958, 11333, 12350, 12511, 12636,
                   12868, 13171, 13246, 13335, 13413, 13574, 13907, 13963, 14192, 14255,
                   14274, 14690-14701

$C_8H_{16}O_4$ 2(2-Ethoxyethoxy)ethyl acetate. B.p. 218.5
                   728, 4351, 11866, 12246, 13103, 13689, 13908, 13964, 14477, 14702-
                   14713

$C_8H_{17}Cl$ 1-Chloro-2-ethylhexane. B.p. 173
                   729

$C_8H_{17}Cl$ 3-(Chloromethyl)heptane.
                   12137, 14714

$C_8H_{17}N$ N-Ethylcyclohexylamine. B.p. 164.9
                   730

$C_8H_{17}N$ 5-Ethyl-2-methylpiperidine. B.p. 163.4
                   731

$C_8H_{17}N$ Methyl(methylcyclohexyl)amine.
                   732

$C_8H_{17}NO$ N,N-Dimethylhexaneamide.
                   11867

$C_8H_{17}NO$ 4-Ethyl-2,6-dimethylmorpholine. B.p. 158.1
                   733

C_8H_{18} 2,2-Dimethylhexane. B.p. 106.54
                   4158, 8626, 8703, 8898

C_8H_{18} 2,3-Dimethylhexane. B.p. 115.8
                   4159, 4536

C_8H_{18} 2,4-Dimethylhexane. B.p. 109.4
                   6612, 12410

C_8H_{18} 2,5-Dimethylhexane. B.p. 109.4
                   164, 1083, 1168, 1501, 1701, 1866, 2112, 2369, 2808, 2928, 3010, 3218,
                   3475, 3727, 3896, 3941, 4160, 4352, 4837, 4893, 5394, 5447, 5641, 5709,
                   5820, 6326, 6417, 6522, 6611, 6712, 7385, 7553, 7723, 7816, 7846, 7916,
                   8193, 8255, 8286, 8376, 8475, 8627, 8866, 8897, 8905, 9240, 9274, 9289,
                   9303, 9398, 9656, 9708, 9807, 9880, 9913, 11699, 12203, 12211, 12301,
                   13038, 13613

C_8H_{18} 3,3-Dimethylhexane. B.p. 111.93
                   8476, 8628

C_8H_{18} 3,4-Dimethylhexane. B.p. 117.9
                   4161

C_8H_{18} 3-Ethyl-3-methylpentane. B.p. 118.26
                   8487

C_8H_{18} Isooctane. B.p. 99.3
                   980, 8254, 11196, 13040, 16167, 16562, 734a

C_8H_{18} 2-Methyl-3-ethylpentane. B.p. 114
                   2809

C_8H_{18} 2-Methylheptane. B.p. 117.2
                   4162, 4537, 7847, 8896, 8904, 13039

C_8H_{18} 3-Methylheptane. B.p. 118.93
                   4163

C_8H_{18} 4-Methylheptane. B.p. 117.7
                   4164

Formula	Name and System No.

$C_8H_{18}O$ Octyl alcohol (*continued*)
12869, 12944, 13172, 13247, 13336, 13414, 13474, 13500, 13519, 13544, 13575, 13606, 13767, 13788, 13909, 13947, 13965, 13990, 14113, 14142, 14165, 14275, 14367, 14395, 14418, 14526, 14579, 14690, 14726-14757

$C_8H_{18}O$ *sec*-Octyl alcohol. B.p. 178.7
742, 1599, 2468, 3731, 4357, 4985, 5126, 5228, 5289, 5998, 6183, 7094, 7160, 7291, 7331, 7753, 9076, 9121, 10161, 10349, 10403, 10472, 10493, 10561, 10686, 10817, 10963, 11201, 11407, 11434, 11493, 11532, 11578, 11645, 12022, 12077, 12103, 12248, 12478, 12551, 12638, 12658, 12684, 12759, 12794, 12816, 13045, 13173, 13248, 13337, 13475, 13501, 13520, 13545, 13607, 13768, 13789, 13818, 13910, 14114, 14125, 14166, 14227, 14368, 14396, 14527, 14586, 14651, 14657, 14758-14786, 16426, 16520, 16525, 16541, 16549, 16550

$C_8H_{18}OS$ 2-Hexylthioethanol.
14787

$C_8H_{18}O_2$ 2-Ethyl-1,3-hexanediol. B.p. 243.1
744, 4172, 14788, 16019

$C_8H_{18}O_2$ Acetaldehyde dipropyl acetal. B.p. 147.7
743, 6526, 16100

$C_8H_{18}O_2$ 1-Butoxy-2-ethoxyethane. B.p. 164.2
745

$C_8H_{18}O_2$ Butyraldehyde diethyl acetal. B.p. 146.3
746

$C_8H_{18}O_2$ 5-Ethoxy-3-methylpentanol. B.p. 211.7
747

$C_8H_{18}O_2$ 2-Ethyl-3-methyl-1,5-pentanediol. B.p. 265.5
748

$C_8H_{18}O_2$ 2-Hexyloxyethanol. B.p. 208.1
749

$C_8H_{18}O_2$ 2(2-Methylpentyloxy)ethanol. B.p. 197.1
750

$C_8H_{18}O_3$ Bis(2-ethoxyethyl) ether. B.p. 186
752, 4359, 10404, 12351, 13174, 13338, 14545, 14789, 14790

$C_8H_{18}O_3$ 2(2-Butoxyethoxy)ethanol. B.p. 231.2
751, 4358, 5142, 6713, 8556, 10133, 10213, 10881, 12870, 12890, 14055, 14193, 14327, 14436, 14791-14802, 16307

$C_8H_{18}O_4$ 2(2(2-Ethoxyethoxy)ethoxy)ethanol.
8557

$C_8H_{18}O_4$ 1,2-Bis(2-methoxyethoxy)ethane.
753

$C_8H_{18}S$ Butyl sulfide. B.p. 185.0
3361, 3732, 4609, 4718, 6184, 7233, 7292, 7481, 8800, 9077, 9354, 10405, 10494, 10562, 10818, 10964, 11435, 11681, 11747, 12174, 12479, 12491, 12836, 12839, 13249, 13339, 13415, 13790, 14167, 14228, 14528, 14587, 14803-14810

$C_8H_{18}S$ Isobutyl sulfide. B.p. 172.0
3479, 3733, 4610, 6185, 6962, 7161, 7234, 7482, 8801, 8968, 9024, 9695, 10406, 10563, 10965, 11436, 12078, 12175, 12249, 12480, 12492, 12817, 13250, 14588, 14639, 14658, 14811-14815

$C_8H_{19}N$ Dibutylamine. B.p. 159.6
754, 8197, 16148, 16193, 16478

$C_8H_{19}N$ Diisobutylamine. B.p. 138.5
11401, 11791, 11807, 13067, 13624, 13665, 14104, 14115, 14725

$C_8H_{19}N$ 2-Ethylhexylamine. B.p. 169.1
755

Formula   Name and System No.

Formula   Name and System No.

$C_9H_{10}O$ p-Methylacetophenone. B.p. 226.3
                   3738, 4366, 5144, 6003, 6187, 6716, 10215, 10606, 10821, 11046, 12444,
                   12588, 12740, 12893, 12947, 13176, 13341, 13557, 13691, 14057, 14089,
                   14144, 14194, 14257, 14278, 14437, 14478, 14504, 14548, 14622, 14702,
                   14827, 14916-14934
$C_9H_{10}O$ Propiophenone. B.p. 217.7
                   3739, 4367, 4774, 5145, 6004, 6188, 10216, 10607, 10756, 10969, 11869,
                   12445, 12741, 12872, 12894, 12948, 13106, 13177, 13253, 13342, 13416,
                   13522, 13547, 13558, 13692, 14058, 14145, 14195, 14258, 14279, 14328,
                   14370, 14397, 14419, 14479, 14505, 14549, 14935-14945
$C_9H_{10}O_2$ Benzyl acetate. B.p. 214.9
                   762, 3740, 4368, 4775, 6005, 6717, 8559, 10053, 10134, 10217, 10608,
                   10757, 10822, 11870, 12446, 12513, 12760, 12895, 12949, 13107, 13178,
                   13254, 13343, 13417, 13693, 13844, 13911, 14059, 14146, 14196, 14259,
                   14280, 14312, 14329, 14359, 14480, 14506, 14572, 14703, 14935, 14946-
                   14962, 16538
$C_9H_{10}O_2$ 1,2-Epoxy-3-phenoxypropane. B.p. 244.4
                   763
$C_9H_{10}O_2$ Ethyl benzoate. B.p. 212.4
                   764, 1502, 2631, 3741, 4369, 4776, 5509, 6006, 6189, 6718, 7755, 8560,
                   10135, 10218, 10496, 10565, 10609, 10758, 10823, 10882, 11335, 11871,
                   12447, 12514, 12873, 12950, 13108, 13179, 13255, 13344, 13418, 13576,
                   13694, 13845, 13912, 13967, 14060, 14147, 14197, 14260, 14281, 14313,
                   14330, 14360, 14481, 14507, 14550, 14704, 14727, 14936, 14946, 14963-
                   14980
$C_9H_{10}O_2$ Methyl α-toluate. B.p. 215.3
                   765, 3742, 6007, 13109, 14061, 14963, 14979-14982
$C_9H_{10}O_3$ Ethyl salicylate. B.p. 233.7
                   3743, 4370, 4777, 6008, 6719, 8561, 8985, 10219, 10260, 10304, 11047,
                   12327, 12365, 12448, 12589, 12896, 12951, 12991, 13440, 13559, 13846,
                   14198, 14282, 14345, 14438, 14456, 14508, 14793, 14828, 14916, 14983-
                   15005
$C_9H_{11}N$ 5-Ethyl-2-vinylpyridine.
                   766
C_9H_{12} Cumene. B.p. 152.8
                   767, 1706, 1869, 2115, 2470, 2604, 2698, 3039, 3224, 3481, 3744, 4173,
                   4371 4612, 4711, 4755, 5087, 5230, 5449, 5647, 6009, 6190, 6422, 6528,
                   6617, 7097, 7484, 7649, 7819, 8199, 8381, 8403, 8482, 8803, 8969, 9356,
                   9549, 9608, 9886, 10006, 10409, 10474, 10970, 11203, 11372, 11534,
                   11669, 11749, 12023, 12079, 12116, 12177, 12251, 12552, 12796, 12819,
                   13069, 13257, 13625, 13640, 13666, 13680, 13729, 13744, 13750, 13819,
                   13834, 14105, 14231, 14676, 14681, 14719, 14759, 14822, 15006-15008,
                   16551, 5510a
C_9H_{12} m-Ethyltoluene. B.p. 161.3
                   10054, 11923, 14892
C_9H_{12} o-Ethyltoluene. B.p. 165.1
                   8483, 10055
C_9H_{12} p-Ethyltoluene. B.p. 162.0
                   10056
C_9H_{12} Mesitylene. B.p. 164.6
                   768, 1707, 1870, 2116, 2471, 2632, 2699, 2930, 3040, 3225, 3482, 3559,
                   3745, 4174, 4372, 4613, 4712, 4763, 5128, 5231, 5291, 5648, 6010, 6191,
                   6423, 6529, 6618, 6720, 7098, 7162, 7235, 7293, 7485, 7650, 7820, 8200,
                   8382, 8484, 8804, 8970, 9025, 9079, 9122, 9357, 9465, 9550, 9609, 9729,

Formula	Name and System No.

Formula	Name and System No.

$C_9H_{13}N$ *N,N*-Dimethyl-*p*-toluidine. B.p. 210.2
3752, 4379, 4618, 6016, 6671, 10062, 10165, 10221, 10761, 12354, 12953, 13113, 13182, 13262, 13347, 13914, 14148, 14199, 14285, 14332, 14765, 14937, 15103-15108

$C_9H_{13}NO$ 5-Ethyl-2-pyridine ethanol.
771

$C_9H_{14}O$ Isophorone. B.p. 215.2
772, 15109

$C_9H_{14}O$ Phorone. B.p. 197.8
3753, 4380, 6017, 6196, 7757, 9469, 10063, 10166, 10690, 10762, 10825, 10977, 11610, 11876, 12554, 12717, 12841, 13183, 13263, 13348, 13420, 13560, 13580, 13969, 13994, 14691, 14733, 14851, 15110-15118

$C_9H_{14}OSi$ Trimethylsiloxybenzene. B.p. 181.9
10978

$C_9H_{15}O$ 1-Methyl-2,5-endomethylenecyclohexane-1-methanol. B.p. 211.2
773

$C_9H_{15}N$ Triallylamine. B.p. 151.1
774

C_9H_{16} *cis*-Hexahydroindan. B.p. 167.7
12256

$C_9H_{16}O$ 5-Ethyl-3-heptene-2-one. B.p. 193.5
775

$C_9H_{16}O_2$ 2,2-Bis(allyloxy)propane.
16048, 16440

$C_9H_{16}O_2$ 2-Ethylbutyl acrylate.
12138

$C_9H_{16}O_2$ Hexyl acrylate.
12183

$C_9H_{16}O_4$ Dimethyl pimelate. B.p. 248.9
776

C_9H_{18} Nonanaphthene. B.p. 136.7
3227

C_9H_{18} Butylcyclopentane. B.p. 156.56
8487

C_9H_{18} Isobutylcyclopentane. B.p. 147.6
8488

C_9H_{18} Isopropylcyclohexane. B.p. 154.5
8489

C_9H_{18} 1-Nonene. B.p. 146.87
777, 1872, 8490, 15888

C_9H_{18} Propylcyclohexane. B.p. 156.72
8491, 15119

C_9H_{18} 1,1,3-Trimethylcyclohexane. B.p. 136.6
6620

$C_9H_{18}O$ 2,6-Dimethyl-4-heptanone. B.p. 168.0
778, 2118, 2474, 3228, 3362, 3754, 4381, 4683, 5234, 6018, 7236, 7490, 7654, 8059, 8971, 9027, 9361, 9553, 10011, 10414, 10979, 11402, 11442, 11755, 11808, 12026, 12118, 12184, 12555, 12799, 12823, 13264, 13645, 13772, 14611, 14638, 14643, 14661, 14805, 14812, 15120, 15121

$C_9H_{18}O$ 2-Ethylheptanal.
14116

Formula   Name and System No.

$C_9H_{18}O_2$ Butyl isovalerate. B.p. 177.6
                    779, 2256, 2702, 4382, 6197, 7294, 8808, 9083, 9470, 10691, 10980, 11494,
                    11538, 11582, 12482, 12688, 12718, 12764, 13265, 13349, 14171, 14236,
                    14766, 14806, 14816, 15035, 15122-15130

$C_9H_{18}O_2$ Ethyl enanthate. B.p. 188.7
                    780, 2703, 3755, 4383, 4525, 6198, 10612, 10981, 11539, 12689, 13266,
                    13350, 14734, 14852, 15131-15133

$C_9H_{18}O_2$ 2-Heptyl acetate. B.p. 176.4
                    781

$C_9H_{18}O_2$ 3-Heptyl acetate. B.p. 173.8
                    782

$C_9H_{18}O_2$ Isoamyl butyrate. B.p. 178.5
                    783, 2257, 2704, 3756, 4384, 4526, 5235, 5292, 6019, 6199, 7164, 7491,
                    7758, 9028, 9084, 9362, 9471, 10064, 10415, 10498, 10568, 10692, 10982,
                    11443, 11540, 11583, 12027, 12257, 12308, 12483, 12556, 12640, 12690,
                    12719, 12765, 12824, 13184, 13267, 13351, 13773, 13794, 14128, 14172,
                    14237, 14532, 14593, 14652, 14767, 14853, 15012, 15036, 15122, 15134-
                    15147

$C_9H_{18}O_2$ Isoamyl isobutyrate. B.p. 168.9
                    784, 2475, 3561, 3757, 4385, 4988, 5236, 6020, 6200, 7101, 7295, 7492,
                    8809, 9612, 10416, 10983, 11444, 11495, 12083, 12258, 12484, 12557,
                    12691, 14173, 14238, 14533, 14612, 14768, 14817, 15013, 15120, 15148-
                    15153

$C_9H_{18}O_2$ Isobutyl isovalerate. B.p. 171.2
                    785, 1600, 2258, 2476, 2705, 3758, 4386, 4989, 5129, 5237, 5293, 6021,
                    6201, 7165, 7296, 7493, 7759, 8810, 9029, 9085, 9124, 9363, 10065,
                    10356, 10417, 10477, 10569, 10984, 11445, 11496, 11541, 11756, 12028,
                    12084, 12104, 12119, 12185, 12259, 12307, 12485, 12558, 12661, 12692,
                    12767, 12800, 12825, 13071, 13268, 13646, 13774, 13821, 14129, 14174,
                    14239, 14534, 14594, 14613, 14644, 14769, 14807, 14818, 14854, 15014,
                    15029, 15121, 15123, 15134, 15148, 15154-15166

$C_9H_{18}O_2$ Isobutyl valerate. B.p. 171.35
                    10985

$C_9H_{18}O_2$ Methyl caprylate. B.p. 192.9
                    786, 2706, 3759, 4387, 6202, 11300, 11446, 11542, 11877, 13269, 13352,
                    13581, 13745, 15110, 15167-15170

$C_9H_{18}O_2$ Pelargonic acid. B.p. 254
                    4778, 10222, 11049, 12898, 15171-15186

$C_9H_{18}O_3$ β-(2-Ethylbutyoxy)propionic acid.
                    787

$C_9H_{18}O_3$ Isobutyl carbonate. B.p. 190.3
                    788, 2259, 2707, 4388, 5238, 6022, 6203, 6724, 10418, 10613, 10693,
                    10986, 11611, 11878, 12559, 12641, 12662, 12693, 12720, 12768, 13114,
                    13185, 13270, 13353, 13582, 13775, 13970, 13995, 14240, 14735, 14770,
                    14855, 15015, 15111, 15187-15194

$C_9H_{19}NO$ N,N-Dimethylheptamide.
                    13695

C_9H_{20} 3,3-Diethylpentane. B.p. 146.17
                    8492

C_9H_{20} 2-Methyloctane. B.p. 135.2
                    3229

Formula   Name and System No.

C_9H_{20} Nonane. B.p. 150.7
                        789, 1873, 2119, 2812, 3230, 5482, 5651, 5722, 6269, 6293, 7023, 8203,
                        8493, 8705, 8870, 11210, 11250, 12433, 13894, 14106, 14118, 14120,
                        14721 15006, 15195, 15889, 16056, 16149, 16209, 16372, 16376, 16382,
                        16494, 16564, 16565, 13462a

C_9H_{20} 2,2,3,3-Tetramethylpentane. B.p. 140.27
                        8494

C_9H_{20} 2,2,4,4-Tetramethylpentane. B.p. 122.28
                        8495

C_9H_{20} 2,3,3,4-Tetramethylpentane. B.p. 141.55
                        8496

C_9H_{20} 2,2,3,4-Tetramethylpentane. B.p. 133.02
                        6621

C_9H_{20} 2,2,3-Trimethylhexane. B.p. 133.60
                        8497

C_9H_{20} 2,2,4-Trimethylhexane. B.p. 126.54
                        8498

C_9H_{20} 2,3,3-Trimethylhexane. B.p. 137.68
                        8499

C_9H_{20} 2,2,5-Trimethylhexane. B.p. 120.1
                        2813, 9788, 9890, 14107, 1168c

C_9H_{20} 2,3,4-Trimethylhexane. B.p. 139.0
                        6622, 13875

C_9H_{20} 2,3,5-Trimethylhexane. B.p. 131.34
                        6623, 8500

C_9H_{20} 2,4,4-Trimethylhexane. B.p. 130.65
                        8501

C_9H_{20} 3,3,4-Trimethylhexane. B.p. 140.46
                        8502

$C_9H_{20}O$ 2,6-Dimethyl-4-heptanol. B.p. 178.1
                        790, 15196

$C_9H_{20}O_2$ Dibutoxymethane. B.p. 181.8
                        791, 8205, 16150

$C_9H_{20}O_2$ Diisobutoxymethane. B.p. 163.8
                        792, 8204, 8384, 9731, 14662, 16189

$C_9H_{20}O_2$ 2-Ethyl-2-butyl-1,3-propanediol.
                        793

$C_9H_{20}O_3$ 1(2-Butoxyethoxy)-2-propanol. B.p. 230.3
                        794

$C_9H_{20}O_3$ 2(2-Isoamyloxyethoxy)ethanol.
                        13876

$C_9H_{20}O_3$ 2-Methoxymethyl-2,4-dimethyl-1,5-pentanediol.
                        795

$C_9H_{20}O_3$ 1,1,3-Triethoxypropane.
                        796

$C_9H_{20}O_4$ Tripropylene glycol.
                        15197-15199

$C_9H_{21}BO_3$ Isopropyl borate. B.p. 140.8
                        6425

$C_9H_{21}N$ N-Methyldibutylamine. B.p. 163.1
                        797

$C_9H_{21}N$ Tripropylamine. B.p. 156
                        798

Formula   Name and System No.

Formula	Name and System No.

$C_{10}H_{10}O_2$ Safrole (*continued*)
13698, 13855, 13879, 13933, 14064, 14091, 14201, 14442, 14461, 14483, 14510, 14552, 14625, 14833, 14874, 14897, 14919, 14984, 15048, 15080, 15175, 15238, 15304, 15340-15364

$C_{10}H_{10}O_4$ Methyl phthalate. B.p. 283.2
805, 3770, 4395, 6031, 6731, 8572, 11106, 12371, 13856, 14023, 15203, 15218, 15289, 15365-15374

$C_{10}H_{12}$ 1,2,3,4-Tetrahydronaphthalene.
806

$C_{10}H_{12}O$ Anethole. B.p. 235.7
807, 3771, 4396, 6032, 6732, 8573, 8988, 10617, 11057, 11270, 12332, 12372, 12596, 13117, 13857, 14024, 14202, 14443, 14462, 14484, 14511, 14626, 14834, 14875, 14920, 15049, 15239, 15340, 15375-15387

$C_{10}H_{12}O$ Estragole. B.p. 215.6
808, 3772, 4397, 6733, 10224, 11337, 12356, 12451, 12737, 13188, 13422, 14965, 15388

$C_{10}H_{12}O_2$ Ethyl α-toluate. B.p. 228.75
809, 3773, 4398, 6033, 6205, 6734, 8574, 10225, 10307, 10499, 10618, 10827, 11058, 12452, 12998, 13189, 13355, 13699, 14065, 14203, 14288, 14485, 14553, 14627, 14921, 14985, 15050, 15240, 15341, 15389-15407

$C_{10}H_{12}O_2$ Eugenol. B.p. 255
3774, 4399, 6034, 6735, 11059, 11107, 11271, 12597, 12997, 13880, 14025, 14835, 15051, 15081, 15176, 15219, 15308, 15324, 15408-15421

$C_{10}H_{12}O_2$ Isoeugenol. B.p. 268.8
3775, 4400, 6736, 11108, 11272, 13881, 14026, 14876, 14898, 15220, 15325, 15422-15433

$C_{10}H_{12}O_2$ Propyl benzoate. B.p. 230.85
810, 3776, 4401, 4780, 6035, 6206, 6737, 8575, 8989, 10226, 10262, 10308, 10500, 10619, 10828, 11060, 12453, 12598, 12901, 12956, 12999, 13356, 13858 14066, 14150, 14204, 14289, 14512, 14922, 14986, 15052, 15241, 15342, 15375, 15389, 15434-15448

$C_{10}H_{12}O_3$ 2-Phenoxyethyl acetate. B.p. 260.6
811

$C_{10}H_{14}$ Butylbenzene. B.p. 183.1
2260, 2391, 2635, 2709, 3042, 3777, 4402, 4620, 5146, 5240, 5652, 6207, 6426, 6532, 6738, 7103, 7297, 9495, 7655, 7761, 8207, 8576, 8812, 9087, 9365, 9554, 9891, 10012, 10067, 10419, 10501, 10694, 10989, 11212, 11302, 11448, 11544, 11651, 11758, 12029, 12186, 12260, 12357, 12561, 12663, 12721, 13190, 13357, 13503, 13525, 13647, 14175, 14535, 14595, 14771, 14808, 15112, 15131, 15135, 15154, 15167, 15449-15458, 16551

$C_{10}H_{14}$ *sec*-Butylbenzene. B.p. 173.1
10068, 10990, 12261

$C_{10}H_{14}$ *tert*-Butylbenzene. B.p. 168.5
10069, 12262

$C_{10}H_{14}$ Cymene. B.p. 176.7
2120, 2261, 2392, 2477, 2636, 2710, 2931, 3231, 3485, 3778, 4176, 4403, 4621, 4765, 4990, 5130, 5147, 5241, 5295, 5653, 6036, 6208, 6427, 6531, 6624, 6739, 7104, 7166, 7298, 7496, 7656, 7823, 8206, 8385, 8503, 8577, 8813, 9030, 9088, 9125, 9366, 9473, 9555, 9892, 10013, 10070, 10420, 10570, 10695, 10991, 11213, 11449, 11497, 11545, 11652, 11658, 11759, 11881, 11964, 12030, 12085, 12120, 12263, 12310, 12562, 12694, 12722, 12769, 12826, 13118, 13191, 13272, 13358, 13479, 13526, 13584, 13648, 13776, 13795, 13822, 13972, 13997, 14176, 14241, 14375, 14536, 14596, 14614, 14663, 14737, 14772, 14856, 15016, 15030, 15037, 15093, 15136, 15149, 15155, 15449, 15459-15468, 16497

Formula   Name and System No.

Formula   Name and System No.

C₁₀H₁₆   Camphene. B.p. 159.6
              821, 1415, 1709, 2121, 2478, 2711, 3943, 3232, 3486, 3784, 4177, 4408,
              4623, 4991, 5242, 5296, 5450, 5654, 6042, 6209, 6428, 6533, 6625, 6744,
              7049, 7105, 7167, 7222, 7497, 7657, 7824, 8208, 8386, 8504, 8814, 8972,
              9031, 9089, 9126, 9367, 9474, 9556, 9613, 9687, 9732, 9893, 10014,
              10421, 10478, 10992, 11138, 11214, 11375, 11450, 11498, 11546, 11613,
              11659, 11670, 11760, 11835, 12031, 12045, 12086, 12121, 12187, 12264,
              12311, 12426, 12563, 12770, 12801, 12827, 13072, 13121, 13274, 13480,
              13585, 13627, 13649, 13730, 13777, 13796, 13824, 13836, 13998, 14130,
              14242, 14376, 14537, 14597, 14615, 14645, 14664, 14677, 14685, 14740,
              14774, 14813, 14819, 15017, 15025, 15094, 15124, 15137, 15150, 15156,
              15187, 15459, 15554-15560, 16601

C₁₀H₁₆   Dipentene. B.p. 177.7
              2479, 3044, 3785, 4626, 5131, 5148, 6043, 7168, 7299, 8822, 9032, 9475,
              10015, 10077, 10357, 10696, 11215, 11303, 11451, 11547, 11585, 11653,
              11965, 12265, 12312, 12359, 12664, 12828, 13481, 13586, 13650, 13778,
              13825, 14243, 14377, 14538, 14598, 14790, 14857, 15038, 15132, 15151,
              15168, 15188, 15460, 15554, 15561-15565

C₁₀H₁₆   d-Limonene. B.p. 177.8
              1710, 2122, 2262, 2393, 2712, 3786, 4178, 4409, 4527, 4992, 5243, 5297,
              5451, 5655, 6044, 6210, 6429, 6534, 6745, 7106, 7498, 7658, 7825, 8209,
              8387, 8815, 9033, 9090, 9368, 9557, 9894, 10422, 10571, 10993, 11216,
              11452, 11548, 11614, 11646, 11761, 11882, 11966, 12087, 12122, 12188,
              12313, 12564, 12695, 12723, 12771, 13122, 13194, 13275, 13361, 13482,
              13587, 13608, 13797, 13949, 13973, 13999, 14244, 14378, 14539, 14555,
              14575, 14599, 14616, 14741, 14775, 14858, 15125, 15138, 15157, 15189,
              15461, 15566-15572, 16350, 16351, 16415, 16427, 16429-16431, 16456,
              16496, 16498, 16504, 16514, 16517, 16521, 16524, 16525, 16527, 16528,
              16530, 16532, 16536, 16539, 16541, 16549

C₁₀H₁₆   Nopinene. B.p. 163.8
              1417, 2481, 2713, 3787, 4410, 5244, 5298, 5656, 6045, 6626, 6747, 7107,
              7169, 7237, 7499, 7659, 7826, 8210, 8388, 8505, 8817, 9091, 9369, 9476,
              9558, 9614, 9688, 9733, 10078, 10358, 10423, 10479, 10994, 11217,
              11376, 11453, 11549, 11836, 11883, 12032, 12046, 12088, 12123, 12189,
              12266, 12565, 12772, 12802, 12829, 13073, 13195, 13276, 13362, 13483,
              13588, 13731, 13779, 13798, 13826, 13837, 14000, 14131, 14245, 14379,
              14617, 14665, 14672, 14686, 14742, 14776, 15018, 15026, 15126, 15139,
              15462, 15556, 15573, 15574, 16457, 16529

C₁₀H₁₆   α-Phellandrene. B.p. 171.5
              2714, 5246, 5657, 7108, 7500, 7661, 9034, 9370, 9895, 10995, 11139,
              11218, 11454, 11762, 12089, 12566, 12773, 13277, 14777, 15575

C₁₀H₁₆   α-Pinene. B.p. 155.8
              1416, 2123, 2480, 2715, 3233, 3788, 4179, 4411, 4624, 4947, 4993, 5088,
              5132, 5247, 5299, 5452, 5658, 6046, 6430, 6535, 6627, 6746, 7050, 7109,
              7170, 7223, 7332, 7501, 7660, 7827, 8211, 8287, 8389, 8506, 8816, 8973,
              9035, 9092, 9127, 9371, 9477, 9559, 9616, 9689, 9734, 9810, 9896, 10016,
              10424, 10480, 10572, 10996, 11140, 11219, 11304, 11377, 11455, 11499,
              11550, 11615, 11660, 11671, 11763, 11837, 12033, 12047, 12090, 12124,
              12190, 12267, 12314, 12567, 12803, 12830, 13074, 13123, 13278 13363,
              13484, 13589, 13628, 13651, 13667, 13674, 13681, 13732, 13751, 13758,
              13799, 14246, 14380, 14540, 14576, 14600, 14618, 14646, 14666, 14673,
              14678, 14720, 14743, 14778, 14814, 14820, 15007, 15019, 15027, 15095,
              15127, 15133, 15152, 15158, 15190, 15555, 15561, 15576-15578, 16452,
              16454, 16458-16460, 16473, 16474, 16500, 16502, 16503, 16526, 16531

Formula | Name and System No.

C₁₀H₁₆ | α-Terpinene. B.p. 173.4
2268, 2482, 2716, 3234, 3487, 3789, 4180, 4412, 4625, 5245, 5300, 5659, 6047, 6211, 6431, 6536, 6748, 7111, 7171, 7238, 7502, 7662, 7828, 8507, 8818, 9093, 9372, 9478, 9615, 10017, 10079, 10359, 10425, 10573, 10697, 10997, 11220, 11305, 11378, 11456, 11551, 11616, 11764, 12034, 12091, 12125, 12191, 12268, 12568, 12665, 12696, 12724, 12774, 12804, 12831, 13075, 13124, 13196, 13279, 13364, 13485, 13590, 13652, 13780, 13827, 13974, 14001, 14132, 14177, 14247, 14381, 14541, 14601, 14619, 14647, 14674, 14688, 14744, 14779, 15020, 15140, 15159, 15450, 15463, 15562, 15573, 15576, 15579-15582

C₁₀H₁₆ | α-Terpinene. B.p. 183
3790, 4413, 5248, 5301, 6212, 7112, 7503, 8819, 9373, 9479, 10426, 11221, 11552, 11617, 11765, 12569, 12697, 12725, 12775, 13365, 13950, 13975, 14002, 14248, 14556, 14602, 14745, 14949, 15141, 15160, 15583-15585

C₁₀H₁₆ | Terpinene. B.p. 181.5
2718, 7110, 9036, 9094, 9374, 10427, 10698, 10998, 11457, 11766, 13125, 13280, 13366, 13591, 13800, 14780, 15566, 16518, 16537, 16540, 16550

C₁₀H₁₆ | Terpinolene. B.p. 185
2264, 2717, 3791, 4414, 5249, 5302, 5660, 6749, 7113, 7300, 7504, 8820, 9095, 9375, 9480, 9560, 9897, 10018, 10428, 10999, 11065, 11222, 11458, 11553, 11618, 11767, 11884, 12035, 12570, 12698, 12726, 12776, 13126, 13281, 13367, 13454, 13486, 13549, 13801, 14178, 14382, 14401, 14781, 15142, 15161, 15451, 15586, 15587

C₁₀H₁₆ | Terpinylene. B.p. 175
7114, 7505

C₁₀H₁₆ | Thymene. B.p. 179.7
1711, 2124, 2265, 3792, 4181, 4415, 5250, 5303, 5661, 6432, 6750, 7115, 7506, 7663, 8212, 8390, 8821, 9096, 9376, 9561, 9898, 10429, 11000, 11223, 11554, 11619, 11768, 11885, 12699, 12727, 12777, 13127, 13197, 13282, 13368, 14003, 14383, 14557, 14746, 14782, 15588-15593

C₁₀H₁₆O | Camphor. B.p. 209.1
3793, 4416, 5847, 6048, 6213, 6673, 8580, 10080, 10139, 10502, 10623, 10766, 10832, 11001, 11224, 11886, 11967, 12516, 12843, 12960, 13128, 13198, 13283, 13369, 13424, 13504, 13527, 13550, 13592, 13951, 13976, 14261, 14292, 14336, 14384, 14402, 14421, 14488, 14558, 14695, 14747, 14950, 14968, 15096, 15246, 15388, 15507, 15538, 15594-15603

C₁₀H₁₆O | Carvenone. B.p. 234.5
3794, 11002, 14447, 14466, 15394, 15437, 15475, 15508, 15604, 15605

C₁₀H₁₆O | Citral. B.p. 226
15247, 15438, 15539, 15606, 15607

C₁₀H₁₆O | Dicyclopentenol.
822

C₁₀H₁₆O | Fenchone. B.p. 193
2266, 3235, 3795, 6214, 10699, 10767, 10833, 11003, 11225, 11459, 12571, 12642, 12728, 12844, 13284, 13370, 13425, 14385, 15039, 15608

C₁₀H₁₆O | Menthenone. B.p. 222.5
15348, 15409, 15476

C₁₀H₁₆O | Pulegone. B.p. 223.8
3796, 4417, 4783, 6049, 10232, 10768, 11066, 12742, 12906, 12961, 13371, 13505, 14070, 14209, 14337, 14448, 14489, 14516, 14559, 14951, 14979, 14990, 15248, 15349, 15380, 15395, 15439, 15477, 15509, 15540, 15609-15617

Formula	Name and System No.

$C_{10}H_{16}O$ Trimethyltetrahydrobenzaldehyde. B.p. 204.5
823

$C_{10}H_{16}O_4$ Diisopropyl maleate. B.p. 228.7
824

$C_{10}H_{17}Cl$ Bornyl chloride. B.p. 207.5
3797, 7507, 10624, 10769, 11620, 11887, 12962, 13285, 13372, 13593, 13977, 14577, 14696, 14952, 14969, 15249, 15609, 15618

$C_{10}H_{18}$ Decahydronaphthalene.
15619

$C_{10}H_{18}$ m-Menthene-8. B.p. 170.8
2483, 4418, 7116, 9037, 9097, 11004, 11141, 11226, 11460, 11555, 11769, 12778, 12832, 15031, 15162, 16533

$C_{10}H_{18}O$ Borneol. B.p. 213.4
3798, 4419, 4784, 6050, 6215, 7762, 10081, 10140, 10233, 10625, 10770, 10834, 11005, 11306, 11338, 12517, 12907, 12963, 13003, 13129, 13199, 13286, 13373, 13426, 13487, 13506, 13528, 13551, 13918, 13952, 13978, 14004, 14071, 14210, 14293, 14314, 14338, 14364, 14386, 14403, 14490, 14560, 14706, 14859, 14942, 14953, 14970, 14980, 14991, 15097, 15250, 15350, 15452, 15493, 15510, 15541, 15567, 15577, 15588, 15595, 15610, 15620-15626

$C_{10}H_{18}O$ Cineol. B.p. 176.35
825, 1601, 2267, 2484, 2719, 3799, 4420, 4627, 5304, 5662, 6051, 6216, 7117, 7172, 7301, 7508, 7664, 7763, 8508, 8823, 8974, 9038, 9098, 9377, 9481, 10019, 10082, 10360, 10430, 11006, 11227, 11307, 11461, 11500, 11556, 11621, 11661, 11770, 11968, 12036, 12092, 12192, 12269, 12315, 12333, 12360, 12427, 12486, 12572, 12666, 12700, 12729, 12779, 12833, 13287, 13374, 13459, 13488, 13529, 13653, 13781, 13802, 13828, 14005, 14179, 14249, 14387, 14542, 14603, 14620, 14648, 14653, 14667, 14748, 14783, 14809, 14860, 15032, 15098, 15128, 15143, 15153, 15163, 15191, 15453, 15464, 15568, 15575, 15579, 15583, 15589, 15627-15630, 16499

$C_{10}H_{18}O$ Citronellal. B.p. 208.0
3800, 4421, 6052, 7764, 9482, 10083, 10771, 11308, 11339, 11888, 12270, 12518, 12643, 12738, 12964, 13130, 13200, 13288, 13375, 13427, 13919, 13953, 13979, 14006, 14133, 14152, 14211, 14294, 14491, 14517, 14697, 14749, 14954, 14971, 15040, 15454, 15596, 15631-15634

$C_{10}H_{18}O$ Geraniol. B.p. 229.6
3801, 4422, 6053, 10084, 10234, 10626, 10772, 11067, 11340, 12908, 12965, 13004, 13201, 13376, 13428, 13887, 14072, 14350, 14404, 14492, 14561, 14707, 14797, 14926, 14992, 15058, 15104, 15251, 15351, 15396, 15440, 15478, 15494, 15511, 15533, 15542, 15606, 15635-15642

$C_{10}H_{18}O$ Linalool. B.p. 199
826, 2268, 3802, 4423, 5251, 6054, 6217, 8824, 9099, 10361, 10627, 10700, 10773, 11009, 11228, 11341, 11557, 11622, 11662, 12271, 12701, 12730, 12780, 12966, 13202, 13377, 13429, 13489, 13530, 13594, 13609, 13803, 13920, 13954, 13980, 14007, 14073, 14295, 14388, 14405, 14698, 14750, 14861, 14972, 15021, 15041, 15074, 15099, 15144, 15192, 15252, 15455, 15465, 15512, 15543, 15557, 15569, 15580, 15590, 15597, 15643-15645

$C_{10}H_{18}O$ Menthone. B.p. 209.5
4424, 6674, 10141, 10774, 10835, 11010, 12519, 12644, 13131, 13378, 13490, 13507, 13552, 14406, 14751, 15479, 15513, 15598, 15620, 15646

Formula   Name and System No.

C$_{10}$H$_{18}$O   α-Terpineol. B.p. 217.8
                    3803, 4425, 6055, 6218, 6751, 8581, 10085, 10235, 10628, 10775, 10836,
                    11011, 11342, 12909, 12967, 13005, 13132, 13290, 13379, 13430, 13508,
                    13531, 13955, 14074, 14153, 14212, 14296, 14315, 14389, 14407, 14449,
                    14493, 14518, 14708, 14839, 14927, 14955, 14973, 14981, 14993, 15253,
                    15352, 15381, 15397, 15441, 15466, 15480, 15514, 15544, 15591, 15611,
                    15621, 15627, 15631, 15635, 15647-15652

C$_{10}$H$_{18}$O   β-Terpineol. B.p. 210.5
                    3804, 4426, 6219, 7341, 10629, 10776, 10837, 11012, 11229, 12968,
                    13006, 13291, 13380, 13491, 13509, 13532, 13921, 13981, 14008, 14862,
                    14956, 14974, 15100, 15254, 15545, 15653-15655

C$_{10}$H$_{18}$O$_2$ Vinyl 2-ethylhexanoate. B.p. 185.2
                    827

C$_{10}$H$_{18}$O$_2$ Vinyl octanoate.
                    828

C$_{10}$H$_{18}$O$_4$ Propyl succinate. B.p. 250.5
                    4427, 11109, 12603, 14027, 15178, 15223, 15255, 15310, 15353, 15481,
                    15515, 15656-15662

C$_{10}$H$_{19}$N   Bornylamine. B.p. 199.8
                    10362, 12702

C$_{10}$H$_{20}$ Butylcyclohexane. B.p. 180.95
                    12272

C$_{10}$H$_{20}$ sec-Butylcyclohexane. B.p. 179.3
                    12273

C$_{10}$H$_{20}$ tert-Butylcyclohexane. B.p. 171.5
                    8509

C$_{10}$H$_{20}$ 1-Decene. B.p. 172.0
                    829, 1874, 2814, 15890

· C$_{10}$H$_{20}$ Isobutylcyclohexane. B.p. 171.3
                    12274

C$_{10}$H$_{20}$ cis-1-Methyl-4-isopropylcyclohexane. B.p. 172.7
                    12275

C$_{10}$H$_{20}$ trans-1-Methyl-4-isopropylcyclohexane. B.p. 170.5
                    12276

C$_{10}$H$_{20}$O   Citranellol. B.p. 224.4
                    3805, 4428, 6057, 8582, 10236, 10630, 10777, 10838, 11013, 11110, 12910,
                    12969, 13381, 13936, 14075, 14390, 14408, 14433, 14519, 14709, 14928,
                    14957, 14994, 15059, 15256, 15354, 15382, 15398, 15410, 15442, 15495,
                    15516, 15534, 15546, 15632, 15663-15667

C$_{10}$H$_{20}$O   2-Ethylhexyl vinyl ether. B.p. 177.7
                    830, 16192

C$_{10}$H$_{20}$O   Menthol. B.p. 216.4
                    3806, 4429, 5149, 6056, 6220, 6752, 7765, 10086, 10142, 10237, 10631,
                    10778, 10839, 11014, 11273, 11623, 12520, 12874, 12911, 12970, 13007,
                    13133, 13203, 13292, 13382, 13431, 13492, 13510, 13533, 13553, 13563,
                    13922, 13956, 13982, 14076, 14213, 14297, 14316, 14339, 14391, 14409,
                    14422, 14450, 14494, 14562, 14840, 14863, 14929, 14958, 14975, 14982,
                    14995, 15257, 15383, 15399, 15443, 15456, 15496, 15517, 15547, 15570,
                    15592, 15599, 15612, 15622, 15633, 15646, 15647, 15668-15674

C$_{10}$H$_{20}$O   Octyl vinyl ether.
                    14752

C$_{10}$H$_{20}$OS   2-Hexylthioethyl vinyl ether.
                    4430, 14787

Formula   Name and System No.

$C_{10}H_{20}O_2$ Capric acid. B.p. 268.8
                      11068, 11111, 15224, 15258, 15311, 15327, 15675-15679

$C_{10}H_{20}O_2$ Ethyl octanoate. B.p. 208.35
                      833, 3807, 4431, 6058, 6221, 6753, 7767, 10632, 10840, 11309, 11624,
                      12645, 12875, 13134, 13293, 13383, 13432, 13983, 14009, 14077, 14298,
                      14340, 14699, 14710, 14753, 15259, 15668, 15680-15682

$C_{10}H_{20}O_2$ 2-Ethylbutyl butyrate. B.p. 199.6
                      831

$C_{10}H_{20}O_2$ 2-Ethylhexyl acetate. B.p. 198.4
                      832, 14722

$C_{10}H_{20}O_2$ Isoamyl isovalerate. B.p. 193.5
                      834, 2269, 2720, 3808, 4432, 4528, 5252, 5305, 5663, 6060, 6222, 6754,
                      7118, 7766, 9039, 9100, 9378, 9483, 10087, 10167, 10363, 10503, 10633,
                      10701, 11015, 11310, 11343, 11462, 11558, 11586, 11625, 11889, 11969,
                      12487, 12521, 12703, 12731, 13135, 13204, 13294, 13384, 13595, 13782,
                      13957, 13984, 14180, 14250, 14262, 14365, 14754, 14864, 15042, 15115,
                      15169, 15260, 15571, 15584, 15586, 15593, 15628, 15643, 15683-15687

$C_{10}H_{20}O_2$ Isoamyl valerate. B.p. 192.7
                      3809

$C_{10}H_{20}O_2$ Methyl pelargonate. B.p. 213.5
                      836, 3810, 4433, 6059, 6223, 6755, 10238, 10841, 11890, 12971, 13205,
                      13433, 13923, 14495, 14563, 14943, 15518, 15688

$C_{10}H_{20}O_2$ 4-Methyl-2-pentyl butyrate. B.p. 182.6
                      835

$C_{10}H_{20}O_3$ 2-Butoxyethyl 2-vinyloxyethyl ether. B.p. 226.7
                      837

$C_{10}H_{20}O_3$ 2,2-Dipropoxy-3-butanone. B.p. 196
                      838

$C_{10}H_{20}O_4$ 2(2-Butoxyethoxy)ethyl acetate. B.p. 245.3
                      839, 12604, 14902, 15060, 15328, 15400, 15444, 15689-15692

$C_{10}H_{21}Cl$ Chlorodecane. B.p. 210.6
                      840

$C_{10}H_{21}N$ N-Butylcyclohexylamine. B.p. 209.5
                      841

$C_{10}H_{21}NO$ N,N-Dimethyloctanamide.
                      14630

$C_{10}H_{22}$ Decane. B.p. 173.3
                      842, 1875, 2126, 2485, 2721, 2815, 3237, 4434, 5396, 5664, 6061, 6628,
                      6756, 7024, 7665, 8213, 8872, 9379, 11016, 11230, 13467, 13534, 14543,
                      15033, 15581, 15619, 15891, 16210, 16373, 16377, 16380

$C_{10}H_{22}$ 2,7-Dimethyloctane. B.p. 160.1
                      843, 1418, 1712, 2125, 2722, 3236, 3488, 3811, 4182, 4435, 4994, 5253,
                      5665, 6224, 6433, 6629, 6757, 7119, 7509, 7666, 7829, 8214, 8391, 8510,
                      8825, 9101, 9380, 9562, 9617, 9735, 9899, 10020, 10481, 11017, 11231,
                      11379, 11463, 11559, 11771, 12037, 12126, 12193, 12316, 12573, 12834,
                      13076, 13295, 13654, 14668, 14689, 14784, 15022, 15164, 15558, 15578

$C_{10}H_{22}$ 3-Ethyl-3-methylheptane. B.p. 163.0
                      8706

$C_{10}H_{22}$ 3,3,5-Trimethylheptane. B.p. 155.5
                      12277, 15693

Formula   Name and System No.

C$_{10}$H$_{22}$O   Amyl ether. B.p. 187.4
                           123, 844, 2723, 3813, 4437, 4628, 6062, 6225, 7302, 7510, 7768, 8826,
                           9040, 9102, 9381, 9484, 9772, 10021, 10088, 10364, 10431, 11018, 11232,
                           11464, 11560, 11587, 11663, 11772, 11970, 12278, 12361, 12488, 12574,
                           13296, 13385, 13460, 13655, 13783, 14392, 14785, 14865, 15129, 15145,
                           15165, 15457, 15563, 16217

C$_{10}$H$_{22}$O   Decyl alcohol. B.p. 232.9
                           845, 3812, 4436, 6063, 10089, 10239, 10779, 10842, 11069, 12912, 12972,
                           13008, 13046, 14078, 14154, 14410, 14451, 14520, 14711, 14798, 14930,
                           14996, 15061, 15105, 15261, 15355, 15384, 15401, 15445, 15482, 15497,
                           15519, 15535, 15548, 15636, 15694-15704, 16241

C$_{10}$H$_{22}$O   2-Ethyloctanol. B.p. 220.5
                           846

C$_{10}$H$_{22}$O   Isoamyl ether. B.p. 172.6
                           124, 847, 1602, 2486, 2724, 3045, 3489, 3562, 3814, 4438, 4629, 4720,
                           5254, 5666, 6064, 6226, 7120, 7173, 7303, 7511, 7667, 7769, 8511, 8827,
                           8894, 9041, 9103, 9128, 9382, 9485, 9900, 10022, 10090, 10365, 10432,
                           10504, 10574, 11019, 11233, 11311, 11465, 11501, 11561, 11588, 11626,
                           11664, 11773, 11971, 12038, 12093, 12127, 12194, 12279, 12317, 12334,
                           12428, 12489, 12575, 12704, 12781, 13206, 13297, 13386, 13461, 13493,
                           13656, 13784, 13804, 13829, 14010, 14181, 14251, 14393, 14544, 14621,
                           14649, 14654, 14669, 14755, 14786, 14810, 14815, 14821, 14866, 15023,
                           15034, 15130, 15146 15166, 15193, 15467, 15572, 15574, 15585, 15585,
                           15629, 15683, 15705, 16228, 152a

C$_{10}$H$_{22}$O   2-Propylheptanol. B.p. 217.9
                           848

C$_{10}$H$_{22}$OS   2(2-Ethylhexylthio)ethanol.
                           15706

C$_{10}$H$_{22}$O$_2$ Acetaldehyde dibutyl acetal. B.p. 188.8
                           849, 8215, 12280, 16151

C$_{10}$H$_{22}$O$_2$ Acetaldehyde diisobutyl acetal. B.p. 171.3
                           850, 8392, 16190

C$_{10}$H$_{22}$O$_2$ 1,2-Dibutoxyethane. B.p. 203.6
                           851

C$_{10}$H$_{22}$O$_3$ Dipropylene glycol butyl ether.
                           6674a

C$_{10}$H$_{22}$O$_3$ 2(2-Hexyloxyethoxy)ethanol. B.p. 259.1
                           852

C$_{10}$H$_{22}$O$_3$ Isoamyl carbonate. B.p. 232.2
                           4439, 12457, 15536

C$_{10}$H$_{22}$O$_4$ 1,2-Bis(2-ethoxyethoxy)ethane. B.p. 246.9
                           853

C$_{10}$H$_{22}$O$_4$ Tripropylene glycol methyl ether. B.p. 243
                           4440, 6675, 15707

C$_{10}$H$_{22}$O$_5$ Bis(2(2-Methoxyethoxy)ethyl) ether.
                           854

C$_{10}$H$_{22}$S   Isoamyl sulfide. B.p. 214.8
                           3815, 6065, 10266, 10505, 10634, 10780, 10843, 11020, 11344, 11891,
                           11972, 12876, 12973, 13136, 13298, 13434, 13924, 14155, 14263, 14299,
                           14317, 14341, 14700, 14756, 15116, 15262, 15600, 15618, 15680, 15708-
                           15710

C$_{10}$H$_{23}$N   Decylamine. B.p. 203.7
                           855

Formula   Name and System No.

C₁₀H₂₃N   Diamylamine. B.p. 190
               856
C₁₀H₂₃N   *N,N*-Dimethyl-2-ethylhexylamine. B.p. 176.1
               857
C₁₀H₂₃N   Diisoamylamine. B.p. 188.2
               14182, 14252, 14867, 15468, 15559, 15564, 15630, 15705
C₁₀H₂₃NO   2-Dibutylaminoethanol. B.p. 228.7
               858
C₁₁H₁₀   1-Methylnaphthalene. B.p. 245.1
               859, 3816, 4441, 4630, 6066, 6227, 6758, 8583, 8992, 10267, 10312,
               10844, 11070, 11112, 11260, 11274, 11892, 12373, 12605, 12913, 12974,
               13009, 13137, 13448, 13859, 14028, 14079, 14214, 14351, 14452, 14467,
               14564, 14631, 14799, 14841, 14879, 14903, 14997, 15062, 15082, 15179,
               15194, 15225, 15275, 15312, 15329, 15356, 15385, 15402, 15411, 15446,
               15483, 15498, 15520, 15637, 15648, 15656, 15669, 15675, 15694, 15711-
               15724
C₁₁H₁₀   2-Methylnaphthalene. B.p. 241.15
               2637, 2725, 3817, 4442, 4631, 4766, 6067, 6228, 6759, 8584, 8993, 10091,
               10285, 10313, 10635, 10845, 11071, 11113, 11131, 11234, 11261, 11345,
               11893, 12335, 12362, 12374, 12606, 12914, 12975, 13010, 13138, 13207,
               13306, 13387, 13449, 13564, 13830, 13860, 14029, 14080, 14095, 14352,
               14434, 14453, 14468, 14496, 14521, 14565, 14632, 14650, 14712, 14800,
               14826, 14842, 14880, 14931, 14959, 14998, 15063, 15083, 15109, 15180,
               15226, 15276, 15313, 15330, 15357, 15386, 15403, 15412, 15422, 15447,
               15499, 15521, 15549, 15649, 15657, 15663, 15670, 15676, 15695, 15707,
               15711, 15725-15731
C₁₁H₁₂O₂   Ethyl cinnamate. B.p. 272
               860, 3818, 4443, 6068, 6760, 8585, 11072, 11114, 12375, 14030, 14904,
               15084, 15204, 15227, 15290, 15365, 15423, 15732-15740
C₁₁H₁₄OS   2(Benzylthio)ethyl vinyl ether.
               8586, 15075
C₁₁H₁₄O₂   1-Allyl-3,4-dimethoxybenzene. B.p. 255
               861, 3819, 4444, 6069, 6761, 8587, 8734, 11073, 11115, 11275, 12336,
               12376, 12607, 13450, 13883, 13937, 14031, 14469, 14843, 14881, 15228,
               15314, 15331, 15358, 15413, 15522, 15658, 15712, 15741-15747
C₁₁H₁₄O₂   Butyl benzoate. B.p. 249.8
               862, 3820, 4445, 6070, 6762, 8589, 10286, 10314, 10636, 11074, 12377,
               12608, 13011, 13861, 13938, 14032, 14320, 14882, 14905, 15085, 15229,
               15315, 15359, 15414, 15484, 15523, 15659, 15713, 15725, 15741, 15748-
               15750
C₁₁H₁₄O₂   1,2-Dimethoxy-4-Propylbenzene. B.p. 270.5
               863, 3821, 4446, 6071, 6763, 8588, 11075, 11116, 11276, 12609, 13884,
               13939, 14033, 14470, 14883, 15181, 15205, 15230, 15278, 15332, 15366,
               15424, 15732, 15751-15757
C₁₁H₁₄O₂   Ethyl β-phenylpropionate. B.p. 248.1
               3822, 6764, 12915, 13012, 14034, 14906, 14999, 15064, 15086, 15360,
               15689, 15714, 15726, 15758-15760
C₁₁H₁₄O₂   Isobutyl benzoate. B.p. 242.15
               864, 3823, 4447, 6072, 6765, 8590, 8994, 10287, 10315, 10637, 11076,
               12610, 12916, 13013, 13451, 13862, 13940, 14300, 14884, 14907, 15000,
               15065, 15263, 15316, 15387, 15404, 15415, 15485, 15500, 15524, 15604,
               15660, 15690, 15696, 15715, 15727, 15742, 15758

Formula   Name and System No.

Formula   Name and System No.

$C_{11}H_{22}O_3$ Isoamyl carbonate. B.p. 232.2
                                     876, 3826, 4452, 4786, 6076, 6231, 8995, 10242, 10268, 10316, 10639,
                                     11079, 12457, 12613, 12918, 12977, 13389, 14082, 14158, 14303, 14524,
                                     14635, 14908, 14933, 15002, 15068, 15266, 15363, 15406, 15448, 15487,
                                     15503, 15527, 15536, 15605, 15615, 15640, 15665, 15691, 15699, 15718,
                                     15729, 15772, 15773

$C_{11}H_{22}O_3$ 4-Methoxy-2,6-dipropyl-1,3-dioxane. B.p. 223.6
                                     877

$C_{11}H_{24}$ Undecane. B.p. 194.5
                                     878, 1876, 2127, 2816, 3238, 5667, 6077, 7025, 7512, 10095, 11236,
                                     13538, 14424, 15892, 16211, 16374, 16378, 16381, 16442, 16468, 11021a,
                                     13462b, 16514a

$C_{11}H_{24}O$ 5-Ethyl-2-nonanol. B.p. 225.4
                                     879, 15719, 15730

$C_{11}H_{24}O_2$ Diamyloxymethane. B.p. 221.6
                                     880, 9773, 16218

$C_{11}H_{24}O_2$ Diisoamyloxymethane. B.p. 210
                                     881, 10143, 10783, 13142, 13598, 13986, 14012, 14414, 14569, 14961,
                                     14977, 15552, 16505

$C_{11}H_{24}O_2$ 2,2-Dibutoxypropane.
                                     882

$C_{11}H_{24}O_2$ 2,6-Dimethyl-4-heptyloxyethanol. B.p. 225.5
                                     883

$C_{11}H_{24}O_4$ 1,1,3,3-Tetraethoxypropane. B.p. 220.1
                                     884

$C_{11}H_{25}N$ 2-Ethylhexylpropylamine.
                                     14724

$C_{11}H_{25}NO$ 1-Dibutylamino-2-propanol. B.p. 225.1
                                     885

$C_{12}F_{27}N$ Tris(perfluorobutyl)amine. B.p. 177
                                     10105, 15008, 15119, 15195, 15693, 16510

$C_{12}H_9N$ Carbazole. B.p. 355
                                     8594, 12378, 15197, 15774, 15775

$C_{12}H_{10}$ Acenaphthene. B.p. 277.9
                                     3827, 4453, 4633, 6078, 6768, 8595, 11080, 11118, 11132, 11277, 12379,
                                     12614, 13865, 14035, 14353, 14885, 14909, 15206, 15231, 15279, 15291,
                                     15298, 15333, 15367, 15427, 15733, 15752, 15761, 15776-15781

$C_{12}H_{10}$ Biphenyl. B.p. 255.9
                                     3828, 4454, 6079, 6769, 8596, 10317, 10640, 11081, 11119, 11133, 11278,
                                     12380, 12615, 13015, 13453, 13704, 13866, 14036, 14218, 14471, 14847,
                                     14886, 14910 15003, 15069, 15182, 15232, 15280, 15292, 15318, 15334,
                                     15368, 15418, 15428, 15488, 15528, 15661, 15700, 15720, 15734, 15744,
                                     15748, 15759, 15767, 15782-15788

$C_{12}H_{10}O$ 1 and 2-Acetylnaphthalene.
                                     10784

$C_{12}H_{10}O$ Phenyl ether. B.p. 259.3
                                     887, 3829, 4455, 6080, 6770, 8597, 8735, 11082, 11120, 11279, 12339,
                                     12381, 12616, 13016, 13867, 13886, 13941, 14037, 14354, 14472, 14887,
                                     14911, 15004, 15087, 15183, 15233, 15281, 15319, 15335, 15369, 15419,
                                     15429, 15529, 15662, 15677, 15721, 15735, 15745, 15749, 15753, 15760,
                                     15768, 15782, 15789-15794

$C_{12}H_{10}O$ o-Phenylphenol.
                                     886

Formula   Name and System No.

Formula   Name and System No.

$C_{12}H_{22}O_2$ 2-Ethylhexyl crotonate. B.p. 241.2
                               897

$C_{12}H_{22}O_2$ Vinyl decanoate isomers.
                               898

$C_{12}H_{22}O_4$ Isoamyl oxalate. B.p. 268.0
                               3836, 7122, 8603, 11086, 11123, 12385, 12619, 14040, 14913, 15185,
                               15209, 15236, 15285, 15321, 15338, 15678, 15738, 15756, 15778, 15786,
                               15804-15805

$C_{12}H_{22}O_4$ Diethyl 2-ethyl-3-methylglutarate. B.p. 255.8
                               899

$C_{12}H_{23}N$ Dicyclohexylamine. B.p. 255.8
                               900

$C_{12}H_{24}$ 2,6,8-Trimethylnonene.
                               15196, 15806

$C_{12}H_{24}O$ 2,6,8-Trimethyl-4-nonanone. B.p. 218.2
                               901

$C_{12}H_{24}OS$ 2(2-Ethylhexylthio)ethyl vinyl ether.
                               8604, 15706

$C_{12}H_{24}O_2$ 2-Ethylbutyl 2-ethylbutyrate. B.p. 222.6
                               902

$C_{12}H_{24}O_2$ 2-Ethylbutyl hexanoate. B.p. 236.2
                               903

$C_{12}H_{24}O_2$ Hexyl 2-ethylbutyrate. B.p. 230.3
                               904

$C_{12}H_{24}O_2$ Hexyl hexanoate. B.p. 245.2
                               905

$C_{12}H_{24}O_3$ 2,2-Dibutoxy-3-butanone. B.p. 228
                               906

$C_{12}H_{24}O_3$ 2,2-Diisobutyoxy-3-butanone. B.p. 214
                               907

$C_{12}H_{26}$ Dodecane. B.p. 216
                               908, 1877, 3240, 4461, 6085, 6676, 10097, 11239, 13541, 14427, 14801,
                               15270, 15807, 15893, 16212

$C_{12}H_{26}$ 2,2,4,4,6-Pentamethylheptane. B.p. 185.6
                               10098

$C_{12}H_{26}$ 2,2,4,6,6-Pentamethylheptane. B.p. 177.9
                               10099

$C_{12}H_{26}O$ 2-Butyl-1-octanol. B.p. 253.4
                               909

$C_{12}H_{26}O$ Dodecyl alcohol.
                               15703

$C_{12}H_{26}O$ Hexyl ether.
                               4462, 8605

$C_{12}H_{26}O$ 2,6,8-Trimethyl-4-nonanol. B.p. 225.5
                               910, 13047, 15806, 16242

$C_{12}H_{26}O_2$ Acetaldehyde diamyl acetal. B.p. 225.3
                               911, 9774, 16219

$C_{12}H_{26}O_2$ Acetaldehyde diisoamyl acetal. B.p. 213
                               912, 9901, 16229

$C_{12}H_{26}O_2$ 2-Ethylhexaldehyde diethyl acetal. B.p. 207.81
                               913

$C_{12}H_{26}O_2$ 3-Ethoxy-4-ethyloctanol. B.p. 249.2
                               914

Formula   Name and System No.

Formula   Name and System No.

$C_{14}H_{14}O$ Benzyl ether. B.p. 297
               3843, 4472, 6090, 6781, 8611, 12342, 12393, 15301, 15303, 15374, 15794,
               15811

$C_{14}H_{22}O$ 2-Ethylhexylphenol. B.p. 297.0
               921

$C_{14}H_{23}N$ N-Ethylhexylaniline.
               922

$C_{14}H_{24}$ 1,3,6,8-Tetramethyl-1,6-cyclodecadiene. B.p. 220.5
               923

$C_{14}H_{26}O_4$ Dibutyl adipate.
               924

$C_{14}H_{28}O$ Trimethylnonyl vinyl ether. B.p. 223.4
               925

$C_{14}H_{28}O_2$ 2-Ethylbutyl 2-ethylhexanoate. B.p. 261.5
               926

$C_{14}H_{28}O_2$ 2-Ethylhexyl 2-ethylbutyrate. B.p. 252.8
               927

$C_{14}H_{28}O_2$ 2-Ethylhexyl hexanoate. B.p. 267.21
               928

$C_{14}H_{28}O_2$ Hexyl 2-ethylhexanoate. B.p. 254.3
               929

$C_{14}H_{29}N$ N(2-Ethylhexyl)cyclohexylamine.
               930

$C_{14}H_{30}$ Tetradecane. B.p. 252.5
               4473, 6677, 11241, 14429

$C_{14}H_{30}O$ 7-Ethyl-2-methyl-4-undecanol. B.p. 264.3
               931

$C_{14}H_{30}O$ Tetradecyl alcohol. B.p. 260.0
               13213, 13394, 15774, 15808, 15815

$C_{14}H_{30}O_2$ 2(2,6,8-Trimethyl-4-nonyloxy)ethanol.
               932

$C_{15}H_{10}N_2O_2$ Bis(p-isocyanatophenyl)methane.
               10366

$C_{15}H_{18}$ 2-Amylnaphthalene. B.p. 292.3
               11984, 13214, 13395

$C_{15}H_{28}O_4$ Dibutyl pimelate.
               933

$C_{15}H_{30}$ Pentadecene.
               14802

$C_{15}H_{30}O_2$ Methyl myristate.
               15813, 15819

$C_{15}H_{32}O$ 2,8-Dimethyl-6-isobutyl-4-nonanol. B.p. 265.4
               934

$C_{15}H_{33}BO_3$ Isoamyl borate. B.p. 255
               15323, 15750

$C_{16}H_{18}O$ Bis(α-methylbenzyl)ether. B.p. 286.7
               935

$C_{16}H_{20}$ Diisopropylnaphthalene. B.p. 305
               13215, 13396

$C_{16}H_{22}O_4$ Dibutyl phthalate.
               6868

$C_{16}H_{28}O_4$ Bis(4-methyl-2-pentyl)maleate.
               936

Formula   Name and System No.

Formula   Name and System No.

$C_{24}H_{38}O_4$ Octyl phthalate.
                             15788
$C_{24}H_{52}O_4Si$ Tetra(2-ethylbutoxy)silane.
                             953
$C_{31}H_{58}O_6$ Tri(2-ethylhexyl)1,2,4-butanetricarboxylate.
                             916

Bibliography

(1)   Ababi and Mihaila *Analele Stiint. Univ. "A. I. Cuza", Iasi, Sect. Ic: Chem.* **11**, 31 (1965); *C.A.* **63**, 14118.

(1)c Ababi and Mihaila, *Analele Stiint. Univ. "A. I. Cuza", Iasi Sect. I* **12**, 115 (1966); *C.A.* **69**, 70438x.

(2)   Ababi, Mihaila, and Cruceanu, *Rev. Roumaine Chim.* **10**, 793 (1965); *C.A.* **64**, 5806.

(2)c Ababi and Balba, *Analele Stiint. Univ. "A. I. Cuza", Iasi, Sect. Ic* **14**, 155 (1968); *C.A.* **71**, 54224k.

(3)   Adelson and Evans, U.S. Patent 2,605,216 (1952).

(4)   Adelson and Evans, U.S. Patent 2,500,596 (1950).

(5)   Agliardi, *Chim. Ind. (Milan)* **28**, 87 (1946).

(6)   Akers and Eubanks, *Proc. Cryogenic Eng. Conf., 2nd, Boulder, 1957*, p. 275; *C.A.* **52**, 14267 (1958).

(7)   Akita and Yoshida, *J. Chem. Eng. Data* **8**, 484 (1963).

(8)   Albanesi, Pasquon, and Genoni, *Chim. Ind. (Milan)* **39**, 814 1957); *C.A.* **52**, 3440 (1958).

(8)c Alfonso, Jimenez, and Zurbano, *An. Real Soc. Espan. Fis. Quim., Ser. B* **63**, 711 (1967); *C.A.* **69**, 70422n.

(9)   Allen and Ellis, U. K. At. Energy Authority, IGR-R/CA, 216 (1957).

(10) Almasy and Barcanfalvi, *Nehezvegyip. Kut. Int. Kozlemen.* **1**, 297 1959); *C.A.* **54**, 6236.

(11) Alpert and Elving, *Ind. Eng. Chem.* **41**, 2864 (1949).

(12) Alpert and Elving, *Ind. Eng. Chem.* **43**, 1174, 1182 1951).

(13) Al'tshuler, Zviadadze, and Chizhikov, *Zh. Neorgan. Khim.* **2**, 1581 (1957); *C.A.* **52**, 7833.

(14) Altsybeeva, Belousov; Ovtrakht and Morachevskii, *Zh. Fiz. Khim.* **38**, 1242 (1964); *C.A.* **61**, 7757.

(15) Altsybeeva and Morachevskii, *Zh. Fiz. Khim.* **38**, 1569, 1574 (1964); *C.A.* **61**, 7753, 7754.

(16) Amer, Paxton, and Van Winkle, *Ind. Eng. Chem.* **48**, 142 (1956).

(17) American Cyanamid Co., *New Prod. Bull.* **13**, March 1950).

(18) Amick and Harney, U.S. Patent 2,487,036 (1949).

(19) Amick, Weiss, and Kirshenbaum, *Ind. Eng. Chem.* **43**, 969 1951).

(20) Andrews and Spence, U.S. Patent 2,061,889 (1936).

(21) Andrews and Spence, U.S. Patent 2,126,600 (1938).

(21)c Anello and Sweeney, U.S. Patent 3,409,512 (1968); *C.A.* **70**, 37177j.

(22) Anon., *Oil, Paint, Drug Reptr.* **156**, No. 18, 4 (Oct. 4, 1949).

(23) Ansul Chemical Co., Ansul Ethers, *Chem. Prod. Bull.*

(24) Arbuzov and Dianovan, *Izu. Acad. Nauk SSR, Ser. Khim.* **1965**, 1584; *C.A.* **64**, 2803.

(25) Aring and Weber, *J. Prakt. Chem.* **30**, 295 (1965); *C.A.* **64**, 11933.

(26) Aristovich, Levin, and Morachevskii, *Tr. Vses. Nauchn-Issled Inst. Neftekhim, Protsessov.* **1962**, 84; *C.A.* **58**, 5096.

(27) Aristovich, Morachevskii, and Sabylin, *Zh. Prikl. Khim.* **38**, 2694 (1965).

(27)c Assal, *Bull. Acad. Polon. Sci., Ser. Sci. Chim.* **14**, 603 (1966); *C.A.* **66**, 22676q.

(27)f Astakhova, Alekseeva, Talanov, and Nisel'son, *Zh. Neorgan. Khim.* **14**, 832 (1969); *C.A.* **70**, 109557h.

(28) Aston, Kennedy, and Messerly, *J. Am. Chem. Soc.* **63**, 2343 (1941).

(29) Atkins, *J. Chem. Soc.* **117**, 218 (1920).

(29)c Aubert, Carles, and Bethuel, *Chim. Ind., Genie Chim.* **98**, 661 (1967); *C.A.* **68**, 53147c.

(29)e Avots, Ekis, and Kuplenieks, *Latv. PSR Zinat. Akad. Vestis, Kim. Ser.* **1968**, 72; *C.A.* **69**, 70356u.

(30) Babcock, U.S. Patent 2,049,486 (1936).

(31) Babcock, U.S. Patent 2,461,191 (1949).

(31)c Babich, Borozdina, Kushner, and Serafimov, *Zh. Prikl. Khim.* **41**, 589 (1968); *C.A.* **69**, 5650w.

(31)f Babich, Ivanchikova, and Serafimov, *Zh. Prikl. Khim.* **42**, 1354 (1969); *C.A.* **71**, 85057n.

(31)h Bachmaier and Jimenez-Barbera, *Deut. Luft-Raumfahrt, Forschungber.* **1967**, DLR-FB-67-21; *C.A.* **69**, 80819m.

(32) Bachman and Simons, *Ind. Eng. Chem.* **44**, 202 (1952).

(33) Bahr and Zieler, *Z. Angew. Allgem. Chem.* **43**, 286 (1930).

(33)c   Baker, Chaddock, Lindsay, and Werner, *Ind. Eng. Chem.* **31**, 1263 (1939).

(34)   Baker *et al.*, *Ind. Eng. Chem.* **31**, 1260, 1263 (1939).

(35)   Baker, Fisher and Roth, *J. Chem. Eng. Data* **9**, 11 (1964).

(36)   Bakowski and Treszczanogicz, *Przemysl Chem.* **22**, 211 (1938); *C.A.* **33**, 6518.

(36)c   Balashov, Grishunin, and Serafimov, *Zh. Fiz. Khim.* **41**, 1210 (1967); *C.A.* **67**, 8053.

(36)e   Balashov and Serafimov, *Izv. Vyssh. Ucheb. Zaved., Khim. Khim. Tekhnol.* **9**, 885 (1966); *C.A.* **67**, 6131y.

(36)g   Balashov and Serafimov, *Izv. Vyssh. Ucheb. Zaved., Khim. Khim. Tekhnol.* **10**, 867 (1967); *C.A.* **68**, 63159h.

(36)i   Balashov, Serafimov, and Bessonova, *Zh. Fiz. Khim.* **40**, 2294 (1966); *C.A.* **66**, 2170.

(36)k   Balasubramanian, Banerjee, and Doraiswamy, *British Chem. Eng.* **11**, 1540 (1966); *C.A.* **66**, 49677c.

(37)   Ballard and Van Winkle, *Ind. Eng. Chem.* **45**, 1803 (1953)

(38)   Bancelin and Rivat, *Bull. Soc. Chim.* **25**, (4) 552 (1919).

(39)   Baney, U.S. Patent **2,425,220** (1947).

(40)   Banks and Musgrave, *J. Chem. Soc.* **1956**, 4682; *C.A.* **51**, 3216.

(40)b   Baradarajan and Satyanarayana, *Indian J. Technol.* **5**, 264 (1967); *C.A.* **68**, 6833z.

(40)c   Baradarajan and Satyanarayana, *J. Chem. Eng. Data* **13**, 148 (1968).

(40)e   Baranaev. Prokhorova, and Suverova, *Khim. Prom.* **44**, 390 (1968); *C.A.* **69**, 30636d.

(41)   Barber and Cady, *J. Am. Chem. Soc.* **73**, 4247 (1951).

(41)f   Bareggi, Mori, Schwarz, and Beltrame, *Chim. Ind. (Milan)* **50**, 1224 (1968); *C.A.* **70**, 32019z.

(42)   Barr-David and Dodge, *J. Chem. Eng. Data* **4**, 107 (1959).

(43)   Barrett Division, Allied Chemical and Dye Corp., *Chem. Industries* **33**, 513 (1933).

(44)   Baud, *Bull. Soc. Chim.* **5**, (4) 1022 (1909).

(45)   Beduwe, *Bull. Soc. Chim. Belges* **34**, 41 1925).

(46)   Bennett and Parmelee, U.S. Patent **2,999,816** (1961).

(47)   Benning, U.S. Patent **2,641,579** (1953).

(48)   Benning, U.S. Patents **2,450,414-15** (1948).

(49)   Benning and Park, U.S. Patent **2,384,449** (1945).

(50)   Berg and Harrison, *Chem. Eng. Progr.* **43**, 487 (1947).

(51)   Berg and Harrison, U.S. Patent **2,442,229** (1948).

(52)   Berg and Harrison, U.S. Patent **2,477,715** (1949).

(53)   Berg, Harrison, and Montgomery, *Ind. Eng. Chem.* **38**, 1149 (1946).

(54)   Berthelot, *Compt. Rend.* **57**, 430 (1863).

(55)   Beyer, Schuberth and Leibnitz, *J. Prakt. Chem.* **27**, 276 (1965); *C.A.* **63**, 9108.

(56)   Bierlein, *Dissertation Abstr.* **24,101**; *C.A.* **52**, 6909.

(57)   Bierlein and Kay, *Ind. Eng. Chem.* **45**, 618 (1953).

(58)   Bigg, Banerjee and Doraiswamy, *J. Chem. Eng. Data* **9**, 17 (1964).

(59)   Binning, U.S. Patent **3,007,985** (1961).

(60)   Birch, Collis and Lowry, *Nature* **158**, 60 (1946).

(61)   Bischoff and Adkins, *J. Am. Chem. Soc.* **46**, 256 (1924).

(62)   Bishop and Denton, *Ind. Eng. Chem.* **42**, 883 (1950).

(63)   Blackford and York, *J. Chem. Eng. Data* **10**, 313 (1965).

(64)   Blinowska, Brzostowski and Magiera, *Bull. Acad. Polon. Sci., Ser. Sci. Chim.* **14**, 467 (1966).

(65)   Bloomer, U.S. Patent **2,381,996** (1945).

(66)   Bludworth and Flower, U.S. Patent **2,381,032** (1945).

(67)   Blyuum, Khainson, Zhadanov, Vasil'eva, *Zh. Fiz. Khim.* **40**, 1779 (1966); *C.A.* **65**, 16120.

(68)   Bol'shakov, Fandeeva, Budnar, Shakova and Vinogradov, *Izv. Akad. Nauk SSSR, Neorgan. Mater.* **2**, 1537 (1966); *C.A.* **66**, 606.

(68)c   Bol'shakov, Plotinskii, and Bardin, *Zh. Neorgan. Khim.* **12**, 189 (1967); *C.A.* **66**, 119303d.

(68)f   Bol'shakov, Kogan, Bodnar, Shakhova, and Kudobkina, *Izv. Akad. Nauk SSSR, Neorgan. Mater.* **4**, 2019 (1968); *C.A.* **70**, 32018y.

(69)   Bonner, Bonner, and Gurney, *J. Am. Chem. Soc.* **55**, 1406 (1933).

(70)   Bonner and Wallace, *J. Am. Chem. Soc.* **52**, 1747 (1930).

(71)   Boublik and Kuchynka, *Chem. Listy* **50**, 1181 (1956); *C.A.* **50**, 16320; *Collection Czech. Chem. Commun.* **21**, 1634 (1956); *C.A.* **51**, 11794.

(72)   Boublik and Kuchynka, *Collection Czech. Chem. Commun.* **25**, 579 (1960); *C.A.* **54**, 16068.

(72)c   Boublikova and Lu, *J. Appl. Chem.* **19**, 89 (1969); *C.A.* **70**, 81405f.

(73)   Bouillon, *Compt. Rend.* **230**, 1290 (1950).

(73)c   Bourgeois, Enjalbert, and Racine, *Chim. Ind., Genie Chim.* **102**, 213 (1969); *C.A.* **71**, 105814n.

(74)   Bouzat and Schmitt, *Compt. Rend.* **198**, 1923 (1934).

(75)   Bower, U.S. Patent **2,999,817** (1961).

(76)   Bozza and Gallarati, *Giorn. Chim. Ind. Appl.* **13**, 163 (1931).
(77)   Bramer, Ruggles and Robinson, U.S. Patent **2,090,652** (1937).
(78)   Brandon, U.S. Patent **2,459,410** (1949).
(79)   Brant, *J. Am. Chem. Soc.* **64**, 2224 (1942).
(80)   Brazauskiene, Miscenko and Ciparis, *Lietuvos TSR Aukstuju Mokyklu Mokslov Darbai, Chem. Ir Chem. Technol.* **6**, 141 (1965); *C.A.* **64**, 1403.
(81)   Bremner, Jones and Coats, British Patent **592,919** (1947).
(82)   Briner and Cardoso, *Compt. Dend.* **144**, 911 (1907).
(83)   Britton, Nutting and Horsley, *Anal. Chem.* **19**, 601 (1947).
(84)   Broick, Dept. of Commerce, *OTS Rept.*, PB **76303**.
(85)   Broich and Hunsmann, German Patent **1,002,321** (1957); *C.A.* **53**, 21663.
(86)   Bromiley and Quiggle, *Ind. Eng. Chem.* **25**, 1136 (1933).
(87)   Brooks, U.S. Patent **2,436,286** (1948).
(88)   Brooks and Nixon, *J. Am. Chem. Soc.* **75**, 480 (1953).
(89)   Brown, *J. Chem. Soc.* **35**, 547 (1879).
(90)   Brown, U.S. Patent **2,286,056** (1942).
(91)   Brown and Smith, *Australian J. Chem.* **7**, 264 (1954); **8**, 62, 501 (1955).
(92)   Brown and Smith, *Australian J. Chem.* **10**, 423 (1957); *C.A.* **52**, 3441.
(93)   Brown and Smith, *Australian J. Chem.* **12**, 407 (1959); *C.A.* **54**, 1003.
(94)   Brown and Smith, *Australian J. Chem.* **13**, 30 (1960); *C.A.* **54**, 10436.
(95)   Bruner and Darden, U.S. Patent **2,609,336** (1952).
(96)   Brunjes and Furnas, *Ind. Eng. Chem.* **27**, 396 (1935).
(97)   Brzostowski, *Bull. Acad. Polon. Sci., Ser. Sci. Chim.* **9**, 471 (1961); *C.A.* **60**, 7512.
(98)   Brzostowski, *Roczniki Chem.* **35**, 291 (1961); *C.A.* **55**, 16109.
(99)   Brzostowski, Malanowski and Zieborak, *Bull. Acad. Polon. Sci., Classe III* **7**, 421 (1959); *C.A.* **54**, 19067.
(100) Brzostowski and Warycha, *Bull. Acad. Polon. Sci., Ser. Sci. Chim.* **11**, 539 (1963); *C.A.* **60**, 4868.
(101) Buchheim, German Patent **616,596**; *Chem. Zentr.*, **1935**, II, 3703.
(101)c Buckman and Clemmons, U.S. Patent **3,406,099** (1968); *C.A.* **70**, 11130p.
(102) Buell, U.S. Patent **2,382,603** (1945).
(103) Burch and Leeds, *Chem. Eng. Data Ser.* **2**, 3 (1957).
(103)c Burd and Braun, *Proc. Div. Refining, Am. Petrol. Inst.* **48**, 464 (1968); *C.A.* **71**, 33826t.
(104) Bures, Cano, and Wirth, *J. Chem. Eng. Data* **4**, 199 (1959).
(104)c Burfield, Richardson, and Guerece, *AIChE J.* **16**, 97 (1970); *C.A.* **70**, 59710w.
(105) Burgin, Hearne and Rust, *Ind. Eng. Chem.* **33**, 385 (1941).
(105)c Burke and Williams, *Trans. Ky. Acad. Sci.* **27**, 29 (1966); *C.A.* **69**, 46411n.
(106) Burn and Din. *Trans. Faraday Soc.* **58**, 1341 (1962); *C.A.* **58**, 971.
(107) Burtle, *Ind. Eng. Chem.* **44**, 1675 (1952).
(108) Bushmakin, Baldangiin and Molodenko, *Zh. Prikl. Khim.* **38**, 1417 (1965); *C.A.* **63**, 7692.
(109) Bushmakin, Begetova and Kuchinskaya, *Sintet. Kauchuk, No.* **4**, 8 (1936); *C.A.* **30**, 6630.
(110) Bushmakin and Kish, *Zh. Prikl. Khim.* **30**, 200 (1957); *C.A.* **51**, 10989-90.
(111) Bushmakin and Kuchinskaya, *Sintet. Kauchuk, No.* **5**, 3 (1936).
(112) Bushmakin and Molodenko, *Zh. Prikl. Khim.* **37**, 2653 (1964); *C.A.* **62**, 10097.
(113) Bushmakin and Voeikova, *Zh. Obshch. Khim.* **19**, 1615 (1949); *C.A.* **44**, 1317.
(114) Butcher and Robinson, *J. Appl. Chem. (London)* **16**, 289 (1966); *C.A.* **66**, 1405.
(115) Butta, *Chem. Listy* **50**, 1646 (1956); *C.A.* **51**, 2349; *Collection Czech. Chem. Commun.* **22**, 1680 (1957); *C.A.* **52**, 8708.
(116) Byk and Sheherbak, *Zh. Fiz. Khim.* **30**, 56 (1956); *C.A.* **50**, 10469.
(117) Bylewski, *Roczniki Chem.* **13**, 322 (1933).
(118) Cadbury, *J. Chem. Educ.* **12**, 292 (1933).
(119) Calder and Fleer, U.S. Patents **2,401,335-6** (1946).
(120) Calfee, Fukuhara and Bigelow, *J. Am. Chem. Soc.* **61**, 3552 (1939).
(121) Calices and Hannotte, *Ingr. Chimist* **20**, 1 (1936).
(122) Calingaert and Wojciechowski, *J. Am. Chem. Soc.* **72**, 5310 (1950).
(122)c Campbell and Chatterjee, *Can. J. Chem.* **48**, 277 (1970); *C.A.* **72**, 71157d.
(123) Campbell and Dulmage, *J. Am. Chem. Soc.* **70**, 1723 (1948).
(124) Campbell and Hickman, *J. Am. Chem. Soc.* **75**, 2879 (1953).
(125) Canjar, Horni and Rothfus, *Ind. Eng. Chem.* **48**, 427 (1956).
(126) Canjar and Lonergan, *A. I. Ch. E. J.* **2**, 280 (1956).
(127) Capkova and Fried, *Collection Czech. Chem. Commun.* **28**, 2235 (1963); *C.A.* **59**, 12240.
(128) Carbide and Carbon Chemicals Corp., Catalog, 12th ed., 1945.
(129) Carbide and Carbon Chemicals Corp., "Cellosolve and Carbitol Solvents," Jan. 1, 1947.

(130)   Carbide and Carbon Chemicals Corp., *Chem. Industries* **33**, 521 (1933).
(131)   Carey and Lewis, *Ind. Eng. Chem.* **24**, 882 (1932).
(131)c   Carles and Aubert, *Chim. Ind., Genie Chim.* **100**, 211 (1968); *C.A.* **69**, 110436d.
(132)   Carleton, *Chem. Eng. Data Ser.* **1**, 21 (1956).
(133)   Carley and Bertelsen, *Ind. Eng. Chem.* **41**, 2806 (1949).
(134)   Carlson, U.S. Patent **2,381,876** (1945).
(135)   Carnell, U.S. Patent **2,430,388** (1947).
(136)   Carpenter, Davis and Wiedeman, U.S. Patent **2,404,163** (1946).
(137)   Carr and Kropholler, *J. Chem. Eng. Data* **7**, 26 (1962).
(138)   Carra and Beltrame, *Chim. Ind. (Milan)* **43**, 1251 (1961); *C.A.* **56**, 9487.
(139)   Carswell and Morrill, *Ind. Eng. Chem.* **29**, 1247 (1937).
(140)   Celanese Chemical Corp., *New Product Bull.* N-08-**1**.
(141)   Ceslak and Karnatz, *British Patent* **580,048** (1946)
(142)   Chabrier de la Saulniere, *Ann. Chim.* **17**, 353 (1942); *C.A.* **38**, 3255.
(143)   Chaiyavech and Van Winkle, *J. Chem. Eng. Data* **4**, 53 (1959).
(144)   Challis, U.S. Patent **2,691,624** (1954).
(145)   Chalov and Aleksandrova, *Gidroliz. Lesokhim. Prom.* **10**, 15 (1957); *C.A.* **51**, 12585.
(145)c   Chandok and McMillan, *J. Chem. Eng. Data* **14**, 286 (1969).
(146)   Chao, *Dissertation Abstr. No.* **19076**; *C.A.* **51**, 9245.
(147)   Chao and Hougen, *Chem. Eng. Sci.* **7**, 246 (1958); *C.A.* **52**, 15219.
(148)   Chavanne, *Bull. Soc. Chim. Belges* **27**, 205 (1913).
(148)a   Cherkasskaya, Tur, Petrenkova, and Lyubomnov, *Zh. Fiz. Khim.* **42**, 2435 (1968); *C.A.* **70**, 51066d.
(148)c   Cherry, *Univ. Microfilms* **66-15,075**; *C.A.* **66**, 108815w.
(148)f   Chevalier, *J. Chim. Phys. Physicochim. Biol.* **66**, 1457 (1969); *C.A.* **72**, 48112b.
(149)   Chevalley, *Bull. Soc. Chim. France* **1961**, 510; *C.A.* **55**, 15093.
(149)c   Chirikov, Galata, Chirikova, and Kofman, *Teor. Osn. Khim. Tekhnol.* **3**, 766 (1969); *C.A.* **71**, 116971n.
(150)   Chirikova, Galata, Kotova and Kofman, *Zh. Fiz. Khim.* **40**, 918 (1966); *C.A.* **65**, 1451.
(151)   Choffe and Asselineau, *Rev. Inst. France Petrole Ann. Combust. Liquides* **11**, 948 (1956); *C.A.* **51**, 3262.
(152)   Choffe and Asselineau, *Rev. Inst. France Petrole Ann. Combust. Liquides* **12**, 565 (1957); *C.A.* **51**, 17383.
(153)   Choffe, Cliquet and Meunier, *Rev. Inst. France Petrole Ann. Combust. Liquides* **15**, 1051 (1960); *C.A.* **55**, 8009.
(154)   Christian, *J. Phys. Chem.* **61**, 1441 (1957).
(155)   Churchill, U.S. Patent **2,527,916** (1950).
(156)   Churchill, Collamore and Katz, *Oil Gas J.* **41**, 33 (Aug. 6, 1942).
(156)c   Cigna and Luizzo, *Ann. Chim. (Rome)* **57**, 38 (1967); *C.A.* **67**, 6168r.
(157)   Cigna and Sebastiani, *Ann. Chim. (Rome)* **54**, 1038 (1964); *C.A.* **63**, 4995.
(158)   Cigna and Sebastiani, *Ann. Chim. (Rome)* **54**, 1048 (1964).
(158)c   Cihova, Vojtko, and Hrusovsky, *Chem. Zvesti* **22**, 599 (1968); *C.A.* **70**, 6966m.
(158)f   Cihova, Vojtko, and Hrusovsky, *Chem. Zvesti* **23**, 270 (1969); *C.A.* **72**, 6626s.
(159)   Cines, U.S. Patent **2,692,227** (1954).
(160)   Cines, U.S. Patent **2,789,087** (1957).
(160)c   Cislak and Karnatz, British Pat. **580,048** (1946).
(161)   Claiborne and Fuqua, *Anal. Chem.* **21**, 1165 (1949).
(162)   Clark, U.S. Patent **2,385,610** (1945).
(163)   Claxton, Physical and Azeotropic Data, Natl. Benzol and Allied Products Assoc. (1958).
(164)   Clough and Johns, *Ind. Eng. Chem.* **15**, 1030 (1923).
(165)   Colburn and Phillips, *Trans. A. I. Ch. E.* **40**, 333 (1944).
(166)   Cole, *Chem. Eng. Data Ser.* **3**, 213 (1958).
(167)   Coles and Popper, *Ind. Eng. Chem.* **42**, 1434 (1950).
(168)   Commercial Solvents Corp., *Tech. Data Sheet No.* **23** (1954).
(168)c   Comtat, Enjalbert, and Mahenc, *Chim. Ind., Genie Chim.* **102**, 225 (1969); *C.A.* **71**, 105815p.
(169)   Conner, Elving, Benischeck, Tobias and Steingiser, *Ind. Eng. Chem.* **42**, 106 (1950).
(170)   Conner, Elving, and Steingiser, *Ind. Eng. Chem.* **40**, 497 (1948).
(171)   Conti, Othmer, and Gilmont, *J. Chem. Eng. Data* **5**, 301 (1960).
(172)   Copenhaver and Bigelow, Acetylene and Carbon Monoxide Chemistry, Reinhold Publ. Corp. (1949), p. 106, 109, 121.
(173)   Cornish, Archibald, Murphy and Evans, *Ind. Eng. Chem.* **26**, 397 (1934).
(174)   Coulson and Jones, British Patent **585,108** (1947); *C.A.* **41**, 4173.
(175)   Coulson and Jones, *J. Soc. Chem. Ind. (London)*, **65**, 169 (1946).

(176) Coulter, Lindsay, and Baker, *Ind. Eng. Chem.* **33**, 1251 (1941).
(177) Cova, *J. Chem. Eng. Data* **5**, 282 (1960).
(178) Craig, paper presented before Div. of Organic Chemistry, 104th Meeting Am. Chem. Soc., Buffalo, N.Y., 1942.
(179) Crawford, Edwards, and Lindsay, *J. Chem. Soc.* **1949**, 1054; *C.A.* **43**, 8835.
(180) Cretcher, Koch, and Pittenger, *J. Am. Chem. Soc.* **47**, 1173 (1925).
(181) Crutzen, Jost and Sieg, *Z. Elektrochem.* **61**, 230 (1957); *C.A.* **51**, 10214.
(182) Cuculo and Bigelow, *J. Am. Chem. Soc.* **74**, 710 (1952).
(183) Curme and Johnson, "Glycols", *ACS Monograph* **114**, Reinhold, New York, 1952.
(184) Dakshinamurty and Rao, *J. Appl. Chem. (London)* **7**, 654 (1957); *C.A.* **52**, 6911 (1958).
(185) Dakshinamurty and Rao, *J. Sci. Ind. Res. (India)* **15B**, 118 (1956); *C.A.* **50**, 11753.
(186) Dakshinamurty and Rao, *J. Sci. Ind. Res. (India)* **17B**, 105 (1958); *C.A.* **52**, 16851.
(187) Dakshinamurty and Rao, *Trans. Indian Inst. Chem. Engrs.* **8**, 57 (1955-6); *C.A.* **51**, 14400.
(188) Dakshinamurty, Rao, Acharya, and Rao, *Chem. Eng. Sci.* **9**, 69 (1958); *C.A.* **53**, 8727.
(189) Dakshinamurty, Rao, Raghavacharya, and Rao, *J. Sci. Ind. Res. (India)* **16B**, 340 (1957); *C.A.* **52**, 2485.
(190) Dakshinamurty, Rao and Rao, *J. Appl. Chem. (London)* **11**, 226 (1961); *C.A.* **55**, 23020.
(190)b Dancui, *Stud. Cercet. Chim.* **16**, 957 (1968); *C.A.* **70**, 118622c.
(190)c Danneil, Toedheide, and Franck, Chem.-Ing.-Tech. **39**, 816 (1967); *C.A.* **67**, 76654v.
(191) Danov and Shinyaeva, *Zh. Fiz. Khim.* **39**, 486 (1965); *C.A.* **62**, 12493.
(192) Daudt, U.S. Patent **2,390,518** (1945).
(193) Davidson, U.S. Patent **2,506,858** (1950).
(194) Davis and Evans, *J. Chem. Data* **5**, 401 (1960).
(194)c Davison, *J. Chem. Eng. Data* **13**, 348 (1968).
(194)f Davison and Smith, *J. Chem. Eng. Data* **14**, 296 (1969).
(195) Deansley, U.S. Patent **1,866,800** (1932).
(196) Deansley, U.S. Patent **2,290,636** (1942).
(197) Deizenrot, Kogan and Fridman, *Zh. Prikl. Khim.* **39**, 1880 (1966); *C.A.* **65**, 16119.
(198) Delzenne, *Chem. Eng. Data Ser.* **3**, 224 (1958).
(199) Delzenne, *J. Chem. Eng. Data* **5**, 413 (1960).
(200) Delzenne, *Bull. Soc. Chim. France* **1961**, 295; *C.A.* **55**, 25441.
(200)a De Mauduit, *Ann. Genie Chim.* **3**, 70 (1967); *C.A.* **69**, 99957b.
(200)c De Mauduit and Gardy, *Compt. Rend. Acad. Sci., Paris, Ser. C* **266**, 946 (1968).
(201) De Mol, *Ingr. Chimiste* **22**, 262 (1938); *C.A.* **34**, 434.
(202) Denyer, Fidler and Lowry, *Ind. Eng. Chem.* **41**, 2726 (1949).
(203) Deshpande and Lu, *Indian J. Technol.* **1**, 403 (1963); *C.A.* **60**, 4867.
(204) Deshpande and Lu, *J. Appl. Chem. (London)* **15**, 136 (1965).
(205) Desty and Fidler, *Ind. Eng. Chem.* **43**, 905 (1951).
(205)c Devyatykh, Agliulov, Feshchenko and Stepanov, *Zh. Fiz. Khim.* **42**, 2071 (1968); *C.A.* **69**, 110387p.
(206) Devyatykh, Odnosevtsev and Umilin, *Zh. Neorgan. Khim.* **7**, 1928 (1962); *C.A.* **57**, 14485.
(206)c Devyatykh, Zorin, Postnikova, and Umilin, *Zh. Neorgan. Khim.* **14**, 1626 (1969); *C.A.* **71**, 64783f.
(207) Di Cave and Giona, *Ric. Sci., Rend. Sez. A* **4**, 645 (1964); *C.A.* **62**, 7168.
(208) Din, *Inst. Intern. Froid, Comm. Intern. Zurich* **1953**, 17; *C.A.* **49**, 5910.
(208)c Dobroserdov and Bagrov, *Zh. Prikl. Khim.* **40**, 875 (1967); *C.A.* **67**, 26297k.
(209) Debroserdov and Il'ina, *Zh. Prikl. Khim,* **34**, 386 (1961); *C.A.* **55**, 13023.
(209)c Dojcansky, Heinrich, and Surovy, *Chem. Zvesti.* **21**, 713 (1967); *C.A.* **68**, 81755d.
(210) Dominik and Wojciechowska, *Przemysl Chem.* **23**, 61 (1939); *C.A.* **33**, 4582.
(211) Donald and Ridgway, *Chem. Eng. Sci.* **5**, 188 (1956).
(212) Donald and Ridgway, *J. Appl. Chem. (London)* **8**, 403, 408 (1958); *C.A.* **53**, 21109.
(213) Donham, *Dissertation Abstr.* Mic. **58-687**; *C.A.* **52**, 14267.
(213)c Doniec, Krauze, Michalowski, and Serwinski, *Zesz. Nauk. Politech. Lodz., Chem.* **17**, 77 (1966); *C.A.* **67**, 68072p.
(213)f Doniec, Krauze, Michalowski, and Serwinski, *Zesz. Nauk. Politech. Lodz., Chem.* **19**, 171 (1969); *C.A.* **72**, 16123q.
(214) Doumas, U.S. Patent **3,212,998** (1962).
(215) Dow Chemical Co., unpublished data.
(216) Drake, U.S. Patent **2,170,854** (1939).
(217) Drake, Duvall, Jacobs, Thompson, and Sonnichsen, *J. Am. Chem. Soc.* **60**, 73 (1958).
(218) Drout, U.S. Patent **2,647,861** (1953).
(219) Dunlap, Bedford, Woodbrey, and Furrow, *J. Am. Chem. Soc.* **81**, 2927 (1959).

(220)   Dunlop and Trimble, *Ind. Eng. Chem.* **32**, 1000 (1940).
(221)   Dunn, U.S. Patent **2,524,899** (1950).
(221)c Duntov, Zernov, and Lyubelskii, *Zh. Prikl. Khim.* **40**, 599 (1967); *C.A.* **67**, 26277d.
(222)   du Pont de Nemours and Co., Wilmington, Del., Polychemicals Dept. Sales Bull. (1959).
(223)   du Pont de Nemours and Co., Wilmington, Del., *New Prods. Bull. No.* **19**.
(224)   du Pont de Nemours and Co., Wilmington, Del., Product Bulletins on THF, DMF, Hexamethyleneimine.
(225)   du Pont de Nemours and Co., Netherland Patent Appl. **6,412,607** (1965); *C.A.* **63**, 14,700.
(225)c Duras, *Chem. Zvesti* **21**, 177 (1967); *C.A.* **67**, 36757z.
(226)   Dykyj, Paulech, and Seprakova, *Chem. Zvesti* **14**, 327 (1960); *C.A.* **54**, 21908.
(227)   Eastman Chemical Products, Inc., unpublished data.
(228)   Efremov and Zel'venskii, *Khim. Prom.* **1964**, 201; *C.A.* **61**, 6448.
(229)   Efremov and Zel'venskii, *Zh. Prikl. Khim.* **38**, 2513 (1965); *C.A.* **64**, 5807.
(229)c Efremov, Zel'venskii, and Alanas'ev, *Poluch. Anal. Veshchestv Osoboi Christ., Mater. Vses. Konf. Gorky, USSR* **1963**, 44; *C.A.* **67**, 26269c.
(230)   Ehrett and Weber, *J. Chem. Eng. Data* **4**, 142 (1959).
(231)   Eiseman, *J. Am. Chem. Soc.* **79**, 6087 (1957).
(231)   Eiseman, U.S. Patent **2,999,815** (1961).
(233)   Eliot, U.S. Patent **2,635,072** (1953).
(234)   Eliot and Weaver, U.S. Patent **2,662,847** (1953).
(235)   Ellis, *U. K. At. Energy Authority, Ind. Group Hdq.* 5197 (1953); *C.A.* **53**, 15681.
(236)   Ellis and Contractor, *Birmingham Univ. Chem. Engr.* **15**, 10 (1964); *C.A.* **61**, 3736.
(237)   Ellis and Forest, *U. K. At. Energy Auth., IGR-TN/CA* 457 (1957); *C.A.* **51**, 9245.
(238)   Ellis and Johnson, *J. Inorg. Nucl. Chem.* **6**, 194, 199 (1958).
(239)   Ellis and Razavipour, *Chem. Eng. Sci.* **11**, 99 (1959); *C.A.* **54**, 10436.
(239)a Elshayal and Lu, *J. Appl. Chem. (London)* **18**, 277 (1968); *C.A.* **69**, 99909n.
(240)   Engel, U.S. Patent **2,363,159** (1944).
(241)   Engel, U.S. Patent **2,376,870** (1945).
(242)   Engel, U.S. Patent **2,404,167** (1946).
(243)   Engel, U.S. Patent **2,445,944** (1948).
(244)   Engel, U.S. Patent **2,465,716** (1949).
(245)   Engel, U.S. Patent **2,465,717** (1949).
(246)   Engel, U.S. Patent **2,465,718** (1949).
(247)   Engel, U.S. Patent **2,481,734** (1949).
(248)   Engelmann and Bittrich, *J. Prakt. Chem.* **19**, 106 (1962); *C.A.* **58**, 7415.
(249)   Engelmann and Bittrich, *Wiss. Z Tech. Hochsch. Chem. Leuna-Merseburg* **8**, 148 (1966); *C.A.* **66**, 606.
(249)c Engelmann and Bittrich, *Wiss. Z. Tech. Hochsch. Chem. Leuna-Merseburg* **8**, 289 (1966); *C.A.* **66**, 98965d.
(250)   England, U.S. Patent **2,802,028** (1957).
(251)   English and Kidwell, *Science* **139**, 341 (1963); *C.A.* **58**, 7430.
(252)   Engs, Wik, and Roberts, U.S. Patent **2,414,639** (1947).
(253)   Ernst and Kaufler, German Patent **486,492** (1926).
(254)   Evans, British Patent **579,675** (1946); *C.A.* **41**, 1695.
(255)   Evans, U.S. Patent **2,140,694** (1938).
(256)   Evans and Edlund, *Ind. Eng. Chem.* **28**, 1186 (1936).
(257)   Evans and Hass, U.S. Patent **2,442,589** (1948).
(257)c Evans and Lin, *J. Chem. Eng. Data* **13**, 14 (1968).
(258)   Evans, Morris and Shokal, U.S. Patent **2,372,941** (1945).
(258)c Evans, Morris, and Shokal, U.S. Patent **2,408,922** (1946).
(259)   Evans, Morris and Shokal, U.S. Patent **2,426,821** (1947).
(260)   Eversole, U.S. Patent **2,160,064** (1939).
(261)   Ewell and Welch, *Ind. Eng. Chem.* **37**, 1224 (1945).
(262)   Ewell and Welch, *J. Am. Chem. Soc.* **63**, 2475 (1941).
(263)   Faerber, U.S. Patent **2,836,546** (1958).
(263)c Fahmy and Assal, *Bull. Acad. Polon. Sci., Ser. Sci. Chim.* **14**, 657 (1966); *C.A.* **66**, 69343u.
(263)e Fahmy and Assal, *Bull. Acad. Polon. Sci., Ser. Sci. Chim.* **14**, 661 (1966); *C.A.* **66**, 69344v.
(263)g Fahmy and Assal, *Bull. Acad. Pol. Sci., Ser. Sci. Chim.* **14**, 667 (1966); *C.A.* **66**, 69345w.
(263)i Fahmy and Assal, *Bull. Acad. Pol. Sci., Ser. Sci. Chim.* **14**, 773 (1966); *C.A.* **66**, 98968g.
(264)   Fahnoe, U.S. Patent **2,527,358** (1950).
(265)   Fairborne, Gibson and Stephens, *J. Chem. Soc.* **1932**, 1965.

(266)   Farbwerke Hoechst AG. unpublished data.
(267)   Farchan Research Laboratory. Data Sheet B, February 1949.
(268)   Fastovskii and Petrovskii, *Zh. Fiz. Khim.* **31**, 836 (1957); *C.A.* **52**, 25.
(269)   Feldman, Julian, private communication.
(269)c Feldman, U.S. Patent **3,338,801** (1967).
(270)   Feldman and Orchin, *Ind. Eng. Chem.* **44**, 2909 (1952).
(271)   Feldman and Orchin, U.S. Patent **2,581,398** (1952).
(272)   Feldman and Orchin, U.S. Patent **2,590,096** (1952).
(273)   Field, U.S. Patent **2,212,810** (1940).
(274)   Field, U.S. Patent **2,265,939** (1941).
(274)c Findlay and Kenyon, *Aust. J. Chem.* **22**, 865 (1969); *C.A.* **71**, 16307g.
(275)   Fischer, Bingle, and Vogel, *J. Am. Chem. Soc.* **78**, 902 (1956).
(276)   Fisher, U.S. Patent **2,341,433** (1944).
(277)   Fisher and Fein, U.S. Patent **2,438,278** (1948).
(278)   Flatt and Benguerel, *Helv. Chim. Acta* **45**, 1765, (1962); *C.A.* **58**, 1952.
(279)   Fleischer, U.S. Patent **2,191,196** (1940).
(280)   Flom, Alpert, and Elving, *Ind. Eng. Chem.* **43**, 1178 (1951).
(281)   Floyd, *Dissertation Abstr. No.* **17,606**; *C.A.* **51**, 14352.
(282)   Fordyce and Simonsen, *Ind. Eng. Chem.* **41**, 104 (1949).
(283)   Forman, U.S. Patent **2,581,789** (1952).
(284)   Fowler, *J. Soc. Chem. Ind. (London)* **69**, *Suppl.* 2, S65 (1950).
(285)   Fowler and Hunt, *Ind. Eng. Chem.* **33**, 90 (1941).
(286)   Fowler and Lim. *J. Appl. Chem. (London)* **6**, 74 (1956); *C.A.* **53**, 11924.
(287)   Fowler and Norris, *J. Appl. Chem. (London)* **5**, 266 (1955).
(288)   Frazer, U.S. Patent **3,013,953** (1961).
(289)   Fredenhagen and Kerck, *Z. Anorg. Chem.* **252**, 280 (1944).
(289)c Fredenslund and Sather, *J. Chem. Eng. Data* **15**, 17 (1970); *C.A.* **72**, 83445j.
(290)   Free and Hutchison, *J. Chem. Eng. Data* **4**, 193 (1959).
(290)c Freshwater and Pike, *J. Chem. Eng. Data* **12**, 179 (1967).
(291)   Frey, U.S. Patent **2,322,800** (1943).
(292)   Frey, Matuszak, and Snow, U.S. Patent **2,186,524** (1940).
(292)c Fried, Gallant, and Schneier, *J. Chem. Eng. Data* **12**, 504 (1967).
(293)   Fried and Pick, *Collection Czech. Chem. Commun.* **26**, 954, (1961).
(294)   Fried, Pick, Hala, and Vilim. *Chem. Listy* **50**, 1039 (1956); *C.A.* **50**, 16320; *Collection Czech. Cem. Commun.* **21**, 1535 (1956); *C.A.* **51**, 11794.
(295)   Friedel, *Bull. Soc. Chim.*, **24**, (2) 160, 241 (1875).
(296)   Fritzsche and Stockton, *Ind. Eng. Chem.* **38**, 737 1946).
(297)   Fritzweiler and Dietrich, *Angnew. Chem.* **45**, 605 1932); **46**, 241 (1933).
(298)   Frolov, Loginova and Kiseleva, *Zh. Fiz. Khim.* **35**, 1784 (1961); *C.A.* **56**, 63.
(298)c Frolov, Loginova, and Nazarova, *Zh. Fiz. Khim.* **43**, 2632 (1969); *C.A.* **72**, 48083t.
(299)   Frolov, Loginova, Saprykina, and Kondakova, *Zh. Fiz. Khim.* **36**, 2282 (1962); *C.A.* **58**, 3957.
(300)   Frolov, Loginova and Shvestova and Ustavschchikov, *Zh. Fiz. Khim.* **38**, 1303 (1964); *C.A.* **61**, 7757.
(301)   Frolov, Loginova and Ustavshchikov, *Neftekhimiya* **2**, 766 (1962); *C.A.* **58**, 12009.
(301)c Frolov, Loginova, Ustavshchikov, and Dmitricheva, *Zh. Fiz. Khim.* **41**, 2088 (1967); *C.A.* **68**, 16534s.
(302)   Frolov and Spiridonova, *Uchenye Zapiski Yaroslacsk. Teknol. Inst.* **5**, 43 (1960); **56**, 9487.
(302)c Fu and Lu, *J. Appl. Chem.* **16**, 324 (1966); *C.A.* **66**, 2170.
(302)e Fu and Lu, *J. Chem. Eng. Data* **13**, 6 (1968).
(303)   Fuchs, *Chem. Ztg.* **51**, 402 (1927).
(304)   Fuqua, U.S. Patent **2,481,211** (1949).
(304)c Furzer and Ho, *Brit. Chem. Eng.* **15**, 80 (1970); *C.A.* **72**, 83426d.
(305)   Galata, Gubskaya, Kinyapina and Kofman, *Tr. po Khim. i. Khim. Teknol.* **1962**, 256; *C.A.* 59, 8573.
(306)   Galska-Krajewska, *Bull. Acad. Polon. Sci., Classe III*, **6**, 257 (1968); C.A. **52**, 15993.
(307)   Galska- *Bull. Acad. Polon. Sci., Ser. Sci. Chim.* **9**, 455 (1961); *C.A.* **60**, 7512.
(308)   Galska-Krajewska, *Roczniki Chem.* **40**, 863; (1966); *C.A.* **65**, 16121.
(308)c Galska-Krajewska, *Roczniki Chem.* **41**, 609 (1967); *C.A.* **67**, 47665u.
(309)   Galska-Krajewska and Zieborak, *Roczniki Chem.* **36**, 119 (1962); *C.A.* **57**, 6680.
(310)   Gandek, *Dissertation Abst.* **64-9159** (1964); *C.A.* **61**, 13935.
(311)   Garber and Bovkun *Zh. Prikl. Khim.* **37**, 831 (1964); *C.A.* **61**, 1319.
(311)c Garber and Bovkun, *Zh. Prikl. Khim.* **41**, 318 (1968); *C.A.* **69**, 22468k.
(312)   Garber Bovkun and Etimova, *Zh. Prikl. Khim.* **35**, 416 (1962); *C.A.* **56**, 14997.
(313)   Garber and Komarova, *Zh. Prikl. Khim.* **39**, 1366 (1966); *C.A.* **65**, 10482.

(313)c Garber and Komarova, *Zh. Prikl. Khim.* **42**, 1347 (1969); *C.A.* **71**, 101613y.
(314)   Garber and Rabukhina, *Zh. Prikl. Khim.* **33**, 2782 (1960); *C.A.* **55**, 9015.
(315)   Garber, Zelenevskaya, and Rabukhina, *Zh. Prikl. Khim.* **33**, 694 (1960); *C.A.* **54**, 20919.
(316)   Garber and Zelenexskaya, *Zh. Prikl. Khim.* **36**, 2306 (1963); *C.A.* **60**, 4867.
(317)   Garner and Hall, *J. Inst. Petrol.* **41**, 1, 18, 24 (1955).
(317)c Garrett and Van Winkle, *J. Chem. Eng. Data* **14**, 302 (1969).
(318)   Gause and Ernsberger, *Chem. Eng. Data Ser.* **2**, 28 (1957).
(319)   Gautier, *Ann. Cheim. Phys.*, **17**, (4) 191 (1869).
(320)   Gaziev, Zel'venskii, and Shalygin, *Zh. Prikl. Khim.* **31**, 1220 (1958); *C.A.* **52**, 19361.
(321)   Geckler and Fragen, U.S. Patent **2,316,126** (1943).
(322)   Gelperin and Novikova, *J. Appl. Chem. U.S.S.R.* **26**, 841 (1953).
(323)   Gelp'erin and Zelenetskii, *Zh. Fiz. Khim.* **34**, 2230 (1960); *C.A.* **55**, 13022.
(324)   Ghysels, *Bull. Soc. Chim. Belges* **33**, 57 (1924).
(325)   Gibson, U.S. Patent **2,347,317** (1944).
(326)   Giguere and Maass, *Can. J. Res.* **18B**, 181 (1940).
(326)c Gilot, Guiglion, and Enjalbert, *Chim. Ind., Genie Chim.* **98**, 1052 (1967); *C.A.* **69**, 60369a.
(326)e Ginzburg, Pikulina, and Litvin, *Zh. Prikl. Khim.* **39**, 2371 (1966).
(327)   Goldberg and Zinov'ev, *Zh. Prikl. Khim.* **33**, 1913 (1960); *C.A.* **54**, 23680.
(328)   Goldblum, Martin, and Young, *Ind. Eng. Chem.* **39**, 1474 (1947).
(328)c Golubkov, Lapidus, and Nisel'son, *Zh. Fiz. Khim.* **41**, 2090 (1967); *C.A.* **68**, 1594.
(329)   Gomberg, *J. Am. Chem. Soc.* **41**, 1414 (1919).
(330)   Gondzik and Stateczny, *Przemysl Chem.* **9**, 132 (1953); *C.A.* **48**, 11759.
(331)   Gorbunova, Lutugina and Malenko, *Zh. Prikl. Khim.* **38**, 374, 622 (1965); *C.A.* **62**, 13908, 15479.
(331)c Gorbunov, Susarev, and Balashova, *Zh. Prikl. Khim.* **41**, 312 (1968); *C.A.* **69**, 13335n.
(332)   Gordon and Benson, *Can. J. Res.* **24B**, 285 (1946).
(333)   Gordon and Bright, U.S. Patent **2,171,549** (1939).
(334)   Gorodetskii, Morachevskii and Olevskii, *Vestnik Lenningrad Univ.* **14**, No. 22, *Ser. Fiz. Khim.* No. 4, 136 (1959); *C.A.* **54**, 8255.
(335)   Gorodetskii and Olevskii, *Vestnik Leningrad Univ.* **15**, No. 16. *Ser. Fiz. Khim.* No. 3, 102 (1960); *C.A.* **55**, 1162.
(335)c Gothard, Iacob, and Brasat, *Rev. Chim. (Bucharest)* **20**, 307 (1969); *C.A.* **71**, 71060w.
(336)   Gothard and Minea, *Rev. Chim (Bucharest)* **14**, 520 (1965); *C.A.* **60**, 4867.
(337)   Gowing-Scopes, *Analyst*, **39**, 4 (1914).
(338)   Grabner and Clump, *J. Chem. Eng. Data* **10**, 13 (1965).
(339)   Granzhan, Semenenko and Kirillova, *Zh. Prikl. Khim.* **39**, 1399 (1966); *C.A.* **65**, 9820.
(340)   Greenburg, U.S. Patent **2,313,536** (1943).
(341)   Greenburg, U.S. Patent **2,405,300** (1946).
(342)   Greenburg, U.S. Patent **2,480,919** (1949).
(342)c Greene and Sonntag, *Advan. Cryog. Eng.* **13**, 357 (1967); *C.A.* **69**, 70343x.
(343)   Grekl, U.S. Patent **2,564,200** (1951).
(344)   Gresham, U.S. Patent **2,395,265** (1946).
(345)   Gresham, U.S. Patent **2,479,068** (1949).
(346)   Gresham and Brooks, U.S. Patent **2,449,470** (1948).
(347)   Griswold and Ludwig, *Ind. Eng. Chem.* **35**, 117 (1943).
(347)c Gromov, Movsumzade, and Sadykov, *Izv. Vyssh. Ucheb. Zaved., Neft Gaz* **12**, 57 (1969); *C.A.* **71**, 25124x.
(348)   Gropsianu,Kyri, and Gropsianu, *Acad. Rep. Populare Romine, Baza Cercetari Stiint. Timisoara, Studii Cercetari Stiint., Ser., Stiinte Chim.* **4**, 73 (1957); *C.A.* **53**, 19501. Gropsianu and Murarescu. *Ibid.* **3**, 81 (1956); *C.A.* **51**, 16028.
(349)   Guinot, U.S. Patent **2,316,860** (1943).
(350)   Guinot and Chassaing, U.S. Patent **2,437,519** (1948).
(351)   Gurukul and Raju, *J. Chem. Eng. Data* **11**, 501, 1966.
(351)c Haccuria and Mathieu, *Ind. Chem. Belges* **1967**, 165; *C.A.* **66**, 88856e.
(352)   Hack and Van Winkle, *Ind. Eng. Chem.* **46**, 2392 (1954).
(353)   Hacker, Lucas and Gelbin, *Chem. Tech. (Berlin)* **16**, 75 (1964); *C.A.* **61**, 1528.
(354)   Hahn, *Brennstoff-Chem.* **35**, 105 (1954).
(354)c Hakuta, Nagahama, and Hirata, *Bull. Jap. Petrol. Inst.* **11**, 10 (1969); *C.A.* **71**, 42799p.
(355)   Hall, Norris and Downs, *Anesthesiology* **21**, 522 (1960); *C.A.* **56**, 12364.
(356)   Hamilton and Cogdell, U.S. Patent **2,831,902** (1958); *C.A.* **52**, 14649.
(357)   Hammond, U.S. Patent **2,356,785** (1944).

(358)   Hands and Norman, *Ind. Chemist* **21**, 307 (1945).
(359)   Hannotte, *Bull Soc. Chim. Belges* **35**, 85 (1926).
(360)   Hansley, U.S. Patent **2,452,460** (1948).
(361)   Hanson, Hogan, Nelson and Cines, *Ind. Eng. Chem.* **44**, 604 (1952).
(362)   Hanson, Hogan, Ruehlen, and Cines, *Chem. Eng. Progr. Symp. Ser.* **49**, No. 6, 37 (1953).
(362)c   Hanson and Van Winkle, *J. Chem. Eng. Data* **12**, 319 (1967).
(363)   Harney and Amick, U.S. Patent **2,454,447** (1948).
(364)   Harper and Moore, *Ind. Eng. Chem.* **49**, 411 (1957).
(365)   Harrison and Berg, *Ind. Eng. Chem.* **38**, 117 (1946).
(366)   Harrison and Somers, U.S. Patent **2,704,271** (1955).
(367)   Hatch and Ballin, *J. Am. Chem. Soc.* **71**, 1039 (1949).
(368)   Haughton, *Chem. Eng. Sci.* **4**, 97 (1955).
(369)   Haughton, *Chem. Eng. Sci.* **15**, 145 (1961); *C.A.* **55**, 26583.
(370)   Haughton, *Chem. Eng. Sci.* **16**, 82 (1961).
(370)c   Haughton, *Brit. Chem. Eng.* **12**, 1102 (1967); *C.A.* **67**, 76646u.
(371)   Haywood, *J. Phys. Chem.* **3**, 317 (1899).
(372)   Heck and Barrick, *Advan. Cryog. Eng.* **11**, 349 (1965); *C.A.* **65**, 8087.
(372)c   Heck and Barrick, *Advan. Cryog. Eng.* **12**, 714 (1967); *C.A.* **67**, 2497.
(372)e   Heck and Hiza, *Am. Inst. Chem. Engrs.* **13**, 593 (1967); *C.A.* **67**, 1472.
(373)   Heinrich, Ilavsky and Surovy, *Chem. Zvesti* **15**, 414 (1961); *C.A.* **55**, 23021.
(374)   Heitz, *Am. J. Enol. Viticult.* **11**, 19 (1960); *C.A.* **54**, 18007.
(375)   Heldman, *J. Am. Chem. Soc.* **66**, 661 (1944).
(376)   Hellwig and Van Winkle, *Ind. Eng. Chem.* **45**, 624 (1953).
(377)   Hennion and Groebner, *J. Am. Chem. Soc.* **70**, 426 (1948).
(378)   Herold, Wustrow, and Wetzel, U.S. Patent **2,091,636** (1937).
(379)   Herz and Rathmann, *Chem. Ztg.* **36**, 1417 (1912).
(380)   Hessel and Geisler, *Z. Physik. Chem.* **229**, 199 (1965); *C.A.* **63**, 14115.
(381)   Hicks-Bruun and Brunn, *J. Res. Natl. Bur. St.* **8**, 525 (1932).
(382)   Hill, *J. Chem. Soc.* **101**, 2467 (1912).
(383)   Hill and Van Winkle, *Ind. Eng. Chem.* **44**, 205, 208 (1952).
(384)   Hipkin, *Am. Inst. Chem. Engr. J.* **12**, 484 (1966); *C.A.* **65**, 1453.
(385)   Hirata and Hirose, *Mem. Fac. Technol., Tokyo Metropol. Univ.* No. **11**, 876 (1961); *C.A.* **58**, 1950.
(386)   Hirata, Hirose, and Yanagawa, *Kagaku Kogaku* **24**, 561 (1960); C.A. **54**, 21908.
(386)c   Hirata, Suda, Hakuta, and Nagahama, *J. Chem. Eng. Japan* **2**, 143 (1969); *C.A.* **72**, 48104a.
(386)f   Hirata, Suda, Hakuta, and Nagahama, *Sekiyu Gakkai Shi.* **12**, 773 (1969); *C.A.* **72**, 71165e.
(386)i   Hirata and Hakuta, *Mem. Fac. Technol., Tokyo Metropol. Univ.* **18**, 1595 (1968); *C.A.* **70**, 100186d.
(386)m   Hiza, Heck, and Kidnay, *Chem. Eng. Progr., Symp. Ser.* **64**, 57 (1968); *C.A.* **70**, 41289q.
(387)   Hogan, Nelson, Hanson, and Cines, *Ind. Eng. Chem.* **47**, 2210 (1955).
(388)   Holley, *J. Am. Chem. Soc.* **24**, 448 (1902).
(389)   Hlolly and Weaver, *J. Am. Chem. Soc.* **27**, 1049 (1905).
(390)   Hollo, Ember, Lengyel, and Wieg, *Acta Chim. Acad. Sci. Hung.* **13**, 307 (1957); *C.A.* **52**, 17862.
(391)   Hollo and Lengyel, *Fette, Seifen, Anstrichmittel* **62**, 913 (1960); *C.A.* **55**, 8009.
(392)   Hollo and Lengyel, *Ind. Eng. Chem.* **51**, 957 (1959).
(393)   Hollo and Lengyel, *Periodica Polytech.* **2**, 173 (1958); *C.A.* **53**, 5799.
(394)   Hollo, Lengyel and Uzonyi, *Periodica Polytech.* **4**, 173 (1960); *C.A.* **55**, 16108.
(395)   Holst and Hamburger, *Z. Physik. Chem.* **91**, 513 (1916).
(396)   Homfray, *J. Chem. Soc.* **87**, 1431 (1905).
(397)   Hopkins, Yerger, and Lynch, *J. Am. Chem. Soc.* **61**, 2460 (1939).
(398)   Hori, *J. Agr. Chem. Soc. Japan* **18**, 155 (1942); *C.A.* **45**, 4202.
(399)   Horsley, *Anal. Chem.* **19**, 508 (1947).
(400)   Horsley, *Anal. Chem.* **21**, 831 (1949).
(401)   Horvitz, U.S. Patent **3,013,076** (1961).
(402)   Horvitz and Pope, U.S. Patent **3,012,948** (1961); *C.A.* **56**, 12741.
(403)   Horyna, *Coll. Czech. Chem. Commun.* **24**, 3253 (1959); *C.A.* **54**, 10436.
(404)   Houser and Van Winkle, *Chem. Eng. Data Ser.* **2**, 12 (1957).
(405)   Houston, *J. Am. Chem. Soc.* **55**, 4131 (1933).
(406)   Howe and Hass, *Ind. Eng. Chem.* **38**, 251 (1946).
(406)a   Hudson and Van Winkle, *J. Chem. Eng. Data* **14**, 310 (1969).
(406)c   Humphrey and Van Winkle, *J. Chem. Eng. Data* **12**, 526 (1967).
(407)   Hunsmann, *Chem.Ingr. Tech.* **33**, 537 (1961); *C.A.* **55**, 25385.
(408)   Hunt, U.S. Patent 2,862,856 (1958).

(409)   Huntress, "Organic Chlorine Compounds", p. 588, 1038, New York, John Wiley & Sons, 1948.
(410)   Huntres and Sanchez-Nieva, *J. Am. Chem. Soc.* **70**, 2813 (1948).
(411)   Hyatt, U.S. Patent **2,176,500** (1939).
(412)   Ibl, Dandliker, and Trumpler, *Chem. Eng. Sci.* **5**, 193 (1956).
(412)c   Igumenoy, Kharchenko, and Mikhailov, *Zh. Prikl. Khim.* **42**, 1662 (1969); *C.A.* **71**, 95486r.
(413)   Imperial Chemical Industries Ltd., unpublished data.
(414)   Ishiguro, Yagyu, Ikushima, and Nakazawa, *J. Pharm. Soc. Japan* **75**, 434 (1955); *C.A.* **50**, 2587.
(415)   Ishiguro, Yagyu, and Takagi, *Yakugaku Zasshi* **79**, 1138 (1959); *C.A.* **54**, 2857.
(416)   Ishiguro, Yagyu, and Takagi, *Yakugaku Zasshi* **80**, 30 (1960); *C.A.* **54**, 11617.
(417)   Izard, U.S. Patent **2,061,732** (1936).
(418)   Jackson and Young, *J. Chem. Soc.* **73**, 922 (1898).
(419)   Jakubicek, *Collection Czech. Chem. Commun.* **26**, 300 (1961); *C.A.* **55**, 10026.
(420)   Jakubicek, *Collection Czech. Chem. Commun.* **28**, 3180 (1962).
(421)   Jakubicek, Fried, and Vehala, *Chem. Listy* **51**, 1422 (1957); *C.A.* **51**, 17382.
(422)   Janecke, *Z. Physik. Chem.* **164**, 401 (1933).
(423)   Jenkins and King, *Chem. Eng. Sci.* **20**, 921 (1965); *C.A.* **64**, 1403.
(424)   Jensen, U.S. Patent **2,360,685** (1944).
(425)   Johannesen, U.S. Patent **2,656,389** (1953).
(426)   Johnson and Spurlin, U.S. Patent **2,459,433** (1949).
(427)   Johnson, Ward and Furter, *Can. J. Technol.* **34**, 514 (1957); *C.A.* **51**, 14351.
(427)c   Johny, Krishnan, and Pai, *Indian J. Technol.* **6**, 278 (1968); *C.A.* **70**, 14850j.
(428)   Jones, *J. Chem. Eng. Data* **7**, 13 (1962).
(429)   Jones, Schoenborn, and Colburn, *Ind. Eng. Chem.* **35**, 666 (1943).
(430)   Jordan, Univ. *Dissertation Abstr. Mic.* **60-1188**; *C.A.* **54**, 14848.
(431)   Jordan and Kay, *Chem. Eng. Progr. Symp. Ser.* **59**, 46 (1963); *C.A.* **59**, 5853.
(432)   Junghaus and Weber, *J. Prakt. Chem.* [4] **2**, 265 (1955); *C.A.* **54**, 17024.
(433)   Kaiser, *Chem. Ingr. Tech.* **38**, 151 (1966); *C.A.* **64**, 15057.
(433)c   Kaminishi, Arai, Saito, and Maeda, *J. Chem. Eng. Japan* **1**, 109 (1968); *C.A.* **70**, 32009w.
(434)   Kaminishi and Toriumi, *Kogyo Kagaku Zasshi* **69**, 175 (1966); *C.A.* **65**, 9801.
(434)c   Kaminishi and Toriumi, *Rev. Phys. Chem. Japan* **38**, 79 (1968); *C.A.* **70**, 23462m.
(434)f   Kandalova, Aleksandrova, Arushan'yants, and Gasparyan, *Uch. Zap., Erevan. Gos. Univ.* **3**, 191 (1968); *C.A.* **72**, 48105b.
(435)   Kaplan and Monakhova, *J. Gen. Chem. (U.S.S.R.)* **7**, 2499 (1937).
(436)   Karpinski and Swietoslawski, *Compt. Rend.* **198**, 2166 (1934).
(437)   Karr, U.S. Patent **2,463,629** (1949).
(438)   Karr, Scheibel, Bowes, and Othmer, *Ind. Eng. Chem.* **43**, 961 (1951).
(438)c   Katunin and Grisha, *Zh. Prikl. Khim.* **41**, 1348 (1968); *C.A.* **69**, 54711a.
(439)   Katz and Newman, *Ind. Eng. Chem.* **48**, 137 (1956).
(440)   Kawalec, *Bull. Acad. Polon. Sci., Ser. Sci. Chim.* **11**, 211 (1963); *C.A.* **59**, 9379.
(441)   Kawalec, *Bull. Acad. Polon. Sci., Ser. Sci. Chim.* **13**, 771 (1965); *C.A.* **64**, 8939.
(442)   Kay, *J. Phys Chem.* **68**, 827 (1964).
(443)   Kay and Brice, *Ind. Eng. Chem.* **45**, 615 (1953).
(444)   Kay and Fisch, *A. I. Ch. E. J.* **4**, 293 (1958).
(445)   Kay and Rambosek, *Ind. Eng. Chem.* **45**, 221 (1953).
(446)   Kay and Warzel, *A. I. Ch. E. J.* **4**, 296 (1958).
(447)   Keistler and Van Winkle, *Ind. Chem.* **44**, 622 (1952).
(448)   Kellogg and Cady, *J. Am. Chem. Soc.* **70**, 3986 (1948).
(449)   Kenttamaa, Lindberg, and Nissema, *Suomen Kemistilehti* **33B**, 189 (1960); *C.A.* **55**, 8009.
(449)c   Kesselman, Hollenbach, Myers, and Humphrey, *J. Chem. Eng. Data* **13**, 34 (1968).
(450)   Kharakhorin, *Inzhener-Fiz. Zhur. Akad. Nauk Beloruss. S.S.R.* **2**, 55 (1959); *C.A.* **54**, 1003.
(450)c   Kharchenko, Mikhailov, and Igumenov, *Izv. Sib. Otd. Akad. Nauk SSSR, Ser. Khim. Nauk* **4**, 32 (1968); *C.A.* **70**, 61665n.
(451)   Kharin and Perelygin, *Izv. Vysshikh Uchebn. Zavedenii, Khim. i. Khim. Tekhnol.* **8**, 564 (1965); *C.A.* **64**, 1402.
(452)   Kharin and Perelygin, *Izv. Vysshikh Uchebn. Zavedenii, Khim. i. Khim. Tekhnol.* **9**, 210 (1966); *C.A.* **65**, 9801.
(452)b   Kharin, Perelygin, and Remizov, *Izv. Vyssh. Uchebn. Zavedenii, Khim. Khim. Tekhnol.* **11**, 871 (1968); *C.A.* **70**, 14858t.
(452)c   Kharin, Perelygin, and Volkov, *Izv. Vyssh. Uchebn. Zavedenii, Pishch. Tekhnol.* **1968**, 136; *C.A.* **69**, 90291x.
(452)e   Kharin, Perelygin, and Remizov, *Izv. Vyssh. Ucheb. Zaved., Khim. Khim. Tekhnol.* **12**, 424 (1969); *C.A.* **71**, 42797m.

(452)g   Kharin, Perelygin, and Volkov, *Izv. Vyssh. Ucheb. Zaved., Pisch. Tekhnol.* **5**, 130 (1968); *C.A.* **70**, 31990x.
(453)   Khazanova, Lesnevskaya and Zakharova, *Khim. Prom.* **42**, 364 (1966); *C.A.* **65**, 16121.
(454)   Kibler and Gusakova, *Gidroliz. i Lesokhim. Prom.* **12**, 14 (1959); *C.A.* **53**, 12776.
(455)   Kieffer and Grabiel, *Ind. Eng. Chem.* **43**, 973 (1951).
(456)   Kieffer and Holroyd, *Ind. Eng. Chem.* **47**, 457 (1955).
(457)   Kilian and Bittrich, *Z. Physik. Chem.* **230**, 383 (1965); *C.A.* **64**, 8951.
(458)   Killgore, Chew and Orr, *J. Chem. Eng. Data* **11**, 535 (1966).
(459)   Kim and Kim, *Chosun Kwahakwon Tongbo* **1964**, 30; *C.A.* **64**, 10456.
(460)   Kimberlin, U.S. Patent **2,275,151** (1943).
(461)   King, Kuck, and Frampton, *J. Am. Oil Chemists Assoc.* **38**, 19 (1961); *C.A.* **55**, 7688.
(461)c   Kirby and Van Winkle, *J. Chem. Eng. Data* **15**, 177 (1970).
(462)   Kireev, Kaplan, and Zlobin, *J. Appl. Chem. (U.S.S.R.)* **7**, 1333 (1934); *C.A.* **29**, 5712.
(463)   Kireev and Monakhova, *J. Phys. Chem. (U.S.S.R.)* **6**, 71 (1936); *C.A.* **31**, 25.
(464)   Kireev and Sitnikov, *J. Phys. Chem. (U.S.S.R.)* **15**, 492 (1941); *C.A.* **36**, 6404.
(465)   Kireev and Skvortsova, *J. Phys. Chem. (U.S.S.R.)* **6**, 63 (1936); *C.A.* **31**, 25.
(466)   Kirk-Othmer, "Encyclopedia of Chemical Technology", Vol. III, p. 794, Interscience, New York, 1949.
(467)   Kirsanova and Byk, *Zh. Prikl. Khim.* **31**, 1610 (1958); *C.A.* **53**, 2721.
(468)   Kirsanova and Byk, *Zh. Prikl. Khim.* **33**, 2784 (1960); *C.A.* **55**, 9017.
(469)   Kirsanova and Byk, *Zh. Prikl. Khim.* **34**, 1373 (1961); **35**, 198 (1962); *C.A.* **55**, 20592; *C.A.* **56**, 12364.
(470)   Kiss and Kules, *Magy. Tud. Akad. Kozp. Fiz. Kut. Int. Kozlemen* **9**, 317 (1961); *C.A.* **57**, 10579.
(471)   Kleinert, *Angew. Chem.* **46**, 18 (1933).
(471)c   Klekers and Scheller, *J. Chem. Eng. Data* **13**, 480 (1968).
(472)   Kliment, Fried and Pick, *Collection Czech. Chem. Commun.* **29**, 2008 (1964).
(473)   Kliment and Vesely, *Tech. Publ. Stredisko Tech. Inform. Potravinar. Prumyslu* No. **161**, 5 (1959-61); *C.A.* **60**, 7512.
(474)   Kobayashi, *et al.*, Japanese Patent **3066** (1952);*C.A.* **48**, 2772.
(475)   Kodak Ltd., British Patent **501,927** (1939).
(476)   Kofman, Matveeva, Mandel'shtam, Kinyapina, Konilspol'skii and Mitrofanova, *Sintez Monomerov dlya Provizv. Sintetich. Kauchuka, Gos. Inst. po Proektu. Zavadov Kauchuk. Prom. i Vses. Nauchn.-Issled. Inst. Sintetich. Kauchuka* **1960**, 42; *C.A.* **57**, 1565.
(477)   Kogan, Fridman, and Deizenrot, *Zh. Prikl. Khim.* **30**, 1339 (1957); *C.A.* **52**, 2486.
(478)   Kogan, Fridman, and Romanova, *Zh. Fiz. Khim.* **33**, 1521 (1959); *C.A.* **54**, 8195.
(479)   Kogan and Ogorodnikov, *Khim. Prom.* **1962**, 660; *C.A.* **59**, 9609.
(480)   Kogan and Tolstova, *Zh. Fiz. Khim.* **33**, 276 (1959); *C.A.* **53**, 20995.
(481)   Kohoutek, *Collection Czech. Chem. Commun.* **25**, 288 (1960); *C.A.* **54**, 16068.
(481)a   Komarov and Kokurina, *Zh. Prikl. Khim.* **42**, 1431 (1969); *C.A.* **71**, 74852s.
(481)c   Komarov and Krichevtsov, *Zh. Prikl. Khim.* **39**, 2834 (1966); *C.A.* **66**, 99000x.
(481)e   Komarov and Krichevtsov, *Zh. Prikl. Khim.* **39**, 2838 (1966); *C.A.* **66**, 99001y.
(481)g   Komatsu, Suzuki, and Ishikawa, *Kogyo Kogaku Zasshi* **72**, 811 (1969); *C.A.* **71**, 74714y.
(482)   Kominek-Szczepanik, *Roczniki Chem.* **33**, 553 (1959); *C.A.* **53**, 21723.
(483)   Konowaloff, *Ann. Physik.* (2) **14**, 34 (1881).
(484)   Korablina, Barinova, Kurakina and Ryabova, *Plasticheskie Massy* **1963**, 18; *C.A.* **60**, 7511.
(485)   Korchemskaya, Shakhparonov, Lel'chuk, Martynova, Baburina, and Voronia, *Zh. Prikl. Khim.* **33**, 2703 (1960); *C.A.* **55**, 11006.
(486)   Kovalenko and Balandina, *Uchenye Zapiski Rostov-na-Donu Univ.* **41**, 39, (1958); *C.A.* **55**, 6118.
(486)c   Kozhitov and Vanyukov, *Zh. Prikl. Khim.* **42**, 1764 (1969); *C.A.* **71**, 105806m.
(487)   Kramer and Reid. *J. Am. Chem. Soc.* **43**, 880 (1921).
(488)   Kretschmer, Nowakowska, and Wiebe, *J. Am. Chem. Soc.* **70**, 1785 (1948).
(489)   Kretschmer and Wiebe, *J. Am. Chem. Soc.* **71**, 1793 (1949).
(490)   Kretschmer and Wiebe, *J. Am. Chem. Soc.* **71**, 3176 (1949).
(491)   Krichevskii, Khazanova, and Linshits, *Zh. Fiz. Khim.* **31**, 2711 (1957); *C.A.* **52**, 8660.
(491)c   Kriebel, *Kaeltetechnik* **19**, 8 (1967); *C.A.* **69**, 80832k.
(491)e   Krishnan and Pai, *Indian J. Technol.* **5**, 196 (1967); *C.A.* **67**, 8895.
(492)   Krishnamurty and Rao, *J. Am. Sci. Ind. Research (India)* **14B**, 55 (1955); *C.A.* **49**, 11379.

(493)   Krishnamurty, Rao and Rao, *J. Sci. Ind. Res (India)* **21D**, 312 (1962); *C.A.* **58**, 1950.
(494)   Krokhin, *Zh. Fiz. Khim.* **39**, 3076 (1965); *C.A.* **64**, 8977.
(494)c   Krokhin, *Zh. Fiz. Khim.* **43**, 442 (1969); *C.A.* **70**, 109535z.
(494)f   Krokhin, *Zh. Fiz. Khim.* **43**, 1323 (1969); *C.A.* **71**, 74698w.
(494)i   Krokhin, *Zh. Fiz. Khim.* **43**, 2389 (1969); *C.A.* **72**, 25636m.
(495)   Krokhin, *Zh. Fiz. Khim.* **40**, 500 (1966); *C.A.* **64**, 16707.
(496)   Krokhin, *Zh. Fiz. Khim.* **40**, 928 (1966); *C.A.* **65**, 1452.
(496)c   Krokhin, *Zh. Fiz. Khim.* **41**, 671 (1967); *C.A.* **66**, 108818z.
(496)e   Krokhin, *Zh. Fiz. Khim.* **41**, 1509 (1967); *C.A.* **67**, 111796s.
(496)g   Kudryavtseva and Eisen, *Eesti NSV Tead. Akad. Toim., Keem., Geol.* **16**, 97 (1967); *C.A.* **67**, 94409y.
(497)   Kudryavtseva, Eisen and Susarev, *Zh. Fiz. Khim.* **40**, 1285, 1652 (1966); *C.A.* **65**, 14497.
(498)   Kudryavtseva and Susarev, *Zh. Prikl. Khim.* **36**, 1231, 1471, 1710, 2025 (1963); *C.A.* **59**, 10809; C.A. **60**, 57, 3545, 7503.
(498)c   Kudryavtseva, Susarev, and Eisen, *Zh. Fiz. Khim.* **40**, 2637 (1966); *C.A.* **66**, 2170.
(498)f   Kudryavtseva, Viit, and Eisen, *Eesti NSV Tead. Akad. Toim., Keem., Geol.* **17**, 242 (1968); *C.A.* **70**, 14834g.
(498)i   Kudryavtseva, Susarev, and Eisen, *Zh. Fiz. Khim.* **43**, 437 (1969); *C.A.* **70**, 109549g.
(498)m   Kudryavtseva, Viit, and Eisen, *Eesti NSV Tead. Akad. Toim., Keem., Geol.* **18**, 346 (1969); *C.A.* **72**, 59701u.
(499)   Kuenen, *Z. Physik. Chem.* **24**, 667 (1897).
(500)   Kuenen, *Z. Physik. Chem.* **37**, 485 (1901).
(501)   Kurmanadharao, Krishnamurty, and Rao, *J. Sci. Ind. Research (India)* **15B**, 682 (1956); *C.A.* **51**, 14352.
(502)   Kurmanadharao, Krishnamurty, and Rao, *Rec. Trav. Chim.* **76**, 769 (1957).
(503)   Kurmanadharao and Rao, *Chem. Eng. Sci.* **7**, 97 (1957); *C.A.* **52**, 15218.
(504)   Kurtyka, *Bull. Acad. Polon. Sci. Classe III* **2**, 291 (1954); 3, 47 (1955).
(505)   Kurtyka, *Bull. Acad. Polon. Sci. Classe III* **4**, 49 (1956); *C.A.* **51**, 1676.
(506)   Kurtyka, *Bull. Acad. Polon. Sci., Ser. Sci. Chim.* **9**, 741, 745 (1961); C.A. **60**, 3918, 4868.
(507)   Kurtyka and Trabczynski, *Roczniki Chem.* **32**, 623 (1958); *C.A.* **53**, 2077.
(507)c   Kushner, Lebedeva, Tatsievskaya, and Serafimov, *Zh. Fiz. Khim.* **42**, 1104 (1968); *C.A.* **69**, 30617y.
(507)e   Kushner, Tatsievskaya, Irzun, Volkova, and Serafimov, *Zh. Fiz. Khim.* **40**, 3010 (1966); *C.A.* **66**, 39364b.
(507)g   Kushner, Tatsievskaya, and Serafimov, *Zh. Fiz. Khim.* **41**, 237 (1967); *C.A.* **66**, 9294.
(507)i   Kushner, Tatsievskaya, and Serafimov, *Zh. Fiz. Khim.* **42**, 2248 (1968); *C.A.* **70**, 23467s.
(507)k   Kushner, Tatsievskaya, Babich, and Serafimov, *Zh. Prikl. Khim.* **42**, 100 (1969); *C.A.* **70**, 81472a.
(508)   Kvalnes, U.S. Patent **3,085,065** (1960).
(509)   Kyle and Tetlow, *J. Chem. Eng. Data* **5**, 275 (1960).
(510)   Lacher, Buck, and Parry, *J. Am. Chem. Soc.* **63**, 2422 (1941).
(511)   Lacher and Hunt, *J. Am. Chem. Soc.* **63**, 1752 (1941).
(512)   Lacourt, *Bull. Soc. Chim. Belges.* **36**, 346 (1927).
(513)   Lake, U.S. Patent **2,432,771** (1947).
(514)   Lake and McDowell, U.S. Patent **2,456,561** (1948).
(515)   Lake and Stribley, U.S. Patent **2,439,777** (1948).
(516)   Lake and Stribley, U.S. Patent **2,477,303** (1949).
(517)   Landwehr, Yerazunis, and Steinhauser, *Chem. Eng. Data Ser.* **3**, 231 (1958).
(518)   Lang, *Z. Physik. Chem. (Leipzig)* **196**, 278 (1950); *C.A.* **45**, 10025.
(519)   Lange, "Handbook of Chemistry", 5th ed., p. 1386, Sandusky, Ohio, Handbook Publishers, Inc., 1944.
(520)   Langer, Connell, and Wender, *J. Org. Chem.* **23**, 50 (1958).
(521)   Lapidus and Nisel'son, *Russ. J. Phys Chem.* **40**, 340 (1956).
(521)c   Lapidus and Nisel'son, *Zh. Fiz. Khim.* **42**, 1406 (1968); *C.A.* **69**, 54703z.
(522)   Lapidus, Nisel'son and Karateeva, *Zh. Fiz. Khim.* **40**, 1630 (1966); *C.A.* **65**, 14495.
(522)c   Lapidus, Nisel'son, and Karateeva, *Zh. Fiz. Khim.* **41**, 482 (1967); *C.A.* **66**, 119312f.
(522)e   Lapidus, Nisel'son, and Karateeva, *Zh. Obshch. Khim.* **37**, 531 (1967); *C.A.* **67**, 6151.
(522)g   Lapidus, Nisel'son and Kulakova, *Zh. Obshch. Khim.* **39**, 2163 (1969); *C.A.* **72**, 48129n.
(523)   Latimer, *A. I. Ch. E. J.* **3**, 75 (1957); *C.A.* **51**, 9245.
(524)   Lebedeva and Khodeeva, *Zh. Fiz. Khim.* **35**, 2602 (1961); *C.A.* **57**, 8456.
(525)   Lebo, *J. Am. Chem. Soc.* **43**, 1005 (1921).
(526)   Lecat, *Acad. Roy. Belges, Sci., Mem.* **23**, 8 (2) (1943-44).

(527) Lecat, *Ann. Chim.*, (12) **2**, 158-202 (1947).
(528) Lecat, *Ann. Soc. Sci. Bruxelles*, **45**, I, 169 (1925).
(529) Lecat, *Ann. Soc. Sci. Bruxelles*, **45**, I, 284 (1925).
(530) Lecat, *Ann. Soc. Sci. Bruxelles*, **47B**, I, 21 (1927).
(531) Lecat, *Ann. Soc. Sci. Bruxelles*, **47B**, I, 63 (1927).
(532) Lecat, *Ann. Soc. Sci. Bruxelles*, **47B**, I, 108 (1927).
(533) Lecat, *Ann. Soc. Sci. Bruxelles*, **47B**, II, 39 (1927).
(534) Lecat, *Ann. Soc. Sci. Bruxelles*, **47B**, II, 87 (1927).
(535) Lecat, *Ann. Soc. Sci. Bruxelles*, **47B**, I, 149 (1927).
(536) Lecat, *Ann. Soc. Sci. Bruxelles*, **48B**, I, 13 (1928).
(537) Lecat, *Ann. Soc. Sci. Bruxelles*, **48B**, II, 54 (1928).
(538) Lecat, *Ann. Soc. Sci. Bruxelles*, **48B**, II, 113 (1928).
(539) Lecat, *Ann. Soc. Sci. Bruxelles*, **48B**, II, 1 (1928).
(540) Lecat, *Ann. Soc. Sci. Bruxelles*, **48B**, II, 105 (1928).
(541) Lecat, *Ann. Soc. Sci. Bruxelles*, **49B**, I, 17 (1929).
(542) Lecat, *Ann. Soc. Sci. Bruxelles*, **49B**, I, 109 (1929).
(543) Lecat, *Ann. Soc. Sci. Bruxelles*, **49B**, I, 28 (1929).
(544) Lecat, *Ann. Soc. Sci. Bruxelles*, **49B**, I, 119 (1929).
(545) Lecat, *Ann. Soc. Sci. Bruxelles*, **50B**, I, 21 (1930).
(546) Lecat, *Ann. Soc. Sci. Bruxelles*, **55B**, 253 (1935).
(547) Lecat, *Ann. Soc. Sci. Bruxelles*, **56B**, 41 (1936).
(548) Lecat, *Ann. Soc. Sci. Bruxelles*, **56B**, 221 (1936).
(549) Lecat, *Ann. Soc. Sci. Bruxelles*, **60**, 155 (1940-46).
(550) Lecat, *Ann. Soc. Sci. Bruxelles*, **60**, 163 (1940-46).
(551) Lecat, *Ann. Soc. Sci. Bruxelles*, **60**, 169 (1940-46).
(552) Lecat, *Ann. Soc. Sci. Bruxelles*, **60**, 228 (1940-46).
(553) Lecat, *Ann. Soc. Sci. Bruxelles*, **61**, 73 (1947).
(554) Lecat, *Ann. Soc. Sci. Bruxelles*, **61**, 79 (1947).
(555) Lecat, *Ann. Soc. Sci. Bruxelles*, **61**, 148 (1947).
(556) Lecat, *Ann. Soc. Sci. Bruxelles*, **61**, 153 (1947).
(557) Lecat, *Ann. Soc. Sci. Bruxelles*, **61**, 255 (1947).
(558) Lecat, *Ann. Soc. Sci. Bruxelles*, **62**, 55 (1948).
(559) Lecat, *Ann. Soc. Sci. Bruxelles*, **62**, 93 (1948).
(560) Lecat, *Ann. Soc. Sci. Bruxelles*, **62**, 128 (1948).
(561) Lecat, *Ann. Soc. Sci. Bruxelles*, **63**, 58 (1949).
(562) Lecat, *Ann. Soc. Sci. Bruxelles*, **63**, 111 (1949).
(563) Lecat, "Azeotropisme," Lamertin, Brussels, 1918.
(564) Lecat, *Bull. Classe Sci. Acad. Roy. Belg.* **29**, 273 (1943).
(565) Lecat, *Bull. Classe Sci. Acad. Roy. Belg.*, **32**, 351 (1946).
(566) Lecat, *Bull. Classe Sci. Acad. Roy. Belg.* **33**, 160 (1947).
(567) Lecat, *Bull. Classe Sci. Acad. Roy. Belg.* **35**, 484 (1949).
(568) Lecat, *Compt. Rend.* **217**, 242 (1943).
(569) Lecat, *Compt. Rend.* **222**, 733 (1946).
(570) Lecat, *Compt. Rend.* **222**, 882, 1488 (1946).
(571) Lecat, *Compt. Rend.* **223**, 286 (1946).
(571)c Lecat, *Compt. Rend.* **223**, 478 (1946).
(572) Lecat, *Rec. Trav. Chem.* **45**, 620 (1926).
(573) Lecat, *Rec. Trav. Chim.* **46**, 240 (1927).
(574) Lecat, *Rec. Trav. Chim.* **47**, 13 (1928).
(575) Lecat, "Tables azeotropiques," l'Auteur, Brussels, July 1949.
(576) Lecat, *Z. Anorg. Algem. Chem.* **186**, 119 (1929).
(577) Ledwock, *Farbe Lack* **62**, 462 (1956); *C.A.* **51**, 17301.
(577)c Lee and Scheller, *J. Chem. Eng. Data* **12**, 497 (1967).
(578) Leitman and Pevzner, Russian Patent **137,907** (1960); *C.A.* **56**, 409.
(579) Lelakowska, *Bull. Acad. Polon. Sci. Classe III* **6**, 645 (1958); *C.A.* **53**, 6719.
(580) Lepingle, *Bull. Soc. Chim. Belges* **39**, 741, 864 (1926).
(580)c Lesbre and Mazerolles, *Compt. Rend.* **240**, 622; *C.A.* **49**, 7903.
(581) Lessells and Corrigan, *Chem. Eng. Data Ser.* **3**, 43 (1958).
(581)c Lesteva, Kachalova, Morozova, Ogorodnikov, and Trenke, *Zh. Prikl. Khim.* **40**, 1808 (1967); *C.A.* **68**, 43653m.
(581)e Lesteva, Ogorodnikov, and Morozova, *Zh. Prikl. Khim.* **40**, 891 (1967); *C.A.* **67**, 6854.
(581)g Lesteva, Morozova, Morozova, Ogorodnikov, and Tyvina, *Zh. Prikl. Khim.* **42**, 553 (1969); *C.A.* **71**, 6952z.
(582) Lewis, U.S. Patent **2,641,580** (1953).
(583) Liberman, Parnes, and Kursanov, *Bull. Acad. Sci. U.R.S.S., Classe Sci. Chim.* **1948**, 101.

(584)   Licht and Denzler, *Chem. Eng. Progr.* **44**, 627 (1948).
(585)   Lidstone, *J. Chem. Soc.* **1940**, 241.
(586)   Linek, Fried and Pick, *Collection Czech. Chem. Commun.* **30**, 1358 (1965); *C.A.* **62**, 15481.
(587)   Lisicki and Galska-Krajewska, *Roczniki Chem.* **40**, 873 (1966); *C.A.* **65**, 16121.
(587)c Liszi, *Magy. Kem. Foly.* **75**, 452 (1969); *C.A.* **72**, 6629v.
(588)   Lloyd and Wyatt, *J. Chem. Soc.* **1955**, 2248.
(589)   Litkenhous, Van Arsdale, and Hutchison, *J. Phys. Chem.* **44**, 377 (1940).
(590)   Lo, Bieber and Karr, *J. Chem. Eng. Data* **7**, 327 (1962).
(591)   Loder, French Patent **814,838** (1937); U.S. Patents **2,135,447-60** (1938).
(591)c Lodl and Scheller, *J. Chem. Eng. Data* **12**, 485 (1967).
(591)e Loginova, Frolov, and Uslavshchikov, *Khim. Prom.* **44**, 275 (1968); *C.A.* **69**, 13326k.
(592)   Long, *J. Chem. Eng. Data* **8**, 174 (1963).
(593)   Long, Martin, and Vogel, *Chem. Eng. Data Ser.* **3**, 28 (1958).
(594)   Lorette and Howard, *J. Org. Chem.* **25**, 1814 (1960).
(595)   Lu, *Can. J. Technol.* **34**, 468 (1957); *C.A.* **55**, 8009.
(596)   Luettringhaus and Dirksen, *Angew. Chem.* **75**, 1059 (1963); *C.A.* **60**, 2748.
(597)   Lumatainen, *U.S. At. Energy Comm. ANL* **6003** (1959); *C.A.* **54**, 2858.
(597)a Lutugina and Kolbina, *Zh. Prikl. Khim.* **41**, 2766 (1968); *C.A.* **70**, 61653g.
(597)c Lutugina, Kolbina, and Podberezkina, *Vestn. Leningrad. Univ.* **22**, *Fiz. Khim.* 68 (1967); *C.A.* **67**, 68080q.
(598)   Lutugina and Kovalichev, *Vestn. Leningr. Univ. Ser. Fiz. i Khim.* **21**, 91 (1966); *C.A.* **65**, 9801.
(598)c Lutugina, Molodenko, and Dement'ev, *Vestn. Leningrad. Univ.* **22**, *Fiz. Khim.* 107 (1967); *C.A.* **69**, 54659q.
(599)   Lutugina, Molodenko and Orlievskaya, *Zh. Prikl. Khim.* **39**, 1774 (1966); *C.A.* **65**, 16119.
(600)   Lyvers and Van Winkle, *Chem. Eng. Data Ser.* **3**, 60 (1958).
(601)   Macarron, *Rev. Real Acad. Cienc. Exact., Fis. Nat (Madrid)* **53**, 357, 607, (1959); *C.A.* **54**, 16969.
(601)c Mackendrick, Heck, and Barrick, *J. Chem. Eng. Data* **13**, 352 (1968).
(602)   MacWood and Paridon, *J. Phys. Chem.* **63**, 1302 (1959); *C.A.* **54**, 1053.
(603)   Maczynski and Maczynska, *Bull. Acad. Polon. Sci., Ser. Sci. Chim.* **13**, 299 (1965); *C.A.* **64**, 2798.
(604)   Madhaven and Murti, *Chem. Eng. Sci.* **21**, 465 (1966); *C.A.* **65**, 9802.
(605)   Magnesco and Ode, *Scientia* **31**, 5 (1964); *C.A.* **63**, 9112.
(606)   Magnus, *Ann. Physik.* **38**, 488 (1836).
(607)   Mainkar and Mene, *Indian J. Technol.* **3**, 228 (1965); *C.A.* **63**, 14112.
(608)   Mainkar and Mene, *Indian Chem. Engr.* **7**, 47 (1965); *C.A.* **64**, 8977.
(608)c Mainkar and Mene, *Trans. Indian Inst. Chem. Eng.* 120 (1967); publ. in *Indian Chem. Eng.* **9**; *C.A.* **70**, 61648j.
(608)f Mainkar and Mene, *Trans. Indian Inst. Chem. Eng.* **10**, 169 (1968); *C.A.* **71**, 42802j.
(609)   Mair, *Anal. Chem.* **28**, 52 (1956).
(610)   Malanowski, *Bull. Acad. Polon. Sci., Ser. Sci. Chim.* **9**, 77, 83 (1961).
(611)   Malesinska, *Bull. Acad. Polon. Sci., Ser. Sci. Chim.* **12**, 853 (1964); *C.A.* **63**, 4995.
(612)   Malesinska and Malesinski, *Bull. Acad. Polon. Sci., Ser. Sci. Chim.* **8**, 191 (1960); *C.A.* **55**, 11047.
(613)   Malesinska and Malesinski, *Bull. Acad. Polon. Sci., Ser. Sci. Chim.* **11**, 469, 475 (1963).
(614)   Malesinski, *Bull. Acad. Polon. Sci., Classe III* **4**, 365 (1956); *C.A.* **51**, 3217.
(615)   Maletskii and Kogan, *Khim. Prom.* **42**, 626 (1966); *C.A.* **65**, 17765.
(616)   Maltese and Valentini, *Chim. Ind. (Milan)* **40**, 548 (1958); *C.A.* **53**, 798.
(617)   Malusov, Malafeev, and Zhavoronkov, *Zh. Fiz. Khim.* **31**, 699 (1957); *C.A.* **52**, 25.
(617)a Manczinger, Radnai, and Tettamanti, *Period. Polytech., Chem. Eng. (Budapest)* **13**, 189 (1969); *C.A.* **72**, 93743u.
(617)c Manczinger and Tettamanti, *Period. Polytech., Chem. Eng.* **10**, 183 (1966); *C.A.* **67**, 26261u.
(618)   Mann, Pardee, Smyth, *J. Chem. Eng. Data.* **8**, 499 (1963).
(619)   Mann and Shemilt, *J. Chem. Eng. Data* **8**, 189 (1963).
(620)   Maretic and Sirocic, *Nafta (Zagreb)* **13**, 126 (1962); *C.A.* **57**, 14485.
(621)   Marinichev and Susarev, *Zh. Prikl. Khim.* **38**, 378 (1965); *C.A.* **62**, 15480.
(622)   Marinichev and Susarev, *Zh. Prikl. Khim.* **38**, 1054 (1965).
(623)   Marinichev and Susarev, *Zh. Prikl. Khim.* **38**, 1619 (1965).
(623)c Marinichev and Susarev, *Zh. Fiz. Khim.* **43**, 1132 (1969); *C.A.* **71**, 42814q.
(624)   Markowska-Majewska, *Bull. Acad. Polon. Sci. Classe III*, **2**, 291 (1954); *C.A.* **49**, 2804.
(625)   Marks and Wingard, *J. Chem. Eng. Data* **5**, 416 (1960).

(626)   Markuzin, *Vestn. Leningrad Univ.* **16**, No. 4 *Ser. Fiz. i Khim.* No. 1, 148 (1961); *C.A.* **55**, 16112.
(627)   Markuzin, *Zh. Prikl. Khim.* **34**, 1175 (1961); *C.A.* **55**, 19444.
(628)   Markuzin, Gomerova and Lesteva, *Zh. Prikl. Khim.* **39**, 1878 (1966); *C.A.* **65**, 16119.
(629)   Markuzin and Sokolova, *Zh. Prikl. Khim.* **39**, 1765 (1966); *C.A.* **65**, 16120.
(629)c Marnac and Enjalbert, *Chim. Ind., Genie Chim.* **98**, 1047 (1967); *C.A.* **69**, 70362t.
(630)   Marschner and Burney, *Ind. Eng. Chem.* **44**, 1406 (1952).
(631)   Marschner and Cropper, *Ind. Eng. Chem.* **38**, 262 (1946).
(632)   Marschner and Cropper, *Ind. Eng. Chem.* **41**, 1357 (1949).
(633)   Marshall, *J. Chem. Soc.* **89**, 1350 (1906).
(634)   Matejicek and Raska, *Chem. Prumysl.* **16**, 82 (1966); *C.A.* **64**, 13448.
(635)   Mathias and Krausz, *Anais Acad. Brasil. Cienc.* **30**, 511 (1958); *C.A.* **58**, 10790.
(635)c Mato and Sanchez, *An. Real Soc. Espan. Fis. Quim., Ser. B* **63**, 1 (1967); *C.A.* **67**, 26265y.
(635)e Mato and Sanchez, *An. Real Soc. Espan. Fis. Quim., Ser. B* **63**, 971 (1967); *C.A.* **68**, 95303s.
(636)   Matuszak and Frey, *Ind. Eng. Chem., Anal. Ed.* **9**, 111 (1937).
(637)   May, Brit. Patent **864,226** (1961); *C.A.* **55**, 19786.
(637)c McBee, Roberts, Judd, and Chao, *Proc. Indiana Acad. Sci.* **65**, 94 (1955); *C.A.* **52**, 10870.
(638)   McCarty, *J. Am. Chem. Soc.* **71**, 1339 (1949).
(638)c McConnell and Van Winkle, *J. Chem. Eng. Data* **12**, 430 (1967).
(639)   McCormack, Walkup, and Rush, *J. Phys. Chem.* **60**, 826 (1956).
(640)   McDermott, F. A., private communication.
(641)   McDonald and McMillan, *Ind. Eng. Chem.* **36**, 1175 (1944).
(642)   McKinnis, U.S. Patent **2,388,429** (1945).
(643)   McKinnis and Flint, U.S. Patent **3,114,679** (1961).
(644)   McKinnis and Webb, U.S. Patent **3,114,680** (1961).
(645)   McMakin and Van Winkle, *J. Chem. Eng. Data* **7**, 9 (1962).
(646)   McMillan, *J. Am. Chem. Soc.* **58**, 1345 (1936).
(647)   McWilliams, *Dissertation Abstr.* 65-617; *C.A.* **63**, 2635.
(648)   Melnikov and Tsirlin, *J. Appl. Chem. U.S.S.R.* **29**, 1573; *C.A.* **51**, 17377.
(649)   Melnikov and Tsirlin, *Zh. Fiz. Khim.* **30**, 2290 (1956); *C.A.* **51**, 9245.
(650)   Merriman, *J. Chem. Soc.* **103**, 1790, 1801 (1913).
(651)   Mervart, Kubinova and Zelikova, *Collection Czech. Chem. Commun.* **26**, 2480 (1961).
(652)   Metyushev, *Tr. Tekhnol. Inst. Pishchevoi Prom.* **15**, 80 (1955); *C.A.* **51**, 14398.
(653)   Metzger and Disteldorf, *J. Chim. Phys.* **50**, 156 (1953).
(654)   Mian and Wingard, *Pakistan J. Sci. Res.* **12**, 53 (1960); *C.A.* **55**, 15093.
(655)   Michalski, Michalowski, Serwinski and Strumillo, *Zeszyty Nauk. Politech. Lodz., Chem.* **12**, 73 (1962); *C.A.* **61**, 6445.
(656)   Mignolte, *Rev. Inst. Franc. Petrole Ann. Combust. Liquides* **18**, 1 (1963); *C.A.* **60**, 1163.
(657)   Miller, *Chem. Eng. Data Ser.* **3**, 239 (1958).
(658)   Miller, *J. Phys. Chem.* **62**, 512 (1958).
(659)   Minnesota Mining and Manufacturing Co., unpublished data.
(659)c Molochnikov, Markova, and Kogan, *Zh. Prikl. Khim.* **40**, 2083 (1967); *C.A.* **68**, 53889g.
(659)f Molochnikov, Kudryavtseva, and Kogan, *Zh. Prikl. Khim.* **42**, 1916 (1969); *C.A.* **71**, 105805k.
(660)   Montjar, Reed, Mulik and Masi, *U.S. At. Energy Comm.* CCC-**1024**-TR273 (1957); *C.A.* **55**, 18273.
(661)   Morachevskii and Ch'eng, *Zh. Fiz. Khim.* **35**, 2335 (1961); *C.A.* **56**, 2941.
(662)   Morachevskii and Kolbina, *Zh. Fiz. Khim.* **35**, 1694 (1961); *C.A.* **56**, 63.
(663)   Morachevskii and Komarova, *Vestn. Leningrad. Univ.* **12**, No. 4, *Ser. Fiz. Khim.* No. 1, 118 (1957); *C.A.* **51**; 11832.
(664)   Morachevskii and Leont'ev, *Zh. Fiz. Khim.* **34**, 2347 (1960); *C.A.* **55**, 13023.
(665)   Morachevskii, Smirnova and Lyzlova, *Zh. Prikl. Khim.* **38**, 1262 (1965).
(666)   Morachevskii and Zharov, *Zh. Prikl. Khim.* **36**, 2771 (1963); *C.A.* **60**, 8702.
(667)   Morris and Snider, U.S. Patent **2,368,597** (1945).
(668)   Morton, "Laboratory Technic in Organic Chemistry", p. 66, New York, McGraw-Hill Book Co., 1938.
(668)c Motina, Tentser, and Aerov, *Zh. Prikl. Khim.* **41**, 1048 (1968); *C.A.* **69**, 39127w.
(668)e Mozzhukhin, Serafimov, Mitropol'skaya, and Sankina, *Zh. Fiz. Khim.* **41**, 1687 (1967); *C.A.* **71**, 111820v.
(668)h Mozzhukhin, Serafimov, Zetkin, Raskina, and Mitropol'skaya, *Zh. Prikl. Khim.* **41**, 2764 (1968); *C.A.* **70**, 71497y.

(669)   Mukherjee and Grunwald, *J. Phys. Chem.* **62**, 1311 (1958).
(670)   Munter, Aepli, and Kossatz, *Ind. Eng. Chem.* **39**, 427 (1947).
(671)   Murakami and Yamada, *Kagaku Kogaku* **26**, 865 (1962); *C.A.* **57**, 14485.
(672)   Murray, *J. Council Sci. Ind. Res.* **17**, 213 (1944).
(672)c   Murthy and Raghavacharya, *Trans. Indian Inst. Chem. Eng.* 26 (1968); *C.A.* **70**, 51094m.
(673)   Murti and Van Winkle, *A. I. Ch. E. J.* **3**, 517 (1957).
(674)   Murti and Van Winkle, *Chem. Eng. Data Ser.* **3**, 72 (1958).
(674)c   Murto and Kivinen, *Suomen Kemistilehti B* **40**, 258 (1967); *C.A.* **68**, 16616v.
(674)f   Murto, Kivinen, Strandman, and Virtalaine, *Suom. Kemistilehti, B* **41**, 375 (1968); *C.A.* **70**, 91227p.
(675)   Myers, *Ind. Eng. Chem.* **47**, 2215 (1955).
(676)   Myers, *Ind. Eng. Chem.* **48**, 1104 (1956).
(677)   Myers, *Petrol. Refiner* **36**, 175 (1957).
(678)   Nadeau and Fisher, U.S. Patent **2,165,293** (1939).
(679)   Nagata, *J. Chem. Eng. Data* **7**, 367 (1962).
(680)   Nagata, *Can. J. Chem. Eng.* **41**, 21 (1963).
(681)   Nagata, *J. Chem. Eng. Data* **7**, 461 (1962).
(682)   Nagata, *Kanayawa Daigak Kogakubu Kiyo* **3**, 1, 197 (1963); *C.A.* **62**, 7165, 8431.
(683)   Nagata, *Can. J. Chem. Eng.* **42**, 82 (1964); *C.A.* **61**, 322.
(684)   Nagata, *J. Chem. Eng. Data* **10**, 106 (1965).
(684)a   Nagata, *J. Chem. Eng. Data* **15**, 213 (1970).
(684)b   Nageshwar and Mene, *J. Appl. Chem.* **19**, 195 (1969); *C.A.* **71**, 74673j.
(684)c   Nageshwar and Mene, *Indian J. Technol.* **6**, 374 (1968); *C.A.* **70**, 81418n.
(684)d   Nageshwar and Mene, *Indian J. Technol.* **7**, 138 (1969); *C.A.* **71**, 42812n.
(684)e   Nakanishi, Shirai and Minamiyama, *J. Chem. Eng. Data* **12**, 591 (1967).
(684)f   Nakanishi, Nakasato, Toba and Shirai, *J. Chem. Eng. Data* **12**, 441 (1967).
(684)g   Nakanishi, Shirai and Nakasato, *J. Chem. Eng. Data* **13**, 188 (1968).
(685)   Narinskii, *Zh. Fiz. Khim.* **40**, 2022 (1966); *C.A.* **66**, 607.
(685)c   Narinskii, *Tr., Vses. Nauch.-Issled. Inst. Kisloror. Mashinostr.* **11**, 3 (1967); *C.A.* **70**, 51083g.
(686)   Narinskii, *Kislorod* **10**, 9 (1957); *C.A.* **52**, 13348.
(686)c   Nataraj and Rao, *Indian J. Technol.* **5**, 212 (1967).
(686)f   Nataraj and Rao, *Trans. Indian Inst. Chem. Eng.* 95 (1968); *C.A.* **70**, 71470j.
(687)   Natradze and Novikova, *Zh. Fiz. Khim.* **31**, 227 (1957); *C.A.* **51**, 15236.
(688)   Natta, U.S. Patent **2,308,229** (1943).
(689)   Naumann, *Ber.* **10**, 1421, 1819, 2099 (1877).
(690)   Navez, *Bull. Soc. Chim. Belges.* **39**, 435 (1930).
(691)   Nelson, U.S. Patent **2,786,804** (1957); *C.A.* **51**, 11704.
(692)   Nelson, U.S. Patent **2,839,452** (1958); *C.A.* **52**, 15890.
(693)   Nelson, U.S. Patent **2,922,753** (1960).
(694)   Nelson and Markham, *J. Am. Chem. Soc.* **72**, 2417 (1950).
(695)   Newcome and Cady, *J. Am. Chem. Soc.* **78**, 5216 (1956).
(696)   Newman, *Bull. Inst. Intern. Froid. Annexe* **1955**, 390; *C.A.* **53**, 15681.
(697)   Newman, *Dissertation Abstr.* No. **16,290**; *C.A.* **50**, 12577 (1956).
(698)   Nielson and Weber, *J. Chem. Eng. Data* **4**, 145 (1959).
(699)   Nisel'son and Lapidus, *Zh. Fiz. Khim.* **39**, 161 (1965); *C.A.* **62**, 9850.
(700)   Nisel'son and Lapidus, *Zh. Fiz. Khim.* **39**, 1756 (1965); *C.A.* **63**, 12385.
(700)c   Nisel'son, Lapidus, Golubkov and Mogucheva, *Zh. Neorgan. Khim.* **12**, 1952 (1967); *C.A.* **68**, 63137z.
(701)   Nisel'son and Mogucheva, *Zh. Neorgan. Khim.* **11**, 144 (1966); *C.A.* **64**, 10456.
(702)   Nisel'son and Sokolova, *uh. Neorgan. Khim.* **6**, 1645 (1961); *C.A.* **56**, 13612.
(703)   Nixon, U.S. Patent **2,604,439** (1952).
(704)   Novak, Matous, and Pick, *Collection Czech. Chem. Commun.* **25**, 2405 (1960); *C.A.* **55**, 3170.
(705)   Novak and Matous, *Tech. Publ. Stredisko Tech. Inform. Potravinar. Prumyslu* No. **161**, 25 (1959-61); *C.A.* **60**, 7512.
(706)   Noyes and Warfel, *J. Am. Chem. Soc.* **23**, 463 (1901).
(707)   Nycander and Gabrielson, *Acta Chem. Scand.* **8**, 1530 (1954); *C.A.* **49**, 6678.
(708)   Oakeson and Weber, *J. Chem. Eng. Data* **5**, 279 (1960).
(709)   Oblad, U.S. Patent **2,440,414** (1948).
(710)   Ocon and Espantoso, *An. Real. Soc. Españ. Fis. Quim. (Madrid)* **54B**, 401 (1958); *C.A.* **53**, 1879.
(711)   Ocon, Espantoso, and Mato, *Publ. Inst. Quim. Fis. "Antonio de Gregorio Roca-solano"* **10**, 214 (1956); *C.A.* **51**, 16028.
(712)   Oddo, *Gazz. Chim. Ital.* **41**, II, 232 (1911).
(713)   Ogawa, Kishida, and Kuyama, *Kagaku Kogaku* **22**, 151 (1958); *C.A.* **52**, 8661.

(714)   Ogorodnikov, Kogan, and Nemtsov, *Zh. Priklad. Khim.* **33**, 1599 (1960); *C.A.* **54**, 21909.
(715)   Ogorodnikov, Kogan, and Nemtsov, *J. Appl. Chem. U.S.S.R.* **33**, 2650 (1960); *C.A.* **55**, 9017.
(716)   Ogorodnikov, Kogan and Nemtsov, *Zh. Priklad. Khim.* **34**, 323, 386, 841, 1096 (1961); *C.A.* **55**, 21772.
(717)   Ogorodnikov, Rabovskaya, Korol and Presman, *Zh. Prikl. Khim.* **37**, 1597, 1786 (1964); *C.A.* **61**, 11379, 12691.
(718)   Olevskii and Golubev, *Tr. Gos. Nauk* **1957**, 42, 58; *C.A.* **53**, 21107.
(718)c Omote and Nakamura, *Kogyo Kagaku Zasshi* **70**, 1599 (1967); *C.A.* **68**, 108422g.
(718)e Oppermann, *Z. Phys. Chem.* **236**, 161 (1967); *C.A.* **68**, 8724.
(719)   Orr and Coates, *Ind. Eng. Chem.* **52**, 27 (1960).
(720)   Orszagh, Lelakowska, and Beldowicz, *Bull. Acad. Polon. Sci. Classe III* **6**, 419 (1958); *C.A.* **52**, 19415.
(721)   Orszagh, Lelakowska and Radecki, *Bull. Acad. Polon. Sci. Classe III* **6**, 605 (1958); *C.A.* **53**, 6719.
(722)   Othmer, *Ind. Eng. Chem.* **35**, 614 (1943).
(723)   Othmer, U.S. Patent **2,050,234** (1936).
(724)   Othmer, U.S. Patent **2,170,834** (1939).
(725)   Othmer, U.S. Patent **2,395,010** (1946).
(726)   Othmer, Chudgar, and Levy, *Ind. Eng. Chem.* **44**, 1872 (1952).
(727)   Othmer and Josefowitz, *Ind. Eng. Chem.* **39**, 1175 (1947).
(728)   Othmer and Morley, *Ind. Eng. Chem.* **38**, 751 (1946).
(729)   Othmer and Savitt, *Ind. Eng. Chem.* **40**, 168 (1948).
(730)   Othmer, Savitt, Krasner, Goldberg, and Markowitz, *Ind. Eng. Chem.* **41**, 572 (1949).
(731)   Othmer, Schlechter, and Koszalka, *Ind. Eng. Chem.* **37**, 895 (1945).
(732)   Padgitt, U.S. Patent **2,531,361** (1950).
(733)   Padgitt, Amis, and Hughes, *J. Am. Chem. Soc.* **64**, 1231 (1942).
(734)   Pahlovouni, *Bull. Soc. Chim. Belges* **36**, 533 (1927).
(735)   Palazzo, *Dissertation Abstr.* Mic **58-1354**; *C.A.* **52**, 13350.
(736)   Pannetier and Mignotte, *Bull. Soc. Chim. France* **1961**, 985; *C.A.* **56**, 2037; *Ibid.*, **1962**, 2141; C.A. **58**, 7430; *Ibid.*, **1963**, 694; *C.A.* **59**, 10810.
(737)   Papousek and Smekal, *Chem. Listy* **52**, 542 (1958); *C.A.* **52**, 19391; Collection *Czech. Chem. Commun.* **24**, 2031 (1957).
(738)   Paquot and Perron, *Bull. Soc. Chim. France* **1957**, 529; *C.A.* **51**, 10156.
(738)a Paremuzov, Vanyukov, and Kernozhitskii, *Zh. Neorgan. Khim.* **14**, 3124 (1969); *C.A.* **72**, 36360t.
(738)c Parisse, U.S. Patent **3,392,090** (1968).
(739)   Patterson, U.S. Patent **2,407,997** (1946).
(740)   Patterson and Ozol, U.S. Patent **2,386,058** (1945).
(741)   Patton, U.S. Patent **2,940,973** (1960).
(742)   Pavlov, Pavlov and Kirnos. *Zh. Prikl. Khim.* **39**, 1555 (1966); *C.A.* **65**, 11404.
(742)c Pavlov, Pavlova, Serafimov, and Kofman, *Prom. Sin. Kauch.* **4**, 6 (1967); *C.A.* **70**, 109551b.
(743)   Pearce and Gerster, *Ind. Eng. Chem.* **42**, 1418 (1950).
(744)   Pennington, *Ind. Eng. Chem.* **44**, 2397 (1952).
(745)   Pennington, private communication.
(746)   Pennington and Reed, *Chem. Eng. Progr.* **46**, 464 (1950).
(747)   Peppel, *Ind. Eng. Chem.* **50**, 767 (1958).
(748)   Perel'shtein, *Ingh.-Fiz. Zh. Akad. Nauk Belorussk. SSR* **5**, 27 (1962); *C.A.* **58**, 6256.
(748)c Perelygin, Remizov, and Kharin, *Izv. Vyssh. Ucheb. Zaved., Pishch. Tekhnol.* **5**, 123 (1969); *C.A.* **72**, 25626h.
(749)   Perugini, *Chim. Ind. (Milan)* **39**, 445 (1957); *C.A.* **51**, 16028.
(750)   Petry, U.S. Patent **2,411,106** (1946).
(751)   Pfann, *J. Am. Chem. Soc.* **66**, 155 (1944).
(751)c Piacentini and Stein, *Chem. Eng. Progr., Symp. Ser.* **63**, 28 (1967); *C.A.* **69**, 54707d.
(752)   Pick, Hala, and Fried, *Chem. Listy* **52**, 561 (1958); *C.A.* **52**, 19393; Collection *Czech. Chem. Commun.* **24**, 1589 (1959).
(753)   Picon and Flahaut, *Compt. Rend.* **230**, 1954 (1950).
(754)   Pierre, *Compt. Rend.* **74**, 224 (1872).
(755)   Pierre and Puchot, *Compt. Rend.* **73**, 599 (1871).
(756)   Piret and Hall, *Ind. Eng. Chim.* **40**, 661 (1948).
(756)c Pitt, *U.S. At. Energy Comm.* **ORNL-TM-1683**; *C.A.* **67**, 47653p.
(757)   Plate and Tarasova, *Bull. Acad. Sci. U.R.S.S., Classe Sci. Chim.* **1941**, 201; Universal Oil Products Library, *Bull.* **17**, 145 (1942); *C.A.* **37**, 6243.

(758)   Podder, Z. Physik. Chem (Frankfurt) 39, 79 (1963); C.A. 60, 7512.
(759)   Politziner, Chem. Eng. Data Ser. 2, 16 (1957).
(760)   Popelier, Bull. Soc. Chim. Belges 32, 179 (1923).
(760)c   Pozhitkova, Simagina, Rozengauz, Bochkarev, and Nisel'son, Zh. Neorgan. Khim.
         14, 2219 (1969); C.A. 71, 105833t.
(761)   Prabhu and Van Winkle, J. Chem. Eng. Data 9, 9 (1964).
(762)   Prahl and Mathes, Angew. Chem. 47, 11 (1934).
(763)   Pratt, Preprint, Trans. Inst. Chem. Engrs. (London) (March 1947).
(764)   Prausnitz and Targovnik, Chem. Eng. Data Ser. 3, 234 (1958).
(765)   Price and Hickman, Proc. West Va. Acad. Sci. 22, 69 (1962).
(766)   Price and Kobayashi, J. Chem. Eng. Data 4, 40 (1959).
(767)   Prill, U.S. Patent 2,599,482 (1952).
(768)   Prokhorova, Serafimov and Tekhtamysheva, Zh. Fiz. Khim. 38, 1005 (1964); C.A.
         61, 2536.
(768)c   Prusakov and Ezhov, At. Energ. 25, 35 (1968); C.A. 69, 70430p.
(769)   Pryanishinkov and Genin. J. Appl. Chem. U.S.S.R. 13, 140 (1940); C.A. 34, 7682.
(770)   Pupezin, Knezevic and Rubnikar, Glasnik Hem. Drustva, Beograd. 28, 523 (1963);
         C.A. 64, 64.
(770)c   Pupezin, Ribnikar, Knezevic, and Dokic, Bull. Boris Kidric Inst. Nucl. Sci. 17, 297
         (1966); C.A. 67, 47783f.
(770)e   Pustil'nik, Gavrilov, Rodin, and Nisel'son, Zh. Neorgan. Khim. 12, 2186 (1967);
         C.A. 67, 120480m.
(771)   Qozati and Van Winkle, J. Chem. Eng. Data 5, 269 (1960).
(771)c   Quitzsch, Kopp, Renker, and Geiseler, Z. Phys. Chem. (Leipzig) 237, 256 (1968);
         C.A. 69, 22457f.
(772)   Quiggle and Fenske, J. Am. Chem. Soc. 59, 1289 (1937).
(773)   Quintanilla, Riv. Quim. Ing. Monterrey 2, 23 (1956); C.A. 51, 12585.
(773)c   Quitzsch, Koehler, Taubert, and Geiseler, J. Prakt. Chem. 311, 429 (1969); C.A. 71,
         42818u.
(774)   Quitzsch, Wunderlich and Geiseler, J. Prakt. Chem. 30, 119 (1956); C.A. 64,
         10458.
(775)   Rabcewicz-Zubkowski, Roczniki Chem. 13, 193 (1933).
(776)   Rabcewicz-Zubkowski, Roczniki Chem. 13, 334 (1933).
(777)   Rabe, Dissertation Abstr. 58-1920; C.A. 52, 16853.
(778)   Rader, Z. Anorg. Allgem. Chem. 130, 325 (1923).
(778)c   Raju and Rao, Indian Chem. Eng. 8, 108 (1966); C.A. 67, 26264x.
(778)f   Raju and Rao, J. Chem. Eng. Data 14, 283 (1969).
(779)   Rall and Schafer, Z. Elektrochem. 63, 1019 (1959); C.A. 54, 6236.
(779)c   Ramalho and Delmas, J. Chem. Eng. Data 13, 161 (1968).
(779)e   Rao, Chiranjivi, and Dasarao, J. Appl. Chem. (London) 17, 118 (1967); C.A. 67,
         6133a.
(779)g   Rao, Chiranjivi, and Dasarao, J. Appl. Chem. (London) 18, 166 (1968); C.A. 69,
         46407r.
(780)   Rao, Dakshinamurty and Rao, J. Sci. Ind. Res. (India) 20B, 218 (1961); C.A. 56,
         14990.
(781)   Rao, Dakshinamurty and Rao, J. Appl. Chem. (London) 12, 274 (1962); C.A. 57,
         11916.
(782)   Rao, Rao, and Rao, J. Appl. Chem. (London) 7, 666 (1957); C.A. 52, 6909.
(783)   Rao, Sarma, Swami, and Rao, J. Sci. Ind. Research (India) 16B, 4 (1957); C.A. 51,
         10196.
(784)   Rao, Swami, and Rao, A. I. Ch. E. J. 3, 191 (1957).
(784)c   Rao, Swami, and Rao, J. Sci. Ind. Res. (India) 16B, 195 (1957).
(785)   Rao, Swami, and Rao, J. Sci. Ind. Research (India) 16B, 233 (1957); C.A. 51,
         17301.
(786)   Rao, Swami, and Rao, J. Sci. Ind. Research (India) 16B, 294 (1957); C.A. 52, 3440.
(787)   Rashkovskaya and Mozharova, Ukr. Khim. Zh. 29, 1023 (1963); C.A. 60, 7514.
(788)   Ray, U.S. Patent 2,498,928 (1950).
(789)   Ray, U.S. Patent 2,623,072 (1952).
(790)   Razniewska, Roczniki Chem. 35, 665 (1961); C.A. 55, 21773.
(791)   Razniewska, Roczniki Chem. 38, 851 (1964); C.A. 62, 3449.
(791)c   Reddy, Krishna, and Venkatarao, Chem. Age India 16, 576 (1965); C.A. 66, 10175.
(792)   Reddy and Rao, Indian J. Technol. 3, 45 (1965).
(793)   Reddy and Rao, J. Chem. Eng. Data 10, 309 (1965); C.A. 64, 1792.
(793)c   Reddy and Rao, Indian J. Technol. 5, 66 (1967); C.A. 67, 6128.
(794)   Reed, U.S. Patent 2,511,993 (1950).
(795)   Reed, Dissertation Abstr. No. 5338; C.A. 47, 11859.
(796)   Reed and Pennington, Modern Refrig. 53, 123 (1950).

(797)   Reeves and Sadle, *J. Am. Chem. Soc.* **72**, 1251 (1950).
(798)   Rehberg, *J. Am. Chem. Soc.* **71**, 3247 (1949).
(799)   Rehberg, U.S. Patent **2,406,561** (1946).
(800)   Rehberg and Fisher, *J. Am. Chem. Soc.* **66**, 1203 (1944).
(801)   Reid, U.S. Patent **2,070,962** (1937).
(802)   Reinders and DeMinjer, *Rec. Trav. Chim.* **59**, 207, 369 (1940).
(803)   Reinders and DeMinjer, *Rec. Trav. Chim.* **66**, 552, 564, 573 (1947).
(804)   Ricard and Guinot, U.S. Patent **1,915,002** (1933).
(805)   Richards and Hargreaves, *Ind. Eng. Chem.* **36**, 805 (1944).
(806)   Riddick, J. A., Commercial Solvents Corp., unpublished data.
(807)   Riddle, "Monomeric Acrylic Esters," p. 9, Reinhold, New York, 1954.
(807)c Ridgway and Butler, *J. Chem. Eng. Data* **12**, 509 (1967).
(808)   Ridley and Ridley, British Patent **795,866** (1958); *C.A.* **53**, 1154.
(809)   Riethof, U.S. Patent **2,383,016** (1945).
(810)   Riethof, U.S. Patent **2,412,649-51** (1946).
(811)   Rius Otero de la Gandara, and Macarron, *Chem. Eng. Sci.* **10**, 105 (1959); *C.A.* **53**, 19501.
(812)   Rivenq, *Bull. Soc. Chim. France* **1961**, 1392; *C.A.* **56**, 2940.
(812)c Rivenq, *Bull. Soc. Chim. France* **9**, 3034 (1969); *C.A.* **72**, 25628k.
(813)   Robinson, "Elements of Fractional Distillation," p. 230, McGraw-Hill, New York, 1930.
(814)   Robinson, Wright, and Bennett, *J. Phys. Chem.* **36**, 658 (1932).
(815)   Rock and Shroder, *Z. Physik. Chem. (Frankfurt)* **11**, 47 (1957).
(815)c Rodger, Hsu, and Furter, *J. Chem. Eng. Data* **14**, 362 (1969).
(816)   Rohm and Haas, *Tech. Data Sheet SP-148* (1958).
(817)   Rohm and Haas Co., *Bull.* on *tert*-Octylamine (1949).
(818)   Rohm and Haas Co., "Physical Properties of the Methylamines," 1949.
(819)   Rohrback and Cady, *J. Am. Chem. Soc.* **73**, 4250 (1951).
(820)   Roscoe, *Ann.* **116**, 203 (1860).
(821)   Roscoe and Dittmar, *Ann.* **112**, 327 (1859).
(822)   Rose Acciarri, and Williams, *Chem. Eng. Data Serv.* **3**, 210 (1958).
(823)   Rose, Papahronis, and Williams, *Chem. Eng. Data Ser.* **3**, 216 (1958).
(824)   Rose and Schrodt, *J. Chem. Eng. Data* **9**, 12 (1964).
(825)   Rose and Supina, *J. Chem. Eng. Data* **6**, 173 (1961).
(826)   Rossini, Mair, and Streiff, "Hydrocarbons from Petroleum," *ACS Monograph* **121**, p. 89, Reinhold, New York, 1953.
(827)   Rother, Steinbrecher and Bittrich, *Z. Physik. Chem.* **220**, 89 (1962); *C.A.* **57**, 14483.
(828)   Rowlinson, *U. K. At. Energy Authority,* Ind. Group R & DB(CA)TN-96D (1959); *C.A.* **53**, 21114.
(829)   Rowlinson and Sutton, *Proc. Roy. Soc. London* A**229**, 396 (1955); *C.A.* **49**, 14443.
(830)   Rudakov and Kalinovskaya, *Gidroliz Lesokhim. Prom.* **10**, 8 (1957); *C.A.* **51**, 10989.
(830)c Rudakovskaya, Timofeev, Tikhonova, and Serafimov, *Zh. Prikl. Khim.* **41**, 583 (1968); *C.A.* **69**, 5649c.
(831)   Rudoi-Kolker and Gregoryan, *Zh. Prikl. Khim.* **37**, 1843 (1964); *C.A.* **61**, 11638.
(832)   Ruhoff and Reid. *J. Am. Chem. Soc.* **59**, 401 (1937).
(833)   Ryabova, Ustyugov and Kudryavtsev, *Tr. Mosk. Khim. Tekhnol Inst.* **49**, 57 (1965); *C.A.* **65**, 1467.
(834)   Ryland, *Am. Chem. J.* **22**, 390 (1899); *Chem. News (London)* **81**, 15, 42, 50 (1908).
(835)   Sakuyama. *J. Soc. Chem. Ind. Japan,* **44**, 266 (1941).
(836)   Sandberg and Patterson, U.S. Patent **2,428,815** (1947).
(837)   Sastry, *J. Soc. Chem. Ind. (London)* **35**, 450 (1916).
(838)   Satapathy, Rao, Anjaneyulu, and Rao, *J. Appl. Chem. (London)* **6**, 261 (1956); *C.A.* **51**, 1677.
(839)   Sauer, *J. Am. Chem. Soc.* **66**, 1707 (1944).
(840)   Sauer, U.S. Patent **2,381,139** (1945).
(841)   Sauer and Hadsell, *J. Am. Chem. Soc.* **70**, 4258 (1948).
(842)   Sauer and Patnode, *J. Am. Chem. Soc.* **67**, 1548 (1945).
(843)   Sauer and Reed, U.S. Patent **2,388,575** (1945).
(844)   Sauer, Scheiber, and Hadsell, *J. Am. Chem. Soc.* **70**, 4254 (1948).
(845)   Saunders and Spaull, *Z. Physik. Chem.* **28**, 332 (1961); *C.A.* **55**, 21714.
(846)   Scatchard and Raymond, *J. Am. Chem. Soc.* **60**, 1278 (1938).
(847)   Scatchard, Wood and Mochel, *J. Am. Chem. Soc.* **61**, 3206 (1939).
(848)   Scatchard, Wood, and Mochel, *J. Phys. Chem.* **43**, (1939).
(848)c Scheller, Schubert, and Koennecke, *J. Prakt. Chem.* **311**, 974 (1969); *C.A.* **72**, 83446k.
(848)e Scheller, Torres-Soto, and Daphtary, *J. Chem. Eng. Data* **14**, 17 (1969).

(849)   Schelling and Anderson, U.S. Patent **2,422,802** (1947).
(850)   Schichtantz, *J. Res. Natl. Bur. Std.* **18**, 129 (1937).
(850)c   Schmidt, *Kaeltetech.* **18**, 331 (1966); *C.A.* **66**, 32357a.
(850)f   Schmidt, Werner, and Schuberth, *Z. Phys. Chem. (Leipzig)* **242**, 381 (1969); *C.A.* **72**, 48123f.
(851)   Schneider, *Z. Physik. Chem. (Frankfurt)* **27**, 171 (1961); *C.A.* **55**, 13022.
(852)   Schopmeyer and Arnold, U.S. Patent **2,350,370** (1944).
(853)   Schreinmakers, *Z. Physik. Chem.* **35**, 459 (1900).
(854)   Schreinmakers, *Z. Physik. Chem.* **39**, 485; **40**, 440; **41**, 331 (1902).
(855)   Schreinmakers, *Z. Physik. Chem.* **47**, 445; **48**, 257 (1904).
(856)   Schuberth, *Monatsber. Deut. Akad. Wiss. Berlin* **4**, 299 (1962); *C.A.* **59**, 2222.
(856)c   Schuberth, *Z. Phys. Chem. (Leipzig)* **235**, 230 (1967); *C.A.* **68**, 43687a.
(857)   Schumaker and Hunt, *Ind. Eng. Chem.* **34**, 701 (1942).
(858)   Schultz and Mallonee, *J. Am. Chem. Soc.* **62**, 1491 (1940).
(859)   Sebba, *J. Chem. Soc.* **1951**, 1975.
(859)c   Sedletskaya and Kogan, *Zh. Prikl. Khim.* **42**, 2551 (1969); *C.A.* **72**, 59729j.
(859)f   Seetharamaswamy, Subrahmanyam, Chiranjivi, and Dakshinamurty, *J. Appl. Chem.* **19**, 258 (1969); *C.A.* **71**, 116983t.
(859)i   Seetharamaswamy, Subrahmanyam, and Dakshinamurty, *J. Appl. Chem.* **19**, 359 (1969); *C.A.* **72**, 59750j.
(860)   Senkus, U.S. Patent **2,406,713** (1946).
(861)   Sense, Stone, and Filbert, *U.S. At. Energy Comm. BMI-1186* (1957); *C.A.* **51**, 15236.
(861)c   Serafimov and Balashov, *Zh. Prikl. Khim.* **39**, 2344 (1966); *C.A.* **66**, 2168.
(862)   Serafimov, Tikhonova and L'vov, *Zh. Fiz. Khim.* **38**, 2065 (1964); *C.A.* **61**, 12689.
(863)   Serafimov, Timofeev and L'vov, *Zh. Fiz. Khim.* **39**, 1890 (1965); *C.A.* **64**, 2790.
(864)   Serafimov, Timofeev, Strukova and L'vov, *Zh. Fiz. Khim.* **38**, 1865 (1964); *C.A.*
(865)   Serafimov, Tyurikov, Rumyantsev and L'vov, *Zh. Fiz. Khim.* **38**, 1326 (1964); *C.A.* **61**, 10092.
(866)   Serebrennaya and Byk, *Zh. Prikl. Khim.* **39**, 1441 (1966); *C.A.* **65**, 9802. **61**, 10096.
(867)   Serebrennaya and Byk, *Zh. Prikl. Khim.* **39**, 1869 (1966); *C.A.* **65**, 16120.
(868)   Seryakov, Vaks and Sidorina, *Titan. Ego Splavy, Akad. Nauk SSSR, Inst. Met.* **1961**, 220; *C.A.* **57**, 13224.
(869)   Seryakov, Vaks, and Sidorina, *Zh. Obshch. Khim.* **30**, 2130 (1960); *C.A.* **55**, 8009.
(870)   Shair and Schurig, *Ind. Eng. Chem.* **43**, 1624 (1951).
(871)   Shakhova and Braude, *Khim. Prom.* **1964**, 906; *C.A.* **62**, 7165.
(872)   Shakhparonov, Lel'chuk, Korchemskaya, Martynova, Baburina, and Voronina, *Zh. Prikl. Khim.* **33**, 2699 (1960); *C.A.* **55**, 11006.
(873)   Shawinigan Chemical, Ltd., Dept. Chem. Development. "Report on Vinyl Crotonate."
(874)   Shcherbak, Byk and Aerov, *Zh. Prikl. Khim.* **28**, 1120 (1955); *C.A.* **50**, 639.
(875)   Shell Chemical Corp., "Allyl Alcohol," 1946.
(876)   Shell Chemical Corp., "Methyl Ethyl Ketone," 1938.
(877)   Shell Chemical Corp., "Organic Chemicals Manufactured by Shell," 1939.
(878)   Shell Development Co., Data Sheet, 1946.
(879)   Shell Development Co., unpublished data.
(879)c   Shnitko, Kogan, and Sheblom, *Zh. Prikl. Khim.* **41**, 1158 (1968); *C.A.* **69**, 39124t.
(879)f   Shnitko, Kogan, and Sheblom, *Zh. Prikl. Khim.* **42**, 2389 (1969); *C.A.* **72**, 36350q.
(880)   Shorr and Segall, *Israel J. Technol.* **1**, 1 (1963); *C.A.* **60**, 2384.
(881)   Shostakovskii, Kuznetsov, Dubovik and Zikherman, *Izv. Akad. Nauk SSSR, Otd. Khim. Nauk* **1961**, 1495; *C.A.* **56**, 308.
(882)   Shostakovskii and Prilezhaeva, *J. Gen. Chem. (U.S.S.R.),* **17**, 1129 (1947); *C.A.* **42**, 3633.
(882)c   Shukla, *Trans. Indian Inst. Chem. Eng.* **48** (1968); *C.A.* **70**,51079k.
(883)   Silina, Pokorskii, Shiryaeva, Ustraikh. *Tr. Vses. Nauchn.-Issled. Inst. Neftekhim. Protsessov.* **1962**, 124; *C.A.* **58**, 3955.
(884)   Simonetta and Barakan, *Gazz. Chim. Ital.* **77**, 105 (1947); *C.A.* **42**, 23.
(885)   Simonetta and Mugnaini, *Chim. Ind. (Milan)* **30**, 73 (1948).
(885)c   Sinka, *J. Chem. Eng. Data* **15**, 71 (1970).
(886)   Sinor and Weber, *J. Chem. Eng. Data* **5**, 243 (1960).
(886)c   Sivtsova, Kogan, and Ogorodnikov, *Zh. Prikl. Khim.* **39**, 2038 (1966); *C.A.* **66**, 3532.
(886)e   Sivtsova, Kogan, and Ogorodnikov, *Zh. Prikl. Khim.* **41**, 461 (1968); *C.A.* **69**, 13359y.
(887)   Sizmann, *Angew. Chem.* **71**, 243 (1959); *C.A.* **53**, 15425.
(888)   Skaates and Kay, *Chem. Eng. Sci.* **19**, 431 (1964); *C.A.* **61**, 8946.
(889)   Slobodyanik and Babushkina, *Zh. Prikl. Khim.* **39**, 1899 (1966); *C.A.* **65**, 16122.
(890)   Smirnova, *Vestn. Leningrad Univ.* **14**, 80 (1959); *C.A.* **54**, 8194.

(891) Smirnova and Morachevskii, *Zh. Fiz. Khim.* **34**, 2546 (1960); *C.A.* **55**, 6117.
(892) Smirnova, Morachevskii, and Storonkin, *Vestn. Leningrad Univ.* **14**, 70 (1959); *C.A.* **54**, 9475.
(893) Smit and Ruyter, *Rec. Trav. Chim.* **79**, 1244 (1960); *C.A.* **55**, 8008.
(894) Smith, *Ind. Eng. Chem.* **34**, 251 (1942).
(895) Smith, U.S. Patent **2,385,546** (1945).
(896) Smith and Bonner, *Ind. Eng. Chem.* **41**, 2867 (1949).
(896)c Smith, Ferris, and Thompson, *U.S. At. Energy Comm.*, **ORNL-4415** (1969); *C.A.* **72**, 25620b.
(897) Smith and LaBonte, *Ind. Eng. Chem.* **44**, 2740 (1952).
(898) Smith and Wojciechowski, *J. Res. Natl. Bur. Std.* **18**, 461 (1937).
(898)c Smolyan, Golubev, and Knyazev, *Khim. Prom.* **44**, 735 (1968); *C.A.* **70**, 14839n.
(899) Smyth and Engel, *J. Am. Chem. Soc.* **51**, 2646 (1929).
(900) SNAM. S.p.A., Netherlands Appl. **6,405,225** (1964); *C.A.* **62**, 9005.
(900)c Sneed, Sonntag, and Van Wylen, *J. Chem. Phys.* **49**, 2410 (1968); *C.A.* **69**, 100225v.
(901) Snyder and Gilbert, *Ind. Eng. Chem.* **34**, 1519 (1942).
(902) Soday and Bennett, *J. Chem. Educ.* **7**, 1336 (1930).
(903) Soday and Bennett, *J. Chem. Educ.* **7**. 1336 (1930).
(904) Sokolov, Sevryugova and Zhavoronkov, *Zh. Fiz. Khim.* **39**, 1008 (1965); *C.A.* **63**, 1253.
(905) Sokolov, Sevryugova and Zhavoronkov, *Zh. Fiz. Khim.* **40**, 1086 (1966); *C.A.* **65**, 11402.
(905)c Sokolov, Sevryugova, and Zhavoronkov, *Teor. Osn. Khim. Tekhnol.* **2**, 139 (1968); *C.A.* **69**, 61986x.
(905)f Sokolov, Sevryugova, and Zhavoronkov, *Rev. Chim. (Bucharest)* **20**, 169 (1969); *C.A.* **71**, 2938b.
(905)i Sokolov, Sevryugova, and Zhavoronkov, *Teor. Osn. Khim. Tekhnol.* **3**, 128 (1969); *C.A.* **70**, 81428r.
(905)m Sokolov, Sevryugova, and Zhavoronkov, *Teor. Osn. Khim. Tekhnol.* **3**, 288 (1969); *C.A.* **70**, 114565p.
(905)p Sokolov, Sevryugova, and Zhavoronkov, *Teor. Osn. Khim. Tekhnol.* **3**, 449 (1969); *C.A.* **71**, 60646r.
(905)r Soldatenko and Tarabrova, *Teplofiz. Svoistva Veshchestv, Akad. Nauk Ukr. SSR Respub. Mezhvedom. Sb.* **1966**, 165; *C.A.* **67**, 36765a.
(905)t Spano, Heck, and Barrick, *J. Chem. Eng. Data* **13**, 168 (1968).
(906) Speck, U.S. Patent **2,449,152** (1948).
(907) Speier, *J. Am. Chem. Soc.* **70**, 4142 (1948).
(908) Spicer and Kruger, *J. Am. Chem. Soc.* **72**, 1855 (1950).
(909) Spicer and Meyer, *J. Am. Chem. Soc.* **73**, 934 (1951).
(910) Spicer and Page, *J. Am. Chem. Soc.* **75**, 3603 (1953).
(911) Stadniki, *Bull. Acad. Polon. Sci., Ser. Sci. Chim.* **10**, 291, 295, 299, 345, 349, 353, 357, 357 (1962); *C.A.* **58**, 65, 3954.
(912) Stadnicki, *Roczniki Chem.* **38**, 827 (1964); *C.A.* **62**, 67.
(912)c Stankova, Vesely, and Pick, *Collection Czech. Chem. Commun.* **35**, 1 (1970); *C.A.* **72**, 59749r.
(913) Stasse, U.S. Patent **2,363,157** (1944).
(914) Stasse, U.S. Patent **2,363,158** (1944).
(915) Steel and Bagster, *J. Chem. Soc.* **97**, 2607 (1910).
(916) Steinbrecher and Bittrich, *Z. Physik. Chem.* **232**, 313 (1966); *C.A.* **65**, 17770.
(917) Steinhauser and White, *Ind. Eng. Chem.* **41**, 2912 (1949).
(918) Steitz, U.S. Patent **2,500,329** (1950).
(919) Steitz, U.S. Patent **2,552,911** (1951).
(920) Stengel and O'Loughlin, U.S. Patent **2,315,139** (1943).
(920)c Stepanishcheva, Burmistrova, and Vil'shau, *Zh. Fiz. Khim.* **43**, 2145 (1969); *C.A.* **71**, 105824r.
(921) Stockhardt and Hull, *Ind. Eng. Chem.* **23**, 1438 (1931).
(922) Stolyarov, Bannikov and Mashendzhinov, *Tr. Gos. Inst. Prikl. Khim.* No. **49**, 283 (1962); *C.A.* **60**, 3544.
(923) Storonkin and Markuzin, *Vestn. Leningrad Univ.* **13**, 100 (1958); *C.A.* **52**, 12493.
(924) Storonkin and Morachevskii, *Zh. Fiz. Khim.* **31**, 42 (1957); *C.A.* **51**, 15236.
(925) Storonkin, Morachevskii, and Belousov, *Vestn. Leningrad Univ.* **13**, 94 (1958); *C.A.* **52**, 17863.
(926) Streett, *Cryogenics* **5**, 27 (1965); *C.A.* **63**, 2434.
(927) Streett and Jones, *J. Chem. Phys.* **42**, 3989 (1965).
(928) Streiff, Murphy, Sedlak, Willingham and Rossini, paper presented before Div. of Petroleum Chemistry, 110th Meeting Am. Chem. Soc., Chicago, 1946.
(929) Stribley and Lake, U.S. Patent **2,463,919** (1949).

(930)   Studenberg and Thomas, *Proc. S. Dakota Acad. Sci.* **36**, 167 (1957); *C.A.* **52**, 15992.
(930)c Subbarao and Ventakarao, *Can. J. Chem. Eng.* **44**, 357 (1966); *C.A.* **67**, 26287g.
(930)e Subbarao and Ventakarao, *J. Appl. Chem.* **18**, 61 (1968); *C.A.* **68**, 7892.
(930)h Subramanian, Nageshwar, and Mene, *J. Chem. Eng. Data* **14**, 421 (1969); *C.A.* **71**, 129380u.
(931)   Subrahmanyam and Murthy, *J. Appl. Chem. (London)* **14**, 500 (1964); *C.A.* **62**, 3451.
(931)c Sukhova, Vlasov, and Kogan, *Zh. Fiz. Khim.* **43**, 1922 (1969); *C.A.* **71**, 105804j.
(932)   Sullivan, U.S. Patent **2,265,220** (1941).
(932)c Surovy and Heinrich, *Sb. Pr. Chem. Fak. SVST* **1966**, 201; *C.A.* **66**, 69368f.
(933)   Suryanarayana and Van Winkle, *J. Chem. Eng. Data* **11**, 7 (1966).
(934)   Susarev, *Zh. Prikl. Khim.* **34**, 412 (1961); *C.A.* **55**, 13023.
(935)   Susarev and Gorbunov, *Zh. Prikl. Khim.* **36**, 459 (1963); *C.A.* **58**, 12010.
(936)   Susarev and Lyzlova, *Zh. Fiz. Khim.* **36**, 437 (1962); *C.A.* **57**, 1617.
(937)   Susarev, Zapol'skaya and Vinichenko, *Zh. Fiz. Khim.* **39**, 2396 (1965); *C.A.* **64**, 1402.
(937)c Suska, Holub, Vonka, and Pick, *Collection Czech. Chem. Commun.* **35**, 385 (1970); *C.A.* **72**, 93753x.
(938)   Sutherland, U.S. Patent **2,290,654** (1942).
(939)   Swami and Rao, *J. Sci. Ind. Research (India)* **18B**, 11 (1959); *C.A.* **53**, 16628.
(940)   Swami, Rao, and Rao, *J. Sci. Ind. Research (India)* **15B**, 550 (1956); *C.A.* **51**, 6252; *Trans. Indian Inst. Chem. Engrs.* **9**, 47 (1956-7); *C.A.* **53**, 14622.
(941)   Swamy and Van Winkle, *J. Chem. Eng. Data* **10**, 214 (1965).
(942)   Swietoslawski, *Bull. Acad. Polon. Sci., Classe III* **7**, 13 (1959); *C.A.* **53**, 19501.
(943)   Swietoslawski and Wajcenblitt, *Roczniki Chem.* **12**, 48 (1932); *C.A.* **26**, 5821.
(944)   Swietoslawski and Kopczynski, *Ibid.*, **11**, 440 (1931); *C.A.* **25**, 5809.
(945)   Swietoslawski and Kreglewski, *Bull. Acad. Polon. Sci., Classe III* **2**, 77 (1954).
(946)   Swietoslawski and Malesinski, *Bull. Acad. Polon. Sci., Classe III* **4**, 159 (1956).
(947)   Swietoslawski and Wajcenblitt, *Compt. Rend.* **193**, 664 (1931); *C.A.* **26**, 1940.
(948)   Swietoslawski and Zieborak, *Bull. Acad. Polon. Sci., Classe Sci. Math. Nat. Ser. A* **1950**, 9, 13; *C.A.* **46**, 410 (1952).
(949)   Swietoslawski, Zieborak, and Galska-Krajewska, *Bull. Acad. Polon. Sci., Classe III* **7**, 43 (1959); *C.A.* **54**, 16068.
(950)   Swietoslawski and Zielenkiewicz, *Bull. Acad. Polon. Sci.* **6**, 111 (1958); *C.A.* **52**, 15169.
(951)   Swinehart and Shenk, "Boron Fluoride and Its Addition Compounds," Harshaw Chemical Co., 1946.
(951)c Synowiec and Zielenski, *Przem. Chem.* **47**, 363 (1968); *C.A.* **70**, 14835h.
(952)   Szapiro, *Zeszyty Nauk Politech. Lodz. Chem.* **7**, 3 (1958); *C.A.* **52**, 19475.
(953)   Tapp and Montagna, U.S. Patent **2,806,884** (1975).
(953)c Taramasso, Spallanzani, and De Malde, *Chim. Ind. (Milan)* **51**, 253 (1969); *C.A.* **70**, 109531v.
(954)   Tarbutton and Deming, *J. Am. Chem. Soc.* **72**, 2086 (1950).
(954)c Tatscheff, Beyer, Thuemmler, and Thinius, *Z. Phys. Chem. (Leipzig)* **237**, 52 (1968); *C.A.* **69**, 22446b.
(954)f Tatsuya, Adachi, Tanaka, and Matsui, Japanese Patent **11,203** (1968); *C.A.* **70**, 46814s.
(955)   Taylor, Ellis and Hands, *J. Appl. Chem. (London)* **16**, 245 (1966); *C.A.* **65**, 14849.
(956)   Taylor and Horsley, U.S. Patent **2,293,317** (1942).
(956)c Taylor and Wingard, *J. Chem. Eng. Data* **13**, 301 (1968).
(957)   Teague and Felsing, *J. Am. Chem. Soc.* **65**, 485 (1943).
(958)   Terry, Kepner, and Webb, *J. Chem. Eng. Data* **5**, 403 (1960); *C.A.* **55**, 8010.
(959)   Teter and Merwin, U.S. Patent **2,388,507** (1945).
(960)   Thayer, *J. Phys. Chem.* **3**, 36 (1899).
(961)   Thornton and Garner, *J. Appl. Chem. (London)* **1**, S61, S68 (1951).
(961)c Thorpe, *Trans. Faraday Soc.* **64**, 2273 (1968); *C.A.* **69**, 80836q.
(962)   Timmermans and Delcourt, *J. Chim. Phys.* **31**, 98 (1934).
(963)   Tolstova, Kogan and Skorokhodova, *Zh. Prikl. Khim.* **38**, 2617 (1965).
(964)   Tomassi, *Roczniki Chem.* **21**, 108 (1947); *C.A.* **42**, 4812.
(965)   Tomkins, Wheat, and Stranks, *Can. J. Res.* **26F**, 168 (1948); *C.A.* **42**, 5848.
(966)   Tongberg and Johnston, *Ind. Eng. Chem.* **25**, 733 (1933).
(966)c Toriume and Kaminishi, *Asahi Garasu Kogyo Gijulsu Shorei-kai Kenkyu Hokoku* **14**, 67 (1968); *C.A.* **72**, 16124r.
(967)   Toyama, Chappelear, Leland and Kobayashi, *Advan. Cryogenic Eng.* **7**, 125 (1961); *C.A.* **57**, 9606.
(968)   Trabczynski, *Bull. Acad. Polon. Sci., Classe III* **6**, 269 (1958); *C.A.* **52**, 15993.
(969)   Treybal, Weber and Daley, *Ind. Eng. Chem.* **38**, 817 (1946).
(970)   Trillat and Cambier, *Compt. Rend.* **118**, 1277 (1894).

(970)c Trofimov, Teskhanskaya, and Shaburov, *Zh. Prikl. Khim. (Leningrad)* **42**, 2556 (1969); *C.A.* **72**, 93773d.
(971) Tsirlin, *Zh. Prikl. Khim.* **35**, 409 (1962); *C.A.* **56**, 14990.
(972) Tsirlin, *Zh. Fiz. Khim.* **36**, 1673 (1962); *C.A.* **57**, 14486.
(973) Tsunoda, *Kogyo Kagaku Zasshi* **61**, 1526 (1958); *C.A.* **56**, 13613.
(974) Tuda, Oguri, and Hukusima, *J. Pharm. Soc. (Japan)* **61**, 74 (1941); *C.A.* **36**, 3077.
(975) Tuerck and Brittain, U.S. Patent **2,405,471** (1946).
(976) Tumova, Prenosil, and Pinkava, *Chem. Prumysl.* **8**, 585 (1958); *C.A.* **54**, 12702.
(977) Tyerman, British Patent **590,713** (1947).
(978) Tyrer, *J. Chem. Soc.* **101**, 81, 1104 (1912).
(979) Udovenko and Aleksandrova, *Zh. Fiz. Khim.* **37**, 52 (1963); *C.A.* **58**, 10790.
(979)c Udovenko and Mazanko, *Zh. Fiz. Khim.* **41**, 1615 (1967); *C.A.* **68**, 9592.
(980) Union Carbide Chemicals Co., "Alkylene Oxides," (1961).
(981) Union Carbide Chemicals Co., unpublished data.
(982) Union Carbide Chemicals Co., "Glycols" (1958); "Alcohols" (1961).
(983) Union Carbide Chemicals Co., *Tech. Inform. Bull.* (July 1959).
(984) Urbancova, *Chem. Zvesti* **13**, 43 (1959); *C.A.* **53**, 14621.
(985) Usines de Melle, French Patent **844,000** (1939).
(985)c Utkin, Balashov, and Serafimov, *Izv. Vyssh. Ucheb. Zaved., Khim. Khim. Tekhnol.* **12**, 1360 (1969); *C.A.* **72**, 71167g.
(986) Uusitalo, *Teknillisen Kemian Aikakausilehti* **18**, 635 (1961); *C.A.* **61**, 3947.
(987) Vaks, Seryakov, Nisel'son, and Sidorina, *Zh. Neorgan. Khim.* **6**, 756 (1961); *C.A.* **56**, 8070.
(988) van de Walle and Henne, *Bull. Soc. Chim. Belges* **34**, 10, 399 (1925).
(989) van Klooster and Douglas, *J. Phys. Chem.* **49**, 67 (1945).
(989)c Van Ness and Kochar, *J. Chem. Eng. Data* **12**, 38 (1967).
(990) Vaughn, U.S. Patent **2,088,935** (1937).
(991) Vdovenko and Kovaleva, *Zh. Prikl. Khim.* **31**, 89 (1958); *C.A.* **52**, 8661.
(991)a Verhoeye, *J. Chem. Eng. Data* **13**, 462 (1968); *C.A.* **69**, 110367g.
(991)c Verhoeye, *J. Chem. Eng. Data* **15**, 222 (1970).
(991)d Verhoeye and Lauwers, *J. Chem. Eng. Data* **14**, 306 (1969).
(992) Vijayaraghavan, Deshpande, and Kuloor, *Indian J. Technol.* **2**, 249 (1964).
(993) Vijayaraghavan, Deshpande, and Kuloor, *Chem. Age (India)* **15**, 1016 (1964); *J. Chem. Eng. Data* **11**, 147 (1966); *J. Chem. Eng. Data* **12**, 13 (1967); *J. Indian Inst. Sci.* **48**, 138 (1966).
(994) Vijayaraghavan, Deshpande, and Kuloor, *Indian J. Technol.* **3**, 267 (1965); *C.A.* **64**, 1402.
(994)c Vijayaraghavan, Deshpande, and Kuloor, *J. Indian Inst. Sci.* **47**, 139 (1965); *C.A.* **66**, 10879r.
(995) Vijayaraghavan, Deshpande, and Kuloor, *J. Chem. Eng. Data* **12**, 15 (1967).
(996) Vilim and Szlaur, *Collection Czech. Chem. Commun.* **29**, 1878 (1964).
(996)c Vitman, Markova, and Zharov, *Zh. Prikl. Khim.* **42**, 2360 (1969); *C.A.* **72**, 25625g.
(996)f Vitman and Zharov, *Zh. Prikl. Khim.* **42**, 2858 (1969); *C.A.* **72**, 71169j.
(996)i Vlasov, Sukhova, and Kogan, *Zh. Fiz. Khim.* **43**, 973 (1969); *C.A.* **71**, 95470f.
(997) Volpicelli and Zizza, *Chim. Ind. (Milan)* **45**, 1502 (1963); *C.A.* **60**, 7512.
(997)c Volpicelli, *Chim. Ind. (Milan)* **49**, 720 (1967); *C.A.* **67**, 85447g; *J. Chem. Eng. Data* **13**, 150 (1968).
(997)f Volpicelli and Campanile, *Corsi Semin. Chim.* **7**, 30 (1967); *C.A.* **70**, 91244s.
(998) Vostrikova, Aerov, Gurovich and Solomatina, *Zh. Prikl. Khim.* **37**, 2210 (1964); *C.A.* **62**, 11201.
(999) Wacker-Chemie G.m.b.H., British Patent **937,550** (1963); *C.A.* **60**, 6505.
(1000) Wade and Finnemore, *J. Chem. Soc.* **85**, 938 (1904).
(1001) Wade and Merriman, *J. Chem. Soc.* **99**, 997 (1911).
(1002) Wagner and Weber, *Chem. Eng. Data Ser.* **3**, 220 (1958).
(1003) Walker and Carlisle, *Chem. Eng. News* **21**, 1250 (1943).
(1004) Wallace and Atkins, *J. Chem. Soc.* **101**, 1179 (1912).
(1005) Wallace and Atkins, *J. Chem. Soc.* **101**, 1958 (1912).
(1006) Walls and Dean, U.S. Patent **2,371,860** (1945).
(1007) Wang, *Proc. Cryogenic Eng. Conf. 2nd, Boulder,* **1957**, 294; *C.A.* **52**, 14267.
(1007)a Washburn, Levens, Albright, and Billig, ADVAN. CHEM. SER. **23**, 134 (1959).
(1008) Watanabe and Conlon, U.S. Patent **2,760,990** (1956); *C.A.* **51**, 3654.
(1009) Weber, *Ind. Eng. Chem.* **48**, 134 (1956).
(1010) Weck and Hunt, *Ind. Eng. Chem.* **46**, 2521 (1954).
(1011) Wehe and Coates, *A. I. Ch. E. J.* **1**, 241 (1955).
(1012) Weismann and Wood, *J. Chem. Phys.* **32**, 1153 (1960); *C.A.* **54**, 18008.
(1013) Welling, U.S. Patent **2,376,104** (1945).
(1014) Welling, U.S. Patent **2,386,375** (1945).

(1015)   Welling, U.S. Patent **2,401,282** (1946).
(1016)   Wen, Mo, Chien, Chung and Kuang, *Hua Kung Hsueh Pao* **1960**, 150; *C.A.* **57**, 14483.
(1017)   Wentworth *et al.*, U.S. Patent **2,038,865** (1936); U.S. Patent **2,041,668** (1936).
(1018)   Werner, *J. Prakt. Chem.* **29**, 26 (1965); *C.A.* **63**, 4998.
(1019)   Whipple, *Ind. Eng. Chem.* **44**, 1664 (1952).
(1020)   White and Rose, *J. Res. Natl. Bur. Std.* **17**, 943 (1936).
(1021)   White and Rose, *J. Res. Natl. Bur. Std.* **21**, 151 (1938).
(1022)   Wichterle, *Collection Czech. Chem. Commun.* **30**, 3388 (1965); *C.A.* **63**, 14116.
(1023)   Willert, U.S. Patent **2,445,738** (1948).
(1024)   Williams, *Trans. A. I. Ch. E.* **37**, 157 (1941).
(1025)   Williams and Meeker, *Anal. Chem.* **20**, 733 (1948).
(1026)   Williams, Rosenberg, and Rothenberg, *Ind. Eng. Chem.* **40**, 1273 (1948).
(1026)c   Willock and Van Winkle, *J. Chem. Eng. Data* **15**, 281 (1970).
(1027)   Wilson and Simons, *Ind. Eng. Chem.* **44**, 2214 (1952).
(1028)   Wingard and Durant, *J. Alabama Acad. Sci.* **27**, 11 (1955); *C.A.* **50**, 10469.
(1029)   Wingard, Durant, Tubbs, and Brown, *Ind. Eng. Chem.* **47**, 1757 (1955).
(1030)   Wingard and Piazza, *Alabama Polytech. Inst. Eng. Expt. Sta. Bull.* No. **32** (1958); *C.A.* **53**, 12776.
(1030)c   Wong and Eckert, *J. Chem. Eng. Data* **14**, 432 (1969); *C.A.* **71**, 129342h.
(1031)   Wood, *J. Am. Chem. Soc.* **59**, 1510 (1937).
(1032)   Woods, *J. Soc. Chem. Ind. (London)* **66**, 26 (1947).
(1032)c   Wu, *Bull. Inst. Chem., Acad. Sinica* **13**, 31 (1967); *C.A.* **68**, 6822v.
(1033)   Wuyts, *Bull. Soc. Chim. Belges* **33**, 178 (1924).
(1034)   Wuyts and Bailleux, *Bull. Soc. Chim. Belges* **29**, 55 (1920).
(1035)   Wuyts and Docquier, *Bull. Soc. Chim. Belges* **44**, 297 (1935).
(1036)   Wyandotte Chemical Corp., Market Development Property Sheet (Feb. 25, 1955).
(1037)   Wyandotte Chem. Corp., British Patent **863,498** (1961); *C.A.* **55**, 18782.
(1038)   Wyrzykowska-Stankiewicz and Zieborak, *Bull. Acad. Polon. Sci., Ser. Sci. Chim.* **8**, 655 (1960); *C.A.* **57**, 4098.
(1039)   Yamamoto and Maruyama, *Kagaku Kogaku* **23**, 635 (1959); *C.A.* **54**, 1004.
(1040)   Yates and Kelly, U.S. Patent **2,752,295** (1956).
(1041)   Yen and Reed, *J. Chem. Eng. Data* **4**, 102 (1959).
(1042)   Yorizane and Yoshimura, *Hiroshima Daigaku Kogakuba Kenkya Hokoku* **13**, 41 (1965); *C.A.* **63**, 1253.
(1042)c   Yorizane and Yoshimura, *Hiroshima Daigaku Kogakubu Kenkyu Hokoku* **14**, 93 (1966); *C.A.* **66**, 22693t.
(1042)e   Yorizane, Yoshimura, and Masuoka, *Kagaku Kogaku* **30**, 1093 (1966); *C.A.* **67**, 47634h.
(1042)g   Yorizane, Yoshimura, and Yamamoto, *Kagaku Kogaku* **31**, 451 (1967); *C.A.* **69**, 51954k.
(1043)   Young, *J. Chem. Soc.* **81**, 707 (1902).
(1044)   Young and Fortey, *J. Chem. Soc.* **81**, 739 (1902).
(1045)   Young and Fortey, *J. Chem. Soc.* **83**, 45, 68, 77 (1903).
(1046)   Young and Fortey, *J. Chem. Soc., Trans.* **81**, 717 (1902).
(1047)   Young and Nelson, *Ind. Eng. Chem., Anal. Ed.* **4**, 67 (1932).
(1048)   Yu and Hickman, *J. Chem. Educ.* **26**, 207 (1949).
(1049)   Yuan, Ho, Keshpande and Lu, *J. Chem. Eng. Data* **8**, 549 (1963).
(1049)c   Zawisza, *Bull. Acad. Pol. Sci., Ser. Sci. Chim.* **15**, 307 (1967); *C.A.* **69**, 22486q.
(1049)e   Zawisza and Glowka, *Bull. Acad. Pol. Sci., Ser. Sci. Chim.* **17**, 373 (1969); *C.A.* **72**, 54701j.
(1050)   Zharov and Morachevskii, *Zh. Prikl. Khim.* **36**, 2397 (1963); *C.A.* **60**, 7513.
(1051)   Zhdanov, *J. Gen. Chem. (U.S.S.R.)* **11**, 471 (1941); *C.A.* **35**, 7275.
(1052)   Zieborak, *Bull. Acad. Polon. Sci., Classe III* **3**, 531 (1955).
(1053)   Zieborak, *Bull. Acad. Polon. Sci., Classe III* **6**, 443, 449 (1958); *C.A.* **52**, 19392.
(1054)   Zieborak, *Bull. Intern. Polon. Sci., Classe Sci., Math., Nat. Ser. A* **1950**, 15; *C.A.* **46**, 410.
(1055)   Zieborak, *Z. Physik. Chem.* **231**, 248 (1966); *C.A.* **65**, 65.
(1056)   Zieborak and Brzostowski, *Bull. Acad. Polon. Sci., Classe III* **5**, 309 (1957); *C.A.* **51**, 14399.
(1057)   Zieborak and Brzostowski, *Bull. Acad. Polon. Sci., Classe III* **6**, 169 (1958); *C.A.* **52**, 13349.
(1058)   Zieborak, Brzostowski, and Kaminski, *Bull. Acad. Polon. Sci., Classe III* **6**, 377, (1958); *C.A.* **52**, 19393.
(1059)   Zieborak and Galska, *Bull. Acad. Polon. Sci., Classe III* **3**, 383 (1955); *C.A.* **50**, 9080.
(1060)   Zieborak and Galska-Krajewska, *Bull. Acad. Polon. Sci., Classe III* **6**, 763 (1958); *C.A.* **53**, 12777.

(1061) Zieborak and Galska-Krajewska, *Bull. Acad. Polon. Sci., Classe III* **7**, 253 (1959); *C.A.* **54**, 16068 (1960).
(1062) Zieborak, Kaczorowna-Badyoczek, and Maczynska, *Roczniki Chem.* **29**, 783 (1955); *C.A.* **50**, 6119 (1956).
(1063) Zieborak, Maczynska, and Maczynski, *Roczniki Chem.* **32**, 85 (1958); *C.A.* **52**, 12493 (1958).
(1064) Zieborak and Maczynska, *Roczniki Chem.* **32**, 295; *C.A.* **52**, 17862 (1958).
(1065) Zieborak and Markowska-Majewska, *Bull. Acad. Polon. Sci., Classe III* **2**, 341 (1954).
(1066) Zieborak and Olszewski, *Bull. Acad. Polon. Sci., Classe III* **4**, 823 (1956); *C.A.* **51**, 7789.
(1067) Zieborak and Wyrzykowska-Stankiewicz, *Bull. Acad. Polon. Sci., Classe III* **6**, 377 (1958); *C.A.* **52**, 19392.
(1068) Zieborak and Wyrzykowska-Stankiewicz, *Bull. Acad. Polon. Sci., Classe III* **6**, 517 (1958); *C.A.* **53**, 3875 (1959).
(1069) Zieborak and Wyrzykowska-Stankiewicz, *Bull. Acad. Polon. Sci., Classe III* **8**, 137 (1960); *C.A.* **55**, 11047.
(1070) Zieborak and Wyrzykowska-Stankiewicz, *Bull. Acad. Polon. Sci., Classe III* **7**, 247 (1959); *C.A.* **54**, 16068.
(1071) Zieborak and Zieborak, *Bull. Acad. Polon. Sci., Classe III* **2**, 287 (1954); *C.A.* **49**, 2803.
(1072) Zilberman, *J. Appl. Chem. U.S.S.R.* **26**, 809 (1954).
(1073) Zmaczynski, *Roczniki Chem.* **11**, 449 (1931); *C.A.* **25**, 5809.
(1074) Zorin, Devyatykh, Krupnova and Krasnova, *Zh. Neorgan. Khim.* **9**, 2280 (1964); *C.A.* **62**, 2261.
(1075) Zvaritskaya and Delarova, *Tr., Vses. Nauchn.-Issled. Alyumin.-Magnievyi Inst.* **1961**, 96; *C.A.* **57**, 2914.
(1076) Zvaritskaya and Delarova, *Tr. Vses. Nauchn.-Issled. Alyumin.-Magnievy Inst.* **1962**, 152; *C.A.* **59**, 12235.

Prediction of Azeotropism and
Calculation of Azeotropic Data

Whether or not a system of two compounds is azeotropic depends essentially on two factors: (1) the difference in boiling point of the two components, and (2) the degree to which the two components form an ideal system. The closer the boiling points of the two components are, the more likely that they will be azeotropic; the more ideal the solution of the two components is, the less likely that they will form an azeotropic system.

The ideality of a two-component system depends largely on the difference in certain physical properties of the components such as polarity, degree of association, tendency toward hydrogen bonding, etc. When the two components are similar—e.g., two hydrocarbons or two alcohols—there is little tendency toward azeotropism. However, if the two components are markedly dissimilar—such as an alcohol and a hydrocarbon—they tend strongly toward azeotropism.

By taking account of the above factors it is sometimes possible to predict azeotropic data fairly well. For example, the departure from ideality of a series of solutions of methanol with hydrocarbons will be relatively constant. As a result the tendency for a methanol–hydrocarbon system to be azeotropic will be related to the other factor, the difference in boiling point. As a result the lowering of the boiling point (designated δ) and the azeotropic composition will be related to this difference in boiling point. Therefore we can predict δ and azeotropic composition for any methanol–hydrocarbon system from the boiling point of the hydrocarbon.

This procedure is the basis for the series of relations shown in Figures 1–6 which can be used to predict azeotropic data for related systems. The procedure also indicates the effect of pressure on azeotropic composition if one determines the difference in boiling point of the system at any desired pressure and applies the data to the appropriate curve. This is illustrated in Figure 6.

In the special case where the two components are immiscible, it is possible to calculate azeotropic boiling point and composition to a high degree of accuracy. For such a system the total vapor pressure is equal to the sum of the vapor pressures of the two components at a given temperature. Therefore, from a plot of vapor pressures of the two com-

615

ponents it is possible to determine the temperature at which the sum of the component vapor pressures is equal to 760 mm. This temperature will be the azeotropic boiling point. The azeotropic boiling point can be determined at any other pressure in a similar manner.

Further, the azeotropic composition can be calculated from the expression:

$$\text{Mole \% A} = \frac{V_A \times 100}{V_A + V_B}$$

where $V_A + V_B$ are vapor pressures of A and B at the azeotropic boiling point.

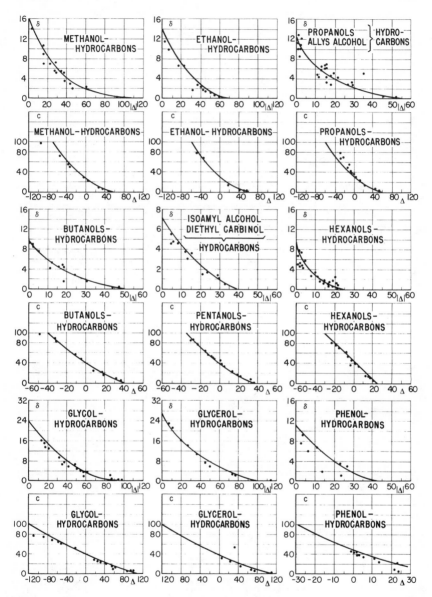

Figure 1. C-Δ and δ-|Δ| *curves for alcohol–hydrocarbon, glycol–hydrocarbon, and phenol–hydrocarbon systems*

C: *Azeotropic composition in weight % first component*
δ: *Boiling point of lower boiling component minus azeotropic boiling point*
|Δ|: *Absolute difference in boiling points of components*
Δ: *Boiling point of first component minus boiling point of second component*

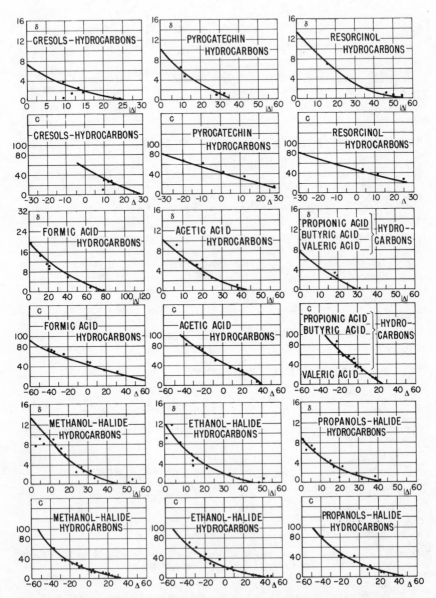

Figure 2. C-Δ and δ-|Δ| *curves for phenol–hydrocarbon, acid–hydrocarbon, and alcohol–halide hydrocarbon systems*

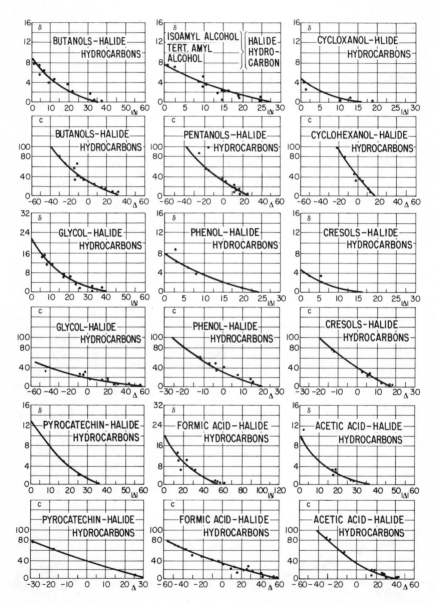

Figure 3. C-Δ and δ-|Δ| curves for alcohol–halide hydrocarbon, glycol–halide hydrocarbon, phenol–halide hydrocarbon, and acid–halide hydrocarbon systems

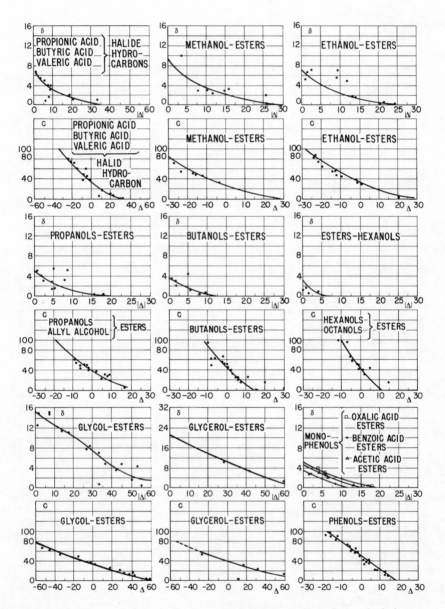

Figure 4.   C-Δ and δ-|Δ| curves for acid–halide hydrocarbon, alcohol–ester, glycol–ester, and phenol–ester systems

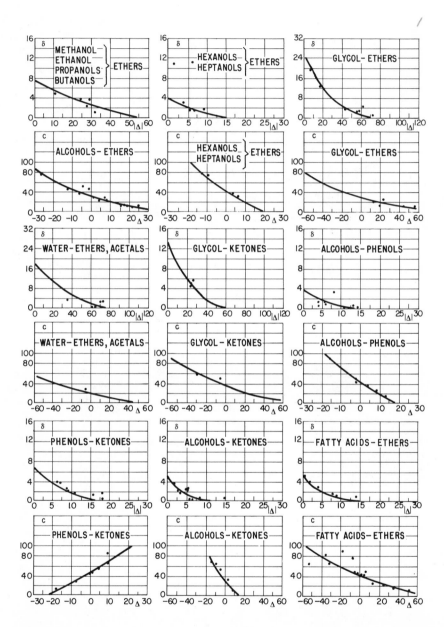

Figure 5. C-Δ and δ-|Δ| curves for alcohols–ethers, glycols–ethers, water–ethers, acids–ethers, alcohols–ketones, glycol–ketones, alcohols–phenols, and phenols–ketones

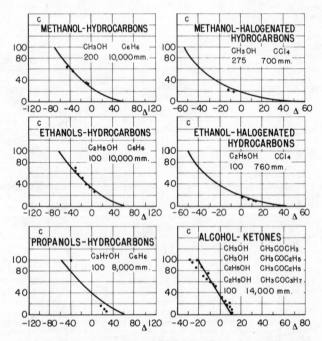

Figure 6. C-Δ *curves for alcohol–hydrocarbons, alcohol–halide hydrocarbons, and alcohols–ketones*

Curves show agreement with experimental data at various
pressures

C: *Weight % alcohol*   *hydrocarbon*
Δ: *Boiling point of alcohol minus*   *halide hydrocarbon*
   boiling water   *ketone*

Vapor-Liquid Equilibrium Diagrams of Alcohol-Ketone Azeotropes as a Function of Pressure

E. C. BRITTON,[1] H. S. NUTTING, and L. H. HORSLEY

The Dow Chemical Co., Midland, Mich.

Pressure has a marked effect on the azeotropic composition and vapor-liquid equilibrium diagrams of alcohol-ketone systems (1). This is due to the fact that the slopes of the vapor pressure curves of alcohols are appreciably greater than for ketones; it results in an unusually large change in the relative boiling points of the components of an alcohol-ketone system with change in pressure.

As a result of the study of these systems, it has been found that the methanol-acetone azeotrope exhibits the unusual phenomenon of becoming nonazeotropic at both low and high pressures—that is, below 200-mm. pressure the system is nonazeotropic with methanol as the more volatile product, while above 15,000 mm. the system is nonazeotropic with acetone the more volatile component.

Some of the equilibrium data for this system and two other alcohol-ketone azeotropes are shown in Figures 1 and 2 on the following pages.

The similarity of the diagrams for the different systems at suitable pressures is of interest. For example, the diagram for methanol-acetone at 10,000 mm. corresponds approximately to the diagram for methanol-methyl ethyl ketone at 1000 mm. and for ethanol-methyl propyl ketone at 100 mm.

Literature Cited

(1) Britton, E. C., Nutting, H. S., Horsley, L. H. U.S. Patent **2,324,255** (July 13, 1943).

---

[1] Deceased.

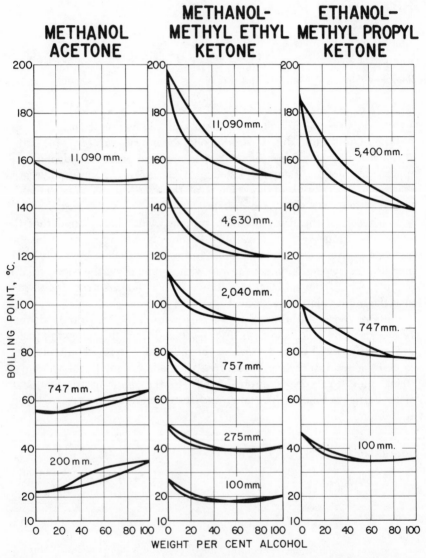

Figure 1. Vapor–liquid equilibrium diagrams of alcohol–ketone systems at various pressures

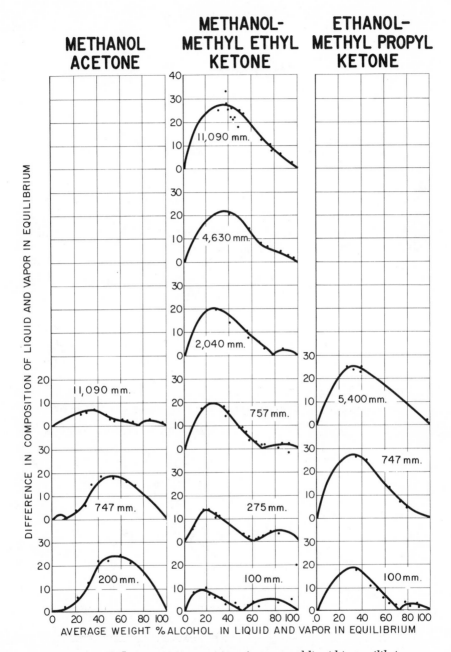

Figure 2. Difference in composition of vapor and liquid in equilibrium

Shown as a function of corresponding average composition of vapor and liquid for alcohol–ketone systems

Graphical Method for Predicting Effect of Pressure on Azeotropic Systems

H. S. NUTTING and L. H. HORSLEY

The Dow Chemical Co., Midland, Mich.

A rapid and easily applicable method has been found for indicating the effect of pressure on the composition and boiling point of an azeotropic system. The method is based on the use of the Cox vapor pressure chart *(1)* on which the log of vapor pressure is plotted as a function of $1/(t°$ C. $+ 230)$ to give a straight line over a wide range of pressures.

Lecat *(2)* has considered the use of the vapor pressure curves of azeotropes to indicate the pressure at which a system would become nonazeotropic. However, he plotted in the conventional manner and could obtain the curves only by detailed experimental work.

It has been found that the vapor pressure curves of azeotropes are straight lines when plotted on a Cox chart which permits determination of the complete vapor pressure curve from the data at two pressures.

Since an azeotrope by definition has either a higher or a lower vapor pressure than that of any of the components, the azeotropic vapor pressure curve will always lie above or below the curves of the components. This is indicated schematically in Figure 1 where A and B are vapor

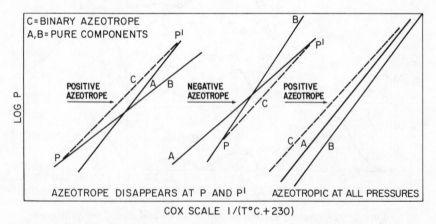

Figure 1.   Schematic of vapor pressure curves of binary azeotropes

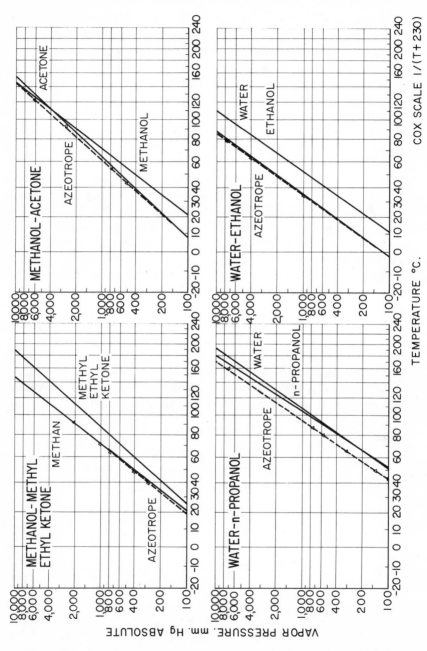

Figure 2. Azeotropic vapor pressure curves of methanol–methyl ethyl ketone, methanol–acetone, water–n-propanol, and water–ethanol

pressure curves of the components and C is the vapor pressure of the azeotrope. If curve C crosses either A or B, the azeotropic vapor pressure is no longer greater or less than any of the components and the system will become nonazeotropic at the point of intersection. On the other hand, if the azeotropic curve is parallel to the other curves the system will be azeotropic up to the critical pressure.

The method has been successfully applied to numerous systems, four of which are shown in Figure 2. The azeotrope methanol-methyl ethyl ketone became nonazeotropic at 3000 mm. of mercury after it was predicted that this would occur at 2000 to 4000 mm. The azeotrope methanol-acetone was studied in detail after it was predicted that the azeotropism would disappear at both low and high pressures. This system is nonazeotropic below 200 mm. of mercury and above 15,000 mm. compared to predicted limits of 200 to 500 mm. and 10,000 to 20,000 mm. While this is the only azeotropic system known to become nonazeotropic at both low and high pressures, there are indications that the phenomenon occurs in several other systems, contrary to the conclusions of Lecat that such systems probably do not exist (3).

Caution should be used in extrapolating curves to very low pressures because of the possibility of curvature in the vapor pressure lines over a manyfold range of pressures.

In cases where only the normal azeotropic boiling point is known, it is possible to predict the effect of pressure on the sytem by drawing the azeotrope curve through the normal boiling point with a slope equal to the average slopes of the component vapor pressure curves. This procedure will permit a fairly accurate prediction of whether the azeotrope will cease to exist below the critical pressure.

Literature Cited

(1) Cox, *Ind. Eng. Chem.*, **15,** 592 (1923).
(2) Lecat, *Ann. soc. sci. Bruxelles,* **49B,** 261–333 (1929).
(3) Lecat, "Traité de Chimie Organique," Vol. 1, p. 139, Paris, Grignard, Mason et Cie., 1935.